WELDING

Principles and Applications

Fourth Edition

WELDING

Principles and Applications

Fourth Edition

LARRY JEFFUS

DELMAR

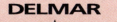

THOMSON LEARNING

Africa • Australia • Canada • Denmark • Japan • Mexico • New Zealand • Philippines
Puerto Rico • Singapore • Spain • United Kingdom • United States

Delmar Staff

Acquisitions Editor: Vernon Anthony
Developmental Editor: Denise Denisoff
Project Editor: Coreen Filson
Production Manager: Mary Ellen Black
Art and Design Coordinator: Mary Beth Vought

Printed in the United States of America
7 8 9 10 XXX 03 02 01

For more information, contact Delmar, 3 Columbia Circle, PO Box 15015, Albany, NY 12212-0515; or find us on the World Wide Web at http://www.delmar.com

International Division List

Asia
Thomson Learning
60 Albert Street, #15-01
Albert Complex
Singapore 189969
Tel: 65 336 6411
Fax: 65 336 7411

Japan:
Thomson Learning
Palaceside Building 5F
1-1-1 Hitotsubashi, Chiyoda-ku
Tokyo 100 0003 Japan
Tel: 813 5218 6544
Fax: 813 5218 6551

Australia/New Zealand:
Nelson/Thomson Learning
102 Dodds Street
South Melbourne, Victoria 3205
Australia
Tel: 61 39 685 4111
Fax: 61 39 685 4199

UK/Europe/Middle East
Thomson Learning
Berkshire House
168-173 High Holborn
London
WC1V 7AA United Kingdom
Tel: 44 171 497 1422
Fax: 44 171 497 1426

Latin America:
Thomson Learning
Seneca, 53
Colonia Polanco
11560 Mexico D.F. Mexico
Tel: 525-281-2906
Fax: 525-281-2656

Canada:
Nelson/Thomson Learning
1120 Birchmount Road
Scarborough, Ontario
Canada M1K 5G4
Tel: 416-752-9100
Fax: 416-752-8102

Library of Congress Cataloging-in-Publication Data:
Jeffus, Larry F.
    [4th ed]
    Welding: principles and applications/Larry Jeffus.
      p. cm.
    Includes index.
    ISBN 0-8273-8240-5 (alk. paper)
    1. Welding. I. Title.
TS227.J418   1997
671.5'2—dc21

97-13165
CIP

CONTENTS

PREFACE

INTRODUCTION

The welding industry today presents a continually growing and changing series of opportunities for skilled welders. Despite economic fluctuations, there is a positive job outlook in welding. Due to a steady growth in the demand for goods fabricated by welding, new welders are needed in every area of welding such as small shops, specialty fabrication shops, and large industries. The student who is preparing for a career in welding will need to:

- have excellent eye-hand coordination,
- work well with tools and equipment,
- know the theory and application of the various welding and cutting processes,
- be able to follow written and verbal instructions,
- work with or without close supervision,
- communicate effectively both verbally and through writing,
- read and interpret welding drawings and sketches, and
- be alert to possible problems and work safely.

A thorough study of WELDING PRINCIPLES AND APPLICATIONS in a classroom/shop setting will help students prepare for the opportunities in modern welding technology. The comprehensive technical content provides the basis for the welding processes. The extensive descriptions of equipment and supplies with in-depth explanations of their operation and function familiarize students with the tools of the trade. The process descriptions, practices, and experiments coupled with actual performance teach the critical manuals skills required on the job. The text also discusses occupational opportunities in welding and explains the training required for certain welding occupations. The skills and personal traits recommended by the American Welding Society for their Certified Welder program are included within the text. Students wishing to become certified under the AWS program must contact the American Welding Society for specific details.

ORGANIZATION

The text is organized to guide the student's learning from an introduction to welding, through critical safety information, to details of specific welding processes, and on to the related areas of welding metallurgy, weldability of metals, fabrication, certification, testing and inspection of welds, and welding joint design, costs, and welding symbols.

Each section of the text introducing a welding process(es) begins with an introduction to the equipment and materials to be used in the process(es), including setup in preparation for welding. The remaining chapters for the specific process concentrate on the actual welding techniques in various applications and positions. The content progresses from basic concepts to the more complex welding technology. Once this technology is understood, the student is able to quickly master new welding tasks or processes.

The welding processes in the text are presented in a manner that allows the student to begin with any section. It is not necessary to learn all of the processes if only one or two are required of your job.

Each chapter begins with a list of *learning objectives* that tell the student and instructor what is to be learned while studying the chapter. A survey of the objectives will show that the student will have the opportunity to develop a full range of welding skills, depending upon the topics selected for the program. Each major process is presented in such a way that the instructor can eliminate processes having little economic value in the market served by the program. However, the student will still learn all essential information needed for a thorough understanding of all processes studied.

In each chapter, *new terms* are highlighted in color and defined. In addition, the new terms are listed at the beginning of the chapter to enable students to recognize the terms when they appear. Terms and definitions used throughout the text are based on the American Welding Society's standards. Industry jargon has also been included where appropriate.

Safety precautions and *notes* for the student are given throughout the text. *Metric equivalents* are listed in parentheses for dimensions. The metric equivalent in most cases has been rounded to the nearest whole number. Numerous photographs, line drawings, and plans illustrate concepts and clarify the discussions.

Most of the chapters contain learning activities in the form of *experiments* and *practices*. The end of the experiments are identified by the (♦) and the end of the practices are identified by the (♦) symbol.

By completing the *experiments,* the student learns the parameters of each welding process. In the experiments, the student changes the parameters to observe the effect on the process. In this way, the student learns to manipulate the variables to obtain the desired welding outcome for given conditions. The experiments provided in the chapters do not have right or wrong answers. They are designed to allow the student to learn the operating limitations or the effects of changes that may occur during the welding process.

A large selection of *practices* are included to enable the student to develop the required manipulative skills, using different materials and material thicknesses in different positions. A sufficient number of practices are provided so that after the basics are learned, the student may choose an area of specialization. Materials specified in the various practices may be varied in both thickness and length to accommodate those supplies that you have in your lab. Changes within a limited range of both

thickness and length will not affect the learning process designed for the practice.

The *review questions* at the end of each chapter can be used as indicators of how well the student has learned the material in each chapter.

REVISION

The fourth edition of WELDING: PRINCIPLES AND APPLICATIONS has been thoroughly revised and reorganized to reflect the latest welding technologies. New chapters added are as follows: Flux Cored Arc Welding, Fabrication, Welder Certification, and Railroad Welding.

In this edition there are many new black-and-white photos and detailed line art. The very unique photographs in this book were taken from the welder's viewpoint. Approximately one-third of the photos were taken from a left-handed view to aid the left-handed students. The use of quality graphics make it much easier for the student to see what is expected to produce a quality weld. The color insert includes many close-up shots of actual welding processes.

The "Success Story" vignettes of real welders provide motivation to students considering welding as a career. Industrial vignettes have been expanded and now appear in every chapter.

SUPPLEMENTS

Accompanying the text are an Instructor's Guide and Study Guide/Lab Manual. The Instructor's Guide contains lesson plan outlines for each chapter. Transparency masters have also been included to assist in the classroom. Answers to the questions in the textbook and the Study Guide/Lab Manual are included along with additional questions and answers for testing. Certification information is also provided in the Instructor's Guide, including samples of typical certification tests from certifying agencies. Supplementary references will direct the instructor to additional sources of information for specific content areas.

The Study Guide/Lab Manual has been thoroughly revised with many new added features. The Study Guide/Lab Manual is designed to test student understanding of the concepts presented in the text. Each chapter starts off with a review of the important topics discussed in the text. Students can then test their knowledge by answering additional questions. Lab exercises are included in those chapters as appropriate to reinforce the primary objectives of the lesson. Artwork and safety precautions are included throughout the manual.

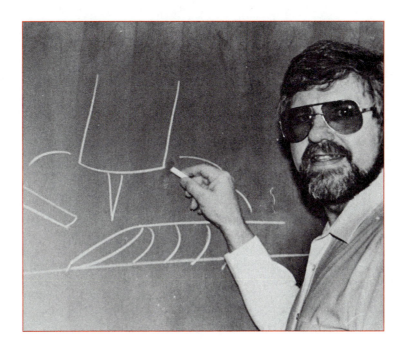

ABOUT THE AUTHOR

During my junior year of high school, I learned to weld in metal shop. There I was taught basic welding principles and applications, and I was able to build a number of projects in shop using oxyacetylene welding, shielded metal arc welding, and brazing.

The practice welds helped me develop welding skills, and building the projects allowed me to start developing some fabrication skills. By the end of my junior year, I had become a fairly skilled welder.

In my sophomore year I joined the Vocational Industrial Clubs of America (VICA). Through VICA I learned how businesses and industries work, developed leadership skills, established goals, and much more. I learned things in VICA that I still use in my life today.

In my senior year at New Bern High School, I was given an opportunity to join Mr. Z.T. Koonce's first class in a new program called Industrial Cooperative Training (ICT). ICT is a cooperative work experience program that coordinates school experiences with real jobs. This allowed me to attend high school in the morning, where I completed my required English, math, and other academic courses for graduation. We were also taught skills that would help us get a job — like how to fill out a job application, interviewing skills, and so on. In the afternoons, I worked as a welder. After graduation, I started a full-time job as a welder at Barbour Boat Works. There my welding skills were refined, and I was allowed to work with the other welders in the shipyard. My first welding assignment was building a barge, making intermittent welds to attach the deck to the barge's ribs.

As my welding skills improved, my supervisor allowed me to apply my new welding skills to more difficult jobs. I welded on barges, military landing crafts, tugboats, PT boats, small tankers, and others. This is how I earned money toward my college education.

With my welding skills, I was able to get a job in a small shop in Madisonville, Tennessee, and attended Hiwassee Junior College. After graduating from Hiwassee, I found other welding jobs that allowed me to continue my education at the University of Tennessee, where I earned a bachelor's degree. After four years, I had both a college degree and four years of industrial experience, which together qualified me for my job as a vocational teacher.

During my career as a welder, I have welded on tanks, pressure vessels, oil well drilling equipment, farm equipment, buildings, race cars, and more. As a vocational teacher, I have taught in high schools, schools for special education, schools for the deaf, and three colleges. I have been a consultant to the welding industry.

Larry Jeffus is a recognized welding instructor with many years of experience teaching welding technology at the community college level. He has been actively involved in the American Welding Society, having served on the General Education Committee and as the Chairman of the North Texas Section of the American Welding Society.

ACKNOWLEDGMENTS

The author would like to thank the following individuals who contributed their time and skills to produce the photographs of the welding in progress: Dewayne Roy, Daniel Evans, David Gibson, Elmo Herrera, and Carl West.

The welding photographs provided in this text are due in large part to the efforts and determination of John L. Chastain, who worked with the author for many long hours to perfect the photographic techniques required to achieve the action photos.

The author also wishes to acknowledge the numerous contributions of Dewayne Roy.

Special thanks are extended to Alton E. Hand and the staff of American Testing Laboratory Inc. for developing the welder qualification and certification, and procedure qualification material which appears in the text. In addition, the author wishes to thank Howard R. Adams, Training Manager, and Alloy Rods Corporation for granting permission to reprint the lesson "ESTIMATING AND COMPARING WELD METAL COSTS" which appears in the text.

The special interest articles were generously contributed by a number of companies, each of which has a long-standing commitment to education.

Special thanks are due to the Miller Electric Manufacturing Co., Hobart Brothers Company, The Lincoln Electric Company, Alloy Rods Corporation, GMF Robotics, ESAB Automation, Martin Marietta Electronics, Silver Engineering, and United States Welding Corporation. Appreciation is also expressed to the E.I. Du Pont De Nemours and Company which granted permission for the reprinting of portions of the Du Pont NDT Welder's Reference. Also to The American Welding Society, Inc. whose *Welding Journal* was a valuable source in providing many of the special interest articles.

Thanks are extended to Jim Sector for permitting location photographs at Albany Calcium Light Co., Inc., Loudonville, NY, and to George Govel, Jr. of Govel Welding, Inc., Albany, NY.

I would like to express my deepest appreciation to Jessica Alvarez and Kathy Cott for all the hours spent typing in the preparation of this edition, and to Kristi Webb, Jennie Rothenberg, and my daughters Wendy and Amy for all of the general office help they provided, and to my wife, Carol, for all of her moral support.

Finally, a special thanks is extended to Samuel L. Evans, Dana Gillenwalters, Robert M. Vaughn, Richard Rowe, Dennis Peabody, and Courtney Newcomer for sharing their success with us. They are valuable contributors to the textbook and an inspiration for those entering the welding industry.

The reviewers of the fourth edition of the text have given their time and expertise to critique the content. Their recommendations have been invaluable to the author:

Edwin Bromley, Lane Community College, Eugene, OR; John Penaz, Dunwoody Institute, Minneapolis, MN; Michael Bost, Cheboygan Area High School, Cheboygan, MI; Ron Samuelson, Southeast Community College, Milford, NE; John Adams, New Castle School of Trades, Pulaski, PA

This book is dedicated to two very special people, my daughters Wendy and Amy.

Index of Experiments and Practices

The following Experiments and Practices are listed in the order in which they appear in the chapter. It should be noted that not all chapters have Experiments and Practices.

CHAPTER 17

CHAPTER 18

CHAPTER 19

CHAPTER 22

RAILROAD WELDING

OBJECTIVES

After completing this chapter, the student will be able to:

- explain the thermite welding process for rails.
- describe the characteristics of austenitic manganese steel.
- list the steps required to repair cracks in rails and rail components.
- explain the reason for keeping thermite welding materials dry.

KEY TERMS

frog
hydrogen embrittlement
flash welding
molecular hydrogen (H_2)
mold

austenitic manganese steel
thermite welding
atomic hydrogen (H)
crucible
slag pan

INTRODUCTION

The railroad was the first mass transit system. Early railroads provided a way for large numbers of immigrants to move out west. Railroads were also instrumental in the development of the agricultural and cattle industries throughout the Midwest, by providing farmers and ranchers ready access to eastern markets.

New high-speed bullet trains and local transit systems are expected to someday provide cities with a more environmentally friendly transportation system than automobiles.

The railroad systems are so large, diverse, and widespread that no matter where you live you can see it in action. Because of this and its impact on our lives, it has been selected for this chapter to serve as an example of the diversity required by welders within an industry.

We all see trains moving about and give little thought to the welding required to keep them moving. This industry, like most others, relies on skilled welders to fabricate and repair essential parts. Welding is used to build locomotives, cars, trussels, rails, switches, and much more for the railroad. Almost every major welding, cutting, and brazing process is used by this expansive industry.

The range of metals and alloys used is extensive. Commonly used metals and alloys include most plain carbon steel, high-strength steels, stainless steels, aluminums, and many others. This diverse group of alloys requires an equally diverse collection of welding procedures. This chapter will cover several different specialty welding needs while concentrating on one area—rails and related components. A wide variety of welding processes and procedures are needed for this one segment of the railroad industry.

Rails are used for many applications other than just the railroad industry. They are used for trolleys, large cranes, overhead cranes, local transit systems, amusement park rides, guard rails, electrical contact rails, and so on. Each type of rail has its own requirements for which rails are specifically designed and alloyed to meet those needs.

Objectives, found at the beginning of each chapter, are a brief list of the most important topics found in each chapter.

Key Terms are the most important technical words you will learn in the chapter. These are listed at the beginning of each chapter following the Objectives and appear in color print where they are first defined. These terms are also given in the glossary at the end of the book.

Chapter 24   Weldability of Metals   627

fourth that of steel and approximately two-thirds that of aluminum. Its melting point is 1,202°F (650°C). Magnesium has considerable resistance to corrosion and compares favorably with some aluminum alloys in this respect.

Magnesium must be alloyed with other elements to provide the necessary strength for most applications. Common alloying elements include zinc, aluminum, manganese, zirconium, and the rare earths. Magnesium alloys may be classified as wrought or casting types. Sheet, plate, and extrusions are part of the first group.

The alloy designations for magnesium consist of one or two letters, which represent the alloying elements. The alloying percentages are next listed in the designation. A letter is also added after the alloying percentages. One example is the ASTM designation AZ91C. This indicates an alloy of 9% aluminum and 1% zinc. The letter following the designation is defined as follows:

- A — Aluminum
- C — Copper
- E — Rare earths
- H — Thorium
- K — Zirconium
- M — Manganese
- Z — Zinc

The temper designation is the same as that used for aluminum. Wrought alloys are used to a great extent because of their properties of high strength, ductility, formability, and toughness.

Magnesium can be welded in somewhat the same manner as aluminum. The most widely used processes are GTA and GMA. Spot, seam, and mechanized resistance welding processes can also be used to weld magnesium on a production basis. Spot and seam welding applications consist of welding sheets and extrusions to thicknesses up to 3/16 in. (approximately 4.8 mm).

REPAIR WELDING

Repair or maintenance welding is one of the most difficult types of welding. Some of the major problems include preparing the part for welding, identifying the material, and selecting the best repair method.

The part is often dirty, oily, and painted, and it must be cleaned before welding. There are many hazardous compounds that might be part of the material on the part. These compounds may or may not be hazardous on the part, but when they are heated or burned during welding they can become life threatening.

■ CAUTION

It's never safe to weld on any part that has not been cleaned before welding. All surface contamination must be removed before welding to prevent the possibility of injuring your health from exposure to materials released during welding. Some chemicals can be completely safe until they are exposed to the welding. The smoke or fumes they produce can be an irritant to the skin or eyes; it can be absorbed through the skin or lungs. If you are exposed to an unknown contamination, get professional help immediately.

Contamination can be removed by sand blasting, grinding, or using solvents. If a solvent is used be sure it does not leave a dangerous residue. Clean the entire part if possible or a large enough area so that any remaining material is not affected by the welding.

Before the joint can be prepared for welding you must try to identify the type of metal. There are several ways to determine metal type before welding. One method is to use a metal identification kit. These kits use a series of chemical analyses to identify the metal. Some kits can not only identify a type of metal but also tell the specific alloy.

Another way to identify metal is to look at its color, test for magnetism, and do a spark test. The spark test should be done using a fine grinding stone. With experience, it's often possible to determine specific types of alloys with great accuracy. The sparks given off by each metal and its alloy are so consistent that the U.S. Bureau of Mines uses a camera connected to a computer to identify metals to aid in recycling. For the beginner it is best if you use samples of a known alloy and compare the sparks to your unknown. The test specimen and the unknown should be tested using the same grinding wheel and the same force against the wheel.

EXPERIMENT 24-1

Identifying Metal Using a Spark Test

In this experiment you will use proper eye safety equipment, a grinder, several different known and unknown samples of metal, and a pencil and paper to identify the unknown metal samples. Starting with the known samples, make several tests and draw the spark patterns as described below. Next test the unknown samples and compare the drawings with the drawings from the known samples. See how many of the unknowns you can identify.

There are several areas of the spark test pattern that you must observe carefully, Figure 24-5. The first

Cautions summarize critical safety rules. They alert you to operations that could hurt you or someone else. They are not only covered in the safety chapter, but you will find them throughout the text when they apply to the discussion, practice, or experiment.

Experiments are designed to allow you to see what effect changes in the process settings, operation, or techniques have on the type of weld produced.

Chapter 13   Flux Cored Arc Welding   301

SIDE BEND

SIDE BEND

45°

7" (178 mm)

9.5 mm)

9.5 mm)

6" (152 mm)   ¾" (19 mm)

Welding Principles and Applications
MATERIAL:
3/4" (9 mm) MILD STEEL PLATE 7 x 5" (178 mm x 76mm)
PROCESS:
FCAW 3G BUTT JOINT
NUMBER:
PRACTICE 13-11   DRAWN BY:
AMY JEFFUS

Figure 13-47

Figure 13-48   45° vertical up fillet weld.

Turn off the welding machine and shielding gas and clean up your work area when you are finished welding.

PRACTICE 13-24

Lap Joint and Tee Joint 3F 100% to Be Tested

Using a properly set up and adjusted FCA welding machine, proper safety protection, 0.035-inch and/or 0.045-inch (0.9-mm and/or 1.2-mm) -diameter E70T-1 and/or E70T-5 electrodes, and one or more pieces of mild steel plate, 12-inch (305-mm) -long and 3/8-inch (9.5-mm) -thick beveled plate, you will make a fillet weld in the vertical position.

Following the same instructions for assembly and welding procedure as outlined in Practice 13-22, repeat each type of joint with both classifications of electrodes as needed until welds can be made with 100% penetration that will pass the test. Turn off the welding machine and shielding gas and clean up your work area when you are finished welding.

PRACTICE 13-25

Tee Joint 3F

Using a properly set up and adjusted FCA welding machine, proper safety protection, 0.035-inch and/or through 1/16-inch (0.9-mm and/or through 1.6-mm) -diameter E70T-1 and/or E70T-5 electrodes, and one or

and/or E70T-5 electrodes, and one or more pieces of mild steel plate, 12-inch (305-mm) -long and 3/8-inch (9.5-mm) -thick beveled plate, you will make a fillet weld in the vertical position.

Following the same instructions for the assembly and welding procedure as outlined in Practice 13-22, repeat each type of joint with both classifications of electrodes as needed until defect-free welds can consistently be made.

Practices are hands-on exercises designed to build your welding skills. Each practice describes in detail what skill you will learn and what equipment, supplies, and tools you will need to complete the exercise.

Real World Features at the end of each chapter present a feature story which describes a real world application of the theory learned in the chapter. You will see how particular knowledge and skills are important to the world.

---

630 ||| Section 6 Related Processes and Technology

T-Boom Control Manifold

Welding operators at a manufacturer of sophisticated geological survey and agricultural equipment were skeptical when the company installed a robot to start automating the welding process. The robot, they thought, would never achieve their high-strength, quality welds or be cost-effective with short production runs.

Experience proved them wrong. Additional advantages gained by the implementation of the welding robot include:

■ relieving the welding operators of monotonous, dangerous, and uncomfortable work,

■ improving the quality of manufactured products by reducing worker inconsistency and fatigue,

■ improving the factory parts flow through reduced set-up times and other economies on short runs,

■ the capability to weld parts that are difficult, undesirable, and uneconomical to handle manually, and

■ reducing parts inventory, handling, and raw material costs.

The company has maintained exceptionally high production standards for more than thirty years of operation. Management considered robots for their welding process for several years. Investigation indicated that arc welding robots could provide improvements in both production and quality. But there were two other requirements to consider — cost-effective short runs and a system capable of handling workpieces with some variation in weld starting points and weld paths.

The arc welding robot selected met all of these requirements and provided many additional benefits, **Figure 1**. This included the long reach (73.8") of the manipulator, the fast bend and twist movement of the wrist — reducing air-cut time — and the fast and easy programming system. In addition, the robot supplier also provided a seam tracking system. This is an adaptive thru-the-arc

Figure 1

WELDING SYSTEM LAYOUT

1. Robot
2. Controller
3. Interface
4. Power Source
5. Wirefeed System and Pneumatic Cleaning System
6. Seam Tracker Arc Sensing Unit
7. Nozzle Reamer
8. Positioner Controller
9. Positioner Control Panel
10. Turntable Positioner
11. Preheat Oven
12. Parts Bin
13. Finished Parts Bin
14. Air Compressor

Figure 2

---

316 ||| Section 4 Gas Shielded Welding

REVIEW

1. Why is it important to make sure that an FCA welding system is set up properly if out-of-position welds are going to be made?

2. What major safety concerns should an FCA welder be cautious of?

3. Why should the FCA welding practice plates be large?

4. Why should the FCA welds be of substantial length?

5. What must be done to a shielding gas cylinder before its cap is removed?

6. What can happen to the wire if the conduit is misaligned at the **feed rollers**?

7. Why is it a good idea for a new student welder to use the gas nozzle even if a shielding gas is not used?

8. Why is the curl in the wire end straightened out?

9. What problems can a high-feed roller pressure cause?

10. Referring to **Table 13-3** answer the following:

    a. What would the range of the feed speed be for an amperage of 150 at 25 volts for E70T-1 0.035-inch (0.9-mm) electrode?

    b. What would the approximate amperage be for E70T-5 0.045-inch (1.2-mm) electrode if it's being fed at a rate of 200 inches per minute (508 cm per minute)?

11. What are the disadvantages of beveling plates for welding?

12. FCA welds with 100% joint penetration can be made in plates up to what thickness?

13. What is the smallest V-groove angle that can be welded using the FCA welding process?

14. What is the purpose of the root pass?

15. Why is the FCA welding process not used for open-root **critical welds**?

16. Why should convex weld faces be avoided?

17. What bead pattern is best for overhead welds?

18. Why is the appearance of the cover pass so important?

19. What is the visual inspection standard's limitations of acceptance?

20. Why is the backing strip 2 inches (50 mm) longer than the test plate?

21. Why should there be a space between the plates when making a fillet weld on thick plates of a tee joint?

22. What would a small notch at the root welds leading edge in a fillet weld mean?

23. What changes should be made in the setup for making a vertical up weld?

24. How can a cold weld be corrected?

25. What problem must be overcome if the amperage and voltage are lowered to make a vertical weld?

26. How can higher welding speeds help control distortion?

27. How can spatter buildup on the welding gun be controlled in the overhead position?

Review Questions will help measure the skills and knowledge you learned in the chapter. Each question is designed to help you apply and understand the information in the chapter.

SUCCESS STORY

Life in Illinois for Samuel L. Evans was like a roller coaster ride of bad decisions and wrong roads. Family and friends tried to help him stop this self-abusive lifestyle. Everyone knew that Sam had promise.

In her efforts to get him away from the streets of Illinois, Sam's sister talked to him about moving to Charlotte, North Carolina. Sam knew if he stayed on the track he was on, his life would be short-lived or lived out in a secured environment. He decided to make the move and start a more productive life for himself.

Outgoing, impressive, and enthusiastic, he landed his first job with a major insurance company in Charlotte. There Sam had the prestige of a white-collar corporate job. This job was great, but it was not the cure or distraction that Sam needed to shake the lifestyle that he had hoped he left behind in Illinois. Then one day he made up his mind to stop his self-destructive activities.

Sam's family had always been involved in the welding trade. Four uncles were welders. His grandfather was cutting and welding in junkyards in the 1960s and 1970s. He decided to continue the tradition and gave welding a try.

Sam enrolled in a welding class at Central Piedmont Community College. It was there that he met his instructor and mentor, Jim Payne. Payne encouraged Sam by letting him know that he had a natural ability that should not go to waste. He also found welding jobs for Sam while he was completing his training.

After completing the welding class at Central Piedmont and being certified in welding, Sam traveled for three years welding his way around the country on the Pipelines and other welding jobs, with General Piping, Inc., Alcar, and the Pipe Welder's Union. He traveled from Massachusetts to Oklahoma, New Mexico to New York, Pennsylvania to Florida.

He stopped traveling in 1992. Now a family man, Sam started doing small welding jobs to support his family. From these small jobs, Evans Welding Services, Inc. was founded. Sam found his niche in life. He enjoys being the company that provides the structural steel for building frames in Charlotte and surrounding areas. Evans Welding Services, Inc. not only provides for the Evans family but at peak construction season provides for up to twelve jobs for other men and women and their families.

Sam continues to learn and attain education in the welding field. He is very involved and respected in the Charlotte business community. Evans Welding Services, Inc. is a member of the Carolinas Association of General Contractors and the Charlotte Chamber of Commerce. It is with great pride that Sam received the first ever "CABWE" Man award from the Carolinas Association of Black Women Entrepreneur's in 1997.

Success Stories are found at the beginning of each of the seven sections in the text. These stories are about real people who have become successful by using their welding skills. Each story is different but one message is repeated by all of our story contributors: welding can be a rich and rewarding career.

Section One

INTRODUCTION

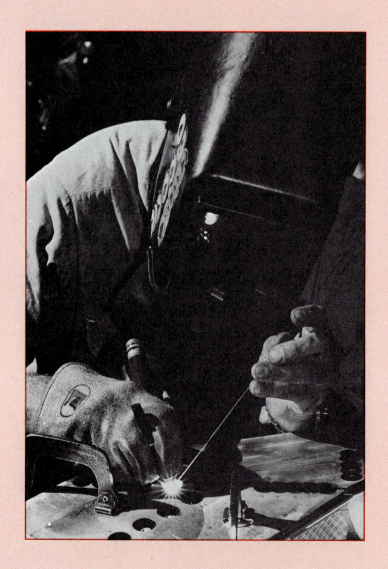

SUCCESS STORY

*L*ife in Illinois for Samuel L. Evans was like a roller coaster ride of bad decisions and wrong roads. Family and friends tried to help him stop this self-abusive lifestyle. Everyone knew that Sam had promise.

*I*n her efforts to get him away from the streets of Illinois, Sam's sister talked to him about moving to Charlotte, North Carolina. Sam knew if he stayed on the track he was on, his life would be short-lived or lived out in a secured environment. He decided to make the move and start a more productive life for himself.

*O*utgoing, impressive, and enthusiastic, he landed his first job with a major insurance company in Charlotte. There Sam had the prestige of a white-collar corporate job. This job was great, but it was not the cure or distraction that Sam needed to shake the lifestyle that he had hoped he left behind in Illinois. Then one day he made up his mind to stop his self-destructive activities.

*S*am's family had always been involved in the welding trade. Four uncles were welders. His grandfather was cutting and welding in junkyards in the 1960s and 1970s. He decided to continue the tradition and gave welding a try.

*S*am enrolled in a welding class at Central Piedmont Community College. It was there that he met his instructor and mentor, Jim Payne. Payne encouraged Sam by letting him know that he had a natural ability that should not go to waste. He also found welding jobs for Sam while he was completing his training.

*A*fter completing the welding class at Central Piedmont and being certified in welding, Sam traveled for three years welding his way around the country on the Pipelines and other welding jobs, with General Piping, Inc., Alcar, and the Pipe Welder's Union. He traveled from Massachusetts to Oklahoma, New Mexico to New York, Pennsylvania to Florida.

*H*e stopped traveling in 1992. Now a family man, Sam started doing small welding jobs to support his family. From these small jobs, Evans Welding Services, Inc. was founded. Sam found his niche in life. He enjoys being the company that provides the structural steel for building frames in Charlotte and surrounding areas. Evans Welding Services, Inc. not only provides for the Evans family but at peak construction season provides for up to twelve jobs for other men and women and their families.

*S*am continues to learn and attain education in the welding field. He is very involved and respected in the Charlotte business community. Evans Welding Services, Inc. is a member of the Carolinas Association of General Contractors and the Charlotte Chamber of Commerce. It is with great pride that Sam received the first ever "CABWE" Man award from the Carolinas Association of Black Women Entrepreneur's in 1997.

Chapter 1

INTRODUCTION TO WELDING

OBJECTIVES

After completing this chapter, the student should be able to

- explain how each one of the major welding processes works.
- list the factors that must be considered before a welding process is selected.
- discuss the history of welding.
- describe briefly the responsibilities and duties of the welder in various welding positions.
- define the terms *weld, forge welding, resistance welding, fusion welding, coalescence,* and *certification.*

KEY TERMS

forge welding

resistance welding

fusion welding

weld

coalescence

welding

oxyfuel gas welding (OFW)

oxyfuel gas cutting (OFC)

oxyfuel brazing (TB)

oxyfuel gas (OF)

shielded metal arc welding (SMAW)

gas tungsten arc welding (GTAW)

gas metal arc welding (GMAW)

flux cored arc welding (FCAW)

manual operation

semiautomatic operation

machine operation

automatic operation

automated operation

certification

qualification

American Welding Society (AWS)

INTRODUCTION

As joining techniques improved through the ages, so did the environment and mode of living for humans. Materials, tools, and machinery improved as civilization developed.

Fastening together the parts of work implements began when an individual attached a stick to a stone to make a spear or axe. Egyptians used applications of stones to create temples and pyramids that were fastened together with a gypsum mortar. Some walls that still exist depict a space-oriented figure that was as appropriate then as now — an ibis-headed god named Thoth who protected the moon and was believed to cruise space in a vessel.

Other types of adhesives were used to join wood and stone in ancient times. However, it was a long time before the ancients discovered a method for joining metals. Workers in the Bronze and Iron Ages began to solve the problems of forming, casting, and alloying metals. Welding metal surfaces was a problem that long puzzled metal-workers of that time period. Early metal-joining methods included such processes as forming a sand mold on top of a piece of metal and casting the desired shape directly on the base metal, so that both parts fused together, forming a single piece of metal, Figure 1-1. Another metal-joining method used in early years was to place two pieces of metal close together and pour molten metal between them. When the edges of the base metal melted, the flow of metal was then dammed and allowed to harden. Figure 1-2.

The Industrial Revolution, from 1750 to 1850, introduced a method of joining pieces of iron together known as forge welding or hammer welding. This process involved the use of a forge to heat the metal to a soft, plastic temperature. The ends of the iron were then placed together and hammered until fusion took place.

Forge welding remained as the primary welding method until Elihu Thomson, in the year 1886, developed the resistance welding technique. This technique provided a more reliable and faster way of joining metal than did previous methods.

As techniques were further developed, riveting was replaced in the United States and Europe by fusion welding to repair ships at the end of World War I. At that time the welding process was considered to be vital to military security. Repairs to the ships damaged during World War I were done in great secrecy. Even today some aspects of welding are closely guarded secrets.

Since the end of World War I, many welding methods have been developed for joining metals. These various welding methods are playing an important role in the expansion and production of the welding industry. Welding has become a dependable, efficient, and economical method for joining metal.

WELDING DEFINED

A weld is defined by the American Welding Society (AWS) as "a localized coalescence (the fusion or growing together of the grain structure of the materials being welded) of metals or nonmetals produced either by heating the materials to the required welding temperatures, with or without the application of pressure, or by the

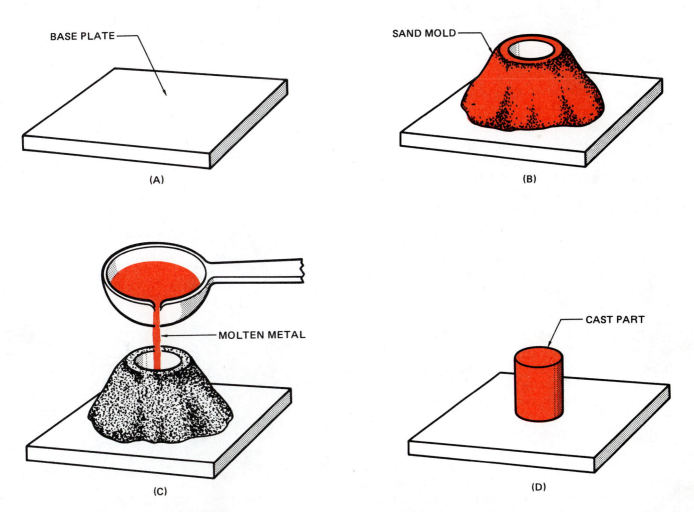

Figure 1-1 Direct casting: (A) base plate to have part cast on it, (B) sand molded into shape desired, (C) pouring hot metal into mold, and (D) part cast is now part of the base plate.

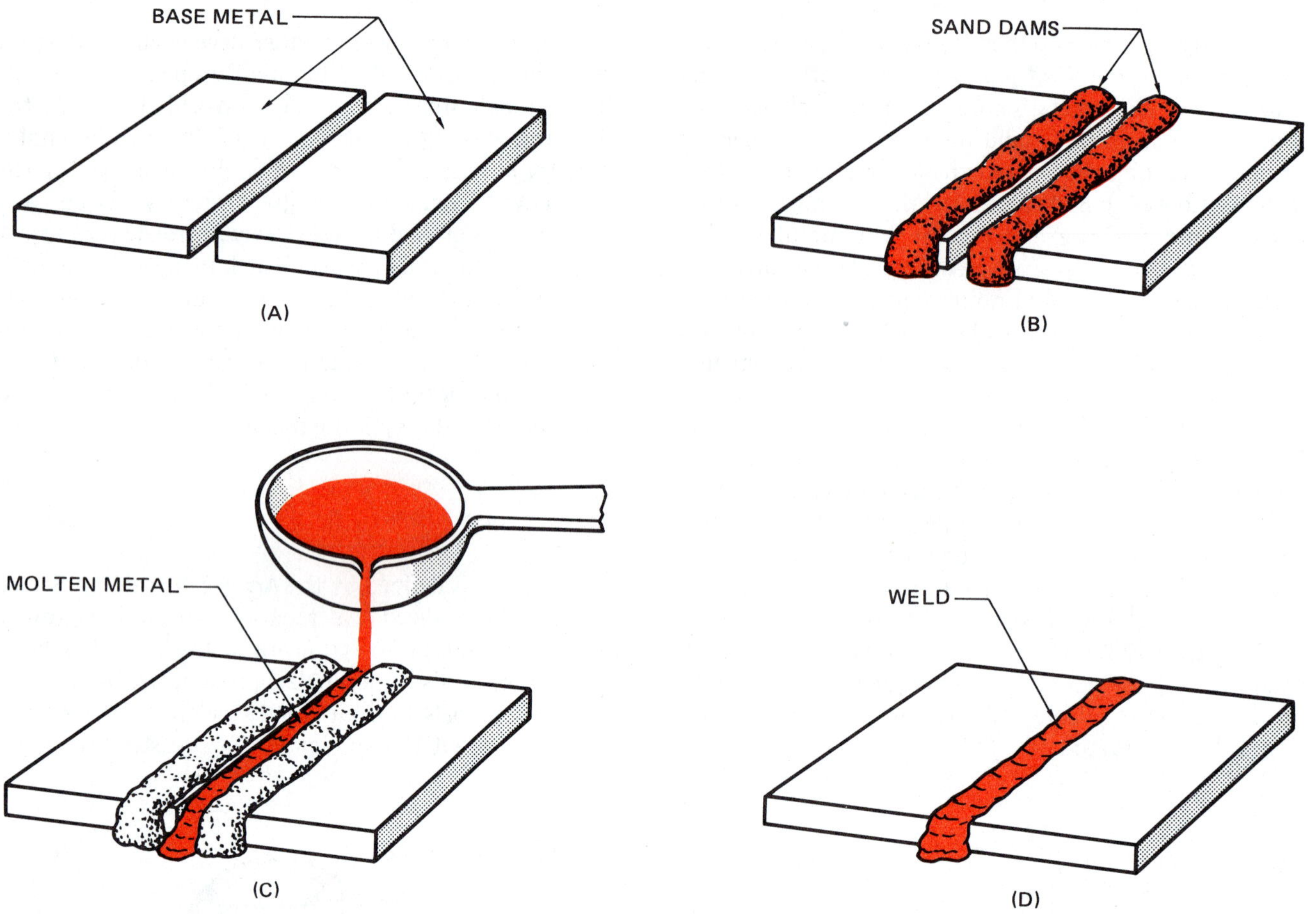

BASE METAL

(A)

SAND DAMS

(B)

MOLTEN METAL

(C)

WELD

(D)

Figure 1-2 Flow welding: (A) two pieces of metal plate, (B) sand dams to hold molten metal in place, (C) molten metal poured between metal plates, and (D) finished welded plate.

application of pressure alone, and with or without the use of filler materials." Welding is defined as "a joining process that produces coalescence of materials by heating them to the welding temperature, with or without the application of pressure or by the application of pressure alone, and with or without the use of filler metal." In less technical language, a weld is made when separate pieces of material to be joined combine and form one piece when heated to a temperature high enough to cause softening or melting and flow together. Pressure may or may not be used to force the pieces together. In some instances, pressure alone may be sufficient to force the separate pieces of material to combine and form one piece. Filler material is added when needed to form a completed weld in the joint. It is also important to note that the word *material* is used because today welds can be made from a growing list of materials such as plastic, glass, and ceramics.

USES OF WELDING

Modern welding techniques are employed in the construction of numerous products, **Figures 1-3** and **1-4**; ships, buildings, bridges, and recreational rides are fabricated by welding processes. Welding is often used to produce the machines that are used to manufacture new products.

Figure 1-3 Space shuttle being made ready for its launch into space. Notice the large welded support structure used to prepare the shuttle for launch. (Photo by Wendy Jeffus.)

(Courtesy of Miller Electric Co.)

(Courtesy of Amoco Corporation)

(Courtesy of AAI Corporation)

(Courtesy of Amoco Corporation)

(Courtesy of Caterpillar Inc.)

Figure 1-4 Welded joints are a critical component of structural integrity.

Welding has made it possible for airplane manufacturers to meet the design demands of strength-to-weight ratios for both commercial and military aircraft.

Man's exploring of space would not be possible without modern welding techniques. From the very beginning of early rockets to today's aerospace industry, welding has played an important role. The space shuttle's construction required the improvement of welding processes. Many of these improvements have helped improve our daily lives.

Many experiments aboard the space shuttle have concerned welding. In the coming years we plan to establish a permanent space station. This will require welding and welders to build such a large structure in the vacuum of space. **Figure 1-5** is a welding machine designed to be used in space. **Figure 1-6** shows a cosmonaut using the welder in open space. The specialized welder was developed at the E. O. Paton Electric Welding Institute in Kiev, Commonwealth of Independent States, the former Soviet Union. As the welding techniques are developed for this major project, we will see them being used here on Earth to improve our world.

Welding is used extensively in the manufacture of automobiles, farm equipment, home appliances, computer components, mining equipment, and earth-moving equipment. Railway equipment, furnaces, boilers, air-conditioning units, and hundreds of other products we use in our daily lives are also joined together by some type of welding process.

Items ranging from dental braces to telecommunication satellites are assembled by welding. Very little in our modern world is not produced using some type of this versatile process.

WELDING PROCESSES

The number of different welding processes has grown in recent years. These processes differ greatly in the manner in which heat, pressure, or both heat and pressure are applied, and in the type of equipment used. Table 1-1 lists various welding and allied processes. Some 67 welding processes are listed, requiring hammering, pressing, or rolling to effect the coalescence in the weld joint. Other methods bring the metal to a fluid state, and the edges flow together.

The most popular welding processes are oxyacetylene welding (OAW), shielded metal arc welding (SMAW), often called stick welding, gas tungsten arc welding (GTAW), gas metal arc welding (GMAW), flux cored arc welding (FCAW), and torch brazing (TB).

The use of regional terms by skilled workers is a common practice in all trade areas, including welding. As an example, oxyacetylene welding is one part of the larger group of processes known as oxyfuel gas welding (OFW). Some of the names used to refer to oxyacetylene welding (OAW) include *gas welding* and *torch welding*. Shielded metal arc welding (SMAW) is often called *stick welding*, *rod welding*, or just *welding*. As you begin your work career you will learn the various names used in your area, but you should always keep in mind and use the more formal terms whenever possible.

Oxyacetylene is the most commonly used fuel gas mixture. It is widely used for welding (OFW), cutting (OFC), and brazing (TB). The oxyfuel (OF) gas processes are the most versatile of the welding processes.

The equipment required is comparatively inexpensive, and the cost of operation is low, **Figure 1-7**.

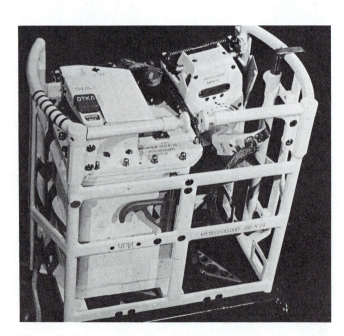

Figure 1-5 Machine designed to be used to weld in space. (Courtesy of E. O. Paton Electric Welding Institute, Commonwealth of Independent States, the former Soviet Union.)

Figure 1-6 A cosmonaut makes a weld outside a space ship. (Courtesy of E. O. Paton Electric Welding Institute, Commonwealth of Independent States, the former Soviet Union.)

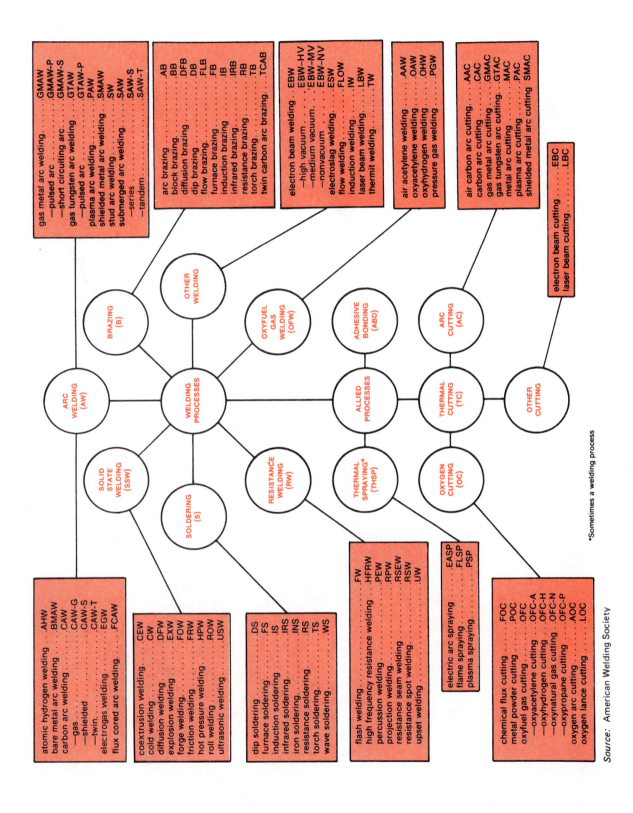

Table 1-1 Master Chart of Welding and Allied Processes.

Source: American Welding Society

*Sometimes a welding process

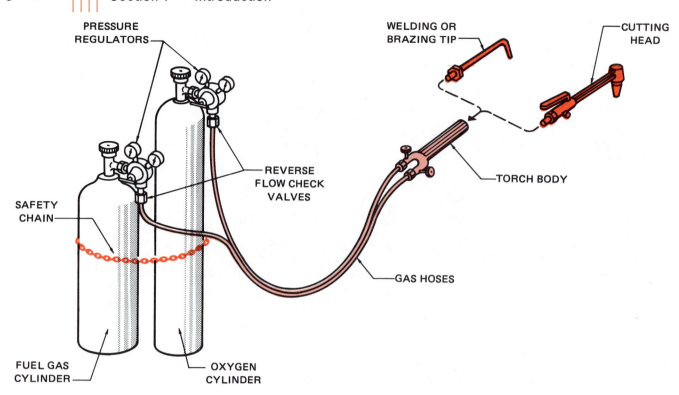

Figure 1-7 Oxyfuel welding and cutting equipment.

Shielded metal arc welding (SMAW) is the most common method of joining metal. High-quality welds can be made rapidly and with excellent uniformity. A variety of metal types and metal thicknesses can be joined with one machine, Figure 1-8.

Gas tungsten arc welding (GTAW) is easily performed on almost any metal. Its clean, high-quality welds often require little or no postweld finishing, Figure 1-9.

Gas metal arc welding (GMAW) is extremely fast and economical. This process is easily used for welding on thin-gauge metal as well as on heavy plate. The high welding rate and reduced postweld cleanup are making gas metal arc welding an outstanding welding process, Figure 1-10.

Flux cored arc welding (FCAW) uses the same type of equipment that is used for the gas metal arc welding process. A major advantage of this process is that with the addition of flux to the center of the filler wire it is often possible to make welds without the use of an external shielding gas. The introduction of smaller wire sizes and the elimination of the shielding gas from some welds has resulted in an increase in the use of FCAW process. Although slag must be cleaned from the welds after completion, the process's advantages of high quality, versatility, and welding speed offset this requirement.

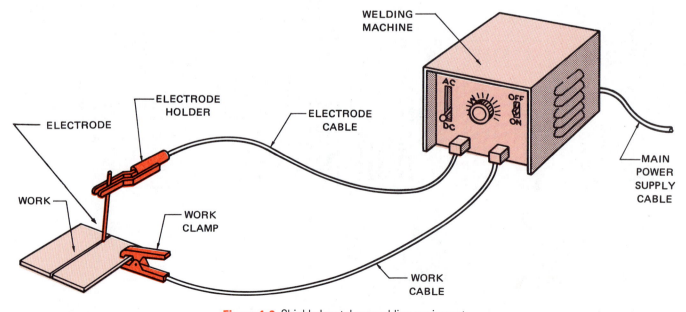

Figure 1-8 Shielded metal arc welding equipment.

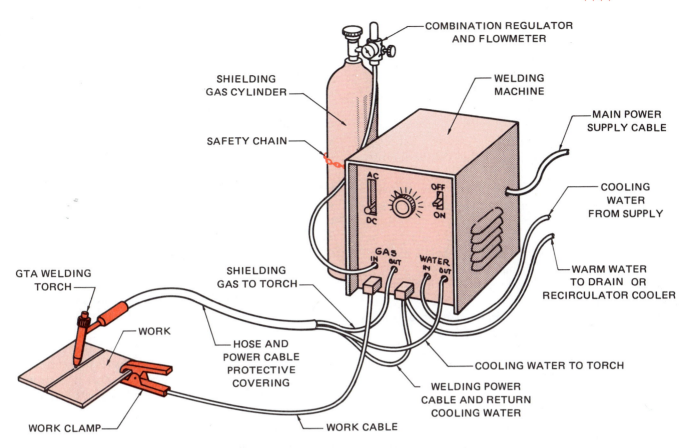

Figure 1-9 Gas tungsten arc welding equipment.

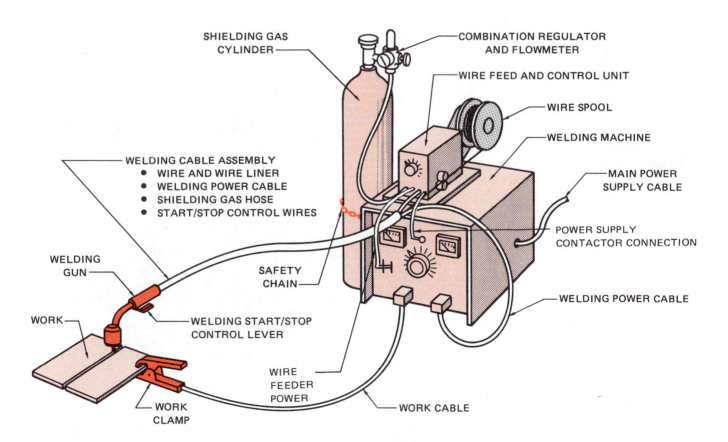

Figure 1-10 Gas metal arc welding equipment.

Selection of the Joining Process

The selection of the joining process for a particular job depends upon many factors. No one specific rule controls the welding process to be selected for a certain job. A few of the factors that must be considered when choosing a joining process include:

- Availability of equipment — The types, capacity, and condition of equipment that can be used to make the welds.
- Repetitiveness of the operation — How many of the welds will be required to complete the job, and are they all the same?
- Quality requirements — Is this weld going to be used on a piece of furniture, to repair a piece of equipment, or to join a pipeline?
- Location of work — Will the weld be in a shop or on a remote job site?
- Materials to be joined — Are the parts made out of a standard metal or some exotic alloy?
- Appearance of the finished product — Will this be a weldment that is only needed to test an idea, or will it be a permanent structure?
- Size of the parts to be joined — Are the parts small, large, or different sizes, and can they be moved or must they be welded in place?
- Time available for work — Is this a rush job needing a fast repair, or is there time to allow for pre- and postweld cleanup?
- Skill or experience of workers — Do the welders have the ability to do the job?
- Cost of materials — Will the weldment be worth the expense of special equipment materials or finishing time?
- Code or specification requirements — Often the selection of the process is dictated by the governing agency, codes, or standards.

The welding engineer and/or the welder must not only decide on the welding process but must also select the method of applying it. The following methods are used to perform welding, cutting, or brazing operations.

- Manual — The welder is required to manipulate the entire process.
- Semiautomatic — The filler metal is added automatically, and all other manipulation is done manually by the welder.
- Machine — Operations are done mechanically under the observation and correction of a welding operator.
- Automatic — Operations are performed repeatedly by a machine that has been programmed to do an entire operation without interaction of the operator.
- Automated — Operations are performed repeatedly by a robot or other machine that is programmed flexibly to do a variety of processes.

OCCUPATIONAL OPPORTUNITIES IN WELDING

The American welding industry has contributed to the widespread growth of the welding and allied processes. Without welding much of what we use on a daily basis could not be manufactured. The list of these products grows every day, thus increasing the number of jobs for people with welding skills. The need to fill these well-paying jobs is not concentrated in major metropolitan areas, but is found throughout the country and the world. Because of the diverse nature of the welding industry, the exact job duties of each skill area will vary. The following are general descriptions of the job classifications used in our profession; specific tasks may vary from one location to another.

Welders perform the actual welding. They are the skilled craftspeople who, through their own labor, produce the welds on a variety of complex products, **Figure 1-11**.

Tack welders, also skilled workers, often help the welder by making small welds to hold parts in place. The tack weld must be correctly applied so that it is strong enough to hold the assembly and still not interfere with the finished welding.

Welding operators, often skilled welders, operate machines or automatic equipment used to make welds.

Welders' helpers are employed in some welding shops to clean slag from the welds and help move and position weldments for the welder.

Welder assemblers, or **welder fitters**, position all the parts in their proper places and make ready for the tack welders. These skilled workers must be able to interpret blueprints and welding procedures. They also must have

Figure 1-11 A welder stitches new leaves to the Swiss Family Robinson Tree House at Walt Disney World. The renovation of this attraction required thousands of new leaves to be welded into place. (Photo by Wendy Jeffus.)

knowledge of the effects of contraction and expansion of the various types of metals.

Welding inspectors are often required to hold a special certification such as the one supervised by the American Welding Society known as Certified Welding Inspector (CWI). To become a CWI, candidates must pass a test covering the welding process, blueprint reading, weld symbols, metallurgy, codes and standards, and inspection techniques. Vision screening is also required on a regular basis once the technical skills have been demonstrated.

Welding shop supervisors may or may not weld on a regular basis, depending on the size of the shop. In addition to their welding skills, they must demonstrate good management skills by effectively planning jobs and assigning workers.

Welding salespersons may be employed by supply houses or equipment manufacturers. These jobs require a broad understanding of the welding process as well as good marketing skills. Good salespersons are able to provide technical information about their products in order to convince customers to make a purchase.

Welding shop owners are often welders who have a high degree of skill and knowledge of small-business management and prefer to operate their own businesses. These individuals may specialize in one field, such as hardfacing, repair and maintenance, or specialty fabrications, or they may operate as subcontractors of manufactured items. A welding business can be as small as one individual, one truck, and one portable welder or as large as a multimillion-dollar operation employing hundreds of workers.

Welding engineers design, specify, and oversee the construction of complex weldments. The welding engineer may work with other engineers in areas such as mechanics, electronics, chemicals, or civil engineering

in the process of bringing a new building, ship, aircraft, or product into existence. The welding engineer is required to know all of the welding process and metallurgy as well as have good math, reading, communication, and design skills. This person usually has an advanced college degree and possesses a professional certification.

In many industries, the welder, welding operator, and tack welder must be able to pass a performance test to specific code or standard.

The highest paid welders are those who have the education and skills to read blueprints and do the required work to produce a weldment to strict specifications.

Large industrial concerns employ workers who serve as support for the welders. These engineers and technicians must have knowledge of chemistry, physics, metallurgy, electricity, and mathematics. Engineers are responsible for research, design, development, and fabrication of a project. Technicians work as part of the engineering staff. These individuals may oversee the actual work for the engineer by providing the engineer with progress reports as well as chemical, physical, and mechanical test results. Technicians may also require engineers to build prototypes for testing and evaluation.

Another group of workers employed by industry does layouts or makes templates. These individuals have had drafting experience and have a knowledge of such operations as punching, cutting, shearing, twisting, and forming, among others. The layout is generally done directly on the material. A template is used for repetitive layouts and is made from sheet metal or other suitable materials.

The flame-cutting process is closely related to welding. Some operators use hand-held torches, and others are skilled operators of oxyfuel cutting machines. These machines range from simple mechanical devices to highly sophisticated, tape controlled, multiple-head machines that are operated by specialists, **Figure 1-12.**

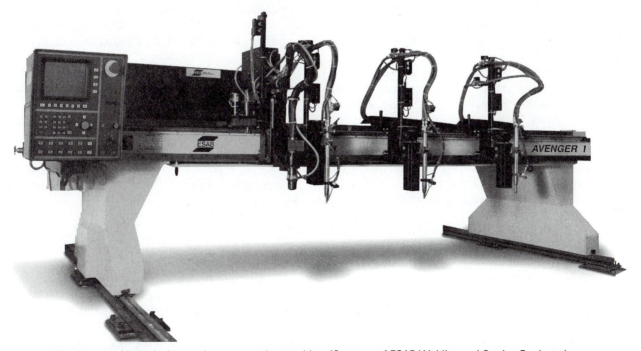

Figure 1-12 Numerical control oxygen cutting machine. (Courtesy of ESAB Welding and Cutting Products.)

TRAINING FOR WELDING OCCUPATIONS

Generally, several months of training are required to learn to weld. To become a skilled welder, both welding school and on-the-job experience are required. Because of the diverse nature of the welding industry, no single list of skills can be used to meet every job's requirement. However, there are specific skills that are required of most entry level welders. This text covers those skill requirements.

In addition to welding skills an entry level welder must possess workplace skills. Workplace skills include a proficiency in reading, writing, math, communication, and science as well as good work habits and an acceptance of close supervision. Some welding jobs may also require a theoretical knowledge of welding, blueprint reading, welding symbols, metal properties, and electricity. A few of the jobs that require less skill can be learned after a few months of on-the-job training. However, the fabrication of certain alloys requires knowledge of metallurgical properties as well as the development of a greater skill in cutting and welding them.

Robotics and computer-aided manufacturing (CAM) both require more than a basic understanding of the welding process; they require that the student be computer literate.

A young person planning a career as a welder needs good eyesight, manual dexterity, hand and eye coordination, and understanding of welding technology. For entry into manual welding jobs, employers prefer to hire young people who have high school or vocational training in welding processes. Courses in drafting, blueprint reading, mathematics, and physics are also valuable.

Beginners in welding who have no training often start in manual welding production jobs that require minimum skill. Occasionally, they first work as helpers and are later moved into welding jobs. General helpers, if they show promise, may be given a chance to become welders by serving as helpers to experienced welders.

A formal apprenticeship is usually not required for general welders. A number of large companies have welding apprenticeship programs. The military, at several of its installations, has programs in welding.

Skill and technical knowledge requirements are higher in some industries. In the fields of atomic energy, aerospace, and pressure vessel construction, high standards for welders must be met to ensure that weldments will withstand the critical forces that they will be subjected to in use.

After two years of training at a vocational school or technical institute, the skilled welder may qualify as a technician. Technicians are generally involved in the interpretation of engineers' plans and instructions. Employment of welders is expected to increase rapidly through the 1990s and into the next century as a result of the wider use of various welding processes, including the use of robots.

Many more skilled welders will be needed for maintenance and repair work in the expanding metalworking industries. The number of welders in production work is expected to increase in plants manufacturing sheet metal products, pressure vessels, boilers, railroads, storage tanks, air-conditioning equipment, ships, and in the field of energy exploration and energy resources. The construction industry will need an ever-increasing number of good welders as the use of welded steel buildings grows.

Employment prospects for welding operators are expected to continue to be favorable because of the increased use of machine, automatic, and automated welding in the manufacture of aircraft, missiles, railroad cars, automobiles, and numerous other products.

Before being assigned a job where service requirements of the weld are critical, welders usually must pass a certification test given by an employer.

In addition, some localities require welders to obtain a license for certain types of outside construction.

After a welder, welding operator, or tack welder has received a certification or qualification by passing a standardized test, he or she is only allowed to make welds covered by that specific test. The welding certification is very restrictive; it allows a welder to perform only code welds covered by that test. Certifications are usually good for a maximum of six months unless a welder is doing code-quality welds routinely. As a student, you should check into the acceptance of a welding qualification test before investing time and possibly money in the test.

The American Welding Society (AWS) has developed three levels of certification for welders. The first level, Entry Level Welder, is for the beginning welder, and Level II and Level III are for the more skilled welders. These certifications are gaining widespread acceptance by industry. Certification allows welders to prove their skills on a standard test.

For additional information about the skills required for various welding job classifications, refer to Table 1-2.

Each year the Vocational Clubs of America, VICA, sponsors a series of welding skill contests for its student members. The contests allow students from around the nation to test their skills against other students. Local contests lead to state or regional contests, then to national, and on to the International Skill Olympics, Figure 1-13. VICA, like other professional organizations, helps to promote quality in the welding profession.

EXPERIMENTS AND PRACTICES

A number of the chapters in this book contain both experiments and practices. These are intended to help you develop your welding knowledge and skills.

The experiments are designed to allow you to see what effect changes in the process settings, operation,

Recommended Instructional Chapters for Welding Occupations*

Category	Skill	Chapter Number
GENERAL	Introduction	1
	Safety	2
OXYFUEL GAS	Oxyfuel Gas	3|4
	Gases	3
	Filler Metals	3
	Equipment and Setup	4
	Plate Welding Flat Position	5
	Plate Welding All Positions	5
	Pipe Welding Flat Position	6
	Pipe Welding All Positions	6
	Soldering, Brazing, and Braze Welding	7
	Filler Metals and Fluxes	7
	Soldering	7
	Brazing and Braze Welding	7
	Oxygen Cutting	8
	Equipment and Setup	8
	Plate and Pipe Oxygen Cutting	8
SHIELDED METAL ARC	Shielded Metal Arc Welding	9/10
	Equipment and Setup	9
	Filler Metals	10
	Plate Welding Flat Position	11
	Plate Welding All Positions	12
	Pipe Welding Flat Position	13
	Pipe Welding All Positions	13
GAS SHIELDED ARC	Gas Tungsten Arc Welding	14
	Equipment and Setup	14
	Filler Metal	14
	Plate Welding	15
	Pipe Welding	16
	Gas Metal Arc Welding	17
	Equipment and Setup	17
	Filler Metal	17
	Plate Welding	18
RELATED INFORMATION	Automation and Robotics	19
	Special Welding Processes	20
	Weldability of Metals	21
	Testing and Inspection	22
	Welding Metallurgy	23
	Welding Symbols	24
	Joint Designs	24
	Special Knowledge or Skill	

Welding Occupations (rows) and associated Special Knowledge or Skill:

Occupation	Special Knowledge or Skill
Tack Welder	
Production SMA Welder	
Production GTA Welder	
Production GMA Welder	
Production Machine Operator	Numerical Control
Construction Welder	
Welder Assembler	
Welder Fitter	
Maintenance Welder	
Owner/Operator Welder	
Welder Craftsperson	Surfacing
Welding Foreman	Business
Welding Technician	Leadership
Welding Engineer	
Welding Sales	Business
RELATED OCCUPATIONS	
Auto Mechanic	
Auto Body	
Machinist	
Air Conditioning/Heating	
Sheet Metal	
General Maintenance	
Farm/Agricultural	
Art/Sculpture	Surfacing

Legend

Principles
- □ Essential
- / General knowledge of
- • Nice to know
- ■ Nonessential

Applications
- □ Pass appropriate weld test
- / Make good welds
- • Nice to know how to
- ■ Nonessential

*Welding occupational job titles and specific responsibilities vary greatly. For a description of these job titles, refer to the glossary in this text.

Table 1-2 Recommended Skills for Various Welding Occupations.

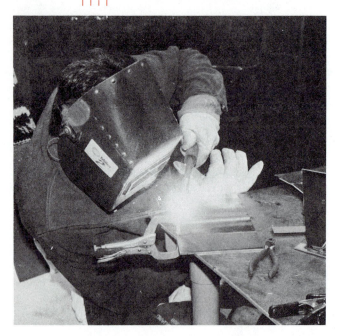

Figure 1-13 A welding student takes part in VICA's United States Skill Olympics. Highly skilled student welders from around the nation compete at the national competition. The winners can then participate in the World Skill Olympics.

Metric Conversions Approximations	
1/4 inch	= 6 mm
1/2 inch	= 13 mm
3/4 inch	= 18 mm
1 inch	= 25 mm
2 inches	= 50 mm
1/2 gal	= 2 L
1 gal	= 4 L
1 lb	= 1/2 K
2 lb	= 1 K
1 psig	= 7 kPa
1°F	= 2°C

Table 1-3 By using an approximation for converting standard units to metric it is possible to quickly have an idea of how large or heavy an object is in the other units. For estimating it is not necessary to be concerned with the exact conversions.

or techniques have on the type of weld produced. When you do an experiment, you should observe and possibly take notes of how the change affected the weld. Often as you make a weld it will be necessary for you to make changes in your equipment settings or your technique

in order to ensure that you are making an acceptable weld. By watching what happens when you make the changes in the welding shop, you will be better prepared to decide on changes required to make good welds on the job.

The practices are designed to build your welding skills. Each practice tells you in detail what equipment, supplies, and tools you will need as you develop the specific skill. In most chapters the practices start off easy and become progressively harder. Welding is a skill that requires that you develop in stages from the basic to the more complex.

Each practice gives the evaluation or acceptable limits for the weld. All welds have some discontinuities; but if they are within the acceptable limits, they are not defects. As you practice your welding, keep in mind the acceptable limits so that you can progress to the next level when you have mastered the process and weld you are working on.

METRIC UNITS

Both standard and metric (SI) units are given in this text. The SI units are in brackets ( ) following the standard unit. When nonspecific values are used—for example, "set the gauge at 2 psig" where 2 is an approximate value—the SI units have been rounded off to the nearest whole number. Round-off occurs in these cases to agree with the standard value and because whole numbers are easier to work with. SI units are not rounded off only when the standard unit is an exact measurement.

Often students have difficulty understanding metric units because exact conversions are used even when the standard measurement was an approximation. Rounding off the metric units makes understanding the metric system much easier, Table 1-3. By using this approximation method, you can make most standard-to-metric conversions in your head without needing to use a calculator.

Once you have learned to use approximations for metric, you will find it easier to make exact conversions whenever necessary. Conversions must be exact in the shop when a part is dimensioned with one system's units and the other system must be used to fabricate the part. For that reason you must be able to make those conversions. Table 1-4 and Table 1-5 are set up to be used with or without the aid of a calculator. Many calculators today have built-in standard–metric conversions. It is a good idea to know how to make these conversions with and without these aids, of course. Practice making such conversions whenever the opportunity arises.

Table of Conversions

TEMPERATURE
Units

°F (each 1° change)	=	0.555°C (change)
°C (each 1° change)	=	1.8°F (change)
32°F (ice freezing)	=	0°Celsius
212°F (boiling water)	=	100°Celsius
–460°F (absolute zero)	=	0°Rankine
–273°C (absolute zero)	=	0°Kelvin

Conversions

°F to °C _____ °F – 32 = _____ × .555 = _____ °C
°C to °F _____ °C × 1.8 = _____ + 32 = _____ °F

LINEAR MEASUREMENT
Units

1 inch	=	25.4 millimeters
1 inch	=	2.54 centimeters
1 millimeter	=	0.0394 inch
1 centimeter	=	0.3937 inch
12 inches	=	1 foot
3 feet	=	1 yard
5280 feet	=	1 mile
10 millimeters	=	1 centimeter
10 centimeters	=	1 decimeter
10 decimeters	=	1 meter
1,000 meters	=	1 kilometer

Conversions

in. to mm _____ in. × 25.4 = _____ mm
in. to cm _____ in. × 2.54 = _____ cm
ft to mm _____ ft × 304.8 = _____ mm
ft to m _____ ft × 0.3048 = _____ m
mm to in. _____ mm × 0.0394 = _____ in.
cm to in. _____ cm × 0.3937 = _____ in.
mm to ft _____ mm × 0.00328 = _____ ft
m to ft _____ m × 32.8 = _____ ft

AREA MEASUREMENT
Units

1 sq in.	=	0.0069 sq ft
1 sq ft	=	144 sq in.
1 sq ft	=	0.111 sq yd
1 sq yd	=	9 sq ft
1 sq in.	=	645.16 sq mm
1 sq mm	=	0.00155 sq in.
1 sq cm	=	100 sq mm
1 sq m	=	1,000 sq cm

Conversions

sq in. to sq mm _____ sq in. × 645.16 = _____ sq mm
sq mm to sq in. _____ sq mm × 0.00155 = _____ sq in.

VOLUME MEASUREMENT
Units

1 cu in.	=	0.000578 cu ft
1 cu ft	=	1728 cu in.
1 cu ft	=	0.03704 cu yd
1 cu ft	=	28.32 L
1 cu ft	=	7.48 gal (U.S.)
1 gal (U.S.)	=	3.737 L
1 cu yd	=	27 cu ft
1 gal	=	0.1336 cu ft
1 cu in.	=	16.39 cu cm
1 L	=	1,000 cu cm
1 L	=	61.02 cu in.
1 L	=	0.03531 cu ft
1 L	=	0.2642 gal (U.S.)
1 cu yd	=	0.769 cu m
1 cu m	=	1.3 cu yd

Conversions

cu in. to L _____ cu in. × 0.01638 = _____ L
L to cu in. _____ L × 61.02 = _____ cu in.
cu ft to L _____ cu ft × 28.32 = _____ L
L to cu ft _____ L × 0.03531 = _____ cu ft
L to gal _____ L × 0.2642 = _____ gal
gal to L _____ gal × 3.737 = _____ L

WEIGHT (MASS) MEASUREMENT
Units

1 oz	=	0.0625 lb
1 lb	=	16 oz
1 oz	=	28.35 g
1 g	=	0.03527 oz
1 lb	=	0.0005 ton
1 ton	=	2,000 lb
1 oz	=	0.283 kg
1 lb	=	0.4535 kg
1 kg	=	35.27 oz
1 kg	=	2.205 lb
1 kg	=	1,000 g

Conversions

lb to kg _____ lb × 0.4535 = _____ kg
kg to lb _____ kg × 2.205 = _____ lb
oz to g _____ oz × 0.03527 = _____ g
g to oz _____ g × 28.35 = _____ oz

PRESSURE AND FORCE MEASUREMENTS
Units

1 psig	=	6.8948 kPa
1 kPa	=	0.145 psig
1 psig	=	0.000703 kg/sq mm
1 kg/sq mm	=	6894 psig
1 lb (force)	=	4.448 N
1 N (force)	=	0.2248 lb

Conversions

psig to kPa _____ psig × 6.8948 = _____ kPa
kPa to psig _____ kPa × 0.145 = _____ psig
lb to N _____ lb × 4.448 = _____ N
N to lb _____ N × 0.2248 = _____ psig

VELOCITY MEASUREMENTS
Units

1 in./sec	=	0.0833 ft/sec
1 ft/sec	=	12 in./sec
1 ft/min	=	720 in./sec
1 in./sec	=	0.4233 mm/sec
1 mm/sec	=	2.362 in./sec
1 cfm	=	0.4719 L/min
1 L/min	=	2.119 cfm

Conversions

ft/min to in./sec _____ ft/min × 720 = _____ in./sec
in./min to mm/sec _____ in./min × .4233 = _____ mm/sec
mm/sec. to in./min _____ mm/sec × 2.362 = _____ in./min
cfm to L/min _____ cfm × 0.4719 = _____ L/min
L/min to cfm _____ L/min × 2.119 = _____ cfm

Table 1-4 Table of Conversions: U.S. Customary (Standard) Units and Metric Units (SI).

Abbreviations and Symbols

U.S. Customary (Standard) Units					
°F	=	degrees Fahrenheit			
°F	=	degrees Rankine			
	=	degrees absolute F			
lb	=	pound			
psi	=	pounds per square inch			
	=	lb per sq in.			
psia	=	pounds per square inch absolute			
	=	psi + atmospheric pressure			
in.	=	inches	=	i	=	"
ft	=	foot or feet	=	f	=	'
sq in.	=	square inch	=	in.		
sq ft	=	square foot	=	ft		
cu in.	=	cubic inch	=	in.		
cu ft	=	cubic foot	=	ft		
ft-lb	=	foot-pound			
ton	=	ton of refrigeration effect			
qt	=	quart			

Metric Units (SI)		
°C	=	degrees Celsius
K	=	kelvin
mm	=	millimeter
cm	=	centimeter
cm²	=	centimeter squared

cm³	=	centimeter cubed
dm	=	decimeter
dm²	=	decimeter squared
dm³	=	decimeter cubed
m	=	meter
m²	=	meter squared
m³	=	meter cubed
L	=	liter
g	=	gram
kg	=	kilogram
J	=	joule
kJ	=	kilojoule
N	=	newton
Pa	=	pascal
kPa	=	kilopascal
W	=	watt
kW	=	kilowatt
MW	=	megawatt

Miscellaneous Abbreviations					
P	=	pressure	sec	=	seconds
h	=	hours	r	=	radius of circle
D	=	diameter	π	=	3.1416 (a constant
A	=	area			used in determining
V	=	volume			the area of a circle)
∞	=	infinity			

Table 1-5 Abbreviations and Symbols.

Even in Its Infancy, Welding Provided the Solution

Figure 1

Welding was not fully developed by 1910. However, it was called upon at that time to solve a problem that could not be solved by any other means. The test resulted in the first welded pipeline in the U.S. The measure of its success is that the pipeline is still used for hydroelectric power generation.

Boulder Hydroelectric Generating Station of the Public Service Company of Colorado started generating electricity in 1910. This happened only after certain engineering ideas that had only been theories were put into practice.

First, Barker Dam had to be constructed to provide a source of power to turn the turbines. The 175-ft-high concrete dam has a peak length of 720 ft and a storage capacity of 11,687 acre-ft. It took eighteen months to build the dam with much of the material hauled by six- and eight-horse teams over rough terrain.

Water is released from the reservoir into a 36-in. concrete gravity line, 11.7 miles in length. There it flows into a reservoir for use by the power plant. This reservoir has a capacity of 165 acre-ft and is located two miles from the power plant at an elevation of 1,835 ft above the plant.

It was this high elevation above the plant that presented a pipeline problem. The pressure line that carries water from the reservoir is 9,340 feet long (Figure 1), but in that stretch, it drops 1,828 ft (Figure 2). The sections of pipe were originally riveted together (Figure 3), but with the drop in elevation, tremendous pressure was

Figure 2

Figure 4

Figure 3

Figure 5

placed on the line, causing some spectacular leaks (**Figure 4**). At 800 psi the hydro plant had the highest pressure of any plant in the world at that time.

Welding was called upon to solve the problem. The oxyacetylene process in use at the time was chosen to weld over the fastened joints (**Figure 5**). Although not much is known about the procedures or parameters used, it is known that peening was used for stress relieving. The welds held (**Figure 6**), and the electric power generated from the plant was used to develop the area around Boulder, Colorado. To this day, the company is still generating power for the surrounding area.

(Adapted from an article in the *Welding Journal*. Courtesy of the American Welding Society, Inc. Information for article supplied by B. Barr, General Cable Company, Westminster, Colorado)

Figure 6

REVIEW

1. Explain how to forge weld.

2. What welding process is Elihu Thomason credited with developing in 1886?

3. During World War I, what process replaced riveting for ship repair?

4. Welding has become a _____, _____, and _____ method of joining metal.

5. What do we call the localized growing together of the grain structure during a weld?

6. Welding can be used to join both _____ and _____ .

7. Some welding processes require both _____ and _____ to make a weld.

8. List six items that use welding in their construction.

9. Why have welding experiments been performed aboard the space shuttle?

10. What three things differ greatly from one welding process to another?

11. Which gases are most commonly used for the OFW process?

12. What is the technically correct name for *gas welding*?

13. Which welding process is the most commonly used to join metal?

14. What is the technically correct name for *stick welding*?

15. GTAW is the abbreviation for which process?

16. What is the ideal process for high welding rates on thin-gauge metal?

17. Flux inside the welding wire gives which process its name?

18. List six items that may be considered in selecting a welding process.

19. Which method of welding application requires a welder to manipulate the entire process?

20. Which method of welding requires the welder to control everything except the adding of filler metal?

21. Which welding process is repeatedly performed by a machine that has been programmed?

22. A _____ works with engineers to produce prototypes for testing.

23. A _____ places parts together in their proper position for the tack welder.

24. A _____ is a piece of sheet metal cut to the shape of a part so that it may be repetitively laid out.

25. What are the approximate standard units for the following SI values?
    a. 13 mm   c. 100°C
    b. 4 L   d. 4 K

26. What are the exact standard units for the following SI values?
    a. 13 mm   c. 100°C
    b. 4 L   d. 4 K

Chapter 2

SAFETY IN WELDING

OBJECTIVES

After completing this chapter, the student should be able to

- describe the type of protection that should be worn for welding.
- describe the proper method of handling, storing, and setting up cylinders.
- discuss the proper way to ventilate a welding area.
- explain how to avoid electric shock.
- describe how to avoid possible health hazards for welding.
- explain how to prevent fires in the welding shop.

KEY TERMS

ultraviolet light
infrared light
visible light
flash burn
safety glasses
goggles
full face shield
flash glasses
welding helmet
earmuffs
earplugs
ventilation
material specification data sheet (MSDS)
natural ventilation

forced ventilation
exhaust pickups
electric shock
electrical resistance
warning label
electrical ground
valve protection cap
acetylene
acetone
Type A fire extinguisher
Type B fire extinguisher
Type C fire extinguisher
Type D fire extinguisher

INTRODUCTION

Accident prevention is the main intent of this chapter. The safety information included in this text is intended as a guide. There is no substitute for caution and common sense. A safe job is no accident; it takes work to make the job safe. Each person working must do what it takes to keep the job safe.

Welding, like other heavy industrial jobs, has a number of potential safety hazards. These hazards need not result in anyone being injured. Learning to work safely with these hazards is as important as learning to be a skilled welder.

You must approach new jobs with your safety in mind. Your safety is your own responsibility, and you must take that responsibility. It's not feasible to anticipate all of the possible dangers in every welding job. There may be some dangers not covered in this text. You can get specific safety information from your supervisor, the shop safety officer or from other workers.

BURNS

Burns are one of the most common and painful injuries that occur in the welding shop. Burns can be caused by ultraviolet light rays as well as by contact with hot welding material. The chance of infection is high with burns because of the dead tissue. It is important that all burns receive proper medical treatment to reduce the chance of infection. Burns are divided into three classifications, depending upon the degree of severity. The three classifications include first-degree, second-degree, and third-degree burns.

First-degree Burns

First-degree burns have occurred when the surface of the skin is reddish in color, tender, and painful and does not involve any broken skin. The first step in treating a first-degree burn is to immediately put the burned area under cold water (not iced) or apply cold water compresses (clean towel, washcloth, or handkerchief soaked in cold water) until the pain decreases. Then cover the area with sterile bandages or a clean cloth. Do not apply butter or grease. Do not apply any other home remedies or medications without a doctor's recommendation. See **Figure 2-1**.

Second-degree Burns

Second-degree burns have occurred when the surface of the skin is severely damaged, resulting in the formation of blisters and possible breaks in the skin. Again, the most important first step in treating a second-degree burn is to put the area under cold water (not iced) or apply cold water compresses until the pain decreases.

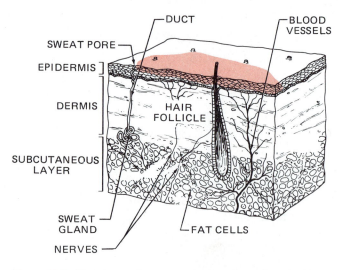

Figure 2-1 First-degree burn — only the skin surface (epidermis) is affected.

Gently pat the area dry with a clean towel, and cover the area with a sterile bandage or clean cloth to prevent infection. Seek medical attention. If the burns are around the mouth or nose, or involve singed nasal hair, breathing problems may develop. Do not apply ointments, sprays, antiseptics, or home remedies. Note: in an emergency any cold liquid you drink— for example, water, cold tea, soft drinks, or milk shake—can be poured on a burn. The purpose is to reduce the skin temperature as quickly as possible to reduce tissue damage. See **Figure 2-2**.

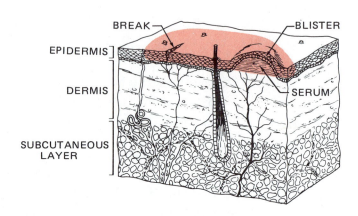

Figure 2-2 Second-degree burn — the epidermal layer is damaged, forming blisters or shallow breaks.

Third-degree Burns

Third-degree burns have occurred when the surface of the skin and possibly the tissue below the skin appear white or charred. Initially, little pain is present because nerve endings have been destroyed. Do not remove any clothes that are stuck to the burn. Do not put ice water or ice on the burns; this could intensify the shock reaction. Do not apply ointments, sprays, antiseptics, or home remedies to burns. If the victim is on fire, smother the flames with a blanket, rug, or jacket. Breathing difficulties are common with burns around the face, neck, and mouth; be sure that the victim is breathing. Place a cold cloth or cool (not iced) water on burns of the face, hands, or feet to cool the burned areas. Cover the burned area with thick, sterile, nonfluffy dressings. Call for an ambulance immediately; people with even small third-degree burns need to consult a doctor. See **Figure 2-3**.

Burns Caused by Light

Some types of light can cause burns. The three types of light include ultraviolet, infrared, and visible. Ultraviolet and infrared are not visible to the unaided human eye. They are the types of light that can cause burns. During welding, one or more of the three types of light may be present. Arc welding produces all three types of light, but gas welding produces visible and infrared light only.

SERUM

EPIDERMIS

DERMIS

SUBCUTANEOUS LAYER

Figure 2-3   Third-degree burn — the epidermis, dermis, and subcutaneous layers of tissue are destroyed.

The light from the welding process can be reflected from walls, ceilings, floors, or any other large surface. This reflected light is as dangerous as the direct welding light. To reduce the danger from reflected light, the welding area, if possible, should be painted flat black. Flat black will reduce the reflected light by absorbing more of it than any other color. When the welding is to be done on a job site, in a large shop or other area that cannot be painted, weld curtains can be placed to absorb the welding light, **Figure 2-4.** These special portable welding curtains may be either transparent or opaque. Transparent welding curtains are made of a special high-temperature, flame-resistant plastic that will prevent the harmful light from passing through.

Figure 2-4   Portable welding curtains. (Courtesy Frommelt Safety Products.)

Welding curtains must always be used to protect other workers in an area that might be exposed to the welding light.

Ultraviolet Light. Ultraviolet light waves are the most dangerous. They can cause first-degree and second-degree burns to a welder's eyes or to any exposed skin. Because a welder cannot see or feel ultraviolet light while being exposed to it, the welder must stay protected when in the area of any of the arc welding processes. The closer a welder is to the arc and the higher the current, the quicker a burn may occur. The ultraviolet light is so intense during some welding processes that a welder's eyes can receive a flash burn within seconds, and the skin can be burned within minutes. Ultraviolet light can pass through loosely woven clothing, thin clothing, light-colored clothing, and damaged or poorly maintained arc welding helmets.

Infrared Light. Infrared light is the light wave that is felt as heat. Although infrared light can cause burns, a person will immediately feel this type of light. Therefore, burns can easily be avoided.

Visible Light. Visible light is the light that we see. It is produced in varying quantities and colors during welding. Too much visible light may cause temporary night blindness (poor eyesight under low light levels). Too little visible light may cause eye strain, but visible light is not hazardous.

Whether burns are caused by ultraviolet light or hot material, they can be avoided if proper clothing and other protection are worn.

EYE AND EAR PROTECTION

Face and Eye Protection

Eye protection must be worn in the shop at all times. Eye protection can be safety glasses, Figure 2-5,

Figure 2-5 Safety glasses with side shields.

goggles, or a full face shield. To give better protection when working in brightly lit areas or outdoors, some welders wear flash glasses, which are special, lightly tinted, safety glasses. These safety glasses provide both protection from flying debris and reflected light.

Suitable eye protection is important because eye damage caused by excessive exposure to arc light is not noticed. Welding light damage occurs often without warning, like a sunburn's effect that is felt the following day. Therefore welders must take appropriate precautions in selecting filters or goggles that are suitable for the process being used, Table 2-1. Selecting the correct shade lens is also important, because both extremes of too light or too dark can cause eye strain. New welders often select too dark a lens, assuming it will give them better protection, but this results in eye strain in the same manner as if they were trying to read in a poorly lit room. In reality, any approved arc welding lenses will filter out the harmful ultraviolet light. Select a lens that lets you see comfortably. At the very least, the welder's eyes must not be strained by excessive glare from the arc.

Ultraviolet light can burn the eye in two ways. This light can injure either the white of the eye or the retina, which is the back of the eye. Burns on the retina are not painful but may cause some loss of eyesight. The whites of the eyes can also be burned by ultraviolet light, Figure 2-6. The whites of the eyes are very sensitive, and burns are very painful. The eyes are easily infected because, as with any burn, many cells are killed. These dead cells in the moist environment of the eyes will promote the growth of bacteria that cause infection. When the eye is burned, it feels as though there is something in the eye. Without a professional examination, however, it is impossible to tell if there is something in the eye. Because there may be something in the eye and because of the high risk of infection, home remedies or other medicines should never be used for eye burns. Anytime you receive an eye injury you should see a doctor.

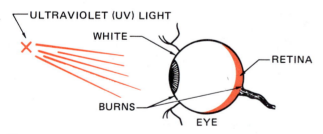

Figure 2-6 The eye can be burned on the white or on the retina by ultraviolet light.

Even with quality welding helmets, like those shown in Figure 2-7, the welder must check for potential problems that may occur from accidents or daily use. Small, undetectable leaks of ultraviolet light in an arc welding helmet can cause a welder's eyes to itch or feel sore after a day of welding. To prevent these leaks, make sure the lens gasket is installed correctly, Figure 2-8. The outer clear lens can be either glass or plastic, but the

Figure 2-7 Typical arc welding helmets used to provide eye and face protection during welding. (Photos courtesy of (top) Thermacote Welco, (bottom) Hornell Speedglas, Inc. [R.H. Blake Inc.].)

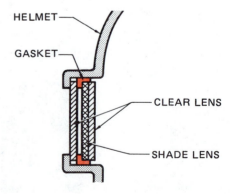

Figure 2-8 The correct placement of the gasket around the shade lens is important because it can stop ultraviolet light from bouncing around the lens assembly.

HUNTSMAN® SELECTOR CHART

Selection Chart for Eye and Face Protectors for Use in Industry and Schools

1
Goggles, flexible fitting, regular ventilation

2
Goggles, flexible fitting, hooded ventilation

3
Goggles, cushioned fitting, rigid body

4
Spectacles

5
Spectacles, eyecup type eyshields

6
Spectacles, semi-flat-fold sideshields

7
Welding goggles, eyecup type, tinted lenses

7A
Chipping goggles, eyecup type, tinted lenses

8
Welding goggles, coverspec type, tinted lenses

8A
Chipping goggles, coverspec type, clear safety lenses

9
Welding goggles, coverspec type, tinted plate lens

10
Face shield, plastic or mesh window (see caution note)

11
Welding helmet

Non-sideshield spectacles are available for limited hazard use requiring only frontal protection

Applications

Operation	Hazards	Protectors
Acetylene-Burning Acetylene-Cutting Acetylene-Welding	Sparks, Harmful Rays, Molten Metal, Flying Particles	7,8,9
Chemical Handling	Splash, Acid Burns, Fumes	2 (for severe exposure add 10)
Chipping	Flying Particles	1,2,4,5,6,7A,8A
Electric (Arc) Welding	Sparks, Intense Rays, Molten Metal	11 (in combination with 4,5,6 in tinted lenses advisable)
Furnace Operations	Glare, Heat, Molten Metal	7,8,9 (for severe exposure add 10)
Grinding-Light	Flying Particles	1,3,5,6 (for severe exposure add 10)
Grinding-Heavy	Flying Particles	1,3,7A,8A (for severe exposure add 10)
Laboratory	Chemical Splash, Glass Breakage	2 (10 when in combination wih 5,6)
Machining	Flying Particles	1,3,5,6 (for severe exposure add 10)
Molten Metals	Heat, Glare, Sparks, Splash	7,8 (10 in combination with 5,6 in tinted lenses)
Spot Welding	Flying Particles, Sparks	1,3,4,5,6 (tinted lenses advisable; for wevere exposure add 10)

CAUTION:
Face shields alone do not provide adequate protection. Plastic lenses are advised for protection against molten metal splash.
Contact lenses, of themselves, do not provide eye protection in the industrial sense and shall not be worn in a hazardous environment without appropriate covering safety eyewear.

Table 2-1 Huntsman Selector Chart. (Courtesy of Kedman Co., Huntsman Product Division.)

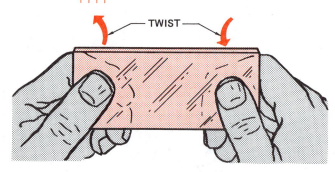

Figure 2-9 To check the shade lens for possible cracks, gently twist it.

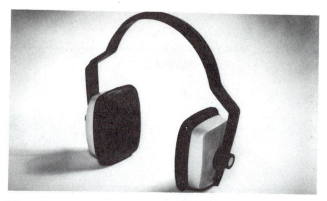

Figure 2-11 Earmuffs provide complete ear protection and can be worn under a welding helmet. (Courtesy of Mine Safety Appliances Company.)

Figure 2-10 Full face shield. (Courtesy of Jackson Products.)

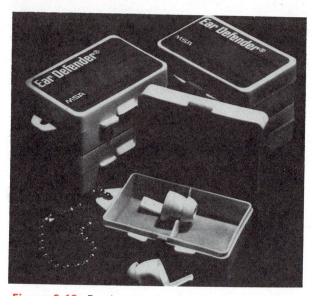

Figure 2-12 Earplugs used as protection from noise only. (Courtesy of Mine Safety Appliances Company.)

inside clear lens must be plastic. As shown in **Figure 2-9**, the lens can be checked for cracks by twisting it between your fingers. Worn or cracked spots on a helmet must be repaired. Tape can be used as a temporary repair until the helmet can be replaced or permanently repaired.

Safety glasses with side shields are adequate for general use, but if heavy grinding, chipping, or overhead work is being done, goggles or a full face shield should be worn in addition to safety glasses, **Figure 2-10**. Safety glasses are best for general protection because they can be worn under an arc welding helmet.

Ear Protection

The welding environment can be very noisy. The sound level is at times high enough to cause pain and some loss of hearing if the welder's ears are unprotected. Hot sparks can also drop into an open ear, causing severe burns.

Ear protection is available in several forms. One form of protection is **earmuffs** that cover the outer ear completely, **Figure 2-11**. Another form of protection is **earplugs** that fit into the ear canal, **Figure 2-12**. Both of these protect a person's hearing, but only the earmuffs protect the outer ear from burns.

■ CAUTION

Damage to your hearing caused by high sound levels may not be detected until later in life, and the resulting loss in hearing is nonrecoverable. Your hearing will not improve with time, and each exposure to high levels of sound will further damage your hearing.

RESPIRATORY PROTECTION

All welding and cutting processes produce undesirable by-products such as fumes and gases. Production of these by-products cannot be avoided. They are created when the temperature of metals and fluxes is raised above the temperatures at which they boil or decompose. Most of the by-products are recondensed in the weld. However, some do escape into the atmosphere, producing the haze that occurs in improperly ventilated welding shops. Some fluxes used in welding electrodes produce fumes that may irritate the welder's nose, throat, and lungs.

■ CAUTION

Welding or cutting must never be performed on drums, barrels, tanks, vessels, or other containers until they have been emptied and cleaned thoroughly, eliminating all flammable materials and all substances (such as detergents, solvents, greases, tars, or acids) that might produce flammable, toxic, or explosive vapors when heated.

Some materials that can cause respiratory problems are used as paints, coating, or plating on metals to prevent rust or corrosion. Other potentially hazardous materials might be used as alloys in metals to give them special properties.

Before welding or cutting, any metal that has been painted or has any grease, oil, or chemicals on its surface must be thoroughly cleaned. This cleaning may be done by grinding, sand blasting, or applying an approved solvent. Metals that are plated or alloyed may not be able to be cleaned before welding or cutting begins.

Most paints containing lead have been removed from the market. But some industries, such as marine or ship applications, still use these lead-based paints. Often old machinery and farm equipment surfaces still have lead-based paint coatings. Solder often contains lead alloys. The welding and cutting of lead-bearing alloys or metals whose surfaces have been painted with lead-based paint can generate lead oxide fumes. Inhalation and ingestion of lead oxide fumes and other lead compounds will cause lead poisoning. Symptoms include metallic taste in the mouth, loss of appetite, nausea, abdominal cramps, and insomnia. In time, anemia and a general weakness, chiefly in the muscles of the wrists, develop.

Both cadmium and zinc are plating materials used to prevent iron or steel from rusting. Cadmium is often used on bolts, nuts, hinges, and other hardware items, and it gives the surface a yellowish-gold appearance. Acute exposure to high concentrations of cadmium fumes can produce severe lung irritation. Long-term exposure to low levels of cadmium in air can result in emphysema (a disease affecting the lung's ability to absorb oxygen) and can damage the kidneys.

Zinc, often in the form of galvanizing, may be found on pipes, sheet metal, bolts, nuts, and many other types of hardware. Zinc plating that is thin may appear as a shiny metallic patchwork or crystal pattern; thicker, hot-dipped zinc appears ruff and may look dull. Zinc is used in large quantities in the manufacture of brass and is found in brazing rods. Inhalation of zinc oxide fumes can occur when welding or cutting on these materials. Exposure to these fumes is known to cause metal fume fever, whose symptoms are very similar to those of common influenza.

Some concern has been expressed about the possibility of lung cancer being caused by some of the chromium compounds that are produced when welding stainless steels.

■ CAUTION

Extreme caution must be taken to avoid the fumes produced when welding is done on dirty or used metal. Any chemicals that might be on the metal will become mixed with the welding fumes, a combination that can be extremely hazardous. All metal must be cleaned before welding to avoid this potential problem.

Despite these fumes and other potential hazards in welding shops, welders have been found to be as healthy as workers employed in other industrial occupations. Rather than take chances, welders should recognize that fumes of any type, regardless of their source, should not be inhaled. The best way to avoid problems is to provide adequate **ventilation**. If this is not possible, breathing protection should be used. Protective devices for use in poorly ventilated or confined areas are shown in **Figures 2-13 and 2-14**.

Potentially dangerous gases also can be present in a welding shop. Proper ventilation or respirators are necessary when welding in confined spaces, regardless of the welding process being used. Ozone is a gas that is produced by the ultraviolet radiation in the air in the vicinity of arc welding and cutting operations. Ozone is very irritating to all mucous membranes, with excessive exposure producing pulmonary edema. Other effects of exposure to ozone include headache, chest pain, and dryness in the respiratory tract.

Phosgene is formed when ultraviolet radiation decomposes chlorinated hydrocarbon. Fumes from chlorinated hydrocarbons can come from solvents such as those used for degreasing metals and from refrigerants from air-conditioning systems. They decompose in the arc to produce a potentially dangerous chlorine acid

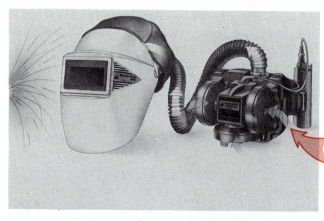

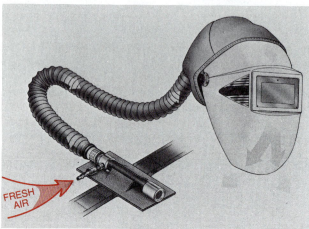

Figure 2-13 Filtered fresh air is forced into the welder's breathing area. The air can come from a belt-mounted respirator or through a hose for a remote location. (Courtesy Hornell Speedglas, Inc. [R.H. Blake Inc.].)

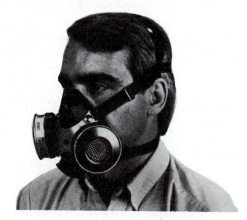

Figure 2-14 Typical respirator for contaminated environments. The filters can be selected for specific types of contaminant. (Courtesy of Mine Safety Appliances Company.)

compound. This compound reacts with the moisture in the lungs to produce hydrogen chloride, which in turn destroys lung tissue. For this reason, any use of chlorinated solvents should be well away from welding operations in which ultraviolet radiation or intense heat is generated. Any welding or cutting on refrigeration or air-conditioning piping must be done only after the

refrigerant has been completely removed in accordance with EPA regulations.

Care also must be taken to avoid the infiltration of any fumes or gases, including argon or carbon dioxide, into a confined working space, such as when welding in tanks. The collection of some fumes and gases in a work area can go unnoticed by the welders. Concentrated fumes or gases can cause a fire or explosion if they are flammable, asphyxiation if they replace the oxygen in the air, or death if they are toxic.

MATERIAL SPECIFICATION DATA SHEETS (MSDS'S)

All manufacturers of potentially hazardous materials must provide to the users of their products detailed information regarding possible hazards resulting from the use of their products. These material specification data sheets are often called MSDS's. They must be provided to anyone using the product or anyone working in the area where the products are in use. Often companies will post these sheets on a bulletin board or put them in a convenient place near the work area.

VENTILATION

The actual welding area should be well ventilated. Excessive fumes, ozone, or smoke may collect in the welding area; ventilation should be provided for their removal. Natural ventilation is best, but forced ventilation may be required. Areas that have 10,000 cubic feet (283 cubic meters) or more per welder, or that have ceilings 16 feet (4.9 meters) high or higher, **Figure 2-15**, may not require forced ventilation unless fumes or smoke begin to collect.

Forced Ventilation

Small shops or shops with large numbers of welders require forced ventilation. Forced ventilation can be general or localized using fixed or flexible exhaust

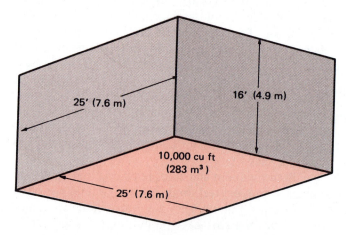

25′ (7.6 m)

16′ (4.9 m)

10,000 cu ft (283 m³)

25′ (7.6 m)

Figure 2-15 A room with a ceiling 16 ft (4.9 m) high may not require forced ventilation for one welder.

Figure 2-16 A flexible exhaust pickup.

WARNING

READ AND UNDERSTAND THIS WARNING　　PROTECT YOURSELF AND OTHERS

ELECTRIC SHOCK can kill.
- Do not permit electrically live parts or electrodes to contact skin · · · or your clothing or gloves if they are wet.
- Insulate yourself from work and ground.

FUMES AND GASES can be dangerous to your health.
- Keep fumes and gases from your breathing zone and general area.
- Keep your head out of fumes.
- Use enough ventilation or exhaust at the arc or both.

ARC RAYS can injure eyes and burn skin.
- Wear correct eye, ear, and body protection.

ENGINE EXHAUST GASES can kill.
- Use in open, well ventilated areas or vent the engine exhaust to the outside.

HIGH VOLTAGE can kill
- Do not touch electrically live parts.
- Stop engine before servicing.

MOVING PARTS can cause serious injury.
- Keep clear of moving parts.
- Do not operate with protective covers, panels, or guards removed.

- Only qualified personnel should install, use, or service this equipment.
- Read operating manual for details.

READ AND UNDERSTAND THE MANUFACTURER'S INSTRUCTIONS AND YOUR EMPLOYER'S SAFETY PRACTICES.

See American National Standard Z49.1, "Safety in Welding and Cutting", published by the American Welding Society, 550 Le Jeune Rd., Miami, Florida 33126; OSHA Safety and Health Standards, 29 CFR 1910, available from U.S. Government Printing Office, Washington, D.C. 20402.

DO NOT REMOVE THIS WARNING　E202

Figure 2-17 Note the warning information for electrical shock and high voltage contained on this typical label, which is attached to welding equipment by the manufacturer. (Courtesy of the Lincoln Electric Company.)

pickups, Figure 2-16. General room ventilation must be at a rate of 2,000 cu ft (56 m³) or more per person welding. Localized exhaust pickups must have a suction strong enough to provide 100 linear feet (30.5 m) per minute velocity of welding fumes away from the welder. Local, state, or federal regulations may require that welding fumes be treated to remove hazardous components before they are released into the atmosphere.

Any system of ventilation should draw the fumes or smoke away before they rise past the level of the welder's face.

Forced ventilation is always required when welding on metals that contain zinc, lead, beryllium, cadmium, mercury, copper, austenetic maganese, or other materials that give off dangerous fumes.

ELECTRICAL SAFETY

Injuries and even death can be caused by electric shock unless proper precautions are taken. Most welding and cutting operations involve electrical equipment in addition to the arc welding power supplies. Grinders, electric motors on automatic cutting machines, and drills are examples. Most electrical equipment in a welding shop is powered by alternating-current (AC) sources having input voltages ranging from 115 volts to 460 volts. However, fatalities have occurred when working with equipment operating at less than 80 volts. Most electric shock in the welding industry does not occur from contact with welding electrode holders, but as a result of accidental contact with bare or poorly insulated conductors. Electrical resistance is lowered in the presence of water or moisture, so welders must take special precautions when working under damp or wet conditions, including perspiration. Figure 2-17 shows a typical warning label attached to welding equipment.

The workpiece being welded, and the frame or chassis of all electrically powered machines, must be connected to a good electrical ground. The work lead from the welding power supply is not an electrical ground and is not sufficient. A separate lead is required to ground the workpiece and power source.

Electrical connections must be tight. Terminals for welding leads and power cables must be shielded from accidental contact by personnel or by metal objects. Cables must be used within their current carrying and duty cycle capacities; otherwise, they will overheat and break down the insulation rapidly. Cable connectors for lengthening leads must be insulated.

■ CAUTION

Welding cables must never be spliced within 10 ft (3 m) of the electrode holder.

Cables must be checked periodically to be sure that they have not become frayed, and, if they have, they must be replaced immediately.

Welders should not allow the metal parts of electrodes or electrode holders to touch their skin or wet coverings on their bodies. Dry gloves in good condition must always be worn. Rubber-soled shoes are advisable. Precautions against accidental contact with bare conducting surfaces must be taken when the welder is required to work in cramped kneeling, sitting, or lying positions. Insulated mats or dry wooden boards are desirable protection from the earth.

Welding circuits must be turned off when the work station is left unattended. The main power supply must be turned off and locked or tagged, to prevent electrocution, when working on the welder, welding leads, electrode holder, torches, wire feeder, guns, or other parts. Since the electrode holder is energized when changing coated electrodes, the welder must wear dry gloves.

GENERAL WORK CLOTHING

Because of the amount and temperature of hot sparks, metal, and slag produced during welding, cutting, or brazing, and the fact that special protective clothing cannot be worn at all times, it is important to choose general work clothing that will minimize the possibility of getting burned.

Wool clothing (100% wool) is the best choice but difficult to find. All-cotton (100% cotton) clothing is a good second choice, and it is the most popular material used. Synthetic materials, including nylon, rayon, and polyester, should be avoided because they are easily burned, produce a hot, sticky ash (because it sticks, burns can be more severe), and some produce poisonous gases. The clothing must also stop ultraviolet light from passing through it. This is accomplished if the material chosen is a dark color, thick, and tightly woven.

The following are some guidelines for selecting work clothing:

■ Shirts must be long sleeved to protect the arms, have a high buttoned collar to protect the neck, **Figure 2-18**, be long enough to tuck into the pants to protect the waist, and have flaps on the pockets to keep sparks out (or have no pockets).

■ Pants must have legs long enough to cover the tops of the boots and must be without cuffs that would catch sparks.

■ Boots must have high tops to keep out sparks, have steel toes to prevent crushed toes, **Figure 2-19**, and have smooth tops to prevent sparks from being trapped in seams.

■ Caps should be thick enough to prevent sparks from burning the top of a welder's head.

All clothing must be free of frayed edges or holes. The clothing must be relatively tight-fitting in order to prevent excessive folds or wrinkles that might trap sparks.

Butane lighters and matches may catch fire or explode if they are subjected to welding heat or sparks. There is no safe place to carry these items when welding. They must always be removed from the welder's pockets and placed a safe distance away before any work is started.

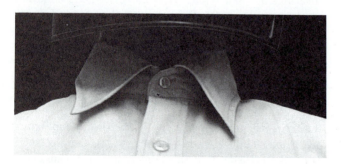

Figure 2-18 The top button of the shirt worn by the welder should always be buttoned in order to avoid severe burns to that person's neck.

STEEL

Figure 2-19 Safety boots with steel toes are required by many welding shops.

■ CAUTION

There is no safe place to carry butane lighters or matches while welding or cutting. They can catch fire or explode if subjected to welding heat or sparks. Butane lighters may explode with the force of 1/4 stick of dynamite. Matches can erupt into a ball of fire. Both butane lighters and matches must always be removed from the welder's pockets and placed a safe distance away before any work is started.

SPECIAL PROTECTIVE CLOTHING

General work clothing is worn by each person in the shop. In addition to this clothing, extra protection is needed for each person who is in direct contact with hot materials. Leather is often the best material to use, as it is lightweight, flexible, resists burning, and is readily available. Synthetic insulating materials are also available. Ready-to-wear leather protection includes capes, jackets, aprons, sleeves, gloves, caps, pants, knee pads, and spats, among other items.

Hand Protection

All-leather, gauntlet-type gloves should be worn when doing any welding, **Figure 2-20**. Gauntlet gloves that have a cloth liner for insulation are best for hot work. Noninsulated gloves will give greater flexibility for fine work. Some leather gloves are available with a canvas gauntlet top, which should be used for light work only.

When a great deal of manual dexterity is required for gas tungsten arc welding, brazing, soldering, oxyfuel gas welding, and other delicate processes, soft leather gloves may be used, **Figure 2-21**. All-cotton gloves are sometimes used when doing very light welding.

Figure 2-20 All leather, gauntlet-type, welding gloves.

Figure 2-22 Full leather jacket. (Courtesy Elliott Glove Co., Inc.)

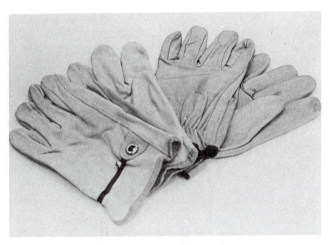

Figure 2-21 For welding that requires a great deal of manual dexterity, soft leather gloves can be worn.

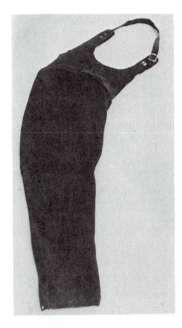

Figure 2-23 Full leather sleeve.

Body Protection

Full leather jackets and capes will protect a welder's shoulders, arms, and chest. Figure 2-22. A jacket, unlike the cape, protects a welder's back and complete chest. A cape is open and much cooler but offers less protection. The cape can be used with a bib apron to provide some additional protection while leaving the back cooler. Either the full jacket or the cape with a bib apron should be worn for any out-of-position work.

Waist and Lap Protection

Bib aprons or full aprons will protect a welder's lap. Welders will especially need to protect their laps if they squat or sit while working and when they bend over or lean against a table.

Arm Protection

For some vertical welding, a full or half sleeve can protect a person's arm, Figure 2-23. The sleeves work best if the work level is not above the welder's chest. Work levels higher than this usually require a jacket or cape to keep sparks off the welder's shoulders.

Leg and Foot Protection

When heavy cutting or welding is being done and a large number of sparks are falling, leather pants and spats should be used to protect the welder's legs and feet. If the weather is hot and full leather pants are uncomfortable, leather aprons with leggings are available. Leggings can be strapped to the legs, leaving the back open. Spats will prevent sparks from burning through the front of lace-up boots.

HANDLING AND STORING CYLINDERS

Oxygen and fuel gas cylinders or other flammable materials must be stored separately. The storage areas must be separated by 20 ft (6.1 m), or by a wall 5 ft high (1.5 m) with at least a 1/2-hour (hr) burn rating, **Figure 2-24**. The purpose of the distance or wall is to keep the heat of a small fire from causing the oxygen cylinder safety valve to release. If the safety valve releases the oxygen, a small fire would become a raging inferno.

Inert gas cylinders may be stored separately or with either fuel cylinders or oxygen cylinders.

Empty cylinders must be stored separately from full cylinders, although they may be stored in the same room or area. All cylinders must be stored vertically and have the protective caps screwed on firmly.

Securing Gas Cylinders

Cylinders must be secured with a chain or other device so that they cannot be knocked over accidentally. Even though more stable, cylinders attached to a manifold or stored in a special room used only for cylinder storage should be chained.

Storage Areas

Cylinder storage areas must be located away from halls, stairwells, and exits so that in case of an emergency they will not block an escape route. Storage areas should also be located away from heat, radiators, furnaces, and welding sparks. The location of storage areas should be such that unauthorized people cannot tamper with the cylinders. A warning sign that reads "Danger — No Smoking, Matches, or Open Lights," or similar wording, should be posted in the storage area, **Figure 2-25**.

Cylinders with Valve Protection Caps

Cylinders equipped with a valve protection cap must have the cap in place unless the cylinder is in use. The protection cap prevents the valve from being broken off if the cylinder is knocked over. If the valve of a full high-pressure cylinder (argon, oxygen, CO_2, and mixed gases) is broken off, the cylinder valve can fly around the shop like a missile if it has not been secured properly.

Never lift a cylinder by the safety cap or the valve. The valve can easily break off or be damaged.

When moving cylinders, the valve protection cap should be replaced, especially if the cylinders are mounted on a truck or trailer for out-of-shop work. The cylinders must never be dropped or handled roughly.

General Precautions

Use warm water (not boiling) to loosen cylinders that are frozen to the ground. Any cylinder that leaks,

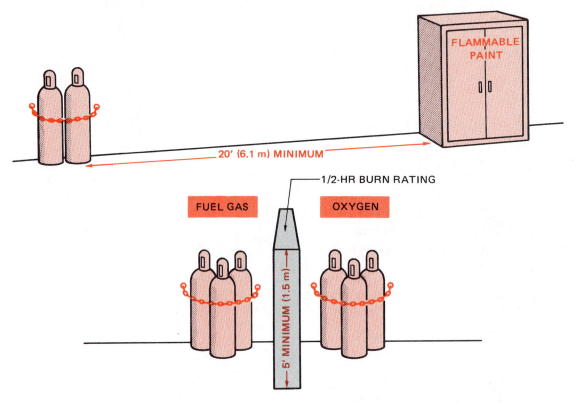

Figure 2-24 The minimum safe distance between stored fuel gas cylinders and any flammable material is 20 ft (6.1 m) or a wall 5 ft (1.5 m) high.

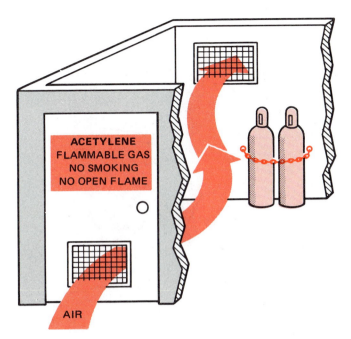

Figure 2-25 A separate room used to store acetylene must have good ventilation and should have a warning sign posted on the door.

Figure 2-26 Move a leaking fuel gas cylinder out of the building or any work area. The pressure should slowly be released after posting a warning of the danger.

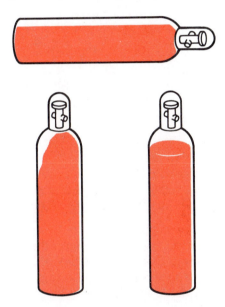

Figure 2-27 The acetone in an acetylene cylinder must have time to settle before the cylinder can be used safely.

has a bad valve, or has gas-damaged threads must be identified and reported to the supplier. A piece of soapstone is used to write the problem on the cylinder. If the leak cannot be stopped by closing the cylinder valve, the cylinder should be moved to a vacant lot or open area. The pressure should then be slowly released after posting a warning sign, Figure 2-26.

Acetylene cylinders that have been lying on their sides must stand upright for 15 minutes or more before they are used. The acetylene is absorbed in acetone, and the acetone is absorbed in a filler. The filler does not allow the liquid to settle back away from the valve very quickly, Figure 2-27. If the cylinder has been in a horizontal position, using it too soon after it is placed in a vertical position may draw acetone out of the cylinder. Acetone lowers the flame temperature and can damage regulator or torch valve settings.

FIRE PROTECTION

Fire is a constant danger to the welder. The possibilities of fires cannot always be removed, but they should be minimized. Highly combustible materials should be 35 ft (10.7 m) or more away from any welding. When it is necessary to weld within 35 ft (10.7 m) of combustible materials, when sparks can reach materials farther than 35 ft (10.7 m) away, or when anything more than a minor fire might start, a fire watch is needed.

Fire Watch

A fire watch can be provided by any person who knows how to sound the alarm and use a fire extinguish-er. The fire extinguisher must be the type required to put out a fire of the type of combustible materials near the welding. Combustible materials that cannot be removed from the welding area should be soaked with water or covered with sand or noncombustible insulating blankets, whichever is available.

Fire Extinguishers

The four types of fire extinguishers are type A, type B, type C, and type D. Each type is designed to put out fires on certain types of materials. Some fire extinguishers can be used on more than one type of fire. However, using the wrong type of fire extinguisher can be dangerous, either causing the fire to spread, causing electrical shock, or causing an explosion.

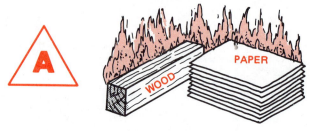

Figure 2-28 Type A fire extinguisher symbol.

Figure 2-29 Type B fire extinguisher symbol.

Figure 2-30 Type C fire extinguisher symbol.

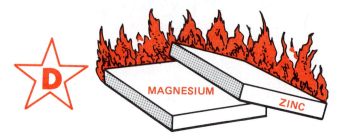

Figure 2-31 Type D fire extinguisher symbol.

Type A Extinguishers. Type A extinguishers are used for combustible solids (articles that burn), such as paper, wood, and cloth. The symbol for a type A extinguisher is a green triangle with the letter A in the center, Figure 2-28.

Type B Extinguishers. Type B extinguishers are used for combustible liquids, such as oil, gas, and paint thinner. The symbol for a type B extinguisher is a red square with the letter B in the center, Figure 2-29.

Type C Extinguishers. Type C extinguishers are used for electrical fires. For example, they are used on fires involving motors, fuse boxes, and welding machines. The symbol for a type C extinguisher is a blue circle with the letter C in the center, Figure 2-30.

Type D Extinguishers. Type D extinguishers are used on fires involving combustible metals, such as zinc, magnesium, and titanium. The symbol for a type D extinguisher is a yellow star with the letter D in the center, Figure 2-31.

Location of Fire Extinguishers

Fire extinguishers should be of a type that can be used on the types of combustible materials located nearby, **Figure 2-32**. The extinguishers should be placed so that they can be easily removed without reaching over combustible material. They should also be placed at a level low enough to be easily lifted off the mounting, **Figure 2-33**. The location of fire extinguishers should be marked with red paint and signs, high enough so that their location can be seen from a distance over people and equipment. The extinguishers should also be marked near the floor so that they can be found even if a room is full of smoke, **Figure 2-34**.

Use

A fire extinguisher works by breaking the fire triangle of heat, fuel, and oxygen. Most extinguishers both cool the fire and remove the oxygen. They use a variety of materials to extinguish the fire. The majority of fire extinguishers found in welding shops use foam, carbon dioxide, a soda-acid gas cartridge, a pump tank, or dry chemicals.

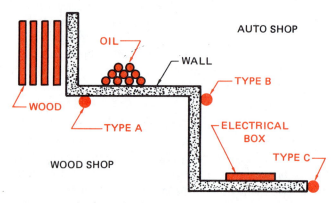

Figure 2-32 The type of fire extinguisher provided should be appropriate for the materials being used in the surrounding area.

Figure 2-33 Mount the fire extinguisher so that it can be lifted easily in an emergency.

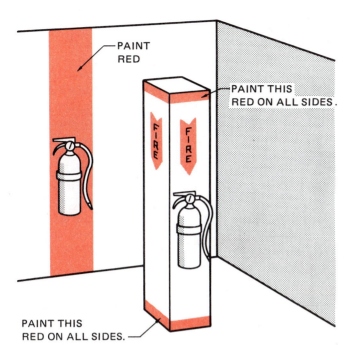

Figure 2-34 The locations of fire extinguishers should be marked so they can be located easily in an emergency.

When using a **foam** extinguisher: Don't spray the stream directly into the burning liquid. Allow the foam to fall lightly on the fire.

When using a **carbon dioxide** extinguisher: Direct the discharge as close to the fire as possible, first at the edge of the flames and gradually to the center.

When using a **soda-acid gas cartridge** extinguisher: Place your foot on the footrest and direct the stream at the base of the flames, **Figure 2-35**.

When using a **dry chemical** extinguisher: Direct the extinguisher at the base of the flames. In the case of type A fires, follow up by directing the dry chemicals at the remaining material still burning. Therefore, the extinguisher must be directed at the base of the fire where the fuel is located, **Figure 2-35**.

Figure 2-35 Point the extinguisher at the material burning, not at the flames.

EQUIPMENT MAINTENANCE

A routine schedule of equipment maintenance will aid in detecting potential problems such as leaking coolant, loose wires, poor grounds, frayed insulation, or split hoses. Small problems, if fixed in time, can prevent the loss of valuable time due to equipment breakdown or injury.

Any maintenance beyond routine external maintenance should be referred to a trained service technician. In most areas, it is against the law for anyone but a licensed electrician to work on arc welders and anyone but a factory-trained repair technician to work on regulators. Electrical shock and exploding regulators can cause serious injury or death.

Hoses

Hoses must be used only for the gas or liquid for which they were designed. Green hoses are to be used only for oxygen, and red hoses are to be used only for acetylene or other fuel gases. Using unnecessarily long lengths of hoses should be avoided. Never use oil, grease, lead, or other pipe-fitting compounds for any joints. Hoses should also be kept out of the direct line of sparks. Any leaking or bad joints in gas hoses must be repaired.

WORK AREA

The work area should be kept picked up and swept clean. Collections of steel, welding electrode stubs, wire, hoses, and cables are difficult to work around and easy to trip over. An electrode caddy can be used to hold the electrodes and stubs, **Figure 2-36**. Hooks can be made to

Figure 2-36 An easy-to-build electrode caddy can be used to hold both electrodes and stubs.

hold hoses and cables, and scrap steel should be thrown into scrap bins.

Arc welding areas should be painted flat black to absorb as much of the ultraviolet light as possible. Portable screens should be used whenever arc welding is to be done outside of a welding booth.

If a piece of hot metal is going to be left unattended, write the word *hot* on it before leaving. This procedure can also be used to warn people of hot tables, vises, firebricks, and tools.

HAND TOOLS

Hand tools are used by the welder to do necessary assembly and disassembly of parts for welding as well as to perform routine equipment maintenance.

The adjustable wrench is the most popular tool used by the welder. When using this wrench, it should be adjusted tightly on the nut and pushed so that most of the force is on the fixed jaw, **Figure 2-37**. When a wrench is being used on a tight bolt or nut, the wrench should be pushed with the palm of an open hand or pulled to prevent injuring the hand. If a nut or bolt is too tight to be loosened with a wrench, obtain a longer wrench. A cheater bar should not be used. The fewer points a box end wrench or socket has, the stronger it is and the less likely it is to slip or damage the nut or bolt, **Figure 2-38**.

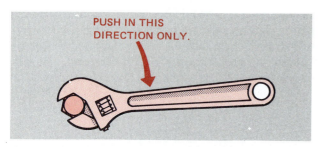

Figure 2-37 The adjustable wrench is stronger when used in the direction indicated.

Figure 2-38 The fewer the points, the less likely the wrench is to slip.

Striking a hammer directly against a hard surface such as another hammer face or anvil may result in chips flying off and causing injury.

The mushroomed heads of chisels, punches, and the faces of hammers should be ground off, **Figure 2-39**. Chisels and punches that are going to be hit harder than a slight tap should be held in a chisel holder or with pliers to eliminate the danger of injuring your hand. A handle should be placed on the tang of a file in order to avoid injuring your hand, **Figure 2-40**. A file can be kept free of chips by rubbing a piece of soapstone on it before it is used.

It is important to remember to use the correct tool for the job. Do not try to force a tool to do a job it was not designed to do.

POWER TOOLS

All power tools must be properly grounded to prevent accidental electrical shock. If even a slight tingle is felt while using a power tool, stop and have the tool checked by a technician. Power tools should never be used with force or allowed to overheat from excessive or incorrect use. If an extension cord is used, it should

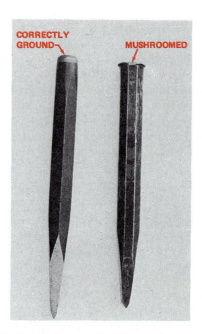

Figure 2-39 Any mushroomed heads must be ground off.

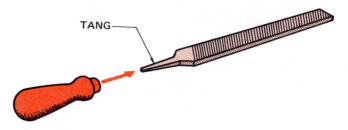

Figure 2-40 To protect yourself from the sharp tang of a file, always use a handle with a file.

Name-plate Amperes	Cord Length in Feet							
	25	50	75	100	125	150	175	200
1	16	16	16	16	16	16	16	16
2	16	16	16	16	16	16	16	16
3	16	16	16	16	16	16	14	14
4	16	16	16	16	16	14	14	12
5	16	16	16	16	14	14	12	12
6	16	16	16	14	14	12	12	12
7	16	16	14	14	12	12	12	10
8	14	14	14	14	12	12	10	10
9	14	14	14	12	12	10	10	10
10	14	14	14	12	12	10	10	10
11	12	12	12	12	10	10	10	8
12	12	12	12	12	10	10	8	8

Note: Wire sizes shown are American Wire Gauge (AWG).

Table 2-2 Recommended Extension Cord Sizes for Use with Portable Electric Tools.

have a large enough current rating to carry the load, Table 2-2. An extension cord that is too small will cause the tool to overheat.

Safety glasses must be worn at all times when using any power tools.

Grinders

Grinding using a pedestal grinder or a portable grinder is required to do many welding jobs correctly. Often it is necessary to grind a groove, remove rust, or smooth a weld. Grinding stones have the maximum revolutions per minute (RPM) listed on the paper blotter, **Figure 2-41**. They must never be used on a machine with a higher rated RPM. If grinding stones are turned too fast, they can explode.

Grinding Stone. Before a grinding stone is put on the machine, it should be tested for cracks. This is done by tapping the stone in four places and listening for a sharp ring, which indicates it is good, **Figure 2-42**. A dull sound indicates that the grinding stone is cracked and should not be used. Once a stone has been installed and has been used, it may need to be trued and balanced by using a special tool designed for that purpose, **Figure 2-43**. Truing keeps the stone face flat and sharp for better results.

Types of Grinding Stones. Each grinding stone is made for grinding specific types of metal. Most stones are for ferrous metals, meaning iron, cast iron, steel, and stainless steel, among others. Some stones are made for nonferrous metals such as aluminum, copper, and brass. If a ferrous stone is used to grind nonferrous metal, the stone will become glazed (the surface clogs with metal) and may explode due to frictional heat building up on the surface. If a nonferrous stone is used to grind ferrous metal, the stone will be quickly worn away.

When the stone wears down, keep the tool rest adjusted to within 1/16 in. (2 mm), **Figure 2-44**, so that the metal being ground cannot be pulled between the tool rest and the stone surface. Stones should not be used when they are worn down to the size of the paper blotter. If small parts become hot from grinding, pliers can be used to hold them. Gloves should never be worn

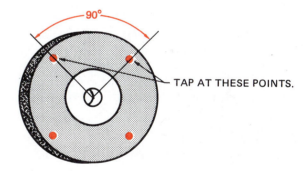

Figure 2-42 Grinding stones should be checked for cracks before they are installed.

Figure 2-41 Always check to be sure that the grinding stone and the grinder are compatible before installing a stone.

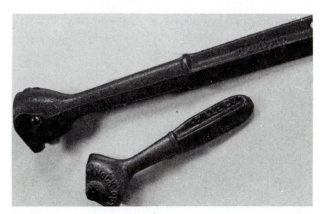

Figure 2-43 Use a grinding stone redressing tool as needed to keep the stone in balance.

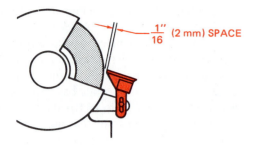

Figure 2-44 Keep the tool rest adjusted.

when grinding. If a glove gets caught in a stone, the whole hand may be drawn in.

The sparks from grinding should be directed down and away from other welders or equipment.

Drills

Holes should be center punched before they are drilled to help stop the drill bit from wandering. If the bit gets caught, stop the motor before trying to remove the bit. All metal being drilled on a drill press should be securely clamped to the table.

The sharp metal shavings should be avoided as they come out of the hole. If they start to become long, stop pushing downward pressure until the shaving breaks. Then the hole can be continued.

METAL CUTTING MACHINES

Many types of metal cutting machines are used in the welding shop, for example, shears, punches, cut-off machines, and band saws. Their advantages include little or no post-cutting cleanup, the wide variety of metals that can be cut, and the fact that the metal is not heated.

■ *CAUTION*

Before operating any power equipment for the first time you must read the manufacturer's safety and operating instructions and should be assisted by someone with experience with the equipment. Be sure your hands are clear of the machine before the equipment is started. Always turn off the power and lock it off before working on any part of the equipment.

Shears and Punches

Welders frequently use shears and punches in the fabrication of metal for welding. These machines can be operated either by hand or by powerful motors. Hand-operated equipment is usually limited to thin sheet stock or small bar stock. Powered equipment can be used on material an inch or more in thickness and several feet wide, depending on its rating. Their power can be a

potential danger if these machines are not used correctly. Both shears and punches are rated by the thickness, width, and type of metal that they can be safely used to work. Failure to follow these limitations can result in damage to the equipment, damage to the metal being worked, and injury to the operator.

Shears work like powerful scissors. The correct placement of the metal being cut is as close to the pivoting pin as possible, **Figure 2-45**. The metal being sheared must be securely held in place by the clamp on the shear before it is cut. If you are cutting a long piece of metal that is not being supported by the shear table, then portable supports must be used. As the metal is being cut it may suddenly move or bounce around; if you are holding on to it, this can cause a serious injury.

Power punches are usually either hydraulic or flywheel operated. Both types move quickly, but only the hydraulic can usually be stopped midstroke. Once the flywheel-type punch has been engaged, in contrast, it will make a complete cycle before it stops. Because punches move quickly or may not be stopped, it is very important that the operator's two hands be clear of the machine and that the metal be held firmly in place by the machine clamps before starting the punching operation.

Cut-off Machines

Cut-off machines may use abrasive wheels or special saw blades to make their cuts, **Figure 2-46**. Most abrasive cut-off wheels spin at high speeds (high RPMs) and are used dry (without coolant). Most saws operate much more slowly and with a liquid coolant. Both types of machines produce quality cuts in a variety of bar or structural shaped metals. The cuts require little or no post-cut cleanup. Always wear eye protection when operating these machines. Before a cut is started the metal must be clamped securely in the machine vise. Even the

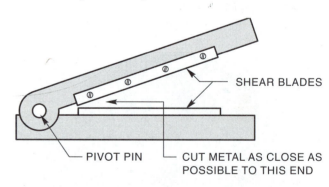

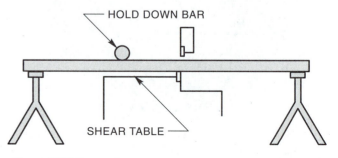

Figure 2-45 Power shear.

Figure 2-46 Cut-off machine.

slightest movement of the metal can bind or break the wheel or blade. If the machine has a manual feed, the cutting force must be applied at a smooth and steady rate. Apply only enough force to make the cut without dogging down the motor. Use only reinforced abrasive cut-off wheels that have an RPM rating equal to or higher than the machine rated speed.

Band Saws

Band saws can be purchased as vertical or horizontal, and some can be used in either position, **Figure 2-47**.

Some band saws can be operated with a cooling liquid and are called *wet saws*; most operate dry. The blade guides must be adjusted as close as possible to the metal being cut. The cutting speed and cutting pressure must be low enough to prevent the blade from overheating. When using a vertical band saw with a manual feed, you must keep your hands away from the front of the blade so that if your hand slips, it will not strike the moving blade. If the blade brakes, sticks, or comes off the track, turn off the power, lock it off, and wait for the band saw drive wheels to come to a complete stop before touching the blade. Be careful of hot flying chips.

MATERIAL HANDLING

Proper lifting, moving, and handling of large, heavy, welded assemblies is important to the safety of the workers and the weldment. Improper work habits can cause serious personal injury as well as cause damage to equipment and materials.

Lifting

When lifting a heavy object, the weight of the object should be distributed evenly between both hands, and your legs should be used to lift, not your back, **Figure 2-48**. Do not try to lift a large or bulky object without help if the object is heavier than you can lift with one hand.

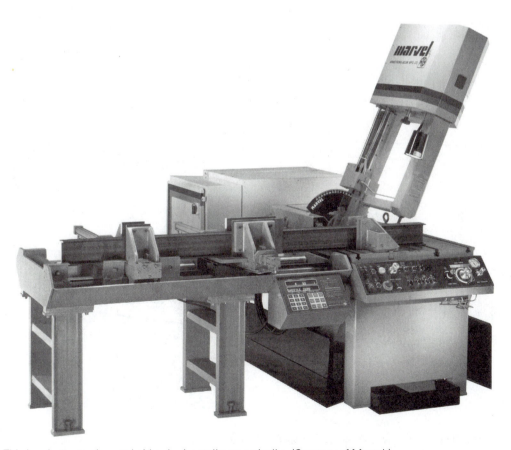

Figure 2-47 This band saw can be used either horizontally or vertically. (Courtesy of Marvel.)

Hoists or Cranes

Hoists or cranes can be overloaded with welded assemblies. The capacity of the equipment should be checked before trying to lift a load. Keep any load as close to the ground as possible while it is being moved. Pushing a load on a crane is better than pulling a load. It is advisable to stand to one side of ropes, chains, and cables that are being used to move or lift a load, **Figure 2-49**. If they break and snap back, they will miss you. If it is necessary to pull a load, use a rope, **Figure 2-50**.

SAFETY FIRST

The safety of the welder working in industry is of utmost importance to the industry. A sizable amount of money is spent for the protection of welders. Usually manufacturers have a safety department with one individual in charge of plant safety. The safety officer's job is to make sure that all welders comply with safety rules during production. The proper clothing, shoes, and eye protection to be worn is emphasized in these plants. Any worker who does not follow established safety rules is subject to dismissal.

If an accident does occur, it's important that appropriate and immediate first aid steps be taken. All welding shops should have established plans for accidents. You should take time to learn the proper procedure for accident response and reporting before you need to respond in an emergency. After the situation has been properly taken care of, you should fill out an accident report.

Equipment is periodically checked to be sure that it is safe and in proper working condition. Maintenance workers are employed to see that the equipment is in proper working condition at all times.

Further safety information is available in *Safety for Welders*, by Larry F. Jeffus, published by Delmar Publishers Inc., and from the American Welding Society or the U.S. Department of Labor (OSHA) Regulations.

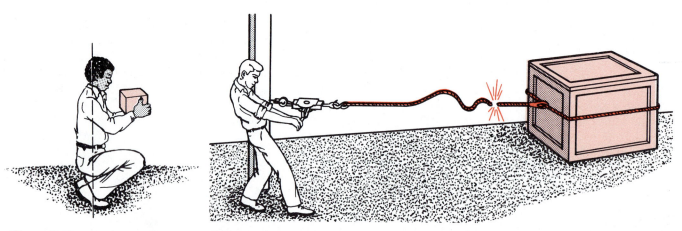

Figure 2-48 Lift with your legs, not your back.

Figure 2-49 Never stand in line with a rope, chain, or cable that is being used to move or lift a load.

Figure 2-50 When moving a load overhead, stay out of the way of the load in case it falls.

Masts Present a Special Welding Problem

Imagine yourself sailing in rough weather somewhere in the south Atlantic, enroute to the Pacific via the Straits of Magellan: not a good place to have your masts give out on you. You have to know that those masts are going to hold up no matter what the conditions. If you are into racing yachts or dinghies, the design and construction of your mast is a key factor in your success or failure.

It makes sense that a high percentage of discerning boat builders and sailing aficionados turn to a company in Southampton, England, that manufactures alloy masts. The founder of the company is a famous boat designer who, 30 years ago, amidst protest and skepticism from wooden mast traditionalists, first used aluminum masts on sailboats, **Figure 1**.

The company manufactures masts ranging from small spars to huge, 160-foot masts for sail-training ships and mere 90 footers used in the 12-meter America's Cup yachts.

Special problems in the manufacture of the aluminum masts required special welding solutions. Ideally, sailing boat masts should taper at the top. This improves performance by cutting down weight and creating less wind disturbance. The company tapers its masts by cutting a long, narrow V in the tube at the top of the mast, **Figure 2**. The taper can be 4 feet, 6 inches or it can be up to 10 feet, depending on the mast. The gap is then closed and welded. Because the way the mast bends under stress is important, the extruded tube is not of an even thickness. The problem is that when the material is cut out of the extruded section there is a variation in thickness between the beginning and the end of the weld. It goes from thick to thin to thick again. At the same time there are different lengths of weld: some are 5 feet and some are 8 feet. The cost of designing and building a fully automatic welding machine to handle these welds was prohibitive, so a semiautomatic unit was selected.

Figure 1

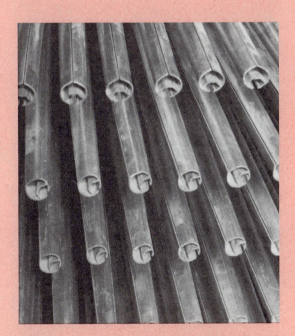

Figure 2

Figure 3

The welding process used is GTAW. The power supply, including a programmer, gives excellent weld repeatability, a controlled start, upslope and downslope, and helps cut down crater problems. The pulsed current reduces warping problems and helps with penetration.

To meet specific production needs, a semiautomatic machine was designed and built. There were so many variables in this particular welding job that an operator was required for part of the welding process, Figure 3. The mast is clamped together. But because the V shape is cut out with a circular saw, there isn't the 100% fit of machined pieces that is needed for an automatic weld. There are slight gaps when the two sides of the cut are joined. In addition, the cut is not always 100% straight but can wander a little over the length. Since the heat of the welding would cause the mast to bend, it has to be pre-bent in the other direction, and the arc has to drop as much as 6 inches from one end of the weld to the other. This was handled with an automatic arc length control, which keeps the torch on track vertically. In addition, the machine is used for small batches of a great variety of masts with different lengths of taper where the variation of material thickness differs from batch to batch.

The operator controls the lateral movement of the torch so that the seam can be followed exactly and varies the amperes as needed to maintain an even weld.

The weld has to be neat because when the profile is ground off and the mast anodized, the welded seam looks darker than the mast on either side, so it is quite visible.

(Courtesy of Miller Electric Mfg. Co., Appleton, Wisconsin 54912.)

REVIEW

1. What is the key to preventing accidents in a welding shop?

2. Describe first-, second-, and third-degree burns and the first aid that would be administered for each type of burn.

3. Which type of light is the most likely to cause burns? Why?

4. What is the name of the eye burn that can occur within seconds?

5. Why must eye protection be worn at all times in the welding shop?

6. Why is it important to seek medical treatment with eye burns?

7. From which material must the inside lens in a welding helmet be made?

8. Why should a welder wear earmuff type protection?

9. List four metals that can cause hazardous fumes during welding.

10. Why is ventilation important when welding in a confined space?

11. What is a MSDS, and how is it used?

12. List three conditions that would require forced ventilation in a weld shop.

13. Why is it easy to become shocked if the welding area is damp?

14. A welding cable must not have a splice within _____ of the welder.

15. Clothing made from what types of materials should never be worn in the welding shop? Why?

16. Describe the best type of clothing to be worn in a welding shop.

17. Ready-to-wear leather protective clothing is available as _____.

18. General welding gloves should be _____.

19. How must oxygen and acetylene cylinders that must be stored near each other be separated?

20. How must gas cylinders be prevented from accidently being knocked over?

21. What can happen if a high-pressure gas cylinder has its valve knocked off?

22. What should be done with a leaking cylinder if the leak cannot be stopped?

23. Why must an acetylene cylinder that was stored horizontally be set in a vertical position for several minutes before it is used?

24. When is a fire watch needed?

25. What type of fire extinguisher would be used on each of the following items?

    a. Paint   c. Trash

    b. Motor   d. Zinc

26. Why is it important to inspect equipment on a regular basis?

27. Why must mushroomed chisels be reground?

28. How should a grinding stone be tested before it is first put on a grinder?

29. How should the tool rest on a grinder be adjusted?

30. Describe how a person should lift heavy objects.

31. Define the following terms:
    (a) infrared light, (b) ultraviolet light,
    (c) forced ventilation, (d) natural ventilation,
    (e) valve protection cap, and (f) acetone.

Section Two
Shielded Metal Arc Welding

Success Story

As a young boy in an agricultural area of Minnesota, Dennis Peabody was always looking for things to design, build, or rebuild—from building mini bikes to restoring cars to fixing farm equipment

In the eighth grade Dennis was allowed to use an old Lincoln arc welder for one hour a day. He was thrilled!

In his junior and senior high school years he took all the welding and metal classes that were available. After graduating from high school he attended technical college for welding and related classes. During his college years he paid for his schooling by welding at night for an agricultural manufacturing company. Between his schooling and night welding jobs he gained a great deal of experience in fabrication and design work.

After graduating from college with a degree in welding and metals, he moved back to his home town and started his welding and design career in a small agricultural company with only four employees and a very small shop. That was twenty-five years ago.

Today, Dennis is still working for the same manufacturing company, but now the company has a manufacturing plant with over forty employees and Dennis is the plant manager. He oversees all plant production and design work for agricultural equipment that is sold both nationally and internationally.

Dennis has taught welding and is on many welding department advisory boards for area high schools and a local technical college.

The changes he has seen over the years in the welding field have predominantly been in the technological improvements made to welding and fabricating equipment and in the strong demand for computer use in the welding trade.

And, yes, Dennis still loves to pick up a welder and strike an arc.

Chapter 3

SHIELDED METAL ARC EQUIPMENT, SETUP, AND OPERATION

OBJECTIVES

After completing this chapter, the student should be able to

■ explain the differences in welding with each of the three types of current.

■ identify welding machines according to their type.

■ demonstrate how to select and set the welding current.

■ describe the proper maintenance of welding equipment.

■ demonstrate how to safely set up an arc welding station.

KEY TERMS

electrons	voltage
amperage	wattage
anode	cathode
output	open circuit voltage
operating voltage	magnetic flux lines
step-down transformer	inverter
rectifier	duty cycle

INTRODUCTION

Shielded Metal Arc Welding (SMAW) is a welding process that uses a flux covered metal electrode to carry an electrical current, **Figure 3-1**. The current forms an arc across the gap between the end of the electrode and the work. The electric arc creates sufficient heat to melt both the electrode and the work. Molten metal from the electrode travels across the arc to the molten pool on the base metal, where they mix together. The end of the electrode and molten pool of metal are surrounded, purified, and protected by a gaseous cloud and a covering of slag produced as the flux coating of the electrode burns or vaporizes. As the arc moves away, the mixture of molten electrode and base metal solidifies and becomes one piece.

SMAW is the most widely used welding process because of its low cost, flexibility, portability, and versatility. The machine and the electrodes are low cost. The machine itself can be as simple as a 110-volt, step-down transformer. The electrodes are available from a large number of manufacturers in packages from 1 lb (0.5 kg) to 50 lb (22 kg).

The SMAW process is very flexible in terms of the metal thicknesses that can be welded and the variety of positions it can be used in. Metal as thin as 1/16 in. (2 mm) thick, or approximately 16 gauge, to several feet thick can be welded using the same machine with different settings. The flexibility of the process also allows metal in this thickness range to be welded in any position.

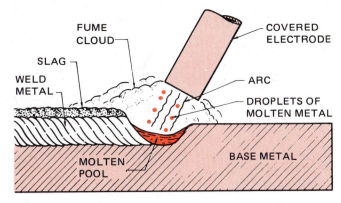

Figure 3-1 Shielded metal arc welding.

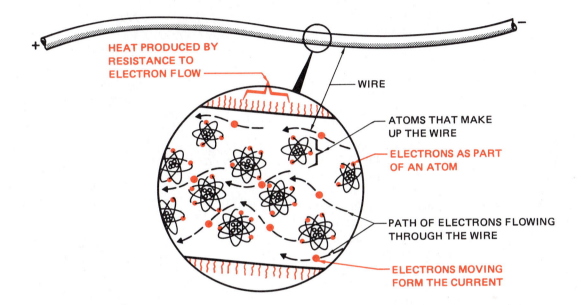

Figure 3-2 Electrons traveling along a conductor.

SMAW is a very portable process because it is easy to move the equipment and engine-driven generator-type welders are available. Also the limited amount of equipment required for the process makes moving easy.

The process is versatile, and it's used to weld almost any metal or alloy, including cast iron, aluminum, stainless steel, and nickel.

WELDING CURRENT

The welding current is an electric current. An electric current is the flow of electrons. Electrons flow through a conductor from negative (-) to positive (+), Figure 3-2. Resistance to the flow of electrons (electricity) produces heat. The greater the resistance, the greater the heat. Air has a high resistance to current flow. As the electrons jump the air gap between the end of the electrode and the work, a great deal of heat is produced. Electrons flowing across an air gap produce an arc.

Measurement

Three units are used to measure a welding current.

The three units are VOLTAGE (V), AMPERAGE (A), and WATTAGE (W). Voltage, or volts (V), is the measurement of electrical pressure, in the same way that pounds per square inch is a measurement of water pressure. Voltage controls the maximum gap the electrons can jump to form the arc. A higher voltage can jump a larger gap. Amperage, or amps (A), is the measurement of the total number of electrons flowing, in the same way that gallons is a measurement of the amount of water flowing. Amperage controls the size of the arc. Wattage,

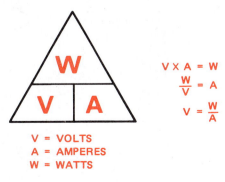

$$V \times A = W$$
$$\frac{W}{V} = A$$
$$V = \frac{W}{A}$$

V = VOLTS
A = AMPERES
W = WATTS

Figure 3-3

or watts (W), are calculated by multiplying voltage (V) times amperes (A), **Figure 3-3**. Wattage is a measurement of the amount of electrical energy or power in the arc. The amount of watts being put into a weld per inch (cm) controls the width and depth of the weld bead, **Figure 3-4**.

Temperature

The temperature of a welding arc exceeds 11,000°F (6,000°C). The exact temperature depends on the resistance to the current flow. The resistance is affected by the arc length and the chemical composition of the gases formed as the electrode covering burns and vaporizes. As the arc lengthens, the resistance increases, thus causing a rise in the arc voltage and temperature. The shorter the arc, the lower the arc temperature produced.

Most shielded metal arc welding electrodes have chemicals added to their coverings to stabilize the arc. These arc stabilizers reduce the arc resistance, making it easier to hold an arc. By lowering the resistance, the arc stabilizers also lower the arc temperature. Other chemicals within the gaseous cloud around the arc may raise or lower the resistance.

The amount of heat produced is determined by the size of the electrode and the amperage setting. Not all of the heat produced by an arc reaches the weld. Some of the heat is radiated away in the form of light and heat waves, **Figure 3-5**. Additional heat is carried away with the hot gases formed by the electrode covering. Heat also is lost through conduction in the work. In total, about 50% of all heat produced by an arc is missing from the weld.

The 50% of the remaining heat the arc produced is not distributed evenly between both ends of the arc. This distribution depends on the composition of the electrode's coating, the type of welding current, and the polarity of the electrode's coating.

Currents

The three different types of current used for welding are alternating current (AC), direct-current electrode negative (DCEN), and direct-current electrode positive (DCEP). The terms DCEN and DCEP have replaced the former terms direct-current straight polarity (DCSP) and direct-current reverse polarity (DCRP). DCEN and DCSP are the same currents, and DCEP and DCRP are the same

currents. Some electrodes can be used with only one type of current. Others can be used with two or more types of current. Each welding current has a different effect on the weld.

DCEN. In direct-current electrode negative, the electrode is negative, and the work is positive, **Figure 3-6**. DCEN welding current produces a high electrode melting rate.

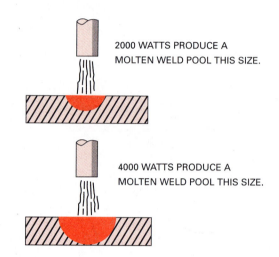

2000 WATTS PRODUCE A MOLTEN WELD POOL THIS SIZE.

4000 WATTS PRODUCE A MOLTEN WELD POOL THIS SIZE.

Figure 3-4 The molten weld pool size depends on the energy (watts), the metal mass, and the thermal conductivity.

Figure 3-5 Energy is lost from the weld in the form of radiation and convection.

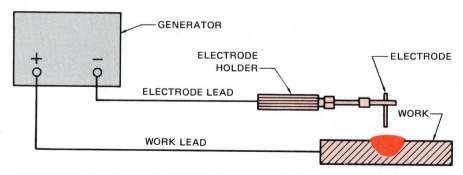

Figure 3-6 Straight polarity (DCSP), electrode negative (DCEN).

DCEP. In direct-current electrode positive, the electrode is positive, and the work is negative, **Figure 3-7**. DCEP current produces the best welding arc characteristics.

AC. In alternating current, the electrons change direction every 1/120 of a second so that the electrode and work alternate from anode to cathode, **Figure 3-8**. The rapid reversal of the current flow causes the welding heat to be evenly distributed on both the work and the electrode; that is, half on the work and half on the electrode. The even heating gives the weld bead a balance between penetration and buildup.

TYPES OF WELDING POWER

Welding power can be supplied as:

- **Constant Voltage (CV)** — The arc voltage remains constant at the selected setting even if the arc length and amperage increase or decrease.

- **Rising Arc Voltage (RAV)** — The arc voltage increases as the amperage increases.

- **Constant Current (CC)** — The total welding current (watts) remains the same. This type of power is also called Drooping Arc Voltage (DAV), because the arc voltage decreases as the amperage increases.

The shielded metal arc welding (SMAW) process requires a constant current arc voltage characteristic, illustrated by the red line in **Figure 3-9**. The shielded metal arc welding machine's voltage output decreases as current increases. This output power supply provides a reasonably high open circuit voltage before the arc is struck. The high open circuit voltage quickly stabilizes the arc. The arc voltage rapidly drops to the lower closed circuit level after the arc is struck. Following this short starting surge, the power (watts) remains almost constant despite the changes in arc length. With a constant voltage output, small changes in arc length would cause the power (watts) to make large swings. The welder would lose control of the weld.

OPEN CIRCUIT VOLTAGE

Open circuit voltage is the voltage at the electrode before striking an arc (with no current being drawn). This voltage is usually between 50 V and 80 V. The higher the open circuit voltage, the easier it is to strike an arc. The higher voltage also increases the chance of electrical shock.

■ CAUTION

The maximum safe open circuit voltage for welders is 80 volts.

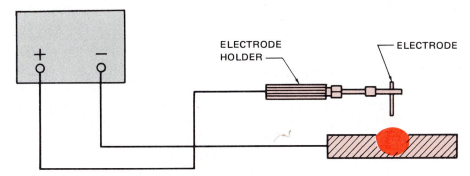

Figure 3-7 Reverse polarity (DCRP), electrode positive (DCEP).

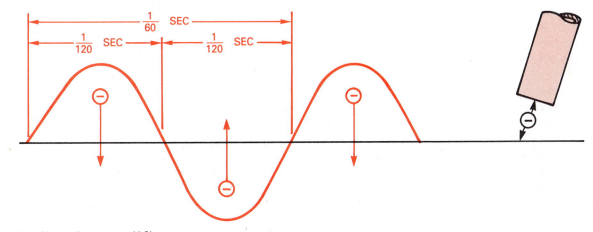

Figure 3-8 Alternating current (AC).

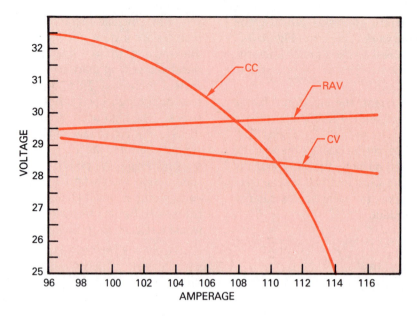

Figure 3-9 CV (constant voltage), RAV (rising arc voltage), CC (constant current).

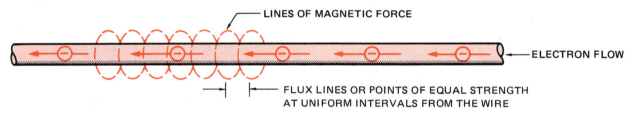

Figure 3-10 Magnetic force around a wire.

OPERATING VOLTAGE

Operating voltage, or closed circuit voltage, is the voltage at the arc during welding. This voltage will vary with arc length, type of electrode being used, type of current, and polarity. The voltage will be between 17 V and 40 V.

ARC BLOW

When electrons flow they create lines of magnetic force that circle around the line of flow, **Figure 3-10**. Lines of magnetic force are referred to as magnetic flux lines. These lines space themselves evenly along a current-carrying wire. If the wire is bent, the flux lines on one side are compressed together, and those on the other side are stretched out, **Figure 3-11**. The unevenly spaced flux lines try to straighten the wire so that the lines can be evenly spaced once again. The force that they place on the wire is usually small. However, when welding with very high amperages, 600 amperes or more, the force may cause the wire to move.

The welding current flowing through a plate or any residual magnetic fields in the plate shill result in uneven flux lines. These uneven flux lines can, in turn, cause an arc to move during a weld. This movement of the arc is

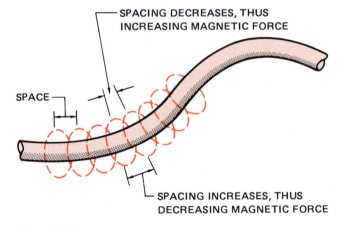

Figure 3-11

called *arc blow*. Arc blow makes the arc drift like a string would drift in the wind. Arc blow is more noticeable in corners, at the ends of plates, and when the work lead is connected to only one side of a plate, **Figure 3-12**. If arc blow is a problem, it can be controlled by connecting the work lead to the end of the weld joint and making the weld in the direction away from the work lead, **Figure 3-13**. Another way of controlling arc blow is to use two work leads, one on each side of the weld. The best way to eliminate arc blow is to use alternating current. AC usually does not allow the flux lines to build long enough to bend the arc before the current changes direction. If it is impossible to move the work connection or to change to AC, a very short arc length can help control arc blow. A large tack weld or a change in the electrode angle can also help control arc blow.

Arc blow may not be a problem as you are learning to weld in the shop, because most welding tables are all steel. If you are using a pipe stand to hold your welding practice plates, arc blow can become a problem. In this case, try reclamping your practice plates.

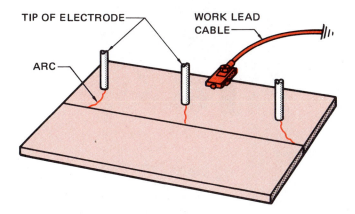

Figure 3-12 Arc blow.

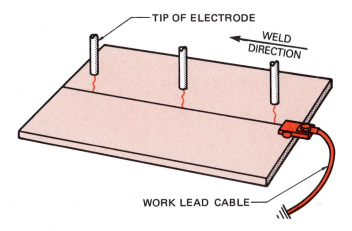

Figure 3-13 Correct current connections to control arc blow.

TYPES OF POWER SOURCES

Two types of electrical devices can be used to produce the low-voltage, high-amperage current combination that arc welding requires. One type uses electric motors or internal combustion engines to drive alternators or generators. The other type uses step-down transformers. Because transformer type welding machines are quieter, more energy efficient, require less maintenance, and are less expensive, they are now the industry standards. However, engine-powered generators are still widely used for portable welding.

Transformers

A welding transformer uses the alternating current (AC) supplied to the welding shop at a high voltage to produce the low voltage welding power. As electrons flow through a wire they produce a magnetic field around the wire. If the wire is wound into a coil the weak magnetic field of each wire is concentrated to produce a much stronger central magnetic force. Because the current being used is alternating or reversing each 1/120 of a second, the magnetic field is constantly being built and allowed to collapse. By placing a second or secondary winding of wire in the magnetic field produced by the first or primary winding a current will be induced in the secondary winding. The placing of an iron core in the center of these coils will increase the concentration of the magnetic field, **Figure 3-14.**

A transformer with more turns of wire in the primary winding than in the secondary winding is known as a step-down transformer. A step-down transformer takes a high-voltage, low-amperage current and changes it into a low-voltage, high-amperage current. Except for some power lost by heat within a transformer, the power (watts) into a transformer equals the power (watts) out because the volts and amperes are mutually increased and decreased.

A transformer welder is a step-down transformer. It takes the high line voltage (110 V, 220 V, 440 V, etc.) and low-amperage current (30 A, 50 A, 60 A, etc.) and changes it into 17 V to 45 V at 190 A to 590 A.

Welding machines can be classified by the method by which they control or adjust the welding current. The

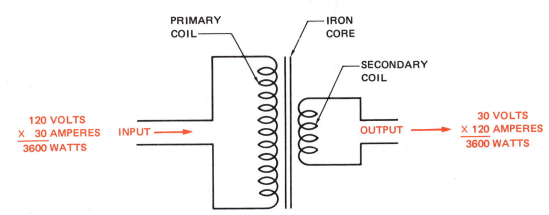

Figure 3-14 Diagram of a step-down transformer.

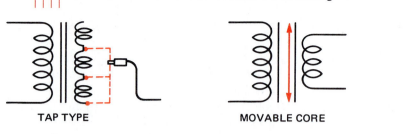

Figure 3-15 Major types of adjustable welding transformers.

major classifications are multiple coil, called taps, movable coil or movable core, **Figure 3-15**, and inverter type.

Multiple Coil

The multiple-coil machine, or tap-type machine, allows the selection of different current settings by tapping into the secondary coil at a different turn value. The greater the number of turns, the higher is the amperage induced in the turns. These machines may have a large number of fixed amperes, **Figure 3-16**, or they may have two or more amperages that can be adjusted further with a fine adjusting knob. The fine adjusting knob may be marked in amperes, or it may be marked in tenths, hundredths, or in any other unit.

EXPERIMENT 3-1

Estimating Amperages

Using a pencil and paper, you will prepare a rough estimate of the amperage setting of a welding machine. **Figure 3-17** shows a welding machine with low, medium, and high tap amperage ranges. A fine adjusting knob is marked with ten equal divisions, and each division is again divided by ten smaller lines.

The machine is set on the medium range, 50 to 250 amperes, and the fine adjusting knob is turned until it points to the line marked 5 (halfway between 0 and 10). This means that the amperage is halfway from 50 to 250, or 150 amperes. If the fine adjusting knob points between 2 and 3, the resulting amperage is 1/4 of the way from 50 to 250, or about 100 amperes. If the knob points between 7 and 8, the amperage is 3/4 of the way from 50 to 250, or about 200 amperes. If the knob points at 4, the amperage is more than 100 but a little less than 150, or about 130 to 140 amperes. What is the amperage if the knob points at 6?

Since this is a method of estimating only, the amperage value obtained is close enough to allow an arc to be struck. The welder can then finish the fine adjusting to obtain a good weld. ◆

EXPERIMENT 3-2

Calculating the Amperage Setting

Using a pencil and paper or calculator, you will calculate the exact value for each space on the fine adjusting knob of a welding machine.

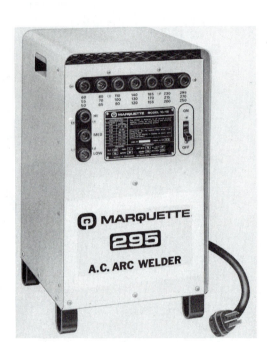

Figure 3-16 Tap-type transformer welding machine. (Courtesy of ESAB Welding and Cutting Products.)

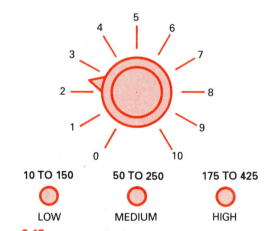

Figure 3-17

With the machine set on the medium range, from 50 to 250 amperes, first subtract the low amperage from the high amperage to get the amperage spread (250 - 50 = 200). Now divide the amperage spread by the number of units shown on the fine adjusting knob (200 ÷ 10 = 20). Each unit is equal to a 20-ampere increase, **Figure 3-18**. When the knob points to 0, the amperage is 50; when the knob points to 1, the amperage is 70; and at 2,

SETTING	VALUE IN AMPERES
0 = 50 + 0,	or 50 A
1 = 50 + 20,	or 70 A
2 = 50 + 40,	or 90 A
3 = 50 + 60,	or 110 A
4 = 50 + 80,	or 130 A
5 = 50 + 100,	or 150 A
6 = 50 + 120,	or 170 A
7 = 50 + 140,	or 190 A
8 = 50 + 160,	or 210 A
9 = 50 + 180,	or 230 A
10 = 50 + 200,	or 250 A

Figure 3-18

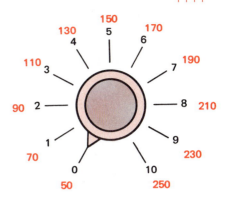

Figure 3-19

the amperage is 90, **Figure 3-19.** There are 100 small units on the fine adjusting knob. Dividing the amperage spread by the number of small units gives the amperage value for each unit (200 ÷ 100 = 2). Therefore, if the knob points to 6.1, the amperage is set at a value of 50 + 120 + 2 = 172 amperes. This method provides a good starting place for the current setting, but if the welding is to be made in accordance to a welding procedure's specific amperage setting it will be necessary to use a calibrated meter to make the correct setting. ◆

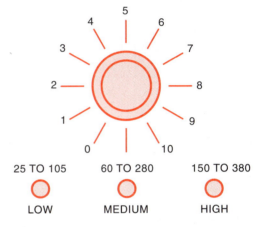

Figure 3-20

PRACTICE 3-1

Estimating Amperages

Using a pencil and paper and the amperage ranges given in this practice (or from machines in the shop), you will estimate the amperage when the knob is at the 1/4, 1/2, and 3/4 settings, **Figure 3-20.** ◆

PRACTICE 3-2

Calculating Amperages

Using a pencil and paper or a calculator, and the amperage ranges given in this practice (or from machines in the shop), you will calculate the amperages for each of the following knob settings: 1, 4, 7, 9, 2.3, 5.7, and 8.5. ◆

Movable Coil or Core

Movable coil or movable core machines are adjusted by turning a handwheel that moves the internal parts closer together or farther apart. The adjustment may also be made by moving a lever, **Figure 3-21.** These machines may have a high and low range, but they do not have a fine adjusting knob. The closer the primary and secondary coils are, the greater is the induced current; the greater the distance between the coils, the smaller is the induced current, **Figure 3-22.** Moving

Figure 3-21 A movable, core-type welding machine. (Courtesy of Lincoln Electric, Cleveland, OH.)

the core in concentrates more of the magnetic force on the secondary coil, thus increasing the current. Moving the core out allows the field to disperse, and the current is reduced, **Figure 3-23.**

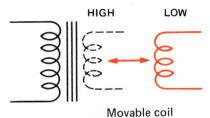

Figure 3-22 Movable coil.

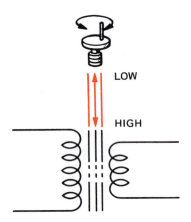

Figure 3-23 Movable core.

Inverter

Inverter welding machines will be much smaller than other types of machines of the same amperage range. This smaller size makes the welder much more portable as well as increasing the energy efficiency. In a standard welding transformer the iron core used to concentrate the magnetic field in the coils must be a specific size. The size of the iron core is determined by the length of time it takes for the magnetic field to build and collapse. By using solid state electronic parts the incoming power in an inverter welder is changed from 60 cycles a second to several thousand cycles a second. This higher frequency allows the use of a transformer that may be as light as 7 pounds and still do the work of a standard transformer weighing 100 pounds. Additional electronic parts remove the high frequency for the output welding power.

The use of electronics in the inverter type welder allows it to produce any desired type of welding power. Before the invention of this machine, each type of welding required a separate machine. Now a single welding machine can produce the specific type of current needed for shielded metal arc welding, gas tungsten arc welding, gas metal arc welding, and plasma arc cutting. Because the machine can be light enough to be carried closer to work, shorter welding cables can be used. The welder does not have to walk as far to adjust the machine. Welding machine power wire is cheaper than welding cables. Some manufacturers produce machines that can be stacked so that when you need a larger machine all you have to do is add another unit to your existing welder, **Figure 3-24**.

Figure 3-24 Typical 300-ampere inverter-type power supply weighing only 70 pounds. (Courtesy of Cyclomatic Industries Inc.)

GENERATORS AND ALTERNATORS

Generators and alternators both produce welding electricity from a mechanical power source. Both devices have an armature that rotates and a stator that is stationary. As a wire moves through a magnetic force field, electrons in the wire are made to move, producing electricity.

In an alternator, magnetic lines of force rotate inside a coil of wire, **Figure 3-25**. An alternator can produce AC only. In a generator, a coil of wire rotates inside a magnetic force. A generator can produce AC or DC. It is possible for alternators and generators to both use diodes to change the AC to DC for welding. In generators, the welding current is produced on the armature and is picked up with brushes, **Figure 3-26**. In alternators, the welding current is produced on the stator, and only the small current for the electromagnetic force field goes across the brushes. Therefore, the brushes in an alternator are smaller and last longer. Alternators can be

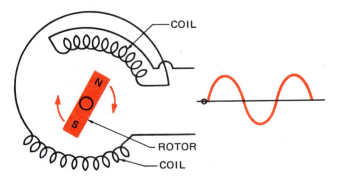

Figure 3-25 Schematic diagram of an alternator.

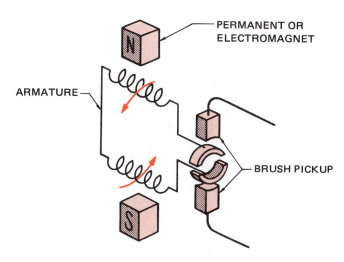

Figure 3-26 Diagram of a generator.

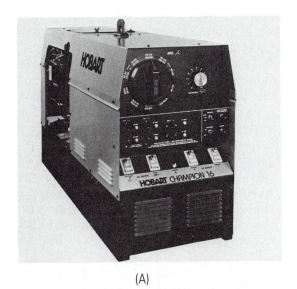

(A)

(B)

Figure 3-27 Portable engine generator welders. (A. Courtesy of Hobart Brothers Company; B. Courtesy of Lincoln Electric, Cleveland, OH.)

smaller in size and lighter in weight than generators and still produce the same amount of power.

Engine-driven generators and alternators may run at the welding speed all the time, or they may have an option that reduces their speed to an idle when welding stops. This option saves fuel and reduces wear on the welding machine. To strike an arc when using this type of welder, stick the electrode to the work for a second. When you hear the welding machine (welder) pick up speed, remove the electrode from the work and strike an arc. In general, the voltage and amperage are too low to start a weld, so shorting the electrode to the work should not cause the electrode to stick. A timer can be set to control the length of time that the welder maintains speed after the arc is broken. The time should be set long enough to change electrodes without losing speed.

Portable welders often have 110-volt or 220-volt plug outlets, which can be used to run grinders, drills, lights, and other equipment. The power provided may be AC or DC. If DC is provided, only equipment with brush-type motors or tungsten light bulbs can be used. If the plug is not specifically labeled 110 volts AC, check the owner's manual before using it for such devices as radios or other electronic equipment. A typical portable welder is shown in **Figure 3-27**.

It is recommended that a routine maintenance schedule for portable welders be set up and followed. By checking the oil, coolant, battery, filters, fuel, and other parts, the life of the equipment can be extended. A checklist can be posted on the welder, **Figure 3-28**.

RECTIFIERS

Alternating welding current can be converted to direct current by using a series of rectifiers. A rectifier allows current to flow in one direction only, **Figure 3-29**.

CHECK EACH DAY BEFORE STARTING
Oil level
Water level
Fuel level
CHECK EACH MONDAY
Battery level
Cables
Fuel line filter
CHECK AT BEGINNING OF MONTH
Air filter
Belts and hoses
Change oil and filter
CHECK EACH FALL
Antifreeze
Test battery
Pack wheel bearings
Change gas filter

Figure 3-28 Portable welder checklist. The owner's manual should be checked for any additional items that might need attention.

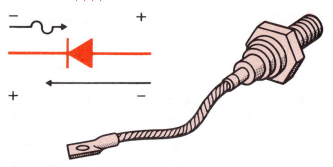

Figure 3-29 Rectifier.

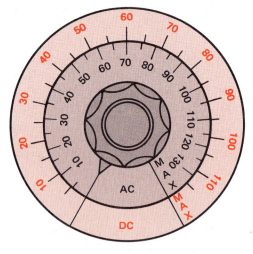

Figure 3-32 Typical dial on an AC–DC transformer rectifier welder.

If one rectifier is added, the welding power appears as shown in **Figure 3-30**. It would be difficult to weld with pulsating power such as this. A series of rectifiers, known as a bridge rectifier, can modify the alternating current so that it appears as shown in **Figure 3-31**.

Rectifiers become hot as they change AC to DC. They must be attached to a heat sink and cooled by having air blown over them. The heat produced by a rectifier reduces the power efficiency of the welding machine. **Figure 3-32** shows the amperage dial of a typical machine. Notice that at the same dial settings for AC and DC, the DC is at a lower amperage. The difference in amperage (power) is due to heat lost in the rectifiers. The loss in power makes operation with AC more efficient and less expensive compared to DC.

A DC adapter for small AC machines is available from manufacturers. For some types of welding, AC does not work properly.

DUTY CYCLE

Welding machines produce internal heat at the same time they produce the welding current. Except for automatic welding machines, welders are rarely used every minute for long periods of time. The welder must

take time to change electrodes, change positions, or change parts. Shielded metal arc welding never continues for long periods of time.

The duty cycle is the percentage of time a welding machine can be used continuously. A 60% duty cycle means that out of every ten minutes, the machine can be used for six minutes at the maximum rated current. When providing power at this level, it must be cooled off for four minutes out of every ten minutes. The duty cycle increases as the amperage is lowered and decreases for higher amperages, **Figure 3-33**. Most welding machines weld at a 60% rate or less. Therefore, most manufacturers list the amperage rating for a 60% duty cycle on the nameplate that is attached to the machine. Other duty cycles are given on a graph in the owner's manual.

The manufacturing cost of power supplies increases in proportion to their rated output and duty cycle. To reduce their price, it is necessary to reduce either their

Figure 3-30 One rectifier in a welding power supply results in pulsating power.

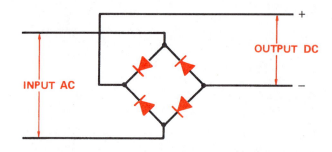

Figure 3-31 Bridge rectifier.

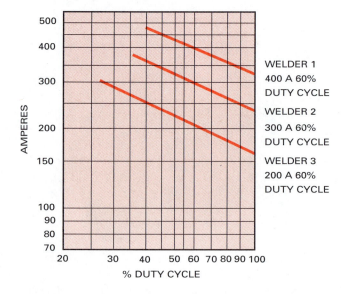

Figure 3-33 Duty cycle of a typical shielded metal arc welding machine.

rating or their duty cycle. For this reason, some home-hobby welding machines may have duty cycles as low as 20% even at a low welding setting of 90 to 100 amperes. The duty cycle on these machines should never be exceeded because a buildup of the internal temperature can cause the transformer insulation to break down, damaging the power supply.

PRACTICE 3-3

Reading Duty Cycle Chart

Using a pencil and paper and the Duty Cycle Chart in **Figure 3-33** (or one from machines in the shop), you will determine the following:

Welder 1: maximum welding amperage
    percent duty cycle at maximum amperage

Welder 2: maximum welding amperage
    percent duty cycle at maximum amperage

Welder 3: maximum welding amperage
    percent duty cycle at maximum amperage

Welder 1: maximum welding amperage at 100% duty cycle

Welder 2: maximum welding amperage at 100% duty cycle

Welder 3: maximum welding amperage at 100% duty cycle

WELDING CABLES

The terms welding cables and welding leads mean the same thing. Cables to be used for welding must be flexible, well-insulated, and the correct size for the job. Most welding cables are made from stranded copper wire. Some manufacturers sell a newer type of cable made from aluminum wires. The aluminum wires are lighter and less expensive than copper. Because aluminum as a conductor is not as good as copper for a given wire size, the aluminum wire should be one size larger than would be required for copper.

The insulation on welding cables will be exposed to hot sparks, flames, grease, oils, sharp edges, impact, and other types of wear. To withstand such wear, only specially manufactured insulation should be used for welding cable. Several new types of insulation are available that will give longer service against these adverse conditions.

As electricity flows through a cable, the resistance to the flow causes the cable to heat up and increase the voltage drop. To minimize the loss of power and prevent overheating, the electrode cable and work cable must be the correct size. **Table 3-1** lists the minimum size cable that is required for each amperage and length. Large welding lead sizes make electrode manipulation difficult. Smaller cable can be spliced to the electrode end of a large cable to make it more flexible. This whip-end cable must not be over 10 ft (3 m) long.

■ CAUTION

A splice in a cable should not be within 10 ft (3 m) of the electrode because of the possibility of electrical shock.

PRACTICE 3-4

Determining Welding Lead Sizes

Using a pencil and paper and Table 3-1, Copper and Aluminum Welding Lead Sizes, you will determine the following:

1. The minimum copper welding lead size for a 200-amp welder with 100-ft (30-m) leads.

2. The minimum copper welding lead size for a 125-amp welder with 225-ft (69-m) leads.

3. The maximum length aluminum welding lead that can carry 300 amps.

Splices and end lugs are available from suppliers. Be sure that a good electrical connection is made whenever splices or lugs are used. A poor electrical connection will result in heat buildup, voltage drop, and poor service from the cable. Splices and end lugs must be well insulated against possible electrical shorting, **Figure 3-34**.

ELECTRODE HOLDERS

The electrode holder should be of the proper amperage rating and in good repair for safe welding. Electrode holders are designed to be used at their maximum amperage rating or less. Higher amperage values will cause the holder to overheat and burn up. If the holder is too large for the amperage range being used, manipulation is hard, and operator fatigue increases. Make sure that the correct size holder is chosen, **Figure 3-35**.

■ CAUTION

Never dip a hot electrode holder in water to cool it off. The problem causing the holder to overheat should be repaired.

A properly sized electrode holder can overheat if the jaws are dirty or too loose, or if the cable is loose. If the holder heats up, welding power is being lost. In addition, a hot electrode holder is uncomfortable to work with.

Replacement springs, jaws, insulators, handles, screws, and other parts are available to keep the holder in good working order, **Figure 3-36**. To prevent excessive damage to the holder, welding electrodes should not be burned too short. A 2-in. (51-mm) electrode stub is short enough to minimize electrode waste and save the holder.

Amperes		Copper Welding Lead Sizes								
		100	150	200	250	300	350	400	450	500
ft	m									
50	15	2	2	2	2	1	1/0	1/0	2/0	2/0
75	23	2	2	1	1/0	2/0	2/0	3/0	3/0	4/0
100	30	2	1	1/0	2/0	3/0	4/0	4/0		
125	38	2	1/0	2/0	3/0	4/0				
150	46	1	2/0	3/0	4/0					
175	53	1/0	3/0	4/0						
200	61	1/0	3/0	4/0						
250	76	2/0	4/0							
300	91	3/0								
350	107	3/0								
400	122	4/0								

Amperes		Aluminum Welding Lead Sizes								
		100	150	200	250	300	350	400	450	500
ft	m									
50	15	2	2	1/0	2/0	2/0	3/0	4/0		
75	23	2	1/0	2/0	3/0	4/0				
100	30	1/0	2/0	4/0						
125	38	2/0	3/0							
150	46	2/0	3/0							
175	53	3/0								
200	61	4/0								
225	69	4/0								

Table 3-1 Copper and Aluminum Welding Lead Sizes.

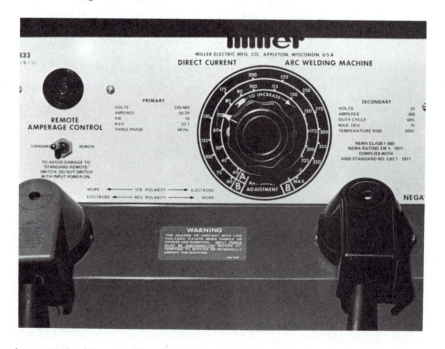

Figure 3-34 Some power lug protection is provided by using add-on boots. (Courtesy of D. Rhodes.)

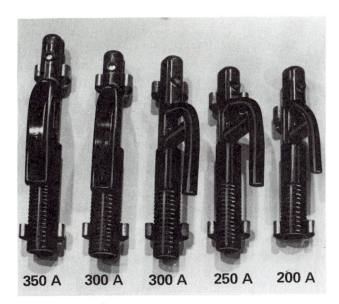

Figure 3-35 Various sizes of electrode holders.

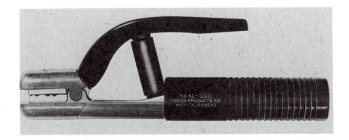

Figure 3-36 Replaceable parts of an electrode holder. (Courtesy of Tweco Products Company.)

PRACTICE 3-5

Repairing Electrode Holders

Using the manufacturer's instructions for your type of electrode holder, required hand tools, and replacement parts, you will do the following:

■ CAUTION

Before starting any work, make sure that the power to the welder is off and locked off or the welding lead has been removed from the machine.

1. Remove the electrode holder from the welding cable.

2. Remove the jaw insulating covers.

3. Replace the jaw insulating covers.

4. Reconnect the electrode holder to the welding cable.

5. Turn on the welding power or reconnect the welding cable to the welder.

6. Make a weld to ensure that the repair was made correctly.

WORK CLAMPS

The work clamp must be the correct size for the current being used, and it must clamp tightly to the material. Heat can build up in the work clamp, reducing welding efficiency, just as was previously described for the electrode holder. Power losses in the work clamp are often overlooked. The clamp should be touched occasionally to find out if it is getting hot.

In addition to power losses due to poor work lead clamping, a loose clamp may cause arcing that can damage a part. If the part is to be moved during welding, a swivel-type work clamp may be needed, Figure 3-37. It may be necessary to weld a tab to thick parts so that the work lead can be clamped to the tab, Figure 3-38.

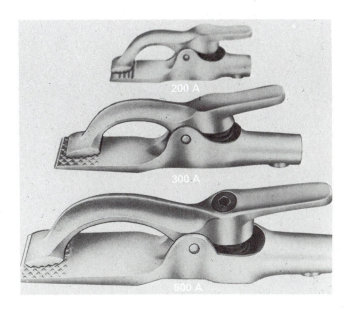

Figure 3-37 A work clamp may be attached to the workpiece. (Courtesy of Lenco, Inc.)

Figure 3-38 Tack welded ground to part.

SETUP

Arc welding machines should be located near the welding site, but far enough away so that they are not covered with spark showers. The machines may be stacked to save space, but there must be enough room between the machines to ensure the air can circulate so as to keep the machines from overheating. The air that is circulated through the machine should be as free as possible of dust, oil, and metal filings. Even in a good location, the power should be turned off periodically and the machine blown out with compressed air, **Figure 3-39**.

The welding machine should be located away from cleaning tanks and any other sources of corrosive fumes that could be blown through it. Water leaks must be fixed and puddles cleaned up before a machine is used.

Power to the machine must be fused, and a power shut-off switch provided. The switch must be located so that it can be reached in an emergency without touching either the machine or the welding station. The machine case or frame must be grounded.

The welding cables should be sufficiently long to reach the work station but not so long that they must always be coiled. Cables must not be placed on the floor in aisles or walkways. If workers must cross a walkway, the cable must be installed overhead, or it must be protected by a ramp, **Figure 3-40**. The welding machine and its main power switch should be off while a person is installing or working on the cables.

The work station must be free of combustible materials. Screens should be provided to protect other workers from the arc light.

The welding cable should never be wrapped around arms, shoulders, waist, or any other part of the body. If the cable was caught by any moving equipment, such as a forklift, crane, or dolly, a welder could be pulled off balance or more seriously injured. If it is necessary to hold the weight off the cable so that the welding can more easily be done, a free hand can be used. The cable should be held so that if it is pulled it can be easily released.

■ CAUTION

The cable should never be tied to scaffolding or ladders. If the cable is caught by moving equipment, the scaffolding or ladder may be upset, causing serious personal injury.

Check the surroundings before starting to weld. If heavy materials are being moved in the area around you, there should be a safety watch. A safety watch can warn a person of danger while that person is welding.

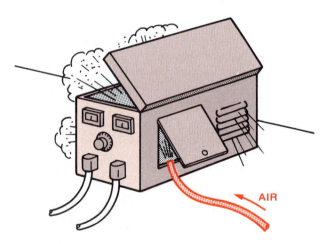

Figure 3-39 Slag, chips from grinding, and dust must be blown out occasionally so that they will not start a fire or cause a short-out or other types of machine failure.

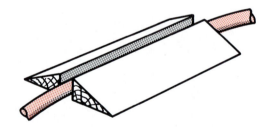

Figure 3-40 To prevent people from tripping when cables must be placed in walkways, lay two blocks of wood beside the cables.

Small Company Uses Welding Robot to Cut Costs

A large manufacturer of robots produces a popular articulated arm robot that has a 6-axis range of motion. The robot is vertically articulated. One customer using this type of robot is a small (30 employees) manufacturer of small-bore hydraulic motors (in addition to agricultural machinery). The company diversified to include the hydraulic motors in its product line when sales of its agricultural products began to decline. The company did not replace any people when it brought in its robot. The robot was purchased for the small bore hydraulics venture; agricultural machine workers who were with the company when the robot came are still at the company.

Even though the company is small, every production machine is computer numerically controlled. The articulated arm robot performs SMA welding of ports on tubing stock. The robot can do three times the welding a human can do, according to the general manager of the company. After watching the human welders, the manager found that they were doing actual welding work about 25% to 30% of the time. The robot, on the other hand, gets actual welding done 75% of the time. The additional welding up-time is a savings for the company and its customers.

The robot performs all necessary welds while working from the middle of a C-shaped workbench, Figure 1. In the cell concept the machined part is not picked up and put down several times before it is finished. It is picked up once and not set down again until the job on that part is completely done. The cells eliminate endless moving, picking, and putting.

The company opted to purchase the robot with a special controller that makes Manufacturing Automation Protocol (MAP) upgrades as simple as plugging in MAP circuit boards and downloading MAP files. The company justified the extra cost of the controller by its plans for future expansion. According to the general manager, "We don't plan to stay small, and a robot that's not equipped with the controller would be obsolete in a year."

The success of the robot in this operation depends inherently on the robot's operator and his/her attitude. The operator is both programmer and operator. He has the attitude that he wants to make the system work. The operator had no programming experience and only a small amount of welding time when the equipment was installed. However, he wanted to learn and had the ability to apply himself. The operator must think, because the robot can't.

The benefits gained from the use of the welding robot include:

- Improved productivity — robotic welding is three times faster than previous manual welding.
- More consistent welding with proper penetration and fewer "leakers."
- Better quality, less workpiece distortion, with only 10% of the weldments requiring followup sizing or honing, compared to 75% of manual welds requiring an after-operation.

(Courtesy of GMF Robotics, Rochester Hills, Michigan 48057.)

Figure 1

REVIEW

1. Describe the welding current.

2. What produces the heat during a shielded metal arc weld?

3. Voltage can be described as _____.

4. Amperage can be described as _____.

5. Wattage can be described as _____.

6. What determines the exact temperature of the shielded metal welding arc?

7. Does all of the heat produced by an SMA weld stay in the weld? Why or why not?

8. What do the following abbreviations mean: AC, DCEN, DCEP, DCSP, and DCRP?

9. Sketch a welding machine, an electrode lead, an electrode holder, an electrode, a work lead, and work connected for DCEN welding.

10. Sketch a welding machine, an electrode lead, an electrode holder, an electrode, a work lead, and work connected for DCEP welding.

11. Why is SMA welding current referred to as *constant current?*

12. What is the higher voltage at the electrode before the arc is struck called? What is its advantage to welding?

13. Referring to the graph in Figure 3-9, page 48, what would the voltage be for the CC power supply at 110 amps? What would the watts be?

14. How does arc blow affect welding?

15. How can arc blow be controlled?

16. How does a welding transformer work?

17. What are *taps* on a welding transformer?

18. What would the approximate amperage setting be if a welder were set to the high range (150 to 350 amps) and the fine adjustment knob was pointing at 5 on a 10-point scale?

19. What are the advantages of the invertor-type welding power supply?

20. What is the difference between the welding current produced by alternators and by generators?

21. What are the advantages of alternators over generators?

22. What must be checked before using the 110-volt power plug on a portable welder?

23. What is meant by a *welder's duty cycle?*

24. Why must a welding machine's duty cycle never be exceeded?

25. Why is copper better than aluminum for welding cables?

26. A splice in a welding cable should never be any closer than _____ to the electrode holder.

27. Why must the electrode holder be correctly sized?

28. What can cause a properly sized electrode holder to overheat?

29. What problem can occur if welding machines are stacked or placed too close together?

30. Why must welding cables never be tied to scaffolding or ladders?

Chapter 4

SHIELDED METAL ARC WELDING OF PLATE

OBJECTIVES

After completing this chapter, the student should be able to

- set the welding amperage correctly.
- explain the effect of changing arc length on a weld.
- control weld bead contour during welding by using the proper weave pattern.
- demonstrate an ability to control undercut, overlap, porosity, and slag inclusions when welding.
- explain the effect of electrode angle on a weld.

KEY TERMS

chill plate	rutile-based flux
arc length	mineral-based electrodes
amperage range	stringer bead
electrode angle	square butt joint
weave pattern	lap joint
cellulose-based fluxes	tee joint

INTRODUCTION

Shielded metal arc welding (SMAW) is the most often used method of joining plate. This method provides a high temperature and concentration of heat, which allow a small molten weld pool to be built up quickly. The addition of filler metal from the electrode adds reinforcement and increases the strength of the weld. SMAW can be performed on almost any type of metal 1/8 in. (3 mm) thick or thicker. A minimum of equipment is required, and it can be portable.

High-quality welds can be consistently produced on almost any type of metal and in any position. The quality of the welds produced depends largely upon the skill of the welder. Developing the necessary skill level requires practice. However, practicing the welds repeatedly without changing techniques will not aid in developing the required skills. Each time a weld is completed it should be evaluated, and then a change should be made in the technique to improve the next weld.

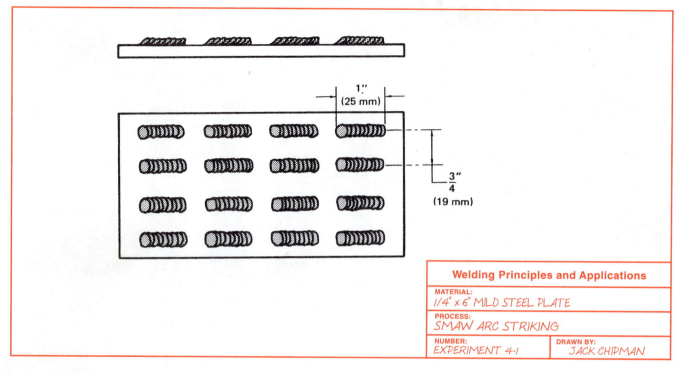

Welding Principles and Applications		
MATERIAL: 1/4" x 6" MILD STEEL PLATE		
PROCESS: SMAW ARC STRIKING		
NUMBER: EXPERIMENT 4-1		**DRAWN BY:** JACK CHIPMAN

Figure 4-1 Striking an arc.

PRACTICE 4-1

Shielded Metal Arc Welding Safety

Using a welding work station, welding machine, welding electrodes, welding helmet, eye and ear protection, welding gloves, proper work clothing, and any special protective clothing that may be required, demonstrate the safe way to prepare yourself and the welding work station for welding. Include in your demonstration appropriate references to burn protection, eye and ear protection, material specification data sheets, ventilation, electrical safety, general work clothing, special protective clothing, and area clean-up.

EXPERIMENT 4-1

Striking the Arc

Using a properly set up and adjusted arc welding machine, the proper safety protection, as demonstrated in Practice 4-1, E6011 welding electrodes having a 1/8-in. (3-mm) diameter, and one piece of mild steel plate, 1/4-in. (6-mm) thick, you will practice striking an arc, **Figure 4-1**.

With the electrode held over the plate, lower your helmet. Scratch the electrode across the plate (like striking a large match), **Figure 4-2**. As the arc is established, slightly raise the electrode to the desired arc length. Hold the arc in one place until the molten weld pool builds to the desired size. Slowly lower the electrode as it burns off and move it forward to start the bead.

If the electrode sticks to the plate, quickly squeeze the electrode holder lever to release the electrode. Break

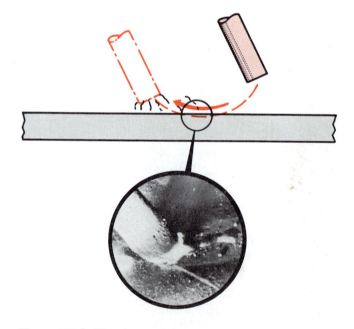

Figure 4-2 Striking the arc.

the electrode free by bending it back and forth a few times. Do not touch the electrode without gloves, because it will still be hot. If the flux breaks away from the end of the electrode, throw out the electrode because restarting the arc will be very difficult, **Figure 4-3**.

Break the arc by rapidly raising the electrode after completing a 1-in. (25-mm) weld bead. Restart the arc as you did before, and make another short weld. Repeat this process until you can easily start the arc each time. Turn off the welding machine and clean up your work area when you are finished welding. ◆

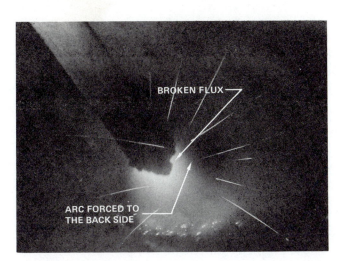

Figure 4-3 If the flux is broken off the end completely or on one side, the arc can be erratic or forced to the side.

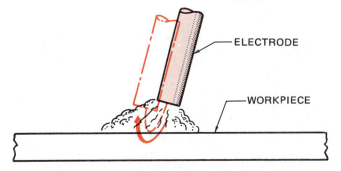

Figure 4-4 Striking the arc on a spot.

EXPERIMENT 4-2

Striking the Arc Accurately

Using the same materials and setup as described in Experiment 4-1, you will start the arc at a specific spot in order to prevent damage to the surrounding plate.

Hold the electrode over the desired starting point. After lowering your helmet, swiftly bounce the electrode against the plate, Figure 4-4. A lot of practice is required to develop the speed and skill needed to prevent the electrode from sticking to the plate.

A more accurate method of starting the arc involves holding the electrode steady by resting it on your free hand like a pool cue. The electrode is rapidly pushed forward so that it strikes the metal exactly where it should. This is an excellent method of striking an arc. Striking an arc in an incorrect spot may cause damage to the base metal.

Practice starting the arc until you can start it within 1/4 in. (6 mm) of the desired location. Turn off the welding machine and clean up your work area when you are finished welding. ◆

EFFECT OF TOO HIGH OR TOO LOW CURRENT SETTINGS

Each welding electrode must be operated in a particular current (amperage) range, Table 4-1. Welding with the current set too low results in poor fusion and

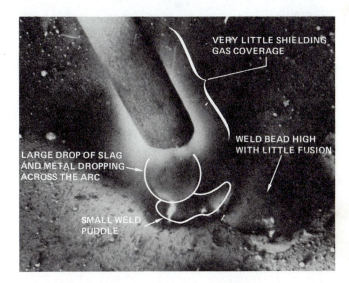

Figure 4-5 Welding with the amperage set too low.

poor arc stability, Figure 4-5. The weld may have slag or gas inclusions because the molten weld pool was not fluid long enough for the flux to react. Little or no penetration of the weld into the base plate may also be evident. With the current set too low, the arc length is very short. A very short arc length results in frequent shortening and sticking of the electrode.

The core wire of the welding electrode is limited in the amount of current it can carry. As the current is increased, the wire heats up because of electrical resistance. This preheating of the wire causes some of the chemicals in the covering to be burned out too early, Figure 4-6. The loss of the proper balance of elements causes poor arc stability. This condition leads to spatter, porosity, and slag inclusions.

An increase in the amount of spatter is also caused by a longer arc. The weld bead made at a high amperage

Electrode	Classification					
Size	E6010	E6011	E6012	E6013	E7016	E7018
3/32 in (2.4 mm)	40–80	50–70	40–90	40–85	75–105	70–110
1/8 in (3.2 mm)	70–130	85–125	75–130	70–120	100–150	90–165
5/32 in (4 mm)	110–165	130–160	120–200	130–160	140–190	125–220

Table 4-1 Welding Amperage Range.

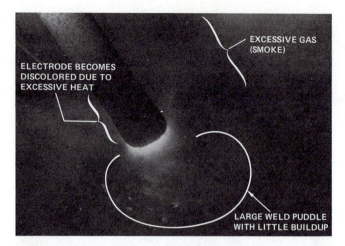

ELECTRODE BECOMES DISCOLORED DUE TO EXCESSIVE HEAT

EXCESSIVE GAS (SMOKE)

LARGE WELD PUDDLE WITH LITTLE BUILDUP

Figure 4-6 Welding with too high an amperage.

setting is wide and flat with deep penetration. The spatter is excessive and is mostly hard. The spatter is called hard because it fuses to the base plate and is difficult to remove, **Figure 4-7**. The electrode covering is discolored more than 1/8 in. (3 mm) to 1/4 in. (6 mm) from the end of the electrode. Extremely high settings may also cause the electrode to discolor, crack, glow red, or burn.

EXPERIMENT 4-3

Effects of Amperage Changes on a Weld Bead

For this experiment, you will need an arc welding machine, welding gloves, safety glasses, welding helmet, appropriate clothing, E6011 welding electrodes having a 1/8-in. (3-mm) diameter, and one piece of mild steel plate, 1/4 in. (6 mm) to 1/2 in. (13 mm) thick. You will observe what happens to the weld bead when the amperage settings are raised and lowered.

Starting with the machine set at approximately 90 A AC or DCRP, strike an arc and make a weld 1 in. (25

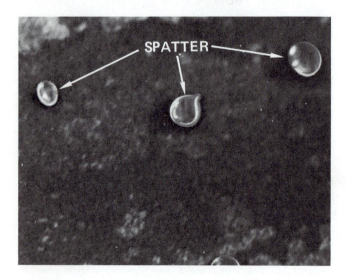

SPATTER

Figure 4-7 Hard weld spatter fused to base metal.

mm) long. Break the arc. Raise the current setting by 10 A, strike an arc, and make another weld 1 in. (25 mm) long. Repeat this procedure until the machine amperage is set at the maximum value.

Replace the electrode and reset the machine to 90 A. Make a weld 1 in. (25 mm) long. Stop and lower the current setting by 10 A. Repeat this procedure until the machine amperage is set at a minimum value.

Cool and chip the plate, comparing the different welds for width, buildup, molten weld pool size, spatter, slag removal, and penetration, **Figure 4-8** (A) and (B). In addition, compare the electrode stubs. Turn off the welding machine and clean up your work area when you are finished welding. ◆

■ CAUTION

Do not change the current settings during welding. A change in the setting may cause arcing inside the machine, resulting in damage to the machine.

ELECTRODE SIZE AND HEAT

The selection of the correct size of welding electrode for a weld is determined by the skill of the welder, the thickness of the metal to be welded, and the size of the metal. Using small-diameter electrodes requires less skill than using large-diameter electrodes. The deposition rate, or the rate that weld metal is added to the weld, is slower when small-diameter electrodes are used. Small-diameter electrodes will make acceptable welds on thick plate, but more time is required to make the weld.

Large-diameter electrodes may overheat the metal if they are used with thin or small pieces of metal. To determine if a weld is too hot, watch the shape of the trailing edge of the molten weld pool, **Figure 4-9**. Rounded ripples indicate the weld is cooling uniformly

(A) WELD BEFORE CLEANING

(B) WELD AFTER CLEANING

Figure 4-8 A weld must be cleaned before it can be accurately inspected.

AMOUNT OF HEAT DIRECTED AT WELD	WELD POOL
TOO LOW	
CORRECT	
TOO HOT	

Figure 4-9 The effect on the shape of the molten weld pool caused by the heat input.

and that the heat is not excessive. If the ripples are pointed, the weld is cooling too slowly because of excessive heat. Extreme overheating can cause a burn-through, which is hard to repair.

To correct an overheating problem, a welder can turn down the amperage, use a shorter arc, travel at a faster rate, use a chill plate (a large piece of metal used to absorb excessive heat), or use a smaller electrode at a lower current setting.

EXPERIMENT 4-4

Excessive Heat

Using a properly set up and adjusted arc welding machine, the proper safety protection, E6011 welding electrodes having a 1/8-in. (3-mm) diameter, and three pieces of mild steel plate, 1/8 in. (3 mm), 3/16 in. (4.8 mm), and 1/4 in. (6 mm) thick, you will observe the effects of overheating on the weld. Make a stringer weld on each of the three plates using the same amperage setting, travel rate, and arc length for each weld. Cool and chip the welds. Then compare the weld beads for width, reinforcement, and appearance.

Using the same amperage settings, make additional welds on the 1/8-in. (3-mm) and 3/16-in. (4.8-mm) plates. Vary the arc lengths and travel speeds for these welds. Cool and chip each weld and compare the beads for width, reinforcement, and appearance. Make additional welds on the 1/8-in. (3-mm) and 3/16-in. (4.8-mm) plates, using the same arc length and travel speed as in the earlier part of this experiment, but at a lower amperage setting. Cool and chip the welds and compare the beads for width, reinforcement, and appearance.

The plates should be cooled between each weld so that the heat from the previous weld does not affect the test results. Turn off the welding machine and clean up your work area when you are finished welding. ◆

ARC LENGTH

The arc length is the distance the arc must jump from the end of the electrode to the plate. As the weld progresses, the electrode becomes shorter as it is consumed. To maintain a constant arc length, the electrode must be lowered continuously. Maintaining a constant arc length is important, as too great a change in the arc length will adversely affect the weld.

As the arc length is shortened, metal transferring across the gap may short out the electrode, causing it to stick to the plate. The weld that results is narrow and has a high buildup, Figure 4-10.

Long arc lengths produce more spatter because the metal being transferred may drop outside of the molten weld pool. The weld is wider and has little buildup, Figure 4-11.

There is a narrow range for the arc length in which it is stable, metal transfer is smooth, and the bead shape is controlled. Factors affecting the length are the type of electrode, joint design, metal thickness, and current setting.

Some welding electrodes, such as E7024, have a thick flux covering. The rate at which the covering melts is slow enough to permit the electrode coating to be rested against the plate. The arc burns back inside the covering as the electrode is dragged along touching the joint,

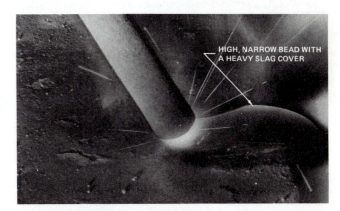

Figure 4-10 Welding with too short an arc length.

Figure 4-11 Welding with too long an arc length.

Figure 4-12 Welding with a drag technique.

Figure 4-12. For this type of welding electrode, the arc length is maintained by the electrode covering.

An arc will jump to the closest metal conductor. On joints that are deep or narrow, the arc is pulled to one side and not to the root, Figure 4-13. As a result, the root fusion is reduced or may be nonexistent, thus causing a poor weld. If a very short arc is used, the arc is forced into the root for better fusion.

Because shorter arcs produce less heat and penetration, they are best suited for use on thin metal or thin-to-thick metal joints. Using this technique, metal as thin as 16 gauge can be arc welded easily. Higher amperage settings are required to maintain a short arc that gives good fusion with a minimum of slag inclusions. The higher settings, however, must be within the amperage range for the specific electrode.

Finding the correct arc length often requires some trial and adjustment. Most welding jobs require an arc length of 1/8 in. (3 mm) to 3/8 in. (10 mm), but this distance varies. It may be necessary to change the arc length when welding to adjust for varying welding conditions.

EXPERIMENT 4-5

Effect of Changing the Arc Length On a Weld

Using an arc welding machine, welding gloves, safety glasses, welding helmet, appropriate clothing, E6011 welding electrodes having a 1/8-in. (3-mm) diameter, and one piece of mild steel plate, 1/4 in. (6 mm) to l/2 in. (13 mm) thick, you will observe the effect of changing the arc length on a weld.

Starting with the welding machine set at approximately 90 A AC or DCRP, strike an arc and make a weld 1 in. (25 mm) long. Continue welding while slowly increasing the arc length until the arc is broken. Restart the arc and make another weld 1 in. (25 mm) long. Welding should again be continued while slowly shortening the arc length until the arc stops. Quickly break the

electrode free from the plate, or release the electrode by squeezing the lever on the electrode holder.

Cool and chip both welds. Compare both welding beads for width, reinforcement, uniformity, spatter, and appearance. Turn off the welding machine and clean up your work area when you are finished welding. ◆

ELECTRODE ANGLE

The electrode angle is measured from the electrode to the surface of the metal. The term used to identify the electrode angle is affected by the direction of travel, generally leading or trailing, Figure 4-14. The relative angle is important because there is a jetting force blowing the metal and flux from the end of the electrode to the plate.

Leading Angle

A leading electrode angle pushes molten metal and slag ahead of the weld, Figure 4-15. When welding in the flat position, caution must be taken to prevent cold lap and slag inclusions. The solid metal ahead of the weld cools and solidifies the molten filler metal and slag before they can melt the solid metal. This rapid cooling prevents the metals from fusing together, Figure 4-16. As the weld passes over this area, heat from the arc may not melt it. As a result, some cold lap and slag inclusions are left.

The following are suggestions for preventing cold lap and slag inclusions:

- Use as little leading angle as possible.
- Ensure that the arc melts the base metal completely, Figure 4-17.
- Use a penetrating-type electrode that causes little buildup.
- Move the arc back and forth across the molten weld pool to fuse both edges.

A leading angle can be used to minimize penetration or to help hold metal in place for vertical welds, Figure 4-18.

Trailing Angle

A trailing electrode angle pushes the molten metal away from the leading edge of the molten weld pool

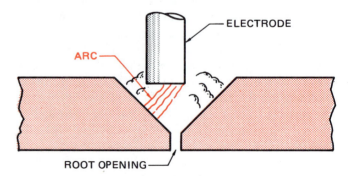

Figure 4-13 The arc may jump to the closest metal, reducing root penetration.

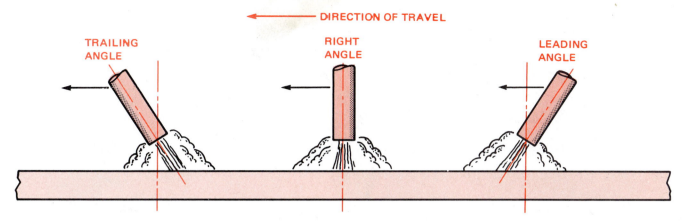

Figure 4-14 Direction of travel and electrode angle.

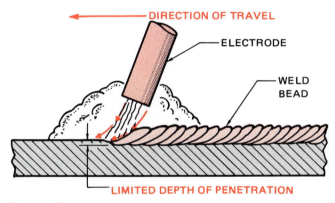

Figure 4-15 Leading electrode angle.

toward the back where it solidifies, **Figure 4-19**. As the molten metal is forced away from the bottom of the weld, the arc melts more of the base metal, which results in deeper penetration. The molten metal pushed to the back of the weld solidifies and forms reinforcement for the weld, **Figure 4-20**.

EXPERIMENT 4-6

Effect of Changing the Electrode Angle On a Weld

Using a properly set up and adjusted arc welding machine, the proper safety protection, E6011 welding electrodes having a 1/8-in. (3-mm) diameter, and one piece of mild steel plate, 1/4 in. (6 mm) to 1/2 in. (13 mm) thick, you will observe the effect of changes in the electrode angle on a weld.

Start welding with a sharp trailing angle. Make a weld about 1 in. (25 mm) long. Closely observe the molten weld pool at the points shown in **Figure 4-21**. Slowly increase the electrode angle and continue to observe the weld.

When you reach a 90° electrode angle, make a weld about 1 in. (25 mm) long. Observe the parts of the weld molten weld pool as shown in Figure 4-21.

Continue welding and change the electrode angle to a sharp leading angle. Observe the weld molten weld pool at the points shown in **Figure 4-22**.

During this experiment, you must maintain a constant arc length, travel speed, and weave pattern if the observations and results are to be accurate.

Cool and chip the weld. Compare the weld bead for uniformity in width, reinforcement, and appearance.

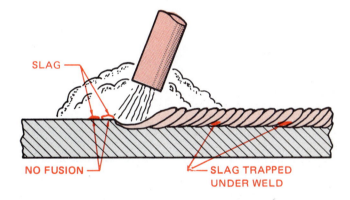

Figure 4-16 Some electrodes, such as E7018, may not remove the deposits ahead of the molten weld pool, resulting in discontinuities within the weld.

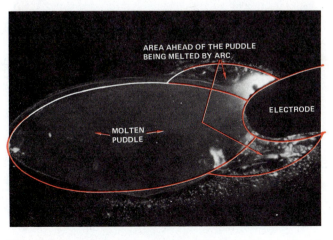

Figure 4-17 Metal being melted ahead of the molten weld pool helps to ensure good weld fusion.

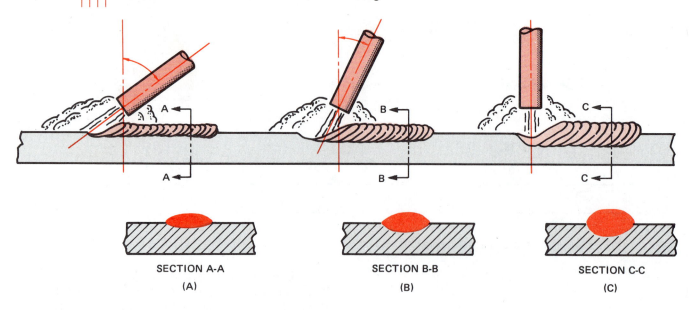

SECTION A-A
(A)

SECTION B-B
(B)

SECTION C-C
(C)

Figure 4-18 Effect of a leading angle on weld bead buildup, width, and penetration. As the angle decreases toward the vertical position (C), penetration increases.

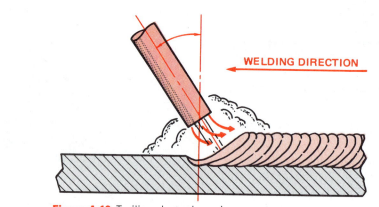

WELDING DIRECTION

Figure 4-19 Trailing electrode angle.

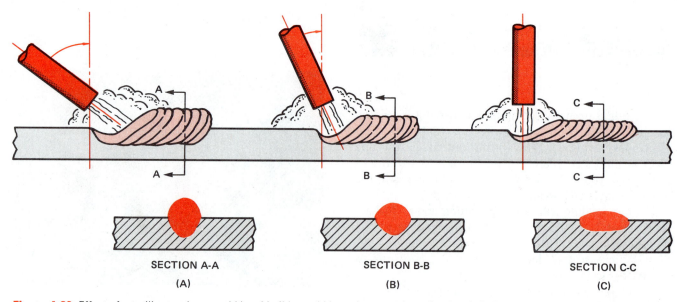

SECTION A-A
(A)

SECTION B-B
(B)

SECTION C-C
(C)

Figure 4-20 Effect of a trailing angle on weld bead buildup, width, and penetration. Section A-A shows a greater weld buildup due to large angle of electrode from a vertical position.

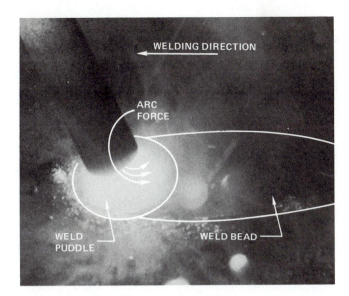

Figure 4-21 Welding with a trailing angle.

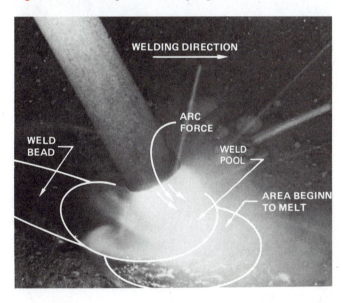

Figure 4-22 Welding with a leading angle.

Turn off the welding machine and clean up your work area when you are finished welding. ◆

ELECTRODE MANIPULATION

The movement or weaving of the welding electrode can control the following characteristics of the weld bead: penetration, buildup, width, porosity, undercut, overlap, and slag inclusions. The exact weave pattern for each weld is often the personal choice of the welder. However, some patterns are especially helpful for specific welding situations. The pattern selected for a flat (1G) butt joint is not as critical as is the pattern selection for other joints and other positions.

Many weave patterns are available for the welder to use. **Figure 4-23** shows ten different patterns that can be used for most welding conditions.

The circular pattern is often used for flat position welds on butt, tee, outside corner joints, and for buildup or surfacing applications. The circle can be made wider or longer to change the bead width or penetration, **Figure 4-24**.

The "C" and square patterns are both good for most 1G (flat) welds, but can also be used for vertical (3G) positions. These patterns can also be used if there is a large gap to be filled when both pieces of metal are nearly the same size and thickness.

The "J" pattern works well on flat (1F) lap joints, all vertical (3G) joints, and horizontal (2G) butt and lap (2F) welds. This pattern allows the heat to be concentrated on

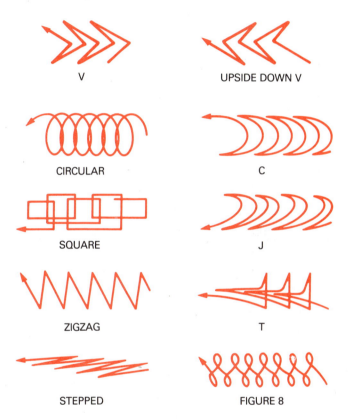

Figure 4-23 Weave patterns.

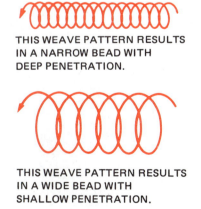

Figure 4-24 Changing the weave pattern size to change the weld bead characteristics.

the thicker plate, Figure 4-25. It also allows the reinforcement to be built up on the metal deposited during the first part of the pattern. As a result, a uniform bead contour is maintained during out-of-position welds.

The "T" pattern works well with fillet welds in the vertical (3F) and overhead (4F) positions, Figure 4-26. It also can be used for deep groove welds for the hot pass. The top of the "T" can be used to fill in the toe of the weld to prevent undercutting.

The straight step pattern can be used for stringer beads, root pass welds, and multiple pass welds in all positions. For this pattern, the smallest quantity of metal is molten at one time as compared to other patterns. Therefore, the weld is more easily controlled. At the same time that the electrode is stepped forward, the arc length is increased so that no metal is deposited ahead of the molten weld pool, Figures 4-27 and 4-28. This action allows the molten weld pool to cool to a controllable size. In addition, the arc burns off any paint, oil, or dirt from the metal before it can contaminate the weld.

The figure 8 pattern and the zigzag pattern are used as cover passes in the flat and vertical positions. Do not weave more than 2 1/2 times the width of the electrode. These patterns deposit a large quantity of metal at one time. A shelf can be used to support the molten weld pool when making vertical welds using either of these patterns, Figure 4-29.

POSITIONING OF THE WELDER AND THE PLATE

The welder should be in a relaxed, comfortable position before starting to weld. A good position is

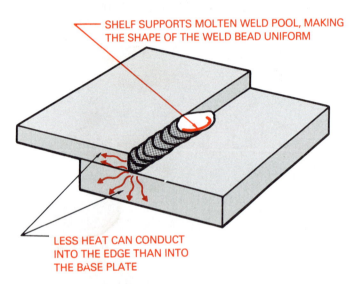

SHELF SUPPORTS MOLTEN WELD POOL, MAKING THE SHAPE OF THE WELD BEAD UNIFORM

LESS HEAT CAN CONDUCT INTO THE EDGE THAN INTO THE BASE PLATE

Figure 4-25 The "J" pattern allows the heat to be concentrated on the thicker plate.

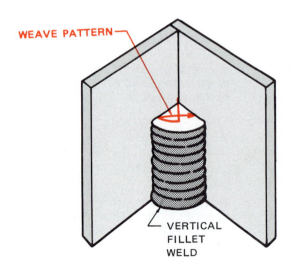

WEAVE PATTERN

VERTICAL FILLET WELD

Figure 4-26 "T" pattern.

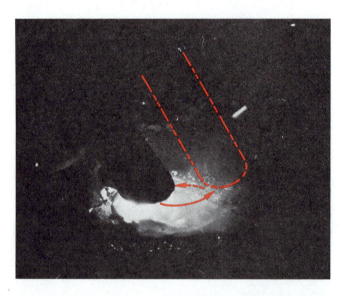

Figure 4-27 The electrode is moved slightly forward and then returned to the weld pool.

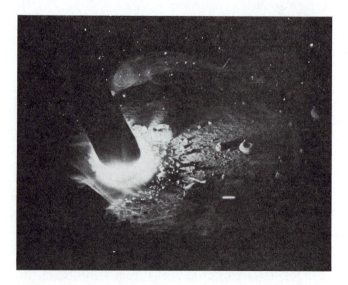

Figure 4-28 The electrode does not deposit metal or melt the base metal.

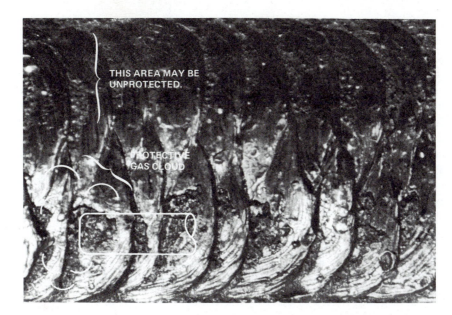

THIS AREA MAY BE
UNPROTECTED.

PROTECTIVE
GAS CLOUD

Figure 4-29 Using the shelf to support the molten pool for vertical welds.

important for both the comfort of the welder and the quality of the welds. Welding in an awkward position can cause welder fatigue, which leads to poor welder coordination and poor-quality welds. Welders must have enough freedom of movement so that they do not need to change position during a weld. Body position changes should be made only during electrode changes.

When the welding helmet is down, the welder is blind to the surroundings. Due to the arc, the field of vision of the welder is also very limited. These factors often cause the welder to sway. To stop this swaying, the welder should lean against or hold on to a stable object. When welding, even if a welder is seated, touching a stable object will make that welder more stable and will make welding more relaxing.

Welding is easier if the welder can find the most comfortable angle. The welder should be in either a seated or a standing position in front of the welding table. The welding machine should be turned off. With an electrode in place in the electrode holder, the welder can draw a straight line along the plate to be welded. By turning the plate to several different angles, the welder should be able to determine which angle is most comfortable for welding, **Figure 4-30.**

PRACTICE WELDS

Practice welds are grouped according to the type of joint and the type of welding electrode. The welder or instructor should select the order in which the welds are made. The stringer beads should be practiced first in each position before the welder tries the different joints in each position. Some time can be saved by starting with the stringer beads. If this is done, it is not necessary to cut or tack the plate together, and a number of beads can be made on the same plate.

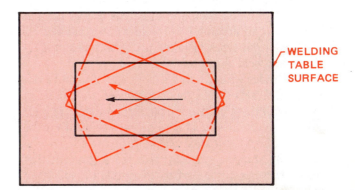

WELDING
TABLE
SURFACE

Figure 4-30 Change the plate angle to find the most comfortable welding position.

Students will find it easier to start with butt joints. The lap, tee, and outside corner joints are all about the same level of difficulty.

Starting with the flat position allows the welder to build skills slowly, so that out-of-position welds become easier to do. The horizontal tee and lap welds are almost as easy to make as the flat welds. Overhead welds are as simple to make as vertical welds, but they are harder to position. Horizontal butt welds are more difficult to perform than most other welds.

Electrodes

Arc welding electrodes used for practice welds are grouped into three filler metal (F number) classes according to their major welding characteristics. The groups are E6010 and E6011, E6012 and E6013, and E7016 and E7018.

F3 E6010 and E6011 Electrodes. Both of these electrodes have cellulose-based fluxes. As a result, these electrodes have a forceful arc with little slag left on the weld bead. See Color Plates 1 and 2.

F2 E6012 and E6013 Electrodes. These electrodes have rutile-based fluxes, giving a smooth, easy arc with a thick slag left on the weld bead. See Color Plates 3 and 4.

F4 E7016 and E7018 Electrodes. Both of these electrodes have a mineral-based flux. The resulting arc is smooth and easy, with a very heavy slag left on the weld bead. See Color Plate 5.

The cellulose- and rutile-based groups of electrodes have characteristics that make them the best electrodes for starting specific welds. The electrodes with the cellulose-based fluxes do not have heavy slags that may interfere with the welder's view of the weld. This feature is an advantage for flat tee and lap joints. Electrodes with the rutile-based fluxes (giving an easy arc with low spatter) are easier to control and are used for flat stringer beads and butt joints.

Unless a specific electrode has been required by a Welding Procedure Specification (WPS), welders can select what they consider to be the best electrode for a specific weld. Without a WPS a recommendation can be made and should be tried, but often the welder has the final choice. An accomplished welder can make defect-free welds on all types of joints using all types of electrodes in any weld position.

Electrodes with mineral-based fluxes should be the last choice. Welds with a good appearance are more easily made with these electrodes, but strong welds are hard to obtain. Without special care being taken during the start of the weld, porosity will be formed in the weld. Figure 4-31 shows a starting tab used to prevent this porosity from becoming part of the finished weld. More information on electrode selection can be found in Chapter 18.

STRINGER BEADS

A straight weld bead on the surface of a plate, with little or no side-to-side electrode movement, is known as a stringer bead. Stringer beads are used by students to practice maintaining arc length, weave patterns, and electrode angle so that their welds will be straight, uniform, and free from defects. Stringer beads, Figure 4-32, are also used to set the machine amperage and for buildup or surfacing applications.

The stringer bead should be straight. A beginning welder needs time to develop the skill of viewing the entire welding area. At first, the welder sees only the arc, Figure 4-33. With practice, the welder begins to see parts of the molten weld pool. After much practice, the welder will see the molten weld pool (front, back, and both sides), slag, buildup, and the surrounding plate, Figure 4-34. Often, at this skill level, the welder may not even notice the arc.

A straight weld is easily made once the welder develops the ability to view the entire welding zone. The welder will occasionally glance around to ensure that the weld is straight. In addition, it can be noted if the weld is uniform and free from defects. The ability of the welder to view the entire weld area is demonstrated by making consistently straight and uniform stringer beads.

After making practice stringer beads, a variety of weave bead patterns should be practiced to gain the ability to control the molten weld pool when welding out-of-position.

Figure 4-32 Stringer bead.

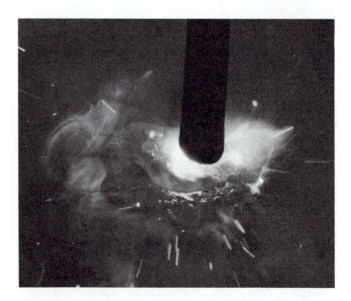

Figure 4-33 New welders frequently see only the arc and sparks from the electrode.

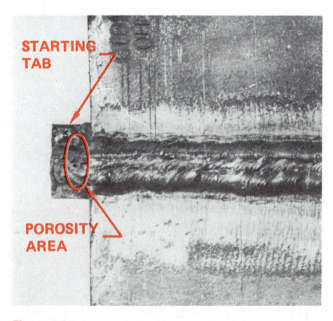

STARTING TAB

POROSITY AREA

Figure 4-31 Porosity is found on the starting tab where it will not affect the weld.

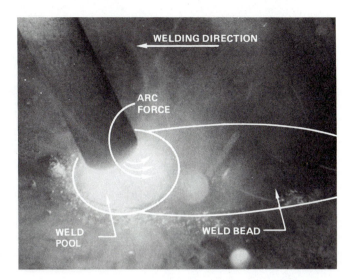

Figure 4-34 More experienced welders can see the molten pool, metal being transferred across the arc, and penetration into the base metal.

PRACTICE 4-2

Straight Stringer Beads in the Flat Position Using E6010 or E6011 Electrodes, E6012 or E6013 Electrodes, and E7016 or E7018 Electrodes

Using a properly set up and adjusted arc welding machine, proper safety protection, as demonstrated in Practice 4-1, arc welding electrodes with a 1/8-in. (3-mm) diameter, and one piece of mild steel plate, 6 in. (152 mm) long x 1/4 in. (6 mm) thick, you will make straight stringer beads.

- Starting at one end of the plate, make a straight weld the full length of the plate.
- Watch the molten weld pool at the points, not the end of the electrode. As you become more skillful, it is easier to watch the molten weld pool.
- Repeat the beads with all three (F) groups of electrodes until you have consistently good beads.
- Cool, chip, and inspect the bead for defects after completing it. Turn off the welding machine and clean up your work area when you are finished welding ◆

PRACTICE 4-3

Stringer Beads in the Vertical Up Position Using E6010 or E6011 Electrodes, E6012 or E6013 Electrodes, and E7016 or E7018 Electrodes

Using the same setup, materials, and electrodes as listed in Practice 4-2, you will make vertical up stringer beads. Start with the plate at a 45° angle.

This technique is the same as that used to make a vertical weld. However, a lower level of skill is required at 45°, and it is easier to develop your skill. After the welder masters the 45° angle, the angle is increased successively until a vertical position is reached, **Figure 4-35**.

Before the molten metal drips down the bead, the back of the molten weld pool will start to bulge, **Figure 4-36**. When this happens, increase the speed of travel and the weave pattern.

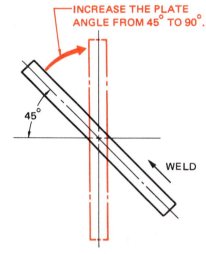

Figure 4-35 Once the 45° angle is mastered, the plate angle is increased successively until a vertical position (90°) is reached.

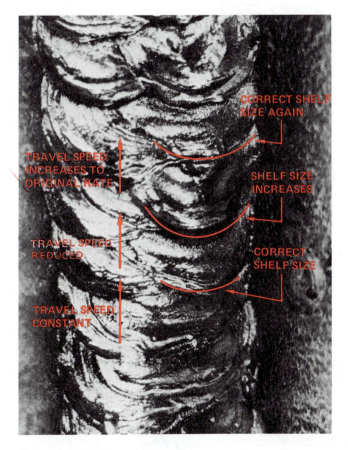

Figure 4-36 E7018 vertical up weld.

Cool, chip, and inspect each completed weld for defects. Repeat the beads as necessary with all three (F) groups of electrodes until consistently good beads are obtained in this position. Turn off the welding machine and clean up your work area when you are finished welding ◆

PRACTICE 4-4

Horizontal Stringer Beads Using E6010 or E6011 Electrodes, E6012 or E6013 Electrodes, and E7016 or E7018 Electrodes

Using the same setup, materials, and electrodes as listed in Practice 4-2, you will make horizontal stringer beads on a plate.

When the welder begins to practice the horizontal stringer bead, the plate may be reclined slightly, Figure

SQUARE BUTT JOINT

The square butt joint is made by tack welding two flat pieces of plate together, Figure 4-39. The space

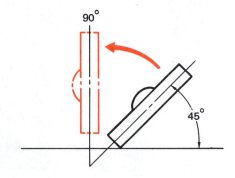

Figure 4-37 Change the plate angle as welding skill improves.

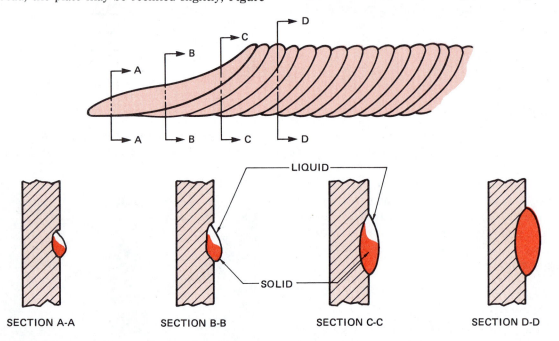

Figure 4-38 The progression of a horizontal bead.

4-37. This placement allows the welder to build the required skill by practicing the correct techniques successfully. The "J" weave pattern is suggested for this practice. As the electrode is drawn along the straight back of the "J," metal is deposited. This metal supports the molten weld pool, resulting in a bead with a uniform contour, Figure 4-38.

Angling the electrode up and back toward the weld causes more metal to be deposited along the top edge of the weld. Keeping the bead small allows the surface tension to hold the molten weld pool in place.

Gradually increase the angle of the plate until it is vertical and the stringer bead is horizontal. Repeat the beads as needed with all three (F) groups of electrodes until consistently good beads are obtained in this position. Turn off the welding machine and clean up your work area when you are finished welding. ◆

Figure 4-39 The tack weld should be small and uniform to minimize its effect on the final weld.

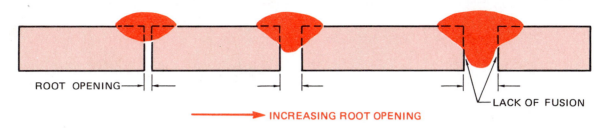

ROOT OPENING

INCREASING ROOT OPENING

LACK OF FUSION

Figure 4-40 Effect of root opening on weld penetration.

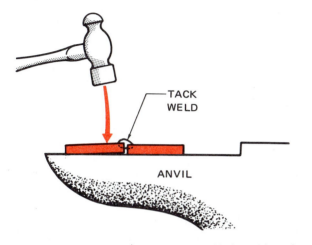

TACK WELD

ANVIL

Figure 4-41 After the plates are tack welded together, they can be forced into alignment by striking them with a hammer.

between the plates is called the root opening or root gap. Changes in the root opening will affect penetration. As the space increases, the weld penetration also increases. The root opening for most butt welds will vary from 0 in. (0 mm) to 1/8 in. (3 mm). Excessively large openings can cause burnthrough or a cold lap at the weld root, **Figure 4-40.**

After a butt weld is completed, the plate can be cut apart so it can be used for rewelding. The strips for butt welding should be no smaller than 1 in. (25 mm) wide. If they are too narrow, there will be a problem with heat buildup.

If the plate strips are no longer flat after the weld has been cut out, they can be tack welded together and flattened with a hammer, **Figure 4-41.**

PRACTICE 4-5

Welded Square Butt Joint in the Flat Position (1G) Using E6010 or E6011 Electrodes, E6012 or E6013 Electrodes, and E7016 or E7018 Electrodes

Using a properly set up and adjusted arc welding machine, proper safety protection, arc welding electrodes having a 1/8-in. (3-mm) diameter, and two or more pieces of mild steel plate, 6 in. (152 mm) long x 1/4 in. (6 mm) thick, you will make a welded square butt joint in the flat position, **Figure 4-42.**

Tack weld the plates together and place them flat on the welding table. Starting at one end, establish a molten weld pool on both plates. Hold the electrode in the molten weld pool until it flows together, **Figure 4-43.** After the gap is bridged by the molten weld pool, start weaving the electrode slowly back and forth across the joint. Moving the electrode too quickly from side to side may result in slag being trapped in the joint, **Figure 4-44.**

Continue the weld along the 6-in. (152-mm) length of the joint. Normally, deep penetration is not required for this type of weld. If full plate penetration is required, the edges of the butt joint should be beveled or a larger than normal root gap should be used. Cool, chip, and inspect the weld for uniformity and soundness. Repeat the welds as needed to master all three (F) groups of electrodes in this position. Turn off the welding machine and clean up your work area when you are finished welding. ◆

PRACTICE 4-6

Vertical (3G) Up-Welded Square Butt Weld Using E6010 or E6011 Electrodes, E6012 or E6013 Electrodes, and E7016 or E7018 Electrodes

Using the same setup, materials, and electrodes as listed in Practice 4-5, you will make vertical up-welded square butt joints.

With the plates at a 45° angle, start at the bottom and make the molten weld pool bridge the gap between the plates, **Figure 4-45.** Build the bead size slowly so that the molten weld pool has a shelf for support. The "C," "J," or square weave pattern works well for this joint.

As the electrode is moved up the weld, the arc is lengthened slightly so that little or no metal is deposited ahead of the molten weld pool. When the electrode is brought back into the molten weld pool, it should be lowered to deposit metal, **Figure 4-46.**

As skill is developed, increase the plate angle until it is vertical. Cool, chip, and inspect the weld for uniformity and defects. Repeat the welds with all three (F) groups of electrodes until you can consistently make welds free of defects. Turn off the welding machine and clean up your work area when you are finished welding. ◆

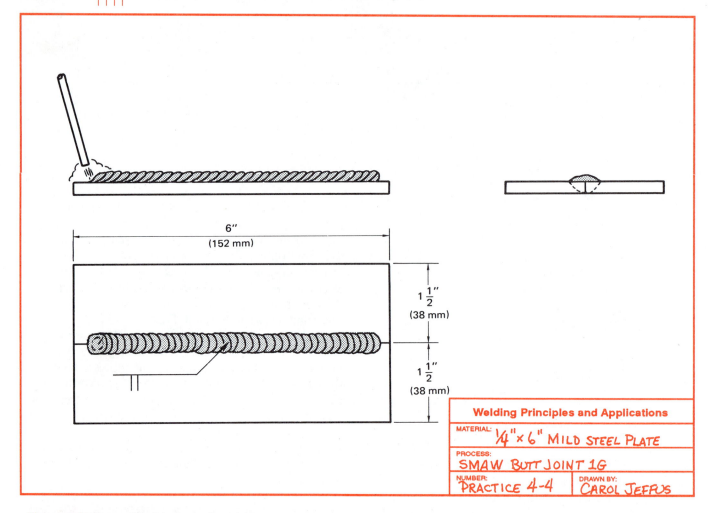

Welding Principles and Applications

MATERIAL: 1/4" x 6" MILD STEEL PLATE

PROCESS: SMAW BUTT JOINT 1G

NUMBER: PRACTICE 4-4

DRAWN BY: CAROL JEFFUS

Figure 4-42 Square butt joint in the flat position.

Figure 4-43 Hold the arc in one area long enough to establish the size of molten weld pool desired.

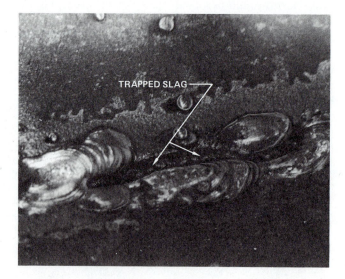

TRAPPED SLAG

Figure 4-44 Moving the electrode from side to side too quickly can result in slag being trapped between the plates.

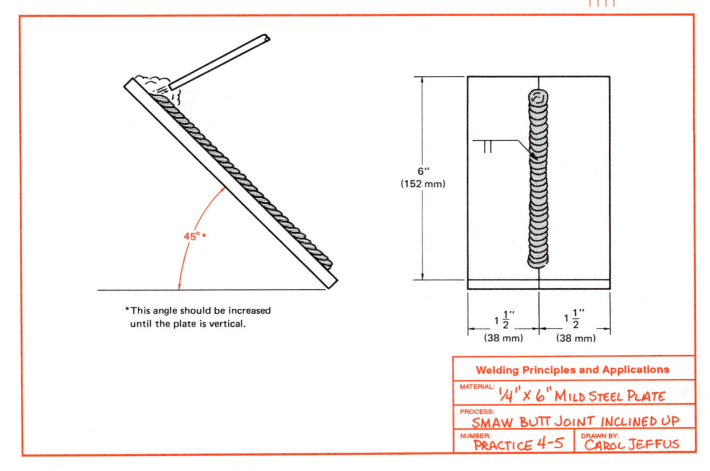

Figure 4-45 Square butt joint in the vertical up position.

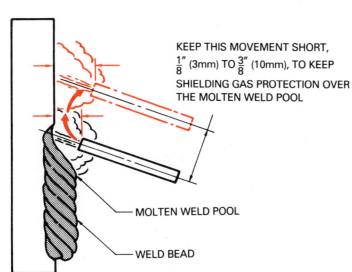

Figure 4-46 Electrode movement for vertical up welds.

PRACTICE 4-7

Welded Horizontal (2G) Square Butt Weld Using E6010 or E6011 Electrodes, E6012 or E6013 Electrodes, and E7016 or E7018 Electrodes

Using the same setup, materials, and electrodes as described in Practice 4-5, you will make a welded horizontal square butt joint.

■ Start practicing these welds with the plate at a slight angle.

■ Strike the arc on the bottom plate and build the molten weld pool until it bridges the gap.

If the weld is started on the top plate, slag will be trapped in the root at the beginning of the weld because of poor initial penetration. The slag may cause the weld to crack when it is placed in service.

The "J" weave pattern is recommended in order to deposit metal on the lower plate so that it can support the bead. By pushing the electrode inward as you cross the gap between the plates, deeper penetration is achieved.

As you acquire more skill, gradually increase the plate angle until it is vertical and the weld is horizontal.

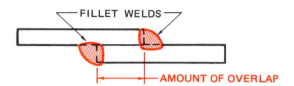

Figure 4-47 Lap joint.

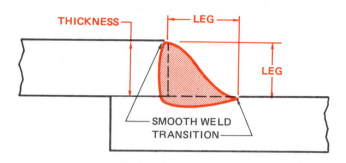

Figure 4-48 The legs of a fillet weld generally should be equal to the thickness of the base metal.

- Cool, chip, and inspect the weld for uniformity and defects.

- Repeat the welds with all three (F) groups of electrodes until you can consistently make welds free of defects. Turn off the welding machine and clean up your work area when you are finished welding. ◆

LAP JOINT

A **lap joint** is made by overlapping the edges of the two plates, Figure 4-47. The joint can be welded on one side or both sides with a fillet weld. In Practice 4-7, both sides should be welded unless otherwise noted.

As the fillet weld is made on the lap joint, the buildup should equal the thickness of the plate, Figure 4-48. A good weld will have a smooth transition from the plate surface to the weld. If this transition is abrupt, it can cause stresses that will weaken the joint.

Penetration for lap joints does not improve their strength; complete fusion is required. The root of fillet welds must be melted to ensure a completely fused joint. If the molten weld pool shows a notch during the weld, Figure 4-49, this is an indication that the root is not being fused together. The weave pattern will help prevent this problem, Figure 4-50.

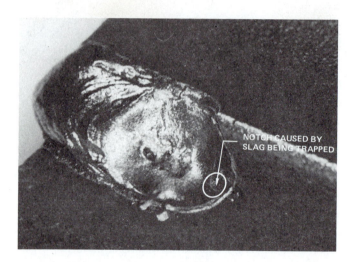

Figure 4-49 Watch the root of the weld bead to be sure there is complete fusion.

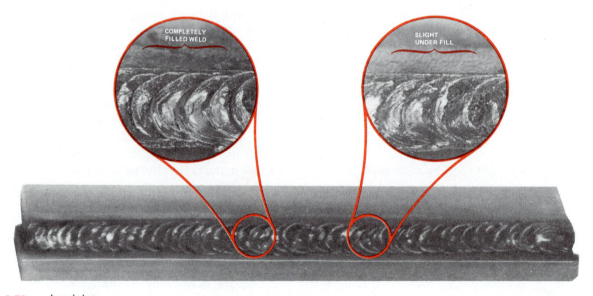

Figure 4-50   Lap joint.

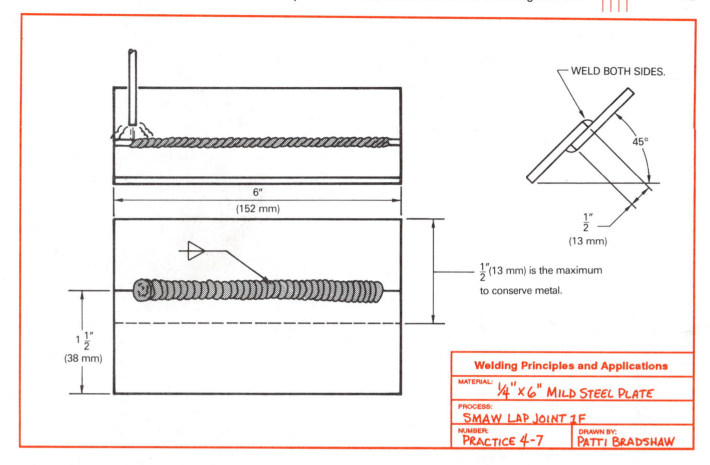

WELD BOTH SIDES.

45°

$\frac{1}{2}$"
(13 mm)

$\frac{1}{2}$"(13 mm) is the maximum
to conserve metal.

6"
(152 mm)

1 $\frac{1}{2}$"
(38 mm)

Welding Principles and Applications

MATERIAL:
$\frac{1}{4}$" X 6" MILD STEEL PLATE

PROCESS:
SMAW LAP JOINT 1F

NUMBER:
PRACTICE 4-7

DRAWN BY:
PATTI BRADSHAW

Figure 4-51 Lap joint in the flat position.

PRACTICE 4-8

Welded Lap Joint in the Flat Position (1F) Using E6010 or E6011 Electrodes, E6012 or E6013 Electrodes, and E7016 or E7018 Electrodes

Using a properly set up and adjusted arc welding machine, proper safety protection, arc welding electrodes having a l/8-in. (3-mm) diameter, and two or more pieces of mild steel plate, 6 in. (152 mm) long x 1/4 in. (6 mm) thick, you will make a welded lap joint in the flat position, **Figure 4-51**.

Hold the plates together tightly with an overlap of no more than l/4 in. (6 mm). Tack weld the plates together. A small tack weld may be added in the center to prevent distortion during welding, **Figure 4-52**. Chip the tacks before you start to weld.

The "J," "C," or zigzag weave pattern works well on this joint. Strike the arc and establish a molten pool

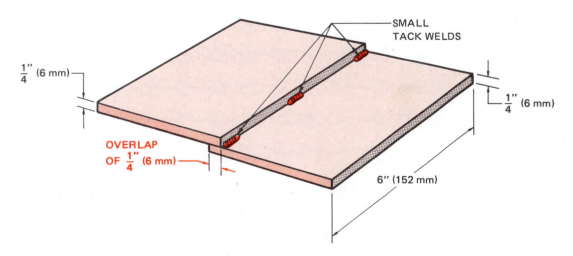

SMALL
TACK WELDS

$\frac{1}{4}$" (6 mm)

$\frac{1}{4}$" (6 mm)

OVERLAP
OF $\frac{1}{4}$" (6 mm)

6" (152 mm)

Figure 4-52 Tack welding the plates together.

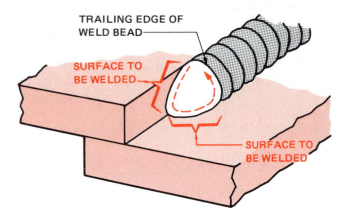

Figure 4-53 Follow the surface of the plate to ensure good fusion.

directly in the joint. Move the electrode out on the bottom plate and then onto the weld to the top edge of the top plate, **Figure 4-53**. Follow the surface of the plates with the arc. Do not follow the trailing edge of the weld bead. Following the molten weld pool will not allow for good root fusion and will also cause slag to collect in the root. If slag does collect, a good weld is not possible. Stop the weld and chip the slag to remove it before the weld is completed. Cool, chip, and inspect the weld for uniformity and defects. Repeat the welds with all three (F) groups of electrodes until you can consistently make welds free of defects. Turn off the welding machine and clean up your work area when you are finished welding ◆

PRACTICE 4-9

Welded Lap Joint in the Horizontal Position (2F) Using E6010 or E6011 Electrodes, E6012 or E6013 Electrodes, and E7016 or E7018 Electrodes

Using the same setup, materials, and electrodes as listed in Practice 4-8, you will make a welded horizontal lap joint.

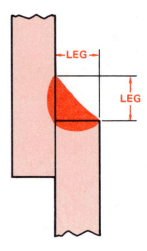

Figure 4-54 The horizontal lap joint should have a fillet weld that is equal on both plates.

The horizontal lap joint and the flat lap joint require nearly the same technique and skill to achieve a proper weld, **Figure 4-54**. Use the "J," "C," or zigzag weave pattern to make the weld. Do not allow slag to collect in the root. The fillet must be equally divided between both plates for good strength. After completing the weld, cool, chip, and inspect the weld for uniformity and defects. Repeat the welds using all three (F) groups of electrodes until you can consistently make welds free of defects. Turn off the welding machine and clean up your work area when you are finished welding. ◆

PRACTICE 4-10

Lap Joint in the Vertical Position (3F) Using E6010 or E6011 Electrodes, E6012 or E6013 Electrodes, and E7016 or E7018 Electrodes

Using the same setup, materials, and electrodes as listed in Practice 4-8, you will make a vertical up welded lap joint.

- Start practicing this weld with the plate at a 45° angle.

- Gradually increase the angle of the plate to vertical as skill is gained in welding this joint. The "J" or "T" weave pattern works well on this joint.

- Establish a molten weld pool in the root of the joint.

- Use the "T" pattern to step ahead of the molten weld pool, allowing it to cool slightly. Do not deposit metal ahead of the molten weld pool.

- As the molten weld pool size starts to decrease, move the electrode back down into the molten weld pool.

- Quickly move the electrode from side to side in the molten weld pool, filling up the joint.

- Cool, chip, and inspect the weld for uniformity and defects.

- Repeat the welds as necessary with all three (F) groups of electrodes until you can consistently make welds free of defects. Turn off the welding machine and clean up your work area when you are finished welding. ◆

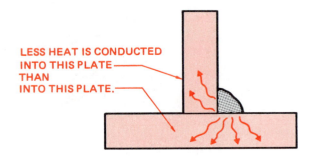

Figure 4-55 Tee joint.

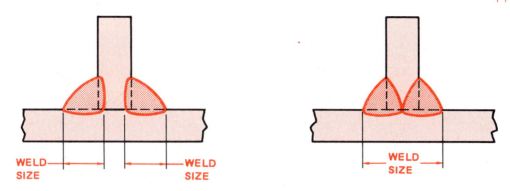

Figure 4-56 If the total weld sizes are equal, then both tee joints would have equal strength.

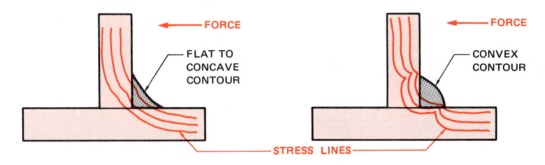

Figure 4-57 The stresses are distributed more uniformly through a flat or concave fillet weld.

TEE JOINT

The tee joint is made by tack welding one piece of metal on another piece of metal at a right angle, **Figure 4-55**. After the joint is tack welded together, the slag is chipped from the tack welds. If the slag is not removed, it will cause a slag inclusion in the final weld.

The heat is not distributed uniformly between both plates during a tee weld. Because the plate that forms the stem of the tee can conduct heat away from the arc in only one direction, it will heat up faster than the base plate. Heat escapes into the base plate in two directions. When using a weave pattern, most of the heat should be directed to the base plate to keep the weld size more uniform and to help prevent undercut

A welded tee joint can be strong if it is welded on both sides, even without having deep penetration, **Figure 4-56**. The weld will be as strong as the base plate if the size of the two welds equals the total thickness of the base plate. The weld bead should have a flat or slightly concave appearance to ensure the greatest strength and efficiency, **Figure 4-57**.

PRACTICE 4-11

Tee Joint in the Flat Position (1F) Using E6010 or 6011 Electrodes, E6012 or E6013 Electrodes, and E7016 or E7018 Electrodes

Using a properly set up and adjusted arc welding machine, proper safety protection, arc welding electrodes

having a 1/8-in. (3-mm) diameter, and two or more pieces of mild steel plate, 6 in. (152 mm) long x l/4 in. (6 mm) thick, you will make a welded tee joint in the flat position, **Figure 4-58**.

After the plates are tack welded together, place them on the welding table so the weld will be flat. Start at one end and establish a molten weld pool on both plates. Allow the molten weld pool to flow together before starting the weave pattern. Any of the weave patterns will work well on this joint. To prevent slag inclusions, use a slightly higher than normal amperage setting.

When the 6-in. (152-mm)-long weld is completed, cool, chip, and inspect it for uniformity and soundness. Repeat the welds as needed for all these groups of electrodes until you can consistently make welds free of defects. Turn off the welding machine and clean up your work area when you are finished welding. ◆

PRACTICE 4-12

Tee Joint in the Horizontal Position (2F) Using E6010 or E6011 Electrodes, E6012 or E6013 Electrodes, and E7016 or E7018 Electrodes

Using the same setup, materials, and electrodes as listed in Practice 4-11, you will make a welded tee joint in the horizontal position.

Place the tack welded tee plates flat on the welding table so that the weld is horizontal and the plates are flat and vertical, **Figure 4-59**. Start the arc on the flat plate and establish a molten weld pool in the root on both

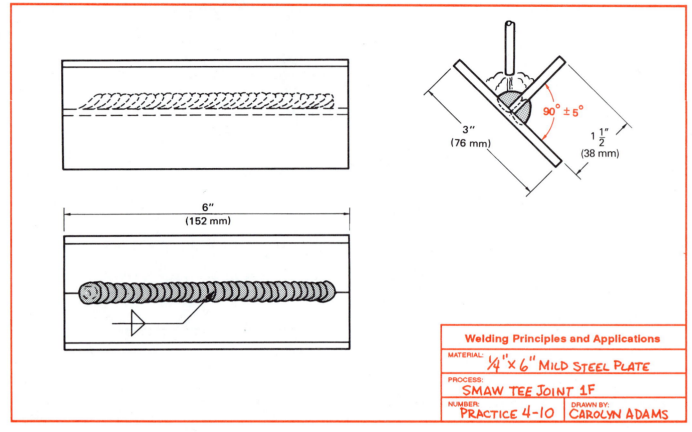

Welding Principles and Applications

MATERIAL: ¼" x 6" MILD STEEL PLATE

PROCESS: SMAW TEE JOINT 1F

NUMBER: PRACTICE 4-10 DRAWN BY: CAROLYN ADAMS

Figure 4-58 Tee joint in the flat position.

plates. Using the "J" or "C" weave pattern, push the arc into the root and slightly up the vertical plate. You must keep the root of the joint fusing together with the weld metal. If the metal does not fuse, a notch will appear on the leading edge of the weld bead. Poor or incomplete root fusion will cause the weld to be weak and easily cracked under a load.

When the weld is completed, cool, chip, and inspect it for uniformity and defects. Undercut on the vertical plate is the most common defect. Repeat the welds with all three (F) groups of electrodes until you can consistently make welds free of defects. Turn off the welding machine and clean up your work area when you are finished welding. ◆

PRACTICE 4-13

Tee Joint in the Vertical Position (3F) Using E6010 or E6011 Electrodes, E6012 or E6013 Electrodes, and E7016 or E7018 Electrodes

Using the same setup, materials, and electrodes as listed in Practice 4-11, you will make a welded tee joint in the vertical position.

Practice this weld with the plate at a 45° angle. This position will allow you to develop your skill for the vertical position. Start the arc and molten weld pool deep in the root of the joint. Build a shelf large enough to support the bead as it progresses up the joint. The

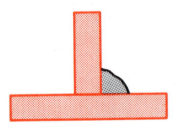

Figure 4-59 Horizontal tee.

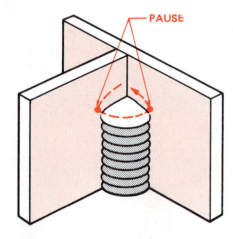

Figure 4-60 Pausing just above the undercut will fill it. This action also causes undercut, but that will be filled on the next cycle.

square, "J," or "C" pattern can be used, but the "T" or stepped pattern will allow deeper root penetration.

For this weld, undercut is a problem on both sides of the weld. It can be controlled by holding the arc on the side long enough for filler metal to flow down and fill it, **Figure 4-60**. Cool, chip, and inspect the weld for uniformity and defects. Repeat the welds as necessary with all three (F) groups of electrodes until you can consistently make welds free of defects. Turn off the welding machine and clean up your work area when you are finished welding. ◆

Tee Joint in the Overhead Position (4F) Using E6010 or E6011 Electrodes, E6012 or E6013 Electrodes, and E7016 or E7018 Electrodes

Using the same setup, materials, and electrodes as listed in Practice 4-10, you will make a welded tee joint in the overhead position.

Start the arc and molten weld pool deep in the root of the joint. Keep a very short arc length. The stepped pattern will allow deeper root penetration.

For this weld, undercut is a problem on both sides of the weld with a high buildup in the center. It can be controlled by holding the arc on the side long enough for filler metal to flow down and fill it. Cool, chip, and inspect the weld for uniformity and defects. Repeat the welds as necessary with all three (F) groups of electrodes until you can consistently make welds free of defects. Turn off the welding machine and clean up your work area when you are finished welding.

Consumable Guide Saves Time in Dragline Assembly

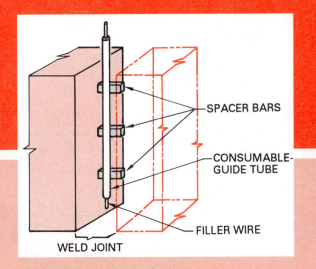

SPACER BARS

CONSUMABLE-GUIDE TUBE

FILLER WIRE

WELD JOINT

Figure 1

Arch of West Virginia is modernizing operations at its recently purchased Amherst Strip Mine outside Logan, West Virginia. When the mine was acquired, bulldozers and small shovel units were used to remove overburden, and mine the underlying seams of coal. Arch, a major mining organization, determined that efficiency at the Amherst mine would be greatly enhanced with the addition of a dragline bucket unit.

GETTING STARTED

F&E Erection, Inc., San Antonio, Texas, was charged with disassembling an existing Marion 8400 dragline in Oakman, Alabama, moving it to the Logan strip mine, and reassembling the parts on the job site. The initial task was to disassemble the parts of the base, revolving frame, counterweight, boom, gantry and legs, bucket, and control house. Each part was

carefully labeled and crated prior to the unit being shipped by rail to the Appalachian mountain site.

The assembly project started in May, 1988, and was completed in May, 1989. A staff of seventy-six people, including forty skilled welders, was assigned to handle the on-site reassembly on the Marion 8400 dragline. Welding needs on a project of this type range from simple shielded metal arc welding to automatic electrogas welding applications, so the versatility of the welding operators is a particularly important efficiency factor — see **Figure 1**.

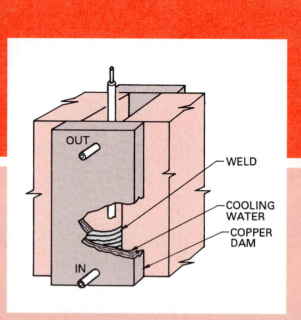

Figure 2

THE RIGHT PROCESS FOR THE RIGHT JOB

Welding the vertical butt joints of the base and the revolving frame presented a special technical challenge. Since the project was assembled on the job site, the plates could not be positioned to permit the use of semiautomatic, high-deposition electrodes. Bob Branson at F&E Erection decided to seek an alternative to using all-position electrodes and the semiautomatic flux-cored arc welding process. Given the nature of the application, Bob soon discovered that using electrogas welding with a consumable-guide tube offered some very significant advantages.

ELECTROGAS CONSUMABLE-GUIDE TUBE WELDING PROCEDURES

Using Lincoln Electric's Vertishield consumable-guide tube process, welding procedures were set. Up to twelve consumable-guide tubes per foot were constructed to complete the longest welds in the revolving base structure. Ceramic insulation rings were installed at regular intervals up and down the tubes to prevent electrical shorts to the base metal. Spacer bars, which were wrapped around the ceramic insulation rings, were welded to the inside of the weld joints — Figure 2. This prevented any movement of the shielded tube during the

welding operation. Periodically, C-braces were welded to the wall to hold up and support the water-cooled copper dams. Each copper dam is about 24 in. long and contains specially designed slag release slots located in the middle of the copper shoes to allow removal of excessive slag buildup from the weld pool. This feature ensures that slag impurities will not be trapped in the molten weld metal.

The welding procedures developed by F&E required an initial preheat of 500°F to remove all moisture from the joint area. Only the first several feet of each joint had to be reheated to this level because the heat generated during welding served to preheat the remaining section of the weld area. Deposition rates were set at 40 lb/h, and each joint was welded in one continuous pass. A572 steel was used, and the weld areas were oxygen cut to provide a precise, square-edged, butt joint.

PROCESS QUALITY AND EFFICIENCY

The use of Vertishield in this application cut the man-hours required to complete one weld to sixteen, versus the sixty hours logged with the semiautomatic flux-cored process. With two operators working together, an entire joint could be completed in one day. Therefore, the consumable-guide tube process yielded a time savings of approximately 75%. X-ray quality results were obtained. All of the welding procedures used in the Amherst Strip Mine dragline project were qualified and approved according to the AWS D1.1 Structural Welding Code — Steel.

(Adapted from an article in the Welding Journal based on a story from The Lincoln Electric Company, Cleveland, Ohio. Courtesy of the American Welding Society, Inc.)

REVIEW

1. Describe two methods of striking an arc with an electrode.

2. Why is it important to strike the arc only in the weld joint?

3. What problems may result by using an electrode at too low a current setting?

4. What problems may result by using an electrode at too high a current setting?

5. According to Table 4-1, page 63, what would the amperage range be for the following electrodes?

   a. 1/8 in. (3.2 mm),   c. 3/32 in. (2.4 mm),
   E6010   E7016

   b. 5/32 in. (4 mm),   d. 1/8 in. (3.2 mm),
   E7018   E6011

6. What makes some spatter "hard"?

7. Why should you never change the current setting during a weld?

8. What factors should be considered when selecting an electrode size?

9. What can a welder do to control overheating of the metal pieces being welded?

10. What effect does changing the arc length have on the weld?

11. What arc problems can occur in deep or narrow weld joints?

12. Describe the difference between using a leading and a training electrode angle.

13. Can all electrodes be used with a leading angle? Why or why not?

14. What characteristics of the weld bead does the weaving of the electrode have?

15. What are some of the applications for the circular pattern in the flat position?

16. Using a pencil, draw two complete lines of the weave patterns you are most comfortable making.

17. Why is it important to find a good welding position?

18. Which electrodes would be grouped in the following F numbers: F3, F2, F4?

19. Give one advantage of using electrodes with cellulose-based fluxes.

20. What are stringer beads?

21. Describe an ideal tack weld.

22. What effect does the root opening or root cap have on a butt joint?

23. What can happen if the fillet weld on a lap joint does not have a smooth transition?

24. Which plate heats up faster on a tee joint? Why?

25. Can a tee weld be strong if the welds on both sides do not have deep penetration? Why or why not?

Chapter 5

SHIELDED METAL ARC WELDING OF PIPE

OBJECTIVES

After completing this chapter, the student should be able to

- explain the difference between low-, medium-, and high-pressure pipe systems.

- describe the differences between pipe and tubing.

- list the different weld passes and explain the purposes of each one.

- demonstrate skill in welding pipe in all positions.

KEY TERMS

pipe
tubing
pressure range
root face
land
root gap
root suck back
concave root surface
icicles

horizontal rolled pipe position
vertical fixed pipe position
horizontal fixed pipe position
welding uphill or downhill
fixed inclined position

INTRODUCTION

The pipe welder is considered by some other welders to be the best welder in the industry. Often pipe welders share a great deal of pride. Some finished welded piping systems are considered works of art. Mastering the skills required to be a master pipe welder often takes a large commitment on the part of the welding student, but this commitment is well rewarded by the industry. The rewards of being a quality pipe welder include better pay, working with the best equipment, and often having a helper to do the less glamorous jobs of cleanup and setup.

The shielded metal arc welding process is very well suited to the fabrication and repair of piping systems. Welded steel pipe is used in factories, power plants, refineries, and buildings to carry liquids and gases for a variety of purposes, **Figure 5-1**. Pipe is used to carry such materials as water, steam, chemicals, gases, petroleum, radioactive materials, and many others. Oil and gas are distributed through cross-country piping systems all over the U.S., from Texas, California, Maine, and the northern slopes of Alaska. Welded pipe is used throughout ships, planes, and spacecraft to carry such liquids as fuels and hydraulic fluids. Pipe and tubing are also used for structures such as handrails, columns in buildings, light posts, bicycle and motorcycle frames, as well as many other items we come in contact with each day.

Figure 5-1 Welded piping system. (Courtesy of MG Technical Products, Inc.)

The purpose for which pipe will be used largely determines how it is welded. This text groups pipe welds into the following three general categories:

■ Low pressure or light service
■ Medium pressure or medium service
■ High pressure or heavy service

Low Pressure — Light Service

Low-pressure or noncritical piping systems may be used to carry water, residential natural gas, noncorrosive or noncombustible chemicals, and other nonhazardous materials used in industry. These types of noncritical piping assemblies are also found on such structural items as handrails, truck racks, columns for residences, and other light-duty products.

These pipe joints must be free from such defects as pinholes, undercut, slag inclusions, or any other defect that may cause the joints to leak or break prematurely. The weld does not require 100% penetration, although penetration should be uniform. Much of the strength of these pipe joints comes from the reinforcement. These welds should always be located so that they are easily repairable if necessary.

Medium Pressure — Medium Service

Medium-pressure piping is used for low-pressure steam heat, corrosive or flammable chemicals, waste disposal, ship plumbing, and for medium- to heavy-service structural items such as highway signs, railings or light posts, trailer axles, and equipment frames or stands. These pipe joints must withstand heavy loads, but

their failure will not be disastrous. Much of their strength is due to weld reinforcement, but there should be 100% root penetration around most of the joint. The root, filler, and reinforcement passes are usually welded with E7018 electrodes for added strength.

High Pressure — Heavy Service

High-pressure or critical piping systems are used for high-pressure steam, radioactive materials, the Alaskan pipeline, fired or unfired boilers, refinery reactor lines, aircraft air frames, offshore oil-rig jackets (legs), motorcycle frames, race car roll cages, truck axles, and for several other critical, heavy-duty applications.

The welds on critical piping systems must be as strong or stronger than the pipe itself. Often, the pipe used for these applications is extra heavy-duty pipe, with heavier wall thicknesses. The weld must have 100% root penetration over 100% of the joint. Root, filler, and reinforcement weld passes are made with an E7018 or stronger electrode. The welds are usually tested, and defects are repaired.

PIPE AND TUBING

Although **pipe** and **tubing** are similar in some aspects, they have different types of specifications and uses. They are only sometimes interchangeable.

The specifications for pipe sizes are given as the inside diameter for pipe 12 in. (305 mm) in diameter or smaller, and as the outside diameter for pipe larger than 12 in. (305 mm) in diameter. Tubing sizes are always given as the outside diameter. The desired shape of tubing, such as square, round, or rectangular, must also be listed with the ordering information, **Figure 5-2**.

The wall thickness of tubing is measured in inches (millimeters) or as U.S. standard sheet metal gauge thickness. The wall thickness for pipe is determined by its schedule, or **pressure range.** The larger the diameter of the pipe, the greater its area. As the area increases, so must the wall thickness in order for the wall to withstand the same pressure range, **Figure 5-3**.

The strength of pipe is given as a schedule. Schedules 10 through 180 are available; schedule 40 is often considered a standard strength. Tubing strength is the ability of tubing to withstand compression, bending, or twisting loads. Tubing should also be specified as rigid or flexible.

Pipe and tubing are both available as welded (seamed) or extruded (seamless).

Most pipe that will be welded into a system is used to carry liquids or gases from one place to another. These systems may be designed to carry large or small quantities or materials having a wide range of pressures. Small-diameter pipe may be used for some structural applications, but usually only on a limited scale.

TUBING

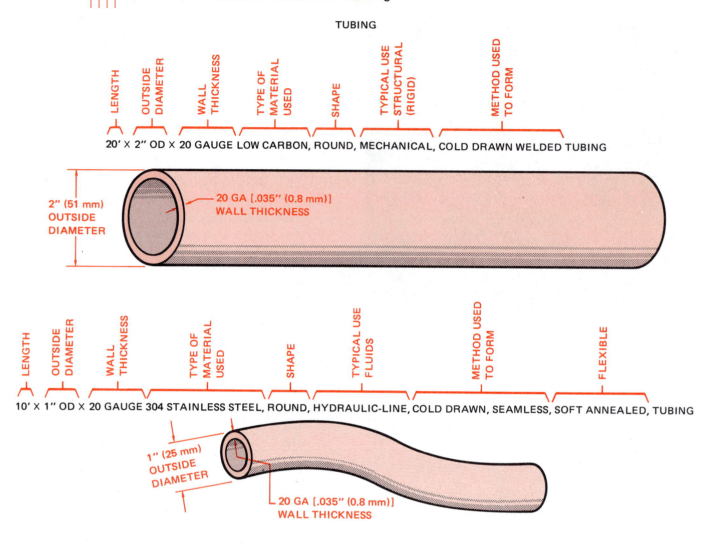

LENGTH | OUTSIDE DIAMETER | WALL THICKNESS | TYPE OF MATERIAL USED | SHAPE | TYPICAL USE STRUCTURAL (RIGID) | METHOD USED TO FORM

20' × 2" OD × 20 GAUGE LOW CARBON, ROUND, MECHANICAL, COLD DRAWN WELDED TUBING

2" (51 mm) OUTSIDE DIAMETER

20 GA [.035" (0.8 mm)] WALL THICKNESS

LENGTH | OUTSIDE DIAMETER | WALL THICKNESS | TYPE OF MATERIAL USED | SHAPE | TYPICAL USE FLUIDS | METHOD USED TO FORM | FLEXIBLE

10' × 1" OD × 20 GAUGE 304 STAINLESS STEEL, ROUND, HYDRAULIC-LINE, COLD DRAWN, SEAMLESS, SOFT ANNEALED, TUBING

1" (25 mm) OUTSIDE DIAMETER

20 GA [.035" (0.8 mm)] WALL THICKNESS

Figure 5-2 Typical specifications used when ordering tubing.

Small-diameter flexible tubing is commonly used to carry pressurized liquids or gases. Ridged tubing is normally used for structural applications. Some tubing is designed for specific purposes, such as electrical mechanical tubing (EMT), which is used to protect electrical wiring. Tubing can be used to replace some standard structural shapes such as I-beams, channels, and angles for buildings. Tubing is also available in sizes that will slide one inside the other to be used in places where telescoping tubing is required, **Figure 5-4.**

In this chapter, the term *pipe* will refer to pipe only. However, it should be understood that the welding sequence, procedures, and skill can also be used on thick-wall round tubing.

Advantages of Welded Pipe

Most pipe 1 1/2 in. (38 mm) in diameter and all steel pipe 2 in. (51 mm) and larger are generally arc welded. Welded piping systems, compared to pipe joined by any other method, are stronger, require less maintenance, last for longer periods of time, allow smoother flow, and weigh less.

Strength. The thickness of the pipe and fitting is the same when they are welded together. Threaded pipe is weakened because the threads reduce the wall thickness of the pipe, **Figure 5-5.**

Less Maintenance Required. Over much time and use, welded pipe joints are resistant to leaks.

Longer Lasting. Welded pipe joints resist corrosion caused by electrochemical reactions because all the parts are made of the same types of metal. Small cracks between the threads on threaded pipe are likely spots for corrosion to start.

Smoother Flow. The inside of a welded fitting is the same size as the pipe itself. As material flows through the pipe, less turbulence is caused by unequal diameters, **Figure 5-6.** Large piping systems may be several miles in length. Lowered resistance to product flow can save on operating energy costs.

Lighter Weight. Threaded fittings are larger and weigh more than welded fittings. The weight savings

PIPE

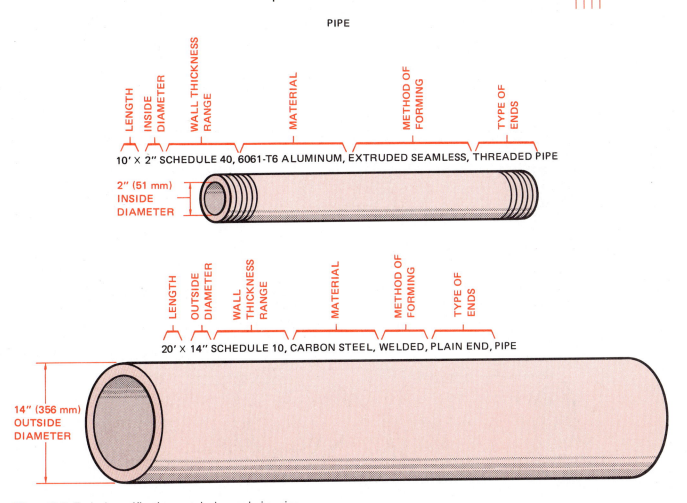

LENGTH INSIDE DIAMETER WALL THICKNESS RANGE MATERIAL METHOD OF FORMING TYPE OF ENDS

10' X 2" SCHEDULE 40, 6061-T6 ALUMINUM, EXTRUDED SEAMLESS, THREADED PIPE

2" (51 mm) INSIDE DIAMETER

LENGTH OUTSIDE DIAMETER WALL THICKNESS RANGE MATERIAL METHOD OF FORMING TYPE OF ENDS

20' X 14" SCHEDULE 10, CARBON STEEL, WELDED, PLAIN END, PIPE

14" (356 mm) OUTSIDE DIAMETER

Figure 5-3 Typical specifications used when ordering pipe.

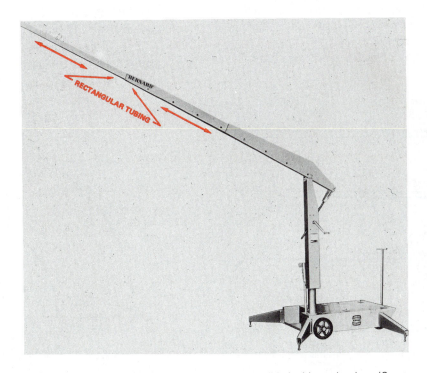

RECTANGULAR TUBING

Figure 5-4 A hand-operated positioner uses pieces of rectangular tubing that slide inside each other. (Courtesy of Bernard Company.)

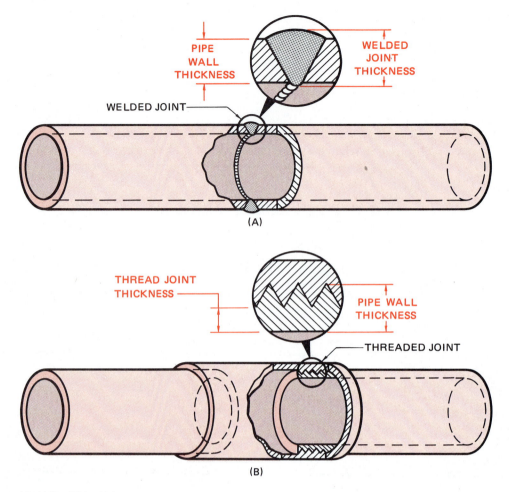

Figure 5-5 The welded joint (A) is thicker than the original pipe; the threaded joint (B) is thinner than the original pipe.

when welded fittings are used for an aircraft means that it can fly longer, faster, and carry a larger load for less money. Buildings and factories will also realize a savings because lighter fittings are less expensive.

Other advantages of welded pipe include the following:

■ An ability to make specially angled fittings by cutting existing fittings, **Figure 5-7.**

■ Odd-shaped parts can be fabricated.

■ A lot of highly specialized equipment is not required for each different size of pipe.

■ Alignment of parts is easier. It is not necessary to

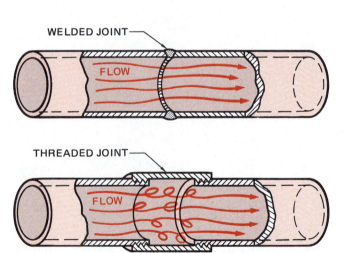

Figure 5-6 The flow along a welded pipe is less turbulent than that in a threaded pipe.

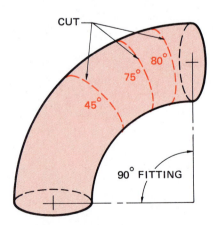

Figure 5-7 A standard 90° fitting can be cut to any special angle that is needed.

Figure 5-8 Pipe beveling machine. (Courtesy of Esco Tool Co.)

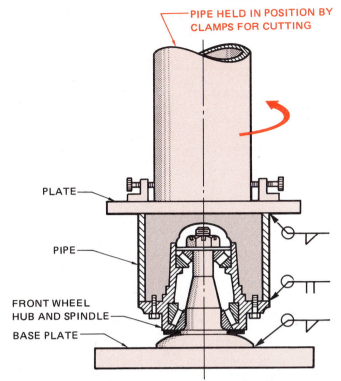

Figure 5-9 Turntable built from a front wheel assembly.

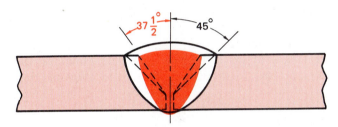

Figure 5-10 The 37 1/2° angled joint may use nearly 50% less filler metal, time, and heat than the 45° angled joint.

overtighten or undertighten fittings so that they will line up.

■ Removing, replacing, or changing parts is easy because special connections are not needed to remove the parts.

PREPARATION AND FITUP

The ends of pipe must be beveled for maximum penetration and high joint strength. The end can be beveled by flame cutting, machining, grinding, or a combination process. It is important that the bevel be at the correct angle, about 37 1/2°, and that the end meet squarely with the mating pipe. The sharp, inner edge of the bevel should be ground flat. This area is called a **root face** or **land.** Final shaping should be done with a grinder so that the **root gap** will be uniform.

The bevel on the end of the pipe can be flame cut using a portable pipe beveling machine, **Figure 5-8,** or a hand-held torch. Chapter 7 described how to set up and operate flame-cutting equipment. A turntable, similar to the one shown in **Figure 5-9,** can be made in the school shop and used for beveling short pieces of pipe. The turntable can be used vertically or horizontally. By turning the table slowly with pipe held between the clamps, a hand torch can be used to produce smooth pipe bevels. Large-production welding shops may also use machines designed specifically for beveling pipe. These machines

will accurately cut a 37 1/2° angle on the pipe.

The 37 1/2° angle allows easy access for the electrode with a minimum amount of filler metal required to fill the groove, **Figure 5-10.**

The root face will help a welder control both penetration and root suck back, **Figure 5-11.** Penetration control is improved because there is more metal near the edge to absorb excessive arc heat. This makes machine adjustments less critical by allowing the molten weld pool to be quickly cooled between each electrode movement. **Root suck back** is caused by the surface tension of the molten metal trying to pull itself into a ball, forming a **concave root surface.** The root face allows a larger weld molten weld pool to be controlled, and because of the increased size of the molten weld pool, it is not so affected by surface tension, **Figure 5-12.**

Fitting pipe together and holding it in place for welding becomes more difficult as the diameter of the pipe gets larger. Devices for clamping and holding pipe in place are available, or a series of wedges and dogs can

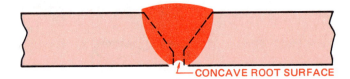

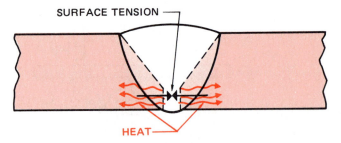

Figure 5-11

Figure 5-12 Heat is drawn out of the molten weld pool, and surface tension holds the pool in place.

be used, **Figure 5-13**. In the practices for this chapter, the pieces of pipe the welder will be using are about 1 1/2 in. (38 mm) wide. However, when welding on larger diameter pipe sizes, the weld specimens must be larger than 1 1/2 in. (38 mm). Welds on larger pipe sizes need more metal to help absorb the higher heat required to make these welds.

A welder can use a vise to hold the pipes in place for tack welding. If the pipe is not round and does not align properly, first tack weld the pipe together and then

quickly hit the tack while holding the pipe over the horn on an anvil, **Figure 5-14**. This action will force the pipe into alignment. For pipe that is too distorted to be forced into alignment in this manner, a welder must grind down the high points to ensure a good fit.

PRACTICE WELDS

One of the major problems to be overcome in pipe welding is learning how to make the transition from one position to another. The rate of change in welding position is slower with large-diameter pipes, but the large-diameter pipes require more time to weld. When a welder first starts welding, a large-diameter pipe should be used in order to make learning this transition easier. As welders develop skill and the technique of pipe welding, they can change to the small-diameter pipe sizes. Pipe as small as 3 in. (76 mm) can be welded quickly. It is large enough for the welder to be able to cut out test specimens.

Pipe used for these practice welds should be no shorter than 1 1/2 in. (38 mm). Pipe that is shorter than 1 1/2 in. (38 mm) rapidly becomes overheated, making welding more difficult.

To progress more quickly with pipe welds, a welder should master grooved plate welds. Once plate welding is mastered in all positions, pipe skills are faster and easier to develop.

Pipe welding is either performed with E6010 or E6011 electrodes for the complete weld, or these electrodes are used for the root pass, and E7018 electrodes

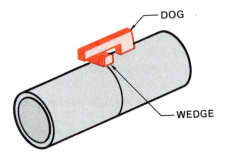

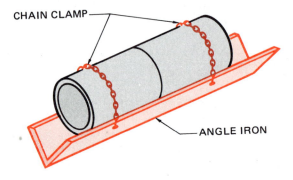

Figure 5-13 Shop fabrications used to align pipe joints.

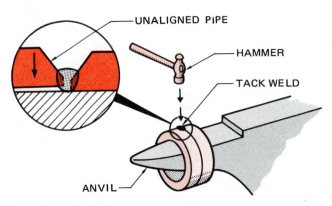

Figure 5-14 Hitting a hot tack weld can align a pipe joint.

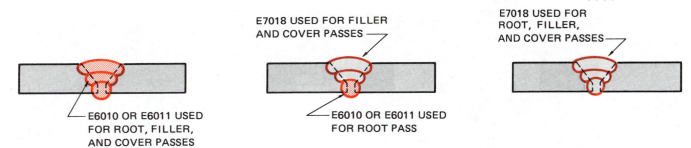

E7018 USED FOR FILLER
AND COVER PASSES

E7018 USED FOR
ROOT, FILLER,
AND COVER PASSES

E6010 OR E6011 USED
FOR ROOT, FILLER,
AND COVER PASSES

E6010 OR E6011 USED
FOR ROOT PASS

Figure 5-15 Single or multiple types of electrodes may be used when producing a pipe weld. The electrode selected is most often controlled by a code or specification.

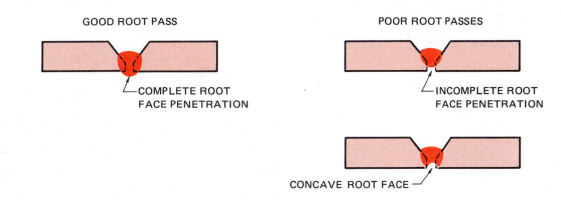

GOOD ROOT PASS

POOR ROOT PASSES

COMPLETE ROOT
FACE PENETRATION

INCOMPLETE ROOT
FACE PENETRATION

CONCAVE ROOT FACE

Figure 5-16 Root pass.

are used to complete the joint. Pipe welding can also be done using the E7018 electrode for the entire weld, Figure 5-15.

The practice pieces of pipe used in the shop are much shorter than the pieces of pipe used in industry. When learning to weld on short pipe, it is a good idea to avoid positioning oneself where longer pipe would eventually be located. Often it is easier to stand at the end of a pipe rather than to one side; however, this cannot be done if the weld was being made on a full length of pipe.

Weld Standards

Weld quality is very important to the pipe welding industry. Like other welds, the major parts of the weld come under a higher level of inspection. However, the surface of the pipe on both sides of the weld is also important. No arc strikes should be made on this surface. Arc strikes outside of the weld groove are considered to be defects by much of the pipe welding industry. Arc strikes form small hardness spots, which if not remelted by the weld, will crack as the pipe expands and contracts with pressure changes. Thus, they are a defect and must be removed and the area repaired. This removal under some standards may simply involve grinding them off, but under some codes they must be treated like any other defect, and a special weld repair procedure must be followed. Because of the importance of not having arc strikes outside the weld groove on pipe welds, you should try to avoid them from the beginning. In Chapter 4, Experiment 4-2, several techniques are described to avoid arc strikes outside of the welding zone. You may

want to refer back to this section if you have difficulty in making arc starts accurately.

Root Weld

A root weld is the first weld in a joint, **Figure 5-16**. It is part of a series of welds that make up a multiple pass weld. The root weld is used to establish the contour and depth of penetration. The most important part of a root weld is the root face, or, in the case of pipe, the inside surface. The face, or outside shape, or contour of the root weld is not so important.

In order for cleaners (pigs) to be used in a pipeline, the inside of each welded joint must be smooth, **Figure 5-17**. Excessive penetration, known as icicles, gets in the way of cleaning, adds resistance to flow, and will cause weakened points on the weld. On some piping systems, consumable inserts or backing rings are used to control penetration and the inside contour. Most pipe welds are made without these devices to control penetration.

The face of a root weld is not important if the root surface is clean, smooth, and uniform. A grinder is used to remove excessive buildup and reshape the face of the root pass. This grinding removes slag along the sides of the weld bead and makes it easier to add the next pass. Not all root passes are ground. Pipe that is to be used in low- and medium-pressure systems is not usually ground. Grinding each root pass takes extra time and does not give the welder the experience of using a hot pass. Most slag must be completely removed by chipping before the hot pass is used.

Figure 5-17 The root face must be uniform.

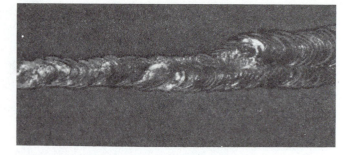

Figure 5-18 Filler pass using stringer beads.

If a welder needs more experience or practice in making an open root weld, refer to Chapter 4.

Hot Pass

The hot pass is used to quickly burn out small amounts of slag trapped along the edge of the root pass. This is slag that cannot be removed easily by chipping or wire brushing. The hot pass can also be used to reshape the root pass by using high current settings and a faster than normal travel speed.

Slag is mostly composed of silicon dioxide, which melts at about 3,100°F (1,705°C). Steel melts at approximately 2,600°F (1,440°C). A temperature of more than 500°F (270°C) hotter than the surrounding metal is required to melt slag. The slag can be floated to the surface by melting the surrounding metal. A high current will quickly melt enough surface to allow the slag to float free; a fast travel speed will prevent burnthrough. The fast travel speed forms a concave weld bead that is easy to clean for the welds that will follow.

Filler Pass

After thoroughly removing slag from the weld groove by chipping, wire brushing, or grinding, it is ready to be filled. The filler pass(es) may be either a series of stringer beads, Figure 5-18, or a wave bead, Figure 5-19. Stringer beads require less welder skill because of the small amount of metal that is molten at one time. Stringer beads are as strong or stronger than weave beads.

The weld bead crater must be cleaned before the next electrode is started. Failure to clean the crater will result in slag inclusions. On high-strength, high-pressure pipe welds, the crater should be slightly ground to ensure its cleanliness, Figure 5-20. When the bead has gone completely around the pipe, it should continue past the starting point so that good fusion is ensured, Figure 5-21. The

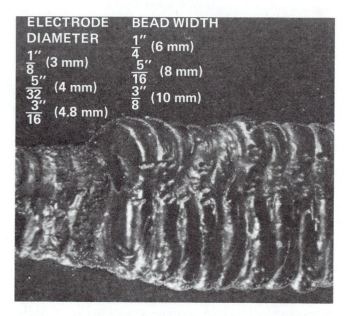

ELECTRODE DIAMETER	BEAD WIDTH
$\frac{1''}{8}$ (3 mm)	$\frac{1''}{4}$ (6 mm)
$\frac{5''}{32}$ (4 mm)	$\frac{5''}{16}$ (8 mm)
$\frac{3''}{16}$ (4.8 mm)	$\frac{3''}{8}$ (10 mm)

Figure 5-19 Filler pass using weave bead. The bead width should not be more than two times the rod diameter.

CRATER FILLED AND PROPERLY CLEANED

Figure 5-20 The weld crater should be filled to prevent cracking and cleaned of slag before restarting the arc.

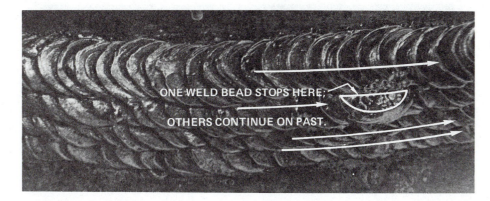

Figure 5-21 Avoid starting and stopping all weld passes in the same area.

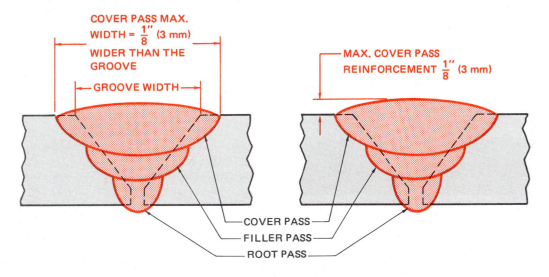

COVER PASS MAX. WIDTH = $\frac{1''}{8}$ (3 mm) WIDER THAN THE GROOVE

GROOVE WIDTH

MAX. COVER PASS REINFORCEMENT $\frac{1''}{8}$ (3 mm)

COVER PASS
FILLER PASS
ROOT PASS

Figure 5-22 Excessively wide or built-up welds restrict pipe expansion at the joint, which may cause premature failure. Check the appropriate code or standard for exact specifications.

locations of starting and stopping spots for each weld pass must be staggered. The weld groove should be filled level with these beads so that it is ready for the cover pass.

Cover Pass

The final covering on a weld is referred to as the cover pass. It may be a weave or stringer bead. The cover pass should not be too wide or have too much reinforcement, **Figure 5-22.** Cover passes that are excessively large will reduce the pipe's strength, not increase it. A large cover pass will cause the stresses in the pipe to be concentrated at the sides of the weld. An oversized weld will not allow the pipe to expand and contract uniformly along its length. This concentration is similar to the restriction a rubber band would have on an inflated balloon if it were put around its center.

The cover pass should be kept as uniform and as neat looking as possible, **Figure 5-23.** A visual checking is often all that low- and medium-pressure welds receive, and a nice looking cover will pass testing each time. A

good cover pass, during a visual inspection, is used to indicate that the weld underneath is sound.

1G HORIZONTAL ROLLED POSITION

The horizontal rolled pipe position is commonly used in fabrication shops where structures or small systems can be positioned for the convenience of the welder, **Figure 5-24.** The consistent high quality and quantity of welds produced in this position make it very desirable for both the welder and the company.

The penetration and buildup of the weld are controlled more easily with the pipe in this position. Weld visibility and welder comfort are improved so that welder fatigue is less of a problem. The pipe can be rolled continuously with some types of positioners, and the weld can be made in one continuous bead.

Because of the ease in welding and the level of skill required, welders who are certified in this position are not qualified to make welds in other positions.

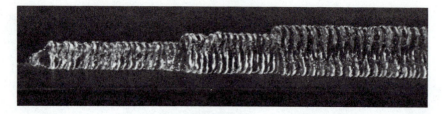

Figure 5-23 Uniformity in each pass shows a high degree of welder skill and increases the probability the weld will pass testing.

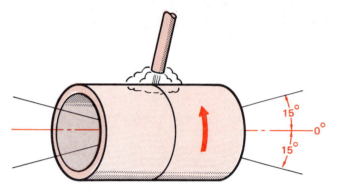

Figure 5-24 1G position. The pipe is rolled horizontally.

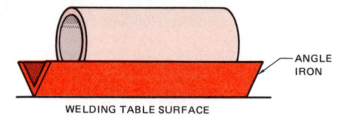

WELDING TABLE SURFACE

ANGLE IRON

Figure 5-25

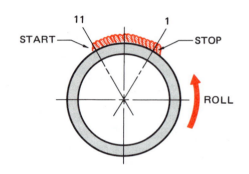

Figure 5-26

PRACTICE 5-1

Beading, 1G Position, Using E6010 or E6011 Electrodes and E7018 Electrodes

Using a properly set up and adjusted arc welding machine, proper safety protection, E6010 or E6011 and E7018 arc welding electrodes, having a 1/8 in. (3 mm) diameter, schedule 40 mild steel pipe, 3 in. (76 mm) or larger in diameter, you will make a straight stringer bead around a horizontally rolled pipe.

Place the pipe horizontally on the welding table in a vee block made of angle iron, **Figure 5-25.** The vee block will hold the pipe steady and allow it to be moved easily between each bead. Strike an arc on the pipe at the 11 o'clock position. Make a stringer bead over the 12 o'clock position, stopping at the 1 o'clock position, **Figure 5-26.** Roll the pipe until the end of the weld is at the 11 o'clock position. Clean the weld crater by chipping and wire brushing.

Strike the arc again and establish a molten weld pool at the leading edge of the weld crater. With the molten weld pool reestablished, move the electrode back on the weld bead just short of the last full ripple. **Figure 5-27.** This action will both reestablish good fusion and keep the weld bead size uniform. Now that the new weld bead is tied into the old weld, continue welding to the 1 o'clock position again. Stop welding, roll the pipe, clean the crater, and resume welding. Keep repeating this procedure until the weld is completely around the pipe. Before the last weld is started, clean the end of the first weld so that the end and beginning beads can be tied together smoothly.

When you reach the beginning bead, swing your electrode around on both sides of the weld bead. The beginning of a weld bead is always high, narrow, and has little penetration, **Figure 5-28.** By swinging the weave pattern (the "C" pattern is best) on both sides of the bead, you can make the bead width uniform. The added heat will give deeper penetration at the starting point. Hold the arc in the crater for a moment until it is built up.

Cool, chip, and inspect the bead for defects. Repeat the beads as needed until they are mastered. Turn off the welding machine and clean up your work area when you are finished welding. ◆

PRACTICE 5-2

Butt Joint, 1G Position, Using E6010 or E6011 Electrodes

Using a properly set up and adjusted arc welding machine, proper safety protection, E6011 or E6011 arc welding electrodes having a 1/8-in. (3-mm) diameter, and two or more pieces of schedule 40 mild steel pipe, 3 in. (76 mm) or larger in diameter, you will make a pipe butt joint in the horizontal rolled position, **Figure 5-29.**

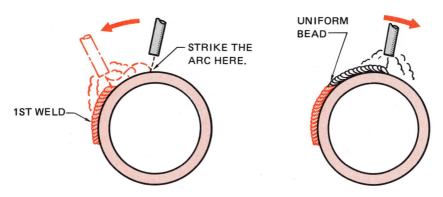

Figure 5-27 Keeping the weld uniform is important when restarting the arc.

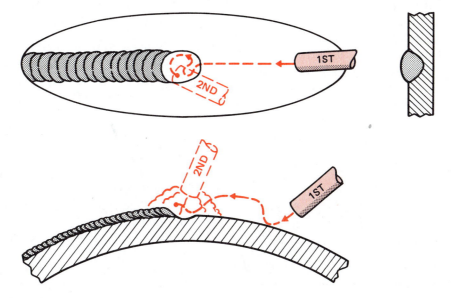

Figure 5-28 Restarting the weld.

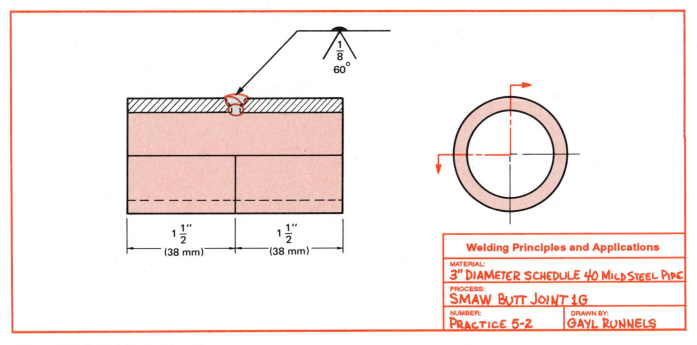

Figure 5-29 Butt joint in the 1G position.

Tack weld two pieces of pipe together as shown in Figure 5-30. Place the pipe horizontally in a vee block on the welding table. Start the root pass at the 11 o'clock position. Using a very short arc and high current setting, weld toward the 1 o'clock position. Stop and roll the pipe, chip the slag, and repeat the weld until you have completed the root pass.

Clean the root pass by chipping and wire brushing. The root pass should not be ground this time. Replace the pipe in the vee block on the table so that the hot pass can be done. Turn up the machine amperage, enough to remelt the root weld surface, for the hot pass. Use a step electrode pattern, moving forward each time the molten weld pool washes out the slag, and returning each time the molten weld pool is nearly all solid, Figure 5-31. Weld from the 11 o'clock position to the 1 o'clock position before stopping, rolling, and chipping the weld. Repeat this procedure until the hot pass is complete.

The filler pass and cover pass may be the same pass on this joint. Turn down the machine amperage. Use a "T," "J," "C," or zigzag pattern for this weld. Start the weld at the 10 o'clock position and stop at the 12 o'clock position. Sweep the electrode so that the molten weld pool melts out any slag trapped by the hot pass. Watch the back edge of the bead to see that the molten weld pool is filling the groove completely. Turn, chip, and continue the bead until the weld is complete. Repeat this weld until you can consistently make welds free of defects. Turn off the welding machine and clean up your work area when you are finished welding. ◆

PRACTICE 5-3

Butt Joint, 1G Position, Using E6010 or E6011 Electrodes for the Root Pass with E7018 Electrodes for the Filler and Cover Passes

Using the same setup, materials, and procedures as described in Practice 5-2, you will make a horizontal rolled butt joint in pipe, Figure 5-32. The root pass is to have 100% penetration over 80% or more of the length of the weld.

Set the pipe in the vee block on the welding table and make the root pass as explained in Practice 5-2. Watch for 100% penetration with no icicles. A hot pass or grinder can be used to clean the face of the root pass. Use an E7018 electrode for the filler and cover passes. The E7018 electrode should not be weaved more than 2 1/2 times the diameter of the electrode. Excessively wide weaving will allow the molten weld pool to become contaminated, Figure 5-33.

After the weld is completed, visually inspect it for 100% penetration around 80% of the root length. Check the weld for uniformity and visual defects on the cover pass. Repeat this weld until you can consistently make welds free of defects. Turn off the welding machine and clean up your work area when you are finished welding. ◆

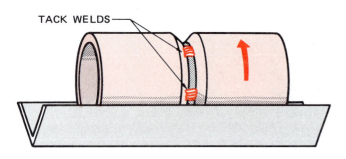

Figure 5-30 The tack welds are to be evenly spaced around the pipe. Use four tacks on small-diameter pipe and six or more on large-diameter pipe.

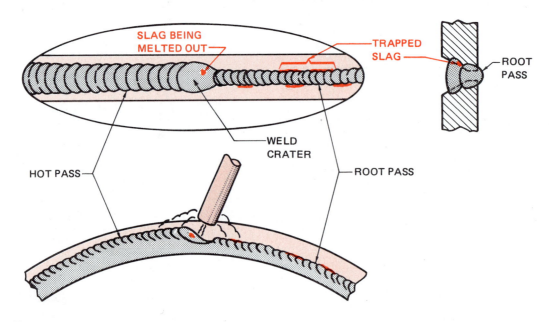

Figure 5-31 Hot pass.

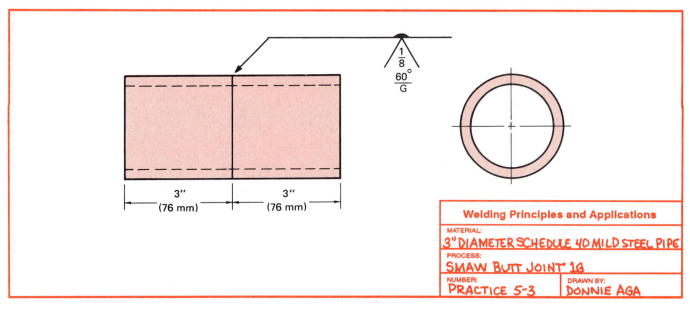

Figure 5-32 Butt joint in the 1G position to be tested.

Figure 5-33 Weave beads more than two and one half times the diameter of the electrode may be nice looking, but the atmosphere may contaminate the unprotected part of the molten weld pool.

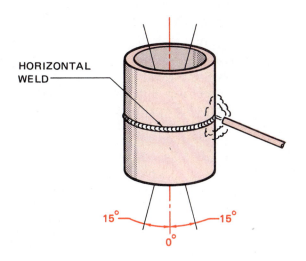

Figure 5-34 2G position. The pipe is fixed vertically, and the weld is made horizontally around it.

2G VERTICAL FIXED POSITION

In the 2G vertical fixed pipe position, the pipe is vertical and the weld is horizontal, Figure 5-34. With these welds, the welder does not need to change welding positions constantly. The major problem that faces welders when welding pipe in this position is that the area to be welded is often located in corners. Because of this location, reaching the back side of the weld is difficult. In the practices that follow, you may turn the pipe between welds. As a welder gains more experience, welds in tight places will become easier.

The welds must be completed in the correct sequence, Figure 5-35. The root pass goes in as with other joints. To reduce the sagging of the bottom of the weld, increase the electrode to work angle. As long as the weld is burned in well and does not have cold lap on the bottom,

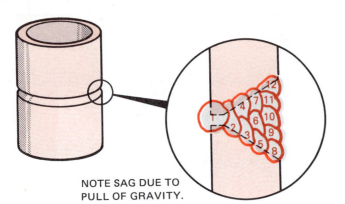

NOTE SAG DUE TO
PULL OF GRAVITY.

Figure 5-35

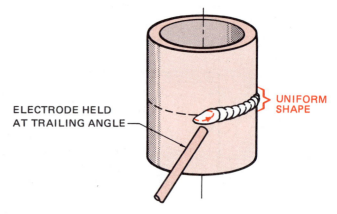

ELECTRODE HELD
AT TRAILING ANGLE

UNIFORM SHAPE

Figure 5-36 Electrode position and weave pattern for a weld on vertical pipe.

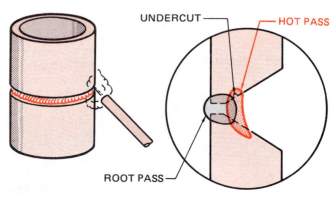

UNDERCUT — HOT PASS

ROOT PASS

Figure 5-37 Hot pass.

the weld is correct. Each of the filler and cover welds that will follow must be supported by the previous weld bead.

PRACTICE 5-4

Stringer Bead, 2G Position, Using E6010 or E6011 Electrodes and E7018 Electrodes

Using the same setup, materials, and electrodes as listed in Practice 5-1, you will make straight stringer beads on a pipe in the vertical position.

The "J" weave pattern should be used so that the molten weld pool will be supported by the lower edge of the solidified metal, **Figure 5-36.** Keep the electrode at an upward and trailing angle so the arc force will help to keep the weld in place.

Repeat these stringer beads as needed, with both groups of electrodes, until you can consistently make welds free of defects. Turn off the welding machine and clean up your work area when you are finished welding. ◆

PRACTICE 5-5

Butt Joint, 2G Position, Using E6010 or E6011 Electrodes

Using a properly set up and adjusted arc welding machine, proper safety protection, E6010 or E6011 arc welding electrodes having a 1/8-in. (3-mm) diameter, and two or more pieces of schedule 40 mild steel pipe, 3 in. (76 mm) or larger in diameter, you will make a butt joint on a pipe in the vertical position.

Place the pipe on the arc welding table. Strike an arc and make a root weld that is as long as possible. If the root gap is close and uniform, a straight, forward movement can be used. For wider gaps, a step pattern must be used. After completing and cleaning the root pass, make a hot pass. The hot pass need only burn the root pass clean, **Figure 5-37.** Undercut on the top pipe is acceptable.

The filler and cover passes should be stringer beads or small weave beads. By keeping the molten weld pool

size small, control is easier. Cool, chip, and inspect the completed weld for uniformity and defects. Repeat this weld until you can consistently make welds free of defects. Turn off the welding machine and clean up your work area when you are finished welding. ◆

PRACTICE 5-6

Butt Joint, 2G Position, Using E6010 or E6011 Electrodes for the Root Pass and E7018 Electrodes for the Filler and Cover Passes

Using the same setup, materials, and procedures as described in Practice 5-4, you will make a vertical fixed pipe weld. The root pass is to have 100% penetration over 80% or more of the length of the weld.

Place the pipe vertically on the welding table. Hold the electrode at a 90° angle to the pipe and with a slight trailing angle, **Figure 5-38.** The electrode should be held tightly into the joint. If a burnthrough occurs, quickly push the electrode back over the burnthrough while increasing the trailing angle. This action forces the weld metal back into the opening. When the root pass is complete, chip the surface slag and then clean out the trapped slag by grinding or chipping, or use a hot pass.

Use E7018 electrodes for the filler and cover passes with a stringer pattern. The weave bead with this elec-

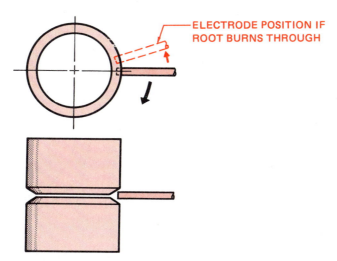

Figure 5-38 Electrode position and movement for the root pass.

trode and position tends to undercut the top and overlap the bottom edge. After the weld is completed, visually inspect it for 100% penetration around 80% of the root length. Check the weld for uniformity and visual defects on the cover pass. Repeat the weld until you can consistently make welds free of defects. Turn off the welding machine and clean up your work area when you are finished welding. ◆

5G HORIZONTAL FIXED POSITION

The 5G horizontal fixed pipe position is the most often used pipe welding position. Welds produced in flat, vertical up or vertical down, and overhead positions must be uniform in appearance and of high quality.

When practicing these welds, mark the top of the pipe for future reference. Moving the pipe will make welding easier, but the same side must stay on the top at all times, Figure 5-39.

The root pass can be welded uphill or downhill. In industry, the method used to weld the root pass is determined by established weld procedures. If there are no procedures requiring a specific direction, the choice is usually made based upon fitup. A close parallel root opening can be welded uphill or downhill. A root opening that is wide or uneven must be welded uphill. In the following practices, the welder can make the choice of direction, but both directions should be tried.

The pipe may be removed from the welding position for chipping, wire brushing, or grinding. The pipe can be held in place by welding a piece of flat stock to it and clamping the flat stock to a pipe stand, Figure 5-40.

The electrode angle should always be upward, Figure 5-41. Changing the angle toward the top and bottom will help control the bead shape. The bead, if welded downhill, should start before the 12 o'clock position and continue past the 6 o'clock position to ensure good fusion and tie-in of the welds. The arc must always be struck inside the joint preparation groove.

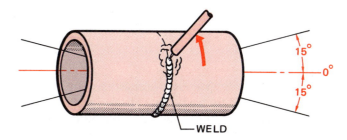

Figure 5-39 5G horizontal fixed position.

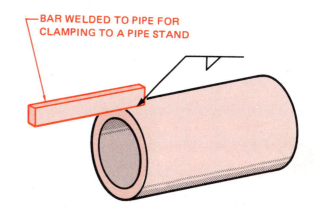

Figure 5-40 Holding the pipe in place by welding a piece of flat stock to the pipe and then clamping the flat stock to a pipe stand.

PRACTICE 5-7

Stringer Bead, 5G Position, Using E6010 or E6011 Electrodes and E7018 Electrodes

Using the same setup, materials, and electrodes as listed in Practice 5-1, you will make straight stringer beads in the horizontal fixed position using both groups of electrodes.

Clamp the pipe horizontally about chest level. Starting at the 11 o'clock position, make a downhill straight stringer bead through the 12 o'clock and 6

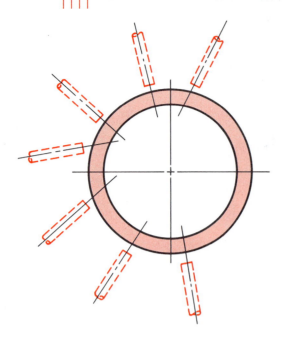

Figure 5-41 Electrode angle.

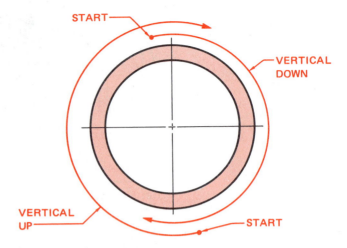

Figure 5-42 Stop at the 7 o'clock position.

o'clock positions. Stop at the 7 o'clock position, **Figure 5-42.** Using a new electrode, start at the 5 o'clock position and make an uphill straight stringer bead through the 6 o'clock and 12 o'clock positions. Stop at the 1 o'clock position. Change the electrode angle to control the molten weld pool.

Repeat these stringer beads as needed, with each group of electrodes, until you can consistently make welds free of defects. Turn off the welding machine and clean up your work area when you are finished welding. ◆

PRACTICE 5-8

Butt Joint, 5G Position, Using E6010 or E6011 Electrodes for the Root Pass and E7018 Electrodes for the Filler and Cover Passes

Using the same setup, materials, and procedures as listed in Practice 5-2, you will make a horizontal fixed pipe weld. The root pass is to have 100% penetration over 80% or more of the length of the weld.

Mark the top of the pipe and mount it horizontally about chest level. Weld the root pass uphill or downhill using E6010 or E6011 electrodes. Either grind the root pass or use a hot pass to clean out trapped slag.

Use E7018 electrodes for the filler and cover passes with stringer or weave patterns. When the weld is completed, visually inspect it for 100% penetration around 80% of the root length. Check the weld for uniformity and visual defects on the cover pass. Repeat the weld until you can consistently make welds free of defects. Turn off the welding machine and clean up your work area when you are finished welding. ◆

PRACTICE 5-9

Butt Joint, 5G Position, Using E6010 or E6011 Electrodes

Using a properly set up and adjusted arc welding machine, proper safety protection, E6010 or E6011 arc welding electrodes having a 1/8-in. (3-mm) diameter, and two or more pieces of schedule 40 mild steel pipe, 3 in. (76 mm) or larger in diameter, you will make a butt joint on a horizontally fixed pipe.

Mark the top of the pipe and mount it about chest level. Depending upon the root gap, make a root weld uphill or downhill using E6010 or E6011 electrodes. Check the root penetration to determine if it is better in one area than in another area. Chip and wire brush the weld and set the machine for a hot pass. Start the hot pass at the bottom and weld upward on both sides. The bead should be kept uniform with little buildup.

Using stringer or weave beads, make the filler and cover welds. If stringer beads are used, downhill welds can be made. Cool, chip, and inspect the weld for uniformity and defects. Repeat this weld until you can consistently make welds free of defects. Turn off the welding machine and clean up your work area when you are finished welding. ◆

6G 45° INCLINED POSITION

The 45° fixed inclined position is thought to be the most difficult pipe position. Qualifying in this position will certify the welder in the other positions using the same size electrodes and pipe sizes. The weld must be uniform and not have defects even though its position is changing in more than one direction at a time, **Figure 5-43.**

During this weld, it is necessary to continuously change the weld weave pattern, electrode angle, and weld speed. Small multipass stringer beads work best. With experience, however, weave beads are possible.

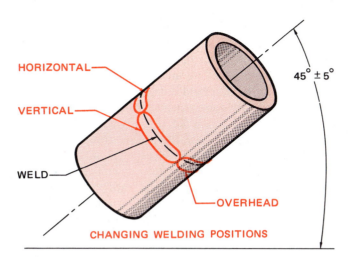

HORIZONTAL

VERTICAL

WELD

OVERHEAD

CHANGING WELDING POSITIONS

45° ± 5°

Figure 5-43 In the 6G position, the pipe is fixed at a 45° angle to the work surface. The effective welding angle changes as the weld progresses around the pipe.

PRACTICE 5-10

Stringer Bead, 6G Position, Using E6010 or E6011 Electrodes and E7018 Electrodes

Using the same setup, materials, and electrodes as listed in Practice 5-1, you will make straight stringer beads on a pipe in the 45° fixed inclined position.

Using the straight step, "T," or zigzag weave pattern, start at the bottom of the pipe with a very short arc. Keep the molten weld pool small and narrow for easier control. As the bead moves up to the side, the electrode angle should stay to the downhill side with a trailing angle. When the weld passes beyond the side, the down-hill and trailing angles are decreased. This is done so that when the weld reaches the top, the electrode is perpendicular to the top of the pipe. Repeat this procedure on the opposite side.

Repeat these stringer beads as needed, with both groups of electrodes, until you can consistently make welds free of defects. Turn off the welding machine and clean up your work area when you are finished welding. ◆

PRACTICE 5-11

Butt Joint, 6G Position, Using E6010 or E6011 Electrodes

Using a properly set up and adjusted arc welding machine, proper safety protection, E6010 or E6011 arc welding electrodes having a 1/8-in. (3-mm) diameter, and two or more pieces of schedule 40 mild steel pipe, 3 in. (76 mm) or larger in diameter, you will make a butt joint on a pipe in the 45° fixed inclined position.

Starting at the top, make a vertical down root pass that ends just beyond the bottom. Repeat this weld on the other side. Chip and wire brush the slag so that an uphill hot pass can be made. The hot pass must be kept small and concave so that more slag will not be trapped along the downhill side. Clean the bead, turn down the machine amperage, and complete the joint with stringer beads, **Figure 5-44**.

Cool, chip, and inspect the completed weld for uniformity and defects. Repeat this weld until you can consistently make welds free of defects. Turn off the welding machine and clean up your work area when you are finished welding. ◆

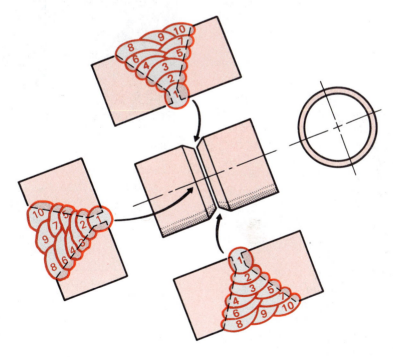

Figure 5-44

Butt Joint, 6G Position, Using E6010 or E6011 Electrodes for the Root Pass and E7018 Electrodes for the Filler and Cover Passes

Using the same setup, materials, and procedures as listed in Practice 5-11, you will make a 45° fixed inclined pipe weld. The root pass is to have 100% penetration over 80% of the root length of the weld.

Make the root pass as either a vertical up or down weld, depending upon the root opening. Chip the slag and clean the weld by grinding or by using a hot pass. If a hot pass is used, chip and wire brush the joint. Using the E7018 electrode, start slightly before the center on the bottom and make a small stringer bead in an upward direction. Keep a trailing and a somewhat uphill electrode angle so that the weld is deposited on the bottom of the lower pipe. The next pass should use a downhill electrode angle so that the bead is on the uphill pipe. Alternate this process until the bead is complete. Clean the weld and inspect it for 100% penetration over 80% of the root length. Check for uniformity and visual defects on the cover pass. Repeat the weld until you can consistently make welds free of defects. Turn off the welding machine and clean up your work area when you are finished welding. ◆

Race of Champions Depends on Welding

Twenty-four-year-old Al Unser, Jr., crossed the finish line triumphant, winning the final contest — and the 1986 title — in the International Race of Champions (IROC). The youngest driver ever to win a major series did so by outrunning eleven of the very best racers from the worlds of stock car, Indy, European, and American road racing. And he did it in a car that was, except for the color and number, identical to all the other cars in the race. Figure 1 shows nine IROC Camaros on a high-bank super speedway. In fact, that's the premise behind this 11-year-old war on wheels.

In 1974, someone decided that the way to find out who was the very best driver was to put the top people from different kinds of racing in identical cars. A dozen Chevrolet Camaros were specially constructed, and drivers drew straws just before each race to see who would drive the car. The sole changes allowed involved seat, safety belt, steering wheel, and pedal adjustment.

The series of four races each season, scheduled around the various auto racing series schedules, was an immediate hit. Though stock car drivers appeared to have an advantage — the sporty, modified Camaro looks, sounds, and handles like a

Figure 1

racing stock car — sports car and Indy car drivers won the first several series. Names such as Mario Andretti and A.J. Foyt dominated the scene as the cars were driven on ovals and twisting road courses. The series has grown in popularity through the years. See **Figure 1**.

Speaking of twists, these competition Camaros are bent, bruised, and battered after a typical IROC race. Trackside spectators and television viewers alike are treated to numerous fender bend-

Figure 2

Figure 3

ing, bump-and-run events. In many cases, welding repairs are needed to restore the racing cars to their original, rigid specifications — just in time for the next high-speed showdown.

The repairs often take place at trackside, with welders frequently working through the night to restore a car. The machines are prepared by one company, which supports the cars with spacious vans filled with parts. The welding equipment is a gas metal arc welding (GMAW) package, mated with an E70S6 wire electrode and a 75-25 gas mixture of argon and CO_2. Other equipment used includes 250- and 300-ampere ac/dc welding power sources. Figure 2 shows the 250-ampere power source at the right foreground. Both are used for gas tungsten arc (GTAW) and are also capable of shielded metal arc (SMAW) welding. Cars are built and rebuilt with a combination of gas metal arc (GMAW) and gas tungsten arc (GTAW) welding processes, Figure 3.

The identical frames are produced in North Carolina, then shipped for assembly to New Jersey. The cars come together as modified Chevrolet V-8s producing 450 horsepower and arrive from a Michigan speed shop to join featherweight fiberglass body panels and space-age metals used in the dry-sump engine and elsewhere. Since the cars must be both safe and strong, they are stripped down and rebuilt after each race. The engines, transmissions, frames, and running gear are minutely inspected.

The welding equipment is also used for fabrication. The engine's dry-sump oil system features a remote tank that is attached to the engine via lines run through the cooling system. These remote oil tanks are created with gas tungsten arc (GTAW) welding.

After safety considerations are met, the first priority of the race organizers is to make sure that all the cars are identical. Every part is the same. Each car is built exactly the same way. When the cars go on the track, they are as identical as it is humanly possible to make them.

(Courtesy of Miller Electric Mfg. Co., Appleton, Wisconsin 54912.)

REVIEW

1. List some applications that use low-pressure piping.

2. List some applications that use medium-pressure piping.

3. List some applications that use high-pressure piping.

4. Describe the differences between pipe and tubing.

5. What are the advantages of welded piping systems over other joining methods?

6. Why must the ends of pipe be beveled before being welded?

7. How can the ends of a pipe be beveled?

8. Why is the end of a pipe beveled at a 37 1/2° angle?

9. What causes root suck back or a concave root face?

10. Why are arc strikes outside the welding zone considered a problem on pipe?

11. What are the purposes of backing rings?

12. What is the purpose of a hot pass?

13. Why must the weld crater be cleaned before starting a new electrode?

14. What is the maximum width of the cover pass? Why?

15. When is the 1G welding position used?

16. How can the start of a weld be made less narrow and more uniform?

17. What supports the welds on a 2G pipe joint?

18. On 5G welds, what usually determines the direction of the root pass?

19. What three welding positions are incorporated in the 6G welding position?

20. Why are small multipass stringer beads best for 6G welds?

Chapter 6

ADVANCED SHIELDED METAL ARC WELDING

OBJECTIVES

After completing this chapter, the student should be able to

■ demonstrate an ability to make multiple pass welds in all positions.

■ explain how to make an open and closed root weld.

■ demonstrate how to use a hot pass to clean a weld.

■ identify the types and parts of weld grooves.

■ explain how to prepare, test, and evaluate guided bend specimens.

■ demonstrate an ability to make code quality welds in plate and pipe.

■ demonstrate an ability to weld according to a welding procedure specification sheet.

KEY TERMS

root pass	multiple pass weld
sound weld	burnthrough
molten weld pool	key hole
hot pass	wagon tracks
filler pass	cover pass
weld groove	guided bend
weld specimen	back gouge
preheating	postheating
interpass temperature	

INTRODUCTION

The SMAW process can be used to produce consistent, high-quality welds. Often it's necessary to make welds in less than ideal conditions. Knowing how to produce a weld of high strength in an out-of-position, difficult situation or in an unusual metal takes both practice and knowledge. A welder is frequently required to make these types of welds to a code or standard. This chapter covers the high-quality welding of plate, pipe, and plate to pipe. The practices are designed to give you the experience of taking code type tests in a variety of materials and positions as well as to develop good workmanship.

Any time a code quality weld requiring 100% joint penetration is to be made on metal thicker than 1/4 in. (6 mm), the metal edges must be prepared before welding. A joint is prepared for welding by cutting a groove in the metal along the edge. The preparation is done to allow deeper penetration into the joint of the weld for improved strength. Prepared joints often require more than one weld pass to complete them. By preparing the joint, metal several feet thick can be welded with great success, **Figure 6-1**. The same welding techniques are used when making prepared welds of any thickness.

The root pass is used to fuse the parts together and seal off possible atmospheric contamination from the filler weld. Once the root pass is completed and cleaned, a hot pass may be used to improve the weld contour and burn out small spots of trapped slag. For high-quality welds, a grinder should be used on the root pass to clean it.

Filler welds and cover welds are often made with high-strength electrodes. These passes are used to fill and cap the weld groove.

Welders are often qualified by passing a required qualification test using a groove weld. The type of joint, thickness of metal, type and size of electrode, and position are all specified by agencies issuing codes and standards. Except for the American Welding Society's Certified Welder program, taking a test according to one company or agency's specifications may not qualify a welder for another company or agency's testing procedures. But, being able to pass one type of test will usually help the welder to pass other tests for the same type of joint, thickness of metal, type and size of electrode, and position. Information about the AWS Certified Welder program is available from the AWS's main office in Miami, Florida.

ROOT PASS

The root pass is the first weld bead of a multiple pass weld. The root pass fuses the two parts together and establishes the depth of weld metal penetration. A good root pass is needed in order to obtain a sound weld. The root may be either open or closed, using a backing strip or backing ring, **Figure 6-2**.

The backing strip used in a closed root may remain as part of the weld, or it may be removed. Because leaving the backing strip on a weld may cause it to fail due to concentrations of stresses along the backing strip, removable backup tapes have been developed. Backup tapes are made of high-temperature ceramics, **Figure 6-3**, that can be used to increase penetration and prevent burnthrough. The tape can be peeled off after the weld is completed. Most welds do not use backing strips.

On plates that have the joints are prepared on both sides, the root face may be ground or gouged clean before another pass is applied to both sides, **Figure 6-4**. This practice has been applied to some large diameter pipes. Welds, however, that can be reached from only one side

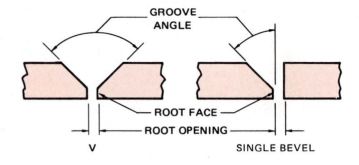
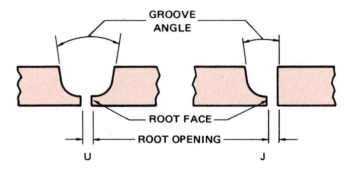

Figure 6-1 Standard grooves used to ensure satisfactory joint penetration.

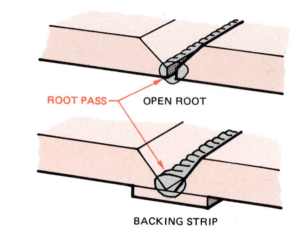

Figure 6-2 Root pass maximum deposit 1/4-in. (6-mm) thick.

must be produced adequately, without the benefit of being able to clean and repair the back side.

The open root weld is widely used in plate and pipe designs. The face side of an open root weld is not so important as the root surface on the back or inside, **Figure 6-5**. The face of a root weld may have some areas of poor uniformity in width, reinforcement, and buildup, or have other defects such as undercut or overlap. As long as the root surface is correct, the front side can be ground, gouged, or burned out to produce a sound weld, **Figure 6-6**. For this reason, during the root pass practices, the weld will be evaluated from the root side only as long as there are not too many defects on the face. To practice the open root welds, the welder will be using

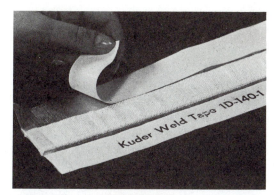

(A) FIBERGLASS

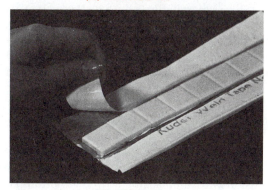

(B) CERAMIC

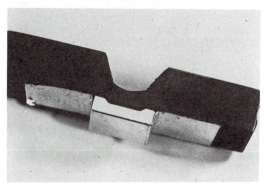

**(C) WELD ROOT PASS MADE USING
CERAMIC BACKING TAPE**

Figure 6-3 Welding backing tapes are available in different materials and shapes. (Courtesy of B.A. Kuder Company.)

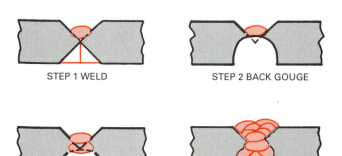

STEP 1 WELD   STEP 2 BACK GOUGE

STEP 3 BACK WELD   STEP 4 COMPLETE WELDING

Figure 6-4 Using back gouging to ensure a sound weld root.

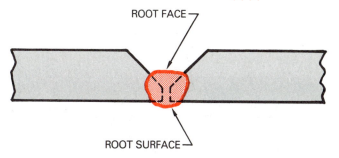

ROOT FACE
ROOT SURFACE

Figure 6-5

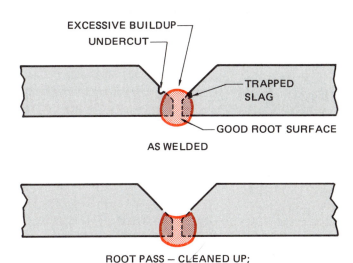

EXCESSIVE BUILDUP
UNDERCUT
TRAPPED SLAG
GOOD ROOT SURFACE
AS WELDED

ROOT PASS — CLEANED UP;
READY FOR NEXT WELD PASS

Figure 6-6 Grinding back the root pass to ensure a sound second pass.

mild steel plate that is 1/8 in. (3 mm) thick. The root face for most grooved joints will be about the same size. This thin plate will help the welder build skill without taking too much time beveling plate just to practice the root pass. Two different methods are used to make a root pass. One method is used only on joints with little or no root gap. This method requires a high amperage and short arc length. The arc length is so short that the electrode flux may drag along on the edges of the joint. The setup for this method must be correct in order for it to work.

The other method can be used on joints with wide, narrow, or varying root gaps. A stepping electrode manipulation and key hole controls the penetration. The electrode is moved in and out of the molten weld pool as the weld progresses along the joint. The edge of the metal is burned back slightly by the electrode just ahead of the molten weld pool, **Figure 6-7.** This is referred to as a key hole, and metal flows through the key hole to the root surface. The key hole must be maintained to ensure 100% penetration. This method requires more welder skill and can be used on a wide variety of joint conditions. The face of the bead resulting from this technique often is defect free.

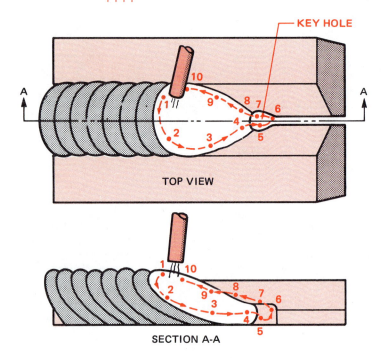

Figure 6-7 Electrode movement to open and use a key.

Root Pass on Plate with a Backing Strip in All Positions

Using a properly set up and adjusted welding machine, proper safety protection, E6010 or E6011 arc welding electrodes having a 1/8-in. (3-mm) diameter, and one or more pieces of mild steel plate, 1/8 in. (3 mm) thick x 6 in. (152 mm) long, and one strip of mild steel, 1/8 in. (3 mm) thick x 1 in. (25 mm) wide x 6 in. (152 mm) long, you will make a root weld in all positions, Figure 6-8. Tack weld the plates together with a 1/16-in. (2-mm) to 1/8-in. (3-mm) root opening. Be sure there are no gaps between the backing strip and plates when the pieces are tacked together, Figure 6-9. If there is a small gap between the backing strip and the plates, it can be removed by placing the assembled test plates on an anvil and striking the tack weld with a hammer. This will close up the gap by compressing the tack welds, Figure 6-10.

Use a straight step or "T" pattern for this root weld. Push the electrode into the root opening so that there is good fusion with the backing strip and bottom edge of

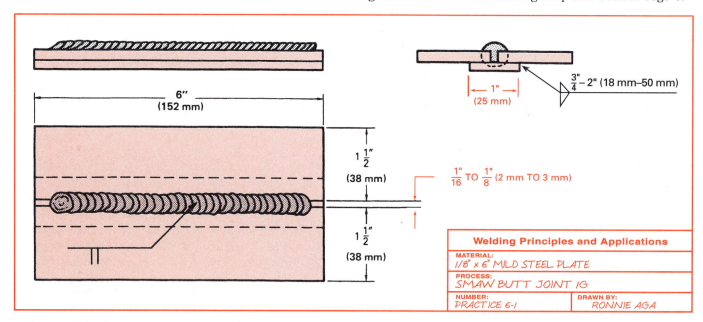

Welding Principles and Applications	
MATERIAL: 1/8" x 6" MILD STEEL PLATE	
PROCESS: SMAW BUTT JOINT 1G	
NUMBER: PRACTICE 6-1	DRAWN BY: RONNIE AGA

Figure 6-8 Square butt joint with a backing strip.

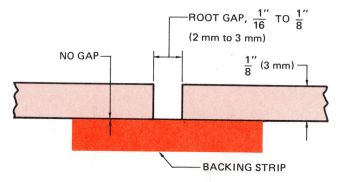

Figure 6-9 Backing strip.

the plates. Failure to push the penetration deep into the joint will result in a cold lap at the root, **Figure 6-11.**

Watch the molten weld pool and keep its size as uniform as possible. As the molten weld pool increases in size, move the electrode out of the weld pool. When the weld pool begins to cool, bring the electrode back into the molten weld pool. Use these weld pool indications to determine how far to move the electrode and when to return to the molten weld pool. After completing the weld, cut the plate and inspect the cross section of the weld for good fusion at the edges. Repeat the welds as necessary until you can consistently make welds free of defects. Turn off the welding machine and clean up your work area when you are finished welding. ◆

Root Pass on Plate with an Open Root in All Positions

Using a properly set up and adjusted arc welding machine, proper safety protection, E6010 or E6011 arc welding electrodes with a 1/8-in. (3-mm) diameter, and two or more pieces of mild steel plate, 6 in. (152 mm) long x 1/8 in. (3 mm) thick, you will make a welded butt joint in all positions with 100% root penetration.

■ Tack weld the plates together with a root opening of 0 in. (0 mm) to 1/16 in. (2 mm).

■ Using a short arc length and high amperage setting, make a weld along the joint.

You can change the electrode angle to control penetration and burnthrough. As the trailing angle is decreased, making the electrode flatter to the plate penetration, depth, and burnthrough decreases, **Figure 6-12,** because both the arc force and heat are directed away from the bottom of the joint back toward the weld. Surface tension holds the metal in place, and the mass of the bead quickly cools the molten weld pool holding it in place. Increasing the electrode angle toward the perpendicular will increase penetration depth and possibly cause more burnthrough. The arc force and heat focused on the gap between the plates will push the molten metal through the joint.

The electrode holder can be slowly rocked from side to side while keeping the end of the electrode in the same spot on the joint, **Figure 6-13.** This will allow the arc force to better tie in the sides of the root to the base metal.

■ When a burnthrough occurs, rapidly move the electrode back to a point just before the burnthrough.

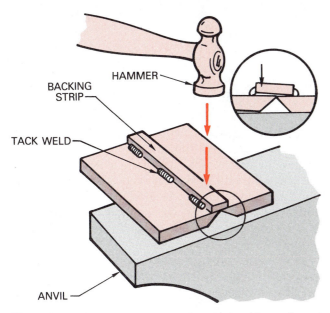

Figure 6-10 Using a hammer to align the backing strip and weld plates.

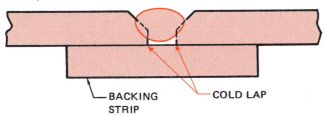

Figure 6-11 Incomplete root fusion.

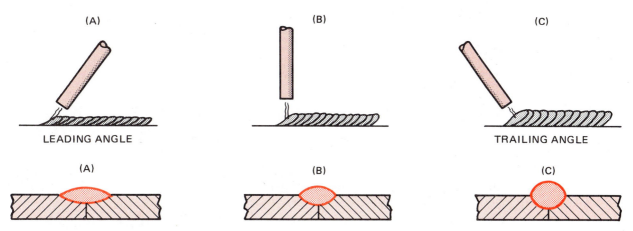

Figure 6-12 Effect of rod angle on weld bead shape.

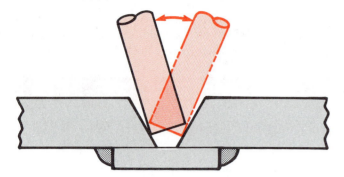

Figure 6-13 Rocking the top of the electrode while keeping the end in the same place helps control the bead shape.

■ Lower the electrode angle and continue welding. If the burnthrough does not close, stop the weld, chip, and wire brush the weld.

■ Check the size of the burnthrough. If it is larger than the diameter of the electrode, the root pass must be continued with the step method described in Practice 6-3. If the burnthrough is not too large, lower the amperage slightly and continue welding.

■ Watch the color of the slag behind the weld. If the weld metal is not fusing to one side, the slag will be brighter in color on one side. The brighter color is caused by the slower cooling of the slag because there is less fused metal to conduct the heat away quickly.

■ After the weld is completed, cooled, and chipped, check the back side of the plate for good root penetration. The root should have a small bead that will look as though it was welded from the back side, **Figure 6-14**. The penetration must be completely free of any drips of metal from the root face, called "icicles."

■ Repeat the welds as necessary until you can consistently make welds free of defects. Turn off the welding machine and clean up your work area when you are finished welding. ◆

PRACTICE 6-3

Open Root Weld on Plate Using the Step Technique in All Positions

Using the same setup, materials, and electrodes as described in Practice 6-2, you will make a welded butt joint in all positions with 100% root penetration.

Tack weld the plates together with a root opening from 0 in. (0 mm) to 1/8 in. (3 mm). Using a medium amperage setting and a short stepping electrode motion, make a weld along the joint.

The electrode should be pushed deeply into the root to establish a key hole that will be used to ensure 100% root penetration. Once the key hole is established, the electrode is moved out and back in the molten weld

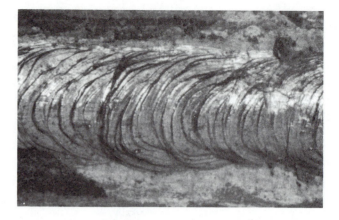

Figure 6-14 The root face (inside) appears uniform.

pool at a steady, rhythmic rate. Watch the molten weld pool and key hole size to determine the rhythm and distance of electrode movement.

If the molten weld pool size decreases, the key hole will become smaller and may close completely. To increase the molten weld pool size and maintain the key hole, slow the rate of electrode movement and shorten the distance the electrode is moved away from the molten weld pool. This will increase the molten weld pool size and penetration because of increased localized heating.

If the molten weld pool becomes too large, metal may drip through the key hole, forming an icicle on the back side of the plate. Extremely large molten weld pool sizes can cause a large hole to be formed or cause burnthrough. Repairing large holes can require much time and skill. To keep the molten weld pool from becoming too large, increase the travel speed, decrease the angle, shorten the arc length, or lower the amperage, **Table 6-1**.

The distance the electrode is moved from the molten weld pool and the length of time in the molten weld pool are found by watching the molten weld pool. The molten weld pool size increases as you hold the arc in the molten weld pool until it reaches the desired size, about twice the electrode diameter, **Figure 6-15**. Move the electrode ahead of the molten weld pool, keeping the arc in the joint but being careful not to deposit any slag or metal. To prevent metal and/or slag from transferring, raise the electrode to increase the arc length, **Figure 6-16**. Keep moving the electrode slowly forward as you watch the molten weld pool. The molten weld pool will suddenly start to solidify. At that time, move the elec-

	Amperage	Travel Speed	Electrode Size	Electrode Angle
To Decrease Puddle Size	Decrease	Increase	Decrease	Leading
To Increase Puddle Size	Increase	Decrease	Increase	Trailing

Table 6-1 Changes Affecting Molten Weld Bead Size.

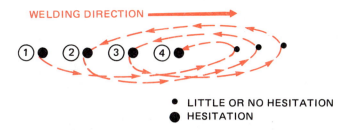

WELDING DIRECTION

① ② ③ ④

● LITTLE OR NO HESITATION
⬤ HESITATION

Figure 6-15 Weave pattern used to control the molten weld bead size.

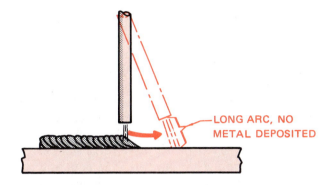

LONG ARC, NO METAL DEPOSITED

Figure 6-16 A long arc prevents metal or slag from being deposited ahead of the weld bead.

trode quickly back to the molten weld pool before it totally solidifies. Moving the electrode in a slight arc will raise the electrode ahead of the molten weld pool and automatically lower the electrode when it returns to the molten weld pool. Metal or slag deposited ahead of the molten weld pool may close the key hole, reduce penetration, and cause slag inclusions. Raising the end of the electrode too high or moving it too far ahead of the molten weld pool can cause all of the shielding gas to be blown away from the molten weld pool. If this happens, oxides can cause porosity. Keeping the electrode movement in balance takes concentration and practice.

Changing from one welding position to another requires an adjustment in timing, amperage, and electrode angle. The flat, horizontal, and overhead positions use about the same rhythm, but the vertical position may require a shorter time cycle for electrode movement. The amperage for the vertical position can be lower than that for the flat or horizontal, but the overhead position uses nearly the same amperage as flat and horizontal. The electrode angle for the flat and horizontal positions is about the same. For the vertical position, the electrode uses a sharper leading angle than does overhead, which is nearly perpendicular and may even be somewhat trailing.

Cool, chip, wire brush, and inspect both sides of the weld. The root surface should be slightly built up and look as though it was welded from that side (refer to Figure 6-14). Repeat the welds as necessary until you can consistently make welds free of defects. Turn off the welding machine and clean up your work area when you are finished welding. ◆

HOT PASS

The surface of a root pass may be irregular, have undercut, overlap, slag inclusions, or other defects, depending upon the type of weld, the code or standards, and the condition of the root pass. The surface of a root pass can be cleaned by grinding or by using a hot pass.

On critical, high-strength code welds it is usually required that the root pass as well as each filler pass be ground (refer to Figure 6-6). This grinding eliminates weld discontinuities caused by slag entrapments. It also can be used to remove most of the E60 series weld metal so that the stronger weld metal can make up most of the weld.) When high-strength, low-alloy welding electrodes are used, this grinding is important to remove most of the low-strength weld deposit. This will leave the weld made up of nearly 100% of the high-strength weld metal, **Figure 6-17.**

The fastest way to clean out trapped slag and make the root pass more uniform is to use a hot pass. The hot pass uses a higher than normal amperage setting and a fast travel rate to reshape the bead and burn out trapped slag. After chipping and wire brushing the root pass to remove all the slag possible, a welder is ready to make the hot pass. The ideal way to apply a hot pass is to rapidly melt a large surface area, **Figure 6-18**, so that the trapped slag can float to the surface. The slag, mostly silicon dioxide (SiO_2), may not melt itself so the surrounding steel must be melted to enable it to float free. The silicon dioxide may not melt because it melts at about 3,100°F (1,705°C), which is more than 500°F (270°C) hotter than the temperature at which the surrounding steel melts, around 2,600°F (1,440°C).

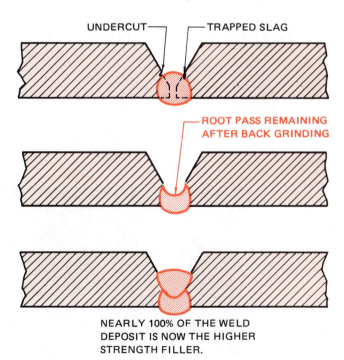

UNDERCUT   TRAPPED SLAG

ROOT PASS REMAINING AFTER BACK GRINDING

NEARLY 100% OF THE WELD DEPOSIT IS NOW THE HIGHER STRENGTH FILLER.

Figure 6-17 Back grinding to remove both discontinuities; filler metal used for the root pass.

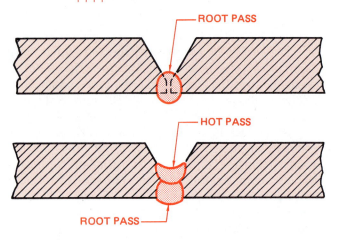

Figure 6-18 Using the hot pass to clean up the face of the root pass.

A very small amount of metal should be deposited during the hot pass so that the resulting weld is concave. A concave weld, compared to a convex weld, is more easily cleaned by chipping, wire brushing, or grinding. Failure to clean the convex root weld will result in a discontinuity showing up on an X ray. Such discontinuities are called wagon tracks. **Figure 6-19.**

The hot pass can also be used to repair or fill most spots of incomplete fusion or pinholes left by the root pass.

The normal weave pattern for a hot pass is the straight step or "T" pattern. The "T" can be used to wash out stubborn trapped slag better than the straight step pattern. The frequency of electrode movement is dependent upon the time required for the molten weld pool to start cooling. As with the root pass, metal or slag should not be deposited ahead of the bead. Do not allow the molten weld pool to cool completely or let the shielding gas covering to be blown away from the molten weld pool.

The hot pass technique can also be used to clean some welds that may first require grinding or gouging for a repair. The penetration of the molten weld pool must be deep enough to free all trapped slag and burn out all porosity.

Hot Pass to Repair a Poor Weld Bead

Using a properly set up and adjusted arc welding machine, proper safety protection, E6010 or E6011 arc welding electrodes having a 1/8-in. (3-mm) diameter, and two or more plates that have welds containing slag inclusions, lack of fusion, porosity, or other defects, you will make a hot pass to burn out the defects.

Chip and wire brush the weld bead. If necessary, use a punch to break apart large trapped slag deposits. The poorer the condition of the weld, the more vertical the joint should be for the hot pass. Large slag deposits tend to float around the molten weld pool and stay trapped in deep pockets in the flat position. With the weld in the vertical position, the slag can run out of the joint and down the face of the weld. Set the amperage as high as possible without overheating and burning up the electrode. Start at the bottom and weld upward using a combination of straight step and "T" patterns to keep the weld deposit uniform. Watch the back edge of the molten weld pool for size and the weld crater for the complete burning out of impurities, **Figure 6-20.**

The plate may start to become overheated because of the high heat input. If you notice that the weld bead is starting to cool too slowly and is growing in length, you should stop welding, **Figure 6-21.** Allow the plate to cool before continuing the weld.

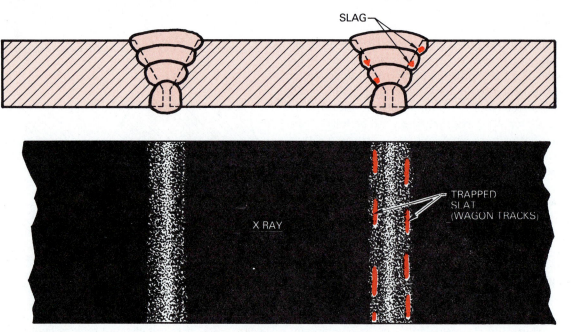

Figure 6-19 Slag trapped between passes will show on an X ray.

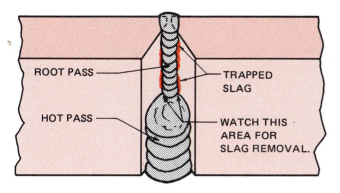

Figure 6-20 Burning out trapped slag by using a hot pass.

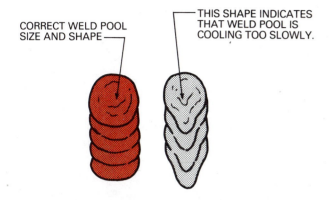

Figure 6-21 The shape of the weld pool can indicate the temperature of the surrounding base metal.

Welds that have large defects in addition to excessive buildup may require some grinding to remove the buildup. Turn off the welding machine and clean up your work area when you are finished welding. ◆

■ *CAUTION*

This hot pass technique is designed to be used on noncritical, noncode welds only. It should not be used to cover bad welds, but as a means of repairing the work of a welder who is less skilled.

After the weld is completed, cool, chip, and inspect it for uniformity. The plate can be cut at places where you know large discontinuities existed before to see if they were repaired or only covered up. If you wish, this experiment can be repeated on other defects and joints.

FILLER PASS

After the root pass is completed and it has been cleaned by grinding or with a hot pass, the groove is filled with weld metal. These weld beads make up the filler pass. More than one pass is often required.

Filler passes are made with stringer beads or weave beads. For multiple pass welds, the weld beads must overlap along the edges. They should overlap enough so

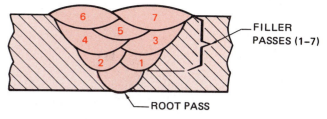

Figure 6-22 Filler passes — maximum thickness 1/8 in. (3 mm) each pass.

that the finished bead is smooth, Figure 6-22. Stringer beads usually overlap about 50%, and weave beads overlap approximately 25%.

Each weld bead must be cleaned before the next bead is started. Slag left on the plate between welds cannot be completely burned out because filler welds should be made with a low amperage setting. Deep penetration will slow the rate of buildup in the joint. Deeply remelting the previous weld metal may weaken the joint. All that is required of a filler weld is that it be completely fused to the base metal.

Chipping, wire brushing, or grinding are the best ways to remove slag between filler weld passes. After the weld is completed, it can be checked by ultrasonic or radiographic nondestructive testing. Most schools are not equipped to do this testing. Therefore, a quick check for soundness can be made by destructive testing. One method of testing the deposited weld metal is by cutting and cross sectioning the weld with an abrasive wheel and inspecting the weld, Figure 6-23. Another fast way to inspect filler passes is to cut a groove through the weld with a gouging tip. Watch the hot metal as it is washed away. The black spots that appear in the cut are slag inclusions. If only a few small spots appear, the weld probably will pass most tests. But, if a long string or large pieces of inclusions appear, the weld will most likely fail.

Figure 6-23 Abrasive cutoff saw used to remove specimens for testing.

Multiple Pass Filler Weld on a V-joint in All Positions

Using a properly set up and adjusted arc welding machine, proper safety protection, E6010 or E6011 arc welding electrodes having a 1/8-in. (3-mm) diameter, and two or more pieces of mild steel plate, 6 in. (152 mm) long x 3 in. (76 mm) wide x 3/8 in. (10 mm) thick, you will make a multiple pass filler weld on a V-joint.

Tack weld the plates together at the corner so that they form a V, **Figure 6-24**. Starting at one end, make a stringer bead along the entire length using the straight step or "T" weave pattern. Thoroughly clean off the slag from the weld before making the next bead. **Figure 6-25** shows the suggested sequence for locating the beads. Continue making welds end cleaning them until the weld is 1 in. (25 mm) or more thick. Both ends of the weld may taper down. If it is important that the ends be square, metal tabs are welded on the ends of the plate for starting and stopping, **Figure 6-26**. The tabs are removed after the weld is completed.

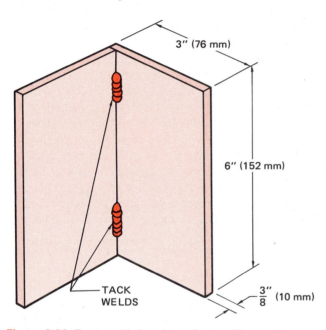

Figure 6-24 Tack weld the plates for the filler weld practice.

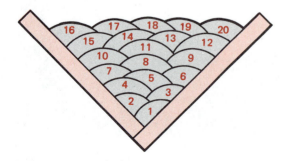

Figure 6-25 Filler pass buildup sequence.

After the weld is completed, it can be visually inspected for uniformity. If nondestructive testing is available, it may be checked for discontinuities. The weld also may be inspected by sectioning it with an abrasive wheel or gouging out with a torch. Repeat these welds until they are mastered. Turn off the welding machine and clean up your work area when you are finished welding. ◆

Multiple Pass Filler Weld on a V-joint in All Positions Using E7018 Electrodes

Using the same setup and materials as in Practice 6-4, and E7018 arc welding electrodes in place of E6010 or E6011 electrodes, you will make a multiple pass filler weld on a V-joint.

- Tack weld the two plates together at the corners so that they form a V.

- Using a slow, straight forward motion, with little or no stepping, "T" or inverted "V" motion, make a stringer bead along the root of the joint.

- Chip the slag and repeat the weld until there is a buildup of 1 in. (25 mm) or more.

If during the weld the buildup should become uneven or large slag entrapments occur, they should be ground out. In industry, groove welds in plate 1 in. (25 mm) or more are normally repaired. Making these repairs now is good experience for the welder. All welders will at some time make a weld that may need repairing.

- After the weld is completed, visually test and nondestructive test the weld for external and internal discontinuities.

- Repeat the welds as necessary until you can consistently make welds free of defects. Turn off the welding machine and clean up your work area when you are finished welding. ◆

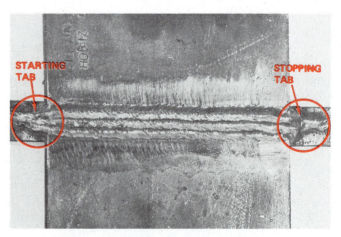

Figure 6-26 Run-off tabs help control possible underfill or burn-back at the starting and stopping points of a groove weld.

COVER PASS

The last weld bead on a multipass weld is known as the cover pass. The cover pass may use a different electrode weave, or it may be the same as the filler beads. Keeping the cover pass uniform and neat looking is important. Most welds are not tested, and often the inspection program is only visual. Thus, the appearance might be the only factor used for accepting or rejecting welds.

The cover pass should be free of any visual defects such as undercut, overlap, porosity, or slag inclusions. It should be uniform in width and reinforcement, **Figure 6-27**. A cover pass should not be more than 1/8 in. (3 mm) wider than the groove opening. Cover passes that are too wide do not add to the weld strength.

<hr>

PRACTICE 6-6

Cover Bead in All Positions

Using a properly set up and adjusted arc welding machine, proper safety protection, E7018 arc welding electrodes having a 1/8-in. (3-mm) diameter, and one or more pieces of mild steel plate, 6 in. (152 mm) long x 1/4 in. (6 mm) thick, you will make a cover bead in each position.

Remember, any time a E7018 low-hydrogen type electrode is to be used, the weave pattern, if used, must not be any larger than 2 1/2 times the diameter of the electrode. This weave cannot be any larger than 5/16 in. (7 mm) wide. Start welding at one end of the plate, and weld to the other end. The weld bead should be about 5/16 in. (7 mm) wide having no more than 1/8 in. (3 mm) of uniform buildup, **Figure 6-28**. The weld buildup should have a smooth transition at the toe, with the plate and the face somewhat convex. Undercut at the toe and a concave or excessively built-up face are the most common problems. Watch the sides of the bead for undercut. When it occurs, keep the electrode just ahead of the spot until it is filled in. There should be a smooth transition between the weld and the plate (refer to Figure 6-27). The shape of the bead face can be controlled by watching the trailing edge of the molten weld pool. That trailing edge is the same as the finished bead, **Figure 6-29**.

Deep penetration is not required with this weld and may even result in some weakening. After the weld is completely cooled, chip and inspect it for uniformity and defects. Repeat the welds as necessary until you can consistently make welds free of defects. Turn off the welding machine and clean up your work area when you are finished welding. ◆

PLATE PREPARATION

When welding on thick plate, it is impossible or impractical for the welder to try to get 100% penetration without preparing the plate for welding. The preparation of the plate is usually in the form of a weld groove. The groove can be cut into one side or both sides of the plate, and it may be cut into either just one plate or both plates of the joint, **Figure 6-30**. The type, depth, angle, and location of the groove is usually determined by a code standard that has been qualified for the specific job.

For SMA welds on plate 1/4 in. (6 mm) or thicker that need to have a weld with 100% joint penetration, the plate must be grooved. The groove may be ground, flame

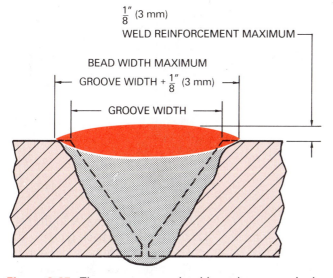

1/8" (3 mm)
WELD REINFORCEMENT MAXIMUM

BEAD WIDTH MAXIMUM
GROOVE WIDTH + 1/8" (3 mm)

GROOVE WIDTH

Figure 6-27 The cover pass should not be excessively large.

Figure 6-28 Practice cover pass.

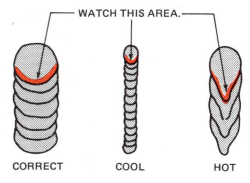

WATCH THIS AREA.

CORRECT   COOL   HOT

Figure 6-29 Watch the back edge of the weld pool to determine the correct current.

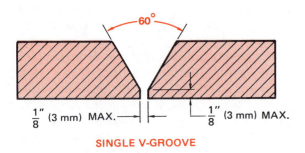

SINGLE V-GROOVE

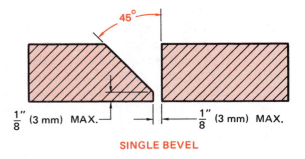

SINGLE BEVEL

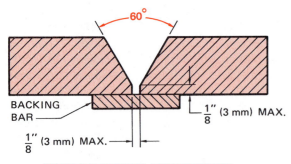

SINGLE V-GROOVE WITH BACKING

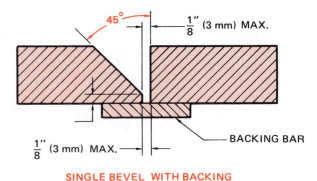

SINGLE BEVEL WITH BACKING

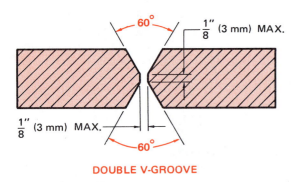

DOUBLE V-GROOVE

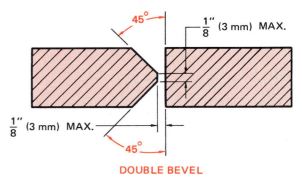

DOUBLE BEVEL

Figure 6-30 Typical butt joint preparations.

cut, gouged, or machined on the edge of the plate before or after the assembly. Bevels and V-grooves are best if they are cut before the parts are assembled. J-grooves and U-grooves can be cut either before or after assembly, **Figure 6-31**. The lap joint is seldom prepared with a groove because little or no strength can be gained by grooving this joint. The only advantage to grooving the lap joint design is to give additional clearance.

Plates that are thicker than 3/8 in. (10 mm) can be grooved on both sides but may be prepared on only one side. The choice to groove one or both sides is most times determined by joint design, position, and application. A tee joint in thick plate is easier to weld and will have less distortion if it is grooved on both sides. Plate in the flat position is usually grooved on only one side unless it can be repositioned. Welds that must have little distortion or that are going to be loaded equally from both sides are usually grooved on both sides. Sometimes plates are either grooved and welded or just welded on one side, and then back gouged and welded, **Figure 6-32**.

Back gouging is a process of cutting a groove in the back side of a joint that has been welded. Back gouging can ensure 100% fusion at the root and remove discontinuities of the root pass. This process can also remove the root pass metal if the properties of the metal are not desirable to the finished weld. **Figure 6-33**. After back gouging, the groove is then welded. See Section 3 for more information of the various methods of gouging.

Heavy plate and pipe sections requiring preparations are often used in products manufactured under a code or standard. The American Welding Society, American Society of Mechanical Engineers, and the American Bureau of Ships are a few of the agencies that issue codes and specifications. The AWS D1.1 and the ASME Boiler and Pressure Vessel (BPV) Section IX standards will be used in this chapter as the standards for multiple pass groove welds that will be tested. The groove depth and angle are determined by the plate or pipe thickness and process. Chapter 19 covers these and other codes for welding plate and pipe.

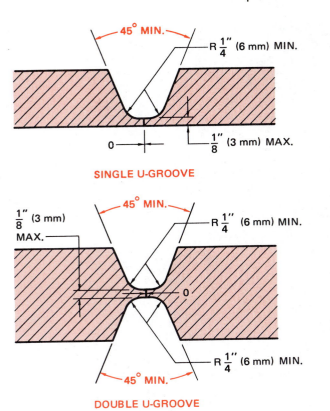

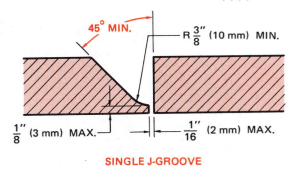

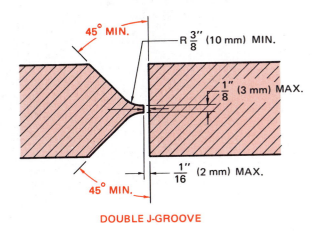

SINGLE U-GROOVE

SINGLE J-GROOVE

DOUBLE U-GROOVE

DOUBLE J-GROOVE

Figure 6-31 Typical butt joint preparations.

Refer to Section 3 for information on beveling. After the plate is beveled, a grinder can be used to clean off oxides and improve the fitup.

PREPARING SPECIMENS FOR TESTING

The detailed preparation of specimens for testing in this chapter are based on the structural welding code AWS D1.1 and the ASME BPV Code, Section IX. The maximum allowable size of fissures (crack or opening) in a

guided bend test specimen are given in codes for specific applications. Some of the standards are listed in ASTM E190 or AWS B4.0, AWS QC10, AWS QC11, and others. Copies of these publications are available from the appropriate organizations. More information on tests and testing can be found in Chapter 20.

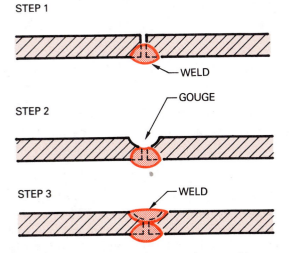

Figure 6-32 Back gouging sequence for a weld to ensure 100% joint penetration.

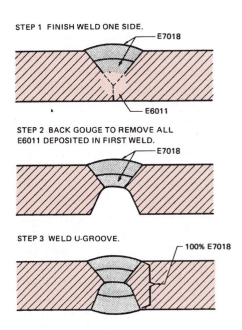

Figure 6-33 Back gouging to remove all weld metal used for the root pass or tacking.

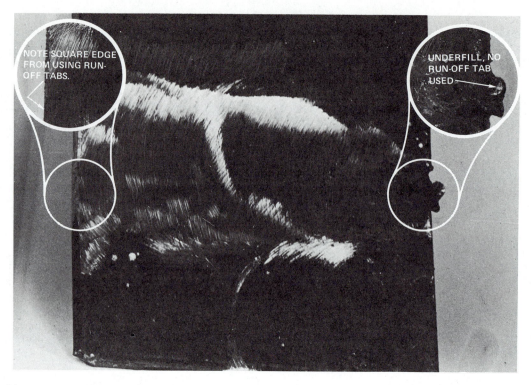

Figure 6-34 Plate ground in preparation for removing test specimens.

Acceptance Criteria for Face Bends and Root Bends

The weld specimen must first pass visual inspection before it can be prepared for bend testing. Visual inspection looks to see that the weld is uniform in width and reinforcement. There should be no arc strikes on the plate other than those on the weld itself. The weld must be free of both incomplete fusion and cracks. The joint penetration must be either 100% or as specified by the specifications. The weld must be free of overlap and undercut must not exceed either 10% of the base metal or 1/32 in. (0.8 mm), whichever is less.

Correct **weld specimen** preparation is essential for reliable results. The weld must be uniform in width and reinforcement and have no undercut or overlap. The weld reinforcement and backing strip, if used, must be removed flush to the surface, **Figure 6-34**. They can be machined or ground off. The plate thickness after removal must be a minimum of 3/8 in. (10 mm), and the pipe thickness must be equal to the pipe's original wall thickness. The specimens may be cut out of the test weldment using an abrasive disc, by sawing, or by cutting with a torch. Flame-cut specimens must have the edges ground or machined smooth after cutting. This procedure is done to remove the heat-affected zone caused by the cut, **Figure 6-35**.

All corners must be rounded to a radius of 1/8 in. (3 mm) maximum, and all grinding or machining marks must run lengthwise on the specimen, **Figures 6-36** and **6-37**. Rounding the corners and keeping all marks run-

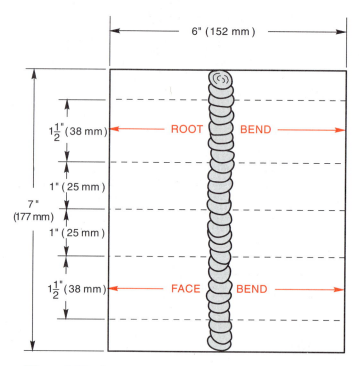

Figure 6-35 Sequence for removing guided bend specimens from the plate once welding is complete.

ning lengthwise reduces the chance of good weld specimen failure due to poor surface preparation.

The weld must pass both the face and root bends in order to be acceptable. After bending there can be no single defects larger than 1/8 in. (3.2 mm), and the sum of all defects larger than 1/32 in. (0.8 mm) but less than 1/8

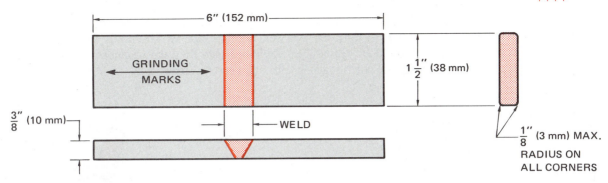

Figure 6-36 Guided bend specimen.

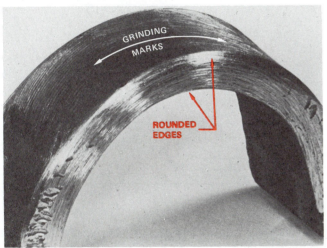

Figure 6-37

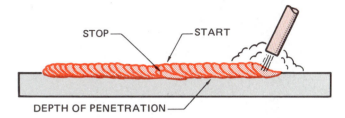

Figure 6-38 Tapering the size of the weld bead helps keep the depth of penetration uniform.

in. (3.2 mm) must not exceed a total of 3/8 in. (9.6 mm) for each bend specimen. An exception is made for cracks that start at the edge of the specimen and do not start at a defect in the specimen.

RESTARTING A WELD BEAD

On all but short welds, the welding bead will need to be restarted after a welder stops to change electrodes. Because the metal cools as a welder changes electrodes and chips slag when restarting, the penetration and buildup may be adversely affected.

When a weld bead is nearing completion, it should be tapered so that when it is restarted the buildup will be more uniform. To taper a weld bead, the travel rate should be increased just before welding stops. A 1/4-in. (6-mm) taper is all that is required. The taper allows the new weld to be started and the depth of penetration re-established without having excessive buildup, **Figure 6-38.**

The slag should always be chipped and the weld crater should be cleaned each time before restarting the weld. This is important to prevent slag inclusions at the start of the weld.

The arc should be restarted in the joint ahead of the weld. The electrodes must be allowed to heat up so

that the arc is stabilized and a shielding gas cloud is re-established to protect the weld. Hold a long arc as the electrode heats up so that metal is not deposited. Slowly bring the electrode downward and toward the weld bead until the arc is directly on the deepest part of the crater where the crater meets the plate in the joint, **Figure 6-39.** The electrode should be low enough to start transferring metal. Next, move the electrode in a semicircle near the back edge of the weld crater. Watch the buildup and match your speed in the semicircle to the deposit rate so that the weld is built up evenly, **Figure 6-40.** Move the electrode ahead and continue with the same weave pattern that was being used previously.

The movement to the root of the weld and back up on the bead serves both to build up the weld and reheat the metal so that the depth of penetration will remain the same. If the weld bead is started too quickly, penetration is reduced and buildup is high and narrow.

Starting and stopping weld beads in corners should be avoided. Tapering and restarting are especially difficult in corners, and this often results in defects, **Figure 6-41.**

PREHEATING AND POSTHEATING

Preheating is the application of heat to the metal before it is welded. This process helps to reduce cracking, hardness, distortion, and stresses. Preheating is most often required on large, thick plates, when the plate is very cold, on days when the temperature is very cold, when small-diameter electrodes are used, on high-carbon or manganese steels, on complex shapes, or with fast welding speeds.

With the practices that are to be tested in this chapter, preheating should be used if the base metal to be welded is very cold. It may also be used to reduce distortion on thick sections and to reduce hardness which may result in weld failure. Preheating the metal will slow the weld cooling rate, which results in a more ductile weld. Table 6-2 lists the recommended preheat temperatures for plain carbon steels.

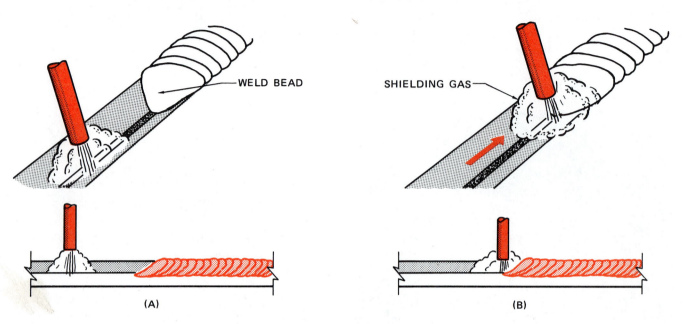

Figure 6-39 When restarting the arc, strike the arc ahead of the weld in the joint (A). Hold a long arc, and allow time for the electrode to heat up, forming the protective gas envelope. Move the electrode so that the arc is focused directly on the leading edge (root) of the previous weld crater (B).

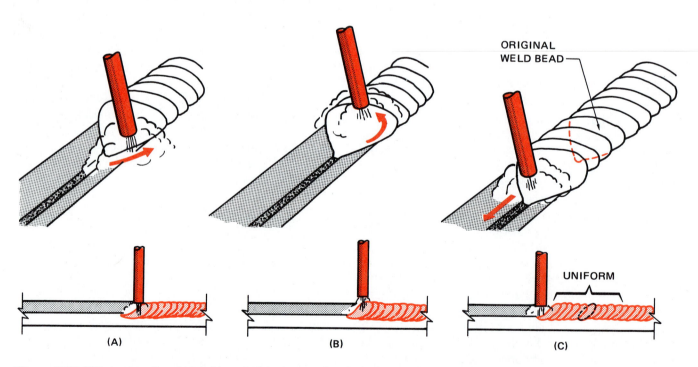

Figure 6-40 When restarting the weld pool after the root has been heated to the melting temperature, move the electrode upward along one side of the crater (A). Move the electrode along the top edge, depositing new weld metal (B). When the weld is built up uniformly with the previous weld, continue along the joint (C).

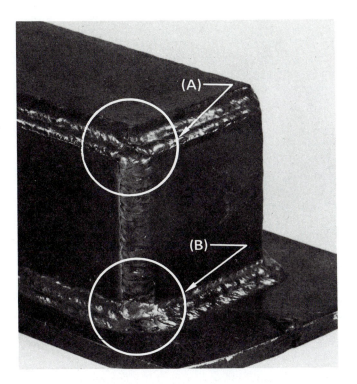

Figure 6-41 Correct method of welding through a corner (A); frequently, leaking is caused by stopping all weld beads at the corner (B).

Plate Thickness in (mm)	Minimum Temperature	
	°F	°C
Up to 1/2 in (13 mm)	70	21
1/2 in (13 mm) to 1 in (25 mm)	100	38
1 in (25 mm) to 2 in (51 mm)	200	95
Over 2 in (51 mm)	300	150

*Metal should be above the dew point.
  Allow 1 hour for each inch in order to provide uniform heating, or for localized preheating. Check to ensure that preheat temperatures are sufficient to melt thermal crayons a minimum of 3 inches (76 mm) in all directions from the area to be welded prior to welding arc application.

Table 6-2 Preheat Temperatures for Arc Welding on Low-Carbon Steels.*

PRACTICE 6-7

WELDING PROCEDURE SPECIFICATION (WPS)

Welding Procedures Specifications No: <u>PRACTICE 6-7</u>
Date: _____

TITLE:
Welding <u>SMAW</u> of <u>plate</u> to <u>plate</u> .

SCOPE:
This procedure is applicable for <u>V-groove plate with a backing strip</u> within the range of <u>3/8 in. (10 mm)</u> through <u>3/4 in. (20 mm)</u> .
Welding may be performed in the following positions <u>1G, 2G, 3G, and 4G</u> .

BASE METAL:
The base metal shall conform to <u>M1020 or A36</u> .
Backing material specification <u>M1020 or A36</u> .

FILLER METAL:
The filler metal shall conform to AWS specification No. <u>E6010 or E6011 root pass and E7018 for the cover pass</u> from AWS specification <u>A5.1</u> . This filler metal falls into F-number <u>F3 and F4</u> and A-number <u>A-1</u> .

SHIELDING GAS:
The shielding gas, or gases, shall conform to the following compositions and purity: <u>N/A</u> .

JOINT DESIGN AND TOLERANCES:

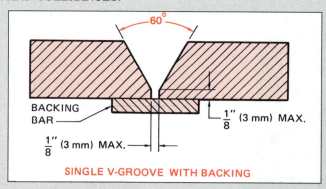

SINGLE V-GROOVE WITH BACKING

PREPARATION OF BASE METAL:

The V-groove is to be ground, flame cut, or machined on the edge of the plate before the parts are assembled. All parts must be cleaned prior to welding of all contaminants, such as paints, oils, grease, or primers. Both inside and outside surfaces within 1 in. (25 mm) of the joint must be mechanically cleaned using a wire brush or grinder.

ELECTRICAL CHARACTERISTICS:

The current shall be AC or DCRP   .
The base metal shall be on the work lead or negative side of the line.

PREHEAT:

The parts must be heated to a temperature higher than 70°F (21°C) before any welding is started.

BACKING GAS:

N/A

WELDING TECHNIQUE:

Tack weld the plates together with the backing strip. There should be about a 1/8-in. (3-mm) root gap between the plates. Use the E6010 or E6011 arc welding electrodes to make a root pass to fuse the plates and backing strip together. Clean the slag from the root pass and use either a hot pass or grinder to remove any trapped slag.

Using the E7018 arc welding electrodes, make a series of filler welds in the groove until the joint is filled. **Figure 6-42** shows the recommended location sequence for the weld beads.

INTERPASS TEMPERATURE:

The plate should not be heated to a temperature higher than 400°F (205°C) during the welding process. After each weld pass is completed, allow it to cool, the weldment must not be quenched in water.

CLEANING:

The slag can be chipped and or ground off between passes but can only be chipped off of the cover pass.

INSPECTION:

Visually inspect the weld for uniformity and other discontinuities, **Figure 6-43**. If the weld passes the visual inspection, then it is to be prepared and guided bend tested according to Chapter 19, "Guided Bend Test," page 427. Repeat each of the welds as needed until you can pass this test.

REPAIR:

No repairs of defects are allowed.

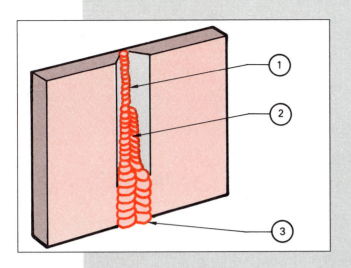

Figure 6-42 Note the sequence in depositing the beads.

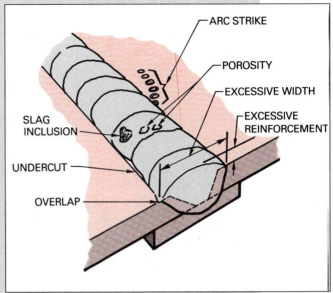

Figure 6-43 Common discontinuities found during a visual examination.

SKETCHES:

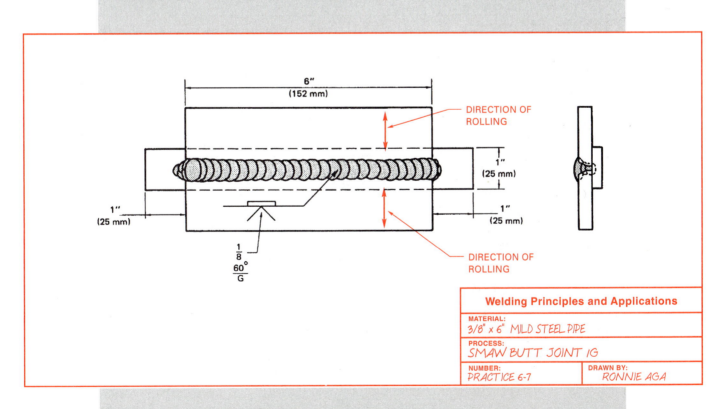

PRACTICE 6-8

WELDING PROCEDURE SPECIFICATION (WPS)

Welding Procedures Specifications No: <u>PRACTICE 6-8</u>
Date: _____

TITLE: Welding <u>SMAW</u> of <u>plate</u> to <u>plate</u> .

SCOPE:
This procedure is applicable for <u>V-groove plate without a backing strip</u> within the range
of <u>3/8 in. (10 mm)</u> through <u>3/4 in. (20 mm)</u> .
Welding may be performed in the following positions <u>1G, 2G, 3G, and 4G</u> .

BASE METAL:
The base metal shall conform to <u>M1020 or A36</u> .
Backing material specification <u>M1020 or A36</u> .

FILLER METAL:
The filler metal shall conform to AWS specification No. <u>E6010 or E6011 root pass and</u>
<u>E7018 for the cover pass</u> from AWS specification <u>A5.1</u> . This filler metal falls into
F-number <u>F3 and F4</u> and A-number <u>A-1</u> .

SHIELDING GAS:
The shielding gas, or gases, shall conform to the following compositions and purity: <u>N/A</u>

JOINT DESIGN AND TOLERANCES:

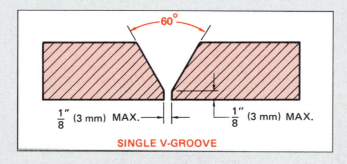

PREPARATION OF BASE METAL:
The V-groove is to be ground, flame cut, or machined on the edge of the plate before the
parts are assembled. All parts must be cleaned prior to welding of all contaminants, such
as paints, oils, grease, or primers. Both inside and outside surfaces within 1 in. (25 mm)
of the joint must be mechanically cleaned using a wire brush or grinder.

ELECTRICAL CHARACTERISTICS:
The current shall be <u>AC or DCRP</u> . The base metal shall be on the <u>work lead or</u>
<u>negative</u> side of the line.

PREHEAT:
The parts must be heated to a temperature higher than 70°F (21°C) before any welding is
started.

BACKING GAS:
N/A

WELDING TECHNIQUE:

Tack weld the plates together; there should be about a 1/8-in. (3-mm) root gap between the plates. Use the E6010 or E6011 arc welding electrodes to make a root pass to fuse the plates together. Clean the slag from the root pass and use either a hot pass or grinder to remove any trapped slag.

Using the E7018 arc welding electrodes, make a series of filler welds in the groove until the joint is filled. **Figure 6-42** shows the recommended location sequence for the weld beads.

INTERPASS TEMPERATURE:

The plate, outside of the heat affected zone, should not be heated to a temperature higher than 400°F (205°C) during the welding process. After each weld pass is completed, allow it to cool; the weldment must not be quenched in water.

CLEANING:

The slag can be chipped and or ground off between passes but can only be chipped off of the cover pass.

INSPECTION:

Visually inspect the weld for uniformity and other discontinuities, **Figure 6-43.** If the weld passes the visual inspection then it is to be prepared and guided bend tested according to Chapter 19, "Guided Bend Test," page 427. Repeat each of the welds as needed until you can pass this test.

REPAIR:

No repairs of defects are allowed.

SKETCHES:

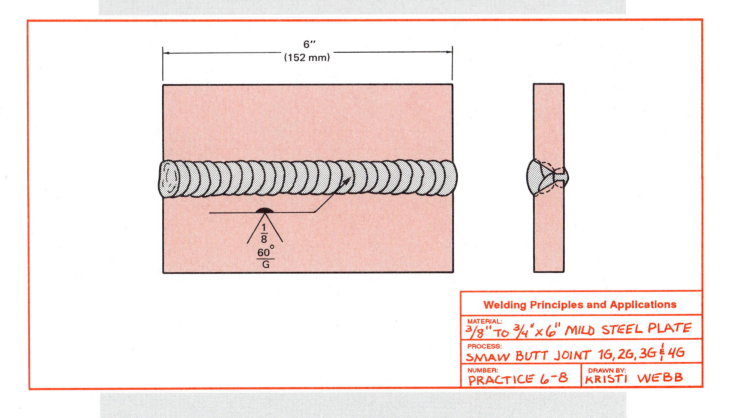

6″
(152 mm)

1/8
60°
G

Welding Principles and Applications

MATERIAL:
3/8″ TO 3/4″ x 6″ MILD STEEL PLATE

PROCESS:
SMAW BUTT JOINT 1G, 2G, 3G & 4G

NUMBER:
PRACTICE 6-8

DRAWN BY:
KRISTI WEBB

PRACTICE 6-9

WELDING PROCEDURE SPECIFICATION (WPS)

Welding Procedures Specifications No: PRACTICE 6-9
Date: _____

TITLE:
Welding SMAW of pipe to pipe .

SCOPE:
This procedure is applicable for V-groove pipe within the range of 3 in. (76 mm) sch 40 through 4 in. (102 mm) schedule 40 .
Welding may be performed in the following positions 1G, 2G, 5G, and 6G .

BASE METAL:
The base metal shall conform to Carbon Steel A52 . Backing material specification N/A.

FILLER METAL:
The filler metal shall conform to AWS specification No. E6010 or E6011 root pass and E7018 for the cover pass from AWS specification A5.1 . This filler metal falls into F-number F3 and F4 and A-number A-1 .

SHIELDING GAS:
The shielding gas, or gases, shall conform to the following compositions and purity: N/A.

JOINT DESIGN AND TOLERANCES:

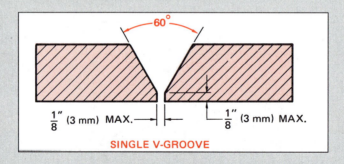

SINGLE V-GROOVE

PREPARATION OF BASE METAL:
The V-groove is to be ground, flame cut, or machined on the end of the pipe before the parts are assembled. All parts must be cleaned prior to welding of all contaminants, such as paints, oils, grease, or primers. Both inside and outside surfaces within 1 in. (25 mm) of the joint must be mechanically cleaned using a wire brush or grinder.

ELECTRICAL CHARACTERISTICS:
The current shall be AC or DCRP . The base metal shall be on the work lead or negative side of the line.

PREHEAT:
The parts must be heated to a temperature higher than 70°F (21°C) before any welding is started.

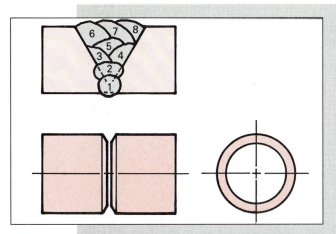

Figure 6-44 Weld bead sequence for 1G and 5G positions.

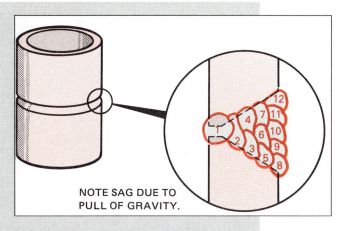

NOTE SAG DUE TO PULL OF GRAVITY.

Figure 6-45 Correct sequence for applying welds to a pipe in the 2G position.

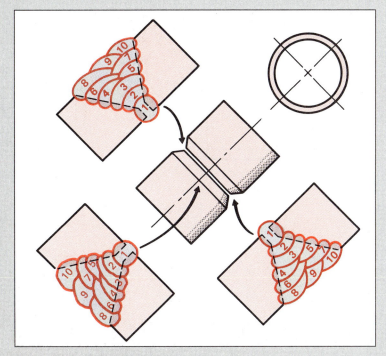

Figure 6-46 6G welding bead sequence.

BACKING GAS:
N/A

WELDING TECHNIQUE:
Tack weld the pipes together; there should be about a 1/8-in. (3-mm) root gap between the pipe ends. Use the E6010 or E6011 arc welding electrodes to make a root pass to fuse the pipe ends together. Clean the slag from the root pass and use either a hot pass or grinder to remove any trapped slag.

Using the E7018 arc welding electrodes, make a series of filler welds in the groove until the joint is filled. **Figure 6-44** shows the recommended location sequence for the weld beads for the 1G and 5G position, **Figure 6-45** for the 2G position and **Figure 6-46** for the 6G position.

INTERPASS TEMPERATURE:
The plate should not be heated to a temperature higher than 400°F (205°C) during the welding process. After each weld pass is completed, allow it to cool; the weldment must not be quenched in water.

CLEANING:
The slag can be chipped and or ground off between passes but can only be chipped off of the cover pass.

INSPECTION:
Visually inspect the weld for uniformity and other discontinuities, Figure 6-43. If the weld passes the visual inspection then it is to be prepared and guided bend tested according to Chapter 19, "Guided Bend Test," page **427**. Repeat each of the welds as needed until you can pass this test.

REPAIR:
No repairs of defects are allowed.

SKETCHES:

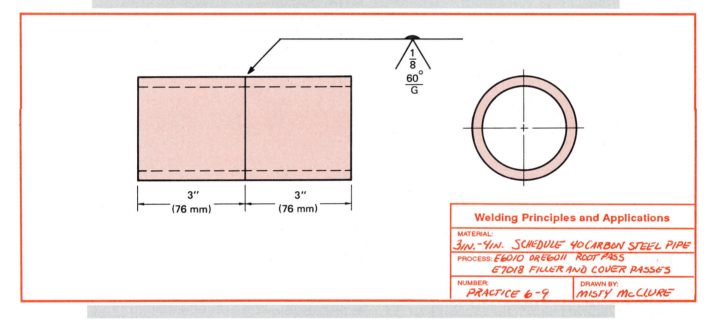

3″	3″
(76 mm)	(76 mm)

Welding Principles and Applications

MATERIAL: *3 IN.– 4 IN. SCHEDULE 40 CARBON STEEL PIPE*

PROCESS: *E6010 OR E6011 ROOT PASS*
E7018 FILLER AND COVER PASSES

NUMBER: *PRACTICE 6-9*   DRAWN BY: *MISTY McCLURE*

Postheating is the application of heat to the metal after welding. Postheating is used to slow the cooling rate and reduce hardening.

Interpass temperature is the temperature of the metal during welding. The interpass temperature is given as a minimum and maximum. The minimum temperature is usually the same as the preheat temperature. If the plate cools below this temperature during welding, it should be reheated. The maximum temperature may be specified to keep the plate below a certain phase change temperature for the mild steel used in these practices. The maximum interpass temperature occurs when the weld bead cannot be controlled because of a slow cooling rate. When this happens, the plate should be allowed to cool down, but not below the minimum interpass temperature.

If, during the welding process, a welder must allow a practice weldment to cool so that the weld can be completed later, the weldment should be cooled slowly and then reheated before starting to weld again. A weld that is to be tested or that is done on any parts other than scrap should not be quenched in water.

POOR FIT

Ideally, all welding will be performed on joints that are properly fitted. Most welds produced to a code or standard are properly fitted. Much of the repair, prototype, and job shop welding, however, is not cut and fitted properly. These welds must be performed under less than ideal conditions, but they still must be strong and have a good appearance.

To make a good weld on a poorly fitted joint requires some special skills. These welds also require a good welder, one whose skill is developed. A skilled welder can watch the molten weld pool and knows how to correct for problems before they develop into disasters. The welder must be able to read the molten weld pool correctly to make needed changes in amperage, current, electrode movement, electrode angle, and timing.

The amperage setting may have to be adjusted up or down by only a few amperes to make the necessary changes in molten weld pool size. Adjusting the machine is often preferable to lengthening the weave pattern

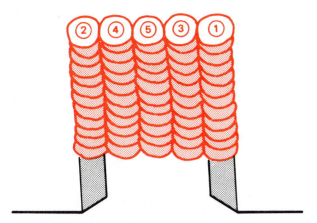

Figure 6-47 Multiple stringer beads used to close a large gap.

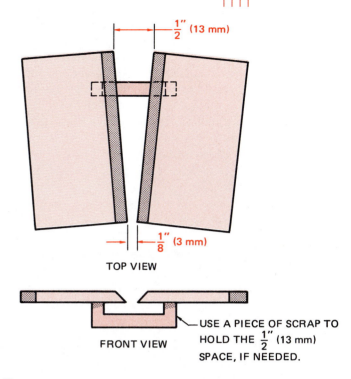

TOP VIEW

FRONT VIEW

USE A PIECE OF SCRAP TO HOLD THE $\frac{1}{2}$" (13 mm) SPACE, IF NEEDED.

Figure 6-48 Welding specimen with poor fitup.

excessively. The current may be changed from AC to DCSP or DCRP to vary the amount of heat input to the molten weld pool. Some electrodes can operate better than other electrodes with lower amperages on some currents. The current also will alter the forcefulness of some electrodes.

The "U," "J," or straight step patterns are usually the best to use, but they should not be moved more than required to close the gap or opening. On some poor-fitting joints, it is necessary to break and restart the arc in order to keep the molten weld pool under control. This will result in a weld with porosity, slag inclusions, and other defects. But it is often better to have a poor weld than to have no weld. A poor root weld can be capped with a sound weld to improve joint strength.

Changing the electrode angle from leading to trailing improves poor fit. Sometimes a very flat angle will also help. The time interval that the electrode is moved into and out of the molten weld pool is critical in maintaining weld control. Returning to a molten weld pool too often or too soon can cause the molten weld pool to drop out of the joint. In most cases, a welder should return to the molten weld pool only after it has started to cool.

On some joints, it is possible to make stringer beads on both sides of the joint until the gap is closed, **Figure 6-47.** Note that the beads are made alternately from the edges of the joint to the center. Welds made in this manner can have good weld soundness and strength, but they require more time to complete.

PRACTICE 6-10

Single V-Groove, Open Root, Butt Joint with an Increasing Root Opening

For this practice, you will need a properly set up and adjusted arc welding machine, proper safety protection, E6010 or E6011 and E7018 arc welding electrodes

having a 1/8-in. (3-mm) diameter, and two or more pieces of mild steel plate, 3/8 in. (10 mm) thick x 4 in. (102 mm) wide x 12 in. (305 mm) long. You will weld a single V-groove open root butt joint that has a poor fitup, starting from the close end.

Tack weld the plates together with a root opening of 1/8 in. (3 mm) at one end and 1/2 in. (13 mm) at the other end, **Figure 6-48.** Using the E6010 or the E6011 electrode, start the root pass at the narrow end and weld to the other end. As the root pass progresses along the widening root gap, care must be taken to maintain molten weld pool control. The "J" or "U" weave pattern works best with low current settings. Long time intervals for electrode movements will give the best weld control.

When the root pass is completed, clean the weld and make a hot pass to burn out any trapped slag. Finish the weld with filler passes using the E7018 electrode. Cool, chip, and inspect the weld for uniformity and defects. Repeat these welds until you can consistently make welds free of defects. Turn off the welding machine and clean up your work area when you are finished welding. ◆

PRACTICE 6-11

Single V-Groove, Open Root, Butt Joint with an Decreasing Root Opening

Using the same setup, equipment, and materials as described in Practice 6-10, you will weld a single

V-groove open root butt joint that has a poor fitup, starting from the wide end.

As in Practice 6-10, tack weld the plates together with a root opening of 1/8 in. (3 mm) at one end and 1/2 in. (13 mm) at the other end. Using the E6010 or E6011 electrode, start welding the root pass at the wide end. Both sides of the joint must be built up until it is possible to get the metal to flow together. The "J" or "U" weave pattern works best to control the bead.

When the root pass is completed, make a hot pass to clean out any trapped slag before making the filler passes with the E7018 electrode. After the weld is completed, cool, chip, and inspect it for uniformity and defects. Repeat these welds until you can consistently make welds free of defects. Turn off the welding machine and clean up your work area when you are finished welding. ◆

Turbine Manufacturer Makes the Move to Robotics

Figure 1

In its quest for greater control of productivity in the fabrication of turbine compressor diaphragms, a manufacturer installed a robotic GTA welding system.

This division produces a wide range of standard and custom units that are sold around the world, including over 100 models in sizes ranging in cost from $40,000 for a small turbine to over $9 million for a large gas turbine. Each turbine must be designed, manufactured, and tested to meet exacting customer requirements for operating efficiency, steam/gas combustion, unit integrity, and control accuracy.

To maintain its high standards of manufacturing, the division is committed to using the latest technologies available. Computer integrated manufacturing (CIM), in which a network is built out of individual islands of automation, has been adopted as an operating policy. Achieving the maximum in productivity and quality is another factor critical to success and requires more than just installing state-of-the-art production equipment — factory layout, materials' flow through the production process, and how tasks are assigned to workers are also important.

A feasibility study was initiated to investigate new dimensions in GTA welding technology. The study revealed some important facts:

1. Robotic GTA welding would improve the division's productivity over manual welding by a factor of about three to one.
2. Weld quality would definitely be improved.
3. Accuracy and reliability would practically eliminate scrap and rework.
4. The system's programming flexibility would allow for frequent and efficient model changeovers.

Although gas tungsten arc welding is inherently a slow process, it offers many advantages, including precision welds, low distortion, high deposition, consistent quality, and eliminates weld spatter, which cannot be tolerated on the workpieces. The process can also be used with various

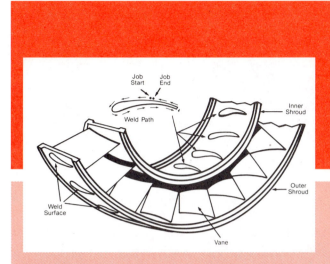

Figure 2

materials, ranging from 403 stainless steel to high-temperature resistant superalloys specified by customers.

The robotic system selected employs a user-friendly teach/playback system, with operator communications conducted through the teach pendant and the control panel. The unit includes a six-axis manipulator and controller, a power source, wire feeder, welding gun assembly, water circulator, and operator stations.

Pulsed gas tungsten arc welding was used for the automatic system. The heat control was excellent because the faster filler wire feed rate was properly matched to current pulsing, producing consistent penetration, low distortion, and high weld quality. The arc start is very reliable, and there is no expulsion to cause contamination. The 2%-thoriated tungsten electrode is prepared each morning and is rarely replaced.

The robot work cell is a unique but simple system consisting of two positioners and a vertical lift mechanism for the robot manipulator. It required about four months each for construction and debugging. For operating efficiency and safety reasons, the robotic work cell is situated in a rectangular area 8 ft below the plant floor level, **Figure 1**. The two 2-axis positioners are located at either end and are automatically placed in a horizontal position at floor level for easy parts' loading and unloading. They swing down below floor level into a vertical position during the welding operation, **Figure 2**. The positioners are integrated into the robot controller so that each vane to the

shroud frame can be manipulated for welding in the flat position. The robot manipulator is bolted to the lift mechanism — a vertical transport system — which is designed to move the manipulator up or down so that welding of the diaphragms can always be accomplished in a flat position.

The parts for the turbine diaphragms are fabricated in two identical semicircular sections. After the parts are fabricated, they are assembled and tacked into position. They are moved to the work cell and loaded onto one of the workstations, which is placed in the horizontal position. The welding fixture, simply designed and operated, consists of two semicircular plates with inserts positioned around the diaphragm to hold the inner and outer shrouds and not to interfere with the gun positioning during welding. The plates also act as heat sinks, reducing possible distortion. To further eliminate any possibility of distortion, the robot is programmed to sequentially weld every other vane. First, the top is welded, then the bottom or inner section, and then the procedure is repeated.

After the system was installed and started up, the company was immediately impressed with the results. The time saving in productivity is substantial. Prior to the robotic installation, all turbine components were manually GTA welded with welding speeds of about 6 ipm and an average of 30% arc time. The robot welding speed varies between 10 and 20 ipm, with 90% arc time.

The metal thickness on the compressor diaphragms averages 5/16 in., and the fillet size varies from 1/16 to 1/32 in., with 50% penetration

Figure 3

required.   Figure 2 shows the shape of the weld path for each weld on the diaphragm. The airfoil-shaped welds can be seen in the center of **Figure 3**. The workpieces average between 48 and 72 welds, and the length is 2 in. The system uses a stainless steel 0.035 in the filler metal.

An extra vane is added to each shroud and removed for testing. Two types of tests are presently being used: dye penetrant inspection and destructive tests. A production engineer says that the quality has been so consistent that the company is planning to remove the extra vane.

(Adapted from information provided by Hobart Brothers Company, Troy, Ohio 45373.)

REVIEW

1. Why are some weld joints grooved?

2. Sketch four of the standard grooves used for welded joints.

3. Why are some backing strips removed from the finished weld?

4. Why are backing tapes used on some joints?

5. Why is it very important to make a weld with a good root surface?

6. What are the two common methods of making a root pass on an open root joint?

7. How can small gaps between the weld plate and backing strip be closed?

8. What effect does changing from a trailing angle to a leading angle have on a weld?

9. What benefit would there be to the root pass if the electrode holder were rocked back and forth while keeping the electrode tip in the joint?

10. What might cause the bright color on the flux as a weld cools?

11. What can happen if the molten weld pool becomes too large on a root weld?

12. What can be done to increase the amount of high-strength welding electrode in the final weld if the root weld was made with low-alloy electrode?

13. What is the purpose of the hot pass?

14. Why should a filler weld pass not have deep penetration?

15. Why is it important to have a good cover pass?

16. What can watching the back edge of a weld pool help you determine?

17. Other than penetration, why would thick butt joints be grooved on both sides?

18. List the things that a weld must be inspected for before it's ground for bend testing.

19. What determines the acceptance or rejection of a bend specimen?

20. What technique can be used to make restarting weld beads easier and more uniform?

21. Why should some weldments be preheated before welding starts?

22. What is postheating used for?

23. How can a wide gap in a joint be closed by welding?

Section Three
Cutting and Gouging

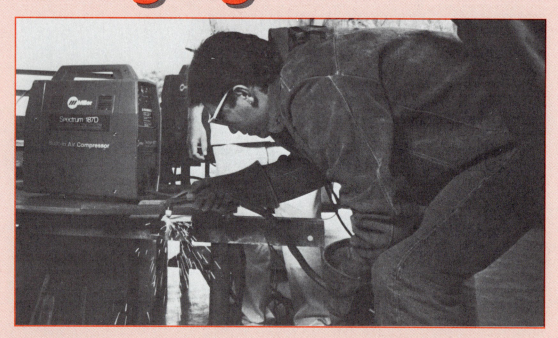

Success Story

Bob Vaughn got his start in the welding field somewhat by chance. Entering his sophomore year of high school, he decided to go into a vocational field. His first choice was drafting, but because all the sections were filled, he needed to select an alternative discipline. His father, a chipper grinder who worked for a pipe fabricating shop, would talk about how the pipe was assembled and welded together. One Saturday morning he got permission to give Bob a tour of the plant. This led to Bob's selection of welding as a second vocational choice.

Shortly after starting the welding program, Bob was told there were some openings in drafting and that he could transfer. By this time he was fascinated by the science of metal joining processes. After completing his high school program, he entered into the college welding program for an additional two years. Bob was encouraged by Bernard Williams, the department head at the time, to continue his education at Penn State in the education field. But like many students, Bob felt he needed a break from the academic environment; besides he didn't see himself as a welding instructor.

Bob became a journeyman welder/fitter and traveled to various construction sites to work on nuclear power plants and chemical refineries. As one of the jobs was coming to an end, an advertisement for a welding instructor was listed in the local newspaper. Wanting to curtail his road traveling, he decided to apply for the job. After fourteen years Bob had returned to the school where he struck his first arc, Pennsylvania College of Technology. It only took him a short time to realize that he thoroughly enjoyed helping others gain skills that lead to good jobs.

For the past fifteen years, while continuing his education, Bob has been helping others become competent welders. This has led many students to excellent financial opportunities. Bob is an Assistant Professor of Welding in the School of Industrial and Engineering Technology. The program at Pennsylvania College of Technology provides an ideal environment for training individuals to become well versed in many of the welding joining processes and technologies.

Bob finds it most gratifying that for years to come he has played a part in shaping the lives and futures of so many. He cannot think of any greater reward than leaving behind a legacy of helping others to achieve their goals and dreams.

Chapter 7

FLAME CUTTING

OBJECTIVES

After completing this chapter, the student should be able to

■ explain how the flame-cutting process works.

■ demonstrate how to properly set up and use an oxyfuel gas cutting torch.

■ safely use an oxyfuel gas cutting torch to make a variety of cuts.

KEY TERMS

oxyfuel gas cutting (OFC)	oxyacetylene hand torch
equal-pressure torches	venturi
cutting lever	machine cutting torch
blowpipe	cutting tips
high-speed cutting tip	preheat flame
preheat holes (orifice)	MPS gases
tip cleaners	hand cutting
coupling distance	drag
soapstone	kindling point
slag	drag lines
soft slag	hard slag

INTRODUCTION

Oxyfuel gas cutting (OFC) is a group of processes that uses a high-temperature oxyfuel gas flame to preheat the metal to a kindling temperature (Color Plates 6A and 6B) at which it will react rapidly with a stream of pure oxygen. See Color Plates 6C and 6D. The kindling temperature for steel, in oxygen, is 1,625°F (884°C). At this temperature, a molten weld pool need not occur to start a cut. The processes in this group are identified by the types of fuel gases used with oxygen to produce the preheat flame. Acetylene is the most commonly used fuel gas. Table 7-1 lists a number of fuel gases, in addition to acetylene, that are used for cutting.

More people use the oxyfuel cutting torch than any other welding process. The cutting torch is used by workers in virtually all areas including manufacturing, maintenance, automotive repair, railroad, farming, and more. It is unfortunately one of the most commonly misused processes. Most workers know how to light the torch and make a cut, but their cuts are very poor quality and often unsafe. A good oxyfuel cut should not only be straight and square, but it should require little or no post-cut cleanup. Excessive post-cutting cleanup results in extra cost, which is an expense that cannot be justified.

Fuel Gas	Flame Fahrenheit	Temperature* Celcius
Acetylene	5,589°	3,087°
MAPP®	5,301°	2,927°
Natural Gas	4,600°	2,538°
Propane	4,579°	2,526°
Propylene	5,193°	2,867°
Hydrogen	4,820°	2,660°

*Approximate neutral Oxyfuel Flame Temperature

Table 7-1 Fuel Gases Used for Flame Cutting.

EYE PROTECTION FOR FLAME CUTTING

The National Bureau of Standards has identified proper filter plates and uses. The recommended filter plates are identified by shade number and are related to the type of cutting operation being performed.

Goggles or other suitable eye protection must be used for flame cutting. Goggles should have vents near the lenses to prevent fogging. Cover lenses or plates should be provided to protect the filter lens. All lens glass should be ground properly so that the front and rear surfaces are smooth. Filter lenses must be marked so that the shade number can be readily identified, **Table 7-2**.

CUTTING TORCHES

The oxyacetylene hand torch is the most common type of oxyfuel gas cutting torch used in industry. The hand torch, as it is often called, may be either a part of a combination welding and cutting torch set, or a cutting torch only, **Figure 7-1**. The combination welding-cutting torch offers more flexibility because a cutting head, welding tip, or heating tip can be attached quickly to the same torch body, **Figure 7-2**. Combination torch sets are often used in schools, automotive repair shops, auto body shops, and small welding shops, or with any job where flexibility in equipment is needed. A cut made with either type of torch has the same quality; however, the

Type of Cutting Operation	Hazard	Suggested Shade Number
Light cutting up to 1 in.	Sparks, harmful rays, molten metal, flying particles	3 or 4
Medium cutting, 1–6 in.		4 or 5
Heavy cutting, over 6 in.		5 or 6

Table 7-2 A General Guide for the Selection of Eye and Face Protection Equipment.

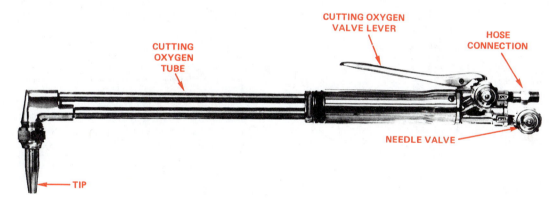

Figure 7-1 Oxyfuel cutting torch. (Courtesy of Victor Equipment Company.)

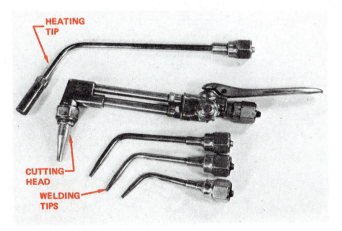

Figure 7-2 The attachments that are used for heating, cutting, welding, or brazing make the combination torch set flexible.

dedicated cutting torches are usually longer and have larger gas flow passages than the combination torches. The added length of the dedicated cutting torch helps keep the operator farther away from the heat and sparks and allows thicker material to be cut.

Oxygen is mixed with the fuel gas to form a high-temperature preheating flame. The two gases must be completely mixed before they leave the tip and create the flame. Two methods are used to mix the gases. One method uses a mixing chamber, and the other method uses an injector chamber.

The mixing chamber may be located in the torch body or in the tip, Figure 7-3. Torches that use a mixing chamber are known as equal-pressure torches because the gases must enter the mixing chamber under the same pressure. The mixing chamber is larger than both the gas inlet and the gas outlet. This larger size causes turbulence in the gases, resulting in the gases mixing thoroughly.

Injector torches will work both with equal gas pressures or low fuel-gas pressures, Figure 7-4. The injector allows the oxygen to draw the fuel gas into the chamber even if the fuel gas pressure is as low as 6 oz/in.^{2} (26 g/cm^2). The injector works by passing the oxygen through a venturi, which creates a low-pressure area that pulls the fuel gases in and mixes them together. An injector-type torch must be used if a low-pressure acetylene generator or low-pressure residential natural gas is used as the fuel gas supply.

The cutting head may hold the cutting tip at a right angle to the torch body or it may be held at a slight angle. Torches with the tip slightly angled are easier for the welder to use when cutting flat plate. Torches with a right-angle tip are easier for the welder to use when cutting pipe, angle iron, I-beams, or other uneven material shapes. Both types of torches can be used for any type of material being cut, but practice is needed to keep the cut square and accurate.

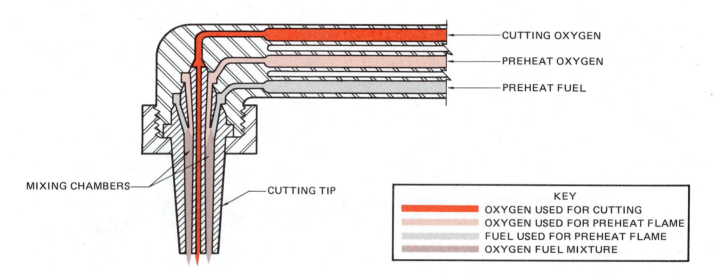

CUTTING OXYGEN
PREHEAT OXYGEN
PREHEAT FUEL

MIXING CHAMBERS
CUTTING TIP

KEY
OXYGEN USED FOR CUTTING
OXYGEN USED FOR PREHEAT FLAME
FUEL USED FOR PREHEAT FLAME
OXYGEN FUEL MIXTURE

Figure 7-3 A mixing chamber located in the tip.

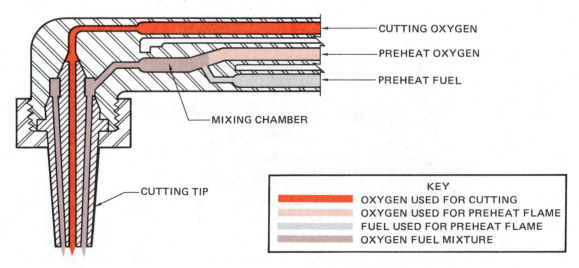

CUTTING OXYGEN
PREHEAT OXYGEN
PREHEAT FUEL

MIXING CHAMBER

CUTTING TIP

KEY
OXYGEN USED FOR CUTTING
OXYGEN USED FOR PREHEAT FLAME
FUEL USED FOR PREHEAT FLAME
OXYGEN FUEL MIXTURE

Figure 7-4 Injector mixing torch.

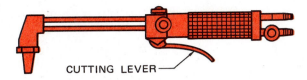

CUTTING LEVER   CUTTING LEVER

Figure 7-5 The cutting lever may be located on the front or back of the torch body.

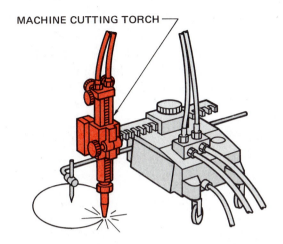

MACHINE CUTTING TORCH

Figure 7-6 Portable flame-cutting machine. (Courtesy of Chemetron Corporation.)

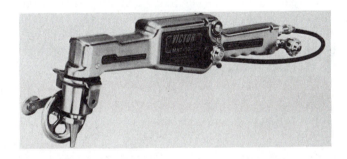

Figure 7-7 Motor-Driver, hand-held cutting torch. (Courtesy of Thermadyne Industries, Inc.)

The location of the cutting lever may vary from one torch to another, **Figure 7-5** Most cutting levers pivot from the front or back end of the torch body. Personal preference will determine which one the welder uses.

A machine cutting torch, sometimes referred to as a blowpipe, operates in a similar manner to a hand cutting torch. The machine cutting torch may require two oxygen regulators, one for the preheat oxygen and the other for the cutting oxygen stream. The addition of a separate cutting oxygen supply allows the flame to be more accurately adjusted. It also allows the pressures to be adjusted during a cut without disturbing the other parts of the flame. Various machine cutting torches are shown in **Figures 7-6, 7-7,** and **7-8.**

CUTTING TIPS

Most cutting tips are made of copper alloy, and some tips are chrome. Chrome plating prevents spatter

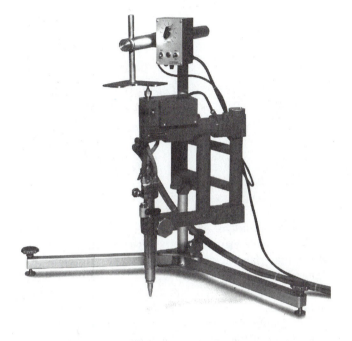

Figure 7-8 Portable cutting machine for highly complex shapes. (Courtesy of ESAB Welding and Cutting Products.)

from sticking to the tip, thus prolonging its usefulness. Tip designs change for the different types of uses and gases, and from one torch manufacturer to another, **Figure 7-9.**

Tips for straight cutting are either standard or high speed, **Figure 7-10.** The high-speed cutting tip is designed to allow a higher cutting oxygen pressure, which allows the torch to travel faster. High-speed tips are also available for different types of fuel gases.

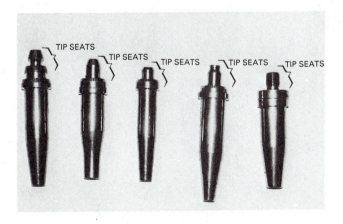

TIP SEATS

Figure 7-9 Five different cutting torch designs for different manufacturers' torches.

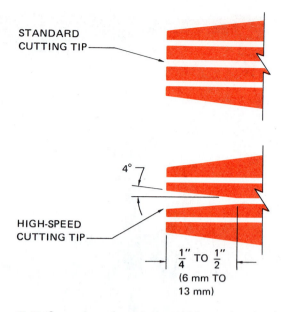

Figure 7-10 Comparison of standard and high-speed cutting tips.

The amount of **preheat flame** required to make a perfect cut is determined by the type of fuel gas used and by the material thickness, shape, and surface condition. Materials that are thick, round, or have surfaces covered with rust, paint, oil, etc., require more preheat flame, **Figure 7-11**.

Different cutting tips are available for each of the major types of fuel gases. The differences in the type or number of **preheat holes** determine the type of fuel gas to be used in the tip. Figure 7-12 lists the fuel gas and

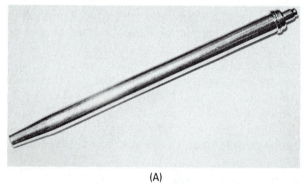

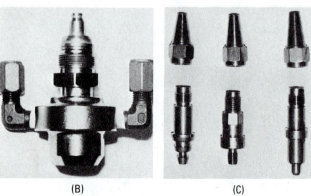

Figure 7-11 Special cutting tips (A) 10-in.-long cutting tip, (B) water-cooled cutting tip, and (C) replaceable end cutting tips.

Gas	Number of Preheat Holes

Figure 7-12 Fuel gas and range of preheat holes.

range of preheat holes or tip designs used with each gas. Acetylene is used in tips having from 1 to 6 preheat holes. Some large acetylene cutting tips may have 8 or more preheat holes.

■ **CAUTION**

If acetylene is used in a tip that was designed to be used with one of the other fuel gases, the tip may overheat, causing a backfire or the tip to explode.

MPS gases are used in tips having eight preheat holes or in a two-piece tip that is not recessed, **Figure 7-13**. These gases have a slower flame combustion rate (see Chapter 26) than acetylene. For tips with less than eight preheat holes, there may not be enough heat to start a cut, or the flame may pop out when the cutting lever is pressed.

■ **CAUTION**

If MPS gases are used in a deeply recessed, two-piece tip, the tip will overheat, causing a backfire or the tip to explode.

Propane and natural gas should be used in a two-piece tip that is deeply recessed, **Figure 7-13**. The flame burns at such a slow rate that it may not stay lit on any other tip.

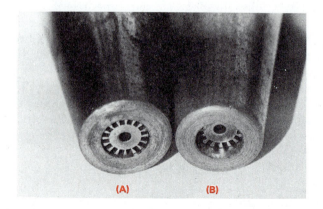

Figure 7-13 Two-piece cutting tips: (A) MAPP®, and (B) propane or natural gas.

Some cutting tips have metal-to-metal seals. When they are installed in the torch head, a wrench must be used to tighten the nut. Other cutting tips have fiber packing seats to seal the tip to the torch. If a wrench is used to tighten the nut for this type of tip, the tip seat may be damaged, Figure 7-14. A torch owner's manual should be checked or a welding supplier should be asked about the best way to tighten various torch tips.

When removing a cutting tip, if the tip is stuck in the torch head, tap the back of the head with a plastic hammer, Figure 7-15. Any tapping on the side of the tip may damage the seat.

To check the assembled torch tip for a good seal, place your thumb over the end of the tip, turn on the oxygen valve, and spray the tip with a leak-detecting solution, Figure 7-16.

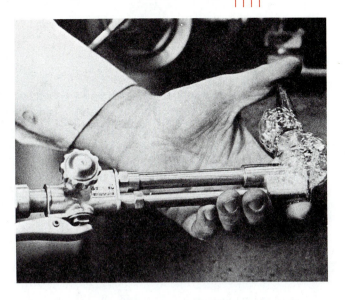

Figure 7-16 Checking a cutting tip for leaks. (Courtesy of Albany Calcium Light Co. Inc., Albany, NY.)

■ *CAUTION*

Carefully handle and store the tips to prevent damage to the tip seats and to keep dirt from becoming stuck in the small holes.

If the cutting tip seat or the torch head seat is damaged, it can be repaired by using a reamer designed for the specific torch tip and head, Figure 7-17, or it can be sent out for repair. New fiber packings are available for

tips with packings. The original leak-checking test should be repeated to be sure the new seal is good.

OXYFUEL CUTTING, SETUP, AND OPERATION

The setting up of a cutting torch system is exactly like setting up oxyfuel welding equipment except for the adjustment of gas pressures. This chapter covers gas pressure adjustments and cutting equipment operations. Chapter 28, Oxyfuel Welding and Cutting, Equipment, Setup, and Operation, gives detailed technical information and instructions for oxyfuel systems. The chapter covers the following topics:

■ Safety

■ Pressure regulator setup and operation

■ Welding and cutting torch design and service

■ Reverse flow and flashback valves

■ Hoses and fittings

■ Types of flames

■ Leak detection

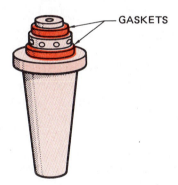

Figure 7-14 Some cutting tips use gaskets to make a tight seal.

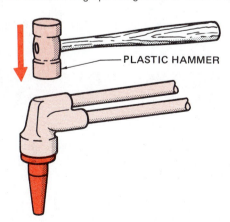

Figure 7-15 Tap the back of the torch head to remove a tip that is stuck. The tip itself should never be tapped.

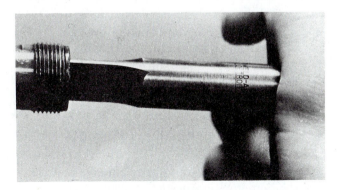

Figure 7-17 Damaged torch seats can be repaired by using a reamer.

Setting Up a Cutting Torch

1. The oxygen and acetylene cylinders must be securely chained to a cart or wall before the safety caps are removed.

2. After removing the safety caps, stand to one side and crack (open and quickly close) the cylinder valves, being sure there are no sources of possible ignition that may start a fire. Cracking the cylinder valves is done to blow out any dirt that may be in the valves.

3. Visually inspect all of the parts for any damage, needed repair, or cleaning.

4. Attach the regulators to the cylinder valves and tighten them securely with a wrench.

5. Attach a reverse flow valve or flashback arrestor, if the torch does not have them built in, to the hose connection on the regulator or to the hose connection on the torch body, depending on the type of reverse flow valve in the set. Occasionally, test each reverse flow valve by blowing through it to make sure it works properly.

6. If the torch you will be using is a combination-type torch, attach the cutting head at this time.

7. Last, install a cutting tip on the torch.

8. Before the cylinder valves are opened, back out the pressure regulating screws so that when the valves are opened the gauges will show zero pounds working pressure.

9. Stand to one side of the regulators as the cylinder valves are opened slowly.

10. The oxygen valve is opened all the way until it becomes tight, but don't over-tighten, and the acetylene valve is opened no more than one-half turn.

11. Open one torch valve and then turn the regulating screw in slowly until 2 psig to 4 psig (14 kPag to 30 kPag) shows on the working pressure gauge. Allow the gas to escape so that the line is completely purged.

12. If you are using a combination welding and cutting torch, the oxygen valve nearest the hose connection must be opened before the flame adjusting valve or cutting lever will work.

13. Close the torch valve and repeat the purging process with the other gas.

14. Be sure there are no sources of possible ignition that may result in a fire.

15. With both torch valves closed, spray a leak-detecting solution on all connections including the cylinder valves. Tighten any connection that shows bubbles, **Figure 7-18.** ◆

Figure 7-18 Leak-check all gas fittings. (Courtesy of Albany Calcium Light Co. Inc., Albany, NY.)

Cleaning a Cutting Tip

Using a cutting torch set that is assembled and adjusted as described in Practice 7-1, and a set of **tip cleaners,** you will clean the cutting tip.

1. Turn on a small amount of oxygen, **Figure 7-19.** This procedure is done to blow out any dirt loosened during the cleaning.

2. The end of the tip is first filed flat, using the file provided in the tip cleaning set, **Figure 7-20.**

3. Try several sizes of tip cleaners in a preheat hole until the correct size cleaner is determined. It should easily go all the way into the tip, **Figure 7-21.**

4. Push the cleaner in and out of each preheat hole several times. Tip cleaners are small, round files. Excessive use of them will greatly increase the **orifice** (hole) size.

5. Next, depress the cutting lever and, by trial and error, select the correct size tip cleaner for the center cutting orifice.

A tip cleaner should never be forced. If the tip needs additional care, refer to the section on tip care in Chapter 25. ◆

Lighting the Torch

Wearing welding goggles, gloves, and any other required personal protective clothing, and with a cutting torch set that is safely assembled, you will light the torch.

1. Set the regulator working pressure for the tip size. If you don't know the correct pressure for the tip, start with the fuel set at 5 psig (35 kPag) and the oxygen set at 25 psig (170 kPag).

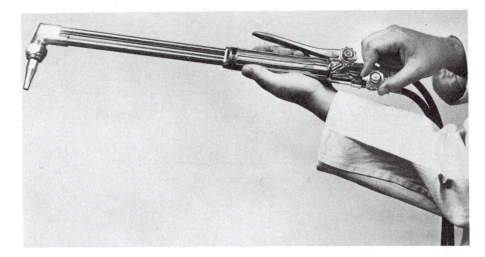

Figure 7-19 Turn on the oxygen valve. (Courtesy of Victor Equipment Company.)

2. Point the torch tip upward and away from any equipment or other students.

3. Turn on just the acetylene valve and use only a sparklighter to ignite the acetylene. The torch may not stay lit. If this happens, close the valve slightly and try to relight the torch.

4. If the flame is small, it will produce heavy black soot and smoke. In this case, turn the flame up to stop the soot and smoke. The welder need not be concerned if the flame jumps slightly away from the torch tip.

5. With the acetylene flame burning smoke free, slowly open the oxygen valve and by using only the oxygen valve adjust the flame to a neutral setting, **Figure 7-22**.

6. When the cutting oxygen lever is depressed, the flame may become slightly carbonizing. This may occur because of a drop in line pressure due to the high flow of oxygen through the cutting orifice.

7. With the cutting lever depressed, readjust the preheat flame to a neutral setting.

The flame will become slightly oxidizing when the cutting lever is released. Since an oxidizing flame is hotter than a neutral flame, the metal being cut will be preheated faster. When the cut is started by depressing the lever, the flame automatically returns to the neutral setting and does not oxidize the top of the plate. Extinguish the flame by first turning off the acetylene and then the oxygen. ◆

■ *CAUTION*

Sometimes with large cutting tips the tip will pop when the acetylene is turned off first. If that happens, turn the oxygen off first.

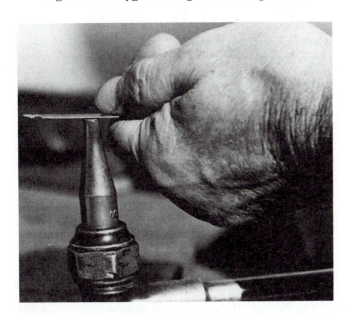

Figure 7-20 File the end of the tip flat.

Figure 7-21 A tip cleaner should be used to clean the flame and center cutting holes.

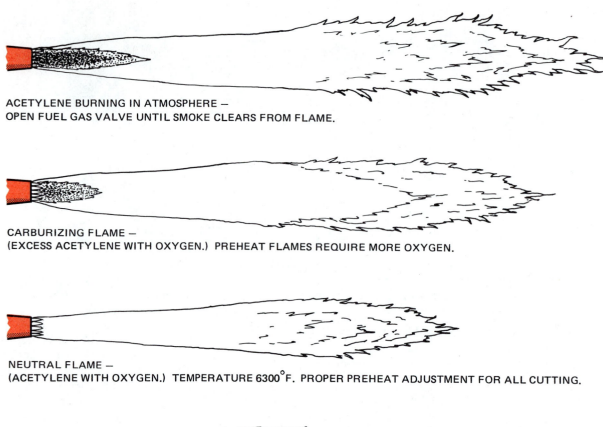

ACETYLENE BURNING IN ATMOSPHERE —
OPEN FUEL GAS VALVE UNTIL SMOKE CLEARS FROM FLAME.

CARBURIZING FLAME —
(EXCESS ACETYLENE WITH OXYGEN.) PREHEAT FLAMES REQUIRE MORE OXYGEN.

NEUTRAL FLAME —
(ACETYLENE WITH OXYGEN.) TEMPERATURE 6300°F. PROPER PREHEAT ADJUSTMENT FOR ALL CUTTING.

NEUTRAL FLAME WITH CUTTING JET OPEN —
CUTTING JET MUST BE STRAIGHT AND CLEAR.

OXIDIZING FLAME —
(ACETYLENE WITH EXCESS OXYGEN.) NOT RECOMMENDED FOR AVERAGE CUTTING.

Figure 7-22 Oxyacetylene flame adjustments for the cutting torch.

HAND CUTTING

When making a cut with a hand torch, it is important for the welder to be steady in order to make the cut as smooth as possible. A welder must also be comfortable and free to move the torch along the line to be cut. It is a good idea for a welder to get into position and practice the cutting movement a few times before lighting the torch. Even when the welder and the torch are braced properly, a tiny movement such as a heartbeat will cause a slight ripple in the cut. Attempting a cut without leaning on the work, to brace oneself, is tiring and causes inaccuracies.

The torch should be braced with the left hand if the welder is right-handed, or with the right hand if the welder is left-handed. The torch may be moved by sliding it toward you over your supporting hand, **Figures 7-23** and **7-24 (A)** and **(B)**. The torch can also be pivoted on the supporting hand. If the pivoting method is used, care must be taken to prevent the cut from becoming a series of arcs.

A slight forward torch angle helps the flame preheat the metal, keeps some of the reflected flame heat off the tip, aids in blowing dirt and oxides away from the cut, and keeps the tip clean for a longer period of time because slag is less likely to be blown back on it, **Figure 7-25**. The forward angle can be used only for a straight line square cut. If shapes are cut using a slight angle, the part will have beveled sides.

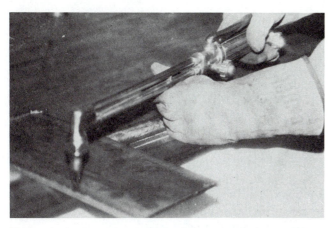

Figure 7-23 For short cuts, the torch can be drawn over the gloved hand.

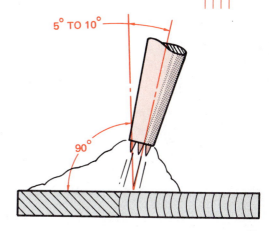

Figure 7-25 A slight forward angle helps when cutting thin material.

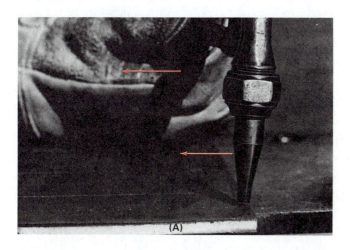

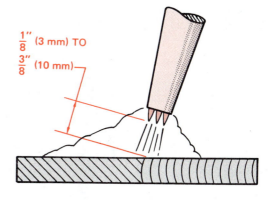

Figure 7-26 Inner cone to work distance.

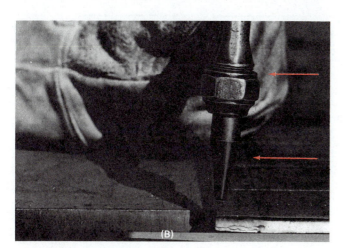

Figure 7-24 For longer cuts, the torch can be moved by sliding your gloved hand along the plate parallel to the cut: (A) start and (B) finish. Always check for free and easy movement before lighting the torch.

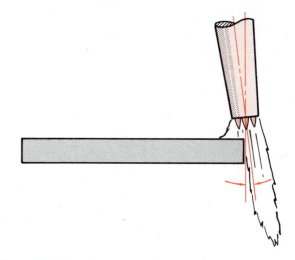

Figure 7-27 Starting a cut on the edge of a plate. Notice how the torch is pointed at a slight angle away from the edge.

When making a cut, the inner cones of the flame should be kept 1/8 in. (3 mm) to 3/8 in. (10 mm) from the surface of the plate, **Figure 7-26.** This distance is known as the coupling distance.

To start a cut on the edge of a plate, hold the torch at a right angle to the surface or pointed slightly away

from the edge, **Figure 7-27.** The torch must also be pointed so that the cut is started at the very edge. The edge of the plate heats up more quickly and allows the cut to be started sooner. Also, fewer sparks will be blown around the shop. Once the cut is started, the torch should be rotated back to a right angle to the surface or to a slight leading angle.

If a cut is to be started in a place other than the edge of the plate, the inner cones should be held as close as possible to the metal. Having the inner cones touch the metal will speed up the preheat time. When the metal is hot enough to allow the cut to start, the torch should be raised as the cutting lever is slowly depressed. When the metal is pierced, the torch should be lowered again, **Figure 7-28**. By raising the torch tip away from the metal, the amount of sparks blown into the air is reduced, and the tip is kept cleaner. If the metal being cut is thick, it may be necessary to move the torch tip in a small circle as the hole goes through the metal. If the metal is to be cut in both directions from the spot where it was pierced, the torch should be moved backward a short distance and then forward, **Figure 7-29**. This prevents slag from refilling the kerf at the starting point, thus making it difficult to cut in the other direction. The kerf is the space produced during any cutting process.

Starts and stops can be made more easily and better if one side of the metal being cut is scrap. When it is necessary to stop and reposition oneself before continuing the cut, the cut should be turned out, a short distance, into the scrap side of the metal, **Figure 7-30**. The extra

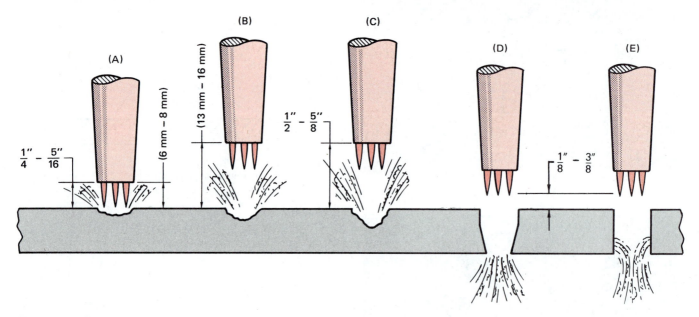

Figure 7-28 Sequence for piercing plate.

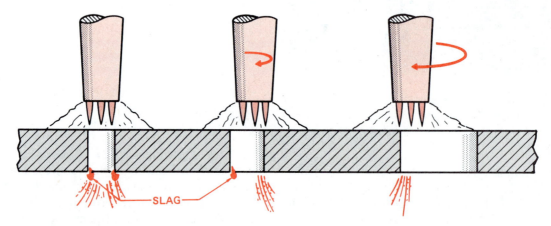

Figure 7-29 A short, backward movement (A), (B) before the cut is carried forward (C) clears the slag from the kerf. Slag left in the kerf may cause the cutting stream to gouge into the base metal, resulting in a poor cut.

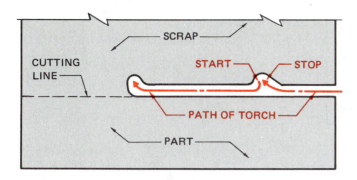

Figure 7-30 Turning out into scrap to make stopping and starting points smoother.

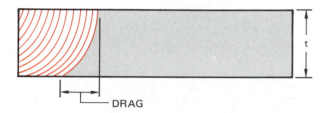

Figure 7-31 Drag is the distance by which the bottom of a cut lags behind the top. (Courtesy of Praxair, Inc.)

space that this procedure provides will allow a smoother and more even start with less chance that slag will block the cut. If neither side of the cut is to be scrap, the forward movement should be stopped for a moment before releasing the cutting lever. This action will allow the drag, or the distance that the bottom of the cut is behind the top, to catch up before stopping, **Figure 7-31.** To restart, use the same procedure that was given for starting a cut at the edge of the plate.

The proper alignment of the preheat holes will speed up and improve the cut. The holes should be aligned so that one is directly on the line ahead of the cut and another is aimed down into the cut when making a straight line square cut, **Figure 7-32.** The flame is di-

rected toward the smaller piece and the sharpest edge when cutting a bevel. For this reason, the tip should be changed so that at least two of the flames are on the larger plate and none of the flames are directed on the sharp edge. **Figure 7-33.** If the preheat flame is directed at the edge, it will be rounded off as it's melted off.

LAYOUT

Laying out a line to be cut can be done with a piece of soapstone or a chalk line. To obtain an accurate line, a scribe or a punch can be used. If a piece of soapstone is used, it should be sharpened properly to increase accuracy, **Figure 7-34.** A chalk line will make a long, straight line on metal and is best used on large jobs. The scribe and punch can both be used to lay out an accurate line, but the punched line is easier to see when cutting. A punch can be held as shown in **Figure 7-35,** with the tip just above the surface of the metal. When the punch is struck with a lightweight hammer, it will make a mark. If you move your hand along the line and rapidly strike the punch, it will leave a series of punch marks for the cut to follow.

SELECTING THE CORRECT TIP AND SETTING THE PRESSURE

Each welding equipment manufacturer uses their own numbering system to designate the tip size. It would be impossible to remember each of the systems. Each manufacturer, however, does relate the tip number to the numbered drill size used to make the holes. On the back of most tip cleaning sets, the manufacturer lists the equivalent drill size of each tip cleaner. By remembering approximately which tip cleaner was used on a particular tip for a metal thickness range, a welder can easily select the correct tip when using a new torch set. Using the tip cleaner that you are familiar with, try it in

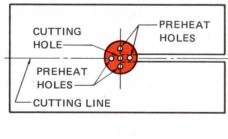

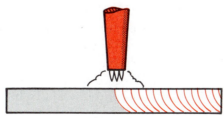

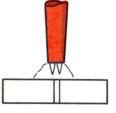

Figure 7-32 Tip alignment for a square cut.

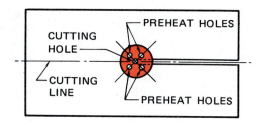

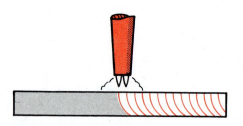

Figure 7-33 Tip alignment for a bevel cut.

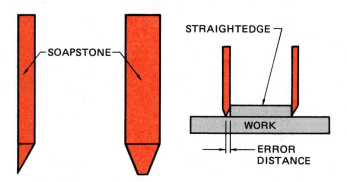

Figure 7-34 Proper method of sharpening a soapstone.

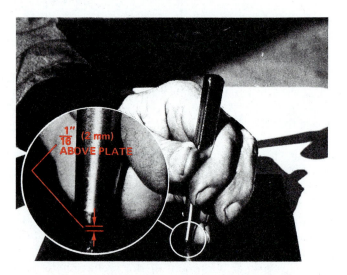

Figure 7-35 Holding the punch slightly above the surface allows the punch to be struck rapidly and moved along a line to mark it for cutting.

Metal Thickness in (mm)	Center Orifice Size No. Drill Size	Tip Cleaner No.*	Oxygen Pressure lb/in² (kPa)	Acetylene lb/in² (kPa)
1/8 (3)	60	7	10 (70)	3 (20)
1/4 (6)	60	7	15 (100)	3 (20)
3/8 (10)	55	11	20 (140)	3 (20)
1/2 (13)	55	11	25 (170)	4 (30)
3/4 (19)	55	11	30 (200)	4 (30)
1 (25)	53	12	35 (240)	4 (30)
2 (51)	49	13	45 (310)	5 (35)
3 (76)	49	13	50 (340)	5 (35)
4 (102)	49	13	55 (380)	5 (35)
5 (127)	45	**	60 (410)	5 (35)

*The tip cleaner number when counted from the small end toward the large end in a standard tip cleaner set

**Larger than normally included in a standard tip cleaner set

Table 7-3 Cutting Pressure and Tip Size.

the various torch tips until you find the correct tip that the tip cleaner fits. **Table 7-3** lists the tip drill size, pressure range, and the metal thickness range for which the tip can be used.

PRACTICE 7-4

Setting the Gas Pressures

Setting the working pressure of the regulators can be done by following a table, or it can be set by watching the flame.

1. To set the regulator by watching the flame, first set the acetylene pressure at 2 psig to 4 psig (14 kPag to 30 kPag) and then light the acetylene flame.

2. Open the acetylene torch valve one to two turns and reduce the regulator pressure by backing out the setscrew until the flame starts to smoke.

3. Increase the pressure until the smoke stops and then increase it just a little more.

This is the maximum fuel gas pressure the tip needs. With a larger tip and a longer hose, the pressure must be set higher. This is the best setting, and it is the safest one to use. With this lowest possible setting, there is less chance of a leak. If the hoses are damaged, the resulting fire will be much smaller than a fire burning from a hose with a higher pressure. There is also less chance of a leak with the lower pressure.

4. With the acetylene adjusted so that the flame just stops smoking, slowly open the torch oxygen valve.

5. Adjust the torch to a neutral flame. When the cutting lever is depressed, the flame will become carbonizing, not having enough oxygen pressure.

6. While holding the cutting lever down, increase the oxygen regulator pressure slightly. Readjust the flame, as needed, to a neutral setting by using the oxygen valve on the torch.

7. Increase the pressure slowly and readjust the flame as you watch the length of the clear cutting stream in the center of the flame, **Figure 7-36**. The center stream will stay fairly long until a pressure is reached that causes turbulence disrupting the cutting stream. This turbulence will cause the flame to shorten in length considerably, **Figure 7-36**.

8. With the cutting lever still depressed, reduce the oxygen pressure until the flame lengthens once again. This is the maximum oxygen pressure that this tip can use without disrupting turbulence in the cutting stream. This turbulence will cause a very poor cut. The lower pressure also will keep the sparks from being blown a longer distance from the work, **Figure 7-37**. ◆

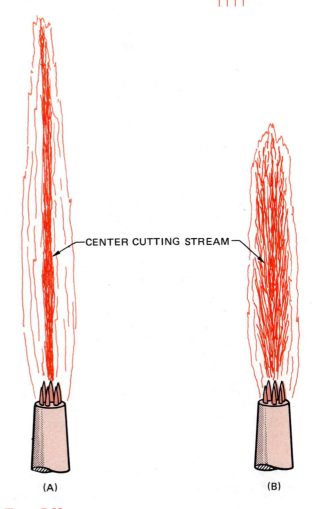

Figure 7-36

THE CHEMISTRY OF A CUT

The oxyfuel gas cutting torch works when the metal being cut rapidly oxidizes or burns. This rapid oxidization or burning occurs when a high-pressure stream of pure oxygen is directed on the metal after it has been preheated to a temperature above its kindling point.

Kindling point is the lowest temperature at which a material will burn. The kindling temperature of iron is 1,600°F (870°C), which is a dull red color. Note that iron is the pure element and cast iron is an alloy primarily of iron and carbon. The process will work easily on any metal that will rapidly oxidize, such as iron, low-carbon steel, magnesium, titanium, and zinc.

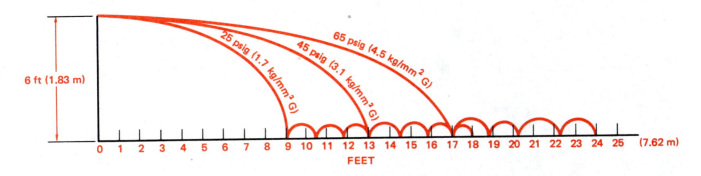

Figure 7-37 The sparks from cutting a mild steel plate, 3/8 in. (10 mm) thick, 6 ft (1.8 m) from the floor, will be thrown much farther if the cutting pressure is too high for the plate thickness. These cuts were made with a Victor cutting tip no. 0-1-101 using 25 psig (1.7 kg/mm^2) as recommended by the manufacturer and by excessive pressures of 45 psig (3.1 kg/mm^2) and 65 psig (4.5 kg/mm^2).

■ *CAUTION*

Some metals release harmful oxides when they are cut. Extreme caution must be taken when cutting used, oily, dirty, or painted metals. They often produce very dangerous fumes when they are cut. You may need extra ventilation and a respirator to be safe.

The process is most often used to cut iron and low-carbon steels, because unlike most of the metals, little or no oxides are left on the metal, and it can easily be welded.

The burning away of the metal is a chemical reaction with iron (Fe) and oxygen (O). The oxygen forms an iron oxide, primarily Fe_3O_4, that is light gray in color. Heat is produced, by the metal, as it burns. This heat helps carry the cut along. On thick pieces of metal, once a small spot starts burning (being cut), the heat generated helps the cut continue quickly through the metal. With some cuts the heat produced may overheat small strips of metal being cut from a larger piece. As an example, the center piece of a hole being cut will quickly become red hot and starts to oxidize with the surrounding air, **Figure 7-38**. This heat produced by the cut makes it difficult to cut out small or internal parts.

EXPERIMENT 7-1

Observing Heat Produced during a Cut

This experiment may require more skill than you have developed by this time. You may wish to observe your instructor performing the experiment or try it at a later time.

Using a properly lit and adjusted cutting torch, welding gloves, appropriate eye protection and clothing,

Figure 7-38 As a hole is cut, the center may be overheated.

and one piece of clean mild steel plate, 6 in. (152 mm) long x 1/4 in. (6 mm) to 1/2 in. (13 mm) thick, you will make an oxyfuel gas cut without the preheat flame.

Place the piece of metal so that the cutting sparks fall safely away from you. With the torch lit, pass the flame over the length of the plate until it is warm, but not hot. Brace yourself and start a cut near the edge of the plate. When the cut has been established, have another student turn off the acetylene regulator. The cut should continue if you remain steady and the plate is warm enough. *Hint: Using a slightly larger tip size will make this easier.* ◆

THE PHYSICS OF A CUT

As a cut progresses along a plate, a record of what happened during the cut is preserved along both sides of the kerf. This record indicates to the welder what was correct or incorrect with the preheat flame, cutting speed, and oxygen pressure.

Preheat

The size and number of preheat holes in a tip has an effect on both the top and bottom edges of the metal. An excessive amount of preheat flame results in the top edge of the plate being melted or rounded off. In addition, an excessive amount of hard-to-remove slag is deposited along the bottom edge. If the flame is too small, the travel speed must be slower. A reduction in speed may result in the cutting stream wandering from side to side. The torch tip can be raised slightly to eliminate some of the damage caused by too much preheat. However, raising the torch tip causes the cutting stream of oxygen to be less forceful and less accurate.

Speed

The cutting speed should be fast enough so that the **drag lines** have a slight slant backward if the tip is held at a 90° angle to the plate, **Figure 7-39**. See Color Plate 6E and 6F. If the cutting speed is too fast, the oxygen stream may not have time to go completely through the metal, resulting in an incomplete cut, **Figure 7-40**. Too slow a cutting speed results in the cutting stream wandering, thus causing gouges in the side of the cut, **Figures 7-41** and **7-42**.

Pressure

A correct pressure setting results in the sides of the cut being flat and smooth. A pressure setting that is too high causes the cutting stream to expand as it leaves the tip, resulting in the sides of the kerf being slightly dished, **Figure 7-43**. When the pressure setting is too low, the cut may not go completely through the metal.

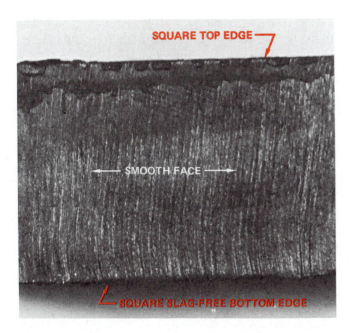

Figure 7-39 Correct cut.

Figure 7-40 Too fast a travel speed resulting in an incomplete cut; too much preheat and the tip is too close, causing the top edge to be melted and removed.

EXPERIMENT 7-2

Effect of Flame, Speed, and Pressure on a Machine Cut

Using a properly lit and adjusted automatic cutting machine, welding gloves, appropriate eye protection and clothing, a variety of tip sizes, and one piece of mild steel plate, 6 in. (152 mm) long x 1/2 in. (13 mm) to 1 in. (25 mm) thick, you will observe the effect of the preheat flame, travel speed, and pressure on the metal being cut.

Using the variety of tips, speeds, and oxygen pressures, make a series of cuts on the plate. As the cut is being made, listen to the sound it makes. Also look at the stream of sparks coming off the bottom. A good cut should have a smooth, even sound, and the sparks should come off the bottom of the metal more like a stream than a spray, Figure 7-44. When the cut is complete, look at the drag lines to determine what was correct or incorrect with the cut, Figure 7-45.

Repeat this experiment until you know a good cut by the sound it makes and the stream of sparks. A good cut has little or no slag left on the bottom of the plate. ◆

Figure 7-41 Too slow a travel speed results in the cutting stream wandering, thus causing gouges in the surface; preheat flame is too close, melting the top edge.

EXPERIMENT 7-3

Effect of Flame, Speed, and Pressure on a Hand Cut

Using a properly lit and adjusted hand torch, welding gloves, appropriate eye protection and clothing, and the same tip sizes and mild steel plate, repeat Experiment 7-2 to note the effects of the preheat flame, travel speed, and pressure on hand cutting. ◆

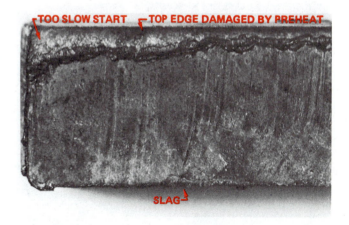

Figure 7-42 Too slow a travel speed at the start; too much preheat.

Slag

The two types of slag produced during a cut are soft slag and hard slag. Soft slag is very porous, brittle, and easily removed from a cut. There is little or no unoxidized iron in it. It may be found on some good cuts. Hard slag may be mixed with soft slag. Hard slag is attached solidly to the bottom edge of a cut, and it requires a lot of chipping and grinding to be removed. There is 30% to 40% or more unoxidized iron in hard slag. The higher the unoxidized iron content, the more difficult the slag is to remove. Slag is found on bad cuts, due to dirty tips, too much pre-

CORRECT CUT

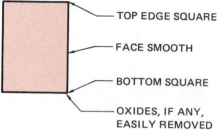

- TOP EDGE SQUARE
- FACE SMOOTH
- BOTTOM SQUARE
- OXIDES, IF ANY, EASILY REMOVED

TRAVEL SPEED TOO SLOW

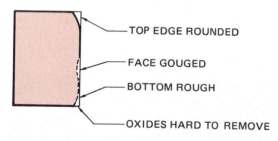

- TOP EDGE ROUNDED
- FACE GOUGED
- BOTTOM ROUGH
- OXIDES HARD TO REMOVE

TRAVEL SPEED TOO FAST

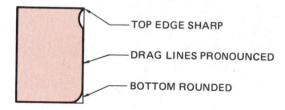

- TOP EDGE SHARP
- DRAG LINES PRONOUNCED
- BOTTOM ROUNDED

PREHEAT FLAMES TOO HIGH ABOVE THE SURFACE

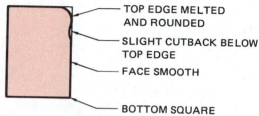

- TOP EDGE MELTED AND ROUNDED
- SLIGHT CUTBACK BELOW TOP EDGE
- FACE SMOOTH
- BOTTOM SQUARE

PREHEAT FLAMES TOO CLOSE TO THE SURFACE

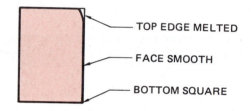

- TOP EDGE MELTED
- FACE SMOOTH
- BOTTOM SQUARE

CUTTING OXYGEN PRESSURE TOO HIGH

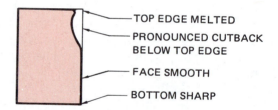

- TOP EDGE MELTED
- PRONOUNCED CUTBACK BELOW TOP EDGE
- FACE SMOOTH
- BOTTOM SHARP

Figure 7-43 Profile of flame-cut plates.

Figure 7-44 A good cut showing a steady stream of sparks flying out from the bottom of the cut.

Figure 7-45 Poor cut. The slag is backing up because the cut is not going through the plate.

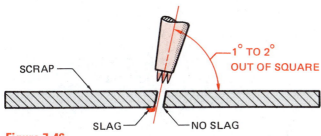

Figure 7-46

heat, too slow a travel speed, too short a coupling distance, or incorrect oxygen pressure.

The slag from a cut may be kept off one side of the plate being cut by slightly angling the cut toward the scrap side of the cut, **Figure 7-46**. The angle needed to force the slag away from the good side of the plate may be as small as 2° or 3°. This technique works best on thin sections; on thicker sections the bevel may show.

PLATE CUTTING

Low-carbon steel plate can be cut quickly and accurately, whether thin-gauge sheet metal or sections more than 4 feet (1.2 m) thick are used. It is possible to achieve cutting speeds as fast as 32 inches per minute (13.5 mm/s), in 1/8-in. (3-mm) plate, and accuracy on machine cuts of ±3/64 in. Some very large hand-cutting torches with an oxygen cutting volume of 600 cfh (2830 L/min) can cut metal that is 4 ft (1.2 m) thick, **Figure 7-47**. Most hand torches will not easily cut metal that is more than 7 in. (178 mm) to 10 in. (254 mm) thick.

The thicker the plate, the more difficult the cut is to make. Thin plate, 1/4 in. (6 mm) or less, can be cut and the pieces separated even if poor techniques and incorrect pressure settings are used. Thick plate, 1/2 in. (13 mm) or thicker, often cannot be separated if the cut is not correct. For very heavy cuts, on plate 12 in. (305

mm) or thicker, the equipment and operator technique must be near perfection or the cut will be faulty.

Plate that is properly cut can be assembled and welded with little or no postcutting cleanup. Poor quality cuts require more time to clean up than is needed to make the required adjustments to make a good weld.

METHODS OF IMPROVING CUTS

Welders can use a variety of techniques to improve the quality of the cuts they make. For example, a piece of angle iron can be clamped to the plate being cut. The angle iron can be used to guide the torch for either square or bevel cuts, **Figure 7-48**.

Devices such as guide rollers for cutting straight lines, circle cutting attachments, and power rollers may be used to reduce operator fatigue and improve the quality of the cuts made, **Figure 7-49**. These devices work best on new stock when repetitive cuts are being made.

PRACTICE 7-5

Flat, Straight Cut in Thin Plate

Using a properly lit and adjusted cutting torch and one piece of mild steel plate, 6 in. (152 mm) long x 1/4 in. (6 mm) thick, you will cut off 1/2-in. (13-mm) strips.

Using a straightedge and soapstone, make several straight lines 1/2 in. (13 mm) apart. Starting at one end, make a cut along the entire length of plate. The strip must fall free, be slag free, and be within ±3/32 in. (2 mm) of a straight line and ±5° of being square. Repeat this procedure until the cut can be made straight and slag free. Turn off the cylinder valves, bleed the hoses, back out the pressure regulators, and clean up your work area when you are finished cutting. ◆

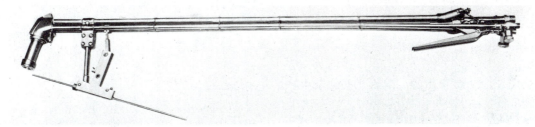

Figure 7-47 Hand torch for thick sections. The wire added to the flame helps to start the cut faster. (Courtesy of Airco Welding Products.)

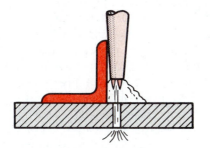

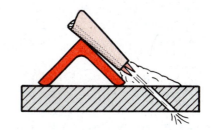

Figure 7-48 Using angle irons to aid in making cuts.

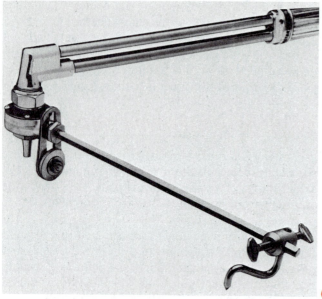

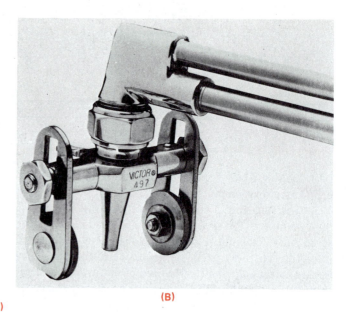

(A)

(B)

Figure 7-49 Devices that are used to improve hand cutting. (Courtesy of Victor Equipment Company.)

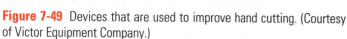

Flat, Straight Cut in Thick Plate

Using a properly lit and adjusted cutting torch and one piece of mild steel plate, 6 in. (152 mm) long x 1/2 in. (13 mm) thick or thicker, you will cut off 1/2-in. (13-mm) strips. *Note: Remember that starting a cut in thick plate will take longer, and the cutting speed will be slower.* Lay out, cut, and evaluate the cut as was done in Practice 7-5. Repeat this procedure until the cut can be made straight and slag free. Turn off the cylinder valves, bleed the hoses, back out the pressure regulators, and clean up your work area when you are finished cutting. ◆

PRACTICE 7-7

Flat, Straight Cut in Sheet Metal

Use a properly lit and adjusted cutting torch and a piece of mild steel sheet that is 10 in. (254 mm) long and 18 gauge to 11 gauge thick. Holding the torch at a very sharp leading angle, **Figure 7-50**, cut the sheet along the line. The cut must be smooth and straight with as little slag as possible. Repeat this procedure until the cut can be made flat, straight, and slag free. Turn off the cylinder valves, bleed the hoses, back out the pressure regulators, and clean up your work area when you are finished cutting. ◆

PRACTICE 7-8

Flame Cutting Holes

Using a properly lit and adjusted cutting torch, welding gloves, appropriate eye protection and clothing, and

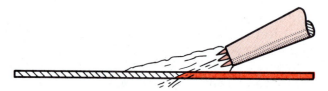

Figure 7-50 Cut the sheet metal at a very sharp angle.

one piece of mild steel plate, 1/4 in. (6 mm) thick, you will cut holes with diameters of 1/2 in. (13 mm) and 1 in. (25 mm). Using the technique described for piercing a hole, start in the center and make an outward spiral until the hole is the desired size, **Figure 7-51**. The hole must be within ±3/32 in. (2 mm) of being round and ±5° of being square. The hole may have slag on the bottom. Repeat this procedure until both small and large sizes of holes can be made within tolerance. Turn off the cylinder valves, bleed the hoses, back out the pressure regulators, and clean up your work area when you are finished cutting. ◆

DISTORTION

Distortion is when the metal bends or twists out of shape as a result of being heated during the cutting process. This is a major problem when cutting a plate. If the distortion is not controlled, the end product might be worthless. There are two major methods of controlling distortion. One method involves making two parallel cuts on the same plate at the same speed and time, **Figure 7-52**. Because the plate is heated evenly, distortion is kept to a minimum, **Figure 7-53**.

The second method involves starting the cut a short distance from the edge of the plate, skipping other short tabs every 2 ft (0.6 m) to 3 ft (0.9 m) to keep the cut from separating. Once the plate cools, the remaining tabs are cut, **Figure 7-54**.

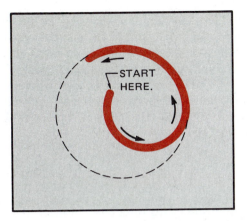

Figure 7-51 Start a cut for a hole near the middle.

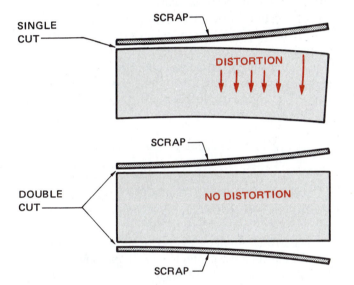

Figure 7-52 Making two parallel cuts at the same time will control distortion.

Figure 7-53 Slitting adaptor for cutting machine. It can be used for parallel cuts from 1 1/2 in. (38 mm) to 12 in. (500 mm). Ideal for cutting test coupons. (Courtesy of ESAB Welding and Cutting Products.)

EXPERIMENT 7-4

Minimizing Distortion

Using a properly lit and adjusted cutting torch, welding gloves, appropriate eye protection and clothing, and two pieces of mild steel, 10 in. (254 mm) long x 1/4 in. (6 mm) thick, you will make two cuts and then compare the distortion. Lay out and cut out both pieces of metal as shown in **Figure 7-55**. Allow the metal to cool,

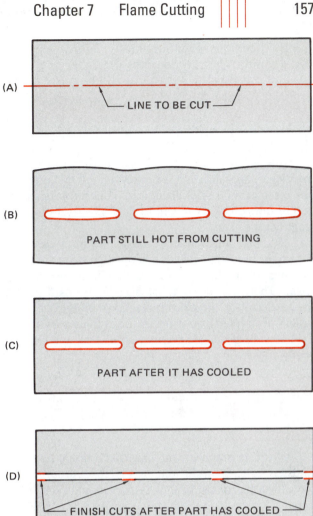

(A)   LINE TO BE CUT

(B)   PART STILL HOT FROM CUTTING

(C)   PART AFTER IT HAS COOLED

(D)   FINISH CUTS AFTER PART HAS COOLED

Figure 7-54 Steps used during cutting to minimize distortion.

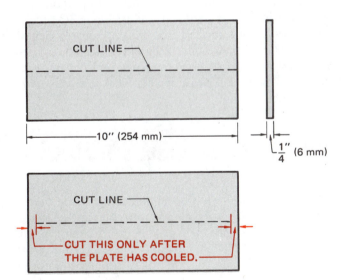

CUT LINE

10″ (254 mm)

1/4″ (6 mm)

CUT LINE

CUT THIS ONLY AFTER THE PLATE HAS COOLED.

Figure 7-55 Making two cuts with minimum distortion. *Note:* Sizes of these and other cutting projects can be changed to fit available stock.

and then cut the remaining tabs. Compare the four pieces of metal for distortion. ◆

Beveling a Plate

Use a properly lit and adjusted cutting torch, welding gloves, appropriate eye protection and clothing, and one piece of mild steel plate, 6 in. (152 mm) long x 3/8 in. (10 mm) thick. You will make a 45° bevel down the length of the plate.

Mark the plate in strips 1/2 in. (13 mm) wide. Set the tip for beveling, and cut a bevel. The bevel should be within ±3/32 in. (2 mm) of a straight line and ±5° of a 45° angle. There may be some soft slag, but no hard slag, on the beveled plate. Repeat this practice until the cut can be made within tolerance. Turn off the cylinder valves, bleed the hoses, back out the pressure regulators, and clean up your work area when you are finished cutting. ◆

Vertical Straight Cut

For this practice, you will need a properly lit and adjusted cutting torch, welding gloves, appropriate eye protection and clothing, and one piece of mild steel plate, 6 in. (152 mm) long x 1/4 in. (6 mm) to 3/8 in. (10 mm) thick, marked in strips 1/2 in. (13 mm) wide and held in the vertical position. You will make a straight line cut. Make sure that the sparks do not cause a safety hazard and that the metal being cut off will not fall on any person or object.

Starting at the top, make one cut downward. Then, starting at the bottom, make the next cut upward. The cut must be free of hard slag and within ±3/32 in. (2 mm) of a straight line and ±5° of being square. Repeat these cuts until they can be made within tolerance. Turn off the cylinder valves, bleed the hoses, back out the pressure regulators, and clean up your work area when you are finished cutting. ◆

Overhead Straight Cut

Using a properly lit and adjusted cutting torch, welding gloves, appropriate eye protection and clothing, and one piece of mild steel plate, 6 in. (152 mm) long x 1/4 in. (6 mm) to 3/8 in. (10 mm) thick, marked in strips 1/2 in. (13 mm) wide, you will make a cut in the overhead position. When making overhead cuts, it is important to be completely protected from the hot sparks. In addition to the standard safety clothing, you should wear a leather jacket, leather apron, cap, ear protection, and a full face shield.

The torch can be angled so that most of the sparks will be blown away. The metal should fall free when the cut is completed. The cut must be within 1/8 in. (3 mm) of a straight line and ±5° of being square. Repeat this practice until the cut can be made within tolerance. Turn off the cylinder valves, bleed the hoses, back out the pressure regulators, and clean up your work area when you are finished cutting. ◆

CUTTING APPLICATIONS

Making practice cuts on a piece of metal that will only become scrap is a good way to learn the proper torch techniques. If a bad cut is made, there is no loss. In a production shop, where each piece of metal is important, scrapped metal due to bad cuts decreases the shop's profits.

A number of factors can affect your ability to make a quality cut on a part that do not exist during practice cuts. The following are some of the things that can become problems when cutting:

■ *Changing positions:* Often, parts are larger than can be cut from one position, so you may have to move to complete the cut. Stopping and restarting a cut can result in a small flaw in the cut surface. If this flaw exceeds the acceptable limits, the cut surface must be repaired before the part can be used. To avoid this problem, always try to stop at corners if the cut cannot be completed without moving.

■ *Sparks:* You will often be making cuts in large plates. Even an ideal cut can create sparks that bounce around the plate surface. These sparks often find their way into your glove, under your arm, or to any other place that will become uncomfortable. Experienced welders will usually keep working if the sparks are not too large or too uncomfortable. With experience you will learn how to angle the torch, direct the cut, and position your body to minimize this problem.

■ *Hot surfaces:* As you continue making cuts to complete the part, it will begin to heat up. Depending on the size of the part, the number of cuts per part, and the number of parts being cut, this heat can become uncomfortable. You may find it necessary to hold the torch farther back from the tip, but this will affect the quality of your cuts, **Figure 7-56.** Sometimes you might be able to rest your hand on a block to keep it off of the plate. Another problem with heat buildup is that it may become high enough to affect the cut quality. Heat becomes a problem when it causes the top edge of the plate to melt during a cut as if the torch tip were too large. This is more of a problem when several cuts are being made in close proximity. Planning your cutting sequence and allowing cooling time will help control this potential problem.

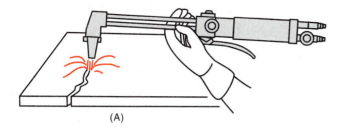

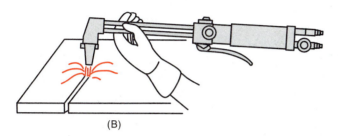

Figure 7-56 It's easier to make straight smooth cuts if you can brace the torch closer to the tip, as in cut B.

- *Tip cleaning:* As with any cutting, the tip will catch small sparks and become dirty or clogged. You must decide how dirty or clogged you will let the tip get before you stop to clean it. Time spent cleaning the tip reduces productivity, unfortunately. On the other hand, if you don't stop occasionally to clean up, the quality of the cut will become so bad that post-cutting cleanup will become excessive. It's your responsibility to decide when and how often to clean the tip.

- *Blow back:* As a cut progresses across the surface of a large plate, it may cross supports underneath the plate. During practice cuts this seldom if ever happens, but, depending on the design of the cutting table, it will occur even under the best of conditions. If the support is small, the blow back may not cover

you with sparks, plug the cutting tip, or cause a major flaw in the cut surface. If the support is large, then one or all of these events can occur. If you see that the blow back is not clearing quickly, it may be necessary to stop the cut. Stopping the cut halts the shower of sparks but leaves you with a problem restart.

PRACTICE 7-12

Cutting Out Internal and External Shapes

Using a properly lit and adjusted cutting torch, welding gloves, appropriate eye protection and clothing, and one piece of plate, 1/4 in. (6 mm) to 3/8 in. (10 mm) thick, you may lay out and cut out one of the sample patterns shown in **Figure 7-57**, one of the projects in Chapter 18, or any other design available.

Choose the pattern that best fits the piece of metal you have and mark it using a center punch. The exact size and shape of the layout is not as important as the accuracy of the cut. The cut must be made so that the center-punched line is left on the part and so that there is no more than 1/8 in. (3 mm) between the cut edge and the line, **Figures 7-58** and **7-59**. Repeat this practice until the cut can be made within tolerance. Turn off the cylinder valves, bleed the hoses, back out the pressure regulators, and clean up your work area when you are finished cutting. ◆

PIPE CUTTING

Freehand pipe cutting may be done in one of two ways. On small-diameter pipe, usually under 3 in. (76 mm), the torch tip is held straight up and down and moved from the center to each side, **Figure 7-60**. This technique can also be used successfully on larger pipe.

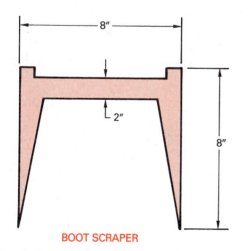

BOOT SCRAPER

Note: Sizes of these and other cutting projects can be changed to fit available stock.

Figure 7-57 Suggested patterns for practice.

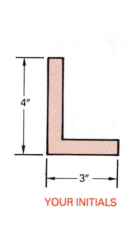

YOUR INITIALS

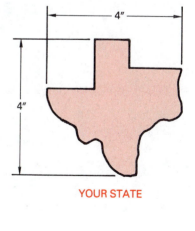

YOUR STATE

For large-diameter pipe, 3 in. (76 mm) and larger, the torch tip is always pointed toward the center of the pipe, Figure 7-61. This technique is also used on all sizes of heavy-walled pipe and can be used on some smaller pipe sizes.

The torch body should be held so that it is parallel to the centerline of the pipe. Holding the torch parallel helps to keep the cut square.

■ **CAUTION**

When cutting pipe, hot sparks can come out of the end of the pipe nearest you, causing severe burns. For protection from hot sparks, plug up the open end of the pipe nearest you, put up a barrier to the sparks, or stand to one side of the material being cut.

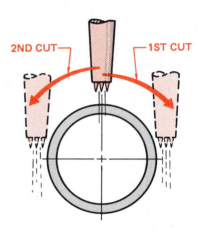

Figure 7-60　Small-diameter pipe can be cut without changing the angle of the torch. After the top is cut, roll the pipe to cut the bottom.

Figure 7-58　Beginning a cut with the torch concentrating the flame on the thin edge to speed starting.

Figure 7-59　The torch is rotated to allow the preheating of the plate ahead of the cut. This speeds the cutting and also provides better visibility of the line being cut.

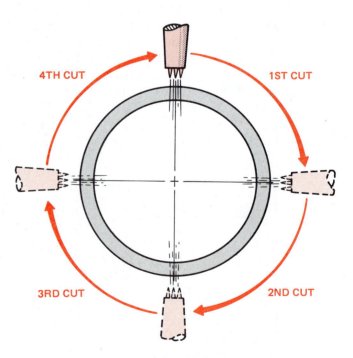

Figure 7-61 On large-diameter pipe, the torch is turned to keep it at a right angle to the pipe. The pipe should be cut as far as possible before stopping and turning it.

PRACTICE 7-13

Square Cut on Pipe, 1G (horizontal rolled) Position

Using a properly lit and adjusted cutting torch, welding gloves, appropriate eye protection and clothing, and one piece of schedule 40 steel pipe with a diameter of 3 in. (76 mm), you will cut off 1/2-in. (13-mm) -long rings.

Using a template and a piece of soapstone, mark several rings, each 1/2 in. (13 mm) wide, around the pipe. Place the pipe horizontally on the cutting table. Start the cut at the top of the pipe using the proper piercing technique. Move the torch backward along the line and then forward; this will keep slag out of the cut. If the end of the cut closes in with slag, this will cause the oxygen to gouge the edge of the pipe when the cut is continued. Keep the tip pointed straight down. When you have gone as far with the cut as you can comfortably, quickly flip the flame away from the pipe. Restart the cut at the top of the pipe and cut as far as possible in the other direction. Stop and turn the pipe so that the end of the cut is on top and the cut can be continued around the pipe. When the cut is completed, the ring must fall free. When the pipe is placed upright on a flat plate, the pipe must stand within 5° of vertical and have no gaps higher than 1/8 in. (3 mm) under the cut. Repeat this procedure until the cut can be made within tolerance. Turn off the cylinder valves, bleed the hoses, back out the pressure regulators, and clean up your work area when you are finished cutting. ◆

PRACTICE 7-14

Square Cut on Pipe, 1G (horizontal rolled) Position

Using the same equipment, materials, and markings as described in Practice 7-13, you will cut off the 1/2-in. (13-mm) -long rings while keeping the tip pointed toward the center of the pipe.

Starting at the top, pierce the pipe. Move the torch backward to keep the slag out of the cut and then forward around the pipe, stopping when you have gone as far as you can comfortably. Restart the cut at the top and proceed with the cut in the other direction. Roll the pipe and continue the cut until the ring falls off freely. Stand the cut end of the pipe on a flat plate. The pipe must stand within 5° of vertical and have no gaps higher than 1/8 in. (3 mm). Repeat this practice until the cut can be made within tolerance. Turn off the cylinder valves, bleed the hoses, back out the pressure regulators, and clean up your work area when you are finished cutting. ◆

PRACTICE 7-15

Square Cut on Pipe, 5G (horizontal fixed) Position

With the same equipment, materials, and markings as described in Practice 7-13, you will cut off 1/2-in. (13-mm) rings, using either technique, without rolling the pipe.

Start at the top and cut down both sides as far as you can comfortably. Reposition yourself and continue the cut under the pipe until the ring falls off freely. Stand the cut end of the pipe on a flat plate. The pipe must stand within 5° of vertical and have no gaps higher than 1/8 in. (3 mm). Repeat this practice until the cut can be made within tolerance. Turn off the cylinder valves, bleed the hoses, back out the pressure regulators, and clean up your work area when you are finished cutting. ◆

PRACTICE 7-16

Square Cut on Pipe, 2G (vertical) Position

With the same equipment, materials, and markings as listed in Practice 7-13, you will cut off 1/2-in. (13-mm) rings, using either technique, from a pipe in the vertical position.

Place a flat piece of plate over the open top end of the pipe to keep the sparks contained, **Figure 7-62**. Start on one side and proceed around the pipe until the cut is completed. Because of slag, the ring may have to be tapped free. Stand the cut end of the pipe on a flat plate. The pipe must stand within 5° of vertical and have no gaps higher than 1/8 in. (3 mm). Repeat this practice until the cut can be made within tolerance. Turn off the cylinder valves, bleed the hoses, back out the pressure regulators, and clean up your work area when you are finished cutting. ◆

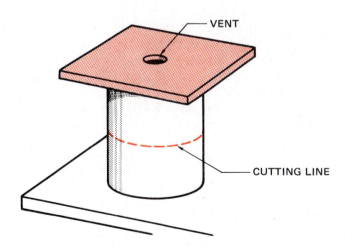

Figure 7-62 Place a plate on top of a short piece of pipe to keep the sparks from flying around the shop.

Programmable Controllers: Production Up, Rejects and Waste Down

By automating the welding of pressure tanks, one company has increased production, cut rejects by as much as 75%, and reduced scrap and welding wire waste. The girth welds on the tanks are produced automatically at separate stations, each equipped with microprocessor-based programmable controllers, Figure 1, and welding power sources. Two operators merely load the tanks, press the start button, and unload the welded tanks.

The circumferential welds, which must withstand 325 psi internal pressure, are produced entirely automatically according to a preprogrammed sequence. The microprocessor controls gas preflow, run-in, wire feed speed, voltage, crater fill, burnback, gas postflow, and starts and stops the rotating fixture. Up to nine different weld programs can be stored in the microprocessor's memory. The microprocessors are easily programmed, Figure 2.

The company started with one programmable control and then added seventeen more, six for each of three plant sites. At least twenty more programmable control and power source combinations will be purchased to further upgrade welding production.

The ability to establish optimum welding parameters for each of five tank sizes saves time and assures a sound weldment from the start. In the past, the first few tanks welded after a setup change would be unacceptable until the system was fine-tuned. Now, the new program is called up from the microprocessor's memory and run without wasting parts.

The controller's accuracy also gives the operator the ability to control wire and gas consumption. For example, a 9-in.-diameter tank requires a certain amount of welding wire. If less wire is used, the weld may not be sound. If more wire is used, it may be wasted. Each automatic station also consumes a predetermined amount of mixed gas per hour, depending on the size of tank being run.

The tanks range from 6 to 22 inches in diameter. They are made of two domes formed from

Figure 1

Figure 2

Figure 3

mild steel in various thickness. After forming, the domes are degreased, and the open, cylindrical end of one of the domes is grooved to form a lap joint when the two halves are mated. After the domes are joined, the joint is welded. **Figure 3** shows the operator GTA welding the cap ring on a tank.

After a component part has been GMA spot welded to the tanks, using a power source equipped with a wire feeder and a spot timer, the tanks are pressure tested and painted.

(Courtesy of Miller Electric Mfg. Co., Appleton, Wisconsin 54912.)

REVIEW

1. Using **Table 7-1**, list the six different fuel gases in rank order according to their temperature.

2. What is a combination welding-cutting torch?

3. State one advantage of owning a combination welding-cutting torch as opposed to just having a cutting torch.

4. State one advantage of owning a dedicated cutting torch as compared to having a combination welding-cutting torch.

5. What is a mixing chamber? Where is it located?

6. Define the term equal-pressure torch. How does it work?

7. How does an injector-type mixing chamber work?

8. State the advantages of having two oxygen regulators on a machine-cutting torch.

9. Why are some copper alloy cutting tips chrome plated?

10. What determines the amount of preheat flame requirements of a torch?

11. What can happen if acetylene is used on a tip designed to be used with propane or other such gas?

12. Why are some propane and natural gas tips made with a deep recessed center?

13. What types of tip seals are used with cutting torch tips?

14. If a cutting tip should stick in the cutting head, how should it be removed?

15. How can cutting torch tip seals be repaired?

16. Why is the oxygen valve turned on before starting to clean a cutting tip?

17. Why does the preheat flame become slightly oxidizing when the cutting lever is released?

18. What causes the tiny ripples in a hand cut?

19. Why is a slight forward torch angle helpful for cutting?

20. Why should cans, drums, tanks, or other sealed containers be opened with a cutting torch?

21. Why is the torch tip raised as the cutting lever is depressed when cutting a hole?

22. Why are the preheat holes not aligned in the kerf when making a bevel cut?

23. Sketch the proper end shape of a soapstone that is to be used for marking metal.

24. Using **Table 7-2**, answer the following:

    a. Oxygen pressure for cutting 1/4-in. (6-mm) thick metal
    b. Acetylene pressure for cutting 1-in. (25-mm) thick metal
    c. Tip cleaner size for a tip for 2-in. (51-mm) thick metal
    d. Drill size for a tip for 1/2-in. (13-mm) thick metal

25. What is the best way to set the oxygen pressure for cutting?

26. What metals can be cut with the oxyfuel gas process?

27. Why is it important to have extra ventilation and/or a respirator when cutting some used metal?

28. What factors regarding a cut can be read from the sides of the kerf after a cut?

29. What is hard slag?

30. Why is it important to make good quality cuts?

31. Describe the methods of controlling distortion when making cuts.

32. How does cutting small-diameter pipe differ from cutting large-diameter pipe?

Chapter 8

PLASMA ARC CUTTING

OBJECTIVES

After completing this chapter, the student should be able to

■ describe plasma and describe a plasma torch.

■ explain how a plasma cutting torch works.

■ list the advantages and disadvantages of using a plasma cutting torch.

■ demonstrate an ability to set up and use a plasma cutting torch.

KEY TERMS

plasma	ionized gas
plasma arc	arc plasma
electrode tip	arc cutting
electrode setback	nozzle insulator
nozzle	nozzle tip
water shroud	cup
jouls	heat affected zone
standoff distance	high frequency alternating current
kerf	pilot arc
dross	stack cutting
plasma arc gouging	water table

INTRODUCTION

The plasma process was originally developed in the mid-1950s as an attempt to create an arc, using argon, that would be as hot as the arc created when using helium gas. The early gas tungsten arc welding process used helium gas and was called 'heliarc.' This early GTA welding process worked well with helium, but helium was expensive. The gas manufacturing companies had argon as a by-product from the production of oxygen. There was no good commercial market for this waste argon gas, but gas manufacturers believed there would be a good market if they could find a way to make argon weld similar to helium.

Early experiments found that by restricting the arc in a fast flowing column of argon a plasma was formed. See Color Plates 7 and 8. The plasma was hot enough to rapidly melt any metal. The problem was that the fast moving gas blew the molten metal away. They could not find a way to control this scattering of the molten metal, so they decided to introduce this as a cutting process, not a welding process. See Figure 8-1.

Several years later, with the invention of the gas lens, plasma was successfully used for welding. Today the plasma arc can be used for plasma arc welding (PAW), plasma spraying (PSP), plasma arc cutting (PAC), and plasma arc gouging. Plasma arc cutting is the most often used plasma process.

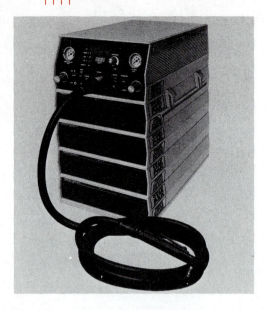

Figure 8-1 Plasma arc cutting machine. This unit can have additional power modules added to the base of its control module to give it more power. (Courtesy of Thermal Dynamics.)

PLASMA

The word plasma has two meanings: it is the fluid portion of blood; and it is a state of matter that is found in the region of an electrical discharge (arc). The plasma created by an arc is an ionized gas that has both electrons and positive ions whose charges are nearly equal to each other. For welding we use the electrical definition of plasma.

A plasma is present in any electrical discharge. A plasma consists of charged particles that conduct the electrons across the gap. Both the glow of a neon tube and the bright fluorescent light bulb are examples of low-temperature plasmas.

A plasma results when a gas is heated to a high enough temperature to convert into positive and negative ions, neutral atoms, and negative electrons. The temperature of an unrestricted arc is about 11,000°F, but the temperature created when the arc is concentrated to form a plasma is about 43,000°F, **Figure 8-2**. This is hot enough to rapidly melt any metal it comes in contact with.

ARC PLASMA

The term arc plasma is defined as gas that has been heated to at least a partially ionized condition, enabling it to conduct an electric current.[1] The term plasma arc is the term most often used in the welding industry when referring to the arc plasma used in welding and cutting processes. The plasma arc produces both the high temperature and intense light associated with all forms of arc welding and arc cutting processes.

PLASMA TORCH

The plasma torch is a device, depending on its design, which allows the creation and control of the plas-

---

[1]ANSI/AWS A3.0-89 An American National Standard, Standard Welding Terms and Definitions

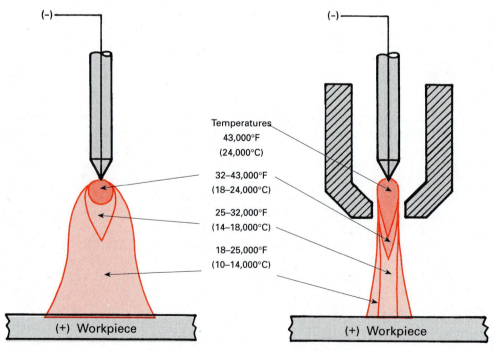

Figure 8-2 Approximate temperature differences between a standard arc and a plasma arc. (Courtesy of the American Welding Society.)

ma for welding or cutting processes. The plasma is created in both the cutting and welding torches in the same basic manner, and both torches have the same basic parts. A plasma torch supplies electrical energy to a gas to change it into the high energy state of a plasma.

Torch Body

The torch body, on a manual type torch, is made of a special plastic that is resistant to high temperatures, ultraviolet light, and impact. The torch body is a place that provides a good grip area and protects the cable and hose connections to the head. The torch body is available in a variety of lengths and sizes. Generally the longer, larger torches are used for the higher capacity machines; however, sometimes you might want a longer or larger torch to give yourself better control or a longer reach. On machine torches the body is often called a barrel and may come equipped with a rack attached to its side. The rack is a flat gear that allows the torch to be raised and lowered manually to the correct height above the work, Figure 8-3.

Torch Head

The torch head is attached to the torch body where the cables and hoses attach to the electrode tip, nozzle tip, and nozzle. The torch and head may be connected at any angle, such as 90°, 75°, 180° (straight), or it can be flexible. The 75° or 90° angles are popular for manual operations, and the 180° straight torch heads are most often used for machine operations. Because of the heat in the head produced by the arc, some provisions for cooling the head and its internal parts must be made. This cooling for low power torches may be either by air or water. Higher power torches must be liquid cooled,

Figure 8-4. It's possible to replace just the torch head on most torches if it becomes worn or damaged.

Power Switch

Most hand-held torches have a manual power switch, which is used to start and stop the power source, gas, and cooling water (if used). The switch most often used is a thumb switch located on the torch body, but it may be a foot control or located on the panel for machine type equipment. The thumb switch may be molded into the torch body or it may be attached to the torch body with a strap clamp. The foot control must be rugged enough to withstand the welding shop environment. Some equipment has an automatic system that starts the plasma when the torch is brought close to the work.

Common Torch Parts

The electrode tip, nozzle insulator, nozzle tip, nozzle guide, and nozzle are the parts of the torch that must be replaced periodically as they wear out or become damaged from use, Figure 8-5.

■ CAUTION

Improper use of the torch or assembly of torch parts may result in damage to the torch body as well as the frequent replacement of these parts.

Figure 8-4 Water-cooled plasma cutting system. The cutting is quieter, and there is less heat distortion of the metal. (Courtesy of Thermal Dynamics Corporation.)

Figure 8-3 Machine plasma torch. (Courtesy of Cerametals Inc.)

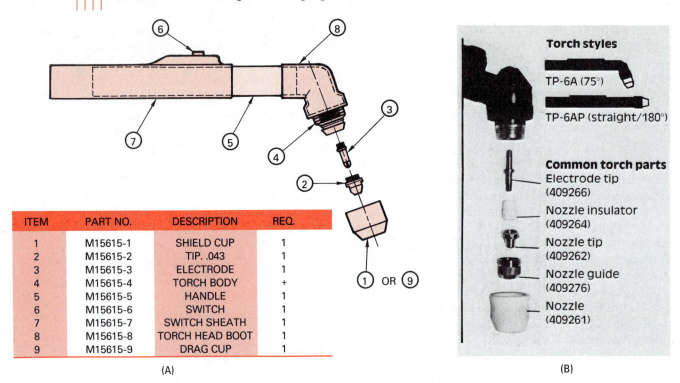

ITEM	PART NO.	DESCRIPTION	REQ.
1	M15615-1	SHIELD CUP	1
2	M15615-2	TIP. .043	1
3	M15615-3	ELECTRODE	1
4	M15615-4	TORCH BODY	+
5	M15615-5	HANDLE	1
6	M15615-6	SWITCH	1
7	M15615-7	SWITCH SHEATH	1
8	M15615-8	TORCH HEAD BOOT	1
9	M15615-9	DRAG CUP	1

(A)   (B)

Figure 8-5 Replaceable torch parts. (Courtesy of A. the Lincoln Electric Company; B. Hobart Brothers Company.)

The metal parts are usually made out of copper, and they may be plated. The plating of copper parts will help them stay spatter-free longer.

Electrode Tip

The electrode tip is often made of copper electrode with a tungsten tip attached. The use of a copper/tungsten tip in the newer torches has improved the quality of work they can produce. By using copper, the heat generated at the tip can be conducted away faster. Keeping the tip as cool as possible lengthens the life of the tip and allows for better quality cuts for a longer time. The newer designed torches are a major improvement over earlier torches, some of which required the welder to accurately grind the tungsten electrode into shape. If you are using a torch that requires the grinding of the electrode tip, you must have a guide to insure that the tungsten is properly prepared.

Nozzle Insulator

The nozzle insulator is between the electrode tip and the nozzle tip. The nozzle insulator provides the critical gap spacing and the electrical separation of the parts. The spacing between the electrode tip and the nozzle tip, called electrode setback, is critical to the proper operation of the system.

Nozzle Tip

The nozzle tip has a small, cone-shaped, constricting orifice in the center. The electrode setback space, between the electrode tip and the nozzle tip, is where the electric current forms the plasma. The preset close-fitting parts provide the restriction of the gas in the presence of the electric current so the plasma can be generated, Figure 8-6. The diameter of the constricting orifice and the electrode setback are major factors in the operation of the torch. As the diameter of the orifice changes, the plasma jet action will be affected. When the setback distance is changed, the arc voltage and current flow will change.

Nozzle

The nozzle, sometimes called the cup, is made of ceramic or any other high-temperature-resistant substance. This helps prevent the internal electrical parts from accidental shorting and provides control of the shielding gas or water injection if they are used, Figure 8-7.

Water Shroud

A water shroud nozzle may be attached to some torches. The water surrounding the nozzle tip is used to control the potential hazards of light, fumes, noise, or other pollutants produced by the process.

POWER AND GAS CABLES

A number of power and control cables and gas and cooling water hoses may be used to connect the power supply with the torch, Figure 8-8. This multi-part cable is usually covered to provide some protection to the cables and hoses inside and to make handling the cable

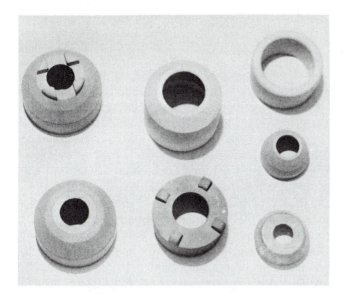

Figure 8-6 Different torches use different types of nozzle tips.

Figure 8-7 Nozzles are available in a variety of shapes for different types of cutting jobs.

easier. This covering is heat resistant but will not prevent damage to the cables and hoses inside if it comes in contact with hot metal or is exposed directly to the cutting sparks.

Power Cable

The power cable must have a high-voltage rated insulation, and it is made of finely stranded copper wire to allow for maximum flexibility of the torch, **Figure 8-9**. For all non-transfer-type torches and those that use a high frequency pilot arc, there are two power conductors, one positive (+) and one negative (–). The size and current carrying capacity of this cable is a controlling factor to the power range of the torch. As the capacity of the equipment increases, the cable must be made large enough to carry the increased current. The larger cables are less flexible and more difficult to manipulate. In order to make the cable smaller on water-cooled torches, the cable is run inside the cooling water return line. By putting the power cable inside the return water line, it allows a smaller cable to carry more current. The water prevents the cable from overheating.

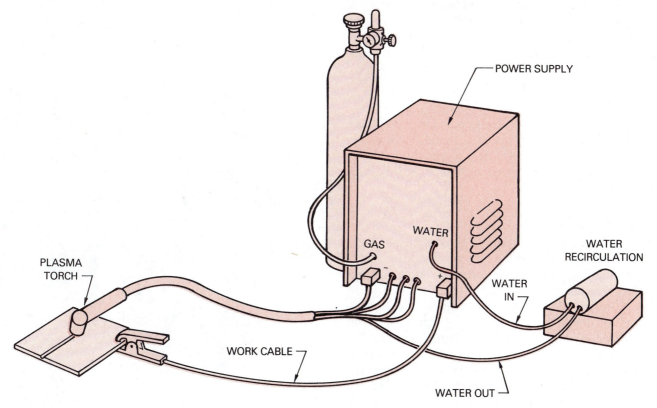

Figure 8-8 Typical manual plasma arc cutting setup.

Figure 8-9 Portable plasma arc cutting machine. (Courtesy of the Lincoln Electric Company.)

Gas Hoses

There may be two gas hoses running the torch. One hose carries the gas used to produce the plasma, and the other provides a shielding gas coverage. On some small amperage cutting torches there is only one gas line. The gas line is made of a special heat-resistant, ultraviolet-light-resistant plastic. If it's necessary to replace the tubing because it is damaged, be sure to use the tubing provided by the manufacturer or a welding supplier. The tubing must be sized to carry the required gas flow rate within the pressure range of the torch, and it must be free from solvents and oils that might contaminate the gas. If the pressure of the gas supplied is excessive, the tubing may leak at the fittings or rupture.

Control Wire

The control wire is a two-conductor, low-voltage, stranded copper wire. This wire connects the power switch to the power supply. This allows the welder to start and stop the plasma power and gas as needed during the cut or weld.

Water Tubing

Medium and high amperage torches may be water cooled. The water for cooling early model torches had to be deionized. Failure to use deionized water on these torches will result in the torch arcing out internally. This arcing may destroy or damage the torch's electrode tip and the nozzle tip. To see if your torch requires this special water, refer to the manufacturer's manual. If cooling water is required, it must be switched on and off at the same time as the plasma power. Allowing the water to circulate continuously might result in condensation in the torch. When the power is reapplied, the water will cause internal arcing damage.

POWER REQUIREMENTS

Voltage

The production of the plasma requires a direct-current (DC), high-voltage, constant-current (drooping arc voltage) power supply. A constant-current-type machine allows for a rapid start of the plasma arc at the high open circuit voltage and a more controlled plasma arc as the voltage rapidly drops to the lower closed voltage level. The voltage required for most welding operations, such as shielded metal arc, gas metal arc, gas tungsten arc, and flux cored arc, ranges from 18 volts to 45 volts. The voltage for a plasma arc process ranges from 50 to 200 volts closed circuit and 150 to 400 volts open circuit. This higher electrical potential is required because the resistance of the gas increases as it's forced through a small orifice. The potential voltage of the power supplied must be high enough to overcome the resistance in the circuit in order for electrons to flow, Figure 8-10.

Amperage

Although the voltage is higher, the current (amperage) flow is much lower than it is with most other welding processes. Some low powered PAC torches will operate with as low as 10 amps of current flow. High powered plasma cutting machines can have amperages as high as 200 amps, and some very large automated cutting machines may have 1,000 ampere capacities. The higher the amperage capacity the faster and thicker they will cut.

Watts

The plasma process uses approximately the same amount of power, in watts, as a similar nonplasma process. Watts are the units of measure for electrical

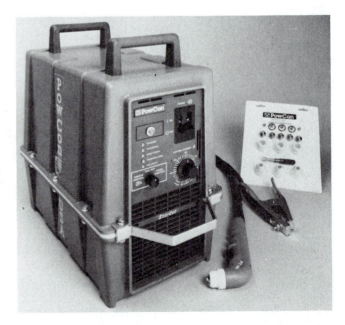

Figure 8-10 Inverter type plasma arc cutting power supply. (Courtesy of Pow Con Inc.)

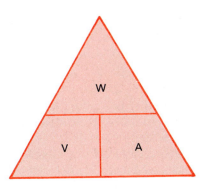

Figure 8-11 Ohm's Law.

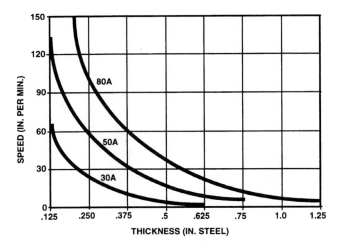

Table 8-1 (Courtesy of ESAB Welding and Cutting Products.)

power. By determining the total watts used for both the nonplasma process and plasma operation, you can make a comparison. Watts used in a circuit are determined by multiplying the voltage times the amperage, **Figure 8-11.** For example, a 1/8-in.-diameter E6011 electrode will operate at 18 volts and 90 amperes. The total watts used would be:

$W = V \times A$
$W = 18 \times 90$
$W = 1620$ watts of power.

A low power PAC torch operating with only 20 amperes and 85 volts would be using a total of:

$W = V \times A$
$W = 85 \times 20$
$W = 1700$ watts of power.

HEAT INPUT

Although the total power used by both plasma and nonplasma processes is similar, the actual energy input into the work per linear foot is less with plasma. The very high temperatures of the plasma process allow much higher traveling rates so that the same amount of heat input is spread over a much larger area. This has the effect of lowering the jouls per inch of heat the weld or cut will receive. **Table 8-1** shows the cutting performance of a typical plasma torch. Note the relationship among amperage, cutting speed, and metal thickness. The lower the amperage, the slower the cutting speed or the thinner the metal that can be cut.

A high travel speed with plasma cutting will result in a heat input that is much lower than that of the oxyfuel cutting process. A steel plate cut using the plasma process may have only a slight increase in temperature following the cut. It's often possible to pick up a part only moments after it's cut using plasma. The same part cut with oxyfuel would be much hotter and require a longer time to cool off.

DISTORTION

Any time metal is heated in a localized zone or spot it expands in that area and, after the metal cools, it

is no longer straight or flat, **Figure 8-12.** If a piece of metal is cut, there will be localized heating along the edge of the cut, and, unless special care is taken, the part will not be usable as a result of its distortion, **Figure 8-13.** This distortion is a much greater problem with thin metals. By using a plasma cutter, an auto body worker can cut the thin, low-alloy sheet metal of a damaged car with little problem from distortion.

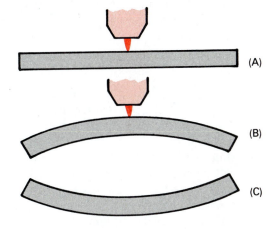

Figure 8-12 (A) When metal is heated, (B) it bends up toward the heat. (C) As the metal cools, it bends away from the heated area.

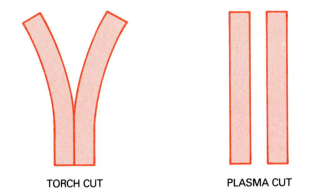

TORCH CUT   PLASMA CUT

Figure 8-13

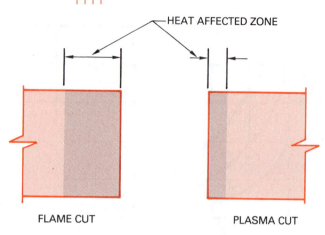

Figure 8-14 A smaller heat-affected zone will result in less hardness or weakening along the cut edge.

On thicker sections, the hardness zone along the edge of a cut will be reduced so small that it is not a problem. When using oxyfuel cutting of thick plate, especially higher alloyed metals, this hardness zone can cause cracking and failure if the metal is shaped after cutting, **Figure 8-14**. Often the plates must be preheated before they are cut using oxyfuel to reduce the heat affected zone. This preheating adds greatly to the cost of fabrication both in time and fuel costs. By being able to make most cuts without preheating, the plasma process will greatly reduce fabrication cost.

APPLICATIONS

Early plasma arc cutting systems required that either helium or argon gas be used for the plasma and shielding gases. As the development of the process improved, it was possible to start the PAC torch using argon or helium and switch to less expensive nitrogen. The use of nitrogen as the plasma cutting gas greatly reduced the cost of operating a plasma system. Because of its operating expense, plasma cutting was limited to metals not easily cut using oxyfuel. Aluminum, stainless steel, and copper were the metals most often cut using plasma.

As the process development improved, less expensive gases and even dry compressed air could be used, and the torches and power supplies improved. By the early 1980s, the PAC process had advanced to a point where it was used for cutting all but the thicker sections of mild steel.

Cutting Speed

High cutting speeds are possible, up to 300 in./min.; that's 25 feet a minute or about 1/4 mile an hour. The fastest oxyfuel cutting equipment could cut at only about one-fourth that speed. A problem with early high-speed machine cutting was that the cutting machines could not reliably make cuts as fast as the PAC torch. That problem has been resolved, and the new machines and robots can operate at the upper limits of the plasma torch capacity. These machines and robots are capable of automatically

maintaining the optimum torch standoff distance to the work. Some cutting systems will even follow the irregular surfaces of preformed part blanks, **Figure 8-15**.

Metals

Any material that is conductive can be cut using the PAC process. In a few applications nonconductive materials can be coated with conductive material so that they can be cut also. Although it's possible to make cuts in metal as thick as 7 inches, it's not cost effective. The most popular materials cut are carbon steel up to 1 inch, stainless steel up to 4 inches, and aluminum up to 6 inches. These are not the upper limits of the PAC process, but beyond these limits other cutting processes may be less expensive. Often a shop may PAC thicker material even if it's not cost effective because they don't have ready access to the alternative process.

Other metals commonly cut using PAC are copper, nickel alloys, high-strength, low-alloy steels, and clad materials. It is also used to cut expanded metals, screens, and other items that would require frequent starts and stops, **Figure 8-16**.

Standoff Distance

The standoff distance is the distance from the nozzle tip to the work, **Figure 8-17**. This distance is very critical to producing quality plasma arc cuts. As the distance increases, the arc force is diminished and tends to spread out. This causes the kerf to be wider, the top edge of the plate to become rounded, and the formation of more dross on the bottom edge of the plate. However, if this distance

Figure 8-15 Plasma arc cut in 2-in.-thick mild steel. Notice how smooth the machine cut edge is.

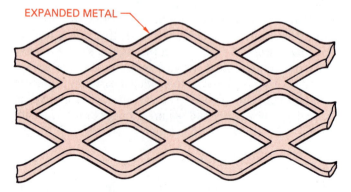

Figure 8-16 Expanded metal.

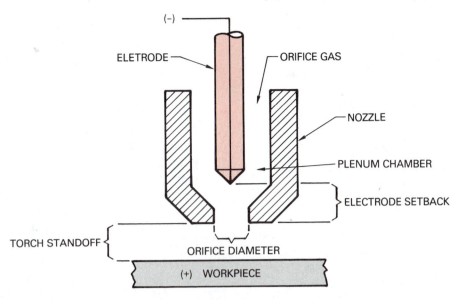

Figure 8-17 Conventional plasma arc terminology. (Courtesy of the American Welding Society.)

Figure 8-18 A castle nozzle tip can be used to allow the torch to be dragged across the surface.

becomes too close, the working life of the nozzle tip will be reduced. In some cases an arc can form between the nozzle tip and the metal that instantly destroys the tip.

On some new torches, it's possible to drag the nozzle tip along the surface of the work without shorting it out. This is a large help when working on metal out of position or on thin sheet metal. Before you use your torch in this manner, you must check the owner's manual to see if it will operate in contact with the work, **Figure 8-18**. This technique will allow the nozzle tip orifice to become contaminated more quickly.

Starting Methods

Because the electrode tip is located inside the nozzle tip, and a high initial resistance to current flow exists in the gas flow before the plasma is generated, it's necessary to have a specific starting method. Two methods are used to establish a current path through the gas.

The most common method uses a high frequency alternating current carried through the conductor, the electrode, and back from the nozzle tip. This high frequency current will ionize the gas and allow it to carry the initial current to establish a pilot arc, **Figure 8-19**.

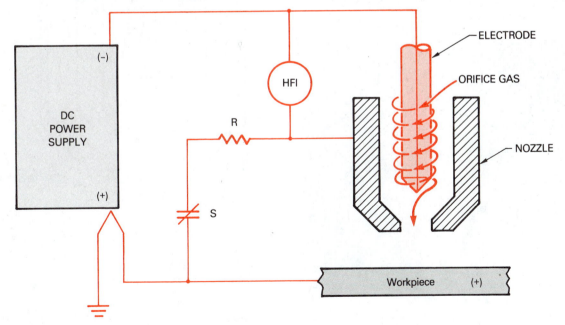

Figure 8-19 Plasma arc torch circuitry. (Courtesy of the American Welding Society.)

After the pilot arc has been started, the high frequency starting circuit can be stopped. A **pilot arc** is an arc between the electrode tip and the nozzle tip within the torch head. This is a nontransfer arc, so the workpiece is not part of the current path. The low current of the pilot arc, although it's inside the torch, does not create enough heat to damage the torch parts. When the torch is brought close enough to the work, the primary arc will follow the pilot arc across the gap, and the main plasma is started. Once the main plasma is started, the pilot arc power can be shut off.

The second method of starting requires the electrode tip and nozzle tip to be momentarily shorted together. This is accomplished by automatically moving them together and immediately separating them again. The momentary shorting allows the arc to be created without damaging the torch parts.

Kerf

The **kerf** is the space left in the metal as the metal is removed during a cut. The width of a PAC kerf is often wider than that of an oxyfuel cut. Several factors will affect the width of the kerf. A few of the factors are as follows:

- standoff distance — The closer the torch nozzle tip is to the work, the narrower the kerf will be, **Figure 8-20.**

- orifice diameter — Keeping the diameter of the nozzle orifice as small as possible will keep the kerf smaller.

- power setting — Too high or too low a power setting will cause an increase in the kerf width.

- travel speed — As the travel speed is increased, the kerf width will decrease; however, the bevel on the sides and the dross formation will increase if the speeds are excessive.

- gas — The type of gas or gas mixture will affect the kerf width as the gas change affects travel speed, power, concentration of the plasma stream, and other factors.

- electrode and nozzle tip — As these parts begin to wear out from use or are damaged, the PAC quality and kerf width will be adversely affected.

- swirling of the plasma gas — On some torches, the gas is directed in a circular motion around the electrode before it enters the nozzle tip orifice. This swirling causes the plasma stream that's produced to be more dense with straighter sides. The result is an improved cut quality, including a narrow kerf, **Figure 8-21.**

- water injection — The injection of water into the plasma stream as it leaves the nozzle tip is not the same as the use of a water shroud. Water injection into the plasma stream will increase the swirl and further concentrate the plasma. This improves the cutting quality, lengthens the life of the nozzle tip, and makes a squarer, narrower kerf, **Figure 8-22.**

Table 8-2 lists some standard kerf widths for several metal thicknesses. These are to be used as a guide for nesting of parts on a plate to maximize the material used and minimize scrap. The kerf size may vary from this depending on a number of variables with your PAC sys-

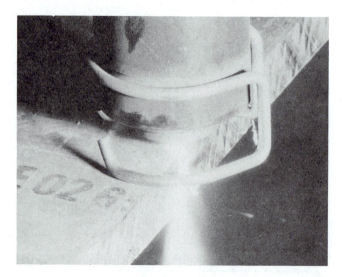

Figure 8-20 A wire adapter can be snapped around some shielding cups, which allows the torch to be slid across the surface of the metal as it is cut.

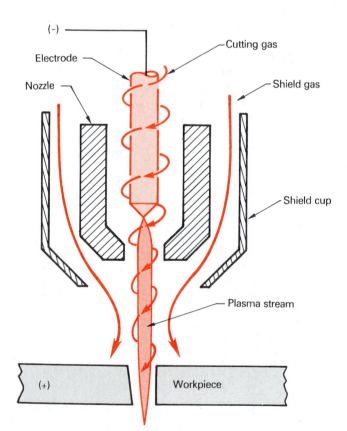

Figure 8-21 The cutting gas can swirl around the electrode to produce a tighter plasma column. (Courtesy of the American Welding Society.)

Standard Kerf Widths for Several Metal Thicknesses			
Plate Thickness		Kerf Allowance	
in.	mm	in.	mm
1/8 to 1	3.2 to 25.4	+3/32	+2.4
1 to 2	25.4 to 51.0	+3/16	+4.8
2 to 5	51.0 to 127.0	+5/16	+8.0

Table 8-2

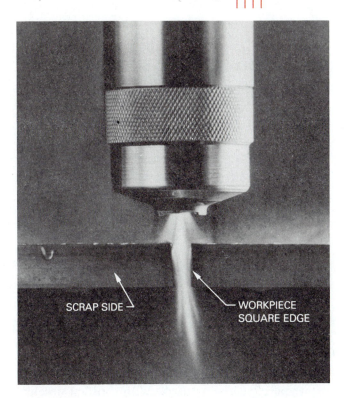

Figure 8-23

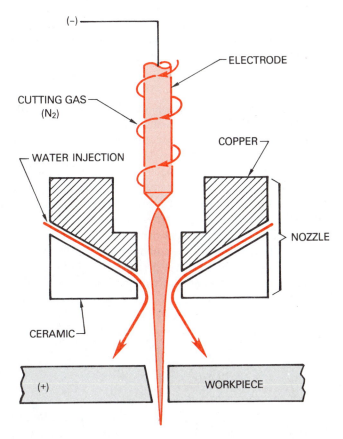

Figure 8-22 Water injection plasma arc cutting. Notice that the kerf is narrow, and one side is square. (Courtesy of the American Welding Society.)

tem. You should make test cuts to verify the size of the kerf before starting any large production cuts.

Because the sides of the plasma stream are not parallel as they leave the nozzle tip, there is a bevel left on the sides of all plasma cuts. This bevel angle is from 1/2° to 3° depending on metal thickness, torch speed, type of gas, standoff distance, nozzle tip condition, and other factors affecting a quality cut. On thin metals, this bevel is undetectable and offers no problem in part fabrication or finishing.

The use of a plasma swirling type torch and the direction the cut is made can cause one side of the cut to be square and the scrap side to have all of the bevel, **Figure 8-23**. This technique is only effective provided that one side of the cut is to be scrap.

GASES

Almost any gas or gas mixture can be used today for the PAC process. Changing the gas or gas mixture is one method of controlling the plasma cut. Although the type of gas or gases used will have a major effect on the cutting performance, it is only one of a number of changes that a technician can make to help produce a quality cut. The following are some of the effects on the cut that changing the PAC gas(es) will have:

- force — The amount of mechanical impact on the material being cut. The density of the gas and its ability to disperse the molten metal.

- central concentration — Some gases will have a more compact plasma stream. This factor will greatly affect the kerf width and cutting speed.

- heat content — As the electrical resistance of a gas or gas mixture changes, it will affect the heat content of the plasma it produces. The higher the resistance, the higher the heat produced by the plasma.

- kerf width — The ability of the plasma to remain in a tightly compact stream will produce a deeper cut with less of a bevel on the sides.

- dross formation — The dross that may be attached along the bottom edge of the cut can be controlled or eliminated.

- top edge rounding — The rounding of the top edge of the plate can often be eliminated by correctly selecting the gas(es) that are to be used.

Gases for Plasma Arc Cutting and Gouging	
Metal	Gas
Carbon and low alloy steel	Nitrogen Argon with 0 to 35% Hydrogen Air
Stainless Steel	Nitrogen Argon with 0°to 35% Hydrogen
Aluminum and aluminum alloys	Nitrogen Argon with 0 to 35% Hydrogen
All plasma arc gouging	Argon with 35% to 40% Hydrogen

Table 8-3

■ metal type — Because of the formation of undesirable compounds on the cut surface as the metal reacts to elements in the plasma, some metals may not be cut with specific gas(es).

Table 8-3 lists some of the popular gases and gas mixtures used for various PAC metals. The selection of a gas or gas mixture for a specific operation to maximize the system performance must be tested with the equipment and set-up being used. With constant developments and improvements in the PAC system, new gases and gas mixtures are continuously being added to the list. In addition to the type of gas, it's important to have the correct gas flow rate for the size tip, metal type, and thickness. Too low a gas flow will result in a cut having excessive dross and sharply beveled sizes. Too high a gas flow will produce a poor cut because of turbulence in the plasma stream and waste gas. A flow measuring kit can be used to test the flow at the plasma torch for more accurate adjustments, **Figure 8-24**.

Stack Cutting

Because the PAC process does not rely on the thermal conductivity between stacked parts, like the oxyfuel process, thin sheets can be stacked and cut efficiently.

With the oxyfuel stack cutting of sheets, it's important that there not be any air gaps between layers. Also, it's often necessary to make a weld along the side of the stack in order for the cut to start consistently.

The PAC process does not have these limitations. It is recommended that the sheets be held together for cutting, but this can be accomplished by using standard C-clamps. The clamping needs to be tight because, if the space between layers is excessive, the sheets may stick together. The only problem that will be encountered is that because of the kerf bevel, the parts near the bottom might be slightly larger if the stack is very thick. This problem can be controlled by using the same techniques as described for making the kerf square.

Dross

Dross is the metal compound that resolidifies and attaches itself to the bottom of a cut. This metal compound is made up mostly of unoxidized metal, metal oxides, and nitrides. It's possible to make cuts dross-free if the PAC equipment is in good operating condition and the metal is not too thick for the size of torch being used. Because dross contains more unoxidized metal than most OFC slag, often it is much harder to remove if it sticks to the cut. The thickness that a dross-free cut can be made is dependent on a number of factors, including the gas(es) used for the cut, travel speed, standoff distance, nozzle tip orifice diameter, wear condition of electrode tip and nozzle tip, gas velocity, and plasma stream swirl.

Stainless steel and aluminum are easily cut dross-free. Carbon steel, copper, and nickel-copper alloys are much more difficult to cut dross-free.

MACHINE CUTTING

Almost any of the plasma torches can be attached to some type of semi-automatic or automatic device to allow it to make machine cuts. The simplest devices are oxyfuel portable flame-cutting machines that run on tracks, **Figure 8-25**. These portable machines are good for mostly straight or circular cuts. Complex shapes can

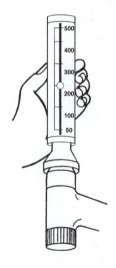

Figure 8-24 Plasma flow measuring kit. (Courtesy of ESAB Welding and Cutting Products.)

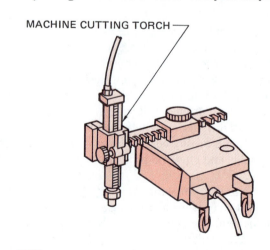

MACHINE CUTTING TORCH

Figure 8-25

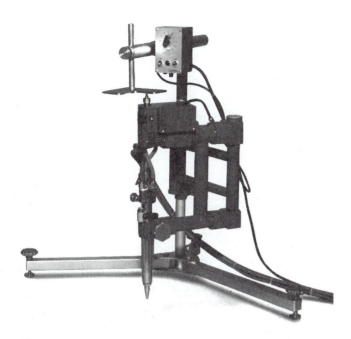

Figure 8-26 Portable pattern cutter can cut shapes, circles, and straight lines. (Courtesy of ESAB Welding and Cutting Products.)

be cut with a pattern cutter that uses a magnetic tracing system to follow the template's shape, **Figure 8-26.**

High powered PAC machines may have amperages up to 1,000 amps. These machines must be used with some semi-automatic or automatic cutting system. The heat, light, and other potential hazards of these machines make them unsafe for manual operations.

Large, dedicated, computer-controlled cutting machines have been built specifically for PAC systems. These machines have the high travel speeds required to produce good quality cuts and have a high volume of production. With these machines, the operator can input the specific cutting instructions such as speed, current, gas flow, location, and shape of the part to be cut, and the machine will make the cut with a high degree of accuracy once or any number of times.

Robotic cutters are also available to perform high-quality, high-volume PAC, **Figure 8-27.** The advantage of using a robot is that, in most cases, the robot is capable of being set up for multitasking. When a robot is programmed, it can cut the part out, change and tool itself and weld the parts together, change the tool and grind, drill, or paint the finished unit.

Water Tables

Machine cutting lends itself to the use of water cutting tables, although they can be used with most hand torches. The water table is used to reduce the noise level, control the plasma light, trap the sparks, eliminate most of the fume hazard and reduce distortion.

Water tables either support the metal just above the surface of the water or they submerge the metal about 3 inches below the water's surface. Both types of water tables must have some method of removing the cut parts, scrap, and slag that build up in the bottom. Often the

surface type tables will have the PAC torch connected to a water shroud nozzle, **Figure 8-28.** By using a water shroud nozzle, the surface table will offer the same advantages to the PAC process as the submerged table offers. In most cases, the manufacturers of this type of equipment have made provisions for a special dye to be added to the water. This dye will help control the harmful light produced by the PAC. Check with the equipment's manufacturer for limitations and application of the use of dyes.

MANUAL CUTTING

Manual plasma arc cutting is the most versatile of the PAC processes. It can be used in all positions, on almost any surface, and on most metals. This process is limited to low power plasma machines; however, even these machines can cut up to 1 1/2-inch-thick metals. The limitation to low power, 100 amperes or less, is primarily for safety reasons. The higher powered machines have extremely dangerous open circuit voltages that can kill a person if accidentally touched.

Setup

The setup of most plasma equipment is similar, but don't ever attempt to set up a system without the manufacturer's owner's manual for the specific equipment.

Be sure all of the connections are tight and that there are no gaps in the insulation on any of the cables.

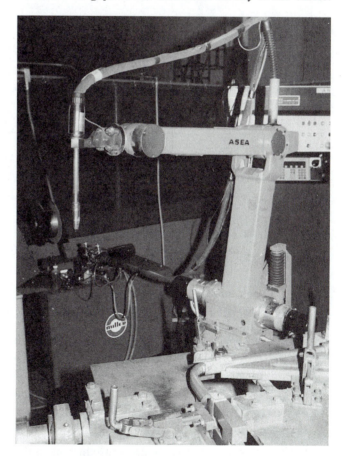

Figure 8-27 Welding Robot. (Courtesy of T.J. Snow Co., Inc.)

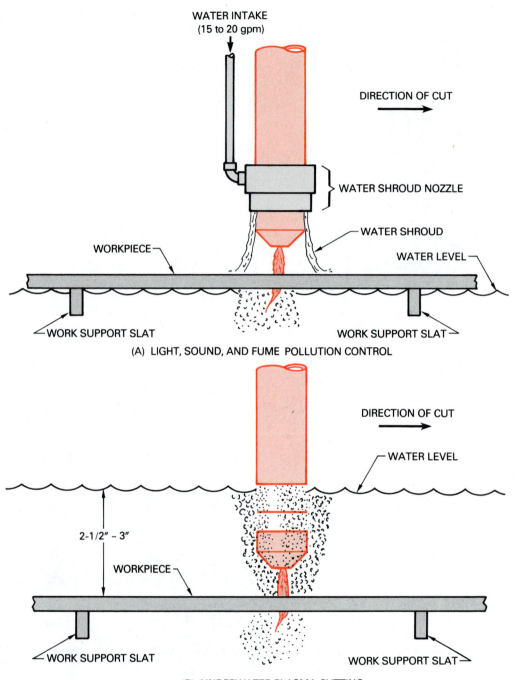

WATER INTAKE
(15 to 20 gpm)

DIRECTION OF CUT

WATER SHROUD NOZZLE

WATER SHROUD

WORKPIECE

WATER LEVEL

WORK SUPPORT SLAT

WORK SUPPORT SLAT

(A) LIGHT, SOUND, AND FUME POLLUTION CONTROL

DIRECTION OF CUT

WATER LEVEL

2-1/2" – 3"

WORKPIECE

WORK SUPPORT SLAT

WORK SUPPORT SLAT

(B) UNDERWATER PLASMA CUTTING

Figure 8-28 A water table can be used either with a water shroud (A) or underwater (B) torches. (Courtesy of the American Welding Society.)

Check the water and gas lines for leaks. Visually inspect the complete system for possible problems.

Before you touch the nozzle tip, be sure that the main power supply is off. The open circuit voltage on even low-powered plasma machines is high enough to kill a person. Replace all parts to the torch before the power is restored to the machine.

PLASMA ARC GOUGING

Plasma arc gouging is a recent introduction to the PAC processes. The process is similar to that of air carbon arc gouging in that a U groove can be cut into the metal's surface. The removal of metal along a joint before the metal is welded or the removal of a defect for repairing can easily be done using this variation of PAC, **Figure 8-29.**

The torch is set up with a less concentrated plasma stream. This will allow the washing away of the molten metal instead of thrusting it out to form a cut. The torch is held at approximately a 30° angle to the metal surface. Once the groove is started it can be controlled by the rate of travel, torch angle, and torch movement.

Plasma arc gouging is effective on most metals. Stainless steel and aluminum are especially good metals

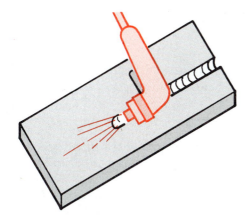

Figure 8-29 Plasma arc gouging a U groove in a plate.

to gouge because there is almost no post cleanup. The groove is clean, bright, and ready to be welded. Plasma arc gouging is especially beneficial with these metals because there is no reasonable alternative available. The only other process that can leave the metal ready to weld is to have the groove machined, and machining is slow and expensive compared to plasma arc gouging.

It's important not to try to remove too much metal in one pass. The process will work better if small amounts are removed at a time. If a deeper groove is required, multiple gouging passes can be used.

SAFETY

PAC has many of the same safety concerns as does most other electric welding or cutting processes. Some special concerns are specific to this process.

- **electrical shock** — Because the open circuit voltage is much higher for this process than for any other, extra caution must be taken. The chance that a fatal shock could be received from this equipment is much higher than from any other welding equipment.

- **moisture** — Often water is used with PAC torches to cool the torch, improve the cutting characteristic, or as part of a water table. Any time water is used it's very important that there be no leaks or splashes. The chance of electrical shock is greatly increased if there is moisture on the floor, cables, or equipment.

- **noise** — Because the plasma stream is passing through the nozzle orifice at a high speed, a loud sound is produced. The sound level increases as the power level increases. Even with low power equipment the decibel (dB) level is above safety ranges. Some type of ear protection is required to prevent damage to operator and other people in the area of the PAC equipment when it's in operation. High levels of sound can have a cumulative effect on one's hearing. Over time, one's ability to hear will decrease unless proper precautions are taken. See the owner's manual for the recommendations for the equipment in use.

- **light** — The PAC process produces light radiation in all three spectrums. This large quantity of visible light, if the eyes are unprotected, will cause night blindness. The most dangerous of the lights is ultraviolet. Like other arc processes, this light can cause burns to the skin and eyes. The third light, infrared, can be felt as heat, and it is not as much of a hazard. Some type of eye protection must be worn when any PAC is in progress. Table 8-4 lists the recommended lens shade numbers for various power level machines.

- **fumes** — This process produces a large quantity of fumes that are potentially hazardous. A specific means for removing them from the work space should be in place. A downdraft table is ideal for manual work, but some special pickups may be required for larger applications. The use of a water table and/or a water shroud nozzle will greatly help to control fumes. Often the fumes cannot be exhausted into the open air without first being filtered or treated to remove dangerous levels of contaminants. Before installing an exhaust system, you must first check with local, state, and federal officials to see if specific safeguards are required.

- **gases** — Some of the plasma gas mixtures include hydrogen; because this is a flammable gas, extra care must be taken to insure that the system is leak-proof.

- **sparks** — As with any process that produces sparks, the danger of an accidental fire is always present. This is a larger concern with PAC because the sparks are often thrown some distance from the work area and the operator's vision is restricted by a welding helmet. If there is any possibility that sparks will be thrown out of the immediate work area, a fire watch must be present. A fire watch is a person whose sole job is to watch for the possible starting of a fire. This person must know how to sound the alarm and have appropriate fire fighting equipment handy. <u>Never cut in the presence of combustible materials.</u>

- **operator check out** — <u>Never operate any PAC equipment until you have read the manufacturer's owner's and operator's manual for the specific equipment to be used first. It's a good idea to have someone who is familiar with the equipment go through the operation after you have read the manual.</u>

Recommended Shade Densities for Filter Lenses		
Current range A	Minimum shade	Comfortable shade
Less than 300	8	9
300 to 400	9	12
400 plus	10	14

Table 8-4

PRACTICE 8-1

Flat, Straight Cuts in Thin Plate

Using a properly set up and adjusted PAC machine, proper safety protection, one or more pieces of mild steel, stainless steel, and aluminum, 6 in. (152 mm) long and 16 gauge and 1/8 in. (3 mm) thick, you will cut off 1/2-inch-wide strips, Figure 8-30.

■ Starting at one end of the piece of metal that is 1/8 in. (3 mm) thick, hold the torch as close as possible to a 90° angle.

■ Lower your hood and establish a plasma cutting stream.

■ Move the torch in a straight line down the plate toward the other end, Figure 8-31.

■ If the width of the kerf changes, speed up or slow down the travel rate to keep the kerf the same size for the entire length of the plate.

Repeat the cut using both thicknesses of all three types of metals until you can make consistently smooth cuts that are within ±3/32 inch of a straight line and ±5° of being square. Turn off the PAC equipment and clean up your work area when you are finished cutting. ◆

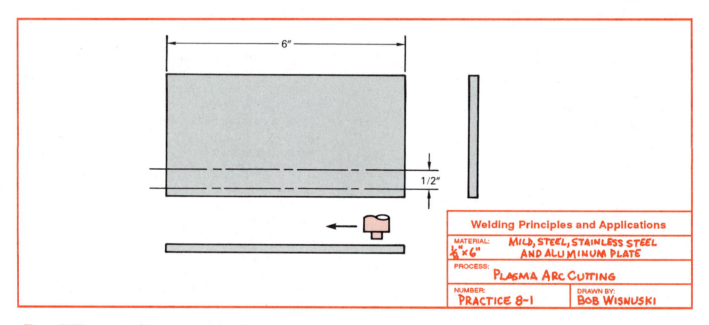

6″

1/2″

Welding Principles and Applications

MATERIAL: ⅛″×6″ MILD, STEEL, STAINLESS STEEL AND ALUMINUM PLATE

PROCESS: PLASMA ARC CUTTING

NUMBER: PRACTICE 8-1

DRAWN BY: BOB WISNUSKI

Figure 8-30

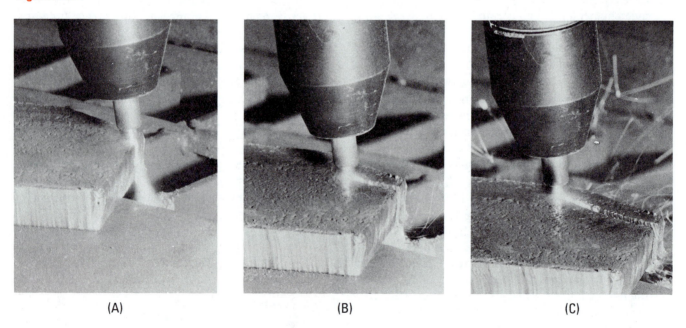

(A)　　　　　　　　(B)　　　　　　　　(C)

Figure 8-31 (A) Starting at the edge of a plate, like starting an oxyfuel cut, (B & C) move smoothly in a straight line toward the other end. Note the roughness left along the side of the plate from a previous cut.

PRACTICE 8-2

Flat, Straight Cuts in Thick Plate

Using a properly set up and adjusted PAC machine, proper safety protection, one or more pieces of mild steel, stainless steel, and aluminum, 6 in. (152 mm) long and 1/4 in. and 1/2 in. thick, you will cut off 1/2-inch-wide strips. Follow the same procedure as outlined in Practice 8-1.

Repeat the cut using both thicknesses of all three types of metals until you can make consistently smooth cuts that are within ±3/32 inch of a straight line and ±5° of being square. Turn off the PAC equipment and clean up your work area when you are finished cutting. ◆

PRACTICE 8-3

Flat Cutting Holes

Using a properly set up and adjusted PAC machine, proper safety protection, one or more pieces of mild steel, stainless steel, and aluminum, 16 gauge, 1/8 in., 1/4 in., and 1/2 in. thick, you will cut 1/2-in. and 1-in. holes.

■ Starting with the piece of metal that is 1/8 in. (3 mm) thick, hold the torch as close as possible to a 90° angle.

■ Lower your hood and establish a plasma cutting stream.

■ Move the torch in an outward spiral until the hole is the desired size, Figure 8-32.

Repeat the hole-cutting process until both sizes of holes are made using all the thicknesses of all three types of metals until you can make consistently smooth cuts that are within ±3/32 inch of being round and ±5° of being square. Turn off the PAC equipment and clean up your work area when you are finished cutting. ◆

PRACTICE 8-4

Beveling of a Plate

Using a properly set up and adjusted PAC machine, proper safety protection, one or more pieces of mild steel, stainless steel, and aluminum, 6 in. (152 mm) long and 1/4 in. and 1/2 in. thick, you will cut a 45° bevel down the length of the plate.

■ Starting at one end of the piece of metal that is 1/4 in. thick, hold the torch as close as possible to a 45° angle.

■ Lower your hood and establish a plasma cutting stream.

■ Move the torch in a straight line down the plate toward the other end, Figure 8-33.

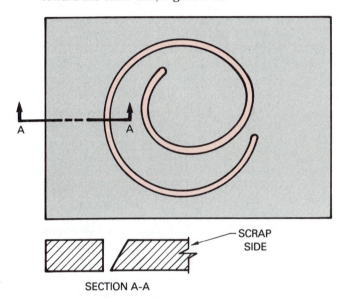

SECTION A-A

Figure 8-32 When cutting a hole, make a test to see which direction to make the cut so that the beveled side is on the scrap piece.

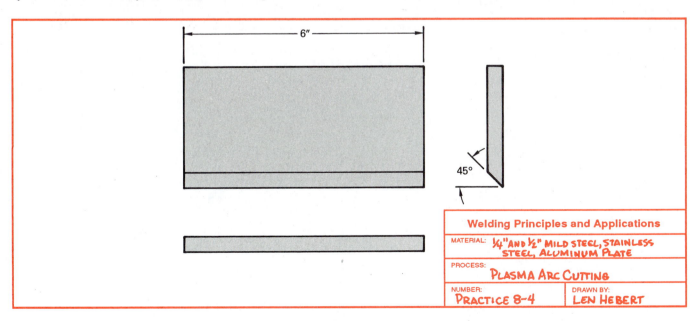

Figure 8-33

Repeat the cut using both thicknesses of all three types of metals until you can make consistently smooth cuts that are within ±3/32 inch of a straight line and ±5° of a 45° angle. Turn off the PAC equipment and clean up your work area when you are finished cutting. ◆

PRACTICE 8-5

U-Grooving of a Plate

Using a properly set up and adjusted PAC machine, proper safety protection, one or more pieces of mild steel, stainless steel, and aluminum, 6 in. (152 mm) long and 1/4 in. or 1/2 in. thick, you will cut a U-groove down the length of the plate.

■ Starting at one end of the piece of metal, hold the torch as close as possible to a 30° angle, **Figure 8-34**.

■ Lower your hood and establish a plasma cutting stream.

■ Move the torch in a straight line down the plate toward the other end.

■ If the width of the U-groove changes, speed up or slow down the travel rate to keep the groove the same width and depth for the entire length of the plate.

Repeat the gouging of the U-groove using all three types of metals until you can make consistently smooth grooves that are within ±3/32 inch of a straight line and uniform in width and depth. Turn off the PAC equipment and clean up your work area when you are finished cutting. ◆

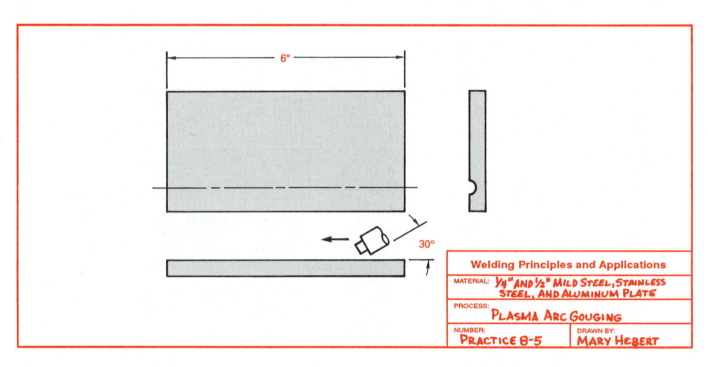

6"

30°

Welding Principles and Applications

MATERIAL: 1/4" AND 1/2" MILD STEEL, STAINLESS STEEL, AND ALUMINUM PLATE

PROCESS: PLASMA ARC GOUGING

NUMBER: PRACTICE 8-5   DRAWN BY: MARY HEBERT

Figure 8-34

Abrasivejet Cutting System Takes on the Competition

For many years, Plasma Cutting Service, Inc., has provided defense, aerospace, and manufacturing companies in the Southwest with cutting services. As capabilities grew from a hand-held plasma arc cutting torch in the '70s to the installation of three CNC plasma arc cutting systems in the early '80s, the company became recognized as one of the few job shops in the Los Angeles area that offered plasma arc cutting services. But by 1985, competition in the area was increasing, and its plasma arc systems had difficulty cutting metals such as titanium and high-nickel alloys. Also, the new composite materials couldn't be cut with plasma arc.

To get an edge on the competition, the company began to search for an additional cutting system that would increase its capabilities, yet provide economy of operation. The system the company determined would give it those features is the Paser II abrasivejet process manufactured by Flow International Corp.

Figure 1

"The abrasivejet cutting system was selected because of its ability to cut all types of material, including composites," said Richard Woolman, president and general manager. Also, since a waterjet does not produce a heat affected zone, heat-sensitive metals can be cut without concern for brittleness in that area.

The system's intensifier pressurizes water up to 55,000 psi and forces it through an orifice at velocities that may reach three times the speed of sound. Abrasives, such as garnet, contained in a hair-thin water stream create the abrasive action needed to cleanly cut hard materials.

The first abrasivejet cutting machine installed by the company was integrated with an existing CNC unit, providing two-axis operation at two cutting stations over a work area of 72 X 180 in. The second unit installed at a later date was incorporated into a gantry robotic system with a working area of 120 X 264 X 36 in. This operation has both five-axis and three-axis cutting stations. Heavy-duty dual cutting applications are handled with this setup. The system also incorporates an interactive graphics program with a nesting program for two-dimensional plotting of a variety of part configurations.

The abrasivejet process is improving the company's cutting capability on a variety of applications. For example, jet engine rotary blade repair patches made of titanium used to be rough cut with plasma arc and then ground and machined to the required dimensions by the customer. Now the part can be cut with a precision that eliminates the customer's grinding and machining operations. Additionally, the abrasivejet cut is five to twenty times narrower than the plasma arc cut, which results in a significant material savings.

In another case, aerospace piping supports made of saw-cut 0.025-in.-thick Inconel Alloy 718 blanks used to be electrical-discharge machine cut at a rate of one part per hour. With the abrasivejet system, a part is now cut to the required tolerances every 20 minutes, including finished profile machining time. This has resulted in a significant increase in productivity and a 75 to 80% reduction in costs.

(Reprinted from the *Welding Journal*. Article based on a story by Flow International Corporation, Kent, Washington. Courtesy of the American Welding Society, Inc.)

REVIEW

1. What is an electrical plasma?

2. Approximately how many times hotter is the plasma arc than an unrestricted arc?

3. Why is the body on a manual plasma torch made from a special type of plastic?

4. How are the torch heads of a plasma torch cooled?

5. What three types of power switches are used on plasma torches?

6. Why are some copper parts plated?

7. Why have copper/tungsten tips helped the plasma torch?

8. What provides the gap between the electrode tip and the nozzle tip?

9. What are the advantages of a water shroud nozzle for plasma cutting?

10. How is water used to control power cable overheating?

11. What factors must be considered when selecting a material for plasma arc gas hoses?

12. Why must the cooling water be turned off when the plasma cutting torch is not being used?

13. What type of voltage is required for a plasma cutting torch?

14. A plasma arc torch that is operating with 90 volts (close circuit) and 25 amps would be using how many watts of power?

15. Why do plasma cuts have little or no distortion?

16. What has limited the upper cutting speed limits of the plasma arc cutting process?

17. What metals can be commonly cut using the PAC process?

18. What happens to torches not using drag nozzle tips if the standoff distance is not maintained?

19. What are the two methods for establishing the plasma path to the metal being cut?

20. List eight items that will affect the quality of a PAC kerf.

21. List seven items that are affected by the choice of PAC gas or gases on the cut.

22. Describe *stack cutting*.

23. How does PAC dross compare with OFC slag?

24. Why must high-power PAC machines be used by a semi-automatic or automatic cutting system?

25. What is the advantage of using a water table for PAC?

26. How is plasma arc gouging performed?

Chapter 9

RELATED CUTTING PROCESSES

OBJECTIVES

After completing this chapter, the student should be able to

■ discuss the various cutting processes.

■ give examples of the types of applications that might be best suited to specific cutting processes.

■ list advantages and disadvantages of using the different cutting processes.

■ explain the safety considerations of each of the different cutting processes.

KEY TERMS

monochromatic	synchronized wave form
YAG laser	solid state laser
gas laser	laser beam welds (LBW)
laser beam cuts (LBC)	laser beam drilling (LBD)
cutting gas assist	exothermic gases
air carbon arc cutting (CAC-A)	carbon electrode
graphite	gouging
washing	oxygen lance cutting
water jet cutting	abrasives
delamination	

INTRODUCTION

The number of specialized cutting processes being developed and improved increases every year. To date there are a number of specialized cutting processes that have been perfected and are in use. The new cutting methods are being used by a much wider group than just the welding industry. These processes can be used to cut a wide variety of different things, such as glass, plastic, printed circuit boards, cloth, and fiberglass insulation, and more things are being added almost daily. Material cutting and even hole drilling using a welding related process have become common in the workplace.

Only a few of the more common cutting processes are covered in this chapter. These are the ones that a welder might be required to either be able to perform or have a working knowledge of.

LASER BEAM CUTTING (LBC) AND LASER BEAM DRILLING (LBD)

Lasers have developed from the first bright red spot generated by the ruby rod to a multi-billion dollar industry. The use of lasers has become very common

Figure 9-1 Bar codes are read by lasers to input information into computers.

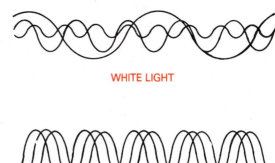

WHITE LIGHT

LASER LIGHT

Figure 9-2 White light is made up of different frequencies (colors) of light waves. Laser light is made up of single frequency light waves traveling parallel to each other.

today. We see their use in our world every day. They help speed our checking out at stores when they are used to read the uniform product code (UPC) on our purchases, **Figure 9-1**. Lasers are used by surveyors to measure distances or to see that a building under construction is level. Doctors can use lasers to make very precise cuts during the most delicate operations. They are even used to entertain us at shows or concerts.

Manufacturers use the laser to do everything from marking products, joining parts, communication between machines, guiding, cutting, or punching machines, to cutting a variety of materials. The industrial laser can be used to guide a machine's cut accurately or to measure distances from a few hundred to thousands of miles.

This versatile tool has helped produce products that could not be produced without the laser's light. A laser can be used to drill holes (LBD) through the hardest materials, such as synthetic diamonds, tungsten carbide cutting tools, quartz, glass, or ceramics. Lasers are used for welding (LBW) materials that are too thin or too hard to be welded with other heat sources. They can be used to cut (LBC) materials that must be cut without overheating delicate parts that might be located just a few thousands of a inch away.

The benefits and capacities of the laser have made it one of the most rapidly growing areas in manufacturing. This is a useful tool that will continue to be developed and improved for years to come.

LASERS

A laser is a form of light that is **monochromatic,** a single color wave length, that travels in parallel waves, **Figure 9-2**. This form of light is generated when certain types of materials are excited either by intense light or with an electric current. The atoms or molecules of the lasing material release their energy in the form of light. The laser light is produced as the atom or molecule slows down from a high energy state to a lower energy state, **Figure 9-3**. The laser light is then bounced back and forth between one fully reflective and one partially reflective surface at the ends of the lasing material. As the light is reflected, it begins to form a **synchronized wave form.** The light that passes through the partially reflective mirror is the laser light, and it's ready to do its job, **Figure 9-4**.

Once the laser beam emerges from the laser rod or chamber, it can be treated as any other type of light. It's possible to focus, reflect, absorb, or defuse the light. By having the light waves traveling in such a uniform manner, they exhibit some specialized characteristics. The light beam tends to remain in a very tight column without

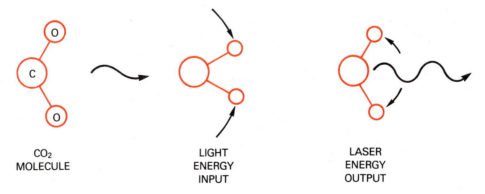

CO_2
MOLECULE

LIGHT
ENERGY
INPUT

LASER
ENERGY
OUTPUT

Figure 9-3 Energy is stored in a carbon dioxide (CO_2) molecule until the molecule can't hold any more. The energy is released suddenly like when a balloon pops. This quick release results in a burst of light energy.

spreading out like ordinary light, **Figure 9-5**. This characteristic allows the laser beam to travel over some distance without significantly being affected by the distance it travels or the air it's traveling through.

Because the beam is not adversely affected by its travel, it can be used to carry information or transmit energy. When the laser wraps around a loaf of bread or a box of crackers, its image is so precise that the data in a uniform product code strip can accurately be received by the store's computer. The military uses lasers to aim weapons or identify targets so that they can be hit by missiles or bombs.

Laser Types

The early lasers all used a synthetic ruby rod as the material that produced the laser light. Today a large number of materials can be used to produce the laser.

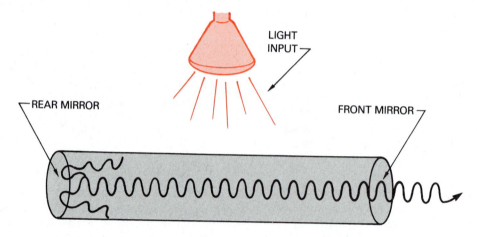

Figure 9-4 Light energy reflects back and forth between the end mirrors until it forms a parallel laser light beam.

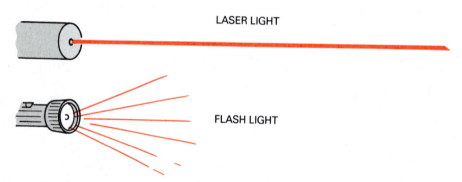

Figure 9-5 Laser light stays in a very tight column unlike most other lights.

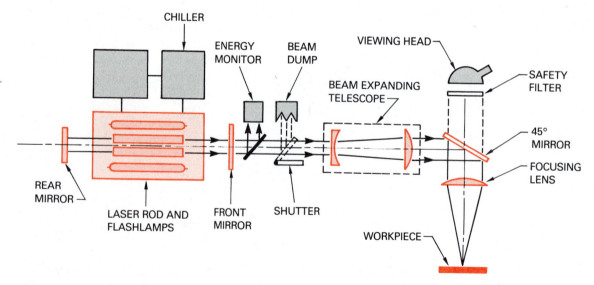

Figure 9-6 Schematic representation of the elements of an Nd: YAG laser. (Courtesy of the American Welding Society.)

These materials include such common items as glass or such exotic items as neodymium-doped, yttrium aluminum garnet (Nd-YAG), often referred to as a YAG laser, Figure 9-6.

Lasers can be divided into two major types; lasers using a solid material for the laser and lasers using a gas. Each of these two types are divided into two groups based on their method of operation; lasers that operate continuously and lasers that are pulsed.

Solid State Lasers

The first lasers were all solid state lasers. They used a ruby rod to produce the laser. Today the most popular solid laser is the neodymium-doped, yttrium aluminum garnet (YAG). This is a synthetic crystal that when exposed to the intense light from the flash tubes can produce high quantities of laser energy. The high-powered solid lasers have a problem because the internal temperatures of the laser rod increase with operation time. These lasers are most often used with low-power continuous or high-powered pulse. The solid state laser is capable of generating the highest powered laser pulses.

Gas Laser

The gas laser uses one or more gases for the laser. Popular gas type lasers use nitrogen, helium, carbon dioxide (CO_2), or mixtures of helium, nitrogen, and CO_2. Gas lasers either use a gas charged cylinder where the gas is static, or they use a chamber that has a means of circulation of the laser gas. The highest continuous power output type lasers are the gas type with a

blower to circulate the gas through a heat exchanger, Figure 9-7.

APPLICATIONS

Laser light beams can be focused into a very compact area. This allows the photo energy that the light possesses to be converted to heat energy as it strikes the surface of a material. The highly concentrated energy can cause the instantaneous melting or vaporization of the material being struck by the laser beam. The equipment used to produce laser beam welds (LBW), laser beam cuts (LBC), or laser beam drilling (LBD) is similar in design and operation. Most laser welding and cutting operations use a gas laser, and most drilling operations use a solid state laser.

In all of the laser applications the ability of the surface of the material to either absorb or reflect the laser greatly affects the operation's efficiency. All materials will reflect some of the laser's light more than others. The absorption rate will increase for any material once the laser beam begins to heat the surface. Once the threshold of the surface temperature is reached, the process will continue at a much higher level of efficiency.

LASER BEAM CUTTING

A laser weld and laser cut both bring the material to a molten state. In the welding process the material is allowed to flow together and cool to form the weld metal. In the cutting process a jet of gas is directed into the molten material to expel it through the bottom of the cut. Although the laser is used primarily to cut very thin materials, it can be used to cut up to 1 inch of carbon steel.

The cutting gas assist can be either a nonreactive gas or an exothermic. Table 9-1 lists the various gases and the materials that they are used to cut. Nonreactive

Figure 9-7 Industrial CO_2 laser beam cuts through a 13-mm-thick titanium plate at a speed of more than 2 cm/s. The invisible 6-kW infrared beam is focused downward on the surface of the plate. The slender tube provides inert gas, which blows molten metal from the narrow (1 mm) cut. (Courtesy of United Technologies Research Center.)

Various Cutting Assist Gases and the Materials They Cut	
Assist Gases	Material
Air	Aluminum
	Plastic
	Wood
	Composites
	Alumina
	Glass
	Quartz
Oxygen	Carbon Steel
	Stainless Steel
	Copper
Nitrogen	Stainless Steel
	Aluminum
	Nickel Alloys
Argon	Titanium

Table 9-1

Figure 9-8 Detailed small part not much larger than a dime made with a laser.

gases do not add any heat to the cutting process; they simply remove the molten material by blowing it out of the kerf. Exothermic gases react with the material being cut, like an oxyfuel cutting torch. The additional heat produced as the exothermic cutting gas reacts with the metal being cut helps blow the molten material out of the kerf.

Some of the advantages of laser cutting include:

■ narrow heat-affected zone — Little or no heating of the surrounding material is observed. It's possible to make very close parallel cuts without damaging the strip that's cut out, **Figure 9-8**.

■ no electrical conductivity required — The part being cut does not have to be electrically conductive, so materials like glass, quartz, and plastic can be cut. There is also no chance that a stray electrical charge might damage delicate computer chips while they are being cut using a laser.

■ noncontact — Nothing comes in contact with the part being cut except the laser beam. Small parts that may have finished surfaces or small surface details can be cut without the danger of disrupting or damaging the surface. It's also not necessary to hold the parts securely as it is when a cutting tool is used.

■ narrow kerf — The width of the kerf is very small, which allows the nesting of parts in close proximity to each other, which will reduce waste of expensive materials, **Figure 9-9**.

■ automation and robotics — The laser beam can easily be directed through an articulated guide to the working end of an automated machine or robot.

■ top edge — The top edge will be smooth and square without being rounded.

LASER BEAM DRILLING

The pulsed laser is the best choice for drilling operations. A very short burst of high laser energy is concentrated on a small spot. The intense heat vaporizes the small spot with enough force to thrust the material out, leaving a small crater. The repeated blasting to the

Figure 9-9 Because of the narrow kerf, small, detailed parts can be nested one inside the other.

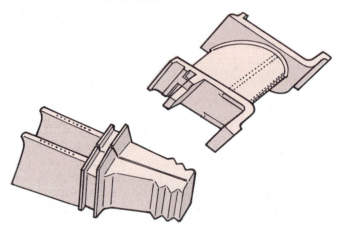

Figure 9-10 Jet engine blades and a rotor component showing laser drilled holes. (Courtesy of the American Welding Society.)

pulsed laser results in the crater becoming increasingly deeper. This process continues until the hole is drilled through the part or to a desired depth, **Figure 9-10**.

A laser can be used to drill holes through the hardest materials, such as tungsten carbide cutting tools, quartz, glass, ceramics, or even synthetic diamonds. No other drilling process can match the precision, speed, or economy of the laser drill.

Holes as small as 0.0001 inch in diameter can be drilled. The limitation on the hole's depth is the laser's focal length. Most holes are less than 1 inch in depth.

LASER EQUIPMENT

Most lasers range from 400 to 1,500 watts in power. Some large machines have as much as 25 KW of power. Although the power of most lasers is relatively small, when compared to other welding processes, it's the laser's ability to concentrate the power into a small area that

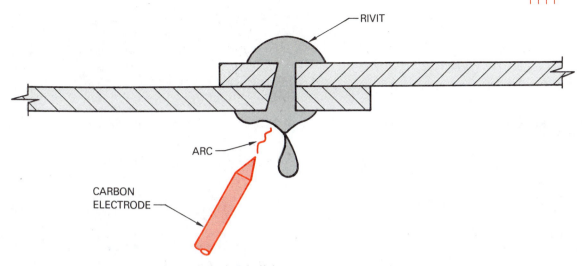

Figure 9-11 A carbon electrode was used to melt the head off rivets.

makes it work so well. The power density of a cutting laser can be equal to 65,000,000 watts per square inch.

Laser equipment is larger than most of the other welding or cutting power supply. A typical unit will require about as much floor space as a large desk, about 10 square feet.

Recent advancements, increased competition, and a rapidly increasing market have helped lower the high cost of the equipment. But even with the current cost of equipment, laser cutting and drilling have one of the lowest average hourly operating costs. The high reliability and long life of the equipment has also helped reduce operating costs.

AIR CARBON ARC CUTTING

The air carbon arc cutting (CAC-A) process was developed in the early 1940s and was originally named air arc cutting (AAC). See Color Plate 9. Air carbon arc cutting was an improvement of the carbon arc process. The carbon arc process was used in the vertical or overhead positions and removed metal by melting a large enough spot so gravity would cause it to drip off the base plate, **Figure 9-11**. This process was slow and could not be accurately controlled. It was found that by using a stream of air the molten metal could be blown away. This greatly improved the speed, quality, and controllability of the process.

In the late 1940s the first air carbon arc cutting torch was developed. Before this development the process required two welders, one to control the carbon arc and the other to guide the air stream. The new torch had both the carbon electrode holder and the air stream in the same unit. This basic design is still in use today, Figure 9-12.

Unlike the oxyfuel process, the air carbon arc cutting process does not require that the base metal be reactive with the cutting stream. Oxyfuel cutting can only be

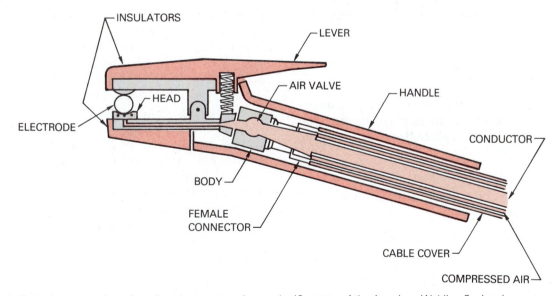

Figure 9-12 Typical cross section of an air carbon arc gouging torch. (Courtesy of the American Welding Society.)

Base Metals	Recommendations
Carbon steel and low alloy steel	Use DC electrodes with DCEP current. AC can be used but with a 50% loss in efficiency.
Stainless steel	Same as for carbon steel.
Cast iron, including malleable and ductile iron	Use of 1/2" or larger electrodes at the highest rated amperage is necessary. There are also special techniques that need to be used when gouging these metals. The push angle should be at least 70 degrees and depth of cut should not exceed 1/2 inch per pass.
Copper alloys (copper content 60% and under)	Use DC electrodes with DCEN (electrode negative) at maximum amperage rating of the electrode.
Copper alloys (copper content over 60%, or size of workpiece is large)	Use DC electrodes with DCEN at maximum amperage rating of the electrode or use AC electrodes with AC.
Aluminum bronze and aluminum nickel bronze (special naval propeller alloy)	Use DC electrodes with DCEN.
Nickel alloys (nickel content is over 80%)	Use AC electrodes with AC.
Nickel alloys (nickel content less than 80%)	Use DC electrodes with DCEP.
Magnesium alloys	Use DC electrodes with DCEP. Before welding, surface of groove should be wire brushed.
Aluminum	Use DC electrodes with DCEP. Wire brushing with stainless wire brushes is mandatory prior to welding. Electrode extension (length of electrode between electrode torch and workpiece) should not exceed 3 inches for good quality work. DC electrodes with DCEN can also be used.
Titanium, zirconium, hafnium, and their alloys	Should not be cut or gouged in preparation for welding or remelting without subsequent mechanical removal of surface layer from cut surface.

Note: Where preheat is required for welding, similar preheat should be used for gouging.

Table 9-2 Recommended Procedures for Air Carbon Arc Cutting of Different Metals.

performed on metals that can be rapidly oxidized by the cutting stream of oxygen. The air stream in this process blows the molten metal away. This greatly increases the list of metals that can be cut, **Table 9-2**.

MANUAL TORCH DESIGN

The air carbon arc cutting torch is designed differently than the shielded metal arc electrode holder. The major differences between an electrode holder and an air carbon arc torch are:

- The lower electrode jaw has a series of air holes.
- The jaw has only one electrode locating groove.
- The electrode jaw can pivot.
- There is an air valve on the torch lead.

By having only one electrode locating groove in the jaw and pivoting the jaw, the air stream will always be aimed correctly. The air must be aimed just under and

behind the electrode and always in the same direction, **Figure 9-13**. This ensures that the air stream will be directed at the spot where the electrode arcs to base the metal.

Torches are available in a number of amperage sizes. The larger torches have greater capacity but are less flexible to use on small parts.

The torch can be permanently attached to a welding cable and air hose or it can be attached to welding power by gripping a tab at the end of the cable with the shielded metal arc electrode holder, **Figure 9-14**. The temporary attachment can be made easier if the air hose is equipped with a quick disconnect. A quick disconnect on the air hose will allow it to be used for other air tools such as grinders or chippers. Greater flexibility for a work station can be achieved with this arrangement.

Electrodes

Air carbon arc cutting electrodes are available as copper coated or plain (without a coating). The copper

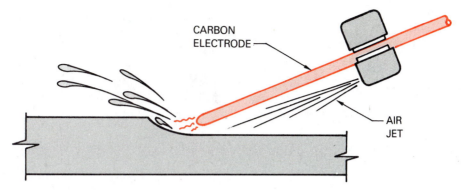

CARBON
ELECTRODE

AIR
JET

Figure 9-13 Air carbon arc gouging.

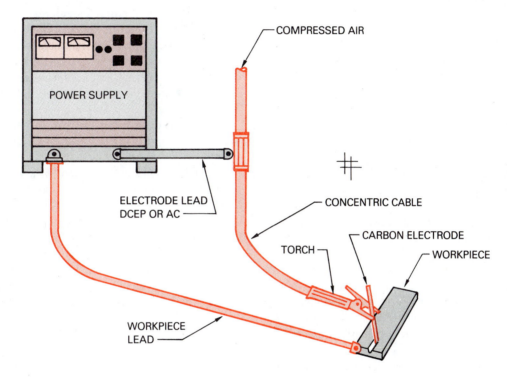

COMPRESSED AIR

POWER SUPPLY

ELECTRODE LEAD
DCEP OR AC

CONCENTRIC CABLE

TORCH

CARBON ELECTRODE

WORKPIECE

WORKPIECE
LEAD

Figure 9-14 Air carbon arc gouging equipment setup. (Courtesy of the American Welding Society.)

coating helps decrease the carbon electrode overheating by increasing its ability to carry higher currents and improves the heat dissipation. The copper coating provides increased strength, to reduce accidental breakage.

Electrodes come in round, flat, and semi-round, **Figure 9-15**. The round electrodes are used for most gouging operations, and the flat electrodes are most often used to scarf off a surface. Round electrodes are also available in sizes ranging from 1/8 to 1 inch in diameter. Flat electrodes are available in 3/8 and 5/8 inch sizes.

Electrodes are available to be used on both direct current electrode positive and alternating current. The DCEP electrodes are the most commonly used, and they are made of carbon in the form of graphite. The AC electrodes are less common; they have some elements added to the carbon to stabilize the arc, which is needed for the AC power.

To reduce waste, electrodes are made so that they can be joined together. The joint consists of a female

ROUND   FLAT   SEMI-ROUND

Figure 9-15 Cross sections of carbon electrodes.

tapered socket at the top end and a matching tang on the bottom end, **Figure 9-16**. The connection of the new electrode to the remaining setup will allow the stub to be consumed with little loss of electrode stock. This connecting of electrodes is required for most track-type air carbon arc cutting operations to allow for longer cuts.

Power Sources

Most shielded metal arc welding power supplies can be used for air carbon arc cutting. The operating voltage required for air carbon arc cutting needs to be 28 volts or

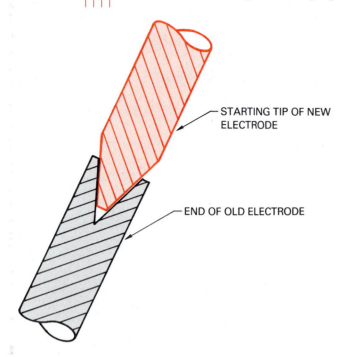

Figure 9-16 Air carbon arc electrode joint.

higher. This voltage is slightly higher than that required for most SMA welding, but most welders will meet this requirement. Check the manufacturer's owner's manual to see if your welder is approved for air carbon arc cutting. If the voltage is lower than the minimum, the arc will tend to sputter out, and it will be hard to make clean cuts.

Because most carbon arc cutting requires a high amperage setting, it may be necessary to stop some cuts so that the duty cycle of the welder is not exceeded. On large industrial welders this is not normally a problem.

Air Supply

Air supplied to the torch must be between 80 to 100 psi. The minimum pressure is around 40 psi. The correct air pressure will result in cuts that are clean, smooth, and uniform. The air flow rate is also important. If the air line is too small or the compressor does not have the required capacity, there will be a loss in air pressure at the torch tip. This line loss will result in a lower than required flow at the tip. The resulting cut will be less desirable in quality.

Application

Air carbon arc cutting can cut a variety of materials. It is a relatively low-cost way of cutting most metals, especially stainless steel, aluminum, nickel alloys, and copper. Air carbon arc cutting is most often used for repair work. Few cutting processes can match the speed, quality, and cost savings of this process for repair or rework. In repair or rework, the most difficult part is the removal of the old weld or cutting a groove so a new weld can be made. The air carbon arc can easily remove the worst welds even if they contain slag inclusions or other

defects. For repairs the arc can cut through thin layers of paint, oil, or rust and make a groove that needs little, if any, cleanup.

■ **CAUTION**

Never cut on any material that might produce fumes that would be hazardous to your health without proper safety precautions, including adequate ventilation.

The highly localized heat results in only slight heating of the surrounding metal. As a result, usually there is no need to preheat hardenable metals to prevent hardness zones. Cast iron is a metal that can be carbon arc gouged to prepare a crack for welding without causing further damage to the part by inputting excessive heat.

Air carbon arc cutting can be used to remove a weld from a part. The removal of welds can be accomplished with such success that often the part needs no post-cutting cleanup. The root of a weld can be back gouged so that a backing weld can be made ensuring 100% weld penetration, **Figure 9-17**.

The electrode should extend approximately 6 inches from the torch when starting a cut; and as the cut progresses, the electrode is consumed. Stop the cut and readjust the electrode when its end is approximately 3 inches from the electrode holder. This will reduce the damage to the torch caused by the intense heat of the operation.

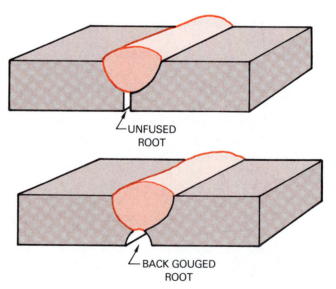

Figure 9-17 Back gouging the root of a weld made in thick metal can ensure that a weld with 100% joint fusion can be made.

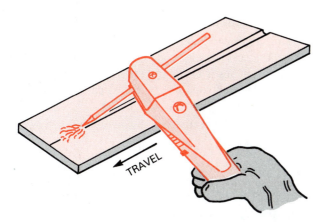

Figure 9-18 Manual air carbon arc gouging operation in the flat position. (Courtesy of the American Welding Society.)

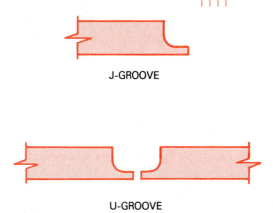

J-GROOVE

U-GROOVE

Figure 9-19

Gouging

Gouging is the most common application of the air carbon arc cutting processes. Arc gouging is the removal of a quantity of metal by completely consuming that portion of the metal removed to form a groove or bevel, **Figure 9-18**. The groove produced along an edge of a plate is usually a J-groove. The groove produced along a joint between plates is usually a U-groove, **Figure 9-19**. Both grooves are used as a means to insure that the weld applied to the joint will have the required penetration into the metal.

Washing

Washing is the process of removing large areas of the surface of the metal, **Figure 9-20**. Washing can be used to remove large areas that contain defects, to reduce the transitional stresses of unequal-thickness plates, or to allow space for the capping of a surface with a wear-resistant material.

SAFETY

In addition to the safety requirements of shielded metal arc, air carbon arc requires several special precautions such as:

■ sparks — The quantity and volume of sparks generated during this process is a major safety hazard. Extra precautions must be taken to ensure that other workers, equipment, materials, or property in the area will not be affected by the spark stream.

■ noise — This process produces a high level of sound. The sound level is high enough to cause hearing damage if proper ear protection is not used.

■ light — The arc light produced is the same as that produced by the shielded metal arc welding process. But because the arc has no smoke to defuse the light and the amperages are usually much higher, the

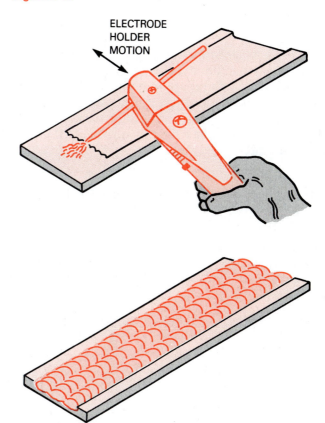

ELECTRODE HOLDER MOTION

Figure 9-20 Hardsurfacing weld applied in the carbon-arc-cut groove. (Courtesy of the American Welding Society.)

chances of receiving arc burns are much higher. Additional protection should be worn, such as thicker clothing, leather jacket, and leather aprons.

■ eyes — Because of the intense arc light, a darker welding filter lens for the helmet should be used.

■ fumes — The combination of the air and the metal being removed result in a high volume of fumes. Special consideration must be made for the removal of these fumes from the work area. Before installing a ventilation system, check with local, state, and federal laws. Some of the fumes may have to be filtered before they can be released into the air.

- surface contamination — Often this process is used to prepare damaged parts so that they can be repaired. If the used parts have paint, oils, or other contamination that might generate hazardous fumes, they must be removed in an acceptable manner before any cutting begins.

- equipment — Check the manufacturer's owner's manuals for specific safety information concerning the power supply and the torch before you start any work with each piece of equipment for your first time.

PRACTICE 9-1

Air Carbon Arc Straight Cut in the Flat Position

Using an air carbon arc cutting torch and welding power supply that has been safely set up in accordance with the manufacturer's specific instructions in the owner's manual and wearing safety glasses, welding helmet, gloves, and any other required personal protection clothing, you will make a 6-inch-long straight U groove gouge in a carbon steel plate.

1. Adjust the air pressure to approximately 80 psi.

2. Set the amperage within the range for the diameter electrode you are using by referring to the box the electrodes came in.

3. Check to see that the stream of sparks will not start a fire or cause any damage to anyone or anything in the area.

4. Make sure the area is safe, and turn on the welder.

5. Using a good dry leather glove to avoid electrical shock, insert the electrode in the torch jaws so that about 6 inches is extending outward. Be sure not to touch the electrode to any metal parts, because it may short out.

6. Turn on the air at the torch head.

7. Lower your arc welding helmet.

8. Slowly bring the electrode down at about a 30° angle so it will make contact with the plate near the starting edge, Figure 9-21. Be prepared for a loud, sharp sound when the arc starts.

9. Once the arc is struck, move the electrode in a straight line down the plate toward the other end. Keep the speed and angle of the torch constant.

10. When you reach the other end, lift the torch so the arc will stop.

11. Raise your helmet and stop the air.

12. Remove the remaining electrode from the torch so it will not accidentally touch anything.

When the metal is cool, chip or brush any slag or dross off of the plate. This material should remove easily. The groove must be within ±1/8 inch of being straight and within ±3/32 inch of uniformity in width and depth. Repeat this cut until it can be made within these tolerances. Turn off the CAC equipment and clean up your work area when you are finished cutting. ◆

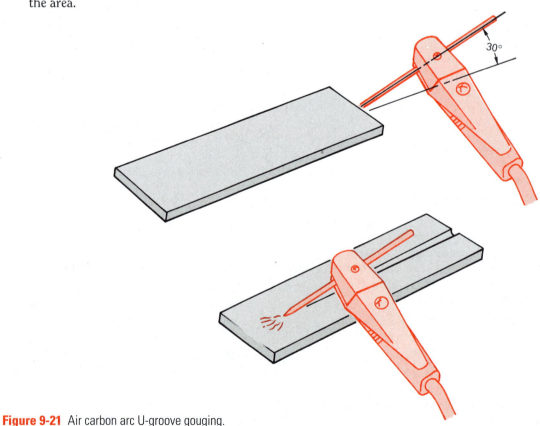

Figure 9-21 Air carbon arc U-groove gouging.

Air Carbon Arc Edge Cut in the Flat Position

Using the same equipment, adjustments, setup, and materials as described in Practice 9-1, you are going to make a J-groove along the edge of the plate.

1. Adjust the air pressure to approximately 80 psi.

2. Set the amperage within the range for the diameter electrode you are using.

3. Check to see that the stream of sparks will not start a fire or cause any damage to anyone or anything in the area.

4. Make sure the area is safe, and turn on the welder.

5. Using a good, dry, leather glove to avoid electrical shock, insert the electrode in the torch jaws so that about 6 inches is extending outward. Be sure not to touch the electrode to any metal parts because it may short out.

6. Turn on the air at the torch head.

7. Lower your arc welding helmet.

8. Slowly bring the electrode down at about a 30° angle so it will make contact with the plate near the starting edge, **Figure 9-22**.

9. Once the arc is struck, move the electrode in a straight line down the edge of the plate toward the other end. Keep the speed and angle of the torch constant.

10. When you reach the other end, lift the torch so the arc will stop.

11. Raise your helmet and stop the air.

12. Remove the remaining electrode from the torch so it will not accidentally touch anything.

When the metal is cool, chip or brush any slag or dross off of the plate. This material should remove easily. The groove must be within ±1/8 inch of being straight and within ±3/32 inch of uniformity in width and depth. Repeat this cut until it can be made within these tolerances. Turn off the CAC equipment and clean up your work area when you are finished cutting. ◆

Air Carbon Arc Back Gouging in the Flat Position

Using the same equipment, adjustments, setup, and materials as described in Practice 9-1, you are going to make a U-groove along the root face of a weld joint on a plate, **Figure 9-23**.

Follow the same starting procedure as you did in Practice 9-1.

1. Start the arc at the joint between the 2 plates.

2. Once the arc is struck, move the electrode in a straight line down the edge of the plate toward the other end. Watch the bottom of the cut to see that it is deep enough. If there is a line along the bottom of the groove, it needs to be deeper. Once the groove depth is determined, keep the speed and angle of the torch constant.

3. When you reach the other end, break the arc off.

4. Raise your helmet and stop the air.

5. Remove the remaining electrode from the torch so it will not accidentally touch anything.

When the metal is cool, chip or brush any slag or dross off of the plate. This material should remove easily.

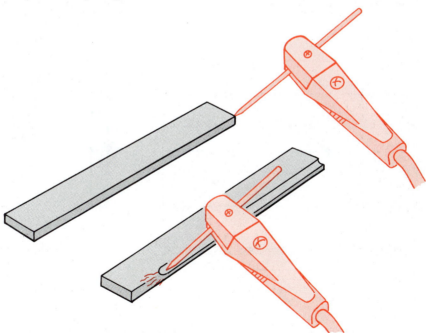

Figure 9-22 Air carbon arc J-groove gouging.

The groove must be within ±1/8 inch of being straight, but it may vary in depth so that all of the unfused root of the weld has been removed. Repeat this cut until it can be made within these tolerances. Turn off the CAC equipment and clean up your work area when you are finished cutting. ◆

PRACTICE 9-4

Air Carbon Arc Weld Removal in the Flat Position

Using the same equipment, adjustments, setup, and materials as described in Practice 9-1, you are going to make a U-groove to remove a weld from a plate, **Figure 9-24**.

Follow the same starting procedure as you did in Practice 9-1.

1. Start the arc on the weld.
2. Once the arc is struck, move the electrode in a straight line down the weld toward the other end. Watch the bottom of the cut to see that it is deep enough. If there is not a line along the bottom of the groove, it needs to be deeper. Once the groove depth is determined, keep the speed and angle of the torch constant.
3. When you reach the other end, break the arc off.
4. Raise your helmet and stop the air.
5. Remove the remaining electrode from the torch so it will not accidentally touch anything.

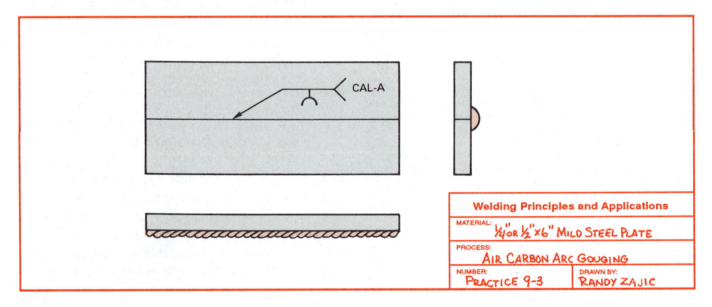

Welding Principles and Applications

MATERIAL: ¼" or ½" x6" Mild Steel Plate

PROCESS: Air Carbon Arc Gouging

NUMBER: Practice 9-3   DRAWN BY: Randy Zajic

Figure 9-23

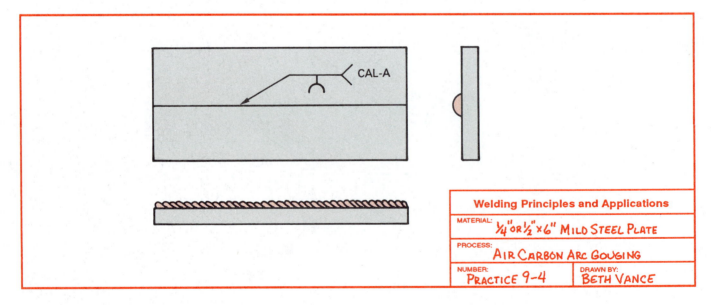

Welding Principles and Applications

MATERIAL: ¼" or ½" x6" Mild Steel Plate

PROCESS: Air Carbon Arc Gouging

NUMBER: Practice 9-4   DRAWN BY: Beth Vance

Figure 9-24

When the metal is cool, chip or brush any slag or dross off of the plate. This material should remove easily. The groove must be within ±1/8 inch of being straight, but it may vary in depth so that all of the weld metal has been removed. Repeat this cut until it can be made within these tolerances. Turn off the CAC equipment and clean up your work area when you are finished cutting. ◆

OXYGEN LANCE CUTTING

The oxygen lance cutting process uses a consumable carbon steel tube. The tip of the tube is heated to its kindling temperature. A high-pressure oxygen flow is started through the lance. The oxygen reacts with the hot lance tip, releasing sufficient heat to sustain the reaction, Figure 9-25.

An oxyfuel torch is usually used to heat the lance tip to a red hot reaction temperature. Other heat sources include electric resistance and electric arcing. Once the oxygen stream is started, it reacts with the lance material, which results in the creation of both a high temperature and heat releasing reaction.

The intense reaction of the lance allows it to be used to cut through a variety of materials. The hot metal leaving the lance tip has not completed its exothermic reaction. As this reactive mass impacts the surface of the material being cut, it releases a large quantity of energy into that surface. Thermal conductivity between the molten metal and the base material is a very efficient method of heat transfer. This, along with the continued burning of the lance material on the surface, causes the base material to become molten.

Once the base material is molten, it may react with the burning lance material, forming fumes or slag, which is then blown from the cut. Any molten material not becoming reactive is carried out of the cut with the slag or blown out with the oxygen stream.

The addition of steel rods or other metals to the center of the oxygen lance tube have increased their productivity. The improved lances last longer and cut faster.

APPLICATION

The oxygen lance's unique method of cutting allows it to be used to cut material not normally cut using a thermal process. Films have portrayed the oxygen lance as a tool used by thieves to cut into safes. In reality, this would result in the valuables in the safe being destroyed. The oxygen lances can be used to cut reinforced concrete.

Cutting concrete has been used in the demolition of buildings. It allows the quick removal of thick sections of the building without the dangerous vibration caused by most conventional methods. This has been a life-saving factor in the use of the oxygen lance for rescue work following earthquakes. The oxygen lance saved thousands of hours and countless lives in Mexico City following the devastation from the city's worst earthquake. Oxygen lances were used to cut large sections of concrete that fell from buildings into manageably sized pieces. Local and national news agencies showed building rubble being cut away by rescue workers using the oxygen lance.

The oxygen lance is also used to cut thick sections of cast iron, aluminum, and steel. Often in the production of these metals thick sections must be cut. Occasionally equipment failure will stop metal production. If the metal in production is allowed to cool, it may need to be cut in sections so it can be removed from the machine. The oxygen lances process is very effective in this type of work.

Safety

It is important to follow all safety procedures when using this process. Manufacturers list specific safety precautions for the oxygen lances they produce. Read and follow those instructions carefully. The major safety concerns are:

■ fumes — The large quantity of fumes generated are often a health hazard. An approved ventilation system must be provided if this work is to be done in a building or enclosed area.

■ heat — This operation produces both high levels of radiant heat and plumes of molten sparks and slag. The operators must wear special heat-resistant clothing.

■ noise — Sound is produced well above safety levels. Ear protection must be worn by anyone in the area.

WATER JET CUTTING

Water jet cutting is not a thermal cutting process. This method of cutting does not put any heat into the material being cut. The cut is accomplished by the rapid erosion of the material by a high pressure jet of water, Figure 9-26. An abrasive powder may be added to the stream of water. Abrasives are added when hard materials such as metals are being cut.

The lack of heat input to the material being cut makes this process unique. Materials that heat might

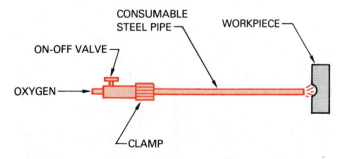

Figure 9-25 Schematic view of oxygen lance cutting. (Courtesy of the American Welding Society.)

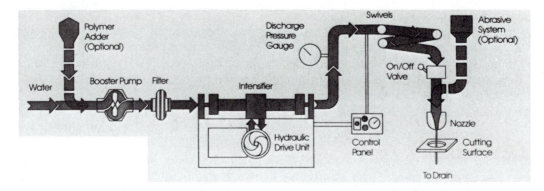

Figure 9-26 Basic diagram illustrating the elements of the water jet cutting system. (Courtesy of Ingersoll-Rand.)

distort, make harder, or cause to delaminate are ideally suited to this process. The lack of heat distortion allows thin material to be cut with the edge quality of a laser cut and as distortion free as a sheer cut. Delamination is not a problem when cutting composite or laminated materials, such as carbon fibers, resins, or computer circuit boards, **Figure 9-27**.

APPLICATIONS

The kerf width does not tend to change unless too high a travel speed is being used. This results in a square, smooth finish on the cut surface. Post-cutting cleanup of the parts is totally eliminated for most materials, and only slight work is needed on a few others, **Figure 9-28**. The quality of the cut surface can be controlled so that even parts for the aerospace industry can often be assembled as cut.

The addition of an abrasive powder can speed up the cutting, allow harder materials to be cut, and improve the surface finish of a cut powder. The powder most often used is garnet. It's also commonly used as an abrasive on sandpaper. If an abrasive is used, the small water jet orifice will wear out faster.

(A)

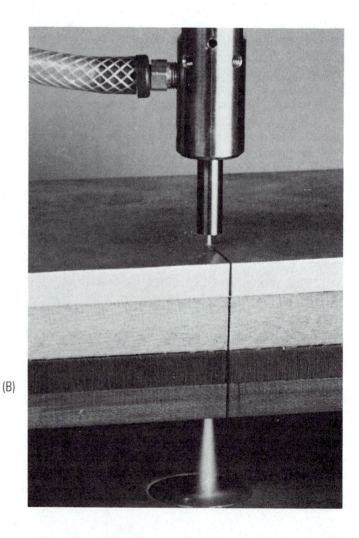

(B)

Figure 9-27 Waterjet cutting is very versatile. (A) A machine with multiple cutting jets is cutting out printed circuit boards. (B) Four different materials, including metal, fiberglass, and plastic, are being cut at one time. (Courtesy of Ingersoll-Rand.)

Figure 9-28 Notice how narrow and clean this cut is. (Courtesy of Ingersoll-Rand.)

Materials that often gum up a cutting blade, such as plexiglass, ABS plastic, and rubber, can be cut easily. There is nothing for the material to adhere to that would disrupt the cut. The lack of heat also reduces the tendency of the material cut surface to become galled.

Most of the water jet cutting is performed by some automated or robotic system. There are a few bandsaw-type, hand-fed cutting machines that are used for single cuts or when limited production is required, **Figure 9-29**.

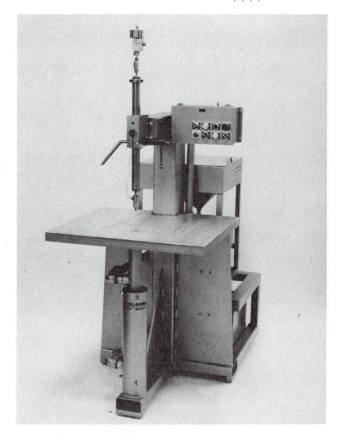

Figure 9-29 Bandsaw-type, hand-fed, water jet cutting machine. (Courtesy of Ingersoll-Rand.)

Cutting Software Speeds Production

Innovative cutting software made it possible for the Elgin Sweeper Company to successfully integrate a gantry shape cutting machine and a CNC unit into its manufacturing operation without months of downtime.

The Elgin Sweeper Company manufactures large, industrial, mechanical and vacuum-style street sweepers (Figure 1) at its 240,000-sq-ft headquarters and plant in Elgin, Illinois. Approximately 200 hourly and 100 salaried employees produce sweepers for municipalities and other customers throughout the world.

When manufacturing engineer Scott Lakari and Manager of Manufacturing Reg Folmar traveled from their plant to ESAB Automation's headquarters in Fort Collins, Colorado, all they planned to do was sign off on a new GXB 1220 shape cutting unit and Auto-Path CNC control — **Figure 2**. They needed the new machine and control in order to take full advantage of the CAM system they purchased several years ago. That system came with a sophisticated nesting package that generates nests in machine language. However, the gantry machine and optical tracer they had been using for

Figure 1

Figure 2

the past eighteen years could only cut traced parts; therefore, the company purchased a new gantry and automated shaped cutting control that could utilize the nesting package.

The GXB is a midsize gantry model, offering many of the same features found on larger machines, such as all-welded steel construction, rack-and-pinion drives, and hardened, ground and polished Thomson rails. The CNC unit combines computer graphics with part programming and process automation. Its color graphics display shows what a shape looks like before it is cut and follows the path of the cutting torch as it moves around the part. It can be used to program new shapes or to transfer part programs from tape readers, floppy disks, or other part programming systems. During production, the control handles all torch and machine movement functions automatically.

While Lakari and Folmar estimated that the new equipment would ultimately save them time and money, they also faced a lengthy process before they could start realizing those savings. "We had 300 parts that ran on our old tracing machine, and had to convert those tracings either to paper tape or some kind of machine language that the new control could read. We had opted to purchase the new machine without a paper tape reader or mini-file because we already has a CAM system and were going to run a DNC link to the machine. In the meantime, we had to figure out how to create readable programs from tracings in a short period of time so that we could get the machine up and running," Lakari explained.

Fortunately, they saw a demonstration of some new digitizing software that supports FastCAM, a PC-based part programming system. The new software, called FastCOPY, allows users to digitize templates and scale drawings with minimal training.

A 16-button cursor pad puts all the geometric definitions of the template or drawing at the operator's fingertips. The operator follows the contour points of the template to produce a CNC file that's ready to go to the machine. FastCOPY also has editing capability, allowing on-screen adjustments. The file can then be transferred back to FastCAM to be verified or geometrically adjusted.

After a demonstration, Folmar and Lakari decided to try the software, digitizing board, and PC for three months. That would give them enough time to digitize parts and get a dc link hooked up so that they could download directly from the CAM system to the control.

Once back at the plant, Lakari set up the software system in two one-hour telephone sessions with ESAB. In another half hour, he learned the system commands, and then was able to teach an operator how to use the system in less than an hour. That operator started working eight hours a day to digitize all of their tracings.

Because they had removed their old machine, they would have been forced to buy all of their cut parts during the several months it took to set up an inter-CAM system if they didn't have the new program. They were cutting their own parts after about four weeks and did not have to out-source them for the entire twelve to fourteen week setup period.

"What was nice about the system," added Lakari, "was that the program interpolated everything,...so you didn't have to know anything about programming to use it." For example, to make a circle, three points on the circle are touched, The program knows a circle is being made by the button being pushed. It knows that only one circle can fit all three of those points, and it draws the circle.

"Whenever engineering asked us to try out new parts, in the past, we'd have to tell them that if our programmers were busy they would have to wait two days to a week for a program. Now, all we needed was a to-scale drawing, which they could plot out off of their CAD system. Then we could lay it on the digitizing board, create a part program, and run the part for them. We found ourselves relying on the digitizing system quite a bit," stated Lakari.

Now that the DNC link is operational, Elgin's programmers use the CAD system to create a source file for their nesting package, and run daily nests on the CNC equipment.

(Reprinted from the *Welding Journal*. Article based on a story from ESAB Automation, Inc., Fort Collins, Colorado. Courtesy of the American Welding Society, Inc. Photos courtesy of ESAB Automation, Inc.)

REVIEW

1. Describe the use of the following processes: LBD, LBW, and LBC.

2. How is laser light formed?

3. Other than the lasing material, what is the difference between solid state lasers and gas lasers?

4. What effect does a material's surface have on the laser beam?

5. Using Table 9-1, list the assist cutting gas and the material it can be used to cut.

6. What are reactive laser assist gases called? And how do they work?

7. What type of laser beams are used for drilling? Why?

8. What is CAC-A?

9. Using Table 9-1, give the recommended procedure for air carbon arc gouging of carbon steel, magnesium alloys, low-alloy copper.

10. Why are some carbon arc electrodes copper coated?

11. What may occur if an SMA welding machine has below minimum arc voltage for air carbon arc gouging?

12. How can carbon arc welding be used to ensure 100% weld penetration?

13. What is washing, and how can it be used?

14. How are oxygen lance cuts usually started?

15. What unusual material can be cut with an oxygen lance?

16. What are the advantages of abrasive powder to the water jet cutting stream?

Section Four
GAS SHIELDED WELDING

SUCCESS STORY

Dana Gillenwaters was not unlike most high school students of his day, more interested in fun than academics. In his senior year, however, he began to realize that life after high school might offer him a challenge.

Dana was always interested in the metal shop classes he attended. The shop instructor he had at Vooresville high school, in upstate New York, was instrumental in developing his interest in the welding field. He also became actively involved in racing cars and snowmobiles. Dana thought that combining two of his interests, sports racing and welding, seemed like a logical choice for a career.

After high school, Dana's father steered him in the direction of a local private trade school. The objective was for Dana to develop more specific welding skills. Dana spent a great deal of time working hands-on in particular processes such as pipe, tig, mig, and electric arc welding. This training turned out to be a turning point in his life.

Soon after graduating from Modern Welding School, Dana gained employment as a New York State Department of Transportation Certified Welder. He spent many hours hanging in safety gear up to 100 feet above water welding on New York State Bridges. Here he became an experienced welder.

Important role models for Dana were his instructors, the owners of Modern Welding School, Jacob Schaad and Edward Percenti. What he learned from these men could never be duplicated. Soon Dana was teaching with them.

An opportunity arose for Dana to purchase Modern Welding School. With the help and support of his family and friends, all the details were worked out. Dana has been the president of Modern Welding School for almost twenty years.

Last year, during the school's 60th anniversary, Dana undertook an immense task. Dana relocated the school to a 10,000 square foot state-of-the-art facility with over 50 welding booths. Major employers such as General Electric, International Paper, BASF, and the United States Postal Service send students to this facility for welding training.

Dana still enjoys teaching welding, as well as consulting for local industry. And the two men who taught him the most about the trade, now in their 70s, still teach at the school on occasion.

Chapter 10

GAS METAL ARC WELDING EQUIPMENT, SETUP, AND OPERATION

OBJECTIVES

After completing this chapter, the student should be able to

- describe the various methods of metal transfer.
- explain the effect of slope and inductance on gas metal arc welding.
- list four variables used to control the gas metal arc welding bead.
- describe the different electrode feed methods.
- name the parts of a gas metal arc welding setup.
- list the advantages of gas metal arc spot welding.

KEY TERMS

axial spray metal transfer

transition current

globular transfer

pulsed-arc metal transfer

synergic system

buried-arc transfer

short-circuiting transfer

electrode extension (stickout)

inductance

slope

pinch effect

INTRODUCTION

In the 1920s, a metal arc welding process using an unshielded wire was being used to assemble the rear axle housings for automobiles. The introduction of the shielded metal arc welding electrode rapidly replaced the bare wire. The shielded metal arc welding electrode made a much higher quality weld. In 1948, the first inert gas metal arc welding (GMAW) process, as it is known today, was developed and became commercially available, **Figure 10-1**. In the beginning, the GMAW process was used to weld aluminum using argon (Ar) gas for shielding. As a result, the process was known as MIG, which stands for metal inert gas welding. The later introduction of CO_2 and O_2 to the shielding gas has resulted in the American Welding Society's preferred term of gas metal arc welding (GMAW). However, it is still commonly referred to as MIG.

The GMAW process may be performed as semiautomatic (SA), machine (ME), or automatic (AU) welding, **Table 10-1**. The GMA welding process is commonly performed as a semiautomatic process and is often mistakenly referred to as "semiautomatic welding." Equipment is available to perform most of the wire-feed processes semiautomatically, and the GMAW process can be fully automated. Robotic arc welding often uses GMAW because of the adaptability of the process in any position, **Figure 10-2**. See Color Plates 10 through 21.

The rising use of all the various types of consumable wire welding processes has resulted in the increased sales of wire. At one time, wire made up less

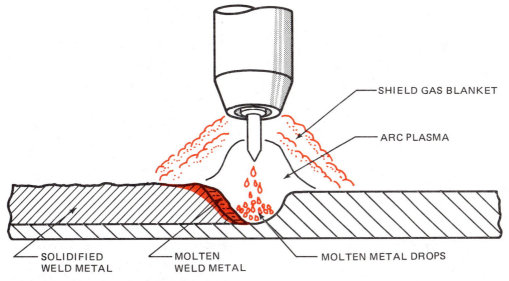

Figure 10-1 Gas shielded metal arc welding (GMAW).

Function	Manual (MA) (Example: SMAW)	Semiautomatic (SA) (Example: GMAW)	Machine (ME) (Example: GMAW)	Automatic (AU) (Example: GMAW)
Maintaining the Arc	Welder	Machine	Machine	Machine
Feeding the Filler Metal	Welder	Machine	Machine	Machine
Provide the Joint Travel	Welder	Welder	Machine	Machine
Provide the Joint Guidance	Welder	Welder	Welder	Machine

Table 10-1 Methods of Performing Welding Processes.

than 1% of the total market of filler metal. The total tonnage of filler metals used has grown and so has the percentage of wire. Today, wire exceeds 50% of the total tonnage of filler metals produced and used.

Much of the increase in the use of the wire welding processes is due to the increases in the quality of the welds produced. This improvement is due to an increased reliability of the wire-feed systems, improvements in the filler metal, smaller wire sizes, faster welding speed, higher weld deposition rates, less expensive shielding gases, and improved welding techniques. Table 10-2 shows the typical weld deposition rates using the GMA welding process. The increased usage has led to a reduction in the cost of equipment. GMA welding equipment is now found even in small shops.

In this chapter, the semiautomatic GMA welding process will be covered. The skill required to set up and

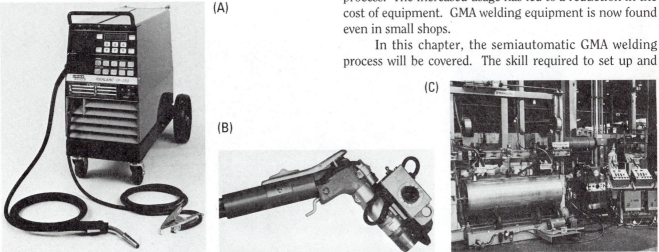

Figure 10-2 (A) Semiautomatic GMA welding setup. (B) Machine GMA welding gun with a friction drive, which provides both uniform nozzle to work distance and travel speed. (C) Automatic GMA welding. (A & B courtesy of Lincoln Electric, Cleveland, OH. C courtesy of The Ransome Co., Division of Big Three Industries, Inc.)

WELD DEPOSITION RATE Pounds per Hour			
Electrode Diameter Amperage	0.35	0.45	0.63
50	2.0		
100	4.8	4.2	
150	7.5	6.7	5.1
200		8.7	7.8
250		12.7	11.1
300			14.4

Table 10-2 GMA Weld Deposition Rates.

operate this process is basic to the understanding and operation of other wire-feed processes. The reaction of the weld to changes in voltage, amperage, feed speed, stickout, and gas is similar to that of most wire-feed processes.

METAL TRANSFER

When first introduced, the GMA process was used with argon as a shielding gas to weld aluminum. Even though argon (Ar) was then expensive, the process was accepted immediately because it was much more produc-

tive than GTA and because it produced higher quality welds than SMA. This new arc welding process required very little post-weld cleanup because it was slag- and spatter-free.

Axial Spray Metal Transfer

The freedom from spatter associated with the argon-shielded GMAW process results from a unique mode of metal transfer called axial spray metal transfer, Figure 10-3. See Color Plates 16 and 17. This process is identified by the pointing of the wire tip from which very small drops are projected axially across the arc gap to the molten weld pool. There are hundreds of drops per second crossing from the wire to the base metal. These drops are propelled by arc forces at high velocity in the direction the wire is pointing. This projection of drops enables welding in the vertical and overhead positions without losing control of transfer. In many cases, the molten weld pool may be too large to be controlled in these positions. Because the drops are very small and directed at the molten weld pool, the process is spatter-free.

This spray transfer process requires three conditions: argon shielding (or argon-rich shielding gas mixtures), DCEP polarity, and a current level above a critical

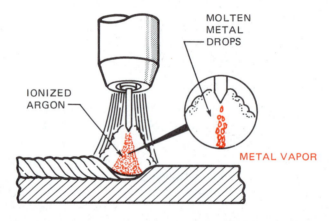

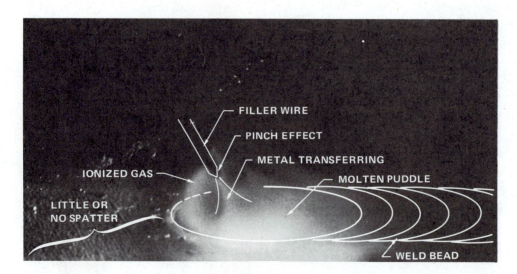

Figure 10-3 Axial spray metal transfer. Note the pinch effect of filler wire and the symmetrical metal transfer column.

amount called the transition current. Figure 10-4 illustrates how the rate of drops transferred changes in relationship to the welding current. At low currents, the drops are large and are transferred at rates below 10 per second. These drops move slowly, falling from the electrode tip as gravity pulls them down. They tend to bridge the gap between the electrode tip end and molten weld pool. This produces a momentary short circuit that throws off spatter. However, the mode of transfer changes very abruptly above the critical current, producing the desirable spray. This change in the rate of transfer as related to current is shown schematically in Figure 10-4.

The transition current depends on the alloy being welded. It also is proportional to the wire diameter, meaning that higher currents are needed with larger diameter wires. The need for high current density imposes some restrictions on the process. The high current hinders welding sheet metal because the high heat cuts through sheet metal. High current also limits its use to the flat, vertical down, and horizontal welding positions. Weld control in the vertical up or overhead positions is very difficult to impossible. Table 10-3 lists the welding parameters for a variety of gases, wire sizes, and metal thicknesses for GMA welding of mild steel.

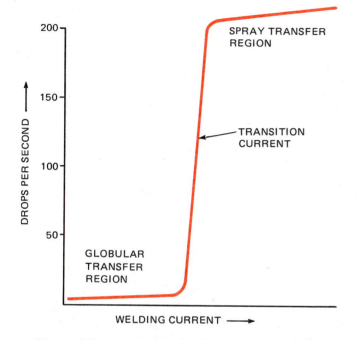

Figure 10-4 Desirable spray transfer shown schematically.

■ CAUTION

The heat produced during axial spray welding using large-diameter wire or high current may be intense enough to cause the filter lens in a welding helmet to shatter. Be sure the helmet is equipped with a clear plastic lens on the inside of the filter lens. Avoid getting your face too close to the intense heat.

GMA WELDING PARAMETERS							
MILD STEEL	Wire-feed speed, in./min		Voltage, V				
Base-material thickness, in.	0.035-in.	0.045-in	CO_2	75 Ar-25 CO_2	Ar	98 Ar-2 O_2	Current,A
0.036	105-115	---	18	16	--	---	50-60
0.048	140-160	70	19	17	--	---	70-80
0.060	180-220	90-110	20	17.7	--	---	90-110
0.075	240-260	120-130	20.7	18	20	--	120-130
1/8	280-300	140-150	21.5	18.5	20.5	---	140-150
3/16	320-340	160-175	22	19	21.5	23.5	160-170
1/4	360-380	185-195	22.7	19.5	22.5	24.5	180-190
5/16	400-420	210-220	23.5	20.5	23.5	25	200-210
3/8	420-520	220-270	25	22	25	26.5	220-250
1/2 and up	---	375	28	26	29	31	300

Table 10-3 GMA Welding Parameters for Mild Steel.

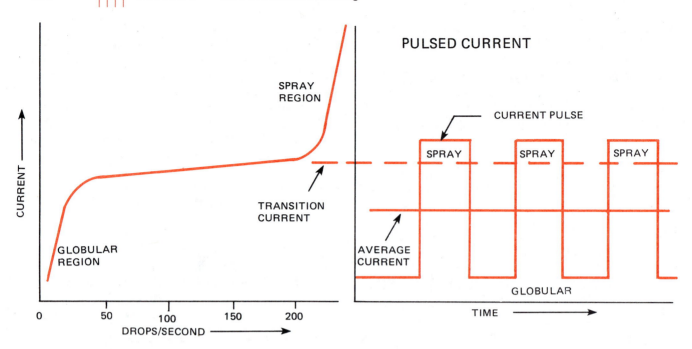

Figure 10-5 Mechanism of pulsed arc spray transfer at a low average current.

Pulsed-arc Metal Transfer

Because the change from spray arc to globular transfer occurs within a very narrow current range and the globular transfer occurs at the rate of only a few drops per second, a controlled spray transfer at significantly lower average currents is achievable. Pulsed-arc metal transfer involves pulsing the current from levels below the transition current to those above it. The time interval below the transition current is short enough to prevent a drop from developing. About 0.1 second is needed to form a globule, so no globule can form at the electrode tip if the time interval at the low base current is about 0.01 second. Actually, the energy produced during this time is very low — just enough to keep the arc alive.

The arc's real work occurs during those intervals when the current pulses to levels above the transition current. The time of that pulse is controlled to allow a single drop of metal to transfer. This is typical of the drops normally associated with spray transfer. In fact, with many power supplies, a few small drops could transfer during the pulse interval. As with conventional spray-arc, the drops are propelled across the arc gap, allowing metal transfer in all positions.

The average current can be reduced sufficiently to reduce penetration enough to weld sheet metal or reduce deposition rates enough to control the molten weld pool in all positions. This level controlling the weld heat input and rate of weld metal deposit is achieved by changing the following variables:

■ frequency — The number of times the current is raised and lowered to form a single pulse. Frequency is measured in pulses per second.

■ amplitude — The amperage or current level of the power at the peak or maximum, expressed in amperage.

■ width of the pulses — The amount of time the peak amperage is allowed to stay on, **Figure 10-5**.

Figure 10-6 shows a typical pulsed arc welding system. Although developed in the mid-1960s, this technol-

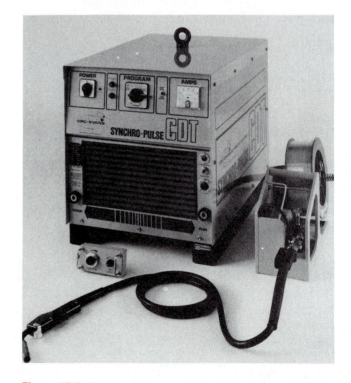

Figure 10-6 GMA pulsed arc welding system. (Courtesy of CRC-Evans Automatic Welding.)

ogy did not receive much attention until solid-state electronics were developed to handle the high power required of welding power supplies. Solid-state electronics provided a better, simpler, and a more economical way to control the pulsing process. The newest generation of pulsed-arc systems interlocks the power supply and wire feeder so that the proper settings of the wire feed end power supply are obtained for any given job by adjusting a single knob. Such systems have been termed synergic. In some respects, these systems are more complex because the correct interrelationships between the wire-feed speeds and power supply settings must be programmed into the equipment, and each wire composition, wire size, and shielding gas requires a special program. The manufacturer generally programs the most common combinations, allowing space in the computer for additional user input.

Shielding Gases for Spray or Pulsed-spray Transfer

Axial spray transfer is impossible without shielding gases containing argon. Pure argon is used with all metals except steels. As much as 80% helium can be added to the argon to increase the power in the arc without affecting the desirable qualities of the spray mode. With more helium, the transfer becomes progressively more globular, forcing the use of a different welding mode, to be described later. Since these gases are inert, they do not react chemically with any metals. This factor makes the GMAW process the only productive, manual, or semiautomatic method for welding metals sensitive to oxygen (essentially all metals except iron or nickel). The cathodic cleaning action associated with argon at DCEP (DCRP) is also very important for fabricating metals such as aluminum, which quickly develops undesirable surface oxides when exposed to air.

This same cleaning action causes problems with steels. Iron oxide in and on the steel surface is a good emitter of electrons that attracts the arc cathode. But these oxides are not uniformly distributed, resulting in very irregular cathode movement and in turn irregular weld deposits. This problem was solved by adding small amounts of oxygen to the argon. The reaction produced a uniform film of iron oxide on the weld pool and provided a stable site for the cathode. This discovery enabled uniform welds in ferrous alloys and expanded the use of GMAW to welding those materials.

The amount of oxygen needed to stabilize arcs in steel varies with the alloy. Generally, 2% is sufficient for carbon and low-alloy steels. In the case of stainless steels, about 1/2% should prevent a refractory scale of chromium oxide. Carbon dioxide can substitute for oxygen. More than 2% is needed, however, and 8% appears to be optimum for low-alloy steels. In many applications, carbon dioxide is the preferred addition because the weld bead has a better contour and the arc appears to be more stable.

Buried-arc Transfer

Carbon dioxide was one of the first gases studied during the development of the GMAW process. It was abandoned temporarily because of excessive spatter and porosity in the weld. After argon was accepted for shielding, further work with carbon dioxide demonstrated that the spatter was associated with globular metal transfer. The large drops are partially supported by arc forces, Figure 10-7. As they become heavy enough to overcome those forces and drop into the pool, they bridge the gap between the wire and the weld pool, producing explosive short circuits and spatter.

Additional work showed that the arc in carbon dioxide was very forceful. Because of this, the wire tip could be driven below the surface of the molten weld pool. With the shorter arcs, the drop size is reduced, and any spatter produced as the result of short circuits was trapped in the cavity produced by the arc. Hence, the name buried-arc transfer, Figure 10-8. The resultant welds tend to be more highly crowned than those produced with open arcs, but they are relatively free of spatter and offer a decided advantage of welding speed. These characteristics make the buried-arc process useful for high-speed mechanized welding of thin sections, such as that found in compressor domes for hermetic air-conditioning and refrigeration equipment or for automotive components.

Because carbon dioxide is an oxidizing gas, its applications to welding carbon steels are restricted. It cannot be used to fabricate most nonferrous materials.

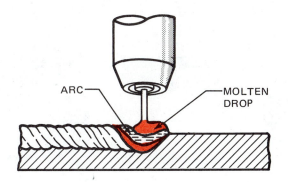

Figure 10-7 Globular metal transfer. Large drop is supported by arc forces.

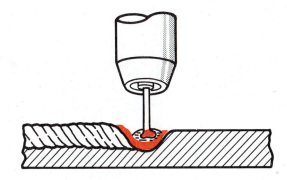

Figure 10-8 Buried-arc transfer. Wire tip is with the weld crater. Spatter is trapped.

Neither should it be used to weld stainless steels because carbon corrodes the weld metal.

Carbon dioxide and helium are similar in that metal transfer in both gases is globular. Helium, too, can be used with the buried-arc technique. It has the advantage of inertness, potentially making it useful for the same types of applications as carbon dioxide but in non-ferrous alloys.

Short-circuiting Transfer GMAW-S

Low currents allow the liquid metal at the electrode tip to be transferred by direct contact with the molten weld pool. This process requires close interaction between the wire feeder and the power supply. This technique is called the short-circuiting transfer.

The transfer mechanisms in this process are quite simple and straightforward, as shown schematically in Figure 10-9. To start, the wire is in direct contact with the molten weld pool (Figure 10-9a). Once the electrode touches the molten weld pool, the arc and its resistance are removed. Without the arc resistance, the welding amperage quickly rises as it begins to flow freely through the tip of the wire into the molten weld pool. The resistance to current flow is highest at the point where the electrode touches the molten weld pool. The resistance is high because both the electrode tip and weld pool are very hot. The higher the temperature the higher the resistance to current flow. A combination of high current flow and high resistance causes a rapid rise in the temperature of the electrode tip.

As the current flow increases, the interface between the wire and molten weld pool is heated until it explodes into a vapor (Figure 10-9b), establishing an arc. This small explosion produces sufficient force to depress the molten weld pool. A gap between the electrode tip and the molten weld pool (Figure 10-9c) immediately opens. With the resistance of the arc reestablished, the voltage increases as the current decreases.

The low current flow is insufficient to continue melting the electrode tip off as fast as it's being fed into the arc. As a result the arc length rapidly decreases (Figure 10-9d) until the electrode tip contacts the molten weld pool (Figure 10-9a). The liquid formed at the wire tip during the arc-on interval is transferred by surface tension to the molten weld pool, and the cycle begins again with another short circuit.

If the system is properly tuned, the rate of short circuiting can reach hundreds per second, causing a characteristic buzzing sound. The spatter is low and the process easy to use. The low heat produced by GMAW-S makes the system easy to use in all positions on sheet metal, low carbon steel, low alloy steel, and stainless steel ranging in thickness from 25 gauge (0.02 in.; 0.5 mm) to 12 gauge (0.1 in.; 2.6 mm). The short-circuiting process does not produce enough heat to make quality welds in sections much thicker than 1/4 in. (6 mm) unless it is used for the root pass on a grooved weld or to fill gaps in joints. Although this technique is highly effective, lack-of-fusion defects can occur unless the process is perfectly tuned and the welder is highly skilled, especially on thicker metal.

Carbon dioxide works well with this short-circuiting process because it produces the forceful arc needed during the arc-on interval to displace the weld pool. Helium can be used as well. Pure argon is not as effective because its arc tends to be sluggish and not very fluid. However, a mixture of 25% carbon dioxide and 75% argon produces a less harsh arc and a flatter, more fluid and desirable weld profile. Although more costly, this gas mixture is preferred.

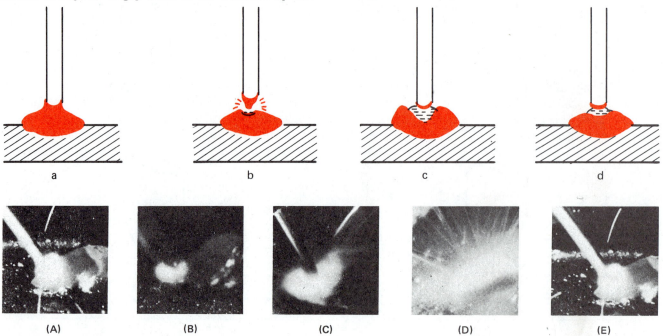

a   b   c   d

(A)   (B)   (C)   (D)   (E)

Figure 10-9 Schematic of short-circuiting transfer.

New technology in wire manufacturing has allowed smaller wire diameters to be produced. These smaller diameters have become the preferred size even though they are more expensive. The short-circuiting process works better with a short electrode stickout.

The power supply is most critical. It must have a constant potential output and sufficient inductance to slow the time rate of current increase during the short-circuit interval. Too little inductance causes spatter due to high current surges. Too much inductance causes the system to become sluggish. The short-circuiting rate decreases enough to make the process difficult to use. Also, the power supply must sustain an arc long enough to pre-melt the electrode tip in anticipation of the transfer at recontact with the weld pool.

FILLER METAL SPECIFICATIONS

GMA welding filler metals are available for a variety of metals, Table 10-4. The most frequently used filler metals are AWS specification A5.18 for carbon steel and AWS specification A5.9 for stainless steel. These filler metals are available in diameter sizes ranging from 0.023 in. (0.6 mm) to 1/8 in. (3.2 mm). Table 10-5 lists the most common sizes and the amperage ranges for these electrodes. The amperage will vary depending on the method of metal transfer, type of shielding gas, and base metal thickness.

For more information on each of the GMA welding electrodes, refer to Chapter 25, "Filler Metal Selection."

Base Metal Type	AWS Filler Metal Specification
Aluminum and aluminum alloys	A5.10
Copper and copper alloys	A5.6
Magnesium alloys	A5.19
Nickel and nickel alloys	A5.14
Stainless steel (austenitic)	A5.9
Steel (carbon)	A5.18
Titanium and titanium alloys	A5.16

Table 10-4 AWS Filler Metal Specifications for Different Base Metals.

Base Metal	Electrode Diameter Inch	Millimeter	Amperage Range
Carbon steel	0.023	0.6	35–190
	0.030	0.8	40–220
	0.035	0.9	60–280
	0.045	1.2	125–380
	1/16	1.6	275–450
Stainless steel	0.023	0.6	40–150
	0.030	0.8	60–160
	0.035	0.9	70–210
	0.045	1.2	140–310
	1/16	1.6	280–450

Table 10-5 Filler Metal Diameters and Amperage Ranges.

WIRE MELTING AND DEPOSITION RATES

The wire melting rates, deposition rates, and wire-feed speeds of the consumable wire welding processes are affected by the same variables. Before discussing them, however, these terms need to be defined. The wire melting rate, measured in inches per minute (mm/sec) or pounds (kg/hr), is the rate at which the arc consumes the wire. The deposition rate, the measure of weld metal deposited, is nearly always less than the melting rate because not all of the wire is converted to weld metal. Some is lost as slag, spatter, or fume. The amount of weld metal deposited in ratio to the wire used is called the deposition efficiency.

Deposition efficiencies depend on the process, on the gas used, and even on how the welder sets welding conditions. With efficiencies of approximately 98%, solid wires with argon shields are best. Some of the self-shielded cored wires are poorest with efficiencies as low as 80%.

Welders can control the deposition rate by changing the current, electrode extension, and diameter of the wire. To obtain higher melting rates, they can increase the current or wire extension, or decrease the wire diameter. Knowing the precise constants is unimportant. However, it is important to know that current greatly affects melting rate and that the extension must be controlled if results are to be reproducible.

WELDING POWER SUPPLIES

To better understand the terms used to describe the different welding power supplies you need to know the following electrical terms:

- Voltage or volts (V) is a measurement of electrical pressure, in the same way that pounds per square inch is a measurement of water pressure.
- Electrical potential means the same thing as voltage and is usually expressed by using the term potential (P). The terms *voltage, volts,* and *potential* can all be interchanged when referring to electrical pressure.
- Amperage or amps (A) is the measurement of the total number of electrons flowing, in the same way that gallons is a measurement of the amount of water flowing.
- Electrical current means the same thing as amperage and is usually expressed by using the term current (C). The terms *amperage, amps,* and *current* can all be interchanged when referring to electrical flow.

GMAW power supplies are the constant-voltage, constant-potential (CV, CP) -type machines, unlike SMAW power supplies, which are the constant-current (CC) -type machines and are sometimes called drooping arc voltage (DAV). It is impossible to make acceptable welds using the wrong type of power supply. Constant-voltage power supplies are available as transformer-rectifiers or as

(A)

(B)

Figure 10-10 (A) Motor-generator welding power supply. (B) Transformer-rectifier welding power supply. (A. Courtesy of Lincoln Electric, Cleveland, OH. B. courtesy of PowCon Incorporated.)

motor-generators, **Figure 10-10.** Some newer machines use electronics, enabling them to supply both types of power at the flip of a switch.

The relationships between current and voltage with different combinations of arc length or wire feed speeds are called volt-ampere characteristics. The volt-ampere characteristics of arcs in argon with constant arc-lengths or constant wire-feed speeds are shown in **Figure 10-11.** To maintain a constant arc length while increasing cur-

rent, it is necessary to increase voltage. For example, with a 1/8-in. (3-mm) arc length, increasing current from 150 to 300 amperes requires a voltage increase from about 26 to 31 volts. The current increase illustrated here results from increasing the wire-feed speed from 200 to 500 inches per minute.

By using **Table 10-6,** the amperage can be determined by knowing the wire-feed speed or the wire-feed speed can be determined by knowing the voltage. This

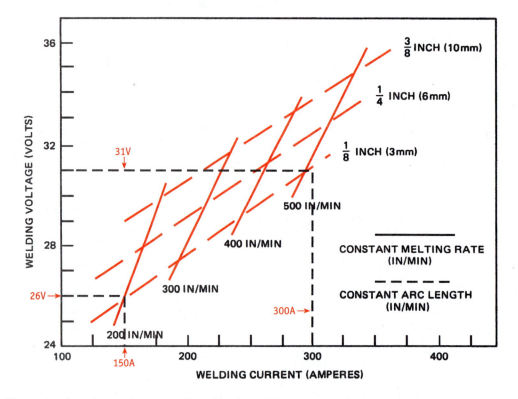

Figure 10-11 The arc length and arc voltage are affected by the welding current and wire-feed speed [0.045-in. (1.43-mm) wire; 1-in. (25-mm) electrode extension.]

Wire-Feed Speed* in/min (m/min)	Wire Diameter Amperages			
	.030 in (0.8 mm)	.035 in (0.9 mm)	.045 in (1.2 mm)	.062 in (1.6 mm)
100 (2.5)	40	65	120	190
200 (5.0)	80	120	200	330
300 (7.6)	130	170	260	425
400 (10.2)	160	210	320	490
500 (12.7)	180	245	365	—
600 (15.2)	200	265	400	—
700 (17.8)	215	280	430	—

*To check feed speed, run out wire for 1 minute and then measure its length.

Table 10-6 Typical Amperages for Carbon Steel.

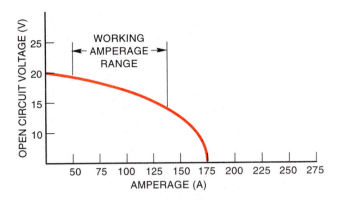

Figure 10-12 Constant potential welder slope.

relationship of amperage to wire feed speed results from the need for a higher amperage to melt the larger quantity of filler metal being introduced to the arc.

Power Supplies for Short-circuiting Transfer

Although the GMA power source is said to have a constant potential (CP), it is not perfectly constant. The graph in **Figure 10-12** shows that there is a slight decrease in voltage as the amperage increases within the working range. The rate of decrease is known as slope. It is expressed as the voltage decrease per 100-ampere increase; for example, 10 V/100 A. For short-circuiting welding, some are equipped to allow changes in the slope by steps or continuous adjustment.

The slope, which is called the *volt-ampere curve,* is often drawn as a straight line because it is fairly straight within the working range of the machine. Whether it is drawn as a curve or a straight line, the slope can be found by finding two points. The first point is the set voltage as read from the voltmeter when the gun switch is activated but no welding is being done. This is referred to as the open circuit voltage. The second point is the voltage and amperage as read during a weld. The voltage control is

not adjusted during the test but the amperage can be changed. The slope is the voltage difference between the first and second readings. The difference can be found by subtracting the second voltage from the first voltage. Therefore, for settings over 100 amperes, it is easier to calculate the slope by adjusting the wire feed so that you are welding with 100 amperes, 200 amperes, 300 amperes, and so on. In other words, the voltage difference can be simply divided by 1 for 100 amperes, 2 for 200 amperes, and so forth.

The machine slope is affected by circuit resistance. Circuit resistance may result from a number of factors, including poor connections, long leads, or a dirty contact tube. A higher resistance means a steeper slope. In short-circuiting machines, increasing the inductance increases the slope. This increase slows the current's rate of change during short circuiting and the arcing intervals, **Figure 10-13**. Therefore, slope and inductance become synonymous in this discussion. As the slope increases, both the short circuit current and pinch effect are reduced. A flat slope has both an increased short circuit current and a greater pinch effect.

The machine slope affects the short-circuiting metal transfer mode more than it does the other modes. Too much current and pinch effect from a flat slope cause a violent short and arc restart cycle, which results in increased spatter. Too little current and pinch effect from a steep slope result in the short circuit not being

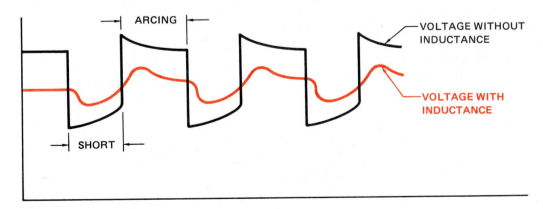

Figure 10-13 Voltage pattern with and without inductance.

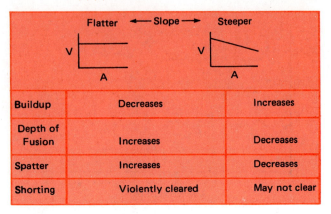

Table 10-7 Effect of Slope.

	Flatter ← Slope →	Steeper
Buildup	Decreases	Increases
Depth of Fusion	Increases	Decreases
Spatter	Increases	Decreases
Shorting	Violently cleared	May not clear

cleared as the wire freezes in the molten pool and piles up on the work, **Table 10-7.**

The slope should be adjusted so that a proper spatter-free metal transfer occurs. On machines that have adjustable slopes, this is easily set. Experiment 11-5 describes a method of adjusting the circuit resistance to change the slope on machines that have a fixed slope. This is done by varying the contact tube-to-work distance. The GMA filler wire is much too small to carry the welding current and heats up due to its resistance to the current flow. The greater the tube-to-work distance, the greater the circuit resistance and the steeper the slope. By increasing or decreasing this distance, a proper slope can be obtained so that the short circuiting is smoother with less spatter.

MOLTEN WELD POOL CONTROL

The GMAW molten weld pool can be controlled by varying the following factors: shielding gas, power settings, weave pattern, travel speed, electrode extension, and gun angle.

Shielding Gas

The shielding gas selected for a weld has a definite effect on the weld produced. The properties that can be affected include the method of metal transfer, welding speed, weld contour, arc cleaning effect, and fluidity of the molten weld pool.

In addition to the effects on the weld itself, the metal to be welded must be considered in selecting a shielding gas. Some metals must be welded with an inert gas such as argon or helium or mixtures of argon and helium.

Other metals weld more favorably with reactive gases such as carbon dioxide or with mixtures of inert gases and reactive gases such as argon and oxygen or argon and carbon dioxide, Table 10-8. The most commonly used shielding gasses are 75% argon + 25% CO_2, argon + 1% to 5% oxygen, and carbon dioxide, Figure 10-14.

- *75% argon + 25% CO_2:* This mixture is used for the short-circuiting metal transfer process on carbon and low-alloy steels. It produces welds with good wetting characteristics, little spatter, high welding speeds, and low distortion.

- *Argon + 1% to 5% oxygen:* This mixture is used for the axial spray transfer method. It produces welds with good wetting, arc stability, little undercut, high welding speeds, and minimum distortion. As the percentage of the oxygen increases, the tendency for oxidation increases. For this reason, only 1% to 2% oxygen is used on stainless steels and low-alloy steels but up to 5% oxygen can be used for carbon steels.

- *Carbon dioxide (CO_2):* This gas is used for the short-circuiting metal transfer process on carbon steel. It produces welds with deep penetration, high welding speeds, and noticeable spatter.

Power Settings

As the power settings, voltage, and amperage are adjusted, the weld bead is affected. Making an acceptable weld requires a balancing of the voltage and amperage. If either or both are set too high or too low, the weld penetration can decrease. A GMA welding machine has no direct amperage settings. Instead, the amperage at the arc is adjusted by changing the wire feed speed. As a result of the welding machine's maintaining a constant voltage when the wire feed speed increases, more amperage flows across the arc. This higher amperage is required to melt the wire so that the same arc voltage can be maintained. The higher amperage is used to melt the filler wire and does not increase the penetration. In fact, the weld penetration may decrease significantly.

Increasing and decreasing the voltage changes the arc length but may not put more heat into the weld. Like changes in the amperage, these voltage changes may decrease weld penetration.

Weave Pattern

The GMA welding process is greatly affected by the location of the electrode tip and molten weld pool.

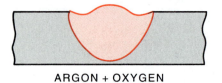
ARGON + OXYGEN

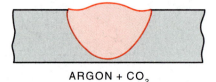
ARGON + CO₂

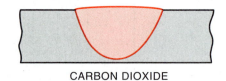
CARBON DIOXIDE

Figure 10-14 Effect of shielding gas on weld bead shape.

During the short-circuiting process if the arc is directed to the base metal and outside the molten weld pool, the welding process may stop. Without the resistance of the hot molten metal, high amperage surges occur each time the electrode tip touches the base metal, resulting in a loud pop and a shower of sparks. It's the something that occurs each time a new weld is started. So when making the weave pattern, you must keep the arc and electrode tip directed into the molten weld pool. Other than the sensitivity to arc location, most of the SMAW weave pattern can be used for GMA welds.

Travel Speed

Because the location of the arc inside the molten weld pool is important, the welding travel speed cannot exceed the ability of the arc to melt the base metal. Too high a travel speed can result in overrunning of the weld pool and an uncontrollable arc. Fusion between the base metal and filler metal can completely stop if the travel rate is too fast. If the travel rate is too slow and the weld pool size increases excessively, it can also restrict fusion to the base plate.

Electrode Extension

The electrode extension (stick-out) is the distance from the contact tube to the arc measured along the wire. Adjustments in this distance cause a change in the wire resistance and the resulting weld bead, Figure 10-15.

GMA welding currents are relatively high for the wire sizes, even for the low current values used in short-circuiting arc metal transfer, Figure 10-16. As the length of wire extending from the contact tube to the work increases, the voltage too should increase. Since this change is impossible with a constant-voltage power supply, the system compensates by reducing the current. In other words, by increasing the electrode extension and maintaining the same wire-feed speed, the current has to change to provide the same resistance drop. This situation

Metal	Shielding Gas	Chemical Reaction
Aluminum	Argon	Inert
	Argon + helium	Inert
Copper and Copper alloys	Argon	Inert
	Argon + helium	Inert
Magnesium	Argon	Inert
	Argon + helium	Inert
Nickel and Nickel alloys	Argon	Inert
	Argon + helium	Inert
Steel, carbon	Argon + oxygen	Slightly oxidizing
	Argon + carbon dioxide	Oxidizing
	Carbon dioxide	Oxidizing
Steel, low-alloy	Argon + oxygen	Slightly oxidizing
	Argon + helium + CO_2	Slightly oxidizing
	Argon + carbon dioxide	Oxidizing
Steel, stainless	Argon + oxygen	Slightly oxidizing
	Argon + helium + CO_2	Slightly oxidizing
Titanium	Argon	Inert

Table 10-8 GMAW Shielding Gases and Base Metals.

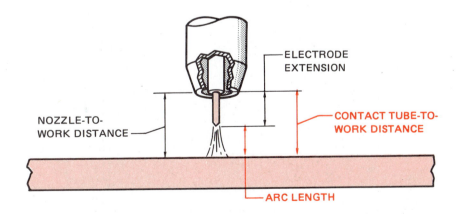

Figure 10-15

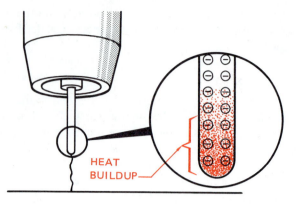

Figure 10-16 Heat buildup due to the extremely high current for the small conductor (electrode).

leads to a reduction in weld heat, penetration, and fusion, and an increase in buildup. On the other hand, as the electrode extension distance is shortened, the weld heats up and penetrates more, and builds up less, **Figure 10-17**.

Experiment 11-3 explains the technique of using varying extension lengths to change the weld characteristics. Using this technique, a welder can make acceptable welds on metal ranging in thickness from 16 gauge to 1/4 in. (6 mm) or more without changing the machine settings. When using this technique, the nozzle-to-work distance should be kept the same so that enough shielding gas coverage is provided. Some nozzles can be extended to provide coverage. Others must be exchanged with the correct length nozzle, **Figure 10-18**.

Gun Angle

The GMA welding gun may be held so that the relative angle between the gun, work, and welding bead being made is either vertical or has a drag angle or a push angle. Changes in this angle will affect the weld bead. The effect is most noticeable during the short-circuiting arc and globular transfer modes.

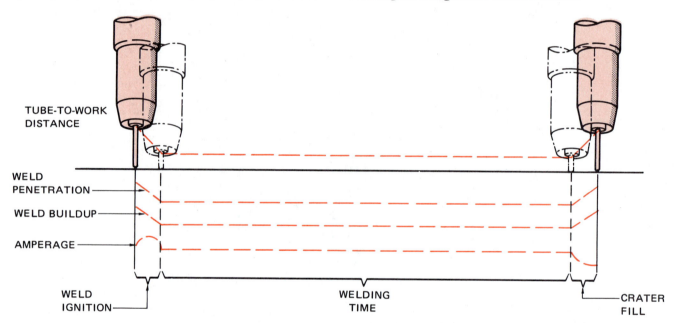

Figure 10-17 Using the changing tube-to-work distance to improve both the starting and stopping points of a weld.

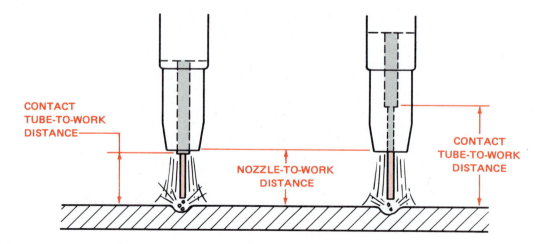

Figure 10-18

Backhand welding is the welding technique that uses a drag angle, Figure 10-19. The welding technique that uses a push angle is known as forehand welding, Figure 10-20.

Backhand Welding

A dragging angle or backhand welding technique directs the arc force into the molten weld pool of metal. This action, in turn, forces the molten metal back onto the trailing edge of the molten weld pool and exposes more of the unmelted base metal, Figure 10-21. The digging action pushes the penetration deeper into the base metal while building up the weld head. If the weld is sectioned, the profile of the bead is narrow and deeply penetrated, with high buildup. See Color Plate 10.

Forehand Welding

Using a pushing angle or forehand welding technique, the arc force pushes the weld metal forward and out of the molten weld pool onto the cooler metal ahead of the weld, Figure 10-22. The heat and metal are spread out over a wider area. The sectional profile of the bead is wide, showing shallow penetration with little buildup. See Color Plate 11.

The greater the angle, the more defined is the effect on the weld. As the angle approaches vertical, the effect is

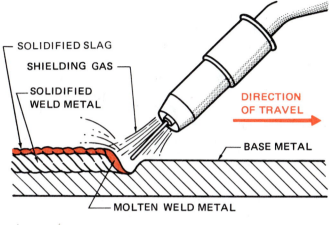

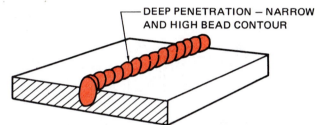

Figure 10-21 Backhand welding or dragging angle.

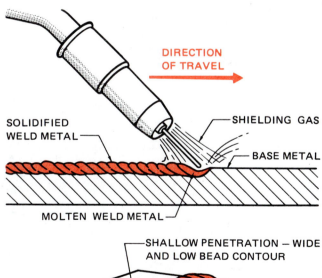

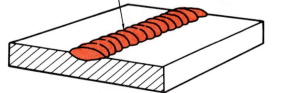

Figure 10-22 Forehand welding or pushing angle.

Figure 10-19 Backhand welding or drag angle.

Figure 10-20 Forehand welding or push angle.

reduced. This allows the welder to change the weld bead as effectively as the changes resulting from adjusting the machine current settings.

EQUIPMENT

The basic GMAW equipment consists of the gun, electrode (wire) feed unit, electrode (wire) supply, power source, shielding gas supply with flowmeter/regulator, control circuit, and related hoses, liners, and cables, **Figures 10-23** and **10-24**. Larger, more complex systems may have water for cooling, solenoids for controlling gas flow, and carriages for moving the work or the gun or both, **Figure 10-25**. The system may be stationary or portable, **Figure 10-26**. In most cases, the system is meant to be used for only one process. Some manufacturers, however, do make power sources that can be switched over for other uses.

Power Source

The power source may be either a transformer rectifier or generator type. The transformers are stationary and commonly require a three-phase power source. Engine generators are ideal for portable use or where sufficient power is not available.

The welding machine produces a DC welding current ranging from 40 amperes to 600 amperes with 10 volts to 40 volts, depending upon the machine. In the past, some GMA processes used AC welding current, but DCRP is used almost exclusively now. Typical power supplies are shown in **Figure 10-27**.

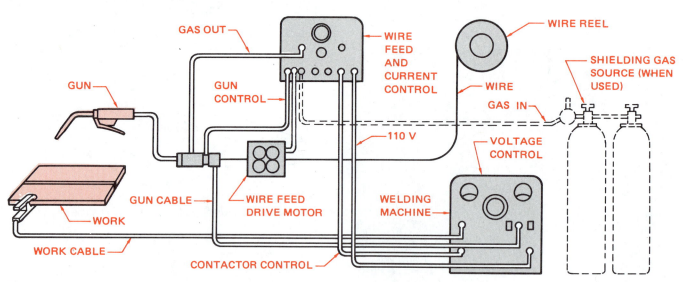

Figure 10-23 Schematic of equipment setup for GMA welding. (Courtesy of Hobart Brothers Company.)

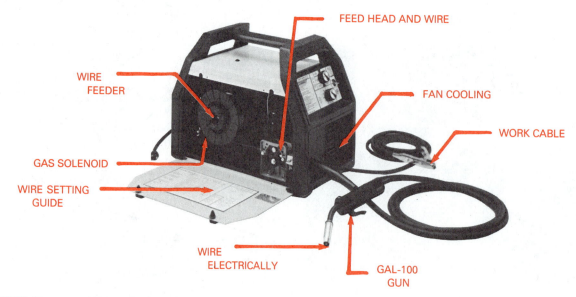

Figure 10-24 (Courtesy of Hobart Brothers Company.)

Figure 10-25 Large stationary GMA multiple welding gun setup. (Courtesy of Caterpillar Inc.)

Figure 10-26 Portable water cooler for GMA welding equipment. (Courtesy of Lincoln Electric, Cleveland, OH.)

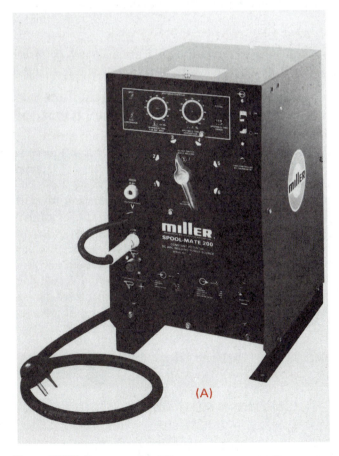

(A)

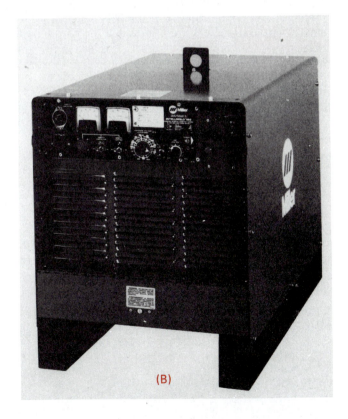

(B)

Figure 10-27 An expensive 200-ampere constant-voltage power supply (A), and a 650-ampere constant-voltage and constant-current power supply (B) for multipurpose GMAW applications. (Courtesy of Miller Electric Manufacturing Company.)

Because of the long periods of continuous use, GMA welding machines have a 100% duty cycle. This allows the machine to be run continuously without damage.

Electrode (Wire) Feed Unit

The purpose of the electrode feeder is to provide a steady and reliable supply of wire to the weld. Slight changes in the rate at which the wire is fed have distinct effects on the weld.

The motor used in a feed unit is usually a DC type that can be continuously adjusted over the desired range. **Figures 10-28** and **10-29** show typical wire feed units and accessories.

Push-type Feed System

The wire rollers are clamped securely against the wire to provide the necessary friction to push the wire through the conduit to the gun. The pressure applied on the wire can be adjusted. A groove is provided in the roller to aid in alignment and to lessen the chance of slippage. Most manufacturers provide rollers with smooth or knurled U-shaped or V-shaped grooves, **Figure 10-30.** Knurling (a series of ridges cut into the groove) helps grip larger diameter wires so that they can be pushed along more easily. Soft wires, such as aluminum, are easy to damage if knurled rollers are used. Soft wires are best used with U-grooved rollers. Even V-grooved rollers can distort the surface of the wire, causing problems. V-grooved rollers are best suited for hard wires, such as mild steel and stainless steel. It is also important to use the correct size grooves in the rollers.

Variations of the push-type electrode wire feeder include the pull type and push-pull type. The difference is in the size and location of the drive rollers. In the push-type system, the electrode must have enough strength to be pushed through the conduit without kinking. Mild steel and stainless steel can be readily pushed 15 ft (4 m) to 20 ft (6 m), but aluminum is much harder to push over 10 ft (3 m).

Pull-type Feed System

In pull-type systems, a smaller but higher speed motor is located in the gun to pull the wire through the conduit. Using this system, it is possible to move even soft wire over great distances. The disadvantages are that the gun is heavier and more difficult to use, rethreading the wire takes more time, and the operating life of the motor is shorter.

Push-Pull-type Feed System

Push-pull-type feed systems use a synchronized system with feed motors located at both ends of the electrode conduit, **Figure 10-31.** This system can be used to move any type of wire over long distances by periodically installing a feed roller into the electrode conduit. Compared to the pull-type system, the advantages of this system include moving wire over longer distances, faster rethreading, and increased motor life due to the reduced load. A disadvantage is that the system is more expensive.

Linear Electrode Feed System

Linear electrode feed systems use a different method to move the wire and change the feed speed.

(A)

(B)

Figure 10-28 (A) A 90-ampere power supply and wire feeder for welding sheet steel with carbon dioxide shielding. (B) Modern wire feeder with digital preset and readout of wire-feed speed and closed-loop control. (A and B courtesy of Miller Electric Manufacturing Company.)

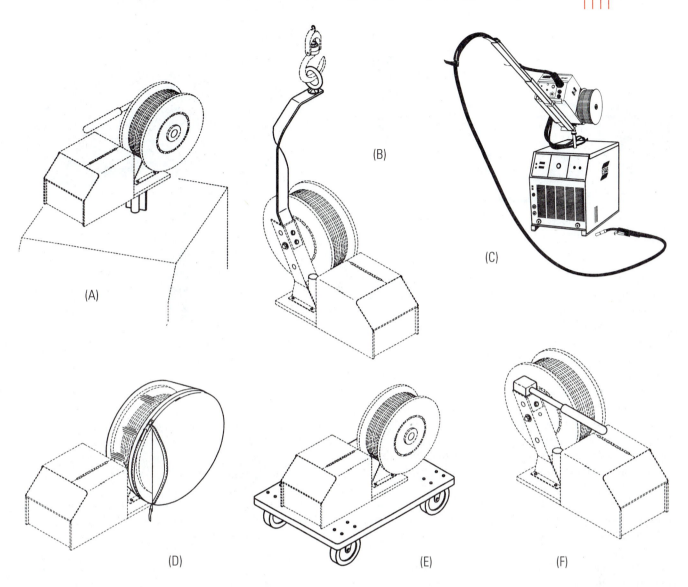

Figure 10-29 (A-F) A variety of accessories are available for most electrode feed systems. (A) Swivel post, (B) boom hanging bracket, (C) counterbalance mini-boom, (D) spool cover, (E) wire feeder wheel cart, (F) carrying handle. (Courtesy of ESAB Welding and Cutting Products.)

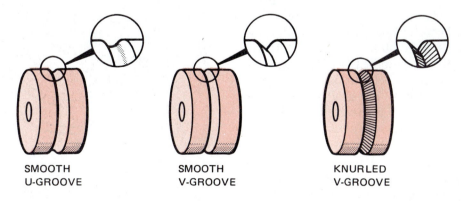

SMOOTH
U-GROOVE

SMOOTH
V-GROOVE

KNURLED
V-GROOVE

Figure 10-30 Feed rollers.

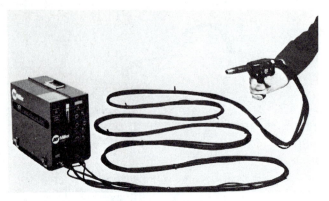

Figure 10-31 Wire-feed system that enables the wire to be moved through a longer cable. (Courtesy of Miller Electric Manufacturing Company.)

Standard systems use rollers that pinch the wire between the rollers. A system of gears is used between the motor and rollers to provide roller speed within the desired range. The linear feed system does not have gears or conventional-type rollers.

The linear feed system uses a small motor with a hollow armature shaft through which the wire is fed. The rollers are attached so that they move around the wire. Changing the roller pitch (angle) changes the speed at which the wire is moved without changing the motor speed. This system works in the same way that changing the pitch on a screw, either coarse threads or fine threads, affects the rate that the screw will move through a spinning nut.

The advantage of a linear system is that the bulky system of gears is eliminated, thus reducing weight, size, and wasted power. The motor operates at a constant high speed where it is more efficient. The reduced size allows the system to be housed in the gun or within an enclosure in the cable. Several linear wire feeders can be synchronized to provide an extended operating range. The disadvantage of a linear system is that the wire may become twisted as it is moved through the feeder.

Spool Gun

A spool gun is a compact, self-contained system consisting of a small drive system and a wire supply, Figure 10-32. This system allows the welder to move freely around a job with only a power lead and shielding gas hose to manage. The major control system is usually mounted on the welder. The feed rollers and motor are found in the gun just behind the nozzle and contact tube. Because of the short distance the wire must be moved, very soft wires (aluminum) can be used. A small spool of welding wire is located just behind the feed rollers. The small spools of wire required in these guns are often very expensive. Although the guns are small, they feel heavy when being used.

Electrode Conduit

The electrode conduit or liner guides the welding wire from the feed rollers to the gun. It may be encased in a lead that contains the shielding gas.

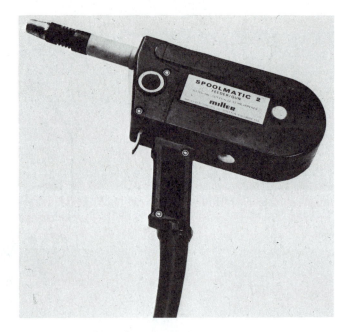

Figure 10-32 Feeder/gun for GMA welding. (Courtesy of Miller Electric Manufacturing Company.)

Power cable and gun switch circuit wires are contained in a conduit that is made of a tightly wound coil having the needed flexibility and strength. The steel conduit may have a nylon or Teflon® liner to protect soft, easily scratched metals, such as aluminum, as they are fed.

If the conduit is not an integral part of the lead, it must be firmly attached to both ends of the lead. Failure to attach the conduit can result in misalignment, which causes additional drag or makes the wire jam completely. If the conduit does not extend through the lead casing to make a connection, it can be drawn out by tightly coiling the lead, Figure 10-33. Coiling will force the conduit out so that it can be connected. If the conduit is too long for the lead, it should be cut off and filed smooth. Too long a

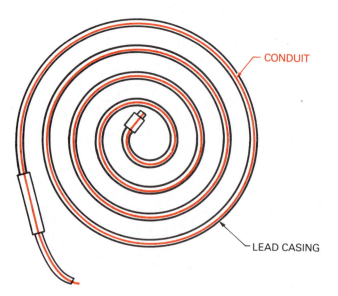

CONDUIT

LEAD CASING

Figure 10-33 Tightly coiled lead casing will force the liner out of the gun.

lead will bend and twist inside the conduit, which may cause feed problems.

Welding Gun

The welding gun attaches to the end of the power cable, electrode conduit, and shielding gas hose, **Figure 10-34.** It is used by the welder to produce the weld. A trigger switch is used to start and stop the weld cycle. The gun also has a contact tube, which is used to transfer welding current to the electrode moving through the gun, and a gas nozzle, which directs the shielding gas onto the weld, **Figure 10-35.**

SPOT WELDING

GMA can be used to make high-quality arc spot welds. Welds can be made using standard or specialized equipment, **Figure 10-36.** The arc spot weld produced by GMAW differs from electric resistance spot welding. The GMAW spot weld starts on one surface of one member and burns through to the other member, **Figure 10-37.** Fusion between the members occurs, and a small nugget is left on the metal surface.

GMA spot welding has some advantages such as the following: (1) welds can be made in thin-to-thick materials; (2) the weld can be made when only one side of the

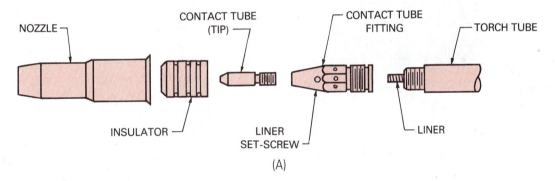

Figure 10-34 A typical GMA welding gun used for most welding processes with a heat shield attached to protect the welder's gloved hand from intense heat generated when welding with high amperages. (Courtesy of Lincoln Electric, Cleveland, OH.)

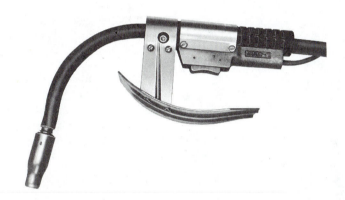

(A)

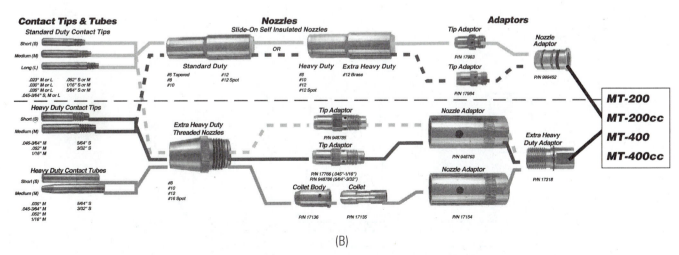

(B)

Figure 10-35 (A) Typical replaceable parts of a GMA welding gun. (Courtesy of Hobart Brothers Company.) (B) Accessories and parts selection guide for a GMA welding gun. (Courtesy of ESAB Welding and Cutting Products.)

materials to be welded is accessible; and (3) the weld can be made when there is paint on the interfacing surfaces. The arc spot weld can also be used to assemble parts for welding to be done at a later time.

Thin metal can be attached to thicker sections using an arc spot weld. If a thin-to-thick butt, lap, or tee joint is to be welded with complete joint penetration, often the thin material will burn back, leaving a hole, or there will not be enough heat to melt the thick section. With an arc spot weld, the burning back of the thin material allows

the thicker metal to be melted. As more metal is added to the weld, the burnthrough is filled, Figure 10-37.

The GMA spot weld is produced from only one side. Therefore, it can be used on awkward shapes and in cases where the other side of the surface being welded should not be damaged. This makes it an excellent process for auto body repair. In addition, because the metals are melted and the molten weld pool is agitated, thin films of paint between the members being joined need not be removed. This is an added benefit for auto body repair work.

■ CAUTION

Safety glasses and/or flash glasses must be worn to protect the eyes from flying sparks.

Specially designed nozzles provide flash protection, part alignment, and arc alignment, Figure 10-38. As a result, for some small jobs it may be is possible to perform the weld with only safety glasses. Welders can shut their eyes and turn their head during the weld.

■ CAUTION

This is not advisable for any work requiring more than just a few spot welds. Prolonged exposure to the reflected ultraviolet light will cause skin burns.

The optional control timer provides weld time and burn-back time. To make a weld, the amperage, voltage, and length of welding time must be set correctly. The burn-back time is a short period at the end of the weld when the wire feed stops but the current does not. This allows the wire to be burned back so it does not stick in the weld, Figure 10-37.

Figure 10-36 GMA spot welding machine. (Courtesy of Miller Electric Manufacturing Co.)

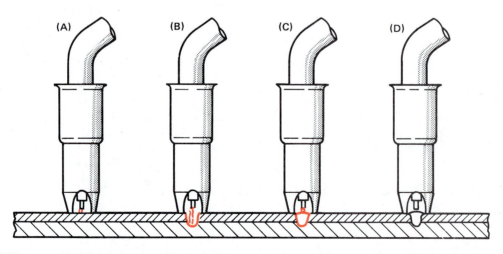

Figure 10-37 GMA spot weld: (A) the arc starts, (B) a hole is burned through the first plate, (C) the hole is filled with weld metal, and (D) the wire feed stops and the arc burns the electrode back.

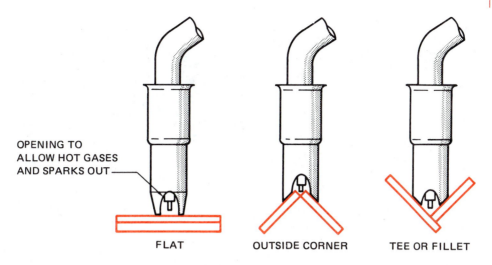

OPENING TO
ALLOW HOT GASES
AND SPARKS OUT

FLAT   OUTSIDE CORNER   TEE OR FILLET

Figure 10-38

Welding Repairs a Worn-Out Giant

Big Muskie, a 26-million-pound walking dragline, is one of the largest machines on earth, **Figure 1**. This giant removes overburden — dirt, rock, and other debris — from the top of coal seams at a large surface mine.

Wired into a local electrical power substation, Big Muskie's eight electric motors draw an average load of 17,000 kW as its 220-cubic-yard-capacity bucket bites out 325-ton loads of earth. Maximum power for the dragline is 63,000 hp. With its 310-foot boom the giant can hoist loads from 180 feet below surface to the height of a 30-story building. Air circulates at 1 million ft³/min to ventilate the dragline house — 140 feet long, 120 feet wide, and 40 feet high. The big machine can move 3 million cubic yards of overburden each month.

However, work had taken its toll on Big Muskie — the dragline base, a cylindrical tub on which the excavating boom turns, needed replacement. In the foreground of **Figure 1**, tractors stand ready to pull the old base from the site. The replacement base is hidden behind the dragline.

The designer and manufacturer of the dragline came to the mine site with 36 prefabricated wedge-shaped segments, **Figure 2**, and a central cylindrical core for the new base. The steel plate segments and core weighed 3 million pounds. Stiffeners honeycombed the segments. The plate

Figure 1

used was mostly 1 to 1-1/4 inches thick, with heavier sections for the base plate (2 inches) and the circular rail pad (12 inches wide, 6 inches thick). The rail pad encircles the top of the base, and the wheels of the boom ride on the rail pad. **Figure 3** shows a typical 6-inch-deep V-joint (indicated by the folding rule), which was filled with weld metal to join the rail pad sections.

Two shifts of field welders, working for a full year, constructed the new base using the flux-

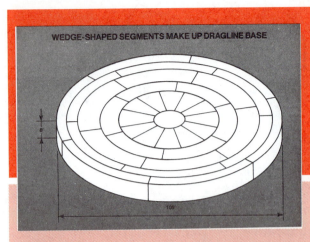

WEDGE-SHAPED SEGMENTS MAKE UP DRAGLINE BASE

Figure 2

Figure 4

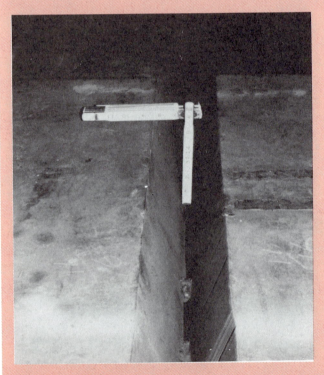

Figure 3

cored arc welding process. They first connected the segments with blocking welds that completely filled the joints for lengths of 18 to 24 inches. Blocking welds accounted for 25% of the welding on the project.

Starting from the central core, welders added segments radially outward. This sequence allowed the material to shrink uniformly around the center as welds cooled, thus reducing stress and distortion. Inspectors checked the large number of welds to assure that beads filled all required joints.

To help maintain a preheat of 250°F and to protect against quenching of welds by rain, welders worked inside an air-inflated tent. Fume extractors cleaned the air in confined work stations inside the segments. Air between the walls of the double-skinned tent insulated the working area for year-round comfort.

For flat position joints, the welders used flux-cored wire. CO_2 gas shielded joints, while guns running at 450 amperes deposited beads. Thinner wire filled out-of-position joints using 75 Ar-25 CO_2 shielding gas at a current of 185 amperes. Sixty percent of the welding was out of position. Most joints were single-bevel or single-V-groove with a 1/8-inch land and 1/4-inch root.

The rail pad and its vertical support beam take the brunt of operational stress. Inspectors checked welds on these components with ultrasonics. Stress relief was 1,150°F for six hours.

After welders completed the new base, nine tractors pulled the assembly from its concrete work pad onto a second pad. While the base spanned the pads, welders in a trench made repairs to the underside of the 3-million-pound base, backgouging and rewelding defective joints.

Little by little, the dragline worked its way over to the site and straddled the original work pad. Using its hydraulically operated walking legs, the dragline lifted itself, allowing workers to remove the worn base from the dragline. The tractors then pulled the old base out, and moved the new base under Big Muskie, in position for installation. Crews attached the new base, **Figure 4.** Big Muskie then walked itself back to work on the coal field.

Engineers selected flux-cored arc welding for its high deposition rate. The repair took 40,000 labor hours compared to an estimated 160,000 hours to do the work with shielded metal arc.

Before welders went to work on Big Muskie, they passed qualifying tests on 1-inch-thick plate in the flat, vertical, and overhead positions, qualifying with field welding equipment and electrodes.

(Adapted from information provided by Alloy Rods Corporation, Hanover, Pennsylvania 17331.)

REVIEW

1. Why is usage of the term *GMAW* preferable to *MIG* for gas metal arc welding?

2. Using Table 10-1, answer the following:
   a. What maintains the arc in machine welding?
   b. What feeds the filler metal in manual welding?
   c. What provides the joint travel in automatic welding?
   d. What provides the joint guidance in semiautomatic welding?

3. What factors have led to increased usage of the GMAW process?

4. In what form is metal transferred across the arc in the axial spray metal transfer method of GMA welding?

5. What three conditions are required for the spray transfer process to occur?

6. Using Table 10-2, answer the following:
   a. What should the wire-feed speed and voltage ranges be to weld 1/8-in. metal with 0.035-in. wire using argon shielding gas?
   b. What should the amperage and voltage range be using 98% Ar 2% O_2 to weld 1/4-in. metal with 0.045-in. wire?

7. What ranges does the pulsed-arc metal transfer shift between?

8. How do frequency, amplitude, and width of the pulses affect the GMAW pulse welding process?

9. How have electronics helped the pulsed-arc process?

10. Why is helium added to argon when making some spray or pulsed-spray transfer welds?

11. Why does DCEP help with welds on metals such as aluminum?

12. Why is CO_2 added to argon when making GMA spray transfer welds?

13. Why should CO_2 not be used to weld stainless steel?

14. How is the metal transferred from the electrode to the plate during the GMAW-S process?

15. Using Figure 10-17, what is the approximate voltage at 100 amps?

16. Using Table 10-6, what would the amperage be for 0.035-in. (0.9-mm) wire at 200 in./min (5 m/min)?

17. Using Table 10-8, what shielding gas should be used for welding on titanium?

18. What may happen if the GMA welding electrode is allowed to strike the base metal outside of the molten weld pool?

19. What effect does shortening the electrode penetration have on weld penetration?

20. Describe the weld produced by a backhand welding angle.

21. Describe the weld produced by a forehand welding angle.

22. What components make up a GMA welding system?

23. Why must GMA welders have a 100% duty cycle?

24. What can happen if rollers of the wrong shape are used on aluminum wire?

25. Where is the drive motor located in a pull-type wire-feed system?

26. How is the wire-feed speed changed with a linear-feed system?

27. What type of liner should be used for aluminum wire?

28. What parts of a typical GMA welding gun can be replaced?

29. Describe the spot welding process using a GMA welder.

Chapter 11

GAS METAL ARC WELDING

OBJECTIVES

After completing this chapter, the student should be able to

■ set up a constant potential, semiautomatic arc welding unit.

■ make satisfactory welds in all positions using the short-circuiting metal transfer method.

■ make satisfactory welds in the 1F, 2F, and 1G positions using the pulsed-arc metal transfer method.

■ make satisfactory welds in the 1F and 1G positions using the axial spray metal transfer method.

KEY TERMS

bird nesting	conduit liner
contact tube	spool drag
cast	flow rate
feed rollers	wire feed speed

INTRODUCTION

Performing a satisfactory GMA weld requires more than just manipulative skill. The setup, voltage, amperage, electrode extension, and welding angle, as well as other factors, can dramatically affect the weld produced. The very best welding conditions are those that will allow a welder to produce the largest quantity of successful welds in the shortest period of time with the highest productivity. Because these are semiautomatic or automatic processes, increased productivity may require only that the welder increase the travel speed and current. This does not mean that the welder will work harder, but rather that the welder will work more productively, resulting in a greater cost efficiency.

The more cost efficient welders can be, the more competitive they and their companies become. This can make the difference between being awarded a bid or a job and having or losing work.

SETUP

The same equipment may be used for semiautomatic GMAW, FCAW, and SAW. Often, FCAW and SAW equipment will have a higher amperage range. In addition, equipment for FCAW and SAW are more likely to be automated than that for GMAW. However, GMA welding equipment can be automated easily.

The basic GMAW installation consists of the following: welding gun, gun switch circuit, electrode conduit-welding contractor control, electrode feed unit, electrode supply, power source, shielding gas supply, shielding gas flowmeter regulator, shielding gas hoses, and both power and work cables. Typical

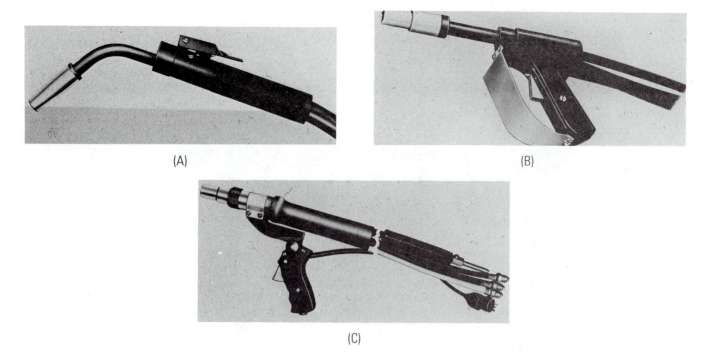

(A)

(B)

(C)

Figure 11-1   (A) Air-cooled GMA welding gun; (B) water-cooled GMA welding gun; (C) water-cooled GMA welding nozzle. (Courtesy of Miller Electric Manufacturing Company.)

water-cooled and air-cooled guns are shown in **Figure 11-1.** The equipment setup in this chapter is similar to equipment built by other manufacturers, which means that any skills developed can be transferred easily to other equipment.

PRACTICE 11-1

GMAW Equipment Setup

For this practice, you will need a GMAW power source, welding gun, electrode feed unit, electrode supply, shielding gas supply, shielding gas flowmeter regulator, electrode conduit, power and work leads, shielding gas hoses, assorted hand tools, spare parts, and any other required materials. In this practice, you will properly set up a GWA welding installation.

If the shielding gas supply is a cylinder, it must be chained securely in place before the valve protection cap is removed, **Figure 11-2.** Standing to one side of the cylinder, quickly crack the valve to blow out any dirt in the valve before the flowmeter regulator is attached, **Figure 11-3.** Attach the correct hose from the regulator to the "gas-in" connection on the electrode feed unit or machine.

Install the reel of electrode (welding wire) on the holder and secure it, **Figure 11-4.** Check the feed roller size to ensure that it matches the wire size, **Figure 11-5.** The conduit liner size should be checked to be sure that it is compatible with the wire size. Connect the conduit to the feed unit. The conduit or an extension should be aligned with the groove in the roller and set as close to

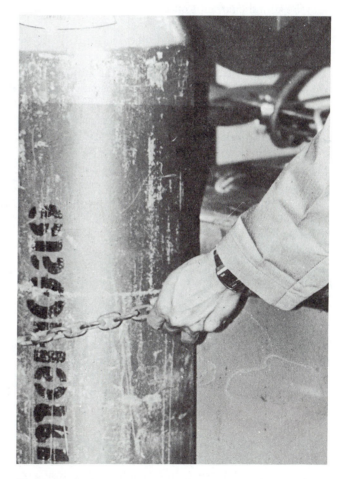

Figure 11-2   Make sure the gas cylinder is chained securely in place before removing the safety cap.

Figure 11-3 Attach the flowmeter regulator. Be sure the tube is vertical.

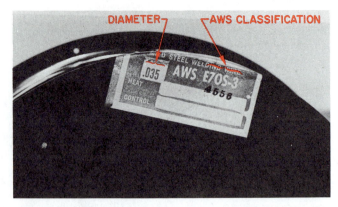

Figure 11-4 When installing the spool of wire, check the label to be sure that the wire is the correct type and size.

the roller as possible without touching, **Figure 11-6.** Misalignment at this point can contribute to a bird's nest, **Figure 11-7.** **Bird nesting** of the electrode wire results when the feed roller pushes the wire into a tangled ball because the wire would not go through the outfeed side conduit, and appears to look like a bird's nest.

Be sure the power is off before attaching the welding cables. The electrode and work leads should be attached to the proper terminals. The electrode lead should be attached to electrode or positive (+). If necessary, it is also attached to the power cable part of the gun lead. The work lead should be attached to work or negative (–).

The shielding "gas-out" side of the solenoid is then also attached to the gun lead. If a separate splice is required from the gun switch circuit to the feed unit, it should be connected at this time, **Figure 11-8A, 11-8B,** and **11-8C.** Check to see that the welding contractor circuit is connected from the feed unit to the power source.

The welding gun should be securely attached to the main lead cable and conduit, **Figure 11-9.** There should be a gas diffuser attached to the end of the conduit liner to ensure proper alignment. A **contact tube** (tip) of the correct size to match the electrode wire size being used should be installed, **Figure 11-10.** A shielding gas nozzle is attached to complete the assembly.

Recheck all fittings and connections for tightness. Loose fittings can leak; loose connections can cause added resistance, reducing the welding efficiency. Some manufacturers include detailed setup instructions with their equipment, **Figure 11-11.** ◆

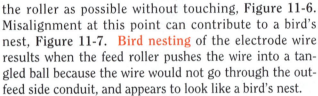

PRACTICE 11-2

Threading GMAW Wire

Using the GMAW machine that was properly assembled in Practice 11-1, you will turn the machine on and thread the electrode wire through the system.

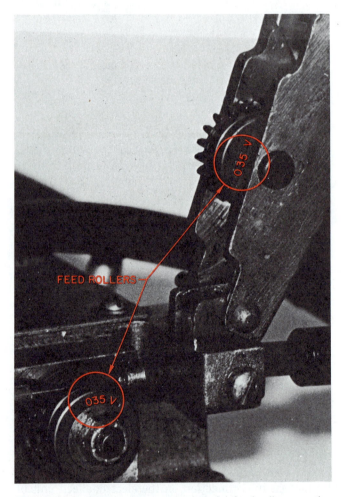

Figure 11-5 Check to be certain that the feed rollers are the correct size for the wire being used.

Check to see that the unit is assembled correctly according to the manufacturer's specifications. Switch on the power and check the gun switch circuit by

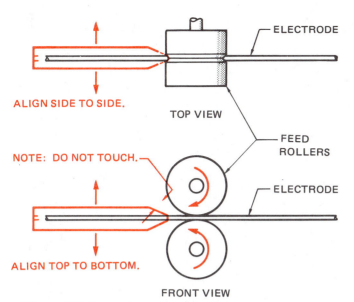

ELECTRODE

ALIGN SIDE TO SIDE.

TOP VIEW

FEED ROLLERS

NOTE: DO NOT TOUCH.

ELECTRODE

ALIGN TOP TO BOTTOM.

FRONT VIEW

Figure 11-6 Feed roller and conduit alignment.

Figure 11-7 "Bird's nest" in the filler wire at the feed rollers.

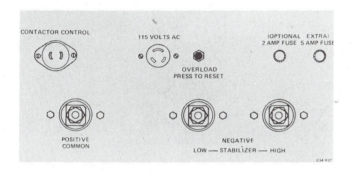

CONTACTOR CONTROL   115 VOLTS AC   (OPTIONAL EXTRA) 2 AMP FUSE 5 AMP FUSE

OVERLOAD PRESS TO RESET

POSITIVE COMMON   NEGATIVE LOW — STABILIZER — HIGH

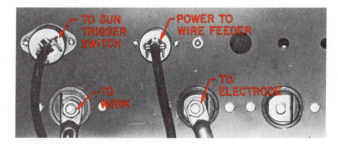

TO GUN TRIGGER SWITCH   POWER TO WIRE FEEDER

TO WORK   TO ELECTRODE

(A)

GUN TRIGGER SWITCH PLUG CONNECTOR

SHIELDING GAS TO GUN

(B)

SHIELDING GAS FROM WIRE FEEDER TO GUN

(C)

Figure 11-8 Connect the leads and other lines as shown in the owner's manual.

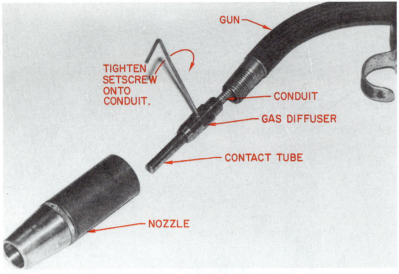

Figure 11-9

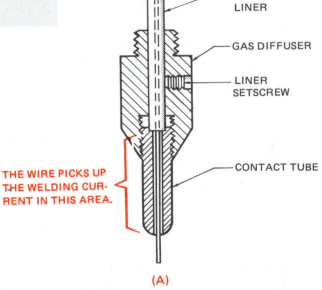

THE WIRE PICKS UP THE WELDING CURRENT IN THIS AREA.

CONDUIT LINER

GAS DIFFUSER

LINER SETSCREW

CONTACT TUBE

(A)

NOTE: ARC SPOTS ON WIRE MAGNIFIED 100 TIMES

(B)

NOTE: CONTACT TUBE TIP MELTED BY HEAT

(C)

Figure 11-10 (A) The contact tube must be the correct size. Too small a contact tube will cause the wire to stick. (B) Too large a contact tube can cause arcing to occur between she wire and tube. (C) Heat from the arcing can damage the tube. (Photo B Courtesy of Brett V. Hahn.)

1
Open side using easy "swell" latches.

2
Remove wire spool nut. Unload wire spool, remove protective packaging, reload wire spool with free end unreeling from bottom, left to right. Reattach wire spool nut.

3
Release upper feed roller.

4
Thread wire through guide between rollers and into wire cable.

5
Clamp upper feed roller.

6
For spec. 7144-1, plug unit into a 120 VAC grounded wall outlet.

7
Turn input switch on. Set weld voltage range switch to "1". Advance wire to end of cable with gun trigger. **WARNING:** Wire is electrically hot when trigger is pulled.

8
Set voltage and wire feed speed using wire settings guide on inside of case.

9
Attach work cable clamp to work to be welded.

10
Connect gas to coupling at rear of case.

11
For gas-shielded solid wire, be sure #B cable is attached to negative (−) terminal and #A cable to positive (+). Unit is shipped this way. To use Handler with self-shielding tubular wire, reverse polarity by connecting #A to negative (−) and #B to positive (+) terminal. Refer to guide inside case. See photo #8 above.

12
ALWAYS WEAR PROPER SAFETY EQUIPMENT. Pull trigger and weld.

Figure 11-11   Example of manufacturer's setup instructions. (Courtesy of Hobart Brothers Company.)

depressing the switch. The power source relays, feed relays, gas solenoid, and feed motor should all activate.

Cut the end of the electrode wire free. Hold it tightly so that it does not unwind. The wire has a natural curve that is known as its cast. The cast is measured by the diameter of the circle that the wire would make if it were loosely laid on a flat surface, Figure 11-12. The cast helps the wire make a good electrical contact as it passes through the contact tube, Figure 11-13. However, the cast can be a problem when threading the system. To make threading easier, straighten about 12 in. (305 mm) of the end of the wire and cut any kinks off.

Separate the wire feed rollers, and push the wire first through the guides, then between the rollers, and finally into the conduit liner, Figure 11-14. Reset the rollers so there is a slight amount of compression on the wire, Figure 11-15. Set the wire feed speed control to a slow speed. Hold the welding gun so that the electrode conduit and cable are as straight as possible.

Press the gun switch. The wire should start feeding into the liner. Watch to make certain that the wire feeds smoothly and release the gun switch as soon as the end comes through the contact tube. ◆

■ CAUTION

If the wire stops feeding before it reaches the end of the contact tube, stop and check the system. If no obvious problem can be found, mark the wire with tape and remove it from the gun. It then can be held next to the system to determine the location of the problem.

With the wire feed running, adjust the feed roller compression so that the wire reel can be stopped easily by a slight pressure. Too light a roller pressure will cause the wire to feed erratically. Too high a pressure can turn a minor problem into a major disaster. If the wire jams at a high roller pressure, the feed rollers keep feeding the wire, causing it to bird nest and possibly short out. With a light pressure, the wire can stop, preventing bird nesting. This is very important with soft wires. The other advantage of a light pressure is that the feed will stop if something like clothing or gas hoses are caught in the reel.

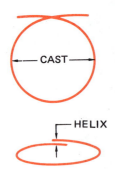

Figure 11-12

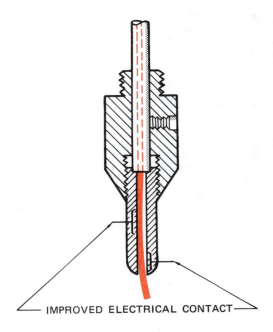

Figure 11-13 Cast forces the wire to make better electrical contact with the tube.

Figure 11-14 Push the wire through the guides by hand.

Figure 11-15 Adjust the wire feed tensioner.

With the feed running, adjust the spool drag so that the reel stops when the feed stops. The reel should not coast to a stop because the wire can be snagged easily. Also, when the feed restarts, a jolt occurs when the slack in the wire is taken up. This jolt can be enough to momentarily stop the wire, possibly causing a discontinuity in the weld.

When the test runs are completed, the wire can either be rewound or cut off. Some wire feed units have a retract button. This allows the feed driver to reverse and retract the wire automatically. To rewind the wire on units without this retract feature, release the rollers and turn them backward by hand. If the machine will not allow the feed rollers to be released without upsetting the tension, you must cut the wire.

■ CAUTION

Do not discard pieces of wire on the floor. They present a hazard to safe movement around the machine. In addition, a small piece of wire can work its way into a filter screen on the welding power source. If the piece of wire shorts out inside the machine, it could become charged with high voltage, which could cause injury or death. Always wind the wire tightly into a ball or cut it into short lengths before discarding it in the proper waste container.

GAS DENSITY AND FLOW RATES

Density is the chief determinant of how effective a gas is for arc shielding. The lower the density of a gas the higher will be the flow rate required for equal arc protection. Flow rates, however, are not in proportion to the densities. Helium, with about one-tenth the density of argon, requires only twice the flow for equal protection.

EXPERIMENT 11-1

Setting Gas Flow Rate

Using the equipment setup as described in Practice 11-1, and the threaded machine as described in Practice 11-2, you will set the shielding gas flow rate.

The exact flow rate required for a certain job will vary depending upon welding conditions. This experiment will help you determine how those conditions affect the flow rate. You will start by setting the shielding gas flow rate at 35 cfh (16 L/min).

Turn on the shielding gas supply valve. If the supply is a cylinder, the valve is opened all the way. With the machine power on and the welding gun switch depressed, you are ready to set the flow rate. Slowly turn in the adjusting screw and watch the float ball as it rises in a tube

on a column of gas. The faster the gas flows, the higher the ball will float. A scale on the tube allows you to read the flow rate. Different scales are used with each type of gas being used. Since various gases have different densities (weights), the ball will float at varying levels even though the flow rates are the same, **Figure 11-16**. The line corresponding to the flow rate may be read as it compares to the top, center, or bottom of the ball, depending upon the manufacturer's instructions. There should be some marking or instruction on the tube or regulator to tell a person how it should be read, **Figure 11-17**.

Release the welding gun switch, and the gas flow should stop. Turn off the power and spray the hose fittings with a leak-detecting solution.

When stopping for a period of time, the shielding gas supply valve should be closed and the hose pressure released. ◆

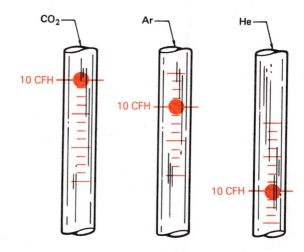

Figure 11-16 Each of these gases is flowing at the same cfh (L/min) rate. Because helium (He) is less dense, its indicator ball is the lowest. Be sure that you are reading the correct scale for the gas being used.

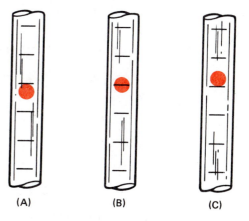

Figure 11-17 Three methods of reading a flowmeter: (A) top of ball, (B) center of ball, and (C) bottom of ball.

ARC-VOLTAGE AND AMPERAGE CHARACTERISTICS

The arc-voltage and amperage characteristics of GMA welding are different from most other welding processes. The voltage is set on the welder, and the amperage is set by changing the wire feed speed. At any one voltage setting the amperage required to melt the wire must change as it is fed into the weld. It requires more amperage to melt the wire the faster it is fed, and less the slower it is fed.

Because changes in the wire feed speed directly change the amperage, it is possible to set the amperage by using a chart and measuring the length of wire fed per minute, Table 11-1. The voltage and amperage required for a specific metal transfer method differ for various wire sizes, shielding gases, and metals.

The voltage and amperage setting will be specified for all welding done according to a welding procedure specification (WPS) or other codes and standards. However, most welding—like that done in small produc-

tion shops, as maintenance welding, for repair work, in farm shops, and the like—is not done to specific code or standard and therefore no specific setting exists. For that reason, it is important to learn to make the adjustments necessary to allow you to produce quality welds.

EXPERIMENT 11-2

Setting the Current

Using a properly assembled GMA welding machine, proper safety protection, and one piece of mild steel plate approximately 12 in. (305 mm) long x 1/4 in. (6 mm) thick, you will change the current settings and observe the effect on GMAW.

On a scale of 0 to 10, set the wire feed speed control dial at 5, or halfway between the low and high settings of the unit. The voltage is also set at a point halfway between the low and high settings. The shielding gas can be CO_2, argon, or a mixture. The gas flow should be adjusted to a rate of 35 cfh (16 L/min).

Hold the welding gun at a comfortable angle, lower your welding hood, and pull the trigger. As the wire feeds and contacts the plate, the weld will begin. Move the gun slowly along the plate. Note the following welding conditions as the weld progresses: voltage, amperage, weld direction, metal transfer, spatter, molten weld pool size, and penetration. Stop and record your observations in Table 11-2. Evaluate the quality of the weld as acceptable or unacceptable.

Reduce the voltage somewhat and make another weld, keeping all other weld variables (travel speed, stickout, direction, amperage) the same. Observe the weld and upon stopping record the results. Repeat this procedure until the voltage has been lowered to the minimum value indicated on the machine. Near the lower end the wire may stick, jump, or simply no longer weld.

Return the voltage indicator to the original starting position and make a short test weld. Stop and compare the results to those first observed. Then slightly increase the voltage setting and make another weld. Repeat the procedure of observing and recording the results as the voltage is increased in steps until the maximum machine

Wire Feed Speed* in/min (m/min)	Wire Diameter			
	.030 in (0.8 mm)	.035 in (0.9 mm)	.045 in (1.2 mm)	.062 in (1.6 mm)
100 (2.5)	40	65	120	190
200 (5.0)	80	120	200	330
300 (7.6)	130	170	260	425
400 (10.2)	160	210	320	490
500 (12.7)	180	245	365	—
600 (15.2)	200	265	400	—
700 (17.8)	215	280	430	—

*To check feed speed, run out wire for 1 minute and then measure its length.

Table 11-1 Typical Amperages for Carbon Steel.

Weld Acceptability	Voltage	Amperage	Spatter	Molten Pool Size	Penetration
Good	20	75	Light	Small	Little

Electrode Diameter .035 in (0.9 mm)
Shielding Gas CO_2
Welding Direction Backhand

Table 11-2 Setting the Current.

capability is obtained. Near the maximum setting the spatter may become excessive if CO_2 shielding gas is used. Care must be taken to prevent the wire from fusing to the contact tube.

Return the voltage indicator again to the original starting position and make a short test weld. Compare the results observed with those previously obtained.

Lower the wire feed speed setting slightly and use the same procedure as before. First lower and then raise the voltage through a complete range and record your observations. After a complete set of test results is obtained from this amperage setting, again lower the wire feed speed for a new series of tests. Repeat this procedure until the amperage is at the minimum setting shown on the machine. At low-amperages and high-voltage settings, the wire may tend to pop violently as a result of the uncontrolled arc.

Return the wire feed speed and voltages to the original settings. Make a test weld and compare the results with the original tests. Slightly raise the wire speed and again run a set of tests as the voltage is changed in small steps. After each series, return the voltage setting to the starting point and increase the wire feed speed. Make a new set of tests.

All of the test data can be gathered into an operational graph for the machine, wire type, size, and shielding gas. Use Table 11-3 to plot the graph. The acceptable welds should be marked on the lines that extend from the appropriate voltages and amperages. Upon completion, the graph will give you the optimum settings for the operation of this particular GMAW setup. The optimum settings are along a line in the center of the acceptable welds.

Experienced welders will follow a much shorter version of this type of procedure any time they are starting to work on a new machine or testing for a new job. This experiment can be repeated using different types of wire, wire sizes, shielding gases, and weld directions. Turn off the welding machine and shielding gas and clean up your work area when you are finished welding. ◆

ELECTRODE EXTENSION

Because of the constant-potential (CP) power supply, the welding current will change as the distance between the contact tube and the work changes. Although this change is slight, it is enough to affect the weld being produced. The longer the electrode extension the greater the resistance to the welding current flowing through the small welding wire. This results in some of the welding current being changed to heat at the tip of the electrode, Figure 11-18. With a standard SMA welding CC power supply this would also reduce the arc voltage, but with a CP power supply the voltage remains constant and the amperage increases. If the electrode extension is shortened, the welding current decreases.

The increase in current does not result in an increase in penetration, because the current is being

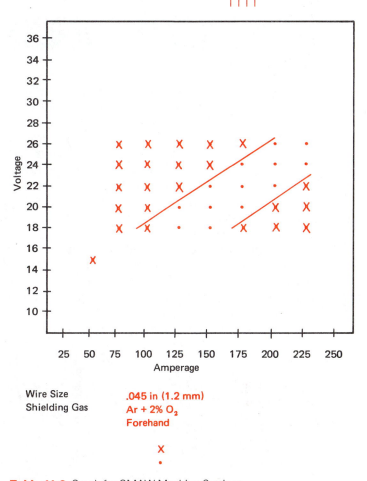

Wire Size   .045 in (1.2 mm)
Shielding Gas   Ar + 2% O_2
                 Forehand

X
•

Table 11-3 Graph for GMAW Machine Settings.

used to heat the electrode tip and not being transferred to the weld metal. Penetration is reduced and buildup is increased as the electrode extension is shortened. Penetration is increased and buildup decreased as the electrode extension is lengthened. Controlling the weld penetration and buildup by changing the electrode will help maintain weld bead shape during welding. It will also help you better understand what may be happening if a weld starts out correctly but begins to change as it progresses along the joint. You may be changing the electrode extension without noticing the change.

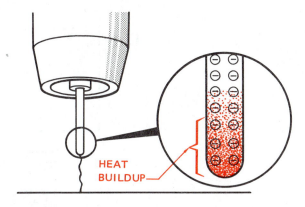

HEAT BUILDUP

Figure 11-18 Heat buildup due to the extremely high current for the small conductor (electrode).

EXPERIMENT 11-3

Electrode Extension

Using a properly assembled GMA welding machine, proper safety protection, and a few pieces of mild steel, each about 12 in. (305 mm) long and ranging in thickness from 16 gauge to 1/2 in. (13 mm), you will observe the effect of changing electrode extension on the weld.

Start at a low current setting. Using the graph developed in Experiment 11-2, set both the voltage and amperage. The settings should be equal to those on the optimum line established for the wire type and size being used with the same shielding gas.

Holding the welding gun at a comfortable angle and height, lower your helmet and start to weld. Make a weld approximately 2 in. (51 mm) long. Then reduce the distance from the gun to the work while continuing to weld. After a few inches, again shorten the electrode extension. Keep doing this in steps until the nozzle is as close as possible to the work. Stop and return the gun to the original starting distance.

Repeat the process just described but now increase the electrode extension in steps of a few inches each. Keep increasing the electrode extension until the weld will no longer fuse or the wire becomes impossible to control.

Change the plate thickness and repeat the procedure. When the series has been completed with each plate thickness, raise the voltage and amperage to a medium setting and repeat the process again. Upon completing this series of tests, adjust the voltage and amperage upward to a high setting. Make a full series of tests using the same procedures as before.

Record the results in Table 11-4 after each series of tests. The final results can be plotted on a graph, as was done in Table 11-5, to establish the optimum electrode extension for each thickness, voltage, and amperage. Turn off the welding machine and shielding gas and clean up your work area when you are finished welding. ◆

Weld Acceptability	Voltage	Amperage	Electrode Extension	Contact Tube-to-Work Distance	Bead Shape
Poor	20	100	1 in (25 mm)	1 1/4 in (31 mm)	Narrow, high with little penetration

Electrode Diameter .035 in (0.9 mm)
Shielding Gas CO_2
Welding Direction Forehand

Table 11-4 Electrode Extension.

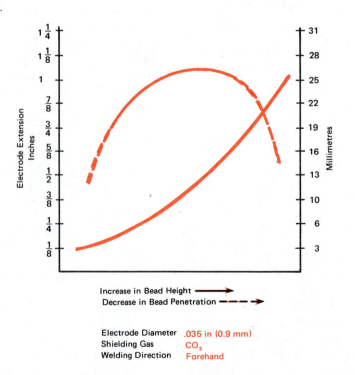

Increase in Bead Height ⟶
Decrease in Bead Penetration ⤏

Electrode Diameter .035 in (0.9 mm)
Shielding Gas CO_2
Welding Direction Forehand

Table 11-5

WELDING GUN ANGLE

The term *welding gun angle* refers to the angle between the GMA welding gun and the work as it relates to the direction of travel. Backhand welding or dragging angle, **Figure 11-19**, produces a weld with deep penetration and higher buildup. Forehand welding or pushing angle, **Figure 11-20**, produces a weld with shallow penetration and little buildup.

Slight changes in the welding gun angle can be used to control the weld as the groove spacing changes. A narrow gap may require more penetration, but as the gap spacing increases a weld with less penetration may be required. Changing the electrode extension and welding gun angle at the same time can result in a quality weld being made with less than ideal conditions.

EXPERIMENT 11-4

Welding Gun Angle

Using a properly assembled GMA welding machine, proper safety protection, and some pieces of mild steel, each approximately 12 in. (305 mm) long and ranging in thickness from 16 gauge to 1/2 in. (13 mm), you will observe the effect of changing the welding gun angle on the weld bead.

Starting with a medium current setting and a plate that is 1/4 in. (6 mm) thick, hold the welding gun at a 30° angle to the plate in the direction of the weld, **Figure 11-21**. Lower your welding hood and depress the trigger. When the weld starts, move in a straight line and slowly pivot the gun angle as the weld progresses. Keep the travel speed, electrode extension, and weave pattern (if used) constant so that any change in the weld bead is caused by the angle change.

The pivot should be completed in the 12 in. (305 mm) of the weld. You will proceed from a 30° pushing angle to a 30° dragging angle. Repeat this procedure using different welding currents and plate thicknesses.

After the welds are complete, note the differences in width and reinforcement along the welds. Turn off the welding machine and shielding gas and clean up your work area when you are finished welding. ◆

EFFECT OF SHIELDING GAS ON WELDING

Shielding gases in the gas metal arc process are used primarily to protect the molten metal from oxidation and contamination. Other factors must be considered, however, in selecting the right gas for a particular application. Shielding gas can influence arc and metal transfer characteristics, weld penetration, width of fusion zone, surface shape patterns, welding speed, and undercut tendency. Inert gases such as argon and helium provide the necessary shielding because they do not form compounds with any other substance and are insoluble in molten

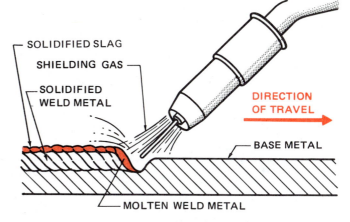

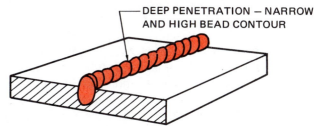

Figure 11-19 Backhand welding or dragging angle.

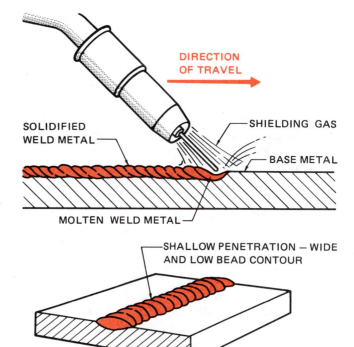

Figure 11-20 Forehand welding or pushing angle.

metal. When used as pure gases for welding ferrous metals, argon and helium may produce an erratic arc action, promote undercutting, and result in other flaws.

It is therefore usually necessary to add controlled quantities of reactive gases to achieve good arc action and metal transfer with these materials. Adding oxygen

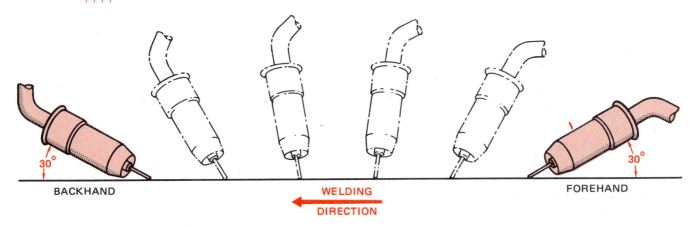

Figure 11-21 Welding gun angle.

or carbon dioxide to the inert gas tends to stabilize the arc, promote favorable metal transfer, and minimize spatter. As a result, the penetration pattern is improved and undercutting is reduced or eliminated.

Oxygen or carbon dioxide are often added to argon. The amount of reactive gas required to produce the desired effects is quite small. As little as 0.5% of oxygen will produce a noticeable change; 1–5% of oxygen is more common. Carbon dioxide may be added to argon in the 20–30% range. Mixtures of argon with less than 10% carbon dioxide may not have enough arc voltage to give the desired results.

Adding oxygen or carbon dioxide to an inert gas causes the shielding gas to become oxidizing. This in turn may cause porosity in some ferrous metals. In this case, a filler wire containing suitable deoxidizers should be used. The presence of oxygen in the shielding gas can also cause some loss of certain alloying elements, such as chromium, vanadium, aluminum, titanium, manganese, and silicon. Again, the addition of a deoxidizer to the filler wire is necessary.

Pure carbon dioxide has become widely used as a shielding gas for GMA welding of steels. It allows higher welding speed, better penetration, and good mechanical properties, and it costs less than the inert gases. The chief drawback in the use of carbon dioxide is the less-steady-arc characteristics and considerable weld-metal-spatter losses. The spatter can be kept to a minimum by maintaining a very short, uniform arc length. Consistently sound welds can be produced using carbon dioxide shielding, provided that a filler wire having the proper deoxidizing additives is used.

EXPERIMENT 11-5

Effect of Shielding Gas Changes

Using a properly assembled GMA welding machine; proper safety protection; a source of CO_2, argon, and oxygen gases or a variety of premixed shielding gases; two flowmeters (or one two-gas mixing regulator); and some pieces of mild steel plate, each about 12 in. (305 mm) long and ranging in thickness from 16 gauge to 1/2 in. (13 mm), you will observe the effect of various shielding gas mixtures on the weld.

Using a mixing flow meter regulator will allow the gases to be mixed in any desired mixture. **Table 11-6** is an example of a mixing ratio chart to be used to arrive at the approximate gas percentages. The exact ratios are not so important to you, as a student, as they are on code work.

With a medium voltage and amperage setting and using a 100% carbon dioxide (CO_2) shielding gas, start making a weld. Either change the mixture after each weld or have another person changing the shielding gas during the weld. Keep the total flow rate the same by adding argon (Ar) while reducing the CO_2 to preserve the same flow rate. During the experiment, change over the shielding gas to 100% argon (Ar).

After the weld is complete, evaluate it for spatter, penetration, undercut, buildup, width, or other noticeable changes along its length. Using **Table 11-7**, record the results of your evaluation.

Repeat the procedure just explained two times more with both low and high power settings. Again, record your observations.

Starting with 100% argon (Ar), add oxygen (O_2) to the shielding gas. The oxygen percentage will range from 0% to 10%, **Table 11-8**. Very slight changes in the percentage will have dramatic effects on the weld. You will make three welds using low, medium, and high power settings. For each weld, you will record your observations.

During some of the welding tests, you will notice a change in the method of metal transfer, weld heat, and general weld performance without a change in the current settings. The shielding gas mixture can have major effects on the rate of metal transfer and the welding speed, as well as other welding variables. Higher speeds and greater production can be obtained by using some gas mixtures. However, the savings can be completely offset by the higher gas cost. Before making a final decision about the gas to be used, all the variables must be compared. **Table 11-9**

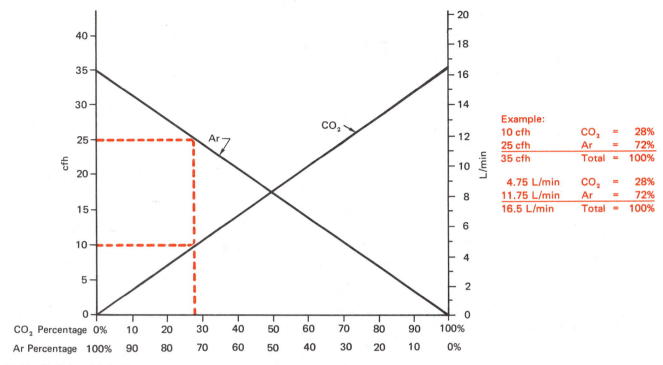

Table 11-6 Gas Mixing Percentages.

Weld Acceptability	Gas Mixture	Spatter	Penetration	Puddle Size	Bead Appearance
Good	75 Ar 25 CO₂	Very little	Deep	Large	Wide with little buildup

Electrode Diameter .035 in (0.9 mm)
Welding Direction Forehand
Voltage 25
Amperage 150

Table 11-7

lists premixed shielding gases and their uses. Turn off the welding machine and shielding gas and clean up your work area when you are finished welding. ◆

PRACTICES

The practices in this chapter are grouped according to those requiring similar techniques and setups. To make acceptable GMA welds consistently, the major skill required is the ability to set up the equipment and weldment. Changes such as variations in material thickness, position, and type of joint require changes both in technique and setup. A correctly set up GMA welding station can, in many cases, be operated with minimum skill. Often the only difference between a welder earning a minimum wage and one earning the maximum wage is the ability to correct machine setups.

Ideally, only a few tests would be needed for the welder to make the necessary adjustments in setup and manipulation techniques to achieve a good weld. The previous welding experiments should have given the welder a graphic set of comparisons to help that welder make the correct changes. In addition to keeping the test data, you may want to keep the test plates for a more accurate comparison.

The grouping of practices in this chapter will keep the number of variables in the setup to a minimum. Often, the only change required before going on to the next weld is to adjust the power settings.

Figures that are given in some of the practices will give the welder general operating conditions, such as voltage, amperage, and shielding gas and/or gas mixture. These are general values, so the welder will have to make some fine adjustments. Differences in the type of machine being used and the material surface condition

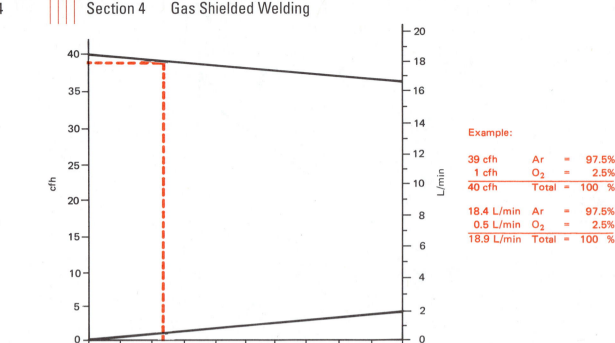

Example:

39 cfh	Ar	=	97.5%
1 cfh	O_2	=	2.5%
40 cfh	Total	=	100 %

18.4 L/min	Ar	=	97.5%
0.5 L/min	O_2	=	2.5%
18.9 L/min	Total	=	100 %

Table 11-8 Ar and O_2 Mixture Percentages.

will affect the settings. For this reason, it is preferable to use the settings developed during the experiments.

METAL PREPARATION

All hot-rolled steel has an oxide layer that is formed during the rolling process called mill scale. *Mill scale* is a thin layer of dark gray or black iron oxide. Some hot-rolled steels that have had this layer removed either mechanically or chemically can be purchased. However, almost all of the hot-rolled steel used today still has this layer because it offers some protection from rusting.

Mill scale is not removed for noncode welding, because it does not prevent most welds from being suitable for service. For practice welds that will be visually inspected mill scale can usually be left on the plate. Filler metals and fluxes usually have deoxidizers added to them so that the adverse effects of the mill scale are reduced or eliminated, Table 11-10. But with GMA welding wire it is difficult to add enough deoxidizers to remove all effects of mill scale. The porosity that mill scale causes is most often confined to the interior of the weld and is not visible on the surface, Figure 11-22. Because it is not visible on the surface, it usually goes unnoticed and the weld passes visual inspection.

Shielding Gas	Chemical Behavior	Uses and Usage Notes
1. Argon	Inert	Welding virtually all metals except steel
2. Helium	Inert	Al and Cu alloys for greater heat and to minimize porosity
3. Ar & He (20–80 to 50–50%)	Inert	Al and Cu alloys for greater heat and to minimize porosity, but with quieter, more readily controlled arc action
4. N_2	Reducing	On Cu, very powerful arc
5. Ar + 25–30% N_2	Reducing	On Cu, powerful but smoother operating more readily controlled arc than with N_2
6. Ar + 1–2% O_2	Oxidizing	Stainless and alloy steels, also for some deoxidized copper alloys
7. Ar + 3–5% O_2	Oxidizing	Plain carbon, alloy, and stainless steels (generally requires highly deoxidized wire)
8. Ar + 3–5% O_2	Oxidizing	Various steels using deoxidized wire
9. Ar + 20–30% O_2	Oxidizing	Various steels, chiefly with short-circuiting arc
10. Ar + 5% O_2 + 15% CO_2	Oxidizing	Various steels using deoxidized wire
11. CO_2	Oxidizing	Plain-carbon and low-alloy steels, deoxidized wire essential
12. CO_2 + 3–10% O_2	Oxidizing	Various steels using deoxidized wire
13. CO_2 + 20% O_2	Oxidizing	Steels

Table 11-9 Shielding Gases and Gas Mixtures Used for Gas Metal Arc Welding.

Deoxidizing Element	Strength
Aluminum (Al)	Very strong
Manganese (Mn)	Weak
Silicon (Si)	Weak
Titanium (Ti)	Very strong
Zirconium (Zr)	Very strong

Table 11-10 Sufficient Deoxidizing Elements Must Be Added to the Filler Wire to Minimize Porosity in the Molten Weld Pool.

If the practices are going to be destructively tested, then all welding surfaces within the weld groove and the surrounding surfaces within 1 in. (25 mm) must be cleaned to bright metal, **Figure 11-23.** Cleaning may be either grinding, filing, sanding, or blasting.

FLAT POSITION, 1G AND 1F POSITIONS

PRACTICE 11-3

Stringer Beads Using the Short-circuiting Metal Transfer Method in the Flat Position

Using a properly set up and adjusted GMA welding machine (see Table 10-3, page 209), proper safety protection, .035-in. and/or .045-in. (0.9-mm and/or 1.2-mm)-diameter wire, and two or more pieces of mild steel sheet, 12 in. (305 mm) long and 16 gauge and 1/8 in. (3 mm) thick, you will make a stringer bead weld in the flat position, **Figure 11-24.**

Starting at one end of the plate and using either a pushing or dragging technique, make a weld bead along the entire 12-in. (305-mm) length of the metal. See Color Plates 14 and 15. After the weld is complete, check its

appearance. Make any needed changes to correct the weld (refer to **Tables 11-3** and **11-5**). Repeat the weld and make additional adjustments. After the machine is set, start to work on improving the straightness and uniformity of the weld.

Keeping the bead straight and uniform can be hard because of the limited visibility due to the small amount of light and the size of the molten weld pool. The welder's view is further restricted by the shielding gas nozzle **Figure 11-25.** Even with limited visibility, it is possible to make a satisfactory weld by watching the edge of the molten weld pool, the sparks, and the weld bead produced. Watching the leading edge of the molten weld pool (forehand welding, push technique) will show you the molten weld pool fusion and width. Watching the trailing edge of the molten weld pool (backhand welding, drag technique) will show you the amount of buildup and the relative heat input, **Figure 11-26.** The quantity and size of sparks produced can indicate the relative location of the filler wire in the molten weld pool. The number of sparks will increase as the wire strikes the solid metal ahead of the molten weld pool. The gun itself will begin to vibrate or bump as the wire momentarily pushes against the cooler, unmelted base metal before it melts. Changes in weld width, buildup, and proper joint tracking can be seen by watching the bead as it appears from behind the shielding gas nozzle.

Repeat each type of bead as needed until consistently good beads are obtained. Turn off the welding machine and shielding gas and clean up your work area when you are finished welding. ◆

PRACTICE 11-4

Flat Position Butt Joint, Lap Joint, and Tee Joint

Using the same equipment, materials, and procedures listed in Practice 11-3, make welded butt joints, lap

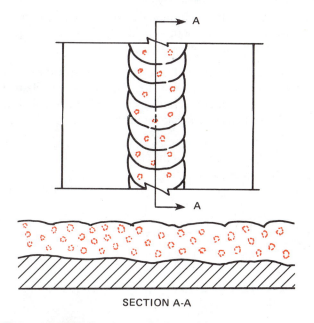

SECTION A-A

Figure 11-22 Uniformly scattered porosities.

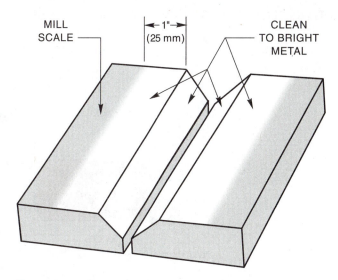

MILL SCALE

1" (25 mm)

CLEAN TO BRIGHT METAL

Figure 11-23 Clean all surfaces to bright metal before welding.

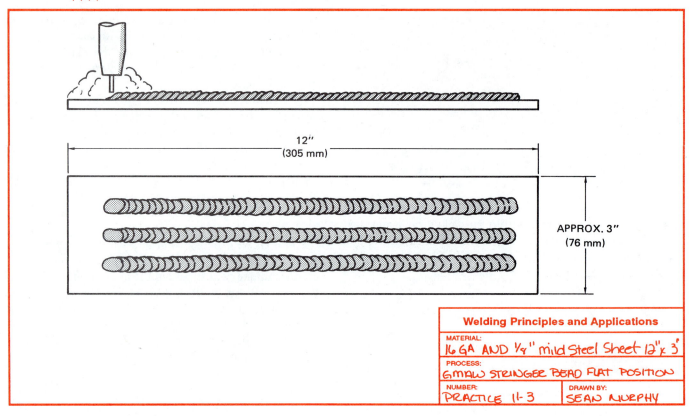

Welding Principles and Applications
MATERIAL:
16 GA AND 1/8" mild Steel Sheet 12" x 3'
PROCESS:
GMAW STRINGER BEAD FLAT POSITION
NUMBER: DRAWN BY:
PRACTICE 11-3   SEAN MURPHY

Figure 11-24 Stringer beads in the flat position.

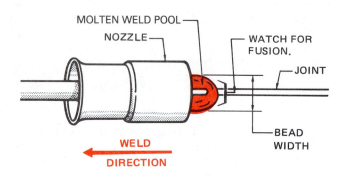

Figure 11-25 The shielding gas nozzle restricts the welder's view.

joints, and tee joints in the flat position, **Figure 11-27A, 11-27B, and 11-27C.** See Color Plate 12.

■ Tack weld the sheets together and place them flat on the welding table, **Figure 11-28.**

■ Starting at one end, run a bead along the joint. Watch the molten weld pool and bead for signs that a change in technique may be required.

■ Make any needed changes as the weld progresses. By the time the weld is complete, you should be making the weld nearly perfectly.

■ Using the same technique that was established in the last weld, make another weld. This time, the entire 12 in. (305 mm) of weld should be flawless.

Repeat each type of joint with both thicknesses of metal as needed until consistently good beads are obtained. Turn off the welding machine and shielding gas and clean up your work area when you are finished welding. ◆

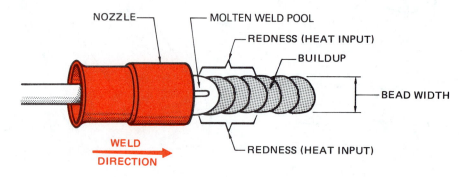

Figure 11-26 Watch the trailing edge of the molten weld pool.

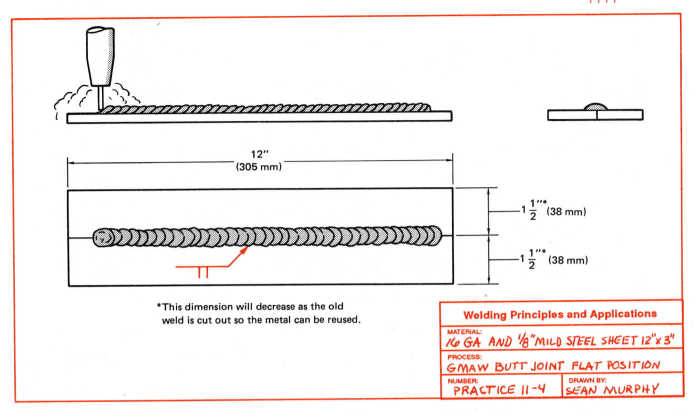

Figure 11-27(A) Butt joint in the flat position.

*This dimension will decrease as the old weld is cut out so the metal can be reused.

Welding Principles and Applications

MATERIAL:
16 GA AND 1/8" MILD STEEL SHEET 12" x 3"

PROCESS:
GMAW BUTT JOINT FLAT POSITION

NUMBER:
PRACTICE 11-4

DRAWN BY:
SEAN MURPHY

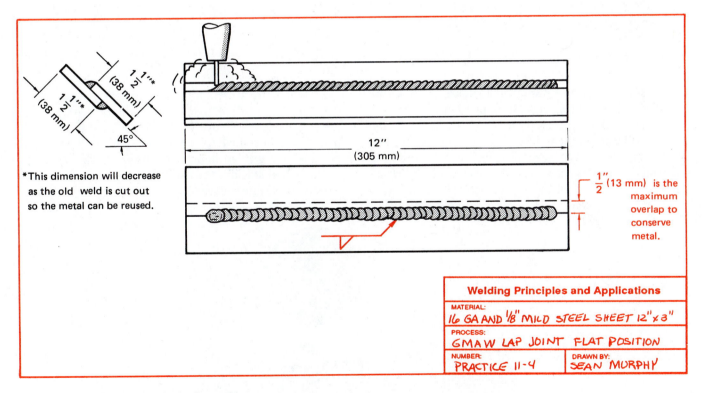

Figure 11-27(B) Lap joint in the flat position.

*This dimension will decrease as the old weld is cut out so the metal can be reused.

1/2" (13 mm) is the maximum overlap to conserve metal.

Welding Principles and Applications

MATERIAL:
16 GA AND 1/8" MILD STEEL SHEET 12" x 3"

PROCESS:
GMAW LAP JOINT FLAT POSITION

NUMBER:
PRACTICE 11-4

DRAWN BY:
SEAN MURPHY

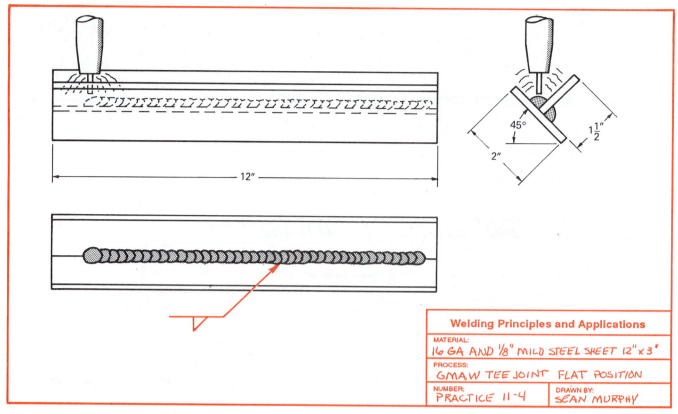

Welding Principles and Applications

MATERIAL:
16 GA AND 1/8" MILD STEEL SHEET 12"x3"

PROCESS:
GMAW TEE JOINT FLAT POSITION

NUMBER:
PRACTICE 11-4

DRAWN BY:
SEAN MURPHY

Figure 11-27(C) Tee joint in the flat position.

PRACTICE 11-5

Flat Position Butt Joint, Lap Joint, and Tee Joint, All with 100% Penetration

Using the same equipment, materials, and setup listed in Practice 11-3, make a welded joint in the flat position with 100% penetration, along the entire 12-in. (305-mm) length of the welded joint. Repeat each type of joint as needed until consistently good beads are obtained. Turn off the welding machine and shielding gas and clean up your work area when you are finished welding. ◆

PRACTICE 11-6

Flat Position Butt Joint, Lap Joint, and Tee Joint, All Welds to Be Tested

Using the same equipment, materials, and setup listed in Practice 11-3, make each of the welded joints in the flat position, **Figure 11-29**. Each weld joint must pass the bend test. Repeat each type of weld joint until all pass the guided bend test. Turn off the welding machine and shielding gas and clean up your work area when you are finished welding. ◆

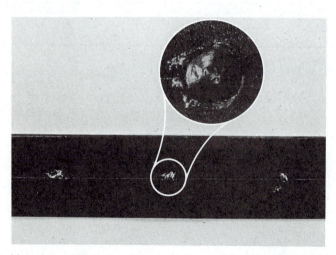

Figure 11-28 Use enough tack welds to keep the joint in alignment during welding. Small tack welds are easier to weld over without adversely affecting the weld.

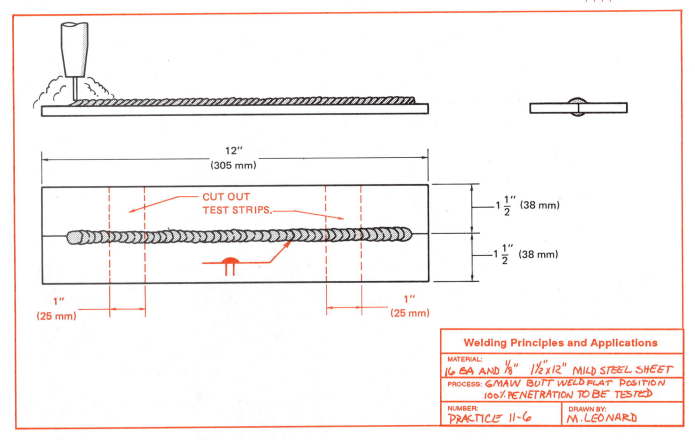

Welding Principles and Applications

MATERIAL:	
16 GA AND ⅛" 1½"x12" MILD STEEL SHEET	
PROCESS: GMAW BUTT WELD FLAT POSITION	
100% PENETRATION TO BE TESTED	
NUMBER: PRACTICE 11-6	DRAWN BY: M.LEONARD

Figure 11-29 Butt joint in the flat position to be tested.

VERTICAL UP 3G AND 3F POSITIONS

PRACTICE 11-7

Stringer Bead at a 45° Vertical Up Angle

Using the same equipment, materials, and setup as listed in Practice 11-3, you will make a vertical up stringer bead on a plate at a 45° inclined angle.

Start at the bottom of the plate and hold the welding gun at a slight angle to the plate, **Figure 11-30**. Brace yourself, lower your hood, and begin to weld.

Depending upon the machine settings and type of shielding gas used, you will make a weave pattern.

If the molten weld pool is large and fluid (hot), use a "C" or "J" weave pattern to allow a longer time for the molten weld pool to cool, **Figure 11-31**. Do not make the weave so long or fast that the wire is allowed to strike the metal ahead of the molten weld pool. If this happens, spatter increases and a spot or zone of incomplete fusion may occur, **Figure 11-32**.

If the molten weld pool is small and controllable, use a small "C", zig zag, or "J" weave pattern to control the width and buildup of the weld. A slower speed can also be used. Watch for complete fusion along the leading

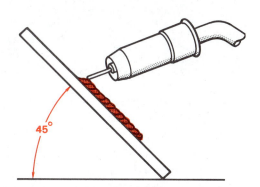

Figure 11-30 Vertical up position.

SHELF

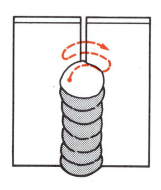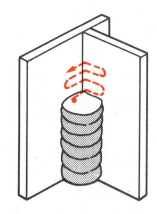

Figure 11-31

edge of the molten weld pool. **Figure 11-33** shows a weld that did not fuse with the plate.

A weld that is high and has little or no fusion is too "cold." Changing the welding technique will not correct this problem. The welder must stop welding and make the needed adjustments.

As the weld progresses up the plate, the back or trailing edge of the molten weld pool will cool, forming a shelf to support the molten metal. Watch the shelf to be sure that molten metal does not run over, forming a drip. When it appears that the metal may flow over the shelf, either increase the weave lengths or stop and start the current for brief moments to allow the weld to cool. Stopping for brief moments will not allow the shielding gas to be lost.

Continue to weld along the entire 12-in. (305-mm) length of plate. Repeat this weld as needed until a straight and uniform weld bead is produced. Turn off the welding machine and shielding gas and clean up your work area when you are finished welding. ◆

PRACTICE 11-8

Stringer Bead in the Vertical Up Position

Repeat Practice 11-7 and increase the angle until you have mastered a straight and uniform weld bead in the vertical up position. Turn off the welding machine and shielding gas and clean up your work area when you are finished welding. ◆

PRACTICE 11-9

Butt Joint, Lap Joint, and Tee Joint in the Vertical Up Position at a 45° Angle

Using the same equipment, materials, and setup as listed in Practice 11-3, you will make vertical up welded joints on a plate at a 45° inclined angle.

Tack weld the metal pieces together shand brace them in position. Check to see that you have free movement along the entire joint to prevent stopping and restarting

Figure 11-32 Burst of spatter caused by incorrect electrode contact with base metal.

Figure 11-33 Weld separated from the plate; there is no fusion between the weld and plate.

during the weld. Avoiding stops and starts both speeds up the welding time and eliminates discontinuities.

The weave pattern should allow for adequate fusion on both edges of the joint. Watch the edges to be sure that they are being melted so that adequate fusion and penetration occurs.

Repeat each type of joint as needed until consistently good beads are obtained. Turn off the welding machine and shielding gas and clean up your work area when you are finished welding. ◆

PRACTICE 11-10

Butt Joint, Lap Joint, and Tee Joint in the Vertical Up Position with 100% Penetration

Using the same equipment, materials, and setup as listed in Practice 11-3, you will increase the plate angle gradually as you develop skill until you are making satisfactory welds in the vertical up position.

Repeat each type of joint as needed until consistently good beads are obtained. Turn off the welding machine and shielding gas and clean up your work area when you are finished welding. ◆

PRACTICE 11-11

Butt Joint, Lap Joint, and Tee Joint in the Vertical Up Position, All Welds to Be Tested

Using the same equipment, materials, and setup as listed in Practice 11-3, you will make the welded joints in the vertical up position. Each weld must pass the bend test. Repeat each type of weld joint until all pass the bend test. Turn off the welding machine and shielding gas and clean up your work area when you are finished welding. ◆

VERTICAL DOWN 3G AND 3F POSITIONS

The vertical down welding technique can be useful when making some types of welds. The major advantages of this technique are the following:

■ Speed — Very high rates of travel are possible.

■ Shallow penetration — Thin sections or root openings can be welded with little burnthrough.

■ Good bead appearance — The weld has a nice width-to-height ratio and is uniform.

Vertical down welds are often used on thin sheet metals or in the root pass in grooved joints. The combination of controlled penetration and higher welding speeds makes vertical down the best choice for such welds. The ease with which welds having a good appearance can be made is deceiving. Generally, more skill is required to make sound welds with this technique than in the vertical up position. The most common problem with these welds is lack of fusion or overlap. To prevent these problems, the arc must be kept at or near the leading edge of the molten weld pool.

PRACTICE 11-12

Stringer Bead at a 45° Vertical Down Angle

Using the same equipment, materials, and setup as listed in Practice 11-3, you will make a vertical down stringer bead on a plate at a 45° inclined angle.

Hold the welding gun at the top of the plate with a slight dragging angle, Figure 11-34, will help to increase penetration, hold back the molten weld pool, and improve visibility of the weld. Be sure that your movements along the 12-in. (305-mm) length of plate are unrestricted.

Lower your hood and start the weld. Watch both the leading edge and sides of the molten weld pool for fusion. The leading edge should flow into the base metal, not curl over it. The sides of the molten weld pool should also show fusion into the base metal and not be flashed (ragged) along the edges.

The weld may be made with or without a weave pattern. If a weave pattern is used, it should be a "C" pattern. The "C" should follow the leading edge of the weld. Some changes on the gun angle may help to increase penetration. Experiment with the gun angle as the weld progresses.

Repeat these welds until you have established a rhythm and technique that work well for you. The welds must be straight and uniform and have complete fusion. Turn off the welding machine and shielding gas and clean up your work area when you are finished welding. ◆

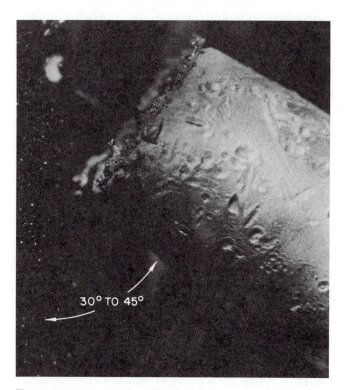

30° TO 45°

Figure 11-34 Vertical down position.

PRACTICE 11-13

Stringer Bead in the Vertical Down Position

Repeat Practice 11-12 and increase the angle of the plate until you have developed the skill to repeatedly make good welds in the vertical down position. The weld bead must be straight and uniform and have complete fusion. Turn off the welding machine and shielding gas and clean up your work area when you are finished welding. ◆

PRACTICE 11-14

Butt Joint, Lap Joint, and Tee Joint in the Vertical Down Position

Using the same equipment, materials, and setup as listed in Practice 11-3, you will make vertical down welded joints.

Tack weld the pieces of metal together and brace them in position. Using the same technique developed in Practice 11-11, start at the top of the joint and weld down the length of the joint. When the weld is complete, inspect it for discontinuities and make any necessary changes in your technique. Repeat each type of joint as needed until consistently good welds are obtained. Turn off the welding machine and shielding gas and clean up your work area when you are finished welding. ◆

PRACTICE 11-15

Butt Joint and Tee Joint in the Vertical Down Position with 100% Penetration

Using the same equipment, materials, and setup as listed in Practice 11-3, you will make welded joints with 100% weld penetration.

It may be necessary to adjust the root opening to meet the penetration requirements. The lap joint was omitted from this practice because little additional skill can be developed with it that is not already acquired with the tee joint. Repeat each type of joint as needed until consistently good welds are obtained. Turn off the welding machine and shielding gas and clean up your work area when you are finished welding. ◆

PRACTICE 11-16

Butt Joint and Tee Joint in the Vertical Down Position, Welds to Be Tested

Using the same equipment, materials, and setup as listed in Practice 11-3, you will make the welded joints in the vertical down position. Each weld must pass the bend test. Repeat each type of weld joint until both pass the bend test. Turn off the welding machine and shield-

ing gas and clean up your work area when you are finished welding. ◆

HORIZONTAL 2G AND 2F POSITIONS

PRACTICE 11-17

Horizontal Stringer Bead at a 45° Angle

Using the same equipment, materials, and setup as listed in Practice 11-3, you will make a horizontal stringer bead on a plate at a 45° reclined angle.

Start at one end with the gun pointed in a slightly upward direction, Figure 11-35. You may use a pushing or a dragging leading or a trailing gun angle, depending upon the current setting and penetration desired. Undercutting along the top edge and overlap along the bottom edge are problems with both gun angles. Careful attention must be paid to the manipulation "weave" technique used to overcome these problems.

The most successful weave patterns are the "C" and "J" patterns. The "J" pattern is the most frequently used. The "J" pattern allows weld metal to be deposited along a shelf created by the previous weave, Figure 11-36. The length of the "J" can be changed to control the weld bead size. Smaller weld beads are easier to control than large ones.

Repeat these welds until you have established the rhythm and technique that work well for you. The weld must be straight and uniform and have complete fusion. Turn off the welding machine and shielding gas and clean up your work area when you are finished welding. ◆

PRACTICE 11-18

Stringer Bead in the Horizontal Position

Repeat Practice 11-7, and increase the angle of the plate until you have developed the skill to repeatedly make good horizontal welds on a vertical surface. The weld bead must be straight and uniform and have com-

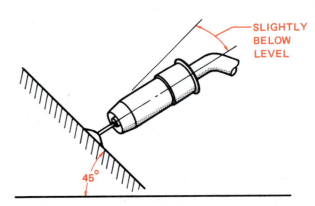

SLIGHTLY BELOW LEVEL

45°

Figure 11-35 45° horizontal position.

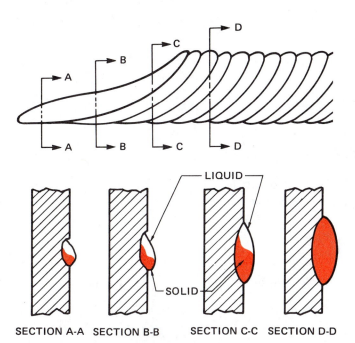

Figure 11-36 The actual size of the molten weld pool remains small along the weld.

SECTION A-A SECTION B-B SECTION C-C SECTION D-D

plete fusion. Turn off the welding machine and shielding gas and clean up your work area when you are finished welding. ◆

PRACTICE 11-19

Butt Joint, Lap Joint, and Tee Joint in the Horizontal Position

Using the same equipment, materials, and setup listed in Practice 11-3, you will make horizontal welded joints.

Tack weld the pieces of metal together and brace them in position using the same skills developed in Practice 11-17. Starting at one end, make a weld along the entire length of the joint. When making the butt or lap joints, it may help to recline the plates at a 45° angle until you have developed the technique required. Repeat each type of joint as needed until consistently good welds are obtained. Turn off the welding machine and shielding gas and clean up your work area when you are finished welding. ◆

PRACTICE 11-20

Butt Joint and Tee Joint in the Horizontal Position with 100% Penetration

Using the same equipment, materials, and setup as listed in Practice 11-3, you will make overhead joints having 100% penetration in the horizontal position.

It may be necessary to adjust the root opening to meet the penetration requirements. Repeat each type of joint as needed until consistently good welds are obtained. Turn off the welding machine and shielding gas and clean up your work area when you are finished welding. ◆

PRACTICE 11-21

Butt Joint and Tee Joint in the Horizontal Position, Welds to Be Tested

Using the same equipment, materials, and setup as listed in Practice 11-3, you will make the welded joints in the horizontal position. Each weld must pass the bend test. Repeat each type of weld joint until both pass the bend test. Turn off the welding machine and shielding gas and clean up your work area when you are finished welding. ◆

OVERHEAD 4G AND 4F POSITIONS

There are several advantages to the use of short-circuiting arc metal transfer in the overhead position, including:

- Small molten weld pool size — The smaller size of the molten weld pool allows surface tension to hold it in place. Less molten weld pool sag results in improved bead contour with less undercut and icicles, **Figure 11-37**.
- Direct metal transfer — The direct metal transfer method does not rely on other forces to get the filler metal into the molten weld pool. This results in efficient metal transfer and less spatter and loss of filler metal.

PRACTICE 11-22

Stringer Bead Overhead Position

Using the same equipment, materials, and setup as listed in Practice 11-3, you will make a welded stringer bead in the overhead position.

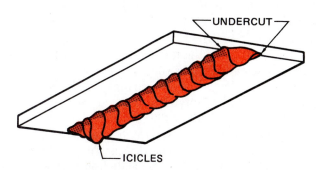

Figure 11-37 Overhead weld.

The molten weld pool should be kept as small as possible for easier control. A small molten weld pool can be achieved by using lower current settings, traveling faster, or by pushing the molten weld pool. The technique used is the welder's choice. Often a combination of techniques can be used with excellent results.

Lower current settings require closer control of gun manipulation to ensure that the wire is fed into the molten weld pool just behind the leading edge. The low power will cause overlap and more spatter if this wire-to-molten weld pool contact position is not closely maintained.

Faster travel speeds allow the welder to maintain a high production rate even if multiple passes are required to complete the weld. Weld penetration into the base metal at the start of the bead can be obtained by using a slow start or quickly reversing the weld direction. Both the slow start and reversal of weld direction put more heat into the meld start to increase penetration, **Figure 11-38**. The higher speed also reduces the amount of weld distortion by reducing the amount of time that heat is applied to a joint.

The pushing or trailing gun angle forces the bead to be flatter by spreading it out over a wider area as compared to the bead resulting from a dragging or backhand gun angle. The wider, shallow molten weld pool cools faster, resulting in less time for sagging and the formation of icicles.

When welding overhead, extra personal protection is required to reduce the danger of burns. Leather sleeves or leather jackets should be worn.

Much of the spatter created during overhead welding falls into the shielding gas nozzle. The effectiveness of the shielding gas is reduced, **Figure 11-39**, and the contact tube may short out to the gas nozzle, **Figure 11-40**. Turbulence caused by the spatter obstructing the gas may lead to weld contamination. The shorted gas nozzle may arc to the work causing damage both to the nozzle and to the plate. To control the amount of spatter, a longer stickout and/or a sharper gun-to-plate angle is required to allow most of the spatter to fall clear of the gas nozzle. The nozzle can be dipped, sprayed, or injected automatically, **Figure 11-41**, with antispatter to help prevent the spatter from sticking. Applying antispatter won't stop the spatter from building up, but it does make its removal much easier.

Make several short weld beads using various techniques to establish the method that is most successful and most comfortable for you. After each weld, stop and evaluate it before making a change. When you have decided on the technique to be used, make a welded stringer bead that is 12 in. (305 mm) long.

Repeat the weld until it can be made straight, uniform, and free from any visual defects. Turn off the welding machine and shielding gas and clean up your work area when you are finished welding. ◆

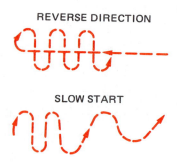

Figure 11-38 Two methods of concentrating heat at the beginning of a weld bead to aid in penetration depth.

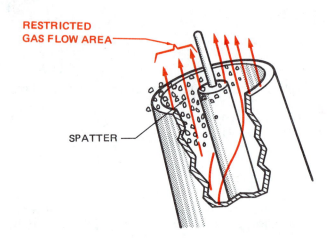

Figure 11-39 Shielding gas flow affected by excessive weld spatter in nozzle.

Figure 11-40 Gas nozzle damaged after shorting out against the work.

Anti-Spatter Unit - Model 4050

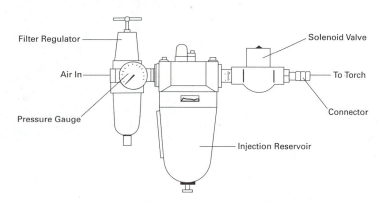

Filter Regulator

Air In

Pressure Gauge

Solenoid Valve

To Torch

Connector

Injection Reservoir

Operating Instructions & Parts Manual

Figure 11-41 Automatic antispatter system that can be added to a GMA welding gun.

PRACTICE 11-23

Butt Joint, Lap Joint, and Tee Joint in the Overhead Position

Using the same equipment, materials, and setup as listed in Practice 11-3, you will make an overhead welded joint.

Tack weld the pieces of metal together and secure them in the overhead position. Be sure you have an unrestricted view and freedom of movement along the joint. Start at one end and make a weld along the joint. Use the same technique developed in Practice 11-22.

Repeat the weld until it can be made straight, uniform, and free from any visual defects. Turn off the welding machine and shielding gas and clean up your work area when you are finished welding. ◆

PRACTICE 11-24

Butt Joint and Tee Joint in the Overhead Position with 100% Penetration

Using the same equipment, materials, and setup as listed in Practice 11-3, you will make overhead welded joints having 100% penetration.

Tack weld the metal together. It may be necessary to adjust the root opening to allow 100% weld metal penetration. During these welds, it may be necessary to use a dragging or backhand torch angle. When used with a "C" or "J" weave pattern, this torch angle helps to achieve the desired depth of penetration. A key hole just ahead of the molten weld pool is a good sign that the metal is being penetrated, **Figure 11-42.**

Repeat the weld until it can be made straight, uniform, and free from any visual defects. Turn off the welding machine and shielding gas and clean up your work area when you are finished welding. ◆

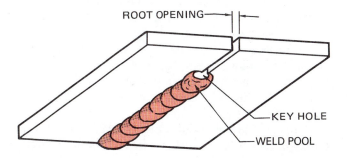

ROOT OPENING

KEY HOLE

WELD POOL

Figure 11-42

PRACTICE 11-25

Butt Joint and Tee Joint in the Overhead Position, Welds to Be Tested

Using the same equipment, materials, and setup as listed in Practice 11-3, you will make a welded joint in the overhead position. Each weld must pass the bend test. Repeat each type of weld joint until both pass the bend test. Turn off the welding machine and shielding gas and clean up your work area when you are finished welding. ◆

PULSED-ARC METAL TRANSFER, 1G POSITION

PRACTICE 11-26

Stringer Bead

Using a properly set up and adjusted GMA welding machine (see Table 10-2, page 208), proper safety protection, .035-in. and/or .045-in. (0.9-mm and/or 1.2-mm)-diameter wire, and two or more pieces of mild steel plate,

12 in. (305 mm) long x 1/4 in. (6 mm) thick, make a stringer bead weld in the flat position.

■ Start at one end of the plate and use either a push or drag technique to make a weld bead along the entire 12-in. (305-mm) length of the metal.

■ After completing the weld, check its appearance and make any changes needed to correct the weld, Figure 11-43.

■ Repeat the weld and make additional adjustments as required in the frequency, amplitude, or pulse width.

■ After the machine is set, start working on improving the straightness and uniformity of the weld.

The location of the arc in the molten weld pool is not as critical in this method of metal transfer as it is in the short-circuiting arc method. If the arc is too far back on the molten weld pool, however, fusion along the leading edge may be reduced, Figure 11-44. Moving the arc to the cool metal ahead of the molten weld pool often causes an increase in spatter.

The weave pattern selected should allow the arc to follow the leading edge of the molten weld pool if deep penetration is needed. A pattern that moves the arc back from the leading edge completely or in part results in reduced penetration.

Repeat the weld until it can be made straight, uniform, and free from any visual defects. Turn off the welding machine and shielding gas and clean up your work area when you are finished welding. ◆

Figure 11-43 Weld bead made with GMAW globular metal transfer mode.

MOLTEN PUDDLE

SPOT WHERE ARC STRUCK OUTSIDE OF THE MOLTEN PUDDLE

Figure 11-44 Momentary break in the arc during globular metal transfer due to improper electrode manipulation.

PRACTICE 11-27

Butt Joint

Using the same equipment, materials, and setup as listed in Practice 11-26, you will make a welded joint in the flat position, Figure 11-45.

The heat produced during GMA welding is not enough to force 100% penetration consistently through metal thicker than 3/16 in. (4.8 mm). On occasion, or with extra effort, it is possible to obtain 100% penetration. However, the method generally is not reliable enough for most industrial applications. To ensure that the joint is completely fused, it is prepared for welding with a groove. Any one of several groove designs can be used, Figure 11-46. The V-groove is most frequently used because of the ease with which it can be produced.

The V-groove should be made with a 45° inclined angle, Figure 11-47. The small-diameter filler metal allows the usual 60% inclined angle for SMAW to be reduced. This reduced groove size requires less filler metal and can be welded in less time, resulting in lower cost.

After the metal has been grooved, tack weld it together. The root opening between the plates should be 1/8 in. (3 mm) or less. Starting at one end, make a single pass weld along the entire joint. A weave pattern that fol-

lows the contour of the groove should be used to ensure complete fusion. Figure 11-48.

The completed weld should be uniform in width and reinforcement and have no visual defects. Repeat the weld until it can be made straight, uniform, and free from any visual defects. Turn off the welding machine and shielding gas and clean up your work area when you are finished welding. ◆

PRACTICE 11-28

Butt Joint With 100% Penetration

Using the same equipment, materials, and setup as listed in Practice 11-16, you will make a groove weld having 100% penetration.

Using the technique developed in Practice 11-27 and making the necessary adjustments in the root gap, the weld metal must fuse 100% of the plate thickness. Watch the molten weld pool. If it appears to sink or does not increase in size, you are probably burning through. This action will cause excessive root reinforcement. To correct this problem, speed up the travel rate and/or increase the contact tube-to-work distance.

After the weld is complete, inspect the root surface for the proper appearance. Repeat the weld until it

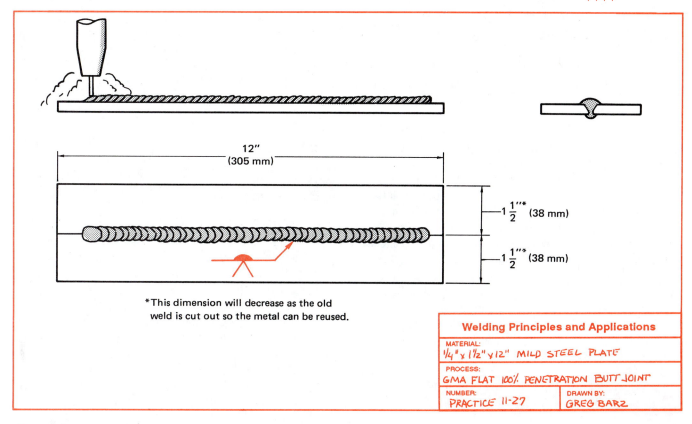

12"
(305 mm)

$1\frac{1}{2}''^*$ (38 mm)

$1\frac{1}{2}''^*$ (38 mm)

*This dimension will decrease as the old weld is cut out so the metal can be reused.

Welding Principles and Applications

MATERIAL:
1/4" x 1 1/2" x 12" MILD STEEL PLATE

PROCESS:
GMA FLAT 100% PENETRATION BUTT JOINT

NUMBER:
PRACTICE 11-27

DRAWN BY:
GREG BARZ

Figure 11-45 Single V-groove butt joint in the flat position.

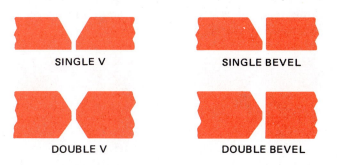

SINGLE V   SINGLE BEVEL   SINGLE U   SINGLE J

DOUBLE V   DOUBLE BEVEL   DOUBLE U   DOUBLE J

Figure 11-46 Typical groove designs.

can be made straight, uniform, and free from any visual defects. Turn off the welding machine and shielding gas and clean up your work area when you are finished welding. ◆

PRACTICE 11-29

Butt Joint to Be Tested

Using the same equipment, materials, and setup as listed in Practice 11-26, you will make a flat groove weld. The weld must pass the bend test. Repeat the weld until the test can be passed. Turn off the welding machine and shielding gas and clean up your work area when you are finished welding. ◆

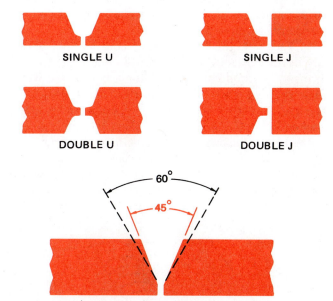

Figure 11-47 A smaller groove angle reduces both weld time and filler metal required to make the weld.

Figure 11-48 Uniform weld produced with a weave pattern.

PRACTICE 11-30

PRACTICE 11-30

Tee Joint and Lap Joint in the IF Position

Using the same equipment, materials, and setup as listed in Practice 11-26, you will make a fillet weld in the flat position.

It is not necessary to groove the plates used for a tee or lap joint to obtain a sound weld. Fillet weld strength can be obtained by making the weld the proper size for the plate thickness.

The face or surface of a fillet weld should be as flat as possible. Welds with excessive buildup waste metal. Welds with too little buildup will be weak, flat, or concave. Fillet weld beads have fewer stress points to cause weld failure during cyclical loading.

A weave pattern that follows the contour of the joint should be used to ensure adequate fusion, **Figure 11-49.** Repeat the weld until it can be made straight, uniform, and free from any visual defects. Turn off the welding machine and shielding gas and clean up your work area when you are finished welding. ◆

PRACTICE 11-31

Tee Joint and Lap Joint in the 2F Position

Using the same equipment, materials, and setup as listed in Practice 11-26, you will make a fillet weld in the horizontal position.

Tack weld the metal together and place the assembly in the horizontal position. The pulsed-arc metal transfer method can be used for a horizontal weld. However, care must be taken to ensure that the legs of the fillet are equal. Because of the size and fluidity of the molten weld pool, undercutting along the top edge and overlap along the bottom edge can also be problems.

To control or eliminate these defects, the beads must be small and quickly made. In addition, a proper weave pattern must be established. The pattern must follow the plate surfaces and also establish a shelf to support the weld. After the weld is complete, inspect it for defects and measure it for uniformity. Repeat the weld until it can be made straight, uniform, and free from any visual defects. Turn off the welding machine and shielding gas and clean up your work area when you are finished welding. ◆

AXIAL SPRAY

PRACTICE 11-32

Stringer Bead, 1G Position

Using a properly set up and adjusted GMA welding machine (see Table 10-2, page 208), proper safety protection, .035-in. and/or .045-in. (0.9-mm and/or 1.2-mm)-diameter wire, and two or more pieces of mild steel plate, 12 in. (305 mm) long x 1/4 in. (6 mm) thick, you will make a welded stringer bead in the flat position.

Start at one end of the plate and use either a push or drag technique to make a weld bead along the entire 12-in. (305-mm) length of the metal. After the weld is complete, check its appearance and make any changes needed to correct the weld, **Figure 11-50.** Repeat the weld and make any additional adjustments required. After the machine is set, start working on improving the straightness and uniformity of the weld. Turn off the welding machine and shielding gas and clean up your work area when you are finished welding. ◆

PRACTICE 11-33

Butt Joint, Lap Joint, and Tee Joint Using the Axial Spray Method

Using the same equipment, materials, and setup as listed in Practice 11-32, you will make a flat and horizontal weld using axial spray metal transfer, **Figure 11-51.**

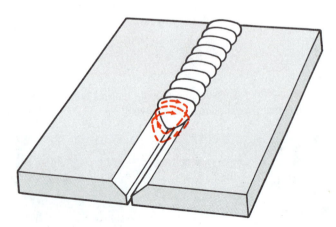

Figure 11-49 The electrode should be moved along the groove contour.

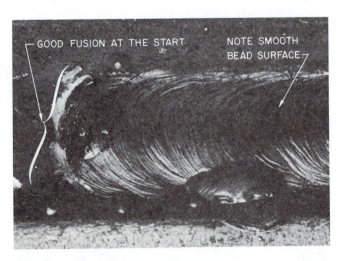

Figure 11-50 Weld bead made with GMAW axial spray metal transfer.

Tack weld the metal together and place the assembly in the flat position on the welding table. Start at one end and make a uniform weld along the entire 12-in. (305-mm) length of the joint. Watch the sides of the fillet weld for signs of undercutting.

Repeat the weld until it can be made straight, uniform, and free from any visual defects. Turn off the welding machine and shielding gas and clean up your work area when you are finished welding. ◆

PRACTICE 11-34

Butt Joint and Tee Joint

Using the same equipment, materials, and setup as listed in Practice 11-32, you will make a flat weld using axial spray metal transfer. Each weld must pass the guided bend test. Repeat each type of weld joint as needed until the bend test can be passed. Turn off the welding machine and shielding gas and clean up your work area when you are finished welding. ◆

Figure 11-51 GMAW axial spray metal transfer.

Self-Shielded, Flux-Cored, Arc Welding on a Major Project

A major international corporation, Standard Oil Company, has a new 1.5 million square foot headquarters building in Cleveland, Ohio, **Figure 1**. The office tower is 45 stories tall, with an 8-floor atrium of shops and restaurants. The building is the work-day home of 2,500 headquarters employees.

Ironworkers spent a year on-site, in all kinds of weather, working on the high steel construction. The architectural/engineering firms awarded the contracts for the building chose welded steel tube construction with braced frame support. The job required the on-site erection of 22,000 tons of steel. The steel fabricator decided to construct in the shop the intermediate and corner tree columns, which constitute 45 percent of the weight of the office tower. Shop welding of the tree columns limited the major on-site welding to splicing the ends of the "trees" to columns above and below. This was a critical consideration, given construction deadlines and the challenging nature of the weather conditions, which during the year of steel frame construction would treat field-welders to all forms of precipitation, high winds, and wind-chill factors reaching −37°F.

For its assigned task of accomplishing a top quality field-welding job on a stringent time schedule, the fabricators elected to use self-shielded, flux-cored arc welding. In this process, all arc and slagging materials reside in the core of the tubular electrode, eliminating the need for external flux or gas-handling equipment. Because it is an open arc process, operators can accurately position the weld metal and visually monitor the weld pool for maximum weld quality. It was the welding foreman's responsibility to see that twenty-six ironworkers completed fifty tons of field-welding as quickly and as well as it could be done.

Steel beam construction began by working two operators to a column, **Figure 2**, in specially designed tents constructed to protect them from

Figure 1

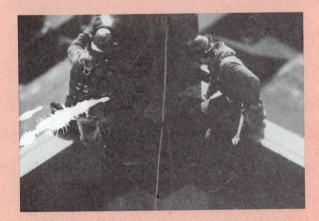

Figure 2

Figure 3

rain, snow, and wind. The operators ultimately completed 1,100 column splices, each requiring about 190 passes. Electric heating bars and supplementary propane torches, **Figure 3**, held the weld joints at the necessary 300°F throughout the welding process. Rain was diverted from the hot welds by miniature gutters of metal flashing tack-welded above the operators' heads.

The two-man approach employed on each column splice joint minimized possible column distortion and helped ensure the plumbness of the member. All welds were visually inspected and then ultrasonically tested to ensure their viability.

On the entire job, only three faulty welds were detected; each was repaired within the day it was found. This extraordinarily low cutout rate may be a record for high-rise projects of the magnitude of this building.

The major construction of the 45 story Standard Oil Company headquarters was completed within a one-year construction deadline. Stringent construction schedules, expert teamwork, and the finest materials available made the erection of the building possible within a short period of time.

(Courtesy of The Lincoln Electric Company, Cleveland, Ohio 44117.)

REVIEW

1. What items make up a basic semiautomatic welding system?

2. What must be done to the shielding gas cylinder before the valve protection cap is removed?

3. Why is the shielding gas valve "cracked" before the flowmeter regulator is attached?

4. What causes the electrode to bird nest?

5. Why must all fittings and connections be tight?

6. What parts should be activated by depressing the gun switch?

7. What benefit does a welding wire's cast provide?

8. What can be done to determine the location of a problem that stops the wire from being successfully fed through the conduit?

9. What are the advantages of using a feed roller pressure that is as light as possible?

10. Why should the feed roller drag prevent the spool from coasting to a stop when the feed stops?

11. Why must you always wind the wire tightly into a ball or cut it into short lengths before discarding it in the proper waste container?

12. Why would the flowmeter ball float at different heights with different shielding gases if the shielding gases are flowing at the same rate?

13. Using Table 11-1, determine the amperage if 400 in. (10.2 m/min) of 0.45-in. (1.2-mm) steel wire is fed in one minute.

14. How is the amperage adjusted on a GMA welder?

15. What happens to the weld as the electrode extension is lengthened?

16. What is the effect on the weld of changing the welding angle from a dragging to a pushing angle?

17. What are the advantages of adding oxygen or CO_2 to argon for welds on steel?

18. What are the advantages of using CO_2 for making GMA welds on steel?

19. What is mill scale?

20. What type of porosity is most often caused by mill scale?

21. What should the welder watch if the view of the weld is obstructed by the shielding gas nozzle?

22. When making a vertical weld and it appears that the weld metal is going to drip over the shelf, what should you do?

23. What are the advantages of making vertical down welds?

24. How can small weld beads be maintained during overhead welds?

25. How can spatter be controlled on the nozzle when making overhead welds?

26. How should the electrode be manipulated for the deepest penetration when using the pulsed-arc metal transfer process?

Chapter 12

FLUX CORED ARC WELDING EQUIPMENT, SETUP, AND OPERATION

OBJECTIVES

After completing this chapter, the student should be able to

- explain the effect of flux on the weld.
- explain the difference between self shielding and gas shielding.
- describe the different ways electrodes are made.
- name the parts of a flux cored arc welding setup.
- list the advantages of flux cored arc spot welding.
- list the major limitations of the flux cored arc welding process.

KEY TERMS

flux cored arc welding (FCAW)
slag
carbon
deoxidizer
self-shielding
rutile
lime
G

INTRODUCTION

Flux cored arc welding (FCAW) is a fusion welding process in which weld heating is produced from an arc between the work and a continuously fed filler metal electrode. Atmospheric shielding is provided completely or in part by the flux sealed within the tubular electrode, **Figure 12-1**. See Color Plate 21. Extra shielding may or may not be supplied through a nozzle in the same way as in GMAW.

Although the process was introduced in the early 1950s, it represented less than 5% of the total amount of welding done in 1965. In 1980, it passed the 20% mark and has been rising rapidly since then. The rapid rise in the use of FCAW has been due to a number of factors. The improvements in the fluxes, smaller electrode diameters, reliability of the equipment, better electrode feed systems, and improved guns have all lead to the increased usage. Guns equipped with smoke extraction nozzles and electronic controls are the latest in a long line of improvements to this process, **Figure 12-2**.

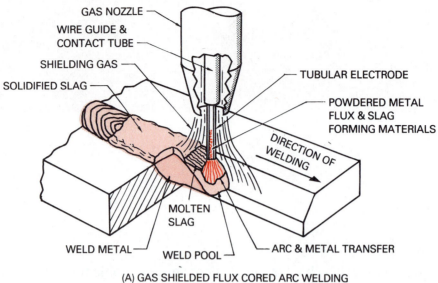

GAS NOZZLE

WIRE GUIDE & CONTACT TUBE

SHIELDING GAS

SOLIDIFIED SLAG

TUBULAR ELECTRODE

POWDERED METAL FLUX & SLAG FORMING MATERIALS

DIRECTION OF WELDING

MOLTEN SLAG

WELD METAL

WELD POOL

ARC & METAL TRANSFER

(A) GAS SHIELDED FLUX CORED ARC WELDING

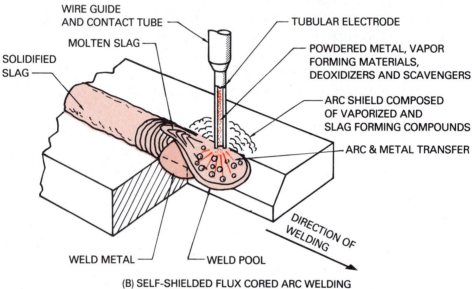

WIRE GUIDE AND CONTACT TUBE

MOLTEN SLAG

SOLIDIFIED SLAG

TUBULAR ELECTRODE

POWDERED METAL, VAPOR FORMING MATERIALS, DEOXIDIZERS AND SCAVENGERS

ARC SHIELD COMPOSED OF VAPORIZED AND SLAG FORMING COMPOUNDS

ARC & METAL TRANSFER

DIRECTION OF WELDING

WELD METAL

WELD POOL

(B) SELF-SHIELDED FLUX CORED ARC WELDING

Figure 12-1 (A) Gas shielded flux cored arc welding; (B) self-shielded flux cored arc welding. (Courtesy of the American Welding Society.)

PRINCIPLES OF OPERATION

FCA welding is similar in a number of ways to the operation of GMA welding, **Figure 12-3.** Both processes use a constant potential (CP) or constant voltage (CV) type power supply. CP power supplies provide a controlled voltage (potential) to the welding electrode. The amperage (current) varies with the speed that the electrode is being fed into the molten weld pool. Just like GMA welding, higher electrode feed speeds produce higher currents and slower feed speeds result in lower currents, assuming all other conditions remain constant.

The effects on the weld of electrode extension, gun angle, welding direction, travel speed, and other welder manipulations are similar to those experienced in GMA welding, **Figure 12-4.** Like GMA welding, having a correctly set

welder does not ensure a good weld. The skill of the welder is an important factor in producing high-quality welds.

The flux inside the electrode provides the molten weld pool with protection from the atmosphere, improves strengths through chemical reactions and alloys, and improves the weld shape.

Atmospheric contamination of molten weld metal occurs as it travels across the arc gap and within the pool before it solidifies. The major atmospheric contaminations come from oxygen and nitrogen, the major elements in air. The addition of fluxing and gas forming elements to the core electrode reduce or eliminate their effects.

Improved strength and other physical or corrosive properties of the finished weld are improved by the flux. Small additions of alloying elements, deoxidizers, and slag agents all can improve the desired weld prop-

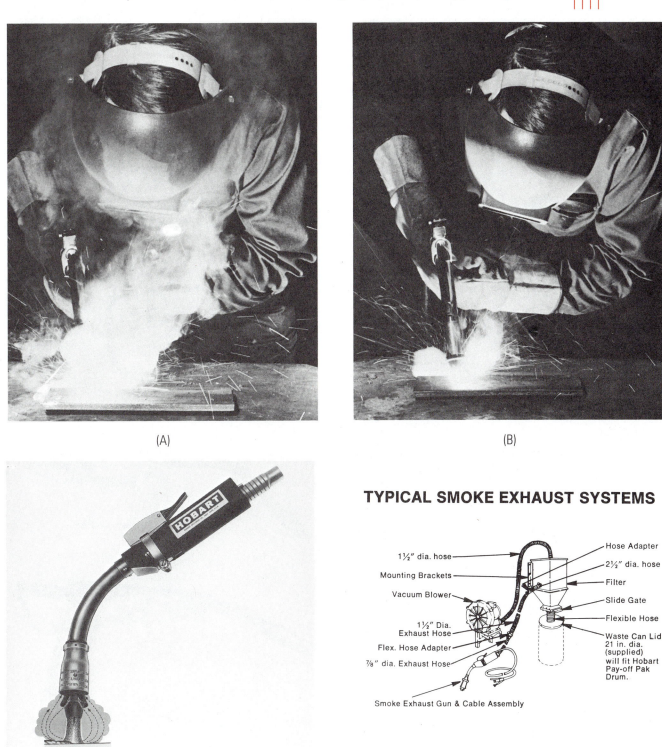

(A)

(B)

TYPICAL SMOKE EXHAUST SYSTEMS

1½" dia. hose

Mounting Brackets

Vacuum Blower

1½" Dia. Exhaust Hose

Flex. Hose Adapter

⅞" dia. Exhaust Hose

Smoke Exhaust Gun & Cable Assembly

Hose Adapter

2½" dia. hose

Filter

Slide Gate

Flexible Hose

Waste Can Lid 21 in. dia. (supplied) will fit Hobart Pay-off Pak Drum.

(C)

(D)

Figure 12-2 (A) FCA welding without smoke extraction and (B) with smoke extraction. (C) Typical FCAW smoke extraction gun. (D) Typical smoke exhaust system. (Courtesy of The Hobart Brothers Company.)

erties. Carbon, chromium, and vanadium can be added to improve hardness, strength, creep resistance, and corrosion resistance. Aluminum, silicon, and titanium all help remove oxidizes and/or nitrides in the weld. Potassium, sodium, and zirconium are added to form slag.

A slag covering of the weld is useful for several reasons. Slag helps the weld by protecting the hot metal from the effects of the atmosphere, controlling the bead shape by serving as a dam or mold, and serving as a blanket to slow the weld's cooling rate, which improves its physical properties, Figure 12-4.

Figure 12-3 Large-capacity wire-feed unit can be used with FCAW or GMAW. (Courtesy of Lincoln Electric, Cleveland, OH.)

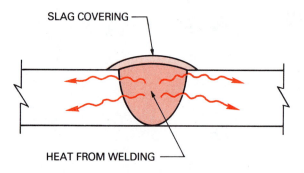

Figure 12-4 The slag covering keeps the welding heat from escaping as quickly, thus slowing the cooling rate.

EQUIPMENT

Power Supply

The FCA welding power supply is the same type that is required for GMAW, called constant-potential, constant-voltage (CP, CV). The words *potential* and *voltage* have the same electrical meaning and are used interchangeably. FCAW machines can be much more powerful than GMAW machines and are available with up to 1500 amperes of welding power.

Guns

Guns are available as either air- or water-cooled types. Air-cooled guns are the most common. Often there is a hand-protecting shield to improve operator comfort when higher amperages are being used, **Figure 12-5.**

Electrode Feed

Electrode feed systems are similar to those used for GMAW. The major difference is that larger FCA machines that can use large-diameter wire most often have two sets of feed rollers. The two sets of rollers help reduce the drive pressure on the electrode. Excessive pressure can distort the electrode wire diameter, which can allow some flux to be dropped inside the electrode guide tube.

ADVANTAGES

FCA welding offers the welding industry a number of important advantages.

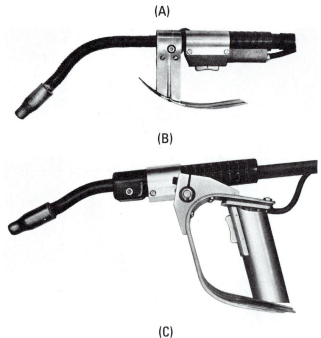

Figure 12-5 Typical FCA welding guns (A) 350 ampere rating self-shielding, (B) 450 ampere rating gas-shielding, and (C) 600 ampere rating gas-shielding. (Courtesy of Lincoln Electric, Cleveland, OH.)

High Deposition Rate

High rates on depositing weld metal are possible. FCA welding deposition rates of more than 25 lb/hr (12 kg/hr) of weld metal are possible. This compares to about 10 lb/hr (6 kg/hr) for SMA welding using a very large diameter electrode of 1/4 in. (6 mm).

Minimum Electrode Waste

The FCA method makes efficient use of filler metal; from 75% to 90% of the weight of the FCA electrode is metal, the remainder being flux. SMA electrodes have a maximum of 75% filler metal; some SMA electrodes have much less. Also a stub must be left at the end of each SMA welding electrode. The stub will average 2 in. (51 mm) in length, resulting in a loss of 11% or more of the SMAW filler electrode purchased. FCA welding has no stub loss, so nearly 100% of the FCAW electrode purchased is used.

Less Edge Preparation

Because of the deep penetration characteristic, no edge beveling preparation is required on some joints in metal up to 1/2 in. (13 mm) in thickness. When bevels are cut, the joint included angle can be reduced to as small as 35°, **Figure 12-6**. The reduced groove angle results in a smaller sized weld. This can save 50% of filler metal with about the same savings in time and weld power used.

Minimum Precleaning

The addition of **deoxidizers** and other fluxing agents permits high-quality welds to be made on plates with light surface oxides and mill scale. This eliminates most of the precleaning required before GMA welding could be performed. Often it's possible to make excellent welds on plates in the "as cut" condition; no cleanup needed.

All Position

Small diameter electrode sizes in combination with special fluxes allow excellent welds in all positions. The slags produced assist in supporting the weld metal. This process is easy to use, and, when properly adjusted, it is much easier to use than other all-position arc welding processes.

Flexibility

Changes in power settings can permit welding to be made on thin-gauge sheet metals or thicker plates using the same electrode size. Multipass welds allow joining unlimited thickness metals. This, too, is attainable with one size of electrode.

High-quality

Many codes permit welds to be made using FCAW. The addition of the flux gives the process the high level of reliability needed for welding on boilers, pressure vessels, or structural steel.

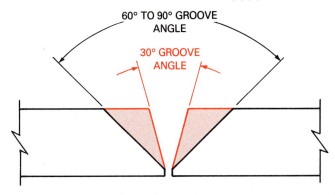

Figure 12-6 The narrower groove angle for FCAW saves on filler metal, welding time, and heat input into the part.

Excellent Control

The molten weld pool is more easily controlled with FCAW than with GMAW. The surface appearance is smooth and uniform even with less operator skill. Visibility is improved by removing the nozzle when using self-shielded electrodes.

LIMITATIONS

The main limitation of flux cored arc welding is that it is confined to ferrous metals and nickle-based alloys. Generally, all low- and medium-carbon steels and some low-alloy steels, cast irons, and a limited number of stainless steels are presently weldable using FCAW.

The equipment and electrodes used for the FCAW process are more expensive. However, the cost is quickly recoverable through higher productivity.

The removal of postweld slag requires another production step. The flux must be removed before the weldment is finished (painted) to prevent crevice corrosion.

With the increased welding output comes an increase in smoke and fume generation. The existing ventilation system in a shop might need to be increased to handle the added volume.

ELECTRODES: METHODS OF MANUFACTURING

The electrodes have flux tightly packed inside. One method used to make them is to first form a thin sheet of metal into a U-shape, **Figure 12-7**. Then a measured quantity of flux is poured into the U-shape before it is squeezed shut. It is then passed through a series of dies to size it and further compact the flux, **Figure 12-8**.

A second method of manufacturing the electrode is to start with a seamless tube. The tube is usually about 1 in. in diameter. One end of the tube is sealed, and the flux powder is poured into the open end. The tube is vibrated during the filling process to insure that it fills completely. Once the tube is full, the open end is sealed. The tube is now sized using a series of dies.

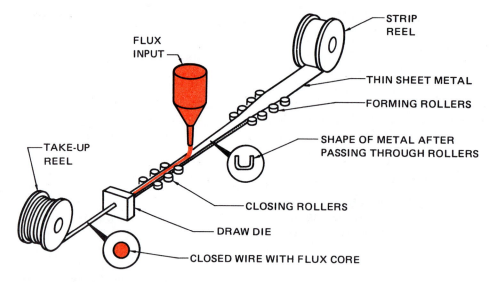

Figure 12-7 Putting the flux in the flux-cored wire.

In both these methods of manufacturing the electrode, the sheet and tube are made up of the desired alloy. Also in both cases the flux is compacted inside the metal skin. This compacting helps make the electrode operate smoother and more consistently.

Electrodes are currently available in sizes from 0.030 in. to 5/32 in. (0.8 mm to 3.9 mm) in diameter, **Table 12.1**. Smaller diameter electrodes are much more expensive per pound than the same type in a larger diameter. Larger diameter electrodes produce such large welds they cannot be controlled in all positions. The most popular diameters range from .045 in. to 3/32 in. (1.2 mm to 2.3 mm).

FCAW filler metal can be purchased on spools weighing around 50 lbs., on coils weighing around 60 lbs., and on reels or in drums up to 750 lbs., **Figure 12-9**.

FLUX

The fluxes used are mainly rutile or lime based. The purpose of the fluxes is the same as in the SMAW process. That is, they can provide all or part of the following to the weld:

■ *Deoxidizers:* Oxygen that is present in the welding zone has two forms. It can exist as free oxygen from the atmosphere surrounding the weld. Oxygen also can exist as part of a compound such as an iron oxide

or carbon dioxide (CO_2). In either case it can cause porosity in the weld if it is not removed or controlled. Chemicals are added that react to the presence of oxygen in either form and combine to form a harmless compound, **Table 12-2**. The new com-

Electrode Diameter	Minimum Metal Thickness
0.035 in. (0.9 mm)	18 gauge 0.047 in. (1.19 mm)
0.045 in. (1.2 mm)	Plate 1/8 in. (3.2 mm)
0.052 in. (1.3 mm)	Plate 1/4 in. (6.3 mm)
5/64 in. (2.0 mm)	Plate 3/8 in. (9.5 mm)
0.035 in. (0.9 mm)	Pipe wall less than 1/2 in. (13 mm)
0.045 in. (1.2 mm)	Pipe wall more than 1/2 in. (13 mm)

Table 12-1 Electrode Size and Minimum Metal Thickness.

Deoxidizing Element	Strength
Aluminum (Al)	Very strong
Manganese (Mn)	Weak
Silicon (Si)	Weak
Titanium (Ti)	Very strong
Zirconium (Zr)	Very strong

Table 12-2 Deoxidizing Elements Added to Filler Wire (to minimize porosity in the molten weld pool).

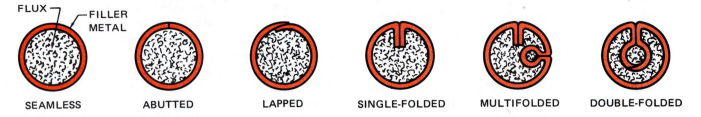

Figure 12-8 Some variations of the typical abutted cored wire. These variations allow higher percentage of filler metal-to-flux without losing the necessary flexibility. The abutted shape is the most common by far.

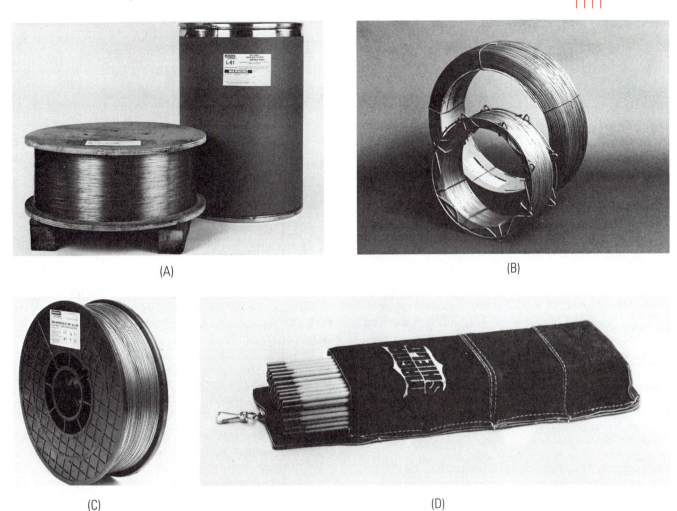

(A)

(B)

(C)

(D)

Figure 12-9 FCAW filler metal weights are approximate. They will vary by alloy and manufacturer. (A, B, C Courtesy of Lincoln Electric, Cleveland, OH.; D Courtesy of Elliott Clove Company.)

pound can become part of the slag that solidifies on top of the weld, or some of it may stay in the weld as very small inclusions. Both methods result in a weld with better mechanical properties with less porosity.

■ *Slag formers:* Slag serves several vital functions for the weld. It can react with the molten weld metal chemically, and it can affect the weld bead physically. In the molten state it moves through the molten weld pool and acts like a magnet or sponge to chemically combine with impurities in the metal and remove them, **Figure 12-10**. Slags can be refractory, become solid at a high temperature, and solidify over the weld molten, helping it hold its shape and slowing its cooling rate.

■ *Fluxing agents:* Molten weld metal tends to have a high surface tension, which prevents it from flowing outward toward the edges of the weld. This causes undercutting along the junction of the weld and the base metal. Fluxing agents make the weld more fluid and allow it to flow outward, filling the undercut.

■ *Arc stabilizers:* Chemicals in the flux affect the arc resistance. As the resistance is lowered the arc voltage drops and penetration is reduced. When the arc resistance is increased, the arc voltage increases and weld penetration is increased. Although the resistance within the ionized arc stream may change, the arc is more stable and easier to control. It also

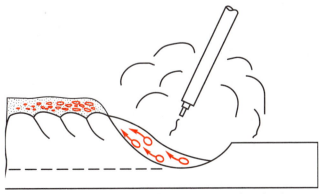

Figure 12-10 Impurities being floated to the surface by slag.

improves the metal transfer by reducing spatter caused by an erratic arc.

■ *Alloying elements:* Because of the difference in the mechanical properties of metal that is formed by rolling or forging and metal that is melted to form a cast weld nugget, the metallurgical requirements of the two also differ. Some elements change the weld's strength, ductility, hardness, brittleness, toughness, and corrosion resistance. Other alloying elements in the form of powder metal can be added to both alloys and add to the deposition rate.

■ *Shielding gas:* As elements in the flux are heated by the arc some of them vaporize and form voluminous gaseous clouds hundreds of times larger than their original volume. This rapidly expanding cloud forces the air around the weld zone away from the molten weld metal, **Figure 12-11.** Without the protection this process affords the molten metal, it would rapidly oxidize. Such oxidization would severely affect the weld's mechanical properties, rendering it unfit for service.

All FCAW fluxes are divided into two groups based on the acid or basic chemical reactivity of the slag. The AWS classifies T-1 as acid and T-5 as basic.

The rutile-based flux is acidic, T-1. It produces a smooth, stable arc and a refractive high temperature slag for out-of-position welding. These electrodes produce a fine drop transfer, a relatively low fume, and an easily removed slag. The main limitation of the rutile fluxes is that their fluxing elements do not produce as high a quality deposit as do the T-5 systems.

The lime-based flux is basic, T-5. It is very good at removing certain impurities from the weld metal, but its low melting temperature slag is fluid, which makes it generally unsuitable for out-of-position welding. These electrodes produce a more globular transfer, more spatter, more fume, and a more adherent slag than do the T-1 systems. These characteristics are tolerated when it is necessary to deposit very tough weld metal and for welding materials having a low tolerance for hydrogen.

Some rutile-based electrodes allow the addition of a shielding gas. With the weld being protected partially by the shielding gas, more elements can be added to the flux, which produces welds with the best of both flux systems, high-quality welds in all positions.

Some fluxes can be used on both single and multiple pass welds, and others are limited to single pass welds only. Using a single pass welding electrode for multipass welds may result in an excessive amount of manganese. The manganese is necessary to retain strength when making large, single pass welds. However, with the lower dilution associated with multipass techniques, it can strengthen the weld metal too much and reduce its ductility. In some cases, small welds that deeply penetrate the base metal can help control this problem.

Table 12-3 lists the shielding gas and polarity for the flux classifications of mild steel FCAW electrodes. The letter G is used to indicate an unspecified classification.

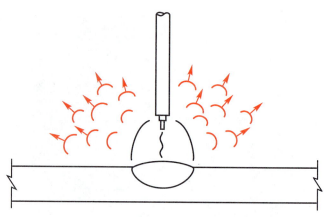

Figure 12-11 Rapidly expanding gas cloud.

The G means that the electrode has not been classified by the American Welding Society. Often the exact composition of fluxes are kept as a manufacturer's trade secret. Therefore, only limited information about the electrode's composition will be given. The only information often supplied is current, type of shielding required, and some strength characteristics.

As a result of the relatively rapid cooling of the weld metal, the weld may tend to become hard and brittle. This factor can be controlled by both adding elements to the flux and the slag formed by the flux, **Table 12-4.** Ferrite is the softer, more ductal form of iron. The addition of ferrite-forming elements can control the hardness and brittleness of a weld. Refractory fluxes are sometimes called "fast-freeze" because they solidify at a higher temperature than the weld metal. By becoming solid first, this slag can cradle the molten weld pool and control its shape. This property is very important for out-of-position welds.

The impurities in the weld pool can be metallic or nonmetallic compounds. Metallic elements that are added to the metal during the manufacturing process in small quantities may be concentrated in the weld. These elements improve the grain structure, strength, hardness, resistance to corrosion, or other mechanical properties in the metal's as-rolled or formed state. But weld nugget is a small casting, and some alloys adversely affect the properties of this casting (weld metal). Nonmetallic compounds are primarily slag inclusions left in the metal from the fluxes used during manufacturing. The welding fluxes form slags that are less dense than the weld metal so that they will float to the surface before the weld solidifies.

Electrode Classification

The American Welding Society classifies FCA welding electrodes as Tubular Wire. **Table 12-5** lists the AWS specifications for filler metals for flux cored arc welding.

Mild Steel. In the AWS classification for mild steel, FCA electrode starts with the letter E, which stands for electrode, **Figure 12-12.** The E is followed by a single one- or two-digit number to indicate the minimum

Classifications	Comments	Shielding Gas
T-1	Requires clean surfaces and produces little spatter. It can be used for single and multiple pass welds in the flat (1G and 1F) and horizontal (2F) positions.	Carbon dioxide (CO_2)
T-2	Requires clean surfaces and produces little spatter. It can be used for single pass welds in the flat (1G and 1F) and horizontal (2F) positions only.	Carbon dioxide (CO_2)
T-3	Used on thin-gauge steel for single pass welds in the flat (1G and 1F) and horizontal (2F) positions only.	None
T-4	Low penetration and moderate tendency to crack for single and multiple pass welds in the flat (1G and 1F) and horizontal (2F) positions.	None
T-5	Low penetration and a thin, easily removed slag, used for single and multiple pass welds in the flat (1G and 1F) position only.	With or without carbon dioxide (CO_2)
T-6	Similar to T-5 without externally applied shielding gas.	None
T-G	The composition and classification of this electrode is not given in the preceding classes. It may be used for single or multiple pass welds.	With or without shielding

Table 12-3 Welding Characteristics of Seven Flux Classifications.

Element	Reaction in Weld
Silicon (Si)	Ferrite former and deoxidizer
Chromium (Cr)	Ferrite and carbide former
Molybdenum (Mo)	Ferrite and carbide former
Columbium (Cb)	Strong ferrite former
Aluminum (Al)	Ferrite former and deoxidizer

Table 12-4 Ferrite-Forming Elements Used in FCA Welding Fluxes.

Metal	AWS Filler Metal Classification
Mild steel	A5.20
Stainless steel	A5.22
Chromium–Molybdenum	A5.29

Table 12-5

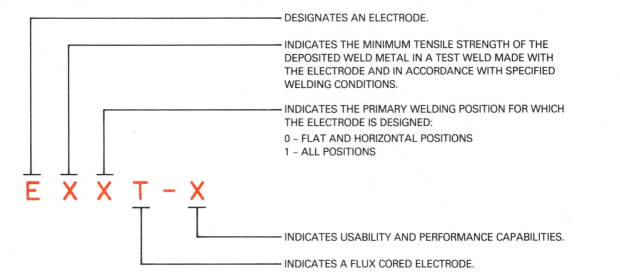

DESIGNATES AN ELECTRODE.

INDICATES THE MINIMUM TENSILE STRENGTH OF THE DEPOSITED WELD METAL IN A TEST WELD MADE WITH THE ELECTRODE AND IN ACCORDANCE WITH SPECIFIED WELDING CONDITIONS.

INDICATES THE PRIMARY WELDING POSITION FOR WHICH THE ELECTRODE IS DESIGNED:
0 – FLAT AND HORIZONTAL POSITIONS
1 – ALL POSITIONS

E X X T - X

INDICATES USABILITY AND PERFORMANCE CAPABILITIES.

INDICATES A FLUX CORED ELECTRODE.

Figure 12-12 Identification system for mild steel FCAW electrodes. (Courtesy of the American Welding Society.)

tensile strength, in pounds per square inch (psi), of a good weld. The actual strength is obtained by adding four zeros to the right of the number given. For example, E6xT-x is 60,000 psi, and E11xT-x is 110,000 psi.

The next number, 0 or 1, indicates the welding positions. Ex0T is to be used in a horizontal or flat position only. Ex1T is an all-position filler metal.

The T located to the right of the tensile strength and weld position numbers indicates that this is a tubular, flux-cored wire. The last number—1, 2, 3, 4, 5, 6, 7, 8, 10, 11—or the letters G or GS are used to indicate the type of flux and whether the filler metal can be used for single- or multiple-pass welds. The electrodes with the numbers ExxT-2, ExxT-3, ExxT-10, and ExxT-GS are intended for single-pass welds only.

Stainless Steel Electrodes.   The AWS classification for stainless steel for FCA electrodes starts with the letter E as its prefix. Following the E prefix, the American Iron and Steel Institute's (AISI) three-digit stainless steel number is used. This number indicates the type of stainless steel in the filler metal.

To the right of the AISI number, the AWS adds a dash followed by a suffix number. The number 1 is used to indicate an all-position filler metal, and the number 3 is used to indicate an electrode to be used in the flat and horizontal positions only.

SHIELDING GAS

FCA welding wire can be manufactured so that all of the required shielding of the molten weld pool is provided by the vaporization of some of the flux within the tubular electrode. When the electrode provides all of the shielding, it is called *self-shielding*. Other FCA welding wire must use an externally supplied shielding gas to provide the needed protection of the molten weld pool. When a shielding gas is added, it is called *dual shield*.

Care must be taken to use the cored electrodes with the recommended gases or not to use gas at all with the self-shielded electrodes. Using a self-shielding, flux-cored electrode with a shielding gas may produce a defective weld. The shielding gas will prevent the proper disintegration of much of the deoxidizers. This results in the transfer of these materials across the arc to the weld. In high concentrations, the deoxidizers can produce slags that become trapped in the welds, causing undesirable defects. Lower concentrations may cause brittleness only. In either case, the chance of weld failure is increased. If these electrodes are used correctly, there is no problem.

The selection of a shielding gas will affect the arc and weld properties. The weld bead width, buildup, penetration, spatter, chemical composition, and mechanical properties are all affected as a result of the shielding gas selection.

Gases used for FCA welding include CO_2 or mixtures of argon and CO_2. Argon gas is easily ionized by the arc.

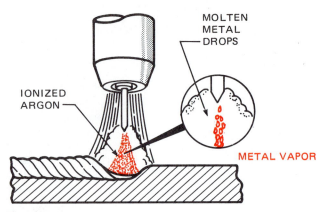

Figure 12-13

Ionization results in a highly concentrated path from the electrode to the weld. This concentration results in a smaller droplet size that is associated with the axial spray mode of metal transfer, **Figure 12-13**. A smooth stable arc results and there is a minimum of spatter. This transfer mode continues as CO_2 is added to the argon until the mixture contains more than 25% of CO_2.

As the percentage of CO_2 increases in the argon mixture, weld penetration increases. This increase in penetration continues until a 100% CO_2 shielding gas is reached. But as the percentage of CO_2 is increased the arc stability decreases. The less stable arc causes an increase in spatter. A mixture of 75% argon and 25% CO_2 works best for jobs requiring a mixed gas.

Straight CO_2 is used for some welding. But the CO_2 gas molecule is easily broken down in the welding arc. It forms carbon monoxide (CO) and free oxygen (O). Both gases are reactive to some alloys in the electrode. As these alloys travel from the electrode to the molten weld pool, some of them form oxides. Silicon and manganese are the primary alloys that become oxidized and lost from the weld metal.

Most FCA welding electrodes are specifically designed to be used with or without shielding gas and for a specific shielding gas or percentage mixture. For example, an electrode designed specifically for use with 100% CO_2 would have higher levels of silicon and manganese to compensate for the losses to oxidization. But if 100% argon or a mixture of argon and CO_2 is used, the weld will have an excessive amount of silicon and manganese. The weld will not have the desired mechanical or metallurgical properties. Although the weld may look satisfactory, it will probably fail prematurely.

■ *CAUTION*

Never use an FCA welding electrode with a shielding gas it is not designated to be used with. The weld it produces may be unsafe.

WELDING TECHNIQUES

Porosity

FCA welding can produce high-quality welds in all positions, although porosity in the weld can be a persistent problem. Porosity can be caused by moisture in the flux, improper gun manipulation, or surface contamination.

The flux used in the FCA welding electrode is subject to picking up moisture from the surrounding atmosphere, so the electrodes must be stored in a dry area. Once the flux becomes contaminated with moisture, it is very difficult to remove. Water (H_2O) breaks down into free hydrogen and oxygen in the presence of an arc, Figure 12-14. The hydrogen can be absorbed into the molten weld metal where it can cause postweld cracking. The oxygen is absorbed into the weld metal also, but it forms bubbles of porosity as the weld begins to solidify.

If a shielding gas is being used, the FCA welding gun gas nozzle must be close enough to the weld to provide adequate shielding gas coverage. If there is a wind or if the nozzle to work distance is excessive, the shielding will be inadequate and cause weld porosity. If welding is to be done outside or in an area subject to drafts, the gas flow rate must be increased or a wind shield must be placed to protect the weld, Figure 12-15.

A common misconception is that the flux within the electrode will either remove or control weld quality problems caused by surface contaminations. That is not true. The addition of flux makes FCA welding more tolerant to surface conditions than GMA welding, although it still is adversely affected by such contaminations.

New hot-rolled steel has a layer of dark gray or black iron oxide called mill scale. Although this layer is very thin, it may provide a source of enough oxygen to cause porosity in the weld. If mill scale causes porosity, it is usually uniformly scattered through the weld, Figure 12-16. Unless severe, uniform scattered porosity is usually not visible in the finished weld. It is trapped under the surface as the weld cools.

Because porosity is under the weld surface, nondestructive testing methods, including X ray, magnetic particle, and ultra sound, must be used to locate it in a weld. It can be detected by mechanical testing such as guided-bend, free-bend, and nick-break testing for establishing weld parameters. Often it is better to remove the mill scale before welding rather than risking the production of porosity.

All welding surfaces within the weld groove and the surrounding surfaces within 1 in. (25 mm) must be cleaned to bright metal, Figure 12-17. Cleaning may be either grinding, filing, sanding, or blasting.

Anytime FCA welds are to be made on metals that are dirty, oily, rusty, wet, or have been painted, the surface must be precleaned. Cleaning can be chemically or mechanically.

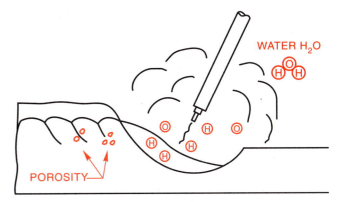

Figure 12-14 Water (H_2O) breaks down in the presence of the arc and the hydrogen (H) is dissolved in the molten weld metal.

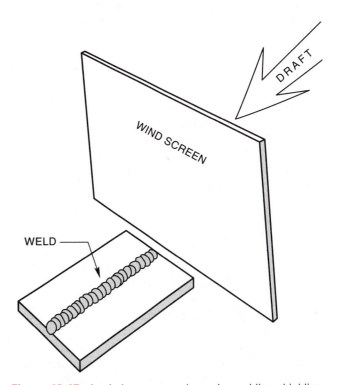

Figure 12-15 A wind screen can keep the welding shielding from being blown away.

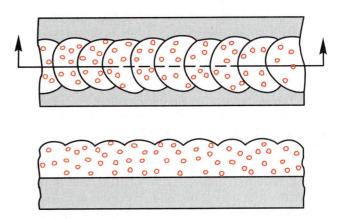

Figure 12-16 Uniformly scattered porosity.

One advantage of chemically cleaning oil and paint is that it's easier to clean larger areas. Both oil and paint smoke easily when heated, and such smoke can cause weld defects. They must be removed far enough from the weld so that weld heat does not cause them to smoke. In the case of small parts the entire part may need to be cleaned.

Gun Angle

The FCA welding gun may be held so that the relative angle between the gun, work, and welding bead being made is either vertical or has a drag angle or a push angle, **Figure 12-18**. Changes in this angle will affect the weld bead.

FCA electrodes have a mineral-based flux, often called low-hydrogen. These fluxes are refractory and become solid at a high temperature. If too steep a forehand or pushing angle is used, slag from the electrode can be pushed ahead of the weld bead and solidify quickly on the cooler plate, **Figure 12-19**. Because the slag remains solid at temperatures above that of the molten weld pool, it can be trapped under the weld by the molten weld metal. To avoid this problem, most flat and horizontal welds should be performed with a backhand or trailing angle.

Vertical-up welds require a forehand gun angle. The forehand angle is needed to direct the arc deep into the groove or joint for better control of the weld pool and deeper penetration, **Figure 12-20**. Slag entrapment associated with most forehand welding is not a problem for vertical welds.

A gun angle around 90° either slightly forehand or backhand works best for overhead welds, **Figure 12-21**. The slight angle aids with visibility of the weld and it helps control spatter buildup in the gas nozzle.

Electrode Extension

The electrode extension is measured from the end of the electrode contact tube to the point the arc begins at the end of the electrode, **Figure 12-22**. Compared to GMA welding, the electrode extension required for FCAW is much greater. The longer extension is required for several reasons. The electrical resistance of the wire causes the wire to heat up, which can drive out moisture from the flux. This preheating of the wire also results in a smoother arc with less spatter.

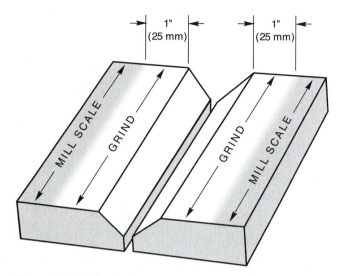

Figure 12-17 Grind mill scales off plates within 1 in. (25mm) of the groove.

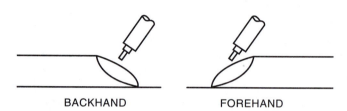

BACKHAND   FOREHAND

Figure 12-18

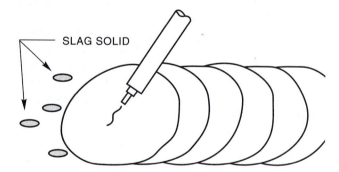

SLAG SOLID

Figure 12-19

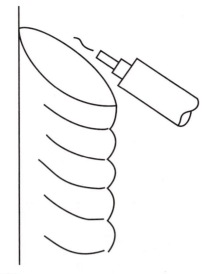

Figure 12-20

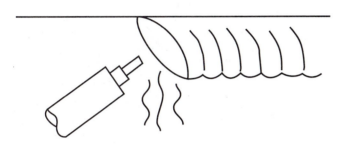

Figure 12-21

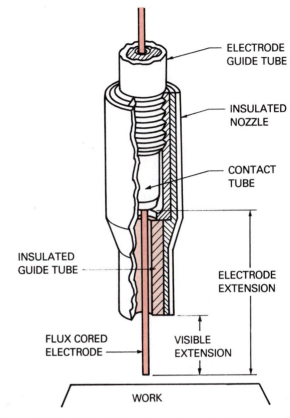

Figure 12-22 Self-shielded electrode nozzle. (Courtesy of the American Welding Society.)

Flux Cored Wire Welds Large Stainless Steel Plenum

Figure 1

Silver Engineering Works, Inc., faced the challenge to custom manufacture a huge drying plenum for the Department of Energy's Rocky Flats plutonium processing plant in Colorado. This plenum is used to decontaminate workers' uniforms and is part of a $2-million program to improve safety conditions at Rocky Flats.

"Actually, the plenum is a stainless steel room, designed for installation within a warehouse building," said Ron Applehans, manufacturing manager at Silver. "It has three stairways on two levels and is built of stainless steel for better corrosion resistance." Figure 1.

FLUX CORED WIRE SPEEDS OUT-OF-POSITION WELDING

To meet the demanding requirements of the nuclear industry and complete the project according to its budget and schedule, Silver selected 0.045-in.-diameter Shield-Bright (AWS E308LT-1) — an all-position, gas shielded, stainless steel, flux cored welding wire produced by Alloy Rods Corporation, Hanover, Pennsylvania.

Applehans said Silver selected the flux-cored welding wire because covered electrodes are too slow and inefficient, and solid wire are very difficult to weld in the overhead position. Welds were mostly T joints, butt joints, and corner joints.

CUTS WELDING TIME

During the three-month production schedule, Silver Engineering's welders reported "outstanding performance" in all positions: horizontal, vertical, and overhead.

"Due to the size of the plenum, welders worked at least 30% to 40% of the time out of position," Applehans explained, "so good penetration in all welding positions was extremely important." Plates were welded in a single pass, and welders used a 75% argon, 25% CO_2 shielding gas with current settings of 25-27 V and 175-200 A, depending on the welding position. Wire feed speed was 400-475 in./min.

Applehans estimated that using the flux cored welding wire instead of covered electrodes cut welding time on the project by "at least two-thirds."

QUALITY WELDS PASS DOE TESTS

Applehans said Silver Engineering randomly x-rayed the welds to confirm they met specifications: no visible inclusions, no porosity, and no cracking.

During the initial visual inspection, he noted that the welders liked the way the flux cored wire "handled" and the appearance of the weld bead. "The appearance of the weld is an indicator of quality," he said, "and welders take special pride in a good job that looks good."

In addition to the visual inspection at Silver, welds in the exhaust plenum were subjected to dye penetrant inspection and ultrasonic testing before delivery to Rocky Flats. At Rocky Flats, the final vacuum testing was conducted by DOE. The plenum met every test; no rework was required, and no leaks or failures were encountered.

(Reprinted from the Welding Journal. Article based on a story from Alloy Rods Corporation, Hanover, Pennsylvania. Courtesy of the American Welding Society, Inc.)

REVIEW

1. List some factors that have led to the increased use of FCA welding.

2. How is FCAW similar to GMAW?

3. What does the FCA flux provide to the weld?

4. What are the major atmospheric contaminations of the molten weld metal?

5. How does slag help an FCA weld?

6. What is the electrical difference between a constant-potential and a constant-voltage power supply?

7. How can FCA welding guns be cooled?

8. What problems does excessive drive roller pressure cause?

9. List the advantages that FCA welding offers the welding industry.

10. Describe the two methods of manufacturing FCA electrode wire.

11. Why are the large-diameter electrodes not used for all position welding?

12. How do deoxidizers remove oxygen from the weld zone?

13. What do fluxing agents do for a weld?

14. Why are alloying elements added to the flux?

15. How does the flux form a shielding gas to protect the weld?

16. What are the main limitations of the rutile fluxes?

17. Why is it more difficult to use lime-based fluxed electrodes on out-of-position welds?

18. What benefit does adding an externally supplied shielding gas have on some rutile-based electrodes?

19. How do excessive amounts of manganese affect a weld?

20. Why are elements added that cause ferrite to form in the weld?

21. Why are some slags called refractory?

22. Why must a flux form a less dense slag?

23. Referring to Table 12-5, what is the AWS classification for FCA welding electrodes for stainless steel?

24. Describe the meaning of each part of this FCA welding electrode identification: E81T-5.

25. What does the number 316 in E316T-1 mean?

26. What is the advantage of using an argon-CO_2 mixed shielding gas?

27. What are the primary alloying elements lost if 100% CO_2 shielding gas is used?

28. What can cause porosity in an FCA weld?

29. What happens to water in the welding arc?

30. What is the thin dark gray or black layer on new hot-rolled steel? How can it affect the weld?

31. Why is uniform scattered porosity hard to detect in a weld?

32. What cautions must be taken when chemically cleaning oil or paint from a piece of metal?

33. What can happen to slag that solidifies on the plate ahead of the weld?

34. How is the electrode extension measured?

Chapter 13

FLUX CORED ARC WELDING

OBJECTIVES

After completing this chapter, the student should be able to

- set up a constant potential, semiautomatic FCA welding system.
- explain the effects that changing from a self-shielded to a dual-shielded electrodes system has on welding.
- control weld bead contour during welding by using the proper weave pattern.
- demonstrate an ability to control undercut, overlap, porosity, and slag inclusions when welding in all positions.
- pass standard weld test using FCA welding.

KEY TERMS

weave bead	lap joint
stringer bead	tee joint
amperage range	feed rollers
wire-feed speed	contact tube
conduit liner	critical weld
root face	

INTRODUCTION

Setup of the FCA weld station is the key to making quality welds. It may be possible, using a poorly set up FCA welder, to make an acceptable weld in the flat position. The FCA welding process is often forgiving; thus welds can often be made even when the welder is not set correctly. However, such welds will have major defects such as excessive spatter, undercut, overlap, porosity, slag inclusions, and poor weld bead contours. Setup becomes even more important for out-of-position welds. Making vertical and overhead welds can be difficult for a student welder with a properly set up system, but it becomes impossible with a system that is out of adjustment.

Learning to set up and properly adjust the FCA welding system will allow you to produce high-quality welds at a high level of productivity.

FCAW is set up and manipulated in a manner similar to that of GMAW. The results of changes in electrode extension, voltage, amperage, and torch angle are essentially the same. Be sure to check the owner's manual for the equipment you are using to be sure it's approved for FCAW and to see whether there are any special instructions for FCAW.

■ *CAUTION*

FCA welding produces a lot of ultraviolet light, heat, sparks, slag, and welding fumes. Proper personal, protective clothing and special protective clothing must be worn to prevent burns from the ultraviolet light and hot weld metal. Eye protection must be worn to prevent injury from flying sparks and slag. Forced ventilation and possibly a respirator must be used to prevent fume-related injuries. Refer to the safety precautions provided by the equipment and electrode manufacturers and to Chapter 2, "Safety in Welding," for additional safety help.

PRACTICES

The practices in this chapter are grouped according to those requiring similar techniques and setups. Plate welds are covered first, then sheet metal. The practices start with 1/4-in. (6-mm) mild steel plates; they are used because they require the least preparation times. The thicker 3/8-in. (9.5-mm) plates provide the basics of practicing groove welding. The 3/4-in. (19-mm) and thicker plates are used to develop the skills required to pass the unlimited thickness test often given to FCA welders. Sheet metal is grouped together because it presents a unique set of learning skills.

The major skill required for making consistently acceptable FCA welds is the ability to set up the welding system. Changes such as variations in material thickness, position, and type of joint require changes both in technique and setup. A correctly set up FCA welding station can, in many cases, be operated by a less-skilled welder. Often the only difference between a welder earning a minimum wage and one earning the maximum wage is the ability to correct machine setups.

For several reasons the FCA welding practice plates will be larger than most other practice plates. Welding heat and welding speed are the major factors that necessitate this increased size. FCA welding is both high energy and fast, and the welding energy (heat) input is so great that small practice plates may glow red by the end of a single weld pass. This would seriously affect the weld quality. To prevent this from happening, wider plates are used. Because of the higher welding speeds, longer plates are usually used.

Plates less than 1/2 in. (13-mm) will be 12 in. (305 mm) long for most practices. In addition to controlling the heat buildup, the longer plates are needed to give the welder enough time to practice welding. Learning to make longer welds is a skill that must also be practiced, because the FCA welding process is used in industry to make long production welds.

Plates thicker than 1/2 in. (13 mm) can be shorter than 12 in. (305 mm). Most codes allow test plates of

"Unlimited thickness" [usually to be as short as 7 in. (178 mm)].

PRACTICE 13-1

FCAW Equipment Setup

For this practice, you will need a semiautomatic welding power source approved for FCA welding, welding gun, electrode feed unit, electrode supply, shielding gas supply, shielding gas flowmeter regulator, electrode conduit, power and work leads, shielding gas hoses (if required), assorted hand tools, spare parts, and any other required materials. In this practice, you will properly set up an FCA welding station. Some manufacturers include detailed setup instructions with their equipment. If such instructions are available for your equipment, follow them. Otherwise, use the following instructions.

If the shielding gas is to be used and it comes from a cylinder, the cylinder must be chained securely in place before the valve protection cap is removed, Figure 13-1. Stand to one side of the cylinder and quickly crack the valve to blow out any dirt in the valve before the flowmeter regulator is attached, Figure 13-2. Attach the cor-

Figure 13-1 Make sure the gas cylinder is chained securely in place before removing the safety cap.

rect hose from the regulator to the "gas-in" connection on the electrode feed unit or machine.

Install the reel of electrode (welding wire) on the holder and secure it, Figure 13-3. Check the feed roller size to ensure that it matches the wire size, Figure 13-4. The conduit liner size should be checked for compatibility with the wire size. Connect the conduit to the feed unit. The conduit or an extension should be aligned with the groove in the roller and set as close to the roller as possible without touching, Figure 13-5. Misalignment at this point can contribute to a bird's nest, Figure 13-6. Bird-nesting of the electrode wire, so-called because it looks like a bird's nest, results when the feed roller pushes the wire into a tangled ball because the wire would not go through the outfeed side conduit.

Be sure the power is off before attaching the welding cables. The electrode and work leads should be attached to the proper terminals. The electrode lead should be attached to electrode or positive (+). If necessary, it is also attached to the power cable part of the gun lead. The work lead should be attached to work or negative (–).

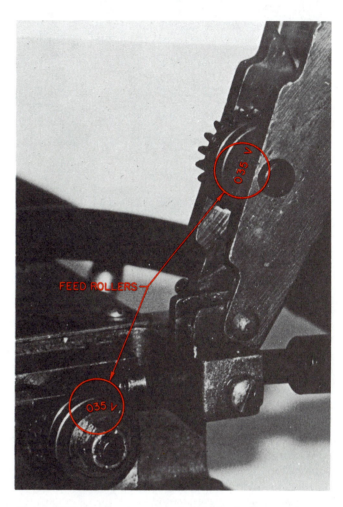

Figure 13-4 Check to be certain that the feed rollers are the correct size for the wire being used.

Figure 13-2 Attach the flowmeter regulator. Be sure the tube is vertical.

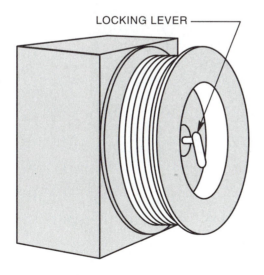

Figure 13-3 Wire reel may be secured by a center nut or locking lever.

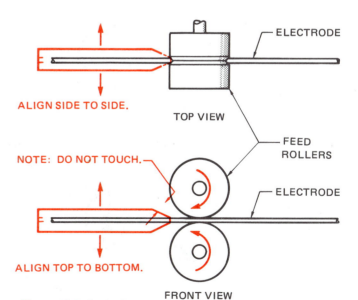

Figure 13-5 Feed roller and conduit alignment.

Figure 13-6 "Bird's nest" in the filler wire at the feed rollers.

The shielding "gas-out" side of the solenoid is then also attached to the gun lead. If a separate splice is required from the gun switch circuit to the feed unit, it should be connected at this time. Check that the welding contractor circuit is connected from the feed unit to the power source.

The welding cable liner or wire conduit must be securely attached to the gas diffuses and **contact tube** (tip), **Figure 13-7**. The contact tube must be the correct size to match the electrode wire size being used. If a shielding gas is to be used, a gas nozzle would be attached to complete the assembly. If a gas nozzle is not needed for a shielding gas, it may still be installed. Because it's easy for a student to touch the work with the contact tube during welding, an electrical short may occur. This short-out of the contact tube will immedi-

ately destroy the tube. Although the gas nozzle may interfere with some visibility, it may be worth the trouble for a new welder. ◆

PRACTICE 13-2

Threading FCAW Wire

Using the FCAW machine that was properly assembled in Practice 11-1, you will turn the machine on and thread the electrode wire through the system.

Check that the unit is assembled correctly according to the manufacturer's specifications. Switch on the power and check the gun switch circuit by depressing the switch. The power source relays, feed relays, gas solenoid, and feed motor should all activate.

Cut off the end of the electrode wire if it's bent. When working with the wire, be sure to hold it tightly. The wire will become tangled if it's released. The wire has a natural curl known as *cast*. Straighten out about 12 in. (300 mm) of the curl to make threading easier.

Separate the wire **feed rollers** and push the wire first through the guides, then between the rollers, and finally into the conduit liner, **Figure 13-8**. Reset the rollers so there is a slight amount of compression on the wire, **Figure 13-9**. Set the **wire-feed speed** control to a slow speed. Hold the welding gun so that the electrode conduit and cable are as straight as possible.

Press the gun switch. The wire should start feeding into the liner. Watch to make certain that the wire feeds smoothly and release the gun switch as soon as the end comes through the contact tube.

If the wire stops feeding before it reaches the end of the contact tube, stop and check the system. If no obvious problem can be found, mark the wire with tape and remove it from the gun. It then can be held next to the system to determine the location of the problem.

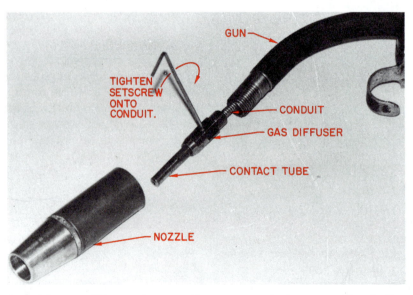

Figure 13-7 Securely attach conduit to gas diffuser and contact tube to prevent wire jams caused by misalignment.

Figure 13-8 Push the wire through the guides by hand.

Figure 13-9 Adjust the wire-feed tensioner.

Figure 13-10 Covered wire reel. (Courtesy of Lincoln Electric Company.)

With the wire feed running, adjust the feed roller compression so that the wire reel can be stopped easily by a slight pressure. Too light a roller pressure will cause the wire to feed erratically. Too high a pressure can crush some wires, causing some flux to be dropped inside the wire liner. If this happens, you will have a continual problem with the wire not feeding smoothly or jamming.

With the feed running, adjust the spool drag so that the reel stops when the feed stops. The reel should not coast to a stop, because the wire can be snagged easily. Also, when the feed restarts, a jolt occurs when the slack in the wire is taken up. This jolt can be enough to momentarily stop the wire, possibly causing a discontinuity in the weld.

When the test runs are completed, the wire can either be rewound or cut off. Some wire feed units have a retract button. This allows the feed driver to reverse and retract the wire automatically. To rewind the wire on units without this retract feature, release the rollers and turn them backward by hand. If the machine will not allow the feed rollers to be released without upsetting the tension, you must cut the wire. Some wire reels have covers to prevent the collection of dust, dirt, and metal filings on the wire, Figure 13-10. ◆

FLAT-POSITION WELDS

PRACTICE 13-3

Stringer Beads Flat Position

Using a properly set up and adjusted FCA welding machine, Table 13-1, proper safety protection, 0.035-in. and/or 0.045-in. (0.9-mm and/or 1.2-mm) diameter E70T-1 and/or E70T-5 electrodes, and one or more pieces of mild steel plate, 12 in. (305 mm) long and 1/4 in. (6-mm) or thicker, you will make a stringer bead weld in the flat position, Figure 13-11.

Starting at one end of the plate and using a dragging technique, make a weld bead along the entire 12-in. (305-mm) length of the metal. After the weld is complete, check its appearance. Make any needed changes to correct the weld. Repeat the weld and make additional adjustments. After the machine is set, start to work on improving the straightness and uniformity of the weld. Use weave patterns of different widths and straight stringers without weaving.

| Electrode | | | Welding Power | | Shielding Gas | | Base Metal | |
Type	Size	Amps	Wire Feed Speed IPM (cm/min)	Volts	Type	Flow	Type	Thickness
E70T-1	0.035 in. (0.9 mm)	130 to 150	288 to 380 (732 to 975)	22 to 25	None	n/a	Low-carbon steel	1/4 in. to 1/2 in. (6 mm to 13 mm)
E70T-1	0.045 in. (1.2 mm)	150 to 210	200 to 300 (508 to 762)	28 to 29	None	n/a	Low-carbon steel	1/4 in. to 1/2 in. (6 mm to 13 mm)
E70T-5	0.035 in. (0.9 mm)	130 to 200	288 to 576 (732 to 1463)	20 to 28	75% argon 25% CO_2	30 cfh	Low-carbon steel	1/4 in. to 1/2 in. (6 mm to 13 mm)
E70T-5	0.045 in. (1.2 mm)	150 to 250	200 to 400 (508 to 1016)	23 to 29	75% argon 25% CO_2	35 cfh	Low-carbon steel	1/4 in. to 1/2 in. (6 mm to 13 mm)

Table 13-1 FCA Welding Parameters for Use If Specific Settings Unavailable from Electrode Manufacturer.

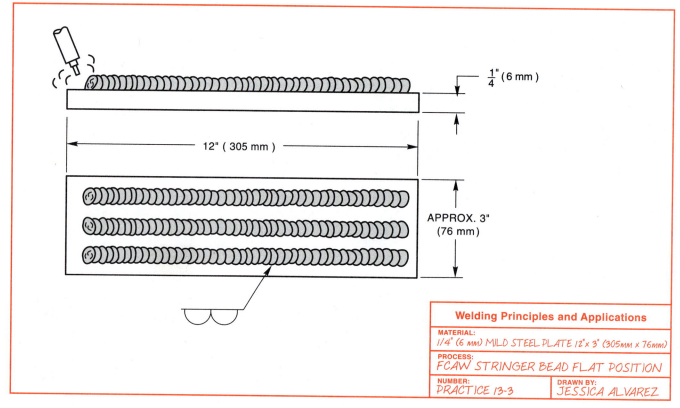

Welding Principles and Applications

MATERIAL:
1/4" (6 mm) MILD STEEL PLATE 12" x 3" (305mm x 76mm)

PROCESS:
FCAW STRINGER BEAD FLAT POSITION

| NUMBER: PRACTICE 13-3 | DRAWN BY: JESSICA ALVAREZ |

Figure 13-11

Repeat with both classifications of electrodes as needed until beads can be made straight, uniform, and free from any visual defects. Turn off the welding machine and shielding gas and clean up your work area when you are finished welding. ◆

SQUARE-GROOVE WELDS

One advantage of FCA welding is the ability to make 100%-joint-penetrating welds without beveling the edges of the plates. These full-joint-penetrating welds can be made in plates that are 1/4 in. (6 mm) or less in thickness. Welding on thicker plates risks the possibility of a lack of fusion on both sides of the root face, Figure 13-12.

There are several disadvantages of having to bevel a plate before welding:

■ Beveling the edge of a plate adds an operation to the fabrication process. Unbeveled plates can be sheared or thermally cut (OFC, PAC, PAC, LBC, etc.) to size,

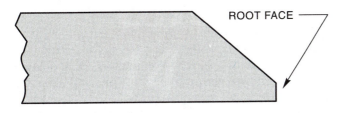

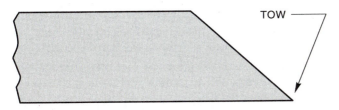

Figure 13-12 A beveled joint may or may not have a flat surface, called a root face.

assembled, and welded; but beveled plates must first be cut to size, then beveled by grinding, machining, or thermally, assembled, and welded.

■ Both more filler metal and welding time are required to fill a beveled joint than are required to make a square jointed weld.

■ Beveled joints have more heat from the thermal beveling and additional welding required to fill the groove. The lower heat input to the square joint means less distortion.

The major disadvantage of making square jointed welds is that as the plate thickness approaches 1/4 in. (6 mm) or the weld is out of position a much higher level of skill is required. The skill required to make quality square welds can be acquired by practicing on thinner metal. It is much easier to make this type of weld in 1/8-in. (3-mm) -thick metal and then move up in thickness as your skills improve.

PRACTICE 13-4

Butt Joint 1G

Using a properly set up and adjusted FCA welding machine, proper safety protection, 0.035-in. and/or 0.045-in. (0.9-mm and/or 1.2-mm) -diameter E70T-1 and/or E70T-5 electrodes, and one or more pieces of mild steel plate, 12 in. (305-mm) long and 1/4 in. (6 mm) or less in thickness, you will make a groove weld in the flat position, **Figure 13-13**.

■ Tack weld the plates together and place them in position to be welded.

■ Starting at one end, run a bead along the joint. Watch the molten weld pool and bead for signs that a change in technique may be required.

■ Make any needed changes as the weld progresses in order to produce a uniform weld.

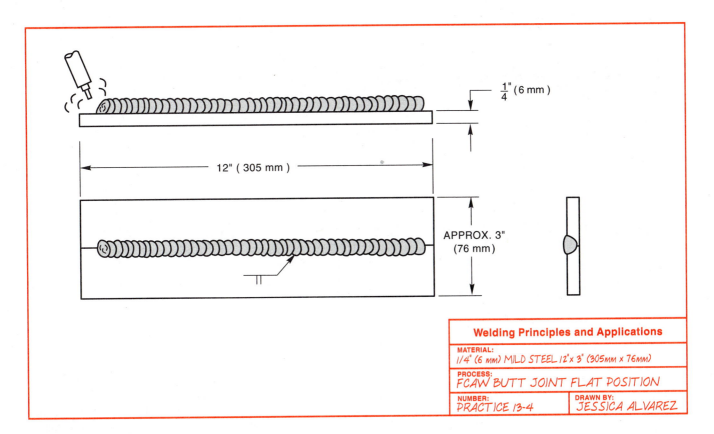

Figure 13-13

Repeat with both classifications of electrodes as needed until defect-free welds can consistently be made in the 1/4-in. (6-mm) -thick plate. Turn off the welding machine and shielding gas and clean up your work area when you are finished welding. ◆

Butt Joint 1G 100% to Be Tested

Using a properly set up and adjusted FCA welding machine, proper safety protection, 0.035-in. and/or 0.045-in. (0.9-mm and/or 1.2-mm) -diameter E70T-1 and/or E70T-5 electrodes, and one or more pieces of mild steel plate, 12 in. (305 mm) long and 1/4 in. (6 mm) thick, you will make a groove weld in the flat position, Figure 13-14.

Following the same instructions for the assembly and welding procedure outlined in Practice 13-4, repeat the weld using each electrode classification until welds using both electrodes can be made with 100% penetration that will pass a bend test. Turn off the welding machine and shielding gas and clean up your work area when you are finished welding. ◆

V-GROOVE AND BEVEL-GROOVE WELDS

Although for speed and economy engineers try to avoid specifying welds that require beveling the edges of plates, it is not always possible. Anytime the metal being welded is thicker than 1/4 in. (6 mm) and a 100% joint penetration weld is required, the edges of the plate must be prepared with a bevel. Fortunately, FCA welding allows a narrower groove to be made and still achieve a thorough thickness weld, Figure 13-15.

All FCA groove welds are made using three different types of weld passes, Figure 13-16:

- ■ *Root pass:* The first weld bead of a multiple pass weld. The root pass fuses the two parts together and establishes the depth of weld metal penetration.

- ■ *Filler pass:* Made after the root pass is completed and used to fill the groove with weld metal. More than one pass is often required.

- ■ *Cover pass:* The last weld pass on a multipass weld. The cover pass may be made with one or more

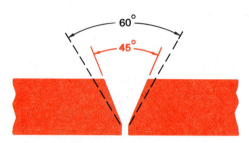

Figure 13-15 A smaller groove angle reduces both weld time and filler metal required to make the weld.

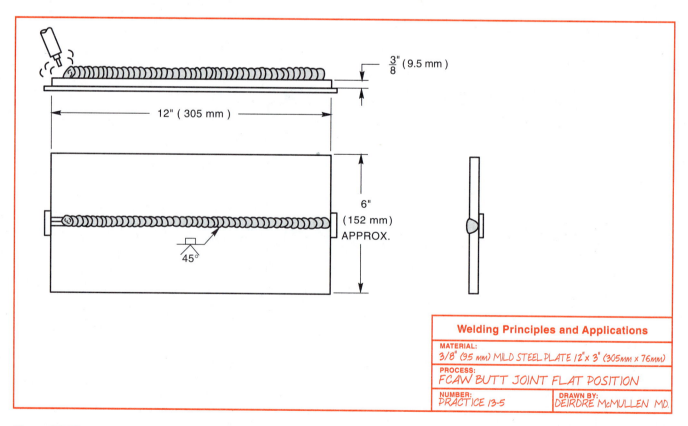

Figure 13-14

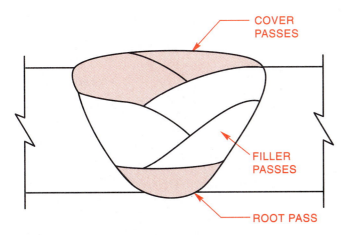

Figure 13-16 The three different types of weld passes that make up a weld.

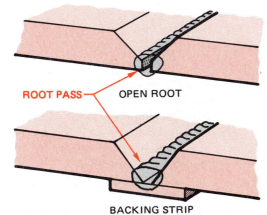

Figure 13-17 Root pass maximum deposit 1/4 in. (6 mm) thick.

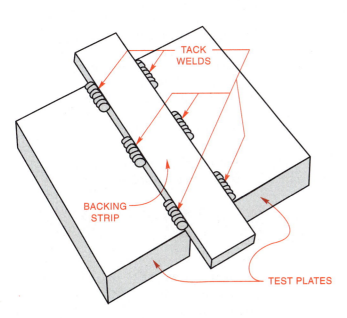

Figure 13-18 Securely tack weld the backing strip to the test plates.

welds. It must be uniform in width, reinforcement, and appearance.

Root Pass

A good root pass is needed in order to obtain a sound weld. The root may be either open or closed and made using a backing strip, **Figure 13-17**.

The backing strips are usually made from a piece of 1/4-in. (6-mm) -thick, 1-in. (25-mm) -wide metal that should be 2 in. (50 mm) longer than the base plates. The strip is attached to the plate by tack welds made on the sides of the strip, **Figure 13-18**.

Most production welds do not use backing strips, so they are made as open root welds. Because of the difficulty in controlling FCA weld's root weld face contours, however, open-root joints are often avoided on **critical welds.** If an open-root weld is needed because of weldment design, the root pass may be put in with an SMA electrode or the root face of the FCA weld can be retouched by grinding and/or back welding.

Care must be taken with any root pass not to have the weld face too convex, **Figure 13-19**. Convex weld faces tend to trap slag along the tow of the weld. FCA weld slag can be extremely difficult to remove in this area, especially if there is any undercutting. To avoid this, adjust the welding power settings, speed, and weave pattern so that a flat or slightly concave weld face is produced, **Figure 13-20**.

Filler Pass

Filler pass(es) are made with either **stringer bead**s or **weave bead**s for flat or vertically positioned welds, but stringer beads work best for horizontal and overhead-positioned welds. When multiple-pass filler welds are

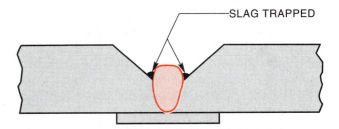

Figure 13-19 Slag trapped beside weld bead is hard to remove.

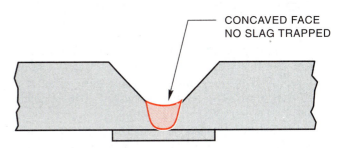

Figure 13-20 Flat or concave weld faces are easier to clean.

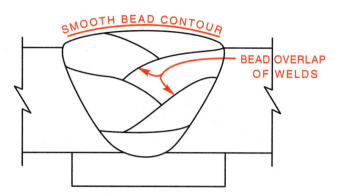

Figure 13-21 The surface of a multipass weld should be as smooth as if it were made by one weld.

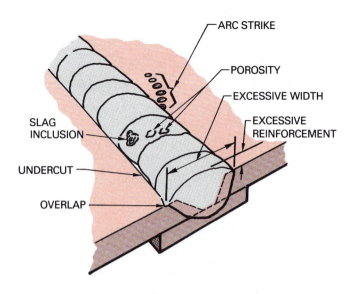

Figure 13-22 Common discontinuities found during a visual examination.

required, each weld bead must overlap the others along the edges. Edges should overlap smoothly enough so that the finished bead is uniform, **Figure 13-21**. Stringer beads usually overlap about 25% to 50%, and weave beads overlap approximately 10% to 25%.

Each weld bead must be cleaned before the next bead is started. The filler pass ends when the groove has been filled to a level just below the plate surface.

Cover Pass

The cover pass may or may not simply be a continuation of the weld beads used to make the filler pass(es). The major difference between the filler pass and the cover pass is the weld face importance. Keeping the face and tow of the cover pass uniform in width, reinforcement, and appearance and defect-free is essential. Most welds are not tested beyond a visual inspection. For that reason the appearance might be the only factor used for accepting or rejecting welds.

The cover pass must meet a strict visual inspection standard. The visual inspection looks to see that the weld is uniform in width and reinforcement. There should be no arc strikes on the plate other than those on the weld itself. The weld must be free of both incomplete fusion and cracks. The weld must be free of overlap, and undercut must not exceed either 10% of the base metal or 1/32 in. (0.8 mm), whichever is less. Reinforcement must have a smooth transition with the base plate and be no higher than 1/8 in. (3 mm), **Figure 13-22**.

PRACTICE 13-6

Butt Joint 1G

Using a properly set up and adjusted FCA welding machine, **Table 13-1**, proper safety protection, 0.035-in. and/or 0.045-in. (0.9-mm and/or 1.2-mm) -diameter E70T-1 and/or E70T-5 electrodes, and one or more pieces of mild steel plate, 12-in. (305-mm) -long and 3/8-in. (9.5-mm) -thick beveled plate, and a 14-in. (355-mm) -long 1-in. (25-mm) -wide and 1/4-in. (6-mm) -thick

backing strip, you will make a groove weld in the flat position, **Figure 13-23**.

Tack weld the backing strip to the plates. There should approximately be an 1/8-in. (3-mm) root gap between the plates. The beveled surface can be made with or without a root face, **Figure 13-24**.

Place the test plates in position at a comfortable height and location. Be sure that you have complete and free movement along the full length of the weld joint. It's often a good idea to make a practice pass along the joint with the welding gun without power to make sure nothing will interfere with your making the weld. Be sure the welding cable is free and will not get caught on anything during the weld.

Start the weld outside the groove on the backing strip tab, **Figure 13-25**. This is done so that the arc is smooth and the molten weld pool size is established at the beginning of the groove. Continue the weld out on to the tab at the outer end of the groove. This process ensures that the end of the groove is completely filled with weld.

Repeat with both classifications of electrodes as needed until consistently defect-free welds can be made. Turn off the welding machine and shielding gas and clean up your work area when you are finished welding. ◆

PRACTICE 13-7

Butt Joint 1G 100% to Be Tested

Using a properly set up and adjusted FCA welding machine, proper safety protection, 0.035-in. and/or 0.045-in. (0.9-mm and/or 1.2-mm) -diameter E70T-1 and/or E70T-5 electrodes, and one or more pieces of mild

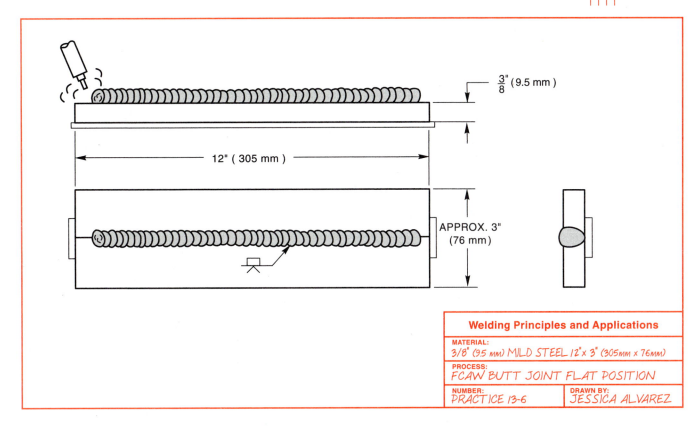

Welding Principles and Applications
MATERIAL:
3/8" (9.5 mm) MILD STEEL 12" x 3" (305mm x 76mm)
PROCESS:
FCAW BUTT JOINT FLAT POSITION
NUMBER: DRAWN BY:
PRACTICE 13-6   JESSICA ALVAREZ

Figure 13-23

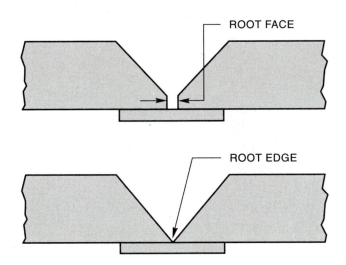

ROOT FACE

ROOT EDGE

Figure 13-24

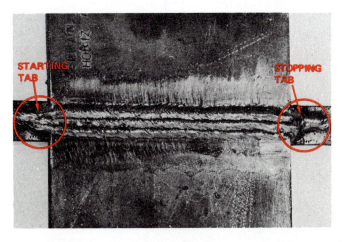

Figure 13-25 Run-off tabs help control possible underfill or burn back at the starting and stopping points of a groove weld.

steel plate, 12-in. (305-mm) -long and 3/8-in. (9.5-mm) -thick beveled plate, and a 14-in. (355-mm) -long 1-in. (25-mm) -wide and 1/4-in. (6-mm) -thick backing strip, you will make a groove weld in the flat position, **Figure 13-26**.

Following the same instructions for the assembly and welding procedure outlined in Practice 13-6, repeat the weld using each electrode classification until welds using both electrodes can be made with 100% penetration that will pass a bend test. Turn off the welding machine and shielding gas and clean up your work area when you are finished welding. ◆

PRACTICE 13-8

Butt Joint 1G

Using a properly set up and adjusted FCA welding machine, **Table 13-2**, proper safety protection, 0.035-in. and/or through 1/16-in. (0.9-mm and/or through 1.6-mm) -diameter E70T-1 and/or E70T-5 electrodes, one or

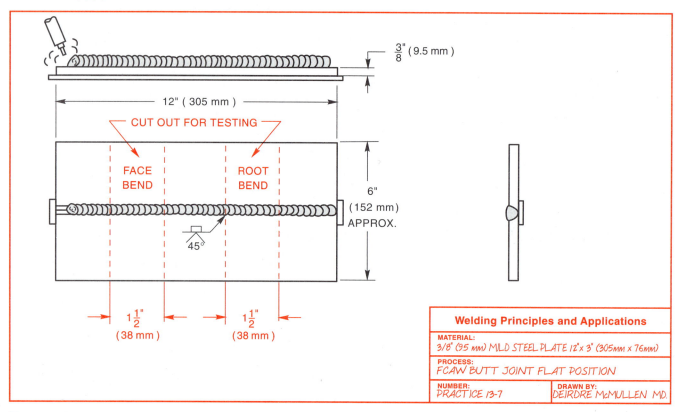

Welding Principles and Applications

MATERIAL:
3/8" (95 mm) MILD STEEL PLATE 12" x 3" (305mm x 76mm)

PROCESS:
FCAW BUTT JOINT FLAT POSITION

| NUMBER: | DRAWN BY: |
| PRACTICE 13-7 | DEIRDRE McMULLEN MD. |

Figure 13-26

Electrode			Welding Power		Shielding Gas		Base Metal	
Type	Size	Amps	Wire Feed Speed IPM (cm/min)	Volts	Type	Flow	Type	Thickness
E70T-1	0.035 in. (0.9 mm)	130 to 150	288 to 380 (732 to 975)	22 to 25	None	n/a	Low-carbon steel	1/2 in. to 3/4 in. (13 mm to 19 mm)
E70T-1	0.045 in. (1.2 mm)	150 to 210	200 to 300 (508 to 762)	28 to 29	None	n/a	Low-carbon steel	1/2 in. to 3/4 in. (13 mm to 19 mm)
E70T-1	.052 in. (1.4 mm)	150 to 300	150 to 350 (381 to 889)	25 to 33	None	n/a	Low-carbon steel	1/2 in. to 3/4 in. (13 mm to 19 mm)
E70T-1	1/16 in. (1.6 mm)	200 to 400	150 to 300 (381 to 762)	27 to 33	None	n/a	Low-carbon steel	1/2 in. to 3/4 in. (13 mm to 19 mm)
E70T-5	0.035 in. (0.9 mm)	130 to 200	288 to 576 (732 to 1463)	20 to 28	75% argon 25% CO_2	30 cfh	Low-carbon steel	1/2 in. to 3/4 in. (13 mm to 19 mm)
E70T-5	0.045 in. (1.2 mm)	150 to 250	200 to 400 (508 to 1016)	23 to 29	75% argon 25% CO_2	35 cfh	Low-carbon steel	1/2 in. to 3/4 in. (13 mm to 19 mm)
E70T-5	.052 in. (1.4 mm)	150 to 300	150 to 350 (381 to 889)	21 to 32	75% argon 25% CO_2	35 cfh	Low-carbon steel	1/2 in. to 3/4 in. (13 mm to 19 mm)
E70T-5	1/16 in. (1.6 mm)	180 to 400	145 to 350 (368 to 889)	21 to 34	75% argon 25% CO_2	40 cfh	Low-carbon steel	1/2 in. to 3/4 in. (13 mm to 19 mm)

Table 13-2 FCA Welding Parameters for Use If Specific Settings Unavailable from Electrode Manufacturer.

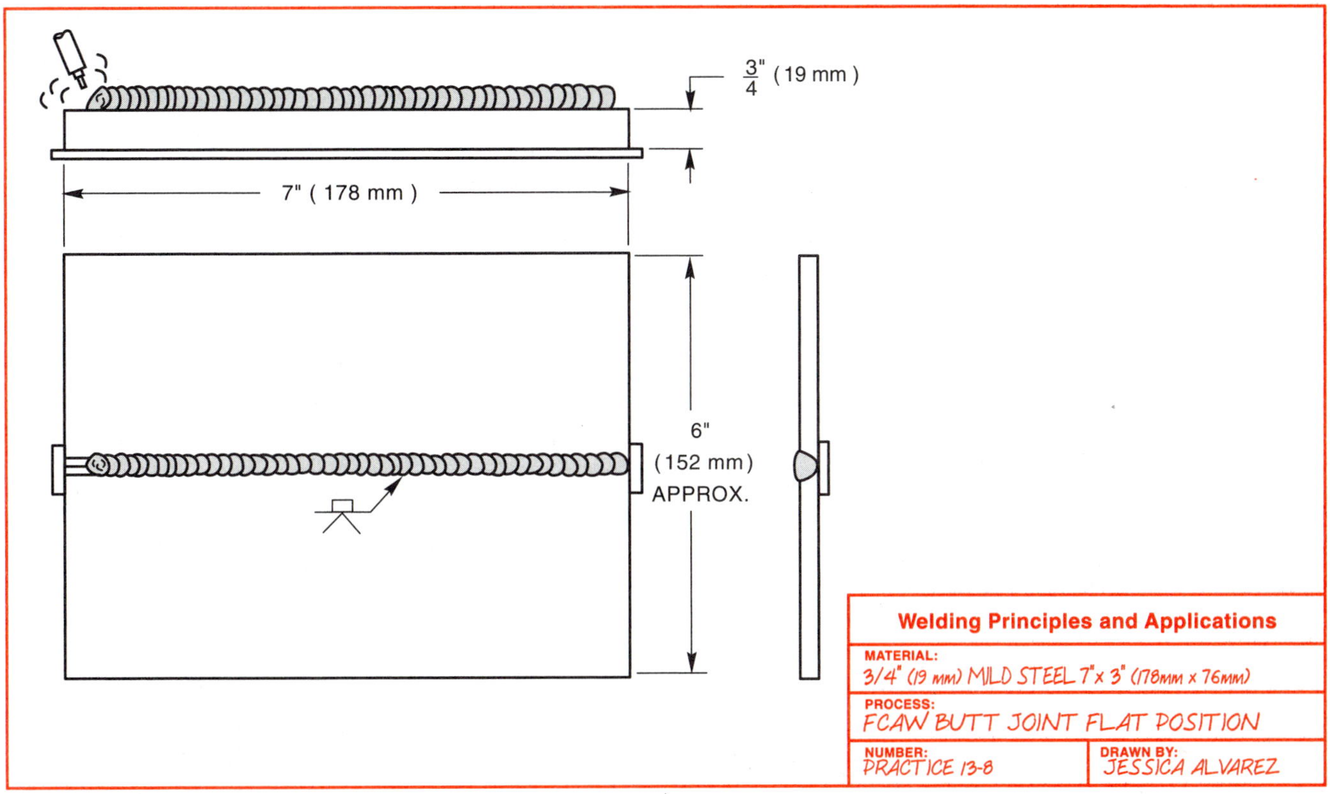

Figure 13-27

more pieces of mild steel plate, 7-in. (178-mm) -long and 3/4-in. (19-mm) or thicker beveled plate, and a 9-in. (230-mm) -long 1-in. (25-mm) -wide and 1/4-in. (6-mm) -thick backing strip, you will make a groove weld in the flat position, **Figure 13-27**.

Following the same instructions for assembly and welding procedure outlined in Practice 13-6, repeat the weld with both classifications of electrodes as needed until consistently defect-free welds can be made. Turn off the welding machine and shielding gas and clean up your work area when you are finished welding. ◆

PRACTICE 13-9

Butt Joint 1G 100% to Be Tested

Using a properly set up and adjusted FCA welding machine, proper safety protection, 0.035-in. and/or through 1/16-in. (0.9-mm and/or through 1.6-mm) -diameter E70T-1 and/or E70T-5 electrodes, one or more pieces of mild steel plate, 7-in. (178-mm) -long and 3/4-in. (19-mm) or thicker beveled plate, and a 9-in. (230-mm) -long 1-in. (25-mm) -wide and 1/4-in. (6-mm) -thick backing strip, you will make a groove weld in the flat position, **Figure 13-28**.

Following the same instructions for the assembly and welding procedure outlined in Practice 13-6, repeat the weld using each electrode classification until welds using both electrodes can be made with 100% penetration that will pass a bend test. Turn off the welding

machine and shielding gas and clean up your work area when you are finished welding. ◆

FILLET WELDS

A fillet weld is the type of weld made on the **lap joint** and **tee joint.** It should be built up equal to the thickness of the plate, **Figure 13-29**. On thick plates the fillet must be made up of several passes as with a groove weld. The difference with a fillet weld is that a smooth transition from the plate surface to the weld is required. If this transition is abrupt, it can cause stresses that will weaken the joint.

The lap joint is made by overlapping the edges of the plates. They should be held together tightly before tack welding them together. A small tack weld may be added in the center to prevent distortion during welding, **Figure 13-30**. Chip the tacks before you start to weld.

The tee joint is made by tack welding one piece of metal on another piece of metal at a right angle, **Figure 13-31**. After the joint is tack welded together, the slag is chipped from the tack welds. If the slag is not removed, it will cause a slag inclusion in the final weld.

Holding thick plates tightly together on tee joints may cause underbead cracking or lamellar tearing, **Figure 13-32**. On thick plates the weld shrinkage can be great enough to pull the metal apart well below the bead or its heat-affected zone. In production welds cracking can be controlled by not assembling the plates tightly together. The space between the two plates can be set by placing a small wire spacer between them, **Figure 13-33**.

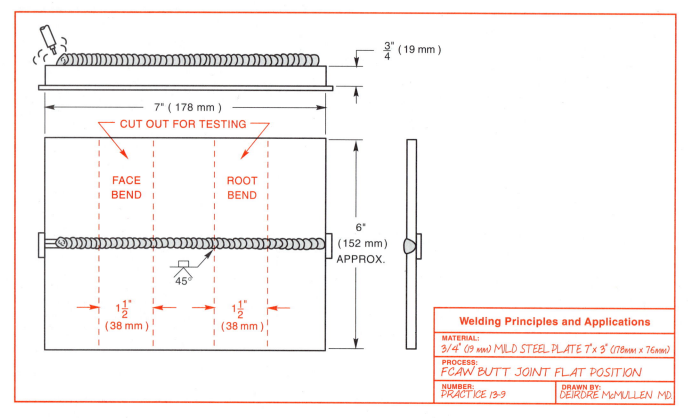

Welding Principles and Applications

MATERIAL:
3/4" (19 mm) MILD STEEL PLATE 7" x 3" (178mm x 76mm)

PROCESS:
FCAW BUTT JOINT FLAT POSITION

NUMBER:	DRAWN BY:
PRACTICE 13-9	DEIRDRE McMULLEN M.D.

Figure 13-28

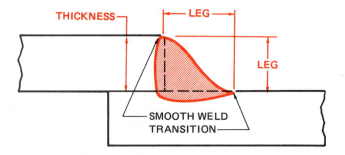

Figure 13-29 The legs of a fillet weld should generally be equal to the thickness of the base metal.

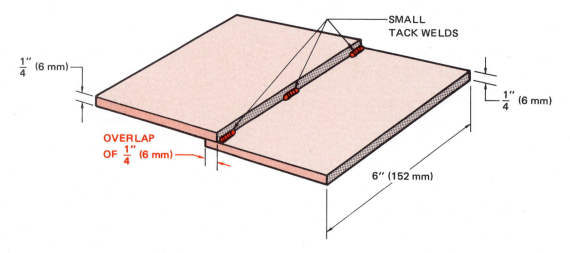

Figure 13-30 Tack welding the plates together.

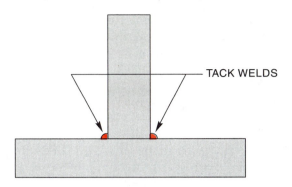

Figure 13-31 Tack welding both sides of a tee joint will help keep the tee square for welding.

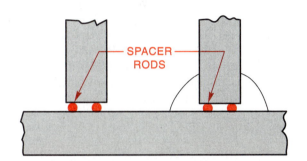

Figure 13-33 Base plate cracking can be controlled by placing spacers in the joint before welding.

A fillet welded lap or tee joint can be strong if it is welded on both sides, even without having deep penetration, **Figure 13-34.** Some tee joints may be prepared for

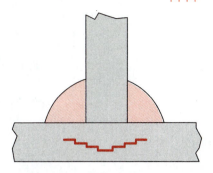

Figure 13-32 Underbead cracking or lamellar tearing of the base plate.

welding by cutting either a bevel or a J-groove in the vertical plate. This cut is not required for strength but may be necessary because of design limitations. Unless otherwise instructed, most fillet welds will be equal in size to the plates welded. A fillet weld will be as strong as the base plate if the size of the two welds equals the total thickness of the base plate. The weld bead should have a flat or slightly concave appearance to ensure the greatest strength and efficiency, **Figure 13-35.**

The root of fillet welds must be melted to ensure a completely fused joint. A notch along the root of the weld pool is indication that the root is not being fused together, **Figure 13-36.** To achieve complete root fusion, move the arc to a point as close as possible to the leading edge of the weld pool, **Figure 13-37.** If the arc strikes the unmelted plate ahead of the molten weld pool, it may become erratic, which will increase weld spatter.

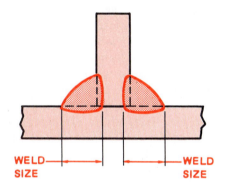

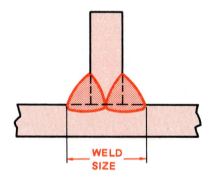

Figure 13-34 If the total weld sizes are equal, then both tee joints would have equal strength.

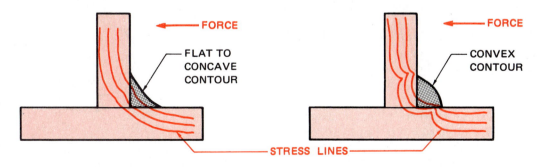

Figure 13-35 The stresses are distributed more uniformly through a flat or concave fillet weld.

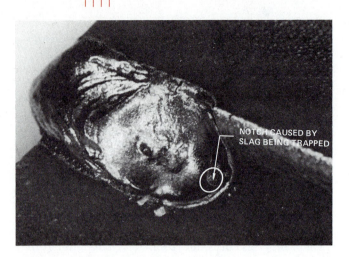

Figure 13-36 Watch the root of the weld bead to be sure there is complete fusion.

NOTCH CAUSED BY SLAG BEING TRAPPED

PRACTICE 13-10

Lap Joint and Tee Joint 1F

Using a properly set up and adjusted FCA welding machine, proper safety protection, 0.035-in. and/or 0.045-in. (0.9-mm and/or 1.2-mm) -diameter E70T-1 and/or E70T-5 electrodes, and one or more pieces of mild steel plate, 12-in. (305-mm) -long and 3/8-in. (9.5-mm) -thick beveled plate, you will make a fillet weld in the flat position.

Tack weld the pieces of metal together and brace them in position. When making the lap or tee joints in the flat position, the plates must be at a 45° angle so that the surface of the weld will be flat, **Figure 13-38(A)** and **(B)**. Starting at one end, make a weld along the entire length of the joint.

Repeat each type of joint with both classifications of electrodes as needed until consistently defect-free welds can be made. Turn off the welding machine and shielding gas and clean up your work area when you are finished welding. ◆

PRACTICE 13-11

Lap Joint and Tee Joint 1F 100% to Be Tested

Using a properly set up and adjusted FCA welding machine, proper safety protection, 0.035-in. and/or 0.045-in. (0.9-mm and/or 1.2-mm) -diameter E70T-1 and/or E70T-5 electrodes, and one or more pieces of mild steel plate, 12-in. (305-mm) -long and 3/8-in. (9.5-mm) -thick beveled plate, you will make a fillet weld in the flat position, **Figure 13-39**.

Following the same instructions for the assembly and welding procedure outlined in Practice 13-10, repeat the weld using each electrode classification until welds using both electrodes can be made with 100% penetra-

tion that will pass a bend test. Turn off the welding machine and shielding gas and clean up your work area when you are finished welding. ◆

PRACTICE 13-12

Tee Joint 1F

Using a properly set up and adjusted FCA welding machine, proper safety protection, 0.035-in. and/or through 1/16-in. (0.9-mm and/or through 1.6-mm) -diameter E70T-1 and/or E70T-5 electrodes, and one or more pieces of mild steel plate, 7-in. (178-mm) -long and 3/4-in. (19-mm) or thicker beveled plate, you will make a fillet weld in the flat position.

Following the same instructions for the assembly and welding procedure outlined in Practice 13-10, repeat each type of joint with both classifications of electrodes as needed until consistently defect-free welds can be made. Turn off the welding machine and shielding gas and clean up your work area when you are finished welding. ◆

PRACTICE 13-13

Tee Joint 1F 100% to Be Tested

Using a properly set up and adjusted FCA welding machine, proper safety protection, 0.035-in. and/or through 1/16-in. (0.9-mm and/or through 1.6-mm) -diameter E70T-1 and/or E70T-5 electrodes, and one or more pieces of mild steel plate, 7-in. (178-mm) -long and 3/4-in. (19-mm) or thicker beveled plate, you will make a fillet weld in the flat position, **Figure 13-40**.

Following the same instructions for the assembly and welding procedure outlined in Practice 13-10, repeat with both classifications of electrodes as needed until welds can be made with 100% penetration that will pass the test. Turn off the welding machine and shielding gas and clean up your work area when you are finished welding. ◆

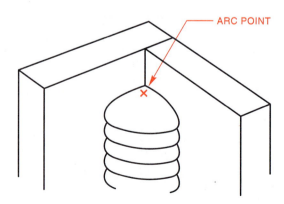

ARC POINT

Figure 13-37 Moving the arc as close as possible to the leading edge of the weld will provide good root fusion.

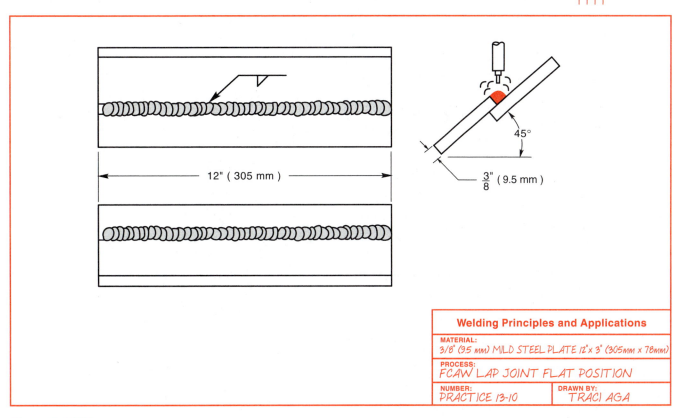

Figure 13-38(A)

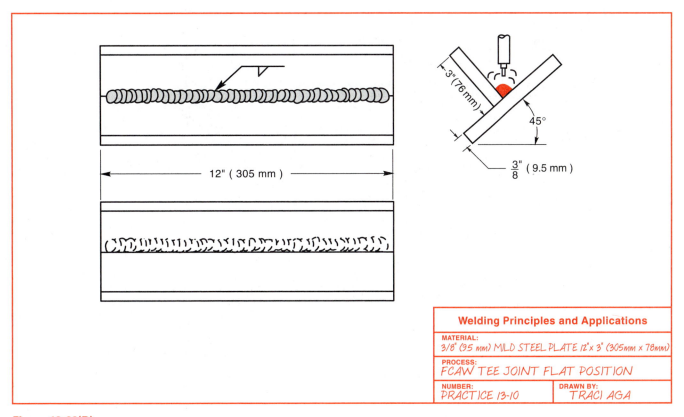

Figure 13-38(B)

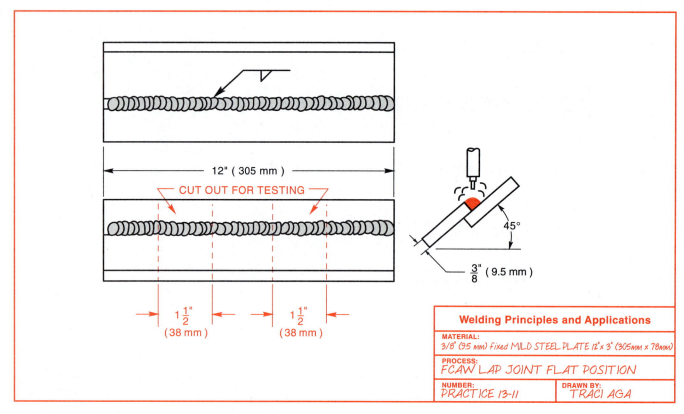

Figure 13-39(A)

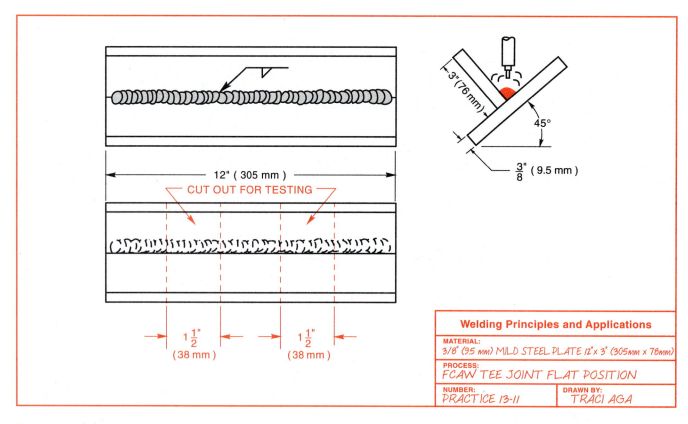

Figure 13-39(B)

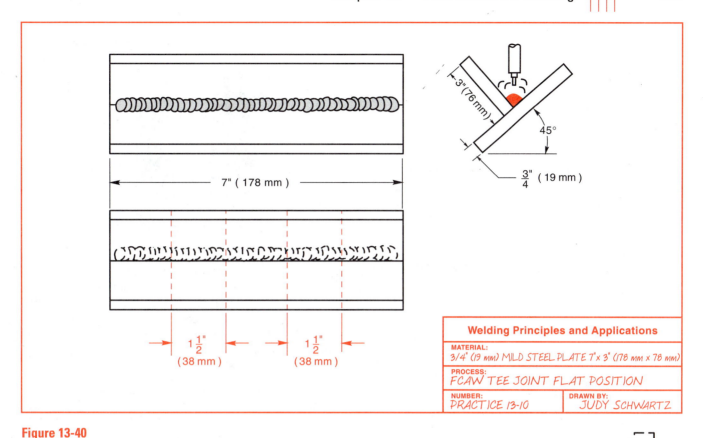

Welding Principles and Applications

MATERIAL:
3/4" (19 mm) MILD STEEL PLATE 7"x 3" (178 mm x 78 mm)

PROCESS:
FCAW TEE JOINT FLAT POSITION

NUMBER:
PRACTICE 13-10

DRAWN BY:
JUDY SCHWARTZ

Figure 13-40

VERTICAL WELDS

PRACTICE 13-14

Butt Joint at a 45° Vertical Up Angle

Using a properly set up and adjusted FCA welding machine, proper safety protection, 0.035-in. and/or 0.045-in. (0.9-mm and/or 1.2-mm) -diameter E70T-1 and/or E70T-5 electrodes, and one or more pieces of mild steel plate, 12 in. (305 mm) long and 1/4 in. (6 mm) thick or thinner, you will increase the plate angle gradually as you develop skill until you are making satisfactory welds in the vertical up position, **Figure 13-41.**

- Start practicing this weld with the plate at a 45° angle.

- Gradually increase the angle of the plate to vertical as skill is gained in welding this joint. A straight stringer bead or slight zigzag will work well on this joint.

- Establish a molten weld pool in the root of the joint.

- Cool, chip, and inspect the weld for uniformity and defects.

It's easier to make a quality weld in the vertical up position if both the amperage and voltage are set at the lower end of their ranges. This will make the molten

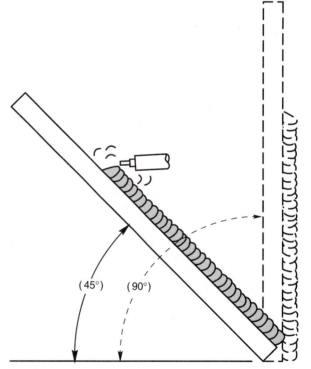

Figure 13-41 Start making welds with the plate at a 45° angle. As your skill develops increase the angle until the plate is vertical.

weld pool smaller, less fluid, and easier to control. A problem with lower power settings is that the weld bead often can be very convex, **Figure 13-42.** Faster travel speed and/or slightly wider weave patterns can be used to control the bead shape.

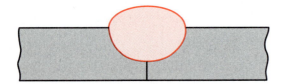

Figure 13-42 Low amperage causes too much buildup, not enough penetration.

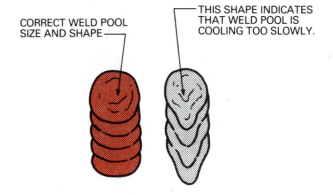

CORRECT WELD POOL SIZE AND SHAPE ——

—— THIS SHAPE INDICATES THAT WELD POOL IS COOLING TOO SLOWLY.

Figure 13-44 The shape of the weld pool can indicate the temperature of the surrounding base metal.

Start at the bottom of the plate and hold the welding gun at a slight upward angle to the plate, **Figure 13-43**. Brace yourself, lower your hood, and begin to weld. Depending on the machine settings and type of electrode used, you will make a weave pattern.

If the molten weld pool is large and fluid (hot), use a C or J weave pattern to allow a longer time for the molten weld pool to cool, **Figure 13-44**. Do not make the weave so long or fast that the electrode is allowed to strike the metal ahead of the molten weld pool. If this happens, spatter increases and a spot or zone of incomplete fusion may occur.

A weld that is high and has little or no fusion is too "cold." Changing the welding technique will not correct this problem. The welder must stop welding and make the needed adjustments to the power supply or electrode feeder. Continue to weld along the entire 12-in. (305-mm) length of plate.

Repeat welds with both electrodes as needed until defect-free welds can be consistently made vertically in the 1/4-in. (6-mm) -thick plate. Turn off the welding machine and shielding gas and clean up your work area when you are finished welding. ◆

Butt Joint 3G

Using a properly set up and adjusted FCA welding machine, proper safety protection, 0.035-in. and/or 0.045-in. (0.9-mm and/or 1.2-mm) -diameter E70T-1 and/or E70T-5 electrodes, and one or more pieces of mild steel plate, 12 in. (305 mm) long and 1/4 in. (6 mm) thick or thinner, you will make a groove weld in the vertical position, **Figure 13-45**.

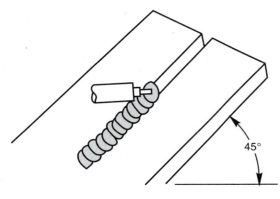

45°

Figure 13-43

Following the same instructions for the assembly and welding procedure outlined in Practice 13-14, repeat with both classifications of electrodes as needed until defect-free welds can be consistently made in the 1/4-in. (6-mm) -thick plate. Turn off the welding machine and shielding gas and clean up your work area when you are finished welding. ◆

Butt Joint 3G 100% to Be Tested

Using a properly set up and adjusted FCA welding machine, proper safety protection, 0.035-in. and/or 0.045-in. (0.9-mm and/or 1.2-mm) -diameter E70T-1 and/or E70T-5 electrodes, and one or more pieces of mild steel plate, 12 in. (305 mm) long and 1/4 in. (6 mm) thick, you will make a groove weld in the vertical position.

Following the same instructions for the assembly and welding procedure as outlined in Practice 13-14, repeat the weld using each electrode classification until welds using both electrodes can be made with 100% penetration that will pass a bend test. Turn off the welding machine and shielding gas and clean up your work area when you are finished welding. ◆

Butt Joint 3G

Using a properly set up and adjusted FCA welding machine, proper safety protection, 0.035-in. and/or 0.045-in. (0.9-mm and/or 1.2-mm) -diameter E70T-1 and/or E70T-5 electrodes, and one or more pieces of mild steel plate, 12-in. (305-mm) -long and 3/8-in. (9.5-mm) -thick beveled plate, and a 14-in. (355-mm) -long 1-in. (25-mm) -wide and 1/4-in. (6-mm) -thick backing strip, you will make a groove weld in the vertical position.

Following the same instructions for assembly and welding procedure as outlined in Practice 13-14, repeat with both classifications of electrodes as needed until defect-free welds can consistently be made. Turn off the

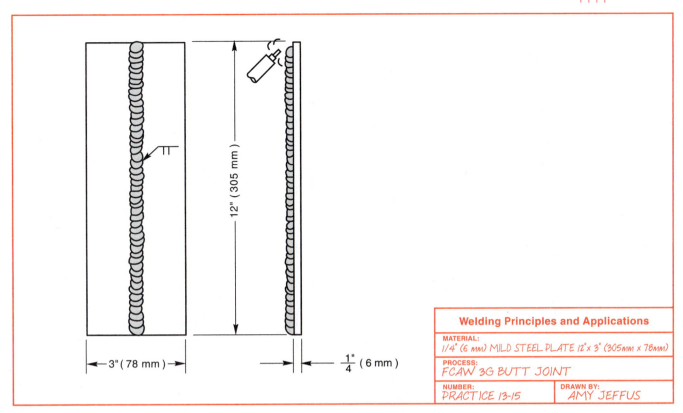

Welding Principles and Applications

MATERIAL:
1/4" (6 mm) MILD STEEL PLATE 12"x 3" (305mm x 78mm)

PROCESS:
FCAW 3G BUTT JOINT

NUMBER:
PRACTICE 13-15

DRAWN BY:
AMY JEFFUS

Figure 13-45

welding machine and shielding gas and clean up your work area when you are finished welding. ◆

PRACTICE 13-18

Butt Joint 3G 100% To Be Tested

Using a properly set up and adjusted FCA welding machine, proper safety protection, 0.035-in. and/or 0.045-in. (0.9-mm and/or 1.2-mm) -diameter E70T-1 and/or E70T-5 electrodes, and one or more pieces of mild steel plate, 12-in. (305-mm) -long and 3/8-in. (9.5-mm) -thick beveled plate, and a 14-in. (355-mm) -long 1-in. (25-mm) -wide and 1/4-in. (6-mm) -thick backing strip, you will make a groove weld in the vertical position, Figure 13-46.

Following the same instructions for the assembly and welding procedure outlined in Practice 13-14, repeat electrodes as needed until welds can be made with 100% penetration that will pass a bend test. Turn off the welding machine and shielding gas and clean up your work area when you are finished welding. ◆

PRACTICE 13-19

Butt Joint at a 45° Vertical Up Angle

Using a properly set up and adjusted FCA welding machine, proper safety protection, 0.035-in. and/or through 1/16-in. (0.9-mm and/or through 1.6-mm) -diameter E70T-1 and/or E70T-5 electrodes, one or more pieces of mild steel plate, 7-in. (178-mm) -long and 3/4-

in. (19-mm) or thicker beveled plate, and a 9-in. (230-mm) -long 1-in. (25-mm) -wide and 1/4-in. (6-mm) -thick backing strip, you will increase the plate angle gradually as you develop skill until you are making satisfactory welds in the vertical up position.

Following the same instructions for the assembly and welding procedure outlined in Practice 13-14, repeat with both classifications of electrodes as needed until defect-free welds can consistently be made. Turn off the welding machine and shielding gas and clean up your work area when you are finished welding. ◆

PRACTICE 13-20

Butt Joint 3G

Using a properly set up and adjusted FCA welding machine, proper safety protection, 0.035-in. and/or through 1/16-in. (0.9-mm and/or through 1.6-mm) -diameter E70T-1 and/or E70T-5 electrodes, one or more pieces of mild steel plate, 7-in. (178-mm) -long and 3/4-in. (19-mm) or thicker beveled plate, and a 9-in. (230-mm) -long 1-in. (25-mm) -wide and 1/4-in. (6-mm) -thick backing strip, you will make a groove weld in the vertical position.

Following the same instructions for the assembly and welding procedure outlined in Practice 13-14, repeat with both classifications of electrodes as needed until defect-free welds can consistently be made. Turn off the welding machine and shielding gas and clean up your work area when you are finished welding. ◆

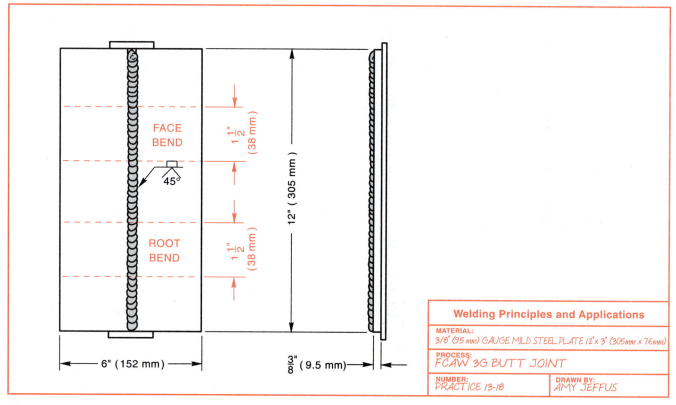

FACE BEND

45°

ROOT BEND

1 1/2" (38 mm)

1 1/2" (38 mm)

12" (305 mm)

6" (152 mm)

3/8" (9.5 mm)

Welding Principles and Applications

MATERIAL:
3/8" (95 mm) GAUGE MILD STEEL PLATE 12" x 3" (305mm x 76mm)

PROCESS:
FCAW 3G BUTT JOINT

NUMBER:
PRACTICE 13-18

DRAWN BY:
AMY JEFFUS

Figure 13-46

PRACTICE 13-21

Butt Joint 3G 100% To Be Tested

Using a properly set up and adjusted FCA welding machine, proper safety protection, 0.035-in. and/or through 1/16-in. (0.9-mm and/or through 1.6-mm) -diameter E70T-1 and/or E70T-5 electrodes, one or more pieces of mild steel plate, 7-in. (178-mm) -long and 3/4-in. (19-mm) or thicker beveled plate, and a 9-in. (230-mm) -long 1-in. (25-mm) -wide and 1/4-in. (6-mm) -thick backing strip, you will make a groove weld in the vertical position, **Figure 13-47.**

Following the same instructions for the assembly and welding procedure outlined in Practice 13-14, repeat the weld using each electrode classification until welds using both electrodes can be made with 100% penetration that will pass a bend test. Turn off the welding machine and shielding gas and clean up your work area when you are finished welding. ◆

PRACTICE 13-22

Fillet Weld Joint at a 45° Vertical Up Angle

Using a properly set up and adjusted FCA welding machine, proper safety protection, 0.035-in. and/or 0.045-in. (0.9-mm and/or 1.2-mm) -diameter E70T-1 and/or E70T-5 electrodes, and one or more pieces of mild steel plate, 12 in. (305 mm) long and 3/8-in. (9.5 mm) thick, you will increase the plate angle gradually as you

develop skill until you are making satisfactory welds in the vertical up position, **Figure 13-48.**

Tack weld the metal pieces together and brace them in position. Check to see that you have free movement along the entire joint to prevent stopping and restarting during the weld. Avoiding stops and starts both speeds up the welding time and eliminates discontinuities.

It's easier to make a quality weld in the vertical up position if both the amperage and voltage are set at the lower end of their ranges. This will make the molten weld pool smaller, less fluid, and easier to control. A problem with the lower power settings is that the weld bead often may be very convex. A convex face on a weld bead often makes it more difficult to remove the slag along the tow of the weld.

The weave pattern should allow for adequate fusion on both edges of the joint. Watch the edges to be sure that they are being melted so that adequate fusion and penetration occurs.

Repeat electrodes as needed until defect-free welds can consistently be made vertically. Turn off the welding machine and shielding gas and clean up your work area when you are finished welding. ◆

PRACTICE 13-23

Lap Joint and Tee Joint 3F

Using a properly set up and adjusted FCA welding machine, proper safety protection, 0.035-in. and/or 0.045-in. (0.9-mm and/or 1.2-mm) -diameter E70T-1

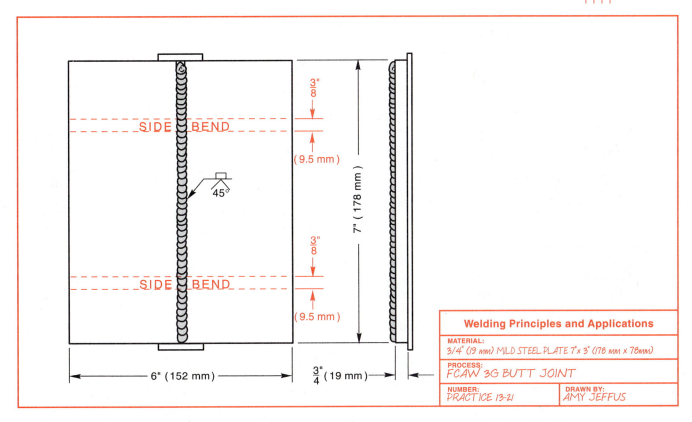

SIDE BEND

45°

SIDE BEND

3/8"
(9.5 mm)

3/8"
(9.5 mm)

7" (178 mm)

6" (152 mm)

3/4" (19 mm)

Welding Principles and Applications

MATERIAL:
3/4" (19 mm) MILD STEEL PLATE 7" x 3" (178 mm x 78mm)

PROCESS:
FCAW 3G BUTT JOINT

NUMBER:
PRACTICE 13-21

DRAWN BY:
AMY JEFFUS

Figure 13-47

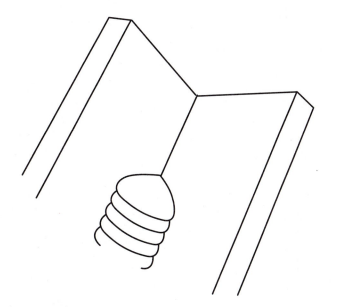

Figure 13-48 45° vertical up fillet weld.

and/or E70T-5 electrodes, and one or more pieces of mild steel plate, 12-in. (305-mm) -long and 3/8-in. (9.5-mm) -thick beveled plate, you will make a fillet weld in the vertical position.

Following the same instructions for the assembly and welding procedure outlined in Practice 13-22, repeat each type of joint with both classifications of electrodes as needed until defect-free welds can consistently be made.

Turn off the welding machine and shielding gas and clean up your work area when you are finished welding. ◆

PRACTICE 13-24

Lap Joint and Tee Joint 3F 100% to Be Tested

Using a properly set up and adjusted FCA welding machine, proper safety protection, 0.035-in. and/or 0.045-in. (0.9-mm and/or 1.2-mm) -diameter E70T-1 and/or E70T-5 electrodes, and one or more pieces of mild steel plate, 12-in. (305-mm) -long and 3/8-in. (9.5-mm) -thick beveled plate, you will make a fillet weld in the vertical position.

Following the same instructions for assembly and welding procedure as outlined in Practice 13-22, repeat each type of joint with both classifications of electrodes as needed until welds can be made with 100% penetration that will pass the test. Turn off the welding machine and shielding gas and clean up your work area when you are finished welding. ◆

PRACTICE 13-25

Tee Joint 3F

Using a properly set up and adjusted FCA welding machine, proper safety protection, 0.035-in. and/or through 1/16-in. (0.9-mm and/or through 1.6-mm) -diameter E70T-1 and/or E70T-5 electrodes, and one or more pieces of mild steel plate, 7-in. (178-mm) -long and 3/4-in. (19-mm) or thicker beveled plate, you will make a fillet weld in the vertical position.

Following the same instructions for the assembly and welding procedure outlined in Practice 13-22, repeat each type of joint with both classifications of electrodes as needed until defect-free welds can consistently be made. Turn off the welding machine and shielding gas and clean up your work area when you are finished welding. ◆

PRACTICE 13-26

Tee Joint 3F 100% to Be Tested

Using a properly set up and adjusted FCA welding machine, proper safety protection, 0.035-in. and/or through 1/16-in. (0.9-mm and/or through 1.6-mm) -diameter E70T-1 and/or E70T-5 electrodes, and one or more pieces of mild steel plate, 7-in. (178-mm) -long and 3/4-in. (19-mm) or thicker beveled plate, you will make a fillet weld in the vertical position, **Figure 13-49**.

Following the same instructions for the assembly and welding procedure outlined in Practice 13-22, repeat with both classifications of electrodes as needed until welds can be made with 100% penetration that will pass the test. Turn off the welding machine and shielding gas and clean up your work area when you are finished welding. ◆

HORIZONTAL WELDS

PRACTICE 13-27

Lap Joint and Tee Joint 2F

Using a properly set up and adjusted FCA welding machine, proper safety protection, 0.035-in. and/or 0.045-in. (0.9-mm and/or 1.2-mm) -diameter E70T-1 and/or E70T-5 electrodes, and one or more pieces of mild steel plate, 12-in. (305-mm) -long and 3/8-in. (9.5-mm) -thick beveled plate, you will make a fillet weld in the horizontal position, **Figure 13-50(A)** and **(B)**.

The root weld must be kept small so that its contour can be controlled. Too large a root pass can trap slag under overlap along the lower edge of the weld, **Figure 13-51**. Clean each pass thoroughly before the weld bead is started. Follow the weld bead sequence shown in **Figure 13-52**. Keeping all of the weld beads small will help control their contour.

Repeat each type of joint with both classifications of electrodes as needed until defect-free welds can consistently be made. Turn off the welding machine and shielding gas and clean up your work area when you are finished welding. ◆

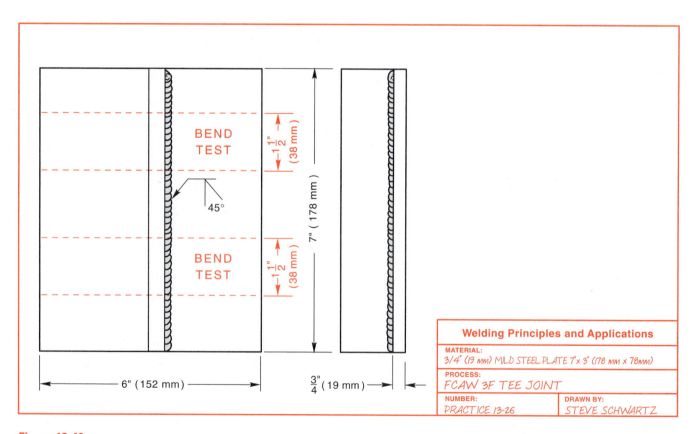

Welding Principles and Applications

MATERIAL:
3/4" (19 mm) MILD STEEL PLATE 7"x 3" (178 mm x 78mm)

PROCESS:
FCAW 3F TEE JOINT

NUMBER:
PRACTICE 13-26

DRAWN BY:
STEVE SCHWARTZ

Figure 13-49

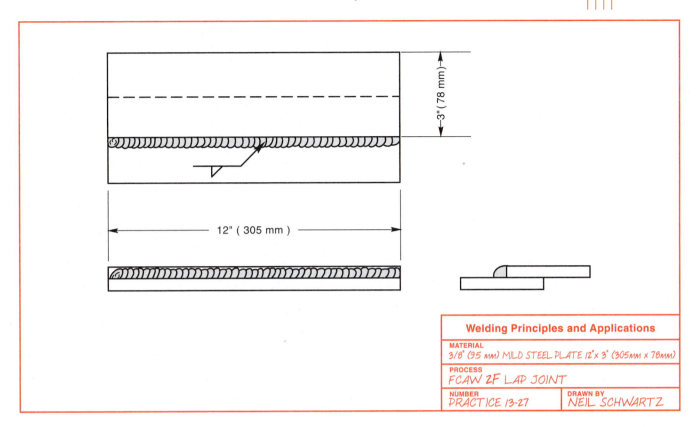

Welding Principles and Applications

MATERIAL
3/8" (9.5 mm) MILD STEEL PLATE 12"x 3" (305mm x 78mm)

PROCESS
FCAW 2F LAP JOINT

NUMBER
PRACTICE 13-27

DRAWN BY
NEIL SCHWARTZ

Figure 13-50(A)

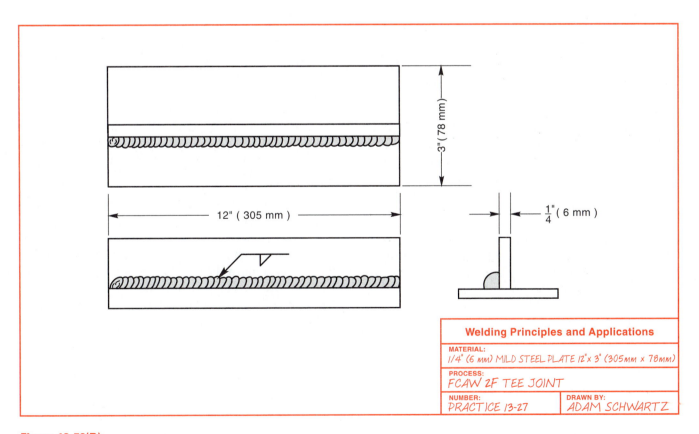

Welding Principles and Applications

MATERIAL:
1/4" (6 mm) MILD STEEL PLATE 12"x 3" (305mm x 78mm)

PROCESS:
FCAW 2F TEE JOINT

NUMBER:
PRACTICE 13-27

DRAWN BY:
ADAM SCHWARTZ

Figure 13-50(B)

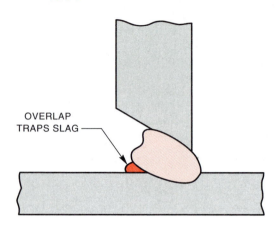

OVERLAP
TRAPS SLAG

Figure 13-51

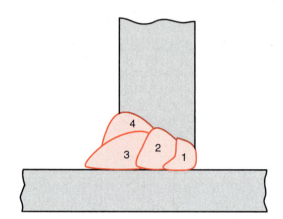

Figure 13-52

through 1/16-in. (0.9-mm and/or through 1.6-mm) -diameter E70T-1 and/or E70T-5 electrodes, and one or more pieces of mild steel plate, 7-in. (178-mm) -long and 3/4-in. (19-mm) or thicker beveled plate, you will make a fillet weld in the horizontal position.

Following the same instructions for the assembly and welding procedure outlined in Practice 13-27, repeat each type of joint with both classifications of electrodes as needed until defect-free welds can consistently be made. Turn off the welding machine and shielding gas and clean up your work area when you are finished welding. ◆

PRACTICE 13-30

Tee Joint 2F 100% to Be Tested

Using a properly set up and adjusted FCA welding machine, proper safety protection, 0.035-in. and/or through 1/16-in. (0.9-mm and/or through 1.6-mm) -diameter E70T-1 and/or E70T-5 electrodes, and one or more pieces of mild steel plate, 7-in. (178-mm) long and 3/4-in. (19-mm) or thicker beveled plate, you will make a fillet weld in the horizontal position.

Following the same instructions for the assembly and welding procedure outlined in Practice 13-27, repeat with both classifications of electrodes as needed until welds can be made with 100% penetration that will pass the test. Turn off the welding machine and shielding gas and clean up your work area when you are finished welding. ◆

PRACTICE 13-28

Lap Joint and Tee Joint 2F 100% to Be Tested

Using a properly set up and adjusted FCA welding machine, proper safety protection, 0.035-in. and/or 0.045-in. (0.9-mm and/or 1.2-mm) -diameter E70T-1 and/or E70T-5 electrodes, and one or more pieces of mild steel plate, 12-in. (305-mm) -long and 3/8-in. (9.5-mm) -thick beveled plate, you will make a fillet weld in the horizontal position.

Following the same instructions for the assembly and welding procedure outlined in Practice 13-27, repeat each type of joint with both classifications of electrodes as needed until welds can be made with 100% penetration that will pass the test. Turn off the welding machine and shielding gas and clean up your work area when you are finished welding. ◆

PRACTICE 13-31

Stringer Bead at a 45° Horizontal Angle

Using a properly set up and adjusted FCA welding machine, proper safety protection, 0.035-in. and/or 0.045-in. (0.9-mm and/or 1.2-mm) -diameter E70T-1 and/or E70T-5 electrodes, and one or more pieces of mild steel plate, 12 in. (305 mm) long and 1/4 in. (6 mm) thick or thinner, you will increase the plate angle gradually as you develop skill until you are making satisfactory horizontal welds across the vertical face of the plate, **Figure 13-53**.

Repeat the weld using each electrode classification until welds using both electrodes can be made horizontally with uniform bead contours when the plates are vertical. Turn off the welding machine and shielding gas and clean up your work area when you are finished welding. ◆

PRACTICE 13-29

Tee Joint 2F

Using a properly set up and adjusted FCA welding machine, proper safety protection, 0.035-in. and/or

PRACTICE 13-32

Butt Joint 2G

Using a properly set up and adjusted FCA welding machine, proper safety protection, 0.035-in. and/or 0.045-in. (0.9-mm and/or 1.2-mm) -diameter E70T-1 and/or E70T-5 electrodes, and one or more pieces of mild steel plate, 12 in. (305 mm) long and 1/4 in. (6 mm) thick or thinner, you will make a groove weld in the horizontal position, **Figure 13-54**.

Following the same instructions for the assembly and welding procedure outlined in Practice 13-31, repeat with both classifications of electrodes as needed until defect-free welds can consistently be made in the 1/4-in. (6-mm) -thick plate. Turn off the welding machine and shielding gas and clean up your work area when you are finished welding. ◆

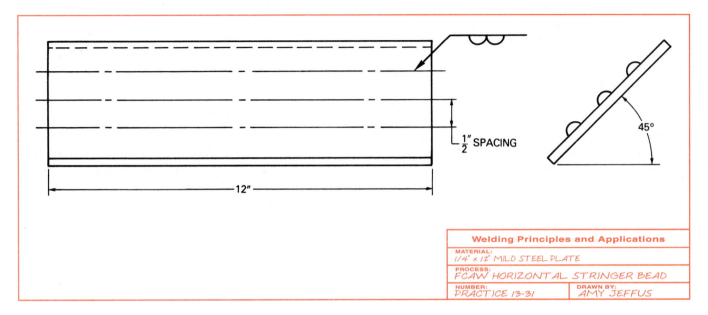

Welding Principles and Applications

MATERIAL:
1/4" x 12" MILD STEEL PLATE

PROCESS:
FCAW HORIZONTAL STRINGER BEAD

NUMBER:	DRAWN BY:
PRACTICE 13-31	AMY JEFFUS

Figure 13-53 Horizontal stringer bead.

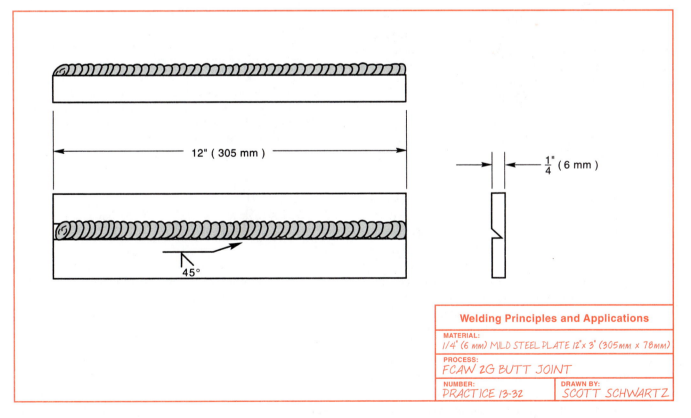

Welding Principles and Applications

MATERIAL:
1/4" (6 mm) MILD STEEL PLATE 12" x 3" (305mm x 78mm)

PROCESS:
FCAW 2G BUTT JOINT

NUMBER:	DRAWN BY:
PRACTICE 13-32	SCOTT SCHWARTZ

Figure 13-54

Butt Joint 2G 100% to Be Tested

Using a properly set up and adjusted FCA welding machine, proper safety protection, 0.035-in. and/or 0.045-in. (0.9-mm and/or 1.2-mm) -diameter E70T-1 and/or E70T-5 electrodes, and one or more pieces of mild steel plate, 12 in. (305 mm) long and 1/4 in. (6 mm) thick, you will make a groove weld in the horizontal position.

Following the same instructions for the assembly and welding procedure outlined in Practice 13-31, repeat the weld using each electrode classification until welds using both electrodes can be made with 100% penetration that will pass a bend test. Turn off the welding machine and shielding gas and clean up your work area when you are finished welding. ◆

Butt Joint 2G

Using a properly set up and adjusted FCA welding machine, proper safety protection, 0.035-in. and/or 0.045-in. (0.9-mm and/or 1.2-mm) -diameter E70T-1 and/or E70T-5 electrodes, and one or more pieces of mild steel plate, 12-in. (305-mm) -long and 3/8-in. (9.5-mm) -thick beveled plate, and a 14-in. (355-mm) -long 1-in. (25-mm) -wide and 1/4-in. (6-mm) -thick backing strip, you will make a groove weld in the horizontal position.

Following the same instructions for the assembly and welding procedure outlined in Practice 13-31, repeat with both classifications of electrodes as needed until defect-free welds can consistently be made. Turn off the welding machine and shielding gas and clean up your work area when you are finished welding. ◆

Butt Joint 2G 100% to Be Tested

Using a properly set up and adjusted FCA welding machine, proper safety protection, 0.035-in. and/or 0.045-in. (0.9-mm and/or 1.2-mm) -diameter E70T-1 and/or E70T-5 electrodes, and one or more pieces of mild steel plate, 12-in. (305-mm) -long and 3/8-in. (9.5-mm) -thick beveled plate, and a 14-in. (355-mm) -long 1-in. (25-mm) -wide and 1/4-in. (6-mm) -thick backing strip, you will make a groove weld in the horizontal position.

Following the same instructions for the assembly and welding procedure outlined in Practice 13-31, repeat the weld using each electrode classification until welds using both electrodes can be made with 100% penetration that will pass a bend test. Turn off the welding machine and shielding gas and clean up your work area when you are finished welding. ◆

Butt Joint 2G

Using a properly set up and adjusted FCA welding machine, proper safety protection, 0.035-in. and/or through 1/16-in. (0.9-mm and/or through 1.6-mm) -diameter E70T-1 and/or E70T-5 electrodes, one or more pieces of mild steel plate, 7-in. (178-mm) -long and 3/4-in. (19-mm) or thicker beveled plate, and a 9-in. (230-mm) -long 1-in. (25-mm) -wide and 1/4-in. (6-mm) -thick backing strip, you will make a groove weld in the horizontal position, **Figure 13-55**.

Following the same instructions for the assembly and welding procedure as outlined in Practice 13-31, repeat with both classifications of electrodes as needed until defect-free welds can consistently be made. Turn off the welding machine and shielding gas and clean up your work area when you are finished welding. ◆

Butt Joint 2G 100% to Be Tested

Using a properly set up and adjusted FCA welding machine, proper safety protection, 0.035-in. and/or through 1/16-in. (0.9-mm and/or through 1.6-mm) diameter E70T-1 and/or E70T-5 electrodes, one or more pieces of mild steel plate, 7-in. (178-mm) -long and 3/4-in. (19-mm) or thicker beveled plate, and a 9-in. (230-mm) -long 1-in. (25-mm) -wide and 1/4-in. (6-mm) -thick backing strip, you will make a groove weld in the horizontal position.

Following the same instructions for the assembly and welding procedure outlined in Practice 13-31, repeat the weld using each electrode classification until welds using both electrodes can be made with 100% penetration that will pass a bend test. Turn off the welding machine and shielding gas and clean up your work area when you are finished welding. ◆

OVERHEAD-POSITION WELDS

Butt Joint 4G

Using a properly set up and adjusted FCA welding machine, proper safety protection, 0.035-in. and/or 0.045-in. (0.9-mm and/or 1.2-mm) -diameter E70T-1 and/or E70T-5 electrodes, and one or more pieces of mild steel plate, 12 in. (305 mm) long and 1/4 in. (6 mm) thick or thinner, you will make a groove weld in the overhead position.

The molten weld pool should be kept as small as possible for easier control. A small molten weld pool can be achieved by using lower current, faster traveling settings.

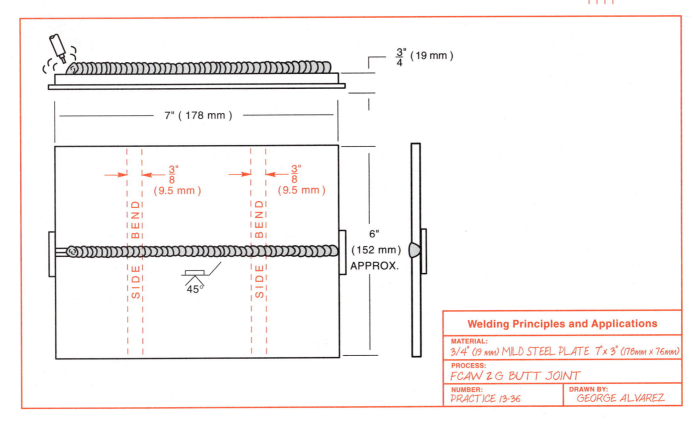

Figure 13-55

Lower current settings require closer control of gun manipulation to ensure that the electrode is fed into the molten weld pool just behind the leading edge. The low power will cause overlap and more spatter if this electrode-to-molten weld pool contact position is not closely maintained.

Faster travel speeds allow the welder to maintain a high production rate even if multiple passes are required to complete the weld. Weld penetration into the base metal at the start of the bead can be obtained by using a slow start or quickly reversing the weld direction. Both the slow start and reversal of weld direction put more heat into the weld start to increase penetration. The higher speed also reduces the amount of weld distortion by reducing the amount of time that heat is applied to a joint.

When welding overhead, extra personal protection is required to reduce the danger of burns. Leather sleeves or leather jackets should be worn.

Much of the spatter created during overhead welding falls into or on the nozzle and contact tube. The contact tube may short out to the gas nozzle. The shorted gas nozzle may arc to the work, causing damage both to the nozzle and to the plate. To control the amount of spatter, a longer stickout and/or a sharper gun-to-plate angle is required to allow most of the spatter to fall clear of the gun or nozzle, **Figure 13-56.**

Make several short weld beads using various techniques to establish the method that is most successful and most comfortable for you. After each weld, stop and evaluate it before making a change. When you have decided on the technique to be used, make a welded stringer bead that is 12 in. (305 mm) long.

Repeat with both classifications of electrodes as needed until defect-free welds can consistently be made in the 1/4-in. (6-mm) -thick plate. Turn off the welding machine and shielding gas and clean up your work area when you are finished welding. ◆

PRACTICE 13-39

Butt Joint 4G 100% to Be Tested

Using a properly set up and adjusted FCA welding machine, proper safety protection, 0.035-in. and/or 0.045-in. (0.9-mm and/or 1.2-mm) -diameter E70T-1 and/or E70T-5 electrodes, and one or more pieces of

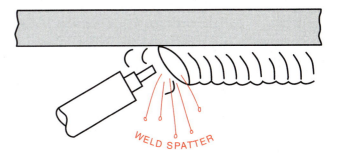

Figure 13-56 Hold the gun so that weld spatter will not fall onto the gun.

mild steel plate, 12 in. (305 mm) long and 1/4 in. (6 mm) thick, you will make a groove weld in the overhead position.

Following the same instructions for the assembly and welding procedure outlined in Practice 13-38, repeat the weld using each electrode classification until welds using both electrodes can be made with 100% penetration that will pass a bend test. Turn off the welding machine and shielding gas and clean up your work area when you are finished welding. ◆

PRACTICE 13-40

Butt Joint 4G

Using a properly set up and adjusted FCA welding machine, proper safety protection, 0.035-in. and/or 0.045-in. (0.9-mm and/or 1.2-mm) -diameter E70T-1 and/or E70T-5 electrodes, and one or more pieces of mild steel plate, 12-in. (305-mm) -long and 3/8-in. (9.5-mm) -thick beveled plate, and a 14-in. (355-mm) -long 1-in. (25-mm) -wide and 1/4-in. (6-mm) -thick backing strip, you will make a groove weld in the overhead position, **Figure 13-57.**

Following the same instructions for the assembly and welding procedure outlined in Practice 13-38, repeat with both classifications of electrodes as needed until defect-free welds can consistently be made. Turn off the welding machine and shielding gas and clean up your work area when you are finished welding. ◆

PRACTICE 13-41

Butt Joint 4G 100% to Be Tested

Using a properly set up and adjusted FCA welding machine, proper safety protection, 0.035-in. and/or 0.045-in. (0.9-mm and/or 1.2-mm) -diameter E70T-1 and/or E70T-5 electrodes, and one or more pieces of mild steel plate, 12-in. (305-mm) -long and 3/8-in. (9.5-mm) -thick beveled plate, and a 14-in. (355-mm) -long 1-in. (25-mm) -wide and 1/4-in. (6-mm) -thick backing strip, you will make a groove weld in the overhead position.

Following the same instructions for the assembly and welding procedure outlined in Practice 13-38, repeat the weld using each electrode classification until welds using both electrodes can be made with 100% penetration that will pass a bend test. Turn off the welding machine and shielding gas and clean up your work area when you are finished welding. ◆

PRACTICE 13-42

Butt Joint 4G

Using a properly set up and adjusted FCA welding machine, proper safety protection, 0.035-in. and/or through 1/16-in. (0.9-mm and/or through 1.6-mm) -diameter E70T-1 and/or E70T-5 electrodes, one or more pieces of mild steel plate, 7-in. (178-mm) -long and 3/4-in. (19-mm) or thicker beveled plate, and a 9-in. (230-

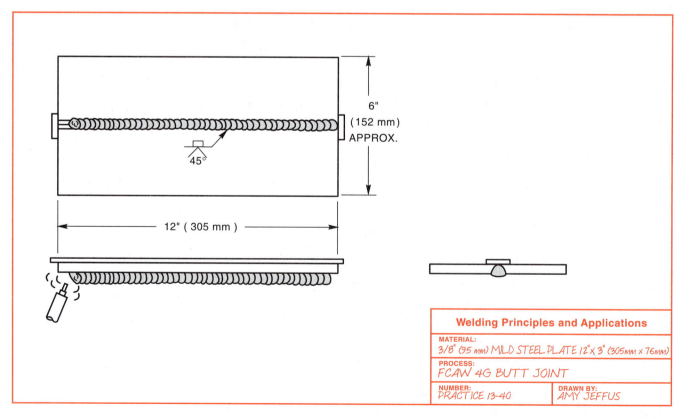

Welding Principles and Applications

MATERIAL:
3/8" (95 mm) MILD STEEL PLATE 12" x 3" (305mm x 76mm)

PROCESS:
FCAW 4G BUTT JOINT

NUMBER:
PRACTICE 13-40

DRAWN BY:
AMY JEFFUS

Figure 13-57

mm) -long 1-in. (25-mm) -wide and 1/4-in. (6-mm) -thick backing strip, you will make a groove weld in the overhead position.

Following the same instructions for the assembly and welding procedure outlined in Practice 13-38, repeat with both classifications of electrodes as needed until defect-free welds can consistently be made. Turn off the welding machine and shielding gas and clean up your work area when you are finished welding. ◆

Butt Joint 4G 100% to Be Tested

Using a properly set up and adjusted FCA welding machine, proper safety protection, 0.035-in. and/or through 1/16-in. (0.9-mm and/or through 1.6-mm) -diameter E70T-1 and/or E70T-5 electrodes, one or more pieces of mild steel plate, 7-in. (178-mm) -long and 3/4-in. (19-mm) or thicker beveled plate, and a 9-in. (230-mm) -long 1-in. (25-mm) -wide and 1/4-in. (6-mm) -thick backing strip, you will make a groove weld in the overhead position.

Following the same instructions for the assembly and welding procedure outlined in Practice 13-38, repeat the weld using each electrode classification until welds using both electrodes can be made with 100% penetration that will pass a bend test. Turn off the welding machine and shielding gas and clean up your work area when you are finished welding. ◆

PRACTICE 13-44

Lap Joint and Tee Joint 4F

Using a properly set up and adjusted FCA welding machine, proper safety protection, 0.035-in. and/or 0.045-in. (0.9-mm and/or 1.2-mm) -diameter E70T-1 and/or E70T-5 electrodes, and one or more pieces of mild steel plate, 12-in. (305-mm) -long and 3/8-in. (9.5-mm) -thick beveled plate, you will make a fillet weld in the overhead position, **Figure 13-58(A)** and **(B)**.

Following the same instructions for the assembly and welding procedure outlined in Practice 13-38, repeat each type of joint with both classifications of electrodes as needed until defect-free welds can consistently be made. Turn off the welding machine and shielding gas and clean up your work area when you are finished welding. ◆

PRACTICE 13-45

Lap Joint and Tee Joint 4F 100% to Be Tested

Using a properly set up and adjusted FCA welding machine, proper safety protection, 0.035-in. and/or 0.045-in. (0.9-mm and/or 1.2-mm) -diameter E70T-1 and/or E70T-5 electrodes, and one or more pieces of mild steel plate, 12-in. (305-mm) -long and 3/8-in. (9.5-mm) -thick beveled plate, you will make a fillet weld in the overhead position.

Following the same instructions for the assembly and welding procedure outlined in Practice 13-38, repeat

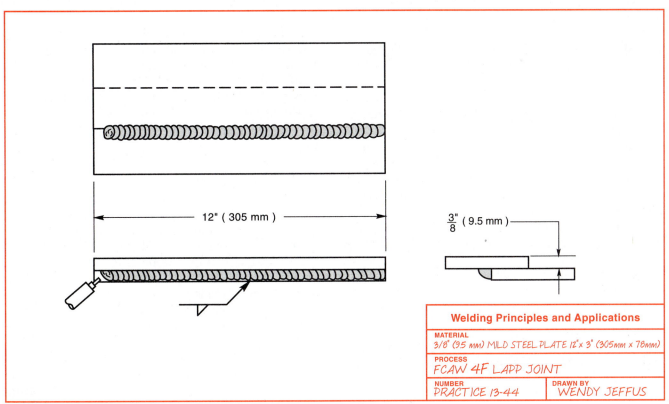

Welding Principles and Applications

MATERIAL
3/8" (9.5 mm) MILD STEEL PLATE 12"x 3" (305mm x 78mm)

PROCESS
FCAW 4F LAPP JOINT

NUMBER
PRACTICE 13-44

DRAWN BY
WENDY JEFFUS

Figure 13-58(A)

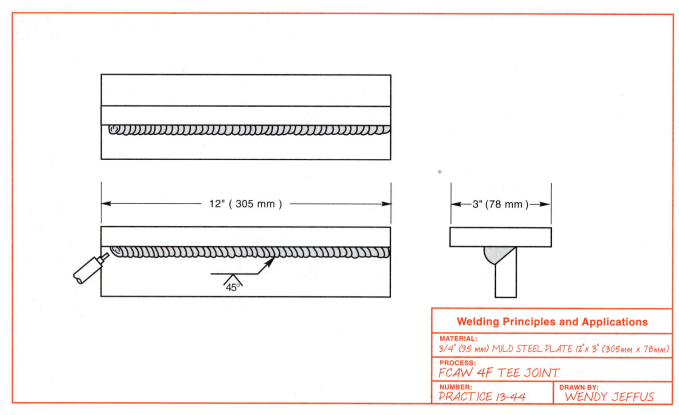

Welding Principles and Applications

MATERIAL:
3/4" (9.5 mm) MILD STEEL PLATE 12"x 3" (305mm x 78mm)

PROCESS:
FCAW 4F TEE JOINT

NUMBER:
PRACTICE 13-44

DRAWN BY:
WENDY JEFFUS

Figure 13-58(B)

each type of joint with both classifications of electrodes as needed until welds can be made with 100% penetration that will pass the test. Turn off the welding machine and shielding gas and clean up your work area when you are finished welding. ◆

PRACTICE 13-46

Tee Joint 4F

Using a properly set up and adjusted FCA welding machine, proper safety protection, 0.035-in. and/or through 1/16-in. (0.9-mm and/or through 1.6-mm) -diameter E70T-1 and/or E70T-5 electrodes, and one or more pieces of mild steel plate, 7-in. (178-mm) -long and 3/4-in. (19-mm) or thicker beveled plate, you will make a fillet weld in the overhead position.

Following the same instructions for the assembly and welding procedure outlined in Practice 13-38, repeat each type of joint with both classifications of electrodes as needed until defect-free welds can consistently be made. Turn off the welding machine and shielding gas and clean up your work area when you are finished welding. ◆

PRACTICE 13-47

Tee Joint 4F 100% to Be Tested

Using a properly set up and adjusted FCA welding machine, proper safety protection, 0.035-in. and/or

through 1/16-in. (0.9-mm and/or through 1.6-mm) -diameter E70T-1 and/or E70T-5 electrodes, and one or more pieces of mild steel plate, 7-in. (178-mm) -long and 3/4-in. (19-mm) or thicker beveled plate, you will make a fillet weld in the overhead position.

Following the same instructions for the assembly and welding procedure as outlined in Practice 13-38, repeat with both classifications of electrodes as needed until welds can be made with 100% penetration that will pass the bend test. Turn off the welding machine and shielding gas and clean up your work area when you are finished welding. ◆

THIN-GAUGE WELDING

The introduction of small electrode diameters has allowed FCA welding to be used on thin sheet metal. Usually these welds will be a fillet type. Filler welds are the easiest weld to make on thin stock. An effort should be taken when possible to design the weld so it is not a butt-type joint. A common use for FCA welding on thin stock is to join it to a thicker member, Figure 13-59. This type of weld is used to put panels in frames.

The following practices include some butt type joints. You will find that the vertical down welds are the easiest ones to make. If it is possible to position the weldment, production speeds can be increased if butt joints are required.

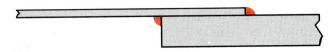

Figure 13-59 FCA welding thin to thick metal.

PRACTICE 13-48

Butt Joint 1G

Using a properly set up and adjusted FCA welding machine, Table 13-3, proper safety protection, 0.030-in. and/or 0.035-in. (0.8-mm and/or 0.9-mm) -diameter E70T-1 and/or E70T-5 electrodes, and one or more pieces of mild steel sheet, 12 in. (305-mm) long and 16 gauge to 18 gauge thick, you will make a butt weld in the flat position, Figure 13-60.

Do not leave a root opening for these welds. Even the slightest opening will result in a burn-through. If a burn-through occurs, the welder can be pulsed off and on so that the hole can be filled. This process will leave a larger than usual buildup. Excessive buildup could be ground off if necessary as part of the postweld cleanup.

Repeat with both classifications of electrodes as needed until defect-free welds can consistently be made. Turn off the welding machine and shielding gas and clean up your work area when you are finished welding. ◆

Electrode			Welding Power			Shielding Gas		Base Metal	
Type	Size	Amps	Wire Feed Speed IPM (cm/min)	Volts	Type	Flow	Type	Thickness
E70T-1	0.030 in. (0.8 mm)	40 to 145	90 to 340 (228 to 864)	20 to 27	None	n/a	Low-carbon steel	16-gauge to 18-gauge
E70T-1	0.035 in. (0.9 mm)	130 to 200	288 to 576 (732 to 1463)	20 to 28	None	n/a	Low-carbon steel	16-gauge to 18-gauge
E70T-5	0.035 in. (0.9 mm)	90 to 200	190 to 576 (483 to 1463)	16 to 29	57% argon 25% CO_2	35 cfh	Low-carbon steel	16-gauge to 18-gauge

Table 13-3 FCA Welding Parameters for Use If Specific Settings Unavailable from Electrode Manufacturer.

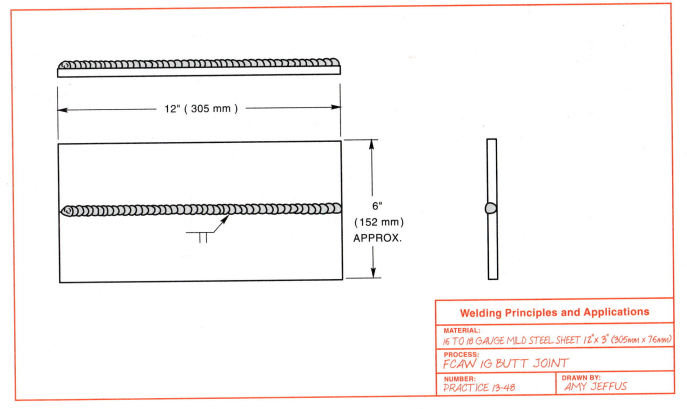

12" (305 mm)

6"
(152 mm)
APPROX.

Welding Principles and Applications

MATERIAL:
16 TO 18 GAUGE MILD STEEL SHEET 12" x 3" (305mm x 76mm)

PROCESS:
FCAW 1G BUTT JOINT

NUMBER:
PRACTICE 13-48

DRAWN BY:
AMY JEFFUS

Figure 13-60

PRACTICE 13-49

Butt Joint 1G 100% to Be Tested

Using a properly set up and adjusted FCA welding machine, proper safety protection, 0.030-in. and/or 0.035-in. (0.8-mm and/or 0.9-mm) -diameter E70T-1 and/or E70T-5 electrodes, and one or more pieces of mild steel sheet, 12 in. (305 mm) long and 16 gauge to 18 gauge thick, you will make a butt weld in the flat position, **Figure 13-61**.

Following the same instructions for the assembly and welding procedure outlined in Practice 13-47, repeat the weld using each electrode classification until welds using both electrodes can be made with 100% penetration that will pass a bend test. Turn off the welding machine and shielding gas and clean up your work area when you are finished welding. ◆

PRACTICE 13-50

Lap Joint and Tee Joint 1F

Using a properly set up and adjusted FCA welding machine, proper safety protection, 0.030-in. and/or 0.035-in. (0.8-mm and/or 0.9-mm) -diameter E70T-1 and/or E70T-5 electrodes, and one or more pieces of mild steel sheet, 12 in. (305 mm) long and 16 gauge to

18 gauge thick, you will make a fillet weld in the flat position.

Following the same instructions for the assembly and welding procedure outlined in Practice 13-47, repeat each type of joint with both classifications of electrodes as needed until defect-free welds can consistently be made. Turn off the welding machine and shielding gas and clean up your work area when you are finished welding. ◆

PRACTICE 13-51

Lap Joint and Tee Joint 1F 100% to Be Tested

Using a properly set up and adjusted FCA welding machine, proper safety protection, 0.030-in. and/or 0.035-in. (0.8-mm and/or 0.9-mm) -diameter E70T-1 and/or E70T-5 electrodes, and one or more pieces of mild steel sheet, 12 in. (305 mm) long and 16 gauge to 18 gauge thick, you will make a fillet weld in the flat position, **Figure 13-62(A)** and **(B)**.

Following the same instructions for the assembly and welding procedure outlined in Practice 13-47, repeat each type of joint with both classifications of electrodes as needed until welds can be made with 100% penetration that will pass the bend test, **Figure 13-63(A)** and **(B)**. Turn off the welding machine and shielding gas and clean up your work area when you are finished welding. ◆

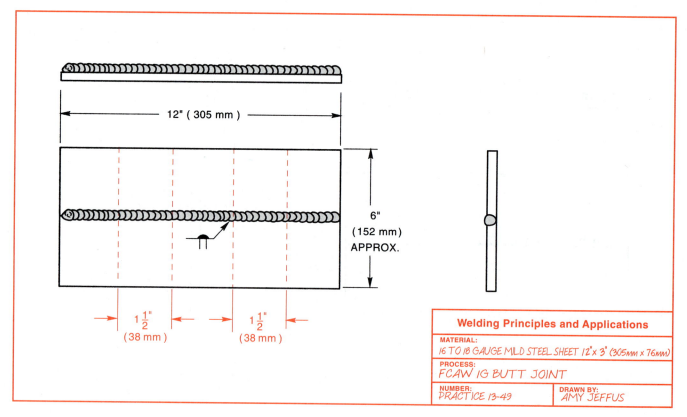

Welding Principles and Applications

MATERIAL:
16 TO 18 GAUGE MILD STEEL SHEET 12"x 3" (305mm x 76mm)

PROCESS:
FCAW 1G BUTT JOINT

NUMBER:
PRACTICE 13-49

DRAWN BY:
AMY JEFFUS

Figure 13-61

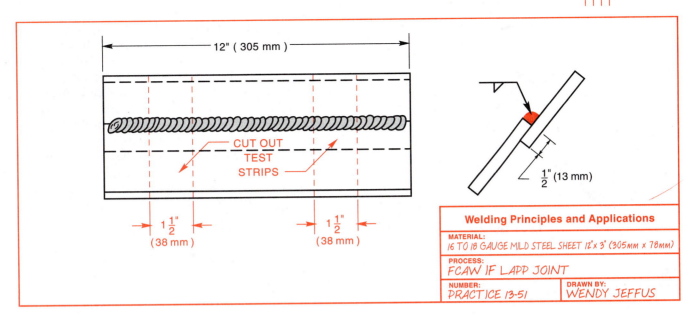

Figure 13-62(A) Cut out test strips.

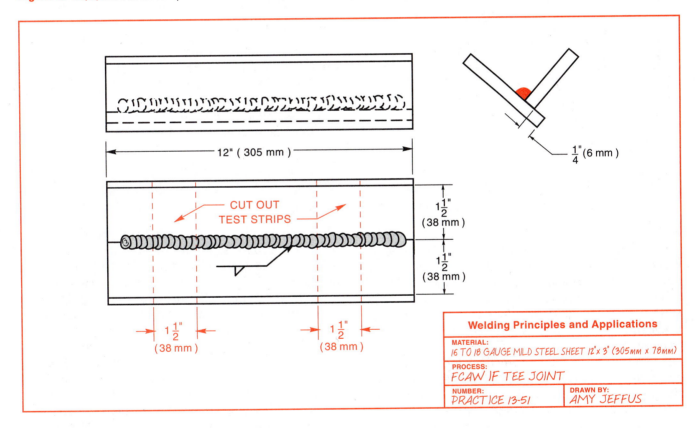

Figure 13-62(B)

PRACTICE 13-52

Butt Joint 3G

Using a properly set up and adjusted FCA welding machine, proper safety protection, 0.030-in. and/or 0.035-in. (0.8-mm and/or 0.9-mm) -diameter E70T-1 and/or E70T-5 electrodes, and one or more pieces of mild steel sheet, 12 in. (305 mm) long and 16 gauge to 18

gauge thick, you will make a butt weld in the vertical up or down position.

Following the same instructions for the assembly and welding procedure outlined in Practice 13-47, repeat with both classifications of electrodes as needed until defect-free welds can consistently be made. Turn off the welding machine and shielding gas and clean up your work area when you are finished welding. ◆

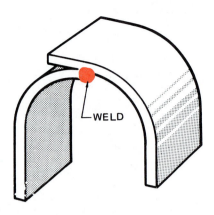

Figure 13-63(A) 180° bend to test lap weld quality.

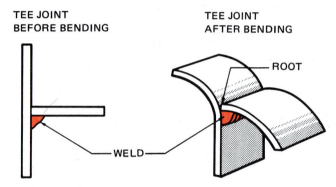

TEE JOINT
BEFORE BENDING

TEE JOINT
AFTER BENDING

ROOT

WELD

Figure 13-63(B) Bend the test strip to be sure the weld has good root fusion.

Butt Joint 3G 100% to Be Tested

Using a properly set up and adjusted FCA welding machine, proper safety protection, 0.030-in. and/or 0.035-in. (0.8-mm and/or 0.9-mm) -diameter E70T-1 and/or E70T-5 electrodes, and one or more pieces of mild steel sheet, 12 in. (305 mm) long and 16 gauge to 18 gauge thick, you will make a butt weld in the vertical up or down position.

Following the same instructions for the assembly and welding procedure outlined in Practice 13-47, repeat the weld using each electrode classification until welds using both electrodes can be made with 100% penetration that will pass a bend test. Turn off the welding machine and shielding gas and clean up your work area when you are finished welding. ◆

Lap Joint and Tee Joint 3F

Using a properly set up and adjusted FCA welding machine, proper safety protection, 0.030-in. and/or 0.035-in. (0.8-mm and/or 0.9-mm) -diameter E70T-1 and/or E70T-5 electrodes, and one or more pieces of mild

steel sheet, 12 in. (305 mm) long and 16 gauge to 18 gauge thick, you will make a fillet weld in the vertical up or down position.

Following the same instructions for the assembly and welding procedure outlined in Practice 13-47, repeat each type of joint with both classifications of electrodes as needed until defect-free welds can consistently be made. Turn off the welding machine and shielding gas and clean up your work area when you are finished welding. ◆

Lap Joint and Tee Joint 3F 100% to Be Tested

Using a properly set up and adjusted FCA welding machine, proper safety protection, 0.030-in. and/or 0.035-in. (0.8-mm and/or 0.9-mm) -diameter E70T-1 and/or E70T-5 electrodes, and one or more pieces of mild steel sheet, 12 in. (305 mm) long and 16 gauge to 18 gauge thick, you will make a fillet weld in the vertical up or down position.

Following the same instructions for the assembly and welding procedure outlined in Practice 13-47, repeat each type of joint with both classifications of electrodes as needed until welds can be made with 100% penetration that will pass the test. Turn off the welding machine and shielding gas and clean up your work area when you are finished welding. ◆

Lap Joint and Tee Joint 2F

Using a properly set up and adjusted FCA welding machine, proper safety protection, 0.030-in. and/or 0.035-in. (0.8-mm and/or 0.9-mm) -diameter E70T-1 and/or E70T-5 electrodes, and one or more pieces of mild steel sheet, 12 in. (305 mm) long and 16 gauge to 18 gauge thick, you will make a fillet weld in the horizontal position.

Following the same instructions for the assembly and welding procedure outlined in Practice 13-38, repeat each type of joint with both classifications of electrodes as needed until defect-free welds can consistently be made. Turn off the welding machine and shielding gas and clean up your work area when you are finished welding. ◆

Lap Joint and Tee Joint 2F 100% to Be Tested

Using a properly set up and adjusted FCA welding machine, proper safety protection, 0.030-in. and/or 0.035-in. (0.8-mm and/or 0.9-mm) -diameter E70T-1 and/or E70T-5 electrodes, and one or more pieces of mild steel sheet, 12 in. (305 mm) long and 16 gauge to 18 gauge thick, you will make a fillet weld in the horizontal position.

Following the same instructions for the assembly and welding procedure outlined in Practice 13-38, repeat each type of joint with both classifications of electrodes as needed until welds can be made with 100% penetration that will pass the bend test. Turn off the welding machine and shielding gas and clean up your work area when you are finished welding. ◆

PRACTICE 13-58

Butt Joint 2G

Using a properly set up and adjusted FCA welding machine, proper safety protection, 0.030-in. and/or 0.035-in. (0.8-mm and/or 0.9-mm) -diameter E70T-1 and/or E70T-5 electrodes, and one or more pieces of mild steel sheet, 12 in. (305 mm) long and 16 gauge to 18 gauge thick, you will make a butt weld in the horizontal position.

Following the same instructions for the assembly and welding procedure outlined in Practice 13-38, repeat with both classifications of electrodes as needed until defect-free welds can consistently be made. Turn off the welding machine and shielding gas and clean up your work area when you are finished welding. ◆

PRACTICE 13-59

Butt Joint 2G 100% to Be Tested

Using a properly set up and adjusted FCA welding machine, proper safety protection, 0.030-in. and/or 0.035-in. (0.8-mm and/or 0.9-mm) -diameter E70T-1 and/or E70T-5 electrodes, and one or more pieces of mild steel sheet, 12 in. (305 mm) long and 16 gauge to 18 gauge thick, you will make a butt weld in the horizontal position.

Following the same instructions for the assembly and welding procedure outlined in Practice 13-38, repeat the weld using each electrode classification until welds using both electrodes can be made with 100% penetration that will pass a bend test. Turn off the welding machine and shielding gas and clean up your work area when you are finished welding. ◆

PRACTICE 13-60

Butt Joint 4G

Using a properly set up and adjusted FCA welding machine, proper safety protection, 0.030-in. and/or 0.035-in. (0.8-mm and/or 0.9-mm) -diameter E70T-1 and/or E70T-5 electrodes, and one or more pieces of mild steel sheet, 12 in. (305 mm) long and 16 gauge to 18 gauge thick, you will make a butt weld in the overhead position.

Following the same instructions for the assembly and welding procedure outlined in Practice 13-38, repeat with both classifications of electrodes as needed until defect-free welds can consistently be made. Turn off the

welding machine and shielding gas and clean up your work area when you are finished welding. ◆

PRACTICE 13-61

Butt Joint 4G 100% to Be Tested

Using a properly set up and adjusted FCA welding machine, proper safety protection, 0.030-in. and/or 0.035-in. (0.8-mm and/or 0.9-mm) -diameter E70T-1 and/or E70T-5 electrodes, and one or more pieces of mild steel sheet, 12 in. (305 mm) long and 16 gauge to 18 gauge thick, you will make a butt weld in the overhead position.

Following the same instructions for the assembly and welding procedure outlined in Practice 13-38, repeat the weld using each electrode classification until welds using both electrodes can be made with 100% penetration that will pass a bend test. Turn off the welding machine and shielding gas and clean up your work area when you are finished welding. ◆

PRACTICE 13-62

Lap Joint and Tee Joint 4F

Using a properly set up and adjusted FCA welding machine, proper safety protection, 0.030-in. and/or 0.035-in. (0.8-mm and/or 0.9-mm) -diameter E70T-1 and/or E70T-5 electrodes, and one or more pieces of mild steel sheet, 12 in. (305 mm) long and 16 gauge to 18 gauge thick, you will make a fillet weld in the overhead position.

Following the same instructions for the assembly and welding procedure outlined in Practice 13-38, repeat each type of joint with both classifications of electrodes as needed until defect-free welds can consistently be made. Turn off the welding machine and shielding gas and clean up your work area when you are finished welding. ◆

PRACTICE 13-63

Lap Joint and Tee Joint 4F 100% to Be Tested

Using a properly set up and adjusted FCA welding machine, proper safety protection, 0.030-in. and/or 0.035-in. (0.8-mm and/or 0.9-mm) -diameter E70T-1 and/or E70T-5 electrodes, and one or more pieces of mild steel sheet, 12 in. (305 mm) long and 16 gauge to 18 gauge thick, you will make a fillet weld in the overhead position.

Following the same instructions for the assembly and welding procedure outlined in Practice 13-38, repeat each type of joint with both classifications of electrodes as needed until welds can be made with 100% penetration that will pass the test. Turn off the welding machine and shielding gas and clean up your work area when you are finished welding. ◆

REVIEW

1. Why is it important to make sure that an FCA welding system is set up properly if out-of-position welds are going to be made?

2. What major safety concerns should an FCA welder be cautious of?

3. Why should the FCA welding practice plates be large?

4. Why should the FCA welds be of substantial length?

5. What must be done to a shielding gas cylinder before its cap is removed?

6. What can happen to the wire if the conduit is misaligned at the feed rollers?

7. Why is it a good idea for a new student welder to use the gas nozzle even if a shielding gas is not used?

8. Why is the curl in the wire end straightened out?

9. What problems can a high-feed roller pressure cause?

10. Referring to Table 13-1, answer the following:

   a. What would the range of the feed speed be for an amperage of 150 at 25 volts for E70T-1 0.035-in. (0.9-mm) electrode?
   b. What would the approximate amperage be for E70T-5 0.045-in. (1.2-mm) electrode if it's being fed at a rate of 200 in. per minute (508 cm per minute)?

11. What are the disadvantages of beveling plates for welding?

12. FCA welds with 100% joint penetration can be made in plates up to what thickness?

13. What is the smallest V-groove angle that can be welded using the FCA welding process?

14. What is the purpose of the root pass?

15. Why is the FCA welding process not used for open-root critical welds?

16. Why should convex weld faces be avoided?

17. What bead pattern is best for overhead welds?

18. Why is the appearance of the cover pass so important?

19. What is the visual inspection standard's limitations of acceptance?

20. Why is the backing strip 2 in. (50 mm) longer than the test plate?

21. Why should there be a space between the plates when making a fillet weld on thick plates of a tee joint?

22. What would a small notch at the root weld's leading edge in a fillet weld mean?

23. What changes should be made in the setup for making a vertical up weld?

24. How can a cold weld be corrected?

25. What problem must be overcome if the amperage and voltage are lowered to make a vertical weld?

26. How can higher welding speeds help control distortion?

27. How can spatter buildup on the welding gun be controlled in the overhead position?

Chapter 14

OTHER CONSTANT POTENTIAL WELDING PROCESSES

OBJECTIVES

After completing this chapter, the student should be able to

- describe the various types of SAW fluxes.
- explain how the SAW process works.
- explain how the ESW and EGW processes work.
- name the parts of an SAW setup.
- list the methods of starting the SAW arc.
- list the major advantages and limitations of the SAW, ESW, and EGW processes.

KEY TERMS

submerged arc welding (SAW)	mechanically mixed flux
granular fluxing	arc starting
twisted wire	electroslag welding (ESW)
strip electrode	electrogas welding (EGW)
fused flux	water-cooled dams
bonded flux	

INTRODUCTION

In addition to GMAW and FCAW there are three other continuous feed electrode processes. As with GMAW and FCAW, a constant potential power supply provides the welding current. These processes are submerged arc welding (SAW), electroslag welding (ESW), and electrogas welding (EGW). Each of these three processes provides the welding industry with significant specialized benefits.

SAW, ESW, and EGW can be considered the workhorses of the fabrication industry. These processes are used to fabricate large, thick sections. They can weld metal ranging from 3/8 in. (10 mm) to more than 20 in. (508 mm) thick in a single operation, depending on the process selected. They are also used to cover large areas on tanks and pressure vessels with welded coverings of special alloys. This allows the unit to be fabricated out of a less expensive metal but have the surface of the expensive alloy. The total cost of the fabrication is less, yet it will provide the same or better service.

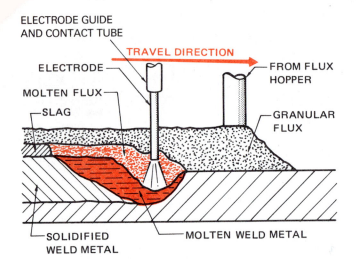

Figure 14-1 Submerged arc welding (SAW). (Courtesy of Hobart Brothers Company.)

Figure 14-2 Hand-held submerged arc welding provides high-quality weld deposits. (Courtesy of Lincoln Electric Company.)

Figure 14-3 Submerged arc welding of 1-in. (25-mm) -thick steel plates for redecking bridges. (Courtesy of Lincoln Electric Company.)

Figure 14-4 Submerged arc welding of pipe done automatically on a field construction site. (Courtesy of Lincoln Electric Company.)

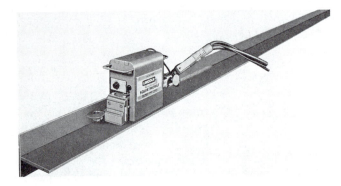

Figure 14-5 Portable gun drive unit. (Courtesy of Lincoln Electric Company.)

SUBMERGED ARC WELDING (SAW)

The early 1930s saw the introduction of the submerged arc welding (SAW) process. The first of many processes to use a continuous wire as the electrode, its acceptance accelerated during World War II. Although the yearly tonnage of SAW electrode wire used has increased, its percentage of the total amount of filler metals used has remained below 10%.

Submerged arc welding (SAW) is a fusion welding process in which the heat is produced from an arc between the work and a continuously fed filler metal electrode. The molten weld pool is protected from the surrounding atmosphere by a thick blanket of molten flux and slag formed from the granular fluxing material preplaced on the work, Figure 14-1. Figures 14-2, 14-3, and 14-4 show several applications of the SAW process.

WELD TRAVEL

Movement along a joint can be provided manually or mechanically. Hand-held travel can be a welder man-ually moving the gun at a constant speed along the joint, or there might be a small gun-mounted motor and friction drive wheel to provide a more consistent travel rate, Figure 14-5.

Mechanical travel can be provided by either moving the gun along the joint or moving the work past a fixed gun, Figure 14-5. Gun travel can be as simple as using a portable tractor type carriage, Figure 14-6, or as complex

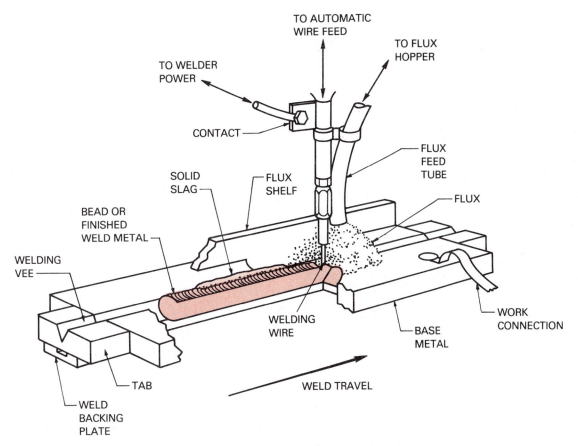

Figure 14-6 Schematic of a SAW setup in which the work is moved past the welding gun. (Courtesy of the American Welding Society.)

as a robotic arm. Work travel can be provided using rollers or other types of positioners, **Figure 14-7.** In some cases a positioner may be used in unison with some type of gun travel. These systems lend themselves to computer controlled automation.

ELECTRODE FEED

SAW electrode feed systems are similar to all of the other CP processes. The filler metal is provided on spools, coils, or in bulk drums. The feed system consists of a variable speed drive motor, motor controller, and drive rollers. The drive rollers are usually the knurled V-groove type, although sometimes the smooth V-groove type rollers are used.

The filler metal is fed at a constant rate to the arc. The CP power provides the arc current to melt the wire at a rate matching the fed rate. The electrode melts and is transferred across the arc to the work through the molten pool of flux. The flux reacts with the metal both during the arc and within the molten weld pool. Weld cooling is somewhat slowed by the heat retaining blanket provided by the slag. If the cooling rate is too fast in large welds, some impurities can be trapped in the center of the weld nugget. The rate of cooling can be slowed if preheat and postheat are used.

Figure 14-7 The large cylinder is being rotated below a sub-arc welding manipulator gantry. (Courtesy of the Ransome Co., Div. of Air Liquide America Corporation.)

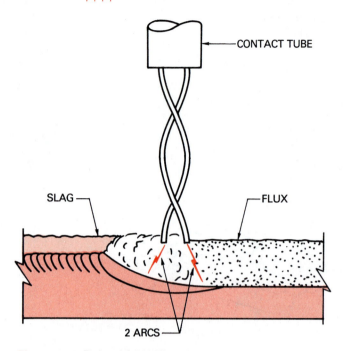

Figure 14-8 Twisted SAW filler wire.

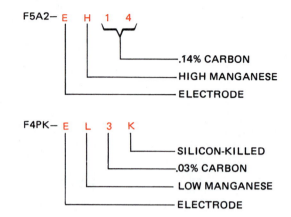

Figure 14-9 AWS filler wire identification system.

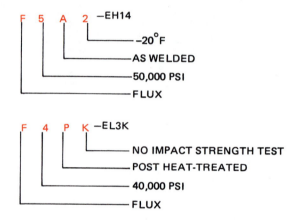

Figure 14-10 AWS flux identification system.

CONTACT TIP

A contact tip provides the source of welding current to the wire. This tip also directs the wire into the joint. The tip is usually made of beryllium-copper so it can withstand the heat of welding.

ELECTRODE

SAW filler metals are available as standard wire and in several special forms. The wire comes in sizes from 1/16 in. to 1/4 in. (1.6 mm to 6 mm). Twisted wire is used to give the arc some oscillating movement as the wire enters the weld, Figure 14-8. This oscillation helps fuse the toe of the weld to the base metal. Strip electrodes are used for surfacing applications. The strips are available up to 3 in. (76 mm) wide and in several thicknesses.

Electrode Classification

Electrode classifications are prefixed with the letter E, designating an electrode, Figure 14-9. The next letter L (low), M (medium), or H (high) refers to the range of manganese. The next one or two digits indicate the normal carbon point of the wire. One carbon point equals 0.01% carbon. The last letter is K and may or may not be used. When it is used, it means the electrode was drawn from a silicon-killed, deoxidized, steel.

FLUX

Fluxes are classified according to the mechanical properties of the weld metal deposited. The same chemically composed flux can have many different classifications, depending upon the classifications of the electrode it is used with and the condition of heat treatment given the weld for testing.

Flux Classification

The basic flux classification is prefixed with the letter F, which designates it as a flux, Figure 14-10. This is followed by one digit, which represents 10,000 psi (69 MPa) minimum tensile strength of the weld. The digit is followed by the letter A or P. The A means the weld was tested and classified in the "as-welded condition." The P means the weld was tested and classified after the prescribed amount of postweld heat treatment. The next item in the classification is a single digit or the letter Z. The digit indicates the lowest temperature in divisions of 10°F, starting at 0°F units, that the weld metal will meet or exceed the required 20 foot-pound (27 J) impact strength test. For example, in the flux identified as F 5 A "1" – xxxx, the "1" is –10°F (–23°C) and in the flux identified as F 5 A "3" – xxxx, the "3" is –30°F (–34°C). The letter Z indicates that no impact strength test is required.

Flux Types

Fluxes are grouped into three types according to their method of manufacture. These types are fused, bonded, and mechanically mixed. Granulated fluxes are available in bags or bulk containers.

Fused Fluxes

Fused fluxes are mixtures that have been heated until they melt into a solid metallic glass. They are then cooled and ground into the desired granular size range. Fused fluxes cannot be alloyed because they are a form of glass and all components in that glass are essentially oxides. They will not dissolve metals without reacting with them, thereby reducing their effectiveness as alloying materials.

Bonded Fluxes

Bonded fluxes are a mixture of fine particles of fluxing agents, deoxidizers, alloying elements, metal compounds, and a suitable binder that holds the mixture together in small, hard granules. Each granule is composed of all the ingredients in the correct proportions.

Mechanically Mixed

Mechanically mixed fluxes are mixtures of fused and bonded fluxes, or a fine mixture of agents in a desired proportion for a certain job.

Flux Storage

To prevent contamination of the weld by hydrogen, the flux must be kept dry and free from oils or other hydrocarbons. If flux becomes damp, it must be redried. Excessive levels of hydrogen in some steels can cause porosity. In hardenable steels, even small amounts of hydrogen can cause underbead cracking. Commercially available dryers are the best method of drying flux. Do not dry flux by using a direct flame. This may fuse the flux together; and, at the same time, the flame produces water that might condense on the flux.

ADVANTAGES OF SAW

The growth in the use of SAW has been due to its major advantages, as follows:

■ Minimum operator protection required — The heavy blanket of granular flux covers all of the arc light except for an occasional flash. The absence of the arc light means that many welders can work close to each other without the problem of arc flash. The welder does not have to wear a welding helmet, so visibility and safety are improved. The granular flux also prevents most of the welding smoke from escaping. Even with a large number of welders operating in confined spaces, forced ventilation is virtually eliminated. The heating and cooling costs of the shop are lessened because the amount of ventilation required is greatly reduced.

■ Highest deposition rate — Using large-diameter wires, more than 40 lb/hr (18 kg/hr) can be deposited. This rate is nearly two times the rate of FCAW and four times that of SMAW. No process other than electroslag welding can come close to this deposition rate.

■ Efficient use of materials — With SAW, there is no spatter to waste metal and cause cleanup problems. All of the electrode is transferred and becomes weld deposit. Only the melted flux needed for the weld is lost. Unfused granular flux can be retrieved and reused. The amount of flux consumed can be controlled by varying the arc length, which is done by changing the arc voltage.

■ Weld size — Flat groove or fillet welds as large as 1 in. (25 mm) can be made in one pass using a single electrode. Larger sizes are possible with multiple electrodes.

■ Easily adapted — With this process, the flux and wire are purchased separately. The flux can be used to change the alloys in the weld metal deposited from the electrode. By changing the flux, the properties of the weld are altered. The composition of the flux is easily changed to meet specific metallurgical properties. Two or more fluxes can be mixed, or granulated metal can be added to a flux or mixture to meet individual needs.

■ High-quality welds — Many codes permit SAW to be used on structural iron, pressure vessels, cryogenic cylinders, and in many other critical applications.

DISADVANTAGES

As with other processes, the submerged arc processes have several disadvantages.

■ Restricted to flat position and horizontal fillets — Because the fluxes needed for submerged arc welding flow easily, welding is restricted to those positions in which the flux can produce a self-supporting blanket.

■ Welding parameters need careful control — Because the flux hides the weld pool, welding conditions must be preset on the basis of experiments or with proven tabular information, including the contact tip-to-work distance, the current, the travel speed, and the voltage. Arc voltage must be carefully controlled to ensure the proper weld profile. Equally important, deviations in arc voltage can cause significant changes in the weld composition when using the fluxes as the source of alloys.

■ Mechanical guidance is necessary — Without some sort of guidance, the arc could easily move away from the joint being welded. To take advantage of the deep penetration possible with the process, accurate positioning is very important. This need can become an expensive complication with other than perfectly straight joints. Even with good guidance systems, positioning problems can develop if the wire does not have a large and uniform cast (or little bend) and a negligible helix (twist). Significant variations in cast or a large helix will cause the arc to wander. This is particularly important when a long wire electrode extension is used to increase deposition rates.

ARC STARTING

There are six commonly used methods of starting the arc. These methods are:

- Steel wool ball starting — A small ball of steel wool is placed between the electrode and the work before the flux is added. The welding current causes the steel wool ball to quickly heat up, and the arc is started, **Figure 14-11**.

- High-frequency starting — A high-frequency current is sent through the electrode, and it establishes the arc when the welding power is turned on.

- Scratch starting — The electrode is dragged along the joint before the welding current is started. When the welding current makes contact with the moving base metal, it begins to spark, which starts the arc.

- Wire retract start — The wire is advanced until it touches the base metal and the welding current is applied. At that moment the electrode is withdrawn slightly to start the arc.

- Sharp wire start — If the end of the wire is cut to a point, that point will quickly arc when it contacts the base metal.

- Molten flux start — The arc will restart on its own if the electrode is lowered into a pool of molten slag and the welding power is restarted.

WELD BACKING

Because the welds made with SA welding are usually very large, it's often necessary to support the root face of the weld. The support can be provided by placing something under the joint, such as a strip of copper, a trough filled with flux, or a backing weld, **Figure 14-12**.

HAND-HELD SAW

Hand-held SA welding is increasing in usage for a variety of reasons. One of the most significant is that there are little if any fumes or smoke to exhaust. New local, state, and federal laws have put restrictions on the material that can be free-vented into the atmosphere. There are requirements in some locations that require

the welding shop ventilation systems to have collectors and filtration equipment in operation. This process does not eliminate all fumes and smoke, but the reduction can significantly reduce the shop's ventilation costs. A shop using SA welding is a much cleaner place to work.

The arc light and spatter are blanketed by the flux covering so the welding technicians can wear lighter protective clothing. This reduces welder fatigue and increases productivity, **Figure 14-13**.

EXPERIMENTS

Most welder training shops are not equipped to make SA welds. The gun and flux handling systems are the parts of the system that are not available. The welding power and wire feeder from a GMA welding station are the same as required for SAW.

With a little adaptation you can make SA welds using your GMAW equipment. This will allow you to experience on a limited basis the hand-held method of SA welding. The welds you produce will not meet any code or standard, but the welding technique you will learn can be applied to SAW equipment in a fabrication shop.

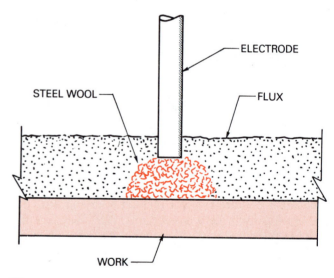

Figure 14-11 Steel wool placed between the electrode and work to aid in starting some SA welding.

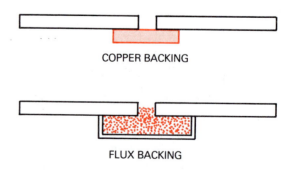

Figure 14-12 Two ways to provide backing for an open root SAW.

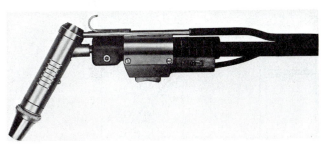

Figure 14-13 Hand-held SA welding gun. (Courtesy of Lincoln Electric Company.)

Because most SA welding wires are relatively large, they require high amperage welders. The higher amperages only produce large welds; they don't help you learn. By using a smaller GMA welding wire, you will be making smaller, more controllable welds. The smaller welds enable you to see how your movements affect the weld.

Setup and Use

For this series of experiments, set up the equipment in the same manner as for GMAW. If you have not completed Practice 11-1, GMAW Equipment Setup, and Practice 11-2, Threading GMAW Wire, you should do them now before you continue. Be sure to check the manufacturer's owner's manual of the equipment to see that it is safely set up.

EXPERIMENT 14-1

Use a properly set up and adjusted GMA welding machine, proper safety protection, 0.035-in. and/or 0.045-in. (0.9-mm and/or 1.2-mm) -diameter electrodes, one or more weld experiment pieces fabricated out of mild steel plate, 3/8 in. (10 mm) or thicker, **Figure 14-14**. You are going to make a fillet weld in the flat position.

- Tack weld the plates together and place them flat on the welding table.
- You won't need a welding helmet but you must wear flash glasses, gloves, and other safety equipment.
- Pour flux in the test specimen to a depth of 1 in.
- Leave the gas nozzle on the gun to prevent accidentally shorting the contact tube to the base metal. Leave the shielding gas off.
- Starting at one end, place the nozzle down to the top of the flux layer.
- Pull the gun trigger to start the welding current. You will feel the wire strike the joint, but the arc may not start. To start the arc if needed, use the scratch start method by moving the gun in a small circle. If the arc fails to start within a few moments, stop and cut the wire off and retry.
- When the arc starts, you will hear a smooth frying sound punctuated by an occasional pop.

- Move the nozzle at a steady speed along the joint until you are at the other end.

■ CAUTION

The unfused flux on the weld is very hot. If not handled with care, it will cause severe burns. Using pliers after the weld has cooled significantly, pour the flux into a metal container so it can continue to cool. Moving the weld before it cools may result in a large quantity of molten slag to be dumped out.

- Once the weld is cooled and flux and slag removed, inspect the weld for uniformity.

Make changes in the current and voltage and try the weld again. Note the effect these changes have on the weld. You may also draw a weave pattern in the surface of the flux with the gun nozzle as the weld progresses along the joint. By the time the weld is finished, you should be making the weld nearly perfectly.

Repeat each type of joint as needed until consistently good beads are obtained. Turn off the power to the welder and wire feeder, then clean up your work area. ◆

ELECTROSLAG WELDING (ESW)

Fluxes used for electroslag welding (ESW) produce slags during the weld that have special characteristics. The slags are designed to be so highly conductive in the molten state that they can replace the arc without interrupting the electrical circuit. These slags are very fluid, and they must be relatively deep to maintain control of the circuit. Because the slags are fluid, some type of dam must be used to contain them.

The electroslag process is designed to contain the pool of fused slag within a cavity whose wells are the edges of the two plates being welded and the two water-cooled dams that bridge those plates, **Figure 14-15**. The wire or wires are fed into the cavity and guided by contact tips using equipment similar to that used with the submerged arc process. The need for a controlled pool size and the use of contact tip or tips within the cavity mean that only relatively heavy plate can be welded with this process.

Heated by its own resistance to the flow of current, the flux temperature is raised well above that of the plate to be welded, generally above 3,200°F (1,760°C). That heat is used to prefuse the plate, which melts at temperatures below 2,790°F (1,532°C) so that metal being cast into it can bond with the plate edges. The flux also serves as a lubricant between the solidified metal and the dams. Additional flux must be added to the cavity to compensate for those losses. The wire is heated to its melting temperature by its internal resistance and remains fluid

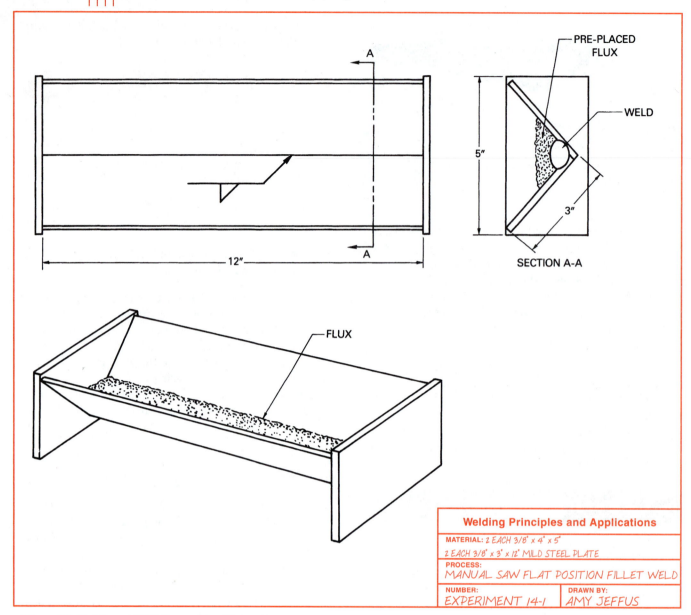

PRE-PLACED FLUX

WELD

5"

3"

SECTION A-A

A

A

12"

FLUX

Welding Principles and Applications

| MATERIAL: 2 EACH 3/8" x 4" x 5" |
| 2 EACH 3/8" x 3" x 12" MILD STEEL PLATE |

| PROCESS: |
| MANUAL SAW FLAT POSITION FILLET WELD |

| NUMBER: | DRAWN BY: |
| EXPERIMENT 14-1 | AMY JEFFUS |

Figure 14-14 Manual SAW flat position fillet weld.

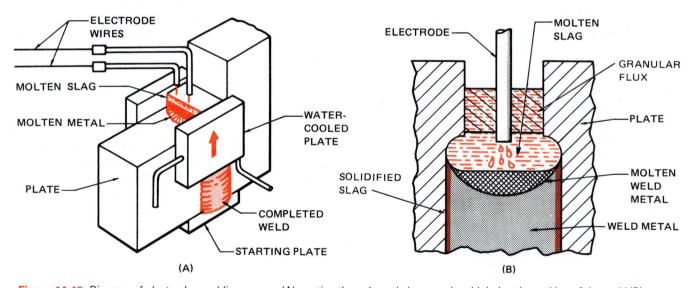

ELECTRODE WIRES

MOLTEN SLAG

MOLTEN METAL

PLATE

WATER-COOLED PLATE

COMPLETED WELD

STARTING PLATE

(A)

ELECTRODE

MOLTEN SLAG

GRANULAR FLUX

PLATE

SOLIDIFIED SLAG

MOLTEN WELD METAL

WELD METAL

(B)

Figure 14-15 Diagram of electroslag welding process (A); section through workpieces and weld during the making of the weld (B).

as it drops through the superheated flux to the molten weld pool.

The welding process is initiated by striking an arc between the wire and a bottom plate. After the arc energy melts enough flux to produce a pool of conductive slag, the arc is extinguished by increasing the distance between the wire tip and plate or by reducing the voltage from the power supply. One drawback of the process is that the metal produced during the starting process is defective and must be removed and repaired manually. This procedure must be followed at any position where welding has been interrupted and restarted.

Water-cooled copper dams confine the molten slag. The copper dams may extend the full length of the joint, or smaller dams may be moved up the joint as the weld metal solidifies. The rate of movement must match the speed at which the electrodes and the base metal are melted. The vertical progression is controlled automatically by sensing voltage changes with constant current power supplies or by sensing current with constant voltage power supplies. The dams and the base metal plates serve as a cooling medium. The lower part of the metal bath solidifies progressively in an upward direction. Slag is added by the operator as needed.

Electroslag welding is suitable for welding joints in thick metal plates because the large molten weld pool is confined and molded by the dams. The joint is welded in one pass.

Electroslag welding tends to produce a large grain size for two reasons: (1) the large mass of metal that is molten at a given time, and (2) the slow cooling of the metal.

Advantages

The electroslag welding process has the following advantages:

- Minimum joint preparation.
- Welds can be made having a thickness of several feet (meters) in plain and alloy steels using multiple electrodes.
- High welding speed.
- Minimum distortion.
- The weld metal is highly purified by the metallurgical slag.

Disadvantages

The electroslag welding process has the following disadvantages:

- Massive, expensive welding equipment and guidance systems are required.
- Lengthy setup times are needed.
- It cannot be used for welding the most commonly used thicknesses of plate.
- It can be used for only vertically positioned joints.

- It produces welds with very pronounced columnar microstructures and poor toughness.

ELECTROGAS WELDING (EGW)

Electrogas welding (EGW) is similar to electroslag welding in a number of ways. In addition to the power supply and wire feed, both processes are welded in the vertical up position using dams to contain the molten weld pool. Both processes can be used to weld on thick sections; however, EGW can weld plates as thin as 3/8 in. (10 mm).

The major difference in these processes is that there is an arc between the electrode and the molten weld pool with EGW. The molten weld pool is protected from the atmosphere by a shielding gas or by a flux core in the filler electrode, Figure 14-16A and B.

When a shielding gas is used, a mixture of carbon dioxide (CO_2) and Argon (Ar) works well. The shielding gas flows through holes in the copper dam to cover the molten weld pool.

Figure 14-17 shows the AWS numbering system for EGW electrodes. There are two systems — one for the solid wire and the other for the tubular, flux-filled electrodes.

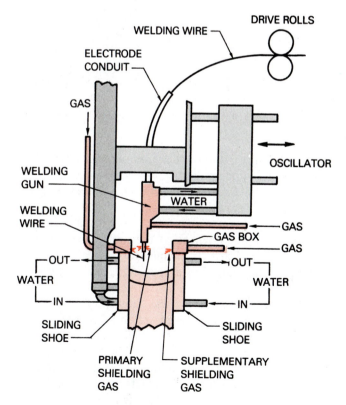

Figure 14-16A Electrogas welding with a solid wire electrode. (Courtesy of the American Welding Society.)

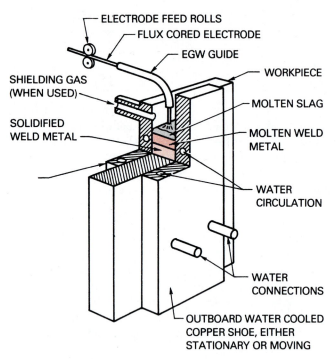

ELECTRODE FEED ROLLS
FLUX CORED ELECTRODE
EGW GUIDE
SHIELDING GAS (WHEN USED)
WORKPIECE
SOLIDIFIED WELD METAL
MOLTEN SLAG
MOLTEN WELD METAL
WATER CIRCULATION
WATER CONNECTIONS
OUTBOARD WATER COOLED COPPER SHOE, EITHER STATIONARY OR MOVING

Figure 14-16B Electrogas welding with a self-shielded flux cored electrode. (Courtesy of the American Welding Society.)

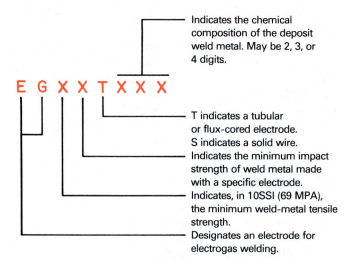

E G X X T X X X

Indicates the chemical composition of the deposit weld metal. May be 2, 3, or 4 digits.

T indicates a tubular or flux-cored electrode. S indicates a solid wire.

Indicates the minimum impact strength of weld metal made with a specific electrode.

Indicates, in 10SSI (69 MPA), the minimum weld-metal tensile strength.

Designates an electrode for electrogas welding.

Figure 14-17 Electrogas Welding Electrode Classification System. (Courtesy of the American Welding Society.)

Barge Construction

Figure 1

A multi-million-dollar contract was awarded to the ship-building division of a company that has serviced the Great Lakes shipping industry for nearly fifty years. The project involves the construction of a 360-ft-long, 8,000 ton, self-unloading cement barge that will travel the Great Lakes. The design of the barge is planned to provide an economical alternative to transporting material by rail or ship.

The cement barge was fabricated using a combination of submerged arc and gas-shielded, flux-cored welding processes. Virtually no stick electrode was used in the assembly of the major structure. The deck, bottom, and side plates were submerged arc welded. Each plate was tack welded in the yard, **Figure 1**, prior to one-pass welding of the seams from each side. Welding procedures were set at 700 amperes (DC+), 33 volts to produce x-ray quality welds at 25 ipm travel speed, all according to ABS specifications. Plates ranged in thickness from 3/8 in. to 1/2 in. thick. The 30-foot subassembled plates were then positioned on the barge with a large overhead crane and welded into place.

Semiautomatic welding was used to reduce costs and improve productivity, **Figure 2**. To per-

Figure 2

Figure 3

form semiautomatic welding, the company used more than twenty wire feeder packages and .052 gas-shielded flux cored electrodes. Welding procedures were set at 240 amperes, 28 volts using a straight CO_2 shielding gas at 35 CFH.

The self-unloading barge will transport cement in eight separate compartments, each of which incorporates a minimum 8 degree angle slope, Figure 3. Small, box-shaped aerators pump air into the cement mixer, which, at an 8 degree

slope, causes the material to flow like a liquid. The mixer is funneled into a large valve opening that drops into an electric-powered pumping station. Two large 250-hp pumping stations remove the material from the cargo bays. Each compartment can be unloaded separately, allowing for eight different cargos.

(Courtesy of the Lincoln Electric Company, Cleveland, Ohio 44117.)

REVIEW

1. What protects the molten SAW pool from the atmosphere?

2. How can manual SA welding gun movement be performed?

3. What are the two methods of mechanical travel for SA welding?

4. How is the weld metal deposited in the molten weld pool of the SA welding process?

5. In what forms can SA welding filler metal be purchased?

6. How is the manganese range of the SA electrode noted in the AWS classification?

7. Why could a single SA welding flux have more than one AWS classification?

8. List the three groupings of SA welding fluxes according to their method of manufacturing.

9. Why are alloys not added to fused SA fluxes?

10. What is in bonded SA fluxes?

11. What must be done with SA fluxes to prevent contamination of the weld by hydrogen?

12. Why does the welder not have to wear a welding helmet?

13. What happens to the unfused SA welding flux?

14. Why is some form of mechanical guidance required with SA welding?

15. List the common methods used to start the SA arc.

16. Why would hand-held SA welding be used?

17. What special characteristics must ES welding slags have?

18. What heats the ES welding flux?

19. How is an ES weld started?

20. Why do ES welds have large grain sizes?

21. List the advantages of ES welding.

22. What is the major difference between ESW and EGW?

Chapter 15

GAS TUNGSTEN ARC WELDING EQUIPMENT, SETUP, OPERATION, AND FILLER METALS

OBJECTIVES

After completing this chapter, the student should be able to

- explain the purposes of the various tungsten and shapes.
- demonstrate the proper method of reshaping tungsten.
- describe the advantages of using different shielding gases.
- explain the effect of alloy oxides on the performance of tungsten.
- demonstrate how to correctly set up a GTA welding machine.

KEY TERMS

inert gas	frequency
tungsten	spark gap oscillator
collet	noble inert gases
flowmeter	prepurge
cleaning action	postpurge

INTRODUCTION

The gas tungsten arc welding (GTAW) process is sometimes referred to as TIG, or heliarc. TIG is short for tungsten inert gas welding, and the term *heliarc* was used because helium was the first gas used for the process. The aircraft industry developed the GTAW process for welding magnesium during the late 1930s and the early 1940s. During that time, helium was the primary shielding gas used, along with DCEP welding current. These caused many of the problems that limited application of the GTA welding process. But improvements in gas composition and a better understanding of the importance of polarity improved the process's effectiveness and reduced its cost.

To use this process, an arc is established between a tungsten electrode and the work. Under the correct welding conditions, the electrode does not melt, although the work does at the point where the arc impacts the work surface and produces a molten weld pool. The filler metal is thin wire that is fed directly into the molten weld pool where it melts. Since hot tungsten is sensitive to oxygen in the air, good shielding with oxygen-free gases is required. The same inert gas provides the needed arc characteristics and protects the molten weld pool. Because fluxes are not used, the welds produced are sound, free of contaminants and slags, and as corrosion-resistant as the parent metal.

Before development of the GTAW process, welding aluminum and magnesium was difficult. The welds produced were porous and corrosion-prone.

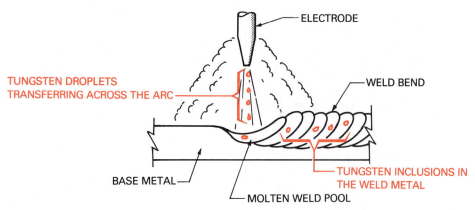

Figure 15-1 Some tungsten will erode from the electrode and be transferred across the arc and become trapped in the weld deposit.

When argon became plentiful and DCEN was recognized as more suitable than DCEP, the GTA process became more common. Until the development of GMAW in the late 1940s, GTAW was the only acceptable process for welding such reactive materials as aluminum, magnesium, titanium, and some grades of stainless steel, regardless of thickness. Although economical for welding sheet metal, the process proved tedious and expensive for joining sections much thicker than 1/4 in. (6 mm). The eye-hand coordination required to make GTA welds is very similar to the coordination required for oxyfuel gas welding. GTA welding is often easier to learn when a person can gas weld.

Tungsten, atomic symbol W, has the following properties:

■ High tensile strength. 500,000 lb/in², or 3,447 kg/mm²

■ Hardness, Rockwell C45

■ High melting temperature, 6,170°F, or 3,410°C

■ High boiling temperature, 10,700°F, or 5,630°C

■ Good electrical conductor

Tungsten is produced mainly by reduction of its trioxide with hydrogen. Powdered tungsten is then purified to 99.95+%, compressed, and sintered (heated to a temperature below melting where grain growth can occur) to make an ingot. The ingot is heated to increase ductility and then is swaged and drawn through dies to produce electrodes. These electrodes are available in sizes varying from .01 in. to .25 in. (0.25 mm to 6 mm) in diameter. The tungsten electrode, after drawing, has a heavy black oxide that is later chemically cleaned or ground off.

The high melting temperature and good electrical conductivity make tungsten the best choice for a nonconsumable electrode. The arc temperature, around 11,000°F (6,000°C), is much higher than the melting temperature of tungsten but not much higher than its boiling temperature of 10,600°F (5,900°C).

As the tungsten electrode becomes hot the arc between the electrode and the work will stabilize. Because electrons are more freely emitted from a hot tungsten, the very highest temperature possible at the tungsten electrode tip is desired. Maintaining a balance between the heat required to have a stable arc and that high enough to melt the tungsten requires an understanding of the GTA torch and electrode.

The thermal conductivity of tungsten and the heat input are prime factors in the use of tungsten as an electrode. In general, tungsten is a good conductor of heat. This conductive property is what allows the tungsten electrode to withstand the arc temperature well above its melting temperature. The heat of the arc is conducted away from the electrode's end so fast that it does not reach its melting temperature. For example, a wooden match burns at approximately 3,000°F (1,647°C). Because aluminum melts at 1,220°F (971°C), a match should easily melt an aluminum wire. However, a match will not even melt a 1/16-in. (2-mm) aluminum wire. The aluminum, like a tungsten electrode, conducts the heat away so quickly that it will not melt.

Because of the intense heat of the arc some erosion of the electrode will occur. This eroded metal is transferred across the arc, **Figure 15-1**. Slow erosion of the electrode results in limited tungsten inclusions in the weld, which are acceptable. Standard codes give the size and amount of tungsten inclusions that are allowable in various types of welds. The tungsten inclusions are hard spots that cause stresses to concentrate, possibly resulting in weld failure. Although tungsten erosion cannot be completely eliminated, it can be controlled. A few ways of limiting erosion include:

■ having good mechanical and electrical contact between the electrode and the collet,

■ using as low a current as possible,

■ using a water cooled torch,

■ using as large a size of tungsten as possible,

■ using DCEN current,

■ using as short an electrode extension from the collet as possible,

■ using the proper electrode end shape,

■ using an alloyed tungsten electrode.

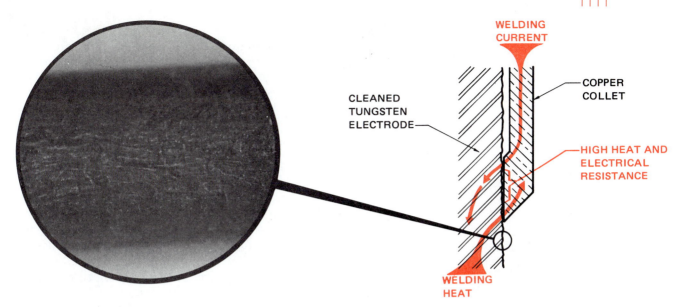

Figure 15-2 Irregular surface of a cleaned tungsten electrode (poor heat transfer to collet).

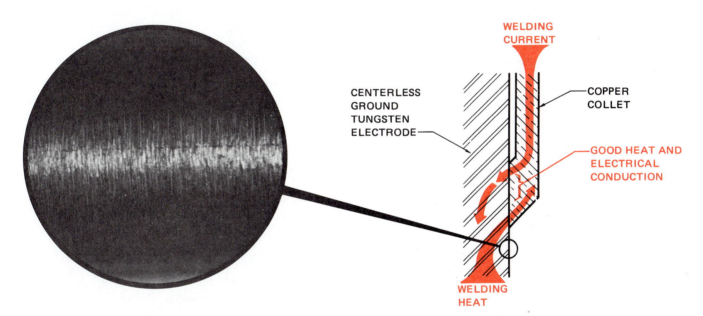

Figure 15-3 Smooth surface of a centerless ground tungsten electrode (good heat transfer to collet).

The torch end of the electrode is tightly clamped in a collet. The collet inside the torch is cooled by air or water. The collet is the cone-shaped sleeve that holds the electrode in the torch. Heat from both the arc and the tungsten electrode's resistance to the flow of current must be absorbed by the collet and torch. To ensure that the electrode is being cooled properly, be sure the collet connection is clean and tight. And for water-cooled torches, make sure water flow is adequate.

Collet-tungsten connection efficiency is shown in Figures 15-2 and 15-3.

Large-diameter electrodes conduct more current because the resistance heating effects are reduced. However, excessively large sizes may result in too low a temperature for a stable arc.

In general, the current-carrying capacity at DCEN is about ten times greater than that at DCEP.

The preferred electrode tip shape impacts the temperature and erosion of the tungsten. With DCEN, a pointed tip concentrates the arc as much as possible and improves arc starting with either a short, high-voltage electrical discharge or a touch start. Because DCEN does not put much heat on the tip it is relatively cool, the point is stable, and it can survive extensive use without damage, **Figure 15-4A.**

With alternating current, the tip is subjected to more heat than with DCEN. To allow a larger mass at the tip to withstand the higher heat the tip is rounded. The melted end must be small to ensure the best arc stability, **Figure 15-4B.**

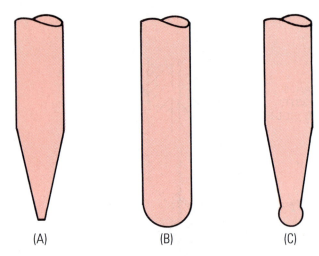

Figure 15-4 Basic tungsten electrode and shapes: pointed (A), rounded (B), and tapered with a balled end (C).

DCEP has the highest heat input to the electrode tip. For this reason a slight ball of molten tungsten is suspended at the end of a tapered electrode tip. The liquid tungsten surface tension with the larger mass above the molten ball holds it in place like a drop of water on your fingertip, **Figure 15-4C**.

TYPES OF TUNGSTEN

Pure tungsten has a number of properties that make it an excellent nonconsumable electrode for the GTA welding process. These properties can be improved by adding cerium, lanthanum, thorium, or zirconium to the tungsten.

For GTA welding, tungsten electrodes are classified as the following:

■ Pure tungsten, EWP
■ 1% thorium tungsten, EWTh-1
■ 2% thorium tungsten, EWTh-2
■ 1/4% to 1/2% zirconium tungsten, EWZr
■ 2% cerium tungsten, EWCe-2
■ 1% lanthanum tungsten, EWLa-1
■ Alloy not specified, EWG

See **Table 15-1**.

The type of finish on the tungsten must be specified as cleaned or ground. More information on composition and other requirements for tungsten welding electrodes is available in the AWS publication A5.12, "Specifications for Tungsten and Tungsten Alloy Electrodes for Arc Welding and Cutting."

Pure Tungsten, EWP

Pure tungsten has the poorest heat resistance and electron emission characteristic of all the tungsten electrodes. It has a limited use with AC welding of metals, such as aluminum and magnesium.

Thoriated Tungsten, EWTh-1 and EWTh-2

Thorium oxide (ThO_2), when added in percentages up to 0.6% to tungsten, improves its current-carrying capacity by improving its thermionic emission. The addition of 1% to 2% does not further improve current-carrying capacity and/or starting, but it does help with electron emission. This percentage also makes tungsten more resistant to contamination.

Zirconium Tungsten, EWZr

Zirconium oxide (ZrO_2) also helps the tungsten to emit electrons freely. The addition of zirconium to the tungsten has the same effect on the electrode characteristics as thorium but to a lesser degree. Zirconium-tungsten electrodes are more easily melted than thorium-tungsten when forming rounded or tapered tungsten with a balled end for AC or DCEP welds. Thoriated tungsten can be melted, but not as easily, and can form a thorium spike on the rounded end, **Figure 15-5**. This thorium spike may result in reduced arc stability for AC welding.

Cerium Tungsten, EWCe-2

Cerium oxide (CeO_2) is added to tungsten to improve the current-carrying capacity in the same manner as thorium does. Cerium does not work as well as thorium, but it is not radioactive like thorium.

AWS Classification	Tungsten Composition	Tip Color
EWP	Pure tungsten	Green
EWTh-1	1% thorium added	Yellow
EWTh-2	2% thorium added	Red
EWZr	1/4% to 1/2% zirconium added	Brown
EWCe-2	2% cerium added	
EWLa-1	1% lanthanum added	
EWG	Alloy not specified	

Table 15-1 Tungsten Electrode Types and Identification.

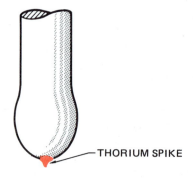

THORIUM SPIKE

Figure 15-5 Thorium spike on a balled end tungsten electrode.

Lanthanum Tungsten, EWLa-1

Lanthanum oxide (La_2O_3) is added to tungsten with about the same result as cerium 2% has on its capacity.

Alloy Not Specified Tungsten, EWG

This classification is for tungsten electrodes whose alloys are being tested by manufacturers. The exact alloy and its ability to affect the tungsten welding characteristics will vary depending on the composition.

SHAPING THE TUNGSTEN

The desired end shape of a tungsten can be obtained by grinding, breaking, remelting the end, or by chemical compounds. Tungsten is brittle and easily broken. Welders must be sure to make a smooth, square break where they want it to be located.

Grinding

A grinder is often used to clean a contaminated tungsten or to point the end of a tungsten. The grinder used to sharpen tungsten should have a fine, hard stone. It should be used for grinding tungsten only. Because of

the hardness of the tungsten and its brittleness, the grinding stone chips off small particles of the electrode. A coarse grinding stone will result in more tungsten breakage and a poorer finish. If the grinder is used for metals other than tungsten, particles of these metals may become trapped on the tungsten as it is ground. The metal particles will quickly break free when the arc is started, resulting in contamination.

EXPERIMENT 15-1

Grinding the Tungsten to the Desired Shape

Using an electric grinder with a fine grinding stone, one piece of tungsten 2 in. (51 mm) long or longer, and safety glasses, you will grind a point on the tungsten electrode.

Because of the hardness of the tungsten, it will become hot. Its high thermal conductivity means that the heat will be transmitted quickly to your fingers. To prevent overheating, only light pressure should be applied against the grinding wheel. This will also reduce the possibility of accidentally breaking the tungsten.

Grind the tungsten so that the grinding marks run lengthwise, Figures 15-6 and 15-7. Lengthwise grinding reduces the amount of small particles of tungsten contaminating the weld. Move the tungsten up and down as it is twisted during grinding. This will prevent the tungsten from becoming hollow-ground. ◆

■ CAUTION

When holding one end of the tungsten against the grinding wheel, the other end of the tungsten must not be directed toward the palm of your hand, Figure 15-8. This will prevent the tungsten from being stuck into your hand if the grinding wheel catches it and suddenly pushes it downward.

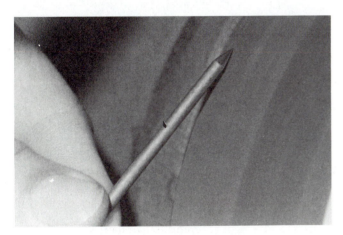

Figure 15-6 Correct method of grinding a tungsten electrode.

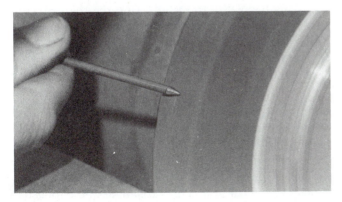

Figure 15-7 Incorrect method of grinding a tungsten electrode.

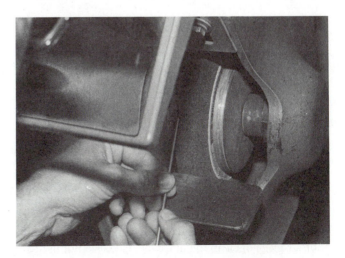

Figure 15-8 Correct way of holding a tungsten when grinding.

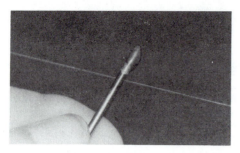

STEP 1

STEP 2

STEP 3

Figure 15-9 Breaking the contaminated end from a tungsten by striking it with a hammer.

Figure 15-10 Correctly breaking the tungsten using two pairs of pliers.

Figure 15-11 Using wire cutters to correctly break the tungsten.

Breaking and Remelting

Tungsten is hard but brittle, resulting in a low impact strength. If tungsten is struck sharply, it will break without bending. When it is held against a sharp corner and hit, a fairly square break will result. **Figures 15-9, 15-10 and 15-11** show how to break the tungsten correctly on a sharp corner using two pliers and using wire cutters.

Once the tungsten has been broken squarely, the end must be melted back so that it becomes somewhat rounded. This is accomplished by switching the welding current to DCEP and striking an arc under argon shielding on a piece of copper. If copper is not available, another piece of clean metal can be used. Do not use carbon as it will contaminate the tungsten.

EXPERIMENT 15-2

Removing a Contaminated Tungsten End by Breaking

Using short scrap pieces of tungsten, pliers, wire cutters, and a light machinist's hammer, you will break the end from the tungsten.

Break about 1/4 in. (6 mm) from the end of the tungsten using the appropriate method, depending upon the diameter of the tungsten. Observe the break; it should be square and relatively smooth, **Figure 15-12.** The end of the tungsten should be broken only if the tungsten is badly contaminated. ◆

Chemical Cleaning and Pointing

The tungsten can be cleaned and pointed using one of several compounds. The tungsten is heated by shorting it against the work. The tungsten is then dipped in the compound, a strong alkaline, which rapidly dissolves the hot tungsten. The chemical reaction is so fast that enough additional heat is produced to keep the tungsten hot, **Figure 15-13.** When the tungsten is removed, cooled, and cleaned, the end will be tapered to a fine

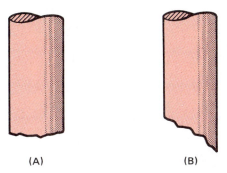

Figure 15-12 Correctly broken tungsten electrode (A); incorrectly broken tungsten electrode (B).

point. If the electrode is contaminated, the chemical compound will dissolve the tungsten, allowing the contamination to fall free.

Pointing and Remelting

The tapered tungsten with a balled end, a shape sometimes used for DCEP welding, is made by first grinding or chemically pointing the electrode. Using DCEP, as in the procedure for the remelted broken end, strike an arc on some copper under argon shielding and slowly increase the current until a ball starts to form on the tungsten. The ball should be made large enough so that the color of the end stays between dull red and bright red. If the color turns white, the ball is too small and should be made larger. To increase the size of the ball, simply apply more current until the end begins to melt. Surface tension will pull the molten tungsten up onto the tapered end. Lower the current and continue welding. DCEP is seldom used for welding. If the tip is still too hot, it may be necessary to increase the size of the tungsten.

EXPERIMENT 15-3

Melting the Tungsten End Shape

Using a properly set up GTA welding machine, proper safety protection, one piece of copper or other clean piece of metal, and the tungsten that was sharpened and broken in Experiments 15-1 and 15-2, you will melt the end of the tungsten into the desired shape.

Properly install the tungsten, set the argon gas flow, switch the current to DCEP, and turn on the machine. Strike an arc on the copper and slowly increase the amperage. Watch the tungsten as it begins to melt, and stop the current when the desired shape has been obtained, **Figure 15-14.** See Color Plates 28, 29, 30, and 31. ◆

GTA WELDING EQUIPMENT

Figures **15-15** and **15-16** show two industrial applications of gas tungsten arc welding.

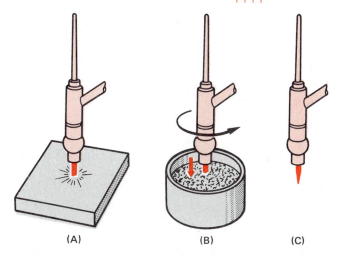

Figure 15-13 Chemically cleaning and pointing tungsten: (A) shorting the tungsten against the work to heat it to red hot, (B) inserting the tungsten into the compound and moving it around, and (C) cleaned and pointed tungsten ready for use.

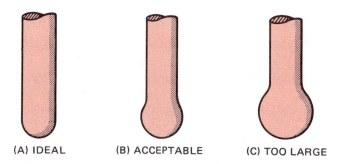

(A) IDEAL   (B) ACCEPTABLE   (C) TOO LARGE

Figure 15-14 Melting the tungsten end shape.

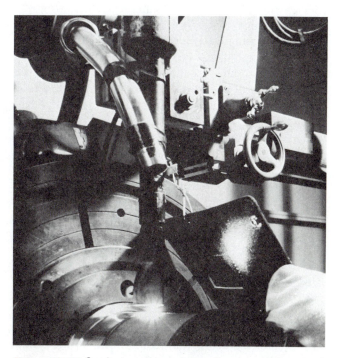

Figure 15-15 Semiautomatic operation allows a stainless steel part to be GTA welded as it's turned past the torch. (Courtesy of Cerametals, Inc.)

Figure 15-16 An operator GTA welds a cap ring on a pneumatic tank. (Courtesy of Miller Electric Manufacturing Company.)

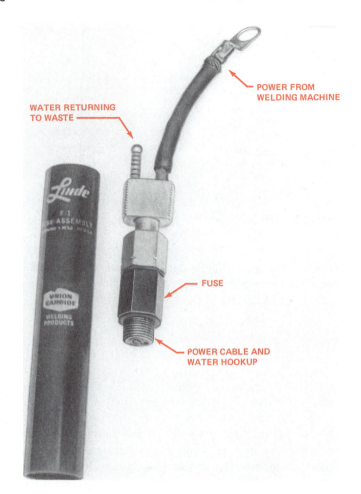

Figure 15-17 Power cable safety fuse.

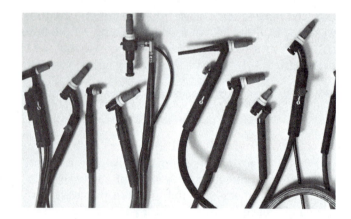

Figure 15-18 GTA welding torches. (Courtesy of Miller Electric Manufacturing Co.)

Torches

GTA welding torches are available water-cooled or air-cooled. The heat transfer efficiency for GTA welding may be as low as 20%. This means that 80% of the heat generated does not enter the weld. Much of this heat stays in the torch and must be removed by some type of cooling method.

The water-cooled GTA welding torch is more efficient than an air-cooled torch at removing waste heat. The water-cooled torch, as compared to the air-cooled torch, operates at a lower temperature, resulting in a lower tungsten temperature and less erosion.

The air-cooled torch is more portable because it has fewer hoses, and it may be easier to manipulate than the water-cooled torch. Also, the water-cooled torch requires a water reservoir or other system to give the needed cooling. The cooling water system should contain some type of safety device, **Figure 15-17,** to shut off the power if the water flow is interrupted. The power cable is surrounded by the return water to keep it cool, so a smaller size cable can be used. Without the cooling water, the cable quickly overheats and melts through the hose.

The water can become stopped or restricted for a number of reasons, such as a kink in the hose, a heavy object set on the hose, or failure to turn on the system. Water pressures higher than 35 psi (241 kg/mm²) may cause the water hoses to burst. When an open system is used, a pressure regulator must be installed to prevent pressures that are too high from damaging the hoses.

GTA welding torch heads are available in a variety of amperage ranges and designs, **Figure 15-18.** The amperage listed on a torch is the maximum rating and cannot be exceeded without possible damage to the torch. The various head angles allow better access in tight places. Some of the heads can be swiveled easily to new angles.

The back cap that both protects and tightens the tungsten can be long or short, **Figures 15-19** and **15-20.**

Hoses

A water-cooled torch has three hoses connecting it to the welding machine. The hoses are for shielding gas to the torch, cooling water to the torch and cooling water return, and housing the power cables to the torch,

Figure 15-21. Air-cooled torches may have one hose for shielding gas attached to the power cable, Figure 15-22.

The shielding gas hose must be plastic to prevent the gas from being contaminated. Rubber hoses contain

Figure 15-19 Short back caps are available for torches when space is a problem.

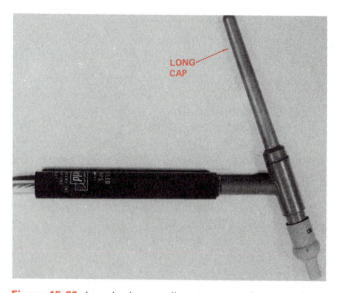

Figure 15-20 Long back caps allow tungstens that are a full 7 in. (177 mm) long to be used. (Courtesy of Hobart Brothers Company.)

oils that can be picked up by the gas, resulting in weld contamination.

The water-in hose may be made of any sturdy material. Water hose fittings have left-hand threads, and gas hose fittings have right-hand threads. This prevents the water and gas hoses from accidentally being reversed when attaching them to the welder. The return water hose also contains the welding power cable. This permits a much smaller size cable to be used because the water keeps it cool.

The water must be supplied to the torch head and return around the cable. This allows the head to receive the maximum cooling from the water before the power cable warms it. Running the water through the torch first has another advantage. That is, when the water solenoid is closed, there is no water pressure in the hoses, which is particularly important. This feature also prevents condensation in the torch. If a water leak should occur during welding, the welding power is stopped, closing the water solenoid and thus stopping the leak.

A protective covering can be used to prevent the hoses from becoming damaged by hot metal, **Figure 15-23**. Even with this protection, the hoses should be supported, **Figure 15-24**, so that they are not underfoot on the floor. By supporting the hoses, the chance of them being damaged by hot sparks is reduced.

Nozzles

The nozzle or cup is used to direct the shielding gas directly on the welding zone. The nozzle size is determined by the diameter of the opening and its length, **Table 15-2**. Nozzles may be made from a ceramic such as alumina or silicon nitride (opaque) or from fused quartz (clear). The nozzle may also have a gas lens to improve the gas flow pattern.

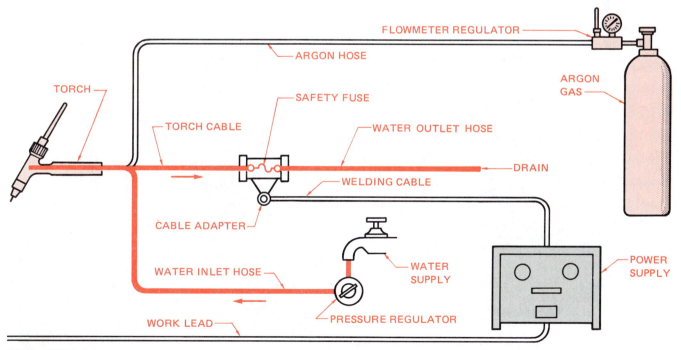

Figure 15-21 Schematic of a GTA welding setup with a water-cooled torch.

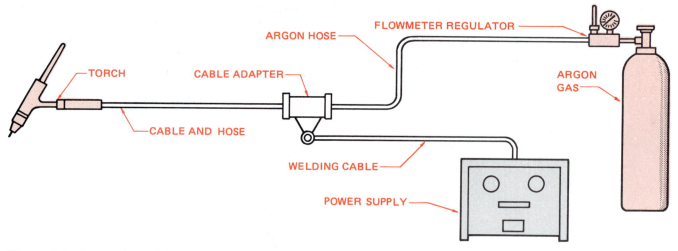

Figure 15-22 Schematic of a GTA welding setup with an air-cooled torch.

Tungsten Electrode Diameter		Nozzle Orifice Diameter	
in	(mm)	in	(mm)
1/16	(2)	1/4 to 3/8	(6 to 10)
3/32	(2.4)	3/8 to 7/16	(10 to 11)
1/8	(3)	7/16 to 1/2	(11 to 13)
3/16	(4.8)	1/2 to 3/4	(13 to 19)

Table 15-2 Recommended Cup Sizes.

The nozzle size, both length and diameter, is often the welder's personal preference. Occasionally, a specific choice must be made based upon joint design or location. Small nozzle diameters allow the welder to better see the molten weld pool and can be operated with lower gas flow rates. Larger nozzle diameters can give better gas coverage, even in drafty places.

Ceramic nozzles are heat resistant and offer a relatively long life. The useful life of a ceramic nozzle is affected by the current level and proximity to the work. Silicon nitride nozzles will withstand much more heat resulting, in a longer useful life.

The fused quartz (glass) used in a nozzle is a special type that can withstand the welding heat. These nozzles are no more easily broken than ceramic ones, but are more expensive. The added visibility with glass nozzles in tight, hard-to-reach places is often worth the added expense.

The longer a nozzle, the longer the tungsten must be extended from the collet. This can cause higher tungsten temperatures, resulting in greater tungsten erosion. When using long nozzles, it is better to use low amperages or a larger sized tungsten.

Flowmeter

The flowmeter may be merely a flow regulator used on a manifold system or it may be a combination flow and pressure regulator used on an individual cylinder, Figures 15-25 and 15-26.

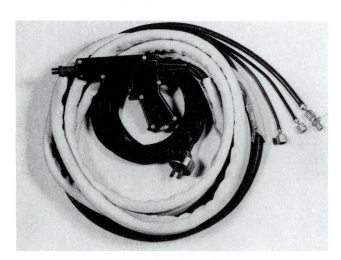

Figure 15-23 Zip-on protective covering also helps keep the hoses neat. (Courtesy of Hobart Brothers Company.)

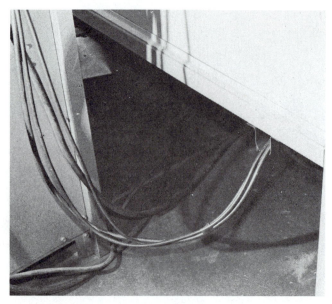

Figure 15-24 A bracket holds the leads off the floor.

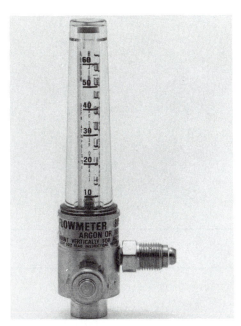

Figure 15-25 Flowmeter. (Courtesy of Concoa Controls Corporation of America.)

Figure 15-26 Flowmeter regulator. (Courtesy of Concoa Controls Corporation of America.)

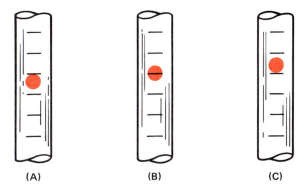

Figure 15-27 Three methods of reading a flowmeter: (A) top of ball, (B) center of ball, and (C) bottom of ball.

The flow is metered or controlled by opening a small valve at the base of the flowmeter. The rate of flow is then read in units of cfh (cubic feet per hour), or L/min (liters per minute). The reading is taken from a fixed scale that is compared to a small ball floating on the stream of gas. Meters from various manufacturers may be read differently. For example, they may read from the top, center, or bottom of the ball, **Figure 15-27**. The ball floats on top of the stream of gas inside a tube that gradually increases in diameter in the upward direction. The increased size allows more room for the gas flow to pass by the ball. If the tube is not vertical, the reading is not accurate, but the flow is unchanged. Also, when using a line flowmeter, it is important to have the correct pressure. Changes in pressure will affect the accuracy of the flowmeter reading. In order to get accurate readings, be sure the gas being used is read on the proper flow scale. Less dense gases, such as helium and hydrogen, will not support the ball on as high a column with the same flow rate as a denser gas, like argon.

TYPES OF WELDING CURRENTS

All three types of welding current, or polarities, can be used for GTA welding. Each current has individual features that make it more desirable for specific conditions or with certain types of metals.

The major differences among the currents are in their heat distributions and the presence or degree of arc cleaning. Figure 15-28 shows the heat distribution for each of the three types of currents.

Direct-current electrode negative (DCEN), which used to be called direct-current straight polarity (DCSP), concentrates about two-thirds of its welding heat on the

work and the remaining one-third on the tungsten. The higher heat input to the weld results in deep penetration. The low heat input into the tungsten means that a smaller sized tungsten can be used without erosion problems. The smaller sized electrode may not require pointing, resulting in a savings of time, money, and tungsten.

Direct-current electrode positive (DCEP), which used to be called direct-current reverse polarity (DCRP), concentrates only one-third of the arc heat on the plate and two-thirds of the heat on the electrode. This type of current produces wide welds with shallow penetration, but it has a strong cleaning action. The high heat input to the tungsten indicates that a large-sized tungsten is required, and the end shape with a ball must be used. The low heat input to the metal and the strong cleaning action on the metal make this a good current for thin, heavily oxidized metals. The metal being welded will not emit electrons as freely as tungsten, so the arc may wander or be more erratic than DCEN.

There are many theories as to why DCEP has a cleaning action. The most probable explanation is that the electrons accelerated from the cathode surface lift

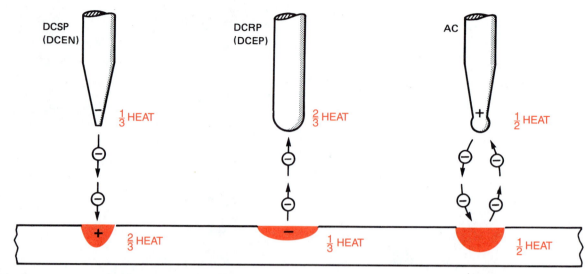

Figure 15-28 Heat distribution between the tungsten electrode and the work with each type of welding current.

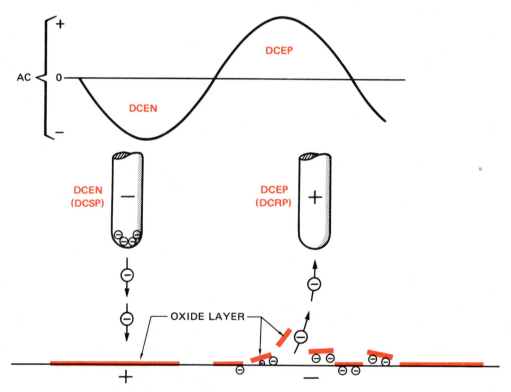

Figure 15-29 Electrons collect under the oxide layer during the DCEP portion of the cycle and lift the oxides from the surface.

the oxides that interfere with their movement. The positive ions accelerated to the metal's surface provide additional energy. In combination, the electrons and ions cause the surface erosion needed to produce the cleaning. Although this theory is disputed, it is important to note that DCEP occurs, that it requires argon-rich shield gases and DCEP polarity, and that it can be used to advantage, **Figure 15-29**.

Alternating current (AC) concentrates about half of its heat on the work and the other half on the tungsten. Alternating current is DCEN half of the time and DCEP the other half of the time. The **frequency** at which the current

cycles is the rate at which it makes a full change in direction, **Figure 15-30**. In the United States, the current cycles at the rate of 60 times per second or 60 hertz (60 Hz). Referring again to **Figure 15-30**, the current is at its maximum peak at points A and B. The rate gradually decreases until it stops at points C and D. The arc at these points is extinguished and, as the current reversal begins, must be re-established. This event requires the emission of electrons from the cathode to ionize the plasma. When the hot, emissive electrode becomes the cathode, re-establishing the arc is easy. However, it is often quite difficult to re-establish the arc when the colder and less emissive work-

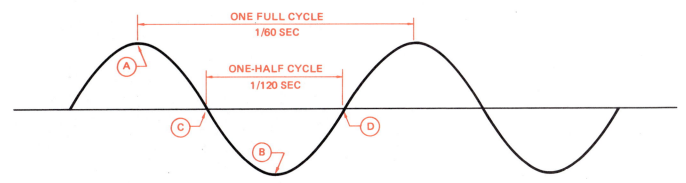

Figure 15-30 Sine wave of alternating current.

piece becomes the cathode. Because voltage from the power supply is designed to support a relatively low voltage arc, it may be insufficient to initiate electron flow. Thus a voltage assist from another source is needed. A high-voltage but low current spark gap oscillator commonly provides the assist at a relatively low cost. The high frequency ensures that a voltage peak will occur reasonably close to the current reversal in the welding arc, creating a low-resistance ionized path for the welding current to follow,

Figure 15-31. This same device is often used to initiate direct-current arcs, a particularly useful technique for mechanized welding. See Color Plates 34, 35, 36, and 37.

The high-frequency current is established by capacitors discharging across a gap set on points inside the machine. Changing the point gap setting will change the frequency of the current. The closer the points are, the higher the frequency; the wider the spacing between the points, the lower the frequency. The voltage is stepped up with a transformer from the primary voltage supplied to the machine. The available amperage to the high-frequency circuit is very low. Thus, when the circuit is complete, the voltage quickly drops to a safe level. The high frequency is induced on the primary welding current in a coil.

The high frequency may be set so that it automatically cuts off after the arc is established, usually with DC. It is used as a continuous current with AC. When used in this manner, it is referred to as alternating current, high-frequency stabilized, or ACHF.

SHIELDING GASES

The shielding gases used for the GTA welding process are argon (Ar), helium (He), hydrogen (H), nitrogen (N), or a mixture of two or more of these gases. The purpose of the shielding gas is to protect the molten weld pool and the tungsten electrode from the harmful effects of air. The shielding gas also affects the amount of heat produced by the arc and the resulting weld bead appearance.

Argon and helium are noble inert gases. This means that they will not combine chemically with any other material. Argon and helium may be found in mixtures but never as compounds. Because they are inert, they will not affect the molten weld pool in any way.

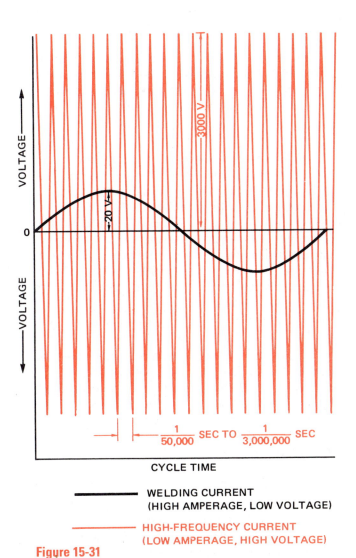

WELDING CURRENT
(HIGH AMPERAGE, LOW VOLTAGE)

HIGH-FREQUENCY CURRENT
(LOW AMPERAGE, HIGH VOLTAGE)

Figure 15-31

■ *CAUTION*

Never allow non-inert gases such as O_2, CO_2, or N to come in contact with your inert gas system. Very small amounts can contaminate the inert gas, which may result in the weld failing.

Argon

Argon is a by-product in air separation plants. Air is cooled to temperatures that cause it to liquify; then its constituents are fractionally distilled. The primary products are oxygen and nitrogen. Before these gases were produced on a tonnage scale, argon was a rare gas. Now it is distributed in cylinders as gas or in bulk as liquid forms.

Because argon is denser than air, it effectively shields welds in deep grooves in the flat position. However, this higher density can be a hindrance when welding overhead because higher flow rates are necessary. The argon is relatively easy to ionize, and thus suitable for alternating-current applications and easier starts. This property also permits fairly long arcs at lower voltages, making it virtually insensitive to changes in arc length. Argon is also the only commercial gas that produces the cleaning discussed earlier. These characteristics are most useful for manual welding, especially with filler metals added as shown in **Figure 15-32**.

Helium

Helium is a by-product of the natural gas industry. It is removed from natural gas as the gas undergoes separation (fractionation) for purification or refinement.

Helium offers the advantage of deeper penetration. The arc force with helium is sufficient to displace the molten weld pool with very short arcs. In some mechanized applications, the tip of the tungsten electrode is positioned below the workpiece surface to obtain very deep and narrow penetration. This technique is especially effective for welding aged aluminum alloys prone to overaging. It also is very effective at high welding speeds, as for tube mills. However, helium is less forgiving for manual welding. With helium, penetration and bead profile are sensitive to the arc length, and the long arcs needed for feeding filler wires are more difficult to control.

Helium has been mixed with argon to gain the combined benefits of cathode cleaning and deeper penetration, particularly for manual welding. The most common of these mixtures is 75% helium and 25% argon.

Although the GTA process was developed with helium as the shielding gas, argon is now used whenever possible because it is much cheaper. Helium also has some disadvantages because it is lighter than air, thus preventing good shielding. Its flow rates must be about twice as high as argon's for acceptable stiffness in the gas stream, and proper protection is difficult in drafts unless high flow rates are used. It is difficult to ionize, necessitating higher voltages to support the arc and making the arc more difficult to ignite. Alternating-current arcs are very unstable. However, helium is not used with alternating current because the cleaning action does not occur.

Hydrogen

Hydrogen is not an inert gas and is not used as a primary shielding gas. However, it can be added to argon when deep penetration and high welding speeds are

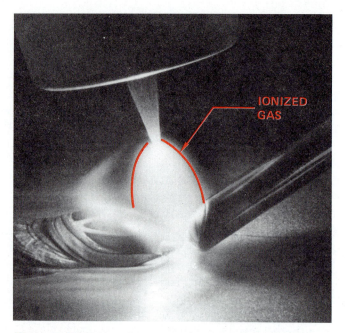

Figure 15-32 Highly concentrated ionized argon gas column.

needed. It also improves the weld surface cleanliness and bead profile on some grades of stainless steel that are very sensitive to oxygen. Hydrogen additions are restricted to stainless steels because hydrogen is the primary cause of porosity in aluminum welds. It can cause porosity in carbon steels and, in highly restrained welds, underbead cracking in carbon and low-alloy steels.

Nitrogen

Like hydrogen, nitrogen has been used as an additive to argon. But it cannot be used with some materials, such as ferritic steels, because it produces porosity. In other cases, such as with austenitic stainless steels, nitrogen is useful as an austenite stabilizer in the alloy. It is used to increase penetration when welding copper. Unfortunately, because of the general success with inert gas mixtures and because of potential metallurgical problems, nitrogen has not received much attention as an additive for GTA welding.

Hot Start

The hot start allows a controlled surge of welding current as the arc is started to establish a molten weld pool quickly. Establishing a molten weld pool rapidly on metals with a high thermal conductivity is often hard without this higher-than-normal current. Adjustments can be made in the length of time and the percentage above the normal current, **Figure 15-33**.

Prepurge and Postpurge

Prepurge is the time during which gas flows to clear out any air in the nozzle or surrounding the weld zone. The operator sets the length of time that the gas flows before the welding current is started, **Figure 15-34**.

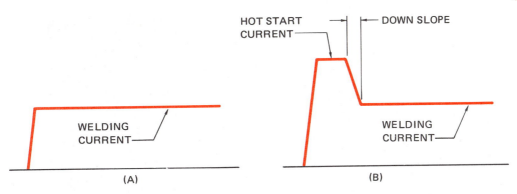

Figure 15-33 Standard method of starting welding current (A); hot start method of starting welding current (B).

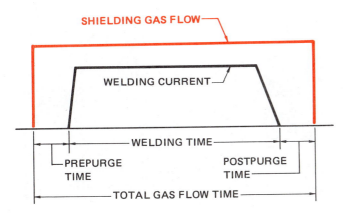

Figure 15-34

Electrode Diameter		Post Gas Flow Time*
in	(mm)	
.01	(0.25)	5 sec
.02	(0.5)	5 sec
.04	(1.0)	5 sec
1/16	(2)	8 sec
3/32	(2.4)	10 sec
1/8	(3)	15 sec
5/32	(4)	20 sec
3/16	(4.8)	25 sec
1/4	(6)	30 sec

*The time may be longer if either the base metal or the tungsten electrode does not cool below the rapid oxidation temperatures within the postpurge times shown.

Table 15-3 Postwelding Gas Flow Times.

Because some machines do not have prepurge, many welders find it hard to hold a position while waiting for the current to start. One solution to this problem is to use the postpurge for prepurging. Switch on the current to engage the postpurge. Now, with the current off, the gas is flowing, and the GTA torch can be lowered to the welding position. The welder's helmet should be lowered and the current restarted before the postpurge stops. This allows welders to have prepurge and to start the arc when they are ready.

The postpurge is the time during which the gas continues flowing after the welding current has stopped. This period serves to protect both the molten weld pool and the tungsten electrode as they cool to a temperature at which they will not oxidize rapidly. The time of the flow is determined by the welding current and the tungsten size, Table 15-3.

Gas Flow Rate

The shielding gas flow rate is measured in cubic feet per hour (CFH) or in metric measure as liters per minute (L/min). The rate of flow should be as low as possible and still give adequate coverage. High gas flow rates waste shielding gases and may lead to contamination. The contamination comes from turbulence in the gas at high flow rates. Air is drawn into the gas envelope by a venturi effect around the edge of the nozzle. Also, the air can be drawn in under the nozzle if the

torch is held at too sharp an angle to the metal, Figure 15-35.

The larger the nozzle size, the higher is the flow rate permissible without causing turbulence. Table 15-4 shows the average and maximum flow rates for most nozzle sizes. A gas lens can be used in combination with the nozzle to stabilize the gas flow, thus eliminating some turbulence. A gas lens will add to the turbulence problem if there is any spatter or contamination on its surface.

REMOTE CONTROLS

A remote control can be used to start the weld, increase or decrease the current, and stop the weld. The remote control can be either a foot-operated or hand-operated device. The foot control works adequately if the welder can be seated. Welds that must be performed away from a welding station must have a hand or thumb control.

Most remote controls have an on-off switch that is activated at the first or last part of the control movement. A variable resistor increases the current as the control is pressed more. A variable resistor works in a manner

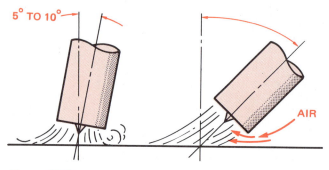

Figure 15-35 Too steep an angle between the torch and work may draw in air.

Nozzle Inside Diameter		Gas Flow*	
in	(mm)	cfh	(L/min)
1/4	(6)	10 – 14	(4.7 – 6.6)
5/16	(8)	11 – 15	(5.2 – 7.0)
3/8	(10)	12 – 16	(5.6 – 7.5)
7/16	(11)	13 – 17	(6.1 – 8.0)
1/2	(13)	17 – 20	(8.0 – 9.4)
5/8	(16)	17 – 20	(8.0 – 9.4)

*The flow rates may need to be increased or decreased depending upon the conditions under which the weld is to be performed.

Table 15-4 Suggested Argon Gas Flow Rate for Given Cup Sizes.

Figure 15-36 A foot-operated device can be used to increase the current.

Figure 15-37 GTA welding unit that can be added to a standard power supply so that it can be used for GTA welding. (Courtesy of Lincoln Electric Company.)

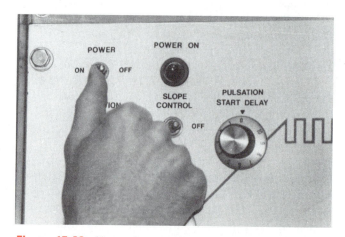

Figure 15-38 Always be sure the power is off when making machine connections.

similar to the accelerator pedal on a car to increase the power (current), **Figure 15-36**. The operating amperage range is determined by the value that has been set on the main controls of the machine.

EXPERIMENT 15-4

Setting Up a GTA Welder

Using a GTA welding machine, remote control welding torch, gas flowmeter, gas source (cylinder or manifold), tungsten, nozzle, collet, collet body, and cap, and any other hoses, special tools, and equipment required, you will set up the machine for GTA welding, **Figure 15-37**.

1. Start with the power switch off, **Figure 15-38**. Use a wrench to attach the torch hose to the machine. The water hoses should have left-hand threads to prevent incorrectly connecting them. Tighten the fittings only as tight as needed to prevent leaks, **Figure 15-39**. Attach the cooling water "in" to the machine solenoid and the water "out" to the power block.

2. The flowmeter or flowmeter regulator should be attached next. If a gas cylinder is used, secure it in place with a safety chain. Then remove the valve protection cap and crack the valve to blow out any dirt, **Figure 15-40**. Attach the flowmeter so that the tube is vertical.

3. Connect the gas hose from the meter to the gas "in" connection on the machine.

4. With both the machine and main power switched off, turn on the water and gas so that the connection to the machine can be checked for leaks. Tighten any leaking fittings to stop the leak.

5. Turn on both the machine and main power switches and watch for leaks in the torch hoses and fittings.

Figure 15-39 Tighten each fitting as it is connected to avoid missing a connection.

Figure 15-40 During transportation or storage, dirt may collect in the valve. Cracking the valve is the best way to remove any dirt.

■ **CAUTION**

Turn off all power before attempting to stop any leaks in the water system.

6. With the power off, switch the machine to the GTA welding mode.
7. Select the desired type of current and amperage range, **Figures 15-41** and **15-42**.
8. Set the fine current adjustment to the proper range, depending upon the size of tungsten used, **Table 15-5.** Refer to Chapter 6 for more information on setting the fine current adjustment.
9. Place the high-frequency switch in the appropriate position, auto for DC or continuous for AC, **Figure 15-43**.
10. The remote control can be plugged in and the selector switch set, **Figure 15-44**.
11. The collet and collet body should be installed on the torch first, **Figure 15-45**.
12. On the Linde or copies of Linde torches, installing the back cap first will stop the collet body from being screwed into the torch fully. A poor connection will result in excessive electrical and thermal resistance, causing a heat buildup in the head.
13. The tungsten can be installed and the end cap tightened to hold the tungsten in place.
14. Select and install the desired nozzle size. Adjust the tungsten length so that it does not stick out more than the diameter of the nozzle, **Figure 15-46**.
15. Check the manufacturer's operating manual for the machine to ensure that all connections and settings are correct.

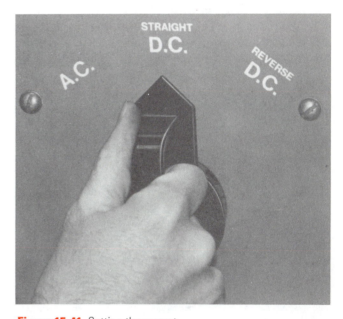

Figure 15-41 Setting the current.

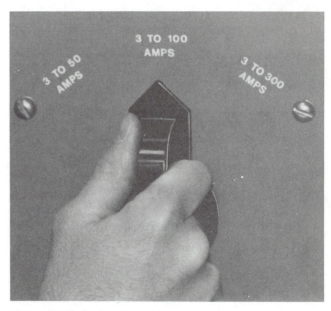

Figure 15-42 Setting the amperage range.

| Electrode Diameter | | DCEN | DCEP | AC |
in	(mm)			
.04	(1)	15–60	Not recommended	10–50
1/16	(2)	70–100	10–20	50–90
3/32	(2.4)	90–200	15–30	80–130
1/8	(3)	150–350	25–40	100–200
5/32	(4)	300–450	40–55	160–300

Table 15-5 Amperage Range for Tungsten Electrodes.

16. Turn on the power, depress the remote control, and again check for leaks.
17. While the postpurge is still engaged, set the gas flow by adjusting the valve on the flowmeter.

The GTA welding system is now ready to be used. ◆

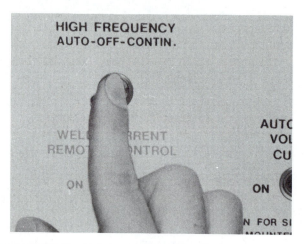

Figure 15-43 The high-frequency switch should be placed in the appropriate position.

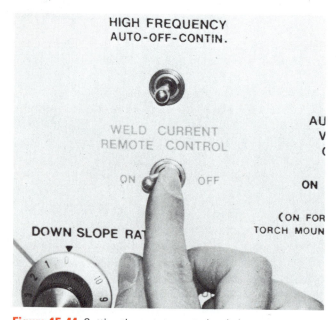

Figure 15-44 Setting the remote control switch.

EXPERIMENT 15-5

Striking an Arc

Using a properly set up GTA welding machine, proper protection, and clean scrap metal, you will strike a GTA welding arc.

1. Position yourself so that you are comfortable and can see the torch, tungsten, and plate while the tungsten tip is held about 1/4 in. (6 mm) above the metal. Try to hold the torch at a vertical angle ranging from 0° to 15°. Too steep an angle will not give adequate gas coverage, Figure 15-47.

■ CAUTION

Avoid touching the metal table with any unprotected skin or jewelry. The high frequency can cause an uncomfortable shock.

2. Lower your arc welding helmet and depress the remote control. A high-pitched, erratic arc should be immediately jumping across the gap between the

Figure 15-45 Inserting collet and collet body.

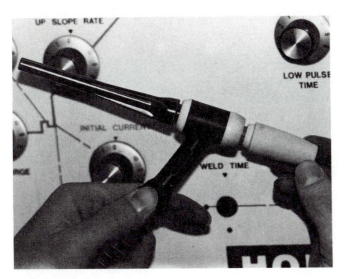

Figure 15-46

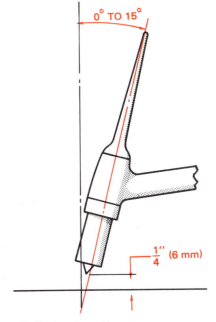

Figure 15-47 GTA torch position.

tungsten and the plate. If the high-frequency arc is not established, lower the torch until it appears.

3. Slowly increase the current until the main welding arc appears, **Figure 15-48**.
4. Observe the color change of the tungsten as the arc appears.
5. Move the tungsten around in a small circle until a molten weld pool appears on the metal.
6. Slowly decrease the current and observe the change in the molten weld pool.
7. Reduce the current until the arc is extinguished.
8. Hold the torch in place over the weld until the post-purge stops.
9. Raise your hood and inspect the weld.

Repeat this procedure until you can easily start the arc and establish a molten weld pool using both AC and DCEN currents. Turn off the welding machine, water, and shielding gas when you are finished, then clean up your work area. ◆

Figure 15-48 Stable gas tungsten arc.

Use of Welding Robots in Short Production Runs

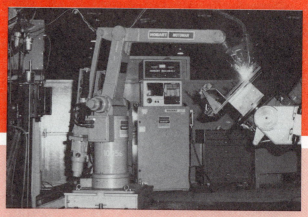

Figure 1

Until recently, an entire area of robotic arc welding has been neglected — short production runs. Now, one company is proving that short arc welding production runs can translate into reduced manufacturing costs.

The company is a leader in finding innovative manufacturing techniques to cut costs and turn-around time. Management continually searches for improved tools that will get a job done better, faster, and with greater cost effectiveness. Sometimes this has meant designing and building their own equipment. Sometimes it has meant purchasing the latest in robotic arc welding equipment.

Recently, the company installed two robotic workcells, **Figure 1.** In addition to their arc welding capabilities, the robots were selected because of their large working range and the availability of a seam tracking system.

The welding robot supplier also provided an adaptive thru-the-arc seam tracking system, which enables the robot weld gun to adjust automatically to changes in weld starting points and welding path. Once the seam tracker (weld gun) locates the start of the seam (path), the seam tracker will maintain the correct relationship between the weld torch and the weld centerline.

Figure 2 shows the layout of the components of the robotic welding system. The combination of the robot and seam tracker allows the company to produce smaller volumes and a wider variety of complex parts with some variation in weld starting points and weld path. The operator can switch rapidly from one workpiece to another without expensive and complex tooling — efficiently and cost-effectively.

This cost-effective short-run production capability also meshes with the company's just-in-time manufacturing program by reducing parts inventory, minimizing storage space and material handling, and reducing raw materials. Just-in-time parts manufacturing can be just as effective internally as with outside vendors.

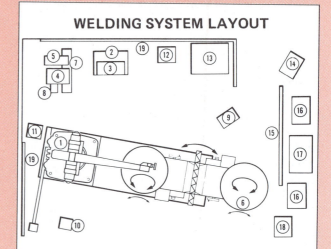

WELDING SYSTEM LAYOUT

1. Robot
2. Controller
3. Interface
4. Power Source
5. Seam Tracker Arc Sensing Unit
6. Turntable-type Work Positioner
7. Positioner Controller
8. Air Compressor
9. Positioner Control Panel
10. Torch Nozzle Cleaner
11. Wirefeed System and Pneumatic Cleaning System
12. Tool Cabinet
13. Work Bench
14. Power Source (for tack welding)
15. Overhead Conveyor for Torch Cables
16. Parts Bins
17. Tack Welding Table
18. Finished Parts Bin
19. Protection Fence

Figure 2

Figure 3A

Figure 3B

To date, the company has selected and programmed some thirty parts for the welding roots. Two views of one typical part (a knife support for a hay baler) are shown in Figure 3. Parts to be welded by the robots are not matched only to the robot's capabilities, but also to the operator's work motions. The idea is to combine the handling time of setting-up on the second work station with the arc-on time on the first work station.

It was found that if the operator can keep ahead of the robot and rest a few seconds, this is the best setup. If the robot continually beats the operator, eventually the operator will become discouraged. So the handling time of setting-up is paired with the arc-on time on the other work station.

(Courtesy of Hobart Brothers Company, Troy, Ohio 45373.)

REVIEW

1. What early advancements made the GTA welding process more effective and reduced its cost?

2. What metals were weldable only by the GTAW process before GMAW was developed?

3. Which two of tungsten's properties make it the best choice for GTA welding?

4. Why must the tip of the tungsten be hot?

5. Why does some tungsten erosion occur?

6. What function regarding tungsten heat do the collet and torch play?

7. What problem can an excessively large tungsten cause?

8. What holds the molten ball of tungsten in place at the tip of the electrode during DCEP welding?

9. Using Table 15-1, answer the following:

   a. What color identifies EWTh-1?

   b. What is the composition of TWCe-2?

10. What does adding thorium oxide do for the tungsten electrode?

11. How can the end of a tungsten electrode be shaped?

12. Why should a grinding stone that is used for sharpening tungsten not be used for other metals?

13. Why should the grinding marks run lengthwise on the tungsten electrode end?

14. What are three ways of breaking off the contaminated end of a tungsten electrode?

15. What is the correct color to use on the balled end of a pointed and remelted tungsten tip on DCEP?

16. Why should the torch be as cool as possible?

17. What will happen to a water-cooled torch cable if the flow of cooling water stops?

18. Why must shielding gas hoses not be made from rubber?

19. Why should the water solenoid be on the supply side of the water system?

20. What materials can be used to make nozzles?

21. What problem can a long nozzle cause to the tungsten?

22. Why must the tube of a flowmeter be vertical?

23. What is the heat distribution with DCEN welding current?

24. What is the heat distribution with DCEP welding current?

25. What is the heat distribution with AC welding current?

26. Why must AC welding power use high frequencies in order to work?

27. Why are argon and helium known as inert gases?

28. Why is argon's ease of ionization a benefit?

29. What makes helium difficult to use for manual welding?

30. What are the benefits of adding hydrogen to argon for welding?

31. What is the purpose of a hot start?

32. Using Table 15-3, determine the post–gas flow time for a 3/23-inch (2.4-mm) tungsten.

33. How can air be drawn into the shielding gas?

34. Using Table 15-4, determine the minimum gas flow rate for a 1/2-inch (13-mm) nozzle.

35. What functions can a remote control provide the welder?

Chapter 16

GAS TUNGSTEN ARC WELDING OF PLATE

OBJECTIVES

After completing this chapter, the student should be able to

- set the correct amperage for each type and size of tungsten.
- set the correct gas flow times and rates.
- make a variety of GTA welds in different positions.
- explain proper rod manipulation techniques.
- demonstrate proper GTA welding torch manipulation techniques.

KEY TERMS

gas coverage
protective zone
contamination
surface tension
oxide layer
chill plate

INTRODUCTION

The gas tungsten arc welding process can be used to join nearly all types and thicknesses of metal. See Color Plates 32, 33, 38, 39, and 40.

The slow travel speed and high cost usually limit GTA welding to thin sections, small parts, or high-integrity joints. In some welds, the root pass may be made with GTA, and the joint then completed with another, more efficient process.

The proper setup of GTA equipment can often affect the quality of the weld performed. Charts and graphs are available that give the correct amperage, gas flow rate, and time. These charts are designed for optimum laboratory or classroom conditions. Actual work in the field will have an effect on these values. The experiments in this chapter are designed to help the welder understand the harmful effects on welding of less than ideal conditions. This will allow the welder to evaluate the appearance of a weld and make the necessary changes in technique or setup to improve the weld.

After a person has learned to weld, troubleshooting welding problems will become much easier. The weld should be watched carefully to pick up changes that indicate a needed adjustment. When welders can do this, they have mastered the GTA process and have made themselves better potential employees. To do a weld is good; to solve a welding problem is better.

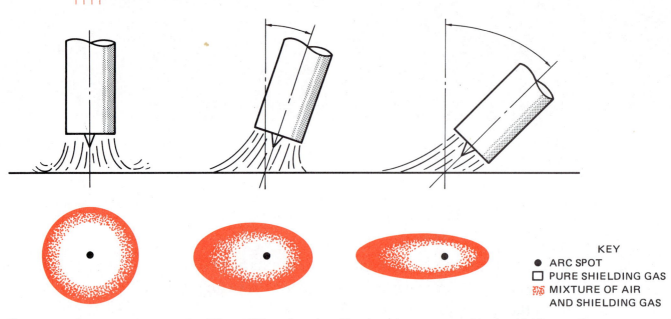

KEY
● ARC SPOT
☐ PURE SHIELDING GAS
▨ MIXTURE OF AIR
     AND SHIELDING GAS

Figure 16-1 Gas coverage patterns for different GTA torch angles. Note how the area covered by the shielding gas becomes narrower and elongates as the angle of the torch increases from the perpendicular.

TORCH ANGLE

The torch should be held as close to perpendicular as possible in relation to the plate. The torch may be angled from 0° to 15° from perpendicular for the proper shielding gas coverage. As the gas flows out it must form a protective zone around the weld. Tilting the torch changes the shape of this protective zone, **Figure 16-1**.

The velocity of the shielding gas also affects the protective zone as the torch angle changes. As the velocity increases, a low-pressure area develops behind the cup. When the low-pressure area becomes strong enough, air is pulled into the shielding gas. The sharper the angle and the higher the flow rate, the possibility of contamination is increased by the onset of turbulence in the gas stream. This causes air to become mixed with the shielding gas. Turbulence caused by the shielding gas striking the work will also cause air to mix with the shielding gas at high velocities.

FILLER ROD MANIPULATION

The filler rod end must be kept inside the protective zone of the shielding gas, **Figure 16-2**. The end of the filler rod is hot, and if it is removed from the gas protection, it will oxidize rapidly. The oxide will then be added to the molten weld pool, **Figure 16-3**. When a weld is stopped so that the welder can change position, the shielding gas must be kept flowing around the rod end to protect it until it is cool. If the end of the rod becomes oxidized, it should be cut off before restarting. The following method can be used both to protect the rod end and reduce the possibility of crater cracking; that is, breaking the arc but keeping the torch over the

crater while, at the same time, sticking the rod in the molten weld pool before it cools, **Figure 16-4**. When the weld is restarted, the rod is simply melted loose again, **Figure 16-5**.

The rod should enter the shielding gas as close to the base metal as possible, **Figure 16-6**. A 15° angle or less prevents air from being pulled into the welding zone behind the rod, **Figure 16-7**. As an example, if a rod is held in a stream of running water, air can be pulled in. The faster the water flows or the sharper the angle at which the rod is held, the more air is pulled in. The same action occurs with the shielding gas as its flow increases or as the rod angle increases.

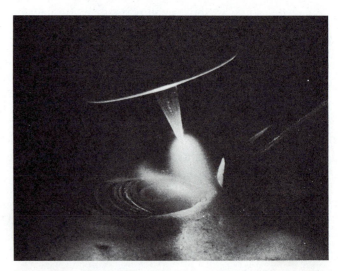

Figure 16-2 The hot filler rod end is well within the protective gas envelope.

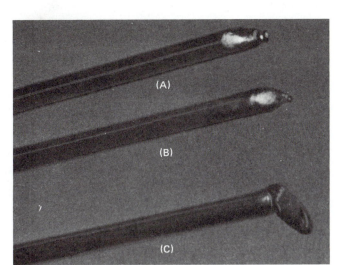

Figure 16-3 Filler properly protected (A), some oxides on filler (B), and excessive oxides caused by improper filler rod manipulation (C).

Figure 16-5 Filler being remelted as the weld is continued.

Figure 16-4 Filler being left in the molten weld pool as the arc is extinguished.

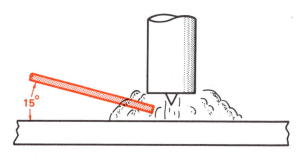

Figure 16-6

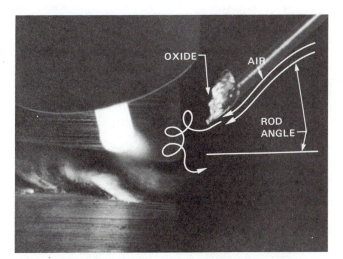

Figure 16-7 Too much filler rod angle has caused oxides to be formed on the filler rod end.

TUNGSTEN CONTAMINATION

For new welding students, the most frequently occurring and most time-consuming problem is tungsten contamination. The tungsten becomes contaminated when it touches the molten weld pool or when it is touched by the filler metal. When this happens, especially with aluminum, surface tension pulls the contamination up onto the hot tungsten, Figure 16-8. The extreme heat causes some of the metal to vaporize and form a large, widely scattered oxide layer. On aluminum, this layer is black. On iron (steel and stainless steel), this layer is a reddish color.

The contamination caused by the tungsten touching the molten weld pool or filler metal forms a weak weld. On a welding job, both the weld and the tungsten must be cleaned before any more welding can be done. The weld crater must be ground or chiseled, and the tungsten end must be reconditioned. Extremely tiny tungsten particles will show up if the weld is x-rayed. Failure to remove the contamination properly will result in the failure of the weld.

When starting to weld, the beginning student will save weld practice time by burning off the contamination. On a scrap plate, strike an arc using a higher-than-normal amperage setting. The arc will be erratic and discolored at first, but, as the contamination vaporizes, the

Figure 16-8 Contaminated tungsten.

arc will stabilize. Contamination can also be knocked off by quickly flipping the torch head.

■ CAUTION

This procedure should never be used with heavy contaminations or when a welder is on the job in the field. It is designed only to help the new student in the first few days of training to save time and increase weld production.

CURRENT SETTING

The amperage set on a machine and the actual welding current are often not the same. The amperage indicated on the machine's control is the same as that at the arc only for the following conditions:

■ The power to the machine is exactly correct.

■ The lead length is very short.

■ All cable connections are perfect with zero resistance.

■ The arc length is exactly the right length.

If any one of these factors changes, the actual welding amperage will change.

In addition to the difference between indicated and actual welding amperage, there is a more significant difference between amperage and welding power. The welding power, in watts, is based on the formula $W = E \times I$, or volts (E) multiplied by amperes (I) equals watts (W). Thus, the indicated power to a weld from two different types of welding machines set at 100 amperes will vary depending upon the voltage of the machine.

The welding machine setting will vary within a range from low to high (cool to hot). The range for one machine may be different from that of another machine. The setting will also be different for various types and sizes of tungstens, polarities, types and thicknesses of metal, joint position or design, and shielding gas used.

A chart, such as the one in **Figure 16-9,** and a series of tests can be used to set the lower and upper limits for the amperage settings. As students' welding skills improve with practice, they will become familiar with the machine settings so that a table for these settings is no longer needed. In the welding industry, some welders will mark a line on the dial of the machine to help in resetting the machine. If a welder is required to make a number of different machine setups, a list or chart can be made and taped to the machine. This practice is more professional than marking the machine dials.

EXPERIMENTS

Experiments are designed to help new welders learn some basic skills that will help them troubleshoot welding problems. If you do the experiments listed in this chapter, you will be better able to determine what is causing a problem with your weld. As you learn more about welding, subtle changes will become more noticeable. Even experienced welders make changes in the setup, current, or welding technique as they try to resolve a problem.

Experiment 16-1 will help the welder determine the correct machine settings for the minimum and maximum welding current for the machine used, the types and sizes of tungstens, and the metal types and thicknesses. Most welding will be performed with a medium-range or mid-range machine setting. The exact setting is more important for machines without remote controls. The remote control allows changes in welding current to be made during the welding without having to stop.

CURRENT AND TUNGSTEN ELECTRODE SIZE	AMPERAGE/MACHINE SETTING				
	TOO LOW	LOW	GOOD	HIGH	TOO HIGH

Figure 16-9 Sample chart used to record GTA welding machine settings.

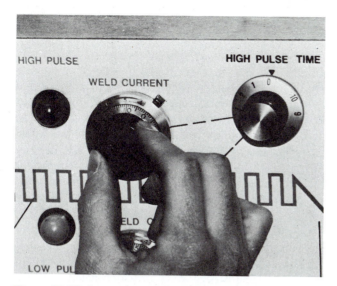

Figure 16-10 Lower the welding current to zero or as low as possible.

Figure 16-11 Melting first occurring.

EXPERIMENT 16-1

Setting the Welding Current

Using a properly set up GTA welding machine and torch, proper safety protection, one of each available tungsten size and type, and 16-gauge mild steel, 1/8 in. (3 mm) and 1/4 in. (6 mm) thick, you will develop a chart of the correct machine current setting for each type and size of tungsten.

Set the machine welding power switch for DCEN (DCSP) and the amperage control to its lowest setting, **Figure 16-10**. Sharpen a point on each tungsten and install one of the smaller diameter tungstens in the GTA torch. Select a nozzle with a 1/2-in. (13-mm) -diameter hole and attach it to the torch head. Set the prepurge time to 0 and postpurge to 10 to 15 seconds. Connect the remote control if it is available. Turn on the main power and hold the torch so that it cannot short out. Depress the remote controls to start the shielding gas so the flow rate can be set at 20 CFH (8 L/min). Switch the high frequency to start. All other functions, such as pulse, hot start, slope, and so on, should be in the off position.

Place the piece of 16-gauge sheet metal flat on the welding table. Hold the torch vertically with the tungsten about 1/4 in. (6 mm) above the metal. Lower your welding hood and fully depress the remote control. Watch the arc to see if it stabilizes and melts the metal. After a short period of time (15 to 30 seconds), stop, raise your hood, and check the plate for a melted spot. If melting occurred, note the size of the spot and depth of penetration, **Figure 16-11**. Increase the amperage setting by 5 or 10 amperes, note the setting on the chart, and repeat the process.

After each test, observe and record the results. The important settings to note are:

1. When the tungsten first heats up and the arc stabilizes.
2. When the metal first melts.
3. When 100% penetration of the metal first occurs.
4. When burnthrough first occurs.
5. When the tungsten starts glowing white hot and/or melts.

The lowest (minimum) acceptable amperage setting is when the molten weld pool first appears on the base metal and the arc is stable. The highest (maximum) amperage setting is when the base metal burnthrough or melting of the tungsten occurs. Any current setting in between the high and low points is within the amperage range for that specific setup.

To establish the range for the next tungsten type or size, repeat the test. After each test, the metal should be cooled to prevent overheating. After each type and size of tungsten has been tested and an operating range established, repeat the procedure using the next thicker metal. Repeat this procedure until you have set up the operating ranges for all of the metals and tungstens you will be using. Turn off the welding machine, shielding gas, and cooling water and clean up your work area when you are finished welding. ◆

GAS FLOW

The gas preflow and postflow times required to protect both the tungsten and the weld depend upon the following factors:

■ Wind or draft speed
■ Nozzle size used
■ Tungsten size
■ Amperage
■ Joint design

- Welding position
- Type of metal welded

The weld quality can be adversely affected by improper gas flow settings. The lowest possible gas flow rates and the shortest preflow or postflow time can help reduce the cost of welding by saving the expensive shielding gas.

In Experiment 16-2, the minimum and maximum gas flow settings for each nozzle size, tungsten size, and amperage setting will be determined. The chart a welder prepares based on experiments is to improve that welder's skill and welding technique. Charts may differ slightly from one welder to another. As a welder's skill improves, the chart may change. As experience is gained, a welder will learn how to set the gas flow effectively without the need for this chart.

The minimum flow rates and times must be increased to weld in drafty areas or for out-of-position welds. The rates and times can be somewhat lower for tee joints or welds made in tight areas. The maximum flow rates must never be exceeded. Exceeding these flow rates causes weld contamination and increases the rejection rate.

EXPERIMENT 16-2

Setting Gas Flow

Using a properly set up GTA welding machine and torch, proper safety protection, one of each available tungsten size, metal that is 16 gauge to 1/4 in. (6 mm) thick, and the welding current chart developed in Experiment 16-1, you will make a chart of the minimum and maximum flow rates and times for each nozzle size, tungsten

size, and amperage setting. An assistant will also be needed to change and record the flow rate while you work.

Set the machine welding power switch for DCEN (DCSP). Set the amperage to the lowest setting for the size of tungsten used. Set the prepurge time to 0 and postpurge at 20 seconds, **Figure 16-12**. Turn on the main power. With the torch held so that it cannot short out, depress the remote control to start the shielding gas flow and set the flow at 20 CFH (9 L/min). Switch the high frequency to start. All other functions, such as pulse, hot start, slope, and so on, should be in the off position.

Starting with the smallest nozzle and tungsten size, strike an arc and establish a molten pool on a piece of metal in the flat position. Watch the molten weld pool and tungsten for signs of oxide formation as another person slowly lowers the gas flow rate. Have that person note this setting (where oxide formation begins), **Figure 16-13**, as the minimum flow rate on the chart next to the nozzle size and current setting. Now slowly increase the flow rate until the molten pool starts to be blown back or oxides start forming. This setting should be noted on the chart as the maximum flow rate for this current and nozzle size, **Figure 16-14**. Lower the flow to a rate of 2 CFH or 3 CFH (1 L/min or 2 L/min) above the minimum value noted on the chart and then stop the arc. Record the length of time from the point when the arc stops and the tungsten stops glowing as the postpurge time. Repeat this test at a medium and then high current setting for this nozzle and tungsten size. When using high current settings, it may be necessary to move the torch or use thicker plate to prevent burnthrough.

Repeat this test procedure with each available nozzle and tungsten size. Stainless steel or aluminum is preferred for this experiment because the oxides are more quickly noticeable than when mild steel is used. If

Figure 16-12 Setting the postpurge timer.

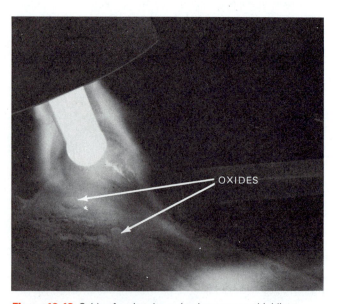

Figure 16-13 Oxides forming due to inadequate gas shielding.

ELECTRODE AND NOZZLE SIZE	FLOW RATE					POST-FLOW TIME		
	TOO LOW	LOW	GOOD	HIGH	TOO HIGH	TOO SHORT	OK	TOO LONG

Figure 16-14 Sample chart for setting shielding gas flow rate and time.

aluminum is used, the welding current must be AC, and the high-frequency switch should be set on continuous.

To establish the minimum prepurge time for each nozzle and tungsten size, set the amperage to a medium-high setting. Hold the torch above the metal so that an arc will be instantly started. Set the prepurge timer to 0 and the gas flow to just above the minimum value noted on the chart. Quickly strike an arc on metal thin enough to cause a weld pool to form instantly at that power setting. Stop the arc and examine the weld pool and tungsten for oxides. Repeat this procedure, increasing the prepurge time until no oxides are formed on either the plate or tungsten. Record this time on the chart as the minimum prepurge time. Repeat this test with each available nozzle and tungsten size. Turn off the welding machine, shielding gas, and cooling water and clean up your work area when you are finished welding. ◆

PRACTICE WELDS

The practice welds are grouped according to the weld position and type of joint and not by the type of metal. The order in which a person decides to do the welds is that person's choice. It is suggested that the stringer beads be done in each metal and position before the different joints are tried. Each metal has its own characteristics that may make one metal easier for a person to work on than another metal.

Mild steel is inexpensive and requires the least amount of cleaning. Slight changes in the metal have little effect on the welding skill required. Stainless steel is somewhat affected by cleanliness, requiring little preweld cleaning. However, the weld pool shows overheating or poor gas coverage. With aluminum, cleanliness is a critical factor. Oxides on aluminum may prevent the molten weld pool from flowing together. The surface tension helps hold the metal in place, giving excellent bead contour and appearance.

The degree of difficulty a welder encounters with each of these metals depends upon the individual's experience. Try each weld with each metal to determine which metal will be easiest to master first. The type of welding machine and materials used will also affect a welder's progress. Practice will help welders overcome any obstacle to their progress.

Mild and Stainless Steels

Both mild and stainless steels are GTA welded with DCEN (DCSP). The manipulation techniques required are nearly the same for each, and the transfer ability of skills is easy. The major differences are that stainless steels show more the effects of poor cleaning, excessive weld heat, poor gas shielding, mismatched filler and base metals, and incorrect torch manipulation. Most welds on mild steel do not show the effects of contamination as easily as do welds on stainless.

The most common sign that there is a problem with a stainless steel weld is the bead color after the weld. The greater the contamination, the darker is the color. The exposure of the weld bead to the atmosphere before it has cooled will also change the bead color. It is impossible, however, to determine the extent of contamination of a weld with only visual inspection. Both light-colored and dark-colored welds may not be free from oxides. Thus, it is desirable to take the time and necessary precautions to make welds that are no darker than dark blue, Table 16-1. Welds with only slight oxide layers are better for multiple passes.

Black crusty spots may appear on both mild and stainless steel weld beads. These spots are often caused by improper cleaning of the filler rod or failure to keep the end of the rod inside the shielding gas.

Surface Color	Approximate Temperature at Which Color Was Formed	
	°F	(°C)
Light straw	400	(200)
Tan	450	(230)
Brown	525	(275)
Purple	575	(300)
Dark blue	600	(315)
Black	800	(425)

Table 16-1 Temperatures at Which Various Colored Oxide Layers Form on Steel.

Base Metal AISI No.			AWS Filler Metal No.	
Low Carbon	1010		RG60	ER70S-3
	1020		RG60	ER70S-3
	1030		RG60	ER70S-3
Stainless Steel	301		ER308	ER308L
	302		ER308	ER308L
	304		ER308	ER308L
	305		ER308	ER308L
	308		ER308	ER308L
	309		ER309	
	310		ER310	
	316		ER316	ER316L
	347		ER347	

Table 16-2 Basic Filler Metal Selection.

Table 16-2 lists the most common types of mild and stainless steels and the recommended filler metals.

Aluminum

Aluminum is GTA welded with AC high frequency stabilized. The alternating current provides good arc cleaning.

The high frequency aids in arc restarting as the current stops to change direction. Aluminum has low density and high surface tension, which allow a large weld bead to be controlled easily. The high thermal conductivity of the metal may make starting a weld on thick sections difficult without preheating. Although aluminum resists oxidation at room temperatures, it rapidly oxidizes at welding temperatures. The oxide that forms reduces the ability of the weld pool to flow together.

The processes of cleaning and keeping the metal clean take a lot of time. Removal of the oxide layer is easy using a chemical or mechanical method. Ten minutes after cleaning, however, the oxide layer may again be thick enough to require recleaning. If the welder's hands or gloves are not clean and oil-free, the metal or filler rods will be recontaminated.

If the filler rod is not kept inside the shielding gas, it will quickly oxidize. But, because of the low melting temperature of the filler rod, the end will melt before it is added to the weld pool if it is held too close to the arc, Figures 16-15 and 16-16. See Color Plates 32 and 33.

Metal Preparation

Both the base metal and the filler metal used in the GTA process must be thoroughly cleaned before welding. Contamination left on the metal will be deposited in the weld because there is no flux to remove it. Oxides, oils, and dirt are the most common types of contaminants. They can be removed mechanically or chemically. Mechanical metal cleaning may be done by grinding, wire brushing, scraping, machining, or filing. Chemical cleaning may be done by using acids, alkalies, solvents, or detergents.

■ CAUTION

The manufacturer's recommendations for using these products must be followed. Failure to do so may result in chemical burns, fires, fumes, or other safety hazards that could lead to serious injury. If anyone should come in contact with any chemicals, immediately refer to the Materials Specification Data Sheet (MSDS) for the proper corrective action.

PRACTICE 16-1

Surfacing Weld (stringer beads), Flat Position, on Mild Steel

Using a properly set up and adjusted GTA welding machine, proper safety protection, and one or more

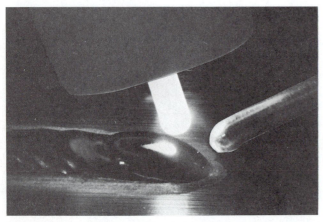

Figure 16-15 Aluminum filler being correctly added to the molten weld pool.

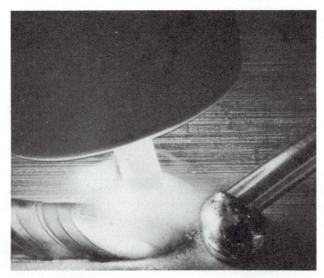

Figure 16-16 Filler rod being melted before it is added to the molten pool.

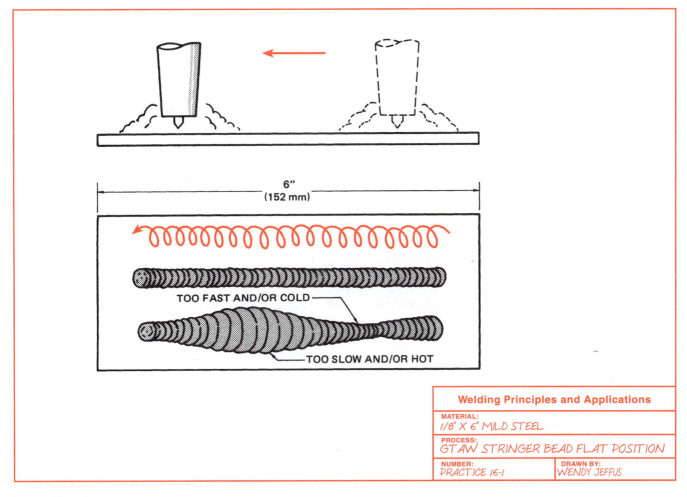

Welding Principles and Applications

MATERIAL:
1/8" X 6" MILD STEEL

PROCESS:
GTAW STRINGER BEAD FLAT POSITION

NUMBER:	DRAWN BY:
PRACTICE 16-1	WENDY JEFFUS

Figure 16-17 Surfacing weld in the flat position.

pieces of mild steel, 6 in. (152 mm) long and 16 gauge and 1/8 in. (3 mm) thick, you will push a weld pool in a straight line down the plate, **Figure 16-17**. Maintain uniform weld pool size and penetration.

■ Starting at one end of the piece of metal that is 1/8 in. (3 mm) thick, hold the torch as close as possible to a 90° angle.

Figure 16-18 Surfacing weld.

■ Lower your hood, strike an arc, and establish a weld pool.

■ Move the torch in a circular oscillation pattern down the plate toward the other end, **Figure 16-18**.

■ If the size of the weld pool changes, speed up or slow down the travel rate to keep the weld pool the same size for the entire length of the plate.

The ability to maintain uniformity in width and keep a straight line increases as you are able to see more than just the weld pool. As your skill improves, you will relax, and your field of vision will increase.

Repeat the process using both thicknesses of metal until you can consistently make the weld visually defect free. Turn off the welding machine, shielding gas, and cooling water and clean up your work area when you are finished welding. ◆

PRACTICE 16-2

Surfacing Weld, Flat Position, on Stainless Steel

Using the same equipment and material thicknesses as listed in Practice 16-1 and one or more pieces of stainless steel, 6 in. (152 mm) long and 1/4 in. (6 mm) thick,

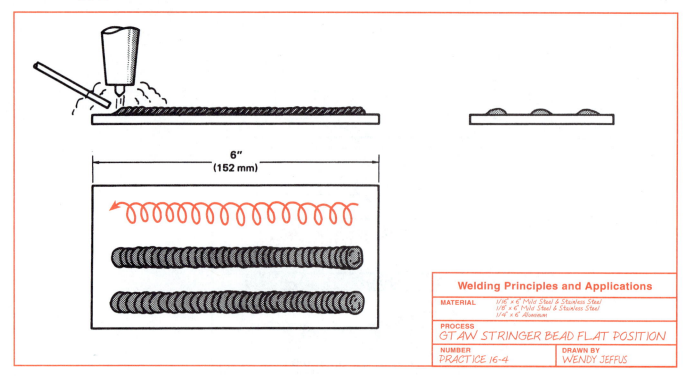

Welding Principles and Applications

MATERIAL	1/16" x 6" Mild Steel & Stainless Steel 1/8" x 6" Mild Steel & Stainless Steel 1/4" x 6" Aluminum

PROCESS
GTAW STRINGER BEAD FLAT POSITION

| NUMBER
PRACTICE 16-4 | DRAWN BY
WENDY JEFFUS |

Figure 16-19 Stringer beads in the flat position.

you will push a molten weld pool in a straight line down the plate, keeping the width and penetration uniform.

To keep the formation of oxides on the bead to a minimum, a chill plate (a thick piece of metal used to absorb heat) may be required. Another method is to make the bead using as low a heat input as possible. When the weld is finished, the weld bead should be no darker than dark blue.

Repeat the process using both thicknesses of metal until you can consistently make the weld visually defect free. Turn off the welding machine, shielding gas, and cooling water and clean up your work area when you are finished welding. ◆

PRACTICE 16-3

Surfacing Weld, Flat Position, on Aluminum

Using the same equipment, setup, and procedure as listed in Practice 16-1, and one or more pieces of aluminum, 6 in. (152 mm) long and 1/16 in. (2 mm), 1/8 in. (3 mm), and 1/4 in. (6 mm) thick, you will push a weld pool in a straight line, maintaining uniform width and penetration for the length of the plate.

A high current setting will allow faster travel speeds. The faster speed helps control excessive penetration. Hot cracking may occur on some types of aluminum after a surfacing weld. This is not normally a problem when filler metal is added. If hot cracking should occur during this practice, do not be concerned.

Repeat the process using all thicknesses of metal until you can consistently make the weld visually defect

free. Turn off the welding machine, shielding gas, and cooling water and clean up your work area when you are finished welding. ◆

PRACTICE 16-4

Flat Position, Using Mild Steel, Stainless Steel, Aluminum

For this practice, you will need a properly set up and adjusted GTA welding machine, proper safety protection, filler rods 36 in. (0.9 m) long x 1/16 in. (2 mm),

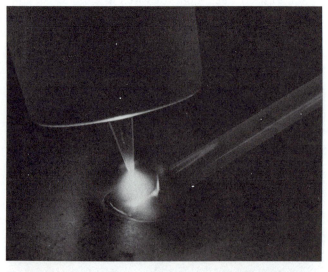

Figure 16-20 Establish a molten weld pool and dip the filler rod into it.

(A)   (B)

Figure 16-21 Note the difference in the weld produced when different size filler rods are used.

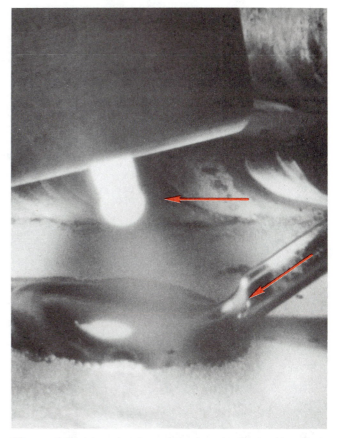

Figure 16-22 Move the electrode back as the filler rod is added.

3/32 in. (2.4 mm), and 1/8 in. (3 mm) in diameter, one or more pieces of mild steel, stainless steel, and aluminum, 6 in. (152 mm) long x 1/16 in. (2 mm) and 1/8 in. (3 mm) thick, and aluminum plate 1/4 in. (6 mm) thick, **Tables 16-3, 16-4,** and **16-5.** In this practice, you will make a straight stringer bead, 6 in. (152 mm) long, that is uniform in width, reinforcement, and penetration, **Figure 16-19.**

Starting with the metal that is 1/8 in. (3 mm) thick and the filler rod having a 3/32-in. (2.4-mm) diameter, strike an arc and establish a weld pool, **Figure 16-20.** Move the torch in a circle as in the practice beading. When the torch is on one side, add filler rod to the other side of the molten weld pool, **Figure 16-21(A)** and **(B).** The end of the rod should be dipped into the weld pool but should not be allowed to melt and drip into the weld pool. **Figure 16-22.** See Color Plates 38 and 39. Change to another size filler rod and determine its effect on the weld pool.

Maintain a smooth and uniform rhythm as filler metal is added. This will help to keep the bead uniform.

See Color Plate 40. Vary the rhythms to determine which one is easiest for you. If the rod sticks, move the torch toward the rod until it melts free.

When the full 6-in. (152-mm) -long weld bead is completed, cool and inspect it for uniformity and defects. Repeat the process using all thicknesses of metal until you can consistently make the weld visually defect free. Turn off the welding machine, shielding gas, and cooling water and clean up your work area when you are finished welding. ◆

Tungsten Electrode			Welding Power			Shielding Gas		Nozzle	Filler Metal	
Type	**Size**	**Tip**	**Amp**	**Current**	**HF**	**Type**	**Flow**	**Size**	**Type**	**Size**
EWTh-1 or EWTh-2	1/16" 2 mm	Point	50 to 100	DCEN DCSP	Start or Auto	Argon	16 cfh 7 L/min	3/8" 10 mm	RG60 or ER70S-3	1/16" 3/32" 2-2.4 mm
EWTh-1 or EWTh-2	3/32" 2.4 mm	Point	70 to 150	DCEN DCSP	Start or Auto	Argon	16 cfh 7 L/min	3/8" 10 mm	RG60 or ER70S-3	1/16" 3/32" 2-2.4 mm
EWTh-1 or EWTh-2	1/8" 3 mm	Point	90 to 250	DCEN DCSP	Start or Auto	Argon	20 cfh 9 L/min	1/2" 13 mm	RG60 or ER70S-3	3/32" 1/8" 2.4-3 mm

Table 16-3 Suggested Setting for GTA Welding of Mild Steel.

Tungsten Electrode			Welding Power			Shielding Gas		Nozzle	Filler Metal	
Type	Size	Tip	Amp	Current	HF	Type	Flow	Size	Type	Size
EWTh-1 or EWTh-2	1/16" 2 mm	Point	70 to 100	DCEN DCSP	Start or Auto	Argon	16 cfh 7 L/min	3/8" 10 mm	ER308 or ER316	1/16" 3/32" 2-2.4 mm
EWTh-1 or EWTh-2	3/32" 2.4 mm	Point	70 to 150	DCEN DCSP	Start or Auto	Argon	16 cfh 7 L/min	3/8" 10 mm	ER308 or ER316	1/16" 3/32" 2-2.4 mm
EWTh-1 or EWTh-2	1/8" 3 mm	Point	90 to 250	DCEN DCSP	Start or Auto	Argon	20 cfh 9 L/min	1/2" 13 mm	ER308 or ER316	3/32" 1/8" 2.4-3 mm

Table 16-4 Suggested Setting for GTA Welding of Stainless Steel.

Tungsten Electrode			Welding Power			Shielding Gas		Nozzle	Filler Metal	
Type	Size	Tip	Amp	Current	HF	Type	Flow	Size	Type	Size
EWP or EWZr	1/16" 2 mm	Round	50 to 90	AC	Continues or On	Argon	17 cfh 8 L/min	7/16" 11 mm	ER1100 or ER4043	1/16" 3/32" 2-2.4 mm
EWP or EWZr	3/32" 2.4 mm	Round	80 to 130	AC	Continues or On	Argon	20 cfh 9 L/min	1/2" 13 mm	ER1100 or ER4043	1/16" 3/32" 2-2.4 mm
EWP or EWZr	1/8" 3 mm	Round	100 to 200	AC	Continues or On	Argon	20 cfh 9 L/min	5/8" 16 mm	ER1100 or ER4043	3/32" 1/8" 2.4-3 mm

Table 16-5 Suggested Setting for GTA Welding of Aluminum.

PRACTICE 16-5

Outside Corner Joint, 1G Position, Using Mild Steel, Stainless Steel, Aluminum

Using the same equipment and materials listed in Practice 16-4, weld an outside corner joint in the flat position, **Figure 16-23.**

■ Place one of the pieces of metal flat on the table and hold or brace the other piece of metal horizontally on it.

■ Tack weld both ends of the plates together, **Figure 16-24.**

■ Set the plates up and add two or three more tack welds on the joint as required, **Figure 16-25.**

■ Starting at one end, make a uniform weld, adding filler metal as needed. In **Figure 16-26,** note the metal areas that are precleaned before the weld is made.

Repeat each weld as needed until all are mastered. Turn off the welding machine, shielding gas, and cooling water and clean up your work area when you are finished welding. ◆

PRACTICE 16-6

Butt Joint, 1G Position, Using Mild Steel, Stainless Steel, Aluminum

Using the same equipment and materials as listed in Practice 16-4, you will weld a butt joint in the flat position, **Figure 16-27.** See Color Plate 22.

Place the metal flat on the table and tack weld both ends together, **Figure 16-28.** Two or three additional tack welds can be made along the joint as needed. Starting at one end, make a uniform weld along the joint. Add filler metal as required to make a uniform weld.

Repeat the process using all thicknesses of metal until you can consistently make the weld visually defect free. Turn off the welding machine, shielding gas, and cooling water and clean up your work area when you are finished welding. ◆

PRACTICE 16-7

Butt Joint, 1G Position, with 100% Penetration, To Be Tested, Using Mild Steel, Stainless Steel, Aluminum

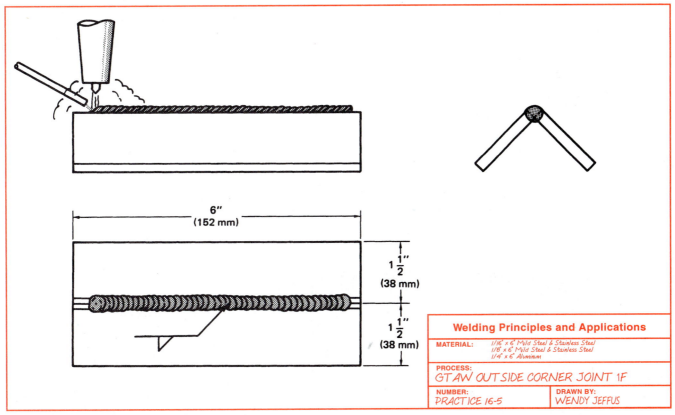

Welding Principles and Applications

MATERIAL: 1/16" x 6" Mild Steel & Stainless Steel
1/8" x 6" Mild Steel & Stainless Steel
1/4" x 6" Aluminum

PROCESS:
GTAW OUTSIDE CORNER JOINT 1F

NUMBER: PRACTICE 16-5   DRAWN BY: WENDY JEFFUS

Figure 16-23 Outside corner joint in the flat position.

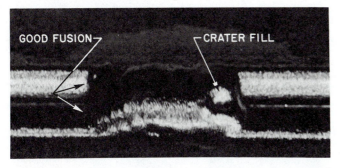

Figure 16-24 Tack weld. Note the good fusion at the start and crater fill at the end.

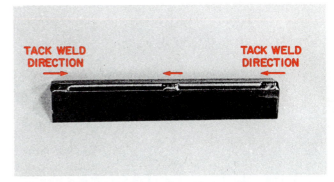

Figure 16-25 Outside corner tack welded together.

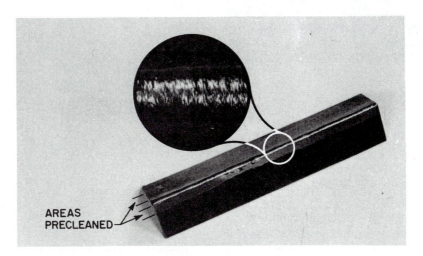

Figure 16-26 Outside corner joint. Note precleaning along weld.

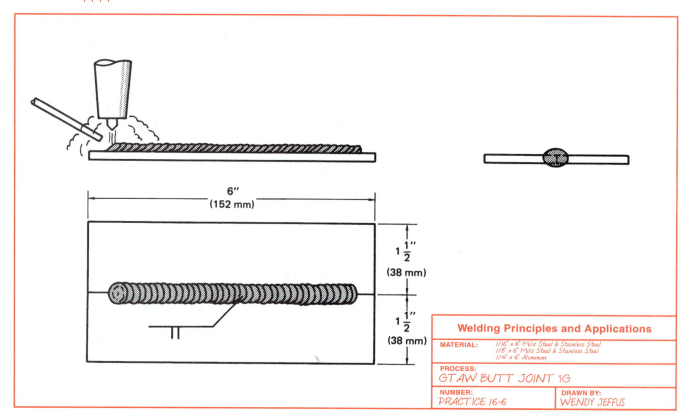

Welding Principles and Applications

MATERIAL:	1/16" x 6" Mild Steel & Stainless Steel 1/8" x 6" Mild Steel & Stainless Steel 1/4" x 6" Aluminum

PROCESS:
GTAW BUTT JOINT 1G

NUMBER:	DRAWN BY:
PRACTICE 16-6	WENDY JEFFUS

Figure 16-27 Square butt joint in the flat position.

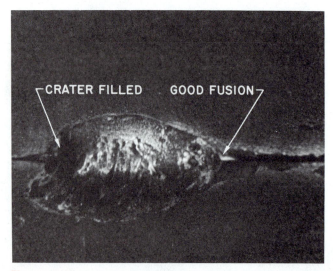

Figure 16-28 Tack weld on butt joint.

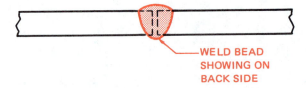

WELD BEAD SHOWING ON BACK SIDE

Figure 16-29 100% weld penetration.

Using the same equipment and materials as listed in Practice 16-4, you will weld a butt joint with 100% penetration, **Figure 16-29**, along the entire 6-in. (152-mm) length of the joint. After the weld is completed, shear out strips 1 in. (25 mm) wide and bend-test them as shown in **Figure 16-30**.

Repeat each weld until all have 100% root penetration. Turn off the welding machine, shielding gas, and cooling water and clean up your work area when you are finished welding. ◆

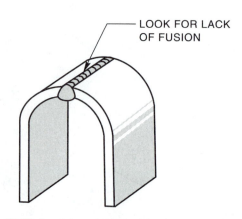

LOOK FOR LACK OF FUSION

Figure 16-30 Bend the 1-in. (25-mm) strip of butt joint backward and look at the root for 100% penetration.

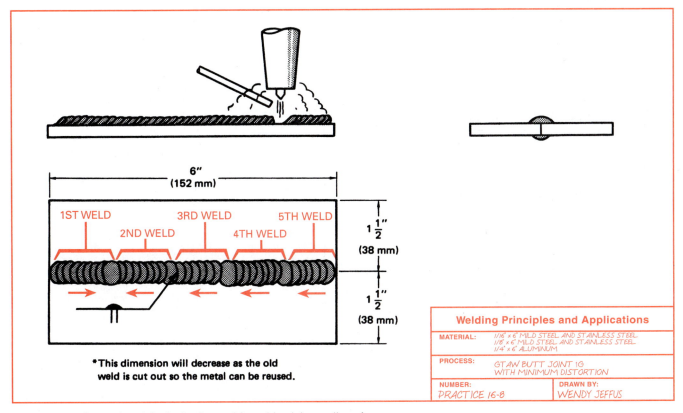

Welding Principles and Applications		
MATERIAL:	1/16" x 6" MILD STEEL AND STAINLESS STEEL 1/8" x 6" MILD STEEL AND STAINLESS STEEL 1/4" x 6" ALUMINUM	
PROCESS:	GTAW BUTT JOINT 1G WITH MINIMUM DISTORTION	
NUMBER: PRACTICE 16-8		DRAWN BY: WENDY JEFFUS

Figure 16-31 Square butt joint in the flat position with minimum distortion.

PRACTICE 16-8

Butt Joint, 1G Position, with Minimum Distortion, Using Mild Steel, Stainless Steel, Aluminum

Using the same equipment and materials as listed in Practice 16-4, you will weld a flat butt joint, while controlling both distortion and penetration, **Figure 16-31**.

Tack weld the plates together as shown in **Figure 16-32**. Using a back stepping weld sequence, make a series of welds approximately 1 in. (25 mm) long along the joint. Be sure to fill each weld crater adequately to reduce crater cracking.

Repeat the process using all thicknesses of metal until you can consistently make the weld visually defect free. Turn off the welding machine, shielding gas, and cooling water and clean up your work area when you are finished welding. ◆

PRACTICE 16-9

Lap Joint, 1F Position, Using Mild Steel, Stainless Steel, Aluminum

Using the same equipment and materials as listed in Practice 16-4, you will weld a lap joint in the flat position, **Figure 16-33**.

Place the two pieces of metal flat on the table with an overlap of 1/4 in. (6 mm) to 3/8 in. (10 mm). Hold the pieces of metal tightly together and tack weld them as shown in **Figures 16-34** and **16-35**. Starting at one end,

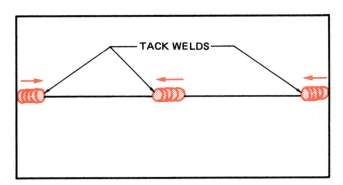

Figure 16-32 Tack welds on a butt joint.

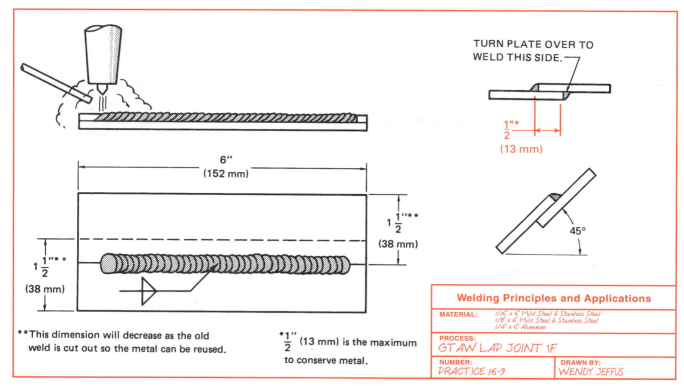

Welding Principles and Applications

MATERIAL:	1/16" x 6' Mild Steel & Stainless Steel
	1/8" x 6' Mild Steel & Stainless Steel
	1/4" x 6' Aluminum

PROCESS:
GTAW LAP JOINT 1F

| NUMBER: | DRAWN BY: |
| *PRACTICE 16-9* | *WENDY JEFFUS* |

6" (152 mm)

1 1/2 "** (38 mm)

1 1/2 "** (38 mm)

**This dimension will decrease as the old weld is cut out so the metal can be reused.

*1/2" (13 mm) is the maximum to conserve metal.

TURN PLATE OVER TO WELD THIS SIDE.

1/2"* (13 mm)

45°

Figure 16-33 Lap joint in the flat position.

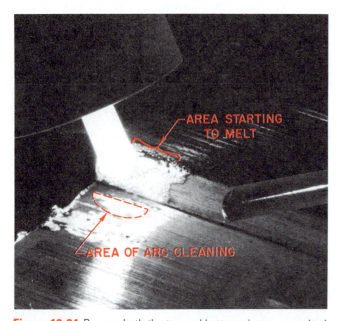

AREA STARTING TO MELT

AREA OF ARC CLEANING

Figure 16-34 Be sure both the top and bottom pieces are melted before adding filler metal.

make a uniform fillet weld along the joint. Both sides of the joint can be welded.

Repeat the process using all thicknesses of metal until you can consistently make the weld visually defect free. Turn off the welding machine, shielding gas, and cooling water and clean up your work area when you are finished welding. ◆

PRACTICE 16-10

Lap Joint, 1F Position, To Be Tested, Using Mild Steel, Stainless Steel, Aluminum

Using the same equipment and materials as listed in Practice 16-4, you will make a fillet weld on one side of a lap joint and test it for 100% root penetration, **Figures 16-36, 16-37,** and **16-38.** After the weld is completed, shear out strips 1 in. (25 mm) wide and bend test them as shown in **Figure 16-39** (page 368).

Repeat each weld until all have 100% root penetration. Turn off the welding machine, shielding gas, and cooling water and clean up your work area when you are finished welding. ◆

PRACTICE 16-11

Tee Joint, 1F Position, Using Mild Steel, Stainless Steel, Aluminum

Using the same equipment and materials as listed in Practice 16-4, you will weld a tee joint in the flat position, **Figure 16-40** (page 369). See Color Plate 24.

Place one of the pieces of metal flat on the table and hold or brace the other piece of metal horizontally on it. Tack weld both ends of the plates together, **Figure 16-41** (page 369). Set up the plates in the flat position and add two or three more tack welds to the joint as required, **Figure 16-42** (page 369).

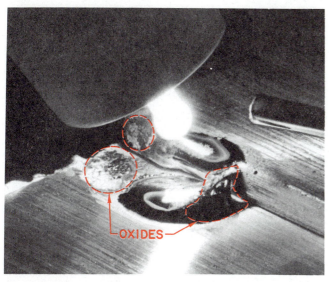

Figure 16-35 Oxides form during tack welding. Do not complete the tack welds. These oxides will become part of the finished weld if the tack is completed.

Figure 16-37 Watch the leading edge of the molten weld pool to ensure that there is complete root fusion.

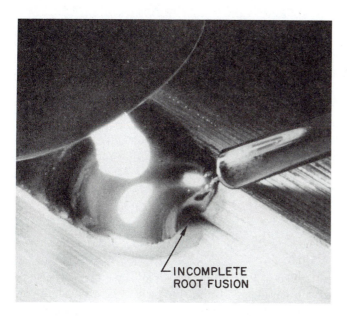

Figure 16-36 A notch indicates that the root was not melted and fused properly.

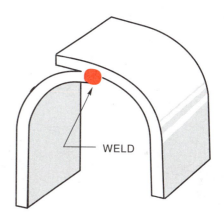

Figure 16-38 Bend the 1-in. (25-mm) strip of lap joint backward and look at the root for 100% penetration.

On the metal that is 1/16 in. (1.5 mm) thick, it may not be possible to weld both sides, but on thicker material a fillet weld can usually be made on both sides. The exception to this is if carbide precipitation occurs on the stainless steel during welding.

Starting at one end, make a uniform weld, adding filler metal as needed.

Repeat the process using all thicknesses of metal until you can consistently make the weld visually defect free. Turn off the welding machine, shielding gas, and cooling water and clean up your work area when you are finished welding. ◆

PRACTICE 16-12

Tee Joint, 1F Position, To Be Tested, Using Mild Steel, Stainless Steel, Aluminum

Using the same equipment and materials as listed in Practice 15-4, you will weld a tee joint and test it for 100% root penetration, **Figure 16-43** (page 369). After the weld is completed, cut or shear out strips 1 in. (25 mm) wide and bend test them as shown in **Figures 16-44** (page 370) and **16-45** (page 370).

Repeat each weld until all have 100% root penetration. Turn off the welding machine, shielding gas, and cooling water and clean up your work area when you are finished welding. ◆

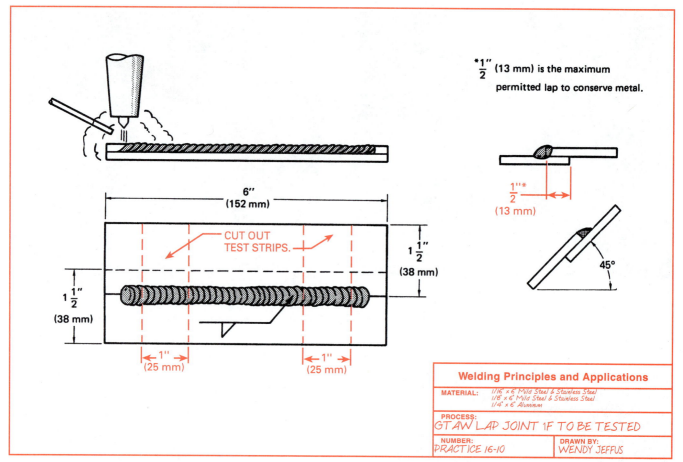

½" (13 mm) is the maximum permitted lap to conserve metal.

6" (152 mm)

CUT OUT TEST STRIPS.

1½" (38 mm)

1½" (38 mm)

1" (25 mm)

1" (25 mm)

½"* (13 mm)

45°

Welding Principles and Applications

MATERIAL:	1/16" x 6" Mild Steel & Stainless Steel 1/8" x 6" Mild Steel & Stainless Steel 1/4" x 6" Aluminum
PROCESS:	GTAW LAP JOINT 1F TO BE TESTED
NUMBER: PRACTICE 16-10	DRAWN BY: WENDY JEFFUS

Figure 16-39 Remove strips and test for root fusion.

PRACTICE 16-13

Stringer Bead at a 45° Vertical Angle, Using Mild Steel, Stainless Steel, Aluminum

Using the same equipment and materials listed in Practice 16-4, you will make a stringer bead in the vertical up position.

■ Starting at the bottom and welding in an upward direction, add the filler metal at the top edge of the weld pool and move the torch in a circle or "C" pattern, **Figure 16-46** (page 371). If the weld pool size starts to increase, the "C" pattern can be increased in length or the power can be decreased.

■ Watch the weld pool and establish a rhythm of torch movement and addition of rod to keep the weld uniform.

Repeat the process using all thicknesses of metal until you can consistently make the weld visually defect free. Turn off the welding machine, shielding gas, and cooling water and clean up your work area when you are finished welding. ◆

PRACTICE 16-14

Stringer Bead, 3G Position, Using Mild Steel, Stainless Steel, Aluminum

Repeat Practice 16-13. Gradually increase the angle as you develop skill until the weld is being made in the vertical up position, **Figure 16-47** (page 371). Repeat the process using all thicknesses of metal until you can consistently make the weld visually defect free. Turn off the welding machine, shielding gas, and cooling water and clean up your work area when you are finished welding. ◆

PRACTICE 16-15

Butt Joint at a 45° Vertical Angle, Using Mild Steel, Stainless Steel, Aluminum

Using the same equipment and materials as listed in Practice 16-4, you will weld a butt joint in the vertical up position.

After tack welding the plates together, start the weld at the bottom and weld in an upward direction.

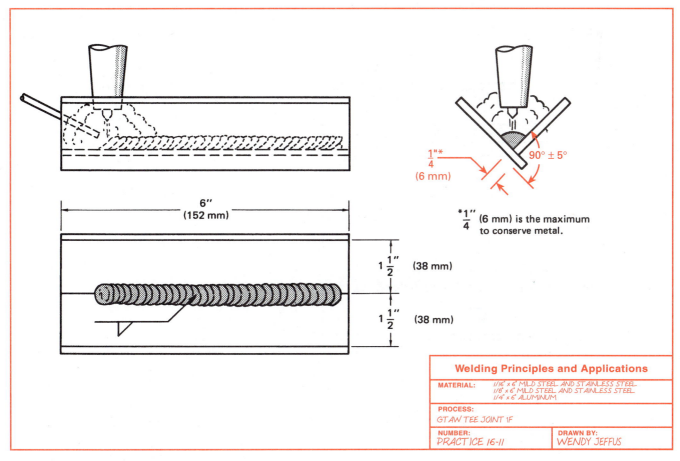

1"*
—
4
(6 mm)

90° ± 5°

*1" (6 mm) is the maximum
—
4
to conserve metal.

6"
(152 mm)

1 1/2" (38 mm)

1 1/2" (38 mm)

Welding Principles and Applications

MATERIAL:	1/16" x 6" MILD STEEL AND STAINLESS STEEL
	1/8" x 6" MILD STEEL AND STAINLESS STEEL
	1/4" x 6" ALUMINUM
PROCESS:	
	GTAW TEE JOINT 1F
NUMBER:	DRAWN BY:
PRACTICE 16-11	WENDY JEFFUS

Figure 16-40 Tee joint in the flat position.

Figure 16-41 Tack weld on a tee joint.

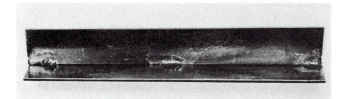

Figure 16-42 Keep the tack welds small so that they will not affect the weld.

The same rhythmic torch and rod movement practiced for the 45° stringer bead should be used to control the weld.

Repeat the process using all thicknesses of metal until you can consistently make the weld visually defect free. Turn off the welding machine, shielding gas, and cooling water and clean up your work area when you are finished welding. ◆

NOTE THE NEARLY FLAT FACE
OF A GOOD FILLET WELD.

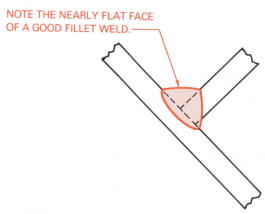

Figure 16-43 100% weld penetration.

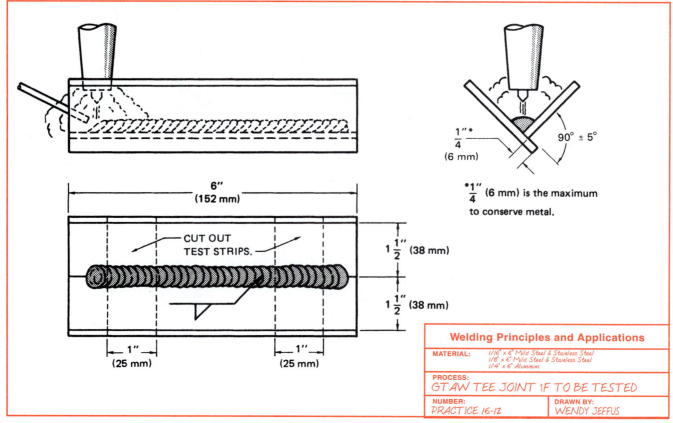

MATERIAL:
1/16" x 6" Mild Steel & Stainless Steel
1/8" x 6" Mild Steel & Stainless Steel
1/4" x 6" Aluminum

PROCESS:
GTAW TEE JOINT 1F TO BE TESTED

NUMBER: PRACTICE 16-12 | **DRAWN BY:** WENDY JEFFUS

Welding Principles and Applications

Figure 16-44 Tee joint in the flat position to be tested.

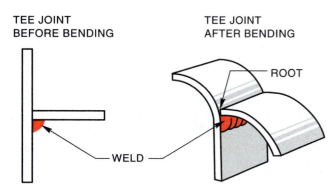

TEE JOINT BEFORE BENDING

TEE JOINT AFTER BENDING

ROOT

WELD

Figure 16-45 Bend the 1-in. (25-mm) strip of the tee joint backward and look at the root for 100% penetration.

PRACTICE 16-16

Butt Joint, 3G Position, Using Mild Steel, Stainless Steel, Aluminum

Repeat Practice 16-15. Gradually increase the plate angle after each weld. As you develop skill, continue increasing the angle until the weld is being made in the vertical up position. Repeat the process using all thicknesses of metal until you can consistently make the weld visually defect free. Turn off the welding machine,

shielding gas, and cooling water and clean up your work area when you are finished welding. ◆

PRACTICE 16-17

Butt Joint, 3G Position, with 100% Penetration, To Be Tested, Using Mild Steel, Stainless Steel, Aluminum

Repeat Practice 16-16. Make the needed changes in the root opening to allow 100% penetration, Figure 16-48 (page 372). It may be necessary to provide a backing gas to protect the root from atmospheric contamination. After the weld is completed, visually inspect it for uniformity and defects. Then shear out strips 1 in. (25 mm) wide and bend test them.

Repeat each weld until all have 100% root penetration. Turn off the welding machine, shielding gas, and cooling water and clean up your work area when you are finished welding. ◆

PRACTICE 16-18

Lap Joint at a 45° Vertical Angle, Using Mild Steel, Stainless Steel, Aluminum

Using the same equipment and materials as listed in Practice 16-4, you will make a vertical up fillet weld on a lap joint.

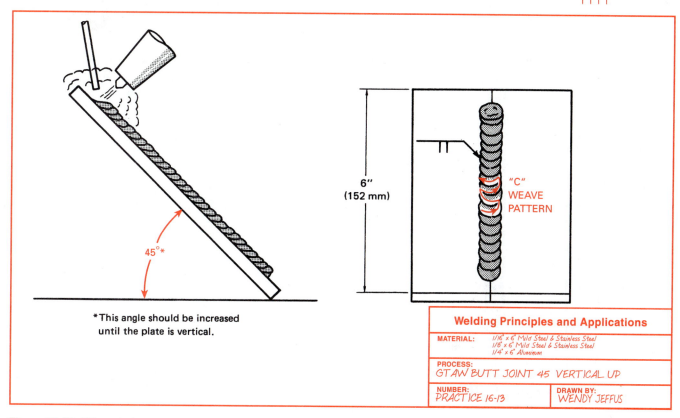

Figure 16-46 45° vertical up.

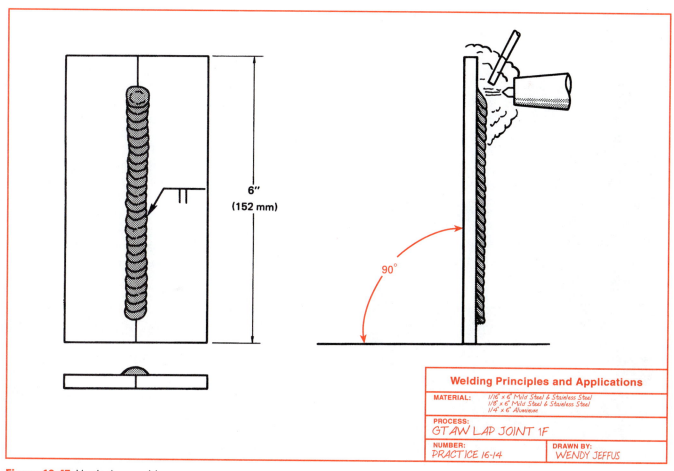

Figure 16-47 Vertical up position.

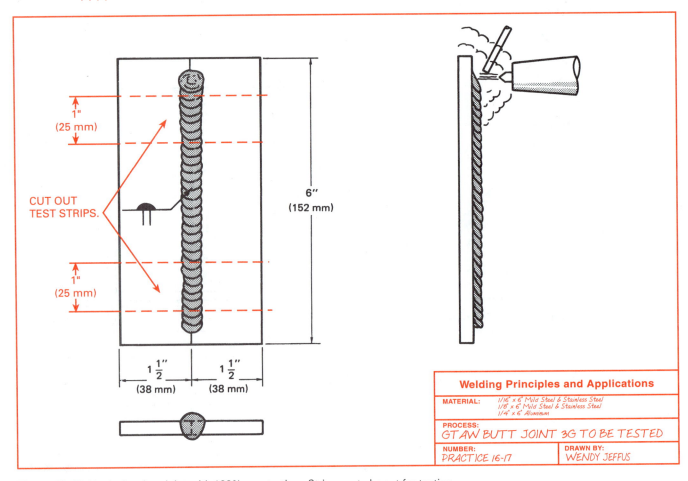

Figure 16-48 Vertical up butt joint with 100% penetration. Strips are to be cut for testing.

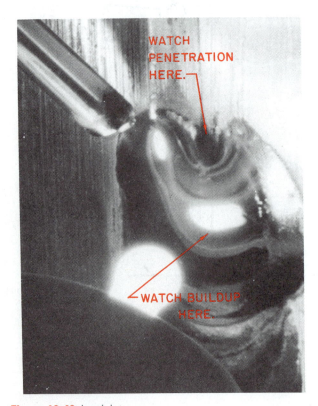

Figure 16-49 Lap joint.

After tack welding the plates together, start the weld at the bottom and weld in an upward direction. It is important to maintain a uniform weld rhythm so that a nice-looking weld bead is formed. It may be necessary to move the torch in and around the base of the weld pool to ensure adequate root fusion, Figure 16-49. The filler metal should be added along the top edge of the weld pool near the top plate.

Repeat the process using all thicknesses of metal until you can consistently make the weld visually defect free. Turn off the welding machine, shielding gas, and cooling water and clean up your work area when you are finished welding. ◆

PRACTICE 16-19

Lap Joint, 3F Position, Using Mild Steel, Stainless Steel, Aluminum

Repeat Practice 16-18. Gradually increase the plate angle after each weld as you develop your skill. Increase the angle until the weld is being made in the vertical up position, Figure 16-50.

Repeat the process using all thicknesses of metal until you can consistently make the weld visually defect

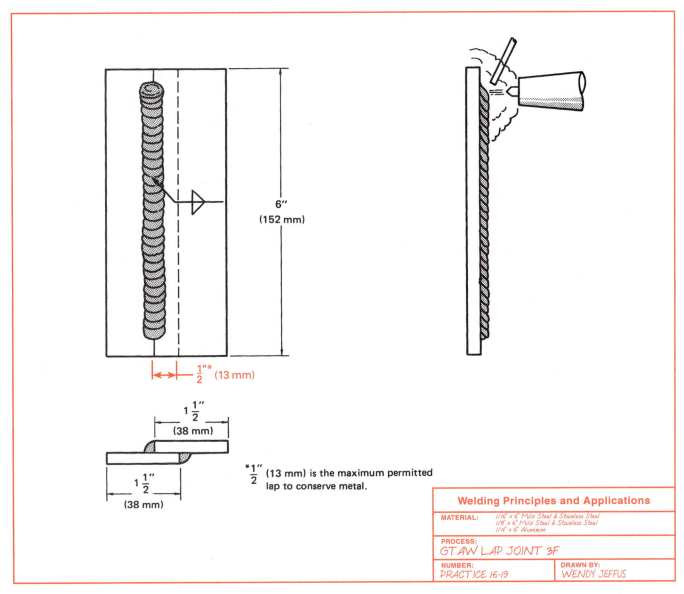

Welding Principles and Applications	
MATERIAL:	1/16" x 6" Mild Steel & Stainless Steel 1/8" x 6" Mild Steel & Stainless Steel 1/4" x 6" Aluminum
PROCESS: GTAW LAP JOINT 3F	
NUMBER: PRACTICE 16-19	DRAWN BY: WENDY JEFFUS

Figure 16-50 Vertical up lap joint.

free. Turn off the welding machine, shielding gas, and cooling water and clean up your work area when you are finished welding. ◆

PRACTICE 16-20

Lap Joint, 3F Position, with 100% Root Penetration, To Be Tested, Using Mild Steel, Stainless Steel, Aluminum

Repeat Practice 16-19 and make the fillet weld in a lap joint with 100% root penetration. After the weld is completed, visually inspect it for uniformity and defects before shearing out strips 1 in. (25 mm) wide and bend test them.

Repeat each weld until all have 100% root penetration. Turn off the welding machine, shielding gas, and cooling water and clean up your work area when you are finished welding. ◆

PRACTICE 16-21

Tee Joint at a 45° Vertical Angle, Using Mild Steel, Stainless Steel, Aluminum

Using the same equipment and materials listed in Practice 16-4, you will make a vertical up fillet weld on a tee joint.

■ After tack welding the plates together, start the weld at the bottom and weld in an upward direction. The edge of the side plate, **Figure 16-51**, will heat up more quickly than the back plate. This rapid heating often leads to undercutting along this edge of the weld.

■ To control undercutting, keep the arc on the back plate and add the filler metal to the weld pool near the side plate.

Repeat the process using all thicknesses of metal until you can consistently make the weld visually defect

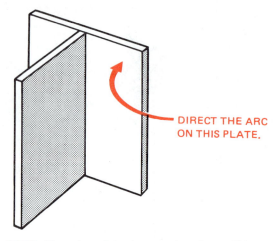

Figure 16-51   The edge of the intersecting plate will heat up faster than the base plate if the heat is not directed away from it.

DIRECT THE ARC ON THIS PLATE.

free. Turn off the welding machine, shielding gas, and cooling water and clean up your work area when you are finished welding. ◆

PRACTICE 16-22

Tee Joint, 3F Position, Using Mild Steel, Stainless Steel, Aluminum

Repeat Practice 16-21. Gradually increase the plate angle after each weld as you develop your skill. Increase the angle until the weld is being made in the vertical up position.

Repeat the process using all thicknesses of metal until you can consistently make the weld visually defect free. Turn off the welding machine, shielding gas, and cooling water and clean up your work area when you are finished welding. ◆

PRACTICE 16-23

Tee Joint, 3F Position, with 100% Root Penetration, To Be Tested, Using Mild Steel, Stainless Steel, Aluminum

Repeat Practice 16-21 and make the fillet weld in a lap joint with 100% root penetration. After the weld is completed, visually inspect it for uniformity and defects before shearing out strips 1 inch (25 mm) wide and bend test them.

Repeat each weld until all have 100% root penetration. Turn off the welding machine, shielding gas, and cooling water and clean up your work area when you are finished welding. ◆

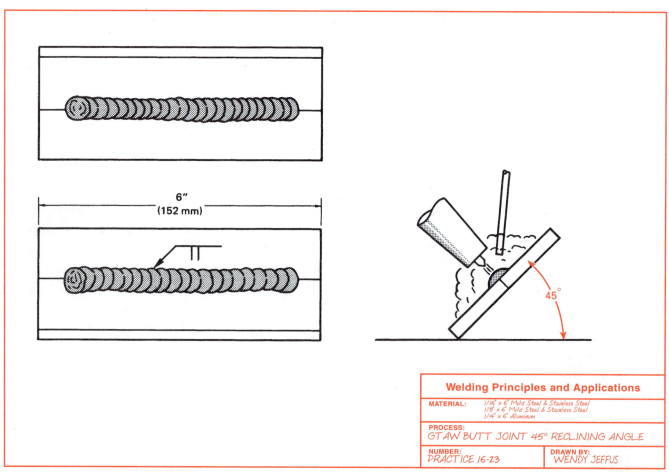

Welding Principles and Applications	
MATERIAL:	1/16" x 6" Mild Steel & Stainless Steel 1/8" x 6" Mild Steel & Stainless Steel 1/4" x 6" Aluminum
PROCESS: GTAW BUTT JOINT 45° RECLINING ANGLE	
NUMBER: PRACTICE 16-23	DRAWN BY: WENDY JEFFUS

Figure 16-52 45° reclining angle.

Stringer Bead at a 45° Reclining Angle, Using Mild Steel, Stainless Steel, Aluminum

Using the same equipment and materials as listed in Practice 16-4, you will make a weld bead on a plate at a 45° reclining angle, **Figure 16-52**. Add the filler metal along the top leading edge of the weld pool. Surface tension will help hold the weld pool on the top if the bead is not too large. The weld should be uniform in width and reinforcement.

Repeat the process using all thicknesses of metal until you can consistently make the weld visually defect free. Turn off the welding machine, shielding gas, and cooling water and clean up your work area when you are finished welding. ◆

Stringer Bead, 2G Position, Using Mild Steel, Stainless Steel, Aluminum

Repeat Practice 16-24. Gradually increase the plate angle as you develop your skill until the weld is being made in the horizontal position on a vertical plate.

Repeat the process using all thicknesses of metal until you can consistently make the weld visually defect free. Turn off the welding machine, shielding gas, and cooling water and clean up your work area when you are finished welding. ◆

Butt Joint, 2G Position, Using Mild Steel, Stainless Steel, Aluminum

Using the same equipment and materials as listed in Practice 16-4, you will weld a butt joint in the horizontal position.

The welding techniques are the same as those used in Practice 16-25. Add the filler metal to the top plate, and keep the bead size small so it will be uniform.

Repeat the process using all thicknesses of metal until you can consistently make the weld visually defect free. Turn off the welding machine, shielding gas, and cooling water and clean up your work area when you are finished welding. ◆

Butt Joint, 2G Position, with 100% Penetration, To Be Tested, Using Mild Steel, Stainless Steel, Aluminum

Repeat Practice 16-26. It may be necessary to increase the root opening to ensure 100% penetration. A backing gas may be required to prevent atmospheric contamination. Using a "J" weave pattern will help to maintain a uniform weld bead. After the weld is completed, visually inspect it for uniformity and defects. Then shear out strips 1 in. (25 mm) wide and bend test them.

Repeat each weld until all have 100% root penetration. Turn off the welding machine, shielding gas, and cooling water and clean up your work area when you are finished welding. ◆

Lap Joint, 2F Position, Using Mild Steel, Stainless Steel, Aluminum

Using the same equipment and materials as listed in Practice 16-4, make a horizontal fillet weld on a lap joint.

■ After tack welding the plates together, start the weld at one end. The bottom plate will act as a shelf to support the molten weld pool, **Figure 16-53**.

■ Add the filler metal along the top edge of the weld pool to help control undercutting.

Repeat the process using all thicknesses of metal until you can consistently make the weld visually defect free. Turn off the welding machine, shielding gas, and cooling water and clean up your work area when you are finished welding. ◆

Lap Joint, 2F Position, with 100% Root Penetration, To Be Tested, Using Mild Steel, Stainless Steel, Aluminum

Repeat Practice 16-28. Be sure the weld is penetrating the root 100%. After the weld is completed, visually inspect it for uniformity and defects before shearing out strips 1 in. (25 mm) wide and bend test them.

Repeat each weld until all have 100% root penetration. Turn off the welding machine, shielding gas, and

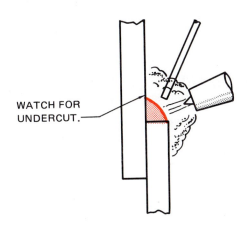

WATCH FOR UNDERCUT.

Figure 16-53 Horizontal lap joint.

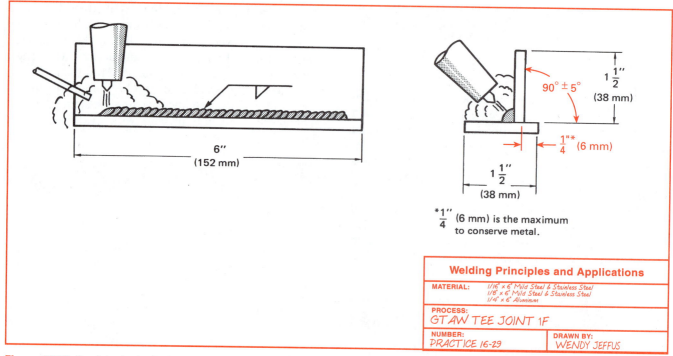

1/4" (6 mm) is the maximum to conserve metal.

Welding Principles and Applications

MATERIAL: 1/16" x 6" Mild Steel & Stainless Steel
1/8" x 6" Mild Steel & Stainless Steel
1/4" x 6" Aluminum

PROCESS: GTAW TEE JOINT 1F

NUMBER: PRACTICE 16-29 | DRAWN BY: WENDY JEFFUS

Figure 16-54 Tee joint in the horizontal position.

cooling water and clean up your work area when you are finished welding. ◆

PRACTICE 16-30

Tee Joint, 2F Position, Using Mild Steel, Stainless Steel, Aluminum

Using the same equipment and materials as listed in Practice 16-4, you will make a horizontal fillet weld on a tee joint. See Color Plates 25, 26, and 27.

After tack welding the plates together, start the weld at one end. The bottom plate will act as a shelf to support the molten weld pool, **Figure 16-54**. As with the horizontal lap joint, add the filler metal along the top leading edge of the weld pool. This will help control undercut.

Repeat the process using all thicknesses of metal until you can consistently make the weld visually defect free. Turn off the welding machine, shielding gas, and cooling water and clean up your work area when you are finished welding. ◆

PRACTICE 16-31

Tee Joint, 2F Position, with 100% Root Penetration, To Be Tested, Using Mild Steel, Stainless Steel, Aluminum

Repeat Practice 16-30. Be sure the weld is penetrating the root 100%. After the weld is completed, visually inspect it for uniformity and defects before shearing out strips 1 in. (25 mm) wide and bend test them.

Repeat each weld until all have 100% root penetration. Turn off the welding machine, shielding gas, and cooling water and clean up your work area when you are finished welding. ◆

PRACTICE 16-32

Stringer Bead, 4G Position, Using Mild Steel, Stainless Steel, Aluminum

Using the same equipment and materials as listed in Practice 16-4, you will make a weld bead on a plate in the overhead position.

The surface tension on the molten metal will hold the welding bead on the plate providing that it is not too large. Add the filler metal along the leading edge of the weld pool, **Figure 16-55**. A wide weld with little buildup will be easier to control and less likely to undercut along the edge.

Repeat the process using all thicknesses of metal until you can consistently make the weld visually defect free. Turn off the welding machine, shielding gas, and cooling water and clean up your work area when you are finished welding. ◆

PRACTICE 16-33

Butt Joint, 4G Position, Using Mild Steel, Stainless Steel, Aluminum

Using the same equipment and materials as listed in Practice 16-4, you will weld a butt joint in the overhead position.

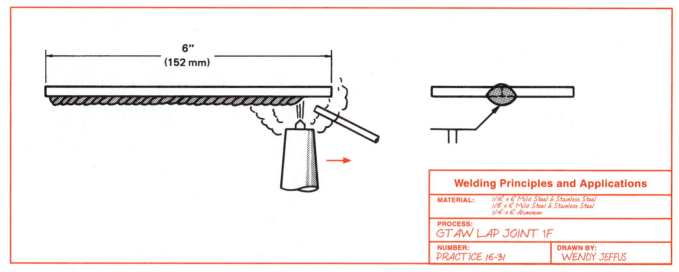

6"
(152 mm)

Welding Principles and Applications	
MATERIAL:	1/16" x 6" Mild Steel & Stainless Steel 1/8" x 6" Mild Steel & Stainless Steel 1/4" x 6" Aluminum
PROCESS: GTAW LAP JOINT 1F	
NUMBER: PRACTICE 16-31	**DRAWN BY:** WENDY JEFFUS

Figure 16-55 Overhead stringer bead.

The same techniques used to make the stringer beads in Practice 16-32 are also used with the butt joint. The size of the bead should be kept small enough so that you can control the weld. The completed weld should be uniform and free from defects.

Repeat the process using all thicknesses of metal until you can consistently make the weld visually defect free. Turn off the welding machine, shielding gas, and cooling water and clean up your work area when you are finished welding. ◆

PRACTICE 16-34

Lap Joint, 4F Position, Using Mild Steel, Stainless Steel, Aluminum

Using the same equipment and materials as listed in Practice 16-4, you will make a fillet weld on a lap joint in the overhead position.

The major concentration of heat and filler metal should be on the top plate. Gravity and an occasional sweep of the torch along the bottom plate will pull the weld pool down. Undercutting along the top edge of the weld can be controlled by putting most of the filler metal along the top edge. The completed weld should be uniform and free from defects.

Repeat the process using all thicknesses of metal until you can consistently make the weld visually defect free. Turn off the welding machine, shielding gas, and cooling water and clean up your work area when you are finished welding. ◆

PRACTICE 16-35

Tee Joint, 4F Position, Using Mild Steel, Stainless Steel, Aluminum

Using the same equipment and materials as listed in Practice 16-4, you will make a fillet weld on a tee joint in the overhead position.

The same techniques used to make the overhead lap weld in Practice 16-34 are used with the tee joint. As with the lap joint, most of the heat and filler metal should be concentrated on the top plate. A "J" weave pattern will help pull down any needed metal to the side plate. The completed weld should be uniform and free from defects.

Repeat the process using all thicknesses of metal until you can consistently make the weld visually defect free. Turn off the welding machine, shielding gas, and cooling water and clean up your work area when you are finished welding. ◆

Robotic Arm Welding

Figure 1

Faced with stiff competition in a fast-changing market, one company emphasizes quality and productivity. An innovator in the automative stamping and casting industry, the future growth of this company depends on its ability to reduce costs and increase productivity in the manufacturing process.

Combining automation with special employee training and motivational programs, the idea of humans and machines working together has taken on a new dimension in the plant.

Management recognized the potential of upgrading plant automation. Arc welding robots were selected to yield the greatest improvement in quality and production costs, **Figure 1**. The robots would also give the company great flexibility and versatility in bidding on a wide range of applications — without the expense of retooling for each job.

The robot arc welding system selected has had eight years' experience in plants with all types of applications. It has proven reliability, dependability, and accuracy.

The robot is designed specifically for arc welding. It is a computer-controlled, electric, servo-driven robot with repeatability of .008 inch. It is fully programmable, with five axes of move-

ment and an exceptionally large working range. Teach/programming with a portable teach pendant is simple and easy to learn and operate with a minimum number of instructions and steps required per program. Programs can be stored and selected for each job-reducing set-up and make-ready.

Management selected an automative brake pedal for the robot's initial application, **Figure 2**. Soon the robot showed savings in labor, scrap, and

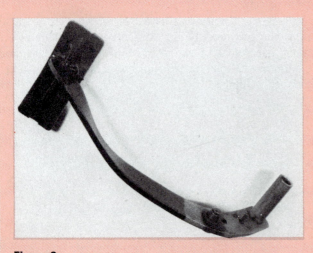

Figure 2

WELDING SYSTEM LAYOUT

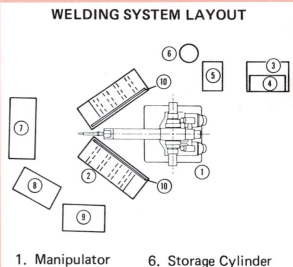

1. Manipulator	6. Storage Cylinder
2. Positioners	7. Parts Supply Bin
3. Controller	8. Parts Supply Bin
4. Interface Unit	9. Finished Parts Bin
5. Power Source	10. Protection Screen

Figure 3

rework pieces, minimizing inspection time and maintenance costs.

The work cell, **Figure 3**, consists of one robot, two stationary table positioners, and one operator for loading and unloading the workpieces. The fix-turing equipment was designed to handle different types and models of brake pedals for quick and easy change-over.

(Adapted from information provided by Hobart Brothers Company, Troy, Ohio 45373.)

REVIEW

1. What effect does torch angle have on the shielding gas protective zone?

2. Why must the end of the filler rod be kept in the shielding gas protective zone?

3. What can cause tungsten contamination?

4. What determines the correct current setting for a GTA weld?

5. What is the lowest acceptable amperage setting for GTA welding?

6. List the factors that affect the gas flow setting for GTA welding.

7. When should the minimum gas flow rates be increased?

8. What is the minimum gas flow rate for a nozzle size?

9. What is the maximum gas flow rate for a nozzle size?

10. Which incorrect welding parameters does stainless steel show clearly?

11. Using Table 16-1, determine the approximate temperature of metal that formed a dark blue color.

12. Using Table 16-2, list the filler metals for the following metals.

    a. 1020 low-carbon steel

    b. 308 stainless steel

13. Why is it possible to control a large aluminum weld bead?

14. What may happen to the end of the aluminum welding rod if it is held too close to the arc?

15. What should be done if someone comes in contact with a cleaning chemical?

16. Using Table 16-3, determine the suggested setting for GTA welding of mild steel using a 3/32-in. (2.4-mm) tungsten.

17. What can be done to limit oxide formation on stainless steel?

18. How should the filler metal be added to the molten weld pool?

19. How can the rod be freed if it sticks to the plate?

20. How is an outside corner joint assembled?

21. What must be done with the weld craters when back stepping a weld? Why?

22. How is the lap joint tested for 100% penetration?

23. What can prevent both sides of a stainless steel tee joint from being welded?

24. How is the filler metal added for a 3F weld?

25. What can cause undercutting on a 3F tee joint?

26. What helps hold the weld in place on a 2F lap joint?

27. What helps hold the weld in place on a 4G weld?

Chapter 17

GAS TUNGSTEN ARC WELDING OF PIPE

OBJECTIVES

After completing this chapter, the student should be able to

- describe how a pipe joint is prepared for welding.
- list the four most common root defects and the causes of each defect.
- discuss when and why a backing gas is used.
- explain the uses of a hot pass.
- sketch a single V-groove and indicate the location and sequence of welds for each position.
- make a single V-groove butt welded joint on a pipe in any position.

KEY TERMS

root	grapes
root penetration	stress point
root reinforcement	backing gas
incomplete fusion	consumable inserts
concavity	

INTRODUCTION

Gas tungsten arc welding of pipe is used when the welded joint must have a high degree of integrity. The welded joint must be strong and free from defects that may cause the premature failure of the welded joint. Many industries, such as oil, gas, nuclear, chemical, and several others, require this kind of joint to be made, Figure 17-1. Failure in a pipe joint can be disastrous in addition to being extremely expensive to repair.

GTA welding gives industry the type of joint it needs. Welders who are skilled in the GTA welding process have the ability to make consistently high-quality welds with a low rejection rate.

GTA pipe welders are among the highest paid workers in the welding industry. The positions available are among the safest and most prestigious and have excellent working conditions. Preparing for such a job requires the development of high levels of skill and technical knowledge through much practice and study.

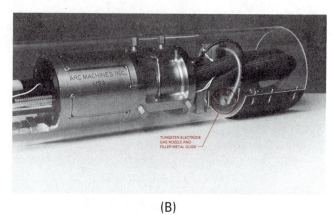

(A)   (B)

Figure 17-1 (A) Automatic GTA outside pipe welding machine, (B) Automatic GTA inside pipe welding machine shown inside a clear plastic pipe. This unit with its remote vision system can work inside pipe as small as 7.189 in. (Courtesy of ARC Machines, Inc.)

PRACTICES

The practices in this chapter are all performed on mild steel. Mild steel is the most readily available material on which to learn. The skill and techniques learned in these practices can be easily transferred to any other piping material. In most cases, the only change is in the composition of the filler material and possibly the current.

JOINT PREPARATION

The end of a pipe must be prepared to ensure a sound weld. The type of groove used in preparing the ends of a pipe for welding will vary depending upon the pipe material, thickness, and application. The most often used grooves on mild steel are the single V-, single U-, single J-, and single bevel-grooves, Figure 17-2. Both the single bevel-groove and single V-groove can be easily

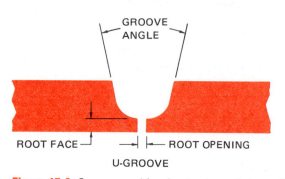
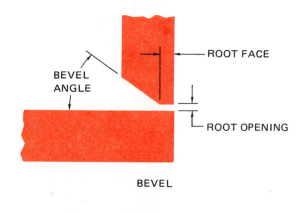
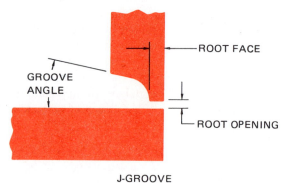

Figure 17-2 Grooves used for pipe-to-pipe and pipe-to-flange welds.

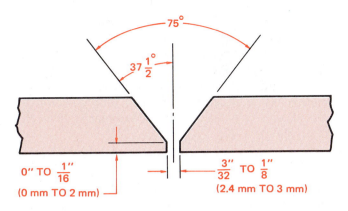

Figure 17-3 Typical V-groove preparation dimensions.

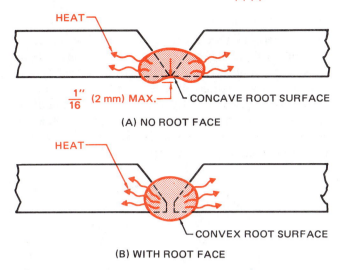

Figure 17-4 (A) This joint cannot conduct enough heat away from the root pass to provide support for the molten root pass. Note the concave surface of the root face caused by suck back. (B) The larger mass of metal helps to cool the puddle quickly and provides for sufficient surface tension, which also helps control the root surface.

flame cut or ground. This factor makes them the most frequently used grooves during training.

The end of the pipe is prepared with a 37 1/2° bevel, leaving a root face of 1/16 in. (2 mm) to 1/8 in. (3 mm), Figure 17-3. When both pipes are prepared in this manner, they form a 75° single V-groove. The groove angle allows good visibility of the weld with a minimum amount of filler metal required to fill the groove. The root face removes the sharp root edge, commonly referred to as a "feather edge." Failure to remove the root edge will result in the weld burning back the material, leaving a key hole that is larger than needed. This also may result in a concave root surface, Figure 17-4.

Before the joint is assembled, the welding surfaces must be cleaned and smoothed so that they are uniform and free of contaminants. To remove any possible sources of contamination to the weld clean a 1-in. (25-mm) wide or wider band both inside and outside of the pipe. The band can be cleaned by grinding, brushing, or filing, Figure 17-5. The prepared edges must also be clean and free from tears, cracks, slivers, or any other defects.

The pipe should be tack welded together with a root opening of 3/32 in. (2.4 mm) to 1/8 in. (3 mm). Four or more tack welds evenly spaced around the pipe should be made. The tack welds should be located so that the root bead will not start against them, Figure 17-6.

ROOT

The root of a weld is the deepest point into the joint where fusion between the base metal and filler metal occurs. The root penetration is the distance measured between the original surface of the joint and the deepest point of fusion. The original surface of the joint may be different from that of the original surface of the metal if the joint was prepared for welding with a groove. The root reinforcement is the amount of metal deposited on the back side of a welded joint, Figure 17-7.

The depth to which a weld must penetrate in a pipe joint, or the amount of root reinforcement, is given in the code or standards being used. Most codes or standards for GTA pipe welding require 100% root penetra-

Figure 17-5 The groove must be cleaned, both outside and inside the pipe joint.

tion and no more than 1/8 in. (3 mm) of root reinforcement. Uniformity and the lack of defects in a weld are also important. The four most common root defects are incomplete fusion, concave root surface (suck back), excessive root reinforcement, and root contamination.

Incomplete Fusion

Sometimes a weld does not completely penetrate the joint, or there may be a lack of fusion on one or both sides of the root. These conditions are caused by not enough heat and temperature reaching the back side of the work, Figure 17-8. The problem with incomplete fusion is that it can form a stress point. Stress points can result in the premature cracking or failure of the weld at a load well under its expected strength.

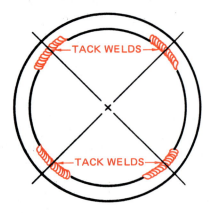

Figure 17-6 Location of tack welds on small-diameter pipe.

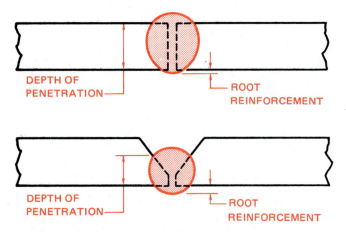

Figure 17-7 Root penetration and reinforcement.

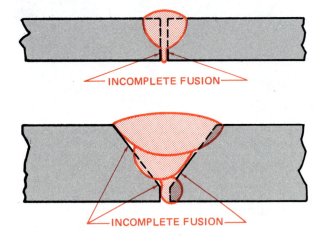

Figure 17-8 Lack of fusion.

Incomplete fusion can be observed if the weld is subjected to a root bend test. A root bend subjects a welded joint to a bending force so that the root of the weld is in tension, Figure 17-9. One test that can be used is a standard test procedure (guided bend or free bend). Another test that can be used is to simply secure the welded joint in a vise and use a hammer to bend the metal, Figure 17-10.

Concave Root Surface

A concave root surface (suck back) occurs when the back side of the root weld is concave in shape, Figure 17-4(A). Common causes for this condition are when insufficient filler metal is added to the joint, or excessive heat is used in the overhead position. The concavity of the root surface results in reduced thickness (actual throat) of the weld, which in turn causes a lower joint strength.

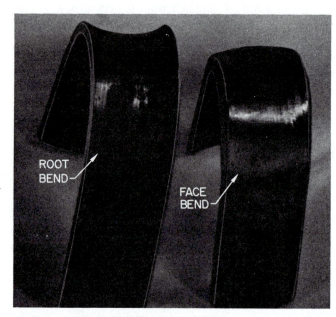

Figure 17-9 Guided bend test specimens.

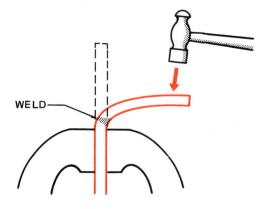

Figure 17-10 Bend test using a hammer and vise.

Concave root surface can be visually inspected (VT) readily without destructive testing. The amount of concavity can be given in the following ways:

■ as a percentage of the total weld length,

■ as the average number and largest size of spots per inch (mm) of weld, or

■ as a percentage of the effective throat. Normally, 1/16 in. (2 mm) is the maximum allowable root concavity.

Excessive Root Reinforcement

Excessive root reinforcement, also known as grapes (or mistakenly as burnthrough), is the excessive buildup of metal on the back side of a weld. This condition results from excessive heat, temperature, and filler metal during welding. Excessive root reinforcement can cause reduced material flow, result in clogged pipes, or form stress points that will result in premature weld failure.

The root of the weld should have some reinforcement (buildup), but this should not exceed 1/8 in. (3 mm). Visual inspection (VT) can be used to verify this condition. The amount of excessive reinforcement is given in the following ways:

■ as a percentage of the total weld length,

■ as the average number and largest size of spots per inch (mm) of weld, or

■ as the maximum distance from the metal surface to the longest spot of excessive reinforcement.

Root Contamination

The back side of the molten weld pool can be contaminated, causing porosity, embrittlement, oxide inclusions, and/or oxide layers. Contamination is caused by overheating, poor cleaning procedures, or improper protection from the surrounding atmosphere, **Figure 17-11**. Root contamination can lead to faster corrosion, oxide flaking, weld brittleness, leaks, stress points, or all of these.

Root surface contamination can be visually inspected. Heavy flakes or large deposits of oxides are signs of root surface contamination. Ideally, the color of the root should be close to that of the base metal. Internal porosity, oxides, or embrittlement can be detected readily by a root bend test. The weld deposit should be as ductile (easily bent) as the surrounding base metal.

BACKING GAS

Root contamination caused by the surrounding atmosphere is not a major problem when welding on mild steel pipe. Some type of protection, however, is needed when welding on low-alloy steel, stainless steel, aluminum, copper, and most other types of pipe. The

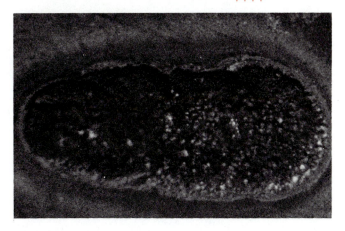

Figure 17-11 Root contamination.

easiest method of protecting the root from atmospheric contamination is to use a backing gas.

The type of gas used for backing will depend upon the type of pipe being welded. Nitrogen and CO_2 are often acceptable, depending upon the code or intended use of the pipe system. Argon, although expensive, can be used to back any type of pipe if a welder is unsure about a less expensive substitution.

There are several methods in common use for containing the backing gas in the pipe. On small diameters or short sections of pipe, the ends of the pipe are capped, **Figure 17-12**. The gas is allowed to purge the complete pipe section. This method requires too much purging time and gas to be practical on large-diameter pipe. For larger diameters, the pipe is plugged on both sides of the joint to be welded so that a smaller area can be purged. If the piping system is complex, consisting of valves and numerous turns, water-soluble plugs or soft plastic gas bags are suggested. They can be blown out with air or water when the system is completed.

When a backing gas is used, the gas must have enough time to purge the pipe completely. When welding on large-diameter pipes or long pipe sections, it is necessary to estimate the length of time required to purge the air. First, estimate the interior volume of the pipe. Second, convert the flow rate into cubic inches per minute (cm^3/min). Third, divide the flow rate by the estimated volume to get the time required. Last, round any fractions of minutes off to the next highest minute.

Standard Units

Volume (in³) = length (in in.) × π × radius² (in in.)
Flow Rate (in³/min) = cfh × 29
Flow Time (min) = Volume ÷ Flow Rate

SI (Metric Units)

Volume (cubic cm) = length (in cm) × π × radius² (in cm)
Flow Rate (cm³/min) = L/min × 1,000
Flow Time (min) = Volume ÷ Flow Rate

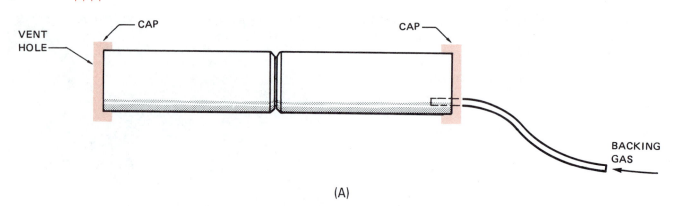

VENT HOLE — CAP

CAP —

BACKING GAS

(A)

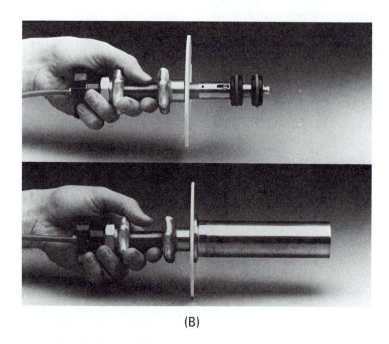

(B)

Figure 17-12 Backing gas can be fed into a pipe for welding by (A) capping the ends externally or (B) by inserting a purge mandrel inside the pipe. (B courtesy of ARC Machines, Inc.)

Standard Unit Example

How long would it take to purge the air out of a 10-ft (305-cm) -long section of 4-in. (10-cm) -diameter pipe if the flow rate is 20 cfh (9.4 L/min)?

Solution

Volume = $120 \times 3.14 \times 2^2$
Volume = 1507 in^3
Flow Rate = 20×29
Flow Rate = 580 in^3/min
Flow Time = $1507 \div 580$
Flow Time = 2.59 min (approximately 3 min)

SI (Metric Unit) Example

How long would it take to purge the air out of a 3-m (10-ft) -long section of 10-cm (4-in.) -diameter pipe if the flow rate is 9 L/min (20 cfh)?

Solution

Volume = $300 \times 3.14 \times 5^2$
Volume = 23,550 cm^3
Flow Rate = 9×1000
Flow Rate = 9,000 cm^3/min
Flow Time = $23,550 \div 9,000$
Flow Time = 2.61 min (approximately 3 min)
$\pi = 3.14$
Radius = 1/2 diameter*
Length in In. = length in ft $\times$ 12
Length in cm = length in m $\times$ 100

*For the purposes of estimating, either the inside diameter (ID) or outside diameter (OD) of the pipe can be used.

Notice in the above examples that the lengths, diameters, and flow rates are about the same and so are the minutes required for purging.

Taping over the joint prevents the gas from being blown out too fast. This will allow a slower flow rate to be used on the purging gas once the pipe has been purged. The tape is removed just ahead of the weld, **Figure 17-13**.

FILLER METAL

The addition of filler metal to the joint can be done by dipping or by preplacement. Dipping can be made easier if the rod is bent to the radius of the pipe, **Figure 17-14**. This will allow the welder to keep the rod angle close to the pipe to minimize contamination. This also makes it possible for welders to reach around the pipe and brace themselves better. The groove can also be used to guide the filler rod, reducing tungsten contamination.

Preplacing the filler rod allows the root pass and some filler passes to be made without rod manipulation. The filler rod, usually 1/8 in. (3 mm) in diameter, is bent in a radius slightly smaller than the pipe being welded. Tension from the wire will hold it in place during the weld. If it does not stay in place, a small tack weld can be made at one end. Both hands can be used to control the torch, making the weld easier and more accurate. This technique is accepted in most codes or standards for welding.

Consumable Inserts

Consumable inserts are preplaced filler metal used for the root pass when consistent, high-quality welds are required. These inserts help to reduce the number of repairs or rejections when welding under conditions that are less than ideal, such as in a limited space. Although most inserts are used on pipe, they are available as strips for flat plate.

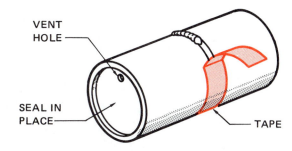

Figure 17-13

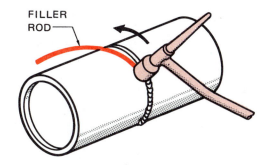

Figure 17-14 Filler rod curved to surface of pipe for ease in welding.

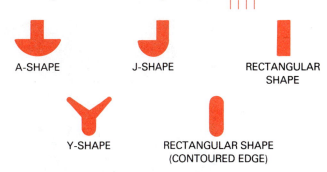

Figure 17-15 Consumable insert standard shapes.

Inserts are classified by their cross-sectional shape, as shown in **Figure 17-15**, and listed as follows:

- Class 1, A-shape
- Class 2, J-shape
- Class 3, rectangular shape
- Class 4, Y-shape
- Class 5, rectangular shape (contoured edges)

These are the most frequently used designs, **Figure 17-16**. Other shapes can be obtained from manufacturers upon request.

When ordering consumable inserts, the welder must specify the classification, size, style, and pipe schedule. Other information is available in the AWS Publication A5.30, "Specification for Consumable Inserts."

PRACTICE 17-1

Tack Welding Pipe

Using a properly set up and adjusted GTA welding machine, proper safety protection, two or more pieces of 3-in. (76-mm) to 10-in. (254-mm) diameter schedule 40 mild steel pipe, prepared with single V-grooved joints, and a few filler rods having 1/16-in. (2-mm), 3/32-in. (2.4-mm), and 1/8-in. (3-mm) diameters, you will tack weld a butt pipe joint.

Bend the rods with the 1/16-in. (2-mm) and 3/32-in. (2.4-mm) diameters into a U-shape. These rods will be used to set the desired root opening, **Figure 17-17**. Lay the pipe in an angle iron cradle. Slide the ends together so that the desired diameter wire is held between the ends. Hold the torch cup against the beveled sides of the groove. The torch should be nearly parallel with the pipe. Bring the filler rod end in close and rest it on the joint just ahead of the torch, **Figure 17-18**.

Lower your helmet and switch on the welding current by using the foot or hand control. Slowly straighten the torch so that the tungsten is brought closer to the groove. This pivoting of the torch around the nozzle will keep the torch aligned with the joint and help prevent arc starts outside of the joint. When the tungsten is close enough, an arc will be established. Increase the current by depressing the foot or hand control until a molten weld pool is established on both root faces. Dip

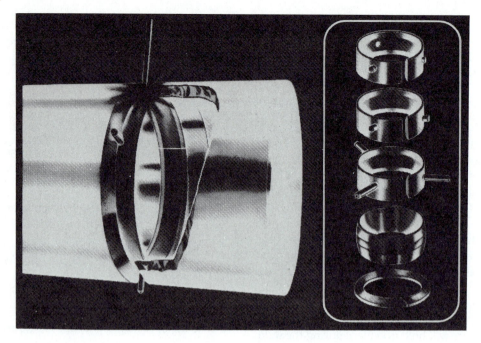

Figure 17-16 Specialized backing rings. (Courtesy of Robvon Backing Ring Company.)

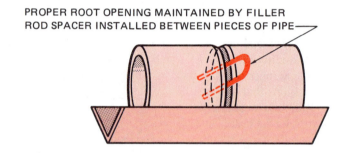

PROPER ROOT OPENING MAINTAINED BY FILLER
ROD SPACER INSTALLED BETWEEN PIECES OF PIPE

Figure 17-17

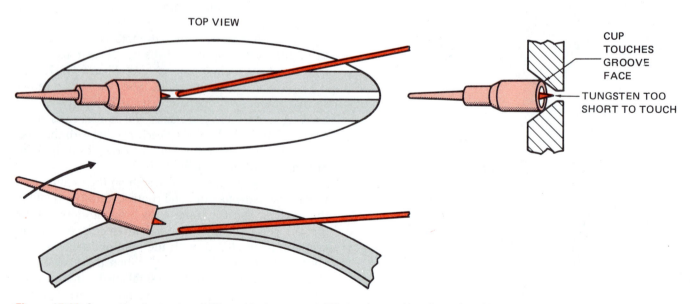

TOP VIEW

CUP
TOUCHES
GROOVE
FACE

TUNGSTEN TOO
SHORT TO TOUCH

Figure 17-18 Supporting the torch and filler rod in the groove will help when making the tack welds.

the filler rod into the molten weld pool as the torch is pivoted from side to side, **Figure 17-19**. Slowly move ahead and repeat this step until you have made a tack weld approximately 1/2 in. (13 mm) long. At the end of the tack weld, slowly reduce the current and fill the weld crater. If the tack weld crater is properly filled and it does not crack, grinding may not be required, **Figure 17-20**.

Figure 17-19

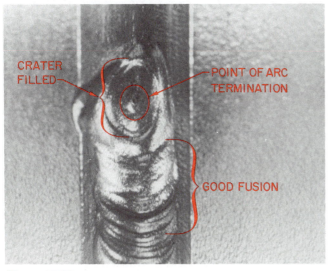

Figure 17-20

Roll the pipe 180° to the opposite side. Check the root opening and adjust the opening if needed. Make a tack weld 1/2 in. (13 mm) long. Roll the pipe 90° to a spot halfway between the first two tacks. Check and adjust the root opening if needed. Make a tack weld. Roll the pipe 180° to the opposite side and make a tack weld halfway between the first two tack welds.

The completed joint should have four tack welds evenly spaced around the pipe (at 90° intervals). The root surfaces of the welds should be flat or slightly convex. The tacks should have good fusion into the base metal with no cold lap at the start, **Figure 17-21**.

Continue making tack welds until this procedure is mastered. Turn off the welding machine, shielding gas, and cooling water and clean up your work area when you are finished welding. ◆

PRACTICE 17-2

Root Pass, Horizontal Rolled Position (1G)

For this practice, you will need a properly set up and adjusted GTA welding machine, proper safety protection, two or more pieces of 3-in. (76-mm) to 10-in. (254-mm) diameter schedule 40 mild steel pipe, prepared with single V-grooved joints and tack welded as described in Practice 16-1, and a few filler rods having 1/16-in. (2-mm), 3/32-in. (2.4-mm) and 1/8-in. (3-mm) diameters. You will make a root welding pass on a pipe in the horizontal rolled (1G) position, **Figure 17-22**.

Place the pipe securely in an angle iron vee block. Using the same procedure as practiced for the tack welds, place the cup against the beveled sides of the joint with the torch at a steep angle (refer to Figure 16-18). Start the weld near the 3 o'clock position and weld toward the 12 o'clock position. Stop welding just past the 12 o'clock position and roll the pipe. Starting the weld above the 3 o'clock position is easier, and starting the weld below this position is harder, **Figure 17-23**. Brace yourself against the pipe and try moving through the length of the weld to determine if you have full freedom of movement.

Hold the filler rod close to the pipe so it is protected by the shielding gas, thus preventing air from being

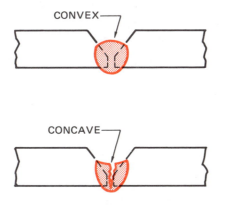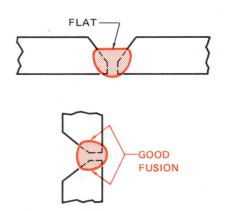

Figure 17-21 Concave tack welds may crack because they cannot withstand the stresses during cooling.

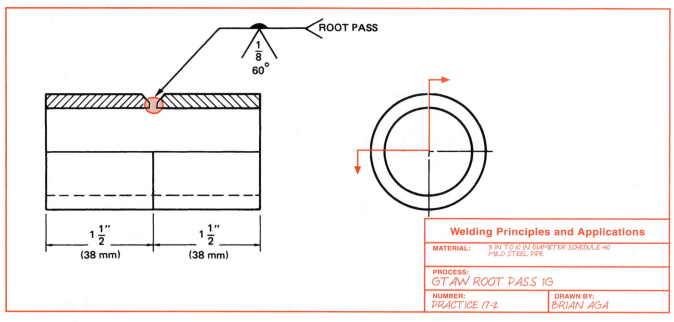

Figure 17-22 Root pass horizontal rolled position butt joint.

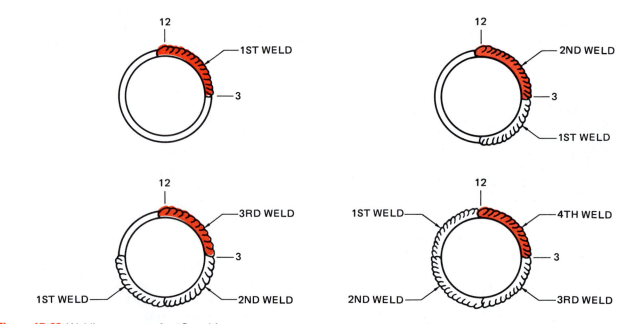

Figure 17-23 Welding sequence for 1G position.

drawn into the inert gas around the weld. The end of the rod must be far enough away from the starting point so it will not be melted immediately by the arc. However, it must be close enough so that it can be seen by the light of the arc.

When you are comfortable and ready to start, lower your helmet. Switch on the welding current by depressing the foot or hand control. Slowly pivot the torch until the arc starts. Increase the current to establish molten weld pools on both sides of the root face. Add the filler rod to the molten weld pool as the torch is rocked from side to side and moved slowly ahead. The side-to-side motion can be used to walk the nozzle along the groove. The walking of the nozzle will ensure good fusion in both root faces and help make a very uniform

weld, **Figure 17-24.** When walking the nozzle along the groove, the tip of the torch cap will make a figure 8 motion. See Color Plate 23.

Note: For the best results with nozzle walking you should use a high-temperature ceramic nozzle such as silicon nitride.

Nozzle walking the torch along a groove can be practiced without welding power so you can see how to move the torch. This technique can be used on most GTA groove welds in pipe or plate, **Figure 17-25.** A skilled welder can also use it on the filler and cover passes.

Watch the molten weld pool to make sure that it is balanced between both sides. If the molten weld pool is not balanced, then one side is probably not as hot nor is it penetrating as well.

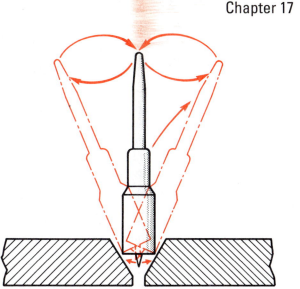

Figure 17-24 Moving the top of the torch in a figure 8 motion will allow you to walk the cup along the joint.

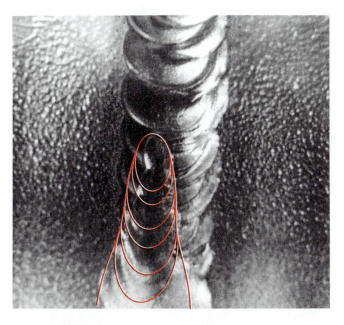

Figure 17-25 Taper down the weld when completing to avoid a large crater.

To ensure good fusion and penetration of the weld at the tack welds, or as the weld is completed, a special tie-in technique is required. Stop adding filler metal to the molten weld pool when it is within 1/16 in. (2 mm) of the tack weld. At this time the key hole will be much smaller or completely closed.

Continue to rock the torch from side to side as you move ahead. Watch the molten weld pool. It should settle down or sink in when the weld has completely tied itself to the tack weld. Failure to do this will cause incomplete fusion or cold lap at the root. If you are crossing a tack weld, carry the molten weld pool across the tack without adding filler and start adding rod at the other end. If you are at the closing end of the weld, slowly decrease the welding current as you add more welding rod to fill the weld crater, **Figure 17-26**.

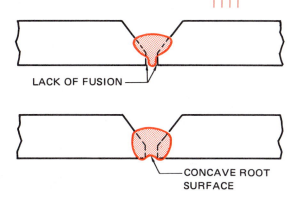

LACK OF FUSION

CONCAVE ROOT SURFACE

Figure 17-26 A hot pass can be used to correct incomplete root fusion or excessive concavity of the root surface.

After the weld is complete around the pipe, visually check the root for any defects. Repeat this practice as needed until it is mastered. Turn off the welding machine, shielding gas, and cooling water and clean up your work area when you are finished welding. ◆

HOT PASS

A hot pass can be used to correct some of the problems caused by a poor root pass. The hot pass may be a hotter-than-normal filler pass or it may be made without adding extra filler. It can be used to correct incomplete root fusion or excessive concavity of the root surface, **Figure 17-27**.

Concave root surfaces are generally caused by insufficient filler metal for the joint. To correct this condition, a hot filler pass is used to add both the needed metal and heat. During the weld, watch the molten weld pool surface to see when it sinks slightly (preferably equal to the needed reinforcement). Move the torch forward. Watch the molten weld pool sink and

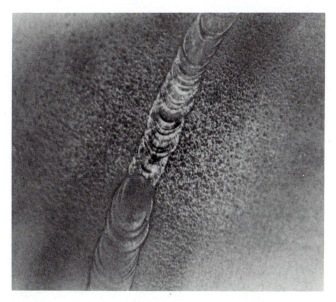

Figure 17-27 Inside root surface with proper penetration after the hot pass.

then add filler. Continue this process along the area of root concavity.

Incomplete fusion generally is caused by insufficient heat and temperature for the joint. To correct this condition, a hot pass is used without adding more filler metal. As before, watch the molten weld pool. When it sinks, move the torch ahead. Continue this process along the area of incomplete root fusion.

EXPERIMENT 17-1

Repairing a Root Pass Using a Hot Pass

Using a properly set up and adjusted GTA welding machine, proper safety protection, two or more pieces of schedule 40 mild steel pipe, 3 in. (76 mm) to 10 in. (254 mm) in diameter, prepared with single V-grooved joints with poor root weld passes, and some filler rods having diameters of 1/16 in. (2 mm), 3/32 in. (2.4 mm), and 1/8 in. (3 mm), you will make a hot pass to correct defects in the root weld.

The first step is to determine the root weld defects and mark the location and type of defect on the outside of the pipe. Most defective root welds will have areas of one type of defect followed by another area of a different type of defect. This occurs most commonly as the student makes over-adjustments to correct a problem or as the root opening and root face dimensions change.

Mark the groove with a punch or die to indicate the type of defect below. Place the pipe with the defect in the 12 o'clock welding position and perform the weld. The hot pass may vary in size as it progresses around the pipe. The weld can go faster across areas that do not require root repair. Turn off the welding machine, shielding gas, and cooling water and clean up your work area when you are finished welding. ◆

PRACTICE 17-3

Stringer Bead, Horizontal Rolled Position (1G)

Using a properly set up and adjusted GTA welding machine, proper safety protection, one or more pieces of mild steel pipe, 3 in. (76 mm) to 10 in. (254 mm) in diameter, and some filler rods having diameters of 1/16 in. (2 mm), 3/32 in. (2.4 mm), and 1/8 in. (3 mm), you will make a straight stringer bead around a pipe in the horizontal rolled position.

Start by cleaning a strip 1 in. (25 mm) wide around the pipe. Brace yourself against the pipe so that you will not sway during the weld, **Figure 17-28(A)** and **(B)**. The torch should have a slight upward slope, about 5° to 10° from perpendicular to the pipe, **Figure 17-29**. The filler wire should be held so that it enters the molten weld pool from the top center at a right angle to the torch.

Start the weld near the 3 o'clock position and weld upward to a point just past the 12 o'clock position. In order to minimize stops and starts, try to make the complete weld without stopping. With the power off and your helmet up, move the torch through the weld to be sure you have full freedom of movement.

With the torch held in position so that the tungsten is just above the desired starting spot, lower your helmet. Start the welding current by depressing the foot or hand control switch. Establish a molten weld pool and dip the filler rod in it. Use a straight forward and back movement with both the torch and rod, **Figures 17-30** and **17-31**. However, never move farther ahead than the leading edge of the molten weld pool. Too large a movement will lead to inadequate molten weld pool coverage by the shielding gas. The short motion is to allow the rod to be dipped into the molten weld pool without the heat from the arc melting the rod back, **Figure 17-32**.

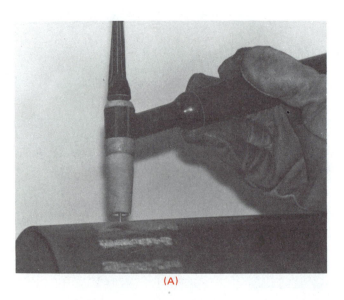
(A)

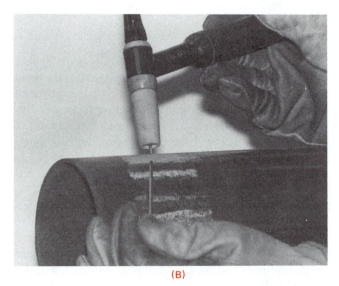
(B)

Figure 17-28 Bracing the gloved hand against the pipe can help you control both the arc (A) for beading or (B) the filler and arc for welding. Note the precleaning of the pipe.

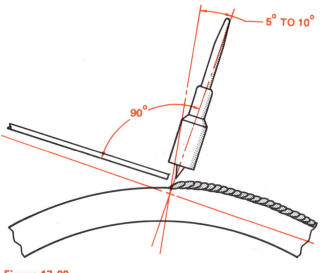

Figure 17-29

Figure 17-32 The filler rod should be melted in the molten weld pool but not allowed to melt and drop into the weld pool.

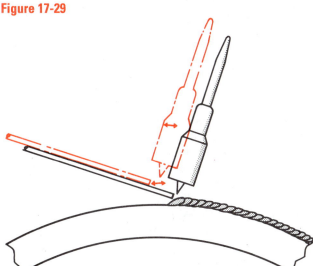

Figure 17-30 The torch and filler rod are moved back and forth together.

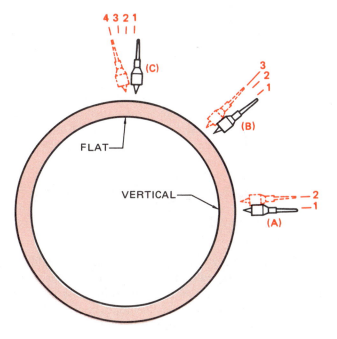

Figure 17-33 Counting to oneself is one way of setting the welding rhythm. As the weld progresses from vertical (A) to flat (C), the rhythm slows.

Figure 17-31 Beading on pipe.

As the weld progresses from a vertical position to a flat position, the frequency of movement and pause times will change. Generally, the more vertical the weld, the faster the movement and the shorter the pause times. As the weld becomes flatter, the movement is slower and the pause times are longer, **Figure 17-33.** In the vertical position, gravity tends to pull the weld toward the back center of the bead. This makes a high crown on the weld with the possibility of undercutting along the edge. Flatter positions cause the weld to be pulled out, resulting in a thinner weld with less apparent reinforcement, **Figure 17-34.**

When the weld reaches the 12 o'clock position, stop and roll the pipe. To stop, slowly decrease the current and add welding rod to fill the weld crater. Keep the torch held over the weld until the postpurge stops. This precaution will prevent the air from forming oxides on the hot weld bead.

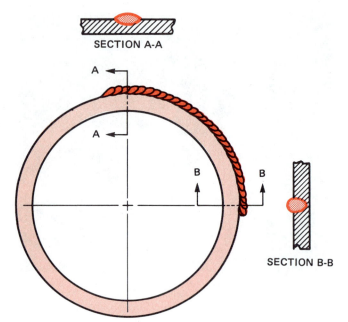

SECTION A-A

SECTION B-B

Figure 17-34 The weld bead contour, if uncontrolled, can change a great deal from the vertical to flat positions.

After the weld is complete around the pipe, it should be visually inspected for straightness, uniformity, and defects. Repeat the process until you can consistently make the weld visually defect-free. Turn off the welding machine, shielding gas, and cooling water and clean up your work area when you are finished welding. ◆

PRACTICE 17-4

Weave and Lace Beads, Horizontal Rolled Position (1G)

Using a properly set up and adjusted GTA welding machine, proper safety protection, one or more pieces of mild steel pipe, 3 in. (76 mm) to 10 in. (254 mm) in diameter, and some filler rods having diameters of 1/16 in. (2 mm), 3/32 in. (2.4 mm), and 1/8 in. (3 mm), you will make a straight weave or lace bead around a pipe in the horizontal rolled position.

Start by cleaning a strip 1 in. (25 mm) wide around the pipe. Brace yourself and check to see that you have enough freedom of movement to make a weld from the 3 to 12 o'clock positions. Hold the torch with a 5° to 10° upward angle. The filler rod is to be held at a right angle to the torch.

Lower your helmet and establish a molten weld pool. Add filler metal as you move the torch to the side. Slowly build a shelf to support the molten weld pool or bead, **Figure 17-35.**

To make a weave bead, move the torch in a "C," "U," or zigzag pattern across the molten weld pool. The pattern should be no wider than one-half the diameter of the cup. Too large a weave means that part of the molten weld pool will not have a cover of the shielding

gas. This results in atmospheric contamination of the molten weld pool.

To make a lace bead slowly, move the torch in a zigzag pattern across the weld bead. The molten weld pool should never be larger than that for a stringer bead. A lace bead can be made any desired width because the small molten weld pool is always protected from atmospheric contamination.

Continue the weave or lace bead up the pipe until you reach the top. To stop slowly, lower the current and add welding rod until the crater is filled. Turn the pipe and continue the bead around the pipe.

After the bead is complete around the pipe, visually inspect it for uniformity in width and reinforcement. The bead must also be free of visible defects. Repeat the process until you can consistently make the weld visually defect free. Turn off the welding machine, shielding gas, and cooling water and clean up your work area when you are finished welding. ◆

FILLER PASS

The filler pass is the next weld pass(es) to be made after the hot pass. The filler pass should be used to fill up the groove quickly. There should be complete fusion but little penetration. Deep penetration serves only to slow the completion of the joint and to subject the pipe to more heat and temperature than needed, **Figure 17-36.**

Often only one GTA filler pass is used to protect the thin root pass from burnthrough. Because of the longer welding time required for the GTA process, pipe joints

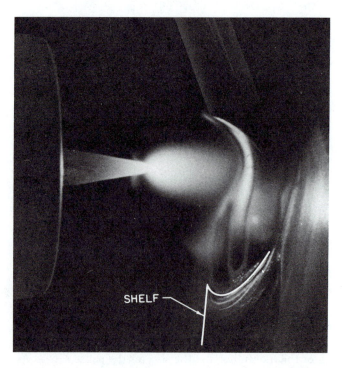

SHELF

Figure 17-35 The shelf will support the molten pool during the vertical portion of the weld.

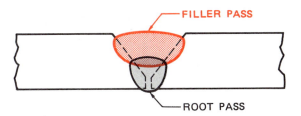

Figure 17-36 The filler pass should fill the groove as much as possible but not more than 1/4 in. (6 mm) at a time.

are often completed using another process. Most other processes, such as SMAW, GMAW, or FCAW, are much hotter and most likely will burn through the root without the added protection of a GTA filler pass.

PRACTICE 17-5

Filler Pass (1G Position)

Using a properly set up and adjusted GTA welding machine, proper safety protection, two or more pieces of 3-in. (76-mm) to 10-in. (254-mm) diameter schedule 40 mild steel pipe, prepared with single V-grooved joints and having a root pass (as described in Experiment 16-1), and some filler rods having 1/16-in. (2-mm), 3/32-in. (2.4-mm), and 1/8-in. (3-mm) diameters, make a filler welding pass in the groove in the horizontal rolled (1G) position

- Place the pipe securely in an angle iron vee block.
- Using the same procedure practiced for the tack welds and root pass, place the cup against the beveled sides of the joint with the torch at a 5° to 10° upward angle.
- Brace yourself and check for full freedom of movement along the welded joint.
- Start the weld near the 1 o'clock position and end just beyond the 12 o'clock position.
- Lower your helmet and establish a molten weld pool. Using the same forward and backward motion that you developed when making the stringer bead in Practice 17-3, add the filler metal at the top center of the molten weld pool until the bead surface is flat or slightly convex. Adding too little filler to this pass may result in burnthrough if another process is used to complete the joint. The filler pass should have as little penetration as possible so that the maximum reinforcement can be added with each pass. Deep penetration beyond the depth required to fuse the weld to the bevel and hot pass will slow the joint fill-up rate.
- To stop the molten weld pool, slowly decrease the current and add rod to fill the weld crater. Another method is to slowly decrease the current and pull the bead up on the beveled side of the joint, **Figure 17-37**.
- Turn the pipe and repeat this bead until the joint is filled up flush with the pipe surface, **Figure 17-38**.

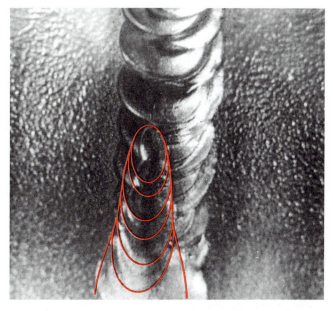

Figure 17-37 Taper down the weld when completing to avoid a large crater.

Figure 17-38 The filler pass should fill the groove flush with the surface.

Turn off the welding machine, shielding gas, and cooling water and clean up your work area when you are finished welding. ◆

COVER PASS

The stringer bead can be continued to cap the weld. If this technique is used, the bead should overlap the pipe surface no more than 1/8 in. (3 mm). Total reinforcement must not exceed 3/32 in. (2.4 mm) on pipe having walls less than 3/8 in. (10 mm) thick, **Figure 17-39**.

After the weld is complete, visually inspect it for uniformity in width and reinforcement and for any defects.

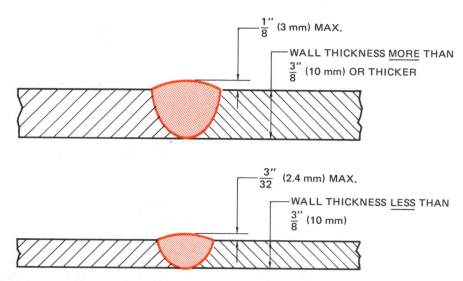

$\frac{1''}{8}$ (3 mm) MAX.

WALL THICKNESS <u>MORE</u> THAN $\frac{3''}{8}$ (10 mm) OR THICKER

$\frac{3''}{32}$ (2.4 mm) MAX.

WALL THICKNESS <u>LESS</u> THAN $\frac{3''}{8}$ (10 mm)

Figure 17-39 Excessively large weld beads prevent the pipe from uniformly expanding under pressure. This restriction from expanding can cause a failure to occur in or near the weld.

PRACTICE 17-6

Cover Pass (1G Position)

Using a properly set up and adjusted GTA welding machine, proper safety protection, two or more pieces of 3-in. (76-mm) to 10-in. (254-mm) diameter schedule 40 mild steel pipe that are single V-grooved and are welded flush with filler passes (as described in Experiment 16-1), and some filler rods having 1/16-in. (2-mm), 3/32-in. (2.4-mm), and 1/8-in. (3-mm) diameters, you will cap the weld with a cover pass made in the horizontal rolled (1G) position, **Figure 17-40**.

Using the same techniques and skill developed in Practice 16-5, make a cover pass. The weld should not be more than 1/8 in. (3 mm) wider than the groove and should have a buildup of no more than 3/32 in. (2.4 mm).

After the weld is complete, visually inspect it for uniformity in width and reinforcement and for any defects. Repeat the process until you can consistently make the weld visually defect free. Turn off the welding machine, shielding gas, and cooling water and clean up your work area when you are finished welding. ◆

PRACTICE 17-7

Single V-groove Pipe Weld, 1G Position, To Be Tested

Using the same equipment, setup, and materials as listed in Practice 17-1, you will make a welded butt joint in the horizontal rolled (1G) position. Then you will test the weld for defects, **Figure 17-41**.

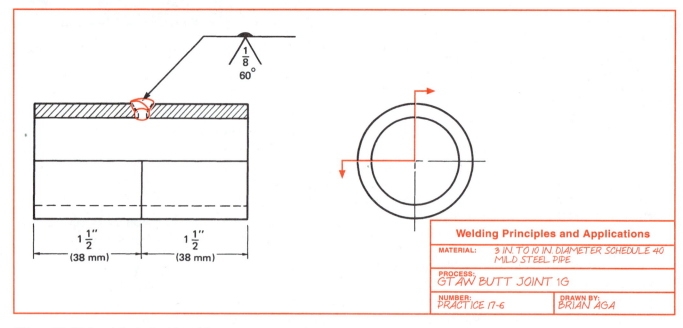

$\frac{1}{8}$ 60°

$1\frac{1}{2}''$ (38 mm)

$1\frac{1}{2}''$ (38 mm)

Welding Principles and Applications

MATERIAL:	3 IN. TO 10 IN. DIAMETER SCHEDULE 40 MILD STEEL PIPE
PROCESS:	GTAW BUTT JOINT 1G
NUMBER: PRACTICE 17-6	DRAWN BY: BRIAN AGA

Figure 17-40 Butt joint in the 1G position.

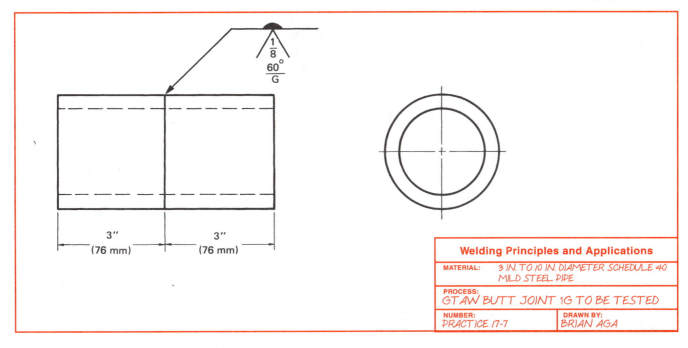

Welding Principles and Applications

MATERIAL:	3 IN. TO 10 IN. DIAMETER SCHEDULE 40 MILD STEEL PIPE
PROCESS:	GTAW BUTT JOINT 1G TO BE TESTED
NUMBER: PRACTICE 17-7	DRAWN BY: BRIAN AGA

Figure 17-41 Butt joint in the 1G position to be tested.

Make the tack welds, root pass, filler pass, and cover pass as described in Practices 17-1, 17-2, 17-5, and 17-6.

When the weld is complete, inspect it for defects. If it passes visual inspection, cut out guided bend test specimens as shown in **Figure 17-42**. Test the specimens for weld defects. Repeat this practice until the test is passed. Turn off the welding machine, shielding gas, and cooling water and clean up your work area when you are finished welding. ◆

PRACTICE 17-8

Stringer Bead, Horizontal Fixed Position (5G)

Using the same equipment, setup, and materials listed in Practice 17-1, make a straight stringer bead around an ungrooved pipe in the horizontal fixed position.

■ Clean a strip 1 in. (25 mm) wide around the pipe.

■ Mark the top for future reference and clamp the pipe at a comfortable work height.

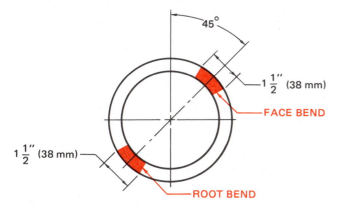

Figure 17-42 Guided bend specimens.

■ Brace yourself so you are steady and move through the weld to make sure that you have freedom of movement.

■ Lower your helmet and establish a molten weld pool at the 6 o'clock position. Keep the molten weld pool size small so that it will stay uniform.

■ Dip the filler metal into the front leading edge of the molten weld pool.

■ Move the torch forward and backward as the rod is added. The frequency of movement will increase as the weld becomes more vertical.

■ Continue the weld without stopping, if possible, to the 12 o'clock position.

■ Repeat the weld up the other side in the same manner.

When the weld is complete, visually inspect (VT) it for straightness and uniformity, and for any defects. Repeat the process until you can consistently make the weld visually defect-free. Turn off the welding machine, shielding gas, and cooling water and clean up your work area when you are finished welding. ◆

PRACTICE 17-9A AND 17-9B

Single-V Butt Joint (5G Position)
A. Root Penetration May Vary
B. 100% Root Penetration To Be Tested

Using the same equipment, setup, and materials listed in Practice 17-1, make a weld on a pipe in the horizontal fixed (5G) position.

The pipes should be tack welded, and the tacks should be made in the following clock positions: 8:30, 4:30, 1:30, and 11:30, **Figures 17-43** and **17-44**.

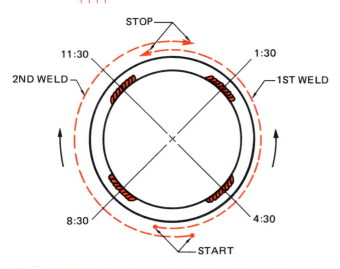

Figure 17-43 Tack welds on pipe in 5G position and the welding sequence.

Figure 17-44

- Start the root weld at the 6:30 o'clock position and weld uphill around one side of the pipe to the 12:30 o'clock position. Start with the cup against the pipe bevels, as was done before, and add the rod at the leading edge of the molten weld pool.
- Repeat this process up the other side of the pipe. The starts and stops of these welds should overlap slightly to ensure a good tie-in.

The filler passes should also start near the 6:30 o'clock position and go up both sides of the pipe to a point near the 12:30 o'clock position, **Figure 17-45**. Staggering the location of these beads will help prevent defects arising from discontinuities caused at the starting and stopping points. Keeping these beads small will help in your ability to control them. A large number of good weld beads is more desirable than a few poorly controlled beads.

The cover pass may be made as either a weave or lace bead. The lace bead is usually more easily controlled because the molten weld pool is smaller.

- After the weld is complete, visually inspect (VT) it for uniformity in width and reinforcement.
- Check also for visible defects and root penetration.
- If the joint passes all visual tests and it has 100% root penetration, guided bend test specimens can be cut out (refer to **Figure 17-9**).
- After bending, evaluate the specimens for defects. Repeat this practice until both parts A and B can be passed. Turn off the welding machine, shielding gas, and cooling water and clean up your work area when you are finished welding. ◆

PRACTICE 17-10

Stringer Bead, Vertical Fixed Position (2G)

Using the same equipment, setup, and materials listed in Practice 17-1, make a straight stringer bead around a pipe in the vertical fixed position.

- Hold the torch at a 5° angle from horizontal and 5° to 10° from perpendicular to the pipe, **Figure 17-46**. This will allow good visibility and simplify the process of adding the rod at the top of the bead.
- Establish a small molten weld pool and filler along the top front edge, **Figure 17-47**. Gravity will tend to pull the metal down, especially if the molten weld pool becomes too large. When the molten weld pool sags, undercut appears along the top edge of the bead. A slight "J" pattern may help to control weld bead sag by making a small shelf to support the molten weld pool while it cools.

When the weld is completed, visually inspect (VT) it for uniformity and straightness, and for any defects.

Repeat the process until you can consistently make the weld visually defect free. Turn off the welding machine, shielding gas, and cooling water and clean up your work area when you are finished welding. ◆

PRACTICE 17-11A AND 17-11B

Single-V Butt Joint (2G Position)
A. Root Penetration May Vary
B. 100% Root Penetration To Be Tested

Using the same equipment, setup, and materials as listed in Practice 17-1, you will weld a butt joint on a pipe in the vertical fixed (2G) position.

Tack weld the pipes and start the root pass at a point between the tack welds. The root weld should be small enough so that surface tension will hold it in place. Too large a root weld bead will cause the top root face to be slightly concave.

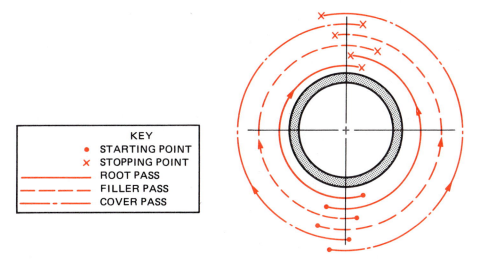

KEY
- • STARTING POINT
- × STOPPING POINT
- —— ROOT PASS
- – – FILLER PASS
- –·– COVER PASS

Figure 17-45 Stagger the starting and stopping points of each bead.

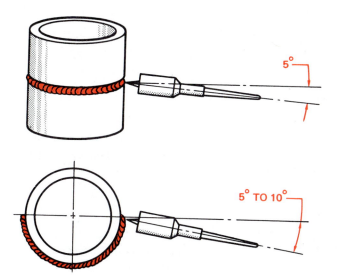

Figure 17-46 When making a straight stringer bead around a pipe in the horizontal fixed position, hold the torch at a 5° angle from horizontal and 5° to 10° from perpendicular to the pipe.

The filler passes can be larger if they are made along the lower beveled surface, **Figure 17-48**. This will support the filler passes and allow a faster joint buildup. The next pass is on the top side of the first. This process of resting the next pass on the last pass continues until the joint is filled.

The cover pass is started around the lower side of the joint, overlapping the pipe surface by no more than 1/8 in. (3 mm), **Figure 17-49**. The next pass covers about one-half of the first cover pass and should be slightly larger. This process of making each pass larger continues until the center weld is complete. Then, each weld is smaller than the last. This process builds a uniform reinforcement having a good profile.

After the weld is complete, visually inspect (VT) it for uniformity and defects. If it passes, inspect the root for 100% penetration. If the root passes, cut out guided bend test specimens. Test the specimens and evaluate the results. Repeat this weld until both parts A and B can

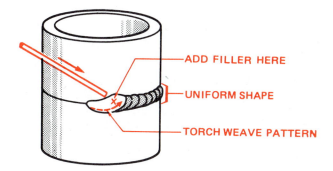

Figure 17-47

be passed. Turn off the welding machine, shielding gas, and cooling water and clean up your work area when you are finished welding. ◆

PRACTICE 17-12

Stringer Bead on a Fixed Pipe at a 45° Inclined Angle (6G Position)

Using the same equipment, setup, and materials listed in Practice 17-1, make a straight stringer bead around a fixed, ungrooved pipe at a 45° inclined angle.

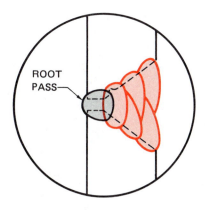

Figure 17-48 Filler pass(es) over root weld on pipe in 2G position.

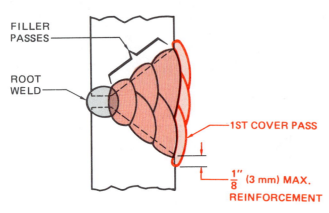

Figure 17-49 Cover passes on pipe in 2G position.

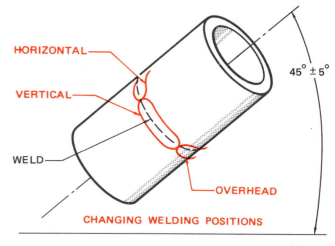

Figure 17-51 In the 6G position, the pipe is at a 45° angle. The effective welding angle changes as the weld progresses around the pipe.

- Starting at the 6:30 o'clock position with the torch at a slight downward angle, establish a small molten weld pool, Figure 17-50.
- Add the rod at the upper leading edge of the molten weld pool. A slight "J" pattern can help control the bead size. As the weld progresses around the pipe, the weld becomes more vertical. The rate of movement should increase.
- Keep the size of the weld small so it can be controlled.

After the weld is complete, visually inspect (VT) it for uniformity and for any defects. Repeat the process until you can consistently make the weld visually defect free. Turn off the welding machine, shielding gas, and cooling water and clean up your work area when you are finished welding. ◆

PRACTICE 17-13A AND 17-13B

Single-V Butt Joint (6G Position)
A. Root Penetration May Vary
B. 100% Root Penetration To Be Tested

Using the same equipment, setup, and materials as listed in Practice 16-1, you will make a welded butt joint on a pipe fixed at a 45° angle (6G) position.

Starting at the bottom, make the root pass upward to the 12:30 o'clock position. The size of the root pass should stay the same; however, it may be off to one side more than the other. As long as the root surface is uniform, this unbalanced appearance is acceptable. The filler passes should correct this problem.

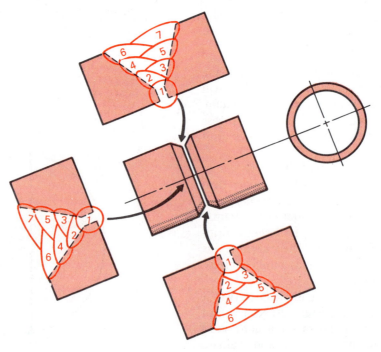

Figure 17-50 6G welding bead sequence.

The filler passes are applied to the downhill side first so that the upper side will be supported, **Figure 17-51.** The starts and stops should be staggered so that they do not increase the possibility of defects at any one point on the weld. A slight "J" weave pattern will help hold the shape of the bead as it changes from overhead to vertical to horizontal.

The cover pass is easy to control if it is a series of stringer beads. Start on the lower side and build up the cover pass as in the 2G position. Each weld should over-lap the preceding weld so that the finished weld is uniform.

After the weld is complete, visually inspect (VT) it for uniformity and for any defects. If the weld passes, inspect the root for 100% penetration. If the root passes this inspection, cut out guided bend test specimens. Test the specimens and evaluate the results. Repeat this practice until both parts A and B can be passed. Turn off the welding machine, shielding gas, and cooling water and clean up your work area when you are finished welding. ◆

GTA Welders Put the Finishing Touches on the Fins for the Patriot Missile

Manual gas tungsten arc welding and automatic gas metal arc welding continue to play critical roles in the manufacture of the now famous Patriot missile, the weapon system that successfully wiped out Iraq's ace-in-the-hole, the SCUD tactical ballistic missile, during the recent war in the Middle East. The two welding processes are being used by the Martin Marietta Electronics, Information & Missiles Group, Orlando, Florida, to fabricate the control fins, or air vanes, for the Patriot.

The welding of the air vanes has been so outstanding that the team representing the weld shop of the Mechanical Fabrication Center was named the Performance Measurement Team of the Year (1990) among more than 70 manufacturing teams engaged in missile fabrication at the Orlando facility. During 1990, the weld shop achieved a "first-time yield" of 99.4%. To date, more than 3,000 Patriot missiles have been produced at Martin Marietta. The total welding on the air vanes at the Florida fabricator has already passed the 44-mile mark.

Figure 1

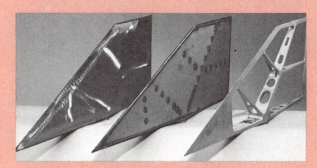

Figure 2

PERFORMANCE MEASUREMENT TEAM

The competition was established by Martin Marietta in 1986 as a part of its Total Quality Management program. Manufacturing groups were organized into Performance Measurement Teams, PMT's. In this arrangement, every discipline that had some impact on the area of manufacture participated in weekly meetings (Figure 1) aimed at identifying ways to improve, then established ways to improve, key issues of concern. According to Martin Marietta, the PMT approach differs from that of quality circles in that the team is empowered to take immediate action to resolve any issue at hand.

The PMT representing the weld shop is made up of twelve welders and six engineers. The team supervisor is James McLean, weld shop supervisor, and the facilitator is Jack Borer, the general foreman. The group's success is best measured, perhaps, in terms of its improvement in performance in 1990 over 1989. Comparing the two years, the weld shop demonstrated a 10.3% improvement in performance and a 6.3% improvement in yield. Scrap and rework, two of the most widely used yardsticks in the determination of quality, were both reduced drastically.

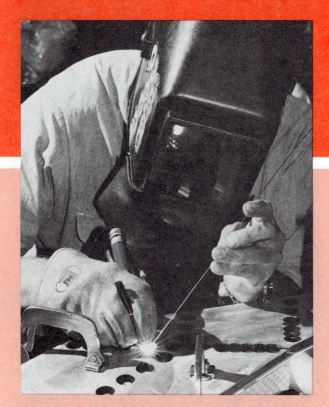

Figure 3

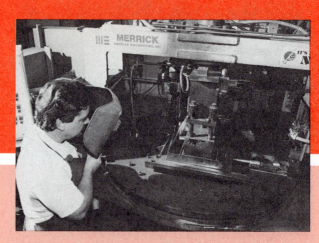

Figure 4

Rework was reduced by 33.3%, while scrap costs were reduced by 74.5%, far greater than the intended goal of 26.4%. The total value of the hardware built for the Patriot in 1990 was slightly more than $5.7 million, yet the value of scrap generated by the weld shop during the same year was only $185, or 0.003% of the total value of hardware built.

Much of the savings obtained in the weld shop during 1990 was generated through improved layout and tooling, as well as improvements in material handling.

In 1989, manpower of 3.6 individuals and 28.6 hours was required to fabricate sixteen units. The following year, the same number of units was cranked out in 14.2 hours with manpower of 1.8 individuals.

When all is said and done, the Martin Marietta weld shop's cost savings through performance improvement in 1990 on the Patriot missile amounted to $68,336.

PRECIPITATION-HARDENING STAINLESS

On the air vanes (there are four to every Patriot missile), matching sets of sheet metal are plug-welded manually to both sides of a similarly shaped investment casting (Figure 2) using the GTA process. The peripheral welds are made in a precision GMA machine. Both the sheet metal and casting alloys are 17-4 PH stainless steels.

John Walvoord, senior welding engineer, said that the sheet metal used in the air vane is 0.040 in. thick. A total of 28 plug welds (Figure 3) is made through each section of sheet, or 56 GTA plug welds for each air vane. The investment-cast part is sandwiched between the two sheets. "The holes for the plug welds are made in a blanking die," he said. "The holes are subsequently run through a countersinking operation. The plug welds are all part of a structural design aimed both at joining the panels to the casting and building up a box structure."

The shielding gas for the GTA process is argon. The welding current ranges from about 75 to 110 A. The filler metal is 0.035-in.-diameter 15-5 PH stainless steel. Walvoord explained that the 15.5 PH stainless filler metal was used to control the crack sensitivity of the 17-4 PH base metal during gas tungsten arc plug welding. "The 15-5 PH stainless filler metal," he said, "enables us to control the hot cracking problem and to obtain the structural integrity needed."

AUTOMATIC WELDING

The automatic GMA welding machine from Merrick Engineering Inc., Nashville, Tennessee (Figure 4), has no joint tracking capability, but follows a joint line for eight weld locations, four on each side. The eight edge welds are made in a sequence designed to minimize distortion. A 17-4 PH filler metal is used there. A lap joint for the GMA welding procedure prevents hot cracking from occurring. The 17-4 PH filler metal, Walvoord said, is about half the cost of the 15-5 PH wire used with GTA welding, and it does the job adequately. A tri-gas consisting of 90% argon-7% helium-3% CO_2 shields the GMA welds.

Walvoord said that tooling holes at precise locations are drilled in the casting structure. A tacking fixture located in front of the plug welding operation fixes the side panels at precise locations in reference to those tooling holes.

The grinding operation follows welding. After that, the air vanes go to a fluorescent-penetrant station. They

then undergo an H-900 aging heat treatment, then go back to penetrant to insure that there are no new problems exposed by the heat-treat operation.

Martin Marietta's Electronics, Information & Missiles Group has been producing the missiles for the U.S. Army under subcontract to the Raytheon Company, Lexington, Massachusetts. The contract with Raytheon calls for the delivery of 7,038 missiles and 734 missile launchers through 1993.

(Reprinted from the *Welding Journal.* Adapted from an article by Bob Irving. Courtesy of the American Welding Society, Inc. Photos by Jon Hammerstein.)

REVIEW

1. What types of industries require GTAW-quality pipe welds?

2. What are the most often used grooves for welding on mild steel pipe?

3. Why is the "feather edge" removed?

4. What condition must the prepared edges of a pipe be in before welding?

5. How is the depth of root penetration measured?

6. What problem can incomplete fusion cause?

7. What are the common causes of a concave root surface?

8. What problems can excessive root reinforcement cause?

9. What can cause root contamination?

10. What gases are used as backing gases?

11. How long would it take to purge a 15-ft section of 6-in.-diameter pipe using a flow rate of 35 cfh?

12. How long would it take to purge a 10-m section of 20-cm-diameter pipe using a flow rate of 17 L/min?

13. How can filler metal be added to make a root pass?

14. What information must be supplied when ordering consumable inserts?

15. How can the proper root opening be maintained for tack welding?

16. How is the tack weld ended so that it will not crack?

17. What is the 1G pipe position?

18. After the weld is set up and ready to begin, how might the welder want to check for freedom of movement?

19. What type of weld can be made using the nozzle walking technique?

20. What two problems can be corrected by a hot pass?

21. Why must you brace yourself when making a weld?

22. What can happen to the weld pool if the torch movement is too long?

23. What technique should be used when stopping a weld?

24. What is the maximum width that a weave pattern should be?

25. Why should the filler pass penetration be limited in depth?

26. How much filler metal should be added to a filler bead?

27. How large should a cover pass be on a pipe with a 1/4-in. (6-mm) wall thickness?

28. What is the 5G pipe position?

29. Why should the starting point of the beads be staggered?

30. How should the torch be held to allow good visibility on a 2G weld?

31. Why should the root bead be small on a 2G weld?

32. Why are stringer beads in the 6G position kept small?

33. Why are the downhill side filler welds put in first on 6G pipe?

Section Five

RELATED PROCESSES

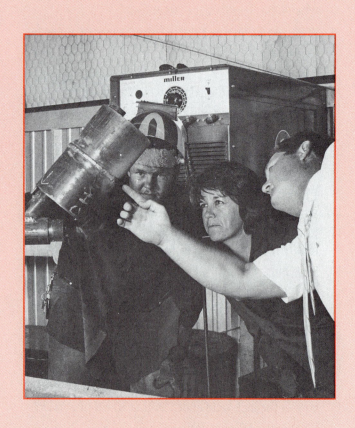

SUCCESS STORY

As a small boy Richard Rowe watched "Dutch," a welder mechanic, working on a coal truck frame. It was 1954 and Rich was looking through Dutch's big black, dirty welder's helmet so that he could see what welding looked like. He was fascinated by what he saw. From that day on Rich was interested in learning to weld.

Twelve years later, Rich began his welding education by enrolling in an adult education course. As his career ambitions grew, so did his desire for continued education. Moorehead Vocational Technical Institute, in Moorehead, Minnesota, granted him a Welding Technology Certificate in 1974. Six years later he completed the A.A. degree at Bismarck College, and in 1989 he earned a B.S. degree from the University of Mary in Bismarck, North Dakota. Just three years ago, he completed the M.S. degree in Technical Teacher Education from Pittsburg State University, Kansas.

Since the very beginning, Rich's work experience has been varied and challenging. He has been a high school teacher, a small-business owner, a college instructor, a community adult education instructor, and a VICA adviser and coach for the North Dakota Welding Skill Olympics.

Along the way he has been actively involved in the welding community. He holds both the AWS Certified Welding Inspector (CWI) and Certified Welding Educator (CWE) certifications and received the North Dakota Outstanding Young Vocational Educator award in 1979. Rich developed and conducted a Comprehensive Employment and Training Assistance program for welding at Bismarck State College, Bismarck, North Dakota. He has been a member of the American Welding Society for 21 years and since 1993 has served as a member of the AWS D1.15c subcommittee on railroad track welding. Currently, Rich serves on the Kansas City AWS local chapter #16 scholarship committee. He is also a member of the American Vocational Association and Kansas Vocational Association.

Rich has compiled many achievements, experiences, and accomplishments in the forty plus years since his encounter with "Dutch." They have all led to today's position, as an instructor of welding at Johnson Community College, Overland Park, Kansas, in the Metal Fabrication Program, which is affiliated with the National Academy of Railroad Sciences. Additionally, he is co-authoring a series of welding textbooks for Delmar Publishers, ITP.

Chapter 18

WELDING JOINT DESIGN, WELDING SYMBOLS, AND FABRICATION

OBJECTIVES

After completing this chapter, the student should be able to:

- understand the basics of joint design.
- identify the major parts of a welding symbol.
- explain the parts of a groove preparation.
- describe how nondestructive test symbols are used.
- list the five major types of joints.
- list seven types of weld grooves.
- lay out a welding project.

KEY TERMS

welding symbol	dimensioning
weld joint	weld types
joint type	weld location
edge preparation	combination symbols
joint dimension	tolerances
G and F	

INTRODUCTION

The joint design affects the quality and cost of the completed weld. Selecting the most appropriate joint design for a welding job requires special attention and skill. The eventual design selection can be influenced by a number of factors.

Every weld joint selected for a job requires some compromises. For example the compromises may be between strength and cost, equipment available and welder skill, or any two, three, or more variables. Because there are so many factors, a good design requires experience. Even with experience, trial welds are necessary before selecting the final joint configuration and welding parameters.

This chapter will familiarize welders with the most important factors and gives them some appreciation of joint design and fabrication. Experience in the welding field will help a welder become a better joint designer and fabricator.

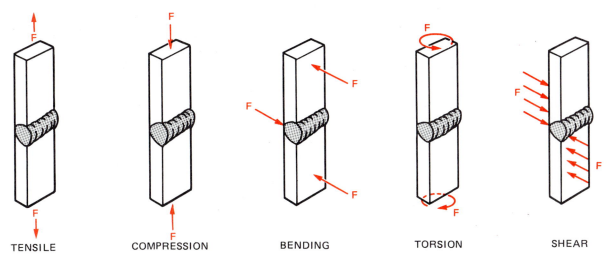

Figure 18-1 Forces on a weld.

Welding symbols are the language used to let the welder know exactly what welding is needed. The welding symbol is used as a shorthand and can provide the welder with all of the required information to make the correct weld. The emphasis in this chapter is on using and interpreting welding symbols so the welder will develop a welder's "vocabulary."

WELD JOINT DESIGN

The selection of the best joint design for a specific weldment requires careful consideration of a variety of factors. Each factor, if considered alone, would result in a part that might not be able to be fabricated. For example, a narrower joint angle requires less filler metal, and that results in lower welding cost. But if the angle is too small for the welding process being used the weld cannot be made.

The purpose of a weld joint is to join parts together so that the stresses are distributed. The forces causing stresses in welded joints are tensile, compression, bending, torsion, and shear, **Figure 18-1**. The ability of a welded joint to withstand these forces depends both upon the joint design and the weld integrity. Some joints can withstand some types of forces better than others.

The basic parts of a weld joint design that can be changed include:

- **Joint type** — The type of joint is considered by the way the joint members come together, **Figure 18-2**.

- **Edge preparation** — The faying surfaces of the mating members that form the joint are shaped for a specific joint. This preparation may be the same on both members of the joint, or each side can be shaped differently, **Figure 18-3** (page 408).

- **Joint dimensions** — The depth and/or angle of the preparation and the joint spacing can be changed to make the weld, **Figure 18-4** (page 409).

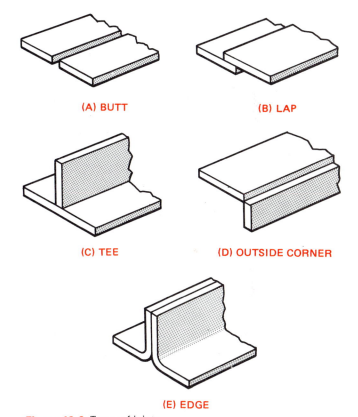

Figure 18-2 Types of joints.

Welding Process

The welding process to be used has a major effect on the selection of the joint design. Each welding process has characteristics that affect its performance. Some processes are easily used in any position; others may be restricted to one or more positions. The rate of travel, penetration, deposition rate, and heat input also affects the welds used on some joint designs. For example, a square butt joint can be made in very thick plates using either electroslag or electrogas welding, but not many other processes can be used on such a joint design.

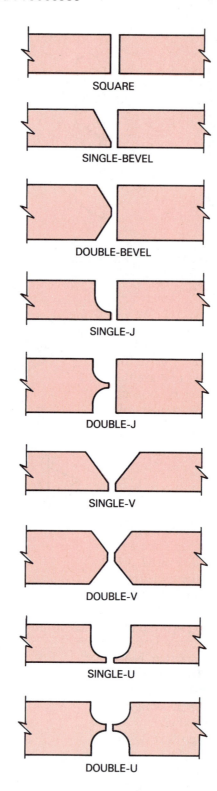

SQUARE

SINGLE-BEVEL

DOUBLE-BEVEL

SINGLE-J

DOUBLE-J

SINGLE-V

DOUBLE-V

SINGLE-U

DOUBLE-U

Figure 18-3 Edge preparation.

Base Metal

Because some metals have specific problems with things like thermal expansion, crack sensitivity, or distortion, the joint selected must control these problems. For example, magnesium is very susceptible to postweld stresses, and the U-groove works best for thick sections.

Plate Welding Positions

The most ideal welding position for most joints is the flat position, because it allows for larger molten weld pools to be controlled. Usually, the larger the weld pool, the faster the joint can be completed. It's not always possible to position the part so that all the welds can be made in

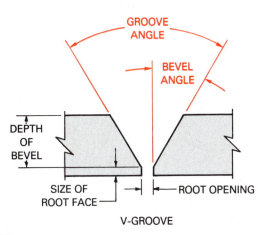

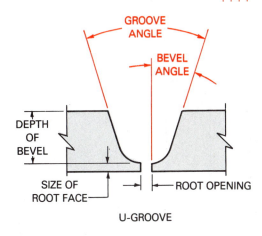

Figure 18-4

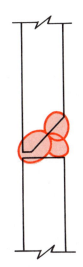

Figure 18-5

the flat position. Some joint design must be used for some out-of-position welding. For example, the bevel joint is often the best choice for horizontal welding, **Figure 18-5**.

The American Welding Society has divided plate welding into four basic positions for grooves **(G)** and fillet **(F)** welds as follows:

- ■ Flat 1G or 1F — When welding is performed from the upper side of the joint, and the face of the weld is approximately horizontal, **Figure 18-6**.

- ■ Horizontal 2G or 2F — The axis of the weld is approximately horizontal, but the type of the weld dictates the complete definition. For a fillet weld, welding is performed on the upper side of an approximately vertical surface. For a groove weld, the face of the weld lies in an approximately vertical plane, **Figure 18-7**.

- ■ Vertical 3G or 3F — The axis of the weld is approximately vertical, **Figure 18-8** (page 410).

- ■ Overhead 4G or 4F — When welding is performed from the underside of the joint, **Figure 18-9** (page 410).

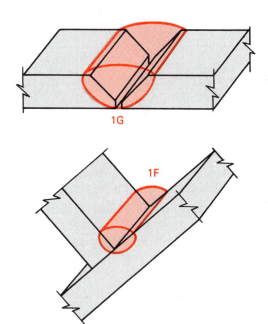

Figure 18-6 Plate flat position.

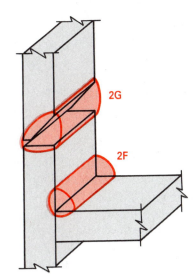

Figure 18-7 Plate horizontal position.

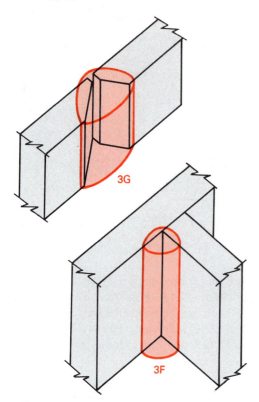

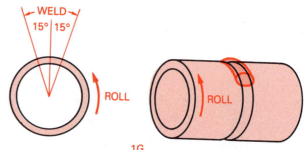

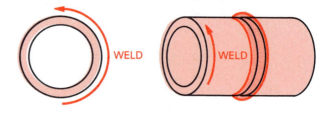

Figure 18-10 Pipe horizontal rolled position.

Figure 18-11 Pipe horizontal fixed position.

Figure 18-8 Plate vertical position.

Figure 18-12 Pipe vertical position.

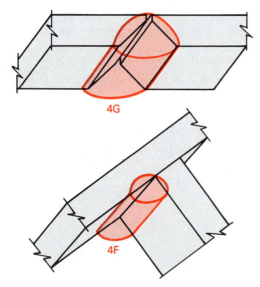

Figure 18-9 Plate overhead position.

Pipe Welding Positions

The American Welding Society Pipe has divided pipe welding into five basic positions:

- Horizontal Rolled 1G — When the pipe is rolled either continuously or intermittently so that the weld is performed within 0° to 15° of the top of the pipe, **Figure 18-10.**

- Horizontal Fixed 5G — When the pipe is parallel to the horizon, and the weld is made vertically around the pipe, **Figure 18-11.**

- Vertical 2G — The pipe is vertical to the horizon, and the weld is made horizontally around the pipe, **Figure 18-12.**

- Inclined 6G — The pipe is fixed in a 45° inclined angle, and the weld is made around the pipe, **Figure 18-13.**

- Inclined with a Restriction Ring 6GR — The pipe is fixed in a 45° inclined angle, and there is a restricting ring placed around the pipe below the weld groove, **Figure 18-14.**

Metal Thickness

As the metal becomes thicker the joint design must change. On thin sections its often possible to make full penetration welds using a square butt joint. But with

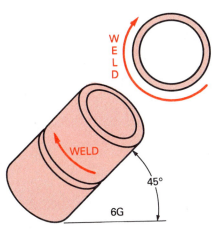

Figure 18-13 Pipe 45° inclined position.

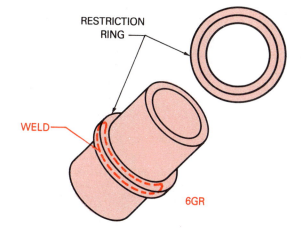

Figure 18-14 Pipe 45° inclined position with a restricting ring.

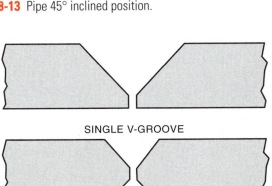

SINGLE V-GROOVE

DOUBLE V-GROOVE

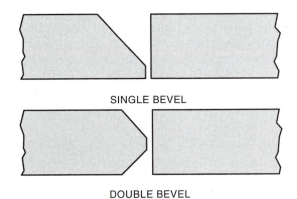

SINGLE BEVEL

DOUBLE BEVEL

Figure 18-15

thicker plates or pipe the edge must be prepared with a groove on one or both sides. The edge may be shaped with either a bevel, V-groove, J-groove, or U-groove. The choice of shape depends on the type of metal, its thickness, and whether it is made before or after assembly.

When welding on thick plate or pipe, it is often impossible for the welder to get 100% penetration without using some type of groove. The groove may be cut into either one of the plates or pipes or both. On some plates it can be cut both inside and outside of the joint, **Figure 18-15**. The groove may be ground, flame cut, gouged, sawed, or machined on the edge of the plate before or after the assembly. Bevels and V-grooves are best if they are cut before the parts are assembled. J-grooves and U-grooves can be cut either before or after assembly, **Figure 18-16**. The lap joint is seldom prepared with a groove, because little or no strength can be gained by grooving this joint.

For most welding processes, plates that are thicker than 3/8 in. (10 mm) may be grooved on both the inside and outside of the joint. Whether to groove one or both sides is most often determined by joint design, position, code, and application. Plates in the flat position are usually grooved on only one side unless they can be repositioned or are required to be welded on both sides. Tee joints in thick plates are easier to weld and have less distortion if they are grooved on both sides.

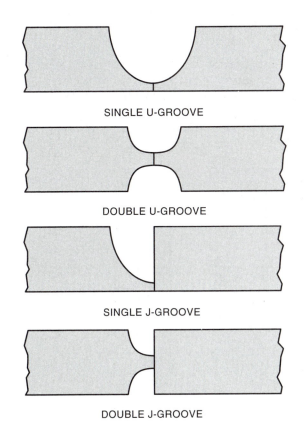

SINGLE U-GROOVE

DOUBLE U-GROOVE

SINGLE J-GROOVE

DOUBLE J-GROOVE

Figure 18-16

Sometimes plates are either grooved and welded or just welded on one side and then back-gouged and welded, **Figure 18-17** (page 412). Back gouging is a process of cutting a groove in the back side of a joint that has been welded. Back gouging can ensure 100% joint fusion at the root and remove discontinuities of the root pass.

Code or Standards Requirements

The type, depth, angle, and location of the groove are usually determined by a code or standard that has been qualified for the specific job. Organizations such as the American Welding Society, the American Society of Mechanical Engineers, and the American Bureau of Ships are among the agencies that issue such codes and specifications. The most common code or standards are the AWS D1.1 and the ASME Boiler and Pressure Vessel (BPV), Section IX.

The joint design for a specific set of specifications often must be what is known as prequalified. These joints have been tested and found to be reliable for the weldments for specific applications. The joint design can be modified, but the cost to have the new design accepted under the standard being used is often prohibitive.

Welder Skill

Often the skills or abilities of the welder are a limiting factor in joint design. A joint must be designed in a manner so that the welders can reliably reproduce it. Some joints have been designed without adequate room for the welder to see the molten weld pool or room to get the electrode or torch into the joint.

Acceptable Cost

Almost any weld can be made in any material in any position. A number of factors can affect the cost of producing a weld. Joint design is one major way to control

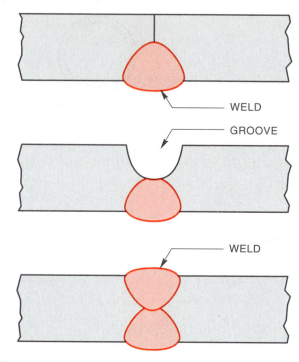

Figure 18-17 Back-gouging a weld joint to ensure 100% joint penetration.

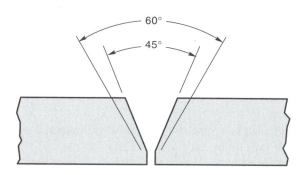

Figure 18-18 A smaller groove angle reduces both weld time and weld metal.

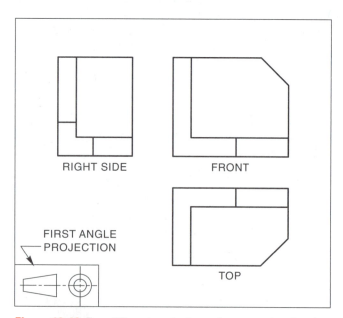

RIGHT SIDE   FRONT

FIRST ANGLE PROJECTION

TOP

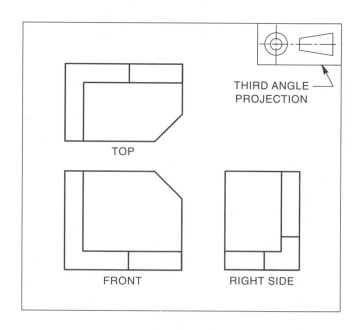

THIRD ANGLE PROJECTION

TOP

FRONT   RIGHT SIDE

Figure 18-19 Two different methods used to rotate drawing views.

welding cost. Changes in the design can reduce cost yet still meet the weldment's strength requirements. Reducing the groove angle can also help, **Figure 18-18.** It will decrease the welding filler metal required to complete the weld as well as decrease the time required to fill the larger groove opening. Joint design must be a consideration in order for any project to be competitive and cost effective.

MECHANICAL DRAWINGS

Mechanical drawings have been around for centuries. Leonardo da Vinci (1452–1519) used mechanical drawings extensively in his inventive works. Many of his drawings still exist today and are as easily understood

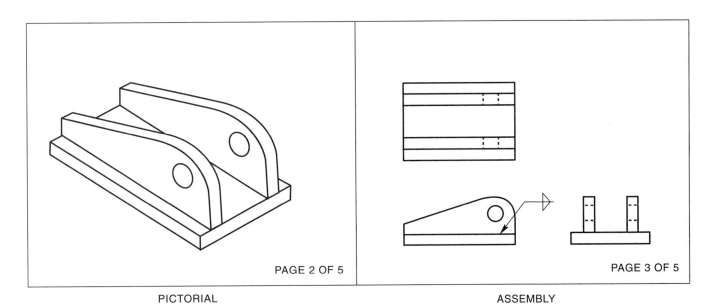

Figure 18-20 Drawings that can make up a set of drawings.

now as when they were drawn. For that reason mechanical drawings have been called the universal language; they are produced in a similar format worldwide. Despite the few differences in how the views are laid out, **Figure 18-19**, the drawings are understandable. Notwithstanding different languages and measuring systems, the basic shape of an object and location of components can be determined from any good drawing.

A group of drawings, known as a *set of drawings*, should contain enough information to enable a welder to produce the weldment. The set of drawings may contain various pages showing different aspects of the project to aid in its fabrication. The pages may include the follow-

ing: title page, pictorial, assembly drawing, detailed drawing, and exploded view, **Figure 18-20**.

In addition to the actual shape as described by the various lines, a set of drawings may contain additional information such as the title box and bill of materials. The *title box*, which appears in one corner of the drawing, should contain the name of the part, the company name, the scale of the drawing, the date of the drawing, who made the drawing, the drawing number, the number of drawings in this set, and tolerances.

A *bill of materials* can also be included in the set of drawings. This is a list of the various items that will be needed to build the weldment, **Figure 18-21**.

		BILL OF MATERIALS		
Part	Number Required	Type of Material	Size Standard Units	Size SI Units
Base	1	Hot Roll Steel	1/2" × 5" × 8"	12.7 mm × 127 mm × 203.2 mm
Cleat	2	Hot Roll Steel	1/2" × 4" × 8"	12.7 mm × 101.6 mm × 203.2 mm

Figure 18-21

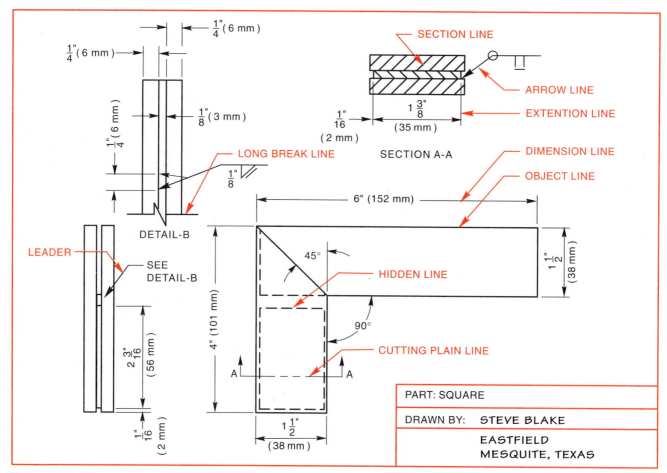

Figure 18-22

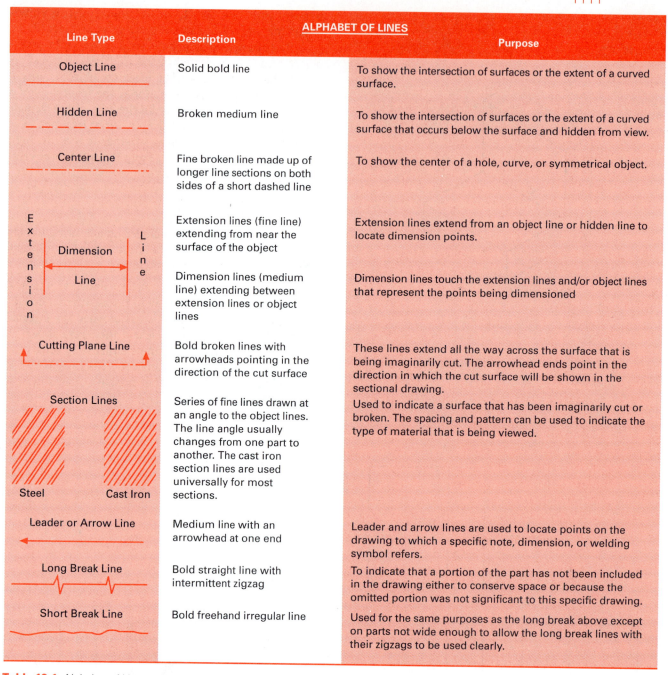

Line Type	Description	Purpose
ALPHABET OF LINES		
Object Line	Solid bold line	To show the intersection of surfaces or the extent of a curved surface.
Hidden Line	Broken medium line	To show the intersection of surfaces or the extent of a curved surface that occurs below the surface and hidden from view.
Center Line	Fine broken line made up of longer line sections on both sides of a short dashed line	To show the center of a hole, curve, or symmetrical object.
Extension Line — Dimension Line	Extension lines (fine line) extending from near the surface of the object	Extension lines extend from an object line or hidden line to locate dimension points.
	Dimension lines (medium line) extending between extension lines or object lines	Dimension lines touch the extension lines and/or object lines that represent the points being dimensioned
Cutting Plane Line	Bold broken lines with arrowheads pointing in the direction of the cut surface	These lines extend all the way across the surface that is being imaginarily cut. The arrowhead ends point in the direction in which the cut surface will be shown in the sectional drawing.
Section Lines (Steel, Cast Iron)	Series of fine lines drawn at an angle to the object lines. The line angle usually changes from one part to another. The cast iron section lines are used universally for most sections.	Used to indicate a surface that has been imaginarily cut or broken. The spacing and pattern can be used to indicate the type of material that is being viewed.
Leader or Arrow Line	Medium line with an arrowhead at one end	Leader and arrow lines are used to locate points on the drawing to which a specific note, dimension, or welding symbol refers.
Long Break Line	Bold straight line with intermittent zigzag	To indicate that a portion of the part has not been included in the drawing either to conserve space or because the omitted portion was not significant to this specific drawing.
Short Break Line	Bold freehand irregular line	Used for the same purposes as the long break above except on parts not wide enough to allow the long break lines with their zigzags to be used clearly.

Table 18-1 Alphabet of Lines.

Lines

To understand drawings, you need to know what the different types of lines represent. The language of drawing uses lines to represent its alphabet and the various parts of the object being illustrated. The various line types are collectively known as the *alphabet of lines,* Table 18-1 and Figure 18-22.

Types of Drawings

Drawings used for most welding projects can be divided into two categories, orthographic projections and pictorial. Projection drawings are made as though one were looking through the sides of a glass box at the object and tracing its shape on the glass, **Figure 18-23** (page 416). If all the sides of the object were traced and the box unfolded and laid out flat, six basic views would be shown, **Figure 18-24** (page 416).

Pictorial drawings present the object in a more realistic or understandable form and usually appear as one of two types, isometric or cavalier, **Figure 18-25** (page 417). The more realistic, perspective drawing form is seldom used for welding projects.

Projection Drawings

Usually not all of the six views are required to build the weldment. Only those needed are normally provided,

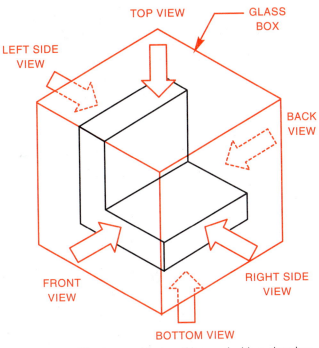

Figure 18-23 Viewing an object as if it were inside a glass box.

usually only the front, right side, and top views. Sometimes only one or two of these views are needed.

The front view is not necessarily the front of the object. A view is selected as the front view because its overall shape is best described when the object is viewed from this direction. As an example, the front view of a car or truck would probably be the side of the vehicle because viewing the vehicle from its front may not show enough detail to let you know whether it is a car, light truck, station wagon, or van. From the front most vehicles may all look very similar.

Special Views

Special views may be included on a drawing to help describe the object so it can be made accurately. Special views on some drawings may include:

■ *Sections view:* The section view is drawn as if part of the object were sawn away to reveal internal details, **Figure 18-26.** This view is useful when the internal details would not be as clear if they were shown as hidden lines. Sections can be either fully across the object or just partially across it. The imaginary cut surface is set off from other noncut surfaces by section lines drawn at an angle on the cut surfaces.

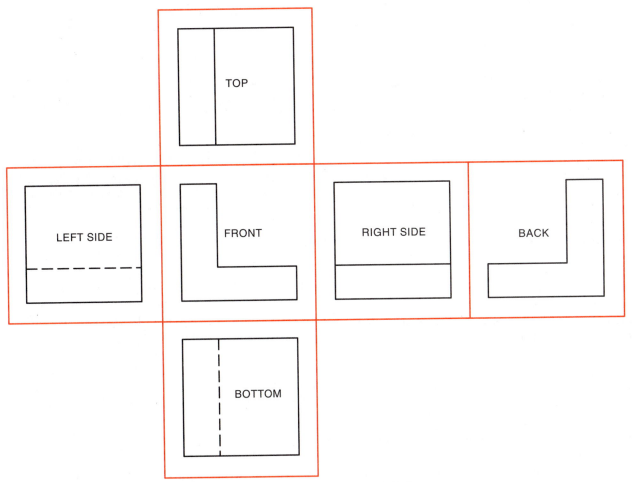

Figure 18-24 The arrangement of views for an object if the glass box were unfolded.

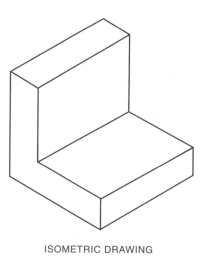

ISOMETRIC DRAWING

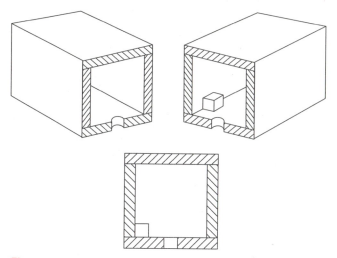

Figure 18-26 Section drawing.

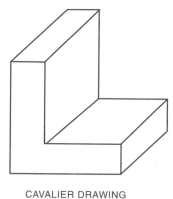

CAVALIER DRAWING

Figure 18-25 Pictorial drawing types.

Some drawings use specific types of section lines to illustrate the type of material the part was made with. The location of this imaginary cut is shown using a cutting plane line, **Figure 18-27.**

■ *Cut-a-ways:* The cut-a-way view is used to show detail within a part that would be obscured by the part's surface. Often a free-hand break line is used to outline the area that has been imaginarily removed to reveal the inner workings.

■ *Detail views:* The detail view is usually an external view of a specific area of a part. Detail views show small details of a part's area and negate the need to draw an enlargement of the entire part. If only a small portion of a view has significance, this area can be shown in a detail view, either at the same scale or larger if needed. By showing only what is needed within the detail, the part drawn can be clearer and not require such a large page.

■ *Rotated views:* A rotated view can be used to show a surface of the part that would not normally be drawn square to any of the six normal view planes. If a surface is not square to the viewing angle, then lines may be distorted. For example, when viewed at an angle, a circle looks like an ellipse, **Figure 18-28** (page 418).

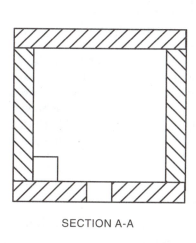

SECTION A-A

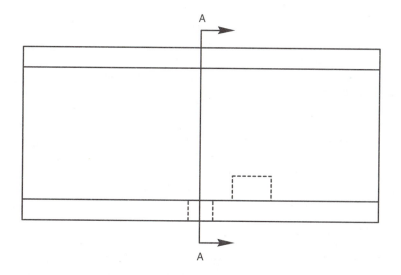

Figure 18-27

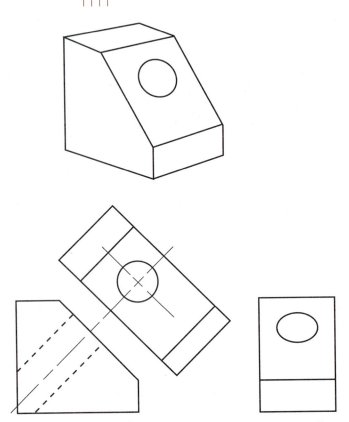

Figure 18-28 Notice that the round hole looks misshapen, elliptical, in the right side view.

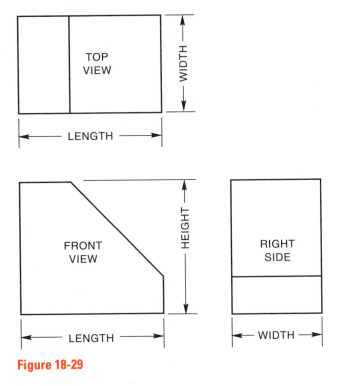

Figure 18-29

Dimensioning

Often it is necessary to look at other views to locate all of the dimensions required to build the object. By knowing how the views are arranged, it becomes easier to locate dimensions. Length dimensions can be found on the front and top views. Height dimensions can be found on the front and right side views. Width dimensions can be found on the top and right side views, Figure 18-29 (page 418). The locations of dimensions on these views is consistent with both the first angle perspective or third angle perspective layouts.

If the needed dimensions cannot be found on the drawings, do not try to obtain them by measuring the drawing itself. Even if the original drawing was made very accurately, the paper it is on changes with changes in humidity. Copies of the original drawing are never the exact same size. The most acceptable way of determining missing dimensions is to contact the person who made the drawing.

Keep the drawing clean and well away from any welding. Avoid writing or doing calculations on the drawing. Often a drawing will be filed following the project for use at a later date. The better care you take with the drawings, the easier it will be for someone else to use them.

WELDING SYMBOLS

The use of welding symbols enables a designer to indicate clearly to the welder important detailed informa-tion regarding the weld. The information in the welding symbol can include the following details for the weld: length, depth of penetration, height of reinforcement, groove type, groove dimensions, location, process, filler metal, strength, number of welds, weld shape, and surface finishing. All this information would normally be included on the welding assembly drawings.

Welding symbols are a shorthand language for the welder. They save time and money and serve to ensure understanding and accuracy. Welding symbols have been standardized by the American Welding Society. Some of the more common symbols for welding are reproduced in this chapter. If more information is desired about symbols or how they apply to all forms of manual and automatic machine welding, these symbols can be found in the complete material, *Standard Symbols for Welding, Brazing and Nondestructive Examination,* ANSI/AWS A2.4, published as an American National Standard by the American Welding Society.

Figure 18-30 shows the basic components of welding symbols, consisting of a reference line with an arrow on one end. Other information relating to various features of the weld are shown by symbols, abbreviations, and figures located around the reference line. A tail is added to the basic symbol as necessary for the placement of specific information.

INDICATING TYPES OF WELDS

Weld types are classified as follows: fillets, grooves, flange, plug or slot, spot or protecting, seam, back or backing, and surfacing. Each type of weld has a specific

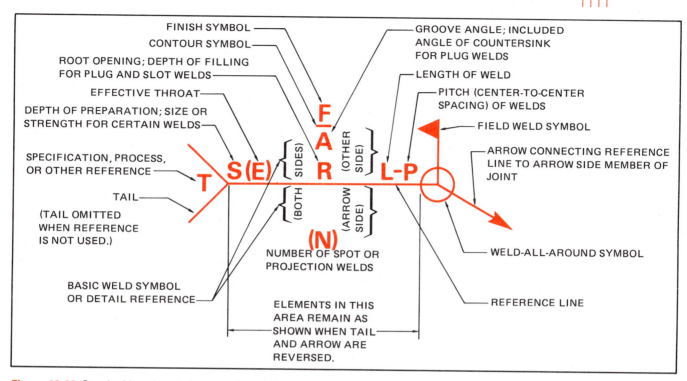

Figure 18-30 Standard location of elements of a welding symbol. (Courtesy of the American Welding Society.)

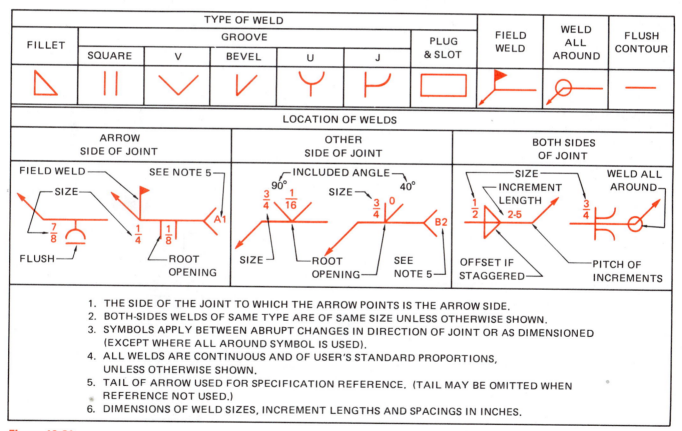

1. THE SIDE OF THE JOINT TO WHICH THE ARROW POINTS IS THE ARROW SIDE.
2. BOTH-SIDES WELDS OF SAME TYPE ARE OF SAME SIZE UNLESS OTHERWISE SHOWN.
3. SYMBOLS APPLY BETWEEN ABRUPT CHANGES IN DIRECTION OF JOINT OR AS DIMENSIONED (EXCEPT WHERE ALL AROUND SYMBOL IS USED).
4. ALL WELDS ARE CONTINUOUS AND OF USER'S STANDARD PROPORTIONS, UNLESS OTHERWISE SHOWN.
5. TAIL OF ARROW USED FOR SPECIFICATION REFERENCE. (TAIL MAY BE OMITTED WHEN REFERENCE NOT USED.)
6. DIMENSIONS OF WELD SIZES, INCREMENT LENGTHS AND SPACINGS IN INCHES.

Figure 18-31

symbol that is used on drawings to indicate the weld. A fillet weld, for example, is designated by a right triangle. A plug weld is indicated by a rectangle. All of the basic symbols are shown in **Figure 18-31**.

WELD LOCATION

Welding symbols are applied to the joint as the basic reference. All joints have an arrow side (near side) and another side (far side). Accordingly, the terms arrow side, other side, and both sides are used to indicate the **weld location** with respect to the joint. The reference line is always drawn horizontally. An arrow line is drawn from one end or both ends of a reference line to the location of the weld. The arrow line can point to either side of the joint and extend either upward or downward.

If the weld is to be deposited on the arrow side of the joint (near side), the proper weld symbol is placed below the reference line, Figure 18-32(A).

If the weld is to be deposited on the other side of the joint (far side), the weld symbol is placed above the reference line, Figure 18-32(B). When welds are to be deposited on both sides of the same joint, the same weld symbol appears above and below the reference line, Figure 18-32(C) and (D).

The tail is added to the basic welding symbol when it is necessary to designate the welding specifications,

procedures, or other supplementary information needed to make the weld, Figure 18-33. The notation placed in the tail of the symbol may indicate the welding process to be used, the type of filler metal needed, whether or not peening or root chipping is required, and other information pertaining to the weld. If notations are not used, the tail of the symbol is omitted.

For joints that are to have more than one weld, a symbol is shown for each weld.

LOCATION SIGNIFICANCE OF ARROW

In the case of fillet and groove welding symbols, the arrow connects the welding symbol reference line to one side of the joint. The surface of the joint the arrow point actually touches is considered to be the arrow side of the joint. The side opposite the arrow side of the joint is considered to be the other (far) side of the joint.

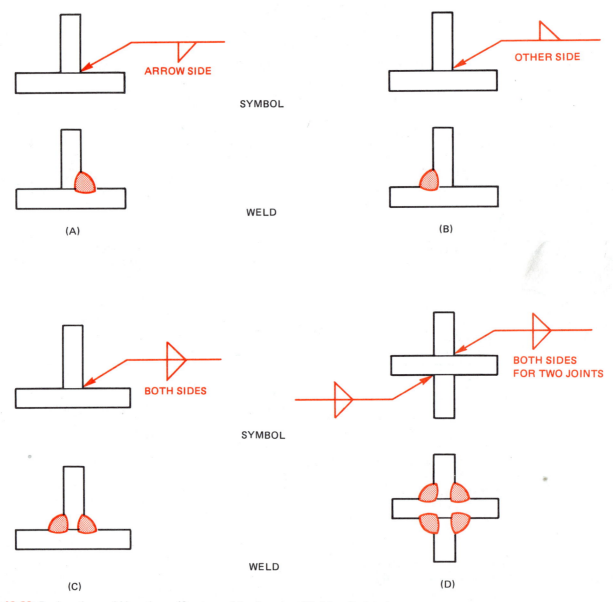

Figure 18-32 Designating weld locations. (Courtesy of the American Welding Society.)

On a drawing, when a joint is illustrated by a single line and the arrow of a welding symbol is directed to the line, the arrow side of the joint is considered to be the near side of the joint.

For welds designated by the plug, slot, spot, seam, resistance, flesh, upset, or projection welding symbols, the arrow connects the welding symbol reference line to the outer surface of one of the members of the joint at the center line of the desired weld. The member to which the arrow points is considered to be the arrow side member. The remaining member of the joint is considered to be the other side member.

FILLET WELDS

Dimensions of fillet welds are shown on the same side of the reference line as the weld symbol and are shown to the left of the symbol, **Figure 18-34(A)**. When both sides of a joint have the same size fillet welds, one or both may be dimensioned as shown in **Figure 18-34(B)**. When both sides of a joint have different size fillet welds, both are dimensioned, **Figure 18-34(C)**. When the dimensions of one or both welds differ from the dimensions given in the general notes, both welds are dimensioned. The size of a fillet weld with unequal legs

is shown in parentheses to the left of the weld symbol, **Figure 18-34(D)**. The length of a fillet weld, when indicated on the welding symbol, is shown to the right of the weld symbol, **Figure 18-34(E)**. In intermittent fillet welds, the length and pitch increments are placed to the right of the weld symbol, **Figure 18-35** (page 422). The first number represents the length of the weld, and the second number represents the pitch or the distance between the centers of two welds.

PLUG WELDS

Holes in the arrow side member of a joint for plug welding are indicated by placing the weld symbol below the reference line. Holes in the other side member of a joint for plug welding are indicated by placing the weld symbol above the reference line, **Figure 18-36** (page 422). Refer to **Figure 18-36** for the location of the dimensions used on plug welds. The diameter or size is located to the left of the symbol (A). The angle of the sides of the hole, if not square, are given above the symbol (B). The depth of buildup, if not completely flush with the surface, is given in the symbol (C). The center-to-center dimensioning or pitch is located on the right of the symbol (D).

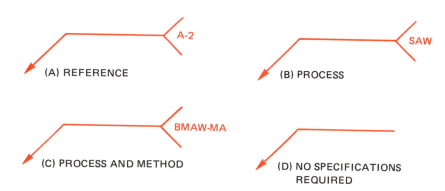

Figure 18-33 Locations of specifications, processes, and other references on weld symbols.

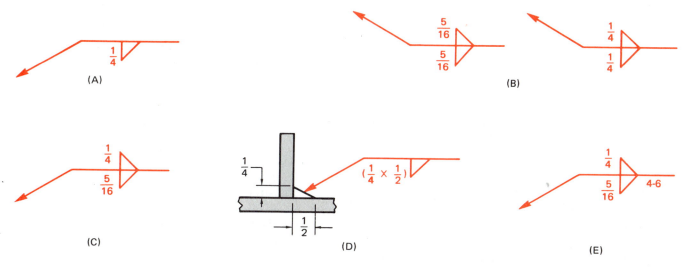

Figure 18-34 Dimensioning the fillet weld symbol.

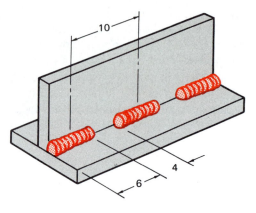

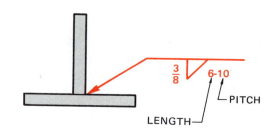

Figure 18-35 Dimensioning intermittent fillet welds.

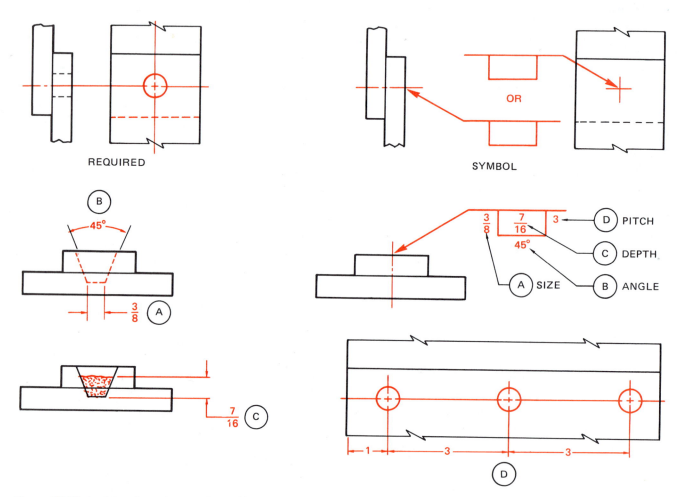

Figure 18-36 Applying dimensions to plug welds.

SPOT WELDS

Dimensions of spot welds are indicated on the same side of the reference line as the weld symbol, **Figure 18-37** (page 423). Such welds are dimensioned either by size or strength. The size is designated as the diameter of the weld expressed in fractions or in decimal hundredths of an inch. The size is shown with or without inch marks to the left of the weld symbol. The center-to-center spacing (pitch) is shown to the right of the symbol.

The strength of spot welds is shown as the minimum shear strength in pounds (newtons) per spot and is shown to the left of the symbol, **Figure 18-38(A)**. When a definite number of spot welds is desired in a certain joint, the quantity is placed above or below the weld symbol in parentheses, **Figure 18-38(B)**.

SHIELDED METAL ARC WELDING (SEE SECTION II)

Plate 1   E6010 electrode.

A   Note the arc tract across the plate to the left.

B   The arc has stabilized and a weld pool has been established. Note the weld is proceeding over the original arc tracts made from the striking.

C   The arc tract has been completely welded over.

Plate 2   E6011 electrode.

A   Note the molten weld pool and slide covering. The electrode is held near the front of the weld pool.

B   The electrode has been moved to the leading edge of the weld pool in a stepping technique frequently used with E6011 and E6010 electrode.

C   Fillet Weld. Note the relatively thin layer of flux covering the solidified weld. Also note the position of the electrode in the weld pool.

Plate 3   E6012 electrode.

The E6012 electrode produces a few more sparks than the E6013 electrode. It also produces slightly less slag.

SHIELDED METAL ARC WELDING (SEE SECTION II)

Plate 4   E6013 electrode.

The E6013 electrode produces a weld with a heavy slag covering. It also produces relatively few sparks.

Plate 5   E7018 electrode.

Note the near lack of spatter and limited smoke. Also the thick smooth slag blanket covering the weld.

CUTTING AND GOUGING (SEE SECTION III)

Plate 6   Oxyfuel Cutting.

A   Oxyfuel torch heating edge of plate before cut is started.

B   Oxyfuel torch heating edge of plate before cut is started.

C   The cut has been started. Note the height of the preheat flame is above the plate.

D   The kerf being made is square to the plate and has parallel sides.

E&F   Note that the top edge of the plate has not been melted as the cut is made and there are few, if any, sparks.

Plate 7   Plasma Arc Cutting — Freehand

A   Plasma arc cutting of 3/4 inch aluminum plate. The plasma arc is started along the edge of the plate in the same manner as with oxyfuel cutting. However there is no need to allow the edge to heat up before the cut is started.

B   Notice the height above the plate that the restricting nozzle is being held.

C   End of the first cut.

Plate 8   Plasma Arc Cutting with a nozzle support guide.

A   Notice how stiff the plasma column is before the cut is started.

B   Cut is started.

C   Cut continues.

D   Notice how the guide keeps the nozzle supported above the plate, which makes cutting much more relaxed.

Plate 9 Air Carbon Arc Gouging.

Air Carbon Arc Gouging of a
U-groove in a square butt joint.

GAS SHIELDED WELDING (SEE SECTION IV)

Plate 10 Gas Metal Arc
Welding, stringer bead using the
backhand or dragging angle.

Plate 11 Gas Metal Arc Welding,
stringer bead using the forehand or
leading angle.

Plate 12 Gas Metal Arc Welding,
fillet weld using the backhand
technique.

Note the silicon slag floating
on the top of the weld pool.
Also note that the arc is directed into the weld near the
leading edge.

Notice how the arc force is pushing
the heat ahead of the weld. Also
notice that the arc is located just
inside of the leading edge of the weld
pool.

The flat weld bead contour can
be seen with its uniform weld
bead ripples.

Plate 13 Gas Metal Arc Welding on pipe.

A&B Notice that
the weld is only
large enough to fill
part of the groove.
Too large of a weld
pool would be
impossible to con-
trol. A forehand or
pushing angle is
used.

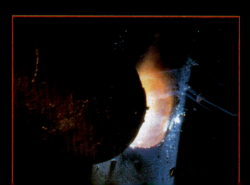

IIII

Plate 14 Gas Metal Arc short circuiting arc metal transfer forehand technique.

The metal is being transferred from the electrode to the weld pool.

Plate 15 Gas Metal Arc short circuiting arc metal transfer backhand technique.

The high weld buildup and shallow weld penetration of this welding technique is shown.

Plate 16 Gas Metal Arc axial spray arc metal transfer forehand technique.

The bright arc is created during this process. The large weld pool with the silicon slag floating in it can easily be seen. Note the smooth weld bead that shows very small ripples.

Plate 17 Gas Metal Arc axial spray arc metal transfer backhand technique.

The arc color is spreading out as it transfers across the gap. The large fluid weld pool can be seen. Note that the weld bead has very little buildup.

Plate 18 Flux Cored Arc Welding, beading using the backhand or dragging angle.

Notice that the weld bead has been covered in a thick layer of slag. There are a large number of sparks and the heavy shielding gas cloud is glowing bright white. This light makes it harder to see details in the weld pool.

Plate 19 Flux Cored Arc Welding, fillet weld using a forehand technique.

Notice how the leading edge of the weld pool has a smooth, even contour as it fused into the base metal. The electrode is entering the weld pool just behind the leading edge.

Plate 20 Flux Cored Arc Welding, fillet weld using a backhand technique.

This welding technique gives a better view on the trailing edge of the weld pool, but you cannot see how the leading edge is fusing into the base metal.

Plate 21 Flux cored arc welding with a leading angle.

A Gas Tungsten arc torch is positioned so that the cup is resting on one side and the tungsten extends to the root opening of the joint.

C From the side you can see the position of the filler metal and the welder as the torch is held in preparation for welding.

Plate 23 Gas Tungsten Arc Welding Pipe, using cup walking technique.

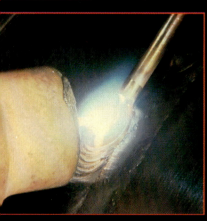

A The filler metal is held in continuous contact with the molten weld pool as the cup is walked along the joint. Notice the area surrounding the molten weld pool that is bright and shiny as it is protected from the atmosphere by shielding gas coverage coming from the torch.

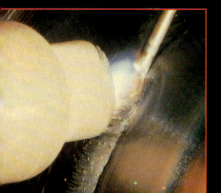

C Less of the weld pool is visible as the torch is continued to be rotated so that the cup will walk up the pipe.

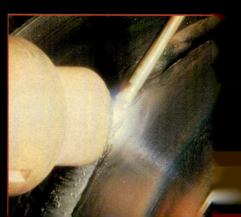

Plate 24 Gas Tungsten Arc
Welding, cup walking, filler pass.

A The tungsten can be seen approximately 1/16 inch above the molten weld pool as the cup is walked up. Notice that in this sequence the root pass is visible and the cover pass can be seen as it solidifies behind the molten weld pool.

B The gas torch has been rotated and the cup edge closest to the right has been raised slightly so it can be moved forward in the next sequence.

C Shows the torch being rotated back toward the welder and stepped forward slightly.

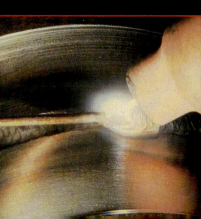

D Now the process is repeated.

Fillet weld has a bright gold color indicating that it was adequately protected by the shielding gas during the welding process. Small line is visible above the weld that was left as the cup walked along that surface during the welding process. Notice that the spacing of the rippled line is exactly in sequence with the weld ripples as the weld pool solidified.

Plate 26 Gas Tungsten Arc Welding, cup walking, cover pass technique, stainless steel.

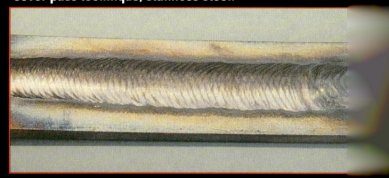

The process using cup walking allows a weld to be made and a filler pass. It is not required that the technique be used in a V-groove or fillet weld to produce an excellent quality weld.

Plate 27 Gas Tungsten Arc Welding, multiple pass, stainless steel using the cup walking technique.

Shows a multiple pass weld in a T-joint. 1 root pass, 3 filler passes and 2 cover passes can distinctly be seen. Note that each weld pass has a bright, clean appearance because of the excellent shielding coverage that this process has.

Plate 28 Forming a balled end on the tungsten by striking a DCEP (DCRP) arc on a copper plate.

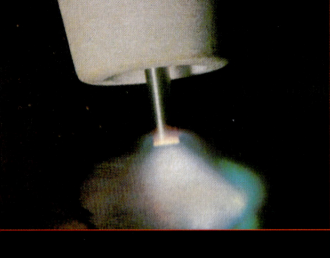

Plate 29 The high frequency has started from the zirconium stripe on the tungsten.

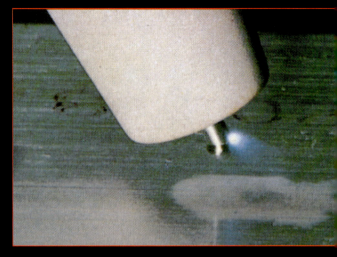

Plate 30 The arc cleaning action can be seen before a molten weld pool is established.

Plate 31 The molten weld pool can be seen as a shin spot in the center of the area cleaned by the arc actio

Plate 32 The end of the filler rod is dipped into the leading edge of the molten weld pool.

Plate 33 The hot end of the filler rod is kept inside the shielding gas envelope to prevent it from being oxidized by the atmosphere.

Plate 34 Welding on stainless steel.

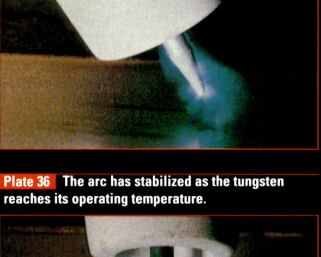

Plate 35 The plasma intensifies at the tungsten tip as the arc starts. The tungsten has not reached its operating temperature.

Plate 36 The arc has stabilized as the tungsten reaches its operating temperature.

Plate 37 A molten weld pool is established.

Plate 38 The filler rod is brought into the arc zone.

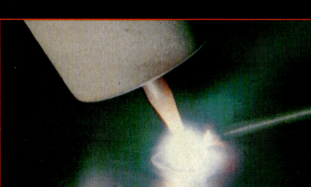

Plate 39 The end of the filler rod is dipped into the leading edge of the molten weld pool.

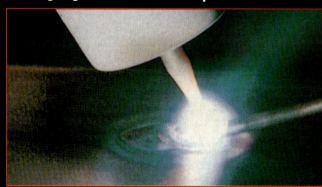

Plate 40 The rod is repeatedly dipped into the molten weld pool to form a weld bead. Note the color change that is taking place in the weld bead as it becomes exposed to the atmosphere.

Plate 41 Manufacturing Steel — Welding Metallurgy.

A   Alloying elements are being added to a melt of steel just prior to pouring the steel.

B   The steel is being tapped from the furnace and a metallurgical sample being taken.

C   Molten steel is transferred from the electric arc furnace into a ladle for transport to the continuous casting machine.

D   Molten steel can be seen flowing from the hopper into the continuous cast forms as the steel is made into bar stock.

E   Below the continuous casting, red hot bars of steel can be seen coming out of the machine.

F   Once the steel has cooled, it is flame-cut into 20-foot-long sections. Each section is identified with a metal tag showing its heat number so its chemical composition can be traced.

Plate 42 Robotic Welding using cold wire feed Gas Tungsten Arc Welding of tubes to tube sheet.

A Cyro robotic arm is shown with a cold wire feed gas tungsten arc welding torch tip for automatically welding tubes to tube sheets frequently used for heat exchangers.

Plate 43 Laser beam welding of a tube to a cap.

A high-powered CO_2 gas laser is being used to audaciously weld a cap onto a thin walled tubing section. The positioner used to rotate the part below the laser beam. The laser beam is being positioned during the weld by a robot.

Plate 44 Thermite welding of railroad track section.

A   Thermite welding of a railroad track section is shown. Railroads are frequently welded in this manner during the construction of new track or to make repairs.

B   A machine used by the railroad for replacing and repairing track sections is shown.

A and B courtesy of Burlington Northern

Plate 45 Plasma arc thermal spraying.

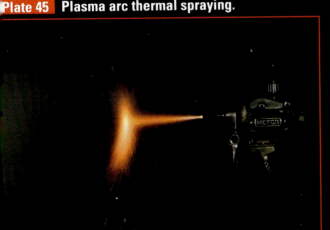

A plasma arc torch is being used to thermally spray

Plate 46 Plasma arc cutting.

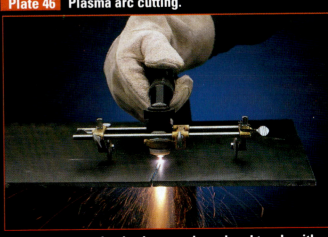

Plasma arc cutting is shown using a hand torch with a

OXYFUEL WELDING (SEE SECTION VI)

Plate 47 Oxyacetylene Welding Flame.

A Excessive oxygen (oxidizing). Observe the distinct, blue inner cone.

B Excessive acetylene flame (carbonizing). In this flame, the inner cone is long and feather-like in appearance.

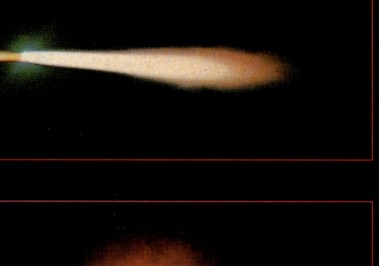

C Equal portions of oxygen and acetylene flame (neutral). Here the inner cone is distinct, but not as blue in color as the oxidizing flame.

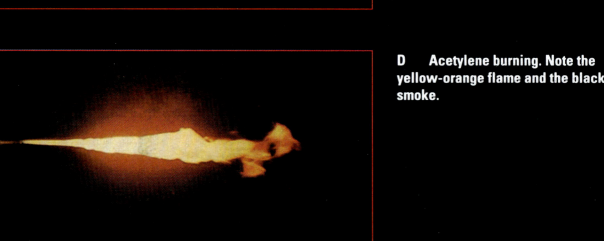

D Acetylene burning. Note the yellow-orange flame and the black smoke.

Plate 48 Oxyacetylene Welding.

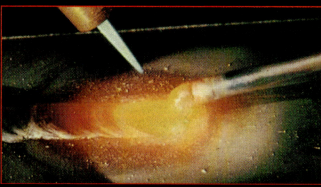

A neutral flame produces a shiny molten weld pool with little or no sparks.

Plate 49 Oxyacetylene Welding.

An excessive acetylene (carbonizing) flame produces a crusty molten weld pool.

Plate 50 Oxyacetylene Welding.

An excessive oxygen (oxidizing) flame produces a foam around molten weld pool and a large amount of sparks.

Plate 51 Oxyacetylene Welding.

Metal is being heated to a melt to start a weld.

Plate 52 Oxyacetylene Welding.

Filler metal is being added to the molten weld pool as it starts melting.

Plate 53 Brazing.

The base metal is heated and the brazing rod is touched to it to determine if the brazing temperature has been reached.

Plate 54 Brazing.

The brazing rod is dipped into the leading edge of the braze pool.

Plate 55 The external fuel tanks of the space shuttle are fabricated by welding. Both gas tungsten and plasma arc welding processes are used to weld the 2219 aluminum tank components.

Courtesy of Ferranti Sciaky, Inc.

Plate 56 This machine is applying a hardfacing to the journal of a rock bit.

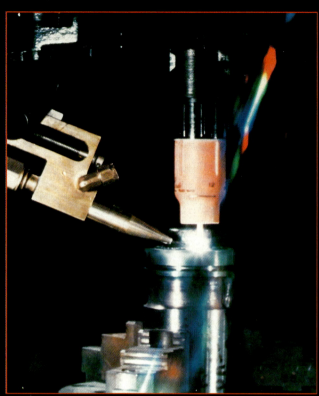

Courtesy of Ferranti Sciaky, Inc.

Plate 57 Semiautomatic flux cored arc welding of box columns on a major construction project.

The Lincoln Electric Company

Plate 58 A robot holds a part and moves it under a stationary plasma torch to cut holes out of manifold parts.

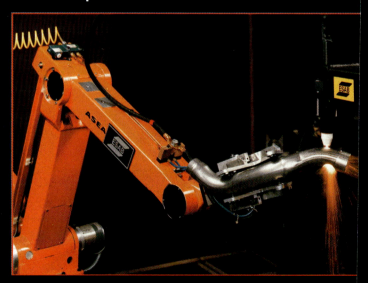

Courtesy of ESAB North America, Inc.

Plate 59 Underwater welding, shown here, is an important part of the many jobs a professional diver may need to be able to perform.

Courtesy of Ocean Corporation

Plate 60 Robot is performing a plasma arc cutting operation on steel pipe.

Courtesy of Cincinnati Milacron Company

Plate 61 Electron beam in vacuum, used in electron beam welding.

Plate 62 Self-shielded flux-cored arc welding system is used to field weld steel beam tree columns. The shop-fabricated columns are then transported to the construction site of a 1.5 million-square-foot building. These columns are then welded to columns at the site. The framework consists of 22,000 pounds of steel. Two welders per column work in specially designed tents to protect them from rain, snow, and wind. the operators ultimately completed 1100 column splices, each requiring about 190 passes.

Plate 63 **Basic Guide to Ferrous Metallurgy.**

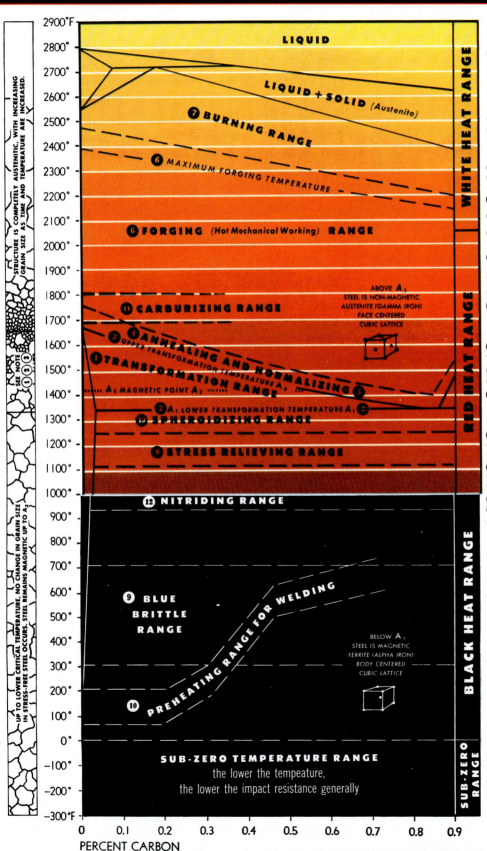

1 **TRANSFORMATION RANGE.** In this range steels undergo internal atomic changes which radically affect the properties of the material.

2 **LOWER TRANSFORMATION TEMPERATURE (A_1).** Termed Ac_1 on heating, Ar_1 on cooling. Below Ac_1 structure ordinarily consists of FERRITE and PEARLITE (see below). On heating through Ac_1 these constituents begin to dissolve in each other to form AUSTENITE (see below) which is non-magnetic. This dissolving action continues on heating through the TRANSFORMATION RANGE until the solid solution is complete at the upper transformation temperature.

3 **UPPER TRANSFORMATION TEMPERATURE (A_3).** Termed Ac_3 on heating, Ar_3 on cooling. Above this temperature the structure consists wholly of AUSTENITE which coarsens with increasing time and temperature. Upper transformation temperature is lowered as carbon increases to 0.85% (eutectoid point).

● **FERRITE** is practically pure iron (in plain carbon steels) existing below the lower transformation temperature. It is magnetic and has very slight solid solubility for carbon.

● **PEARLITE** is a mechanical mixture of FERRITE and CEMENTITE.

● **CEMENTITE** or IRON CARBIDE is a compound of iron and carbon, Fe_3C.

● **AUSTENITE** is the non-magnetic form of iron and has the power to dissolve carbon and alloying elements.

4 **ANNEALING,** frequently referred to as FULL ANNEALING, consists of heating steels to slightly above Ac_3, holding for AUSTENITE to form, then *slowly* cooling in order to produce small grain size, softness, good ductility and other desirable properties. On cooling slowly the AUSTENITE transforms to FERRITE and PEARLITE.

5 **NORMALIZING** consists of heating steels to slightly above Ac_3, holding for AUSTENITE to form, then followed by cooling (in still air). On cooling, AUSTENITE transforms giving somewhat higher strength and hardness and slightly less ductility than in annealing.

6 **FORGING RANGE** extends to several hundred degrees above the UPPER TRANSFORMATION TEMPERATURE.

7 **BURNING RANGE** is above the FORGING RANGE. Burned steel is ruined and *cannot be cured* except by remelting.

8 **STRESS RELIEVING** consists of heating to a point below the LOWER TRANSFORMATION TEMPERATURE, A_1, holding for a sufficiently long period to relieve locked-up stresses, then slowly cooling. This process is sometimes called PROCESS ANNEALING.

9 **BLUE BRITTLE RANGE** occurs approximately from 300° to 700° F. Peening or working of steels should not be done between these temperatures, since they are more brittle in this range than above or below it.

10 **PREHEATING FOR WELDING** is carried out to prevent crack formation. See TEMPIL° PREHEATING CHART for recommended temperature for various steels and non-ferrous metals.

11 **CARBURIZING** consists of dissolving carbon into surface of steel by heating to above transformation range in presence of carburizing compounds.

12 **NITRIDING** consists of heating certain *special steels* to about 1000° F for long periods in the presence of ammonia gas. Nitrogen is absorbed into the surface to produce extremely hard "skins".

13 **SPHEROIDIZING** consists of heating to just below the lower transformation temperature, A_1, for a sufficient length of time to put the CEMENTITE constituent of PEARLITE into globular form. This produces softness and in many cases good machinability.

● **MARTENSITE** is the hardest of the transformation products of AUSTENITE and is formed only on cooling below a certain temperature known as the M_s temperature (about 400° to 600° F for carbon steels). Cooling to this temperature must be sufficiently rapid to prevent AUSTENITE from transforming to softer constituents at higher temperatures.

● **EUTECTOID STEEL** contains approximately 0.85% carbon.

● **FLAKING** occurs in many alloy steels and is a defect characterized by localized micro-cracking and "flake-like" fracturing. It is usually attributed to hydrogen bursts. Cure consists of cycle cooling to at least 600° F before air-cooling.

● **OPEN OR RIMMING STEEL** has not been completely deoxidized and the ingot solidifies with a sound surface ("rim") and a core portion containing blowholes which are welded in subsequent hot rolling.

● **KILLED STEEL** has been deoxidized at least sufficiently to solidify without appreciable gas evolution.

● **SEMI-KILLED STEEL** has been partially deoxidized to reduce solidification shrinkage in the ingot.

● **A SIMPLE RULE:** Brinell Hardness divided by two, times 1000, equals approximate Tensile Strength in pounds per square inch. (200 Brinell ÷ 2 × 1000 = approx. 100,000 Tensile Strength, p.s.i.)

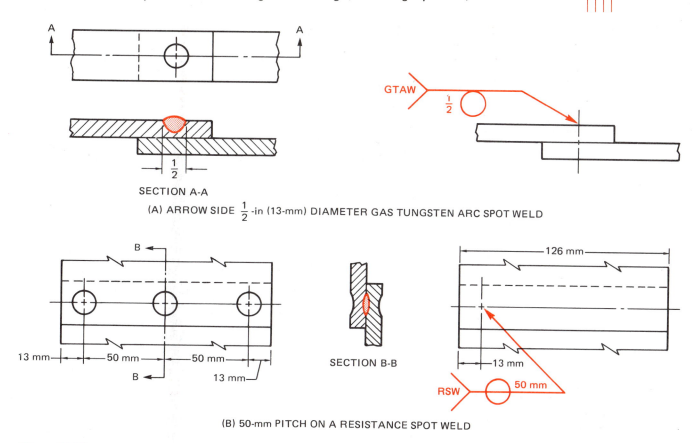

SECTION A-A

(A) ARROW SIDE $\frac{1}{2}$ -in (13-mm) DIAMETER GAS TUNGSTEN ARC SPOT WELD

SECTION B-B

(B) 50-mm PITCH ON A RESISTANCE SPOT WELD

Figure 18-37

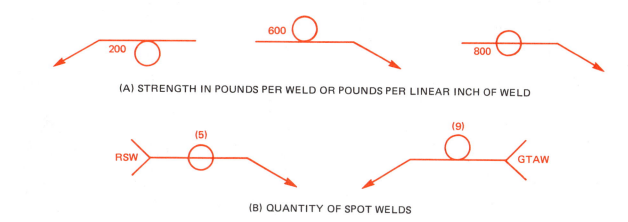

(A) STRENGTH IN POUNDS PER WELD OR POUNDS PER LINEAR INCH OF WELD

(B) QUANTITY OF SPOT WELDS

Figure 18-38 Designating strength and number of spot welds.

SEAM WELDS

Dimensions of seam welds are shown on the same side of the reference line as the weld symbol. Dimensions relate to either size or strength. The size of seam welds is designated as the width of the weld expressed in fractions or decimal hundredths of an inch. The size is shown with or without the inch marks to the left of the weld symbol, Figure 18-39(A) (page 424). When the length of a seam weld is indicated on the symbol, it is shown to the right of the symbol, Figure 18-39(B) (page 424). When seam welding extends for the full distance between abrupt changes in

the direction of welding, a length dimension is not required on the welding symbol.

The strength of seam welds is designated as the minimum acceptable shear strength in pounds per linear inch. The strength value is placed to the left of the weld symbol, Figure 18-40 (page 424).

GROOVE WELDS

Joint strengths can be improved by making some type of groove preparation before the joint is welded. There are seven types of grooves. The groove can be

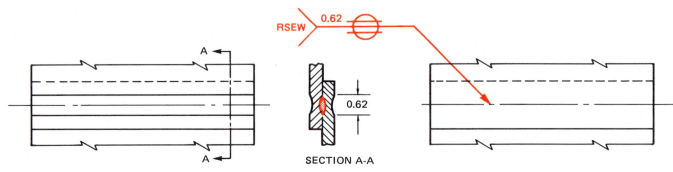

Figure 18-39 Designating the size of a seam weld.

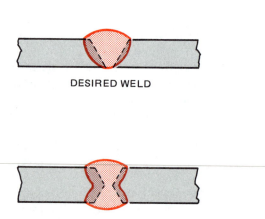

ARROW SIDE 200 LB/IN MINIMUM ACCEPTABLE SHEAR STRENGTH ELECTRON BEAM SEAM WELD

Figure 18-40 Strength of seam weld.

made in one or both plates or on one or both sides. By cutting the groove in the plate, the weld can penetrate deeper into the joint. This helps to increase the joint strength without restricting flexibility.

The grooves can be cut in base metal in a number of different ways. The groove can be cut using an oxyfuel cutting torch, air carbon arc cutting, plasma arc cutting, machined, or saws.

The various types of groove welds are classified as follows:

■ Single-groove and symmetrical double-groove welds that extend completely through the members being joined. No size is included on the weld symbol, **Figure 18-41.**

■ Groove welds that extend only part way through the parts being joined. The size as measured from the top of the surface to the bottom (not including reinforcement) is included to the left of the welding symbol, **Figure 18-42.**

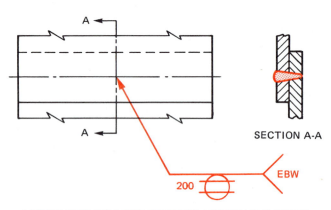

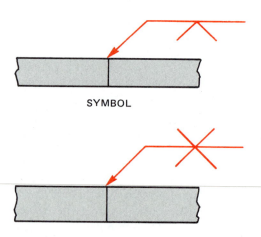

Figure 18-41 Designating single- and double-groove welds with complete penetration. (Courtesy of the American Welding Society.)

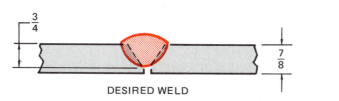
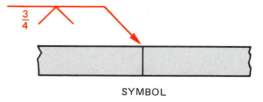

Figure 18-42 Designating the size of grooved welds with partial penetration. (Courtesy of the American Welding Society.)

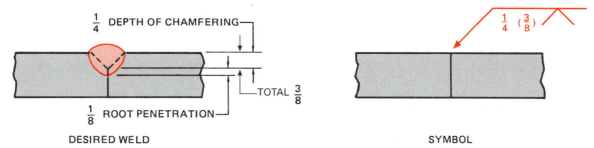

Figure 18-43 Showing size and root penetration of grooved welds. (Courtesy of the American Welding Society.)

■ The size of groove welds with a specified effective throat is indicated by showing the depth of groove preparation with the effective throat appearing in parentheses and placed to the left of the weld symbol, **Figure 18-43**. The size of square groove welds is indicated by showing the root penetration. The depth of chamfering and the root penetration is read in that order from left to right along the reference line.

■ The root face's main purpose is to minimize the burnthrough that can occur with a feather edge. The size of the root face is important to ensure good root fusion, **Figure 18-44**.

■ The size of flare groove welds is considered to extend only to the tangent points of the members, **Figure 18-45** (page 426).

■ The root opening of groove welds is the user's standard unless otherwise indicated. The root opening of groove welds, when not the user's standard, is shown inside the weld symbol, **Figure 18-46** (page 426).

BACKING

A backing (strip) is a piece of metal that is placed on the back side of a weld joint. The backing must be thick enough to withstand the heat of the root pass as it is burned-in. A backing strip may be used on butt joints, tee joints, and outside corner joints, **Figure 18-47** (page 426).

The backing may be either left on the finished weld or removed following welding. If the backing is to be removed the letter "R" is placed in the backing symbol, **Figure 18-48** (page 427). The backing is often removed for a finished weld because it can be a source of stress concentration and a crevice to promote rusting.

FLANGED WELDS

The following welding symbols are used for light-gauge metal joints where the edges to be joined are bent to form flange or flare welds.

■ Edge flange welds are shown by the edge flange weld symbol.

■ Corner flange welds are indicated by the corner flange weld symbol.

■ Dimensions of flange welds are shown on the same side of the reference line as the weld symbol and are placed to the left of the symbol, **Figure 18-49** (page 427). The radius and height above the point of tangency are indicated by showing both the radius and the height separated by a plus sign.

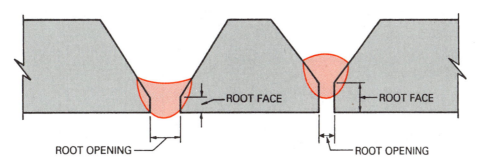

Figure 18-44

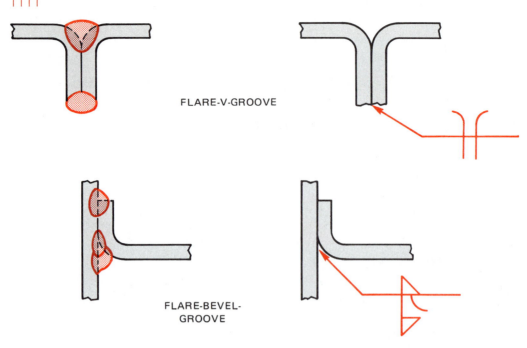

Figure 18-45 Designating flare-V- and flare-bevel-groove welds. (Courtesy of the American Welding Society.)

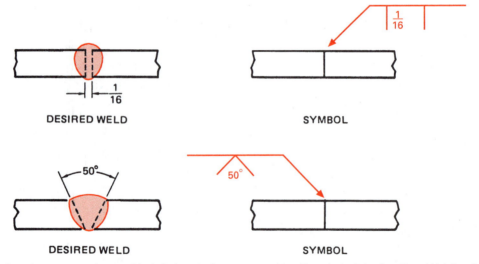

Figure 18-46 Designating root openings and included angle for groove welds. (Courtesy of the American Welding Society.)

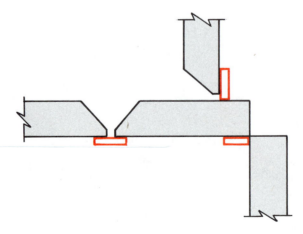

Figure 18-47 Backing strips.

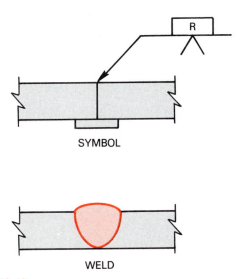

Figure 18-48

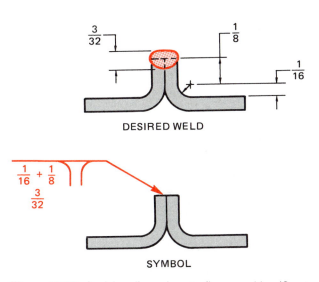

Figure 18-49 Applying dimensions to flange welds. (Courtesy of the American Welding Society.)

Type of Nondestructive Test		Symbol
Visual		VT
Penetrant		PT
Dye penetrant		DPT
Fluorescent penetrant		FPT
Magnetic particle		MT
Eddy current		ET
Ultrasonic		UT
Acoustic emission		AET
Leak		LT
Proof		PRT
Radiographic		RT
Neutron radiographic		NRT

Table 18-2 Standard Nondestructive Testing Symbols.

■ The size of the flange weld is shown by a dimension placed outward from the flanged dimensions.

NONDESTRUCTIVE TESTING SYMBOLS

The increased use of nondestructive testing (NDT) as a means of quality assurance has resulted in the development of standardized symbols. They are used by the designer or engineer to indicate the area to be tested and the type of test to be used. The inspection symbol uses the same basic reference line and arrow as the welding symbol, **Figure 18-50**.

The symbol for the type of nondestructive test to be used, **Table 18-2**, is shown with a reference line. The location above, below, or on the line has the same significance as it does with a welding symbol. Symbols above the line indicate other side, symbols below the line indicate arrow side, and symbols on the line indicate no preference for the side to be tested, **Figure 18-51** (page 428). Some tests may be performed on both sides; therefore, the symbol appears on both sides of the reference line.

Two or more tests may be required for the same section of weld. **Figure 18-52** (page 428) shows methods of combining testing symbols to indicate more than one type of test to be performed.

The length of weld to be tested or the number of tests to be made can be noted on the symbol. The length may be given either to the right of the test symbol, usually in inches, or can be shown by the arrow line, **Figure 18-53** (page 428). The number of tests to be made is given in parentheses ( ) above or below the test symbol, **Figure 18-54** (page 428). The welding symbols and nondestructive testing symbols both can be combined into one symbol, **Figure 18-55** (page 428). The combination symbol may help both the welder and inspector to identify welds that need special attention. A special symbol can be used to show the direction of radiation used in a radiographic test, **Figure 18-56** (page 429).

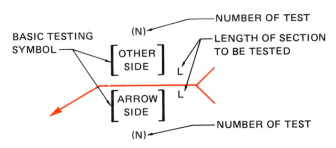

Figure 18-50 Basic nondestructive testing symbol.

Figure 18-51 Testing symbols used to indicate what side is to be tested.

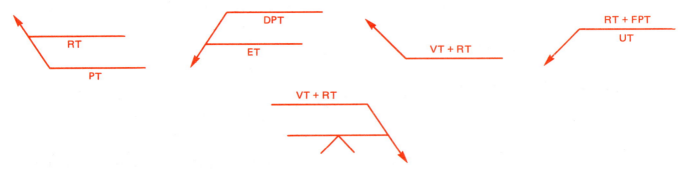

Figure 18-52 Methods of combining testing symbols.

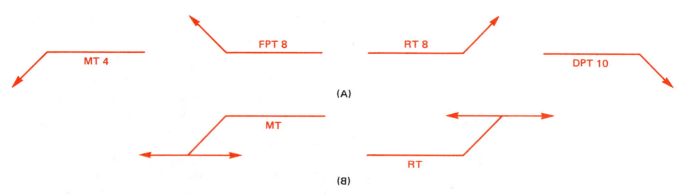

Figure 18-53 Two methods of designating the length of weld to be tested.

Figure 18-54 Method of specifying the number of tests to be made.

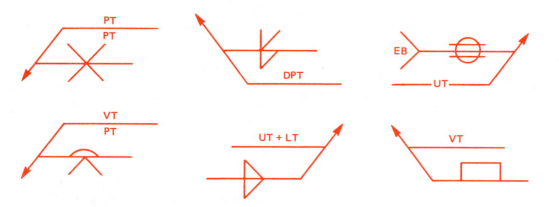

Figure 18-55 Combination welding and nondestructive testing symbols.

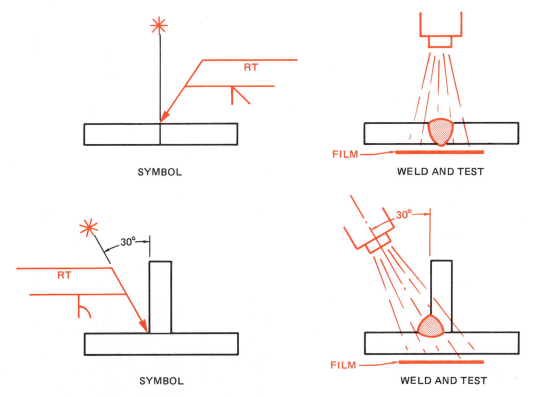

SYMBOL

WELD AND TEST

FILM

30°

RT

SYMBOL

30°

WELD AND TEST

FILM

Figure 18-56 Combination symbol for welding and radiation location for testing.

Figure 18-57 Large welded oil platform. (Courtesy of Amoco Corporation.)

FABRICATION

In addition to straight welding, welders are often required to assemble parts to form a weldment. The weldment may form a completed project or may only be part of a larger structure. Some weldments are composed of two or three parts; others may have hundreds or even thousands of individual parts, Figure 18-57 (page 429). Even the largest weldments start by placing two parts together.

The number and type of steps required to take a plan and create a completed project vary depending on the complexity and size of the finished weldment. All welding projects start with a plan. This plan can range from a simple one that exists only in the mind of the welder to a very complex plan comprising a set of drawings. As a beginning welder, you must learn how to follow a set of drawings to produce a finished weldment.

Soon we will be fabricating large structures in space, Figure 18-58. In fact, work has already begun on the International Space Station, which will be assembled in space from large sections built here on earth. Most of these assemblies required some type of welding. Someday we expect to be welding in space. Research for welding in space dates back to the 1960s with experiments done on board the *U.S. Sky Lab*. Today that research continues with experiments on the Space Shuttle program and in conjunction with the Soviet Mir Space Station.

Safety

As with any welding, safety is of primary concern for fabrication of weldments. Fabrication may present some safety problems not normally encountered in straight welding. Unlike most practice welding, much of the larger fabrication work may need to be performed outside an enclosed welding booth. Additionally, several welders may be working simultaneously on the same structure, so extra care must be taken to prevent burns to you or the other welders from the arc or hot sparks. Ventilation is also important because the normal shop ventilation may not extend to the fabrication area. Often you will be working in an area with welding cables and torch hoses lying scattered on the floor. To prevent accidental tripping, these lines must be flat on the floor and should be covered if they are in a walkway.

Figure 18-58 (A) The neutral buoyancy tank allows divers to work in space suits under water to simulate the microgravity of space. (Courtesy of NASA.)

Figure 18-58 (B) Large structure that could be fabricated in space someday. (Courtesy of NASA.)

These and other safety concerns are covered in Chapter 2, "Welding Safety." You should also read any safety booklets supplied with the equipment before starting any project.

Shop Math

Measuring. Measuring for most welded fabrications does not require accuracies greater than what can be obtained with a steel rule or a steel tape, **Figure 18-59.** Both steel rules and steel tapes are available in standard and metric units. Standard unit rules and tapes are available in fractional and decimal units, **Figure 18-60.**

Tolerances. All measuring, whether on a part or on the drawing, is in essence an estimate, because no matter how accurate the measurement is there will always be a more accurate way of taking it. The more accurate the measurement, the more time it takes. To save time while

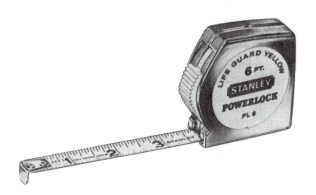

Figure 18-59 (A) Steel tape measures are available in lengths ranging from 6 ft. to 100 ft.. (The Stanley Tool Div.)

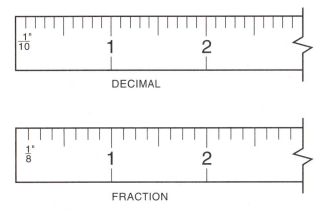

Figure 18-60

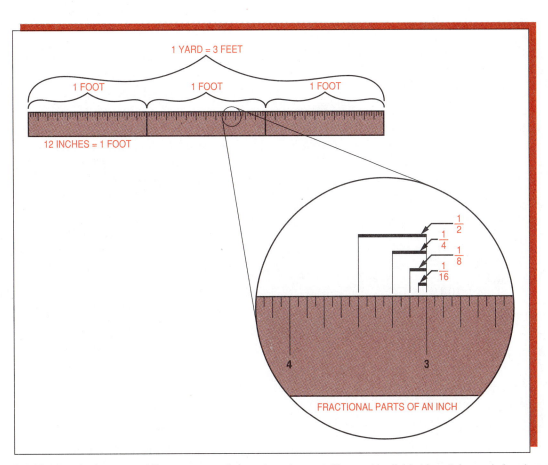

Figure 18-59 (B) The standard system of linear measure is based on the yard. The yard is divided into 3 ft., each foot into 12 in., and each inch into fractional parts. (Courtesy of Mark Huth.)

Dimension	Tolerance	Acceptable Dimensions	
		Minimum	Maximum
10"	±1/8"	9 7/8"	10 1/8"
2' 8"	±1/4"	2' 7 3/4"	2' 8 1/4"
10'	±1/8"	9' 11 7/8"	10' 1/8"
11"	±0.125	10.875"	11.125"
6'	±0.25	5' 11.75"	6' 0.25"
250 mm	±5 mm	245 mm	255 mm
300 mm	±5 mm–0 mm	300 mm	305 mm
175 cm	±10 mm	174 cm	176 cm

Figure 18-61 Dimension tolerances.

still making an acceptable part, dimensioning tolerances have been established. Most drawings will usually state a dimensioning tolerance, the amount by which the part can be larger or smaller than the stated dimensions and still be acceptable. Tolerances are usually expressed as plus (+) and minus (–). If the tolerance is the same for both the plus and the minus, it can be written using this symbol ±, **Figure 18-61** (page 432). In addition to the tolerance for a part, there may be an overall tolerance for the completed weldment. This dimension ensures that if all the parts are either too large or too small, their cumulative effect will not make the completed weldment too large or too small.

Adding and Subtracting. Although most drawings are made with as many dimensions as possible, the welder may have to do some basic math to complete the project. Adding and subtracting fractions and mixed numbers can be accomplished quickly by following a simple rule.

RULE: Fractions that are to be added or subtracted must have the same denominator, or bottom number.

Add 1/2 + 1/4 + 3/8
$$1/2 = 4/8$$
$$1/4 = 2/8$$
$$+ \ 3/8 = 3/8$$
$$9/8 = 1 \ 1/8$$

Add 6 1/2 + 3 3/4
$$6 \ 1/2 = 6 \ 2/4$$
$$+ \ 3 \ 3/4 = 3 \ 3/4$$
$$9 \ 5/4 = 10 \ 1/4$$

Subtract 5/8 from 3/4
$$3/4 = 6/8$$
$$- \ 5/8 = 5/8$$
$$1/8$$

Subtract 7 7/8 from 9 1/4
$$9 \ 1/4 = 9 \ 2/8 = 8 \ 10/8$$
$$- \ 7 \ 7/8 \ \ \ \ = \ \ \ \ \ 7 \ 7/8$$
$$1 \ 3/8$$

Reducing Fractions. Some fractions can be reduced to a lower denominator. For example, 2/4 is the same as 1/2.

When working with a drawing and making measurements, you can locate either one easily on the scale. Usually, such reductions are necessary only when you are working with several different dimensions or various fractional units. For reducing fractions in the shop, it is often easiest to divide both the numerator and denominator by 2. This method will simplify the reduction because all the fractional units found on shop rules and tapes are divisible by 2, for example, halves, fourths, eighths, sixteenths, and thirty-seconds. Using this method may require more than one reduction, but the simplicity of dividing by 2 offsets the time needed to repeat the reduction. Reduction of fractions will become easier with practice.

To reduce 4/8 inch:
$$\frac{4}{8} = \frac{4 \div 2}{8 \div 2} = \frac{2}{4}$$

The new fraction is 2/4 inch.
2/4 can be reduced again.
$$\frac{2}{4} = \frac{2 \div 2}{4 \div 2} = \frac{1}{2}$$

The new fraction is 1/2 inch, the lowest form.

Rounding Numbers. When multiplying or dividing numbers, we often get a whole number followed by a long decimal fraction. When we divide 10 by 3, for example, we get 3.3333333. For all practical purposes, we need not lay out weldments to an accuracy greater than the second decimal place. We would therefore round off this number to 3.33, a dimension that would be easier to work with in the welding shop.

RULE: When rounding off a number, look at the number to the right of the last significant place to be used. If this number is less than 5, drop it and leave the remaining number unchanged. If this number is 5 or greater, increase the last significant number by one and record the new number.

Round off 15.6549 to the second decimal place.
15.6549. Because the number in the third place is less than 5, the new number would be 15.65

Round off 8.2764 to the second decimal place.
8.2764. Because the number in the third place is 5 or more, the new number would be 8.28.

Round off 0.8539 to the third decimal place.
0.8539. Because the number in the fourth place is 5 or more, the new number would be 0.854.

Round off 156.8244 to the first decimal place.
156.8244. Because the number in the second place is less than 5, the new number would be 156.8.

Converting Fractions to Decimals. From time to time it may be necessary to convert fractional numbers to decimal numbers. A fraction-to-decimal conversion is needed before most calculators can be used to solve problems containing fractions. There are some calculators that will allow the inputting of fractions without converting them to decimals.

RULE: To convert a fraction to a decimal, divide the numerator (top number in the fraction) by the denominator (bottom number in the fraction).

To convert 3/4 to a decimal:
$$3 \div 4 = 0.75$$

To convert 7/8 to a decimal:
$$7 \div 8 = 0.875$$

Converting Decimals to Fractions. This process is less exact than the conversion of fractions to decimals. Except for specific decimals, the conversion will leave a remainder unless a small enough fraction is selected. For example, if you are converting 0.765 to the nearest 1/4 in., 3/4 in. would be acceptable and this conversion would leave a remainder of 0.015 in. (0.765 - 0.75 = 0.015). If you are working to a ±1/8-in. tolerance that has up to a 1/4-in. difference from the minimum to maximum dimensions, a measurement of 3/4 is acceptable. More accurately, 0.765 can be converted to 49/64 in., a dimension that would be hard to lay out and impossible to cut using a hand torch.

RULE: To convert a decimal to a fraction, multiply the decimal by the denominator of the fractional units desired; that is, for 8ths (1/8) use 8, for 4ths (1/4) use 4, and so on. Place the whole number (dropping or rounding off the decimal remainder) over the fractional denominator used.

To convert 0.75 to 4ths:
$$0.75 \times 4 = 3.0 \text{ or } 3/4$$

To convert 0.75 to 8ths:
$$0.75 \times 8 = 6.0 \text{ or } 6/8, \text{ which will reduce to } 3/4$$

To convert 0.51 to 4ths:
$$0.51 \times 4 = 2.04 \text{ or } 2/4, \text{ which will reduce to } 1/2$$

To convert 0.47 to 8ths:
$$0.47 \times 8 = 3.76 \text{ or } 3/8$$

(Note that the 0.76 of the 3.76 is more than 0.5, so it could be rounded up, giving 4 or 4/8, which will reduce to 1/2.)

Conversion Charts. Occasionally a welder must convert the units used on the drawing to the type of units used on the layout rule or tape. Fortunately, charts that can easily be used to convert between fractions, decimals, and metric units are available. To use these charts, **Figure 18-62** (page 434), locate the original dimension and then look at the dimension in the adjacent column(s) of the new units needed.

To convert 1/16 inch to millimeters:
$$1/16 \text{ in.} = 1.5875 \text{ mm}$$

To convert 0.5 inch to a fraction:
$$0.5 \text{ in.} = 1/2 \text{ in.}$$

To convert 0.375 inch to millimeters:
$$0.375 \text{ in.} = 9.525 \text{ mm}$$

To convert 25 millimeters to a decimal inch:
$$25 \text{ mm} = 0.98425 \text{ in.}$$

To convert 19 millimeters to a fractional inch:
$$19 \text{ mm} = 3/4 \text{ in. (approximately)}$$

Both metric-to-standard conversions and standard-to-metric conversions result in answers that often contain long decimals number strings. Often this new converted number, because of the decimals now attached to it, cannot be easily located on the rule or tape. In addition, most of the layout and fabrication work welders perform will not require such levels of accuracy. These small decimal fractions, in inches or millimeter scales, represent such a small difference that they cannot be laid out with a steel rule or tape. Such small differences can be important to some weldments, but in these cases some machining is required to obtain that level of accuracy. Because these small units are not normally included in a layout, they can be rounded off. Round off millimeter units to the nearest whole number; for example, 19.050 mm would be 19 mm and 1.5875 mm would be 2 mm, and so on. Round off decimal inch units to the nearest 1/16-in. fractional unit; 0.47244 in. would become 0.5 in. (1/2"), and 0.23622 in. would become 0.25 in. (1/4"). In both cases of rounding, the whole number obtained is well within most welding layout and fabrication drawing tolerances, which are usually ±1/16 in. or ±1/8 in.

Using the rounding off method of conversions with the conversion chart makes the following converted units easier to locate on rules and tapes.

To convert 1/2 inch to millimeters:
$$1/2 \text{ in.} = 13 \text{ mm}$$

To convert 0.625 inch to millimeters:
$$0.625 \text{ in.} = 16 \text{ mm}$$

To convert 2 3/4 inch to millimeters:
$$2 \times 25.4 = 50.8$$
$$3/4 = \underline{19.0}$$
$$69.8 \text{ rounded to 70 mm}$$

To convert 5.5 inch to millimeters:
$$5 \times 25.4 = 127.0$$
$$0.5 = \underline{12.7}$$
$$139.7 \text{ rounded to 140 mm}$$

To convert 10 millimeters to fraction inch:
$$10 \text{ mm} = 3/8 \text{ in.}$$

To convert 14 millimeters to decimal inch:
$$14 \text{ mm} = 0.5625 \text{ in.}$$

To convert 300 millimeters to fraction inch:
$$300 \div 25.4 = 11.81 \text{ in. rounded to 11 13/16 in.}$$

Inches dec	mm	Inches dec	mm
0.01	0.2540	0.51	12.9540
0.02	0.5080	0.52	13.2080
0.03	0.7620	0.53	13.4620
0.04	1.0160	0.54	13.7160
0.05	1.2700	0.55	13.9700
0.06	1.5240	0.56	14.2240
0.07	1.7780	0.57	14.4780
0.08	2.0320	0.58	14.7320
0.09	2.2860	0.59	14.9860
0.10	2.5400	0.60	15.2400
0.11	2.7940	0.61	15.4940
0.12	3.0480	0.62	15.7480
0.13	3.3020	0.63	16.0020
0.14	3.5560	0.64	16.2560
0.15	3.8100	0.65	16.5100
0.16	4.0640	0.66	16.7640
0.17	4.3180	0.67	17.0180
0.18	4.5720	0.68	17.2720
0.19	4.8260	0.69	17.5260
0.20	5.0800	0.70	17.7800
0.21	5.3340	0.71	18.0340
0.22	5.5880	0.72	18.2880
0.23	5.8420	0.73	18.5420
0.24	6.0960	0.74	18.7960
0.25	6.3500	0.75	19.0500
0.26	6.6040	0.76	19.3040
0.27	6.8580	0.77	19.5580
0.28	7.1120	0.78	19.8120
0.29	7.3660	0.79	20.0660
0.30	7.6200	0.80	20.3200
0.31	7.8740	0.81	20.5740
0.32	8.1280	0.82	20.8280
0.33	8.3820	0.83	21.0820
0.34	8.6360	0.84	21.3360
0.35	8.8900	0.85	21.5900
0.36	9.1440	0.86	21.8440
0.37	9.3980	0.87	22.0980
0.38	9.6520	0.88	22.3520
0.39	9.9060	0.89	22.6060
0.40	10.1600	0.90	22.8600
0.41	10.4140	0.91	23.1140
0.42	10.6680	0.92	23.3680
0.43	10.9220	0.93	23.6220
0.44	11.1760	0.94	23.8760
0.45	11.4300	0.95	24.1300
0.46	11.6840	0.96	24.3840
0.47	11.9380	0.97	24.6380
0.48	12.1920	0.98	24.8920
0.49	12.4460	0.99	25.1460
0.50	12.7000	1.00	25.4000

Inches frac	dec	mm	Inches frac	dec	mm
1/64	0.015625	0.3969	33/64	0.515625	13.0969
1/32	0.031250	0.7938	17/32	0.531250	13.4938
3/64	0.046875	1.1906	35/64	0.546875	13.8906
1/16	0.062500	1.5875	9/16	0.562500	14.2875
5/64	0.078125	1.9844	37/64	0.578125	14.6844
3/32	0.093750	2.3812	19/32	0.593750	15.0812
7/64	0.109375	2.7781	39/64	0.609375	15.4781
1/8	0.125000	3.1750	5/8	0.625000	15.8750
9/64	0.140625	3.5719	41/64	0.640625	16.2719
5/32	0.156250	3.9688	21/32	0.656250	16.6688
11/64	0.171875	4.3656	43/64	0.671875	17.0656
3/16	0.187500	4.7625	11/16	0.687500	17.4625
13/64	0.203125	5.1594	45/64	0.703125	17.8594
7/32	0.218750	5.5562	23/32	0.718750	18.2562
15/64	0.234375	5.9531	47/64	0.734375	18.6531
1/4	0.250000	6.3500	3/4	0.750000	19.0500
17/64	0.265625	6.7469	49/64	0.765625	19.4469
9/32	0.281250	7.1438	25/32	0.781250	19.8437
19/64	0.296875	7.5406	51/64	0.796875	20.2406
5/16	0.312500	7.9375	13/16	0.812500	20.6375
21/64	0.328125	8.3344	53/64	0.828125	21.0344
11/32	0.343750	8.7312	27/32	0.843750	21.4312
23/64	0.359375	9.1281	55/64	0.859375	21.8281
3/8	0.375000	9.5250	7/8	0.875000	22.2250
25/64	0.390625	9.9219	57/64	0.890625	22.6219
13/32	0.406250	10.3188	29/32	0.906250	23.0188
27/64	0.421875	10.7156	59/64	0.921875	23.4156
7/16	0.437500	11.1125	15/16	0.937500	23.8125
29/64	0.453125	11.5094	61/64	0.953125	24.2094
15/32	0.468750	11.9062	31/32	0.968750	24.6062
31/64	0.484375	12.3031	62/64	0.984375	25.0031
1/2	0.500000	12.7000	1	1.000000	25.4000

For converting decimal-inches in "thousandths," move decimal point in both columns to left.

Figure 18-62

To convert 240 millimeters to decimal inch:

$$240 \div 25.4 = 9.44 \text{ in. rounded to } 9\ 7/16 \text{ in.}$$

LAYOUT

Parts for fabrication may require that the welder lay out lines and locate points for cutting, bending, drilling, and assembling. Lines may be marked with a soapstone or a chalkline, scratched with a metal scribe, or punched with a center punch. If a piece of soapstone is used, it should be sharpened properly to increase accuracy, **Figure 18-63** (page 435). A chalk line will make a long, straight line on metal and is best used on large jobs, **Figure 18-64** (page 435). Both the scribe and punch can be used to lay out an accurate line, but the punched line is easier to see when cutting. A punch can be held as shown in **Figure 18-65** (page 436), with the tip just above the surface of the metal. When the punch is struck with a lightweight hammer, it will make a mark. If you move

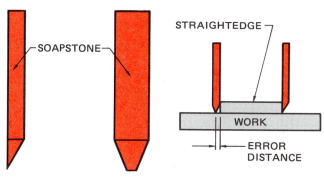

Figure 18-63 Proper method of sharpening a soapstone.

your hand along the line and rapidly strike the punch, it will leave a series of punch marks for the cut to follow.

Always start a layout as close to a corner of the material as possible. By starting in a corner or along the edge, you can take advantage of the preexisting cut as well as reduce wasted material.

It is easy to cut the wrong line. In welding shops one person may lay out the parts and another may make the cuts. Even when one person does both jobs, it is easy to cut the wrong line, either because of the restricted view through cutting goggles or because of the large number of lines on a part. To avoid making a cutting

Figure 18-64 (A) Pull the chalk line tight and then snap it

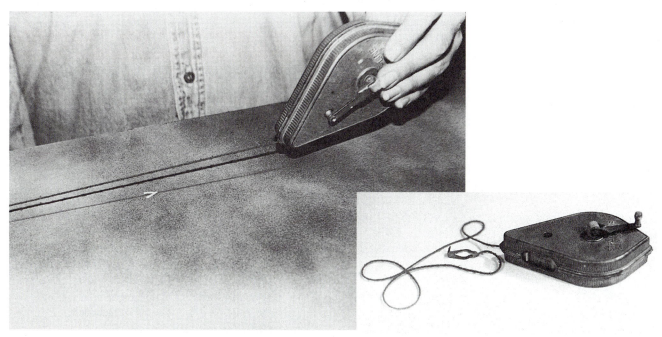

Figure 18-64 (B) Check to see that the line is dark enough.

Figure 18-64 (C) Chalk line reel.

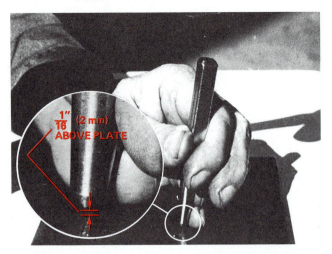

Figure 18-65 Holding the punch slightly above the surface allows the punch to be stuck rapidly and moved along a line to mark it for cutting.

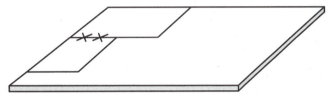

Figure 18-66

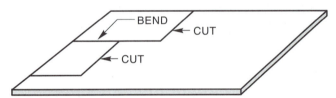

Figure 18-67 Identifying layout lines to avoid mistakes during cutting.

mistake, always identify whether lines are being used for cutting, for locating bends, as drill centers, or as assembly locations. The lines not to be cut may be marked with an X, or they may be identified by writing directly on the part, **Figure 18-66** (page 436). Mark the side of the line that is scrap so that when the kerf is removed from that side the part will be the proper size, **Figure 18-67** (page 436). Any lines that have been used for constructing the actual layout line or to locate points for drilling or are made in error must be erased completely or clearly marked to avoid confusion during cutting and assembly.

Some shops have their own shorthand methods for identifying layout lines, or you can develop your own system. Failure to develop and use a system for identifying

lines will ultimately result in a mistake. In a welding shop you will find only those who have made the wrong cut and those who will make the wrong cut. When it does happen, check with the welding shop supervisor to see what corrective steps can be taken. One advantage for most welding assemblies is that many cutting errors can be repaired by welding. Prequalified procedures are often established for just such an event, so check before deciding to scrap the part.

The process of laying out a part may be affected by the following factors:

■ *Material shape:* **Figure 18-68** (page 437) lists the most common metal shapes used for fabrication. Flat stock such as sheets and plates are easiest to lay out, and pipes and round tubing are the most difficult shapes to work with.

■ *Part shape:* Parts with square and straight cuts are easier to lay out than are parts with angles, circles, curves, and irregular shapes.

■ *Tolerance:* The smaller or tighter the tolerance that must be maintained, the more difficult the layout.

■ *Nesting:* The placement of parts together in a manner that will minimize the waste created is called *nesting*.

Parts with square or straight edges are the easiest to lay out. Simply measure the distance, and use a square or straight edge to lay out the line to be cut, **Figure 18-69** (page 438). Straight cuts that are to be made parallel to an edge can be drawn by using a combination square and a piece of soapstone. Set the combination square to the correct dimension and drag it along the edge of the plate while holding the soapstone at the end of the combination square's blade, **Figure 18-70** (page 438).

PRACTICE 18-1

Layout Square, Rectangular, and Triangular Parts

Using a piece of metal or paper, soapstone or pencil, tape measure, and square, you will lay out the parts shown in **Figure 18-71** (page 439). The parts must be laid out within ±1/16 in. of the dimensions. Convert the dimensions into S.I. metric units of measure.

Circles, arcs, and curves can be laid out by using either a compass or a circle template, **Figure 18-72** (page 439). The diameter is usually given for a hole or round part, and the radius is usually given for arcs and curves, **Figure 18-73** (page 440). The center of the circle, arc, or curve may be located using dimension lines and center lines. Curves and arcs that are to be made tangent to another line may be dimensioned with only their radiuses, **Figure 18-74** (page 440). ◆

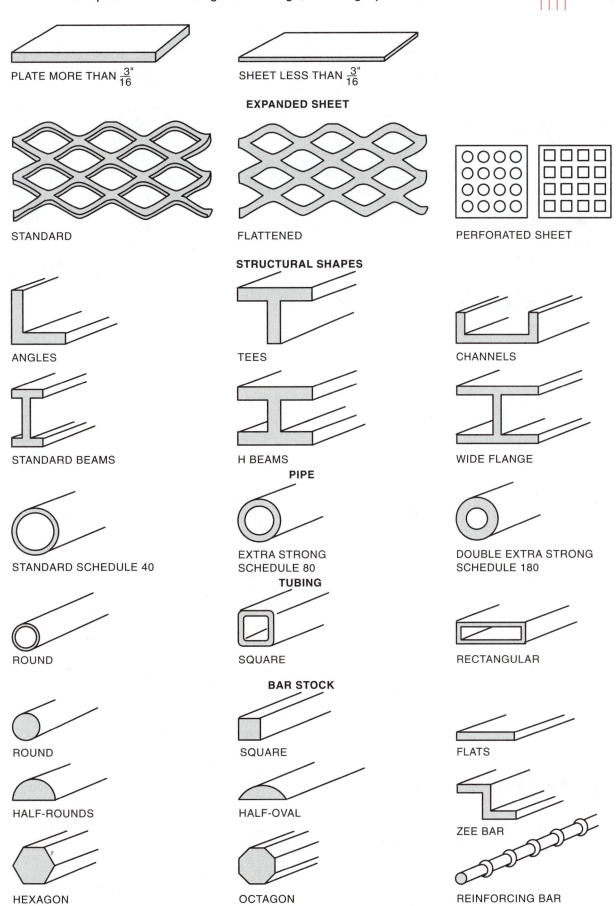

PLATE MORE THAN $\frac{3"}{16}$

SHEET LESS THAN $\frac{3"}{16}$

EXPANDED SHEET

STANDARD

FLATTENED

PERFORATED SHEET

STRUCTURAL SHAPES

ANGLES

TEES

CHANNELS

STANDARD BEAMS

H BEAMS

WIDE FLANGE

PIPE

STANDARD SCHEDULE 40

EXTRA STRONG
SCHEDULE 80

DOUBLE EXTRA STRONG
SCHEDULE 180

TUBING

ROUND

SQUARE

RECTANGULAR

BAR STOCK

ROUND

SQUARE

FLATS

HALF-ROUNDS

HALF-OVAL

ZEE BAR

HEXAGON

OCTAGON

REINFORCING BAR

Figure 18-68 Standard metal shapes, most available with different surface finishes such as hot-rolled, cold-rolled, or galvanized.

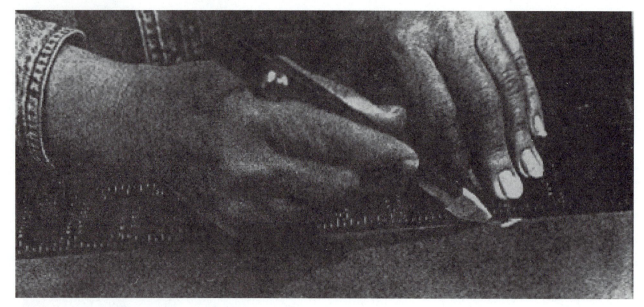

Figure 18-69 Using a square to draw a straight line.

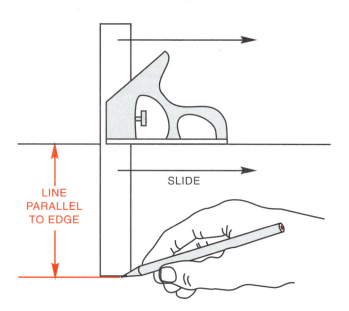

Figure 18-70 Using a combination square to lay out a strip of metal.

Laying Out Circles, Arcs, and Curves

Using a piece of metal or paper, soapstone or pencil, tape measure, compass, or circle template and square, you will lay out the parts shown in **Figure 18-75** (page 441). The parts must be laid out within ±1/16 in. of the dimensions. Convert the dimensions into S.I. metric units of measure. ◆

Nesting

Laying out parts so that the least amount of scrap is produced is important. Odd-shaped and unusual-sized parts often produce the largest amount of scrap. Computers can be used to lay out nested parts with a minimum of scrap. Some computerized cutting machines can also be programmed to nest parts.

Manual nesting of parts may require several tries at laying out the parts in order to achieve the lowest possible scrap.

Nesting Layout

Using metal or paper that is 8 1/2 in. × 11 in., soapstone or pencil, tape measure, and square, you will lay out the parts shown in **Figure 18-76** (page 441) in a manner that will result in the least scrap, assume a 0" kerf width. Use as many 8 1/2 in. × 11 in. pieces of stock as would be necessary to produce the parts using your layout. The parts must be laid out within ±1/16 in. of the dimensions. Convert the dimensions into S.I. metric units of measure. ◆

Bill of Materials

Using the parts laid out in Practice 18-3 and paper and pencil, you will fill out the bill of materials form shown in **Figure 18-77** (page 442). ◆

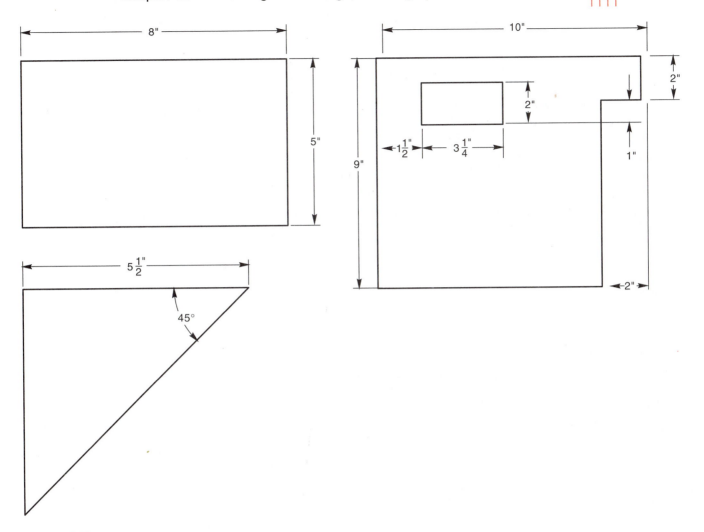

Figure 18-71

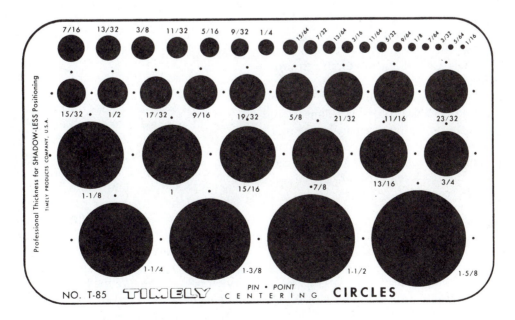

Figure 18-72 (A) Circle template. (Courtesy Timely Products Co.)

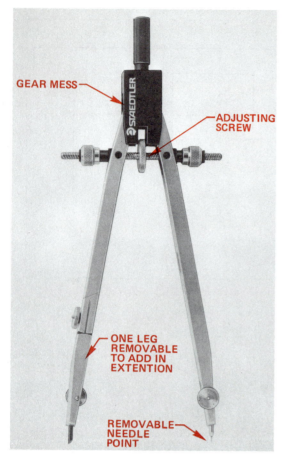

Figure 18-72 (B) Compass. (Courtesy of J.S. Staedtler, Inc.)

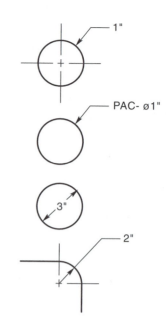

Figure 18-73

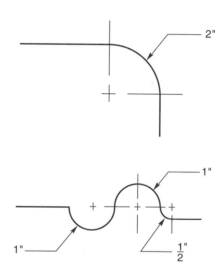

Figure 18-74

Kerf Space

Because all cutting processes, except shearing, produce a kerf during the cut, this space must be included in the layout when parts are laid out side by side. The *kerf* is the space created as material is removed during a cut. The width of a kerf varies depending on the cutting process used. Of the cutting processes used in most shops, the metal saw will produce one of the smallest kerfs and the hand-held oxyfuel cutting torch can produce one of the widest.

When only one or two parts are being cut, the kerf width may not need to be added to the part dimension. This space may be taken up during assembly by the root gap required for a joint. If a large number of parts are being cut out of a single piece of stock, the kerf width can add up and increase the stock required for cutting out the parts, **Figure 18-78 (page 442)**.

PRACTICE 18-5

Allowing Space for the Kerf

Using a pencil, 8 1/2-in. × 11-in. paper, measuring tape or rule, and square, you will lay out four rectangles 2 1/2 in. × 5 1/4 in. down one side of the paper, leaving 3/32 in. for the kerf.

Two ways can be used to provide for the kerf spacing. One method is to draw a double line on the side of the part where the kerf is to be made, **Figure 18-79 (page 443)**. The other way is to lay out a single line and place an X on the side of the line that the cut is to be made, **Figure 18-80 (page 443)**. Note that no kerf space need be left along the sides made next to the edge of the paper or next to the scrap. What is the total length and width of material needed to lay out these four parts?

Parts can be laid out by tracing either an existing part or a template, **Figure 18-81 (page 443)**. When using either process, be sure the line you draw is made as tight as possible to the part's edge, **Figure 18-82 (page 444)**. The inside edge of the line is the exact size of the part. Make the cut on the line or to the outside so that the part will be the correct side once it is cleaned up, **Figure**

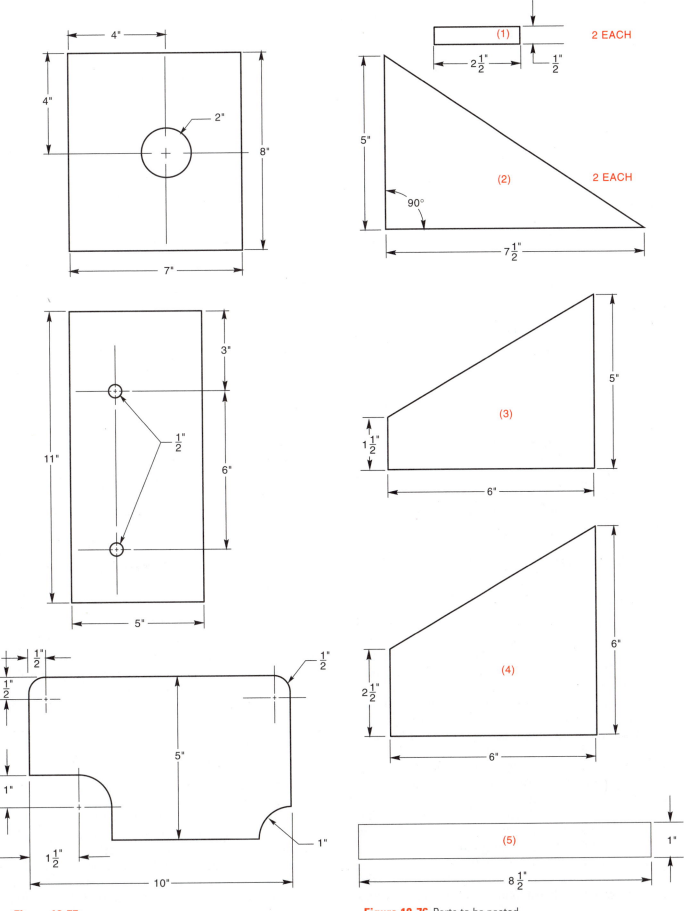

Figure 18-75

Figure 18-76 Parts to be nested.

BILL OF MATERIALS

Part	Number Required	Type of Material	Size Standard Units	SI Units

Figure 18-77 Bill of materials form.

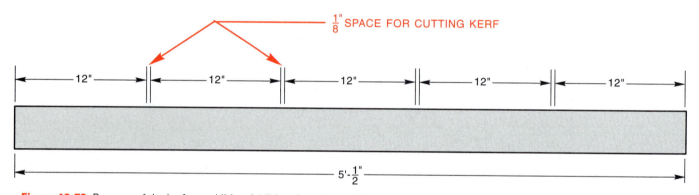

Figure 18-78 Because of the kerf, an additional 1/2 in. of stock would be required to make these five 1-ft pieces.

18-83 (page 444). Sometimes a template may be made of a part. Templates are useful if the part is complex and needs to fit into an existing weldment. They are also helpful when a large number of the same part are to be made or when the part is only occasionally used. The advantage of using templates is that once the detailed layout work is completed exact replicates can be made any time they are needed. Templates can be made out of heavy paper, cardboard, wood, sheet metal, or other appropriate material. The sturdier the material, the longer the template will last.

Special tools have been developed to aid in laying out parts; one such tool is the *contour marker,* Figure 18-84 (page 444). These markers are highly accurate when properly used, but they do require a certain amount of practice. Once familiar with this tool, the user can lay out an almost infinite variety of joints within the limits of the tool being used. One advantage of tools like the contour marker is that all sides of a cut in structural shapes and pipe can be laid out from one side without relocating the tool, **Figure 18-85** (page 444). ◆

MATERIAL SHAPES

Metal stock can be purchased in a wide variety of shapes, sizes, and materials. Weldments may be constructed from combinations of sizes and/or shapes of metals. Only a single type of metal is usually used in most weldments unless a special property such as corrosion resistance is needed. In those cases dissimilar metals may be joined into the fabrication at such locations as needed. The most common metal used is carbon

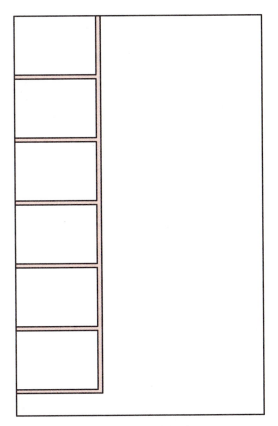

Figure 18-79 Kerf is made between the lines.

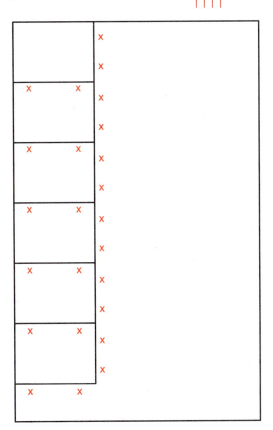

Figure 18-80 X's mark the side of the line on which the kerf is to be made.

Figure 18-81 (A) Tracing a part.

Figure 18-81 (B) Tracing a template.

steel, and the most common shapes used are plate, sheet, pipe, tubing, and angles. For that reason, most of the fabrication covered in this chapter will concentrate on carbon steel in those commonly used shapes. Transferring the fabrication skills learned in this chapter to the other metals and shapes should require only a little practice time.

Plate is usually 3/16 in. (4.7 mm) or thicker and measured in inches and fractions of inches. Plates are available in widths ranging from 12 in. (305 mm) up to 96 in. (2438 mm) and lengths from 8 ft. (2.4 m) to 20 ft. (6 m). Thickness ranges up to 12 in. (305 mm).

Sheets are usually 3/16 in. (4.7 mm) or less and measured in gauge or decimals of an inch. Several

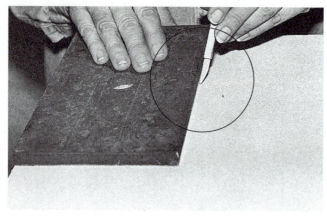

Figure 18-82 Be sure that the soapstone is held tightly into the part being traced.

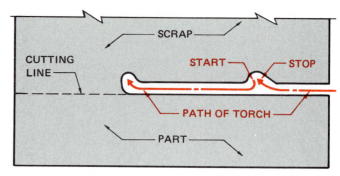

Figure 18-83 Turning out into scrap to make stopping and starting points smoother.

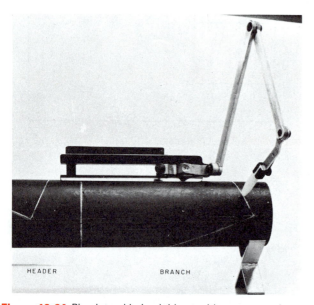

Figure 18-84 Pipe lateral being laid out with contour marker.

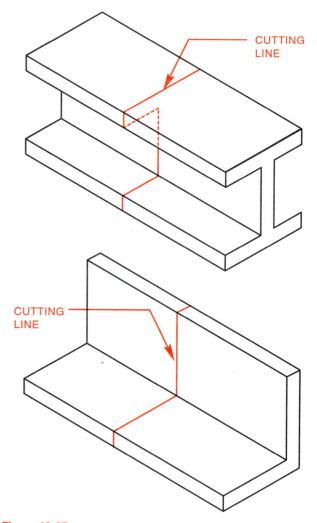

Figure 18-85

different gauge standards are used. The two most common are the Manufacturer's Standard Gauge for Sheet Steel, used for carbon steels, and the American Wire Gauge, used for most nonferrous metals such as aluminum and brass.

Pipe is dimensioned by its diameter and schedule or strength. Pipe that is smaller than 12 in. (305 mm) is dimensioned by its inside diameter, and the outside diameter is given for pipe that is 12 in. (305 mm) in diameter and larger, Figure 18-86. The strength of pipe is given as a schedule. Schedules 10 through 180 are available; schedule 40 is often considered a standard strength. The wall thickness for pipe is determined by its schedule (pressure range). The larger the diameter of the pipe, the greater its area. Pipe is available as welded (seamed) or extruded (seamless).

Tubing sizes are always given as the outside diameter. The desired shape of tubing, such as square, round, or rectangular, must also be listed with the ordering information.

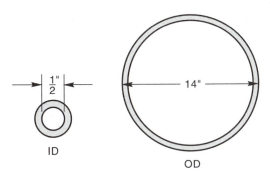

Figure 18-86 ID inside diameter; OD outside diameter.

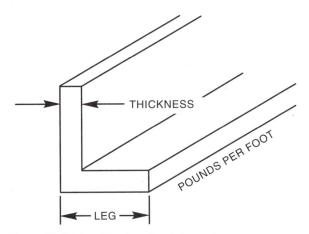

Figure 18-87 Specifications for sizing angles.

The wall thickness of tubing is measured in inches (millimeters) or as Manufacturer's Standard Gauge for Sheet Metal. Tubing should also be specified as rigid or flexible. The strength of tubing may also be specified as the ability of tubing to withstand compression, bending, or twisting loads.

Angles, sometimes referred to as angle iron, are dimensioned by giving the length of the legs of the angle and their thickness, **Figure 18-87**. Stock lengths of angles are 20 ft., 30 ft., 40 ft., and 60 ft. (6 m, 9.1 m, 12.2 m, and 18.3 m).

ASSEMBLY

The assembling process, bringing together all the parts of the weldment, requires a proficiency in several areas. You must be able to read the drawing and interpret the information provided there to properly locate each part. An assembly drawing has the necessary information, both graphically and dimensionally, to allow the various parts to be properly located as part of the weld-

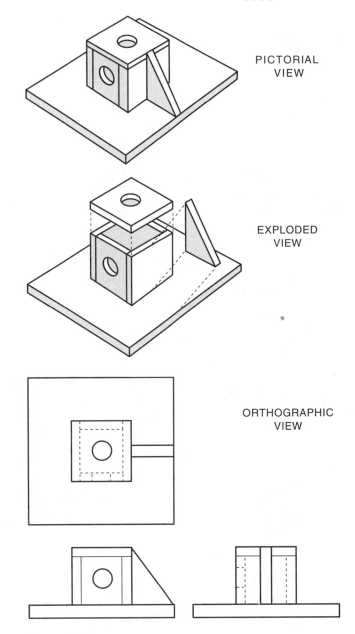

PICTORIAL VIEW

EXPLODED VIEW

ORTHOGRAPHIC VIEW

Figure 18-88

ment. If the assembly drawings include either pictorial or exploded views, this process is much easier for the beginning assembler; however, most assembly drawings are done as two, three, or more orthographic views, **Figure 18-88**. Orthographic views will be more difficult to interpret until you have developed an understanding of their various elements.

On very large projects such as buildings or ships, a corner or centerline is established as a baseline. This is the point where all measurements for all part location begins. When working with smaller weldments, a single part may be selected as such a starting point. Often, selecting the base part is automatic because all other parts are to be joined to this central part. On other weldments, however, the selection of a base part is strictly up to the assembler.

To start the assembly, select the largest or most central part to be the base for your assembly. All other parts will then be aligned to this one part. Using a base also helps to prevent location and dimension errors. Otherwise, a slight misalignment of one part, even within tolerances, will be compounded by the misalignment of other parts, resulting in an unacceptable weldment. Using a baseline or base part will result in a more accurate assembly.

Identify each part of the assembly and mark each piece for future reference. If needed, you can hold the parts together and compare their orientation to the drawing. Locate points on the parts that can be easily identified on the drawing such as holes and notches, **Figure 18-89**. Now mark the location of these parts—top, front, or other such orientation—so you can locate them during the assembly.

Layout lines and other markings can be made on the base to locate other parts. Using a consistent method of marking will help prevent mistakes. One method is to draw parallel lines on both parts where they meet, **Figure 18-90**.

After the parts have been identified and marked, they can be either held or clamped into place. Holding the parts in alignment by hand for tack welding is fast but often leads to errors and thus is not recommended for beginning assemblers. Experienced assemblers rec-

ognize that clamping the parts in place before tack welding is a much more accurate method, **Figure 18-91**.

ASSEMBLY TOOLS

Clamps

A variety of clamps can be used to temporarily hold parts in place so that they can be tack welded.

■ *C-clamps,* one of the most commonly used clamps, come in a variety of sizes, **Figure 18-92**. Some C-clamps have been specially designed for welding. Some of these clamps have a spatter cover over the screw, and others have their screws made of spatter-resistant materials such as copper alloys.

■ *Bar clamps* are useful for clamping larger parts. Bar clamps have a sliding lower jaw that can be positioned against the part before tightening the screw clamping end, **Figure 18-93**. They are available in a variety of lengths.

■ *Pipe clamps* are very similar to bar clamps. The advantage of pipe clamps is that the ends can be attached to a section of standard 1/2-in. pipe. This feature allows for greater flexibility in length, and the pipe can easily be changed if it becomes damaged.

■ *Locking pliers* are available in a range of sizes with a number of various jaw designs, **Figure 18-94** (page 448). The versatility and gripping strength make locking pliers very useful. Some locking pliers have a self-adjusting feature that allows them to be moved

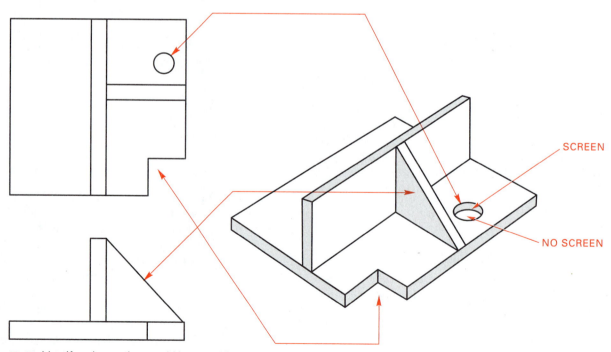

SCREEN

NO SCREEN

Figure 18-89 Identify unique points to aid in assembly.

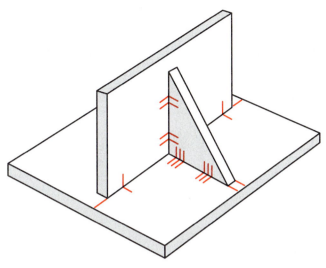

Figure 18-90 Lay out markings to help locate the parts for tack welding.

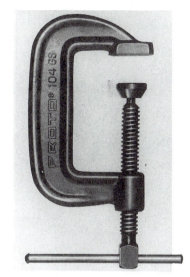

Figure 18-92 C-clamps. (Courtesy of Stanley-Proto Industrial Tools, Covington, GA.)

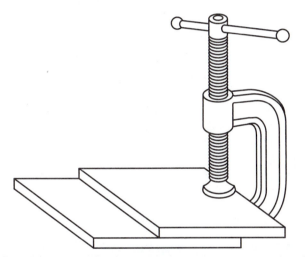

Figure 18-91 C-clamp being used to hold plates for tack welding (Mike Gellerman).

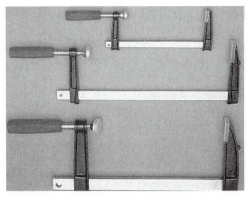

Figure 18-93 Bar clamps. (Courtesy of Woodworker's Supply Inc.)

between different thicknesses without the need to readjust them.

■ *Cam-lock clamps* are specialty clamps that are often used in conjunction with a jig or a fixture. They can be preset, allowing for faster work, **Figure 18-95** (page 448).

■ *Specialty clamps* such as these for pipe welding, **Figure 18-96** (page 449), are available for many different types of jobs. Such specialty clamps make it possible to do faster and more accurate assembling.

Fixtures

Fixtures are devices that are made to aid in assemblies and fabrication of weldments. When a number of similar parts are to be made, fixtures are helpful. They can increase speed and accuracy in the assembly of parts. Fixtures must be strong enough to support the weight of the parts, be able to withstand the rigors of repeated assemblies, and remain in tolerance. They may have clamping devices permanently attached to speed up their use. Often, locating pins or other devices are used to ensure proper part location. A well-designed fixture allows adequate room for the welder to make the necessary tack welds. Some parts are left in the fixture through the entire welding process to reduce distortion. Making fixtures for every job is cost prohibitive and not necessary for a skilled assembler.

FITTING

Not all parts fit exactly as they were designed. There may be slight imperfections in cutting or distortion of parts due to welding, heating, or mechanical

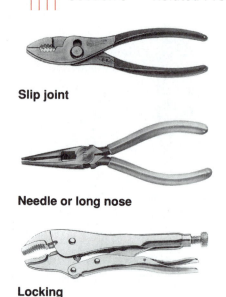

Figure 18-94 Three common types of pliers. (Courtesy of Stanley-Proto Industrial Tools, Covington, GA.)

damage. Some problems can be solved by grinding away the problem area. Hand grinders are most effective for this type of problem, **Figure 18-97**. Other situations may require that the parts be forced into alignment.

A simple way of correcting slight alignment problems is to make a small tack weld in the joint and then use a hammer and possibly an anvil to pound the part into place, **Figure 18-98** (page 450). Small tacks applied in this manner will become part of the finished weld. Be sure not to strike the part in a location that will damage the surface and render the finished part unsightly or unusable.

More aligning force can be applied using cleats or dogs with wedges or jacks. Cleats or dogs are pieces of metal that are temporarily attached to the weldment's parts to enable them to be forced into place. Jacks will do a better job if the parts must be moved more than about a 1/2 in. (13 mm), **Figure 18-99** (page 450). Anytime cleats or dogs are used, they must be removed and the area ground smooth.

Some codes and standards will not allow cleats or dogs to be welded to the base metal. In these cases more expensive and time-consuming fixtures must be constructed to help align the parts if needed.

TACK WELDING

Tack welding is a temporary method of holding the parts in place until they can be completely welded. Usually, all of the parts of a weldment should be assembled before any finishing welding is started. This will help reduce distortion. Tack welds must be strong enough to withstand any pounding or forcing during assembly and any forces caused by weld distortion during final welding. They must also be small enough to be incorporated into the final weld without causing a discontinuity in its size or shape, **Figure 18-100** (page 451).

Tack welds must be made with an appropriate filler metal, in accordance with any welding procedure. They must be located well within the joint so that they can be refused during the finish welding. Post-tack welding cleanup is required to remove any slag or impurities that may cause flaws in the finished weld. Sometimes the ends of a tack weld must be ground

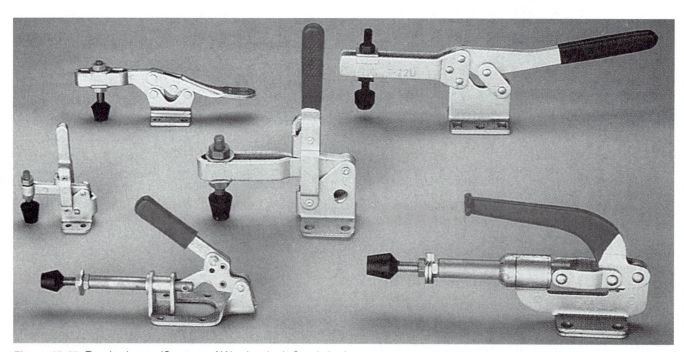

Figure 18-95 Toggle clamps. (Courtesy of Woodworker's Supply Inc.)

Figure 18-96 (A) Pipe alignment clamps.

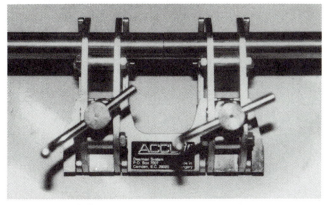

Figure 18-96 (B) Pipe clamps.

down to a taper to improve their tie-in to the finished weld metal.

A good tack weld is one that does its job by holding parts in place yet is undetectable in the finished weld.

WELDING

Good welding requires more than just filling up the joints with metal. The order and direction in which welds are made can significantly affect distortion in the weldment. Generally, welding on an assembly should be staggered from one part to another. This will allow both the welding heat and welding stresses to dissipate.

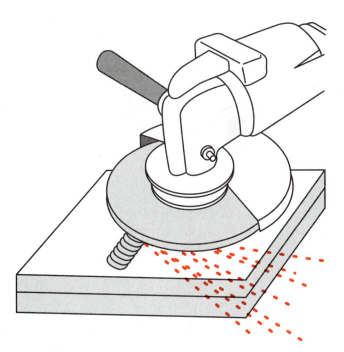

Figure 18-97 (A) (Courtesy of Mike Gellerman.)

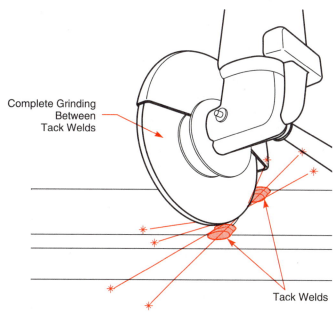

Complete Grinding
Between
Tack Welds

Tack Welds

Figure 18-97 (B) (Courtesy of Mike Gellerman.)

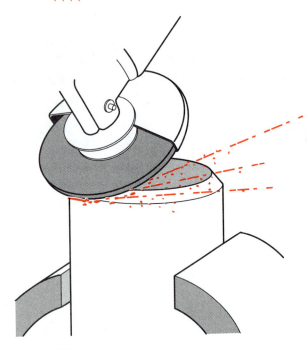

Figure 18-97 (C) (Courtesy of Mike Gellerman.)

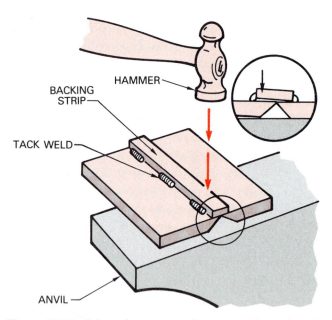

Figure 18-98 Using a hammer to align the backing strip and weld plates.

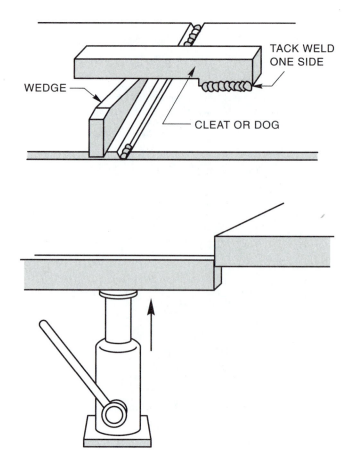

Figure 18-99 Jack, cleat, and wedge used in realignment. (Courtesy of Mike Gillerman.)

Keep the arc strikes in the welding joint so that they will be remelted as the weld is made. This will make the finished weldment look neater and reduce post-weld cleanup. Some codes and standards do not allow arcs to be made outside of the welding joint.

Striking the arc in the correct location on an assembly is more difficult than working on a welding table. When working on an assembly, you will often be in an awkward position, which makes it harder to strike the arc correctly. Several techniques will help you improve your arc starting accuracy. You can use your free hand to guide the electrode in to the correct spot. Resting your arm, shoulder, or hip against the weldment can also help. Practicing starting the weld with the power off sometimes is helpful.

Be sure that you have enough freedom of movement to complete the weld joint. Check to see that your welding leads will not snag on anything that would prevent you from making a smooth weld. If you are welding out of position, be sure that welding sparks will not fall on you or other workers. If the weldment is too large to fit into a welding booth, portable welding screens should be used to protect other workers in the area from sparks and welding light.

Follow all safety and setup procedures for the welding process. Practice the weld to be sure that the machine is set up properly before starting on the weldment.

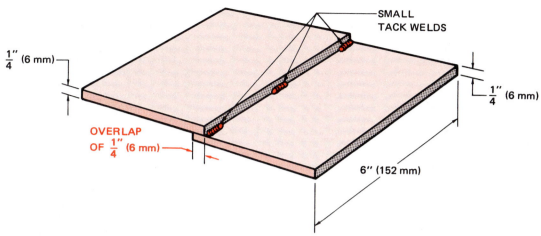

Figure 18-100 (A) Make tack welds as small as possible.

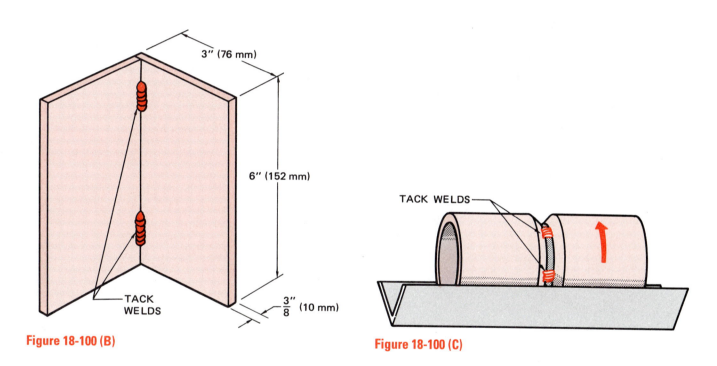

Figure 18-100 (B)

Figure 18-100 (C)

FINISHING

Depending on the size of the shop, the welder may be responsible for some or all of the finish work. Such work may vary from chipping, cleaning, or grinding the welds to applying paint or other protective surfaces.

Grinding of welds should be avoided if pieceable by properly sizing the weld as it is made. Grinding can be an expensive process, adding significant cost to the finished weldment. Sometimes it is necessary to grind for fitting purposes or for appearance, but even in these cases it should be minimized if pieceable.

■ *CAUTION*

When using a portable grinder, be sure that it is properly grounded and that the sparks will not fall on others, cause damage, or start a fire. Always maintain control to prevent the stone from catching and gouging the part or yourself.

Most grinding is done with a hand angle grinder, Figure 18-101. These grinders can be used with a flat or cupped grinding stone or sandpaper. As the grinder is used the stone will wear down and must be discarded once it has worn down to the paper backing. It is a good practice to hold the grinder at an angle so that if anything is thrown off the stone or metal surface, it will not strike you or others in the area. Because of the speed of the grinding stone, any such object can cause serious injury.

The grinder must be held securely so that there is a constant pressure on the work. If the pressure is too great, the grinder motor will overheat and may burn out. If the pressure is too light, the grinder may bounce, which could crack the grinding stone. Move the grinder in a smooth pattern along the weld. Watch the weld surface as it begins to take the desired shape and change your pattern as needed.

Painting and other finishes release fumes such as volatile organic compounds (VOC), which are often regulated by local, state, and national governments. Special ventilation is required for most paints. Such a ventilation system will remove harmful fumes from the air

Figure 18-101 Wire brushes and grinding stones used to clean up welds.

before it is released back into the environment. Check with your local, state, or national regulating authority before using such products. Read and follow all manufacturer's instructions for the safe use of their product.

Robotic Welding Mass-Produces Golf Carts

The world's largest selling golf cart line is mass-produced in Augusta, Georgia, the home of the prestigious Masters Professional Golf Tournament.

The modern production facilities cover over 400,000 square ft. and turn out several thousand golf carts and industrial vehicles each year. Welding equipment is an important component of the total manufacturing facility. The plant uses more than sixty power sources for welding. The recent trend has been to high-tech robotic welding systems that increase productivity, improve weld quality, and save the company both time and money.

Two robotic welding cells were installed in the 4,500 square foot small parts welding shop. The company investigated several systems while seeking a solution to the relatively small lot runs (under 3,000 parts) welding of up to 23 different parts within critical tolerances. Most robotic welding system applications were dedicated to one or two parts, with long run lots. The turnkey robotic systems chosen consisted of robots, controllers,

and four station work tables. Figure 1 shows the robot performing complex welding tasks on one of the four work tables while the operator removes welded parts and loads pieces to be welded into the other workstation fixtures. Behind the robot is a 450-ampere constant voltage dc power source that supports the welding operation. A key element in the proper functioning of the system was the design and installation of customized fixturing and work holders. Figure 2 shows the robot arm striking an arc and welding a part on one of the four fixtured work tables.

The production engineer performed all programming and welding parameter settings of the welding robot systems. On a typical part welded by the system, such as the engine frame assembly, the robot makes twenty-seven different critical welds preprogrammed into the robot controller. Programming in the pulsing parameters, at 60 or 120 pulses/second, helps eliminate product warpage while keeping extremely close tolerances.

With mostly mild steel parts, the robot systems are capable of GMA welding up to 40 in./min

Figure 1

Figure 2

using 1,000 lb spools of .035 wire under 75/25 argon/CO_2 shielding. Random welded samples are fully inspected in the quality control department on a coordinate measuring machine to insure that the high production welding meets strict dimensional standards.

The industrial engineer on the project did an investment justification study on the robotic welding systems that projected a payback period of just 2.7 years through the saving of nearly 10,000 man-hours per year — a reduction of 47% in direct labor in the small parts welding department.

While the robotic welders produce precision welded parts, other welders are busy producing parts and assemblies on welding equipment. **Figure 3** shows a welder in the frame shop checking the settings of a welding power source, one of three surrounding the frame workpiece, which requires 58 different welds.

(Courtesy of Miller Electric Mfg. Co., Appleton, Wisconsin 54912.)

Figure 3

REVIEW

1. What stresses must a welded joint withstand?

2. List the five joint types used in welding.

3. Sketch and label five edge preparations used for welding joints.

4. Sketch a V-grooved butt joint, and label all of the joint's dimensions.

5. Sketch a weld on plates in the 1G and 1F positions.

6. Sketch a weld on plates in the 2G and 2F positions.

7. Sketch a weld on plates in the 3G and 3F positions.

8. Sketch a weld on plates in the 4G and 4F positions.

9. Sketch a weld on a pipe in the 1G position.

10. Sketch a weld on a pipe in the 5G position.

11. Sketch a weld on a pipe in the 2G position.

12. Sketch a weld on a pipe in the 6G position.

13. Sketch a weld on a pipe in the 6GR position.

14. Why is it usually better to make a weld in the flat position?

15. Why are some joints back-gouged?

16. What is a prequalified joint?

17. Why is cost a consideration in joint design?

18. What is contained in a set of drawings?

19. What information can be included in the title box of a drawing?

20. How is the front view of an object selected?

21. Why are sections and cut-a-ways used in drawings?

22. What types of information can be included on a welding symbol?

23. Why are welding symbols used?

24. What types of information may appear on the reference line of a welding symbol?

25. What are the different classifications of welds that a symbol can indicate?

26. How is the reference line always drawn?

27. Why is a tail added to the basic welding symbol?

28. What is meant if the weld symbol is placed below the reference line?

29. How are the dimensions for a fillet weld given?

30. What dimensions can be given for a plug weld?

31. What two units are used to show the minimum shear strength of a spot weld?

32. How is the strength of a seam weld specified?

33. How can the groove be cut on the edge of a plate?

34. Sketch and dimension a V-groove weld symbol for a weld on the arrow side, with 1/8-in. root opening, 3/4 in. in size having a groove angle of 45°.

35. How is the removal of the backing strip noted on a welding symbol?

36. How are flanged edges formed?

37. Sketch two NDT symbols that illustrate different methods that can be used to indicate multiple test requirements for the same section of weld.

38. What safety concerns must be addressed before welding on a large assembly outside of a welding booth?

39. Why do some parts have ± tolerances?

40. What is the rule that must be followed for adding or subtracting fractions?

41. Reduce the following fractions to their lowest denominator: 4/8, 16/32, 8/32, 10/16, 12/8, 8/4, and 10/2.

42. Round off the following numbers to two decimal places: 38.973, 7.976, 0.0137, 100.062, and 12.124.

43. Convert the following fractions to decimal equivalents: 1/2, 3/8, 9/16, 11/32, 15/16, and 1/32.

44. Convert the following decimals to the appropriate fractional units: 0.25 to 4ths, 0.375 to 8ths, 0.956 to 16ths, 0.79 to 4ths, and 1.29 to 32ths.

45. Using the conversion chart in **Figure 18-62,** convert the following standard dimensions to metric units: 1/8 in., 5/32 in., 11/16 in., 7/8 in., and 1 9/16 in.

46. Using the conversion chart in **Figure 18-62,** convert the following metric dimensions to standard fractional units: 6.35 mm, 19.05 mm, 12.7 mm, 24.6062 mm, 42.0688 mm.

47. Using the conversion chart in **Figure 18-62,** convert the following metric dimensions to the nearest standard fractional units ±1/16 in.: 13 mm, 16 mm, 5 mm, 22 mm, and 3 mm.

48. How can layout lines be drawn on a metal surface?

49. Why should some layout lines be marked with an X?

50. How would a small or tight tolerance affect the layout?

51. How can circles, arcs, and curves be laid out?

52. Which cutting process does not produce a kerf space?

53. What can a template be made of?

54. What is the difference between plate and sheet material?

55. Why should you identify a part for assembly as the base?

56. What makes some C-clamps better for welding than others?

57. What precautions should be taken if a hammer is used to shift a part into alignment?

58. Why should the entire weldment be assembled and tack welded in place before finished welding is started?

59. Why is it important not to strike the arc outside of the weld joint?

60. What can happen if a grinding stone with a lower rated RPM than the grinder is used?

Chapter 19

WELDING CODES, STANDARDS AND WELDING COST

OBJECTIVES

After completing this chapter, the student should be able to:

- explain the difference between qualification and certification.
- list the major considerations for selecting a code or standard.
- write a welding procedure and specification.
- identify the three most common codes and describe their major uses.
- outline the steps required to certify a weld and welder.
- explain how a tentative WPS becomes a certified WPS.

KEY TERMS

code	AWS D1.1
standard	ASME Section IX
specification	API Standard 1104
Welding Procedure Specification (WPS)	thickness range
Welding Schedule	P-number
procedure qualification record (PQR)	F-number
welder performance qualification	

INTRODUCTION

It is important to know that any weld produced is going to be the best one for the job. A method is also needed to ensure that each weld made in the same plant or on the same type of equipment in another plant will be of the same quality.

To meet these requirements various agencies have established codes and standards. These detailed written outlines explaining exactly how a weld is to be laid out, performed, and tested have made consistent quality welds possible. By having the required information skilled welders in shops all around the city, state, country, or world can make the same weld to the same level of safety, strength, and reliability.

A testing procedure to certify the welder ensures that the welder has the skills to make the weld. Passing a weld test is much easier when all of the detailed information is provided.

Selecting the code or standard to be used to judge a weld is equally as important as having a skilled welder. Not every product welded needs to be manufactured to the same level. The decision on the appropriate code or standard can be one of the most important aspects of welding fabrication. If the wrong one is selected, the cost of fabrication can be too high, or the parts might not stand up to the service.

To enable a welding business to operate profitably, the owner or manager must be able to make cost-effective welding decisions. A number of factors affect the cost of producing weldments. Some of these factors include the following:

- Material
- Weld design
- Welding processes
- Finishing
- Labor
- Overhead

CODES, STANDARDS, PROCEDURES, AND SPECIFICATIONS

A number of organizations publish codes or specifications that cover a wide variety of welding conditions and applications. The selection of the specific code to be used is made by the engineers, designers, or governmental requirements. Codes and specifications are intended to be guidelines only and must be qualified for specific applications by testing.

A welding code or standard is a detailed listing of the rules or principles that are to be applied to a specific classification or type of product.

A welding specification is a detailed statement of the legal requirements for a specific classification or type of weld to be made on a specific product. Products manufactured to code or specification requirements commonly must be inspected and tested to assure compliance.

A number of agencies and organizations publish welding codes and specifications. The selection of the particular code or specification to a weldment can be the result of one or more of the following requirements:

- Local, state, or federal government regulations — Many governing agencies require that a specific code or standard be followed.

- Bonding or insuring company — The weld must be shown to be fit for service requirements as established through testing. A bonding or insuring company must feel that the product is the safest that can be produced.

- End user (customer) requirements — The manufacturer considers cost and reliability; that is, as stricter standards are applied to the welding, the cost of the weldments increases. The more lax the standard, the lower the cost, but also the reliability and possibly the safety decrease.

- Standard industrial practices — The code or standard used is considered to be the standard one for the industry and has been in use for some time.

The three most commonly used codes are:

- Standard 1104, American Petroleum Institute — Used for pipelines

- Section IX, American Society of Mechanical Engineers — Used for pressure vessels and nuclear components

- D1.1, American Welding Society — Used for bridges, buildings, and other structural steel

The following organizations publish welding codes and/or specifications. Most can be contacted for additional information and current price list either directly or through the World Wide Web.

AASHT

American Association of State Highway and Transportation Officials
444 North Capitol Street, NW
Washington, DC 20001

AIAA

Aerospace Industries Association of America
1725 DeSales Street, NW
Washington, DC 20036

AISC

American Institute of Steel Construction
400 North Michigan Avenue
Chicago, IL 60611

ANSI

American National Standards Institute
1430 Broadway
New York, NY 10018

API

American Petroleum Institute
2101 L Street, NW
Washington, DC 20037

AREA

American Railway Engineering Association
Suite 403
2000 L Street, NW
Washington, DC 20036

ASME

American Society of Mechanical Engineers
345 East 47th Street
New York, NY 10017

AWWA

American Water Works Association
6666 West Quincy Avenue
Denver, CO 80235

AWS

American Welding Society
550 NW LeJeune Road
Miami, FL 33126

AAR

Association of American Railroads
1920 L Street, NW
Washington, DC 20036

MIL

Department of Defense
Washington, DC 20301

SAE

Society of Automotive Engineers
400 Commonwealth Drive
Warrendale, PA 15096

WELDING PROCEDURE QUALIFICATION

Welding Procedure Specification (WPS)

A welding procedure and specification is a set of written instructions by which a sound weld is made. Normally the procedure is written in compliance with a specific code, specification, or definition.

Welding Procedure Specification (WPS) is the standard terminology used by the American Welding Society (AWS) and the American Society of Mechanical Engineers (ASME). Welding Schedule is the standard federal government, military, or aerospace terminology denoting a WPS. The shortened term welding procedures is the most common term used by industry to denote a WPS.

The WPS lists all of the parameters required to produce a sound weld to the specific code, specifications, or definition. Specific parameters such as welding process, technique, electrode or filler, current, amperage, voltage, preheat, and postheat should also be included. The procedure should list a range or set of limitations on each, such as amps = 110–150, voltage = 17–22, etc., with the more essential or critical parameters more closely defined or limited.

The WPS should give enough detail and specific information so that any qualified welder could follow it and produce the desired weld. The WPS should always be prepared as a tentative document until it is tested and qualified.

Qualifying the Welding Procedure Specification

The WPS must be qualified to prove or verify that the list of variables — amperage, voltage, filler, etc. —

will provide a sound weld. Sample welds are prepared using the procedure and specifications listed in the tentative WPS. A record of all of the parameters used to produce the test welds must be kept; be specific for the parameters with limits, like voltage, amperage, etc. This information should be recorded on a form called the Procedure Qualification Record (PQR).

In most cases, the inspection agency, inspector, client, or customer will request a copy of both the WPS and the PQR before allowing production welding to begin.

Qualifying and Certifying

The process of qualifying and then certifying both the WPS and welders requires a number of specific requirements. The requirements may vary from one code or standard to another but the general process is the same for most. Before you invest in the testing required to qualify and certify process and welders under a code, you must first obtain a copy of the code you are planning to use. The requirements of codes and standards change from time to time, and it's important that your copy is the most recent version.

The following is a generic schedule of required activities you might follow when qualifying and certifying the welding process, the welder(s), and/or welding operator.

1. A tentative welding procedure is prepared by a person knowledgeable of the process and technique to be used and the code or specification to be satisfied.

2. Test samples are welded in accordance with the tentative WPS, and the welding parameters are recorded on the PQR. The test must be witnessed by an authorized person from an independent testing lab, the customer, an insurance company, or other individual(s) as specified by the code or listing agency.

3. The test samples are tested under the supervision of the same individuals or group that witnessed the test by the applicable requirements, codes, or specifications.

4. If the test samples pass the applicable test, the procedure has completed qualification. It is then documented as qualified/finalized and is released for use in production.

5. If the test samples do not pass the applicable test, the tentative WPS value parameters are changed as deemed feasible. Test samples are then rewelded and retested to determine if they do or do not meet applicable requirements. This process is repeated until the test samples pass applicable requirements, and the procedure is finalized and released.

6. The welder making the test samples to be used in qualifying the procedure is normally considered qualified and is then certified in the specific procedure.

7. Other welders to be qualified weld test samples per the WPS, and the samples are tested per applicable requirements. If the samples pass, the welder is qualified to the specific procedure and certified accordingly.

8. A qualified WPS is usable for an indefinite length of time, usually until a process considered more efficient for a particular production weld is found.

9. The welder's qualification is normally considered effective for an indefinite period of time, unless the welder is not engaged in the specific process of welding for which he or she is qualified for a period exceeding 6 months; then the welder must requalify. Also a welder will need to requalify if for some reason the qualification is questioned.

Figures 19-1 and 19-2 (page 462) are two examples of test records used to qualify a WPS and a welder for plate, PQR.

GENERAL INFORMATION

Normally, the format of the WPS is not dictated by the code or specification. Any format is acceptable as long as it lists the parameters or variables (essential or nonessential, amps, volts, filler identification, etc.) listed by the code or specification. Most codes or specifications appear in an acceptable or recommended format.

Ideally, the WPS should include all of the information required to make the weld. A welder should be able to be given the WPS without additional instructions and produce the weld. To help with this, it is often a good idea to include supplementary information with each WPS. The information might be basic instructions for the process. With some WPS you might include several pages as attachments that can give the welder a little review of the setup, operation, testing, inspecting, etc, that will help to ensure accuracy and uniformity in the welds.

Essential variables are those parameters in which a change is considered to affect the mechanical properties of the weldment to the point of requiring requalification of the procedure. Nonessential variables are those parameters in which a change may be made without requiring requalification of the procedure. However, a change in nonessential variables usually requires a revision to be made.

There are large differences between various codes. The AWS D1.1, *Structural Welding Code,* allows some prequalified weld joints for specific processes (SMAW, SAW, FCAW, and GMAW). A written procedure is required for these joints; but since the procedure is tentative, it does not require support via a written PQR, Figure 19-3 (pages 463–467).

The Procedure Qualification Requirements regarding positions for groove welds in plate differ between codes. The AWS D1.1 requires a written procedure for each position. The ASME Section IX, however, qualifies

a welder for the 1G position when the welder qualifies for 2G, 3G, or 4G.

Ordinarily, the welder must be qualified/certified in accordance with a specific WPS. The welder's qualifying test plate may be examined radiographically or ultrasonically in place of bend tests. Specific codes or specifications must be referenced for details of the number of actual tensile, bend, or other type test specimens and tests to be performed. For example, AWS D1.1 and ASME Section IX do not require a "Nick-Break Test Specimen," but API Standard 1104 does require it.

PRACTICE 19-1

Writing a Welding Procedure Specification (WPS)

Using the form provided and following the example, **Figure 19-3,** you will write a welding procedure specification. **Figure 19-4** (page 468) is a form that is a composite of sample WPS forms provided by AWS, ASME, and API codes. You may want to obtain a copy of one of the codes or standards and compare a weld you made to the standard. Most of the unique information is provided in this short outline. Additional information that may be required for this form can be found in figures in this chapter. You may need to refer to some of the chapters on welding or to your notes to establish the actual limits of the welding variables (voltage, amperage, gas flow rates, nozzle size, etc.).

NOTE: Not all of the blanks will be filled in on the forms. The forms are designed to be used with a large variety of weld procedures, so they have spaces that will not be used each time.

1. The WPS number is usually made up following a system established by the company. This number may or may not include coded information relating to the date it was written, who wrote it, material or process data, and so on.

2. Date the WPS was written or effective.

3. The welding process(es) that will be used to perform the weld SMAW, GMAW, GTAW, and so on.

4. The actual material type and thickness or pipe type and diameter and/or wall thickness. If all the material or pipe being joined are the same, then the same information will appear before and after "to."

5. Fillet or groove weld and the joint type butt, lap, tee, and so on.

6. Thickness range qualified (or) diameter range qualified: For both plate and pipe, a weld performed successfully on one thickness qualifies a welder to weld on material within that range. See **Table 19-1** (page 467) for a list of thickness ranges.

7. Material position: 1G, 2G, 3G, 4G, 1F, 2F, 3F, 4F, 5G, 6G, 6GR, **Figure 19-3.**

WELDING PROCEDURE SPECIFICATION (WPS)

Welding Procedures Specifications No: _____   Date: _____

TITLE:
Welding _____ of _____ to _____

SCOPE:
This procedure is applicable for _____
within the range of _____ through _____

Welding may be performed in the following positions _____

BASE METAL:
The base metal shall conform to _____

Backing material specification _____

FILLER METAL:
The filler metal shall conform to AWS classification No. _____ from
AWS specification _____. This filler metal falls into F-number
_____ and A-number _____

SHIELDING GAS:
The shielding gas, or gases, shall conform to the following compositions and purity:
_____

JOINT DESIGN AND TOLERANCES:

PREPARATION OF BASE METAL:

ELECTRICAL CHARACTERISTICS:
The current shall be _____

The base metal shall be on the _____ side of the line.

PREHEAT:

BACKING GAS:

WELDING TECHNIQUE:

INTERPASS TEMPERATURE:

CLEANING:

INSPECTION:

REPAIR:

SKETCHES:

Figure 19-1 Welding procedure specification (WPS).

PROCEDURE QUALIFICATION RECORD (PQR)

Welding Qualification Record No: _____(1)_____ WPS No: _____(2)_____ Date: _____(3)_____

Material specification _____(4)_____ to _____

P-No. _____(5)_____ to P-No. _____ Thickness and O.D. _____(6)_____

Welding process: Manual _____(7)_____ Automatic _____(8)_____

Thickness Range _____(9)_____

Filler Metal

Specification No. _____(10)_____ Classification _____(11)_____ F-number _____(12)_____

A-number _____(13)_____ Filler Metal Size _____(14)_____ Trade Name _____(15)_____

Describe filler metal (if not covered by AWS specification) _____(16)_____

_____

Flux or Atmosphere

Shielding Gas _____(17)_____ Flow Rate _____(18)_____ Purge _____(19)_____

Flux Classification _____(20)_____ Trade Name _____(21)_____

Welding Variables

Joint Type _____(22)_____	Position _____(29)_____
Backing _____(23)_____	Preheat _____(30)_____
Passes and Size _____(24)_____	Bead Type _____(31)_____
No. of Arcs _____(25)_____	Current _____(32)_____
Ampere _____(26)_____	Volts _____(33)_____
Travel Speed _____(27)_____	Oscillation _____(34)_____

Interpass Temperature Range _____(28)_____

Weld Results

Appearance _____(35)_____ Weld Size _____(36)_____

Guided Bend Test

Type	Result	Type	Result
(37)	(38)		

Tensile Test

Specimen No.	Dimensions Width|Thickness	Area	Ultimate Total Load, lb.	Ultimate Unit Stress, psi.	Character of Failure And Location
(39)	(40)	(41)	(42)	(43)	(44)

Welder's Name _____(45)_____ Identification No. _____(46)_____ Laboratory Test No. _____

By virtue of these test meets welder performance requirements.

Test Conducted by _____(47)_____ Address _____

per _____(48)_____ Date _____(49)_____

We certify that the statements in this record are correct and that the test welds performed and tested are in accordance with the WPS

Manufacture _____(50)_____

Signed by _____

Date _____

Figure 19-2 Procedure qualification record (PQR).

WELDING PROCEDURE SPECIFICATION (WPS)
FOR
ABC, INC.

Welding Procedures Specifications No:_____WPS-1A_____   Date:____7-11-91____

TITLE:
Welding _____GTAW_____ of _____3" SCHEDULE 80 316 SEAMLESS PIPE_____ to __3" SCHEDULE 80 316 PIPE FLANGE____ .

SCOPE:
This procedure is applicable for _____V-GROOVED WELDS IN PIPING_____
within the range of _____3" SCHEDULE 80_____ through _____3" SCHEDULE 80_____
Welding may be performed in the following positions ____1G____ .

BASE METAL:
The base metal shall conform to ASTM A376 GRADE TP-316 SEAMLESS PIPE SCHEDULE 80 P8
Backing material specification _____NA_____ .

FILLER METAL:
The filler metal shall conform to AWS classification No. ____ER 316L____ from AWS
specification ____A5.9____ . This filler metal falls into F-number _6_ and A-number _8_ .

SHIELDING GAS:
The shielding gas, or gases, shall conform to the following compositions and purity:
_____WELDING GRADE ARGON_____ .

JOINT DESIGN AND TOLERANCES:

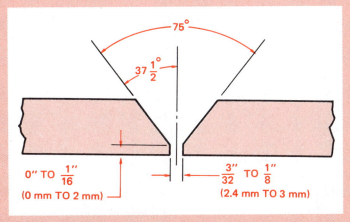

PREPARATION OF BASE METAL:
The edges of parts to be joined shall be prepared by machining. All parts to be joined must be cleaned prior to welding of all hydrocarbons and other contaminants, such as cutting fluids, grease, oil and primers, by suitable solvents. Both the inside and outside surfaces within 2" of the joint must be mechanically cleaned by using a stainless steel wire brush that has not been used for other purposes or pickled with 10 to 20 percent nitric acid solution. Joint alignment shall be maintained by 4 tack welds equally spaced around the joint. The tack welds are to be made using the same GTA Welding process used for the root pass. Both ends of each tack weld are to be ground to a taper prior to the beginning of the root pass.

Figure 19-3(A)

ELECTRICAL CHARACTERISTICS:
The current shall be_____DIRECT CURRENT STRAIGHT POLARITY_____ .
The base metal shall be on the _____NEGATIVE_____ side of the line.

PREHEAT:
The parts shall not be welded if they are below 70° F.

BACKING GAS:
To protect the inside of the root surface from the formation of oxides during welding, a continuous flow of argon into the part is required. The open end of the part must be capped (figure 1) and the unwelded joint must be taped prior to the beginning of any welding. The backing gas must have a flow rate of 10 CFH to 15 CFH, and the flow must begin 2 minutes before welding starts and continue until the part has cooled to room temperature. The backing gas may be stopped between welds only if the part is allowed to cool to room temperature.

WELDING TECHNIQUE:
The GTA welding process is to be used for making the weld.

Electrode	1/8 inch diameter EWTh-2
Electrode tip geometry	Tapered 2 to 3 times length to diameter
Nozzle	1/2 inch diameter
Shielding gas	Argon
Shielding gas flow rate	20 to 45 CFH
Current, A	90 to 150
Polarity	DCEN
Arc voltage, V	12
Filler metal type	ER 316L
Filler metal size	3/32 to 1/8 inch diameter
Backing gas	Argon
Backing gas flow rate	10 to 15 CFH
Preheat	70 degrees min.
Interpass temp.	500 degrees max.
Travel speed	As required

TACK WELDS:
With the pipe securely clamped in to welding jig and the flange fitting properly located with the correct root gap, the four tack welds are to be performed (figure 2). Holding the electrode so that it is very close to the root face but not touching (figure 3), slowly increase the current until the arc starts and a molten weld pool is formed. Adding filler metal to maintain a slight convex weld face and a flat or slightly concave root face (figure 4). When it's time to end the tack weld, lower the current slowly so that the molten weld pool can be tapered down in size (figure 5). When all four tack welds are complete, allow the pipe to cool. Using a grinding wheel that has never been used on any metal other than 316 stainless steel, feather the ends of the tack welds.

Figure 19-3(B)

ROOT WELD:

Holding the electrode so that it is very close to the root face but not touching, slowly increase the current until the arc starts and a molten weld pool is formed. As the weld progresses, add filler metal as required to maintain a slightly convex weld face and a flat or slightly concave root face. When it's necessary to stop the weld, to reposition the part, or the weld is completed, the current must be lowered slowly so that the molten weld pool can be tapered down in size.

FILLER AND COVER WELDS:

Position the pipe so that the weld is in the 1G position. Connect the backing gas and start the flow 2 minutes before welding begins. A 10 to 15 CFH flow must be maintained during the remainder of the welding. The backing gas may be stopped if welding is to be discontinued for more than 15 minutes. A slightly higher welding current level will be required for the remaining welds. Start each of the filler welds so that the starting and ending points will be staggered (Figure 6). The size of each of the filler passes and the cover passes must not be larger than 1/4 inch wide. The weld bead size must be small so as to minimize the heat input in the weld joint and so as to allow the weld to stress relieve itself. The finished weld contour must be within tolerance (Figure 7).

INSPECTION:

The weld is to be inspected visually in accordance to AWS B1. Guide for Nondestructive Inspection of Welds. It must be found to be within tolerance as stated herein.

REPAIR:

Only slight surface discontinuities may be repaired with the approval of the QC department.

Figure 19-3(C)

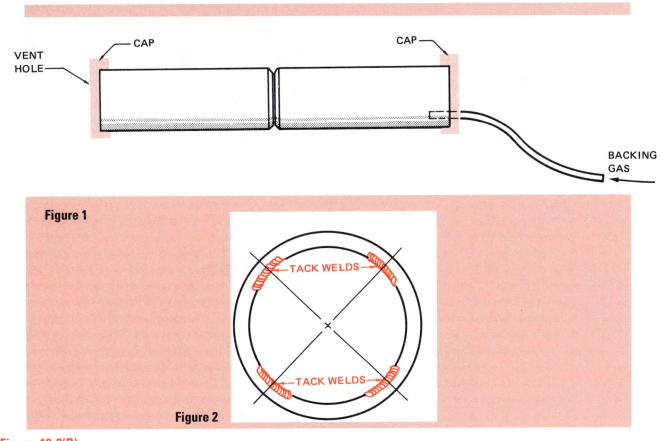

VENT HOLE —
CAP
CAP
BACKING GAS

Figure 1

TACK WELDS

TACK WELDS

Figure 2

Figure 19-3(D)

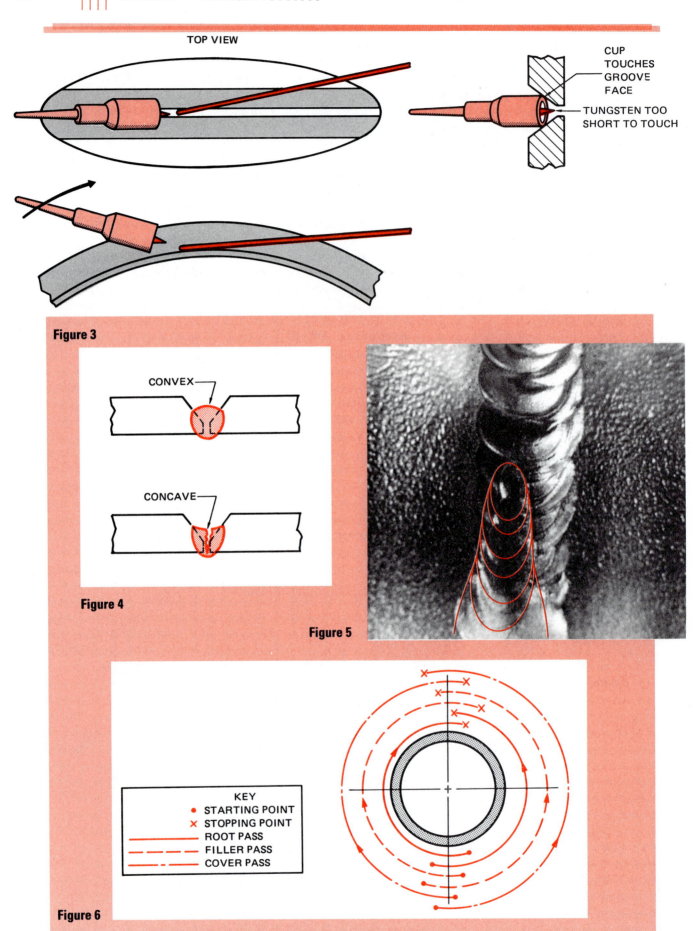

TOP VIEW

CUP TOUCHES GROOVE FACE

TUNGSTEN TOO SHORT TO TOUCH

Figure 3

CONVEX

CONCAVE

Figure 4

Figure 5

KEY
• STARTING POINT
✕ STOPPING POINT
— ROOT PASS
— – FILLER PASS
— · — COVER PASS

Figure 6

Figure 19-3(E)

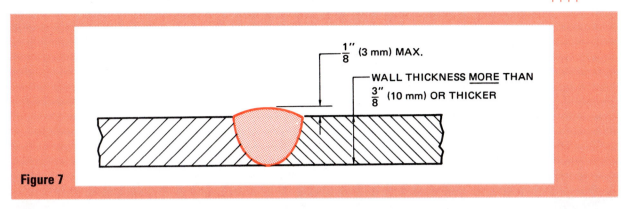

Figure 7

Figure 19-3(F)

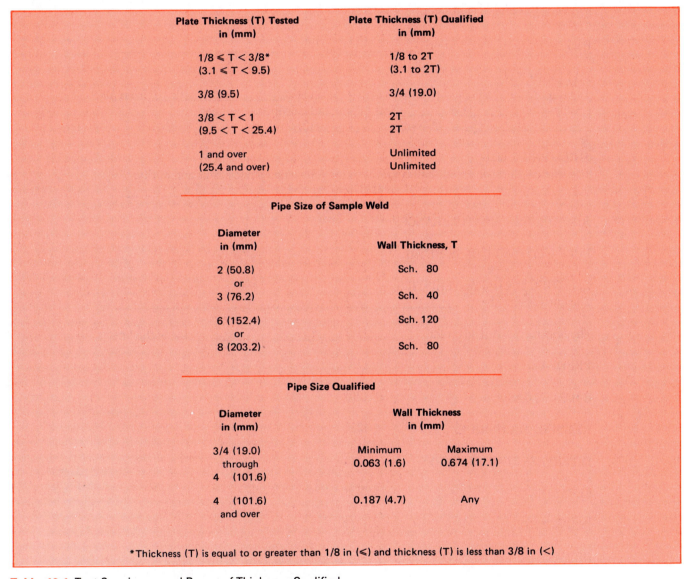

Plate Thickness (T) Tested in (mm)	Plate Thickness (T) Qualified in (mm)
1/8 ≤ T < 3/8* (3.1 ≤ T < 9.5)	1/8 to 2T (3.1 to 2T)
3/8 (9.5)	3/4 (19.0)
3/8 < T < 1 (9.5 < T < 25.4)	2T 2T
1 and over (25.4 and over)	Unlimited Unlimited

Pipe Size of Sample Weld

Diameter in (mm)	Wall Thickness, T
2 (50.8) or	Sch. 80
3 (76.2)	Sch. 40
6 (152.4) or	Sch. 120
8 (203.2)	Sch. 80

Pipe Size Qualified

Diameter in (mm)	Wall Thickness in (mm) Minimum	Maximum
3/4 (19.0) through 4 (101.6)	0.063 (1.6)	0.674 (17.1)
4 (101.6) and over	0.187 (4.7)	Any

*Thickness (T) is equal to or greater than 1/8 in (≤) and thickness (T) is less than 3/8 in (<)

Table 19-1 Test Specimens and Range of Thickness Qualified.

8. Base metal specification: This is the ASTM specification for the type and grade of material including the P-number, **Table 19-2.**

9. If a backing material is used, then its ASTM or other specification information must be included here.

10. Classification number: This is the standard number found on the electrode or electrode box, such as E6010, E7018, E316-15, ER70S-3, or E70T-1.

11. Filler metal specification number: The AWS has specifications for chemical composition and physical

WELDING PROCEDURE SPECIFICATION (WPS)

Welding Procedures Specifications No: _____(1)_____   Date: _____(2)_____

TITLE:
Welding _____(3)_____ of _____(4)_____ to _____(4)_____

SCOPE:
This procedure is applicable for _____(5)_____
within the range of _____(6)_____ through _____(6)_____

Welding may be performed in the following positions _____(7)_____

BASE METAL:
The base metal shall conform to _____(8)_____

Backing material specification _____(9)_____

FILLER METAL:
The filler metal shall conform to AWS classification No. _____(10)_____ from AWS
specification _____(11)_____. This filler metal falls into F-number
_____(12)_____ and A-number _____(13)_____

SHIELDING GAS:
The shielding gas, or gases, shall conform to the following compositions and purity:
_____(14)_____

JOINT DESIGN AND TOLERANCES:
                                                (15)
PREPARATION OF BASE METAL:
                                                (16)
ELECTRICAL CHARACTERISTICS:
The current shall be _____(17)_____

The base metal shall be on the _____(18)_____ side of the line.

PREHEAT: (19)

BACKING GAS: (20)

WELDING TECHNIQUE: (21)

INTERPASS TEMPERATURE: (22)

CLEANING: (23)

INSPECTION: (24)

REPAIR: (25)

SKETCHES: (26)

Figure 19-4 Welding procedure specifications (WPS).

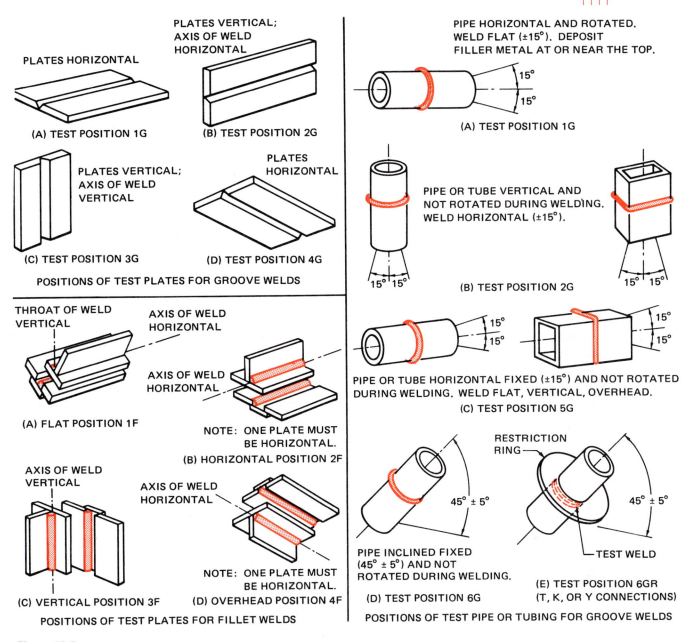

Figure 19-5

Type of Material	
P-1	Carbon Steel
P-3	Low Alloy Steel
P-4	Low Alloy Steel
P-5	Alloy Steel
P-6	High Alloy Steel — Predominently Martensitic
P-7	High Alloy Steel — Predominently Ferritic
P-8	High Alloy Steel — Austenitic
P-9	Nickel Alloy Steel
P-10	Specialty High Alloy Steels
P-21	Aluminum and Aluminum-Base Alloys
P-31	Copper and Copper Alloy
P-41	Nickel

Table 19-2 P-Numbers.

A5.10	Aluminum — bare electrodes and rods
A5.3	Aluminum — covered electrodes
A5.8	Brazing filler metal
A5.1	Steel, carbon, covered electrodes
A5.20	Steel, carbon, flux-cored electrodes
A5.17	Steel-carbon, submerged arc wires and fluxes
A5.18	Steel-carbon, gas metal arc electrodes
A5.2	Steel — oxyfuel gas welding
A5.5	Steel — low-alloy covered electrodes
A5.23	Steel — low-alloy electrodes and fluxes — submerged arc
A5.28	Steel — low-alloy filler metals for gas shielded arc welding
A5.29	Steel — low-alloy, flux-cored electrodes

Table 19-3 Specification Numbers.

properties for electrodes. Some of these specifications are listed in **Table 19-3**.

12. F-number: A specific grouping number for several classifications of electrodes having similar composition and welding characteristics. **Table 19-4** lists the F-number corresponding to the electrode used.

13. A-number: The classification of weld deposit analysis. **Table 19-5** lists the A-numbers.

14. Shielding gas(es) and flow rate for GMAW, FCAW, or GTAW. ◆

PRACTICE 19-2

Procedure Qualification Record (PQR)

Following the procedure you wrote in Practice 19-1, you are going to make the weld to see if your tentative welding procedure and specification can be certified. Use the form provided to record all of the appropriate information, **Figure 19-2** (page 462). ◆

1. The PQR number is usually made up following a system established by the company. This number may or may not include coded information relating to the product being welded, material or process data, and the like.

2. The WPS number on which the PQR is based.

3. The date on which the welding took place.

4. Base metal specification: This is the ASTM specification for the type and grade of material.

5. **Table 19-2** lists some commonly used metals and their P-numbers.

6. Test material thickness (or) test pipe outside diameter (OD) (and) wall thickness.

7. Manual welding processes are used to qualify a welder. Specify the process GMAW, FCAW, SMAW, GTAW, and so on.

8. Automatic welding processes are used to qualify a welding operator. Specify the process (SAW, ESW, etc.).

9. Thickness range qualified (or) diameter range qualified: For both plate and pipe, a weld performed successfully on one thickness qualifies a welder to weld on material within that range. See **Table 19-1** (page 467) for a list of thickness ranges.

10. Filler metal specification number: The AWS has specifications for chemical composition and physical properties for electrodes. Some of these specifications are listed in **Table 19-3**.

11. Classification number: This is the standard number found on the electrode or electrode box, such as E6010, E7018, E316-15, or ER1100.

12. F-number: A specific grouping number for several classifications of electrodes having similar composition and welding characteristics. See **Table 19-4** for the F-number corresponding to the electrode used.

13. A-number: The classification of weld deposit analysis. See **Table 19-5** for a list of A-numbers.

14. Give the diameter of electrode used.

15. Give the manufacturer's identification name or number.

16. List the manufacturer's chemical composition and physical properties as provided if the filler metal is not covered by an AWS specification.

17. Shielding gas or gas mixture for GMAW, FCAW, or GTAW.

18. Flow rate in cubic feet per hour (CFH).

19. The amount of time that the shielding gas is to flow to purge air from the welding zone.

20. SAW flux classification.

21. The manufacturer's identification name or number for the SAW flux.

22. Butt, lapp, tee, or other joint type.

23. Backing strip material specification: This is the ASTM specification number.

24. The number of passes and the size.

25. Usually 1 except for some automatic SAW process that may use multiple electrodes with multiple arcs.

26. The amount of power in amps used to make the weld. If the machine being used for the weld does not have an amp meter, a meter must be attached to the welding leads to get this reading.

Group Designation	Metal Types	AWS Electrode Classification
F1	Carbon Steel	EXX20, EXX24, EXX27, EXX28
F2	Carbon Steel	EXX12, EXX13, EXX14
F3	Carbon Steel	EXX10, EXX11
F4	Carbon Steel	EXX15, EXX16, EXX18
F5	Stainless Steel	EXXX15, EXXX16
F6	Stainless Steel	ERXXX
F22	Aluminum	ERXXXX

Table 19-4 F-Numbers.

A No.	Types of weld deposit	Analysis					
		C %	Mn %	Si %	Mo %	Cr %	Ni %
1	Mild Steel	0.15	1.6	1.0	-	-	-
2	Carbon-Moly	0.15	1.6	1.0	0.4-0.65	0.5	-
3	Chrome (0.4 to 2%)-Moly	0.15	1.6	1.0	0.4-0.65	0.4-2.0	-
4	Chrome (2 to 6%)-Moly	0.15	1.6	2.0	0.4-1.5	2.0-6.0	-
5	Chrome (6 to 10.5%)-Moly	0.15	1.2	2.0	0.4-1.5	6.0-10.5	-
6	Chrome-Martensitic	0.15	2.0	1.0	0.7	11.0-15.0	-
7	Chrome-Ferritic	0.15	1.0	3.0	1.0	11.0-30.0	-
8	Chromium-Nickel	0.15	2.5	1.0	4.0	14.5-30.0	7.5-15.0
9	Chromium-Nickel	0.30	2.5	1.0	4.0	25.0-30.0	15.0-37.0
10	Nickel to 4%	0.15	1.7	1.0	0.55	-	0.8-4.0
11	Manganese-Moly	0.17	1.25-2.25	1.0	0.25-0.75	-	0.85
12	Nickel-Chrome-Moly	0.15	0.75-2.25	1.0	0.25-0.8	1.5	1.25-2.25

Table 19-5 A-Number Classification of Ferrous Metals.

27. The travel speed in inches per minute is usually given only for machine or automatic welds.

28. This is the maximum temperature that the base metal is allowed to reach during the weld. Welding must stop and the part allowed to cool if this temperature is reached.

29. Test position: 1G, 2G, 3G, 4G, 1F, 2F, 3F, 4F, 5G, 6G, 6GR, **Figure 19-5** (page 469).

30. This is the minimum temperature that the base metal must be before welding can start.

31. Groove of fillet weld.

32. AC, DCEP or DCEN

33. The voltage is usually only included for GMAW, FCAW, and SMA welding.

34. The type of electrode movement used when making the weld.

35. Visually inspect the weld and record any flaws.

36. Record the legs and reinforcement dimensions. Measure and record the depth of the root penetration.

37. Four (4) test specimens are used for 3/8-in. or thinner metal. Two (2) will be root bent and two (2) face bent. For thicker metal all four (4) will be side bent.

38. Visually inspect the specimens after testing and record any discontinuities.

39. Identification number that was marked on the specimen.

40. Width and thickness of test section of the specimen.

41. Cross-sectional area of specimen in the test area.

42. The force at which the specimen failed.

43. The force divided by the specimen's area converts the ultimate total load for the specimen to the force that would have been required to break a one square in. section of the material.

44. The type of failure whether it was ductal or brittle and where the failure occurred relative to the weld.

45. Welder's name: The person who performed the weld.

46. Identification no.: On a welding job, every person has an identification number that is used on the time card and paycheck. In this space, you can write the class number or section number, since you do not have a clock number.

47. The name of the person who interpreted the results.

48. Qualifications of the test interpreter: This is usually a Certified Welding Inspector or other qualified person.

49. Date the results of the test were completed.

50. The name of the company that requested the test.

WELDING COSTS

Estimating the costs of welding can be a difficult task because of the many variables involved. One approach is to have a welding engineer specify the type and size of weld to withstand the loads that the weldment must bear. Then the welding engineer must select the welding process and filler metal that will provide the required welds at the least possible cost. This method is used in large shops, where welding on a product can range from a few thousand dollars to well over a million dollars. With competition for work resulting in small profit margins, each job must be analyzed carefully.

The second approach, used by smaller shops, is to get a price for the materials and then estimate the production time. Process and filler metal costs are considered, but little thought is given to the hidden costs of equipment depreciation, joint efficiency, power, and so on. With the majority of the welding jobs in these shops taking from a few hours to a few days and with costs ranging from a few hundred dollars to a few thousand dollars, extensive cost analysis cannot be justified. Extensive estimating may take more time than the job itself. To remain profitable *and* competitive, some cost estimation is required. Only the cost considerations that a small shop should consider when estimating a job will be covered in this section.

COST ESTIMATION

A number of factors affect welding cost. These factors can be divided into two broad categories: fixed and variable. *Fixed costs* are those expenses that must be paid each and every day, week, month, or year, regardless of work or production. Examples of fixed costs include rent, taxes, insurance, and advertising. *Variable costs* are those expenses that change with the quantity of work being produced. Examples of variable cost include supplies, utilities, labor, and equipment leases. Expenses within both categories must be considered when making welding job estimates. These cost areas include the following:

■ *Material cost:* The cost of new stock required to produce the weldment is fixed by the supplier. It is often possible to help control these costs by getting bids from several suppliers and combining as many jobs as possible in order to get any discount for bulk purchases.

■ *Scrap cost:* Scrap is an inevitable part of any project. It costs a company in two ways: by wasting expensive resources and by requiring cleanup and removal. Reducing scrap production through proper planning will result in a direct saving for any project.

■ *Process cost:* The major welding processes—SMAW, GMAW, and FCAW—differ widely in their cost of equipment, operating supplies, and production efficiency. SMAW has the lowest initial cost and has excellent flexibility but also a higher total cost for large jobs.

■ *Filler metal:* The cost of filler metal per pound is only a small part of its actual cost. The major welding processes (SMAW, GMAW, and FCAW) have widely varying deposition and efficiency rates.

■ *Labor cost:* Total labor cost includes wages and benefits. Insurance, sick leave, vacation, social security, retirement, and other benefits can range from 25% to 75% of the total labor cost. Because labor costs are figured on an hourly basis, they can be controlled only by increasing productivity.

■ *Overhead costs:* Overhead costs are often intangible costs related to doing business. These costs include building rent or mortgage, advertising, insurance, utilities, taxes, licenses, governmental fees, accounting, loan payments, and property upkeep.

■ *Finishing cost:* Post-welding cleanup, painting, or other finishing add to the weldment's final production cost. Many of these finishing processes can produce some level of health hazard, and a major concern to everyone is the environment. Complying with local, state, and federal environmental laws can add significantly to the cost of painting, dipping, and plating. New environmentally friendly paints, low-pressure spray guns, and water-based products are a few of the advancements in finishing that have helped reduce environmental compliance costs.

Joint Design

Joint design is an important consideration when estimating weld cost. The root opening, root face thickness, and bevel angles must be studied carefully when making design decisions. All these factors affect the weld dimensions, **Figure 19-6**, which in turn determine the amount of weld metal needed to fill the joint. Plate thickness is a major factor that, in most cases, cannot be changed. Increasing the root opening increases cost. But larger root openings often allow the bevel angle to be reduced while maintaining good access to the weld root.

To keep welding costs down, the joints should have the smallest possible root opening and the smallest reasonable bevel angle. These conditions are more easily achievable with welding processes that provide deep penetration. The other benefit of deep penetration is the option of a deeper root face and its significant effect on the volume of filler metal needed. In addition, the amount of reinforcement affects welding costs. Some reinforcement is unavoidable in order to ensure a full thickness weld. However, too much reinforcement requires extra time and material, which can reduce the weld's strength and fatigue life.

The weld should be approximately the same size as the metal is thick. Welds that are undersized or oversized can cause joint failure. Welds that are undersized do not have enough area to hold the parts under load. Welds that are oversized can make the joint too stiff. The lack of flexibility of the weld causes the metal near the joint to be highly stressed, **Figure 19-7**. The same will happen to a piece of wire bent between two pliers: It will break. But a piece of wire bent between your fingers will withstand more bending before it breaks.

Welds made on parts of unequal size should allow enough joint flexibility to prevent a crack from forming along the edge of the weld. It is possible to taper the thicker metal to reduce the thickness or to build up the thinner metal.

Overwelding also contributes to welding costs. Welders often believe "if a little is good, a lot is better, and too much is just right." Too often it is assumed that a large reinforcement means greater strength, but that is never true.

Cutting weld volume also reduces the labor cost. It should be remembered that the labor content of welds

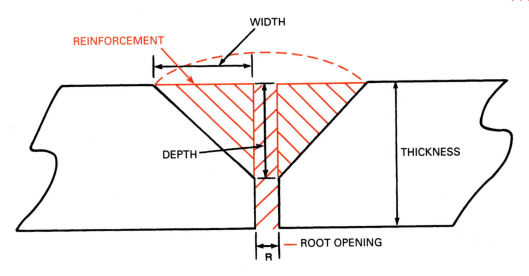

Figure 19-6 Calculation of weld requirements depending on joint design.

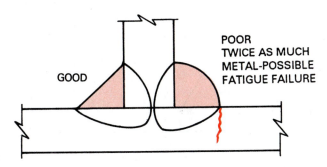

Figure 19-7 Overwelding can be harmful as well as costly.

almost always exceeds the cost of the expendable. Reducing the amount of filler metal needed also cuts the time needed to make the welds. This significant effect on labor costs, **Figure 19-8**, justifies the price of more expensive filler metals or welding processes. This assumes that the metals and processes provide enough increase either in penetration, to allow joint designs requiring less filler metal, or in deposition rates, to reduce welding times.

Groove Welds

For groove welds the bevel angle greatly affects the filler metal volume. As the groove angle increases, more filler metal is required to fill it during the weld. Because the volume is in proportion to the angle, the change in volume can be calculated. The first thing in determining the weld volume is to find the cross-sectional area of the weld groove. First find the rectangular area formed by the root opening:

Root Area = root opening × plate thickness

Next find the area of the triangular space of a bevel, which is equal to 1/2 of the width times the depth. The area of a V-groove (if both sides are at the same bevel angle) is equal to the width times the depth:

$$\text{Bevel area} = \frac{\text{bevel width} \times \text{bevel depth}}{2}$$

$$\text{V-Groove area} = \text{bevel width} \times \text{bevel depth}$$

The cross-sectional area in **Figure 19-6** is the sum of the small rectangular area formed by the root opening and plate thickness, plus the area of the triangular space formed by the bevel angle:

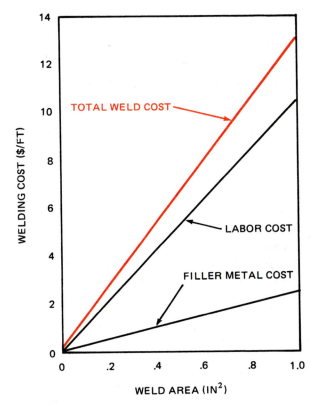

Figure 19-8 Increasing the weld size increases cost. Note the labor is 80% of the total cost (based on typical modern welding rates and efficiencies).

Total cross-sectional area = root area + bevel area(s)

The total groove volume is then determined by multiplying the groove area by the weld length.

Total groove volume = cross-sectional area × weld length

PRACTICE 19-3

Finding Weld Groove Volume

Using a pencil, paper, and calculator, determine the total volume of the following groove welds, **Figure 19-9.**

1. V-groove joint with the following dimensions:
   Width, 3/8 in.
   Depth, 3/8 in.
   Root opening, 1/8 in.
   Thickness, 1/2 in.
   Weld length, 144 in.

2. Bevel joint with the following dimensions:
   Width, 0.25 in.
   Depth, 0.375 in.
   Root opening, 0.062 in.
   Thickness, 0.5 in.
   Weld length, 96 in.

3. V-groove joint with the following dimensions:
   Width, 12 mm
   Depth, 12 mm
   Root opening, 2 mm
   Thickness, 15 mm
   Weld length, 3600 mm

4. Bevel joint with the following dimensions:
   Width, 15 mm
   Depth, 15 mm
   Root opening, 3 mm
   Thickness, 18 mm
   Weld length, 3 M

Fillet Welds

The deep penetration processes of fillet welds offer lower costs and improved weld quality. Smaller fillet welds with deeper penetration also have the potential for yielding much stronger welds, **Figure 19-10.** The cross-sectional area of a fillet weld is equal to 1/2 of the weld leg height times the weld leg width:

$$\text{Fillet area} = \frac{\text{leg height} \times \text{leg width}}{2}$$

The fillet weld volume is determined in the same manner as the groove weld, by multiplying the area times the length:

Total fillet volume = cross-sectional area × weld length

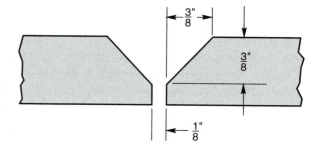

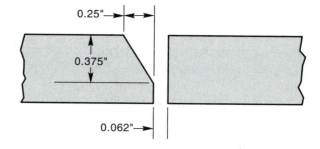

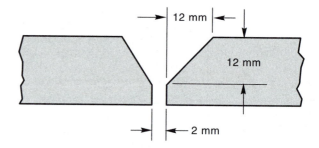

Figure 19-9

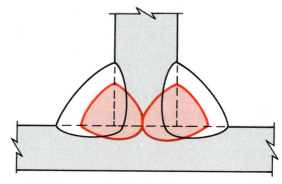

Figure 19-10 The red 1/4-in. (6-mm) fillet weld is stronger than the other fillet weld and contains about one-half the amount of filler metal.

WELD METAL COST

In their technical data sheets, manufacturers of filler metal provide information regarding the welding metal. The number of electrodes per pound or the length of wire per pound can be used to determine the pounds of electrodes needed to produce a weld. To make this determination, the weight of filler metal required to fill the groove or make the fillet weld must be determined. The weight of filler metal is determined by multiplying the weld volume times the density of the metal, **Table 19-6**:

Weight of filler metal = volume × metal density

Material	Weight lb/in₃	Weight g/cm₃
Aluminum	0.096	2.73
Steel	0.287	7.945

Table 19-6 Density of Metals.

PRACTICE 19-4

Using a pencil, paper, and a calculator, determine the weight of metal required for each of the welds described in Practice 19-3. Calculate the weight for both steel and aluminum base and filler metals.

Using weight of weld metal deposited allows for better comparisons when a number of different welds are being made. The weight of weld metal can either be determined for the welding prints or measured as the welder uses up supplies, **Figure 19-11**.

Cost of Electrodes, Wires, Gases, and Flux

You must obtain the current cost per pound of electrode or welding wire plus the cost of shielding gas of flux (if applicable from the supplier). The shielding gas flow rate varies slightly with the kind of gas used. The flow rates in **Figure 19-12** are average values whether the shielding gas is an argon mixture or pure CO_2. Use these rates in your calculations if the actual flow rate is not available.

In the submerged arc process (SAW) the ratio of flux to wire consumed in the weld is approximately 1 to 1

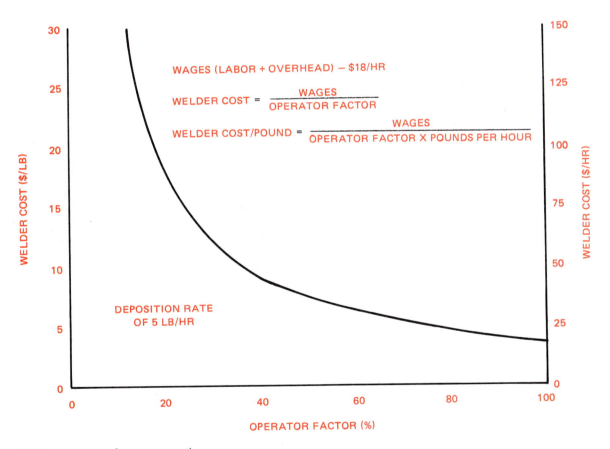

WAGES (LABOR + OVERHEAD) — $18/HR

$$\text{WELDER COST} = \frac{\text{WAGES}}{\text{OPERATOR FACTOR}}$$

$$\text{WELDER COST/POUND} = \frac{\text{WAGES}}{\text{OPERATOR FACTOR} \times \text{POUNDS PER HOUR}}$$

DEPOSITION RATE OF 5 LB/HR

Figure 19-11 Low operator factors are costly.

	GMAW		FCAW		
WIRE DIAMETER	.035"	.045"-1/16"	.045"	1/16"	5/64"-1/8"
CFH	30	35	35	40	45

Figure 19-12 Approximate shielding gas flow rate cubic feet per hour.

by weight. When the loss due to flux handling and flux recovery systems is considered, the average ratio of flux to wire is approximately 1.4 pounds of flux for each pound of wire consumed. If the actual flux-to-wire ratio is unknown, use 1.4 for cost estimating.

Deposition Efficiency

Not every pound of electrode filler metal purchased is converted into weld metal. Some portion of every electrode is lost as slag, spatter, and/or fume. Some, such as SMAW electrode stub ends, are unused. The amount of raw electrode deposited as weld metal is measured as deposition efficiency. If all is deposited, the deposition efficiency is 100%. If half is lost, the deposition efficiency is 50%. The argon-shielded GMA process is about 95% efficient. The SMAW process is between 40% and 60% efficient, depending on the electrode and the welder using it. Thus, a relatively costly filler metal with high efficiency can be as cost effective as one that appears to be cheaper, **Figure 19-13**. The efficiency can then be calculated by the following formula:

$$\text{Deposition efficiency} = \frac{\text{weight of weld metal}}{\text{weight of electrode used}} \quad \text{or}$$

$$\frac{\text{deposition rate (lb/hr)}}{\text{burn-off rate (lb/hr)}}$$

The deposition efficiency tells us how many pounds of weld metal can be produced from a given weight of the electrode of welding wire. As an example, 100 pounds of a flux cored electrode with an efficiency of 85% will produce approximately 85 pounds of weld metal. One hundred pounds of coated electrode with an efficiency of 65% will produce approximately 65 pounds of weld metal (less the weight of the stubs discarded, as described below).

Note that electrodes priced at $.65 actually cost about $1.30 per pound as weld metal because about 50% is wasted. The more expensive GMA wire at $.85 costs about $.90 per pound as weld metal when deposited because only 5% is lost as spatter and fume. Even with the added $.15 for shield gas, the final cost of $1.05 is less than that of welds made with covered electrodes.

Deposition Rate

The deposition rate is the rate at which weld metal can be deposited by a given electrode or welding wire, expressed in pounds per hour. It is based on continuous operation, with no time allowed for stops and starts for inserting a new electrode, cleaning slag, termination of the weld, or other reasons. The deposition rate will increase as the welding current is increased. When using solid or flux cored wires, the deposition rate will increase as the electrical stickout is increased and the same welding current is maintained. True deposition rates for each welding filler metal, whether it is a coated electrode or a solid or flux cored wire, can only be established by an actual test. The weldment is weighed before and after welding as the whole process is timed. The tables in **Figures 19-14** to **19-17** (pages 477–480) contain average values for the deposition rate of the various welding filler metals based on welding laboratory tests and published data.

Deposition Data Tables

Coated Electrodes. The deposition efficiency of coated electrodes (by AWS definition and in published data) does not subtract the unused electrode stub that is discarded. This is understandable, since the stub length can vary with the operator and the application. Long continuous welds are usually conducive to short stubs, while on short intermittent welds the stub length tends to be longer. **Figure**

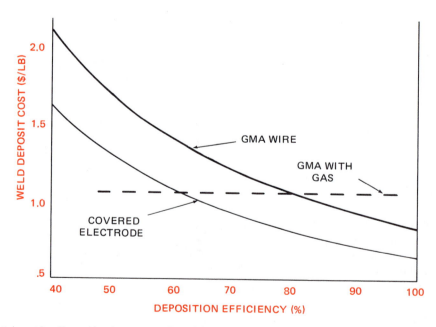

Figure 19-13 Weld metal cost is affected by the process deposition efficiency.

SHIELDED METAL ARC WELDING
DEPOSITION DATA — COATED ELECTRODES

E6010

Electrode Diameter	Amperes	Deposition Rate lbs/hr	Efficiency %
1/8	100	2.1	76.3
	130	2.3	68.8
5/32	140	2.8	73.6
	170	2.9	64.1
3/16	160	3.3	74.9
	190	3.5	69.7
7/32	190	4.5	76.9
	230	5.1	73.1
1/4	220	5.9	77.9
	260	6.2	76.2

E6011

Electrode Diameter	Amperes	Deposition Rate lbs/hr	Efficiency %
1/8	120	2.3	70.7
5/32	150	3.7	77.0
3/16	180	4.1	73.4
7/32	210	5.0	74.2
1/4	250	5.6	71.9

E6012

Electrode Diameter	Amperes	Deposition Rate lbs/hr	Efficiency %
1/8	130	2.9	81.8
5/32	165	3.2	78.8
	200	3.4	69.0
3/16	220	4.0	77.0
	250	4.2	74.5
7/32	320	5.6	69.8
1/4	320	5.6	70.0
	360	6.6	67.7
	380	7.1	66.0
5/16	400	8.1	70.2

E6013

Electrode Diameter	Amperes	Deposition Rate lbs/hr	Efficiency %
5/32	140	2.6	75.6
	160	3.0	74.1
	180	3.5	71.2
3/16	180	3.2	73.9
	200	3.8	71.1
	220	4.1	72.9
7/32	250	5.3	71.3
	270	5.7	73.0
	290	6.1	72.7
1/4	290	6.2	75.0
	310	6.5	73.5
	330	7.1	72.1
5/16	360	8.6	70.7
	390	9.4	71.8
	450	10.3	71.3

E7014

Electrode Diameter	Amperes	Deposition Rate lbs/hr	Efficiency %
1/8	120	2.4	63.9
	150	3.1	61.1
5/32	160	3.0	71.9
	200	3.7	67.0
3/16	230	4.5	70.9
	270	5.5	73.2
7/32	290	5.8	67.2
	330	7.1	70.3
1/4	350	7.1	68.7
	400	8.7	69.9
5/16	440	8.9	62.2
	500	11.1	65.4

Figure 19-14 Deposition data.

19-18 (page 480) illustrates how the stub loss influences the electrode efficiency when using coated electrodes.

In **Figure 19-18**, a 14-in.-long, 5/32-in.-diameter E7018 electrode at 140 amperes is considered. It is 75% efficient and a 2-in. stub loss is assumed. The 75% efficiency applies only to the 12 inches of the electrode consumed in making the weld, not to the 2-in. stub. When the 2-in. stub loss and the 25% lost to slag, spatter, and fumes are considered, the efficiency minus stub loss is lowered to 64.3%. This means that for each 100 pounds of electrodes, you can expect an actual deposit of approximately 64.3 pounds of weld metal if all electrodes are used to a 2-in. stub length.

The formula for efficiency including stub loss is important. It must always be used when estimating the cost of depositing weld metal by the SMAW method. **Figure 19-19** (page 481) shows the formula used to establish the efficiency of coated electrodes including stub loss. It is based on the electrode length and is slightly inaccurate. That is, it does not consider that electrode weight is not evenly distributed because of flux removed from the electrode holder end. (Indicated by the dotted lines in **Figure 19-18**.) Use of the formula will result in a 1.5–2.3% error that varies with electrode size, coating thickness, and stub length. The formula is acceptable for estimating purposes, however.

E7016

Electrode Diameter	Amperes	Deposition Rate lbs/hr	Efficiency %
5/32	140	3.0	70.5
	160	3.2	69.1
	190	3.6	66.0
3/16	175	3.8	71.0
	200	4.2	71.0
	225	4.4	70.0
	250	4.8	65.8
1/4	250	5.9	74.5
	275	6.4	74.1
	300	6.8	73.2
	350	7.6	71.5
5/16	325	8.0	77.3
	375	9.0	76.3
	425	10.2	76.7

E7024

Electrode Diameter	Amperes	Deposition Rate lbs/hr	Efficiency %
1/8	140	4.2	71.8
	180	5.1	70.7
5/32	180	5.3	71.3
	210	6.3	72.5
	240	7.2	69.4
3/16	245	7.5	69.2
	270	8.3	70.5
	290	9 1	68.0
7/32	320	9.4	72.4
	360	11.6	69.1
1/4	400	12.6	71.7

LOW ALLOY, IRON POWDER ELECTRODES
OF THE TYPES E7018, E8018,
E9018, E10018, E11018 & E12018

Electrode Diameter	Amperes	Deposition Rate lbs/hr	Efficiency %
3/32	70	1.37	70.5
	90	1.65	66.3
	110	1.73	64.4
1/8	120	2.58	71.6
	140	2.74	70.9
	160	2.99	68.1
5/32	140	3.11	75.0
	170	3.78	73.5
	200	4.31	73.0
3/16	200	4.85	76.4
	250	5.36	74.6
	300	5.61	70.3
7/32	250	6.50	75.0
	300	7.20	74.0
	350	7.40	73.0
1/4	300	7.72	78.0
	350	8.67	77.0
	400	9.04	74.0

Figure 19-15 Deposition data.

For the values given in **Figure 19-18** the formula is:

$$\text{Efficiency} - \text{stub loss} = \frac{(14 - 2) \times 0.75}{14}$$

$$= \frac{(12 \times 0.75)}{14}$$

$$= \frac{9}{14}$$

$$= 0.6429 \quad \text{or} \quad 64.3\%$$

In this example, the electrode length is known, the stub loss must be estimated, and the efficiency is taken from the tables in **Figures 19-14** and **19-15**. Use an average stub loss and three inches for coated electrodes if the actual shop practices concerning stub loss are not known.

Efficiency of Flux Cored Wires. Flux cored wires have a lower power flux-to-metal ratio than coated electrodes and therefore a higher deposition efficiency. Stub loss need not be considered, since the wire is continuous. The gas shielded wires of the E70T-1 and E70T-2 types have efficiencies of 83%–88%. The gas shielded basic slag wire (E70T-5) is 85%–90% efficient with CO_2 as the shielded gas. The efficiency can reach 92% when a 75%

argon–25% CO_2 gas mixture is used. Use the efficiency figures in **Figure 19-16** for your calculations if the actual values are not known.

The efficiency of self-shielded flux cored wires has more variation because of the large assortment of available types designed for specific applications. The efficiency of the high-deposition, general-purpose type, such as E70T-4, is 81%–86%, depending on wire size and electrical stickout. The chart in **Figure 19-16** shows the optimum conditions for each wire size and may be used in your calculations.

Efficiency of Solid Wire for GMAW. The efficiency of solid wires in GMAW is very high and will vary with the shielding gas or gas mixture used. Using CO_2 will produce the most spatter, and the average efficiency will be about 93%. Using a 75% argon–25% CO_2 gas mixture will result in somewhat less spatter and an efficiency of approximately 96%. A 98% argon–2% oxygen mixture will produce even less spatter and the average efficiency will be about 98%. Stub loss need not be considered, since the wire is continuous. Figure 19-20 (page 481) shows the average efficiencies to use in your calculations if the actual efficiency is not known.

FLUX CORED ELECTRODES—GAS SHIELDED TYPES E70T-1, E71T-1, E70T-2, AND ALL LOW ALLOY TYPES.

Electrode Diameter	Amperes	Deposition Rate lbs/hr	Efficiency %
.045	180	5.3	85
	200	5.5	86
	240	6.9	84
	280	13.0	83
.052	190	4.8	85
	210	5.3	83.5
	270	7.6	83
	300	9.8	85
1/16	200	5.2	85
	275	10.1	85
	300	11.5	85
	350	13.3	86
5/64	250	6.4	85
	350	10.5	85
	450	14.8	85
3/32	400	12.7	85
	450	15.0	86
	500	18.5	86
7/64	550	17.1	85
	625	19.6	86
	700	23.0	86
1/8	600	16.2	86
	725	22.5	86
	850	29.2	85

FLUX CORED ELECTRODES — SELF SHIELDED

Type & Diameter	Amperes	Deposition Rate lbs/hr	Efficiency %
E70T-3 3/32	450	14.0	88
E70T-4 3/32	400	18.0	85
.120	450	20.0	81
E70T-6 5/64	350	11.9	86
3/32	480	14.7	81
E70T-7 3/32	325	11.4	80
7/64	450	18.0	86
E71T-7 .068	200	4.2	76
5/64	300	8.0	84
E71T-8 5/64	220	4.4	77
3/32	300	6.7	77
EG1T8-K6 5/64	235	4.3	76
E71T8-Ni1 5/64	235	4.3	77
3/32	345	8.2	84
E70T-10 3/32	400	13.0	69
E71T-11 5/64	240	4.5	87
3/32	250	5.0	91
E70T4-K2 3/32	300	14.0	83

NOTE: Values shown are optimum for each type and size.

Figure 19-16 Deposition data.

Efficiency of Solid Wires for SAW. In submerged arc welding there is no spatter loss and an efficiency of 99% may be assumed. The only loss during welding is the short piece the operator must clip off the end of the wire to remove the fused flux that forms at the termination of each weld. This is done to ensure a good start on the next weld.

Operating Factor

Operating factor is the percentage of a welder's working day actually spent on welding. It is the arc time in hours divided by the total hours worked. A 45% (0.45) operating factor means that only 45% of the welder's day is actually spent on welding. The rest of this time is spent installing a new electrode or wire, cleaning slag, positioning the weldment, cleaning spatter from the welding gun, and so on.

When using coated electrodes (SMAW), the operating factor can range from 15%–40% depending on material handling, fixturing, and operator dexterity. If the actual operating factor is not known, an average of 30% may be used for cost estimates involving the shielded metal arc welding process.

When welding with solid wires (GMAW) using the semiautomatic method, operating factors ranging from 45%–55% are easily attainable. For cost estimating purposes, use a 45% operating factor. The estimated operating factor of FCAW is about 5% lower than that of GMAW to allow for slag removal time.

GAS METAL ARC WELDING SOLID WIRES

Diameter Inches	Amperes	Melt-Off Rate lbs/hr	Efficiency %
.030	75	2.0	
	100	2.7	98% A
	150	4.2	2% O_2
	200	7.0	98%
.035	80	3.2	
	100	2.8	
	150	4.3	75% Argon
	200	6.3	25% CO_2
	250	9.2	96%
.045	100	2.1	
	125	2.9	
	150	3.7	
	200	5.7	
	250	7.4	Straight
	300	10.4	CO_2
	350	13.5	93%
1/16	250	6.7	
	275	8.6	
	300	9.2	
	350	11.5	
	400	14.3	
	450	17.8	

SUBMERGED ARC WIRES (1″ STICKOUT)

Diameter Inches	Amperes	Melt-Off Rate lbs/hr	Efficiency %
5/64	300	7.0	
	400	10.2	
	500	15.0	
3/32	400	9.4	
	500	13.0	
	600	17.2	
1/8	400	8.5	
	500	11.5	
	600	15.0	
	700	19.0	
5/32	500	11.3	ASSUME 99% EFFICIENCY
	600	14.6	
	700	18.4	
	800	22.0	
	900	26.1	
3/16	600	13.9	
	700	17.5	
	800	21.0	
	900	25.0	
	1,000	29.2	
	1,100	34.0	

NOTE: Values for 1″ stickout.

NOTE: For GMAW and SAW, melt-off rate may be used as the deposition rate in the cost formulas. Using the proper deposition efficiency will account for losses due to spatter, clipping the wire end, etc.

Figure 19-17 Deposition data.

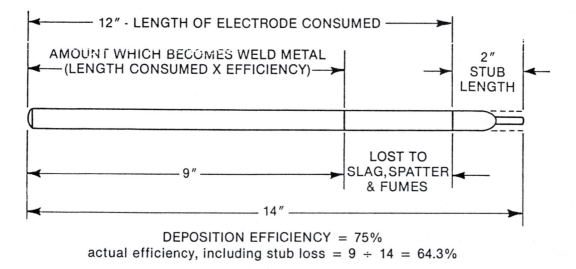

DEPOSITION EFFICIENCY = 75%
actual efficiency, including stub loss = 9 ÷ 14 = 64.3%

Figure 19-18 Deposition efficiency and stub loss.

In semiautomatic submerged arc welding, slag removal and loose flux handling must be considered. A 40% operating factor is typical for this process.

Automatic welding using the GMAW, FCAW, and SAW processes requires that each application be studied individually. Operating factors ranging from 50% to 100% may be obtained depending on the degree of automation.

The chart in **Figure 19-21** shows average operating factor values for the various welding processes. These

$$\text{EFFICIENCY MINUS STUB LOSS} = \frac{(\text{ELECTRODE LENGTH} - \text{STUB LENGTH}) \times \text{DEPOSITION EFFICIENCY}}{\text{ELECTRODE LENGTH}}$$

Figure 19-19 Efficiency minus stub loss formula.

Shielding Gas	Efficiency Range	Average Efficiency
Pure CO_2	88-95%	93%
75% A-25% CO_2	94-98%	96%
98% A-2% O_2	97-98.5%	98%

Figure 19-20 Deposition efficiencies — gas metal arc welding carbon and low alloy steel wires.

figures may be used for cost estimating when the actual operating factor is not known.

Knowing the productivity of welders is necessary when determining the cost of the finished part. In some shops this cost is passed directly on to the customer in the form of cost plus. It may also be used to determine at what level the shop can bid on new work and still make a profit. It is not often used to promote or penalize welders.

The number of parts produced is useful if there are a number of welders making the same or similar parts in a production shop. A comparison of the productivity can be made because there should be little difference in the average time compared with each welder's actual time.

The length of weld produced is a useful tool when a lot of the same welding is required. Welding on items like ships, tanks, or large vessels may take weeks, months, or years. The length that a welder produces in this type of production can be recorded on a regular basis.

The deposition rates of the process also affect costs. Processes with high deposition rates can be very cost-effective, **Figure 19-22** (page 482). A compelling reason for replacing covered electrodes with small-diameter cored wires is that the deposition rates in the vertical position can be increased from about 2 pounds per hour to over 5 pounds per hour. Thus the welder cost in this example drops from over $30 per pound to about $12 per pound of weld metal deposited.

Since operating factors and deposition rates interact strongly, their effects on weld costs are examined together. The time spent preparing and positioning weld joints for submerged arc welding is costly, but that process's high deposition rates justify the time, **Figure 19-23** (page 482). Covered electrodes, however, cannot

compete with most other processes unless the setup time and other factors can be reduced. The speed with which alloys and electrode types can be changed explains why covered electrodes have remained competitive in small job shops, especially when typical welds are quite short. Changeover times with GMA processes can be lengthy, and if the welding jobs are small, the operator factor can drop below 15%. Excessively high deposition rates may be needed to compensate for that deficit.

FACTORS FOR COST FORMULAS

Labor and Overhead

Labor and overhead may be considered jointly in your calculations. Labor is the welder's hourly rate of pay including wages and benefits. Overhead includes allocated portions of plant operating and maintenance costs. Weld shops in manufacturing plants normally have established labor and overhead rates for each department. Labor and overhead rates can vary greatly from plant to plant and also with location. **Figure 19-24** (page 483) shows how labor and overhead can vary and suggests an average value to use in your calculations when the actual value is unknown.

Cost of Power

Cost of electrical power is a very small part of the cost of depositing weld metal and in most cases is less than 1% of the total. It will be necessary for you to know the power cost expressed in dollars per kilowatt-hour ($/kWh) if required for a total cost estimate.

Calculating the Cost per Pound of Deposited Weld Metal

Example 1

Calculate the cost of welding 1,280 feet of a single-bevel butt joint as shown in **Figure 19-25(A)** (page 483), using the following data.

WELDING PROCESS			
SMAW	*GMAW	*FCAW	*SAW
30%	50%	45%	40%

Figure 19-21 Approximate operating factor.

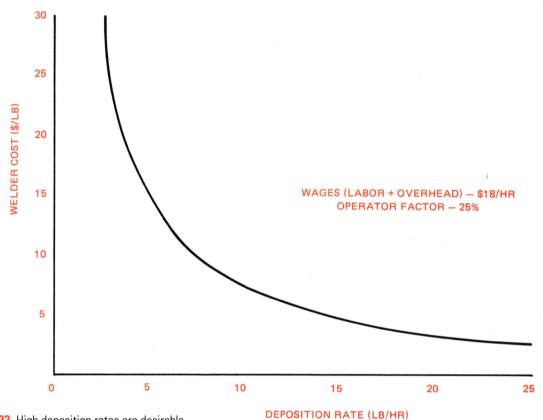

Figure 19-22 High deposition rates are desirable.

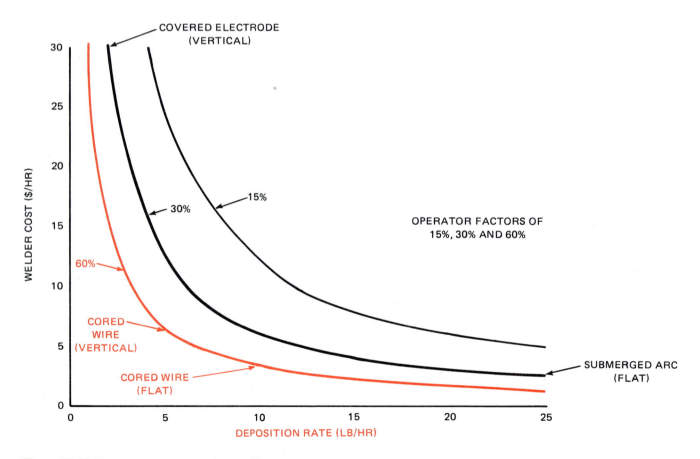

Figure 19-23 Newer processes can reduce welder costs.

HOURLY WELDING LABOR AND OVERHEAD RATES	
Small Shops	$7.50 to $15.00/hr.
Large Shops	$15.00 to $35.00/hr.
Average	$20.00/hr.

Figure 19-24 Approximate labor and overhead rates.

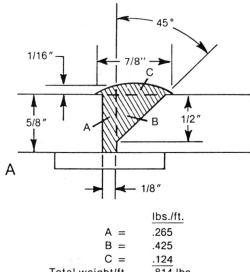

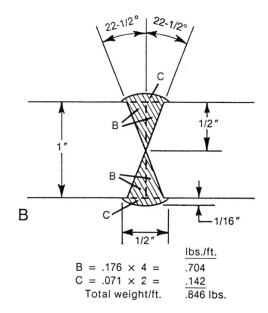

	lbs./ft.
A =	.265
B =	.425
C =	.124
Total weight/ft.	.814 lbs.

	lbs./ft.
B = .176 × 4 =	.704
C = .071 × 2 =	.142
Total weight/ft.	.846 lbs.

Figure 19-25 Estimating weld metal weight.

Electrode, 13/16-in. diameter, 14 in. long, E7018, operated at 25 volts, 250 amps.

Stub loss, 2 in.

Labor overhead, $20.00/hr

Electrode cost, $.57/lb

Power cost, $0.045/kWh

   The formula for the calculations is shown on the Weld Metal Cost Worksheet in **Figure 19-26** (page 484). The following explains each step in the calculations:

Line 1: Labor and Overhead—$20.00/hr (given)

   Deposition rate: From shielded metal arc welding deposition data chart in **Figure 19-15** (page 478); 5.36 lb/hr.

   Operating factor: Since it is not stated above, use the average value of 30% (0.30) shown in **Figure 19-21** (page 481).

   The cost of labor and overhead per pound of deposited weld metal can now be calculated as $12.44/lb.

Line 2: Electrode Cost Per Pound—$.57 (given)

   Deposition efficiency: From shielded metal arc welding deposition table in **Figure 19-15**; 74.6%. Since this is a coated electrode, the efficiency must be adjusted for stub loss by the formula shown in **Figure 19-19** (page 481). We know that the electrode length is 14 in. and the stub loss is 2 in. (given). The formula becomes:

$$\text{Efficiency} - \text{stub loss} = \frac{(14 - 2) \times 0.746}{14}$$

$$= 0.639 \quad \text{or} \quad 63.9\%$$

   63.9% is the adjusted efficiency to be used in line 2. The cost of the electrode per pound of deposited weld metal can now be calculated as $.89/lb.

Lines 3 & 4: Not applicable for coated electrodes.

Line 5: Cost of Power—$.045/kWh (given).

   Volts and amperes: 25 V and 250 A (given).

   Constant: The 1,000 already entered is a constant necessary to convert to watt-hours.

   Deposition rate: 5.36 lbs/hr are used in line 1. The cost of electrical power to deposit one pound of weld metal can now be calculated as $.052.

Line 6: Total lines 1, 2, and 5 to find the total cost of depositing one pound weld metal.

   The total is $13.38.

Calculating the Cost per Foot of Deposited Weld Metal

   Calculating the weight of weld metal requires that we consider the following items:

1. Area of the weld's cross section.

2. Length of the weld.

EXAMPLE 1
WELD METAL COST WORKSHEET
COST PER POUND OF DEPOSITED WELD METAL

1	LABOR & OVERHEAD	$\dfrac{\text{LABOR \& OVERHEAD COST/HR}}{\text{DEPOSITION RATE (LBS/HR)} \times \text{OPERATING FACTOR}}$ =	$\dfrac{20.00}{5.36 \times .30} = \dfrac{20.00}{1.608} = \underline{12.44}$
2	ELECTRODE	$\dfrac{\text{ELECTRODE COST/LB}}{\text{DEPOSITION EFFICIENCY}}$ =	$\dfrac{.57}{.639} = \underline{.89}$
3	GAS	$\dfrac{\text{GAS FLOW RATE (CU FT/HR)} \times \text{GAS COST/CU FT}}{\text{DEPOSITION RATE (LBS/HR)}}$ =	$\dfrac{\_\_\_ \times \_\_\_}{\_\_\_} = \_\_\_ = \underline{NA}$
4	FLUX	$\dfrac{\text{FLUX COST/LB} \times 1.4}{\text{DEPOSITION EFFICIENCY}}$ =	$\dfrac{\_\_\_ \times 1.4}{\_\_\_} = \_\_\_ = \underline{NA}$
5	POWER	$\dfrac{\text{COST/kWh} \times \text{VOLTS} \times \text{AMPS}}{1000 \times \text{DEPOSITION RATE}}$ =	$\dfrac{.045 \times 25 \times 250}{1000 \times 5.36} = \dfrac{281.25}{5,360} = \underline{.052}$
6	TOTAL COST PER LB. OF DEPOSITED WELD METAL	SUM OF 1 THROUGH 5, ABOVE	$ \underline{13.38}$

COST PER FOOT OF DEPOSITED WELD METAL

7	Cost per pound of deposited weld metal	×	pounds per foot of weld joint	=	$\underline{13.38} \times \underline{.814} = \$\underline{10.89}$

COST OF WELD METAL — TOTAL JOB

8	Total feet of weld	×	Cost per foot	=	$\underline{1,280} \times \underline{10.89} = \$\underline{13,939}$

Figure 19-26

3. Volume of the weld in cubic inches.
4. Weight of the weld metal per cubic inch.

In the fillet weld shown in **Figure 19-27**, we know that the area of the cross section (the triangle) is equal to one-half the base times the height. The volume of the weld is equal to the area times the length. The weight of the weld is then the volume times the weight of the material (steel) per cubic inch.

We can then write the formula:

Weight of weld metal = 1/2 × base × height × length × weight of material

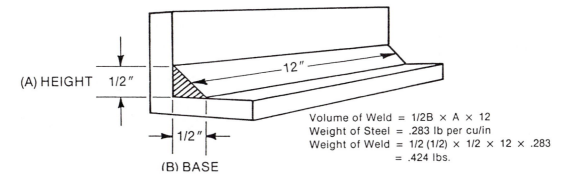

Volume of Weld = 1/2B × A × 12
Weight of Steel = .283 lb per cu/in
Weight of Weld = 1/2 (1/2) × 1/2 × 12 × .283
                          = .424 lbs.

Figure 19-27 Calculating the weight per foot of a fillet weld.

Substituting the values from **Figure 19-27**, we have:

Wt/ft = 0.5 × 0.5 × 0.5 × 12 × 0.283 = 0.4245 lb

The chart in **Figure 19-28** eliminates the need for these calculations for steel fillet and butt joints, since it lists the weight per foot directly.

Estimating the weight per foot of a weld using the chart requires that you make a drawing of the weld joint to exact scale. Dimension the leg lengths, root gap, thickness, angles, and other pertinent measurements as shown in **Figure 19-25** (page 483). Divide the cross section of the weld into right triangles and rectangles as shown. Where required, sketch in the reinforcement, which is the domed portion above or below the surface of the plate. The reinforcement should extend slightly beyond the edges of the joint. Measure the length and height of the reinforcement and note them on your drawing. The reinforcement is only an approximation because the contour cannot be exactly controlled in welding. Refer to the weight tables in **Figure 19-28** (page 486) for the weights per foot of each of the component parts of the weld, as sketched. The sum of the weights of all the components is the total weight of the weld, per foot, as shown in **Figure 19-25(A)** (page 483).

Line 7: The total cost per pound as determined in line 6 is entered and multiplied by the weight per foot as determined in **Figure 19-25(A)**.

Line 8: The cost of the weld for the total job is determined by multiplying the total feet of weld (given) by the cost per foot as determined in line 7.

Example 2

Calculate the total cost of depositing 1,280 feet of weld metal using the CO_2 shielded, flux cored welding process in the double V-groove joint shown in **Figure 19-25(B)** (page 483) using the following data.

1. Electrode: 3/32 in., E70T-1 @ 31 volts, 450 amps.
2. Labor and overhead: $20.00/hour.
3. Deposition rate: 15 pounds/hour. From table in **Figure 19-16** (page 479).
4. Operating factor: 45% (0.45) average from **Figure 19-21** (page 481).
5. Electrode cost: $.80/pound (from supplier).
6. Deposition efficiency: 86% (0.86) from table in **Figure 19-16**.
7. Gas flow rate: 45 cubic feet per hour. From **Figure 19-12** (page 475).
8. Gas cost: $.03/cubic foot (from supplier)
9. Cost of power: $.045/kWh
10. Wt/ft of weld: From **Figure 19-25(B)** @ 0.846 lb/ft.

These values are shown inserted into the formulas on the weld metal cost worksheet in **Figure 19-29** (page 487).

COMPARING WELD METAL COSTS

Note that the amount of weld metal deposited in Example 1 and Example 2 is almost the same, while the total cost of depositing the weld metal is three times higher in Example 1, as shown here:

Example 1: 1,280 ft × 0.814 lb/ft = 1,041.9 lb at $13,939

Example 2: 1,280 ft × 0.846 lb/ft = 1,082.9 lb at $4,342

This is because the flux cored process has a higher deposition rate, efficiency, and operating factor and also allows a tighter joint due to the deep penetrating characteristics of the process.

When comparing welding processes, all efforts should be made to use the proper welding current for the electrode or wire in the position in which the weld must be made. As an example, consider depositing a given size fillet weld in the vertical up position by the GMAW process and FCAW process semiautomatically. In both processes the welding current and voltage must be lowered to weld out of position. In GMAW, the short circuiting arc transfer must be used. Example 3 compares the weld metal cost per pound deposited by these processes, using the proper current and voltage for depositing a 1/4" fillet weld on 1/4" plate, vertically up.

We can eliminate the cost of electrical power when comparing processes because the difference is very small.

WEIGHT PER FOOT OF WELD METAL FOR FILLET WELDS
AND ELEMENTS OF COMMON BUTT JOINTS (lbs/ft)
STEEL

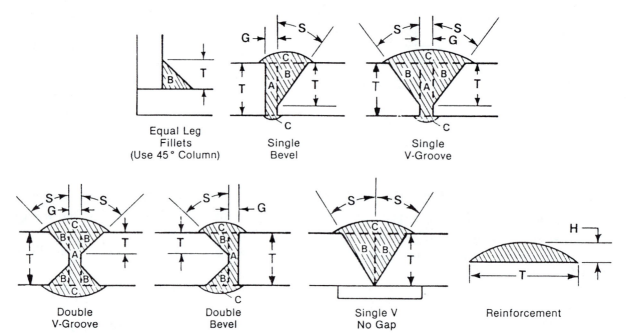

Equal Leg Fillets (Use 45° Column) · Single Bevel · Single V-Groove · Double V-Groove · Double Bevel · Single V No Gap · Reinforcement

T Inches	Weight/ft. of Rectangle A G						Weight per foot of Triangle B S						Weight/ft. Reinforcement C H			
	1/16"	1/8"	3/16"	1/4"	3/8"	1/2"	5°	10°	15°	22½°	30°	45°	1/16"	1/8"	3/16"	1/4"
1/8	.027	.053	.080	.106	.159	.212	.002	.005	.007	.011	.015	.027				
3/16	.040	.080	.119	.159	.239	.318	.005	.011	.016	.025	.035	.060	.027			
1/4	.053	.106	.159	.212	.318	.425	.009	.019	.028	.044	.061	.106	.035			
5/16	.066	.133	.199	.265	.390	.531	.015	.029	.044	.069	.096	.166	.044	.084		
3/8	.080	.159	.239	.318	.478	.637	.021	.042	.064	.099	.138	.239	.053	.106		
7/16	.091	.186	.279	.371	.557	.743	.028	.057	.087	.129	.188	.325	.062	.124		
1/2	.106	.212	.318	.425	.637	.849	.037	.075	.114	.176	.245	.425	.071	.141	.212	
9/16	.119	.239	.358	.478	.716	.955	.047	.095	.144	.223	.311	.451	.080	.159	.239	
5/8	.133	.265	.398	.531	.796	1.061	.058	.117	.178	.275	.383	.664	.088	.177	.265	.354
11/16	.146	.292	.438	.584	.876	1.167	.070	.142	.215	.332	.464	.804	.097	.195	.292	.389
3/4	.159	.318	.478	.637	.995	1.274	.084	.169	.256	.396	.552	.956	.106	.212	.318	.424
13/16	.172	.345	.517	.690	1.035	1.380	.098	.198	.301	.464	.648	1.121	.115	.230	.345	.460
7/8	.186	.371	.557	.743	1.114	1.486	.114	.230	.349	.538	.751	1.300	.124	.248	.371	.495
15/16	.199	.398	.597	.796	1.194	1.592	.131	.263	.400	.618	.863	1.493	.133	.266	.398	.530
1	.212	.425	.637	.849	1.274	1.698	.149	.300	.456	.703	.981	1.698	.141	.283	.424	.566
1 1/8	.239	.478	.716	.955	1.433	1.910	.188	.379	.577	.890	1.241	2.149	.159	.318	.477	.637
1 1/4	.265	.531	.796	1.061	1.592	2.123	.232	.468	.712	1.099	1.532	2.653	.177	.354	.531	.707
1 3/8	.292	.584	.876	1.167	1.751	2.335	.281	.567	.861	1.330	1.853	3.210	.195	.389	.584	.777
1 1/2	.318	.637	.955	1.274	1.910	2.547	.334	.674	1.023	1.582	2.206	3.821	.212	.424	.637	.849
1 5/8	.345	.690	1.035	1.380	2.069	2.759	.393	.792	1.201	1.857	2.589	4.484	.230	.460	.690	.920
1 3/4	.371	.743	1.114	1.486	2.229	2.972	.455	.918	1.393	2.154	3.002	5.200	.248	.495	.743	.990
1 7/8	.390	.796	1.194	1.592	2.388	3.184	.523	1.053	1.599	2.473	3.447	5.970	.266	.531	.796	1.061
2	.425	.849	1.274	1.698	2.547	3.396	.594	1.197	1.820	2.813	3.921	6.792	.283	.566	.849	1.132
2 1/4	.478	.955	1.433	1.910	2.865	3.821	.752	1.516	2.303	3.561	4.963	8.596	.318	.637	.955	1.273
2 1/2	.530	1.061	1.592	2.123	3.184	4.245	.928	1.871	2.844	4.396	6.127	10.613	.354	.707	1.061	1.415
2 3/4	.584	1.167	1.751	2.335	3.502	4.669	1.123	2.264	3.441	5.319	7.414	12.841	.389	.778	1.167	1.556
3	.636	1.274	1.910	2.547	3.821	5.094	1.337	2.695	4.095	6.330	8.823	15.282	.424	.849	1.273	1.698

Figure 19-28

EXAMPLE 2
WELD METAL COST WORKSHEET
COST PER POUND OF DEPOSITED WELD METAL

1	LABOR & OVERHEAD	$\dfrac{\text{LABOR \& OVERHEAD COST/HR}}{\text{DEPOSITION RATE (LBS/HR)} \times \text{OPERATING FACTOR}}$ =	$\dfrac{20.00}{15 \times .45} = \dfrac{20.00}{6.75} = \underline{2.96}$
2	ELECTRODE	$\dfrac{\text{ELECTRODE COST/LB}}{\text{DEPOSITION EFFICIENCY}}$ =	$\dfrac{.80}{.86} = \underline{.93}$
3	GAS	$\dfrac{\text{GAS FLOW RATE (CU FT/HR)} \times \text{GAS COST/CU FT}}{\text{DEPOSITION RATE (LBS/HR)}}$ =	$\dfrac{45 \times .03}{15} = \dfrac{1.35}{15} = \underline{.09}$
4	FLUX	$\dfrac{\text{FLUX COST/LB} \times 1.4}{\text{DEPOSITION EFFICIENCY}}$ =	$\dfrac{ \times 1.4}{} = = \underline{NA}$
5	POWER	$\dfrac{\text{COST/kWh} \times \text{VOLTS} \times \text{AMPS}}{1000 \times \text{DEPOSITION RATE}}$ =	$\dfrac{.045 \times 31 \times 450}{1000 \times 15} = \dfrac{627.75}{15,000} = \underline{.042}$
6	TOTAL COST PER LB. OF DEPOSITED WELD METAL	SUM OF 1 THROUGH 5, ABOVE	\$ $\underline{4.02}$

COST PER FOOT OF DEPOSITED WELD METAL

7	Cost per pound of deposited weld metal	×	pounds per foot of weld joint	=	$\underline{4.02} \times \underline{.846} = \$ \underline{3.40}$

COST OF WELD METAL — TOTAL JOB

8	Total feet of weld	×	Cost per foot	=	$\underline{1,280} \times \underline{3.40} = \$ \underline{4,352}$

Figure 19-29

Example 3

	FCAW
Electrode type	0.045 in. dia. E71T-1
Labor and overhead	$20.00/hour
Welding current	180 amp
Deposition rate	5.3 lb/hr (Figure 19-16)
Operation factor	45% (Figure 19-21)
Electrode cost	$1.44/lb
Deposition efficiency	85% (Figure 19-16)
Gas flow rate	35 cfh (Figure 19-12)
Gas cost per cu ft	$.03 CO_2

Subtracting the values in Example 1:
GMAW

0.045 in. dia. ER70S-3
$20.00/hour
125 amp
2.9 lb/hr (Figure 19-17)
50% (Figure 19-21)
$.66/lb.
96% (Figure 19-20)
35 cfh (Figure 19-12)
$11 75% Ar 25% CO_2

The tabulated data are shown in **Figure 19-30**.

As you can see, the cost of depositing the weld metal is about 33% less using the flux cored arc welding process. Since there is no slag to help hold the vertical weld puddle in the GMAW process, the welding current with solid wire must be lowered considerably. This, of course, lowers the deposition rate, and since labor and overhead is the largest factor involved, it substantially raises deposition costs. In the flat or horizontal position, where the welding current on the solid wire would be much higher, the cost difference would be considerably less.

OTHER USEFUL FORMULAS

The following formulas will assist you in making other useful calculations:
Total Pounds of Electrodes Required
(Ref. Example 1)

$$\text{Total pounds} = \frac{\text{wt/ft of weld} \times \text{no. of ft of weld}}{\text{deposition efficiency}}$$

Subtracting the values from Example 1:

$$\frac{0.814 \times 1,280}{0.639} = 1,631 \text{ lb}$$

Welding Time Required (Ref. Example 1)

$$\text{Welding time} = \frac{\text{wt/ft of weld} \times \text{ft of weld}}{\text{deposition rate} \times \text{operating factor}}$$

$$\frac{0.814 \times 1,280}{5.36 \times 0.30} = \frac{1,042}{1,608} = 648 \text{ hr}$$

Amortization of Equipment Costs

Calculations show that you can save $7.00 per pound of deposited weld metal by switching from E7018 electrodes and the SMAW process to an ER70S-3 solid wire using the GMAW process. However, the cost of the necessary equipment (power source, wire feeder, and gun) is $2,800. How long will it take to amortize or regain the cost of the equipment knowing that the deposition rate of the ER70S-3 is 7.4 lb/hr and the operating factor of the GMAW process is 50%? The formula is:

Equipment cost ÷ (deposition rate × operation factor) = man-hour $ savings/lb

Subtracting the values in the formula:

$$\frac{2,800}{7.00} \div (7.4 \times .50) = \text{man-hours}$$

$$400 \div 3.7 = 108 \text{ man-hours}$$

If we divide 108 into eight-hour days (108/8 = 13.5), the deposited weld metal savings of one man working an eight-hour day for 13 1/2 days will pay for the cost of the equipment.

FORMULAS FOR CALCULATING COST PER POUND DEPOSITED WELD METAL	FLUX CORED ARC WELDING E71T-1 .045" DIAMETER 180 Amps.	GAS METAL ARC WELDING ER70S-3 .045" DIAMETER 125 Amps.
LABOR & OVERHEAD = LABOR & OVERHEAD COST/HR / (DEPOSITION RATE (LBS/HR) X OPERATING FACTOR)	20.00 / (5.3 × .45) = 20.00 / 2.385 = 8.39	20.00 / (2.9 × .50) = 20.00 / 1.45 = 13.79
ELECTRODE: = ELECTRODE COST/LB / DEPOSITION EFFICIENCY	1.44 / .85 = 1.69	.66 / .96 = .69
GAS: = GAS FLOW RATE (CU FT/HR) X GAS COST/CU FT / DEPOSITION RATE (LBS/HR)	CO_2 35 × .03 = 1.05 / 5.3 = .20	75/25 30 × .11 = 3.3 / 2.9 = 1.14
SUM OF THE ABOVE	TOTAL VARIABLE COST/LB DEPOSITED WELD METAL $ 10.28	TOTAL VARIABLE COST/LB DEPOSITED WELD METAL $ 15.62

Figure 19-30

Special Welding Wires Speed Up Maintenance and Repair

Figure 1

The EnDOtec process brings the advantages of gas metal arc welding (GMAW) to maintenance welding and wear protection applications.

The process, developed by Eutectic Corporation, uses ten different metal cored wires specifically engineered for maintenance and repair GMAW. They provide welding characteristics needed for difficult maintenance environments, such as high hardness, corrosion resistance, elevated temperature, ability to weld dissimilar steels, etc. Most are all-position wires, which usually eliminate the need for teardown and setup when maintenance welding is needed.

Labor costs are reduced in comparison to shielded metal arc welding (SMAW) since the alloys can be deposited at rates up to 20 lb/h. With deposition efficiency approaching 96%, these alloys also have greatly reduced waste. This is due to their metal core design, very low spatter, and absence of stub loss inherent in SMAW.

Low heat input reduces detrimental effects on the base metals; low dilution maintains consistent performance characteristics throughout the deposit.

MINE DIGS UP SAVINGS

In Missouri, 1,800 ft below the surface, several 20,000-lb crusher heads work around the clock, pulverizing ore in a lead and copper mine. After approximately one year of this punishing work, the ribs of these heads wear smooth and lose their efficiency. They are then taken out of service and rebuilt.

Originally, this was done every eight months. The welding company's maintenance welding specialist suggested two ways to speed up the repair procedure. First, the old, fatigued metal was removed using the ExOtec process. This involves a specially developed continuous gouging wire and the ExO GUN, and provides high-speed metal removal, up to 30 lb/h. Noise and fumes are less than with other metal removal processes.

The special GMA process is then used to rebuild the heads — **Figure 1.** New steel bars are first welded on using an iron-base alloy, specially formulated for toughness, ductility, and strength. The assembly is capped off with a different alloy designed for high-impact-resistance work, with hardening to Rc 45. This alloy greatly reduces the amount of metal that wears off and eventually must be rewelded.

The mine operations can now rebuild four heads, per year, thanks to the high deposition rate of the alloys. This process also greatly reduces waste. Alloy consumption originally amounted to 2,000 lb per head, but now that has been reduced to 800 lb. Combined labor and material savings amount to $44,000 per year.

ENGINE REBUILDER REVS UP FOR LESS

Valves in big marine engines lead a hard life. Operating temperatures are in excess of 930°F, and the valves suffer abrasion and adhesive wear. This can reduce cylinder compression and cause a power loss. At that point, the owner of the boat brings it in for an engine overhaul.

A large East Coast boat population means that engine rebuilders on the shores of the Atlantic are kept pretty busy. One such firm was looking for ways to speed up the process and reduce their costs. They determined that they could attack both objectives by rebuilding the valves in-house. This would avoid the cost of buying expensive reconditioned valves and waiting for delivery from their suppliers.

The part in question is a valve used on slow-speed diesel engines — **Figure 2.** Each cylinder

Figure 2

Figure 3

has one valve, due to the engine's two-stroke function, and generates approximately 3,500 hp. The cast steel valves have an 18-in.-diameter seat area.

The engine rebuilder consulted with a specialist from the welding company, who recommended that they rebuild the valves using the GMA process developed for maintenance and repair, in a two-step procedure. First the face and seat areas are built up using a stainless steel alloy. Then one or two passes of a Tero-Cote alloy are applied for high resistance to heat and wear. The company's welding advisory service provided guidance on the selection of proper procedures and shielding gas, and on-site training was conducted by the local representative.

They are now rebuilding an average of 35 valves per year. Cost, including labor and materials, is $687.80. The price of rebuilt valves is $1,400 each, so their annual savings amounts to $25,000.

STEEL MILL OPERATION BENEFITS

When steel rods leave the rolling operation at a southern mini-mill, conveyor rolls move them to the straightening operation and then to a warehouse. The rolls, made of 1045 carbon steel, have a 16-in.-diameter. Grooves are gradually worn into them due to the high temperature of the rods and the abrasive and adhesive wear, and eventually they lose their effectiveness.

The rolls are rebuilt for the mill by a machine shop. On average, they had a one year service life, which was expensive for the mill since it meant that 125 rolls per year had to be reconditioned.

The welding company demonstrated a wire specifically engineered for tooling applications with service temperatures up to 900°F. This special alloy provides balanced abrasion and impact resistance. The application development manager from the welding company assisted with inplant training of the welders to be sure optimum results were achieved.

The new welded coating provides a harder, tougher surface, with a projected service life of four years — see **Figure 3.** This means the mill now only needs to recondition 30 rolls per year, instead of 125, which results in a net savings of $33,900.

(Reprinted from the *Welding Journal.* Article based on a story from Eutectic Corporation, Flushing, New Jersey. Courtesy of the American Welding Society, Inc.)

REVIEW

1. What are codes and standards?

2. Why is it important to select the correct welding code or standard?

3. What is the difference between welding codes or standards and welding specifications?

4. What might influence the selection of a particular code or specification for welding?

5. What information should be included in a WPS?

6. What is the purpose of the PQR?

7. Who should witness the test welding being performed for a tentative WPS?

8. Ideally, a WPS should be written with enough information to allow a good welder to _____.

9. List examples of fixed and variable costs that must be considered when estimating a job.

10. List examples of overhead costs that a welding shop might have.

11. What effect on the weld cost does increasing the groove angle have?

12. What potential problems can be caused by having too small or too large a weld bead?

13. What would be the cross-sectional area of a single bevel-groove weld that is 1/2 in. wide and 5/8 in. deep on a 3/4-in.-thick plate having a 1/8-in. root opening? What would be the SI area ?

14. What would be the cross-sectional area of a V-groove weld that is 6 mm wide and 8 mm deep on a 10-mm-thick plate having a 2-mm root opening? What would the area be in square inches?

15. What would be the cross-sectional area of a fillet weld that has an equal leg of 1/2 in.? What would be the SI area ?

16. What would be the cross-sectional area of a fillet weld that has a 10-mm leg and an 8-mm leg? What would be the area in square inches?

17. How many pounds of steel electrode would be required to make a weld that has a volume of 18 in^3? What would be the SI weight?

18. What are the advantages of calculating the weight of filler metal needed to make a weld?

19. Using Figure 19-12, what would be the flow rate for a 1/16-in. FCAW electrode?

20. What is the approximate ratio of flux to filler wire for SAW?

21. Why must deposition efficiency be used when determining how much electrode will be needed for a job?

22. Approximately how many pounds of weld will a 30-pound spool of 3/32-in. E70T-4 FCA welding wire produce? (Refer to Figure 19-16, page 479.)

23. What is the melt-off rate for 0.035 GMAW filler wire at the 100 ampere setting? (Refer to Figure 19-17, page 480.)

24. What is the approximate length of the unused electrode stub of a 1/8-in. E7018 SMAW electrode?

25. What does the flux-to-metal ratio have to do with FCA welding deposition efficiency?

26. Which shielding gas or gas mixture provides GMAW with the highest average efficiency? (Refer to Figure 19-20, page 481.)

27. What is the operation factor?

28. Of the major processes, which has the highest and which has the lowest operation factor, and what are their percentages? (Refer to Figure 19-21, page 481.)

29. Referring to Figure 19-22 (page 482), what would be the welder cost in $/lb if the deposition rate were 15 lb/hr?

30. Why is SMAW still used in many shops on small jobs if its operating factors and deposition rates are so low compared to most other processes?

Chapter 20

TESTING AND INSPECTION OF WELDS

OBJECTIVES

After completing this chapter, the student should be able to:

■ describe the difference between mechanical and nondestructive testing.

■ list the twelve most common discontinuities and the nondestructive methods of locating them.

■ discuss how both the mechanical and nondestructive testing are performed.

■ explain why welds are tested.

■ evaluate a weld according to a given standard or code.

KEY TERMS

quality control	etching
discontinuity	radiograph inspection (RT)
defect	ultrasonic inspection (UT)
tolerance	eddy current inspection (ET)
mechanical testing (DT)	Rockwell hardness tester
nondestructive testing (NDT)	Brinell hardness tester
shearing strength	

INTRODUCTION

It's important to know that a weld will meet the requirements of the company and/or codes or standards. It's also necessary to ensure the quality, reliability, and strength of a weldment. To meet these demands, an active inspection program is needed. The extent to which a welder and product are subjected to testing and inspection depends upon the intended service of the product. Items that are to be used in light, routine-type service, such as ornamental iron, fence posts, gates, and so forth, are not inspected as critically as products in critical use. Some of the items in critical use include a main nuclear reactor containment vessel, oil refinery high-pressure vessels, aircraft airframes, bridges, and so on. The type of inspection required is then very much dependent upon the type of service the welded part will be required to withstand. The quality of the weld that will pass or be acceptable for one welding application may not meet the needs of another.

QUALITY CONTROL (QC)

Once a code or standard has been selected, a method is chosen for ensuring that the product meets the specifications. The two classifications of methods used in product quality control are destructive or mechanical testing and nondestructive testing. These methods can be used individually, or a combination of the two methods can be used. Mechanical testing (DT) methods, except for hydrostatic testing, result in the product being destroyed. Nondestructive testing (NDT) does not destroy the part being tested.

Mechanical testing is commonly used to qualify welders or welding procedures. It can be used in a random sample testing procedure in mass production. In many cases, a large number of identical parts are made, and a chosen number are destroyed by mechanical testing. The results of such tests are valid only for welds made under the same conditions because the only weld strengths known are the ones resulting from the tested pieces. It is then assumed that the strengths of the non-tested pieces are the same.

Nondestructive testing is used for welder qualification, welding procedure qualification, and product quality control. Since the weldment is not damaged, all the welds can be tested. Because the parts are not destroyed, more than one testing method can be used on the same part. Frequently only part of the welds are tested to save time and money. The same comparison of random sampling applies to these tests as it does for mechanical testing. Critical parts or welds are usually 100% tested.

DISCONTINUITIES AND DEFECTS

A discontinuity is an interruption of the typical structure of a weldment. It may be a lack of uniformity in the mechanical, metallurgical, or physical characteristics of the material or weldment. A discontinuity is not necessarily a defect.

A defect, according to AWS, is "a discontinuity or discontinuities, which by nature or accumulated effect (for example, total porosity or slag inclusion length that renders a part or product unable to meet minimum applicable acceptance standards or specifications). This term designates rejectability."

In other words many acceptable products may have welds that contain discontinuities. But no products may have welds that contain defects. The only difference between a discontinuity and a defect is when the discontinuity becomes so large or when there are so many small discontinuities that the weld is not acceptable under the standards of the code for that product. Some codes are more strict than others, so that the same weld might be acceptable under one code but not under another.

Ideally, a weld should not have any discontinuities, but that is practically impossible. The difference between what is acceptable, fit for service, and perfection is known as tolerance. In many industries, the tolerances for welds have been established and are available as codes or standards. Table 20-1 lists a few of the agencies that issue codes or standards. Each code or standard gives the tolerance that changes a discontinuity to a defect.

When evaluating a weld, it is important to note the type of discontinuity, the size of the discontinuity, and the location of the discontinuity. Any one of these factors or all three can be the deciding factor that, based on the applicable code or standard, might change a discontinuity to a defect.

The twelve most common discontinuities are as follows:

- Porosity
- Inclusions
- Inadequate joint penetration
- Incomplete fusion
- Arc strikes
- Overlap
- Undercut
- Cracks
- Underfill
- Laminations
- Delaminations
- Lamellar tears

Porosity

Porosity results from gas that was dissolved in the molten weld pool, forming bubbles that are trapped as the metal cools to become solid. The bubbles that make up porosity form within the weld metal; for that reason they cannot be seen as they form. These gas pockets form in the same way that bubbles form in a carbonated drink as it warms up or as air dissolved in water forms bubbles in the center of a cube of ice. Porosity forms either spherical (ball-shaped) or cylindrical (tube- or tunnel-shaped). The cylindrical porosity is called wormhole. The rounded edges tend to reduce the stresses around them. Therefore, unless porosity is extensive, there is little or no loss in strength.

Porosity is most often caused by improper welding techniques, contamination, or an improper chemical balance between the filler and base metals.

Improper welding techniques may result in shielding gas not properly protecting the molten weld pool. For example, the E7018 electrode should not be weaved wider than two and one-half times the electrode diameter, because very little shielding gas is produced. As a result, parts of the weld are unprotected. Nitrogen from

American Bureau of Shipping
American Petroleum Institute
American Society of Mechanical Engineers
American Society for Testing and Materials
American Welding Society
British Welding Institute
United States Government

*A more complete listing of agencies with addresses is included in the Appendix.

Table 20-1 Major Code Issuing Agencies.*

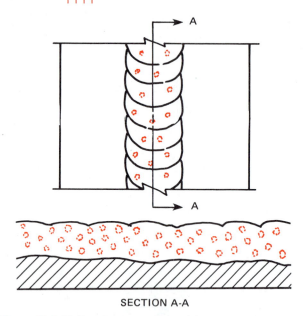

Figure 20-1 Uniformly scattered porosities.

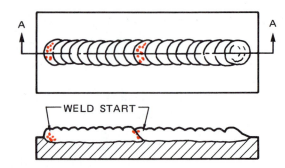

Figure 20-2 Clustered porosity.

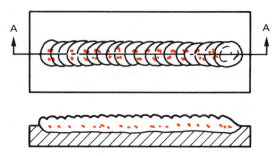

Figure 20-3 Linear porosity.

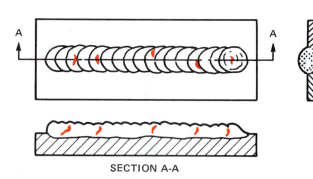

SECTION A-A

Figure 20-4 Piping or wormhole porosity.

the air that dissolves in the weld pool and then becomes trapped during escape can produce porosity.

The intense heat of the weld can decompose paint, dirt, or oil from machining, and rust or other oxides, producing hydrogen. This gas, like nitrogen, can also become trapped in the solidifying weld pool, producing porosity. When it causes porosity, hydrogen can also diffuse into the heat affected zone, producing underbead cracking in some steels. The level needed to crack welds is below that necessary to produce porosity.

Porosity can occur as a result of welds produced by the following processes:

- Plasma arc welding (PAW)
- Submerged arc welding (SAW)
- Gas tungsten arc welding (GTAW)
- Gas metal arc welding (GMAW)
- Flux cored arc welding (FCAW)
- Shielded metal arc welding (SMAW)
- Carbon arc welding (CAW)
- Oxyacetylene welding (OAW)
- Oxyhydrogen welding (OHW)
- Pressure gas welding (PGW)
- Electron beam welding (EBW)
- Electroslag welding (ESW)
- Laser beam welding (LBW)
- Thermit welding (TW)

Porosity can be grouped into the following four major types:

- Uniformly scattered porosity is most frequently caused by poor welding techniques or faulty materials, Figure 20-1.

- Clustered porosity is most often caused by improper starting and stopping techniques, Figure 20-2.
- Linear porosity is most frequently caused by contamination within the joint, root, or interbead boundaries, Figure 20-3.
- Piping porosity, or wormhole, is most often caused by contamination at the root, Figure 20-4. This porosity is unique because its formation depends on the gas escaping from the weld pool at the same rate as the pool is solidifying.

Inclusions

Inclusions are nonmetallic materials, such as slag and oxides, that are trapped in the weld metal, between weld beads, or between the weld and the base metal. Inclusions sometimes are jagged and irregularly shaped. Also, they can form in a continuous line. This causes stresses to concentrate and reduces the structural integrity (loss in strength) of the weld.

Although not visible, their development can be expected if prior welds were improperly cleaned or had a poor contour. Unless care is taken in reading radiographs, the presence of slag inclusions can be interpreted as other defects.

Linear slag inclusions in radiographs generally contain shadow details, otherwise they could be interpreted as lack-of-fusion defects. These inclusions result from a lack of slag control caused by poor manipulation that allows the slag to flow ahead of the arc; by not removing all the slag from previous welds; or by welding highly crowned, incompletely fused welds.

Scattered inclusions can resemble porosity but, unlike porosity, they are generally not spherical. These inclusions can also result from inadequate removal of earlier slag deposits and poor manipulation of the arc. Additionally, heavy mill scale or rust serves as their source, or they can result from unfused pieces of damaged electrode coatings falling into the weld. In radiographs some detail will appear, unlike linear slag inclusions.

Nonmetallic inclusions, Figure 20-5, are caused under the following conditions:

■ Slag and/or oxides do not have enough time to float to the surface of the molten weld pool.

■ There are sharp notches between weld beads or between the weld bead and the base metal that trap the material so that it cannot float out.

■ The joint was designed with insufficient room for the correct manipulation of the molten weld pool.

Nonmetallic inclusions are often found in welds produced by the following processes:

■ Submerged arc welding (SAW)

■ Gas metal arc welding (GMAW)

■ Flux cored arc welding (FCAW)

■ Shielded metal arc welding (SMAW)

■ Carbon arc welding (CAW)

■ Electroslag welding (ESW)

■ Thermit welding (TW)

Inadequate Joint Penetration

Inadequate joint penetration occurs when the depth that the weld penetrates the joint, Figure 20-6, is less than that needed to fuse through the plate or into the preceding weld. A defect usually results that could reduce the required cross-sectional area of the joint or become a source of stress concentration that leads to fatigue failure. The importance of such defects depends on the notch sensitivity of the metal and the factor of safety to which the weldment has been designed. Generally, if proper welding procedures were developed and followed, such defects will not occur.

The major causes of inadequate joint penetration are:

■ Improper welding technique — The most common cause is a misdirected arc. Also the welding technique may require that both starting and run-out tabs be used so that the molten weld pool is well-established before it reaches the joint. Sometimes, a failure to back gouge the root sufficiently provides a deeper root face than allowed for, Figure 20-7 (page 496).

■ Not enough welding current — Metals that are thick or have a high thermal conductivity are often preheated so that the weld heat is not drawn away so quickly by the metal that it cannot penetrate the joint.

■ Improper joint fitup — This problem results when the weld joints are not prepared or fitted accurately. Too small a root gap or too large a root face will keep the weld from penetrating adequately.

■ Improper joint design — When joints are accessible from both sides, back gouging is often used to ensure 100% root fusion.

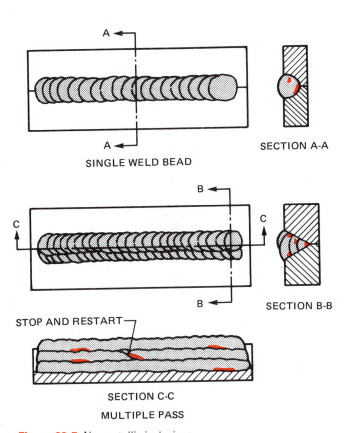

Figure 20-5 Nonmetallic inclusions.

A

SECTION A-A

SINGLE WELD BEAD

B

C   C

B

SECTION B-B

STOP AND RESTART

SECTION C-C

MULTIPLE PASS

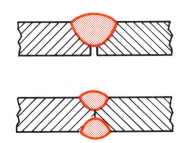

Figure 20-6 Inadequate joint penetration.

Inadequate joint penetration can occur as a result of welds produced by the following processes:

- Plasma arc welding (PAW)
- Submerged arc welding (SAW)
- Gas metal arc welding (GMAW)
- Flux cored arc welding (FCAW)
- Shielded metal arc welding (SMAW)
- Carbon arc welding (CAW)
- Electroslag welding (ESW)
- Oxyacetylene welding (OAW)
- Electron beam welding (EBW)

Incomplete Fusion

Incomplete fusion is the lack of coalescence between the molten filler metal and previously deposited filler metal and/or the base metal, **Figure 20-8**. The lack of fusion between the filler metal and previously deposited weld metal is called **Interpass Cold Lap**. The lack of fusion between the weld metal and the joint face is called **Lack of Sidewall Fusion**. Both of these problems usually travel along all or most of the weld's length. This discontinuity is not as detrimental to the weld's strength in service if it is near the center of the weld and is not open to the surface.

Some major causes of lack of fusion are:

- Inadequate agitation — Lack of weld agitation to break up oxide layers. The base metal or weld filler metal may melt, but a thin layer of oxide may prevent coalescence from occurring.
- Improper welding techniques — Poor manipulation, such as moving too fast or using an improper electrode angle.

- Wrong welding process — For example, the use of short-circuiting transfer with GMAW to weld plate thicker than 1/4 in. (6 mm) can cause the problem because of the process's limited heat input to the weld.
- Improper edge preparation — Any notches or gouges in the edge of the weld joint must be removed. For example if a flame cut plate has notches along the cut, they could result in a lack of fusion in each notch, **Figure 20-9**.
- Improper joint design — Incomplete fusion may also result from not enough heat to melt the base metal, or too little space allowed by the joint designer for correct molten weld pool manipulation.
- Improper joint cleaning — Failure to clean oxides for the joint surfaces resulting from the use of an oxyfuel torch to cut the plate, or failure to remove slag from a previous weld.

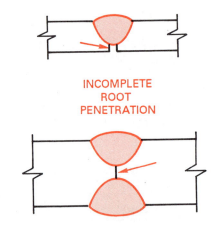

INCOMPLETE ROOT PENETRATION

Figure 20-7 Incomplete root penetration.

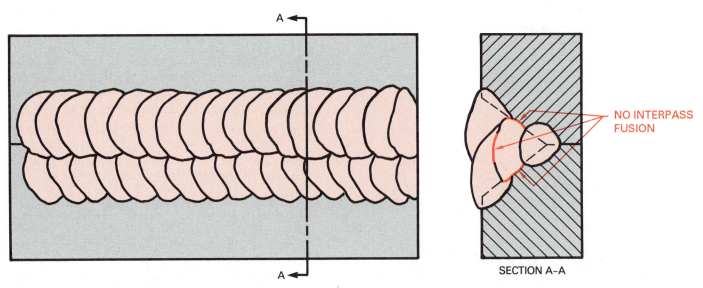

NO INTERPASS FUSION

SECTION A-A

Figure 20-8 Incomplete fusion.

Incomplete fusion can be found in welds produced by all major welding processes.

Arc Strikes

Figure 20-10 shows arc strikes that are small, localized points where surface melting occurred away from the joint. These spots may be caused by accidentally striking the arc in the wrong place and/or by faulty ground connections. Even though arc strikes can be ground smooth, they cannot be removed. These spots will always appear if an acid etch is used. They also can be localized hardness zones or the starting point for cracking. Arc strikes, even when ground flush for a guided bend, will open up to form small cracks or holes.

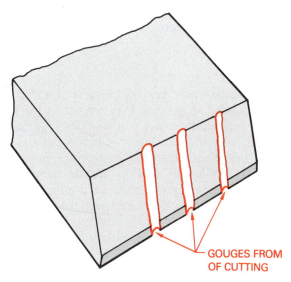

Figure 20-9 Remove gouges along the surface of the joint before welding.

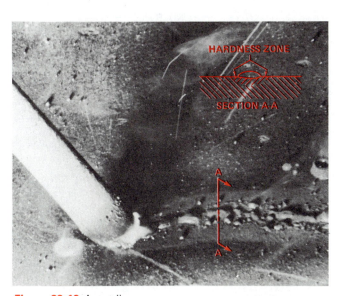

Figure 20-10 Arc strikes.

Overlap

Overlap occurs in fusion welds when weld deposits are larger than the joint is conditioned to accept. The weld metal then flows over a surface of the base metal without fusing to it, Figure 20-11. It generally occurs on the horizontal leg of a horizontal fillet weld under extreme conditions. It can also occur on both sides of flat-positioned capping passes. With GMA welding, overlap occurs when using too much electrode extension to deposit metal at low power. Misdirecting the arc into the vertical leg and keeping the electrode nearly vertical will also cause overlap. To prevent overlap, the fillet weld must be correctly sized to less than 3/8 in. (9.5 mm), and the arc must be properly manipulated.

Undercut

Undercut is the result of the arc plasma removing more metal from a joint face than is replaced by weld metal, Figure 20-12. It can result from excessive current. It is a common problem with GMA welding when insufficient oxygen is used to stabilize the arc. Incorrect welding technique, such as incorrect electrode angle or

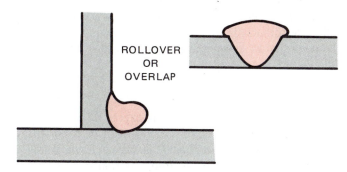

Figure 20-11 Rollover or overlap.

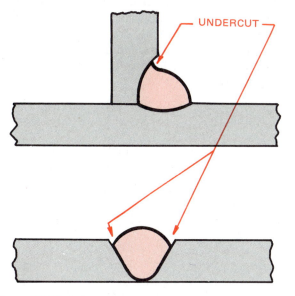

Figure 20-12 Undercut.

excessive weave, can also cause undercut. To prevent undercutting, the welder can weld in the flat position by using multiple instead of single passes, change the shield gas, and improve manipulative techniques.

Crater Cracks

Crater cracks are the tiny cracks that develop in the weld craters as the weld pool shrinks and solidifies, **Figure 20-13.** Low-melting materials are rejected toward the crater center while freezing. Since these materials are the last to freeze, they are pulled apart or separated, as a result of the weld metal's shrinking as it cools. The high shrinkage stresses aggravate crack formation. Crater cracks can be minimized, if not prevented, by not interrupting the arc quickly at the end of a weld. This allows the arc to lengthen, the current to drop gradually, and the crater to fill and cool more slowly. GMAW equipment has a crater filling control that automatically and gradually reduces the wire feed speed at the end of a weld.

Underfill

Underfill on a groove weld is when the weld metal deposited is inadequate to bring the weld's face or root surfaces to a level equal to that of the original plane. For a fillet weld it is when the weld deposit has an insufficient effective throat, **Figure 20-14.** This problem can usually be corrected by slowing the travel rate down or making more weld passes.

Plate Generated Problems

Not all welding problems are caused by weld metal, the process, or the welder's lack of skill in depositing that metal. The material being fabricated can be at fault too. Some problems result from internal plate defects that the welder cannot control. Others are the result of improper welding procedures that produce undesirable hard metallurgical structures in the heat affected zone, as discussed in other chapters. The internal defects are the result of poor steelmaking practices. Steel producers try to keep their steels as sound as possible, but the mistakes that occur are blamed, too frequently, on the welding operation.

Lamination

Laminations differ from lamellar tearing because they are more extensive and involve thicker layers of nonmetallic contaminants. Located toward the center of the plate, **Figure 20-15,** laminations are caused by insufficient cropping (removal of defects) of the pipe in ingots. The slag and oxidized steel in the pipe is rolled out with the steel, producing the lamination.

Delamination

When laminations intersect a joint being welded, some laminations may open up and become delaminate. Contaminating of the weld metal may occur if the lamination contained large amounts of slag, mill scale, dirt, or other undesirable materials. Such contamination can cause wormhole porosity or lack-of-fusion defects.

The problems associated with delaminations are not easily corrected. An effective solution for thick plate is to weld over the lamination to seal it. A better solution is to replace the steel.

Lamellar Tears

These tears appear as cracks parallel to and under the steel surface. In general, they are not in the heat affected zone, and they have a steplike configuration. They result from the thin layers of nonmetallic inclusions that lie beneath the plate surface and have very poor ductility. Although barely noticeable, these inclusions separate when severely stressed, producing laminated cracks. These cracks are evident if the plate edges are exposed, **Figure 20-16.**

A solution to the problem is to redesign the joints in order to impose the lowest possible strain throughout the plate thickness. This can be accomplished by making smaller welds so that each subsequent weld pass heat-treats the previous pass to reduce the total stress in the

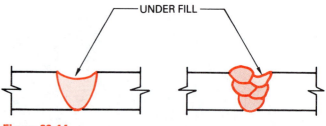

Figure 20-14

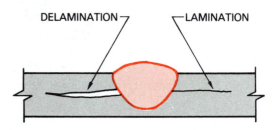

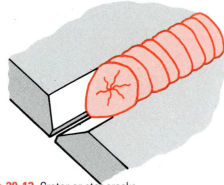

Figure 20-13 Crater or star cracks.

Figure 20-15

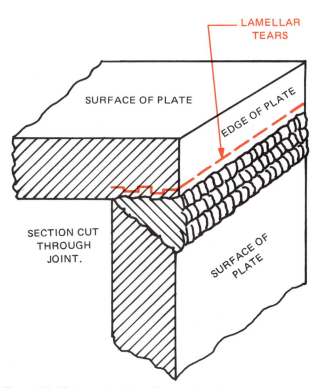

Figure 20-16 Example of lamellar tearing.

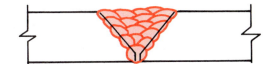

Figure 20-17 Using multiple welds to reduce weld stresses.

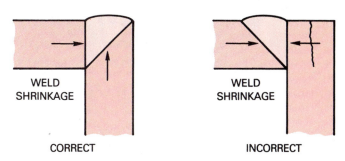

Figure 20-18 Correct joint design to reduce lamellar tears.

finished weld, Figure 20-17. The joint design can be changed to reduce the stress on the through thickness of the plate, Figure 20-18. Also refer to Chapter 18, Welding Joint Design, Welding Symbols, and Fabrication.

DESTRUCTIVE TESTING (DT)

Tensile Testing

Tensile tests are performed with specimen prepared as round bars or flat strips. The simple round bars are often used for testing only the weld metal, sometimes called "all weld metal testing." Round specimens are cut from the center of the weld metal. The flat bars are often used to test both the weld and the surrounding metal. Flat bars are usually cut at a 90° angle to the weld, Figure 20-19. Table 20-2 (page 500) shows how a number of standard smaller size bars can be used, depending on the thickness of the metal to be tested. Bar size also depends on the size of the tensile testing equipment available for the testing, Figure 20-20 (page 500).

Two flat specimens are used, commonly for testing thinner sections of metal. When testing welds, the specimen should include the heat affected zone and the base plate. If the weld metal is stronger than the plate, failure occurs in the plate; if the weld is weaker, failure occurs in the weld. This test, then, is open to interpretation.

After the weld section is machined to the specified dimensions, it is placed in the tensile testing machine

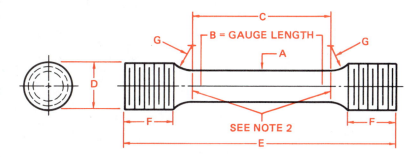

NOTE 1: DIMENSION A, B, AND C SHALL BE AS SHOWN, BUT ALTERNATE SHAPES OF ENDS MAY BE USED AS ALLOWED BY ASTM SPECIFICATION E-8.

NOTE 2: IT IS DESIRABLE TO HAVE THE DIAMETER OF THE SPECIMEN WITHIN THE GAUGE LENGTH SLIGHTLY SMALLER AT THE CENTER THAN AT THE ENDS. THE DIFFERENCE SHALL NOT EXCEED 1 PERCENT OF THE DIAMETER.

Figure 20-19 Tensile testing specimen. (Courtesy of Hobart Brothers Company.)

SPECIMEN	DIMENSIONS OF SPECIMEN						
	in/mm A	in/mm B	in/mm C	in/mm D	in/mm E	in/mm F	in/mm G
C-1	.500/12.7	2/50.8	2.25/57.1	.750/19.05	4.25/107.9	.750/19.05	.375/9.52
C-2	.437/11.09	1.750/44.4	2/50.8	.625/15.8	4/101.6	.750/19.05	.375/9.52
C-3	.357/9.06	1.4/35.5	1.750/44.4	.500/12.7	3.500/88.9	.625/15.8	.375/9.52
C-4	.252/6.40	1.0/25.4	1.250/31.7	.375/9.52	2.50/63.5	.500/12.7	.125/3.17
C-5	.126/3.2	.500/12.7	.750/19.05	.250/6.35	1.750/44.4	.375/9.52	.125/3.17

Table 20-2 Dimensions of Tensile Testing Specimens.

and pulled apart. A specimen used to determine the strength of a welded butt joint for plate is shown in **Figure 20-21**.

The tensile strength, in pounds per square inch, is obtained by dividing the maximum load, required to brake the specimen, by the cross-sectional area of the specimen at the middle. The cross-sectional area is obtained by using either formula $A = \pi \cdot r^2$ or $A = D^2 \cdot 0.785$ [(r^2 is the same as multiplying the radius times itself) (D^2 is the same as multiplying the diameter times itself). Using either formula will give you the same answer.

The elongation is found by fitting the fractured ends of the specimen together, measuring the distance between gauge marks, and subtracting the gauge length. The percent of elongation is found by dividing the elongation by the gauge length and multiplying by 100.

$$E_1 = \frac{L_f - L_o}{L_o} \times 100$$

Where:
E_1 = % of elongation
L_f = Final gauge length
L_o = Original gauge length

Fatigue Testing

Fatigue testing is used to determine how well a weld can resist repeated fluctuating stresses or cyclic loading. The maximum value of the stresses is less than the tensile strength of the material. Fatigue strength can be lowered by improperly made weld deposits, which may be caused by porosity, slag inclusions, lack of penetration, or cracks. Any one of these discontinuities can act as points of stress, eventually resulting in the failure of the weld.

In the fatigue test, the part is subjected to repeated changes in applied stress. This test may be performed in one of several ways, depending upon the type of service the tested part must withstand. The results obtained are usually reported as the number of stress cycles that the part will resist without failure and the total stress used.

In one type of test, the specimen is bent back and forth. This test subjects the part to alternating compres-

sion and tension. A fatigue testing machine is used for this test, **Figure 20-22**. The machine is turned on, and, as it rotates, the specimen is alternately bent twice for each revolution. In this case, failure is usually rapid.

Shearing Strength of Welds

The two forms of **shearing strength** of welds are transverse shearing strength and longitudinal shearing strength. To test transverse shearing strength, a specimen is prepared as shown in **Figure 20-23**. The width of the specimen is measured in inches or millimeters. A tensile

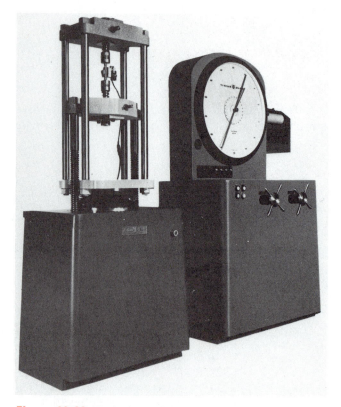

Figure 20-20 Typical tensile tester used for measuring the strength of welds (60,000-lb universal testing machines). (Courtesy of Tinius Olsen Testing Machine Co., Inc.)

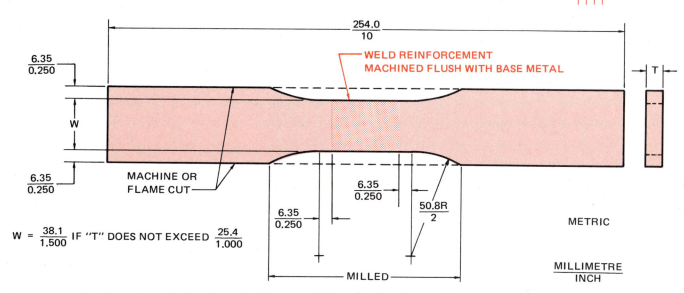

Figure 20-21 Tensile specimen for flat plate weld. (Courtesy of Hobart Brothers Company.)

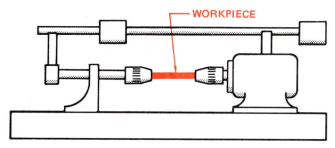

Figure 20-22 Fatigue testing. The specimen is placed in chucks of the machine. The machine is turned on, and, as it rotates, the specimen is alternately bent twice for each revolution.

CONVERSION TABLE — MILLIMETRES TO INCHES

DIM-mm	TOL	DIM-in
9.52		0.375
9.52	± 1.58	0.375
12.7		0.500
19.05		0.750
50.8		2.000
63.5		2.500
228.6		9.000
114.3		4.500

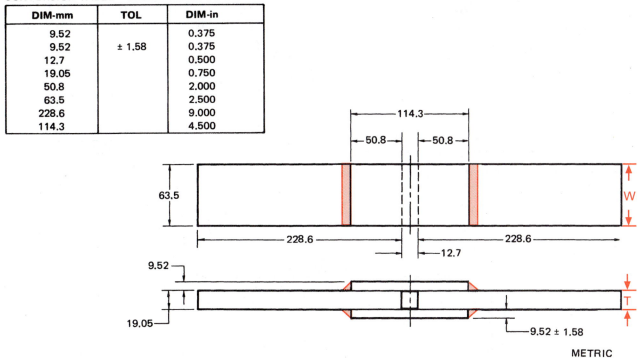

Figure 20-23 Transverse fillet weld shearing specimen after welding. (Courtesy of Hobart Brothers Company.)

load is applied, and the specimen is ruptured. The maximum load in pounds or kilograms is then determined.

The shearing strength of the weld, in pounds per linear inch, is obtained by dividing the maximum force by twice the width of the specimen.

$$\text{Shearing strength lb/in. (kg/mm)} = \frac{\text{Maximum force}}{}$$

To test longitudinal shearing strength, a specimen is prepared as shown in **Figure 20-24**. The length of each weld is measured in inches or millimeters. The specimen is then ruptured under a tensile load, and the maximum force in pounds or kilograms is determined.

The shearing strength of the weld in pounds per linear inch or kilograms/millimeter is obtained by dividing the maximum force by the sum of the length of welds which ruptured.

$$\text{Shearing strength lb/in. (kg/mm)} = \frac{\text{Maximum force}}{\text{Length of ruptured weld}}$$

Welded Butt Joints

The three methods of testing welded butt joints are (1) the nick-break test, (2) the guided bend test, and (3) the free bend test. It is possible to use variations of these tests.

Nick-break Test. A specimen for this test is prepared as shown in **Figure 20-25(A)**. The specimen is supported as shown in **Figure 20-25(B)**. A force is then applied, and the specimen is ruptured by one or more blows of a hammer. The force may be applied slowly or suddenly. Theoretically, the rate of application could affect how the specimen breaks, especially at a critical temperature. Generally, however, there is no difference in the appearance of the fractured surface due to the method of applying the force. The surfaces of the fracture should be checked for soundness of the weld.

Guided Bend Test. To test welded, grooved butt joints on metal that is 3/8 in. (10 mm) thick or less, two specimens are prepared and tested — one face bend and one root bend, **Figure 20-26(A)** and **(B)**. If the welds pass this test, the welder is qualified to make groove welds on plate having a thickness range of from 3/8 in. to 3/4 in. (10 mm to 19 mm). These welds need to be machined as shown in **Figure 20-27(A)** and **(B)** (page 504). If these specimens pass, the welder will also be qualified to make fillet welds on materials of any (unlimited) thicknesses. For welded, grooved butt joints on metal 1 in. (25 mm) thick, two side bend specimens are prepared and tested, **Figure 20-27(C)** (page 504). If the welds pass this test, the welder is qualified to weld on metals of unlimited thickness.

When the specimens are prepared, caution must be taken to ensure that all grinding marks run longitudinally to the specimen so that they do not cause stress cracking. In addition, the edges must be rounded to reduce cracking that tends to radiate from sharp edges.

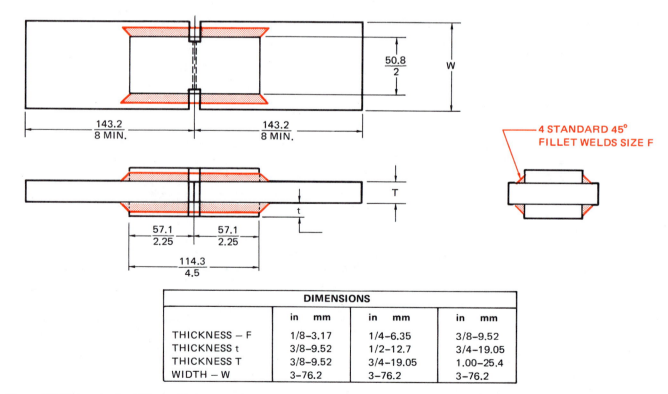

DIMENSIONS						
	in	mm	in	mm	in	mm
THICKNESS – F	1/8	3.17	1/4	6.35	3/8	9.52
THICKNESS t	3/8	9.52	1/2	12.7	3/4	19.05
THICKNESS T	3/8	9.52	3/4	19.05	1.00	25.4
WIDTH – W	3	76.2	3	76.2	3	76.2

Figure 20-24 Longitudinal fillet weld shear specimen.

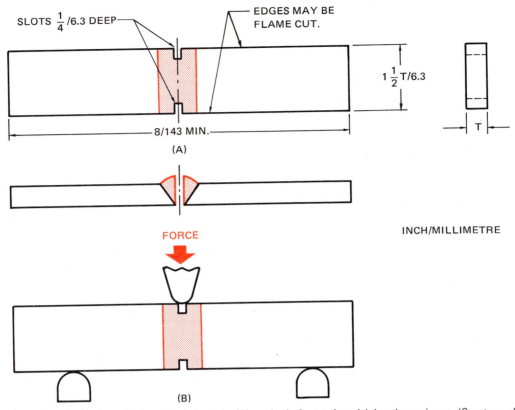

Figure 20-25 (A) Nick-break specimen for butt joints in plate; (B) method of rupturing nick-break specimen. (Courtesy of Hobart Brothers Company.)

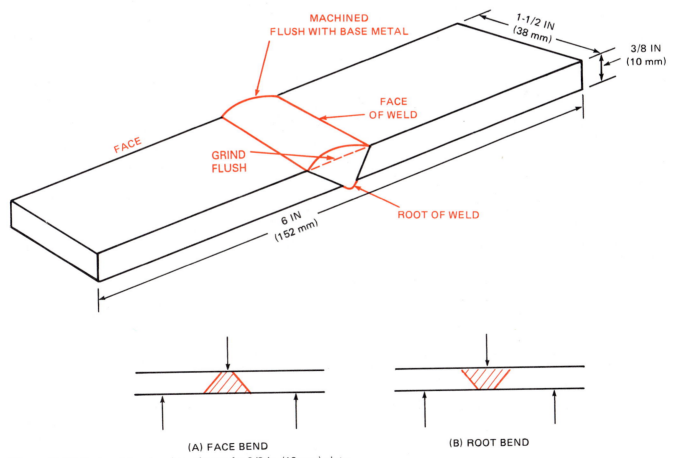

Figure 20-26 Root and face bend specimens for 3/8-in. (10-mm) plate.

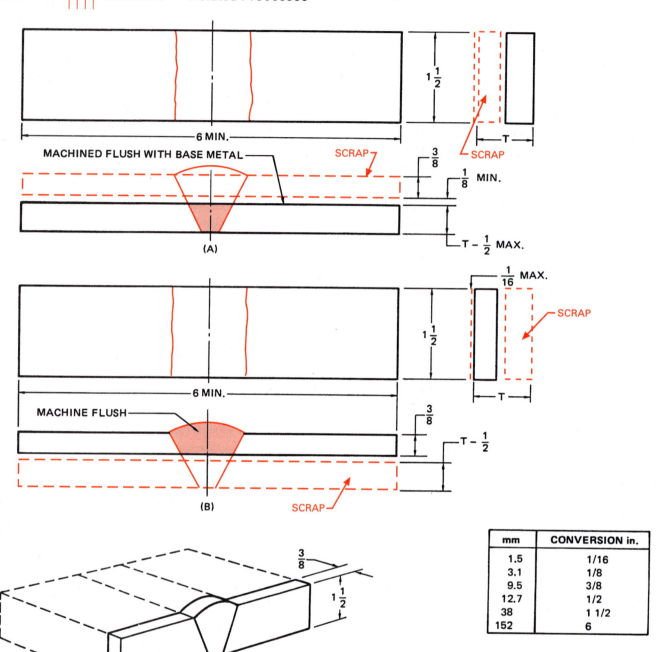

mm	CONVERSION in.
1.5	1/16
3.1	1/8
9.5	3/8
12.7	1/2
38	1 1/2
152	6

Figure 20-27 (A) Root bend specimen; (B) specimen for face bend test; and (C) side bend specimen for plates thicker than 3/8 in. (10 mm).

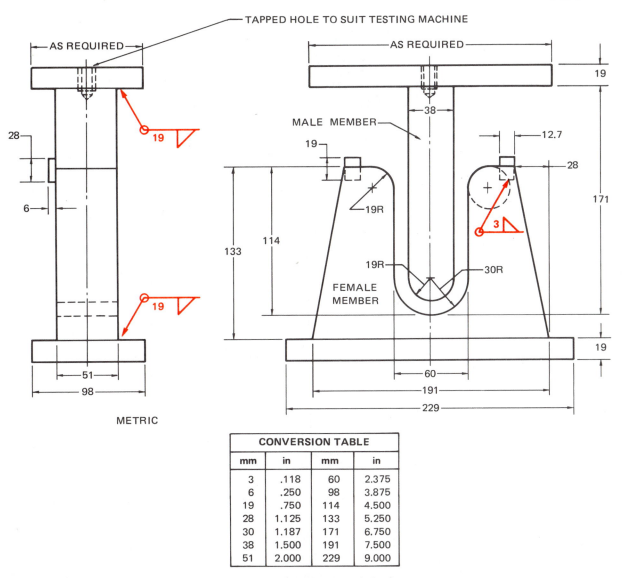

TAPPED HOLE TO SUIT TESTING MACHINE

CONVERSION TABLE			
mm	in	mm	in
3	.118	60	2.375
6	.250	98	3.875
19	.750	114	4.500
28	1.125	133	5.250
30	1.187	171	6.750
38	1.500	191	7.500
51	2.000	229	9.000

Figure 20-28 Fixture for guided bend test. (Courtesy of The Aluminum Association.)

Procedure. The jig shown in **Figure 20-28** is commonly used to bend most specimens. The radius that the specimens are bent will vary for some metals such as aluminum. Place the specimens in the jig with the weld in the middle. Face bend specimens should be placed with the face of the weld toward the gap. Root bend specimens should be positioned so that the root of the weld is directed toward the gap. Side bend specimens are placed with either side facing up. The two parts of the jig are forced together.

Once the test is completed, the specimen is removed. The convex surface is then examined for cracks or other discontinuities.

Free-bend Test. The free-bend test is used to test welded joints in plate. A specimen is prepared as shown in **Figure 20-29** (page 506). Note that the width of the specimen is 1.5 multiplied by the thickness of the specimen. Each corner lengthwise should be rounded in a radius not exceeding 1/10 the thickness of the specimen. Tool marks should run the length of the specimen.

Gauge lines are drawn on the face of the weld, **Figure 20-30** (page 506). The distance between the gauge lines is 1/8 in. (3.17 mm) less than the face of the weld. The initial bend of the specimen is completed in the device illustrated in **Figure 20-31** (page 507). The gauge line surface should be directed toward the supports. The weld is located in the center of the supports and loading block.

Alternate Bend. The initial bend may be made by placing the specimen in the jaws of a vise with one-third the length projecting from the jaws. The specimen is then bent away from the gauge lines through an angle of 30 to 45 degrees by blows of a hammer. The pres-

CONVERSION TABLE							
in	mm	in	mm	in	mm	in	mm
1/4	6.35	9/16	14.2	1 1/8	28.5	2	50.8
3/8	9.52	3/4	19.05	1 1/4	31.7	6	152.4
1/2	12.7	15/16	23.8	1 1/2	38.1	8	203.2
5/8	15.8	1	25.4	1 7/8	47.6	9	228.6

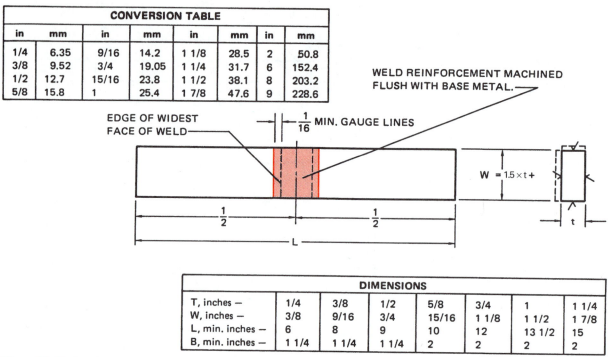

DIMENSIONS							
T, inches —	1/4	3/8	1/2	5/8	3/4	1	1 1/4
W, inches —	3/8	9/16	3/4	15/16	1 1/8	1 1/2	1 7/8
L, min. inches —	6	8	9	10	12	13 1/2	15
B, min. inches —	1 1/4	1 1/4	1 1/4	2	2	2	2

Figure 20-29 Free-bend test specimen. (Courtesy of Hobart Brothers Company.)

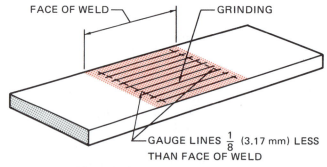

Figure 20-30 Gauge lines are drawn on the weld face of a free-bend specimen.

sure is continued until a crack or depression appears on the convex face of the specimen. The load is then removed.

The elongation is determined by measuring the minimum distance between the gauge lines along the convex surface of the weld to the nearest .01 in. (0.254 mm) and subtracting the initial gauge length. The percent of elongation is obtained by dividing the elongation by the initial gauge length and multiplying by 100.

Fillet Weld Break Test

The specimen for this test is made as shown in Figure 20-32(A). In Figure 20-32(B), a force is applied to the specimen until the rupture breaking of the specimen occurs. Any convenient means of applying the force may be used, such as an arbor press, testing machine, or hammer blows. The break surface should then be examined for soundness; i.e., slag inclusions, overlap, porosity, lack of fusion, or other discontinuities.

Testing by Etching

Specimens to be tested by **etching** are etched for two purposes: (1) to determine the soundness of a weld, or (2) to determine the location of a weld.

A test specimen is produced by cutting a portion from the welded joint so that a complete cross section is obtained. The face of the cut is then filed and polished with fine abrasive cloth. The specimen can then be placed in the etching solution. The etching solution or reagent makes the boundary between the weld metal and base metal visible, if the boundary is not already distinctly visible.

The most commonly used etching solutions are hydrochloric acid, ammonium persulphate, or nitric acid.

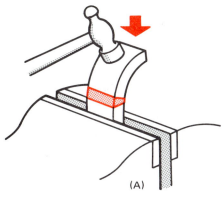

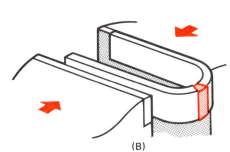

CONVERSION TABLE	
mm	in
12.7	.500
20	0.787
32	1.25
76	3.000

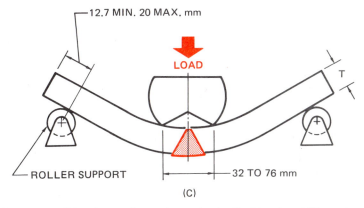

Figure 20-31 Free-bend test: (A) initial bend can be made in this manner, (B) a vise can be used to make the final bend, and (C) another method used to make the bend. (Courtesy of Hobart Brothers Company.)

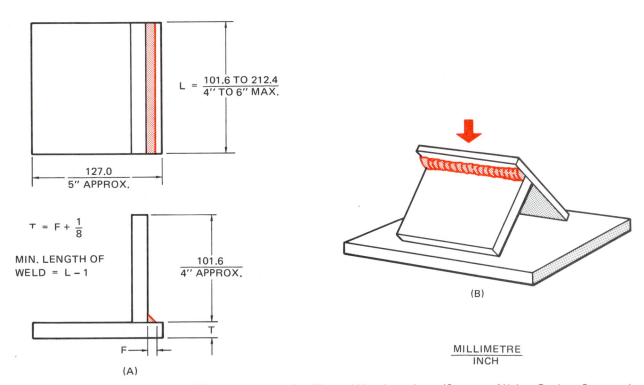

Figure 20-32 (A) Fillet weld break test and (B) method of rupturing fillet weld break specimen. (Courtesy of Hobart Brothers Company.)

Hydrochloric Acid. Equal parts by volume of concentrated hydrochloric (muriatic) acid and water are mixed. The welds are immersed in the reagent at or near the boiling temperature. The acid usually enlarges gas pockets and dissolves slag inclusions, enlarging the resulting cavities.

■ CAUTION

When mixing the muriatic acid into the water, be sure safety glasses and gloves are worn in order to prevent any injuries from occurring.

Ammonium Persulphate. A solution is prepared consisting of one part of ammonium persulphate (solid) to nine parts of water by weight. The surface of the weld is rubbed with cotton saturated with this reagent at room temperature.

Nitric Acid. A great deal of care should be exercised when using nitric acid because severe burns can result if it is used carelessly. One part of concentrated nitric acid is mixed with nine parts of water by volume.

■ CAUTION

When diluting an acid, always pour the acid slowly into the water while continuously stirring the water. Carelessly handling this material or pouring water into the acid can result in burns, excessive fuming, or explosion.

The reagent is applied to the surface of the weld with a glass stirring rod at room temperature. Nitric acid has the capacity to etch rapidly and should be used on polished surfaces only.

After etching, the weld is rinsed in clear, hot water. Excess water is removed, and the etched surface is then immersed in ethyl alcohol and dried.

Impact Testing

A number of tests can be used to determine the impact capability of a weld. One common test is the Izod test, Figure 20-33(A), in which a notched specimen is struck by an anvil mounted on a pendulum. The energy in footpounds required to break the specimen is an indication of the impact resistance of the metal. This test compares the toughness of the weld metal with the base metal.

Another type of impact test is the Charpy test. This test is similar to the Izod test. The difference is in the manner in which specimens are held. A typical impact tester is shown in Figure 20-33(B).

NONDESTRUCTIVE TESTING (NDT)

Nondestructive testing of welds is a method used to test materials for surface defects such as cracks, arc strikes, undercuts, and lack of penetration. Internal or subsurface defects can include slag inclusions, porosity, and unfused metal in the interior of the weld.

Visual Inspection (VT)

Visual inspection is the most frequently used nondestructive testing method. The majority of welds receive only visual inspection. In this method, if the

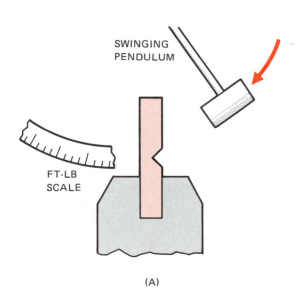

(A)

(B)

Figure 20-33 Impact testing: (A) specimen mounted for Izod impact toughness and (B) typical impact tester used for measuring the toughness of metals. (Photo courtesy of Tinius Olsen Testing Machine Co., Inc.)

weld looks good it passes; if it looks bad it is rejected. This procedure is often overlooked when more sophisticated nondestructive testing methods are used. However, it should not be overlooked.

An active visual inspection schedule can reduce the finished weld rejection rate by more than 75%. Visual inspection can easily be used to check for fitup, interpass acceptance, welder technique, and other variables that will affect the weld quality. Minor problems can be identified and corrected before a weld is completed. This eliminates costly repairs or rejection.

Visual inspection should be used before any other nondestructive or mechanical tests are used to eliminate (reject) the obvious problem welds. Eliminating welds that have excessive surface discontinuities that will not pass the code or standards being used saves preparation time.

Penetrant Inspection (PT)

Penetrant inspection is used to locate minute surface cracks and porosity. Two types of penetrants are now in use, the color-contrast and the fluorescent versions. Color-contrast, often red, penetrants contain a colored dye that shows under ordinary white light. Fluorescent penetrants contain a more effective fluorescent dye that shows under black light.

The following steps outline the procedure to be followed when using a penetrant.

1. Precleaning. The test surface must be clean and dry. Suspected flaws must be cleaned and dried so that they are free of oil, water, or other contaminants.

2. The test surface must be covered with a film of penetrant by dipping, immersing, spraying, or brushing.

3. The test surface is then wiped, washed, or rinsed free of excess penetrant. It is dried with cloths or hot air.

4. A developing powder applied to the test surface acts like a blotter to speed the tendency of the penetrant to seep out of any flaws on the test surface.

5. Depending upon the type of penetrant applied, visual inspection is made under ordinary white light or near-ultraviolet black light when viewed under this light, **Figure 20-34**. The penetrant fluoresces to a yellow-green color, which clearly defines the defect.

Magnetic Particle Inspection (MT)

Magnetic particle inspection uses finely divided ferromagnetic particles (powder) to indicate defects on magnetic materials.

A magnetic field is induced in a part by passing an electric current through or around it. The magnetic field is always at right angles to the direction of current flow. Ferromagnetic powder registers an abrupt change in the

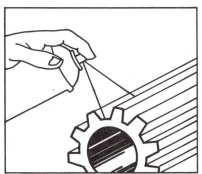

1. Preclean inspection area. Spray on Cleaner/Remover —wipe off with cloth.

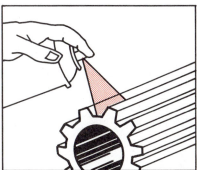

2. Apply Penetrant. Allow short penetration period.

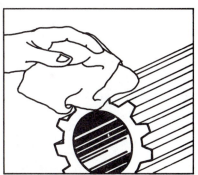

3. Spray Cleaner/Remover on wiping towel and wipe surface clean.

4. Shake Developer can and spray on a thin, uniform film of Developer.

5. Inspect. Defects will show as bright red lines in white Developer background.

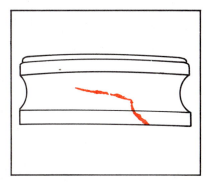

Figure 20-34 Penetrant testing. (Adapted from Magnaflux Corporation.)

resistance in the path of the magnetic field, such as would be caused by a crack lying at an angle to the direction of the magnetic poles at the crack. Finely divided ferromagnetic particles applied to the area will be attracted and outline the crack.

In **Figure 20-35**, the flow or discontinuity interrupting the magnetic field in a test part can be either longitudinal or circumferential. A different type of magnetization is used to detect defects that run down the axis, as opposed to those occurring around the girth of a part. For some applications you may need to test in both directions.

In **Figure 20-36(A)**, longitudinal magnetization allows detection of flaws running around the circumference of a part. The user places the test part inside an electrified coil. This induces a magnetic field down the length of the test part. In **Figure 20-36(B)**, circumferential magnetization allows detection of flaws occurring down the length of a test part. An electric current is sent down the length of the part to be inspected. The magnetic field thus induced allows defects along the length of the part to be detected.

Radiographic Inspection (RT)

Radiographic inspection (RT) is a method for detecting flaws inside weldments. Instead of using visible light rays, the operator uses invisible, short-wavelength rays developed by X-ray machines, radioactive isotopes (gamma rays), and variations of these methods. These rays are capable of penetrating solid materials and reveal most flaws in a weldment on an X-ray film or fluorescent screen. Flaws are revealed on films as dark or light areas against a contrasting background after exposure and processing, **Figure 20-37**.

The defect images in radiographs measure differences in how the X rays are absorbed as they penetrate the weld. The weld itself absorbs most X rays. If something less dense than the weld is present, such as a pore or lack of fusion defect, fewer X rays are absorbed, darkening the film. If something more dense is present, such as heavy ripples on the weld surface, more X rays will be absorbed, lightening the film.

Therefore, the foreign material's relative thickness (or lack of it) and differences in X-ray absorption determine the radiograph image's final shape and shading. Skilled readers of radiographs can interpret the significance of the light and dark regions by their shape and shading. The X-ray image is a shadow of the flaw. The farther the flaw is from the X-ray film, the fuzzier and larger the image appears. When X-raying thick material, flaws near the top surface may appear much larger than the same sized flaw near the back surface. Those skilled at interpreting weld defects in radiographs must also be very knowledgeable about welding.

Figure 20-38 (pages 511–515) shows samples of common weld defects and a representative radiograph for each.

Four factors affect the selection of the radiation source:

■ Thickness and density of the material

■ Absorption characteristics

■ Time available for inspection

■ Location of the weld

Portable equipment is available for examining fixed or hard-to-move objects. The selection of the correct

PARTICLES CLING
TO THE DEFECT
LIKE TACKS TO A
SIMPLE MAGNET

SPECIMEN DEFECT

MAGNETIC FIELD

BY INDUCING A MAGNETIC FIELD WITHIN THE PART TO BE TESTED, AND APPLYING A COATING OF MAGNETIC PARTICLES, SURFACE CRACKS ARE MADE VISIBLE, THE CRACKS IN EFFECT FORMING NEW MAGNETIC POLES.

Figure 20-35 (Adapted from Magnaflux Corporation.)

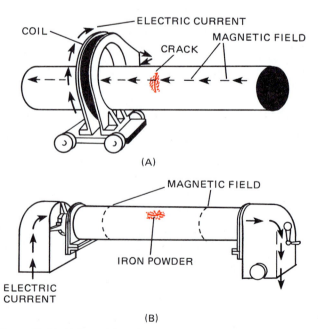

Figure 20-36 (Adapted from Magnaflux Corporation.)

equipment for a particular application is determined by specific voltage required, degree of utility, economics of inspection, and production rates expected, **Figures 20-39 and 20-40 (page 516)**.

Ultrasonic Inspection (UT)

Ultrasonic inspection (UT) is a fast and relatively low-cost nondestructive testing method. This inspection method employs electronically produced high-frequency sound waves (roughly 1/4 to 25 million cycles per second), which penetrate metals and many other materials at speeds of several thousand feet (meters) per second. A portable ultrasonic inspection unit is shown in **Figure 20-41 (page 516)**.

The two types of ultrasonic equipment are pulse and resonance. The pulse-echo system, most often employed in the welding field, uses sound generated in short bursts or pulses. Since the high-frequency sound used is at a relatively low power, it has little ability to travel through air, so it must be conducted from the probe into the part through a medium such as oil or water.

Sound is directed into the part with a probe held in a preselected angle or direction so that flaws will reflect some energy back to the probe. These ultrasonic devices operate very much like depth sounders, or "fish finders." The speed of sound through a material is a known quantity. These devices measure the time required for a pulse to return from a reflective surface. Internal computers calculate the distance and present the information on a cathode ray tube where an operator can interpret the results. The signals can be "monitored" electronically to operate alarms, print systems, or recording equipment. Sound not reflected by flaws continues into the part. If the angle is correct, the sound energy will be reflected back to the probe from the opposite side. Flaw size is determined by plotting the length, height, width, and shape using trigonometric rules.

Figure 20-42 (page 517) shows the path of the sound beam in butt welding testing. The operator must know the exit point of the sound beam, the exact angle of the refracted beam, and the thickness of the plate.

Leak Checking

Leak checking can be performed by filling the welded container with either a gas or liquid. There may or

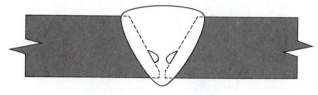

WELDING DEFECT: Lack of sidewall fusion (LOF). Elongated voids between the weld beads and the joint surfaces to be welded.

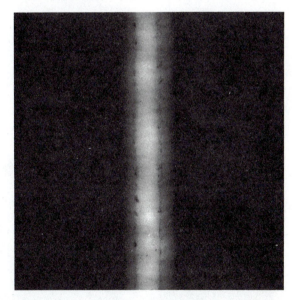

RADIOGRAPHIC IMAGE: Elongated parallel, or single, darker density lines sometimes with darker density spots dispersed along the LOF lines which are very straight in the lengthwise direction and not winding like elongated slag lines. Although one edge of the LOF lines may be very straight like LOP, lack of sidewall fusion images will not be in the center of the width of the weld image.

Figure 20-38 Welding defects with radiographic images. (Reprinted with permission of E. I. DuPont de Nemours & Co., Inc.)

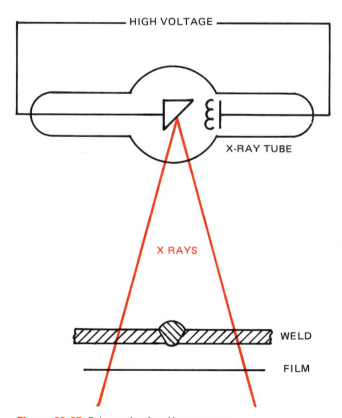

Figure 20-37 Schematic of an X-ray system.

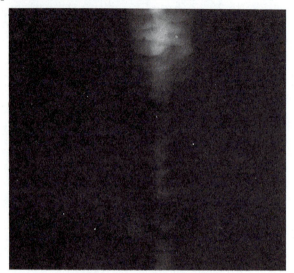

WELDING DEFECT: External concavity or insufficient fill. A depression in the top of the weld, or cover pass, indicating a thinner than normal section thickness.

RADIOGRAPHIC IMAGE: A weld density darker than the density of the pieces being welded, and extending across the full width of the weld image.

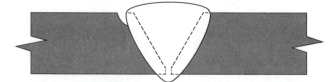

WELDING DEFECT: External undercut. A gouging out of the piece to be welded, alongside the edge of the top or "external" surface of the weld.

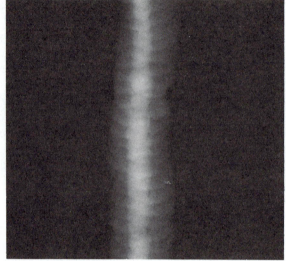

RADIOGRAPH IMAGE: An irregular darker density along the edge of the weld image. The density will always be darker than the density of the pieces being welded.

WELDING DEFECT: Excessive Penetration (Icicles, Drop-thru). Extra metal at the bottom (root) of the weld.

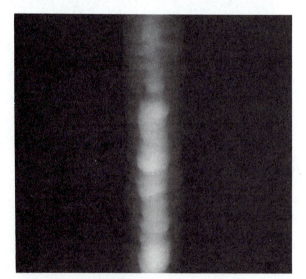

RADIOGRAPHIC IMAGE: A lighter density in the center of the width of the weld image, either extended along the weld or in isolated circular "drops".

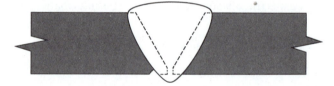

WELDING DEFECT: Internal (root) undercut. A gouging out of the parent metal, alongside the edge of the bottom or "internal" surface of the weld.

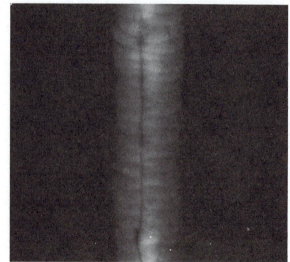

RADIOGRAPHIC IMAGE: An irregular darker density near the center of the width of the weld image and along the edge of the root pass image.

Figure 20-38 Welding defects with radiographic images (continued). (Reprinted with permission of E. I. DuPont de Nemours & Co., Inc.)

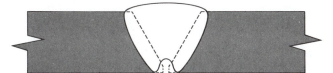

WELDING DEFECT: Internal concavity (suck back). A depression in the center of the surface of the root pass.

WELDING DEFECT: Interpass slag inclusions. Usually nonmetallic impurities that solidified on the weld surface and were not removed between weld passes.

RADIOGRAPHIC IMAGE: An elongated irregular darker density with fuzzy edges, in the center of the width of the weld image.

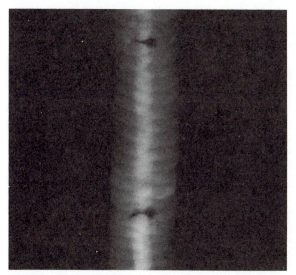

RADIOGRAPHIC IMAGE: An irregularly shaped darker density spot, usually slightly elongated and randomly spaced.

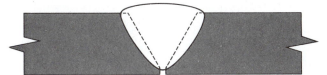

WELDING DEFECT: Incomplete or lack of penetration (LOP). The edges of the pieces have not been welded together, usually at the bottom of single V-groove welds.

WELDING DEFECT: Elongated slag lines (wagon tracks). Impurities that solidify on the surface after welding and were not removed between passes.

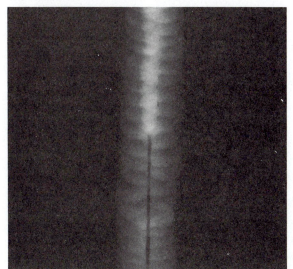

RADIOGRAPHIC IMAGE: A darker density band, with very straight parallel edges, in the center of the width of the weld image.

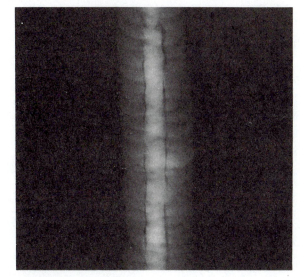

RADIOGRAPHIC IMAGE: Elongated, parallel, or single darker density lines, irregular in width and slightly winding in the lengthwise direction.

Figure 20-38 Welding defects with radiographic images (continued). (Reprinted with permission of E. I. DuPont de Nemours & Co., Inc.)

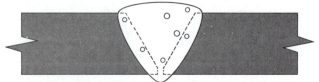

WELDING DEFECT: Scattered porosity. Rounded voids random in size and location.

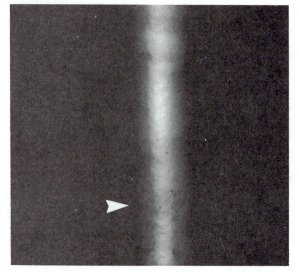

RADIOGRAPHIC IMAGE: Rounded spots of darker densities random in size and location.

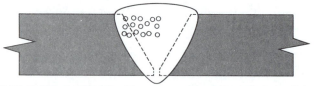

WELDING DEFECT: Cluster porosity. Rounded or slightly elongated voids grouped together.

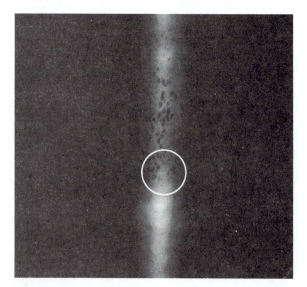

RADIOGRAPHIC IMAGE: Rounded or slightly elongated darker density spots in clusters with the clusters randomly spaced.

WELDING DEFECT: Root pass aligned porosity. Rounded and elongated voids in the bottom of the weld aligned along the weld centerline.

RADIOGRAPHIC IMAGE: Rounded and elongated darker density spots, that may be connected, in a straight line in the center of the width of the weld image.

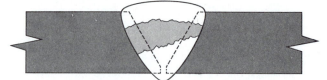

WELDING DEFECT: Transverse crack. A fracture in the weld metal running across the weld.

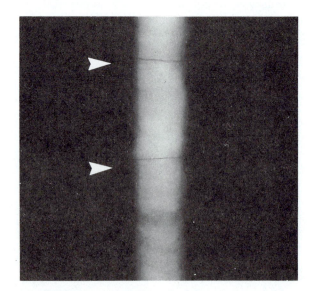

RADIOGRAPHIC IMAGE: Feathery, twisting line of darker density running across the width of the weld image.

Figure 20-38 Welding defects with radiographic images (continued). (Reprinted with permission of E. I. DuPont de Nemours & Co., Inc.)

WELDING DEFECT: Longitudinal root crack. A fracture in the weld metal at the edge of the root pass.

WELDING DEFECT: Tungsten inclusions. Random bits of tungsten fused, but not melted, into the weld metal.

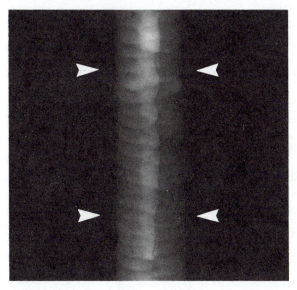

RADIOGRAPHIC IMAGE: Feathery, twisting lines of darker density along the edge of the image of the root pass. The "twisting" feature helps to distinguish the root crack from incomplete root penetration.

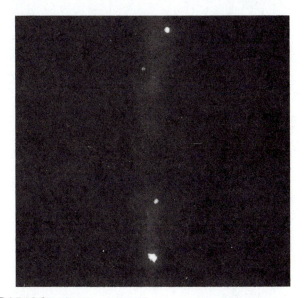

RADIOGRAPHIC IMAGE: Irregularly shaped lower density spots randomly located in the weld image.

Figure 20-38 Welding defects with radiographic images (continued). (Reprinted with permission of E. I. DuPont de Nemours & Co., Inc.)

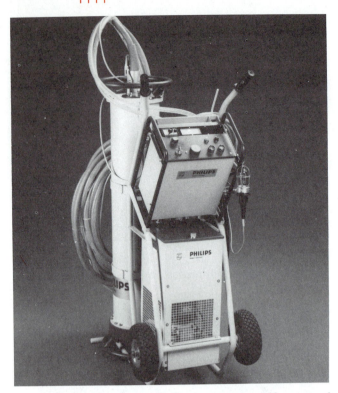

Figure 20-39 Typical mobile X-ray equipment. (Courtesy of Philips GmbH.)

Figure 20-40 Preparing to test the quality of a weld in a pressure vessel using X-ray equipment. (Courtesy of Philips GmbH.)

may not be additional pressure applied to the material in the weldment. Water is the most frequently used liquid, although sometimes a liquid with a lower viscosity is used. If gas is used, it may be either a gas that can be detected with an instrument when it escapes through a flaw in the weld, or it can be an air leak that is checked with bubbles.

Eddy Current Inspection (ET)

Eddy current inspection (ET) is another nondestructive test. This method is based upon magnetic induction in which a magnetic field induces eddy currents within the material being tested. An eddy current is an induced electric current circulating wholly within a mass of metal. This method is effective in testing nonferrous and ferrous materials for internal and external cracks, slag inclusions, porosity, and lack of fusion that are on or very near the surface. Eddy current cannot locate flaws that are not near the surface.

A coil carrying high-frequency alternating current is brought close to the metal to be tested. A current is

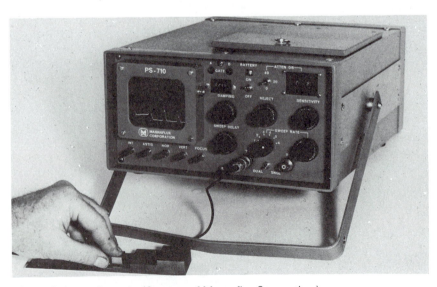

Figure 20-41 Portable ultrasonic inspection unit. (Courtesy of Magnaflux Corporation.)

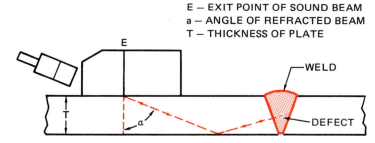

E — EXIT POINT OF SOUND BEAM
a — ANGLE OF REFRACTED BEAM
T — THICKNESS OF PLATE

Figure 20-42 Ultrasonic testing.

Figure 20-44 Brinell hardness tester. (Courtesy of Tinius Olsen Testing Machine Company.)

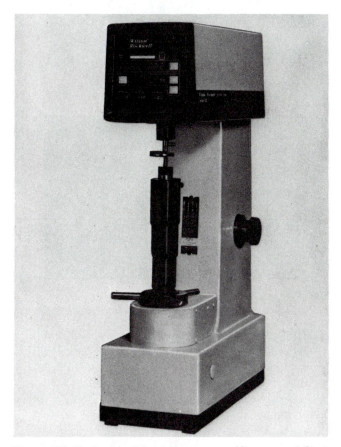

Figure 20-43 Rockwell hardness tester. (Courtesy of Clark Instrument, Inc.)

produced in the metal by induction. The part is placed in or near high-frequency, alternating current coils. The magnitude and phase difference of these currents is, in turn, indicated in the actual impedance value of the pick-up coil. Careful measurement of this impedance is the revealing factor in detecting defects in the weld.

Hardness Testing

Hardness is the resistance of metal to penetration and is an index of the wear-resistance and strength of the metal. Hardness tests can be used to determine the relative hardness of the weld with the base metal. The two types of hardness testing machines in common use are the Rockwell and the Brinell testers.

The Rockwell hardness tester, Figure 20-43, uses a 120-degree diamond cone for hard metals and a 1/16-in. (1.58-mm) hardened steel ball for softer metals. The method is based upon resistance-to-penetration measurement. The hardness is read directly from a dial on the tester. The depth of the impression is measured instead of the diameter. The tester has two scales for reading hardness, known as the B-scale and the C-scale. The C-scale is used for harder metals, and the B-scale for softer metals.

The Brinell hardness tester measures the resistance of material to the penetration of a steel ball under constant pressure (about 3,000 kilograms) for a minimum of approximately 30 seconds, Figure 20-44. The diameter is measured microscopically, and the Brinell number is checked on a standard chart. Brinell hardness numbers are obtained by dividing the applied load by the area of the surface indentation.

Welding Plays Center Field in Ballpark Construction

Figure 1

Old baseball stadiums, unlike the memories they house, don't last forever. This year, sometime after the Chicago White Sox's final game, 80-year-old Comiskey Park, the only park left where Cy Young pitched, will be torn down.

This is the place where Babe Ruth smashed a home run to win the first All-Star Game; where Bob Feller pitched a no-hitter on Opening Day in 1940. Once, a little boy looked up to Shoeless Joe Jackson, who played in this fine old stadium and said, "Say it ain't so, Joe!"

It *is* so. But before the wrecking ball makes its final swing, a new Comiskey Park will be completed on land adjacent to the old one. Hobart Brothers Company welding products are being used to make sure Chicago's baseball fans will have a place to host a new era of memories.

The new Comiskey (Figure 1) was designed by HOK Sports Facilities Group, Kansas City, Missouri. The 43,000-seat stadium will cost millions — $119.4-million, to be exact. The new park will contain elements of the past and some of the future. The dirt from the original stadium will be laid down in the new one, and each of the 43,000 seats will be computer-designed so that each one points to the infield.

Chicago's winds and subzero temperatures this past winter gave the forty-four people building the stadium plenty of challenges to overcome during the building process. Two dozen workers alone were needed to build the handrails that are welded with steel strong enough to support the prestressed concrete. International Erectors, Inc., located in DePere, Wisconsin, the firm responsible for building the stadium's steel structures, selected the Hobart Champion 16 and 18 engine-driven welding power sources combined with 418 filler metal — Figure 2.

International Erectors have only nine months to build this massive structure. That's why they appreciate the efficiency and adaptability of the engine-driven welding machines. The 8500 W of auxiliary power allows them to operate drills and grinders, while welding under the maximum load.

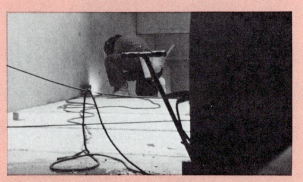

Figure 2

The 8-gallon gas tank provides eight hours of continuous, 100% duty cycle operation, without refueling.

Power losses, typically caused by Chicago's subzero temperatures, were not a problem for the welding power sources. These machines start without being jumped. This is because of the autobattery combination that is built into the devices.

The builder chose Hobart 418 (AWS E7018) electrodes designed for welding a variety of steel where low-hydrogen weld metal with a 70,000-psi tensile strength is required. The 7018 electrode also provided good arc strikes and restrikes. The electrode's ability to produce good welds with less spatter in the uphill and overhead positions made it attractive for welding at the job site.

Thanks to reliable welding equipment and quality filler metals, White Sox fans will have a place to cheer in spring 1991.

(Reprinted from the Welding Journal. Article based on a story from Hobart Brothers Company, Troy, Ohio. Courtesy of the American Welding Society, Inc.)

REVIEW

1. Why are all welds not inspected to the same level or standard?

2. Why is the strength of all production parts not known if a sample number of parts are mechanically tested?

3. Why is it possible to do more than one nondestructive test on a weldment?

4. What is a discontinuity?

5. What is a defect?

6. What is tolerance?

7. What are the 12 most common discontinuities?

8. How can porosity form in a weld and not be seen by the welder?

9. What welding process can cause porosity to form?

10. How is piping porosity formed?

11. What are inclusions, and how are they caused?

12. When does inadequate joint penetration usually become defective?

13. How can a notch cause incomplete fusion?

14. How can an arc strike appear on a guided bend test?

15. What is overlap?

16. What is undercut?

17. What causes crater cracks?

18. What is underfill?

19. What is the difference between a lamination and a delamination?

20. How can stress be reduced through a plate's thickness to reduce lamellar tearing?

21. What would be the tensile strength in pounds per square inch of a specimen measuring 0.375 in. thick and 1.0 in. wide if it failed at 27,000 pounds?

22. What would be the elongation for a specimen for which the original gauge length was 2 in. and final gauge length was 2.5 in.?

23. How are the results of a stress test reported?

24. What would be the transverse shear strength per inch of weld if a specimen that was 2.5 in. wide withstood 25,000 pounds?

25. What would be the longitudinal shearing strength per millimeter of a specimen that was 50.8 mm wide and 116 mm long and withstood 50.0 kg/mm?

26. What are the three methods of distractive testing of a welded butt joint?

27. How are the specimens bent for a guided bend root, face, and side bend test?

28. How wide should a specimen be if the material thickness is
a) 0.375 in.?
b) 6.35 mm?

29. Why are guide lines drawn on the surface of a free-bend specimen?

30. What part of a fillet weld break test is examined?

31. What can happen if acids are handled carelessly?

32. What information about the weld does an impact test provide?

33. Which nondestructive test is the most commonly used test?

34. List the five steps to be followed when using a penetrant test.

35. What properties must metal have before it can be tested with the MT process.

36. Why will some flaws appear larger on the X-ray than they are in the weld?

37. How is the size of a flaw determined using ultrasonic inspection?

38. What is the major limitation of eddy current inspection?

39. What information does a hardness test reveal?

Chapter
21

WELDER CERTIFICATION

OBJECTIVES

After completing this chapter, the student should be able to:

■ explain the difference between qualification and certification.

■ outline the steps required to certify a weld and a welder.

■ make welds that meet a standard.

■ explain what is meant by the terms *qualified welder* and *certified welder*.

KEY TERMS

Entry Level Welder

Welder Performance Qualification

acceptable criteria

transverse face-bend

Welder Certification

certified welder

transverse root-bend

weld test

INTRODUCTION

Welding, in most cases, is one of the few professions that require job applicants to demonstrate their skills even if they are already certified. Some other professions—doctor, lawyer, pilot, for example—do give a written test or require a license initially. But welders are often required to demonstrate their knowledge and their skills before being hired, since welding, unlike most other occupations, requires a high degree of eye-hand coordination.

A method commonly used to test welders' ability is the qualification or certification test. Welders who have passed such a test are referred to as *qualified welders;* if proper written records are kept of the test results, they are referred to as certified welders. Not all welding jobs require that the welder be certified. Some merely require that a basic weld test be passed before applicants are hired.

Welder certification can be divided into two major areas. The first area covers the traditional welder certification that has been used for years. This certification is used to demonstrate welding skills for a specific process on a specific weld, to qualify for a welding assignment or as a requirement for employment.

The American Welding Society has developed the second, newer area of certification. This certification has three levels. The first level is primarily designed for the new welder needing to demonstrate Entry-Level Welder skills. The other levels cover Advanced Welders and Expert Welders. This chapter covers the traditional certification and the AWS QC10 *Specification for Qualification and Certification for Entry Level Welder.*

QUALIFIED AND CERTIFIED WELDERS

Welder qualification and welder certification are often misunderstood. Sometimes it is assumed that a qualified or certified welder can weld anything. Being certified does not mean that a welder can weld everything, nor does it mean that every weld that is made is acceptable. It means that the welder has demonstrated the skills and knowledge necessary to make good welds. To ensure that a welder is consistently making welds that meet the standard, welds are inspected and tested. The more critical the welding, the more critical the inspection and the more extensive the testing of the welds.

All welding processes can be tested for qualification and certification. The testing can range from making spot welds with an electric resistant spot welder to making electron beam welds on aircraft. Being qualified or certified in one area of welding does not mean that a welder can make quality welds in other areas. Most qualifications and certifications are restricted to a single welding process, position(s), metal, and thickness range.

Changes in any one of a number of essential variables can result in the need to recertify. For example:

- *Process:* Welders can be certified in each welding process such as SMAW, GMAW, FCAW, GTAW, EBW, and RSW. Therefore, a new test is required for each process.

- *Material:* The type of metal—such as steel, aluminum, stainless steel, and titanium—being welded will require a change in the certification. Even a change in the alloy within a base metal type can require a change in certification.

- *Thickness:* Each certification is valid on a specific range of thickness of base metal. This range is dependent on the thickness of the metal used in the test. For example, if a 3/8-in. (9.5-mm) plain carbon steel plate is used, then under some codes the welder would be qualified to make welds in plate thickness ranges from 3/16-in. to 3/4-in. (4.7 mm to 19 mm).

- *Filler metal:* Changes in the classification and size of the filler metal can require recertification.

- *Shielding gas:* If the process requires a shielding gas, then changes in gas type or mixture can affect the certification.

- *Position:* In most cases, a weld test taken in the flat position would limit certification to flat and possibly horizontal welding. A test taken in the vertical position, however, would usually allow the welder to work in the flat, horizontal, and vertical positions.

- *Joint design:* Changes in weld type such as groove or fillet welds require a new certification. Additionally, variations in joint geometry, such as groove type, groove angle, and number of passes, can also require retesting.

- *Welding current:* In some cases changing from AC to DC or changes such as to pulsed power and high frequency can affect the certification.

Any welder qualification or certification process must include the specific welding skill level to be demonstrated. The detailed information for a welding test is often given as part of a Welding Procedure Specification (WPS) or similar set of welding specifications or schedules (see Chapter 19). Such standards inform everyone about which skills are required, enabling the welder to prepare for the welding test and to demonstrate welding skills to the company. Varying from the strict limitations in the WPS usually requires that a different test be taken.

Welder Performance Qualification is the demonstration of a welder's ability to produce welds meeting very specific prescribed standards. The form used to document this test is called the *Welding Qualification Test Record.* The detailed written instructions to be followed by the welder are called the *Welder Qualification Procedure.* Welders passing this certification are often referred to as being a *Qualified Welder* or as *qualified.*

Welder Certification is the written verification that a welder has produced welds meeting a prescribed standard of welder performance. A welder holding such a written verification is often referred to as being *Certified* or as a *Certified Welder.*

An *AWS Certified Welder* is one who has complied with all the provisions, requirements, and specifications of the AWS regarding certification. Very specific requirements must be followed by any school or organization before it can offer this certification. Under the AWS program the welder must pass a closed book exam covering specific knowledge areas and a performance test. Written documentation, including the welder's name and social security number and test results, must be sent to the AWS, where such records will be entered into the AWS National Registry for welders. The certification record will expire after one year and will be automatically deleted from the registry.

AWS ENTRY-LEVEL WELDER QUALIFICATION AND WELDER CERTIFICATION

The AWS **Entry Level Welder** qualification and certification program specifies a number of requirements not normally found in the traditional welder qualification and certification process. The additions to the AWS program have broadened the scope of the test. Areas such as practical knowledge have long been an assumed part of most certification programs but have not been a formal part of the process. Most companies have assumed that welders who could produce code quality welds could understand enough of the technical aspects of welding.

Today, however, greater importance is placed on the technical knowledge of the process, code, and other aspects of the complete welding process. This change is due to the greater complexity of many welding processes and an increased responsibility of companies and their welders to ensure the quality and reliability of weldments. It is important not only that the weld be correctly performed but also that the welder know why it must be performed in such a specific manner. This is all intended to increase accuracy and reduce rejection of welds.

Practical Knowledge

A written test must be passed with a minimum grade of 75% on all areas except safety. The safety questions must be answered with a minimum accuracy of 90%. The following subject areas, covered in the given chapters of this text, are included in the test:

- Welding and cutting theory (Chapters 3, 7, 8, 9, 10, 12, and 15).
- Welding and cutting inspection and testing (Chapters 6, 7, 8, 9, and 20).
- Welding and cutting terms and definitions (Glossary).
- Base and filler metal identification (Chapters 23, 24, and 25).
- Common welding process variables (Chapters 3, 10, 12, and 15).
- Electrical fundamentals (Chapters 3, 10, and 15).
- Drawing and welding symbol interpretation (Chapter 18).
- Fabrication principles and practices (Chapter 18).
- Safe practices (Chapter 2).

Refer to the specific chapter(s) that relate to each of the required knowledge areas for the information required in order to pass that area of the test.

Performance Test

Some welders mistakenly believe that passing a welding certification test is a matter of luck. Passing a weld test is both a function of welder skill and knowing the detailed information regarding the acceptance level of the weld. Without both of these elements, passing the test would indeed be just luck.

You will need to practice taking a weld test in order to prepare yourself for passing such a test for a job. As part of this practice you will be required to lay out, cut out, fit up, and weld several different weldments following written and verbal instructions using a variety of processes. It will also be your responsibility to keep the welding area clean and safe.

This text provides the necessary instructions required to perform quality welds, and the AWS material provides the necessary codes and standards. Learning to read and interpret the AWS materials is an important part

of the certification process, since the AWS is the leading authority on the codes. Together, this text and the AWS material provide all of the knowledge and skill standards necessary to become a qualified and certified welder.

Layout. This section requires that you read and interpret simple drawings and sketches including welding symbols. As such, you will fill out a bill of materials that will be required to fabricate the weldment. The material specifications must be given in both standard and S.I. units, requiring that you make the necessary conversions.

Once the bill of material is complete, you must lay out on appropriate metal stock the individual parts that are to be cut out. The parts must be laid out to within a fractional tolerance of ±1/16 in. (1.6 mm) with an angular tolerance of +10° –5°. Be sure to leave an appropriate amount of space between parts for the torch kerf.

Written Procedures. Each weldment drawing includes a written list of notes that must be followed. You must also follow a WPS for each weldment. Specific welding information can be found in this text in the appropriate chapter as it relates to the cutting or welding process. Additional information is available in the AWS EG2.0 publication and in the references listed in **Table 21-1**. The material in this text follows the AWS guidelines but

ANSI/ AWS B2.1.001	Standard Welding Procedure Specification for Shielded Metal Arc Welding of Carbon Steel, (M-1/P-1, Group 1 or 2), 3/16 through 3/4 inch in the As-Welded Condition, with Backing
ANSI/ AWS B2.1.008	Standard Welding Procedure Specification for Gas Tungsten Arc Welding of Carbon Steel (M-1, Group 1), 10 through 18 Gauge, in the As-Welded Condition, with or without Backing
ANSI/ AWS B2.1.009	Standard Welding Procedure Specification for Gas Tungsten Arc Welding of Austenitic Stainless Steel, (M-8/P-8), 10 through 18 Gauge, in the As-Welded Condition, with or without Backing
ANSI/ AWS B2.1.015	Standard Welding Procedure Specification for Gas Tungsten Arc Welding of Aluminum, (M-22 or P-22), 10 through 18 Gauge, in the As-Welded Condition, with or without Backing
ANSI/ AWS B2.1.019	Standard Welding Procedure Specification for CO_2 Shielded Flux Cored Arc Welding of Carbon Steel (M-1/P-1/S-1, Group 1 or 2), 1/8 through 1-1/2 inch Thick, E70T-1 and E71T-1, As-Welded Condition
ANSI/ AWS B2.1.020	Standard Welding Procedure Specification for 75% Argon 25% CO_2 Shielded Flux Cored Arc Welding of Carbon Steel, (M-1/P-1/S-1, Group1 or 2), 1/8 through 1-1/2 inch Thick, E70T-1 and E71T-1, As-Welded Condition

Table 21-1 Standard Welding Procedure Specifications.

TIME CARD

Name _____   Date _____   Job _____

Starting Time	Ending Time	Total Time
Sign		Total

Figure 21-1A Useful forms for keeping track of weldments.

INSPECTION REPORT   JOB: _____

INSPECTION	PASS/FAIL	INSPECTOR'S INITIALS	DATE
Layout			
Cutout			
Assembly			
Welding			
Tack			
Interpass			
Finish			
Overall Rating			
Accuracy			
Appearance			

Welder _____   Date _____

Figure 21-1B (Continued).

does not include all of the AWS material as provided in their publications.

Written Records. Fill out time or job cards, reports, and/or other records as needed, **Figure 21-1.** Written records must be complete, neat, and legible. These records must be turned in with the completed weldment and will be considered as part of the overall evaluation of your skills. Similar records are required by most large welding companies, to determine the productivity of welders and to ensure that each job is charged correctly for time and materials. These records help the company to stay profitable.

Verbal Instructions. In any working shop verbal instructions must be given from time to time. These instructions are as important as written ones. Many welding shops assemble their welders at the beginning of a shift so that special instructions can be given verbally. Instructions may include things such as which stock to use, where scrap is to be deposited, and which welder to use. In some cases critical information such as safety

BILL OF MATERIALS

Name ————————————————————— Date ————————— Job —————————

Part ID	Size Determination	SI Determination

Material Specification —————————————————————————

Figure 21-1C (Continued).

concerns are given verbally. Your safety and the safety of others could depend on your ability to remember and follow verbal instructions.

Safety A score of 90% on safety is acceptable for the knowledge portion of the certification; however, prior to doing any work in the welding shop the student must pass a safety test with 100% accuracy. The test may be repeated as needed until 100% accuracy is obtained. The test must include questions related to protection of the persons in the area and the welding shop in general, including precautionary information, shop and welding area ventilation, fire prevention and protection, and other general aspects of welding safety. Refer to Chapter 2 of *Safety for Welders* by Larry Jeffus and ANSI Z49.1 for specific safety information.

Housekeeping Cleaning the welder's work area is usually the responsibility of the welder. By its nature the welding and cutting process produces large quantities of scrap, including electrode stubs, scrap metal, welding and cutting slag, and grinding dust. It is also necessary to keep welding leads, electrical cords, hoses, guns, torches or electrode holders, hand tools, and power tools up and out of harms way, **Figure 21-2**.

Some housekeeping chores may be provided by other support staff, but the actual cleaning and picking up of the welding area and equipment is up to the welder. As with our industry, the welder or welding stu-

dent must keep the work area clean. Cleanliness is as important to safety as are many other responsibilities. Keeping the area picked up can also improve welding productivity.

Routine repairs of equipment may also fall under housekeeping chores. Most welders are expected to change their own welding hood lenses as needed, and some are expected to make minor repairs and adjustments to the guns, torches, and electrode holders, **Figure 21-3**. Before making any such repairs or adjustments, you must first check with the manufacturer's literature for the equipment and your instructor.

Cutting out Parts Parts will be cut out using the manual oxyfuel gas cutting (OFC), machine oxyfuel gas cutting (OFC—Track Burner), Air Arc Cutting (CAC-A), and manual Plasma Arc Cutting (PAC) processes.

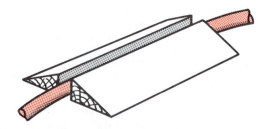

Figure 21-2 To prevent people from tripping if cables must be placed in walkways, lay two blocks of wood beside the cables.

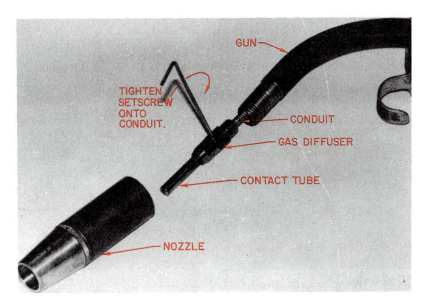

Figure 21-3 Minor repairs of equipment may be the welder's responsibility.

To cut out the various parts, you will need to make straight, angled, and circular cuts. In some cases you will decide which method and process to use. On other cuts the method and process will be specified on the drawing. Inspect all cut surfaces for flaws and defects and repair if necessary, **Figure 21-4** (page 527).

Fit-Up and Assembly of the Parts Putting the parts of a weldment together in preparation of welding requires special skills. The more complex the weldment the more difficult the assembly. Each part must be located and squared to other parts, **Figure 21-5** (page 527). To complicate this process, clamping may be necessary because not all parts may be flat and straight.

The cutting process may distort the metal. Any such distortion or bend that may have been in the metal must be corrected during the fit-up process. Sometimes grinding can be required to make a correct fit-up, **Figure 21-6** (page 527). Other times the parts can be fitted together correctly by using C-clamps or pliers, **Figure 21-7** (page 527). In more difficult cases, tack welds and cleats or dogs must be used to achieve the proper fit-up, **Figure 21-8** (page 527).

Workmanship Standards for Welding Positioning of the weldment must be made so that the welds are being made within 5° of the specified position, **Figure 21-9** (page 528).

All arc strikes must be within the groove. Arc strikes outside the groove will be considered defects.

Tack welds must be small enough so that they do not interfere with the finished weld. They must, however, be large enough to withstand the shrinkage forces from the welds as they are being made. Sometimes it is a good idea to use several small tacks on the same joint to

ensure that the parts are held in place, **Figure 21-10** (page 529).

The weld bead size is important. Beads must be sized in accordance to the WPS or drawing for each specific weld. All weld bead starts and stops must be smooth, **Figure 21-11** (page 529). For bend specimens the starts and stops should be made outside of the areas from which the weld test specimens will be taken.

The weld beads must be cleaned, either with a hand wire brush, a hand chipping hammer, a punch and hammer, or a needle-scaler. All weld cleaning must be performed with the test plate in the welding position. A grinder may not be used to remove weld control problems such as undercut, overlap, or trapped slag.

Because of the size of the test plate and the quantity of weld metal being deposited, the plate may have a tendency to become overheated. If overheating occurs, allow the test plate to air cool, but do not quench it. A good practice is not to waste this time. You can start on another weld test position, fit up the next set of test plates, or find another productive use of your time while the plate air cools.

Weld Inspection Each weld and/or weld pass is to be inspected visually:

- There shall be no cracks or incomplete fusion.

- There shall be no incomplete joint penetration in groove welds except as permitted for partial joint penetration groove welds.

- The Test Supervisor shall examine the weld for acceptable appearance and shall be satisfied that the welder is skilled in the process and procedure specified for the test.

CORRECT CUT

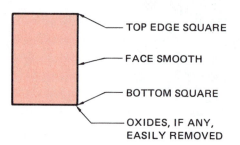

- TOP EDGE SQUARE
- FACE SMOOTH
- BOTTOM SQUARE
- OXIDES, IF ANY, EASILY REMOVED

TRAVEL SPEED TOO SLOW

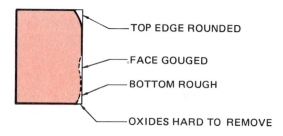

- TOP EDGE ROUNDED
- FACE GOUGED
- BOTTOM ROUGH
- OXIDES HARD TO REMOVE

TRAVEL SPEED TOO FAST

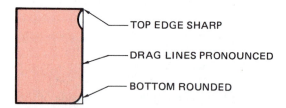

- TOP EDGE SHARP
- DRAG LINES PRONOUNCED
- BOTTOM ROUNDED

SQUARE TOP EDGE

SMOOTH FACE

SQUARE SLAG-FREE BOTTOM EDGE

Figure 21-4 Flame-cut profiles and standards.

PREHEAT FLAMES TOO HIGH ABOVE THE SURFACE

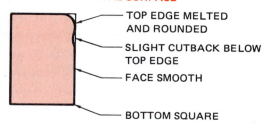

- TOP EDGE MELTED AND ROUNDED
- SLIGHT CUTBACK BELOW TOP EDGE
- FACE SMOOTH
- BOTTOM SQUARE

PREHEAT FLAMES TOO CLOSE TO THE SURFACE

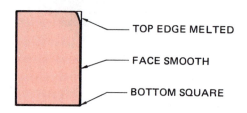

- TOP EDGE MELTED
- FACE SMOOTH
- BOTTOM SQUARE

CUTTING OXYGEN PRESSURE TOO HIGH

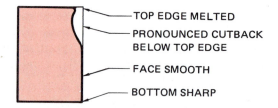

- TOP EDGE MELTED
- PRONOUNCED CUTBACK BELOW TOP EDGE
- FACE SMOOTH
- BOTTOM SHARP

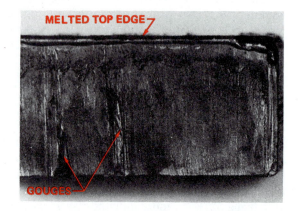

MELTED TOP EDGE

GOUGES

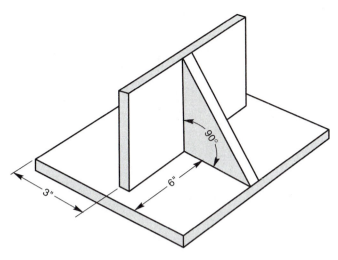

Figure 21-5 Locate and square parts to be welded.

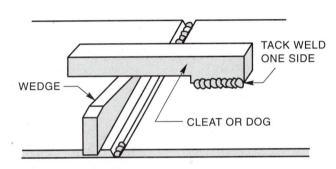

Figure 21-6 A portable grinder can be used to correct cutting or fitting problem.

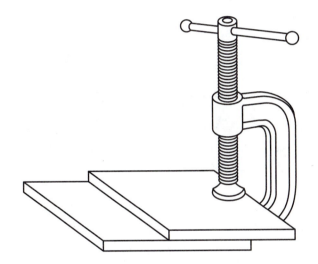

Figure 21-7 C-clamp being used to hold plates for tack welding.

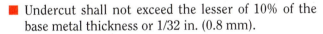

TACK WELD ONE SIDE

WEDGE

CLEAT OR DOG

Figure 21-8A Wedges and cleats can be used on heavier metal to pull the joint into alignment.

- Undercut shall not exceed the lesser of 10% of the base metal thickness or 1/32 in. (0.8 mm).
- Where visual examination is the only criterion for acceptance, all weld passes are subject to visual examination at the direction of the Test Supervisor.
- The frequency of porosity shall not exceed one in each 4 in. (100 mm) of weld length, and the maximum diameter shall not exceed 3/32 in. (2.4 mm).
- Welds shall be free from overlap.*

*American Welding Society *AWS QC10-95 Specification for Qualification and Certification for Entry Level Welders.*

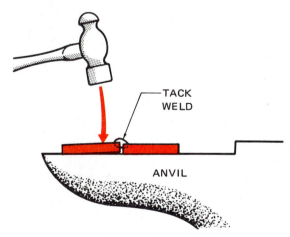

TACK WELD

ANVIL

Figure 21-8B Also the plates can be forced into alignment by striking them with a hammer.

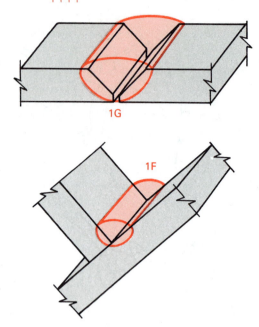

Figure 21-9A Plate flat position.

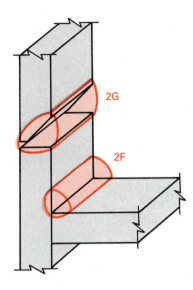

Figure 21-9B Plate horizontal position.

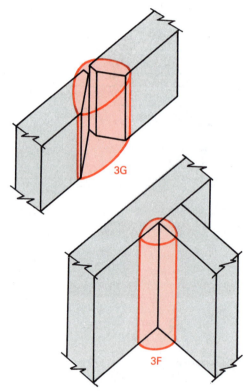

Figure 21-9C Plate vertical position.

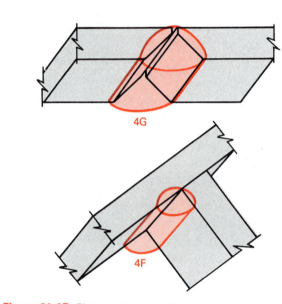

Figure 21-9D Plate overhead position.

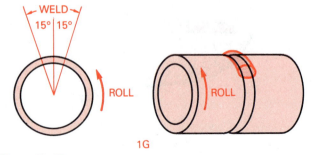

Figure 21-9E Pipe horizontal rolled position.

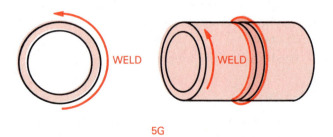

Figure 21-9F Pipe horizontal fixed position.

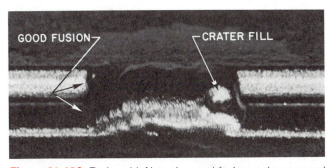

Figure 21-10A Tack weld. Note the good fusion at the start and crater fill at the end.

Figure 21-10B The tack weld should be small and uniform to minimize its effect on the final weld.

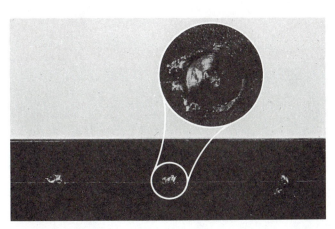

Figure 21-10C Use enough tack welds to keep the joint in alignment during welding. Small tack welds are easier to weld over without adversely affecting the weld.

WELDING SKILL DEVELOPMENT

The *AWS EG2.0-95 Guide for the Training and Qualification of Welding Personnel Entry Level Welder* lists safety, related knowledge, and welding skills that must be mastered as part of a training program. Such welding skills are included in various chapters throughout this text. Table 21-2 (page 530) lists these specific welding skills and cross-references them to the text practices in which they are taught.

Some of the AWS practices require skills that will best be developed by following the skill development listed in this text. For example, it is easier for a new welder to develop vertical up skills by starting welding on a plate that is at a 45° inclined angle. As you develop the skills to control the weld at this angle, slowly increase the angle until you are making welds in the vertical position.

Table 21-3 (page 531) cross-references the list of cutting skills that the AWS states you should develop with their location in the text.

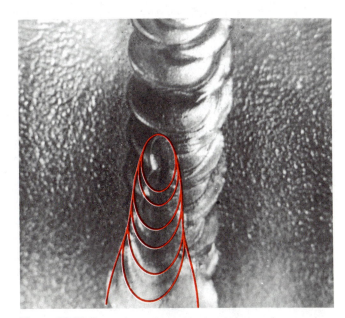

Figure 21-11 Taper down the weld when it is complete or when stopping to reposition. This will make the end smoother and ease restarting the next weld bead if necessary.

WELDING PROCESS	FILLET				GROOVE			
	1F FLAT	2F HORIZONTAL	3F VERTICAL	4F OVERHEAD	1G FLAT	2G HORIZONTAL	3G VERTICAL	4G OVERHEAD
Shielded Metal Arc Welding (SMAW) E6010 or E6011 and E7018 Electrodes on Plain Carbon Steel	P4-7	P4-8	P4-9	P4-13				
Shielded Metal Arc Welding (SMAW) E7018 Electrodes on Plain Carbon Steel					P6-5	P6-5	P6-5	P6-5
Shielded Metal Arc Welding (SMAW) E7018 Electrodes on Plain Carbon Steel to be Bend Tested							P6-6	P6-6
Gas Metal Arc Welding (GMAW-S) Short Circuit Transfer on Plain Carbon Steel using .035 or .045 Diameter E70S-X Electrode with CO_2 or 75% Ar/25% CO_2 Shielding Gas	P11-3	P11-19	P11-10	P11-23	P11-3	P11-19	P11-10	P11-23
Gas Metal Arc Welding (GMAW) Spray Transfer on Plain Carbon Steel using .035 or .045 Diameter E70S-X Electrode with Ar + 2% to 5% O_2 Shielding Gas	P11-33	P11-33			P11-33			
Flux Cored Arc Welding (FCAW) Self-Shielded on Plain Carbon Steel using .035 or .045 Diameter E71T-11 Electrode	P13-10	P13-27	P13-54	P13-62	P13-4	P13-32	P13-52	P13-60
Flux Cored Arc Welding (FCAW-G) Gas-Shielded on Plain Carbon Steel using .035 or .045 Diameter E71T-1 Electrode with CO_2 or 75% Ar/25% CO_2 Shielding Gas	P13-10	P13-27	P13-54	P13-62	P13-4	P13-32	P13-52	P13-60
Gas Tungsten Arc Welding (GTAW) Plain Carbon Steel with ER70S-X Filler Metal using EWTh-2 or EWCe-2 Tungsten Electrodes	P16-9	P16-27	P16-22	P16-33	P16-5	P16-24	P16-16	P16-31
Gas Tungsten Arc Welding (GTAW) Aluminum with ER4043 Filler Metal using EWP or EWZr Tungsten Electrodes	P16-9	P16-27	P16-22		P16-5	P16-24		
Gas Tungsten Arc Welding (GTAW) Stainless Steel with ER3XX Filler Metal using EWTh-2 or EWCe-2 Tungsten Electrodes	P16-9	P16-27			P16-5			

Table 21-2 AWS Process Certification Options for Entry Level Welder and their location in this text.

WELDER QUALIFICATION AND CERTIFICATION TEST INSTRUCTIONS FOR PRACTICES

After you have mastered the welding skills as listed in **Table 21-2** and the cutting skills as listed in **Table 21-3**, you should be ready to start the required assemblies and welding. In this part of the entry-level certification you will need to pass two bend tests and seven workmanship assembly weldments.

CUTTING PROCESS	STRAIGHT CUTTING	BEVELING	SHAPE CUTTING	WELD REMOVAL (WASHING)	METAL REMOVAL (GOUGING)
Oxyfuel Gas Cutting (OFC) Plain Carbon Steel (Manual)	X	X	X	X	
Oxyfuel Gas Cutting (OFC) Plain Carbon Steel [Machine (Track Burner)]	X	X			
Air Carbon Arc Cutting (CAC-A) Plain Carbon Steel (Manual)					X
Plasma Arc Cutting (PAC) Plain Carbon Steel (Manual)			X		
Plasma Arc Cutting (PAC) Stainless Steel (Manual)			X		
Palsma Arc Cutting (PAC) Aluminum (Manual)			X		

Table 21-3 AWS Entry Level Welder Cutting Process.

PRACTICE 21-1

WELDER QUALIFICATION TEST PLATE FOR LIMITED THICKNESS HORIZONTAL 2G POSITION WITH E7018 ELECTRODES

DIMENSIONAL TOLERANCES

Test plates: two (2); each 3/8 in. thick, 3 in. wide, and 7 in. long, both having a 22 1/2° bevel along one edge.
Backing strip: one (1); each either 1/4 in. or 3/8 in. thick, 1 in. wide, and 9 in. long.

WELDING PROCEDURE SPECIFICATION (WPS)
Welding Procedures Specifications No: _Practice 21-1_ Date:_____

TITLE:
Welding _SMAW_ of _plate_ to _plate_ .

SCOPE:
This procedure is applicable for _V-groove plate with a backing strip_
within the range of _3/16 in. (4.7 mm)_ through _3/4 in. (19 mm)_ .
Welding may be performed in the following positions: _2G_ .

BASE METAL:
The base metal shall conform to Carbon Steel M-1 or P-1, Group 1 or 2   .
Backing material specification Carbon Steel M-1 or P-1, Group 1, 2, or 3   .

FILLER METAL:
The filler metal shall conform to AWS specification No. E7018   from AWS specification
 A5.1   . This filler metal falls into F-number F-4   and A-number A-1   .

SHIELDING GAS:
The shielding gas, or gases, shall conform to the following compositions and purity:
 N/A   .

JOINT DESIGN AND TOLERANCES:

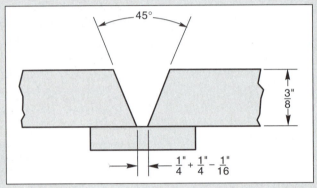

Figure 21-12

PREPARATION OF BASE METAL:
The bevel is to be flame cut on the edge of the plate before the parts are assembled. The
beveled surface must be smooth and free of notches. Any roughness or notches that are
deeper than 1/64 in. (0.4 mm) must be ground smooth.
     All hydrocarbons and other contaminations, such as cutting fluids, grease, oil, and
primers, must be cleaned off all parts and filler metals before welding. This cleaning can be
done with any suitable solvents or detergents. The backing strip, groove face, and inside
and outside plate surface within 1 in. (25 mm) of the joint must be mechanically cleaned of
slag, rust, and mill scale. Cleaning must be done with a wire brush or grinder down to
bright metal.

ELECTRICAL CHARACTERISTICS:
The current shall be Direct Current Electrode Positive (DCEP) . The base metal shall be on
the negative   side of the line.

Welds	Filler Metal Dia.	Current	Amperage Range
Tack	3/32 in. (2.4 mm)	DCEP	70 to 115
Root	1/8 in. (3.2 mm)	DCEP	115 to 165
Filler	5/32 in. (4 mm)	DCEP	150 to 220

PREHEAT:
The parts must be heated to a temperature higher than 50°F (10°C) before any welding is
started.

BACKING GAS:
N/A

SAFETY:

Proper protective clothing and equipment must be used. The area must be free of all hazards that may affect the welder or others in the area. The welding machine, welding leads, work clamp, electrode holder, and other equipment must be in safe working order.

WELDING TECHNIQUE:

Tack weld the plates together with the backing strip. There should be about a 1/4-in. (6-mm) root gap between the plates. Use the E7018 arc welding electrodes to make a root pass to fuse the plates and backing strip together. Clean the slag from the root pass, being sure to remove any trapped slag along the sides of the weld.

Using the E7018 arc welding electrodes, make a series of stringer or weave filler welds, no thicker than 1/4 in. (6.4 mm), in the groove until the joint is filled.

INTERPASS TEMPERATURE:

The plate should not be heated to a temperature higher than 350°F (175°C) during the welding process. After each weld pass is completed, allow it to cool but never to a temperature below 50°F (10°C). The weldment must not be quenched in water.

CLEANING:

The slag must cleaned off between passes. The weld beads may be cleaned by a hand wire brush, a hand chipping hammer, a punch and hammer, or a needle-scaler. All weld cleaning must be performed with the test plate in the welding position. A grinder may not be used to remove weld control problems such as undercut, overlap, or trapped slag.

INSPECTION:

Visually inspect the weld for uniformity and discontinuities. There shall be no cracks, no incomplete fusion, and no overlap. Undercut shall not exceed the lesser of 10% of the base metal thickness or 1/32 in. (0.8 mm). The frequency of porosity shall not exceed one in each 4 in. (100 mm) of weld length, and the maximum diameter shall not exceed 3/32 in. (2.4 mm).

REPAIR:

No repairs of defects are allowed.

SKETCHES:

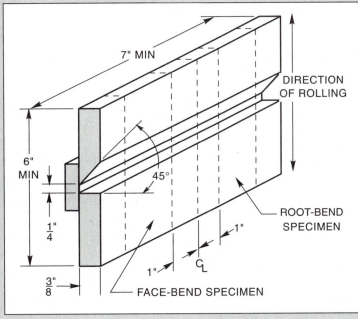

Figure 21-13

BEND TEST:

The weld is to be mechanically tested only after it has passed the visual inspection. Be sure that the test specimens are properly marked to identify the welder, the position, and the process.

Specimen Preparation

1. For 3/8-in. test plates two specimens are to be located in accordance with the requirements of **Figure 21-14.** One is to be prepared for a "Transverse Face-Bend," and the other is to be prepared for a "Transverse Root-Bend."

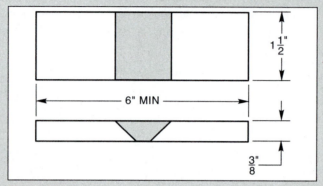

Figure 21-14

Transverse Face-Bend. The weld is perpendicular to the longitudinal axis of the specimen and is bent so that the weld face becomes the tension surface of the specimen. **Transverse face-bend** specimens shall comply with the requirements of **Figure 21-15.**

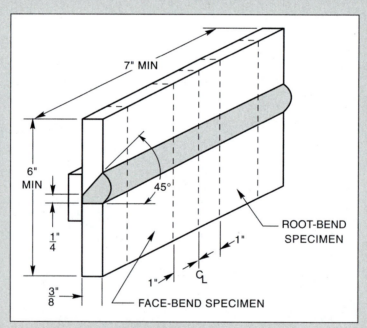

Figure 21-15

Transverse Root-Bend. The weld is perpendicular to the longitudinal axis of the specimen and is bent so that the weld root becomes the tension surface of the specimen. Transverse root-bend specimens shall comply with the requirements of **Figure 21-15.**

Acceptance Criteria for Bend Test:

No single surface indication shall exceed 1/8 in. (3.2 mm) measured in any direction.

The sum of the greatest dimensions of all indications on the surface that exceed 1/32 in. (0.8 mm) but are less than or equal to 1/8 in. (3.2 mm) shall not exceed 3/8 in. (9.6 mm).

PRACTICE 21-2

WELDER QUALIFICATION TEST PLATE FOR LIMITED THICKNESS VERTICAL 3G POSITION WITH E7018 ELECTRODES

DIMENSIONAL TOLERANCES

Test plates: two (2); each 3/8 in. thick, 3 in. wide, and 7 in. long, both having a 22 1/2° bevel along one edge.
Backing strip: one (1); each either 1/4 in. or 3/8 in. thick, 1 in. wide, and 9 in. long.

WELDING PROCEDURE SPECIFICATION (WPS)
Welding Procedures Specifications No: _Practice 21-2_   Date:_____

TITLE:
Welding _SMAW_   of _plate_   to _plate_ .

SCOPE:
This procedure is applicable for _V-groove plate with a backing strip_____
within the range of _3/16 in. (4.7 mm)_ through _3/4 in. (19 mm_____ .
Welding may be performed in the following positions: _3G_____ .

BASE METAL:
The base metal shall conform to _Carbon Steel M-1 or P-1, Group 1 or 2_____ .
Backing material specification _Carbon Steel M-1 or P-1, Group 1, 2 or 3_____

FILLER METAL:
The filler metal shall conform to AWS specification No. _E7018_   from AWS specification
_A5.1_   . This filler metal falls into F-number _F-4_   and A-number _A-1_   .

SHIELDING GAS:
The shielding gas, or gases, shall conform to the following compositions and purity:
_N/A_____ .

JOINT DESIGN AND TOLERANCES:

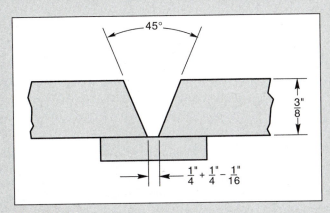

Figure 21-16

PREPARATION OF BASE METAL:
The bevel is to be flame cut on the edge of the plate before the parts are assembled. The beveled surface must be smooth and free of notches. Any roughness or notches that are deeper than 1/64 in. (0.4 mm) must be ground smooth.

All hydrocarbons and other contaminations, such as cutting fluids, grease, oil, and primers, must be cleaned off all parts and filler metals before welding. This cleaning can be done with any suitable solvents or detergents. The backing strip, groove face, and inside and outside plate surface within 1 in. (25 mm) of the joint must be mechanically cleaned of slag, rust, and mill scale. Cleaning must be done with a wire brush or grinder down to bright metal.

ELECTRICAL CHARACTERISTICS:

The current shall be Direct Current Electrode Positive (DCEP) . The base metal shall be on the negative   side of the line.

Welds	Filler Metal Dia.	Current	Amperage Range
Tack	3/32 in. (2.4 mm)	DCEP	70 to 115
Root	1/8 in. (3.2 mm)	DCEP	115 to 165
Filler	5/32 in. (4 mm)	DCEP	150 to 220

PREHEAT:

The parts must be heated to a temperature higher than 50°F (10°C) before any welding is started.

BACKING GAS:

N/A

SAFETY:

Proper protective clothing and equipment must be used. The area must be free of all hazards that can affect the welder or others in the area. The welding machine, welding leads, work clamp, electrode holder, and other equipment must be in safe working order.

WELDING TECHNIQUE:

Tack weld the plates together with the backing strip. There should be about a 1/4-in. (6-mm) root gap between the plates. Use the E7018 arc welding electrodes to make a root pass to fuse the plates and backing strip together. Clean the slag from the root pass, being sure to remove any trapped slag along the sides of the weld.

Using the E7018 arc welding electrodes, make a series of stringer or weave filler welds, no thicker than 1/4 in. (6.4 mm), in the groove until the joint is filled.

INTERPASS TEMPERATURE:

The plate should not be heated to a temperature higher than 350°F (175°C) during the welding process. After each weld pass is completed, allow it to cool but never to a temperature below 50°F (10°C). The weldment must not be quenched in water.

CLEANING:

The slag must cleaned off between passes. The weld beads may be cleaned by a hand wire brush, a hand chipping, a punch and hammer, or a needle-scaler. All weld cleaning must be performed with the test plate in the welding position. A grinder may not be used to remove weld control problems such as undercut, overlap, or trapped slag.

INSPECTION:

Visually inspect the weld for uniformity and discontinuities. There shall be no cracks, no incomplete fusion, and no overlap. Undercut shall not exceed the lesser of 10% of the base metal thickness or 1/32 in. (0.8 mm). The frequency of porosity shall not exceed one in each 4 in. (100 mm) of weld length, and the maximum diameter shall not exceed 3/32 in. (2.4 mm).

REPAIR:
No repairs of defects are allowed.

SKETCHES:

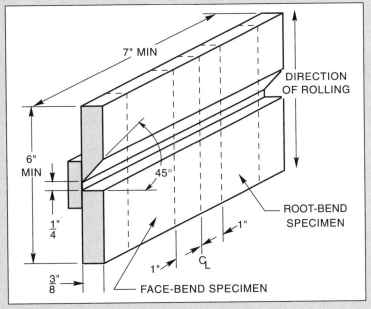

Figure 21-17

BEND TEST:
The weld is to be mechanically tested only after it has passed the visual inspection. Be sure that the test specimens are properly marked to identify the welder, the position, and the process.

SPECIMEN PREPARATION
For 3/8-in. test plates two specimens are to be located in accordance with the requirements of Figure 21-18. One is to be prepared for a "Transverse Face-Bend," and the other is to be prepared for a "Transverse Root-Bend."

■ *Transverse Face-Bend.* The weld is perpendicular to the longitudinal axis of the specimen and is bent so that the weld face becomes the tension surface of the specimen. Transverse face-bend specimens shall comply with the requirements of Figure 21-19.

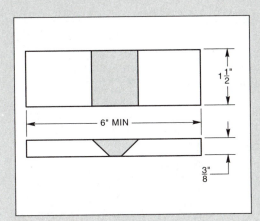

Figure 21-18

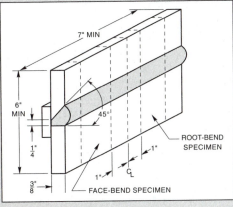

Figure 21-19

■ *Transverse Root-Bend.* The weld is perpendicular to the longitudinal axis of the specimen and is bent so that the weld root becomes the tension surface of the specimen. Transverse face-bend specimens shall comply with the requirements of Figure 21-19.

ACCEPTANCE CRITERIA FOR BEND TEST:
No single surface indication shall exceed 1/8 in. (3.2 mm) measured in any direction.

The sum of the greatest dimensions of all indications on the surface that exceed 1/32 in. (0.8 mm) but are less than or equal to 1/8 in. (3.2 mm) shall not exceed 3/8 in. (9.6 mm).

LAYOUT, ASSEMBLY, AND FABRICATION OF WELDMENTS

Practices 21-3 through 21-9 require that you lay out the parts, assemble the parts, and weld them to form a fabrication. The practices have specific sets of specifications regarding each step in the process. The purpose of these practices is to challenge the student with realistic welding projects like those they will be asked to do on a welding job. These practices tie together all of the skills that have been learned during your training.

You will have to read a set of drawings in order to know how to lay out the parts. Following the instructions, you will cut out the parts with the appropriate process. The parts will then be assembled according to the drawings and written specifications. Finally, the assembly will be welded. The welding must be done in place, which means some of it will be out-of-position.

After post-weld cleanup, the weldment will be visually inspected for acceptability. Each weld must meet the following criteria: There shall be no cracks, no incomplete fusion, and no overlap. Undercut shall not exceed the lesser of 10% of the base metal thickness or 1/32 in. (0.8 mm). The frequency of porosity shall not exceed one in each 4 in. (100 mm) of weld length, and the maximum diameter shall not exceed 3/32 in. (2.4 mm).

PRACTICE 21-3

GAS METAL ARC WELDING–SHORT CIRCUIT METAL TRANSFER (GMAW-S)

WELDING PROCEDURE SPECIFICATION (WPS)
Welding Procedures Specifications No: _Practice 21-3_ Date:_____

TITLE:
Welding _GMAW-S_ of _plate_ to _plate_ .

SCOPE:
This procedure is applicable for _V-groove, bevel, and fillet welds_ .
within the range of _1/8 in. (3.2 mm)_ through _1 1/2 in. (38 mm)_ .
Welding may be performed in the following positions: _All_ .

BASE METAL:
The base metal shall conform to _Carbon Steel M-1, P-1 and S-1 Group 1 or 2_ .
Backing material specification: _None_ .

FILLER METAL:
The filler metal shall conform to AWS specification No. _E70S-3_ from AWS specification _A5.18_ . This filler metal falls into F-number _F-6_ and A-number _A-1_ .

SHIELDING GAS:
The shielding gas, or gases, shall conform to the following compositions and purity:
<u>CO_2 at 30 to 50 CFH or 75% Ar/CO_2 25% at 30 to 50 CFH</u>   .

JOINT DESIGN AND TOLERANCES:

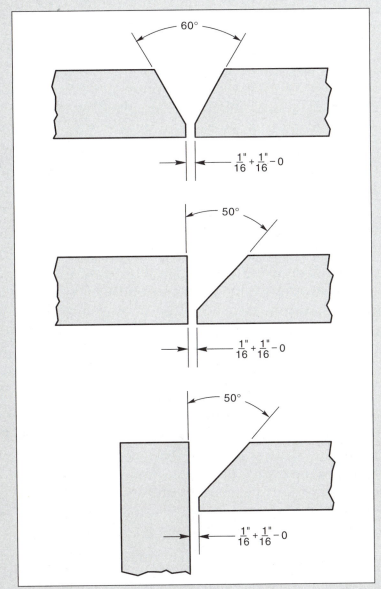

Figure 21-20

PREPARATION OF BASE METAL:
The bevels are to be flame cut on the edges of the plate before the parts are assembled. The beveled surface must be smooth and free of notches. Any roughness or notches deeper than 1/64 in. (0.4 mm) must be ground smooth.

All hydrocarbons and other contaminations, such as cutting fluids, grease, oil, and primers, must be cleaned off all parts and filler metals before welding. This cleaning can be done with any suitable solvents or detergents. The groove face and inside and outside plate surface within 1 in. (25 mm) of the joint must be mechanically cleaned of slag, rust, and mill scale. Cleaning must be done with a wire brush or grinder down to bright metal.

ELECTRICAL CHARACTERISTICS:
The current shall be <u>Direct Current Electrode Positive (DCEP)</u> . The base metal shall be on the <u>negative</u> side of the line.

Electrode		Welding Power			Shielding Gas		Base Metal	
Type	Size	Amps	Wire Feed Speed IPM (cm/min)	Volts	Type	Flow	Type	Thickness
E70S-3	0.035 in. (0.9 mm)	90 to 120	180 to 300 (457 to 762)	15 to 19	CO_2 or 75%Ar/$CO_2$25%	30 to 50	Low-carbon steel	1/4 in. to 1/2 in. (6 mm to 13 mm)
E70S-3	0.045 in. (1.2 mm)	130 to 200	125 to 200 (318 to 508)	17 to 20	CO_2 or 75%Ar/$CO_2$25%	30 to 50	Low-carbon steel	1/4 in. to 1/2 in. (6 mm to 13 mm)

PREHEAT:

The parts must be heated to a temperature higher than 50°F (10°C) before any welding is started.

BACKING GAS:

N/A

SAFETY:

Proper protective clothing and equipment must be used. The area must be free of all hazards that may affect the welder or others in the area. The welding machine, welding leads, work clamp, electrode holder, and other equipment must be in safe working order.

WELDING TECHNIQUE:

Using a 1/2-in. (13-mm) or larger gas nozzle for all welding, first tack weld the plates together according to the drawing. There should be about a 1/16-in. (1.6-mm) root gap between the plates with V-grooved or beveled edges. Use the E70S-3 arc welding electrodes to make a root pass to fuse the plates together. Clean any silicon slag from the root pass, being sure to remove any trapped silicon slag along the sides of the weld.

Using the E70S-3 arc welding electrodes, make a series of stringer or weave filler welds, no thicker than 1/4 in. (6.4 mm), in the groove until the joint is filled. The 1/4- in. (6.4-mm) fillet welds are to be made with one pass.

INTERPASS TEMPERATURE:

The plate should not be heated to a temperature higher than 350°F (175°C) during the welding process. After each weld pass is completed, allow it to cool but never to a temperature below 50°F (10°C). The weldment must not be quenched in water.

CLEANING:

Any slag must be cleaned off between passes. The weld beads may be cleaned by a hand wire brush, a hand chipping, a punch and hammer, or a needle-scaler. All weld cleaning must be performed with the test plate in the welding position. A grinder may not be used to remove weld control problems such as undercut, overlap, or trapped slag.

INSPECTION:

Visually inspect the weld for uniformity and discontinuities. There shall be no cracks, no incomplete fusion, and no overlap. Undercut shall not exceed the lesser of 10% of the base metal thickness or 1/32 in. (0.8 mm). The frequency of porosity shall not exceed one in each 4 in. (100 mm) of weld length, and the maximum diameter shall not exceed 3/32 in. (2.4 mm).

REPAIR:

No repairs of defects are allowed.

SKETCHES: (Figure 21-21)

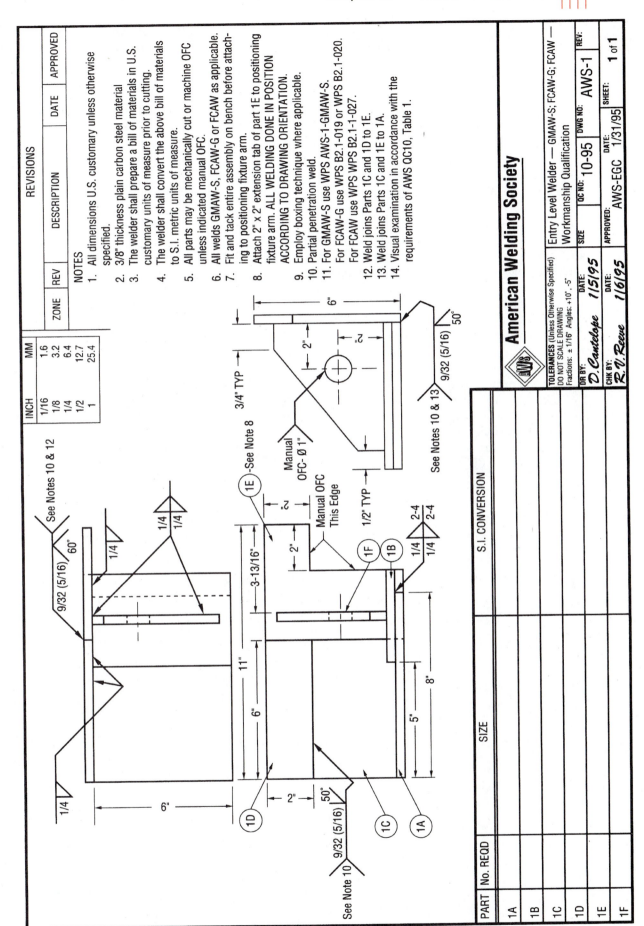

Figure 21-21 GMAW-S Workmanship Qualification Test. (Courtesy of the American Welding Society.)

PRACTICE 21-4

FLUX CORED ARC WELDING–GAS SHIELDED (FCAW-G)

WELDING PROCEDURE SPECIFICATION (WPS)

Welding Procedures Specifications No: <u>Practice 21-4</u>   Date: _____

TITLE:

Welding <u>FCAW</u>   of <u>plate</u>   to <u>plate</u>   .

SCOPE:

This procedure is applicable for <u>V-groove, bevel, and fillet welds</u>
within the range of <u>1/8 in. (3.2 mm)</u>   through <u>1 1/2 in. (38 mm)</u>   .
Welding may be performed in the following positions: <u>All</u>   .

BASE METAL:

The base metal shall conform to <u>Carbon Steel M-1, P-1, and S-1, Group 1 or 2</u>   .
Backing material specification: <u>None</u>   .

FILLER METAL:

The filler metal shall conform to AWS specification No. <u>E71T-1</u>   from AWS specification
<u>A5.20</u>   . This filler metal falls into F-number <u>F-6</u>   and A-number <u>A-1</u>   .

SHIELDING GAS:

The shielding gas, or gases, shall conform to the following compositions and purity:
<u>CO_2 at 30 to 50 CFH or 75% Ar/CO_2 25% at 30 to 50 CFH</u>   .

JOINT DESIGN AND TOLERANCES:

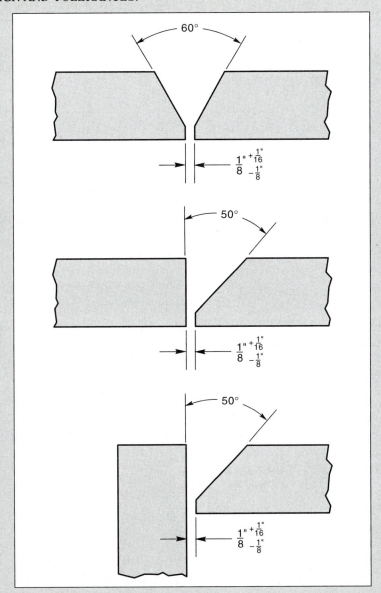

Figure 21-22

PREPARATION OF BASE METAL:

The bevels are to be flame cut on the edges of the plate before the parts are assembled. The beveled surface must be smooth and free of notches. Any roughness or notches deeper than 1/64 in. (0.4 mm) must be ground smooth.

All hydrocarbons and other contaminations, such as cutting fluids, grease, oil, and primers, must be cleaned off all parts and filler metals before welding. This cleaning can be done with any suitable solvents or detergents. The groove face and inside and outside plate surface within 1 in. (25 mm) of the joint must be mechanically cleaned of slag, rust, and mill scale. Cleaning must be done with a wire brush or grinder down to bright metal.

ELECTRICAL CHARACTERISTICS:

The current shall be <u>Direct Current Electrode Positive (DCEP)</u> . The base metal shall be on the <u>negative</u> side of the line.

Electrode		Welding Power			Shielding Gas		Base Metal	
Type	Size	Amps	Wire Feed Speed IPM (cm/min)	Volts	Type	Flow	Type	Thickness
E71T-1	0.035 in (0.9 mm)	130 to 150	288 to 380 (732 to 975)	22 to 25	CO_2 or 75%Ar/$CO_2$25%	30 to 50	Low-carbon steel	1/4 in. to 1/2 in. (6 mm to 13 mm)
E71T-1	0.045 in. (1.2 mm)	150 to 210	200 to 300 (508 to 762)	28 to 29	CO_2 or 75%Ar/$CO_2$25%	30 to 50	Low-carbon steel	1/4 in. to 1/2 in. (6 mm to 13 mm)

PREHEAT:
The parts must be heated to a temperature higher than 50°F (10°C) before any welding is started.

BACKING GAS:
N/A

SAFETY:
Proper protective clothing and equipment must be used. The area must be free of all hazards that may affect the welder or others in the area. The welding machine, welding leads, work clamp, electrode holder, and other equipment must be in safe working order.

WELDING TECHNIQUE:
Using a 1/2-in. (13-mm) or larger gas nozzle and a contact tube approximately 3/4 in. (19 mm) to work distance for all welding, first tack weld the plates together according to the drawing. There should be about a 1/8-in. (3.2-mm) root gap between the plates with V-grooved or beveled edges. Use the E71T-1 arc welding electrodes to make a root pass to fuse the plates together. Clean the slag from the root pass, being sure to remove any trapped slag along the sides of the weld.

Using the E71T-1 arc welding electrodes, make a series of stringer or weave filler welds, no thicker than 1/4 in. (6.4 mm), in the groove until the joint is filled. The 1/4-in. (6.4-mm) fillet welds are to be made with one pass.

INTERPASS TEMPERATURE:
The plate should not be heated to a temperature higher than 350°F (175°C) during the welding process. After each weld pass is completed, allow it to cool but never to a temperature below 50°F (10°C). The weldment must not be quenched in water.

CLEANING:
The slag must be cleaned off between passes. The weld beads may be cleaned by a hand wire brush, a hand chipping, a punch and hammer, or a needle-scaler. All weld cleaning must be performed with the test plate in the welding position. A grinder may not be used to remove weld control problems such as undercut, overlap, or trapped slag.

INSPECTION:
Visually inspect the weld for uniformity and discontinuities. There shall be no cracks, no incomplete fusion, and no overlap. Undercut shall not exceed the lesser of 10% of the base metal thickness or 1/32 in. (0.8 mm). The frequency of porosity shall not exceed one in each 4 in. (100 mm) of weld length, and the maximum diameter shall not exceed 3/32 in. (2.4 mm).

REPAIR:
No repairs of defects are allowed.

SKETCHES: (Figure 21-23)

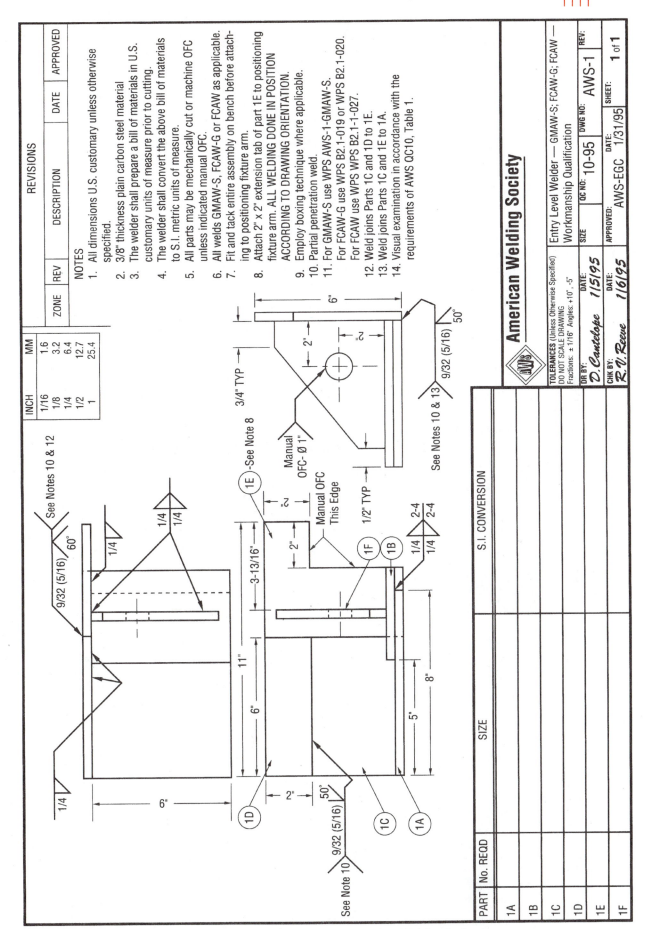

Figure 21-23 FCAW-G Workmanship Qualification Test. (Courtesy of the American Welding Society.)

PRACTICE 21-5

FLUX CORED ARC WELDING SELF-SHIELDED (FCAW)

WELDING PROCEDURE SPECIFICATION (WPS)

Welding Procedures Specifications No: <u>Practice 21-5</u>   Date: _____

TITLE:

Welding <u>FCAW</u> of <u>plate</u> to <u>plate</u> .

SCOPE:

This procedure is applicable for <u>V-groove, bevel, and fillet welds</u>
within the range of <u>1/8 in. (3.2 mm)</u> through <u>1 1/2 in. (38 mm)</u> .
Welding may be performed in the following positions: <u>All</u> .

BASE METAL:

The base metal shall conform to <u>Carbon Steel M-1, P-1, and S-1, Group 1 or 2</u> .
Backing material specification: <u>None</u> .

FILLER METAL:

The filler metal shall conform to AWS specification No. <u>E71T-11</u> from AWS specification
<u>A5.20</u> . This filler metal falls into F-number <u>F-6</u> and A-number <u>A-1</u> .

SHIELDING GAS:

The shielding gas, or gases, shall conform to the following compositions and purity:
<u>None</u> .

JOINT DESIGN AND TOLERANCES:

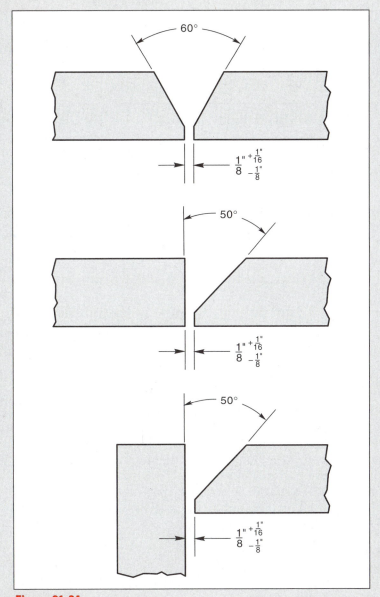

Figure 21-24

PREPARATION OF BASE METAL:
The bevels are to be flame cut on the edges of the plate before the parts are assembled. The beveled surface must be smooth and free of notches. Any roughness or notches deeper than 1/64 in. (0.4 mm) must be ground smooth.

All hydrocarbons and other contaminations, such as cutting fluids, grease, oil, and primers, must be cleaned off all parts and filler metals before welding. This cleaning can be done with any suitable solvents or detergents. The groove face and inside and outside plate surface within 1 in. (25 mm) of the joint must be mechanically cleaned of slag, rust, and mill scale. Cleaning must be done with a wire brush or grinder down to bright metal.

ELECTRICAL CHARACTERISTICS:
The current shall be _Direct Current Electrode Positive (DCEP)_ . The base metal shall be on the _negative_ side of the line.

Electrode		Welding Power			Shielding Gas		Base Metal	
Type	Size	Amps	Wire Feed Speed IPM (cm/min)	Volts	Type	Flow	Type	Thickness
E71T-1	0.035 in. (0.9 mm)	130 to 150	288 to 380 (732 to 975)	22 to 25	CO_2 or 75%Ar/$CO_2$25%	30 to 50	Low-carbon steel	1/4 in. to 1/2 in. (6 mm to 13 mm)
E71T-1	0.045 in. (1.2 mm)	150 to 210	200 to 300 (508 to 762)	28 to 29	CO_2 or 75%Ar/$CO_2$25%	30 to 50	Low-carbon steel	1/4 in. to 1/2 in. (6 mm to 13 mm)

PREHEAT:
The parts must be heated to a temperature higher than 50°F (10°C) before any welding is started.

BACKING GAS:
N/A

SAFETY:
Proper protective clothing and equipment must be used. The area must be free of all hazards that may affect the welder or others in the area. The welding machine, welding leads, work clamp, electrode holder, and other equipment must be in safe working order.

WELDING TECHNIQUE:
Using a 1/2-in. (13-mm) or larger gas nozzle and a contact tube approximately 3/4 in. (19 mm) to work distance for all welding, first tack weld the plates together according to the drawing. There should be about a 1/8-in. (3.2-mm) root gap between the plates with V-grooved or beveled edges. Use the E71T-1 arc welding electrodes to make a root pass to fuse the plates together. Clean the slag from the root pass, being sure to remove any trapped slag along the sides of the weld.

Using the E71T-1 arc welding electrodes, make a series of stringer or weave filler welds, no thicker than 1/4 in. (6.4 mm), in the groove until the joint is filled. The 1/4-in. (6.4-mm) fillet welds are to be made with one pass.

INTERPASS TEMPERATURE:
The plate should not be heated to a temperature higher than 350°F (175°C) during the welding process. After each weld pass is completed, allow it to cool but never to a temperature below 50°F (10°C). The weldment must not be quenched in water.

CLEANING:
The slag must cleaned off between passes. The weld beads may be cleaned by a hand wire brush, a hand chipping, a punch and hammer, or a needle-scaler. All weld cleaning must be performed with the test plate in the welding position. A grinder may not be used to remove weld control problems such as undercut, overlap, or trapped slag.

INSPECTION:
Visually inspect the weld for uniformity and discontinuities. There shall be no cracks, no incomplete fusion, and no overlap. Undercut shall not exceed the lesser of 10% of the base metal thickness or 1/32 in. (0.8 mm). The frequency of porosity shall not exceed one in each 4 in. (100 mm) of weld length, and the maximum diameter shall not exceed 3/32 in. (2.4 mm).

REPAIR:
No repairs of defects are allowed.

SKETCHES: (Figure 21-25)

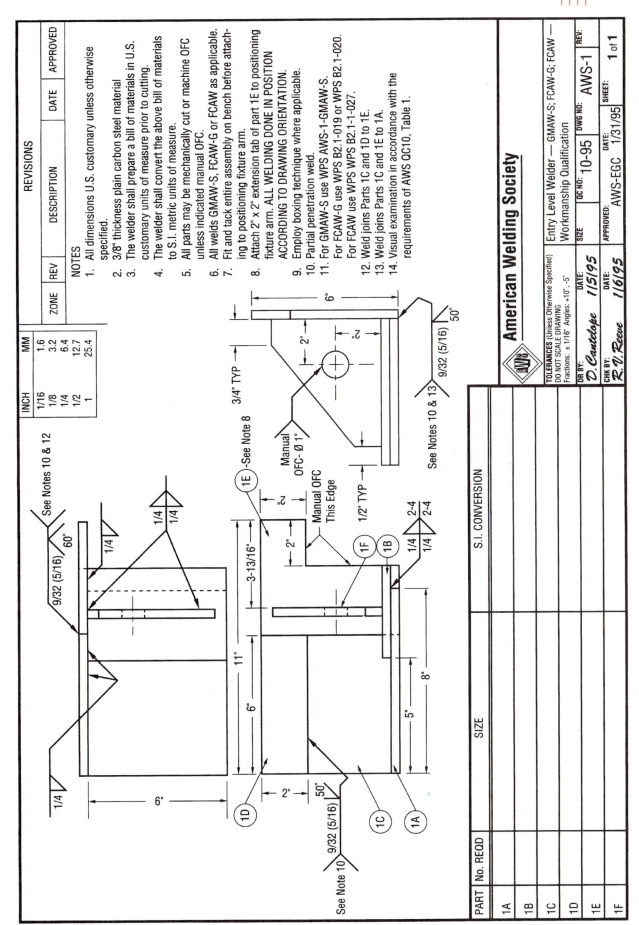

Figure 21-25 FCAW-G Workmanship Qualification Test. (Courtesy of the American Welding Society.)

PRACTICE 21-6

GAS METAL ARC WELDING SPRAY TRANSFER (GMAW)

WELDING PROCEDURE SPECIFICATION (WPS)

Welding Procedures Specifications No: __Practice 21-6__ Date:_____

TITLE:

Welding _GMAW_ of _plate_ to _plate_ .

SCOPE:

This procedure is applicable for _V-groove and fillet welds_____

within the range of _1/8 in. (3.2 mm)_ through _1 1/2 in. (38 mm)_ .

Welding may be performed in the following positions: _1G and 2F_____ .

BASE METAL:

The base metal shall conform to _Carbon Steel M-1, P-1, and S-1, Group 1 or 2_ .

Backing material specification: _None_____ .

FILLER METAL:

The filler metal shall conform to AWS specification No. _E70S-3_ from AWS specification

_A5.18_ . This filler metal falls into F-number _F-6_ and A-number _A-1_ .

SHIELDING GAS:

The shielding gas, or gases, shall conform to the following compositions and purity:

_CO_2 at 30 to 50 CFH or 75% Ar/CO_2 25% at 30 to 50 CFH_____ .

JOINT DESIGN AND TOLERANCES:

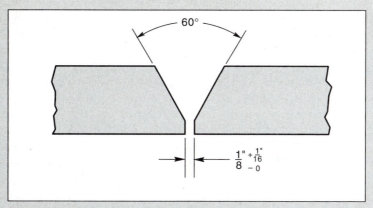

Figure 21-26

PREPARATION OF BASE METAL:
The bevels are to be flame cut on the edges of the plate before the parts are assembled. The beveled surface must be smooth and free of notches. Any roughness or notches deeper than 1/64 in. (0.4 mm) must be ground smooth.

All hydrocarbons and other contaminations, such as cutting fluids, grease, oil, and primers, must be cleaned off all parts and filler metals before welding. This cleaning can be done with any suitable solvents or detergents. The groove face and inside and outside plate surface within 1 in. (25 mm) of the joint must be mechanically cleaned of slag, rust, and mill scale. Cleaning must be done with a wire brush or grinder down to bright metal.

ELECTRICAL CHARACTERISTICS:
The current shall be Direct Current Electrode Positive (DCEP) . The base metal shall be on the negative side of the line.

Electrode		Welding Power			Shielding Gas		Base Metal	
Type	Size	Amps	Wire Feed Speed IPM (cm/min)	Volts	Type	Flow	Type	Thickness
E70S-3	0.035 in (0.9 mm)	180 to 230	400 to 550 (1016 to 1397)	25 to 27	Ar plus 2% to 5% O_2	30 to 50	Low-carbon steel	1/4 in. to 1/2 in. (6 mm to 13 mm)
E70S-3	0.045 in. (1.2 mm)	260 to 340	300 to 500 (762 to 1270)	25 to 30	Ar plus 2% to 5% O_2	30 to 50	Low-carbon steel	1/4 in. to 1/2 in. (6 mm to 13 mm)

PREHEAT:
The parts must be heated to a temperature higher than 50°F (10°C) before any welding is started.

BACKING GAS:
N/A

SAFETY:
Proper protective clothing and equipment must be used. The area must be free of all hazards that may affect the welder or others in the area. The welding machine, welding leads, work clamp, electrode holder, and other equipment must be in safe working order.

WELDING TECHNIQUE:
Using a 3/4-in. (19-mm) or larger gas nozzle for all welding, first tack weld the plates together according to the drawing. There should be about a 1/16-in. (1.6-mm) root gap between the plates with V-grooved or beveled edges. Use the E70S-3 arc welding electrodes to make a root pass to fuse the plates together. Clean any silicon slag from the root pass, being sure to remove any trapped silicon slag along the sides of the weld.

Using the E70S-3 arc welding electrodes, make a series of stringer or weave filler welds, no thicker than 1/4 in. (6.4 mm), in the groove until the joint is filled. The 1/4-in. (6.4-mm) fillet welds are to be made with one pass.

INTERPASS TEMPERATURE:
The plate should not be heated to a temperature higher than 350°F (175°C) during the welding process. After each weld pass is completed, allow it to cool but never to a temperature below 50°F (10°C). The weldment must not be quenched in water.

CLEANING:
Any slag must be cleaned off between passes. The weld beads may be cleaned by a hand wire brush, a hand chipping, a punch and hammer, or a needle-scaler. All weld cleaning must be performed with the test plate in the welding position. A grinder may not be used to remove weld control problems such as undercut, overlap, or trapped slag.

INSPECTION:
Visually inspect the weld for uniformity and discontinuities. There shall be no cracks, no incomplete fusion, and no overlap. Undercut shall not exceed the lesser of 10% of the base metal thickness or 1/32 in. (0.8 mm). The frequency of porosity shall not exceed one in each 4 in. (100 mm) of weld length, and the maximum diameter shall not exceed 3/32 in. (2.4 mm).

REPAIR:
No repairs of defects are allowed.

SKETCHES: (Figure 21-27)

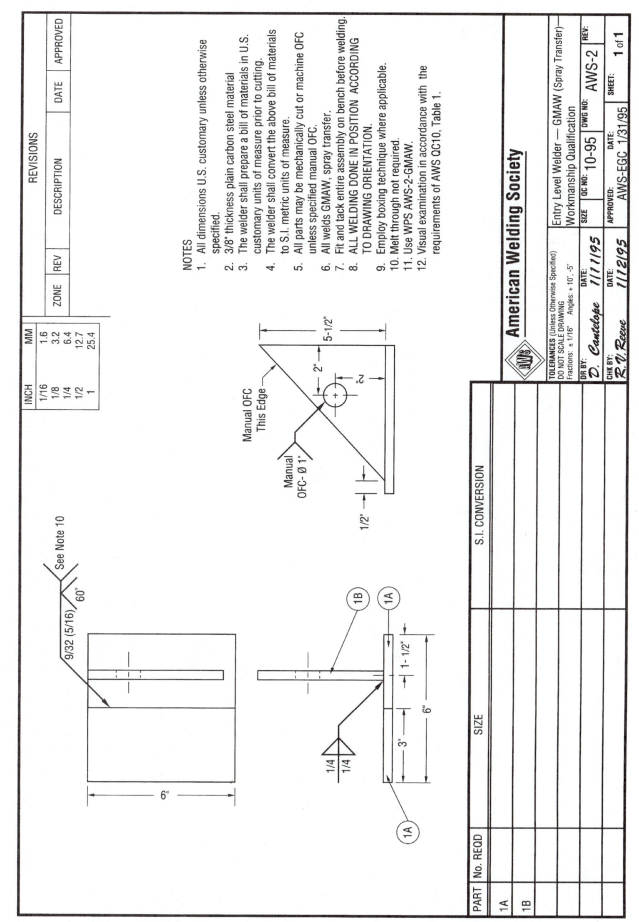

Figure 21-27 GMAW (Spray Transfer) Workmanship Qualification Test. (Courtesy of the American Welding Society.)

PRACTICE 21-7

GAS TUNGSTEN ARC WELDING ON PLAIN CARBON STEEL (GTAW)

WELDING PROCEDURE SPECIFICATION (WPS)
Welding Procedures Specifications No: _Practice 21-7_   Date:_____

TITLE:
Welding _GTAW_ of _sheet_ to _sheet_ .

SCOPE:
This procedure is applicable for _square groove and fillet welds_____
within the range of _18 gauge_____ through _10 gauge_____ .
Welding may be performed in the following positions: _1G and 2F_____ .

BASE METAL:
The base metal shall conform to _Carbon Steel M-1, Group 1_____ .
Backing material specification: _None_____ .

FILLER METAL:
The filler metal shall conform to AWS specification No. _E70S-3_ from AWS specification
_A5.18_ . This filler metal falls into F-number _F-6_ and A-number _A-1_ .

ELECTRODE:
The tungsten electrode shall conform to AWS specification No. _EWTh-2_ from AWS specification _A5.12_ . The tungsten diameter shall be _1/8 in. (3.2 mm) maximum_ .
The tungsten end shape shall be _tapered at 2 to 3 times its length to its diameter_ .

SHIELDING GAS:
The shielding gas, or gases, shall conform to the following compositions and purity:
_Welding Grade Argon_____ .

JOINT DESIGN AND TOLERANCES:

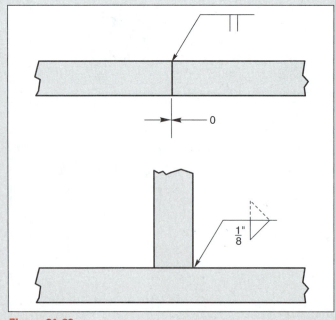

Figure 21-28

PREPARATION OF BASE METAL:

All hydrocarbons and other contaminations, such as cutting fluids, grease, oil, and primers, must be cleaned off all parts and filler metals before welding. This cleaning can be done with any suitable solvents or detergents. The joint face, inside and outside plate surface within 1 in. (25 mm) of the joint, must be mechanically cleaned of slag, rust, and mill scale. Cleaning must be done with a wire brush or grinder down to bright metal.

ELECTRICAL CHARACTERISTICS:

The current shall be _Direct Current Electrode Negative (DCEN)_ . The base metal shall be on the _positive_ side of the line.

Metal Specifications		Gas Flow			Nozzle Size in. (mm)	Amperage Min. Max.
		Rates CFM (L/min)	Purging Times			
Thickness	Dia. of E70S-3*		Prepurging	Postpurging		
18 ga.	1/16 in. (1.6 mm)	15 to 20 (7 to 9)	10 to 15 sec.	10 to 25 sec.	1/4 to 3/8 (6 to 10)	45 to 65
17 ga.	1/16 in. (1.6 mm)	15 to 20 (7 to 9)	10 to 15 sec.	10 to 25 sec.	1/4 to 3/8 (6 to 10)	45 to 70
16 ga.	1/16 in. (1.6 mm)	15 to 20 (7 to 9)	10 to 15 sec.	10 to 25 sec.	1/4 to 3/8 (6 to 10)	50 to 75
15 ga.	1/16 in. (1.6 mm)	15 to 20 (7 to 9)	10 to 15 sec.	10 to 25 sec.	1/4 to 3/8 (6 to 10)	55 to 80
14 ga.	3/32 in. (2.4 mm)	20 to 25 (9 to 12)	10 to 20 sec.	10 to 30 sec.	3/8 to 5/8 (10 to 16)	60 to 90
13 ga.	3/32 in. (2.4 mm)	20 to 25 (9 to 12)	10 to 20 sec.	10 to 30 sec.	3/8 to 5/8 (10 to 16)	60 to 100
12 ga.	3/32 in. (2.4 mm)	20 to 25 (9 to 12)	10 to 20 sec.	10 to 30 sec.	3/8 to 5/8 (10 to 16)	60 to 110
11 ga.	3/32 in. (2.4 mm)	20 to 25 (9 to 12)	10 to 20 sec.	10 to 30 sec.	3/8 to 5/8 (10 to 16)	65 to 120
10 ga.	3/32 in. (2.4 mm)	20 to 25 (9 to 12)	10 to 20 sec.	10 to 30 sec.	3/8 to 5/8 (10 to 16)	70 to 130

*Other E70S-X filler metal may be used.

PREHEAT:

The parts must be heated to a temperature higher than 50°F (10°C) before any welding is started.

BACKING GAS:

N/A

SAFETY:

Proper protective clothing and equipment must be used. The area must be free of all hazards that may affect the welder or others in the area. The welding machine, welding leads, work clamp, electrode holder, and other equipment must be in safe working order.

WELDING TECHNIQUE:

   TACK WELDS: With the parts securely clamped in place with the correct root gap, the tack welds are to be performed. Holding the electrode so that it is very close to the root face but not touching, slowly increase the current until the arc starts and a molten weld pool is formed. Add filler metal as required to maintain a slight convex weld face and a flat or slightly concave root face. When it is time to end the tack weld, lower the current slowly so that the molten weld pool can be tapered down in size. When all tack welds are complete, allow the parts to cool as needed before assembling the remaining parts. Repeat the tack welding procedure until the entire part is assembled.

SQUARE GROOVE AND FILLET WELDS: Holding the electrode so that it is very close to the metal surface but not touching, slowly increase the current until the arc starts and a molten weld pool is formed. As the weld progresses, add filler metal as required to maintain a flat or slightly convex weld face. If it is necessary to stop the weld or to reposition yourself or if the weld is completed, the current must be lowered slowly so that the molten weld pool can be tapered down in size.

INTERPASS TEMPERATURE:
The plate should not be heated to a temperature higher than 120°F (49°C) during the welding process. After each weld pass is completed, allow it to cool but never to a temperature below 50°F (10°C). The weldment must not be quenched in water.

CLEANING:
Recleaning may be required if the parts or filler metal becomes contaminated or reoxides to a degree that the weld quality will be affected. Reclean using the same procedure used for the original metal preparation. Any slag must be cleaned off between passes.

INSPECTION:
Visually inspect the weld for uniformity and discontinuities. There shall be no cracks, no incomplete fusion, and no overlap. Undercut shall not exceed the lesser of 10% of the base metal thickness or 1/32 in. (0.8 mm). The frequency of porosity shall not exceed one in each 4 in. (100 mm) of weld length, and the maximum diameter shall not exceed 3/32 in. (2.4 mm).

REPAIR:
No repairs of defects are allowed.

SKETCHES: (Figure 21-29)

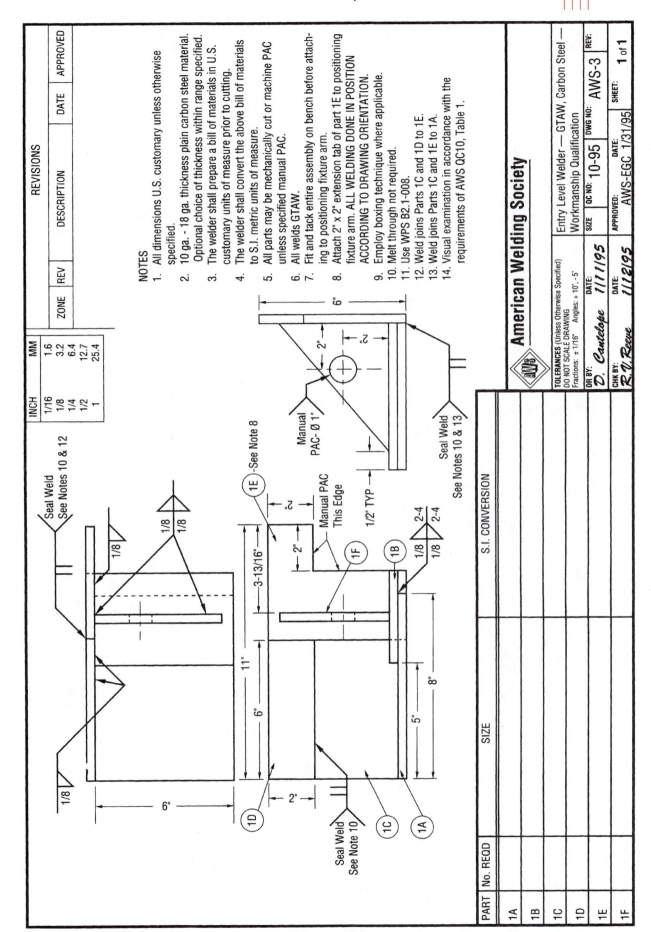

Figure 21-29 GTAW Plain Carbon Steel Workmanship Qualification Test. (Courtesy of the American Welding Society.)

PRACTICE 21-8

GAS TUNGSTEN ARC WELDING ON STAINLESS STEEL (GTAW)

WELDING PROCEDURE SPECIFICATION (WPS)

Welding Procedures Specifications No: _Practice 21-8_   Date:_____

TITLE:
Welding _GTAW_   of _sheet_   to _sheet_   .

SCOPE:
This procedure is applicable for _square groove and fillet welds_
within the range of _18 gauge_   through _10 gauge_
Welding may be performed in the following positions: _1G and 2F_   .

BASE METAL:
The base metal shall conform to _Austenitic Stainless Steel M-8 or P-8_   .
Backing material specification: _None_   .

FILLER METAL:
The filler metal shall conform to AWS specification No. _ER3XX_   from AWS specification
_A5.9_   . This filler metal falls into F-number _F-6_   and A-number _A-8_   .

ELECTRODE:
The tungsten electrode shall conform to AWS specification No. _EWTh-2_   from AWS
specification _A5.12_   . The tungsten diameter shall be _1/8 in. (3.2 mm) maximum_   .
The tungsten end shape shall be _tapered at 2 to 3 times its length to its diameter_   .

SHIELDING GAS:
The shielding gas, or gases, shall conform to the following compositions and purity:
Welding Grade Argon   .

JOINT DESIGN AND TOLERANCES:

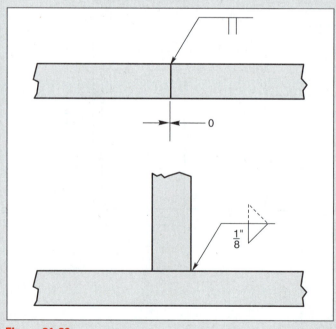

Figure 21-30

PREPARATION OF BASE METAL:
All hydrocarbons and other contaminations, such as cutting fluids, grease, oil, and primers, must be cleaned off all parts and filler metals before welding. This cleaning can be done with any suitable solvents or detergents. The joint face and inside and outside plate surface within 1 in. (25 mm) of the joint must be cleaned of slag, oxide, and scale. Cleaning can be mechanical or chemical. Mechanical metal cleaning can be done by grinding, stainless steel wire brushing, scraping, machining, or filing. Chemical cleaning can be done by using acids, alkalies, solvents, or detergents. Cleaning must be done down to bright metal.

ELECTRICAL CHARACTERISTICS:
The current shall be Direct Current Electrode Negative (DCEN) . The base metal shall be on the positive side of the line.

Metal Specifications		Gas Flow			Nozzle Size in. (mm)	Amperage Min. Max.
		Rates CFM (L/min)	Purging Times			
Thickness	Dia. of ER3XX*		Prepurging	Postpurging		
18 ga.	1/16 in. (1.6 mm)	15 to 20 (7 to 9)	10 to 15 sec.	10 to 25 sec.	1/4 to 3/8 (6 to 10)	35 to 60
17 ga.	1/16 in. (1.6 mm)	15 to 20 (7 to 9)	10 to 15 sec.	10 to 25 sec.	1/4 to 3/8 (6 to 10)	40 to 65
16 ga.	1/16 in. (1.6 mm)	15 to 20 (7 to 9)	10 to 15 sec.	10 to 25 sec.	1/4 to 3/8 (6 to 10)	40 to 75
15 ga.	1/16 in. (1.6 mm)	15 to 20 (7 to 9)	10 to 15 sec.	10 to 25 sec.	1/4 to 3/8 (6 to 10)	50 to 80
14 ga.	3/32 in. (2.4 mm)	20 to 25 (9 to 12)	10 to 20 sec.	10 to 30 sec.	3/8 to 5/8 (10 to 16)	50 to 90
13 ga.	3/32 in. (2.4 mm)	20 to 25 (9 to 12)	10 to 20 sec.	10 to 30 sec.	3/8 to 5/8 (10 to 16)	55 to 100
12 ga.	3/32 in. (2.4 mm)	20 to 25 (9 to 12)	10 to 20 sec.	10 to 30 sec.	3/8 to 5/8 (10 to 16)	60 to 110
11 ga.	3/32 in. (2.4 mm)	20 to 25 (9 to 12)	10 to 20 sec.	10 to 30 sec.	3/8 to 5/8 (10 to 16)	65 to 120
10 ga.	3/32 in. (2.4 mm)	20 to 25 (9 to 12)	10 to 20 sec.	10 to 30 sec.	3/8 to 5/8 (10 to 16)	70 to 130

*Any RE3XX stainless steel A5.9 filler metal may be used.

PREHEAT:
The parts must be heated to a temperature higher than 50°F (10°C) before any welding is started.

BACKING GAS:
N/A

SAFETY:
Proper protective clothing and equipment must be used. The area must be free of all hazards that may affect the welder or others in the area. The welding machine, welding leads, work clamp, electrode holder, and other equipment must be in safe working order.

WELDING TECHNIQUE:
　　　　TACK WELDS: With the parts securely clamped in place with the correct root gap, the tack welds are to be performed. Holding the electrode so that it is very close to the root

face but not touching, slowly increase the current until the arc starts and a molten weld pool is formed. Add filler metal as required to maintain a slight convex weld face and a flat or slightly concave root face. When it is time to end the tack weld, lower the current slowly so that the molten weld pool can be tapered down in size. When all tack welds are complete, allow the parts to cool as needed before assembling the remaining parts. Repeat the tack welding procedure until the entire part is assembled.

SQUARE GROOVE AND FILLET WELDS: Holding the electrode so that it is very close to the metal surface but not touching, slowly increase the current until the arc starts and a molten weld pool is formed. As the weld progresses, add filler metal as required to maintain a flat or slightly convex weld face. If it is necessary to stop the weld, to reposition yourself, or the weld is completed, the current must be lowered slowly so that the molten weld pool can be tapered down in size.

INTERPASS TEMPERATURE:

The plate should not be heated to a temperature higher than 350°F (180°C) during the welding process. After each weld pass is completed, allow it to cool but never to a temperature below 50°F (10°C). The weldment must not be quenched in water.

CLEANING:

Recleaning may be required if the parts or filler metal become contaminated or reoxide to a degree that the weld quality will be affected. Reclean using the same procedure used for the original metal preparation.

INSPECTION:

Visually inspect the weld for uniformity and discontinuities. There shall be no cracks, no incomplete fusion, and no overlap. Undercut shall not exceed the lesser of 10% of the base metal thickness or 1/32 in. (0.8 mm). The frequency of porosity shall not exceed one in each 4 in. (100 mm) of weld length, and the maximum diameter shall not exceed 3/32 in. (2.4 mm).

REPAIR:

No repairs of defects are allowed.

SKETCHES: (Figure 21-31)

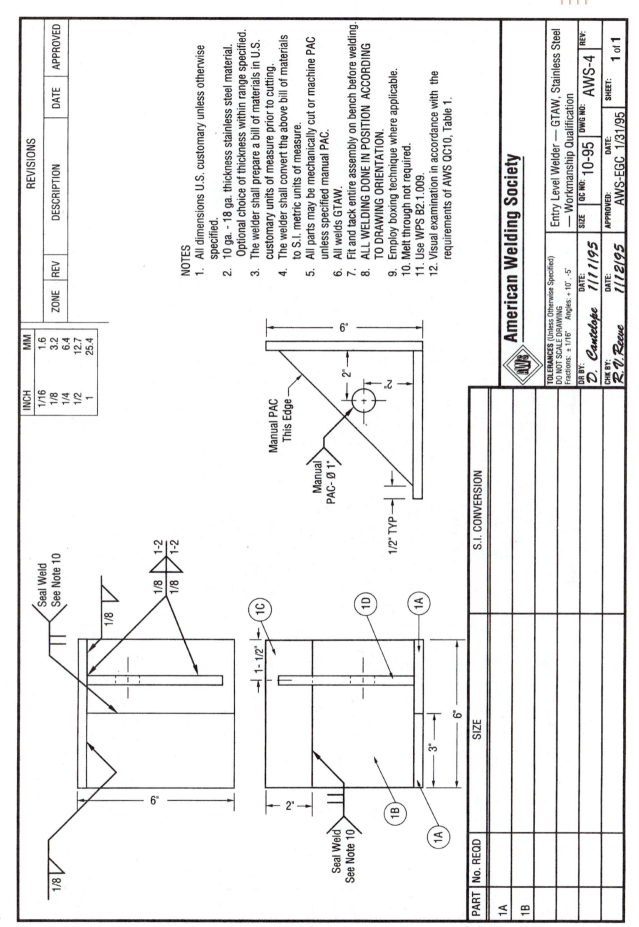

Figure 21-31 Stainless Steel Workmanship Qualification Test. (Courtesy of the American Welding Society.)

PRACTICE 21-9

GAS TUNGSTEN ARC WELDING ON ALUMINUM (GTAW)

WELDING PROCEDURE SPECIFICATION (WPS)
Welding Procedures Specifications No: _Practice 21-9_ Date:_____

TITLE:
Welding _GTAW_ of _sheet_ to _sheet_ .

SCOPE:
This procedure is applicable for _square groove and fillet welds_____
within the range of _18 gauge_____ through _10 gauge_____ .
Welding may be performed in the following positions: _1G and 2F_____ .

BASE METAL:
The base metal shall conform to _Aluminum M-22 or P-22_____ .
Backing material specification: _None_____ .

FILLER METAL:
The filler metal shall conform to AWS specification No. _ER4043___ from AWS specification
_A5.10___ . This filler metal falls into F-number _F-23___ and A-number ___ .

ELECTRODE:
The tungsten electrode shall conform to AWS specification No. _EWP___ from AWS specification _A5.12____ . The tungsten diameter shall be _1/8 in. (3.2 mm) maximum____ . The
tungsten end shape shall be _rounded_____ .

SHIELDING GAS:
The shielding gas, or gases, shall conform to the following compositions and purity:
_Welding Grade Argon_____ .

JOINT DESIGN AND TOLERANCES:

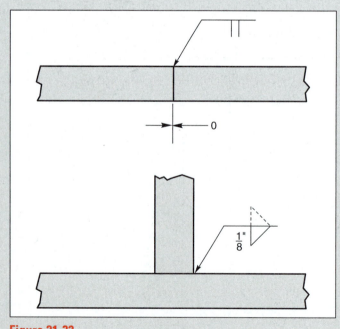

Figure 21-32

PREPARATION OF BASE METAL:
All hydrocarbons and other contaminations, such as cutting fluids, grease, oil, and primers, must be cleaned off all parts and filler metals before welding. This cleaning can be done with any suitable solvents or detergents. The joint face, inside and outside plate surface within 1 in. (25 mm) of the joint, must be mechanically or chemically cleaned of oxides. Mechanical cleaning may be done by stainless steel wire brushing, scraping, machining, or filing. Chemical cleaning may be done by using acids, alkalies, solvents, or detergents. Because the oxide layer may reform quickly and affect the weld, welding should be started within 10 minutes of cleaning.

ELECTRICAL CHARACTERISTICS:
The current shall be <u>Alternating Current High-Frequency Stabilized (balanced wave preferably)</u>. The base metal shall be on the <u>N/A</u> side of the line.

Metal Specifications		Gas Flow			Nozzle Size in. (mm)	Amperage Min. Max.
		Rates CFM (L/min)	Purging Times			
Thickness	Dia. of ER4043*		Prepurging	Postpurging		
18 ga.	3/32 in. (2.4 mm)	20 to 30 (9 to 14)	10 to 15 sec.	10 to 25 sec.	1/4 to 3/8 (6 to 10)	40 to 60
17 ga.	3/32 in. (2.4 mm)	20 to 30 (9 to 14)	10 to 15 sec.	10 to 25 sec.	1/4 to 3/8 (6 to 10)	50 to 70
16 ga.	3/32 in. (2.4 mm)	20 to 30 (9 to 14)	10 to 15 sec.	10 to 25 sec.	1/4 to 3/8 (6 to 10)	60 to 75
15 ga.	3/32 in. (2.4 mm)	20 to 30 (9 to 14)	10 to 15 sec.	10 to 25 sec.	1/4 to 3/8 (6 to 10)	65 to 85
14 ga.	3/32 in. (2.4 mm)	20 to 30 (9 to 14)	10 to 20 sec.	10 to 30 sec.	3/8 to 5/8 (10 to 16)	75 to 90
13 ga.	1/8 in. (3.2 mm)	25 to 40 (12 to 19)	10 to 20 sec.	10 to 30 sec.	3/8 to 5/8 (10 to 16)	85 to 100
12 ga.	1/8 in. (3.2 mm)	25 to 40 (12 to 19)	10 to 20 sec.	10 to 30 sec.	3/8 to 5/8 (10 to 16)	90 to 110
11 ga.	1/8 in. (3.2 mm)	25 to 40 (12 to 19)	10 to 20 sec.	10 to 30 sec.	3/8 to 5/8 (10 to 16)	100 to 115
10 ga.	1/8 in. (3.2 mm)	25 to 40 (12 to 19)	10 to 20 sec.	10 to 30 sec.	3/8 to 5/8 (10 to 16)	100 to 125

*Other aluminum A5.10 filler metal may be used if needed.

PREHEAT:
The parts must be heated to a temperature higher than 50°F (10°C) before any welding is started.

BACKING GAS:
N/A

SAFETY:
Proper protective clothing and equipment must be used. The area must be free of all hazards that may affect the welder or others in the area. The welding machine, welding leads, work clamp, electrode holder, and other equipment must be in safe working order.

WELDING TECHNIQUE:
The welder's hands or gloves must be clean and oil-free to prevent recontaminating the metal or filler rods.

TACK WELDS: With the parts securely clamped in place with the correct root gap, the tack welds are to be performed. Holding the electrode so that it is very close to the root face but not touching, slowly increase the current until the arc starts and a molten weld pool is formed. Add filler metal as required to maintain a slight convex weld face and a flat or slightly concave root face. When it is time to end the tack weld, lower the current slowly so that the molten weld pool can be tapered down in size. When all tack welds are complete, allow the parts to cool as needed before assembling the remaining parts. Repeat the tack welding procedure until the entire part is assembled.

SQUARE GROOVE AND FILLET WELDS: Holding the electrode so that it is very close to the metal surface but not touching, slowly increase the current until the arc starts and a molten weld pool is formed. As the weld progresses, add filler metal as required to maintain a flat or slightly convex weld face. If it is necessary to stop the weld or to reposition yourself or the weld is completed, the current must be lowered slowly so that the molten weld pool can be tapered down in size.

INTERPASS TEMPERATURE:
The plate should not be heated to a temperature higher than 120°F (49°C) during the welding process. After each weld pass is completed, allow it to cool but never to a temperature below 50°F (10°C). The weldment must not be quenched in water.

CLEANING:
Recleaning may be required if the parts or filler metal becomes contaminated or reoxides to a degree that the weld quality will be affected. Reclean using the same procedure used for the original metal preparation.

INSPECTION:
Visually inspect the weld for uniformity and discontinuities. There shall be no cracks, no incomplete fusion, and no overlap. Undercut shall not exceed the lesser of 10% of the base metal thickness or 1/32 in. (0.8 mm). The frequency of porosity shall not exceed one in each 4 in. (100 mm) of weld length, and the maximum diameter shall not exceed 3/32 in. (2.4 mm).

REPAIR:
No repairs of defects are allowed.

SKETCHES: (Figure 21-33)

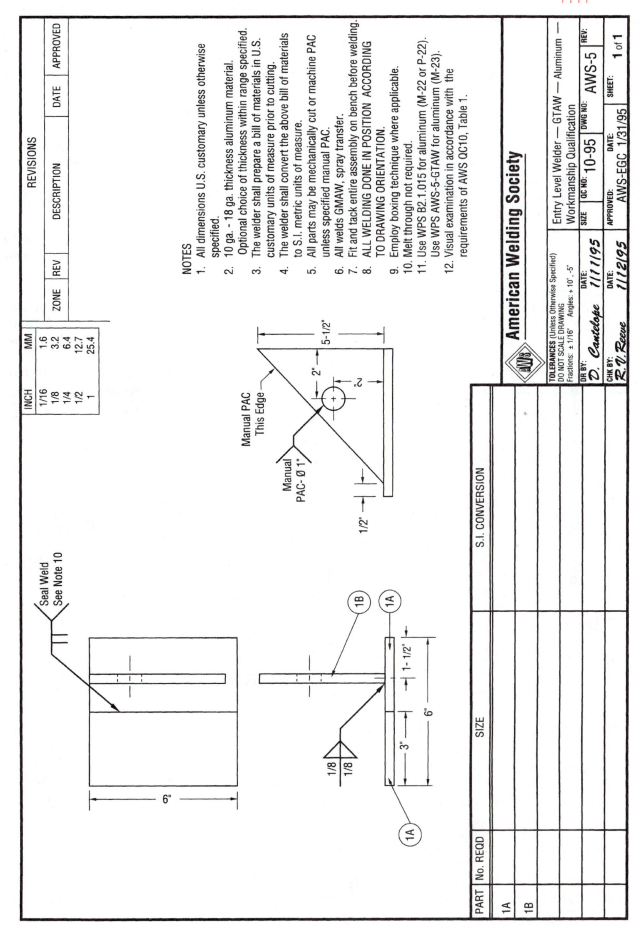

Figure 21-33 GTAW Aluminum Workmanship Qualification Test. (Courtesy of the American Welding Society.)

PRACTICE 21-10

WELDER AND WELDER OPERATOR QUALIFICATION TEST RECORD (WPS)

Using a completed weld such as the ones from Practice 21-1 or 21-2, you will follow the steps listed below and complete the test record shown in **Figure 21-34**. This form is a composite of sample test recording forms provided by AWS, ASME, and API codes. You may want to obtain a copy of one of the codes or standards and compare a weld you made to the standard. This form is useful when you are testing one of the practice welds in this text.

Note: Not all of the blanks will be filled in on the forms. The forms are designed to be used with a large variety of weld procedures, so they have spaces that will not be used each time.

1. Welder's name: The person who performed the weld.

2. Identification no.: On a welding job, every person has an identification number that is used on the time card and paycheck. In this space, you can write the class number or section number since you do not have a clock number.

3. Welding process(es): Was the weld performed with SMAW, GMAW, or GTAW?

4. How was the weld accomplished: manually, semiautomatically, or automatically?

5. Test position: 1G, 2G, 3G, 4G, 1F, 2F, 3F, 4F, 5G, 6G, 6GR, **Figure 21-35**.

6. What WPS was used for this test?

7. Base metal specification: This is the ASTM specification number, **Table 21-4** (page 568).

8. Test material thickness (or) test pipe diameter (and) wall thickness: The actual thickness of the welded material or pipe diameter and wall thickness.

9. Thickness range qualified (or) diameter range qualified: For both plate and pipe, a weld performed successfully on one thickness qualifies a welder to weld on material within that range. See **Table 21-5** (page 568) for a list of thickness ranges.

10. Filler metal specification number: The AWS has specifications for chemical composition and

WELDER AND WELDING OPERATOR QUALIFICATION TEST RECORD (WQR)

Welder or welding operator's name _____(1)_____ Identification no. ___(2)___

Welding process ___(3)___ Manual ___(4)___ Semiautomatic ___(4)___ Machine ___(4)___
Position ___(5)___

(Flat, horizonal, overhead or vertical - if vertical, state whether up or down
In accordance with welding procedure specification no. ___(6)___

Material specification ___(7)___

Diameter and wall thickness (if pipe) - otherwise, joint thickness ___(8)___

Thickness range this qualifies ___(9)___

FILLER METAL

Specification No. ___(10)___ Classification ___(11)___ F-number ___(12)___

Describe filler metal (if not covered by AWS specification) ___(13)___

Is backing strip used? ___(14)___

Filler metal diameter and trade name ___(15)___ Flux for submerged arc or gas
for gas metal arc or flux cored arc welding ___(16)___

Figure 21-34

physical properties for electrodes. Some of these specifications are listed in **Table 21-6** (page 569).

11. Classification number: This is the standard number found on the electrode or electrode box, such as E6010, E7018, E316-15, ER1100, etc.

12. F-number: A specific grouping number for several classifications of electrodes having similar composition and welding characteristics. See **Table 21-7** (page 569) for the F-number corresponding to the electrode used.

13. Give the manufacturer's chemical composition and physical properties as provided.

14. Backing strip material specification: This is the ASTM specification number.

15. Give the diameter of electrode used and the manufacturer's identification name or number.

16. Flux for SAW or shielding gas(es) and flow rate for GMAW, FCAW, or GTAW.

If the weld is a groove weld, follow steps 17 through 22 and then skip to step 27. If the weld test was a fillet weld, skip to step 22.

17. Visually inspect the weld and record any flaws.

18. Record the weld face, root face, and reinforcement dimensions.

19. Four (4) test specimens are used for 3/8-in. (10-mm) or thinner metal. Two (2) will be root bent and two (2) face bent. For thicker metal all four (4) will be side bent.

20. Visually inspect the specimens after testing and record any discontinuities.

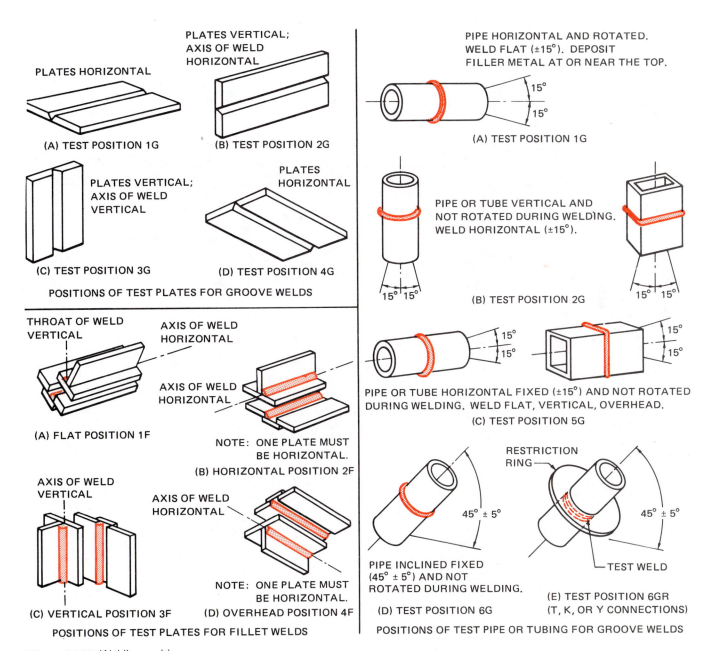

Figure 21-35 Welding positions.

Type of Material	
P-1	Carbon Steel
P-3	Low Alloy Steel
P-4	Low Alloy Steel
P-5	Alloy Steel
P-6	High Alloy Steel — Predominently Martensitic
P-7	High Alloy Steel — Predominently Ferritic
P-8	High Alloy Steel — Austenitic
P-9	Nickel Alloy Steel
P-10	Specialty High Alloy Steels
P-21	Aluminum and Aluminum-Base Alloys
P-31	Copper and Copper Alloy
P-41	Nickel

Table 21-4 P Numbers.

21. Who witnessed the welding for verification that the WPS was followed?

22. The identification number assigned by the testing.

If the weld is a filler weld, follow steps 23 through 28.

23. Visually inspect the weld and record any flaws.

24. Record the legs and reinforcement dimensions.

25. Measure and record the depth of the root penetration.

26. Polish the side of the specimens and apply an acid to show the complete outline of the weld.

27. Who witnessed the welding for verification that the WPS was followed?

28. The identification number assigned by the testing.

If a radiographic test is used, follow steps 29 through 33. If this test is not used, go to step 34.

29. The number the lab placed on the X-ray film before it was exposed on the weld.

30. Record the results of the reading of the film.

Plate Thickness (T) Tested in (mm)	Plate Thickness (T) Qualified in (mm)
1/8 ≤ T < 3/8* (3.1 ≤ T < 9.5)	1/8 to 2T (3.1 to 2T)
3/8 (9.5)	3/4 (19.0)
3/8 < T < 1 (9.5 < T < 25.4)	2T 2T
1 and over (25.4 and over)	Unlimited Unlimited

Pipe Size of Sample Weld

Diameter in (mm)	Wall Thickness, T
2 (50.8) or	Sch. 80
3 (76.2)	Sch. 40
6 (152.4) or	Sch. 120
8 (203.2)	Sch. 80

Pipe Size Qualified

Diameter in (mm)	Wall Thickness in (mm)	
	Minimum	Maximum
3/4 (19.0) through 4 (101.6)	0.063 (1.6)	0.674 (17.1)
4 (101.6) and over	0.187 (4.7)	Any

*Thickness (T) is equal to or greater than 1/8 in (≤) and thickness (T) is less than 3/8 in (<)

Table 21-5 Test Specimens and Range of Thickness Qualified.

A5.10	Aluminum — bare electrodes and rods
A5.3	Aluminum — covered electrodes
A5.8	Brazing filler metal
A5.1	Steel, carbon, covered electrodes
A5.20	Steel, carbon, flux-cored electrodes
A5.17	Steel-carbon, submerged arc wires and fluxes
A5.18	Steel-carbon, gas metal arc electrodes
A5.2	Steel — oxyfuel gas welding
A5.5	Steel — low-alloy covered electrodes
A5.23	Steel — low-alloy electrodes and fluxes — submerged arc
A5.28	Steel — low-alloy filler metals for gas shielded arc welding
A5.29	Steel — low-alloy, flux-cored electrodes

Table 21-6 Specification Numbers.

Group Designation	Metal Types	AWS Electrode Classification
F1	Carbon Steel	EXX20, EXX24, EXX27, EXX28
F2	Carbon Steel	EXX12, EXX13, EXX14
F3	Carbon Steel	EXX10, EXX11
F4	Carbon Steel	EXX15, EXX16, EXX18
F5	Stainless Steel	EXXX15, EXXX16
F6	Stainless Steel	ERXXX
F22	Aluminum	ERXXXX

Table 21-7 F Numbers.

31. Whether the test passed or failed the specific code.
32. Who witnessed the welding for verification that the WPS was followed?
33. The number assigned by the testing.
34. The name of the company that requested the test.
35. The name of the person who interpreted the results. This is usually a Certified Welding Inspector (CWI) or other qualified person.
36. Date the results of the test were completed.

PERFORMANCE QUALIFICATION TEST RECORD

PERFORMANCE QUALIFICATION TEST RECORD

Identification No. _____   Weld Stamp No. _____

Name _____   Social Security # _____ - ___ - _____

Eye correction required   Yes ☐   No ☐   Type of Eye Correction   Eye glasses ☐
Contact lenses ☐
Magnifiers ☐

Qualified with WPS No. _____   Process _____   Manual ☐   Semi-Automatic ☐

Position Flat (1G, 1F ☐), Horizontal (2G, 2F ☐), Vertical (3G, 3F ☐), Overhead (4G, 4F ☐)

Test base metal specification _____   Thickness _____

AWS filler metal classification _____   Size _____

Shielding Gas _____   Flow Rate _____

GUIDED BEND-TEST RESULTS

Visual Test Results   Pass ☐   Fail ☐   Weld size _____

Type of Bend			Result	
Face ☐	Root ☐	Side ☐	Pass ☐	Fail ☐
Face ☐	Root ☐	Side ☐	Pass ☐	Fail ☐

The above-named person is qualified for the welding process used in this test within the limits of essential variables including materials and filler metal as provided by the AWS D1.1 _____ Code. We, the undersigned, certify that the statements in this record are correct and that the test welds were prepared, welded, and tested in accordance with the requirements of the AWS D1.1 _____ Code.

Date Tested _____   Signed by _____

Test Supervisor

Signed by _____

Corporate Representative   Title

REVIEW

1. How does applying for a welding job differ from most other types of jobs?

2. What is the major difference between a qualified welder and a certified welder?

3. What process can a welder be certified for?

4. List the variables that, if changed, would require that a new certification test be given.

5. What is required to become an AWS Certified Welder?

6. What is the tolerance allowed for laying out parts for the AWS Entry Level Welder Certification test?

7. Why are written records about actual welding important to companies?

8. What types of information can be given to welders verbally?

9. What score is needed on the safety test prior to welding in the shop, and what areas must be included on the test?

10. What types of scrap do welding and cutting typically produce?

11. What must you check before making any repairs or adjustments to equipment as part of your routine housekeeping chores?

12. What cutting processes must you be able to use in order to pass the AWS Entry Level Welder Certification test?

13. Why may it be necessary to clamp parts together during the fit-up process?

14. Where must all arc strikes be made during any testing?

15. Why may it be necessary to allow the part to cool during the weld test?

16. What are the visual inspection tolerances of the welds for the AWS Entry Level Welder Certification test?

17. What minimum and maximum temperatures are allowed for the test plates during the SMAW welding test?

18. What is the acceptance criterion for the bend test on the SMAW welding test?

19. How smooth must the flame cut surface of a bevel be before starting a welding test?

20. What shielding gases are required by the SMAW-S WPS?

21. What is the maximum thickness allowed by the WPS for any one FCAW weld pass?

22. What are the minimum and maximum amperages and wire feed speeds allowed for 0.35-in. (0.9-mm) electrodes on the GMAW by the WPS?

23. Sketch the weld joint design, including joint tolerances, for the GTAW plain carbon steel WPS test.

CHAPTER
22

RAILROAD WELDING

OBJECTIVES

After completing this chapter, the student will be able to:

■ explain the thermite welding process for rails.

■ describe the characteristics of austenitic manganese steel.

■ list the steps required to repair cracks in rails and rail components.

■ explain the reason for keeping thermite welding materials dry.

KEY TERMS

frog	austenitic manganese steel
hydrogen embrittlement	thermite welding
flash welding	atomic hydrogen (H)
molecular hydrogen (H_2)	crucible
mold	slag pan

INTRODUCTION

The railroad was the first mass transit system. Early railroads provided a way for large numbers of immigrants to move out west. Railroads were also instrumental in the development of the agricultural and cattle industries throughout the Midwest, by providing farmers and ranchers ready access to eastern markets.

New high-speed bullet trains and local transit systems are expected to someday provide cities with a more environmentally friendly transportation system than automobiles.

The railroad systems are so large, diverse, and widespread that no matter where you live you can see it in action. Because of this and its impact on our lives, it has been selected for this chapter to serve as an example of the diversity required by welders within an industry.

We all see trains moving about and give little thought to the welding required to keep them moving. This industry, like most others, relies on skilled welders to fabricate and repair essential parts. Welding is used to build locomotives, cars, trussels, rails, switches, and much more for the railroad. Almost every major welding, cutting, and brazing process is used by this expansive industry.

The range of metals and alloys used is extensive. Commonly used metals and alloys include most plain carbon steel, high-strength steels, stainless steels, aluminums, and many others. This diverse group of alloys requires an equally diverse collection of welding procedures. This chapter will cover several different specialty welding needs while concentrating on one area—rails and related components. A wide variety of welding processes and procedures are needed for this one segment of the railroad industry.

Rails are used for many applications other than just the railroad industry. They are used for trolleys, large cranes, overhead cranes, local transit systems, amusement park rides, guard rails, electrical contact rails, and so on. Each type of rail has its own requirements for which rails are specifically designed and alloyed to meet those needs.

RAIL TYPES

The first type of rails that come to mind are used by the railroads. Within this single usage are three major classifications of composition: soft carbon rails, standard carbon rails, and premium steel rails, **Figure 22-1**. Within each of these classifications the composition of the rail can be changed to meet a specific requirement of a particular railroad.

All rails used as tracks must have good mechanical properties, among them the following

- *Tensile strength* refers to the property of a rail to resist forces applied to pull it apart, **Figure 22-2**.

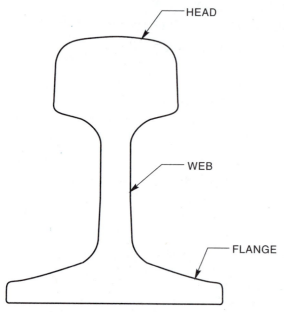

Figure 22-1 Parts of standard T rail.

Tensile strength has two parts: yield strength and ultimate strength. *Yield strength* is the amount of strain needed to permanently deform a test specimen. The *yield point* is the point during tensile loading when the metal stops stretching and begins to be permanently made longer by deforming, **Figure 22-3**. *Ultimate strength* is a measure of the load that breaks a specimen. Steels used for rails become work-hardened as they are stretched during a tensile test. These metals will actually become stronger and harder as a result of being tested. Tensile strength is usually measured in pounds per square inch (psi) (MPa).

- *Hardness* is a rail's resistance to highly localized deformation. The ends of track must resist being deformed by the repeated pounding of the train wheels, **Figure 22-4**. The hardness property of rails is increased by cold working as the rail is reshaped under the rolling stock loads. Hardness is also affected by heating from welds and heat-treatment methods. Since hardness is proportional to strength, it is a quick way to determine strength. It is also useful in determining whether the rail received the proper heat treatment, since heat treatment also affects strength. Rail hardness is usually measurable on the Brinell scale using a 3000-Kg load on a 10-mm diameter ball or the Rockwell C scale.

- *Toughness* is the property that allows a rail to withstand impact loads, sudden shock, cyclic loading, or bends without fracturing. Rails are subjected to impact loads and sudden shock every time the train crosses a joint or enters a switch. Cyclic loading and bending occurs every time a

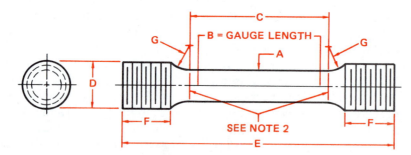

NOTE 1: DIMENSION A, B, AND C SHALL BE AS SHOWN, BUT ALTERNATE SHAPES OF ENDS MAY BE USED AS ALLOWED BY ASTM SPECIFICATION E-8.

NOTE 2: IT IS DESIRABLE TO HAVE THE DIAMETER OF THE SPECIMEN WITHIN THE GAUGE LENGTH SLIGHTLY SMALLER AT THE CENTER THAN AT THE ENDS. THE DIFFERENCE SHALL NOT EXCEED 1 PERCENT OF THE DIAMETER.

Figure 22-2 Tensile test specimen. (Courtesy of Hobart Brothers Company.)

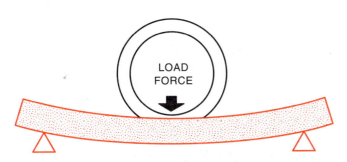

Figure 22-3 Rail deflects under wheel load but returns to straight, unbent, if load is below yield point.

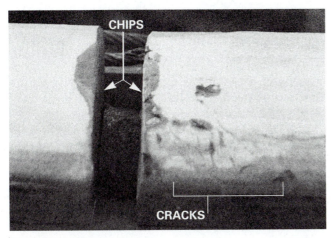

Figure 22-4 Rail end damage.

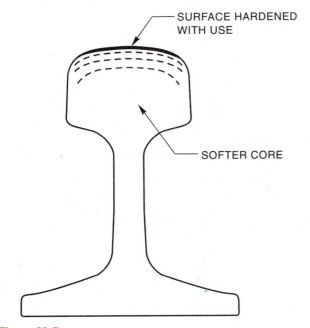

Figure 22-5

loaded wheel passes over a rail section between ties. Rails are subjected to billions of load cycles during their lifetime. Toughness is measured most often with the Charpy impact test at $-100°F$ ($-73°C$). The test is performed at this low temperature because steel's toughness is adversely affected by the cold.

Rails come in many sizes. Rail sizes are expressed as pounds per yard, a standard that dates back to the earliest railroads. However, almost every other stock metal shape, such as channels, H-beams, and I-beams, has its size expressed in pounds per foot.

Soft Carbon Rails

The term *soft carbon rails* is misleading. These rails have strengths in the 100,000-psi (690 MPa) range, which makes them stronger than most plain carbon steels. Soft carbon rails are often used as electrical contact rails, sometimes called *third rails*. They are one way that electric power can be supplied for electric trains (the other way is by overhead wires). Many subways, commuter lines, and other public transit systems use third rail systems.

Standard Carbon Rails

Standard Carbon rails have been widely used throughout the rail industry for many years. Standard carbon rails have provided years of outstanding service, as seen in the thousands of miles of it still in use today. These rails, although primarily used as jointed sections, can be welded.

Standard carbon rails have strengths in the range of 135,000 psi (930 MPa), making them much stronger than standard structural steels.

Premium Steel Rails

Premium steel rails are made from austenitic manganese steel which is used for much of the railroad rails today. It has high-impact strengths even down to sub-zero temperatures, an important property for track laid in cold regions. It also has very good wear resistance in a metal-to-metal application such as between the rail and train wheels. Its wear resistance is increased with use because of work-hardening. The top surface of the track hardens as it is plastically deformed with use while the underlying metal remains unaffected, **Figure 22-5**. This hard surface and tough underlying metal make austenitic manganese steel rail and related components very suited for its job.

Austenitic manganese steel rails have a nominal hardness number of 341 or more on the Brinell hardness scale. This hardness is achieved by alloying elements, heat treatment, or a combination of both. Heat treatment can be full where the entire rail is treated or partially treated where only the head is heat treated. If the rail achieves its

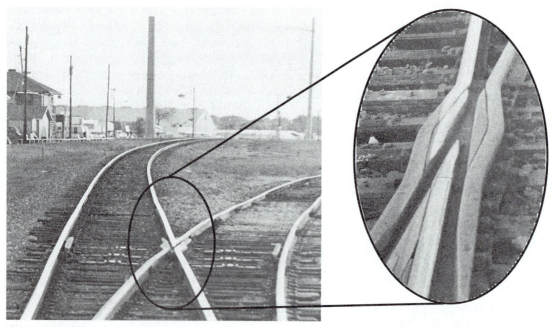

Figure 22-6 See attatched insert.   Frog

hardness as a result of heat treatment, special welding procedures must be followed. You must contact the rail manufacturer for such a specialized procedure.

Rail Components

Rail components are often made from forgings and castings. The same alloys of austenitic manganese steels are used for these parts as are used for rails. The preheating and interpass temperature ranges are similar, but you must check with the specific WPA before welding.

Common components that must be repaired are switches, crossings, and frogs. **Frogs** are the large structures, usually castings, that make up the center of a crossing or the rail crossing point of a switch, **Figure 22-6**. A frog can be either a single casting or made up of several parts including more than one casting.

Because they must have openings so that the train wheels can switch from one rail to another, frogs are subjected to extensive wear. Every time a train passes over a switch each wheel on the frog's side impacts the frog. The frog point, the smallest part of the frog, **Figure 22-7**, receives most of that wear. The wear the point receives results in its being worn down, chipped, cracked, and broken.

Repair welding the frog is a very time-consuming job. A typical WPS repair procedure might consist of the following steps, **Figure 22-8**:

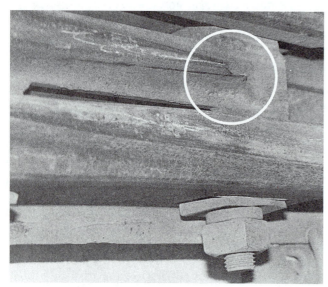

Figure 22-7 Frog point.

Figure 22-8 Frog point being built up.

- Visually inspect the frog for damage.
- Clean all oil, grease, dirt, and other debris from around the damaged area.
- Preheat the frog as needed for air arc gouging.
- Allow the frog to cool down between gouges if the air arc gouging overheats the metal.
- Conduct post-gouging cool down as needed.
- Conduct post-gouging cleanup by grinding, chipping, and wire brushing.
- Inspect the area damaged with dye penetrate to make sure that all cracks have been completely removed.
- Preheat for welding if needed.
- Build up weld in 5-in. (250-mm) -long beads in a skip pattern.
- Do post-weld pass cleanup.
- Repeat welding procedure until the total buildup is complete.
- Allow post-welding cool-down as required.
- Do post-weld cleanup.
- Conduct post-weld inspection with dye penetrate or ultrasonic.
- Repair as needed.
- Do contour grinding and finish work.

The actual welding time is not unusually long. It is the skip welding, cleaning, heating, cooling, and inspections that extend the total repair time. Following the steps of the WPS will ensure a satisfactory weld. Grinding to achieve the desired shape following welding should be kept to a minimum.

CONTINUOUS WELDED TRACK

Before the introduction of continuous welded track, rail sections were joined every 39 feet (11.8 m). These joints produced the familiar sound of clickety-click as the trail wheels rolled along. The track joint spaces were needed to provide for the expansion of the rail sections due to changes in temperature. If the tracks were laid properly, the space would be closed completely on a hot summer day. During a cold winter day, however, the space would be at its widest. Without this expansion space, track would have either pulled itself apart or buckled.

Welded rail provided a number of advantages for the railroads, among them:

- It provides a smoother ride; less rail end damage; and reduced wear and tear on wheels, axles, and running gear.
- It is stronger, provides greater safety, and allows increased axle loads for heavier trains.

- Fuel economy is increased.
- Higher speeds can safely be reached. Some trains in Europe can travel nearly 200 miles per hour (320 kilometers per hour).
- Service life of the rail is increased and maintenance cost is reduced.

Improvements in technology have resulted in continuous welded rail sections that are approximately 1/4 mile in length. The two most commonly used methods for joining rail sections into longer track are flash welding (FW) and thermite welding (TW), **Figure 22-9**.

Flash welding is used in large plants to produce long rail sections that will be hauled to the construction location. **Thermite welding** is a portable process that can be used either at a plant or on location for track construction and repair.

RAIL REPAIRS

Repairing austenitic manganese steel rail and components is difficult because it has a high degree of heat distortion. This steel has a higher rate of expansion than do plain carbon steels. It also has a lower rate of thermal conductivity than plain carbon steels. This characteristic causes the heat of welding or cutting to be localized. The welding can cause a concentration of force from localized heating and localized expansion. Both of these factors can result in the rail being distorted, **Figure 22-10** (page 576).

Distortion can be controlled by proper heating and welding techniques:

- Limit heat input by making short narrow welds. Welds must not be longer than 5 in. (250 mm) and applied in a skip weld pattern, **Figure 22-11** (page 576). The weld beads should be slightly convex and no wider than 5/8 in. (16 mm).

Figure 22-9 Weld point in continuous welded track.

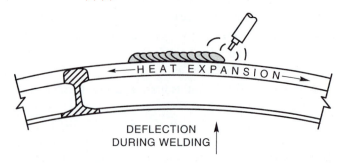

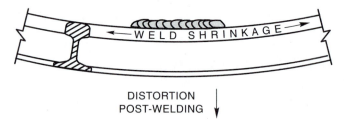

Figure 22-10 Heat deflection during welding results in track distortion after welding.

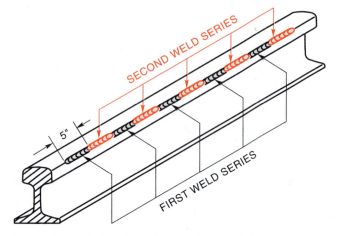

Figure 22-11 Series of 5-in. (250-mm) -long skip welds.

■ Uniform heating and controlled cooling rates allow the rail to expand and contract uniformly.

■ Offset the parts before welding. In **Figure 22-12** the rail ends are slightly raised before welding. Without this offset the weld would bend downward as it cooled. Even though the heating during the thermite process is uniform, the difference in the cross-sectional area of the rail is uneven, causing the top to shrink more than the base. Therefore, offsetting the rail ends correctly will result in a straight section after welding.

■ Jigs and fixtures can be used to restrain the weld throughout the entire welding process. The rail will remain straight if it is not allowed to bend during the welding process, **Figure 22-13**.

Rail alignment is very important. It must be aligned so that the rail head is even on the top and sides, **Figure 22-14**. Slight differences in size between the new rail and the old rail can be expected due to wearing of the old rail. This misalignment can be compensated for by grinding. The length and angle of the taper resulting from the offset must meet the American Railway Engineering Association (AREA) standard and that of the rail's owner. Check with the project engineer if needed for such specifications.

Preheating and Postheating

Preheating is required before any welding or thermal cutting and gouging. Preheating is needed to prevent hardness zones from forming in the repair area. The highly alloyed steels used for rails have a tendency to form hard brittle zones next to any welding or cutting unless properly pre- and postheated. In most cases the rail must be preheated to 800–1,000°F (427°C to 540°C) before welding. Usually the preheating should extend 4 to 6 in. (100 mm to 150 mm) along the track beyond the actual weld area. **Figure 22-15** shows a typical rail heater used for preheating. Different preheating temperatures may be required both for heat treated rails or by WPS supplied by the rail's owner.

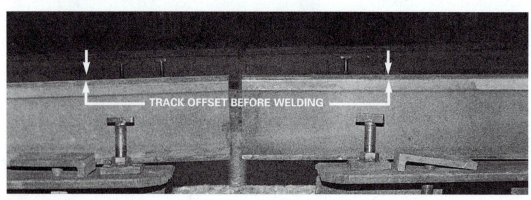

Figure 22-12

Figure 22-13 Clamp for preventing frog from distorting during the weld repair.

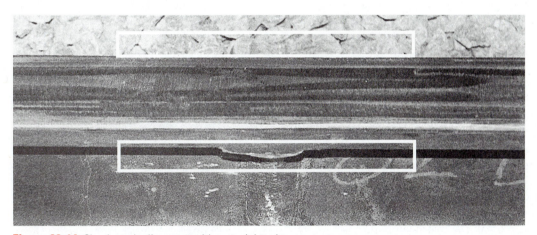

Figure 22-14 Check track alignment with a straight edge.

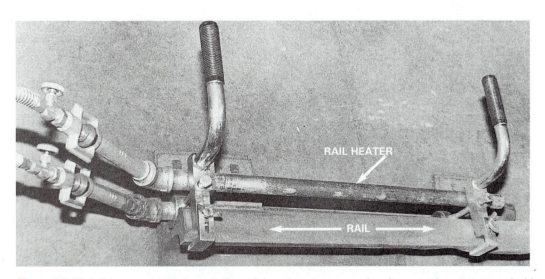

RAIL HEATER

← RAIL →

Figure 22-15 Rail heater attached to rail. Two of these heaters are used at the same time, one on each side.

Weld interpass temperatures must be maintained within the required ranges for preheating. In some cases, because of low heat input from the welding process, additional heat is needed to maintain the proper temperature range for welding or cutting. At other times the rail may need to cool down if the temperature rises above the upper limit of the preheat range due to welding or cutting heat. Overheating of austenitic manganese steels can cause carbide precipitation.

Carbide precipitation occurs when at high temperatures carbon combines with chromium to form chromium carbides in the grain boundaries. The formation of chromium carbides depletes the steel of free chromium, which is a detrimental condition that seriously reduces the mechanical properties of the rail.

Post-weld heat treatment usually consist of controlling the cooling rate. Too rapid a cooling rate will cause embrittleness of the weld and base metal.

Because of the thermal sensitivity of rails, welding must never be performed if the rails will be cooled too quickly. This restricts most repair welding to fair weather or to controlled environmental enclosures. Additionally, if the track is below freezing, 32°F (0°C), or the air temperature is below 0°F (–18°C), welding must not be performed.

Hydrogen Embrittlement

Hydrogen embrittlement is a potential problem for all highly alloyed steels. A bounded pair of hydrogen atoms is called **molecular hydrogen (H$_2$)**, Figure 22-16. In nature, hydrogen atoms exist only if they are bonded together or to other types of atoms, such as oxygen with which it forms water (H$_2$O). Hydrogen atoms are also present in many other substances such as grease and oil. The heat of a welding arc can break the hydrogen bond, forming free, unbonded hydrogen atoms called **atomic hydrogen (H)**. Atomic hydrogen is freely dissolved into the molten weld pool. As the weld metal cools, however, most of the dissolved hydrogen comes out of solution to form bubbles of porosity. Because the hydrogen atom is so small, some of the dissolved atomic hydrogen atoms will move into the base metal through the tiny spaces between the grain of solid metal. Sometimes much later this atomic hydrogen can cause under-bead cracking, Figure 22-17. This hydrogen-caused cracking can appear hours, days, or weeks following welding. To control post-weld hydrogen cracking–related problems, welding must not be performed under the following conditions:

- All welding filler metal and fluxes must be classified as low hydrogen.

- The weld may be contaminated by water, such as when the tracks are wet or it is precipitating (raining, snowing, or sleeting).

- Proper storage or handling procedures for filler metal and flexes must be followed to prevent them

from becoming contaminated with sources of hydrogen such as moisture, oil, grease, or hydrocarbons.

- Strong winds, which may blow the shielding gas cloud away from the molten weld pool, resulting in atmospheric moisture contaminating the weld, are present, Figure 22-18.

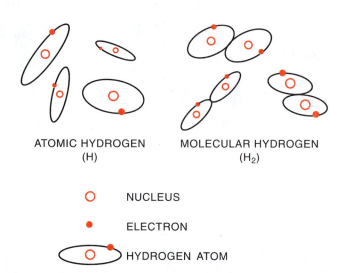

ATOMIC HYDROGEN (H)

MOLECULAR HYDROGEN (H$_2$)

O   NUCLEUS

•   ELECTRON

O•   HYDROGEN ATOM

Figure 22-16 Hydrogen atoms are made up of one proton, in the nucleus, and one electron, in the outer shell.

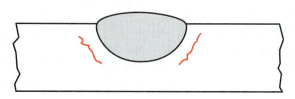

Figure 22-17 Underbead cracking.

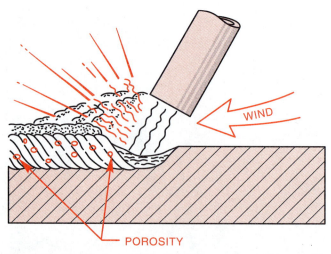

WIND

POROSITY

Figure 22-18 Wind can blow shielding away from molten weld pool, allowing porosity to form in the weld.

GOUGING

Gouging is the removal of metal by a thermal process, usually either oxyfuel gouging or air arc gouging, Figure 22-19. Due to the high heat input produced by the oxyfuel process, it is not recommended for rail repair.

When using air arc gouging, the cuts should be shallow with as high a travel speed as possible. Trying to remove too much metal at a time will slow the process down, possibly resulting in overheating of the base metal. The base metal must be at least 70°F (21°C) before any gouging begins and not more than 500°F (260°C) as measured adjacent to the gouge.

Check the gouged groove with dye penetrant to determine when the entire depth of the crack or defect has been removed. Make the gouge just wide enough to allow good electrode manipulation. Removing excessive metal will make the repair more costly in time and materials. Remove all oxides and other debris from the gouged groove before welding.

Larger areas may be removed by gouging if there are numerous cracks or there is a need to remove a severely damaged area such as frog point, Figure 22-20. The same procedures that are used on single cracks must be used. Make fast shallow narrow cuts. Remove only as much metal as is required to remove the cracks or damage down to sound base metal.

CRACKS

Cracks can be formed in the work-harded surface of rails and rail components. These cracks can be transverse or longitudinal, Figure 22-21. Transverse cracks can go completely across the rail from side to side, start at one edge and continue toward the other side, or not touch either side. Longitudinal cracks usually start and end without extending to an edge, but in some cases they can start at the rail end and extend for some distance along the track.

Because some cracks are so small and hard to see visually, dye penetrant or magnetic particle inspection should be used to locate cracks. Knowing the exact starting and ending points of the crack is important in order to completely remove it before welding repairs are started. If a small portion of the crack remains following the repair, it will quickly spread.

Cracks can be removed by air arc gouging, oxyfuel torch gouging, or grinding. Air arc gouging is the most desirable, and oxyfuel torch gouging is the least desirable.

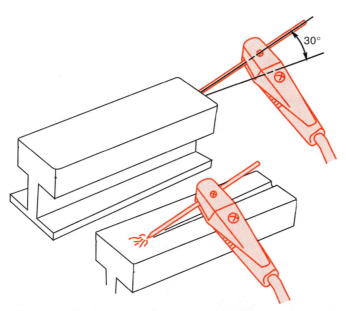

Figure 22-19 Air arc gouging.

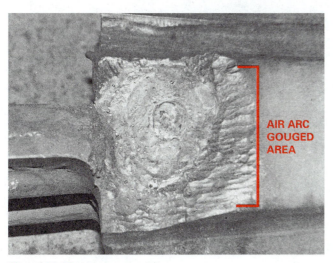

Figure 22-20 Air arc gouged frog.

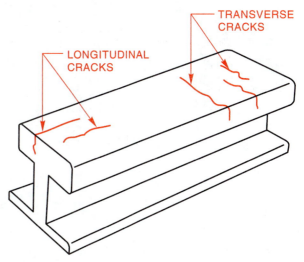

Figure 22-21 Types of cracks that may be found in rails and rail components.

Cracks that start at an edge should be removed by starting at the end of the crack and gouging to the edge. Cracks that do not propagate from an edge must be gouged out by starting at one end and working to the center, stopping and restarting at the other end, and working back until both gouges meet, Figure 22-22.

RAIL ENDS

Rail ends become damaged as the result of repeated impacts with the wheels of rolling stock. The rail ends are beaten down or chipped by these impacts and must be rebuilt, Figure 22-23. The repair weld should be made by starting the welding along the outside edge of the track. Continue welding toward the end, then across to the inside of the rail. Follow the inside edge of the rail to the end of depressed area, Figure 22-24. Continue the weld on around the backside of the depression and back to the starting point. At the weld's beginning point, move to the inside to begin an inward spiral, ending at the center of the depression. Each weld pass should overlap the preceding weld bead by approximately 35% to 50%.

ARC WELDING (AW)

Shielded metal arc welding (SMAW), gas metal arc welding (GMAW), and flux cored arc welding (FCAW) are all approved processes for rail repair.

Shielded Metal Arc Welding

The flexibility of the shielded metal arc welding process enables its extensive use throughout the rail industry. It can be used to repair broken parts or rebuild worn surfaces, Figure 22-25.

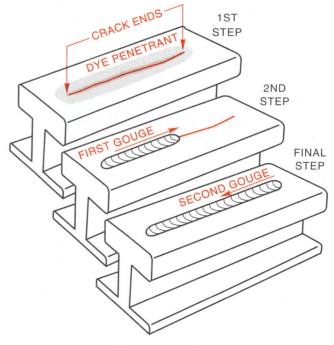

Figure 22-22 Steps for locating and gouging out a crack. Gouge from both ends of the crack toward the center.

Figure 22-23 Damage and wear at the track ends.

Because of the sensitivity of the rail metal to hydrogen embrittlement, all filler metal and fluxes must be kept dry. Once the electrodes are removed from their containers, they must be placed in a suitable electrode oven, **Table 22-1** (page 582).

Arc strikes must not be made outside of the welding zone, **Figure 22-26** (page 582). The arc should be struck just ahead of the starting point of the weld. Once the arc has stabilized, quickly move it to the desired starting point, depositing little or no filler metal. Hold the arc at the starting point until the desired weld size is established, never wider than 5/8 in. (16 mm). Slowly weld back over the starting point while maintaining a consistent weld bead width, **Figure 22-27** (page 582).

There must be as short an arc length as possible. The electrode must be angled into the molten weld pool by using a trailing angle. This angle concentrates most of the gaseous cloud formed as the flux burns over the molten weld pool. Using a leading angle leaves the molten weld pool poorly protected by the gaseous cloud. It also tends to push slag and spatter ahead of the molten weld pool. Poor weld protection, spatter, and slag ahead of the weld can result in weld defects such as porosity and slag inclusions.

Before breaking the arc, stop the forward motion long enough to fill the weld crater. Once the crater has been filled, break the arc back over the weld, **Figure 22-28** (page 583). Both filling the weld crater and breaking the arc over the weld will reduce crater cracks, **Figure 22-29** (page 583).

When multiple weld passes are required to complete the buildup, the first pass must be cleaned and inspected. If any defects are detected in the first pass, they must be corrected before the next pass is started.

To control both distortion and base metal overheating, each weld's bead must be no longer than 5 in. (250 mm) and applied in a skip weld pattern. To improve weld bead tie-ins, each layer should be offset from the previous layer, **Figure 22-30** (Figure 583).

As with any low-hydrogen electrode, open root welds are not recommended. A backing bar of a similar material to the rail must be used for any welding that would be considered an open root weld without such backing bar.

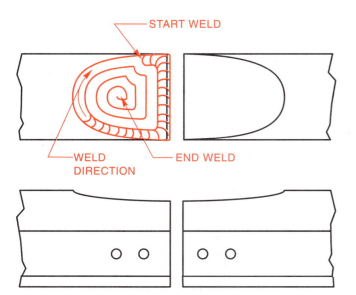

Figure 22-24   Repairing rail end damage.

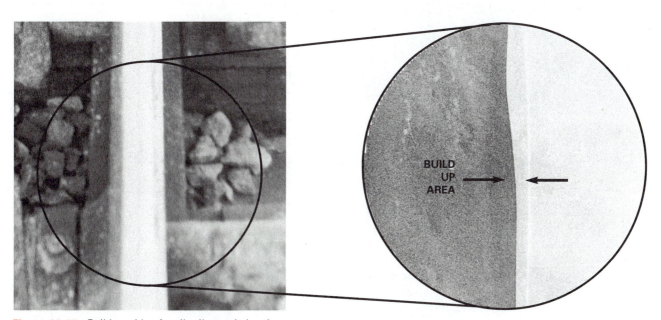

Figure 22-25   Build up side of mail rail at switch point.

| Filler Metal Type | Pre-Drying | | Storage | Maximum Exposure Time Hrs. | Redrying Times After Maximum Exposure* |
	Hermetically Sealed	Non-Hermetically Sealed			
E70XX	Not Required	2 hrs at 450°F to 500°F (230°C to 260°C)	250°F (120°C)	4	2 hrs at 450°F to 500°F (230°C to 260°C)
E80XX	Not Required	1 hr at 700°F to 800°F (370°C to 425°C)	250°F (120°C)	2	1 hr at 700°F to 800°F (370°C to 425°C)
E90XX	Not Required	1 hr at 700°F to 800°F (370°C to 425°C)	250°F (120°C)	1	1 hr at 700°F to 800°F (370°C to 425°C)
E100XX	Not Required	1 hr at 700°F to 800°F (370°C to 425°C)	250°F (120°C)	1/2	1 hr at 700°F to 800°F (370°C to 425°C)
E110XX	Not Required	1 hr at 700°F to 800°F (370°C to 425°C)	250°F (120°C)	1/2	1 hr at 700°F to 800°F (370°C to 425°C)
E120XX	Not Required	1 hr at 700°F to 800°F (370°C to 425°C)	250°F (120°C)	1/2	1 hr at 700°F to 800°F (370°C to 425°C)

*Redrying is allowed a maximum of one time but no redrying is allowed if the electrodes were wet.

Table 22-1 Electrode Handling Requirements to Reduce the Possibility of Moisture Contamination.

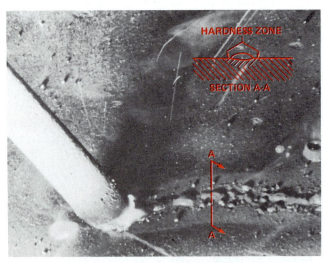

Figure 22-26 Arc strikes.

Gas Metal Arc Welding

Gas metal arc welding is very sensitive to drafts and wind. The shielding gas covering the molten weld pool is easily disturbed. To prevent this, a wind break or enclosure must be set up around the weld area. As with SMA welding, distortion is a major concern; so short skip welds should be used. Excessively long, large, and hot welds can result in the parts being rendered unusable.

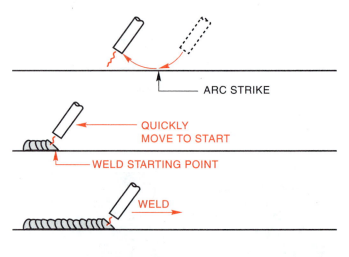
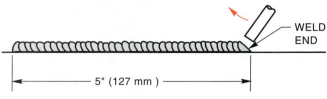

Figure 22-27 Weld starting.

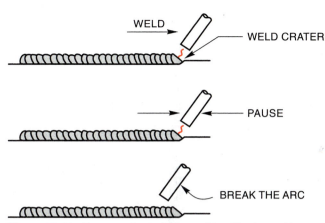

WELD

WELD CRATER

PAUSE

BREAK THE ARC

Figure 22-28 Pause at the end of a weld to file the weld crater before breaking the arc back over the weld bead.

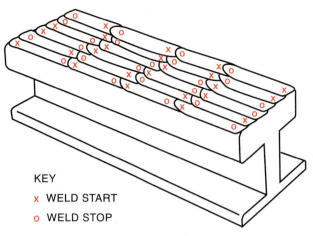

KEY

x   WELD START

o   WELD STOP

Figure 22-30 Stagger starts and stops between welds and weld passes.

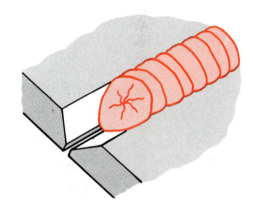

Figure 22-29 Crater cracks.

One of the major advantages of using GMAW is the absence of or small post-weld bead cleanup. It is also easier to build up wear areas, since the deposition rate can be controlled. If water-cooled equipment is used, care must be taken to avoid water leaks because they are a source of hydrogen.

Flux Cored Arc Welding

Flux cored arc welding has many major advantages. It has a high deposition rate for fast buildups. The flux core, like the flux on an SMA welding electrode, provides weld with shielding, metal purification, alloys, weld bead contour control, and arc stabilizing, but it is not as susceptible to atmospheric moisture contamination. Care still must be taken to protect the filler metal from moisture, but it is less likely to develop a problem under normal situations.

The addition of a shielding gas can help produce a tougher weld. Less alloys are lost to the atmosphere when gas shielding is used. There is little effect on the portability of this process with or without shielding gas. The decision to use shielding gas must be based on the WPS and your specific needs.

FLASH WELDING (FW)

Flash welding is a resistance welding process. The heat for fusion is produced when a low-voltage, high-amperage current is passed across the interfacing joint surfaces. For track welding the rail ends are preheated before the welding current is applied. Once the *faying* surfaces have melted sufficiently, powerful 50- to 100-ton cylinders force the ends together and the welding current is turned off, **Figure 22-31** (page 584). Post-weld treatment can include shaping, finishing, and post-weld heat treatment.

These long sections are loaded on special trains and hauled to the construction sight. Each train can haul enough track for approximately 6 miles of rail. Large machines (see Color Plate 44B) are used to remove the old track and replace new sections.

THERMITE WELDING (TW)

Thermite welding is a process that employs an exothermic reaction to develop a high temperature. This process is based on the great attraction of aluminum for oxygen. Aluminum can be used as a reducing agent for many oxides.

Thermite welding of rails is used throughout the world to join lengths of rails into continuous track.

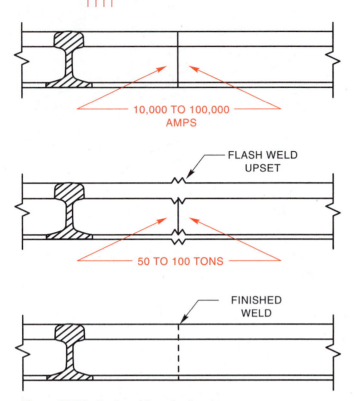

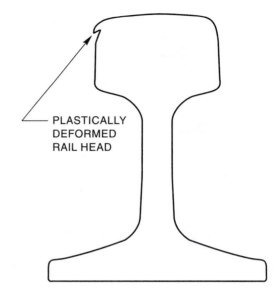

Figure 22-31 Flash welding of rail ends.

Figure 22-32 Plastic deformation of rail head occurs over time with use.

None of the other methods of joining and welding rails offer the advantages and service provided by thermite welding. The thermite welding process is widely used because of its high-quality welds, relative simplicity, portability, and economy.

Several companies make thermite welding products. The exact welding procedure must contain preheat temperature, chemical reaction time, charge size, joint spacing, and other essential welding variables and must be obtained from the manufacturer's WPS for each product being used and rail type being welded.

Thermite compound consists of finely divided aluminum and iron oxide mixed in a ratio of 1 to 3 by weight. Alloys are added to the mixture to obtain the desired weld metal properties. The mixture is not explosive and can be ignited only at a temperature of 2800°F (1537°C). A special ignitor is used to start the reaction. After chemical reaction has taken place, the melt attains a temperature of 4500°F (2482°C).

The products of the reaction are superheated iron alloy and alumina slag. The slag floats to the top and is not used in making the weld (see Color Plate 44A).

The rail ends must be cleaned, prepared, and aligned in preparation for welding. Cleaning must include the removal of all dirt, rust, oil, grease, paint, or other sources of possible contamination. Metal flaws such as burrs, chips, and rolled-over edges must be removed, **Figure 22-32**.

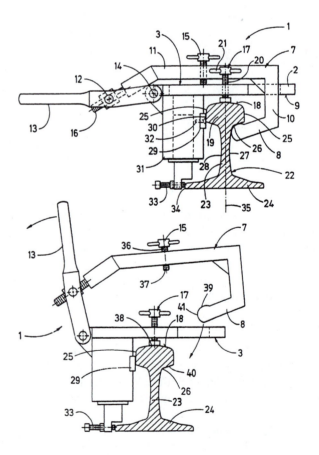

Figure 22-33 Rail clamp used to align and hold rail ends for welding.

Rail end spacing, smoothness, and squareness are all important. The rail ends must be properly prepared for welding. The rail end face must be cut square, smooth, and parallel. They can be cut by sawing, abrasive disk, or oxyfuel gas cutting.

The rails must be secured in proper alignment to prevent their movement during the welding process. Vertical, horizontal, and joint gap alignments as well as the correct offset of the rails can be provided by special clamping and alignment devices, **Figure 22-33**.

Thermite welds are essentially steel castings. The hardware parts used in making a thermite weld consist of a crucible, mold, and slag pan and are all made from a refractory material. The crucible is a container designed to hold the thermite powder while the chemical reaction heats up before welding occurs. The mold is placed around the rail ends and sealed to hold the weld metal in place until it cools. The slag pan provides a container for the overflow of slag and any excess weld metal. The crucible can be designed as either reusable or for a one-time use. **Figure 22-34** illustrates the arrangement of a thermite welding setup.

Moisture of any kind such as rain, snow, ice, or even the slightest dampness can make the thermite process hazardous to use. If the powder, mold, crucible, slag pan, or track are damp, a steam explosion may occur during the weld.

■ CAUTION

A steam explosion can cause serious injury or even death.

Refractory materials are designed to withstand the heat of welding but have relatively low mechanical strength. Care must be taken during their installation so as not to crack or break them by forcing, overtightening, or binding.

■ CAUTION

Never use damaged or improperly fitted molds. Damaged molds present a serious hazard to everyone in the area. If the damaged mold fails during the thermite welding process, superheated metal and slag will be released. Because of the temperature and quantity of this material, it can cause steam explosions even from relatively dry ballast or soil. Such explosions can throw large quantities of molten material several feet from the weld.

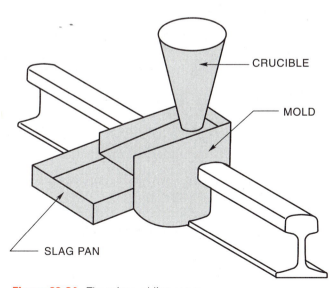

Figure 22-34 Thermite welding setup.

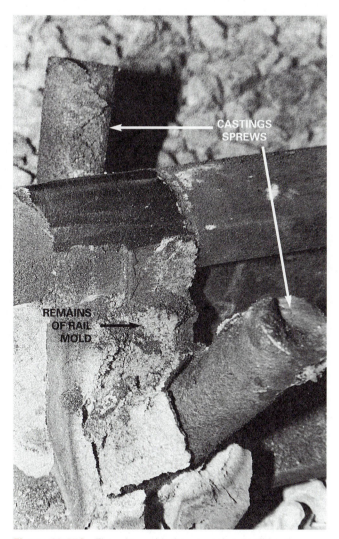

Figure 22-35A Thermite weld after removing crucible, slag pan, and mold.

Following the preheating of the rail ends to WPS recommended temperature, the thermite mixture in the crucible is ignited. When the metal is molten, the plug in the crucible collapses, and the metal flows into the mold. The weld metal temperature is about two times the melting temperature of the base metal. When the weld metal flows into the joint, therefore, fusion takes place. After the deposited metal has cooled, the mold is removed and the riser and sprews are cut off, leaving the desired sound weld, **Figure 22-35** (page 585).

Thermite welding is also used to weld together ends of large reinforcement rods in concrete and to make welds in large sections where other methods of welding prove difficult.

The most unique application of thermite welding is used by the military. It uses a process to sabotage moving parts of equipment that were captured or that were going to be abandoned during battles.

Figure 22-35B Thermite weld after grinding.

REVIEW

1. What are some of the uses of rails?

2. What are the three major classifications of compositions for rails?

3. What does the term *tensile strength* refer to?

4. What does a metal's hardness represent?

5. Why is the Charpy impact test done at −100°F (−73°C)?

6. What are soft carbon rails used for?

7. Why is austenitic manganese steel used for tracks?

8. What gives austenitic manganese steel its hardness?

9. Why do frog points wear so severely?

10. Why did early tracks have joints every 39 feet (11.8 m)?

11. When welding on rails, why do you have to be careful not to have them distort?

12. How can welding distortion be controlled when welding in rails?

13. Why must rails be properly pre- and postheat treated?

14. What can happen to austenitic manganese steels if they are overheated?

15. What will happen to rails if they are cooled too quickly?

16. What is the difference between molecular hydrogen and atomic hydrogen?

17. Why should welding not be done in areas unprotected from strong winds?

18. Why is the gouged groove checked with dye penetrant?

19. Why is it important to know exactly where the starting and ending point of a crack are before starting a repair?

20. Sketch the weld bead pattern that must be used for rebuilding the damaged end of a rail.

21. Why must electrodes and fluxes be kept dry?

22. How can crater cracking be reduced?

23. How is the heat for fusion produced in flash welding?

24. Why is thermite welding used so extensively for welding of rail ends?

25. How can rail ends be secured for thermite welding?

26. Why must thermite welding supplies and equipment be kept dry?

Section
Six

RELATED PROCESSES AND TECHNOLOGY

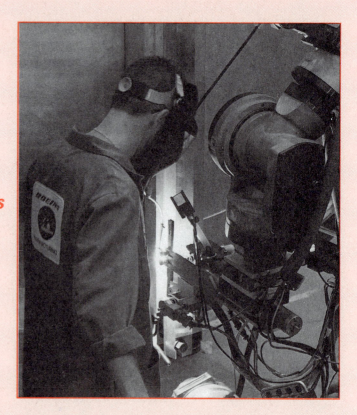

SUCCESS STORY

When Dixie Szumowski found herself a divorced mother with two children, she realized she needed a well-paying job in order to provide for her family's needs. She happened to read a story in a Las Vegas newspaper about a new program in mechanical technology offered by Community College of Southern Nevada.

Always interested in challenges and not too enamored with traditional women's jobs, Dixie enrolled in the program. Since classes were offered at night, she was able to work full time while in training.

She decided that her major emphasis would be on welding. It was with some trepidation that she entered her first class in welding theory; here she was, an attractive young lady with a heartwarming smile, surrounded by classmates who were not only twice her size, but many of them also twice her age. Her fears quickly disappeared, however, when she met the teacher. Bill Sowle, a retired Navy welding instructor, is widely known for his easygoing manner, which belies his demanding standards.

Before long, Dixie became one of the best welding students the college had ever seen. "It was a pleasure to see her work," said Bill Sowle, "I have seldom seen any student so determined to get things right the first time. She's a real perfectionist."

Her amazing success in welding gave her the courage to enter classes in other areas of mechanical technology. Before long she was troubleshooting, repairing, and fabricating electrical systems, became an expert in mechanical power transmission, and could talk gears, belts, sprockets, and chain drives with the best of them.

Soon the repair of hydraulic and pneumatic systems, blueprint reading, and knowledge of structure and behavior of ferrous and nonferrous materials had no secrets for her.

Dixie was graduated in 1990 with high honors as one of the first two students in the mechanical technology program and is presently employed as a maintenance technician with an international clothing company, where she is considered a rising star.

"Welding is my true love, though," said Dixie. "My success in welding gave me the courage to continue in my studies, and I'm still most happy with a welding torch in my hand."

Chapter 23

WELDING METALLURGY

OBJECTIVES

After completing this chapter, the student should be able to:

- list the crystalline structures of metals and explain how grains form.
- work with phase diagrams.
- list the five mechanisms used to strengthen metals.
- explain why steels are such versatile materials.
- describe the types of weld heat affected zones.
- discuss the problems hydrogen causes when welding steel.
- discuss the heat treatments used in welding.
- explain the cause of corrosion in stainless steel welds.

KEY TERMS

metallurgy	solid solution
lattice	allotropic
unit cell	precipitation hardening
crystalline structure	tempering
phase diagram	martensite
eutectic composition	(needle-like) acicular structure
body-centered cubic (BCC)	spheroidized microstructure
face-centered cubic (FCC)	recrystallization temperature
ferrite	grain refinement
pearlite	heat treatments
cementite	heat-affected zone (HAZ)
austenite	allotropic transformation

INTRODUCTION

Skilled welders need to know more than just how to establish an arc and manipulate the electrode in order to consistently deposit uniform high-quality welds. It is important for a competent welder to understand the materials being welded. With this knowledge, the welder can select the best processes and procedure to produce a weldment that is as strong and tough as possible. The welder needs to learn metallurgy to recognize that special attention might be needed when welding certain types of steel and to understand the kind of care required.

Metals gain their desirable mechanical and chemical properties as the result of alloying and heat-treating. Welding operations heat the metals, and that heating will certainly change not only the metal's initial structure but its properties as well. A skilled welder can minimize the effects these changes will have on the metal and its properties. See Color Plate 41.

HEAT, TEMPERATURE, AND ENERGY

Heat and temperature are both terms used to describe the quantity and level of thermal energy present. To better comprehend what takes place during a weld you must understand the differences between heat and temperature. Heat is the quantity of thermal energy, and temperature is the level of thermal activity.

Although both heat and temperature are used to describe the thermal energy in a material, they are independent values. A material can have a large quantity of heat energy in it but the material can be at a low temperature. Conversely, a material can be at a high temperature and have very little heat.

Heat

Heat is the amount of thermal energy in matter. All matter contains heat down to absolute zero (-460°F or -273°C). The basic U.S. unit of measure for heat is the British Thermal Unit (BTU). One BTU is defined as the amount of heat required to raise one pound of water one degree fahrenheit, and the SI unit of heat is the joule (J).

There are two forms of heat. One is called sensible because as it changes a change in temperature can be sensed or measured. The other form of heat is called latent. Latent heat is the heat required to change matter from one state to another, and it does not result in a temperature change. For example, if a pot containing water is heated on a stove, the water picks up heat from the burner and the water's temperature increases. The more heat put into the water the higher its temperature until it begins to boil. Once the water reaches 212°F (100°C), its temperature stops rising. As long as the pot is on the burner, heat is being put into the water; but there is no increase in sensible heat. The heat is all going into the latent heat required to change the water from the liquid state to a gaseous state, Figure 23-1.

Latent heat is absorbed by a material as it changes from a solid to a liquid state and from a liquid to a gaseous state. When matter changes from a gaseous to a liquid state or from a liquid to a solid state latent heat must be removed, Figure 23-2. A change in a material's latent heat also occurs when there's a change in the structure of the material. For example, when the crystal lattice of a metal changes, a change in latent heat occurs.

EXPERIMENT 23-1

Using 2 beakers, 2 hot plates, 2 thermometers, a cup of ice, a cup of ice water, gloves, safety glasses, and any other required safety protection, you are going to observe the effect of latent heat on the temperature rise of water, Figure 23-3 (page 592).

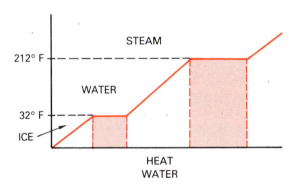

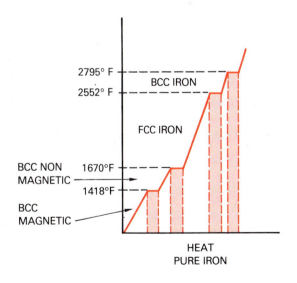

KEY
☐ AREA OF SENSIBLE HEAT CHANGE
▨ AREA OF LATENT HEAT CHANGE

Figure 23-2

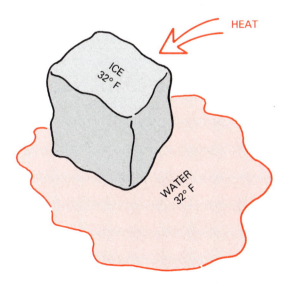

Figure 23-1 There is no change in temperature when there is a change in state.

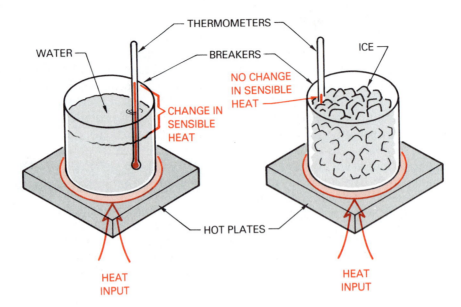

Figure 23-3 Both beakers have heat being added, but only one has a change in temperature.

Put one cup of ice water, 32°F (0°C) in one beaker and one cup of ice, 32°F (0°C) in the other beaker. Place both beakers on a hot plate. Slowly stir the contents of each beaker using the thermometers. Observe the change in temperatures each thermometer measures. Record the following information:

1. What was the temperature when you started?
2. What was the temperature after one minute?
3. What was the temperature of the water without the ice when the ice in the other cup was all gone?
4. How long did it take for all of the ice to melt? ◆

Temperature

Temperature is a measurement of the vibrating speed or frequency of the atoms in matter. The atoms in all matter vibrate down to absolute zero. The basic unit of measure is the degree. The U.S. unit is degrees Fahrenheit, and the SI unit is the degree Celsius.

As matter becomes warmer, its atoms vibrate at a higher frequency. As matter cools, the vibrating frequen-cy slows. This vibration of the atoms is what gives off the infrared light that comes from all objects. When the object becomes hot enough, the vibrating frequency of the atoms gives off visible light. We see that light as a dull red glow when the surface reaches a little above 1,000°F. See Color Plate 63. As the surface becomes even hotter, we can see the color light it gives off change until it glows "white hot," **Figure 23-4**.

We can tell the temperature of an object by the frequency of the light its vibrating atoms produces. That's how scientists tell the temperature of distant stars.

EXPERIMENT 23-2

Using a piece of mild steel, a safely setup oxyfuel torch, gloves, safety glasses, and required personal pro-

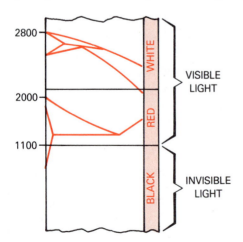

Figure 23-4 See Color Plate 63.

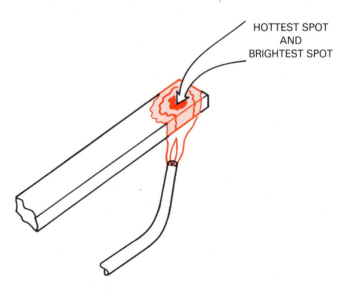

Figure 23-5 As the temperature of the metal increases, it begins to glow.

tection, you are going to observe the changing temperature's effect on the color of the steel, **Figure 23-5**.

Light the torch and hold it near one side of the mild steel. Observe the other side of the metal to see when it begins to change color. Record the following information:

1. What was the first color that you could see?
2. What was the second, third, fourth, etc., colors that you could see?
3. What was the last color that appeared before the metal melted?
4. Compare the colors you saw with the colors of the "Basic Guide to Ferrous Metallurgy." See Color Plate 63.

Repeat this experiment using different metals and see what colors they produce. ◆

MECHANICAL PROPERTIES OF METAL

In learning welding skills, an understanding of the mechanical properties of metals is most important. The mechanical properties of a metal can be described as those quantifiable properties that enable the metal to resist externally imposed forces without failing. If the mechanical properties of a metal are known, a product can be constructed that will meet specific engineering specifications. As a result, a safe and sound structure can be constructed.

All of a metal's properties interact with one another. Some properties are similar and complement each other, but others tend to be opposites. For example, a metal cannot be both hard and ductile. Some metals are hard and brittle, and others are hard and tough. Probably the most outstanding property of metals is the ability to have their properties altered by some form of heat treatment. Metals can be soft and then made hard, brittle, or strong by the correct heat treatment, yet other heat treatments can return the metal back to its original, soft form.

It is the responsibility of the metallurgist or engineer to select a metal that has the best group of properties for any specific job. Except in very unusual cases, a metallurgist would not create a new alloy for a job but merely select one from the tens of thousands of metal alloys already available. Often metallurgists must make difficult choices when designs call for properties that are usually not found together. Additionally the more unique the alloy, the greater its cost and often the more difficult it is to weld.

This section describes some of the significant mechanical properties of metals. The next section describes how various heat treatments can be used to change a metal's properties.

The sections that follow describe some of the significant mechanical properties that the welder should be familiar with in order to do a successful job of welding fabrication.

Hardness

Hardness may be defined as resistance to penetration. Files and drills are made of metals that rank high in hardness when properly heat treated. The hardness property may in many metals be increased or decreased by heat-treating methods, and increased in other metals by cold working. Since hardness is proportional to strength, it is a quick way to determine strength. It is also useful in determining whether the metal received the proper heat treatment, since heat treatment also affects strength. Hardness is measurable in a number of ways. Most methods quantify a metal's resistance to highly localized deformation.

Brittleness

Brittleness refers to the ease with which a metal will crack or break apart without noticeable deformation. Glass is brittle, and when broken, all of the pieces fit back together because it did not bend (deform) before breaking. Some types of cast iron are brittle and once broken will fit back together like a puzzle's parts. Brittleness is related to hardness in metals. Generally, as the hardness of a metal is increased, the brittleness is also increased. Brittleness is not measured by any testing method. It is the absence of ductility.

Ductility

Ductility is the ability of a metal to be permanently twisted, drawn out, bent, or changed in shape without cracking or breaking. Ductile metals include aluminum, copper, zinc, and soft steel. Ductility is measurable in a number of ways. Ductility in tensile tests is usually measured as a percentage of elongation and also as a percentage of reduction in area. It also can be measured with bend tests.

Toughness

Toughness is the property that allows a metal to withstand forces, sudden shock, or bends without fracturing. Toughness may vary considerably with different methods of load application and is commonly recognized as resistance to shock or impact loading.

Toughness is measured most often with the Charpy test. This test yields information about the resistance of a metal to sudden loading in the presence of a severe notch. Because only a small specimen is required, the test is faulted for not providing a general picture of a component's toughness. Unfortunately, tests on a larger scale require very expensive equipment and are very time-consuming.

Strength

Strength is the property of a metal to resist deforming. Common types of strength measurements are tensile, compressive, shear, and torsional, **Figure 23-6** (page 594).

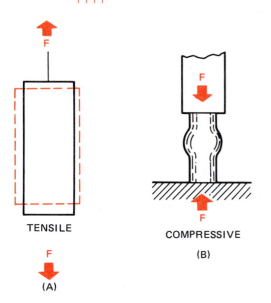
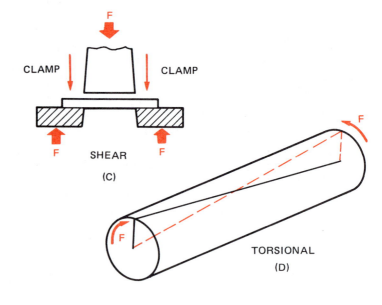

Figure 23-6 Types of forces (F) applied to metal.

■ **Tensile strength:** *Tensile strength* refers to the property of a material that resists forces applied to pull metal apart. Tension has two parts: yield strength and ultimate strength. *Yield strength* is the amount of strain needed to permanently deform a test specimen. The *yield point* is the point during tensile loading when the metal stops stretching and begins to be permanently made longer by deforming. Like a rubber band that stretches and returns to its original size, metal that stretches before the yield point is reached will return to its original shape. After the yield point is reached the metal is usually longer and thinner. Some metals stretch a great deal before they yield, and others stretch a great deal before and after the yield point. These metals are considered to have high ductility. *Ultimate strength* is a measure of the load that breaks a specimen. Some metals may become work-hardened as they are stretched during a tensile test. These metals will actually become stronger and harder as a result of being tested. Other metals lose strength once they pass the yield point and fail at a much lower force. Metals that do not stretch much before they break are brittle. The tensile strength of a metal can be determined by a tensile testing machine.

■ **Compressive strength:** Compressive strength is the property of a material to resist being crushed. The compressive strength of cast iron, rather brittle material, is three to four times its tensile strength.

■ **Shear strength:** Shear strength of a material is a measure of how well a part can withstand forces acting to cut or slice it apart.

■ **Torsional strength:** Torsional strength is the property of a material to withstand a twisting force.

Other Mechanical Concepts

Strain is deformation caused by stress. The part shown in **Figure 23-7** is under stress and was strained

(deformed) by the external load. The deformation is in the form of a bend.

Elasticity is the ability of a material to return to its original form after removal of the load. The yield point of a material is the limit to which the material can be loaded and still return to its original form after the load has been removed, **Figure 23-8**.

Elastic limit is defined as the maximum load, per unit of area, to which a material will respond with a deformation directly proportional to the load. When the force on the material exceeds the elastic limit, the material will be deformed permanently. The amount of permanent deformation is proportional to the stress level above the elastic limit. When stressed below its elastic limit, the metal returns to its original shape.

Impact strength is the ability of a metal to resist fracture under a sudden load. An example of a material that is ductile and yet has low-impact strength is Silly Putty. If it is pulled slowly, it stretches easily, but if it is pulled quickly, it forms a brittle fracture.

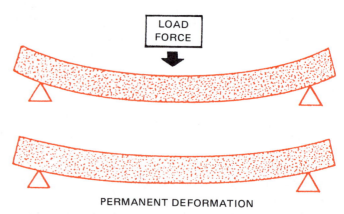

Figure 23-7 Effect of excessive stress causing permanent strain in a beam.

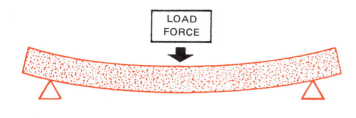

BEAM RETURNS TO ORIGINAL FORM

Figure 23-8 Reaction of an elastic beam to a force.

STRUCTURE OF MATTER

All solid matter exists in one of two basic forms. Solid matter is either crystalline or amorphous in form. Solids that are crystalline in form have an orderly arrangement of their atoms. Each crystal making up the solid can be very small, too small to be seen without a microscope. Examples of materials that are crystalline in form include metals and most minerals like table salt, **Figure 23-9.** Amorphic materials have no orderly arrangement of their atoms into crystals. Examples of amorphic materials include glass and silicon, **Figure 23-10.** Both crystalline and amorphic materials look and feel like solids, so without sophisticated testing equipment, you cannot tell the difference between them.

CRYSTALLINE STRUCTURES OF METAL

The fundamental building blocks of all metals are atoms arranged in very precise three-dimensional patterns called crystal lattices. Each metal has a characteristic pattern that forms these crystal lattices. The smallest identifiable group of atoms is the unit cell. The unit cells that characterize all commercial metals are illustrated in **Figures 23-11, 23-12,** and **23-13** (page 596). It may take millions of these individual unit cells to form one crystal. Although some metals have identical atomic arrangements, the dimensions between individual atoms vary from metal to metal. Some metals change their lattice structure when heated above a specific temperature.

Crystals develop and grow by the attachment of atoms to the submicroscopic unit cells forming the metal's characteristic crystal structure. The final individual crystal size within the metal depends on the length of time the material is at the crystal forming temperature and any obstructions to its growth. In most cases, solid

Metal	Crystal Type
Aluminum	fcc
Chromium	bcc
Copper	fcc
Gold	fcc
Iron (alpha)	fcc
Iron (gamma)	fcc
Iron (delta)	bcc
Lead	fcc
Nickel	fcc
Silver	fcc
Tungsten	bcc
Zinc	hpc

KEY
fcc = Face-centered cubic
bcc = Body-centered cubic
hpc = Hexagoral close-packed

Figure 23-9 Crystal structure of common metals.

Materials		Crystal Type
Copper Acelylide	Cu_2C_2	none
Gadolinium Oxide	Gd_2O_3	none
Iron Hydroxide	$Fe(OH)_2$	none
Lead Oxide	Pb_2O	none
Nickel Monosulfide	NiS	none
Silicane Dioxide	SiO_2	none

Figure 23-10 Amorphic materials.

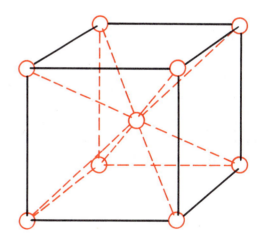

Figure 23-11 Body-centered cubic unit cell.

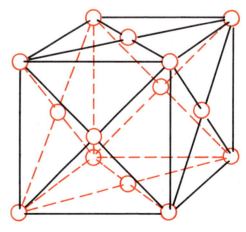

Figure 23-12 Face-centered cubic unit cell.

crystals continue to grow larger randomly from a liquid as it cools, until they encounter other crystals growing in a similar fashion. The result is solid metal composed of microscopic crystals with the same structure but different orientations. Their crystalline shapes depend on how they grew and how other crystals interrupted that growth. Crystals growing from liquid develop a dendritic or columnar structure, Figures 23-14 and 23-15.

The crystal structures are studied by polishing small pieces of the metals, etching them in dilute acids, and examining the etched structures with a microscope. Such examinations enable metallurgists to determine how the metal formed and to observe changes caused by heat treating and alloying. These microscopic examinations reveal various combinations of three different phases: pure metal, solid solutions of two or more metals dissolved in one another, and intermetallic compounds.

PHASE DIAGRAMS

If the metal-working industries used only pure metals, the only required information about their crystalline structure could be a list of their melting temperatures and crystalline structures. However, most engineering materials are alloys, not pure metals. An *alloy* is a metal with one or more elements added to it, resulting in a significant change in the metal's properties. It is inconvenient, if not impossible, to list all phases and temperatures at which alloys exist. That kind of information is summarized in graphs called phase diagrams. Phase diagrams are also known as equilibrium or constitution diagrams, and the terms are used interchangeably. These diagrams do not necessarily describe what happens with rapid changes in temperature since metals are sluggish in response to temperature fluctuations. They do describe the constituents present at temperature equilibrium.

Lead-Tin Phase Diagram

The chart in **Figure 23-16** is a phase diagram representing the changes brought about by alloying lead (Pb) with tin (Sn). Although this phase diagram is simpler than the iron carbon phase diagram used for steel, it has many similarities. On the charts temperature is vertical and alloy percent; in this case, lead and tin are horizontal.

Metallurgy uses Greek letters to identify different crystal structures. On this chart the Greek letter α (alpha, or a) is used for one crystal form and the Greek letter β (beta, or b) is used to represent the other crystal form. The chart has four different areas identified.

1. *Liquid phase:* The area at the top with the highest temperatures is where all the metal is a molten liquid.
2. *Solid phase:* The area in the lower center with the lowest temperatures is a solid mixture of α- and β-type crystals.

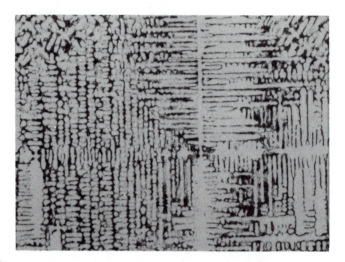

Figure 23-14 Dendritic structure. Weld metal in stellite 6B. (100X magnification.) (Courtesy of General Electric Company.)

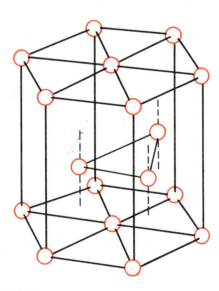

Figure 23-13 Hexagonal close-packed cubic unit cell.

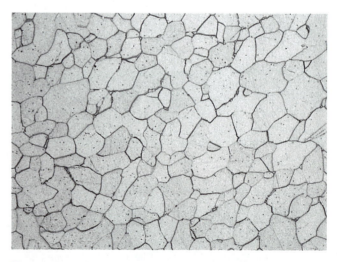

Figure 23-15 Columnar structure. Nickel alloy. (100X magnification.) (Courtesy of LECO Corp.)

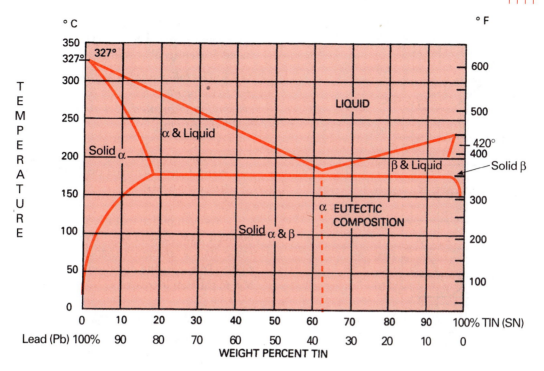

Figure 23-16 Lead-tin phase diagram.

3. *Liquid-solid phase:* The two triangular-shaped areas that touch in the center contain a slurry or paste made up of liquid and a specific type of solid crystals.

4. *Solid-solution phase:* The two triangular-shaped areas that stand vertical, one on each side, next to the temperature scales are solid crystals in either α form on the left side or β form on the right side.

Notice on the chart that although 100% lead becomes a liquid at 620°F (327°C) and 100% tin becomes a liquid at 420°F (232°C), a mixture of 38.1% lead and 61.9% tin becomes a liquid at 362°F (183°C). The mixture has a lower melting temperature than either of the two metals in the mixture. The mixture melts at a temperature 258°F (144°C) below 100% lead and 58°F (49°C) below 100% tin. This mixture is called the *eutectic composition* of lead and tin. A eutectic composition is the lowest possible melting temperature of an alloy.

On the chart the broadest temperature for a slurry or paste is a mixture of approximately 80% (80.5%) lead and 20% (19.5%) tin. This mixture remains partially liquid and solid during a 173°F (78°C) temperature change. While a metal is in the liquid-solid phase it is very weak and any movement will cause cracks to form. For this reason, some aluminum alloys crack when a GTA weld is made without adding filler metal and many other metals form crater cracks at the end of a weld. In both cases as the metal cools it shrinks and pulls itself apart in the center of the weld. The addition of filler metal changes the alloy so it is not as subject to hot cracking. The only metals that don't go through the liquid-solid phase are pure metals and those eutectic composition alloys.

As a 90% lead and 10% tin mixture cools from the 100% liquid phase, it forms an α solid crystal in a liquid.

As cooling continues all of the liquid forms into the α crystal. But as solid α crystals cool to around 300°F (150°C), some of them change into the β crystal form. This type of solid solution phase change occurs in many metals. Steel goes through several such changes as it's heated and cooled even though it never melted. It is these changes in steel that allow it to be hardened and softened by heating, *quenched, tempered,* and *annealed.*

Phase diagrams for other metal alloys are more complicated but they are used in exactly the same way. They describe the effects of changes in temperature or alloying on different phases.

*NOTE: The following information relates to **Figure 23-17** and to Color Plate 63.*

Iron-Carbon Phase Diagram

The iron-carbon phase diagram is illustrated in both Color Plate 63 and **Figure 23-17** (page 598). The iron-carbon phase diagram is more complex than that for the lead-tin—it has more lines with more solid solution phase changes—but they are read in the same way. In the iron-carbon diagram used here the percentage of iron starts at 100% and goes down to 99.1% while the percentage of carbon goes from 0.0% to 0.9%. Unlike the lead-tin alloys that go from 0.0% to 100% mixtures, very small changes in the percentage of carbon produce major changes in the alloys properties.

Iron is a pure metal containing no measurable carbon and is relatively soft. This soft metal when alloyed with as little as 0.80% carbon can become tool steel, **Table 23-1** (page 598). Other alloying elements are added to iron to enhance its properties. No other alloying element has such a dramatic effect as carbon.

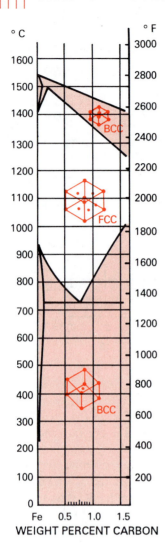

Figure 23-17 Iron iron-carbon phase diagram.

body-centered cubic form of iron is called *alpha ferrite*, abbreviated α-Fe. The face-centered cubic form of iron is called Austenite, abbreviated γ-Fe.

*NOTE: The following information relates to **Figure 23-18** and to **Color Plate 63**.[1] The boxed numbered paragraphs can be found printed in a column on the right side of both illustrations.*

1. **Transformation Range** — In this range steels undergo internal atomic changes that radically affect the properties of the material.[1]

You will notice that the Transformation Range (**1**) is between lines (**2**) and (**3**). This area (**1**) on **Figure 23-19** (page 200) is triangular shaped with the small end to the right. This is in the direction that the percentage of carbon in the iron increases.

2. **Lower Transformation Temperature** (A_1) — Termed Ac_1 on heating, Ar_1 on cooling. Below Ac_1 structure ordinarily consists of Ferrite and Pearlite (see 2a and 2b below). On heating through Ac_1 these constituents begin to dissolve in each other to form Austenite (see 2d below), which is non-magnetic. This dissolving action continues on heating through the Transformation Range until the solid solution is complete at the upper transformation temperature.[1]

We do not think of solids being able to dissolve into each other, but above the lower transformation temperature (A_1) the carbon rich Cementite (see 2c below) and the carbon lean Ferrite, both solids, dissolve into each other. The dissolving of one material into another as the mixture is heated is similar to dissolving salt into water, Experiment 23-3.

Ferrite, cementite, and austenite are all crystalline forms of iron and iron-carbon alloys. Their formation is based on three factors; the carbon content, temperature, and time.

1. *Carbon content:* Before any of the crystal forms can be created, the correct carbon percentage must exist in the alloy.

2. *Temperature:* With the correct carbon content, an iron-carbon alloy will form a specific crystal at the

Iron is called an **allotropic** metal, because it exists in two different crystal forms in the solid state. It changes between the different crystal forms as its temperature changes. The changes in the crystal structure occur at very precise temperatures. Pure iron forms the **body-centered cubic (BCC)** crystal below a temperature of 1675°F (913°C). Iron changes to form the **face-centered cubic (FCC)** crystal above 1675° (913°). The

Alloy Name	% Carbon*	Major Properties
Iron	0.0% to 0.03%	Soft, Easily Formed, Not Hardinable
Low-Carbon	0.03% to 0.30%	Strong, Formable
Medium-Carbon	0.30% to 0.50%	High Strength, Tough
High-Carbon	0.50% to 0.90%	Hard, Tough
Tool Steel	0.80% to 1.50%	Hard, Brittle
Cast Iron	2% to 4%	Hard, Brittle, Most Types Resist Oxidation

Table 23-1 Iron-Carbon Alloys.*Carbon is not the only alloying element added to iron.

[1]*Basic Guide to Ferrous Metallurgy,* Copyright 1954 by Tempil, Hamilton Boulevard, South Plainfield, NJ 07080.

TEMPIL°
BASIC GUIDE TO FERROUS METALLURGY

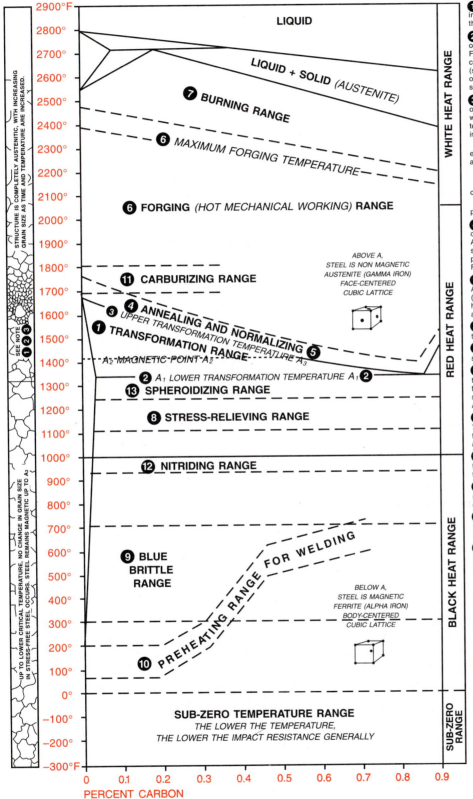

1 TRANSFORMATION RANGE. In this range steels undergo internal atomic changes which radically affect the properties of the material.

2 LOWER TRANSFORMATION TEMPERATURE (A₁). Termed Ac₁ on heating, Ar₁ on cooling. Below Ac₁ structure ordinarily consists of FERRITE and PEARLITE (see below). On heating through Ac₁ these constituents begin to dissolve in each other to form AUSTENITE (see below) which is non-magnetic. This dissolving action continues on heating through the TRANSFORMATION RANGE until the solid solution is complete at the upper transformation temperature.

3 UPPER TRANSFORMATION TEMPERATURE (A₃). Termed Ac₃ on heating, Ar₃ on cooling. Above this temperature the structure consists wholly of AUSTENITE which coarsens with increasing time and temperature. Upper transformation temperature is lowered as carbon increases to 0.85% (eutectoid point).

• FERRITE is practically pure iron (in plain carbon steels) existing below the lower transformation temperature. It is magnetic and has very slight solid solubility for carbon.

• PEARLITE is a mechanical mixture of FERRITE and CEMENTITE.

• CEMENTITE or IRON CARBIDE is a compound of iron and carbon, Fe₃C.

• AUSTENITE is the non-magnetic form of iron and has the power to dissolve carbon and alloying elements.

4 ANNEALING, frequently referred to as FULL ANNEALING, consists of heating steels to slightly above Ac₃, holding for AUSTENITE to form, then slowly cooling in order to produce small grain size, softness, good ductility and other desirable properties. On cooling slowly the AUSTENITE transforms to FERRITE and PEARLITE.

5 NORMALIZING consists of heating steels to slightly above Ac₃, holding for AUSTENITE to form, then followed by cooling (in still air). On cooling, AUSTENITE transforms giving somewhat higher strength and hardness and slightly less ductility than in annealing.

6 FORGING RANGE extends to several hundred degrees above the UPPER TRANSFORMATION TEMPERATURE.

7 BURNING RANGE is above the FORGING RANGE. Burned steel is ruined and cannot be cured except by remelting.

8 STRESS RELIEVING consists of heating to a point below the LOWER TRANSFORMATION TEMPERATURE, A₁, holding for a sufficiently long period to relieve locked-up stresses, then slowly cooling. This process is sometimes called PROCESS ANNEALING.

9 BLUE BRITTLE RANGE occurs approximately from 300° to 700°F. Peening or working of steels should not be done between these temperatures, since they are more brittle in this range than above or below it.

10 PREHEATING FOR WELDING is carried out to prevent crack formation. See TEMPIL° PREHEATING CHART for recommended temperatures for various steels and non-ferrous metals.

11 CARBURIZING consists of dissolving carbon into surface of steel by heating to above transformation range in presence of carburizing compounds.

12 NITRIDING consists of heating certain special steels to about 1000°F for long periods in the presence of ammonia gas. Nitrogen is absorbed into the surface to produce extremely hard "skins".

13 SPHEROIDIZING consists of heating to just below the lower transformation temperature, A₁, for a sufficient length of time to put the CEMENTITE constituent of PEARLITE into globular form. This produces softness and in many cases good machinability.

• MARTENSITE is the hardest of the transformation products of AUSTENITE and is formed only on cooling below a certain temperature known as the M₃ temperature (about 400° to 600°F for carbon steels). Cooling to this temperature must be sufficiently rapid to prevent AUSTENITE from transforming to softer constituents at higher temperatures.

• EUTECTOID STEEL contains approximately 0.85% carbon.

• FLAKING occurs in many alloy steels and is a defect characterized by localized micro-cracking and "flake-like" fracturing. It is usually attributed to hydrogen bursts. Cure consists of cycle cooling to at least 600°F before air-cooling.

• OPEN OR RIMMING STEEL has not been completely deoxidized and the ingot solidifies with a sound surface ("rim") and a core portion containing blowholes which are welded in subsequent hot rolling.

• KILLED STEEL has been deoxidized at least sufficiently to solidify without appreciable gas evolution.

• SEMI-KILLED STEEL has been partially deoxidized to reduce solidification shrinkage in the ingot.

• A SIMPLE RULE: Brinell Hardness divided by two, times 1000, equals approximate Tensile Strength in pounds per square inch. (200 Brinell ÷ 2 x 1000 = approx. 100,000 Tensile Strength, p.s.i.)

Figure 23-18 See Color Plate 63.

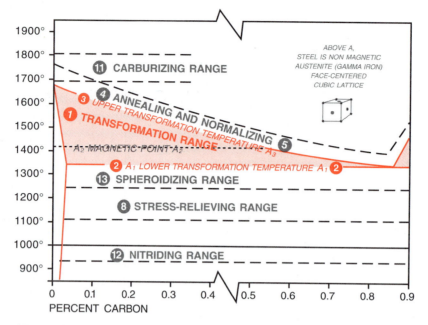

Figure 23-19 Courtesy of Tempil Division, Big Three Industries, Inc.

approximate temperature for that specific alloy given enough time.

3. *Time:* Crystal formation requires time. The more time at the correct temperature, the larger the crystals can grow. Heating is a relatively slow process compared to quenching, which can occur in a fraction of a second. Therefore, crystals tend to grow larger or change forms during heating and stay frozen in the larger size or form when cooled.

2a. Ferrite is practically pure iron (in plain carbon steels) existing below the lower transformation temperature. It is magnetic and has very slight solid solubility, less than 0.02%, of carbon. Very low carbon content alloys produce more ferrite crystals than they do other crystal forms. For this reason, these alloys are considered to be soft. Without enough carbon, these low-carbon alloys will not become very hard and brittle even if they are quenched. The inability to become hard even alongside a weld makes them easily machinable after welding.

2b. Pearlite is a mechanical mixture of Ferrite and Cementite, **Figure 23-20.** Pearlite is a mechanical mixture of ferrite and cementite crystals in much the same way that one might mix two different colors of sand. Each grain of sand still retains its unique color even though it has been mixed with the other colored grains. Unlike sand, each grain of ferrite and cementite fit together like a puzzle's parts, because they grew together.

2c. Cementite, or iron carbide, is a compound of iron and carbon, Fe_2C.

2d. Austenite is the non-magnetic form of iron and has the power to dissolve carbon and alloying elements. An iron-carbon alloy at low temperature can dissolve or absorb very little additional alloys. Like water and salt in Experiment 23-3, however, the higher the temperature and the longer the time the more salt that is

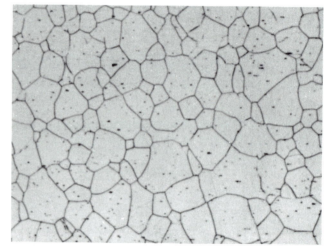

Figure 23-20 Pearlite: a mechanical mixture of ferrite and cementite (Fe_3C). (100X magnification.) (Courtesy of Oak Ridge National Laboratories.)

devolved. In the same manner iron will accept more alloys at higher temperatures. These new alloys may stay trapped within the iron-carbon crystal when they are cooled or they may separate. Once devolved at high temperatures, it is usually the cooling rate or time that affects the final makeup of the iron-carbon crystals.

3. **Upper Transformation Temperature (A_3)** — Termed Ac_3 on heating, Ar_3 on cooling. Above this temperature the structure consists wholly of Austenite, which coarsens with increasing time and temperature. Upper transformation temperature is lowered as carbon increases to 0.85% (eutectoid point).[1]

[1]*Basic Guide to Ferrous Metallurgy,* Copyright 1954 by Tempil, Hamilton Boulevard, South Plainfield, NJ 07080.

The upper transformation temperature (A_3) line slopes downward to the right. That is because alloying carbon to iron lowers the temperature for the allotropic crystal transformation from BCC to FCC. This temperature reduction is similar to what is observed as the mixture of lead-tin approached the 80.5% lead and 19.5% tin ratio, **Figure 23-16** (page 597). **Figure 23-17** (page 598), which represents carbon content ranging from 0.0% up to 1.50%, shows over a 200°F (100°C) drop in the melting temperature of iron-carbon alloys. The transition range starts at about 1,675°F (912°C) and lowers to about 1,335°F (724°C) when 0.85% carbon is present.

The temperature at which the allotropic crystal transition of ferrite to austenite occurs is so important in all heat treatments of steel that metallurgists call it the "critical temperature."

STRENGTHENING MECHANISMS

Perhaps the most important physical characteristic of a metal is strength. Most pure metals are relatively weak. Structures built of pure metals would be more massive and heavier than those built with metals strengthened by alloying and heat treating. Welders must understand numerous methods used to strengthen metals. Some of the strengthening methods used and how welding affects them are described next.

Solid-solution Hardening

It is possible to replace some of the atoms in the crystal lattice with atoms of another metal in a process called solid-solution hardening. In such cases, the lattices of the solid solution and the pure metal are the same except that the lattice of the solid solution is strained as more foreign atoms dissolve. These alloyed prestressed crystals are stronger and less ductile. The amount of change in these properties depends on the number of second atoms introduced, **Figure 23-21**.

Although an important metallurgical tool, solid-solution hardening has its drawbacks. Not all metals have lattice dimensions that allow significant substitution of other atoms. The amount that can be introduced this way is thus limited. Solid-solution hardening does have the advantage of not changing lattice structure as the result of thermal treatments. Thus solid-solution-strengthened alloys are generally considered very weldable, **Figure 23-22** (page 602). Many aluminum alloys are strengthened in this way.

Precipitation Hardening

Alloy systems that show a partial solubility in the solid phase generally have very low solubility at room temperature. Solubility increases with temperature until the alloy system reaches its solubility limit. For example, the aluminum-copper system has a solid-solution solubility of 0.25% copper at room temperature that increases to a

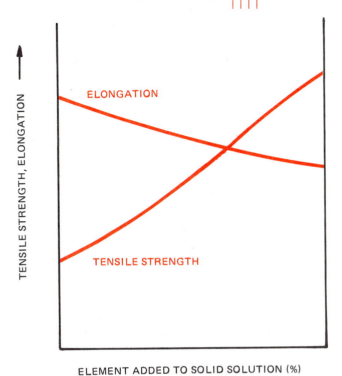

Figure 23-21 The general change in properties caused by the addition of other atoms in alloys, producing solid solutions.

maximum of 5.6% copper at 1,019°F (548°C). For this reason, very few, if any, aluminum alloys have more than 4.5% copper as an alloying element. These alloys reach their saturation just like a sponge saturated with water; when more water is added, it just runs off. When more alloy is added to metals, it will not be combined with the base metal.

Precipitation hardening is a heat treatment involving three steps. First, heating the alloy enough to dissolve the second phase and form a single solid solution. Second, quenching the alloy rapidly from the solution temperature to keep the second phase in solution, thus producing a supersaturated solution. Third, reheating the alloy with careful control of time and temperature to precipitate the second phase as very fine crystals that strengthen the lattice in which they had dissolved. This process, called precipitation hardening or age hardening, is the heat treatment used to strengthen many aluminum alloys.

EXPERIMENT 23-3

Using a beaker, water, salt, spoon, a hot plate, gloves, safety glasses, and any other required safety protection, you are going to observe the ability of salt to be dissolved into water as the water is heated.

The dissolving of salt in water shows an increase in solubility as the temperature increases. Add a spoonful of salt to the beaker of cold water and stir vigorously until the salt is dissolved. This is an example of a liquid

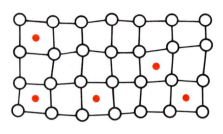

KEY

○   ATOMS OF THE BASE METAL

●   ATOMS OF THE HARDENING ALLOY

(A)

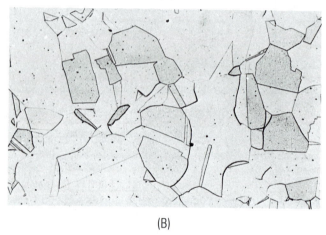

(B)

Figure 23-22 Solid solution austenite (B). (300X magnification.) (Courtesy of Beuhler Ltd.)

the solution. If more salt is slowly added and stirring is continued at some point, the salt will no longer dissolve into the solution. The water has now reached its solubility limit. No more salt will dissolve no matter how long the mixture is stirred.

If the beaker is heated, the salt crystals that did not dissolve at room temperature soon disappear. Now even more salt could be added, and it would dissolve. The increasing of the solubility limits by heating has resulted in a supersaturated solution. A supersaturated solution is one that, because of heating, a higher percentage of another material can be dissolved in it.

When the beaker of water is cooled to room temperature, with time the reverse action occurs. The supersaturated solution can no longer hold the salt in solution and rejects it from the solution. In time, salt crystals reappear in the beaker. ◆

Mechanical Mixtures of Phases

Two phases or constituents may exist in equilibrium, depending on the alloy's temperature and composition. The mixture may consist of two different crystals, which are solid solutions of the two metals of the alloy

(see the α & β area in Figure 23-16), or the mixture may consist of a single solid solution and an intermetallic compound, such as the pearlite in **Figure 23-23**. The properties of alloys that are mechanical mixtures of two phases are generally related linearly to the relative amounts of the metal's two constituents, **Figure 23-24**.

At room temperature, the iron-carbon alloy has two forms: alpha iron ferrite and cementite. The ferrite is very ductile but weak; the cementite is very strong but

Figure 23-23 Pearlite.

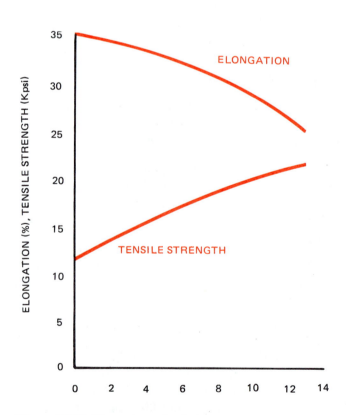

Figure 23-24 Change in mechanical properties caused by beta (silicon phase) in mechanical mixture with alpha (aluminum phase).

brittle. In a careful combination, the cementite stiffens the soft ferrite crystals, increasing its strength without losing too much ductility.

Quench Temper and Anneal

Quenching is the process of rapidly cooling a metal by one of several methods. The quicker the metal cools, the greater the quenching affect. The most common methods of quenching are listed from the slowest to the fastest:

■ Air quenching — Air is blown across the part, cooling it only slightly faster than it would cool in still air.

■ Oil quenching — The part is immersed in a bath of oil; and because oil is a poor conductor of heat, the part cools slower that it would in water.

■ Water quenching — The part is immersed in a bath of water and cools rapidly.

■ Brine quenching — The part is immersed in a salt water solution. The salt does not allow a steam pocket to form around the part as happens with straight water. This results in the fastest quenching time, **Figure 23-25**.

NOTE: Agitating the metal (moving it around rapidly) when it is immersed in a cooling liquid will speed up the cooling rate.

Tempering is the process of reheating a part that has been hardened through heating and quenched. The reheating reduces some of the brittle hardness caused by the quenching replacing it with toughness and a increased tensile strength.

EXPERIMENT 23-4

Using 3 or more pieces of mild steel approximately 1/4 in. thick, 1 in. wide, and 6 in. long, a vice with a hammer or tensile tester, hacksaw, file, water for quenching, two or more fire bricks, a safely assembled and properly lit oxyfuel torch, pliers, safety glasses, gloves, and any other required safety equipment, you are going to observe the effect that quenching and tempering has on metal.

Heat the pieces of metal one at a time to a bright red color. Place one of them, while still red hot, between hot fire bricks. Immerse the other two into the water while they are still glowing bright red. Moving the metal in the water will insure a faster quench.

■ CAUTION

The steam given off from the quenching of the hot metal can cause severe burns. Use your gloves and be careful not to allow the steam to burn you.

Set one of the pieces aside. File a smooth clean area on the other part. Slowly heat this piece on the opposite side from the filed area using the oxyfuel torch, **Figure 23-26**. Watch the clear spot, and it will begin to change colors. It will first become a very pale yellow and gradually change to a dark blue. When the surface is dark blue, stop heating it and allow it to cool as slowly as possible between hot fire bricks. The surface colors formed are called temper colors, and they indicate the temperature of the metal's surface, **Table 23-2**. The color comes from the layer of oxide formed on the hot metal's surface. The bluing on a gun barrel is the same thing, and it's used to both make the barrel stronger and to protect it from rusting. You could also use a 800°F (425°C) colored temperature indicating crayon like a Tempil® Temperature Indicator.

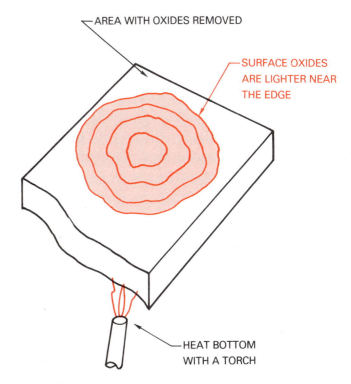

AREA WITH OXIDES REMOVED
SURFACE OXIDES ARE LIGHTER NEAR THE EDGE
HEAT BOTTOM WITH A TORCH

Figure 23-26

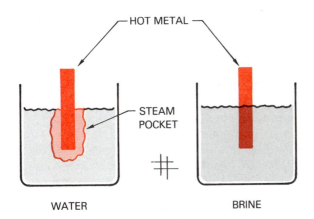

HOT METAL
STEAM POCKET
WATER   BRINE

Figure 23-25

Degrees Fahrenheit	Color of Steel
430	Very pale yellow
440	Light yellow
450	Pale straw-yellow
460	Straw-yellow
470	Deep straw-yellow
480	Dark yellow
490	Yellow-brown
500	Brown-yellow
510	Spotted red-brown
520	Brown-purple
530	Light purple
540	Full purple
550	Dark purple
560	Full blue
570	Light blue
640	Dark blue

Table 23-2 Temperatures Indicated by the Colors of Mild Steel.

steels). Cooling to this temperature must be sufficiently rapid to prevent Austenite from transforming to softer constituents at higher temperatures.[1] As formed, martensite is very hard and brittle and useless for most engineering applications.

When martensite is viewed under a microscope, it has a (needle-like) acicular structure, **Figure 23-29.** When welding medium and high carbon steels, the cooling rates can be fast enough to produce the undesirable martensite. Martensite formation can be minimized by preheating the steel in order to slow the cooling rates. If the surrounding metal is warmed by preheating, the weld does not lose its temperature as quickly, **Figure 23-30.**

Martensite can be tempered to a more useful structure at a temperature below the lower critical temperature, see line A$_1$ on **Figure 23-31** (page 606). The exact temperature, determined by the carbon content and other alloying elements, is generally furnished by the steel maker.

When martensite is tempered, some of the carbon atoms that were trapped interstitially (small holes in the crystalline lattice) between the iron atoms are allowed to defuse out, **Figure 23-32** (page 606). The precipitation of the trapped carbon atoms, allows the body-centered tetragonal crystal to reform into the body-centered cube crystal shape. This relieves the internal strain. The microstructures of tempered and untempered martensite look very much alike. In this condition, the tempered martensite is very strong and tough.

As the tempering time and/or temperature is increased, the structure changes to a spheroidized microstructure. In this state, the steel is in the soft or annealed condition. For example, when a chisel, drill bit or slag hammer is overground so that the sharp edge turned blue, this would indicate that a temperature of about 600°F (316°C) was reached. A temperature this high causes overtempering and softens the edge. Thus time and temperature are critical variables in the loss of strength and hardness, **Figure 23-33** (page 606).

NOTE: For all practical purposes plain carbon steels with less than 0.30% carbon cannot be hardened through heat-treating and quenching.

After both specimens have cooled to room temperature, they can be tested. If you have a tensile testing machine, test each specimen and record their failure strength. If you don't have a tensile tester, make a 1/4-in. (6-mm) -deep saw cut on both edges of each specimen, **Figure 23-27.** Place the specimens in a vise and break them. Note how they broke.

Look at the fractured surface and record which has the light-colored surface and which has the darker surface. Also note which one has the smallest grain sizes shown.

This experiment can be repeated using different metals and alloys. ◆

Martensitic Reactions

Heating a carbon-iron alloy above the temperature at which austenite forms, see Line (A$_3$), **Figure 23-28,** and then quenching it rapidly in water is certainly not an equilibrium condition. The FCC crystals are unable to change to BCC since the rapid cooling was too fast to allow change.

Martensite is the hardest of the transformation products of Austenite and is formed only on cooling below a certain temperature known as the M. temperature (about 400° to 600°F (200°C to 315°C) for carbon

Cold Work

When metals are deformed at room temperature by cold rolling, drawing, or swaging, the grains are flattened and elongated. Complex movements occur within and between the grains. These movements distort and disrupt the crystalline structure of the metal by markedly increasing its strength and decreasing its ductility. The presence of impurities or alloying elements in metals causes them to work harder more quickly. Sheets, bars, and tubes are intentionally cold-worked to increase their

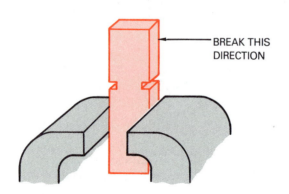

BREAK THIS DIRECTION

Figure 23-27

[1]*Basic Guide to Ferrous Metallurgy,* Copyright 1954 by Tempil, Hamilton Boulevard, South Plainfield, NJ 07080.

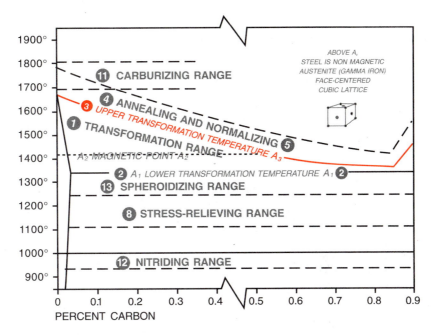

Figure 23-28 Courtesy of Tempil Division, Big Three Industries, Inc.

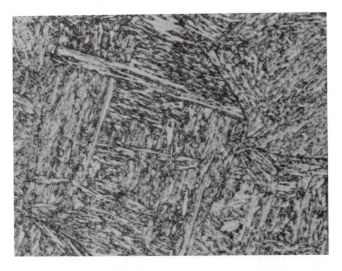

Figure 23-29 Martensite. Note the acicular needlelike structure. (500X magnification.) (Courtesy of Oak Ridge National Laboratories.)

WITHOUT PREHEAT — The temperature starts close together and spreads out quicker (a large temperature change between weld and sides of plate).

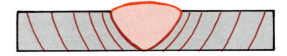

WITH PREHEAT — The temperature starts further apart and spacing changes slowly (little temperature difference between weld and sides of plate).

Figure 23-30 WITHOUT PREHEAT: The temperature starts close together and spreads out quickly (a large temperature change between weld and sides of plate). WITH PREHEAT: The temperature starts farther apart and spacing changes slowly (little temperature difference between weld and sides of plate).

strength, since cold working will strengthen almost all metals and their alloys.

The cold-worked structure can be annealed by heating the metal above the **recrystallization temperature.** Above that temperature new crystals grow. The size of the new grains depends on the severity of the prior cold working and the time above the recrystallization temperature. The final annealed structure is weaker and more ductile than the cold-worked structure. The final properties depend primarily on the alloy and on the grain size. The coarse-grained metals are weaker.

Grain Size Control

One of the few metallurgical effects common to all metals and their alloys is grain growth. When metals are heated, grain growth is expected. The rate of growth increases with temperature and the length of time at that temperature. There are charts that show the effect of time and temperature on the austenite grain growth. These are called time-temperature-transformation (TTT) charts. The process involves larger grains devouring smaller grains. Coarse grains are weaker and tend to be more ductile than fine grains.

The longer the metal is held at a high temperature, above A_3, the larger the grain size. (See the left vertical strip on Figure 23-18, page 599.) The Austenite crystals

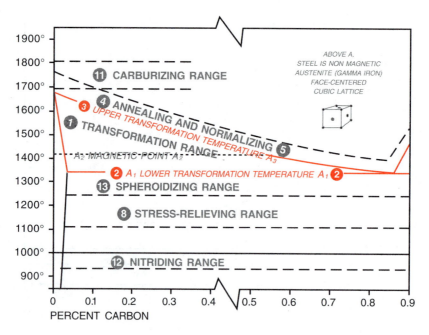

Figure 23-31 Courtesy of Tempil Division, Big Three Industries, Inc.

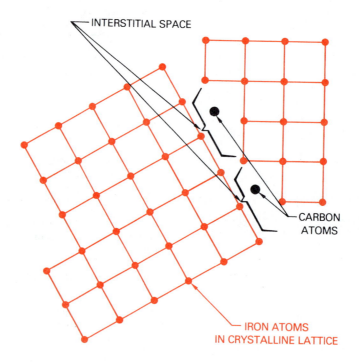

Figure 23-32

will grow so large that when they are cooled and transformed into Pearlite their large size can be easily seen in a fracture. Some welders know that this large grain is detrimental to the metal's strength and often refer to it as having been "crystallized."

The production of fine grains is not as simple because large grains cannot be shrunk. Reducing the grain size requires creation of fresh grains by cold working and recrystallizing the metal, such as that of ferrite to austenite or austenite to ferrite. This technique is common to all metals and alloys. The same results can be obtained with an allotropic transformation.

Allotropic transformation also requires the creation of fresh grains. Since the temperatures of allotropic transformation are high, the new grains begin to grow almost immediately. To obtain a fine-grained structure, the metal must be heated quickly above the critical temperature, line A_3 on **Figure 23-28**, and cooled quickly in a

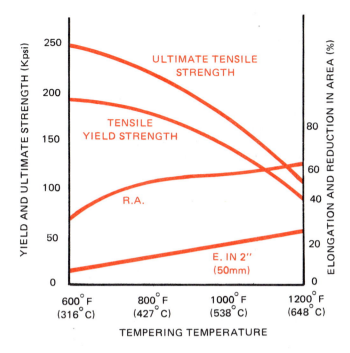

Figure 23-33 Changes in properties of SAE 1050 martensite caused by tempering.

process called grain refinement. Not all metals exhibit allotropic transformation. Fortunately, iron does, and this is one reason why steels are such versatile materials.

HEAT TREATMENTS ASSOCIATED WITH WELDING

Welding specifications frequently call for heat treating joints before welding or after fabrication. To avoid mistakes in their application, welders should understand the reasons for these heat treatments.

Preheat

Preheat is used to reduce the rate at which welds cool. Generally, it provides two beneficial effects — lower residual stresses and reduced cracking. The lowest possible temperature should be selected because preheat increases the size of the heat-affected zone and can damage some grades of quenched and tempered steels. The amount of preheat generally is increased when welding stronger plates or in response to higher levels of hydrogen contamination.

Area 10, "Preheating for Welding," Figure 23-34 (page 608), shows the temperature range for preheating various iron-carbon alloys. As the preheat area moves to the higher carbon percentage side, it rises in temperature. This is because as the percentage of carbon increases the alloy becomes more susceptible to hardening and cracking due to rapid cooling. Preheating charts are available for recommended temperatures for various steels and nonferrous metals, Table 23-3.

The most commonly used preheat temperature range is between 250°F and 400°F (121°C and 204°C) for

	PLATE THICKNESS (INCHES)			
ASTM/AISI	1/4 OR LESS	1/2	1	2 OR MORE
A36	70°F	70°F	70°F	150°-225°F
A572	70°	70°	150°	225°-300°
1330	95°	200°	400°	450°
1340	300°	450°	500°	550°
2315	ROOM TEMP.	ROOM TEMP.	250°	400°
3140	550°	625°	650°	700°
4068	750°	800°	850°	900°
5120	600°	200°	400°	450°

Table 23-3 Recommended Minimum Preheat Temperatures for Carbon Steel.

structural steel. The preheat temperatures can be as high as 600°F (316°C) when welding cast irons.

Care must be taken to soak heat into the region of the intended weld. Superficial heating is not enough because the purpose is to affect the rate at which a relatively large weld cools. To prevent problems and assure uniform heating, the temperature of the preheated section should be measured at least 10 and 20 minutes after heating.

Stress Relief, Process Annealing

Residual stresses are unsuitable in welded structures, and their effects can be significant. The maximum stresses generally equal the yield strength of the weakest material associated with a specific weld. Such stresses can cause distortion, especially if the component is to be machined after welding. They can also reduce the fracture strength of welded structures under certain conditions.

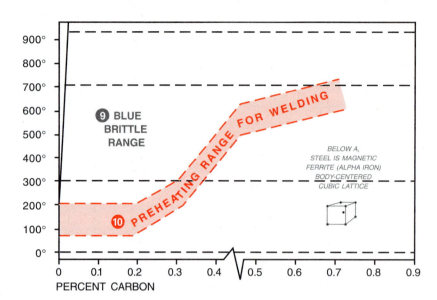

Figure 23-34 Courtesy of Tempil Division, Big Three Industries, Inc.

Area number 8, **Stress Relieving**, consists of heating to a point below the Lower Transformation Temperature, A_1, holding for a sufficiently long period to relieve locked-up stresses, then slowly cooling. This process is sometimes called Process Annealing, **Figure 23-35.**[1]

The yield strength of steels decreases at higher temperatures. When heated, the residual stresses will drop to conform to the lower yield strength; thus, the higher the temperature, the better. But significant changes caused by overheating, overtempering, grain growth, or even a phase change must be avoided. Therefore, the temperatures selected must be less than the tempering temperature used to heat treat the plate. Regardless of the plate's metallurgical structure, these temperatures must be kept below the critical temperature.

The most commonly used temperature range for stress relief steel is between 1,100°F (593°C) and 1,150°F (620°C). This range is high enough to drop the yield residual stresses by 80% and low enough to prevent any harmful metallurgical changes in most steels. While risky, heating to just under the critical temperature does

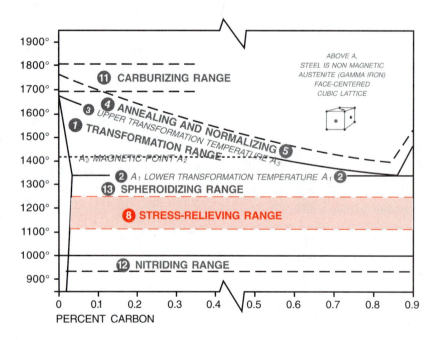

Figure 23-35 Courtesy of Tempil Division, Big Three Industries, Inc.

[1]*Basic Guide to Ferrous Metallurgy*, Copyright 1954 by Tempil, Hamilton Boulevard, South Plainfield, NJ 07080.

offer a stress reduction of about 90%. Caution should be exercised, however, because some steel will become brittle after thermal stress relief.

Time at temperature is also an important factor since it takes time to bring a weldment to temperature. One hour at temperature is the minimum, while six hours offers an additional costly 10% drop in stress. Before cooling, the component must be uniformly heated. That requirement alone generally ensures the component will be at temperature long enough to relieve the stresses. The rate of cooling must also be considered. Rapid cooling results in uneven cooling (as in welding) and causes a new set of residual stresses. Ideally, the cooling rate is slow enough to cool the entire mass of the metal uniformly.

Annealing

Annealing, frequently referred to as Full Annealing, involves heating the structure of a metal to a high enough temperature, slightly above Ac_3, to turn it completely austenitic. See **Figure 23-36**, area 4. After soaking to equalize the temperature throughout the part, it is cooled in the furnace at the slowest possible rate. On cooling slowly the austenite transforms to ferrite and pearlite. The metal is now its softest with small grain size, good ductility, excellent machinability, and other desirable properties.

Normalizing

Area 5, **Normalizing**, consists of heating steels to slightly above Ac_3, holding for Austenite to form, then

followed by cooling (in still air). On cooling, Austenite transforms, giving somewhat higher strength and hardness and slightly less ductility than in annealing, **Figure 23-37** (page 610).[1]

THERMAL EFFECTS CAUSED BY ARC WELDING

During arc welding, liquid weld metal is deposited on the base metal, which is at or near room temperature. In the process, some of the base metal melts from contact with the liquid weld metal and arc, flame, and so on. Conduction, convection, and radiation pull the heat out of the weld metal. This causes the temperature of the deposited weld metal to cool until it becomes solid. Within a few seconds the weld and base metal have gone from being a solid at room temperature to a liquid and back to a solid near room temperature.

Metallurgical changes in the heated region are inevitable. The lowest temperature at which any such changes occur defines the outer extremity of the zone of change. That zone of change is called the **heat-affected zone,** abbreviated **HAZ**, **Figure 23-38** (page 611). The exact size and shape of the HAZ is affected by a number of factors:

■ *Type of metal or alloy:* Some metals are easily affected even by small temperature changes while others are more resistant. In work-hardened metals, the heat-affected zone is defined by the recrystallization temperature, **Figure 23-39** (page 612). In age-

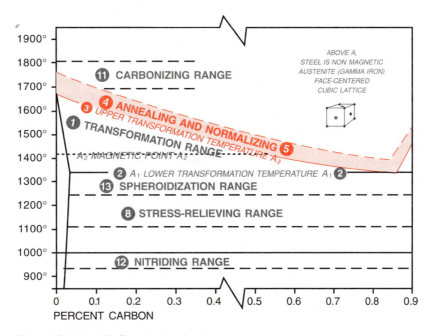

Figure 23-36 Courtesy of Tempil Division, Big Three Industries, Inc.

[1]*Basic Guide to Ferrous Metallurgy,* Copyright 1954 by Tempil, Hamilton Boulevard, South Plainfield, NJ 07080.

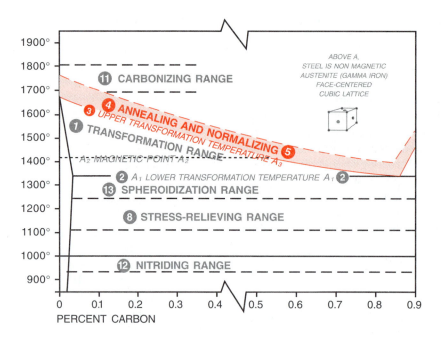

Figure 23-37

hardened alloys, this temperature is the lowest temperature at which evidence of overaging is seen, **Figure 23-40** (page 612). In quenched and tempered steels, it is the temperature at which overtempering is seen.

- *Method of applying the welding heat:* Some heat sources, such as plasma arc welding (PAW), are very concentrated. This high-intensity welding process can have an HAZ area that is only a few thousandths of an inch wide. Oxyacetylene welding (OFW) is a much less intense heating source, and the resulting HAZ will be very large.

- *Mass of the part:* The larger the piece of metal being welded, the greater its ability to absorb heat without a significant change in temperature. Very large weldments may not have a noticeable temperature rise while small parts may almost completely reach the melting temperature. The more metal that gets hot, the larger the HAZ.

- *Pre- and postheating:* As the temperature of the base metal increases—whether from pre- or postheating or the welding process itself—the larger the HAZ. A cold plate may have an extremely narrow HAZ.

The HAZ need not always be small. A larger HAZ is desirable for welding steels that can produce martensite at the same cooling rates as those produced when welding heavy plate. The welder can slow the cooling rates to tolerable levels by preheating the plate and thus increasing the size of the HAZ. In this case, a large HAZ is safer than contending with a brittle martensitic structure.

An important feature of the HAZ, caused by high temperatures, is the severe grain growth at the fusion line. With steels at this critical temperature, the HAZ also produces fine grains as a result of the allotropic transformation. **Figure 23-41** (page 612) is an example of a weld that shows the HAZ in cross section. Regardless of the alloy, the welder must control the HAZ, whether that control is to make it large or small.

GASES IN WELDING

Many welding problems and defects result from undesirable gases that can dissolve in the weld metal, **Figure 23-42** (page 612). Except for the inert gases argon and helium discussed in the Section 4, Gas Shielded Welding, the other common gases either react with or dissolve in the molten weld pool. Those that dissolve have a very high solubility in liquid metal but a very low solubility in solidified weld metal. Thus, during the freezing process, the dissolved gases try to escape. With very slow solidification rates, the gases will escape. With very high solidification rates, they become trapped in the metal as supersaturated solutes. At intermediate rates of solidification, they become trapped as gas bubbles, causing porosity. Some, such as hydrogen, produce other problems as well.

Hydrogen

Hydrogen has many sources, including moisture in electrode coatings, fluxes, very humid air, damp weld joints, organic lubricants, rust on wire or on joint surfaces or in weld joints, organic items such as paint, cloth fibers, dirt in weld joints, and others.

Hydrogen is the principle cause of porosity in aluminum welds and with GMAW welds on stainless steels. It can cause random porosity in most metals and their alloys.

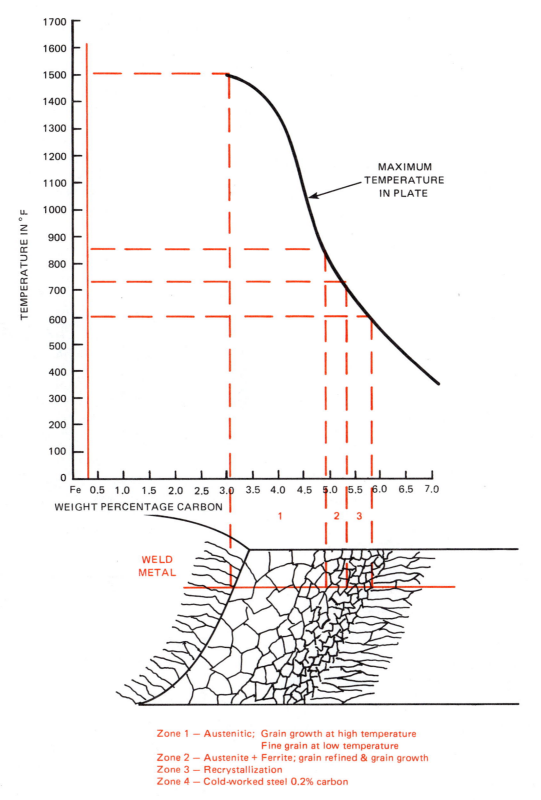

Figure 23-38 Changes in grain structure caused by heating steel plate into different zones of the iron–iron-carbon phase diagram.

Hydrogen can be troublesome in steels. Even in amounts as low as 5 parts per million, it can cause underbead cracking in high-strength steels. This type of cracking, called delayed cracking, cold cracking, or hydrogen-induced cracking, requires three conditions: (1) a high-stress state, (2) a martensitic microstructure, and (3) a critical level of hydrogen. The first two conditions are typical of quenched and tempered steels. With them, the critical amount of hydrogen is inversely proportional to the stress state; that is, the stronger steels can tolerate less hydrogen. The welder must use dry electrodes and dry submerged arc fluxes to avoid the

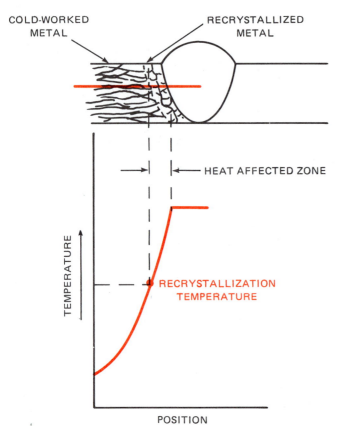

Figure 23-39   A heat-affected zone in cold-worked metal. The zone is identified by a change from the cold-worked grains to the recrystallized grains.

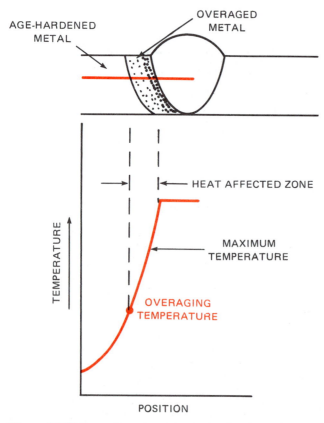

Figure 23-40   Heat-affected zone in age-hardened metal.

hydrogen, in the form of moisture, when welding steels with strength levels above 80,000 psi (5,624 kg/cm²).

Hydrogen problems are avoidable by keeping organic materials away from weld joints, keeping the welding consumables dry, and preheating the components to be welded. Preheating slows the cooling rate of weldments, allowing more time for hydrogen to escape. Generally, the recommended preheat temperatures range between 250°F and 350°F (121°C and 176°C). In the case of very high-strength materials, the acceptable preheat temperatures will exceed 400°F (204°C). Low-temperature preheating of materials before welding can also remove the moisture condensed on the weld joint.

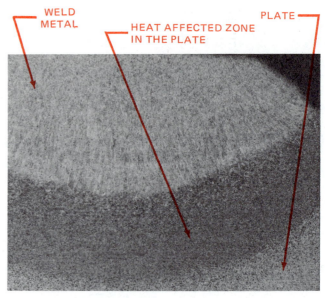

Figure 23-41   Weld heat-affected zone. (25X magnification.) (Courtesy of LECO Corp.)

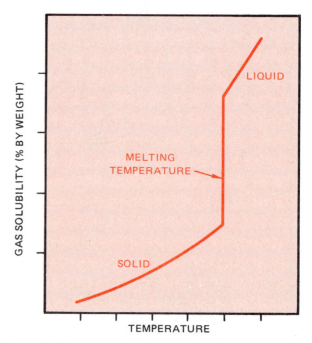

Figure 23-42   Typical change of solubility of "active" gases in metals.

Nitrogen

Nitrogen comes from air drawn into the arc stream. In GMAW processes, it results from poor shielding or strong drafts that disrupt the shield. In SMAW processes, nitrogen can result from carrying an excessively long arc. The primary problem with nitrogen is porosity. In high-strength steels, it can cause a degree of embrittlement that results in marginal toughness.

Some alloys, such as austenitic stainless steels and metals such as copper, have a high solubility for nitrogen in the solid state. This high solubility does not cause porosity in those materials. Since nitrogen improves the strength of stainless steels, it sometimes is added intentionally to shield gases used when making GTA or GMAW welds in them. In Europe it is used for welding copper to increase the arc energy with GTA welding.

Oxygen

As with nitrogen, the common source of oxygen contamination, air, reaches the weld because of poor shielding or excessively long arcs. But metallurgical changes, not porosity, cause most of the effects of oxygen. Oxygen causes the loss of oxidizable alloys such as manganese and silicon, which reduces strength, produces inclusions in weld metals, which reduce their toughness and ductility, and causes an oxide formation on aluminum welds, which affects appearance and complicates multipass welding.

About 2% of oxygen is added intentionally to stabilize the GMAW process when welding steels with argon shielding. At this concentration, oxygen does not cause the metallurgical problems listed previously. Nevertheless, the amount used must be very carefully controlled because the amount needed varies, depending on the alloys being welded. Mixtures containing only enough oxygen to do an effective stabilizing job should be used. Any more could cause problems, particularly with sensitive alloys.

Carbon Dioxide

Carbon dioxide is an oxygen substitute for stabilizing GMAW processes using argon shields, although about 5% to 8% carbon dioxide is usually added to produce the same effects achieved with 2% oxygen. However, the carbon in carbon dioxide is a potential contaminant that can cause problems with corrosion resistance in the low-carbon grades of stainless steels. Even straight carbon dioxide shielding has this problem. In most cases, carbon dioxide levels below 5% do not seem to increase the carbon content of stainless steels enough to cause difficulty.

METALLURGICAL DEFECTS

Cold Cracking

Cold cracking is the result of hydrogen dissolving in the weld metal and then diffusing into the heat-affected zone. The cracks develop long after the weld metal

solidifies. For that reason, it is also known as hydrogen-induced cracking and delayed cracking. This cracking is most commonly found in the course grains of the HAZ, just under the fusion zone. For this reason, it is also called underbead cracking. Generally, these cracks do not surface and can be seen only in radiographs or by sectioning welds. Sometimes they surface in a region of the weld running parallel to the fusion zone. With high-strength steels, hydrogen-induced cracking occurs as transverse cracks in the weld metal that are seen easily with very low magnification, **Figure 23-43**.

Hydrogen-induced cracking requires (1) a high stress, (2) a microstructure sensitive to hydrogen, and, of course, (3) hydrogen. The first two are almost always satisfied with high-strength quenched and tempered steels. The third depends on the welding process, the filler metals, and the preheat, as discussed previously.

Another factor is time. Under very severe conditions, cracking can occur in minutes. With very marginal conditions, cracks might not appear for weeks. For this reason, welds often are not radiographed for weeks to allow time for any potential cracks to develop.

Figure 23-44 illustrates how stress, time, and hydrogen interact to produce such cracks. These cracks

Figure 23-43

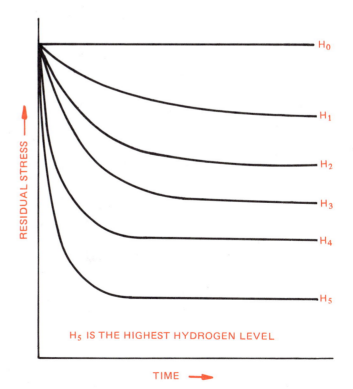

Figure 23-44 The time needed to develop hydrogen-induced cracking decreases at higher stresses and at higher hydrogen levels.

never occur with 0% hydrogen, even with very high yield strength steels. To prevent cracking at hydrogen level H_5, the residual stresses must be reduced to the level indicated by the horizontal section of the curve H_5. Low residual stresses are possible with either a very weak steel or a strong steel that had a stress relief anneal. (The stress relief anneal also will help by dropping the hydrogen level to level H_0.)

Hot Cracking

Hot cracks differ from cold cracks discussed previously. The hot cracks are caused by tearing the metal along partially fused grain boundaries of welds that have not completely solidified. Unlike those caused by hydrogen, these longitudinal cracks are located in the centers of weld beads. They develop immediately after welding, unlike the delayed nature of hydrogen-induced cracking. In the process of freezing, low-melting materials in the weld metal are rejected as the columnar grains solidify, leaving a high concentration of the low-melting materials where the grains intersect at the center of the weld. These partially melted and weak grain boundaries are stressed while the weld metal shrinks, causing them to rupture.

A high sulfur content is most often responsible for hot cracking in steel. It forms a low-melting iron sulfide on the grain boundaries. Hot cracking is more likely to occur in steels that contain higher levels of carbon and

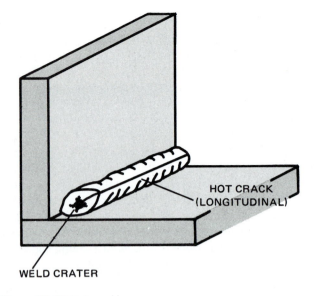

Figure 23-45 Hot cracking.

phosphorous and in steels that are high in sulfur and low in manganese, **Figure 23-45**.

Severely concave welds may also cause hot cracking, because the welds are not as strong. As welds cool, they shrink. If the weld is not thick enough, when it shrinks it cannot pull the metal in, and so it cracks. Even if the weld is larger, it may crack if the weldment is

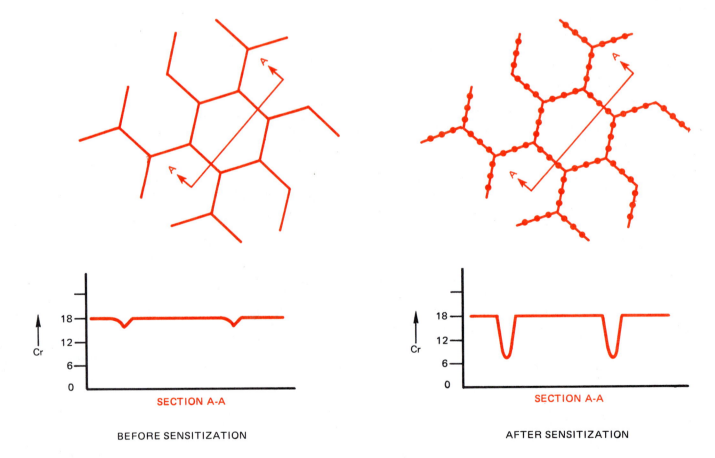

BEFORE SENSITIZATION   AFTER SENSITIZATION

Figure 23-46 Depletion of chromium in grain boundaries due to the precipitation of chromium carbide.

rigid and cannot be pulled inward by the weld. In these situations, even large welds can have hot cracks.

Carbide Precipitation

Stainless steels rely on free chromium for their resistance to corrosion. When carbon is present and the steel is heated to temperatures between about 800°F (427°C) and 1,500°F (816°C), carbon combines with the chromium to form chromium carbides in the grain boundaries. The formation of chromium carbides depletes the steel of the free chromium needed for protection. Thus, for steels, every effort is made to use low-carbon stainless steel or a special stabilized grade made for welding.

Figure 23-46 illustrates how the chromium in solution can drop from the nominal 18% in the most commonly used stainless steels to levels well under the 12% needed for even minimal protection. When in contact with corrosive materials, the low-chromium grain boundaries can dissolve and the weld fail, **Figure 23-47**.

Problems are minimized by using very low-carbon steel called extra-low-carbon (ELC) steels. Without carbon, the chromium carbides cannot form. Another way is to tie up the carbon by forming very stable carbides of titanium or columbium. This also prevents chromium carbides from forming. Weld metals are alloyed similarly with low carbon levels or by stabilizing the carbides. Titanium is not generally used for that purpose in electrodes because it is lost in the transfer. Instead, most

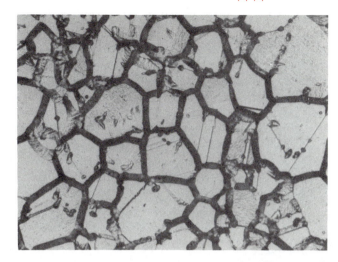

Figure 23-47 Grain boundary corrosion in sensitized type 304 stainless steel. Note that the grain boundaries have dissolved completely. (150X magnification.) (Courtesy of General Electric Company.)

electrodes are alloyed with columbium to avoid the corrosion problems associated with carbide precipitation.

Carbon dioxide shield gases can cause a similar problem, especially with the ELC grades. These gases supply the carbon that impairs corrosion resistance by depleting the free chromium. The small amounts used in argon to stabilize the arc are often tolerable. But carbon dioxide additions should be avoided unless used with caution and an awareness that problems can develop.

Air Conditioning Leader Helps the World Stay Cool

Started as a family business a century ago, the Trane Company is today the world's second-largest manufacturer of air-conditioning equipment.

Since the refrigerant side of most air-conditioning systems is a sealed hermetic device where containment of the refrigerant charge is absolutely essential to proper operation, the air-conditioning industry uses welding extensively in its manufacturing process.

At the company, all welding production from preliminary process development to welder training and plant production followup is the responsibility of the Welding Engineering Group. The employees in this group have access to a fully equipped and up-

Figure 1

to-date 3,000-square-foot welding laboratory. Recently, new GMAW (gas metal arc) and FCAW (flux cored) welding equipment was added for manufacturing. The company ordered eighty-nine power sources, each equipped with a movable boom and digital wire

Figure 2

Figure 3

feed system. Figure 1 shows some of these units prior to installation. The units have outstanding arc characteristics. Arc starting was excellent without any stubbing, and the arc was smooth and consistent with minimal spatter. Weld sample test results showed superior penetration to the root of the weld samples in all welding positions.

In Figure 2, a centrifugal water chiller for a 700-ton capacity air-conditioning unit is spray transfer GMAW welded. The 12-foot counter-balanced boom has a 360° turning radius and provides the operator with greater access to the workpiece. Figure 3 shows a 1,100-ton-capacity absorption cold generator being spray transfer GMAW welded

using steam or hot water to generate chilled water. Again note that the boom simplifies access to the weld site.

Finally, Figure 4 shows how short arc GMAW and FCAW welding is used in the construction for water boxes for the ends of centrifugal air conditioning equipment. Fixed slope and non-variable inductance characteristics of the welding power source are ideal for this welding application. A thermostatically-controlled fan on the power source limits the amount of dirt ingested by the unit and greatly reduces noise in the welding area.

(Courtesy of Miller Electric Mfg. Co., Appleton, Wisconsin 54912.)

Figure 4

REVIEW

1. What gives metals their desirable properties?

2. What is heat?

3. What are the basic units of measure for heat?

4. What is sensible heat?

5. What is latent heat?

6. What does the color of light given off from a hot object indicate?

7. What is a crystal lattice?

8. Why doesn't metal form into one large single crystal?

9. What is an alloy?

10. Using **Table 23-3** (page 607), answer the following questions:
    a. At approximately what temperature does a 70% lead and 30% tin mixture become solid?
    b. What crystal structure is formed first when a 20% lead and 80% tin mixture cools down from a 100% liquid phase?

11. What is an eutectic composition?

12. Using **Table 23-1** (page 598), what is the lowest and highest carbon contents of iron-carbon alloys?

13. Approximately how many degrees wide is the transition range at the 0.1% carbon alloy?

14. Referring to Color Plate 63, what color would low-carbon steel appear when it is in the transition range?

15. Referring to **Figure 23-16** (page 597), what is the approximate temperature of the *Lower Transformation Temperature?*

16. What three factors affect the size and type of crystals formed in an iron-carbon alloy?

17. Referring to Color Plate 63, above what temperature is a 0.1% iron-carbon alloy nonmagnetic?

18. What is the most significant factor that affects how much of an alloying element remains trapped in the iron-carbon crystals?

19. What is known as the critical temperature of iron-carbon alloys?

20. Can a metal have all the mechanical properties at ideal levels? Why or why not?

21. What other properties can a metal's hardness reveal?

22. Which property, brittleness or ductility, will let a metal deform without breaking? Why?

23. What is toughness?

24. What are the common types of strength measurements?

25. What is a major advantage of solid-solution-strengthened alloys?

26. What are the three steps in precipitation hardening?

27. How do ferrite and cementite work together to form a strong ductile steel?

28. Why is brine quenching faster than water quenching?

29. What can be done to speed up the quenching rate in any liquid?

30. Why is the formation of martensite a problem when welding on high-carbon steels?

31. How can the effects of cold-working be removed?

32. How can large-grain crystals be made into smaller sizes?

33. Referring to Color Plate 63, what would the preheat temperature range be for an iron-carbon alloy with 0.60% carbon?

34. Why must the stress-relieving temperature be kept below the critical temperature of the plate?

35. What properties can annealing produce in metals?

36. How long does it take the weld metal to go through its thermal cycle during a weld.

37. What are some sources of hydrogen that can contaminate a weld?

38. How can nitrogen get into an SMAW weld?

39. What are some of the problems that oxygen can cause in welds?

40. When do cold cracks develop?

41. What is carbide precipitation?

Chapter 24

WELDABILITY OF METALS

OBJECTIVES

After completing this chapter, the student should be able to:

■ list the methods used to weld most ferrous metals.

■ list the methods used to weld four nonferrous metals.

■ explain the precautions that must be taken when welding various metals and alloys.

■ describe the effects of preheating and postheating on welding.

KEY TERMS

weldability	chromium-molybdenum steel
steel classification	cast iron
carbon steel	malleable cast iron
alloy steels	copper and copper alloys
tool steel	aluminum
high-manganese steel	titanium
low-alloy steels	magnesium
stainless steels	

INTRODUCTION

All metals can be welded, although some require far more care and skill to produce acceptably strong and ductile joints. The term *weldability* has been coined to describe the ease with which a metal can be welded properly. Good weldability means that almost any process can be used to produce acceptable welds and that little effort is needed to control the procedures. Poor weldability means that the processes used are limited and that the preparation of the joint and the procedure used to fabricate it must be controlled very carefully or the weldment will not function as intended.

Weldability is defined by the AWS as "the capacity of a metal to be welded under the fabrication conditions imposed into a specific, suitably designed structure and to perform satisfactorily in the intended service." Books have been written about this subject. Weldability involves the metallurgy of both the metal to be welded and the filler metal, the welding processes, the joint design, the weld preparation, the heat treatments before and after welding, and many other factors, depending on the complexity of the welded system. This chapter does not discuss the selection of methods and procedures for joining difficult-to-weld metals. Those with no previous experience should seek expert help before attempting to weld these metals. Otherwise, they run the risk of failures in service that could physically harm nearby people.

Therefore, an attempt will not be made in this text to cover the weldability of all metals. The welding done in the average school shop is limited to a small number of metals. However, information will be given in this chapter on a number of other metals that will be found in common use in industry.

Most welding processes produce a thermal cycle in which the metals are heated over a range of temperatures up to, and including, fusion. Cooling of the metal to ambient temperatures then follows. The heating and cooling cycle can set up stresses and strains in the weldment. Whatever the welding process used, certain metallurgical, physical, and chemical changes also take place in the metal. A wide range of welding conditions can exist for welding methods when joining metals with good weldability. However, if weldability is a problem, adjustments usually will be necessary in one or more of the following factors:

- Shielding atmosphere
- Filler metal
- Fluxing material
- Welding method
- Use of preheat, postheat, and/or interpass temperature
- Welding procedures

Table 24-1 (page 620) is a summary of acceptable welding methods for joining a variety of ferrous metals.

STEEL CLASSIFICATION AND IDENTIFICATION

SAE and AISI Classification Systems

Two primary numbering systems have been developed to classify the standard construction grades of steel, including both carbon and alloy steels. These systems classify the types of steel according to their basic chemical composition. One classification system was developed by the Society of Automotive Engineers (SAE). The other system is sponsored by the American Iron and Steel Institute (AISI).

The numbers used in both systems are now just about the same. However, the AISI system uses a letter before the number to indicate the method used in the manufacture of the steel.

Both numbering systems usually have a four-digit series of numbers. In some cases, a five-digit series is used for certain alloy steels. The entire number is a code to the approximate composition of the steel.

In both steel classification systems, the first number often, but not always, refers to the basic type of steel, as follows:

1XXX Carbon
2XXX Nickel
3XXX Nickel chrome
4XXX Molybdenum
5XXX Chromium
6XXX Chromium vanadium
7XXX Tungsten
8XXX Nickel chromium vanadium
9XXX Silicomanganese

The first two digits together indicate the series within the basic alloy group. There may be several series within a basic alloy group, depending upon the amount of the principle alloying elements. The last two or three digits refer to the approximate permissible range of carbon content. For example, the metal identified as 1020 would be 1XXX carbon steel with a XX20 0.20% range of carbon content, and 5130 would be 5XXX chromium steel with an XX30 0.030% range of carbon content.

The letters in the AISI system, if used, indicate the manufacturing process as follows:

- C — Basic open-hearth or electric furnace steel and basic oxygen furnace steel
- E — Electric furnace alloy steel

Table 24-2 (page 621) shows the AISI and SAE numerical designations of alloy steels.

Unified Numbering System (UNS)

Presently being promoted is a Unified Numbering System for all metals. This system eventually will replace the AISI and other systems.

CARBON AND ALLOY STEELS

Steels alloyed with carbon and only a low concentration of silicon and manganese are known as plain carbon steels. These steels can be classified as low-carbon, medium-carbon, and high-carbon steels. The division is based upon the percentage of carbon present in the material.

Plain carbon steel is basically an alloy of iron and carbon. Small amounts of silicon and manganese are added to improve their working quality. Sulfur and phosphorus are present as undesirable impurities. All steels contain some carbon, but steels that do not include alloying elements other than low levels of manganese or silicon are classified as plain carbon steels. Alloy steels contain specified larger proportions of alloying elements.

The AISI has adopted the following definition of carbon steel: "Steel is classified as carbon steel when no minimum content is specified or guaranteed for aluminum, chromium, columbium, molybdenum, nickel, titanium, tungsten, vanadium, or zirconium; and when the minimum content of copper which is specified or guaranteed does not exceed 0.40%; or when the maximum content which is specified or guaranteed for any of

Table of Joining Processes:

Material	Thickness	SMAW	SAW	GMAW	FCAW	GTAW	PAW	ESW	EGW	RW	FW	OFW	DFW	FRW	EBW	LBW	TB	FB	IB	RB	DB	IRB	DFB	S
Carbon Steel	S	X	X	X		X				X	X	X			X	X	X	X	X	X	X	X		X
	I	X	X	X	X	X				X	X	X		X	X	X	X	X	X				X	X
	M	X	X	X	X					X	X	X		X	X	X	X						X	
	T	X	X	X	X			X	X	X	X			X	X		X						X	
Low Alloy Steel	S	X	X	X		X				X	X	X	X		X	X	X	X	X	X	X	X	X	X
	I	X	X	X	X	X				X	X		X	X	X	X	X	X	X				X	X
	M	X	X	X	X						X		X	X	X	X	X	X	X				X	
	T	X	X	X	X			X		X	X		X	X	X	X					X		X	
Stainless Steel	S	X	X	X		X	X			X	X	X			X	X	X	X	X	X	X	X	X	X
	I	X	X	X	X	X	X			X	X			X	X	X	X	X	X				X	X
	M	X	X	X	X						X			X	X	X	X	X	X				X	
	T	X	X	X	X			X			X			X	X	X					X		X	
Cast Iron	I	X										X					X	X	X				X	X
	M	X	X	X	X							X					X	X	X				X	X
	T	X	X	X	X							X					X						X	
Nickel and Alloys	S	X		X		X	X			X	X	X			X	X	X	X	X	X	X	X	X	X
	I	X	X	X		X	X				X			X	X	X	X	X	X	X			X	X
	M	X	X	X			X				X			X	X	X		X					X	
	T	X		X				X			X			X	X			X					X	
Aluminum and Alloys	S	X		X		X	X			X	X	X			X	X	X	X	X	X			X	X
	I	X		X		X				X	X		X		X	X	X	X	X			X	X	X
	M	X		X		X					X			X	X	X	X	X			X		X	
	T	X		X				X	X		X			X		X		X					X	
Titanium and Alloys	S			X		X	X			X	X			X		X	X	X				X		X
	I			X		X	X				X		X	X	X	X		X						X
	M			X		X	X				X		X	X	X	X		X						X
	T			X							X			X		X	X	X						X
Copper and Alloys	S			X		X	X				X				X		X	X	X	X	X			X	X
	I			X			X				X			X	X		X	X	X		X			X	X
	M			X							X			X	X		X	X	X					X	
	T			X							X				X		X	X	X					X	
Magnesium and Alloys	S			X		X				X					X	X	X	X	X			X		X
	I			X		X				X	X			X	X	X	X	X				X		X
	M			X							X			X	X	X		X						X
	T			X							X			X										
Refractory Alloys	S			X		X	X			X	X			X		X	X	X	X	X			X	X
	I			X			X				X			X		X	X	X						X
	M									X	X													
	T																							

*This table presented as a general survey only. In selecting processes to be used with specific alloys, the reader should refer to other appropriate sources of information.

**See legend below:

LEGEND

Process Code

SMAW–Shielded Metal Arc Welding
SAW–Submerged Arc Welding
GMAW–Gas Metal Arc Welding
FCAW–Flux Cored Arc Welding
GTAW–Gas Tungsten Arc Welding
PAW–Plasma Arc Welding
ESW–Electroslag Welding
EGW–Electrogas Welding
RW–Resistance Welding
FW–Flash Welding
OFW–Oxyfuel Gas Welding
DFW–Diffusion Welding

FRW–Friction Welding
EBW–Electron Beam Welding
LBW–Laser Beam Welding
B–Brazing
  TB–Torch Brazing
  FB–Furnace Brazing
  IB–Induction Brazing
  RB–Resistance Brazing
  DB–Dip Brazing
  IRB–Infrared Brazing
  DFB–Diffusion Brazing
S–Soldering

Thickness

S–Sheet: up to 3 mm (1/8 in.)
I–Intermediate: 3 to 6 mm (1/8 to 3/4 in.)
M–Medium: 6 to 19 mm (1/4 to 3/4 in.)
T–Thick: 19 mm (3/4 in.) and up

X–Commercial Process

Table 24-1 Overview of Joining Processes. (Courtesy of the American Welding Society.)

13XX	Manganese 1.75
23XX**	Nickel 3.50
25XX**	Nickel 5.00
31XX	Nickel 1.25; Chromium 0.65
E33XX	Nickel 3.50; Chromium 1.55; Electric furnace
40XX	Molybdenum 0.25
41XX	Chromium 0.50 or 0.95; Molybdenum 0.12 or 0.20
43XX	Nickel 1.80; Chromium 0.50 or 0.80; Molybdenum 0.25
E43XX	Same as above, produced in basic electric furnace
44XX	Manganese 0.80; Molybdenum 0.40
45XX	Nickel 1.85; Molybdenum 0.25
47XX	Nickel 1.05; Chromium 0.45; Molybdenum 0.20 or 0.35
50XX	Chromium 0.28 or 0.40
51XX	Chromium 0.80, 0.88, 0.93, 0.95, or 1.00
E5XXXX	High carbon; High chromium; Electric furnace bearing steel
E50100	Carbon 1.00; Chromium 0.50
E51100	Carbon 1.00; Chromium 1.00
E52100	Carbon 1.00; Chromium 1.45
61XX	Chromium 0.60, 0.80, or 0.95; Vanadium 0.12, or 0.10, or 0.15 minimum
7140	Carbon 0.40; Chromium 1.60; Molybdenum 0.35; Aluminum 1.15
81XX	Nickel 0.30; Chromium 0.40; Molybdenum 0.12
86XX	Nickel 0.55; Chromium 0.50; Molybdenum 0.20
87XX	Nickel 0.55; Chromium 0.50; Molybdenum 0.25
88XX	Nickel 0.55; Chromium 0.50; Molybdenum 0.35
92XX	Manganese 0.85; Silicon 2.00; 9262-Chromium 0.25 to 0.40
93XX	Nickel 3.25; Chromium 1.20; Molybdenum 0.12
98XX	Nickel 1.00; Chromium 0.80; Molybdenum 0.25
14BXX	Boron
50BXX	Chromium 0.50 or 0.28; Boron
51BXX	Chromium 0.80; Boron
81BXX	Nickel 0.33; Chromium 0.45; Molybdenum 0.12; Boron
86BXX	Nickel 0.55; Chromium 0.50; Molybdenum 0.20; Boron
94BXX	Nickel 0.45; Chromium 0.40; Molybdenum 0.12; Boron

Note: The elements in this table are expressed in percent.

*Consult current AISI and SAE publications for the latest revisions.

**Nonstandard steel

Table 24-2 AISI and SAE Numerical Designation of Alloy Steels.*

the following elements does not exceed the respective percentages hereinafter stated: manganese 1.65%, silicon 0.60%, copper 0.60%." Under this classification will be steels of different composition for various purposes.

Many special alloy steels have been developed and sold under various trade names. These alloy steels usually have special characteristics, such as high tensile strength, resistance to fatigue, corrosion resistance, or the ability to perform at high temperatures. Basically, the ability of carbon steel to be welded is a function of the carbon content, Table 24-3 (page 622). (Other factors to be considered include thickness and the geometry of the joint.) All carbon steels can be welded by at least one method. However, the higher the carbon content of the metal, the more difficult it is to weld the steel. Special precautions must be followed in the welding process.

Low-carbon and Mild Steel

Low-carbon steel has a carbon content of 0.15% or less, and mild steel has a carbon content range of 0.15% to 0.30%. Both steels can be welded easily by all welding processes. The resulting welds are of extremely high quality. Oxyacetylene welding of these steels can be done by using a neutral flame. Joints welded by this process are of high quality, and the fusion zone is not hard or brittle.

Both low-carbon and mild steels can be welded readily by the shielded metal arc method. The selection of the correct electrode for the particular welding application helps to ensure high strength and ductility in the weld.

The gas metal arc and flux cored arc welding processes are used for welding both low- and medium-carbon steels due to the ease of welding and because they prevent contamination of the weld. The high productivity and lower cost make them increasingly popular welding processes.

The gas tungsten arc process is slow and will cause severe porosity in the weld if the steel is not fully degassed (degassed metal that has had all the gas normally dissolved in metal during manufacturing removed). GTAW can make superior welds with a very clear x-ray quality.

Medium-carbon Steel

The welding of medium-carbon steels, having 0.30% to 0.50% carbon content, is best accomplished by the various fusion processes, depending upon the carbon

Common Name	Carbon Content	Typical Use	Weldability
Ingot Iron	0.03% max.	Enameling, galvanizing and deep drawing sheet and strip	Excellent
Low-Carbon Steel	0.15% max.	Welding electrodes, special plate and shapes, sheet, strip	Excellent
Mild Steel	0.15-0.30%	Structural shapes, plate and bar	Good
Medium-Carbon Steel	0.30-0.50%	Machinery parts	Fair[1]
High-Carbon Steel	0.50-1.00%	Springs, dies, and railroad rails	Poor[2]

[1]Preheat and frequently postheat required.
[2]Difficult to weld without adequate preheat and postheat.

Table 24-3

content of the base metal. The welding technique and materials used are dictated by the metallurgical characteristics of the metal being welded. For steels containing more than 0.40% carbon, preheating and subsequent heat treatment generally are required to produce a satisfactory weld. Shielded arc electrodes of the type used on low-carbon steels can be used for welding this type of steel. The use of an electrode with a special low-hydrogen coating may be necessary to reduce the tendency toward underbead cracking.

Medium-carbon steels can be resistance welded. However, special techniques may be required. Other welding methods that produce sound welds on medium-carbon steel include submerged arc welding, thermit welding, pressure gas welding, and spot, flash, and seam welding.

High-carbon Steel

High-carbon steels usually have a carbon content of 0.50% to 0.90%. These steels are much more difficult to weld than either the low- or medium-carbon steels. Because of the high carbon content, the heat affected zone can transform to very hard and brittle martensite. The welder can avoid this by using preheat and by selecting procedures that produce high-energy inputs to the weld. Refer to Color Plate 63 and **Figure 23-32 (page 606)** for the preheat temperature for the specific carbon content. The martensite that does form is tempered by postweld heat treatments such as stress relief anneal. Refer to Color Plate 63 and **Figure 23-33 (page 607)** for the temperature for stress relief annealing between 1125°F and 1250°F (600° to 675°C).

In arc welding high-carbon steel, mild-steel shielded arc electrodes are generally used. However, the weld metal does not retain its normal ductility because it absorbs some of the carbon from the steel. Austenitic steel electrodes can sometimes be used to obtain better ductility in the weld.

Welding on high-carbon steels is often done to build up a worn surface to original dimensions or to develop a hard surface. In this type of welding, preheating or heat treatment may not be needed because of the

way in which heat builds up in the part during continuous welding.

TOOL STEEL

Because tool steel has a carbon content of from 0.8% to 1.50%, it is very difficult to weld. Gas welding can be employed if the material is in the lower carbon ranges. However, when welding tool steel by the gas welding method, the rods selected for welding repairs should have a carbon content matching that of the base metal. Gas welding requires correct flame adjustment and careful manipulation of the flame and rod. Recommended practice is to preheat the metal, followed by a slow annealing after the welding. A carburizing or neutral flame is desirable in obtaining a strong weld.

In arc welding tool steels, one of the following procedures should be observed:

■ Anneal the parts, preheat and weld with a proper electrode, and then heat treat to restore the original properties to the metal.

■ Preheat the parts and deposit one or more layers on the kerf surfaces of the joint with a covered electrode before depositing the weld beads that tie the joint together. These procedures dilute the carbon content, and the surfaces of the joint can then be more readily welded together.

HIGH-MANGANESE STEEL

High-manganese steel contains 12% or more manganese and a carbon content ranging from 1% to 1.4%. These steels are used for wear resistance in applications involving impact. They are used for such items as power shovels, rock crushers, mine equipment, augers, switch frogs, and so forth.

These alloys are austenitic and, therefore, tough. When inspected, they form a hard martensitic layer on the surface, which provides wear resistance. As martensite develops during work hardness it reaches a stage where it starts to (roll over) or (mushrooms) as in the head of a chisel, or on the edges of a railroad track. As the components wear in service, they are repaired with electrodes having similar compositions. Care must be taken not to inhale the fumes produced because they are high in manganese.

Martensite is very hard and brittle and must be checked for cracks periodically. If cracks form in the hard martensite layer they can move into the softer austenitic below, **Figure 24-1**. There cracks like cracks in glass will continue moving through the part until it breaks. Removing the cracks by welding them up or cutting them out is the only way to prevent them from causing the part to fail.

LOW-ALLOY, HIGH-TENSILE STRENGTH STEELS

Low-alloy steels are used increasingly because of requirements for high strength with less weight. These types of steel are readily weldable by all of the common welding processes. The methods and types of electrodes used depend upon the end results required. Steel manufacturers can usually supply welding information.

The following types of steel can be found in this class:

- Austenitic manganese
- Chromium alloy
- Carbon molybdenum
- Chromium molybdenum
- Chromium vanadium
- Manganese alloy
- Nickel molybdenum
- Manganese molybdenum
- Nickel chromium
- Nickel chromium molybdenum
- Nickel copper

Other high-strength, low-alloy steels are available under a variety of trade names and are listed in the *Metals Handbook,* published by the American Society for Metals.

STAINLESS STEELS

Stainless steels consist of four groups of alloys: austenitic, ferritic, martensitic, and precipitation hardening. The austenitic group is by far the most common. Its chromium content provides corrosion resistance, while its nickel content produces the tough austenitic microstructure. These steels are relatively easy to weld, and a large variety of electrode types are available.

The most widely used stainless steels are the chromium-nickel austenitic types. They are used for

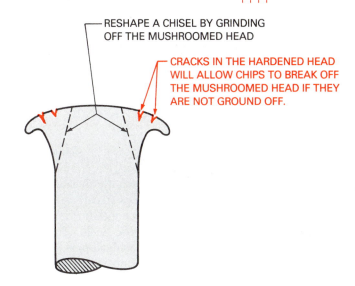

RESHAPE A CHISEL BY GRINDING OFF THE MUSHROOMED HEAD

CRACKS IN THE HARDENED HEAD WILL ALLOW CHIPS TO BREAK OFF THE MUSHROOMED HEAD IF THEY ARE NOT GROUND OFF.

Figure 24-1

items such as chemical equipment, cooking utensils, food processing equipment, and furnace parts and are usually referred to by their chromium-nickel content as 18/8, 25/12, 25/20, and so on. For example, 18/8 contains 18% chromium and 8% nickel, with 0.08% to 0.20% carbon. To improve weldability, the carbon content should be as low as possible. Carbon should not be more than 0.03%, with the maximum being less than 0.10%.

Keeping the carbon content of stainless steel will also help reduce carbide precipitation. Carbide precipitation occurs when alloys containing both chromium and carbon are heated. The chromium and carbon combine to form chromium carbide (Cr_3C_2).

The combining of chromium and carbon lowers the chromium that is available to provide corrosion resistance in the metal. This results in a metal surrounding the weld that will oxidize or rust. The amount of chromium carbide formed is dependent on the percentage of carbon, the time that the metal is in the critical range, and the presence of stabilizing elements.

If the carbon content of the metal is very low, little chromium carbide can form. Some stainless steels types have a special low carbon variation. These low carbon stainless steels are the same as the base type but with much lower carbon content. To identify the low carbon from the standard AISI number the "L" is added as a suffix. See examples 304 and 304L, **Table 24-4 (page 624)**.

Chromium carbides form when the metal is between 800°F and 1,500°F (625°C and 815°C). The quicker the metal is heated and cooled through this range the less time that chromium carbides can form. Since austenitic stainless steels are not hardenable by quinching, the weld can be cooled using a chill plate. The chill plate can be water cooled for larger welds.

Some filler metals have stabilizing elements added to prevent carbide precipitation. Columbium and titanium are both commonly found as chromium stabilizers. Examples of the filler metals are E310Cb and ER309Cb.

| Nominal Composition of Stainless Steels | | | | | | |
AISI Type	C	Mn Max	Si Max	Cr	Ni	Other
			Nominal Composition %			
304	0.08 max	2.0	1.0	18-20	8-12	
304L	0.03 max	2.0	1.0	18-20	8-12	
316	0.08 max	2.0	1.0	16-18	10-14	2.0-3.0 Mo
316L	0.03 max	2.0	1.0	16-18	10-14	2.0-3.0 Mo

Table 24-4 Comparison of Standard-Grade and Low-Carbon Stainless Steels.

In fusion welding, stainless austenitic steels may be welded by all of the methods used for plain carbon steels.

Since *ferritic stainless* steels contain almost no nickel, they are cheaper than austenitic steels. They are used for ornamental or decorative applications such as architectural trim and at elevated temperatures such as used for heat exchanging. However, ferritic stainless steels also tend to be brittle unless specially deoxidized. Special high-purity, high-toughness ferritic stainless steels have been developed, but careful welding procedures must be used with them to prevent embrittlement. This means very careful control of nitrogen, carbon, and hydrogen, and limiting the welding choices to GMAW and GTAW processes.

Martensitic stainless steels are also low in nickel but contain more carbon than the ferritic. They are used in applications requiring both wear resistance and corrosion resistance. Items such as surgical knives and razor blades are made of them. Quality welding requires very careful control of both preheat and tempering immediately after welding.

Precipitation hardening stainless steels can be much stronger than the austenitic, without losing toughness. Their strength is the result of a special heat treatment used to develop the precipitate. They can be solution treated prior to welding and given the precipitation treatment after welding.

The closer the characteristics of the deposited metal match those of the material being welded, the better is the corrosion resistance of the welded joint. The following precautions should be noted:

■ Any carburization or increase in carbon must be avoided, unless a harder material with improved wear resistance is actually desired. In this case, there will be a loss in corrosion resistance.

■ It is important to prevent all inclusions of foreign matter, such as oxides, slag, or dissimilar metals.

In welding with the metal arc process, direct current is more widely used than alternating current. Generally, reverse polarity is preferred where the electrode is positive and the workpiece is negative.

The diameter of the electrode used to weld steel that is thinner than 3/16 in. (4.8 mm) should be equal to, or slightly less than, the thickness of the metal to be welded.

When setting up for welding, material .050 in. (1.27 mm) and less in thickness should be clamped firmly to

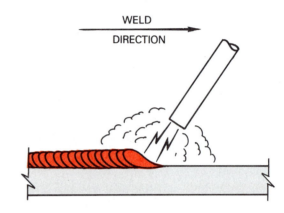

WELD DIRECTION

Figure 24-2 Backhand or drag angle.

prevent distortion or warpage. The edges should be butted tightly together. All seams should be accurately aligned at the start. It is advisable to tack weld the joint at short intervals as well as to use clamping devices.

The electrode should always point into the weld in a backhand or drag angle, **Figure 24-2**. Avoid using a figure-8 pattern or excessive side weaving motion such as that used in welding carbon steel. Best results are obtained with a stringer bead with little or very slight weaving motion and steady forward travel, with the electrode holder leading the weld pool at about 60° in the direction of travel.

To weld stainless steels, the arc should be as short as possible. **Table 24-5** can be used as a guide.

CHROMIUM-MOLYBDENUM STEEL

Chromium-molybdenum steel is used for high-temperature service and for aircraft parts. It can be welded by the following processes: shielded metal arc, gas tungsten arc, gas metal arc, flux cored, and submerged arc. The welds have a tensile strength of 60,000 psi (49 kg/cm²) to 80,000 psi (5,625 kg/cm²) as welded. This type of material lends itself to joints in thin sections.

CAST IRON

In welding **cast iron,** ductility of the weld is critical because of the brittleness of the cast iron itself.

All grades of cast iron have a high carbon content, usually ranging from 2% to 4%. The most common grades of cast iron contain about 2.5% to 3.5% total car-

Metal Thickness			Electrode Diameter		Current (Amperes)	Voltage Open Circuit
in	(mm)		in	(mm)		
.050	(1.27)		5/64	(1.98)	25–50	30–35
.050 – .0625	(1.27–1.58)		3/32	(2.38)	30–90	35–40
.0625 – .1406	(1.58–3.55)		3/32–1/8	(2.38–3.17)	50–100	40–45
.1406 – .1875	(3.55–4.74)		1/8–5/32	(3.17–3.96)	80–125	45–50
.250 and up	(6.35 and up)		3/16	(4.76)	100–175	55–60

Table 24-5 Shielded Metal Arc Welding Electrode Setup for Stainless Steel.

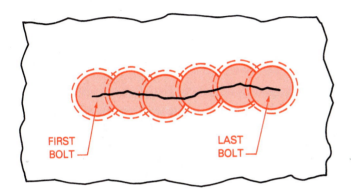

Figure 24-3 A crack in cast iron can be plugged by drilling and tapping overlapping holes. Each bolt overlaps the previous bolt. Only the last one must have a locking compound to prevent it from loosening.

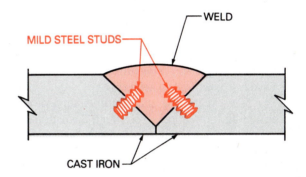

Figure 24-4 Mild steel studs can be threaded into holes drilled and tapped in the side of the groove. The studs help the weld hold the cast iron together.

bon. Part of this carbon is in a combined form, and part is in a free state as graphite. Gray cast iron, malleable cast iron, and nodular cast iron are weldable. White cast iron is practically not weldable, **Figure 24-3.**

Cast iron essentially has no ductility. Its tensile strength may be as low as 20,000 psi (1,406 kg/cm²). For this reason, it is important to guard against cracks due to expansion and contraction during the welding process.

Successful arc welding of cast iron depends upon the skill of the operator. Cast iron should not be welded as rapidly as mild steel. Most gray malleable and nodular cast iron welding jobs can be welded using the shielded metal arc welding process after extensive preheating. The finished weld must then be cooled slowly. The preheating temperature should be at least 600°F (315°C), and preferably around 900° (482°C) to 1,200°F (650°C). Preheating should be applied as uniformly as possible, and cooling should be controlled. It is important in welding cast iron to keep the depth of fusion or penetration to a minimum to prevent the transformation of the metal into some undesirable structure.

Cast iron can be welded using the oxyacetylene method or the shielded metal arc method. Other welding methods can be used but are not considered to be so effective as these. The shielded metal arc process has high deposition rates, but cooling between passes is necessary to prevent heat buildup. The two types of electrodes that can be used when welding cast iron with the

shielded metal arc process are machinable and non-machinable. Machinable types deposit high nickel and nickel-iron weld metal or a nickel-copper composition. The deposited metals are soft and readily machinable. Non-machinable electrodes have a mild steel core covered with a special flux. The deposit is very hard and is used only when machining is not necessary. All of these electrodes are useful in producing leak-tight joints. They are used mainly for repairing water jackets, motor blocks, transmission cases, and other similar assemblies, **Figure 24-4.**

In using an oxyacetylene flame for welding cast iron, a special cast-iron filler rod having a high silicon content should be used. With this type of rod, sufficient silicon is left in the weld area after welding. A proper flux is called for to keep the molten pool fluid and to avoid inclusions and blowholes in the weld area.

Braze welding may be used for the repair of broken castings. The advantage of braze welding is that it is not necessary to heat the metal to a molten condition. This limits the size of the brittle, heat-affected zone that is forming in the base metal. After this method of welding, however, the part cannot be subjected to service temperatures higher than 500°F (260°C).

Malleable Cast Iron

Malleable cast iron can be welded in much the same manner as gray and nodular cast irons. It is a tough and ductile material. It is white cast iron that has been annealed to convert the brittle graphite flakes to nodules of carbon. During the welding process, some of the carbon is reconverted to martensite, thus reducing ductility. After the welding operation is completed, the malleable casting must go through another annealing. The annealing treatment leaves the metal soft but with considerable strength and toughness. Malleable irons have a tensile strength between 40,000 psi (2,812 kg/cm^2) and 100,000 psi (7,031 kg/cm^2).

COPPER AND COPPER ALLOYS

There are many different types of copper alloys. Copper is often alloyed with other metals such as tin, zinc, nickel, silicon, aluminum, and iron. Copper and copper alloys can be joined by most of the commonly used methods such as gas welding, arc welding, resistance welding, brazing, and soldering.

For many years, the successful welding of copper was considered impractical. Too much distortion resulted when using a gas torch. That heat source was barely able to melt the metal due to its high thermal conductivity. The more highly intense electric arc overcomes that difficulty.

When welding copper, the welding current should be considerably higher than when welding steel. On copper 1/8 in. (3 mm) or more in thickness, the current should not be more than 140 A. The preheat should be 500°F (260°C). By using this method of welding copper, satisfactory results can be obtained for the following factors: economy, speed, ductility, strength, and freedom from distortion. However, the copper must not be electrolytic because excessive amounts of gas in the metal increase porosity.

ALUMINUM WELDABILITY

One of the characteristics of aluminum and its alloys is that it has a great affinity for oxygen. Aluminum atoms combine with oxygen in the air to form a high melting point oxide that covers the surface of the metal. This feature, however, is the key to the high resistance of aluminum to corrosion. It is because of this resistance that aluminum can be used in applications where steel is rapidly corroded.

Pure aluminum melts at 1,200°F (650°C). The oxide that protects the metal melts at 3,700°F (2,037°C). This means that the oxide must be cleaned from the metal before welding can begin.

When the GMA or GTA welding processes are used, the stream of inert gas covers the weld pool, excluding all air from the weld area. This serves to prevent reoxidation of the molten aluminum. Neither of these welding processes requires a flux.

Aluminum has high thermal conductivity. Aluminum and its alloys can rapidly conduct heat away

from the weld area. For this reason, it is necessary to apply the heat much faster to the weld area to bring the aluminum to the welding temperature. Therefore, the intense heat of the electric arc makes this method best suited for welding aluminum.

When aluminum welds solidify from the molten state, they will shrink about 6% in volume. The stress that results from this shrinkage may create excessive joint distortion unless allowances are made before joining the metal. Cracking can occur because the thermal contraction is about two times that of steel. The heated parent metal expands when welding occurs. This expansion of the metal next to the weld area can reduce the root opening on butt joints during the process. The contraction that results upon cooling, plus the shrinkage of the weld metal, creates a tension and serves to increase cracking.

The shape of the weld groove and the number of beads can affect the amount of distortion. Less distortion occurs with two-pass square butt welds. Other factors that have an influence on the weld are the speed of welding, the use of properly designed jigs and fixtures to support the aluminum while it is being welded, and tack welding to hold parts in alignment.

TITANIUM

Titanium is a silvery-gray metal weighing about half as much as steel or about one and one-half times as much as aluminum. Two of the most important properties of titanium are its extremely high strength-to-weight ratio (in alloy form) and its generally excellent corrosion resistance. Titanium alloys, unlike most other light metals, retain their strength at temperatures up to about 800°F (426°C).

Pure titanium is comparatively soft and weak. It is very difficult to refine. Commercially pure titanium contains trace impurities that increase the strength considerably, while at the same time causing a loss in ductility. Alloy combinations currently in use contain assortments of tin, chromium, iron, aluminum, and vanadium.

The success of welding titanium depends upon complete shielding from the atmospheric gases, oxygen, nitrogen, and hydrogen, and from sources of carbon. The inert gas welding process seems to be the best method of welding this metal. Shielding gases used are argon, helium, or a combination of the two. It is preferable to perform the welding in a closed chamber.

When welding titanium, the joint must be clean with no traces of contamination. Clean filler metal and perfect shielding are required to eliminate the porosity and embrittlement that can be produced during welding.

MAGNESIUM

Magnesium is an extremely light metal having a silvery-white color. The weight of magnesium is one-fourth that of steel and approximately two-thirds that of aluminum. Its melting point is 1,202°F (650°C). Magnesium has considerable resistance to corrosion and compares favorably with some aluminum alloys in this respect.

Magnesium must be alloyed with other elements to provide the necessary strength for most applications. Common alloying elements include zinc, aluminum, manganese, zirconium, and the rare earths. Magnesium alloys may be classified as wrought or casting types. Sheet, plate, and extrusions are part of the first group.

The alloy designations for magnesium consist of one or two letters, which represent the alloying elements. The alloying percentages are next listed in the designation. A letter is also added after the alloying percentages. One example is the ASTM designation AZ91C. This indicates an alloy of 9% aluminum and 1% zinc. The letter following the designation is defined as follows:

- A — Aluminum
- C — Copper
- E — Rare earths
- H — Thorium
- K — Zirconium
- M — Manganese
- Z — Zinc

The temper designation is the same as that used for aluminum. Wrought alloys are used to a great extent because of their properties of high strength, ductility, formability, and toughness.

Magnesium can be welded in somewhat the same manner as aluminum. The most widely used processes are GTA and GMA. Spot, seam, and mechanized resistance welding processes can also be used to weld magnesium on a production basis. Spot and seam welding applications consist of welding sheets and extrusions to thicknesses up to 3/16 in. (approximately 4.8 mm).

REPAIR WELDING

Repair or maintenance welding is one of the most difficult types of welding. Some of the major problems include preparing the part for welding, identifying the material, and selecting the best repair method.

The part is often dirty, oily, and painted, and it must be cleaned before welding. There are many hazardous compounds that might be part of the material on the part. These compounds may or may not be hazardous on the part, but when they are heated or burned during welding they can become life threatening.

■ CAUTION

Its never safe to weld on any part that has not been cleaned before welding. All surface contamination must be removed before welding to prevent the possibility of injuring your health from exposure to materials released during welding. Some chemicals can be completely safe until they are exposed to the welding. The smoke or fumes they produce can be an irritant to the skin or eyes; it can be absorbed through the skin or lungs. If you are exposed to an unknown contamination, get professional help immediately.

Contamination can be removed by sand blasting, grinding, or using solvents. If a solvent is used be sure it does not leave a dangerous residue. Clean the entire part if possible or a large enough area so that any remaining material is not affected by the welding.

Before the joint can be prepared for welding you must try to identify the type of metal. There are several ways to determine metal type before welding. One method is to use a metal identification kit. These kits use a series of chemical analyses to identify the metal. Some kits can not only identify a type of metal but also tell the specific alloy.

Another way to identify metal is to look at its color, test for magnetism, and do a spark test. The spark test should be done using a fine grinding stone. With experience, it's often possible to determine specific types of alloys with great accuracy. The sparks given off by each metal and its alloy are so consistent that the U.S. Bureau of Mines uses a camera connected to a computer to identify metals to aid in recycling. For the beginner it is best if you use samples of a known alloy and compare the sparks to your unknown. The test specimen and the unknown should be tested using the same grinding wheel and the same force against the wheel.

EXPERIMENT 24-1

Identifying Metal Using a Spark Test

In this experiment you will use proper eye safety equipment, a grinder, several different known and unknown samples of metal, and a pencil and paper to identify the unknown metal samples. Starting with the known samples, make several tests and draw the spark patterns as described below. Next test the unknown samples and compare the drawings with the drawings from the known samples. See how many of the unknowns you can identify.

There are several areas of the spark test pattern that you must observe carefully, **Figure 24-5 (page 628).**

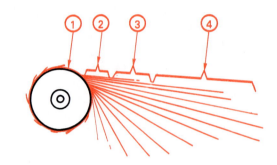

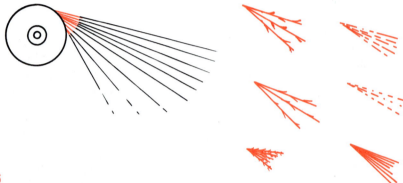

Figure 24-5

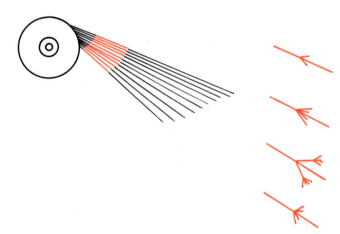

Figure 24-6

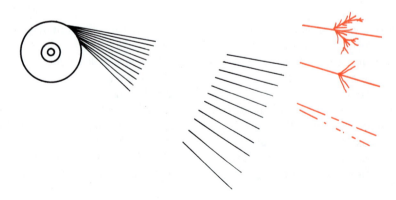

Figure 24-7

The first area is the grinding stone: are there sparks that are being carried around the wheel, or are all of the sparks leaving the wheel, **Figure 24-6.**

The next area is immediately adjacent to the wheel where the spark stream leaves the wheel. Note the color of this area; it may very from white to dull red. Also note the stream to see if the sparks are small, medium, or large and whether the column of sparks is tightly packed or spread out. Draw a sketch of what the spark stream looks like here, **Figure 24-7.**

As the spark stream moves away from the wheel a few inches it will begin to change. This change may be in color, speed of the sparks, or size of the sparks; the sparks may divide into smaller, separate streams, explode in a burst of tiny fragments, or just stop glowing, **Figure 24-8.** Sketch these changes you see in the sparks.

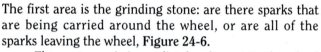

Figure 24-8

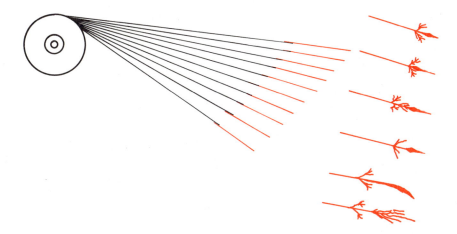

Figure 24-9

When the sparks end they may simply stop glowing, change color, change shape, explode, or divide into smaller parts, **Figure 24-9**. Sketch these changes you see in the sparks

Repeat this experiment with other types of metal as they become available.

Once the type of metal to be repaired is determined, to the best of your ability decide on the type of weld groove that's needed. Some breaks may need to be ground into a V- or U-groove and others may not need to be grooved at all, **Table 24-6**.

Often thin sections of most metals can be repaired without the need for the break to be grooved. Thicker sections of most metals will be easier to repair if the crack is ground into a groove. To help in the realignment of the part, it is a good idea to leave a small part of the crack unground so it can be used to align the parts, **Figure 24-10**.

Metal	Thickness			
	1/8 - 1/4	1/4 - 1/2	1/2 - 1	Over 1
Mild Steel	Square	V	V	V
Aluminum	Square	U	U	U
Magnesium	U	U	U	U
Stainless Steel	Square	V	V	V
Cast Iron	V	V	U	U

Table 24-6

After the part is ready to be welded, a test tab can be welded onto the part. This test tab should be welded in a spot that will not damage the part. Once the tab is welded on, it can be broken off to see if the weld will hold, **Figure 24-11**. If the test tab shows good strength, the repair should continue. If the test tab fails, a new welding procedure should be tried. ◆

Figure 24-10 A small section of the bottom of the break is left so the parts can be realigned.

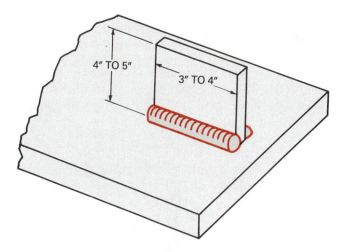

Figure 24-11 Tab test.

T-Boom Control Manifold

Figure 1

Welding operators at a manufacturer of sophisticated geological survey and agricultural equipment were skeptical when the company installed a robot to start automating the welding process. The robot, they thought, would never achieve their high-strength, quality welds or be cost-effective with short production runs.

Experience proved them wrong. Additional advantages gained by the implementation of the welding robot include:

■ relieving the welding operators of monotonous, dangerous, and uncomfortable work,

■ improving the quality of manufactured products by reducing worker inconsistency and fatigue,

■ improving the factory parts flow through reduced set-up times and other economies on short runs,

■ the capability to weld parts that are difficult, undesirable, and uneconomical to handle manually, and

■ reducing parts inventory, handling, and raw material costs.

The company has maintained exceptionally high production standards for more than thirty years of operation. Management considered robots for their welding process for several years. Investigation indicated that arc welding robots could provide improvements in both production and quality. But there were two other requirements to consider — cost-effective short runs and a system capable of handling workpieces with some variation in weld starting points and weld paths.

The arc welding robot selected met all of these requirements and provided many additional benefits, Figure 1. This included the long reach (73.8") of the manipulator, the fast bend and twist movement of the wrist — reducing air-cut time — and the fast and easy programming system. In addition, the robot supplier also provided a seam tracking system. This is an adaptive thru-the-arc

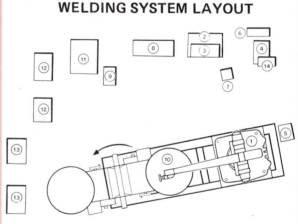

WELDING SYSTEM LAYOUT

1. Robot
2. Controller
3. Interface
4. Power Source
5. Wirefeed System and Pneumatic Cleaning System
6. Seam Tracker Arc Sensing Unit
7. Nozzle Reamer
8. Positioner Controller
9. Positioner Control Panel
10. Turntable Positioner
11. Preheat Oven
12. Parts Bin
13. Finished Parts Bin
14. Air Compressor

Figure 2

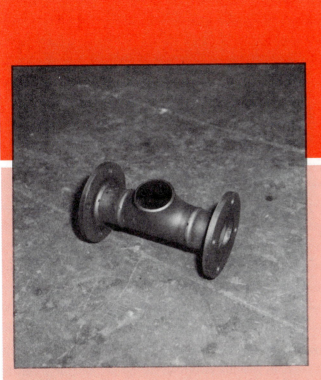

Figure 3

seam tracking system that enables the robot weld gun to adjust automatically to changes in weld starting points and weld path.

Because of the programmability and versatility of the robot, the company has the capability of arc welding a wide variety of parts and material types — including coated and uncoated metals. The workcell, **Figure 2**, has been equipped with a two-station, five-axis work positioner that additionally maximizes short-run efficiency by:

■ providing for the set-up of two different fixtures — thus reducing changeover time,

■ allowing the operator to quickly and easily set up workpieces of various types and sizes,

■ expanding the robot's working range and optimizing the weld path with constant horizontal positioning,

■ helping to reduce air-cut time and increasing welding speeds, and

■ allowing the robot controller to automatically select the proper job program.

During the first four months of operation, the company selected and programmed thirty-five parts for the robot. They all have short run schedules of 25 to 200 parts. A typical control manifold produced by the robotic system is shown in **Figure 3**. The efficiency and reliability of the robot is justifying the company's investment in the system by reducing manufacturing costs, increasing production rates, and improving weld quality and penetration (weld strength).

(Adapted from information provided by Hobart Brothers Company, Troy, Ohio 45373.)

REVIEW

1. What is meant when a metal is said to have good weldability?

2. What does the term *weldability* involve?

3. What properties of a metal can be affected by the choice of welding process?

4. Referring to **Table 24-1 (page 620)**, what commercial joining process can be used to weld 1-inch (25-mm) -thick cast iron?

5. Referring to **Table 24-1**, what commercial joining process can be used to braze sheet stainless steel?

6. Referring to **Table 24-1**, what types of metals can elector gas welding be used to join?

7. What two organizations have developed systems for classifying standard construction grades of steel?

8. Explain the steel classification number 1030.

9. Referring to **Table 24-1**, what is the composition of the metal identified as 44XX?

10. What is the maximum allowable percentage of manganese in carbon steel as defined by the AISI classification for carbon steel?

11. According to **Table 24-3 (page 622)**, at what level of carbon content does weldability become poor?

12. What other factors other than carbon affect a steel's ability to be welded?

13. Why would some low-carbon steels have severe porosity when welded with the GTA welding process?

14. What must be done to steels before welding and after welding if they contain more than 0.40% carbon?

15. Why are high-carbon steels preheated before welding?

16. Explain how to weld tool steel with the oxyfuel process.

17. Why must cracks in the held martensite layer of high-manganese steels be removed?

18. What properties does chromium and nickel produce in stainless steels?

19. What problems can occur to stainless as it is allowed to form carbide precipitation during welding?

20. Why should stainless steel not be held at a temperature between 800°F and 1500°F (425°C and 815°C)?

21. What are the different uses for the three types of stainless steels?

22. Referring to **Table 24-5 (page 625)**, what diameter SMA welding electrode should be used for 1/4-inch (6-mm)-thick stainless steel?

23. Why must cast iron weld metal be ductile?

24. What composition of SMA electrode should be used to make machinable-type weld deposits on cast iron?

25. Why should copper be welded with high currents and preheated?

26. Why can't aluminum oxide be melted off of aluminum?

27. Why are aluminum welds likely to cause distortion and cracking?

28. What alloying elements are added to titanium to give it its strength?

29. What do the letters and numbers in a magnesium identification stand for?

30. How can a part be cleaned before it is welded?

31. How can a spark test be used to identify metals?

Chapter 25

FILLER METAL SELECTION

OBJECTIVES

After completing this chapter, the student should be able to:

- explain how and when to use each type of filler metal.
- select the best filler metal to fit a specific welding job.
- list the forms filler metals come in.
- explain significance of the filler metal prefixes.
- explain how to interpret the standard filler metal numbering systems.
- describe the effects alloys have on ferrous metals.

KEY TERMS

filler metal	fast freezing
carbon equivalent	alloying elements
core wire	arc blow
flux covering	

INTRODUCTION

Manufacturers of filler metals may use any one of a variety of identification systems. There is not a mandatory identification system for filler metals. Manufacturers may use their own numbering systems, trade names, color codes, or a combination of methods to identify filler metal. They may voluntarily choose to use any one of several standardized systems.

The most widely used numbering and lettering system is the one developed by the American Welding Society (AWS). Other numbering and lettering systems have been developed by the American Society for Testing and Materials (ASTM) and the American Iron and Steel Institute (AISI). A system of using colored dots has also been developed by the National Electrical Manufacturers Association (NEMA). Some manufacturers have produced systems that are similar to the AWS system. Most major manufacturers include both the AWS identification and their own identification on the box, package, or directly on the filler metal.

Information that pertains directly to specific filler metal is readily available from most electrode manufacturers. The information given in charts, pamphlets, or pocket electrode guides is specific to their products and may or may not include standard AWS tests, terms, or classifications within their identification systems.

The AWS publishes a variety of books, pamphlets, and charts showing the minimum specifications for filler metal groups, Table 25-1 (A) through (E). They also publish comparison charts that include all of the information manufacturers

See AWS A5.1-81, Specifications for Carbon
Steel Covered Arc Welding Electrodes

Manufacturers	AWS Classification	E6010	E6011	E6012	E6013	E6020	E6022	E6027
Aga de Mexico, S.A.		AGA C10 C12	AGA C11	AGA R12	AGA R10 R11	—	—	—
Airco Welding Products		AIRCO 6010	AIRCO 6010 6011C 6011LOC	AIRCO 6012 6012C	AIRCO 6013 6013C 6013D	AIRCO 6020	—	EASY ARC 6027
Air Products and Chemicals, Inc.		AP6010W	AP6011W	AP6012W	AP6013W	—	—	—
A-L Welding Products, Inc.		AL6010	AL6011	AL6012	AL6013	—	—	—
Alloy Rods, Allegheny International, Inc.		AP100 SW610	SW14	SW612	SW15	—	—	—
Applied Power, Inc.		—	No. 130 Red-Rod	—	No. 140 Production Rod	—	—	—
Arcweld Products Limited		Easyarc 10	Arcweld 230 Easyarc 11	Arcweld 387	Arcweld 90 Satinarc	—	—	—
Bohler Bros. of America, Inc.		Fox CEL	—	—	Fox OHV Fox ETI	Fox UMZ	—	—
Brite-Weld		Brite-Weld E6010	Brite-Weld E6011	Brite-Weld E6012	Brite-Weld E6013	—	—	Brite-Weld E6027
Canadian Liquid Air Ltd.		LA6010	LA6011P	LA6012P	LA6013 LA6013P	—	—	—
Canadian Rockweld Ltd.		R60	R61	R62	R63 R63A	R620	—	R627
C-E Power Systems, Combustion Engineering, Inc.		—	—	—	—	—	—	—
Century Mfg. Co.		—	331 324	—	331 313	—	—	—
Champion Hobart, S.A. de C.V.		6010	6011	6012 Ducto P60	6013 Versa-T	—	—	—
CONARCO, Alambres y Soldaduras, S.A.		CONARCO 10 CONARCO 10P	CONARCO 11	CONARCO 12 CONARCO 12D	CONARCO 13A CONARCO 13	—	—	—
Cronatron Welding Systems, Inc.		—	Cronatron 6011	—	Cronatron 6013	—	—	—
Electromanufacturas, S.A.		West Arco XL-610M ZIP 10T	West Arco ACP611	West Arco FP-612 ZIP 12	West Arco SW-613M SUPER SW-613 SW-10	—	—	West Arco ZIP 27
ESAB		OK 22.45	OK 22.65	—	OK 46.00 OK 43.32 OK 50.10 OK 50.40	—	—	—
Eureka Welding Alloys		Eureka 6010	Eureka 6011	Eureka 6012	Eureka 6013	—	—	Eureka 6027
Eutectic Corporation		—	—	—	DynaTrode 666	—	—	—

Table 25-1 (A) Carbon Steel Covered Arc Welding Electrodes. (Courtesy of the American Welding Society.)

Manufacturers	AWS Classification E6010	E6011	E6012	E6013	E6020	E6022	E6027
Hobart Brothers Company	Hobart 10 60AP	Hobart 335A	Hobart 12 12A	Hobart 413 447A	—	Hobart 1139	Hobart 27H
International Welding Products, Inc.	INWELD 6010	INWELD 6011	INWELD 6012	INWELD 6013	INWELD 6020	—	INWELD 6027
Kobe Steel, Ltd.	KOBE 6010	KOBE 6011	KOBE TB-62 RB-26D	ZERODE 44 RB-26 B-33	—	—	ZERODE 27 AUTOCON 27, B-27
Latamer Company, Inc.	Latco E6010	Latco E6011	Latco E6012	Latco E6013	Latco E6020	—	Latco E6027
The Lincoln Electric Company	Fleetweld 5P	Fleetweld 35 35LS 180	Fleetweld 7	Fleetweld 37 57	—	—	Jetweld 2
Liquid Carbonic, Inc.	RD 704D RD 610P	RD 504D RD 611P	RD 604	RD 613	—	—	—
Murex Welding Products	SPEEDEX 610	Type A 611C 611LV	Type N 13 GENEX M	Type U U13	F.H.P.	—	SPEEDEX 27
Oerlikon, Inc.	OERLIKON E6010	—	OERLIKON E6012	OERLIKON E6013	OERLIKON E6020	—	—
Sodel	—	SODEL 11	—	SODEL 31	—	—	—
Techalloy Maryland, Inc. Reid-Avery Div.	RACO 6010	RACO 6011	RACO 6012	RACO 6013	RACO 6020	—	RACO 6027
Teledyne Canada HARFAC	6010	6011	6012	6013	—	—	—
Teledyne McKay	6010	6011	6012	6013	—	—	—
Thyssen Draht AG	Thyssen Cel 70	—	Thyssen Blau Thyssen A5 Thyssen Grun	Thyssen Grun T SH Blau SH Gelb B SH Gelb T SH Gelb S SH Gelb R SH Grun TB SH Lila R Union 6013	SH Gelb	SH Tiefbrand	—
WASAWELD	WASAWELD E6010	WASAWELD E6011	WASAWELD E6012	WASAWELD E6013	WASAWELD E6020	—	WASAWELD E6027
Weld Mold Company	—	—	—	—	—	—	—
Weldwire Co., Inc.	WELDWIRE 6010	WELDWIRE 6011	WELDWIRE 6012	WELDWIRE 6013	WELDWIRE 6020	WELDWIRE 6022	WELDWIRE 6027
Westinghouse Electric Corporation	ZIP 10 XL 610A	ACP ZIP-11R	ZIP 12	SW	DH 620	—	ZIP 27 ZIP 27M

Table 25-1 (B) Carbon Steel Covered Arc Welding Electrodes (continued). (Courtesy of the American Welding Society.)

Manufacturers	AWS Classification E7014	E7015	E7016	E7016-1	E7018	E7018-1	E7024
Aga de Mexico, S.A.	—	—	—	—	AGA B10	—	AGA RH10
Airco Welding Products	EASY ARC 7014	—	7016 7016M	—	EASY ARC 7018C 7018MR CODE ARC 7018MR	—	EASY ARC 7024 7024D
Air Products and Chemicals, Inc.	AP7014W	—	AP7016W	—	AP7018W	—	AP7024W
A-L Welding Products, Inc.	AL7014	—	—	—	AL7018 Nuclearc 7018	—	AL7024
Alloy Rods, Allegheny International, Inc.	SW15IP	—	70LA-2	Atom Arc 7016-1	Atom Arc 7018 SW-47	Atom Arc 7018-1	7024
Applied Power, Inc.	No. 146 Hy-Pro-Rod	—	—	—	No. 7018 Marq-Rod	—	—
Arcweld Products Limited	Easyarc 14	—	Arcweld 312	—	Easyarc 328 Easyarc 7018MR	—	Easyarc 12
Bohler Brothers of America, Inc.	—	—	Fox EV47	—	Fox EV50	—	Fox HL 180 Ti
Brite-Weld	Brite-Weld 7014	—	—	—	Brite-Weld E7018	—	Brite-Weld E7024
Canadian Liquid Air, Ltd.	LA7014	—	—	—	Super Arc 18 LA7018, AA7018, LA7018B	—	LA7024 LA24HD
Canadian Rockweld, Ltd.	R74	—	Tensilarc 76	—	Hyloarc 78	—	R724
C-E Power Systems, Combustion Engineering, Inc.	—	—	—	—	CE7018	—	—
Century Mfg. Co.	331 363	—	—	—	331 327	—	—
Champion Hobart, S.A. de C.V.	—	—	724	—	7018	718	MULTI-T
CONARCO, Alambres y Soldaduras, S.A.	CONARCO 14	CONARCO 15	CONARCO 16	—	CONARCO 18	CONARCO 18-1	CONARCO 24
Cronatron Welding Systems, Inc.	—	—	—	—	Cronatron 7018	—	—
Electromanufacturas, S.A.	West Arco ZIP 14	—	West Arco WIZ 16	—	West Arco WIZ 18	—	West Arco ZIP 24
ESAB	OK 46.16	—	OK 53.00 OK 53.05	OK 53.68	OK 48.00 OK 48.04 OK 48.15	OK 48.68 OK 55.00	OK Femax 33.65 OK Femax 33.80
Eureka Welding Alloys	Eureka 7014	—	—	—	Eureka 7018	—	Eureka 7024
Eutectic Corporation	—	—	—	—	EutecTrode 3018	—	—
Hobart Brothers Company	Hobart 14A	—	—	—	Hobart 718 718LMP	—	Hobart 24 24H
International Welding Products, Inc.	INWELD 7014	INWELD 7015	INWELD 7016	—	INWELD 7018	—	INWELD 7024

Table 25-1 (C)   Carbon Steel Covered Arc Welding Electrodes (continued). (Courtesy of the American Welding Society.)

Manufacturers	AWS Classification	E7014	E7015	E7016	E7016-1	E7018	E7018-1	E7024
Kobe Steel, Ltd.		KOBE RB-14	—	KOBE LB-52 LB-52U LB-26Vu ZERODE-52	—	KOBE LB-52-18 LTB-52A	—	KOBE RB-24 ZERODE 50F FB-24
Latamer Company, Inc.		Latco E7014	Latco E7015	Latco E7016	—	Latco E7018	—	Latco E7024
The Lincoln Electric Company		Fleetweld 47	—	—	—	Jetweld LH70 LH 73 LH 78 LH 75	Jetweld LH75	Jetweld 1 3
Liquid Carbonic, Inc.		RD 714	—	—	—	RD 718	—	RD 724
Murex Welding Products		SPEEDEX U	—	HTS HTS-18 HTS-180	—	SPEEDEX HTS-MR HTS-M-MR-718	—	SPEEDEX 24 24D
Oerlikon, Inc.		—	—	—	OERLIKON Spezials Extra Tenacito WZ	OERLIKON E7018	OERLIKON E7018-1	OERLIKON Ferromatic Ferrocito R
Sodel		—	—	—	—	—	SODEL 328	SODEL 314
Techalloy Maryland, Inc. Reid-Avery Div.		RACO 7014	RACO 7015	RACO 7016	—	RACO 7018	RACO 7018-1	RACO 7024
Teledyne Canada HARFAC		7014	—	7016	—	—	7018-1	7024
Teledyne McKay		7014	—	7016	—	7018 XLM	7018-1 XLM	7024
Thyssen Draht AG		SH Multifer 130	Thyssen K50 SH Grun K45 (6015)	SH Kb F Thyssen K50R SH Grun K50W Thyssen Kb Spezial SH Grun K70W Thyssen K90S	—	SH Grun K70	Thyssen 120K	Thyssen Rot R160 Thyssen Rot R160S SH Multifer 180 Thyssen Rot AR160
WASAWELD		WASAWELD E7014	—	—	—	WASAWELD E7018	—	WASAWELD E7024
Weld Mold Company		—	—	—	—	WELD MOLD 7018	—	—
Weldwire Co., Inc.		WELDWIRE 7014	WELDWIRE 7015	WELDWIRE 7016	WELDWIRE 7016-1	WELDWIRE 7018	WELDWIRE 7018-1	WELDWIRE 7024
Westinghouse Electric Corporation		ZIP 14	—	LOH 2	—	WIZ 18	—	ZIP 24

Manufacturers	AWS Classification	E7024-1	E7027	E7028	E7048
Aga de Mexico, S.A.		—		—	—
Airco Welding Products		—	—	EASY ARC 7028	—
Air Products and Chemicals, Inc.		—	—	—	—
A-L Welding Products, Inc.		—		—	—
Alloy Rods, Allegheny International, Inc.		7024-1	—	—	—
Applied Power, Inc.		—	—	—	—
Arcweld Products Limited		—	—	Super 28	—

Table 25-1 (D) Carbon Steel Covered Arc Welding Electrodes (continued). (Courtesy of the American Welding Society.)

Manufacturers	AWS Classification E7024-1	E7027	E7028	E7048
Bohler Bros. of America, Inc.	—	—	Fox HL 180 Kb	—
Brite-Weld	—	—	—	—
Canadian Liquid Air, Ltd.	LA7024	—	LA7028 LA7028B	LA7048B
Canadian Rockweld, Ltd.	—	R727	Hyloarc 728	—
C-E Power Systems, Combustion Engineering, Inc.	—	—	—	—
Century Mfg. Co.	—	—	—	—
Champion Hobart, S.A. de C.V.	—	—	—	—
CONARCO, Alambres y Soldaduras, S.A.	—	—	CONARCO 28	CONARCO 48
Cronatron Welding Systems, Inc.	—	—	—	—
Electromanufacturas, S.A.	—	—	—	—
ESAB	—	—	OK 38.48 OK 38.65 OK 38.85 OK 38.95	OK 53.35
Eureka Welding Alloys	—	—	—	—
Eutectic Corporation	—	—	—	—
Hobart Brothers Company	—	—	Hobart 728	—
International Welding Products, Inc.	—	—	INWELD 7028	—
Kobe Steel, Ltd.	—	—	—	KOBE LB-26V LB-52V ZERODE 6V
Latamer Company, Inc.	—	—	Latco E7028	—
The Lincoln Electric Company	Jetweld 1	—	Jetweld LH3800	—
Liquid Carbonic, Inc.	—	—	RD 728	—
Murex Welding Products	—	—	SPEEDEX 28	—
Oerlikon, Inc.	—	—	OERLIKON E7028	—
Sodel	—	—	—	—
Techalloy Maryland, Inc. Reid-Avery Div.	—	—	RACO 7028	RACO 7048
Teledyne Canada HARFAC	—	—	—	—
Teledyne McKay	—	—	—	—
Thyssen Draht AG	—	SH Multifer 200	SH Multifer 150 K11 Thyssen Rot BR160	—
WASAWELD	—	—	—	—
Weld Mold Company	—	—	—	—
Weldwire Co., Inc.	WELDWIRE 7024-1	WELDWIRE 7027	WELDWIRE 7028	WELDWIRE 7048
Westinghouse Electric Corporation	—	—	—	—

Source: Reprinted from AWS A5.0-83, Filler Metal Comparison Charts

Table 25-1 (E) Carbon Steel Covered Arc Welding Electrodes (continued). (Courtesy of the American Welding Society.)

provide the AWS regarding their filler metals. Both the literature on filler metal specifications and filler metal comparisons may be obtained directly from the AWS.

The AWS classification system is for minimum requirements within a grouping. Filler metals manufactured within a grouping may vary but still be classified under that grouping's classification.

A manufacturer may add elements to the metal or flux, such as more arc stabilizers. When one characteristic is improved, another characteristic may also change. The added arc stabilizer may make a smoother weld with less penetration. Other changes may affect the strength and ductility or other welding characteristics.

Because of the variables within a classification, some manufacturers make more than one type of filler metal that is included in a single classification. This and other information may be included in the data supplied by manufacturers.

Manufacturers' Electrode Information

The type of information given by different manufacturers ranges from general information to technical, chemical, and physical information. A mixture of different types of information may be given.

General information given by manufacturers may include some or all of the following: welding electrode manipulation techniques, joint design, prewelding preparation, postwelding procedures, types of equipment that can be welded, welding currents, and welding positions.

Understanding the Electrode Data

Technical procedures, physical properties, and chemical analysis information given by manufacturers may include some or all of the following:

- Number of welding electrodes per pound
- Number of inches of weld per welding electrode
- Welding amperage setting for each size of welding electrode
- Welding codes for which the electrode can be used
- Types of metal that can be welded
- Ability to weld on rust, oil, or paint
- Weld joint penetration characteristics
- Preheating and postheating temperatures
- Weld deposit physical strengths: ultimate tensile strength, yield point, yield strength, elongation, impact strength
- Percentages of such alloys as carbon, sulfur, phosphorus, manganese, silicon, chromium, nickel, molybdenum, and other alloys

The information supplied by the manufacturer can be used for a variety of purposes including the following:

- Estimates of the pounds of electrodes needed for a job
- Welding conditions under which the electrode can be used, for example, on clean or dirty metal

- Welding procedure qualification information regarding amperage, joint preparation, penetration, and welding codes
- Physical and chemical characteristics effecting the weld's strengths and metallurgical properties

Data Resulting from Mechanical Tests

Most of the technical information supplied is self-explanatory and easily understood. The mechanical properties of the weld are given as the results of standard tests. The following are some of the standard tests and the meaning of each test.

- Tensile strength, psi — the load in pounds that would be required to break a section of soundweld that has a cross-sectional area of one square inch.
- Yield Point, psi — the point in low- and medium-carbon steels at which the metal begins to stretch when force (stress) is applied after which it will not return to its original length.
- Elongation, percent in two inches — the percentage that a two-inch piece of weld will stretch before it breaks.
- Charpy V notch, ft-lb — the impact load required to break a test piece of weld metal. This test may be performed on metal below room temperature at which point it is more brittle.

Data Resulting from Chemical Analysis

Chemical analysis of the weld deposit may also be included in the information given by manufacturers. It is not so important to know what the different percentages of the alloys do, but it is important to know how changes in the percentages of the alloys affect the weld. Chemical composition can easily be compared from one electrode to another. The following are the major elements and the effects of their changes on the iron in carbon steel.

- Carbon (C) — As the percentage of carbon increases, the tensile strength increases, the hardness increases, and ductility is reduced. Carbon also causes austenite to form.
- Sulfur (S) — It is usually a contaminant, and the percentage should be kept as low as possible below 0.04%. As the percentage of carbon increases, sulfur can cause hot shortness and porosity.
- Phosphorus (P) — It is usually a contaminant, and the percentage should be kept as low as possible. As the percentage of phosphorus increases, it can cause weld brittleness, reduced shock resistance, and increased cracking.
- Manganese (Mn) — As the percentage of manganese increases, the tensile strength, hardness, resistance to abrasion, and porosity all increase; hot shortness is reduced. It's also a strong austenite former.

- Silicon (Si) — As the percentage of silicon increases, tensile strength increases, and cracking may increase. Used as a deoxidizer and ferrite former.

- Chromium (Cr) — As the percentage of chromium increases, tensile strength, hardness, and corrosion resistance increase with some decrease in ductility. It's also a good ferrite and carbide former.

- Nickel (Ni) — As the percentage of nickel increases, tensile strength, toughness, and corrosion resistance increase. It's also an austenite former.

- Molybdenum (Mo) — As the percentage of molybdenum increases, tensile strengthens at elevated temperatures; creep resistance and corrosion resistance all increase. It's also a ferrite and carbide former.

- Copper (Cu) — As the percentage of copper increases the corrosion resistance and cracking tendency increases.

- Columbium (Cb) — As the percentage of columbium (niobium) increases, the tendency to form chromecarbides is reduced in stainless steels. It's also a strong ferrite former.

- Aluminum (Al) — As the percentage of aluminum increases, the high temperature scaling resistance improves. It's also a good oxidizer and ferrite former.

Carbon Equivalent (CE)

The weldability of an iron alloy is affected by the combination of the alloys used with carbon. To determine the weldability of a specific alloy formulas have been developed to calculate the carbon equivalence. There are several different formulas, some more complex than others. All of the formulas produce the same product, a carbon equivalent (CE) value that can be used to determine the weldability of the alloy. The CE value and the percentage of carbon are used similarly. Carbon equivalence or carbon content should be used to determine whether any special procedures are needed to make an acceptable weld. They are used for selecting such conditions as pre- or postheat temperatures and stress releasing. Carbon and its effects on welding are covered in more depth in Chapter 24.

The CE of an alloy is an indication of how the weld will affect the surrounding metal. This area is called the heat-affected zone (HAZ), Figure 25-1. The higher the CE the more effect the weld can have on the surrounding base metal.

By knowing the CE of the metal the welder can adjust the welding procedure to control problems with the HAZ. Some of the most common adjustments to the welding procedure are: pre and/or post heating, electrode selection, electrode size, electrode type, and current settings.

Using the following CE formula a welder can make the proper adjustments in the welding procedure for plain carbon steels.

$$CE = \%C + \frac{\%Mn}{6} + \frac{\%Mo}{4} + \frac{\%Cr}{5} + \frac{\%Ni}{15} + \frac{\%Cu}{15} + \frac{\%P}{3}$$

- CE= 0.40% or less — no special welding requirements.

- CE= 0.40% to 0.60% — low-hydrogen welding electrode and the related procedure required.

- CE= 0.60% or more — low-hydrogen welding electrode, higher welding heat inputs, pre-heating, and controlled cooling rates.

SMAW OPERATING INFORMATION

Shielded metal arc welding electrodes, sometimes referred to as welding rods, stick electrodes, or simply electrodes, have two parts. These two parts are the inner core wire and a flux covering, Figure 25-2.

The functions of the core wire include the following:

- To carry the welding current
- To serve as most of the filler metal in the finished weld

The functions of the flux covering include the following:

- To provide some of the alloying elements
- To provide an arc stabilizer (optional)
- To serve as an insulator
- To provide a slag cover to protect the weld bead and slow cooling rate
- To provide a protective gaseous shield during welding

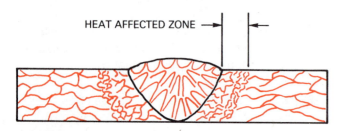

Figure 25-1

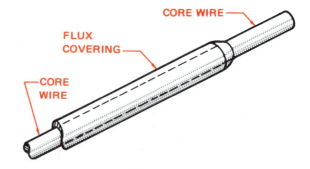

Figure 25-2 The two parts of a welding electrode.

CORE WIRE

A core wire is the primary metal source for a weld. For fabricating structural and low-alloy steels, the core wires of the electrode use inexpensive rimmed or low-carbon steel. For more highly alloyed materials, such as stainless steel, high-nickel alloys, or nonferrous alloys, the core wires are of the approximate composition of the material to be welded. The core wire also supports the coating that carries the fluxing and alloying materials to the arc and weld pool.

FUNCTIONS OF THE FLUX COVERING

Provides Shielding Gases

Heat generated by the arc causes some constituents in the flux covering to decompose and others to vaporize, forming shielding gases. These gases prevent the atmosphere from contaminating the weld metal as it transfers

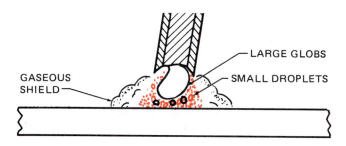

Figure 25-3 Methods of metal transfer during an arc.

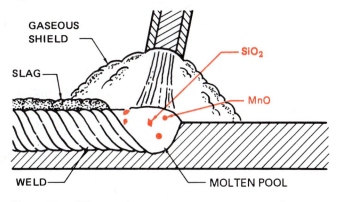

Figure 25-4 Silicon and manganese act as scavengers that combine with contaminants and float to the slag on top of the weld.

across the arc gap. They also protect the molten weld pool as it cools to form solid metal. In addition, shielding gases and vapors greatly affect both the drops that form at the electrode tip and their transfer across the arc gap, **Figure 25-3**. They also cause the spatter from the arc and greatly determine arc stiffness and penetration. For example, the E6010 electrode contains cellulose. Cellulose decomposes into the hydrogen responsible for the deep electrode penetration so desirable in pipeline welding.

Alloying Elements

Elements in the flux are mixed with the filler metal. Some of these elements stay in the weld metal as alloys. Other elements pick up contaminants in the molten weld pool and float them to the surface. At the surface, these contaminants form part of the slag, **Figure 25-4**.

Effect on Weld

Welding fluxes can affect the penetration and contour of the weld bead. Penetration may be pushed deeper if the core wire is made to melt off faster than the flux melts. This forms a small chamber or crucible at the end of the electrode that acts like the combustion chamber of a rocket. As a result, the molten metal and hot gases are forced out very rapidly. The effect of this can be seen on the surface of the molten weld pool as it is blown back away from the end of the electrode. Some electrodes do not use this jetting action, and the resulting molten weld pool is much calmer (less turbulent) and may be rounded in appearance. In addition, the resulting bead may have less penetration, **Figure 25-5**.

Weld bead contour can also be affected by the slag formed by the flux. Some high-temperature slags, called refractory, solidify before the weld metal solidifies, forming a mold that holds the molten metal in place. These electrodes are sometimes referred to as **fast freezing** and are excellent for vertical, horizontal, and overhead welding positions.

FILLER METAL SELECTION

Selecting the best filler metal for a job is seldom delegated to the welder in large shops. The selection of

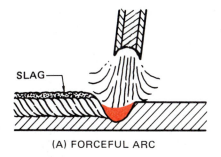

(A) FORCEFUL ARC

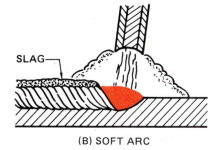

(B) SOFT ARC

(C) HIGH-TEMPERATURE SLAG

Figure 25-5

the correct process and filler metal is a complex process. If the choice is given to the welder, it is one of the most important decisions the welder will make.

Covering all of the variables for selecting a filler metal would be well beyond the scope of this text. A sample of the types of things that must be considered for the selection of a SAW electrode are given below. To further complicate things, welding electrodes have more than one application, and many welding electrodes may be used for the same type of work.

The following conditions that the welder should consider when choosing a welding electrode are not in order of importance. They are also not all of the factors that must be considered.

Shielded Metal Arc Welding Electrode Selection

- Type of electrode — What electrode has been specified in the blueprints or in the contract for this job?

- Type of current — Can the welding power source supply AC only, DC only, or both AC and DC?

- Power range — What is the amperage range on the welder and its duty cycle? Different types of electrodes require different amperage settings even for the same size welding electrode. For example, the amperage range for a 1/8-in. (3-mm)-diameter E6010 electrode is 75 A to 125 A, and the amperage range for a 1/8-in. (3-mm)-diameter E7018 electrode is 90 A to 165 A.

- Type of metal — Some welding electrodes may be used to join more than one similar type of metal. Other electrodes may be used to join together two different types of metal. For example, an E309-15 electrode can be used to join 305 stainless steel to 1020 mild steel.

- Thickness of metal — The penetration characteristics of each welding electrode may differ. Selecting one electrode that will weld on a specific thickness of material is important. For example, E6013 has very little penetration and is therefore good for welding on sheet metal.

- Weld position — Some welding electrodes can be used to make welds in all positions. Other electrodes may be restricted to making flat, horizontal, and/or vertical position welds; a few electrodes may be used to make flat position welds only.

- Joint design — The type of joint and whether it is grooved or not may affect the performance of the welding electrode. For example, the E7018 electrode does not produce a large gaseous cloud to protect the molten metal. For this reason, the electrode movement is restricted so that the molten weld pool is not left unprotected by the gaseous cloud.

- Surface condition — It is not always possible to work on new, clean metal. Some welding electrodes will weld on metal that is rusty, oily, painted, dirty, or galvanized.

- Number of passes — The amount of reinforcement needed may require more than one welding pass. Some welding electrodes will build up faster, and others will penetrate deeper. The slag may be removed more easily from some welds than from others. For example, E6013 will build up a weld faster than E6010, and the slag is also more easily removed between weld passes.

- Distortion — Welding electrodes that will operate on low-amperage settings will have less heat input and cause less distortion. Welding electrodes that have a high rate of deposition (fills the joint rapidly) and can travel faster will also cause less distortion. For example, the flux on an E7024 has 50% iron powder, which gives it a faster fill rate and allows it to travel faster, resulting in less distortion of the metal being welded.

- Preheat or postheat — On low-carbon steel plate 1 in. (25 mm) thick or more, thick preheating is required with most welding electrodes. Postheating may be required to keep a weld zone from hardening or cracking when using some welding electrodes. However, no postheating may be required when welding low-alloy steel using E310-15.

- Temperature service — Weld metals react differently to temperature extremes. Some welds become very brittle and crack easily in cryogenic (low-temperature) service. A few weld metals resist creep and oxidation at high temperatures. For example, E310Mo-15 can weld on most stainless and mild steels without any high-temperature problems.

- Mechanical properties — Mechanical properties such as tensile strength, yield strength, hardness, toughness, ductility, and impact strength can be modified by the selection of specific welding electrodes.

- Postwelding cleanup — The hardness or softness of the weld greatly affects any grinding, drilling, or machining. The ease with which the slag can be removed and the quantity of spatter will affect the time and amount of cleanup required.

- Shop or field weld — The general working conditions such as wind, dirt, cleanliness, dryness, and accessibility of the weld will affect the choice of welding electrode. For example, the E7018 electrode must be kept very dry, but the E6010 electrode is not greatly affected by moisture.

- Quantity of welds — If a few welds are needed, a more expensive welding electrode requiring less skill may be selected. For a large production job requiring a higher skill level, a less expensive welding electrode may be best.

After deciding the specific conditions that may affect the welding, the welder has most likely identified more than one condition that needs to be satisfied. Some of the conditions will not interfere with others. For example, the type of current and whether a welder makes one or more weld passes have little or no effect on each other. However, if a welder needs to machine the finished weld, hardness is a consideration. When two or more conditions conflict, the welder is seldom the person who will make the decision. It may be necessary to choose more than one welding electrode. When welding pipe, E6010 and E6011 are often used for the root pass because of their penetration characteristics, and E7018 is used for the cover pass because of its greater strength and resistance to cracking.

Each AWS electrode classification has its own welding characteristics. Some manufacturers have more than one welding electrode in some classifications. In these cases, the minimum specifications for the classification have been exceeded. An example of more than one welding electrode in a single classification is Lincoln's Fleetweld 35®, Fleetweld 35LS®, and Fleetweld 180R®. These electrodes are all in AWS classification E6011. For the manufacturer's complete description of these electrodes, consult Table 25-2 (page 644).

The characteristics of each manufacturer's filler metals can be compared to one another by using data sheets supplied by the manufacturer. General comparisons can be made easily using an electrode comparison chart.

When making an electrode selection, many variables must be kept in mind; and the performance characteristics must be compared before making a final choice.

AWS FILLER METAL CLASSIFICATIONS

The AWS classification system uses a series of letters and numbers in a code that gives the important information about the filler metal. The prefix letter is used to indicate the filler's form, or a type of process the filler is to be used with, or both. The prefix letters and their meanings are as follows:

- **E** — Indicates an arc welding electrode. The filler carries the welding current in the process. We most often think of the E standing for an SMA "stick" welding electrode. It also is used to indicate wire electrodes used in GMAW, FCAW, SAW, ESW, EGW, etc.

- **R** — Indicates a rod that is heated by some other source other than electric current flowing directly through it. Welding rods are sometimes referred to as being "cut length" or "welding wire." It is often used with OFW and GTAW.

- **ER** — Indicates a filler metal that is supplied for use as either a current carrying electrode or in rod forms. The same alloys are used to produce the electrodes and the rods. This filler metal may be supplied as a wire on a spool for GMAW or as a rod for OFW or GTAW.

- **EC** — Indicates a composite electrode. These electrodes are used for SAW. Don't confuse an ECu, copper arc welding electrode, for an ECNi2, which is a composite nickel submerged arc welding wire.

- **B** — Indicates a brazing filler metal. This filler metal is usually supplied as a rod, but it can come in a number of other forms. Some of the forms it comes in are powder, sheets, washers, rings, etc.

- **RB** — Indicates a filler metal that is used as a current carrying electrode or as a rod or both. The form the filler is supplied in for each of the applications may be different. The composition of the alloy in the filler metal will be the same for all of the forms supplied. This filler can be used for processes like arc braze welding or oxyfuel brazing.

- **RG** — Indicates a welding rod used primarily with oxyfuel welding. This filler can be used with all of the oxyfuels, and some of the fillers are used with the GTAW process.

- **IN** — Indicates a consumable insert. These are most often used for welding on pipe. They are preplaced in the root of the groove to supply both filler metal and support for the root pass. The inserts may provide for some joint alignment and spacing.

The next two classifications are not filler metal. They are classified under the same system because they are welding consumables. The GTA welding tungsten is not a filler, but it is consumed, very slowly, during the welding process.

- **EW** — Indicates a nonconsumable tungsten electrode. The GTAW electrode is obviously not a filler metal, but it falls under the same classification system.

- **F** — Indicates a flux used for SAW. The composition of the weld metal is influenced by the flux. There are alloys and agents in the flux used for SAW that are dissolved into the weld metal. For this reason, the filler metal and flux are specified together with the filler metal identification first and the flux second.

In addition to the prefix, there are some suffix identifiers. The suffix may be used to indicate a change in the alloy in a covered electrode or the type of welding current to be used with stainless steel covered electrodes.

CARBON STEEL

Carbon and Low Alloy Steel Covered Electrodes

The AWS specification for carbon steel covered arc electrodes is A5.1, and for low alloy steel covered arc electrodes it is A5.5. Filler metal classified within these specifications are identified by a system that uses the letter E followed by a series of numbers to indicate the minimum tensile strength of a good weld, the position(s) in which the electrode can be used, the type of flux coating, and the type(s) of welding current, **Figure 25-6 (page 645)**.

The tensile strength is given in pounds per square inch (psi). The actual strength is obtained by adding three zeros to the right of the number given. For example, E60XX is 60,000 psi, E110XX is 110,000 psi, etc.

The next number located to the right of the tensile strength, either 1, 2, or 4, designates the welding position capable. For example:

ELECTRODE IDENTIFICATION AND OPERATING DATA

COATING COLOR	AWS Number on Coating	(L)LINCOLN	ELECTRODE	ELECTRODE POLARITY	SIZES AND CURRENT RANGES (Amps.) (electrodes are manufactured in these sizes for which current ranges are given)					
					3/32" SIZE	1/8" SIZE	5/32" SIZE	3/16" SIZE	7/32" SIZE	1/4" SIZE
Brick Red	6010		FLEETWELD 5P	DC+	40–75	75–130	90–175	140–225	200–275	220–325①
Gray	6011		FLEETWELD 35	AC	50–85	75–120	90–160	120–200	150–260	190–300
				DC+	40–75	70–110	80–145	110–180	135–235	170–270
Red Brown	6011	Green	FLEETWELD 35LS	AC		80–130	120–160			
				DC±		70–120	110–150			
Brown	6011		FLEETWELD 180	AC	40–90	60–120	115–150			
				DC±	40–80	55–110	105–135			
Pink	7010-A1		SHIELD-ARC 85	DC+	50–90	75–130	90–175	140–225		
Pink	7010-A1	Green	SHIELD-ARC 85P	DC+				140–225		
Tan	7010-G		SHIELD-ARC HYP	DC+		75–130	90–185	140–225	160–250	
Gray	8010-G		SHIELD-ARC 70+	DC+		75–130	90–185	140–225		

AWS Number   Lincoln Trademark

① Range for 5/16" size is 240–400 amps.   DC+ is Electrode Positive. DC– is Electrode Negative.

All tests were performed in conformance with specifications AWS A5.5 and ASME SFA.5.5 in the aged condition for the E7010-G and E8010-G electrodes, and in the stress relieved condition for Shield-Arc 85 & 85P. Tests for the other products were performed in conformance with specifications AWS A5.1 and ASME SFA.5.1 for the as-welded condition

TYPICAL MECHANICAL PROPERTIES

Low figures in the stress-relieved tensile and yield strength ranges below for Shield-Arc 85 and 85P are AWS minimum requirements.
Low figures in the as-welded tensile and yield strength ranges below for the other products are AWS minimum requirements.

	FLEETWELD 5P	FLEETWELD 35	FLEETWELD 35LS	FLEETWELD 180	SHIELD-ARC 85	SHIELD-ARC 85P	SHIELD-ARC HYP	SHIELD-ARC 70+
As Welded Tensile Strength—psi	62–69,000	62–68,000	62–67,000	62–71,000	70–78,000	70–78,000	70–84,000	80–92,000
Yield Point—psi Ductility—% Elong. in 2"	52–62,000 22–32	50–62,000 22–30	50–60,000 22–31	50–64,000 22–31	60–71,000 22–26	57–63,000 22–27	60–77,000 22–23	67,000–83,000 19–24
Charpy V-Notch Toughness —ft. lbs.	20–60 @ –20°F	20–90 @ –20°F	20–57 @ –20°F	20–54 @ –20°F	68 @ 70°F	68 @ 70°F	30 @ 20°F	40 @ 50°F
Hardness, Rockwell B (avg) ⑤	76–82	76–85	73–86	75–85			83–92	88–93
Stress Relieved @ 1150°F Tensile Strength—psi	60–69,000	60–66,000	60–65,000		70–83,000	70–74,000	80–82,000	80–84,000
Yield Point—psi Ductility—% Elong. in 2"	46–56,000 28–36	46–56,000 28–36	46–51,000 28–33		57–69,000 22–28	57–65,000 22–27	72–76,000 24–27	71–76,000 22–26
Charpy V-Notch Toughness —ft. lbs.	71 @ 70°F		120 @ 70°F		64 @ 70°F	68 @ 70°F	30 @ –20°F	30 @ –50°F
Hardness, Rockwell B (avg) ⑤					80–89	80–87		

CONFORMANCES AND APPROVALS

See Lincoln Price Book for certificate numbers, size and position limitations and other data

	FW-5P	FW-35	FW-35LS	FW-180	SA-85	SA-85P	SA-HYP	SA-70+
Conforms to Test Requirements of AWS—A5.1 and ASME—SFA5.1 AWS—A5.5 and ASME—SFA5.5	E6010	E6011	E6011	E6011	E7010-A1④	E7010-A1	E7010-G	E8010-G
ASME Boiler Code Group Analysis	F3 A1	F3	F3 A1	F3 A1	F3 A2	F3 A2	F3 A2	F3
American Bureau of Shipping & U. S. Coast Guard	Approved	Approved	Approved		Approved			
Conformance Certificate Available ④	Yes	Yes	Yes	Yes	Yes	Yes	Yes	Yes
Lloyds	Approved	Approved						
Military Specifications	MIL-QQE-450	MIL-QQE-450			MIL-E-22200/7			

③ Also meets the requirements for E7010-G and E6010 in 3/32" size.

④ "Certificate of Conformance" to AWS classification test requirements is available. These are needed for Federal Highway Administration projects.

⑤ Hardness values obtained from welds made in accordance with AWS A5.1.

Source: The Lincoln Electric Company

Table 25-2 Fleetweld 35®, Fleetweld 35LS®, and Fleetweld 180® Lincoln Electrodes.

Figure 25-6 AWS numbering system for A5.1 and A5.5 carbon and low-alloy steel-covered electrodes.

■ 1 — In an E601X means all positions flat, horizontal, vertical, and overhead.

■ 2 — In an E602X means horizontal fillets and flat.

■ 3 — Is an old term no longer used; it meant flat only.

■ 4 — In an E704X means flat, horizontal, overhead, and vertical-down.

The last two numbers together indicate the major type of covering and the type of welding current. For example, EXX10 has an organic covering and uses DCEP polarity. The AWS classification system for A5.1 and A5.5 covered arc welding electrodes is shown in Table 25-3. The type of welding current for any electrode may be expanded to include currents not listed if a manufacturer adds additional arc stabilizers to the electrode covering, Figure 25-7 (page 646).

On some covered arc electrodes, a suffix may be added to indicate the approximate alloy in the deposit as welded. For example, the letter A indicates a 1/2% molybdenum addition to the weld metal deposited. Table 25-4 is a complete list of the major alloying elements in electrodes.

Some of the more popular arc welding electrodes and their uses in these specifications are as follows:

AWS Classification	Type of Covering	Capable of Producing Satisfactory Welds in Positions Shown[a]	Type of Current[b]
		E60 Series Electrodes	
E6010	High cellulose sodium	F,V,OH,H	dc, reverse polarity
E6011	High cellulose potassium	F,V,OH,H	ac or dc, reverse polarity
E6012	High titania sodium	F,V,OH,H	ac or dc, straight polarity
E6013	High titania potassium	F,V,OH,H	ac or dc, either polarity
E6020		H-fillets	ac or dc, straight polarity
E6022[c]	High iron oxide	F	ac or dc, either polarity
E6027	High iron oxide, iron powder	H-fillets, F	ac or dc, straight polarity
		E70 Series Electrodes	
E7014	Iron powder, titania	F,V,OH,H	ac or dc, either polarity
E7015	Low hydrogen sodium	F,V,OH,H	dc, reverse polarity
E7016	Low hydrogen potassium	F,V,OH,H	ac or dc, reverse polarity
E7018	Low hydrogen potassium, iron powder	F,V,OH,H	ac or dc, reverse polarity
E7024	Iron powder, titania	H-fillets, F	ac or dc, either polarity
E7027	High iron oxide, iron powder	H-fillets, F	ac or dc, straight polarity
E7028	Low hydrogen potassium, iron powder	H-fillets, F	ac or dc, reverse polarity
E7048	Low hydrogen potassium, iron powder	F,OH,H,V-down	ac or dc, reverse polarity

[a]The abbreviations F,V,V-down,OH,H, and H-fillets indicate the welding positions as follows:

F = Flat
H = Horizontal
H-fillets = Horizontal fillets
V-down = Vertical down
V = Vertical
OH = Overhead

[b]Reverse polarity means the electrode is positive; straight polarity means the electrode is negative.

[c]Electrodes of the E6022 classification are for single-pass welds.

For electrodes 3/16 in (4.8 mm) and under, except 5/32 in (4.0 mm) and under for classifications E7014, E7015, E7016, and E7018.

Source: Reprinted from AWS A5.1-81, "Specification for Covered Carbon Steel Arc Welding Electrodes," with permission from the American Welding Society.

Table 25-3 Electrode Classification. (Courtesy of the American Welding Society.)

EXXX0	— DCRP only
EXXX1	— AC and DCRP
EXXX2	— AC and DCSP
EXXX3	— AC and DC
EXXX4	— AC and DC
EXXX5	— DCRP only
EXXX6	— AC and DCRP
EXXX8	— AC and DCRP

Figure 25-7 Welding currents.

E6010

The E6010 electrodes are designed to be used with DCEP polarity and have an organic-based flux (cellulose, $C_6H_{10}O_5$). They have a forceful arc that results in deep penetration and good metal transfer in the vertical and overhead positions, **Figure 25-8**. The electrode is usually used with a whipping or stepping motion. This motion helps remove unwanted surface materials such as paint, oil, dirt, and galvanizing. Both the burning of the organic compound in the flux to form CO_2, which protects the molten metal, and the rapid expansion of the hot gases force the atmosphere away from the weld. A small amount of slag remains on the finished weld, but it is difficult to remove, especially along the weld edges. E6010 electrodes are commonly used for welding on fired and unfired pressure vessels, pipe, and in construction jobs, shipyards, and repair work.

E6011

The E6011 electrodes are designed to be used with AC or DCEP reverse polarity and have an organic-based flux. These electrodes have many of the welding characteristics of E6010 electrodes, **Figure 25-9**. In most applications, the E6011 is preferred. The E6011 has added arc stabilizers, which allow it to be used with AC. Using this welding electrode on AC only slightly reduces its penetration but will help control any arc blow problem. **Arc blow** or arc wander is the magnetic deflection of the arc from its normal path. When welding with either E6010 or E6011, the weld pool may be slightly concave from the forceful action of the rapidly expanding

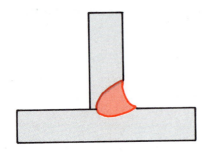

Figure 25-8   E6010.

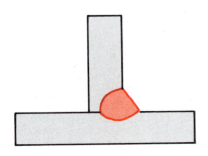

Figure 25-9   E6011.

Suffix Symbol	Molybdenum (Mo)%	Nickel (Ni)%	Chromium (Cr)%	Manganese (Mn)%	Vanadium (Va)%
A 1	0.5				
B 1	0.5		0.5		
B 2	0.5		1.25		
B 3	1.0		2.25		
B 4	0.5		2.0		
C 1		2.5			
C 2		3.5			
C 3		1.0			
D 1	0.3			1.5	
D 2	0.3			1.75	
G	0.2	0.5	0.3	1.0	0.1*
M					

*Only one of these alloys may be used.

Table 25-4 Major Alloying Elements in Electrodes.

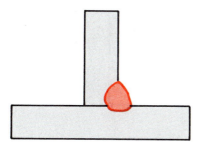

Figure 25-10 E6012.

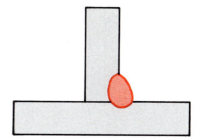

Figure 25-11 E6013.

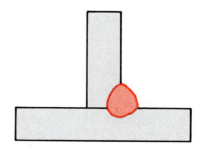

Figure 25-12 E7014.

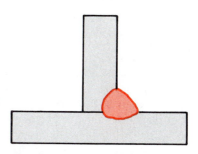

Figure 25-13 E7024.

gas. This forceful action also results in more spatter and sparks during welding.

E6012

The E6012 electrodes are designed to be used with AC or DCEN polarity and have rutile-based flux (titanium dioxide TiO_2). This electrode has a very stable arc that is not very forceful, resulting in a shallow penetration characteristic, **Figure 25-10.** This limited penetration characteristic helps with poor-fitting joints or thin materials. Thick sections can be welded, but the joint must be grooved. Less smoke is generated with this welding electrode than with E6010 or E6011, but a thicker slag layer is deposited on the weld. If the weld is properly made, the slag can be removed easily and may even free itself after cooling. Spatter can be held to a minimum when using both AC and DC. E6012 electrodes are commonly used for all new work including storage tanks, machinery fabrication, ornamental iron, and general repair work.

E6013

The E6013 electrodes are designed to be used with AC or DC, either polarity. They have a rutile-based flux. The E6013 electrode has many of the same characteristics of the E6012 electrode, **Figure 25-11.** The slag layer is usually thicker on the E6013 and is easily removed. The arc of the E6013 is as stable, but there is less penetration, which makes it easier to weld very thin sections. The weld bead will also be built up slightly higher than the E6012. E6013 electrodes are commonly used for sheet metal fabrication, metal buildings, surface buildup, truck and automotive body work, and farm equipment.

E7014

The E7014 electrodes are designed to be used with AC or DC, either polarity. They have a rutile-based flux with iron powder added. The E7014 electrode has many arc and weld characteristics that are similar to those of the E6013 electrode, **Figure 25-12.** Approximately 30% iron powder is added to the flux to allow it to build up a weld faster or have a higher travel speed. The penetration characteristic is light. This welding electrode can be used on metal with a light coating of rust, dirt, oil, or paint. The slag layer

is thick and hard but can be completely removed with chipping. E7014 electrodes are commonly used for welding on heavy sheet metal, ornamental iron, machinery, frames, and general repair work.

E7024

The E7024 electrodes are designed to be used with AC or DC, either polarity. They have a rutile-based flux with iron powder added. This welding electrode has a light penetration and fast fill characteristic, **Figure 25-13.** The flux contains about 50% iron powder, which gives the flux its high rate of deposition. The heavy flux coating helps control the arc and can support the electrode so that a drag technique can be used. The drag technique allows this electrode to be used by welders with less skill. The slag layer is heavy and hard but can easily be removed. If the weld is performed correctly, the slag may remove itself. Because of the large, fluid, molten weld pool, this electrode is equally used in the flat and horizontal position only although it can be used on work that is slightly vertical. E7024 electrodes are commonly used for welding earthmoving, mining, and railroad equipment.

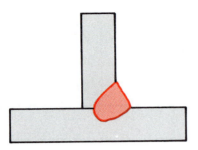

Figure 25-14 E7016.

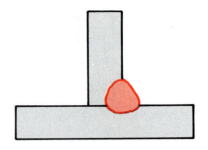

Figure 25-15 E7018.

E7016

The E7016 electrodes are designed to be used with AC or DCEP polarity. They have a low-hydrogen-based (mineral) flux. This electrode has moderate penetration and little buildup, **Figure 25-14**. There is no iron powder in the flux, which helps when welding in the vertical or overhead positions. Welds on high sulfur and cold rolled metals can be made with little porosity. Low-alloy and mild steel heavy plates can be welded with minimum preheating. E7016 electrodes are commonly used for construction, earthmoving and mining equipment, and shipbuilding.

E7018

The E7018 electrodes are designed to be used with AC or DCEP polarity. They have a low-hydrogen-based flux with iron powder added. The E7018 electrodes have moderate penetration and buildup, **Figure 25-15**. The slag layer is heavy and hard but can be removed easily by chipping. The weld metal is protected from the atmosphere primarily by the molten slag layer and not by rapidly expanding gases. For this reason, these electrodes should not be used for open root welds. The atmosphere may attack the root, causing a porosity problem. The E7018 welding electrodes are very susceptible to moisture, which may lead to weld porosity. These electrodes are commonly used for pipe, heavy sections of plate, shipbuilding, boiler work, and low-temperature equipment. E7018 electrodes are sometimes referred to as Lo-Hi rods, because they allow very little hydrogen into the weld pool.

Wire Type Carbon Steel Filler Metals

Solid Wire. The AWS specifications for carbon steel filler metals for gas shielded welding wire is A5.18. Filler metal classified within these specifications can be used for GMAW, GTAW, and PAW process. Because in GTAW and PAW the wire does not carry the welding current, the letters ER are used as a prefix. The ER is followed by two numbers to indicate the minimum tensile strength of a good weld. The actual strength is obtained by adding three zeros to the right of the number given. For example, ER70S-x is 70,000 psi.

The S located to the right of the tensile strength indicates that this is a solid wire. The last number 2, 3, 4, 5, 6, 7, or the letter G are used to indicate the filler metal composition and the weld's mechanical properties, **Figure 25-16**.

ER70S-2

This is a deoxidized mild steel filler wire. The deoxidizers allow this wire to be used on metal that has

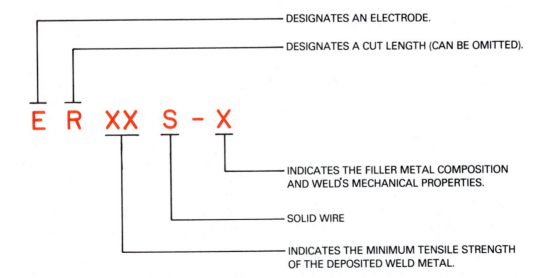

Figure 25-16 AWS numbering system for carbon steel filler metal for GMAW.

light coverings of rust or oxides. There may be a slight reduction in the weld's physical properties if the weld is made on rust or oxides, but this reduction is only slight, and the weld will usually still pass the classification test standards. This is a general purpose filler that can be used on killed, semi-killed, and rimmed steels. Argon-oxygen, argon-CO_2, and CO_2 can be used as shielding gases. Welds can be made in all positions.

ER70S-3

This is a popular filler wire. It can be used in single or multiple pass welds in all positions. ER70S-3 does not have the deoxidizers required to weld over rust, oxides, or on rimmed steels. It produces high-quality welds on killed and semikilled steels. Argon-oxygen, argon-CO_2, and CO_2 can be used as shielding gases.

ER70S-6

This is a good general purpose filler wire. It has the highest levels of manganese and silicon. The wire can be used to make smooth welds on sheet metal or thicker sections. Welds over rust, oxides, and other surface impurities will lower the mechanical properties, but not normally below the specifications of this classification. Argon-oxygen, argon-CO_2, and CO_2 can be used as shielding gases. Welds can be made in all positions.

Tubular Wire. The AWS specifications for carbon steel filler metals for flux cored arc welding wire is A5.20. Filler metal classified within this specification can be used for the FCAW process. The letter E, for electrode, is followed by a single number to indicate the minimum tensile strength of a good weld. The actual strength is obtained by adding four zeros to the right of the number given. For example, E6xT-x is 60,000 psi, and E7xT-x is 70,000 psi.

The next number, 0 or 1, indicates the welding positions. Ex0T is to be used in a horizontal or flat position only. Ex1T is an all-position filler metal.

The T located to the right of the tensile strength and weld position numbers indicates that this is a tubular, flux cored wire. The last number 2, 3, 4, 5, 6, 7, 8, 10, 11, or the letters G or GS are used to indicate if the filler metal can be used for single- or multiple-pass welds. The electrodes with the following numbers ExxT-2, ExxT-3, ExxT-10, and ExxT-GS are intended for single-pass welds only, Figure 25-17.

E70T-1, E71T-1

E70T-1 and E71T-1 have a high-level deoxidizer in the flux cored. It has high levels of silicon and manganese, which allow it to weld over some surface contaminations such as oxides or rust. This filler metal can be used for single- or multiple-pass welds. Argon 75% with 25% CO_2 or 100% CO_2 can be used as the shielding gas. It can be used on ASME A36, A106, A242, A252, A285, A441, and A572, or similar metals. Applications include railcars, heavy equipment, earth-moving equipment, shipbuilding, and general fabrication. The weld metal deposited has a chemical and physical composition similar to that of E7018 low-hydrogen electrodes.

E70T-2, E71T-2

E70T-2 and E71T-2 are highly deoxidized flux cored filler metal that can be used for single-pass welds only. The high levels of deoxidizers allow this electrode to be used over mill scale and light layers of rust and still produce sound welds. Because of the high level of manganese, if the filler is used for multiple-pass welds, there might be manganese-caused center line cracking of the weld; 100% CO_2 can be used as the shielding gas. E70T-2 can be used on ASME A36, A106, A242, A252, A285, A441, and A572, or similar metals. Applications include repair and maintenance work and general fabrication.

E70T-4, E71T-4

E70T-4 and E71T-4 are self-shielding, flux-cored filler metal. The fluxing agents produce a slag,

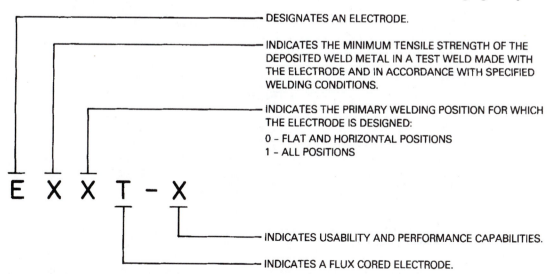

Figure 25-17 Identification system for mild steel FCAW electrodes. (Courtesy of the American Welding Society.)

which allows a larger than usual molten weld pool. The large weld pool permits high deposition rates. Weld deposits are ductile and have a high resistance to cracking. E70T-4 can be used to weld joints that have larger than usual root openings. Applications include large weldments and earth-moving equipment.

E71T-7

E71T-7 is a self-shielding, all position, flux-cored filler metal. The fluxing system allows the control of the molten weld pool required for out-of-position welds. The high level of deoxidizers reduce the tendency for cracking in the weld. It can be used for single- or multiple-pass welds.

Stainless Steel Electrodes

The AWS specifications for stainless steel covered arc electrodes is A5.4 and for stainless steel bare, cored, and stranded electrodes and welding rods is A5.9. Filler metal classified within the A5.4 uses the letter E as its prefix, and the filler metal within the A5.9 uses the letters ER as its prefix, **Figure 25-18.**

Following the prefix, the American Iron and Steel Institute's (AISI), three-digit stainless steel number is used. This number indicates the type of stainless steel in the filler metal.

To the right of the AISI number, the AWS adds a dash followed by a suffix number. The number 15 is used to indicate there is a lime base coating, and the DCEP polarity welding current should be used. The number 16 is used to indicate there is a titania-type coating, and AC or DCEP polarity welding currents can be used. Examples of this classification system are E308-15 and E308-16 electrodes.

The letter L may be added to the right of the AISI number before the dash and suffix number to indicate a low-carbon stainless welding electrode. E308L-15 and E308L-16 arc welding electrodes and ER308L and ER316L are examples of the use of the letter L, **Figure 25-19.**

AISI TYPE NUMBER	442 446	430F 430FSE	430 431	501 502	416 416SE	403 405 410 420 414	321 348 347	317	316L	316	314	310 310S	309 309S	304 L	303S E	201 202 301 302 302B 304 305 308	MILD STEEL
201-202-301 302-302B-304 305-308	310 312 309	310 312 309	310 312 309	310 312 309	309 310 312	309 310 312	308	308	308	308	308	308	308	308	308	308	312 310 309
303 303SE	310 309 312	310 309 312	310 309 312	310 309 312	309 310 312	309 310 312	308	308	308	308	308	308	308	308	308-15	308	312 310 309
304L	310 309 312	310 309 312	310 309 312	310 309 312	309 310 312	309 310 312	308	308	308-L	308	308	308	308	308 -L	308	308	312 310 309
309 309S	310 309 312	310 309 312	310 309 312	310 309 312	309 310 312	309 310 312	308	317 316 309	316	316	309	309	309	308	308	308	309 310 312
310 310S	310 309 312	310 309 312	310 309 312	310 309 312	309 310 312	310 309 312	308	317 316 309	316	316	310	310	309 310	308	308	308	310 309 312
314	310 312 309	310 312 309	310 312 309	310 312 309	310 312 309	309 310 308	309 310	309 310	309 310	310-15	310		309 310	309 310	309 310	309 310	310 309 312
316	310 309 312	310 309 312	310 309 312	310 309 312	309 310 312	309 310 312	308	316	316	316	309 310 316	310 309 316	309 310 316	308 316	308 316	308 316	309 310 312
316L	310 309 312	310 309 312	310 309 312	310 309 312	309 310 312	309 310 312	308	316 317 308	316-L	316	309 310 316	310 309 316	316 316	308 316	308 316	308 316	309 310 312
317	310 309 312	310 309 312	310 309 312	310 309 312	309 310 312	309 310 312	308	317	316 308	316 308	309 310 317	317 316 309	317 316 309	308 316 317	308 316 317	308 316 317	309 310 312
321 348 347	310 309 312	310 309 312	310 309 312	310 309 312	309 310 312	309 310 312	347	308 347	347 308	347 308	309 347	347 308	347 308	347 308 -L	347 308	347 308	309 310 312
403-405 410-420 414	310 309 312	310 309 312	310 309 312	310 309 312	309 310	410† 309††	309 310	309 310	309 310	309 310	310 309	310 309	309 310	309 310	309 310	309 310	309 310 312
416 416SE	310 309	310 309	310 309	310	410-15†	410-15† 309†† 310††	309 310 312	309 310 312	309 310 312	309 310 312	310 312	309 310 312	309 310 312	309 310 312	309 310 312	309 310 312	309 310 312
501 502	310	310	310	502† 310††	310	310	310 309	310 309	310 309	310 309	310 309	310 309	310 309	310 309	310 309	310 309	310 312 309
430 431	310 309	310 309	430-15† 310†† 309††	310	310	310 309	310 309	310 309	310 309	310 309	310 309	310 309	310 309	310 309	310 309	310 309	310 309 312
430F 430FSE	310 309	410- 15†	310 309	310 309	310 309	310 309 312	309 310 312	309 310 312	310 309 312	310 309 312	310 309 312	310 309 312	310 309 312	310 309 312	310 309 312	310 309 312	310 309 312
442 448	309 310	309 310 312	310 309 312	310 309 312	310 309 312	310 309 312	310 309 312	310 309 312	310 309 312	310 309 312	310 309 312	310 309 312	310 309 312	310 309 312	310 309 312	310 309 312	310 309 312

†Preheat
††No Preheat Necessary
Bold numbers indicate first choice, light numbers indicate second and third choice. This choice can vary with specific applications and individual job requirements.

Figure 25-18 Filler metal selector guide for joining different types of stainless to the same type or another type of stainless. (Courtesy of Thermacote Welco.)

UTP Designation	AWS/SFA5.4 Covered	AWS/SFA5.9 TIG and MIG	Description and Applications
6820	E 308-16	ER 308	For welding conventional 308 type SS.
68 Kb	E 308-15		Low hydrogen coating.
6820 Lc	E 308 L-16	ER 308L	Low carbon grade, prevents carbide precipitation adjacent to weld.
308L Fe Hp	E 308 L-16		Fast depositing for maintenance and production coating.
68 LcHL	E 308 L-16		High-performance electrode with rutile-acid coating, core wire alloyed. For stainless and acid-resisting CrNi steels.
68 LcKb	E 308 L-15		Low-carbon electrode for stainless, acid-resisting CrNi-steels.
6824	E 309-16	ER 309	For welding 309 type SS and carbon steel to SS.
6824 Kb	E 309-15		Special lime-coated electrode for corrosion and heat-resistant 22/12 CrNi-steels.
6824 Lc	E 309 L-16	ER 309L	Same as 309, but with low carbon content
6824 Nb	E 309 Cb-16		Corrosion and heat resistant 22/12 CrNi-steels.
6824 MoNb	E309MoCb-16		Corrosion and heat resistant 22/12 CrNi-steels.
309L Fe Hp	E 309 L-16		High deposition rate, easy to use.
6824 Mo Lc	E 309 L-16		For welding similar and dissimilar SS.
68H	E 310-16	ER 310	For high temperature service and cladding steel.
6820 Mo	E 316-16	ER 316	For welding acid resistant Stainless steels.
6820 Mo Lc	E 316L-16	ER 316L	Low carbon grade. Prevents intergranular corrosion.
68 TI Mo	E 316 L-16		Most efficient type. For maintenance and production. High Performance.
68 MoLcHL	E 316 L-16		High-performance electrode with rutile-acid coating, core wire alloyed, for stainless and acid-resisting CrNi-Mo-steels
68 MoLcKb	E 316L-15		Low-carbon electrode for stainless and acid-resisting CrNiMo-steels.
317 Lc Titan	E 317 L-16	ER 317L	Deposit resist sulphuric acid corrosion.

UTP Designation	AWS/SFA5.4 Covered	AWS/SFA5.9 TIG and MIG	Description and Applications
317LFe Hp	E 317 L16		Fast melt off rate, excellent for overlays, easy to use. High Performance.
68 Mo	E 318-16		Versatile stainless all position electrode.
320 Cb	E 320-15		For welding similar acid resistant ss.
3320 Lc	E 320		Is a rutile-coated electrode for welding in all positions except vertical down.
6820 Nb	E 347-16	ER 347	Stabilized grade, prevents carbide precipitation.
347 FeHp	E 347-16		High performance stainless steel electrode of class E 347-16 for welding stabilized Cr Ni alloys.
66	E 410-15		Low-hydrogen electrode for corrosion and heat-resistant 14% Cr-steels.
1915 HST			Low-hydrogen fully austenitic electrode with 0% ferrite content.
1925			Extremely corrosion resistant to phosphoric and sulfuric acids.
68 Hcb	E 310 Cb	ER 310 Cb	For high heat applications, and joining steels to stainless steel.
2535 NbSn			Electrode is a lime-type special electrode and is used for surfacing and joining heat resistant base metals, especially cast steel.
E 330-16	E 330-16		Excellent for welding furnace parts.
6805			For welding of base material 17-4 Ph.
6808 Mo			Rutile lime-type austenitic-ferritic electrode with low carbon content is suited for joining and surfacing on corrosion-resistant steels and cast steel type with austenitic-ferritic structure (Duplex-steels).
6809 Mo			Rutile-basic austenitic-ferritic electrode with low carbon content is suited for joining and surfacing on corrosion-resistant steels and cast steel types with an austenitic-ferritic structure (Duplex-steels).

Figure 25-19 Stainless steel electrodes, filler metals, and wires. (Courtesy of UTP Welding Materials, Inc.)

Stainless steel may be stabilized by adding columbium (Cb) as a carbide former. The designation Cb is added after the AISI number for these electrodes, such as E309Cb-16. Stainless steel filler metals are stabilized to prevent chromium-carbide precipitation.

E308-15, E308-16, E308L-15, E308-16, ER308 and ER308L

All are filler metals for 308 stainless steels. 308 stainless steels are used for food or chemical equipment, tanks, pumps, hoods, and evaporators. All E308 and ER308 filler metals can be used to weld on all 18-8-type stainless steels such as 301, 302, 302B, 303, 304, 305, 308, 201, and 202.

E309-15, E309-16, E309Cb-15, E309Cb-16, ER309, and ER309L

All are filler metals for 309 stainless steels. 309 stainless steels are used for high-temperatures service, such as furnace parts and mufflers. All E309 filler metals can be used to weld on 309 stainless or to join mild steel to any 18-8-type stainless steel.

E310-15, E310-16, E309Cb-15, E309Cb-16, E310Mo-15, E310Mo-16, and ER310

All are filler metals for 310 stainless steels. 310 stainless steels are used for high-temperature service where low creep is desired, such as for jet engine parts, valves, and furnace parts. All E310 filler metals can be used to weld 309 stainless or to join mild steel to stainless or to weld most hard-to-weld carbon and alloy steels. E310Mo-15 and 16 electrodes have molybdenum added to improve their strength at high temperatures and to resist corrosive pitting.

E316-15, E316-16, E3116L-15, E316L-16, ER316, ER316L, and ER316L-Si

All are filler metals for 316 stainless steels. 316 stainless steels are used for high-temperature service where high strength with low creep is desired. Molybdenum is added to improve these properties and to resist corrosive pitting. E316 filler metals are used for welding tubing, chemical pumps, filters, tanks, and furnace parts. All E316 filler metals can be used on 316 stainless steels or when weld resistance to pitting is required.

Nonferrous Electrode

The AWS identification system for covered nonferrous electrodes is based on the atomic symbol or symbols of the major alloy(s) or the metal's identification number. The alloy having the largest percentage appears first in the identification. The atomic symbol is prefixed by the letter E. For example, ECu is a covered copper arc welding electrode, and ECuNiAl is a copper-nickel-aluminum alloy covered arc welding electrode. A letter, number, or letter-number combination may be added to the right of the atomic symbol to indicate some special alloys. For example, ECuAl-A2 is a copper-aluminum welding electrode that has 1.59% iron added.

ALUMINUM AND ALUMINUM ALLOYS

The AWS specifications for aluminum and aluminum alloy filler metals are A5.3 for covered arc welding electrodes and A5.10 for bare welding rods and electrodes. Filler metal classified within the A5.3 uses the atomic symbol Al, and in the A5.10 the prefix ER is used with the Aluminum Association number for the alloy, Figure 25-20.

Aluminum Covered Arc Welding Electrodes

Al-2 and Al-43

The aluminum electrodes do not use the letter E before the electrode number. Aluminum covered arc welding electrodes are designed to weld with DCEP polarity. These electrodes can be used on thin or thick sections, but thick sections must be preheated to between 300° F (150° C) and 600° F (315° C). The preheating of these thick sections allows the weld to penetrate immediately when the weld starts. Aluminum arc welding electrodes can be used on 2024, 3003, 5052, 5154, 5454, 6061, and 6063 aluminum. When welding on aluminum, a thin layer of surface oxide may not prevent welding. Thicker oxide layers must be removed mechanically or chemically. Excessive penetration can be supported by carbon plates or carbon paste. Most arc welding electrodes can also be used for oxyfuel gas welding of aluminum.

Aluminum Bare Welding Rods and Electrodes

ER1100

1100 aluminum has the lowest percentage of alloy agents of all of the aluminum alloys, and it melts at 1,215° F. The filler wire is also relatively pure. ER1100 produces welds that have good corrosion resistance, high ductility, with tensile strengths ranging from 11,000 to 17,000 psi. The weld deposit has a high resistance to cracking during welding. This wire can be used with OFW, GTAW, and GMAW. Preheating to 300° to 350° F is required for GTA welding on plate or pipe 3/8 in. and thicker to insure good fusion. Flux is required for OFW. 1100 aluminum is commonly used for items such as food containers, food processing equipment, storage tanks, and heat exchangers. ER1100 can be used to weld 1100 and 3003 grade aluminum.

ER4043

ER4043 is a general purpose welding filler metal. It has 4.5 to 6.0% silicon added, which lowers its melting temperature to 1,155° F. The lower melting temperature helps promote a free flowing molten weld pool. The welds have high ductility and a high resistance to cracking during welding. This wire can be used with OFW, GTAW, and

BASE METAL	319 355	43 356	214	6061 6063 6151	5456	5454	5154 5254	5086	5083	5052 5652	5005 5050	3004	1100 3003	1060
1060	4145 4043 4047	4043 4047 4145	4043 5183 4047	4043 4047	5356	4043 4043	4043 5183 4047	5356 4043	5356 4043	4043 4047	1100 4043	4043	1100 4043	1260 4043 1100
1100 3003	4145 4043 4047	4043 4047 4145	4043 5183 4047	4043 4047	5356	4043	4043 5183 4047	5356 4043	5356 4043	4043 5183 4047	4043 5183 5356	4043 5183 5356	1100 4043	
3004	4043 4047	4043 4047	5654 5183 5356	4043 5183 5356	5356 5183 5556	5654 5183 5356	5654 5183 5356	5356 5183 5556	5356 5183 5556	4043 5183 4047	4043 5183 5356	4043 5183 5356		
5005 5050	4043 4047	4043 4047	5654 5183 5356	4043 5183 5356	5356 5183 5556	5654 5183 5356	5654 5183 5356	5356 5183 5556	5356 5183 5556	4043 5183 4047	4043 5183 5356			
5052 5652	4043 4047	4043 5183 4047	5654 5183 5356	5356 5183 5356	5356 5183 4043	5654 5183 5556	5654 5183 5356	5356 5183 5556	5356 5183 5556	5654 5183 4043				
5083	NR	5356 4043 5183	5356 5183 5556	5356 5183 5556	5183 5356 5556	5356 5183 5556	5356 5183 5556	5356 5183 5556	5183 5356 5556					
5086	NR	5356 4043 5183	5356 5183 5556	5356 5183 5556	5356 5183 5556	5356 5183 5554	5356 5183 5554	5356 5183 5556						
5154 5254	NR	4043 5183 4047	5654 5183 5356	5356 5183 4043	5356 5183 5554	5654 5183 5356	5654 5183 5356							
5454	4043 4047	4043 5183 4047	5654 5183 4047	5356 5183 4043	5356 5183 5554	5554 4043 5183								
5456	NR	5356 4043 5183	5356 5183 5556	5356 5183 5556	5556 5183 5356									
6061 6063 6151	4145 4043 4047	4043 5183 4047	5356 5183 4043	4043 5183 4047										
214	NR	4043 5183 4047	5654 5183 5356											
43 356	4043 4047	4145 4043 4047												
319 355	4145 4043 4047													

Note: *First filler alloy listed in each group is the all-purpose choice. NR means that these combinations of base metals are not recommended for welding.*

Figure 25-20 Recommended filler metals for joining different types of aluminum to the same type or a different type of aluminum. (Courtesy of Thermacote Welco.)

GMAW. Preheating to 300° to 350° F is required for GTA welding on plate or pipe 3/8 in. and thicker to insure good fusion. Flux is required for OFW. ER4043 can be used to weld on 2014, 3003, 3004, 4043, 5052, 6061, 6062, 6063, and cast alloys 43, 355, 356, and 214.

ER5356

ER5356 has 4.5 to 5.5% magnesium added to improve the tensile strength. The weld has high ductility but only an average resistance to cracking during welding. This wire can be used for GTAW and GMAW. Preheating to 300° to 350° F is required for GTA welding on plate or pipe 3/8 in. and thicker to insure good fusion. ER5356 can be used to weld on 5050, 5052, 5056, 5083, 5086, 5154, 5356, 5454, and 5456.

ER5556

ER5556 has 4.7 to 5.5% magnesium and 0.5 to 1.0% manganese added to produce a weld with high strength. The weld has high ductility and

only average resistance to cracking during welding. This wire can be used for GTAW and GMAW. Preheating to 300° to 350° F is required for GTA welding on plate or pipe 3/8 in. and thicker to insure good fusion. ER5556 can be used to weld on 5052, 5083, 5356, 5454, and 5456.

Special Purpose Filler Metals

ENi

The nickel arc welding electrodes are designed to be used with AC or DCEP polarity. These arc welding electrodes are used for cast iron repair. The carbon in cast iron will not migrate into the nickel weld metal, thus preventing cracking and embrittlement. The cast iron may or may not be preheated. A very short arc length and a fast travel rate should be used with these electrodes.

ECuAl

The aluminum bronzed welding electrodes are designed to be used with DCEP polarity. This welding electrode has copper as its major alloy. The aluminum content is at a much lower percentage. Iron is usually added but at a percentage that is very low. These electrodes are sometimes referred to as arc brazing electrodes, although this is not an accurate description. Stringer beads and a short arc length should be used with these electrodes. Aluminum bronze welding electrodes are used for overlaying bearing surfaces, welding on castings of manganese, bronze, brass, or aluminum bronze, or assembling dissimilar metals.

Surface and Buildup Electrode Classification

Hardfacing or wear-resistant electrodes are the most popular special-purpose electrodes; however, there are also cutting and brazing electrodes. Specialty electrodes may be identified by manufacturers' trade names. Most manufacturers classify or group hardfacing or wear-resistant electrodes according to their resistance to impact, abrasion, or corrosion. Occasionally, electrode resistance to wear at an elevated temperature may be listed. One electrode may have more than one characteristic or type of service listed.

EFeMn-A

The EFeMn-A electrodes are designed to be used with AC or DCEP polarity. This electrode is an impact-resistant welding electrode. It can be used on hammers, shovels, spindles, and in other similar applications.

ECoCr-C

The ECoCr-C electrodes are designed to be used with AC or DCEP polarity. This electrode is a corrosion and abrasion-resistant welding electrode. It also maintains its resistance to elevated temperatures. ECoCr-C is commonly used for engine cams, seats and valves, chain saw bars, bearings, or dies.

Magnesium Alloys

The joining of magnesium alloys by torch welding or brazing is possible without a fire hazard because the melting point of magnesium is 1,202° F (651° C) to 858° F (459° C) below its boiling point where magnesium may start to burn.

ER AZ61A

The ER AZ61A filler metal can be used to join most magnesium wrought alloys. This filler has the best weldability and weld strength for magnesium alloys AZ31B, HK31A, and HM21A.

ER AZ92A

The ER AZ92A filler metal can be used on cast alloys, Mg-Al-Zn and AM 100A. This filler metal has a somewhat higher resistance to cracking.

Filler Metal Considerations for Critical Weldments

Figure 1

When selecting filler metal for weldments that will undergo nondestructive testing, the right choice can be the determining factor in the success or failure of a project or other endeavor. It is important to keep in mind that filler metal cost is normally less than 1% of the weldment cost, but rework of failed welds can cost up to ten times the total amount budgeted for the welding operation. Proper filler metal selection can nonetheless simplify the troubleshooting process.

Initially, the welding engineer should ensure that the properties in the heat-affected zone will conform to the specified properties. This is not

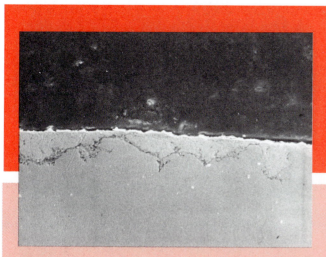

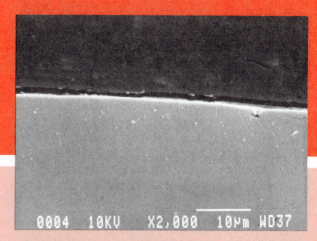

Figure 2A

Figure 2B

just an issue of "weldability" of the base metal. The solution should address the location of the weld in the part, the welding techniques, facilities, and type of tests that will be performed on the weldments or the finished parts.

When some concerns about the base metal have not been eliminated, filler metal selection may make a difference. For example, in nickel-based alloys that may be subject to microfissuring, improvements in filler metal composition and cleanliness may promote "healing" of microfissures. However, improperly prepared weld joints can cause the same problem, so it is important to check the other potential sources when troubleshooting.

Filler metals may be selected on the basis of matching chemistry, matching properties, or special metallurgical considerations. In general, chemistry match is the best choice for most applications. Chemistry match is also best for casting repair and most applications that require a postweld heat treatment. Property selection is normally used in applications in which heat treatment is either partially or fully completed prior to the welding operation.

Whatever the criteria for the selection of filler metal, the reduction of residual elements, tramp elements, and interstitials may be beneficial. A filler metal specification with stricter chemical requirements than the base metal specification is normally used in most nondestructively tested welding applications. However, most specifications are intended for broad applications. For special applications, a quality level higher than the minimum specification is normally advantageous.

In the specification of welding wire, a usage test can determine the difference between "good wires" and "bad wires." Operations with extensive quality control procedures normally require a wipe test for filler metal cleanliness and a test weldment. While these tests are implicit in most filler metal specifications, standards for acceptance have a wide variation; therefore, a standard test has not been developed. However, strict chemistry control eliminates the need for weldment property tests as a filler metal acceptance test (subsequent to weldment qualification), so most filler metal "weldability" tests are normally based on quality rather than mechanical property testing.

In critical weldments, a value engineering study will often show a justification for the best filler metal available. One example is on the Standard missile program — **Figure 1**. Using copper-coated wire, the rejection rate of weldments was over 50%. Upon switching to metallurgically controlled MC-GRADE[tm1] welding wire, the rejection rate dropped to less than 1%. The low-alloy welding wire has proven to exceed military specification requirements. A scanning electron micrograph of the wire cross section shows the contamination in the copper-coated welding wire, as compared to the metallurgically controlled wire — **Figure 2**.

It is important to work with a filler metal supplier who is up to date on the latest advances in welding metallurgy and adapts to the needs of customer quality control procedures. The use of proper filler metal selection and good filler metal specifications is an important start for production of quality weldments.

[1]MC-GRADE[tm] is a trademark of United States Welding Corp.

(Adapted from an article by M.J. Simon appearing in the Welding Journal. Courtesy of American Welding Society, Inc. Figure 1 photo provided courtesy of Kaiser Electroprecision.)

REVIEW

1. What groups have developed electrode identification systems?

2. Referring to Table 25-1, determine the following:
   a. The Lincoln Electric Company E6010 electrode
   b. Hobart Brothers Company E7018 electrode
   c. Teledyne McKay E7024 electrode

3. What types of general information about electrodes may be given by different electrode manufacturers?

4. Define tensile strength.

5. What chemicals alloys are
   a. considered to be contaminants to the weld metal?
   b. used to increase tensile strength in the weld metal?
   c. used to increase corrosion resistance?
   d. used to reduce creep?

6. What should CE be used for?

7. What welding parameters should be used of a metal that has a CE of more than 0.60%?

8. What functions can the flux covering of a an SMA electrode provide to the weld?

9. How does an SMA welding electrode's flux covering produce the shielding gas to protect the weld?

10. What fluxing agents act as scavengers in the molten weld pool?

11. How can an SMA welding electrode's flux help with deeper penetration?

12. What are the advantages of refractory type stages?

13. List the things that must be considered before selecting an electrode for a specific job.

14. Why can there be more than one electrode for each classification manufactured by the same company?

15. What do the following filler metal designations stand for?
    a. E
    b. ER
    c. RG
    d. IN

16. Explain the parts of this AWS electrode classified as E7018.

17. Which SMA welding electrode(s) can be used to weld on metal that has a light covering of paint?

18. Which SMA welding electrode(s) are commonly used to weld on sheet metal fabrications?

19. Which SMA welding electrodes can be used with a dig technique?

20. Referring to Figures 25-8 through 25-15 (pages 646–648), which electrode has the deepest penetration?

21. How is the E7018 molten weld pool protected?

22. What is the purpose of the deoxidizers in ER70S-2?

23. What alloying element used in FCA welding electrodes causes the electrode to produce centerline cracks in the weld if it is used for multipass welds?

24. What does the 15 and 16 stand for in SMA stainless steel welding electrodes?

25. What stainless steel(s) would have
    a. low creep at high temperatures?
    b. food service equipment?

26. Referring to Figure 25-18 (page 650), what stainless steel filler metal would be selected for welding
    a. 304L stainless to 314 stainless?
    b. 321 stainless to 348 stainless?
    c. 316 stainless to mild steel?

27. Referring to Figure 25-19 (page 651), what would be an excellent filler metal for weld on furnace parts?

28. What forms the basis for the AWS identification system for nonferrous-covered electrodes?

29. Why must thick sections of aluminum be preheated before an Al-2 electrode is used for welding?

30. For what types of items would the purest aluminum alloy be used?

31. Referring to Figure 25-20 (page 653), what aluminum filler metal would be selected for welding
    a. 5052 aluminum to 5456 aluminum?
    b. 5454 aluminum to 6061 aluminum?
    c. 1100 aluminum to 4043 aluminum?

32. What are aluminum *arc brazing electrodes* used for?

33. How do most manufacturers classify or group hard-facing or wear-resistant electrodes?

Chapter 26

WELDING AUTOMATION AND ROBOTICS

OBJECTIVES

After completing this chapter, the student should be able to:

- explain the difference between manual (MA), semiautomatic (SA), machine (ME), automatic (AU), and automated welding.
- list the major factors to be considered in establishing a robotic welding station.
- list robotic safety considerations.
- explain the need for interaction between various components of robotic work stations.

KEY TERMS

pick and place	reprogrammable
computer-aided design (CAD)	multifunctional
computer-aided manufacturing (CAM)	manipulator
manual joining	X,Y,Z axis
semiautomatic joining	cycle time
machine joining	work cell
automatic joining	sensors
automated joining	future automation
industrial robots	

INTRODUCTION

The use of technology to reduce production time, increase output, and reduce costs began as early as the 1920s. The automotive industry first used the automatic welding process to produce high-quality welded swing-arm supports and other parts. The process used a continuous feed, unshielded bare wire. Since that time, advances in automatic welding technology have continued to take place.

In the 1960s, American technology produced the first industrial robots. The first industrial robots were used by the nuclear industry to handle radioactive materials. These early units were mainly pick and place robots used to move material with little repetitive accuracy required. During the 1970s computers were applied in increasing numbers by large industry to serve as intelligent controllers for automation. By the 1980s, the decreasing cost of computers and advancements in robotics had made this technology possible, even for small businesses.

More and more modern businesses are using computer-aided design (CAD) to improve products. CAD technology can help in selecting materials, specifying thicknesses, and locating supports to ensure good engineering design of products. Computer-aided manufacturing (CAM) can then be used to improve production. CAM technology can aid in selecting assembly methods, planning the product flow through the manufacturing steps, and scheduling the various operations to reduce the actual production costs.

Computers and microprocessors assist modern industry in producing high-quality products with a minimum waste of materials and time. The use of high technology to monitor part flow through production alerts management to potential problem areas before they cause slowdowns, lost manufacturing time, and increased costs, as well as possible hazardous situations.

Automation is revolutionizing industry worldwide. Some manufacturing plants using robots and other automatic equipment require only supervisory personnel to manage large portions of the production. These facilities are only the beginning.

Many workers whose jobs are modernized by automation find that there are new and more technical opportunities for operating, maintaining, designing, and installing automated equipment and robots. Skilled technicians are needed to set up and operate such automatic manufacturing equipment. The technician needs to understand the manufacturing processes, including welding, to ensure that the operating guidelines established will work consistently. Most robot manufacturers recommend that a highly skilled welder be selected to operate the robot. It is much easier to train a welder to operate the robot than it is to train a computer technician to make good welds.

In welding applications, the increased use of robots, however, will never replace the tremendous need for skilled welders. Welders will always be needed to produce high-quality welds in production and maintenance in locations too restricting for robotics.

Students who have mastered hands-on welding techniques and those who understand the various automated welding procedures will have the advantage when seeking either type of employment opportunity.

This chapter will give the welding student a general overview of automatic welding processes and robotics with guidelines on implementing automation in welding applications. See Color Plates 58 through 60.

MANUAL JOINING PROCESS

A manual joining process is one that is completely performed by hand. The welder controls all of the manipulation, rate of travel, joint tracking, and, in some cases, the rate at which filler metal is added to the weld. The manipulation of the electrode or torch in a straight line or oscillating pattern affects the size and shape of the weld, Figure 26-1. The manipulation pattern may also be used to control the size of the weld pool during out-of-position welding. The rate of travel or speed at which the weld progresses along the joint affects the width, reinforcement, and penetration of the weld, Figure 26-2. The placement or location of the weld bead within the weld joint affects the strength, appearance, and possible acceptance of the joint. The rate at which filler metal is added to the weld affects the reinforcement, width, and appearance of the weld, Figure 26-3.

The most commonly used manual (MA) arc welding process is shielded metal arc welding (SMAW). This process was described in Section II of this text. The flexibility the welder has in performing the weld makes this process one of the most versatile. By changing the manipulation, rate of travel, or joint tracking, the welder can make an acceptable weld on a variety of material thicknesses, Figure 26-4.

The most commonly used manual arc welding, gas welding, and brazing processes are listed in Table 26-1.

SEMIAUTOMATIC JOINING PROCESSES

A semiautomatic joining process is one in which the filler metal is fed into the weld automatically. All other functions are controlled manually by the welder. The addition of filler metal to the weld by an automatic wire feeder system enables the welder to increase the uniformity of welds, productivity, and weld quality. The distance of the welding gun or torch from the work remains constant. This gives the welder better manipulative control as compared to, for example, shielded metal arc welding, in which the electrode holder starts at a dis-

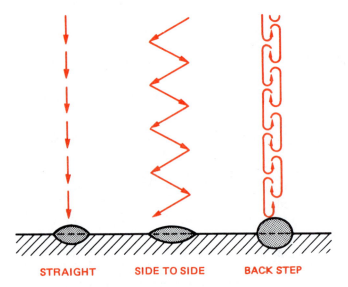

STRAIGHT   **SIDE TO SIDE**   **BACK STEP**

Figure 26-1 The electrode manipulation affects the size and shape of the weld bead.

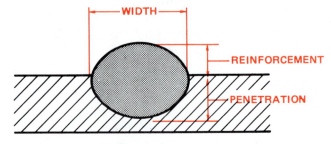

Figure 26-2 The travel rate of the weld affects width, reinforcement, and penetration of the weld bead.

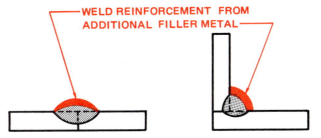

Figure 26-3 Addition of filler metal.

FAST, STRAIGHT TRAVEL   SLOW WEAVE TRAVEL   OFF-CENTER FAST TRAVEL

Figure 26-4

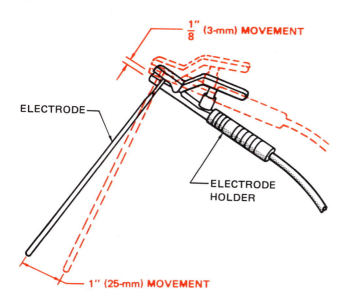

Figure 26-5 A 1/8-in. (3-mm) movement of the electrode holder results in a 1-in. (25-mm) movement of the electrode at the surface of the work.

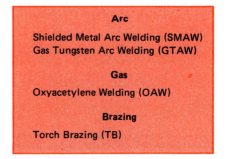

Arc
Shielded Metal Arc Welding (SMAW)
Gas Tungsten Arc Welding (GTAW)
Gas
Oxyacetylene Welding (OAW)
Brazing
Torch Brazing (TB)

Table 26-1 Manual Joining Processes.

tance of 14 in. (356 mm) from the work. This distance exaggerates the slightest accidental movement made during the first part of the weld, **Figure 26-5**. In the SMAW process, the electrode holder must be lowered steadily as the weld progresses to feed the electrode and maintain the correct arc length, **Figure 26-6** (page 660). This constant changing of the distance above the work causes the welder to shift body position frequently. This change, too, may affect the consistency of the weld.

Because the filler metal is being fed from a large spool, the welder does not have to stop welding to change filler electrodes or filler metal. SMA electrodes cannot be used completely as they have a waste stub of approximately 2 in. (51 mm). This waste stub represents approximately 15% of the filler metal that must be discarded. The frequent stopping for rod and electrode changes, followed by restarting, wastes time and increases the number of weld

craters. These craters are often a source of cracks and other discontinuities, **Figure 26-7** (page 660). In some welding procedures each weld crater must be chipped and ground before the weld can be restarted. These procedures can take up to ten minutes — time that can be used for welding in a semiautomatic welding process.

The most commonly used semiautomatic (SA) arc welding process is gas metal arc welding (GMAW). This process was fully described in Section 4. **Table 26-2** (page 660) lists several other semiautomatic processes.

MACHINE JOINING PROCESSES

A machine joining process is one in which the joining is performed by equipment requiring the welding operator to observe the progress of the weld and make adjustments as required. The parts being joined may or may not be loaded and unloaded automatically. The operator may monitor the joining progress by watching it directly, observing instruments only, or a combination of both methods. Adjustments in travel speed, joint tracking, work-to-gun or work-to-torch distance, and current settings may be needed to ensure that the joint is made according to specifications.

The work may move past a stationary welding or joining station, **Figure 26-8** (page 661), or it may be held

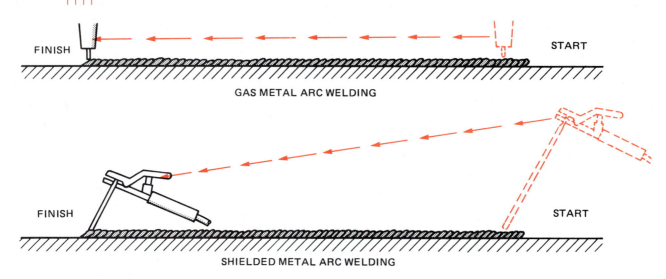

FINISH   START

GAS METAL ARC WELDING

FINISH   START

SHIELDED METAL ARC WELDING

Figure 26-6 In GMAW, the torch height remains constant above the work surface. In SMAW, the height of the electrode holder steadily decreases from the beginning to end of the weld.

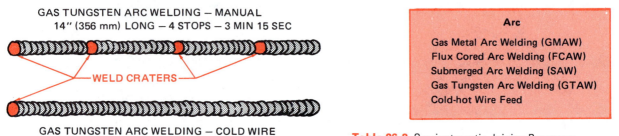

GAS TUNGSTEN ARC WELDING — MANUAL
14" (356 mm) LONG — 4 STOPS — 3 MIN 15 SEC

WELD CRATERS

GAS TUNGSTEN ARC WELDING — COLD WIRE
14" (356 mm) LONG — 1 STOP — 1 MIN 30 SEC

Figure 26-7 Fewer weld craters occur in continuous welding.

Arc
Gas Metal Arc Welding (GMAW)
Flux Cored Arc Welding (FCAW)
Submerged Arc Welding (SAW)
Gas Tungsten Arc Welding (GTAW)
Cold-hot Wire Feed

Table 26-2 Semiautomatic Joining Processes.

stationary and the welding machine moves on a beam or track along the joint, **Figure 26-9** (page 662). On some large machine welds, the operator may ride with the welding head along the path of the weld. During the assembly of the external fuel tanks used for the space shuttle, two operators were required for a few of the machine welds. One operator watched the root side of the weld while the other observed the face side of the weld. They were able to communicate with each other so that any needed changes could be made.

To minimize adjustments during machine welds, a test weld is often performed just before the actual weld is produced. This practice weld helps increase the already high reliability of machine welds.

AUTOMATIC JOINING PROCESSES

An automatic joining process is a dedicated process (designed to do only one type of welding on a specific part) that does not require adjustments to be made by the operator during the actual welding cycle. All operating guidelines are preset, and parts may or may not be loaded or unloaded by the operator. Automatic equipment is often dedicated to one type of product or part. A large investment is usually required in jigs and fixtures used to hold

the parts to be joined in the proper alignment. The operational cycle can be controlled mechanically or numerically (computer). The cycle may be as simple as starting and stopping points, or it may be more complex. A more complex cycle may include such steps as prepurge time, hot start, initial current, pulse power, downslope, final current, and postpurge time, **Figure 26-10** (page 662).

Automatic welding or brazing is best suited to large-volume production runs because of the expense involved in special jigs and fixtures.

AUTOMATED JOINING

Automated joining processes are similar to automatic joining except that they are flexible and more easily adjusted or changed. Unlike automatic joining, there is no dedicated machine for each product. The equipment can be easily adapted or changed to produce a wide variety of high-quality welds.

The industrial robot is rapidly becoming the main component in automated welding or joining stations. The welding or joining cycles are often controlled by computers or microprocessors. The flexibility provided by automated work stations makes it possible for even small companies with limited production runs to invest in automated equipment. The equipment is controlled by programs, or a series of machine commands expressed

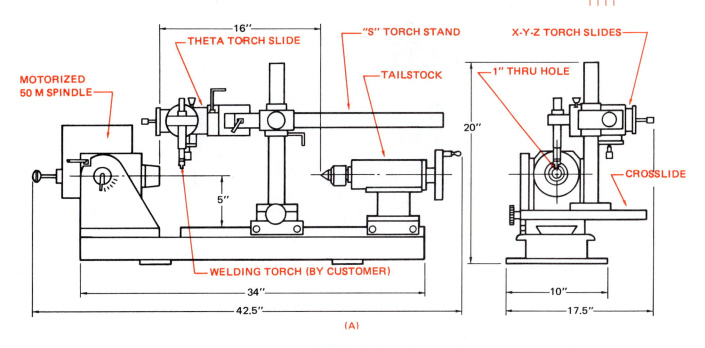

MOTORIZED 50 M SPINDLE

16"

THETA TORCH SLIDE

"S" TORCH STAND

TAILSTOCK

X-Y-Z TORCH SLIDES

1" THRU HOLE

20"

5"

CROSSLIDE

WELDING TORCH (BY CUSTOMER)

34"

42.5"

10"

17.5"

(A)

(B)

Figure 26-8 Two precision bench-welding systems for moving the welding torch, part, or both the torch and part to make repetitive machine welds. (Drawing A courtesy of M Engineering, Inc.; photo B courtesy of Radix Corporation.)

in numerical codes, that direct the welding, cutting, assembling, or any other activities. The programs can be stored and quickly changed. Some systems can store and retrieve many different programs internally. Other systems are controlled by a host computer. Both types of systems can speed up production when frequent changes are required.

INDUSTRIAL ROBOTS

An **industrial robot** is a "**reprogrammable, multifunctional manipulator** designed to move material, parts, tools, or specialized devices through variable programmed motions for the performance of a variety of tasks." Industrial robots are primarily powered by

Figure 26-9 Automatic GTA machine welding along the seam of a stationary pipe.

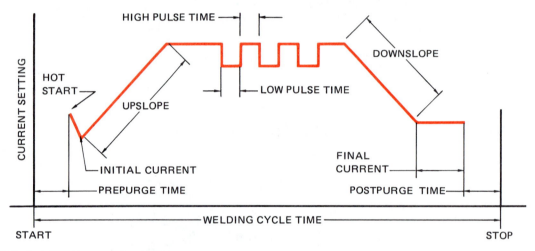

Figure 26-10 Typical GTAW automatic welding program.

electric stepping motors, hydraulics, or pneumatics, and are controlled by a program, **Figure 26-11.**

Robots can be used to perform a variety of industrial functions, including grinding, painting, assembling, machining, inspecting, flame cutting, product handling, and welding. See Color Plate 60.

Robots range in size and complexity from small desk-top units capable of lifting only a few ounces (grams) to large floor models capable of lifting tons. Most robots can perform movements in three basic directions: longitudinal **(X),** transverse **(Y),** and vertical **(Z), Figure 26-12.** The tool end of the robot arm may also be jointed so that it can tilt and rotate, **Figure 26-13.**

The robot may be used with other components to increase production and the flexibility of the system. A computer or microprocessor can synchronize the robot's operation to positioners, conveyors, automatic fixtures, and other production machines. Parallel or multiple work stations increase the duty cycle (the fraction of time during which welding or work is being done) and reduce **cycle time** (the period of time from starting one operation to starting another), **Figure 26-14.** Parts can be

Figure 26-11 Robot control unit. (Courtesy of Welders Supply Inc., Dallas, Texas; photographed by Terry Gentry.)

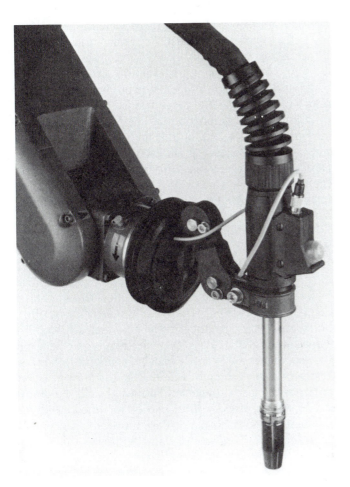

Figure 26-12 Machine axes. (Courtesy of Alexander Binzel Corp.)

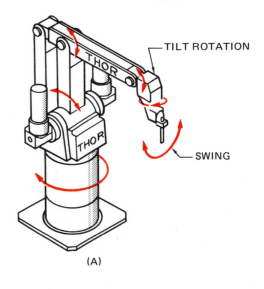

(A)

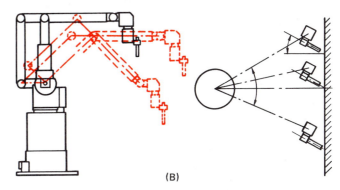

(B)

Figure 26-13 The tool end may be jointed so that it can tilt and rotate. (Courtesy of Merrick Engineering, Inc.)

loaded or unloaded by the operator at one station while the robot welds at another station.

Robot Programming

The program that directs the robot through the desired tasks can be created by several methods. One method of programming the robot is to teach it to move its arm along the desired path. The arm can be moved through the complete cycle by physically directing its movements by hand or by using a remote control panel, **Figure 26-15** (page 663). The remote control panel allows the operator to direct the arm to the desired points along the weld path. In both cases, the robot will follow a straight line between the points established during the training session, **Figure 26-16** (page 663). A circular path or curve can be obtained by using touch points, **Figure 26-17** (page 663). Some units can be given the radius to be followed. A weave pattern can be added to the basic straight line weld at this time.

After the robot has been programmed by the teaching session, a test run should be made. Using a properly located part, with the welding current off, watch the robot arm to see if it follows the correct path. If the test

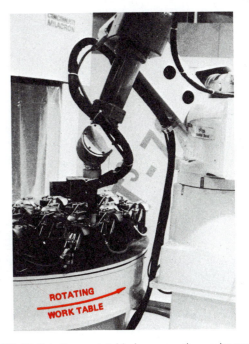

Figure 26-14 Rotating work table increases the work zone.

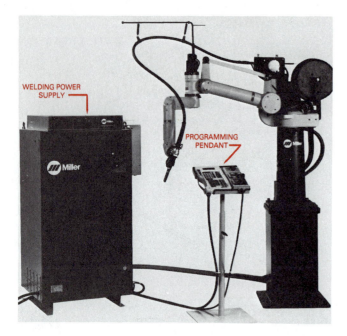

Figure 26-15 Remote teaching pendant for programming robot. (Courtesy of Miller Electric Mfg. Co.)

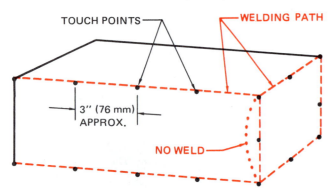

Figure 26-16 Straight line welds.

run appears to be satisfactory, the welding parameters are set. A part should then be welded and the welds inspected for acceptability.

Another method of programming a robot is to use a computer keyboard to produce the program. These programs can be longer and more detailed than the teaching program. Information that can be given to the robot, such as the total length of weld and the number of welding parameter changes, is restricted by the memory capacity of the controller. Programmed robots use less internal memory (memory required for the teaching machines) because they understand basic programming language.

After the program has been tested and determined to be acceptable, it can be stored for future use. The paper tape, cassette tape, and disc storage are the three most common storage methods. The paper tape is a strip 1 in. (25 mm) wide that is punched with a series of holes, Figure 26-18. Paper tapes are easily soiled or torn. But, unlike magnetic tape, they cannot be erased accidentally, damaged, or changed by other electrical equipment. If the paper tapes are properly handled and stored, they can be a permanent program record. The cassette tape is not as fast as the disc storage, but both work well. Programs stored on cassettes or discs can be modified easily for production design changes. Care must be taken to prevent accidental erasure or damage due to the strong magnetism found around welding machines.

System Planning

Evaluating the need for the use of a robot and selecting a robot to meet that need is a complex process. Some of the factors that must be considered are the following:

- Present and future needs
- Parts design
- Equipment selection
- Safety

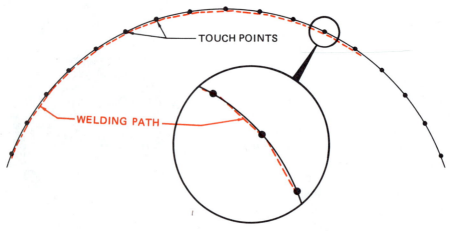

Figure 26-17 Curve welding path.

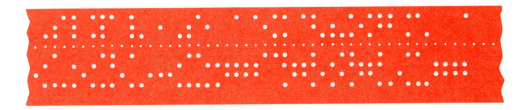

Figure 26-18 Paper tape.

Present and Future Needs. A system using a robot can be 80% more productive than a system using manual welders. This higher rate of productivity means that there must be a market for the higher volume of product produced. Premature investment in automation can be more costly than it can be productive. A comprehensive market analysis of present and future needs must be the first step to ensure that your business is not overextended.

Parts design. The design of the weldment must be compatible with a robotic system. The components should be selected and assembled so that the accessibility of the robot's arm and torch is not restricted, Figure 26-19. Although the torch can be moved along several axes, it cannot reach restricted areas as a welder can using manual processes. Proper design and assembly sequences can reduce the amount of manual pickup welding required to complete the weldment. Mockups and CAD/CAM can be used before the robot is installed to reduce possible problems.

The parts design must also take into consideration the higher heat input from the almost constant welding. This heat could result in distortion of the weldment if some compensation is not provided. One method of controlling this distortion is to stagger the

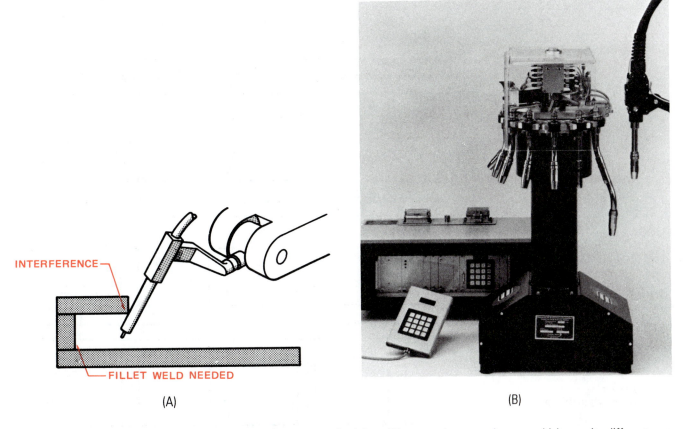

INTERFERENCE

FILLET WELD NEEDED

(A)

(B)

Figure 26-19 (A) Parts should be designed to permit access to the joints; (B) automatic gun exchanger, which permits different gun designs to be used when needed. (Photo courtesy of Alexander Binzel Corp.)

weld locations, Figure 26-20. However, this practice means that more arm articulation (movement) is required, resulting in a slowdown in production. The best method of eliminating distortion is to use a combination of jigs, tack welds, and preset angles, Figures 26-21. This technique can be very successful, but it requires more initial design and setup time than other methods.

When possible, all welds should be performed in the flat position. Large weld sizes are less critical than small weld sizes in their exact location, Figure 26-22. Small weld sizes require better tolerances since even the slightest mislocation may be unacceptable.

Equipment Selection. The robot is the prime component in an automated **work cell** (a manufacturing unit consisting of two or more work stations), Figure 26-23. The selection of the system must be considered in connection with the other pieces of equipment being used, Figure 26-24. To obtain the most effective system, the robot, welding power supply, positioners, conveyors, and other equipment should all be computer controlled.

The robot itself must be evaluated for its ability to work in a welding environment. Other factors of equipment selection to be considered include speeds, work zone configuration, accuracy, programmability, weave capability, and interaction capability with other

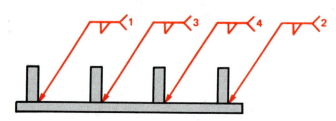

Figure 26-20 Staggering weld positions to prevent distortion of the workpiece.

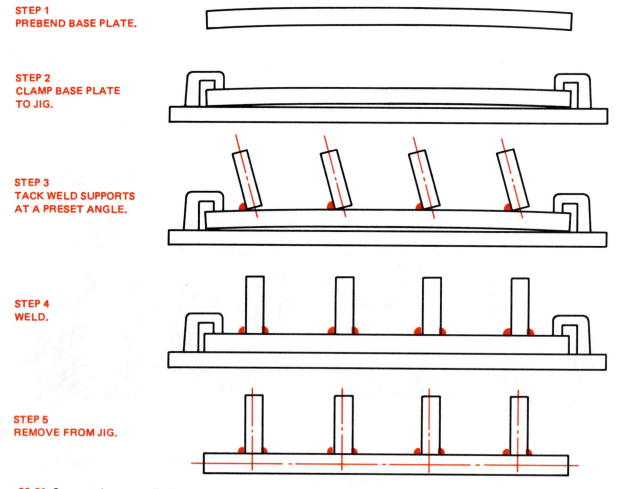

STEP 1
PREBEND BASE PLATE.

STEP 2
CLAMP BASE PLATE
TO JIG.

STEP 3
TACK WELD SUPPORTS
AT A PRESET ANGLE.

STEP 4
WELD.

STEP 5
REMOVE FROM JIG.

Figure 26-21 Suggested steps to eliminate heat distortion in completed part.

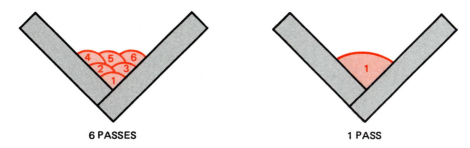

6 PASSES   **1 PASS**

Figure 26-22 One large weld can be made faster and is less critical to produce.

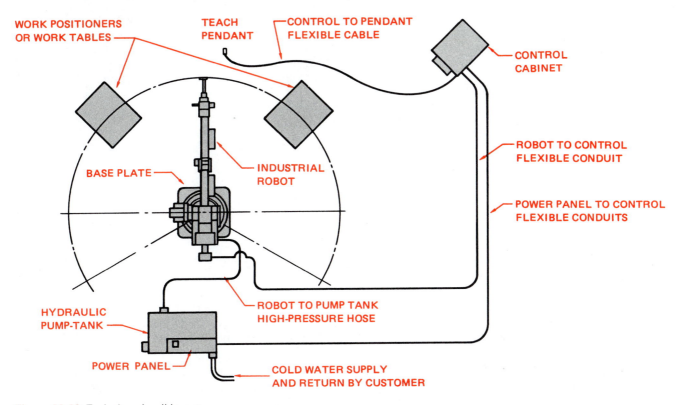

Figure 26-23 Typical work cell layout.

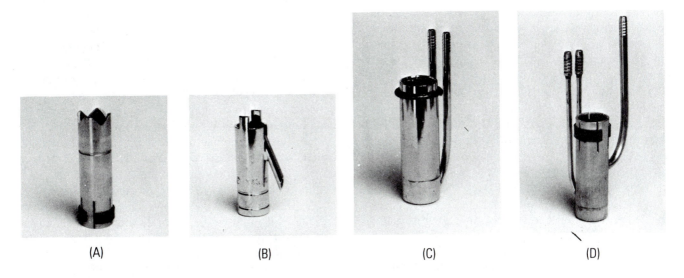

(A)   (B)   (C)   (D)

Figure 26-24 Specially designed nozzles. (Courtesy of Alexander Binzel Corp.)

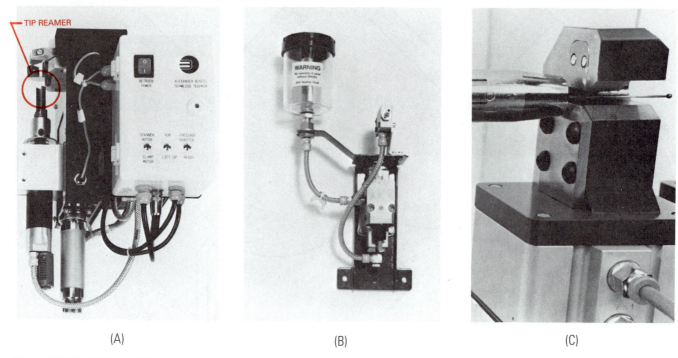

(A) (B) (C)

Figure 26-25 (A) Tip servicing station, (B) automatic antispatter unit, (C) automatic wire cutter. (Courtesy of Alexander Binzel Corp.)

equipment. The welding environment, unlike other manufacturing areas, is hot. The parts and machines in the welding environment are frequently exposed to spatter and subject to electronic noise (RF) from high-frequency starts. Weld spatter can stick to unprotected machine surfaces, causing the arm to jam or resulting in excessive wear. Electronic noise can damage the program, interrupt input or output signals, and cause erratic operations.

The speeds at which the robot arm moves must be compatible with the welding process. The rate of movement must be constant when making a specific size weld, but travel speed will have to be varied to permit different weld sizes and positions. A higher speed during arm articulation to a new position can increase productivity.

A welding tip service station can be included to prevent spatter buildup from clogging the welding torch. The service station can include a reamer to remove spatter and an automatic spray of antispatter, **Figure 26-25**.

The size and configuration of the work zone depends upon the number of axes and reach of the robot, **Figure 26-26(A)** through **(D)**. The articulation of the torch may limit the overall reach of the robot, **Figure 26-27(A)** and **(B)** (page 670). The use of the robot together with a rotating table or positioner can increase its effective work zone (refer to Figure 26-14, page 663).

The accuracy with which a work cycle can be repeated and the joint tolerances that can be accepted must be considered when selecting equipment. To make acceptable welds, it is important that the arm follows the same path within a few thousandths of an inch (millimeters) each time. Some units have **sensors** that track the

joint even if it varies from the programmed path. This feature permits the welding of parts that otherwise might have been rejected.

A variety of programming options is available from robot manufacturers. The greater the programming flexibility available, the more versatile is the unit. A typical welding program cycle can include the components shown in **Figure 26-28** (page 671). Some typical industrial welding robots are shown in **Figures 26-29** and **26-30** (page 672).

A signal from the positioner indicates that the parts are in place. The arm moves to the weld starting point and waits as prepurging gas starts to flow. Both the current and wire feed are started, and a molten weld pool is established before the arm begins to move along the weld joint. At the end of the joint, the arm stops, and the current and wire feed continue to fill the weld crater. The wire feed stops, followed in one second by the current to burn back the wire. Gas flow continues for postpurging. During the welding cycle, the program may perform checks to ensure that the gas flow, voltage, amperage, and other external factors are operating correctly.

The program may also allow the torch to be manipulated in several different weave patterns, **Figure 26-31** (page 672). This feature means that large, single pass welds or out-of-position welds can be performed successfully.

To ensure that the system achieves maximum productivity, it is important that all components be able to be remotely controlled. Preferably, it should be adjustable to various power settings. The wire feeder should have various speed settings in addition to remote starting and stopping. Sensors should provide the robot

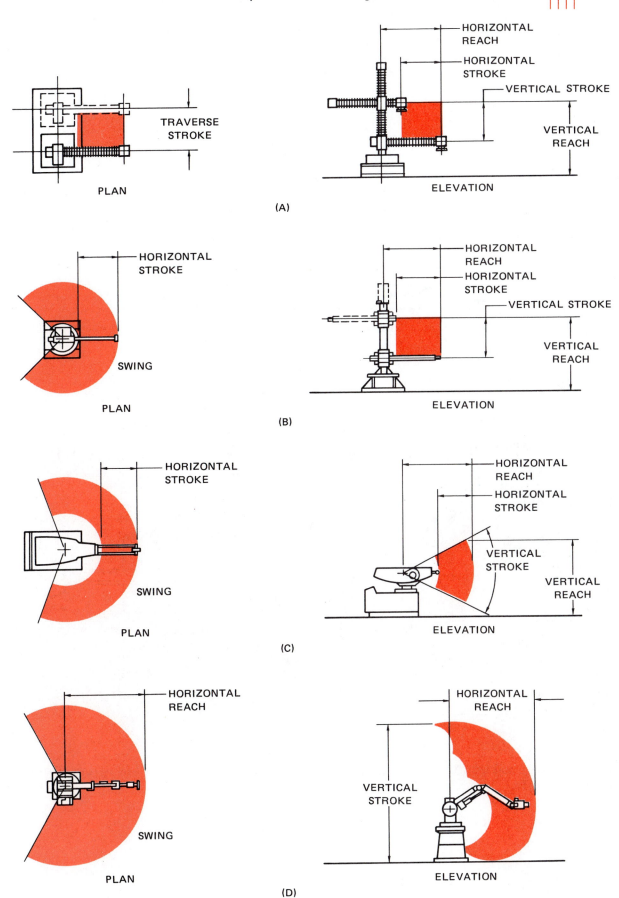

Figure 26-26 The work envelope for various types of robots is different: (A) rectangular coordinate robot, (B) cylindrical coordinate robot, (C) spherical coordinate robot, and (D) jointed arm robot.

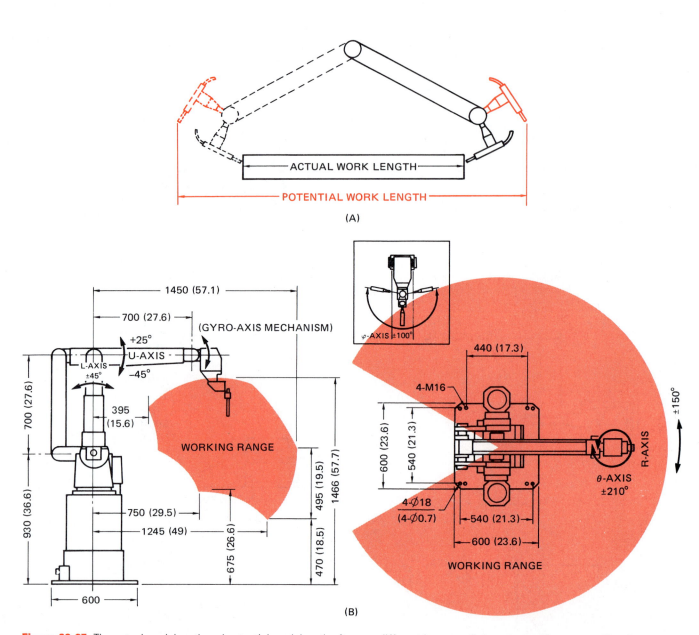

Figure 26-27 The actual work length and potential work length often are different but must fit into the working range. (Part B courtesy of Merrick Engineering, Inc.)

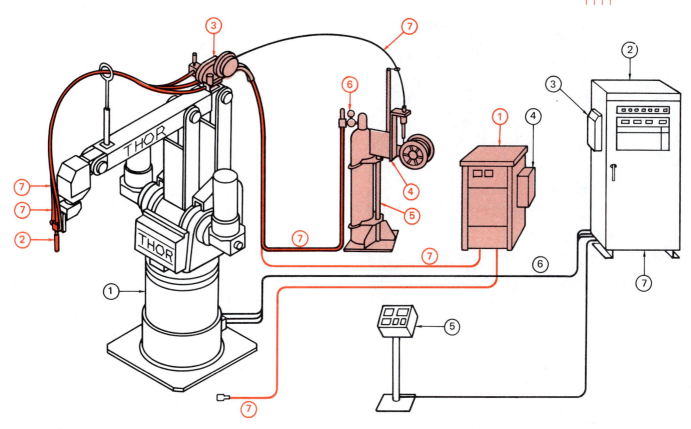

♦ Computerized Arc Welding Robot "THOR-K"

Symbol	Welding Robot
1	Arc Welding Robot
2	Robot Control Unit
3	Teaching Box
4	Welding Signal Unit
5	Control Box
6	Control Cables
7	External Memory Unit (Located inside of the Robot Control Unit.)

Symbol	Welding Equipment
1	Welding Power Supply
2	Welding Torch
3	Wire Feeder
4	Wire Reel Unit
5	Stand for Gas Cylinder
6	CO_2 Gas Flow Regulator
7	Cable and Hose Assembly

Combination with OTC CO_2 Gas Shielded Arc Welding Machine

Figure 26-28 All components must interact. (Courtesy of Merrick Engineering, Inc.)

with an "arc start" signal in addition to continuous monitoring of voltage and amperage readings. Product moving and positioning equipment should have sensors to indicate that the part is "in position" and ready for welding. Safety sensors should immediately stop all movement of equipment if unauthorized persons are in the work zone.

Safety. The following precautions are recommended for the use of automatic welding equipment and robots.

■ All personnel should be instructed in the safe operation of the robot.

■ All personnel should be instructed in the location of an emergency power shutoff.

■ The work area should be restricted to authorized persons only.

■ The work area should have fences, gates, or other restrictions to prevent access by unauthorized personnel.

■ Sensors should be mounted around the floor and work area to stop all movement when unauthorized personnel are detected during the operation.

■ The arc welding light should be screened from other work areas.

Figure 26-29 A robot welds an air cooler. (Courtesy of Arvin Industries, Inc., Columbus, IN.)

- A breakaway toolholder should be used in case of accidental collision with the part, **Figure 26-32**.
- A signal should sound or flash before the robot starts moving.

FUTURE AUTOMATION

The industrial robot today is in its infancy. Advancements we expect in the future are only limited by our imagination. Robots are getting smarter. They can "see" using fiber optics and are beginning to process what they see to make program adjustments. They can also use magnetic sensors to locate parts and touch sensors to pick them up without damaging them. Advancements in welding processes will allow greater weld control. Improved and faster computers will allow more complex programming, and improved sensors will give better feedback. The outcome of these advancements will be increased production at lower costs with higher quality.

Figure 26-30 Dual robot spot welding work cell incorporating transgun-type spot welders. (Courtesy of T.J. Snow Co., Inc.)

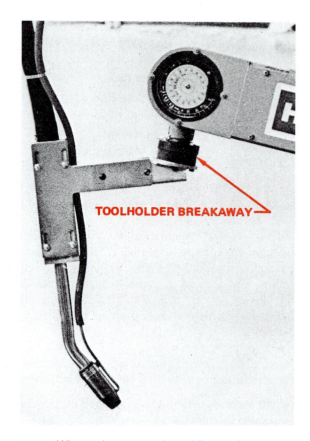

TOOLHOLDER BREAKAWAY

Figure 26-32 When using automatic welding equipment and robots, a breakaway toolholder should be used. (Courtesy of Welders Supply Inc., Dallas, Texas; photographed by Terry Gentry.)

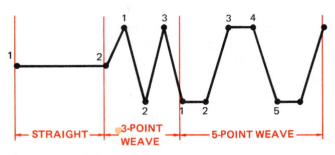

STRAIGHT — 3-POINT WEAVE — 5-POINT WEAVE

Figure 26-31 Weave patterns.

Microprocessor Control Yields Consistent Weld Quality

For many years, the end caps of the hydraulic and pneumatic cylinders for rack-and-pinion rotary actuators have been secured to the actuator housing with threaded, external tie-rods. Recently the design engineers at one company began a switch to welded end caps to eliminate several manufacturing operations and reduce material consumption.

The welds are critical because the cylinders must withstand test pressures up to 4,500 psi. When welds of manually welded prototypes proved to be of extremely inconsistent quality, the plant superintendent realized that the degree of welding control required could probably never be consistently achieved in production by a manual operation.

The solution to the problem is an automatic system controlled by a programmable microprocessor. A special welding fixture was designed for the operation. The resulting automatic GMAW (gas metal arc) welding system provides precision control of every welding variable from preflow to postflow and including start-stop fixture controls. Up to nine welding programs can be stored in the microprocessor's memory, and additional programs can be stored on tape and entered into memory at any time.

The automatic system consists of a microprocessor-controlled wire feeder with a wire drive, power source, water-cooled torch, and a cooling system. In Figure 1, the operator observes as the weld is made automatically under the control of the microprocessor mounted overhead. The power source at the right provides the weld power as called for by the microprocessor, and the cooling system mounted on the power source keeps the torch cool.

The fixture has a hinged torch holder mounted on cross slides for horizontal and vertical adjustment of the stationary welding torch. The workpiece is mounted vertically in a variable speed rotating positioner. The only weld parameter not controlled by the microprocessor is the positioner rotational speed, which is adjusted manually.

Figure 1

The cylinders of the actuators, for which the process is designed, are made from 1026 steel, have 0.375-in. walls, and range in diameter from 1-1/2 to 6 inches. They are welded with .045 diameter AWS ER70S6 wire using 95% argon/5% oxygen shielding gas.

A typical weld program consists of 2 seconds gas preflow and a .25 second run-in at 22 volts and 225 IPM before the fixture begins to rotate. With the work rotating, weld parameters are 28 volts, 350 IPM with a 1-in. electrical stick-out. Following a .5 second crater fill at 20 volts and 200 IPM, after which the fixture stops rotating, there is a 0.05 second burnback and a 4-second gas postflow. After completing the sequence, the microprocessor resets itself for the next weld. Total sequence time varies with the diameter of the cylinder. Typically, the end cap of a 1-1/2-in. diameter cylinder is welded in 45 seconds. Figure 2 shows the operator loading a cylinder with end cap in place on the rotating fixture for welding. The hinged torch positioner swings aside to permit work to be loaded and unloaded. The counterbalance compensates for the weight of the torch and the wire feeder, which is also mounted on the torch positioner.

Compared to the use of external tie-rods, welded cylinder end caps save both material and labor. Welding eliminates eight prestressed tie-rods, eight nuts, and the flanges on two end caps per actuator. In terms of manufacturing operations, it eliminates cutting threads in both ends of eight tie-rods, drilling four holes in each of the two

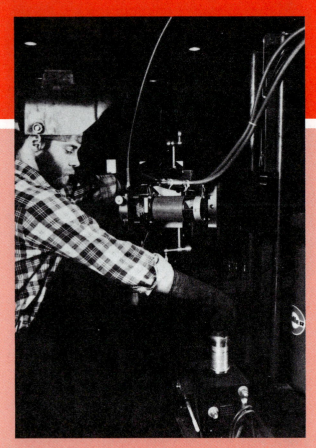

Figure 2

end cap flanges and drilling and tapping eight holes in the actuator housing. But consistent weld quality is essential to achieve these savings. The actuator cylinders operate at from 2,000 to 3,000 psi, and every one is tested at 150 percent operating pressure to assure their safety.

The actuators, as well as specially designed rotary actuators, are used in a variety of industrial equipment. The company has built actuators to provide torques up to 50,000,000 lb in and through rotations up to 1,800 degrees, or 5 full turns. The company is the primary supplier of rotary actuators that operate valves, hatches, and other devices on the United States' fleet of nuclear submarines. The company also designed and built the 70-in.-long rotary actuators that swing aside the gantry arm on the space shuttle launch tower at Cape Canaveral, Florida.

(Courtesy of Miller Electric Mfg. Co., Appleton, Wisconsin 54912.)

REVIEW

1. What were the first industrial robots in America used for?

2. How can CAM technology aid in manufacturing?

3. Why do most robot manufacturers recommend that a skilled welder operate the robot?

4. What must a welder control for a process to be considered manual?

5. List the commonly used manual processes.

6. What must a welder control for a process to be considered semiautomatic?

7. List the commonly used semiautomatic processes.

8. What makes a process a machine process?

9. What makes a process an automatic process?

10. What makes a process an automated process?

11. What types of power provides the industrial robot with movement?

12. In what axes or directions can a robot move?

13. What is meant by "teach" a robot?

14. What is the advantage of storing a robot's program on a disk or cassette tape?

15. Why is it important to reduce restricted areas in a design that will be welded using a robot?

16. Why is it important to make small welds more accurately than large ones?

17. Why must a robot be evaluated for its fitness to work in a welding area?

18. Why is joint tracking important for a robot?

19. Why is it important for a robot to weave the torch?

20. List the safety considerations that must be followed when operating a robotic work station.

Chapter 27

OTHER WELDING PROCESSES

OBJECTIVES

After completing this chapter, the student should be able to:
- explain the operating principles for the different special welding processes.
- list the reasons why a particular process should be selected to make a special weld.
- list the operational limitations of each special welding process explained in this chapter.

KEY TERMS

resistance welding (RW)	ultrasonic welding (USW)
flash welding (FW)	inertia welding
upset welding (UW)	laser welding (LBW)
percussion welding (PEW)	plasma-arc welding (PAW)
electrical resistance	thermit welding (TW)
electron beam welding (EBW)	stud welding (SW)
evacuated chamber	hardfacing
optical viewing system	thermal spraying (THSP)

INTRODUCTION

Over eighty different welding and allied processes are listed by the American Welding Society. This text covers nine of the most commonly used processes that require the welder to have a special skill. This chapter covers seventeen additional processes that call for special equipment and techniques. Some of these processes require less skill or knowledge to set up and operate, such as resistance spot welding (RSW). Others demand a great deal of technical information and training, such as electron beam welding (EBW).

The actual operating procedures vary greatly from one manufacturer's machine to another. The specific settings also change from one material to another. Because of these factors, only the general theory, procedures, and applications are discussed in this chapter. More information can be obtained from the AWS or directly from the manufacturer of the equipment being operated. The skill needed to operate this equipment can be learned quickly on the job or in classes taught by the specific equipment manufacturer.

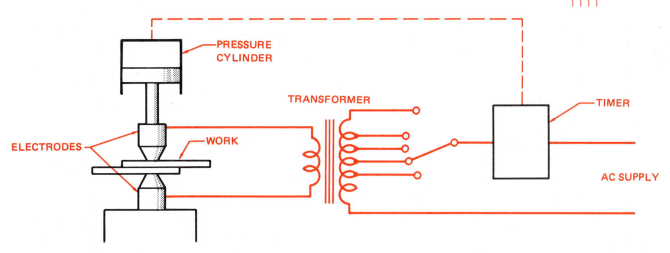

Figure 27-1 Fundamental resistance welding machine circuit.

RESISTANCE WELDING (RW)

Resistance welding (RW) is defined as a process wherein coalescence is produced by the heat obtained from the resistance of the workpiece to the flow of low-voltage, high-density, electric current in a circuit of which the workpiece is a part, **Figure 27-1**. Pressure is always applied to ensure a continuous electrical circuit and to forge the heated parts together. Heat is developed in the assembly to be welded, and pressure is applied by the welding machine through the electrodes. During the welding cycle the mating surfaces of the parts do not have to melt in order for the weld to occur. The parts are usually joined as the result of heat and pressure and not their being melted together. Fluxes or filler metals are not needed for this welding process. The heat produced in the weld may be expressed in the following manner:

$$H = I^2 RT$$

where
H = Heat
I = Current
R = Electrical resistance to the circuit
T = Duration of the current

The current for resistance welding is usually supplied by either a transformer or transformer capacitor. The transformer, in both power supplies, is used to convert the high line voltage (low-amperage) power to the welding high-amperage current at a low voltage. A capacitor, when used, stores the welding current until it is used. This storage capacity allows such machines to use a smaller size transformer. The required pressure, or electrode force, is applied to the workpiece by pneumatic, hydraulic, or mechanical means.

Most resistance welding machines consist of the following three components:

- The mechanical system to hold the workpiece and to apply the electrode.

- The electrical circuit made up of a transformer and, if needed, a capacitor, a current regulator, and a secondary circuit to conduct the welding current to the workpiece.

- The control system to regulate the time of the welding cycle.

There are several basic resistance welding processes. These processes include spot (RSW), seam (RSEW), high-frequency seam (RSEW-HF), projection (PW), flash (FW), upset (UW), and percussion (PEW).

Resistance welding is one of the most useful and practical methods of joining metal. This process is ideally suited to production methods. It also requires workers who are less skilled.

Spot Welding (RSW)

Spot welding is the most common of the various resistance welding processes. In this process, the weld is produced by the heat obtained at the interface between the workpieces. This heat is due to the resistance to the flow of electric current through the workpieces, which are held together by pressure from the electrode, **Figure 27-2**. The size and shape of the formed welds are controlled somewhat by the size and contour of the electrodes.

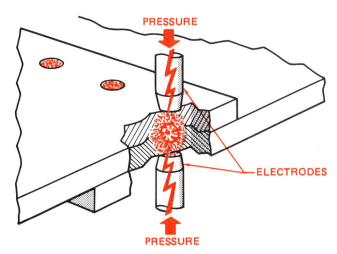

Figure 27-2 Heat resulting from resistance of the current through the metal held under pressure by the electrodes creates fusion of the two workpieces during spot welding.

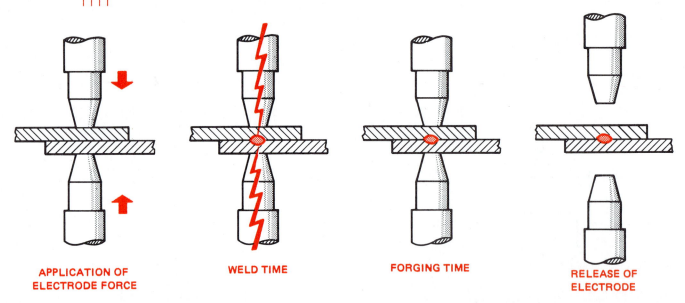

APPLICATION OF
ELECTRODE FORCE

WELD TIME

FORGING TIME

RELEASE OF
ELECTRODE

Figure 27-3 Basic periods of spot welding.

The welding time is controlled by a timer built into the machine. The timer controls four different steps, **Figure 27-3**. The steps are as follows:

- Squeeze time, or the time between the first application of electrode force and the first application of welding current.

- Weld time, or the actual time the current flows.

- Hold time, or the period during which the electrode force is applied and the welding current is shut off.

- Off Period, or the time during which the electrodes are not contacting the workpieces.

Tables supplied by the machine manufacturer provide information for the exact time for each stage for different types and thicknesses of metal.

Material from .001 in. (0.025 mm) to 1 in. (25 mm) thick may be joined by spot welding.

Spot Welding Machines. The three types of spot welding machines commonly used are rocker-arm, press-type, and portable-type.

In rocker-arm spot welders, the lower electrode is stationary while movement is transferred to the upper electrode by means of an arm, which moves about a pivot point. The rocker arm can be moved by one of three methods: foot pedal, air cylinder, or an electric motor. Rocker-arm welders, **Figure 27-4**, are available with throat depths from 12 in. to 48 in. (30 cm to 122 cm), and transformer capacities from 10 kVA to 20 kVA.

In press-type spot welders, **Figure 27-5**, the movement of the upper electrode is controlled by a hydraulic cylinder. This type of welder is used for welding heavy sections. Throats having a depth of 60 in. (152 cm) are available with capacities up to 500 kVA.

Portable spot welders are used where work is too large to be moved. These machines are complete with a

Figure 27-4 Rocker-arm spot welder. The operator depresses the foot switch, tongs drop automatically into position on the work, and the weld cycle stops through a predetermined time adjustment. (Courtesy of Miller Electric Manufacturing Company.)

pressure cylinder, transformers, electrode holders, electrodes, and the necessary controls.

Portable spot-welding guns are used in the mass production of automobiles, aircraft, railroad cars, and similar products where relatively thin sheets are to be welded, **Figure 27-6**.

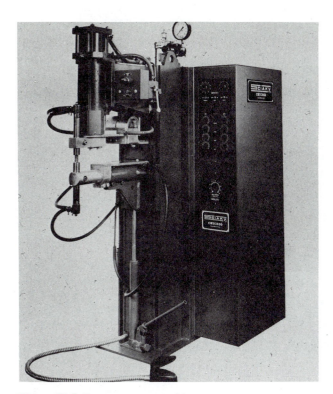

Figure 27-5 Press-type spot welder.

Multiple-spot Welders. Multiple-spot welders are used where a high production rate is a requirement. The multiple-spot welder shown in **Figure 27-7** (page 680) is especially designed for auto instrument panels. It welds 20 projection-type studs at one time. The assembly is hand loaded and then automatically moves into the weld position where it is welded and ejected. The fixture then returns to the load position.

Most multiple-spot welders have a series of air-operated or hydraulically operated guns mounted on a header and use a common bar for the lower electrode. The welding guns are connected by flexible wires to individual transformers or to a common bus bar attached to the transformer, **Figure 27-8** (page 680).

All guns make contact with the workpiece at the same time. They can be fired either in a certain sequence or all at the same time. These types of welders are widely used in the automobile industry.

Seam Welding (RSEW)

Seam welding is similar in some ways to spot welding except that the spots are spaced so close together that they actually overlap one another to make a continuous seam weld.

Seam welding is accomplished by using roller-type electrodes in the form of wheels that are 6 in. (152 mm) to 9 in. (229 mm) or more in diameter, **Figure 27-9** (page 680). These roller-type electrodes are usually copper alloy discs 3/8 in. (10 mm) to 5/8 in. (16 mm) thick. Cooling is achieved by a constant stream of water directed to the electrode near the weld, **Figure 27-10** (page 681).

Welding is done either with the roller electrodes in motion or while they are stopped for an instant. If continual motion is used, the rate of welding usually varies between 1 ft and 5 ft (30 cm to 152 cm) per minute. The greatest welding speed is obtained on the thinnest materials. An indexing mechanism can be used when the wheels are to be stopped for each weld.

Figure 27-6 Small welder for fine, detailed work. (Courtesy of ESAB Welding and Cutting Products.)

Figure 27-7 Multiple-spot welder, specially designed as an auto instrument panel welder. (Courtesy of T.J. Snow Co., Inc., Chattanooga, TN.)

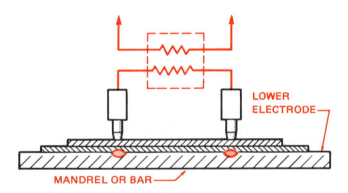

Figure 27-8 The arrangement of electrodes in multiple-spot welding.

Electric timing equipment is useful when it is necessary to provide the precise control required for highest quality welds. Spot welds can be positioned at almost any interval desired by simply adjusting the timing and rate of electrode motion.

It is possible to change the welding current and electrode pressure to control the surface condition and width of the weld. The seam width should be about two times the thickness of the sheet plus 1/16 in. (2 mm). Figure 27-11 shows a tweezer-type seam welder with automatic cycling.

Types of Seam Welds. Figure 27-12 illustrates the type of seams used in most seam welding processes. The lap seam is the most common of these seams. The tops and bottoms of containers are usually fastened together with a flanged seam. For a metal thickness of 1/16 in. (2 mm),

Figure 27-9 Seam welder. (Courtesy of Sciaky Brothers, Inc.)

the mash seam is often used. The electrode face should be wide enough to cover the overlap by approximately one and one-half times the thickness of the sheet.

High Frequency Resistance Seam Welding (RSEW-HF)

This process is similar to RSEW in some ways. The major differences are that the welding current is supplied

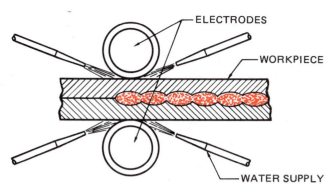

Figure 27-10 Schematic illustration of the seam welding process.

Figure 27-11 Tweezer-type seam welder with automatic cycling for circular flat pack. (Courtesy of Tweezer Weld®, A Division of GEBO Corp. USA.)

as a high frequency 200 to 500 kHz as opposed to DC or 60 cycle AC. The high frequency power provides very localized heating. This heating is a result of the resistance to the current induced in the metal and not to the resistance between parts. As a consequence of this localized heating, pipe and tubing can be welded with little loss of power flowing around the back side of the joint.

RSEW-HF is used in the production of welded pipe, tubing, and structural shapes. It works very well for the fabrication of I-beams, H-beams, channel, etc. These welded structural shapes are lighter and stronger than their roll formed counterparts.

Resistance Projection Welding (RPW)

Projection welding is somewhat similar to spot welding in that it involves joining parts by a resistance welding process.

To make a projection weld, projections are formed on at least one of the workpieces at the points where welds are desired. The projections, small raised areas, can be any shape, such as round, oval, circular, oblong, or diamond. They can be formed by embossing, casting, stamping, or machining, **Figure 27-13** (page 682).

The workpieces that have the projections are placed between plain, large-area electrodes in the welding machine, **Figure 27-14** (page 682). The current is turned on, and pressure is applied. Since nearly all of the resistance is in the projections, most of the heating occurs at the points where welds are desired. The heat causes the projections and the opposing metal to soften. The pressure from the electrodes causes the softened projections to flatten, resulting in fusion of the workpieces.

With this type of welding, a complicated electrode shape is not required. In cases where the weld area does not have a regular shape, the electrode can be shaped to fit the surface.

There are many variables in this type of welding. Some of these variables are thickness of metal, number of projections, and kind of material. These variables make it hard to predetermine the current and electrode pressure required.

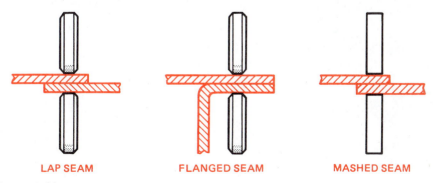

LAP SEAM   FLANGED SEAM   MASHED SEAM

Figure 27-12 Types of seam welds.

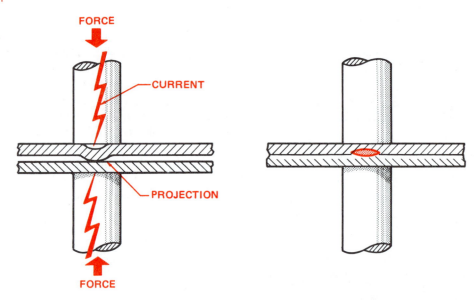

Figure 27-13 Projection welding.

Steel plate, galvanized sheet steel, and stainless steels can be joined using projection welding. Aluminum is not readily welded by this process.

Flash Welding (FW)

Flash welding (FW) may be considered a resistance welding process. Fusion is produced over the ends of stock by heat produced from the resistance to the flow of electric current between the two surfaces. Pressure is applied after heating is completed. As a result, the material is forced together, and fusion takes place. The basic steps in flash welding are as follows:

1. Clamp the parts together in dies (these dies conduct the electric current to the workpieces).
2. Move one part toward the other part until an arc is established.
3. Upset by applying pressure when flashing has caused the parts to reach a plastic temperature.
4. Cut off the welding current when fusion and upset are complete, **Figure 27-15**.

A flash welding machine is shown in **Figure 27-16**. Flash welding may be used to join dissimilar aluminum alloys and to join aluminum to other metals, **Figure 27-17**.

Figure 27-14 Projection welding machine. (Courtesy of Sciaky Brothers, Inc.)

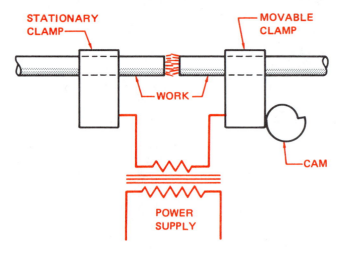

Figure 27-15 Schematic diagram of the flash welding process.

Figure 27-16 Flash welding is a high-speed production method for making miter and butt joints. (Courtesy of Aluminum Company of America.)

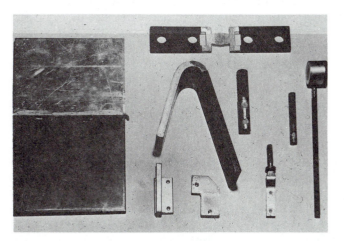

Figure 27-17 Aluminum and copper are flash welded together for electrical connections. (Courtesy of Aluminum Company of America.)

Flash welding is a high-volume production process. Unless a large number of pieces are to be welded, it is very costly to prepare the workholding dies. The time and materials involved in setting up the job are also costly factors. The production rate is high since the welding time is short.

Upset Welding (UW)

Welding small areas is usually done by the **upset welding** method. In this form of welding, **Figure 27-18(A)** through **(C)**, two pieces of metal having the same cross section are gripped and pressed together. Heat is generated in the contact surfaces by **electrical resistance.** The upset at the interface extrudes the contaminated contact area to the outside, where it is trimmed, **Figure 27-18(A).**

Flash welding has an arcing action at the contacting surfaces. No flashing occurs in the upset process. The main difference between the upset and flash processes is

(A)

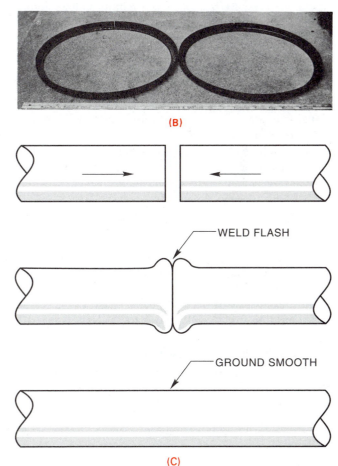

(B)

WELD FLASH

GROUND SMOOTH

(C)

Figure 27-18 (A) Upset welding machine for welding circular parts, and (B) parts welded in this machine. (C) The weld flash is trimmed off after an upset weld. (Courtesy of National Machine Company.)

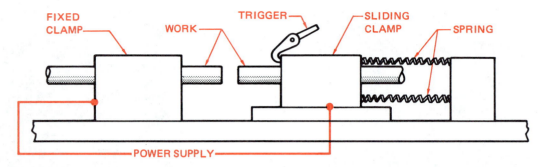

Figure 27-19 Principle of percussion welding.

that less current is needed in the upset process and welding time is extended.

Percussion Welding (PEW)

Percussion welding (PEW) is a resistance welding process and is actually a variation of flash welding. Fusion is obtained simultaneously over the entire area of the contacting surfaces. The heat that causes fusion is obtained from an electric arc produced by the rapid discharge of stored energy. Immediately after or during the electrical discharge, pressure is applied as a hammer blow, **Figure 27-19**. A percussion weld is similar to a flash weld. A short application of high-intensity energy instantly heats the surfaces to be joined. This rapid heating is almost immediately followed by quick blow to make the weld.

The action of this process is very rapid. There is little heating effect upon the metal adjacent to the weld. Thus, it is possible for heat-treated parts to be welded without becoming annealed. Examples of this type of welding include welding copper to aluminum or stainless steel, and adding Stellite® tips to tools, silver contact tips to copper, cast iron to steel, and zinc to steel.

ELECTRON BEAM WELDING (EBW)

Some years ago, electron beam welding was introduced to the metal fabrication field on a limited basis for difficult applications. The applications usually involved highly reactive or oxidizable metals. Basically, electron beam welding produced a strong, very clean, deep, and narrow weld.

Recent technological advances have increased the capability of electron beam welding equipment as a research and production tool. The commercial and aerospace industries have extensively and successfully applied this method of welding. However, it is felt that the ultimate potential of this method of welding has yet to be reached.

Despite the high initial cost, electron beam welders are used in the auto, electronics, pipeline, shipbuilding, aerospace, and high-speed welded tubing industries. In industrial uses, the initial cost of the welding equipment

is offset by the decrease in or elimination of the use of filler metal, by faster welding speeds, and by the elimination of joint preparation.

Temperatures up to 180,000°F (100,000°C) are generated by electron beam welding machines. At such high temperatures, the machine can accurately vaporize metals and ceramics. Deep, strong welds with no heat deformation are produced between close-fitting parts. The welds have high strength and purity. This process can be used to join dissimilar metals.

Electron beam welding utilizes the energy from a fast-moving beam of electrons focused on the base material. See Color Plate 61. The electrons strike the metal surface, giving up their kinetic energy almost completely in the form of heat. Welds are made in a vacuum (10^{-3} mm Hg to 10^{-5} mm Hg), which practically eliminates contamination of the weld material by the gases left in the vacuum chamber. The high vacuum is necessary to produce and focus a stable, uniform electron beam. Welds produced by this process are coalesced from vacuum-melted material, which eliminates the usual fusion weld contaminants caused by water vapor, oxygen, nitrogen, hydrogen, and slag.

Early electron beam welders needed hard vacuums for both the beam and welding chambers. New units need only a soft vacuum in the welding chamber (attained in only a few seconds), or, in the case of some new units, welds can be made 1/4 in. (6 mm) outside the evacuated (vacuum) chamber. Nonvacuum welding is thought of as a supplement to, not as a replacement for, conventional welding.

The Welding Gun

The electron gun consists of a filament, cathode, anode, and focusing coil. These parts are mounted above the work chamber, as shown in **Figure 27-20**. The electrons from the heated filament carry a negative charge and are repelled by the cathode and attracted by the anode. The anode has an opening through which the electrons pass.

The electrons then move through a magnetic field, which is produced by an electromagnetic focusing coil.

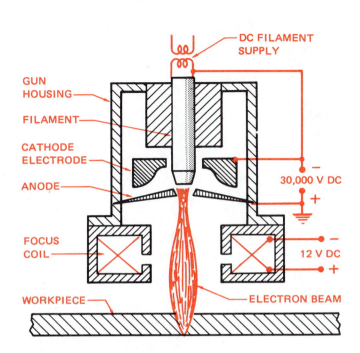

Figure 27-20 Electron beam gun produces an accurately controlled heat source adjustable for length and point of focus.

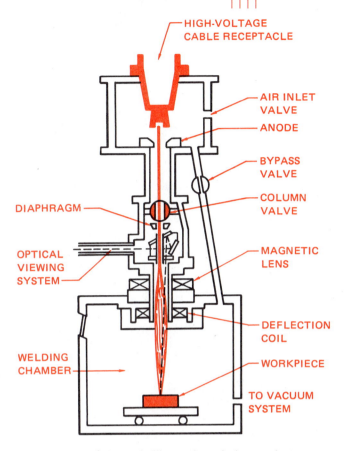

Figure 27-22 Schematic illustration of electron beam gun column.

The machine is equipped with an optical viewing system, Figure 27-21. This system provides a line of sight down the path of the electron beam centerline to the weld area when the beam is off, Figure 27-22.

It is possible to vary the current to the focusing coil so that the operator can focus the beam from a sharp focus to a beam 1/4 in. (6 mm) in diameter. Focal points above, on, or below the work surface are provided to modify the weld pattern.

Seam Tracking

Electron beam equipment is designed and manufactured so that the work table can be rotated beneath the gun for circular welds or driven along a path that corresponds to, or is parallel to, the centerline of the

Figure 27-21 7.5-kW electron beam welder. (Courtesy of Leybold-Heraeus Vacuum Systems, Inc.)

chamber for linear welds. Sometimes the gun itself is moved.

Automatic seam tracking is available to track any seam. The amount of misalignment of the placement of the joint may be a few thousandths of an inch or a few inches. The seam training device continuously checks the actual position of the seam during the welding operation and precisely corrects for the degree of misalignment, Figure 27-23.

Assuming the piece part is driven along the X-axis, Figure 27-24, electromagnetic deflection will move the electron beam transversely to the welding motion. As the misaligned seam moves in a linear direction along the axis, the seam tracker interprets the amount of deviation from the theoretical centerline and corrects the deflection of the beam to match the exact position of the seam at any instant.

Basically, the work is moving on the welding axis, and the beam is scanning on the Y-axis. However, all motions are interchangeable as selected, so that the welding motion may be on the gun, and correction may still be accomplished by moving the beam.

ULTRASONIC WELDING (USW)

Ultrasonic welding is a process for joining similar and dissimilar metals by introducing high-frequency vibrations into the overlapping metals in the area to be joined. Fluxes and filler metals are not required, electrical current does not pass through the weld metal, and only localized heating is generated. The temperature produced is below the melting point of the materials being joined. Thus, no melting occurs during the welding cycle.

Figure 27-23 Contour follower system. (Courtesy of Sciaky Brothers, Inc.)

The most common welds made with this process include spot, overlapping spot, continuous seam, and ring welds. The pieces to be welded are clamped between the welding tip and a hard surface, which serves as an anvil.

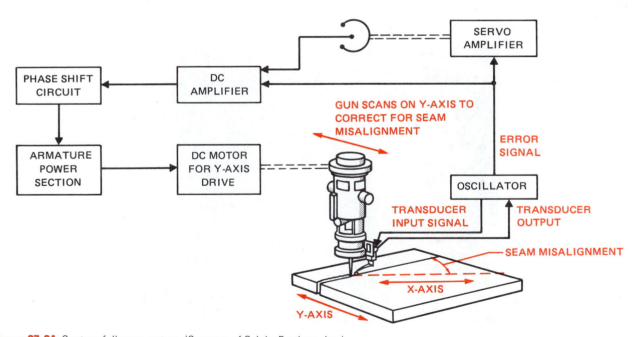

Figure 27-24 Contour follower system. (Courtesy of Sciaky Brothers, Inc.)

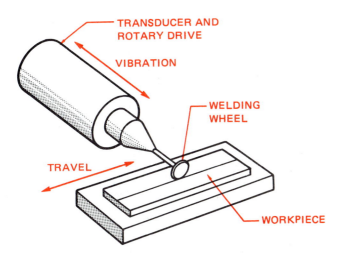

Figure 27-25 Ultrasonic seam welding.

High-frequency energy is fed to a transducer, which converts it to vibrations. The welding tip is attached to the transducer. The coupling system and welding tip form a unit that is referred to as a sonotrode, **Figure 27-25**. The welding tip oscillates in a plane essentially parallel to the joint interface. Transverse (shear) waves in the material produce the weld. Vertical (compression) waves are not effective in ultrasonic welding of metals.

The ultrasonic vibrations combined with the static clamping force cause dynamic shear stress in workpieces. When these shear stresses are great enough, local plastic deformation of the metal occurs at the interface. Surface films of oxide coatings are shattered and dispersed, and pure base metal contact is achieved. A true metallurgical bond is then formed in the solid state without melting the parent metal.

Continuous seam welding is done with a rotating, disk-shaped sonotrode tip. Either a counter rotating roller anvil or table-type anvil is also used. The entire tip assembly rotates so that the peripheral speed of the sonotrode tip matches the traversing speed of the workpiece.

The equipment used in this type of welding consists of the following: (1) a power source or frequency converter that converts 60-hertz (60 cycles per second) line power into high-frequency electrical power, and (2) a transducer that changes high-frequency electrical power to rapid vibrations. Welding speeds can range from a few feet per minute to 400 feet per minute (0.3 meters per minute to 131 meters per minute) for continuous seam welding. Times range from 0.005 second to 1.0 second for spot welding.

Ultrasonic Welding Applications

Ultrasonic welding has many applications in the assembly of electrical products. Typical applications include the following:

- Attaching oxide-resistant contact surface buttons to switches.
- Attaching leads to coils of foil, sheet, or wire made of aluminum.
- Attaching very fine wire leads and elements to other components.

Figure 27-26 shows an ultrasonic spot welder used to perform the types of welds just listed.

In the plastics field (packaging), ultrasonic welding is used for both spot and continuous seam fabrication and for closures on various types of foil or plastic envelopes and pouches.

Various positive drive tip configurations can be designed for specific joint requirements. Applications of the process to aluminum welding include attaching thin ribs to cylinders, welding two pairs of leads at once, and joining two gears and one spacer in a single weld. Attachment of aluminum parts or brackets to stainless steel walls is another job readily performed by ultrasonic welding.

At the present stage of development, ultrasonic welding can be used only for materials having thicknesses up to about 1/8 in. (3 mm). However, equipment is being developed to extend this range.

INERTIA WELDING PROCESS

In **inertia welding,** one workpiece is fixed in a stationary holding device. The other is clamped in a spindle chuck, which is accelerated rapidly. At a predetermined speed, power is cut, as shown in **Figure 27-27(A)** (page 688). As a result, one part is then thrust against the other piece. Friction between the parts causes the spindle to decelerate, converting stored energy to frictional

Figure 27-26 Ultrasonic spot welder. (Courtesy of Sonobond Ultrasonics.)

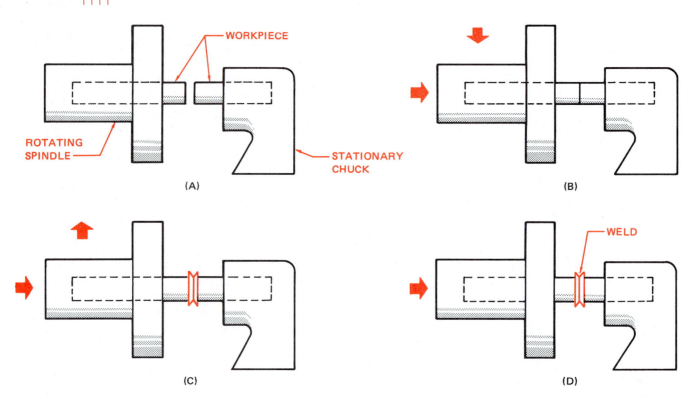

Figure 27-27 Inertia welding process. (Courtesy of Caterpillar Inc.)

heat. Enough heat is formed to soften, but not melt, the faces of the part, **Figure 27-27(B)**.

An accurate heating rate can be obtained since inertia can be controlled to supply whatever energy is needed

Until rotation ceases, the two parts cannot bond. The compression force upsets the metal interface, forcing out any impurities or voids. The heat-affected zone is narrow, **Figure 27-27(C)**. The flash may or may not be cut off. The weld is formed when the spindle stops, as shown in **Figure 27-27(D)**.

A graph of speed, torque, and upset (change in workpiece length) during the weld period will show what happens, **Figure 27-28**. The curves start when the two pieces come together, after flywheel acceleration.

At first, there is a small torque peak and a corresponding change in the length of the parts. This is when initial temperature buildup occurs.

Torque then drops back and is fairly constant. During this period, a state of near equilibrium exists when energy from the rotating mass is being converted to heat at the same approximate rate as it is being conducted away. Little upset occurs during this time.

Finally, the speed drops to a point where heat penetration does not keep up with heat dissipation and the faces cool slightly. The torque peaks sharply as the weld bonds are formed and broken. Most of the upset occurs just before the spindle stops. A solid-state weld is created as the now stationary parts are pressed together.

The inertia welding process produces a superior-quality, complete interface weld. The welding conditions must be consistent so that human judgment is removed in production work. The technique has been applied successfully to super alloys as well as to standard metals.

Bond Characteristics

Microscopic examination of the bond resulting from the inertia welding process shows the following three metallurgical characteristics:

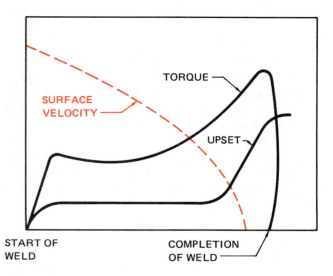

Figure 27-28 Chart shows what happens during inertia welding process. (Courtesy of Caterpillar Inc.)

■ The weld zone is very narrow and has a fine-grained structure with no melt product or grain growth. These conditions indicate a hot-worked structure. When dissimilar metals are welded, often there are streaks of intermixed material near the outside diameter.

■ Hardening phases from the rapid chilling are seen throughout the structure. The degree of hardening can be controlled but is often about the same as that achieved with a mild water quench.

■ There is a zone of varying grain structure between the heat-affected zone and the parent structure, Figure 27-29.

Table 27-1 lists some of the metals that can be joined by inertia welding.

Some parts that have been joined by the inertia welding process are shown in Figures 27-31 to 27-34.

Figure 27-30 shows a hydraulic piston rod assembly. The material is prechromed, cold-drawn tubing with the eye cut from tubing. In the past, this part was manufactured as a one-piece forging. Inertia welding of the parts produced an estimated savings of 50% per rod.

A fan bracket assembly is shown in Figure 27-31 (page 690) with an SAE 1045 shaft welded to an SAE 1020 bracket using the inertia welding process. The steel shaft is pressed into a shrink-fit forging, and then welded on the back side. A considerable savings per assembly is realized.

Figure 27-32 (page 690) shows an automotive engine valve with a 21-4-N head welded to an SAE 8645 stem. This part was formerly welded by the flash butt process. Inertia welding production methods produced 600 welds per hour.

The lift link shown in Figure 27-33 (page 690) is made from SAE 1045 steel. Formerly, it was manufactured as a one-piece forging and now is joined by inertia welding.

Similar Metals	
Carbon steels	Molybdenum
Sintered steels	Waspalloy
Stainless steels	Cobalt alloys
Tool steels	Titanium
Alloy steels	Zircalloy
Aluminum alloys	Inconel®
Copper	Nickel alloys
Brasses, bronzes[1]	

Dissimilar Metals
High-speed steel to various steels
Sintered steels to wrought steels
316 stainless steel to Inconel®
Stainless steel to medium- and low-carbon steels
1100 or 6061 aluminum to medium-carbon steel
Cobalt-base alloys to steel
347 stainless steel to 17-4 PH
Pure titanium to 302 stainless steel
Copper to 1100 or 6061 aluminum
Copper to medium-carbon steel
Copper to various brasses
Aluminum bronze to medium-carbon steel
Nickel-base alloys to steel

[1] Except bearing types

Source: Caterpillar Tractor Company

Table 27-1 Partial List of Metals Joined by Inertia Welding. (Courtesy of Caterpillar Tractor Company.)

Advantages of the Process

The advantages of the inertia welding process are many. A few of these advantages are listed as follows:

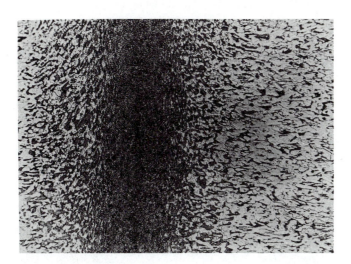

Figure 27-29 Mechanical mixing and grain refinement are evident in this 50X photomicrograph of inertia weld between two pieces of ASE 51201 steel. (Courtesy of Caterpillar Inc.)

Figure 27-30 Hydraulic piston rod inertia welded assembly. (Courtesy of Caterpillar Inc.)

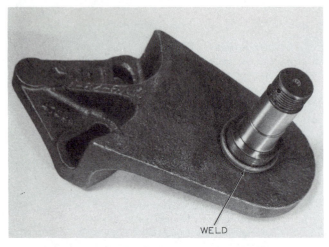

Figure 27-31 Fan bracket. (Courtesy of Caterpillar Inc.)

Figure 27-32 Automotive engine valves. (Courtesy of Caterpillar Inc.)

- Superior weld.
- A very narrow heat-affected zone adjacent to the weld.
- Uniform production welds.
- Fast production welds.
- Clean operation.
- Lowest cost of energy.
- Minimum skill required to operate the welder.

- Amount of upset of parts can be controlled to close tolerances.
- A complete interface weld can be obtained.
- Safe operation.

Applications of this process include welding dry bearing materials such as oxides and leaded bronzes. Metals that cannot be hot forged can also be joined by inertia welding.

LASER WELDING (LBW)

In **laser welding,** Figure 27-34, fusion is obtained by directing a highly concentrated beam of coherent light to a very small spot. Laser beams combine low-heat input (0.1 joule to 10 joules) with high-power intensity of more than 10,000 watts per square centimeter (considerably more than the electron beam). Due to the fact that the heat is provided by a beam of light, there is no physical contact between the workpiece and the welding equipment. It is possible to make welds through transparent materials.

The ease with which the beam can be directed to any area of the work makes laser welding very flexible. In the manufacture of complex forms, for example, it is possible to move the focused laser beam under digital control to seam weld any desired shape. The proven high-quality laser welds, along with the flexibility and the comparatively moderate cost of laser welding equipment, indicates that laser welding plays an increasing role in microelectronics and other light-gauge metal welding applications.

Laser welding of high-thermal-conductivity materials, such as copper, is not difficult to do. The extremely concentrated laser heat will melt the metal locally to make a weld or will vaporize the metal to drill a hole before it can be conducted away by the copper, which happens in most other forms of welding.

Advantages and Disadvantages of Laser Welding

Laser welding has some distinct advantages and disadvantages when compared to other welding processes. Electron beam welding is the only method that rivals the heat output of a laser. Generally, however, electron beam welding must operate in a vacuum. Since the laser beam is a light beam, it can operate in air or any transparent material, and the source need not even be close to the work. The material being welded need not be an

BEFORE

AFTER

Figure 27-33 Lift link before and after inertia welding. (Courtesy of Caterpillar Inc.)

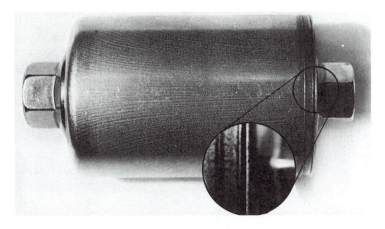

Figure 27-34 Gas filter, laser welded.

electrical conductor that limits most other processes or even part of an electrical or mechanical circuit. However, the light may be diffused by the welding vapors, so techniques to bypass the vapors have been developed.

Laser welds are small, sometimes being less than 0.001 in. (0.0254 mm). Laser welding is used to connect leads to elements in integrated circuitry for electronics. Lead wires insulated with polyurethane can be welded without removing the insulation.

Using the laser welding process, it is possible to weld heat-treated alloys without undoing the heat-treatment.

This method of welding can be used to join dissimilar metals. Metals that are difficult to weld, such as tungsten, stainless steels, titanium alloys, Kovar®, nickel alloy, aluminum, and tin-plated steels, can also be successfully welded by this process.

An optical system is used to focus the beam on the workpiece. The actual control of the welding energy is done by a switch.

The Laser Beam

The laser is based on the principle that atoms in certain crystals and gases can be made to release a coherent, monochromatic (single wavelength) energy when they are excited. The output is self-amplified in the laser because the excited atoms release their energy much more rapidly when stimulated by light emitted by neighboring atoms, Figure 27-35.

When a ruby laser is used in the laser welding process, chromium atoms in the ruby are excited to produce a parallel, nondivergent light beam. The ruby rod is generally 3/8 in. (10 mm) in diameter. It has an almost perfect crystal structure with about 0.05% of its weight consisting of chromium oxide. The chromium atoms contained in the rod give the ruby laser its red color because the atoms absorb green light from external light sources. The chromium atoms are pumped to a higher energy state by the green light.

During the welding process, the atoms return to their original state and surrender a portion of the extra energy they absorbed in the form of red fluorescent light. Collisions between atoms and photons cause other atoms to give off red light exactly in phase with the stimulating radiation. These collisions are increased by using a very intense green light, which serves to excite the chromium atoms in the ruby rod. The mirrored ends of the rod bounce the red light back and forth inside the rod. When the energy builds up to a certain threshold value, the chain-reaction collisions cause a burst of red light. Due to the fact that the mirror at the front end of the rod only partially reflects the light, red light is emitted through the mirror.

PLASMA-ARC WELDING PROCESS (PAW)

The term *plasma* should be defined in its electrical sense. A gas, or plasma, is present in any electrical discharge. The plasma consists of charged particles that transport the charge across the gap.

The two outstanding advantages of plasmas are higher temperature and better heat transfer to other objects. The higher the temperature differential between

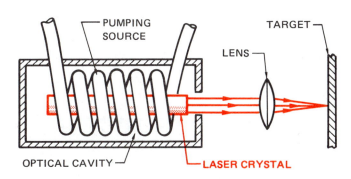

Figure 27-35 Schematic diagram of laser welder.

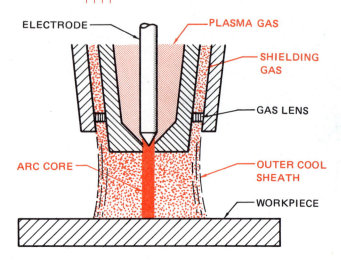

Figure 27-36 Schematic diagram of plasma welding process.

the heating fluid and the object to be heated, the faster the object can be heated.

In plasma-arc welding, a plasma jet is produced by forcing gas to flow along an arc restricted electromagnetically as it passes through a nozzle, **Figure 27-36**. The stiffness of the arc is increased by its decreased cross-sectional area. As a result, the welder has better control of the weld pool. By forcing the gas into the arc stream, it is heated to its ionization temperature, where it forms free electrons and positively charged ions. The plasma jet produced is a brilliant flame. The tip of the electrode is situated above the opening in the torch nozzle that serves to constrict the arc. A plasma welding installation is shown in **Figure 27-37**. The plasma jet passing through the restraining orifice has an accelerated velocity.

When this intense arc is directed on the workpiece, it is possible to make a welded butt joint in metal having a thickness of up to 1/2 in. (13 mm) in a single pass. No edge preparation or filler metal is required.

Any known metal can be melted, even vaporized, by the plasma jet process, making it useful for many welding operations. This process can be used to weld carbon steels, stainless steels, Monel®, Inconel®, aluminum, copper, and brass alloys.

Plasma Generator

A plasma generator produces an electric arc within a cooling system that utilizes gas or liquid. This system keeps the electrode and orifice from vaporizing.

The generator consists of a plasma gun containing a restrictive orifice in a water-cooled copper anode. Power is supplied by a rectifier-type DC power source with a control console. Plasma generators are used as cutting torches, in vaporizing solids, for welding, and in high-temperature chemistry. The plasma process is most often used for cutting; see Chapter 8.

STUD WELDING (SW)

Stud welding is a semiautomatic or automatic arc welding process. An arc is drawn between a metal stud and the surface to which it is to be joined. When the end of the stud and the underlying spot on the surface of the work have been properly heated, they are brought together under pressure.

The process uses a pistol-shaped welding gun, which holds the stud or fastener to be welded. When the trigger of the gun is pressed, the stud is lifted to create an arc and is then forced against the molten pool by a backing spring. The operation is controlled by a timer. The arc is shielded by surrounding it with a ceramic ferrule, which confines the metal to the weld area.

In the welding operation, a stud is loaded in the chuck of the gun, and the ferrule is fastened over the stud. The gun is then placed on the workpiece. The action of the gun when the trigger is squeezed causes

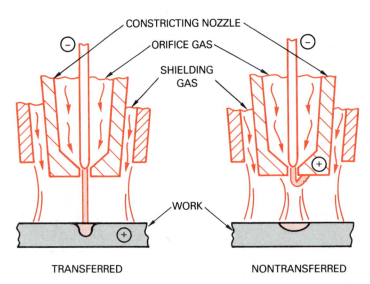

Figure 27-37 Transferred and nontransferred plasma arc modes. (Courtesy of the American Welding Society.)

the stud to pull away from the workpiece, resulting in an arc. The arc melts the end of the stud and an area on the workpiece. At the correct moment, a timing device shuts off the current and causes the spring to plunge the stud into the molten pool, which freezes instantly. The gun is then released from the stud and the ferrule knocked off.

HARDFACING

Hardfacing is defined as the process of obtaining desired properties or dimensions by applying, using oxy-fuel or arc welding, an integral layer of metal of one composition onto a surface, edge, or point of a base metal or another composition. The hardfacing operation makes the surface highly resistant to abrasion.

There are various techniques of hardfacing. Some apply a hard surface coating by fusion welding. In other techniques, no material is added, but the surface metal is changed by heat treatment or by contact with other materials.

Several properties are required of surfaces that will be subjected to severe wearing conditions, including hardness, abrasion resistance, impact resistance, and corrosion resistance.

Hardfacing may involve building up surfaces that have become worn. Therefore, it is necessary to know how the part will be used and the kind of wear to expect. In this way, the proper type of wear-resistant material can be selected for the hardfacing operation.

When a part is subjected to rubbing or continuous grinding, it undergoes abrasion wear. When metal is deformed or lost by chipping, crushing, or cracking, impact wear results.

Selection of Hardfacing Metals

Many different types of metals and alloys are available for hardfacing applications. Most of these materials can be deposited by any conventional manual or automatic arc or oxyfuel welding method. Deposited layers may be as thin as 1/32 in. (0.79 mm) or as thick as necessary. The proper selection of hardfacing materials will yield a wide range of characteristics.

Steel or special hardfacing alloys should be used where the surface must resist hard or abrasive wear. Where surfacing is intended to withstand corrosion-type or friction-type wear, bronze or other suitable corrosion-resistant alloys may be used.

Most hardfacing metals have a base of iron, nickel, copper, or cobalt. Other elements that can be added include carbon, chromium, manganese, nitrogen, silicon, titanium, and vanadium. The alloying elements have a tendency to form carbides. Hardfacing metals are provided in the form of rods for oxyacetylene welding, electrodes for shielded metal arc welding, or in hard wire form for automatic welding. Tubular rods containing a powder metal mixture, powdered alloys, and fluxing ingredients can be purchased from various manufacturers.

Many hardfacing materials are designated by manufacturers' tradenames. Some of the materials have AWS designations. AWS materials are classified into the following designations:

- High-speed steel
- Austenitic manganese steel
- Austenitic high-chromium iron
- Cobalt-base metals
- Copper-base alloy metals
- Nickel-chromium boron metal
- Tungsten carbides

The coding system serves to identify the important elements of the hardfacing metal. The prefix R is used to designate a welding rod, and the prefix E indicates an electrode. Certain materials are further identified by the addition of digits after a suffix.

Hardfacing Welding Processes

Oxyfuel Welding. In hardfacing operations, oxyfuel welding permits the surfacing layer to be deposited by flowing molten filler metal into the underlying surface. This method of surfacing is called sweating or tinning, **Figure 27-38.**

With the oxyacetylene flame, small areas can be hardfaced by applying thin layers of material. In addition, the alloy can be easily flowed to the corners and edges of the workpiece without overheating or building up deposits that are too thick. Placement of the metal can be controlled accurately.

The size of the weld is affected by many factors. These factors include the rate of travel, degree of preheat, type of metal being deposited, and thickness of the work.

Figure 27-39 (page 694) shows the approximate relationship of the tip, rod, molten pool, and base metal during the hardfacing operation.

Iron, nickel, and cobalt-base alloys require an excess acetylene flame. Copper alloys and bronze call for a neutral or slightly oxidizing flame. Laps, blowholes, and poor adhesion of deposits can be prevented by a flame characteristic that is soft and quiet.

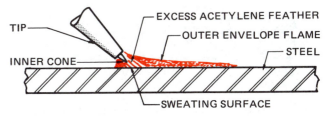

Figure 27-38 An example of how to produce sweating.

In all types of surfacing operations, the metal should be cleaned of all loose scale, rust, dirt, or other foreign substances before the alloy is applied. The best method of removing these impurities is by grinding or machining the surface. Fluxes may be used to maintain a clean surface. They also help to overcome oxidation that may develop during the operation.

Conventional methods may be used in holding the torch and rod. **Figure 27-40** shows the backhand method of hardfacing.

If the base metal is cast iron, it will not "sweat" like steel. Therefore, slightly less acetylene should be used. Alloys do not flow so readily on cast iron as they do on steel. Usually, it is necessary to break the surface crust on the metal with the end of the rod. A cast-iron welding flux is generally necessary. The best method is to apply a thin layer of the alloy and then build on top of it.

The oxyacetylene process is preferred for small parts. Cracking can be minimized by using adequate preheat, postheat, and slow cooling. Metal arc welding is preferred for large parts.

Arc Welding. Hardfacing by arc welding may be accomplished by shielded metal arc, gas metal arc, gas tungsten arc, submerged arc, plasma arc, or other processes.

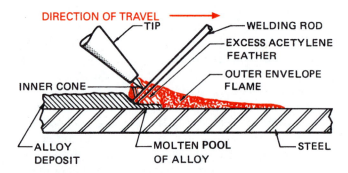

Figure 27-39 Approximate relationship of the tip, rod, and molten weld pool for forehand hardfacing.

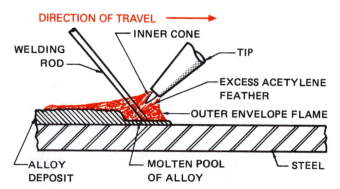

Figure 27-40 Backhand method of hardfacing. (Courtesy of Praxair, Inc.)

The techniques employed for any one of these processes are similar to those used in welding for joining. The factor of dilution must be carefully considered because the composition of the added metal will differ from the base metal. The least amount of dilution of filler metal with base metal is an important goal, especially where the two metals differ greatly. Little dilution means that the deposited metal maintains its desired characteristics. When using high-melting-point alloys, dilution of the weld metal is usually kept well below 15%.

Hardfacing by the arc welding method has many advantages, including high rates of deposition, flexibility of operation, and ease of mechanization.

Hardfacing may be applied to many types of metals, including low- and medium-carbon steels, stainless steels, manganese steels, high speed steels, nickel alloys, white cast iron, malleable cast iron, gray and alloy cast iron, brass, bronze, and copper.

Quality of Surfacing Deposit

The type of service to which a part is to be exposed governs the degree of quality required of the surfacing deposit. Some applications may require that the deposited metal contain no pinholes or cracks. For other applications, these requirements are of little importance. In most cases, the quality of the deposited metals can be very high. Steel-base alloys do not tend to crack, while other materials, such as high-alloy cast steels, are subject to cracking and porosity.

Hardfacing Electrodes

The proper type of surfacing electrode must be selected as one type of electrode will not meet all requirements. Most electrodes are sold under manufacturers' trade names.

Electrodes may be classified into the following three general groups:

- Resistance to severe abrasion
- Resistance to both impact and moderate abrasion
- Resistance to severe impact and moderately severe abrasion

Tungsten carbide and chromium electrodes are included in the first group. The material deposited is very hard and abrasive-resistant. These electrodes can be one of two types, either coated tubular or regular coated cast alloy. The tubular types contain a mixture of powdered metal, powdered ferroalloys, and fluxing materials. The tubes are the coated type. These electrodes are used with the metallic arc.

Electrodes contain small tungsten carbide crystals embedded in the steel alloy. After this material is applied to a surface, the steel wears away with use, leaving the very hard tungsten carbide particles exposed. This wearing away of steel results in a self-sharpening ability of the surfacing material. Cultivator sweeps and scraping tools are among parts that are surfaced with this material, Figure 27-41.

Chromium carbide electrodes are tougher than tungsten carbide-type electrodes. However, chromium carbide electrodes are not as hard and are less abrasion-resistant. This material is too hard to be machined. But it has good corrosion-resistant qualities.

The electrodes in the second group are the high-carbon type. When used for surfacing, these electrodes leave a tough and very hard deposit. Examples of hard-faced products in this group include gears, tractor lugs, and scraper blades.

The third group of electrodes is used for surfacing rock-crusher parts, links, pins, railroad track components, and parts where severe abrasion resistance is a requirement, Figures 27-42 and 27-43. Deposits from these electrodes are very tough but not hard. It is this quality that seems to work-harden the hardfacing material but leaves the material underneath in a softened condition. Therefore, cracking generally is not a problem.

Shielded Metal Arc Method

1. Start the process by cleaning the surface.
2. Since most hardfacing electrodes are too fluid for out-of-position welding, the work should be arranged in the flat position.
3. Set the amperage so that just enough heat is provided to maintain the arc. Too much heat will cause excessive dilution.
4. Hold a medium-long arc, using either a straight or welding pattern. When a thin bead is required, use

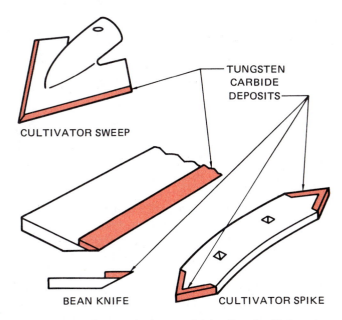

Figure 27-41 Farm tools that can be hardfaced with tungsten carbide electrodes to increase the life of the tools.

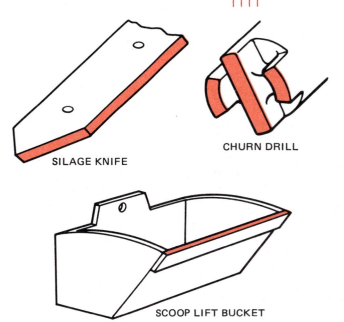

Figure 27-42 Products that are hardfaced to produce moderate impact resistance and severe abrasion resistance.

Figure 27-43 Bearing surfaces are built up on this crankshaft. (Courtesy of Hobart Brothers Company.)

the weave pattern and keep the weave to a width of 3/4 in. (19 mm).

5. If more than one layer is required, remove all slag before placing other layers.

Hardfacing with Gas Shielded Arc

GTA, GMA, and FCA welding processes may be used in hardfacing operations. These three processes, in many instances, are better methods of hardfacing because of the ease with which the metal can be deposited. In addition, the hardfacing materials may be deposited to form a porosity-free, smooth, and uniform surface.

Where the job calls for cobalt-base alloys, the GTA method does an effective job. Very little preheating of the base is required. The GMA and FCA welding processes are somewhat faster than surfacing by GTA due to the fact that continuous wire is used.

Care must be exercised when using the GMA, FCA, and GTA welding processes for hardfacing in order to avoid dilution of the weld. Helium or a mixture of helium-argon normally produces a higher arc voltage than pure argon. For this reason, the dilution of the weld metal increases. An argon and oxygen mixture should be used for surfacing with the gas metal arc processes, and argon with gas tungsten arc processes. When using FCAW, shielding may be provided as either shielding gas or self-shielding. The self-shielding hardsurfacing process is used when working outdoors because of its ability to better resist the effect of light winds.

Carbon Arc Method

1. Clean the surface of all rust, scale, dirt, or other foreign particles. Place the workpiece in a flat position.

2. Using a commercial hardfacing paste, spread the paste evenly over the area to be surfaced. Allow the paste to dry somewhat.

3. Regulate the heat carefully so that just enough heat is provided to obtain a free-flowing molten pool. Too high a welding current tends to create an excessive dilution of the base metal, thus lowering the hardness of the finished surface.

4. The carbon electrode should be moved in a circular motion. When sufficient heat is obtained, the paste will melt and fuse with the base metal.

THERMAL SPRAYING (THSP)

Thermal spraying is the process of spraying molten metal onto a surface to form a coating. Pure or alloyed metal is melted in a flame or electric arc and atomized by a blast of compressed air. The resulting fine spray builds up on a previously prepared surface to form a solid metal coating. Because the molten metal is accompanied by a blast of air, the object being sprayed does not heat up very much. Therefore, thermal spraying is known as a "cold" method of building up metal, **Figure 27-44.**

Thermal Spraying Equipment

A thermal spraying installation requires, at a minimum, the following equipment: air compressor, air control unit, air flowmeter, oxyfuel gas or arc equipment, and exhaust equipment, **Figure 27-45.**

Figure 27-44 Rebuilding a worn crankshaft bearing with a Mogul Turbo-jet thermal spraying gun. (Courtesy of Metallizing Company of America, Inc.)

The three main types of guns available for use in the thermal spraying process are wire guns, powder guns, and crucible guns.

Wire Guns. The wire gun uses metal in the form of wire. Wire sizes range in diameter from 20 gauge to 3/16 in. (4.8 mm). These guns can spray from 4 lb (2 kg) to 12 lb (6 kg) of metal per hour.

The flame gun consists of the following parts: (1) an oxyfuel gas torch with a hole for wire through the center of the tip, (2) a high-speed turbine that drives a pair of knurled rolls equipped with reduction gears to feed wire into the flame at the correct rate, and (3) an "air cap" that encloses the tip of the torch and directs a blast of air to pick up and project the fine molten particles against the workpiece, as shown in **Figure 27-46.** Oxyacetylene gas is most commonly used for the oxyfuel flame. However, other fuel gases such as hydrogen, propane, or natural gas may be used.

The Thermospray (Powder) Process

During the past several years, the development of equipment, materials, and methods has greatly broadened the scope of powder spraying. Now a wide range of alloys and ceramics can be applied at speeds and costs that are economically feasible.

The wire type of flame spray equipment is limited to those materials that can be formed into wires or rods. In contrast, the thermospray process permits the use of metals, alloys, and ceramics in powder form.

The Thermospray Gun. This type of gun, **Figure 27-47** (page 698), usually requires no air. Two lightweight hoses are used to supply oxygen and fuel gas. The powder is fed from a reservoir attached directly to the gun, thus eliminating separate hoppers and hoses. A small reservoir

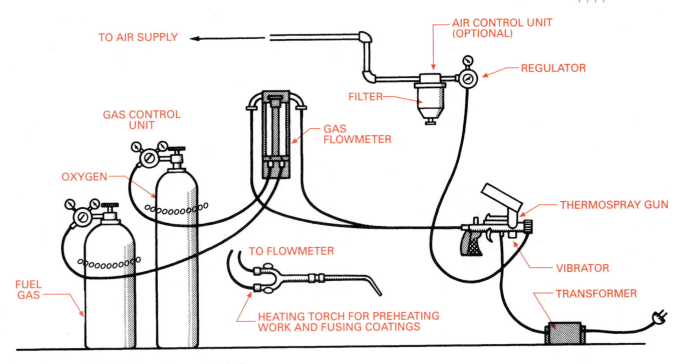

Figure 27-45 Complete thermal spray installation.

is attached to the gun for hand use, and a larger one is provided for lathe-mounted guns or for large-scale production work. An air cooler may be attached to the gun to reduce overheating of small work or thin sections. Extensions are available for coating inside diameters. A trigger-actuated vibrator, which is attached to the gun, is used with ceramic powders and with some metal powders.

By changing powder orifices, any required powder feed may be obtained. This permits spraying the entire range of metals, alloys, ceramics, and cements that can be applied by the thermospray process. Several different types of nozzles can be used for various purposes.

Acetylene is generally used, but hydrogen is required for some applications. The gun may be used as a torch for preheating light work. For large work and for the fusing of coatings of self-fusing alloys, high-capacity acetylene torches are used.

Torch Spraying. A spray metal torch is used for spraying small parts. The torch is made up of a hopper and a spray control mechanism that can be attached to any conventional acetylene torch. Metal powder is placed in the hop-

Figure 27-46 Wire-type thermal spray gun used to resurface a worn roller.

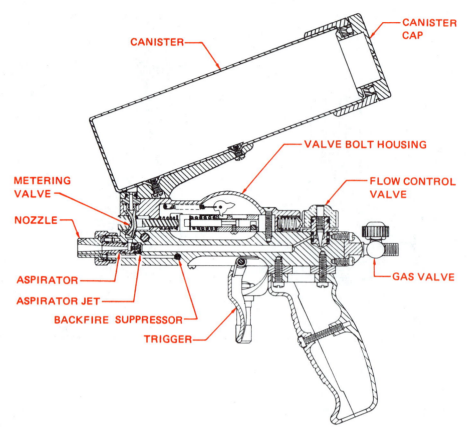

Figure 27-47 Schematic of Thermospray Gun for applying metals, ceramics, and hardfacing alloys.

per. The torch is lit and is then moved over the part to create the overlay, **Figure 27-48**. The powder passes into the acetylene flame, which converts it to a fluid state.

Applying Sprayed Metal. Metal that is sprayed is usually applied in layers less than .010 in. (0.25 mm) thick. Each layer is applied to a surface and bonds to the preceding layer. Greater reliability can be obtained by not trying to lay a heavy coating in one single pass.

Mechanized operation can usually be accomplished by mounting the workpiece in a lathe. The gun is then mounted on the tool post and traverses the workpiece by mechanized operation.

Plasma Spraying Process

Plasma is the term used to describe vapors of materials that are raised to a higher energy level than the ordinary gaseous state. Whereas gases consist of separate molecules, plasma consists of these same gases, which have been broken up and dissociated into ions and electrons.

The plasma spraying process (PSP), **Figure 27-49**, makes use of a spray gun that uses an electric arc contained within a water-cooled jacket. The plasma flame permits the selection of an inert or chemically inactive gas for the flame medium so that oxidation can be controlled during the application of the spray material.

The powder is fed into the plasma flame through the side of the nozzle. The high velocity of the flame propels the powder toward the surface to be coated. As this occurs, the ions and electrons of the plasma are recombining into atoms and releasing energy as heat. This heat is sufficient to melt the powder.

Coatings resulting from this process add extra life and superior resistance to heat, wear, and erosion on parts and products of almost any base material.

Thermal spraying techniques are used in applications that involve wear and high-temperature problems. Typical applications include missile nose cones, rocket nozzles, jet turbine cases, electrical contacts, jet engine burner cam clamps, and many aircraft parts.

COLD WELDING (CW)

This is a solid state process of welding. There is no heating or melting of the metal that forms the bond in this process. The weld takes place at room temperature.

The coalescence of the metal surfaces occurs as a result of the force applied. It's possible to join most soft, ductile metals using this method. Also dissimilar metals such as aluminum to copper, iron to copper, etc. can be joined.

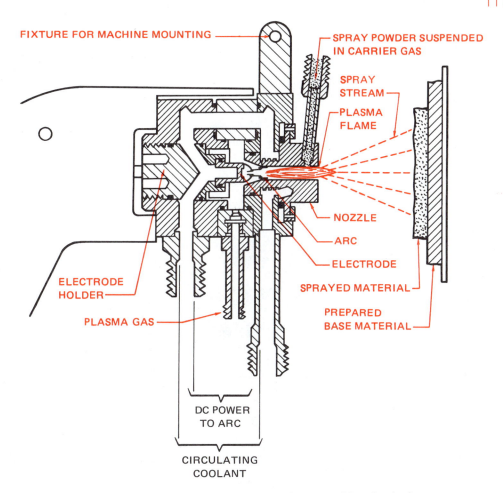

FIXTURE FOR MACHINE MOUNTING

SPRAY POWDER SUSPENDED IN CARRIER GAS

SPRAY STREAM

PLASMA FLAME

NOZZLE

ARC

ELECTRODE

SPRAYED MATERIAL

PREPARED BASE MATERIAL

ELECTRODE HOLDER

PLASMA GAS

DC POWER TO ARC

CIRCULATING COOLANT

Figure 27-48 Schematic of thermal spraying gun. (Courtesy of Metallizing Company of America, Inc.)

Figure 27-49 Internal ceramic coating with Mogul Rokide gun, extension, and angular air cap. (Courtesy of Metallizing Company of America, Inc.)

Early cold welding was primarily done as soft welds, but today it's possible to make lapp, edge, and butt joints. Spot welds can be made using portable, hand-operated tools, **Figure 27-50** (page 700). Other types of joints require special machines.

A major factor in the success of the cold weld is that the surface oxides and other contaminations be removed before welding. The best method to clean the surface is with a power wire brush. The surface must not be touched after it's cleaned, and the welding should be done as soon as possible.

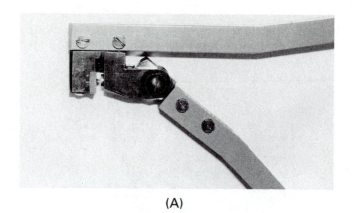

(A)

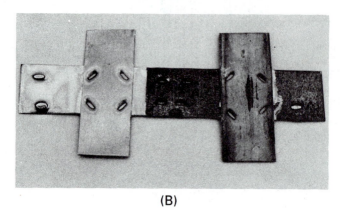

(B)

Figure 27-50 (A) Cold welding tool, (B) cold welds made between aluminum copper and mild steel.

Small Diameter Flux Cored Wire Saves Over-welding Costs at a Boatyard

Figure 1

A boat-building company operates three yards totaling 50 acres. Commercial fishing vessels, Coast Guard ice breakers, high-speed naval craft, ocean incinerator ships, and semisubmersible drilling rigs are but a few of the projects that have passed through the 225,000 sq ft enclosed shops and the found end launchways at the company.

A contract for two high-performance gunboats gave the welding manager a chance to look into the problem almost every fabricator has run into — overwelding (laying down more weld metal than is structurally required).

The 250-ft (76.2-m) vessels are designed for multi-mission purposes, which include long-range patrol, interdiction, or pursuit. Two of the requirements for such vessels are light weight and speed. That is why the design specifications called for steel not exceeding 1/4 in. (6.4 mm) in thickness. The material used is 1/8, 3/16, and 1/4 in. (3.2, 4.8, and 6.4 mm) A36 mild steel with limited amounts of ABS Class EH36 in the hull. Because of the thinness of the steel, 1/8-in. (3.2-mm) fillets were specified. The problem was to keep the weld size down to 1/8 in. (3.2 mm).

A 0.045-in. (1.14-mm) wire was used. Although the welders did their best, the welds were consistently overwelded up to 1/4 to 5/16 in. (6.4 to 7.9 mm).

The welding manager began to check into the negative aspects of overwelding. His findings reinforced his intuition that overwelding cuts deeply into productivity and adds greatly to the cost of welding. The weight of the 1/4-in. (6.4-mm) fillet is four times greater than that of 1/8-in. (3.2-mm) fillet. Not only is weld metal wasted, but it takes longer to lay down the unnecessarily large fillet. An additional waste factor to be considered is that of distortion. Especially with the thinner material required on the gunboat project, the overwelded fillets with their greater heat input were causing plate distortion. Time was required to flame-straighten the distorted material.

Feeling the FCAW process is faster, easier to use, and results in better workmanship, the welding manager converted over 90% of the ferrous metal joining to that method. The solution to the overwelding problem was the use of a small diameter (0.035 in., 0.89 mm) flux cored wire. It is a mild steel flux cored wire that combines both strength and low temperature toughness in an ultra small diameter. **Figure 1** shows a typical bulkhead weldment.

The welding manager estimated that although the smaller diameter welding wire is more expensive, money would be saved by cutting overwelding, **Figure 2**. It is estimated that 150,000 linear feet (45,720 m) of welds will be needed for each boat. One foot of 1/4-in. (6.4-mm) fillet weighs 0.106 lb (48.2 g), while a 1/8-in. (3.2-mm) fillet weighs only 0.027 lb (12.3 g) per foot. Simple multiplication shows that 15,900 lb (7,212 kg) of weld metal is deposited with a 1/4-in. (6.4-mm) fillet and only 4,050 lb (1,837 kg) with the 1/8-in. (3.2-mm) fillet. The difference demonstrates that for one boat alone, 11,850 lb (5,375 kg) of excess weld metal would be purchased and deposited for no reason, if overwelding were allowed.

Before using the 0.035-in. (0.89-mm) flux cored wire in the fabrication of the vessels, it was qualified to ABS requirements. In the course of this qualification, the wire had to maintain soundness when welded over primer. Good mechanical properties were an additional advantage of the wire.

Potential Savings Through Reduced Overwelding*

**Estimated 150,000 Linear
Feet of Welding for Each Boat**

¼ in. fillet = 0.106 lb/ft
⅛ in. fillet = 0.027 lb/ft

150,000 × 0.106 = 15,900 lb of weld metal per boat
150,000 × 0.027 = 4050 lb of weld metal per boat

**Flux Cored Wire
Estimated 85% Efficient**

$15,900 = 0.85(x)$

$\dfrac{15,900}{0.85} = x$

x = 18,705 total lb of filler metal needed to deposit ¼ in. fillet

$\dfrac{4050}{0.85} = x$

x = 4765 total lb of filler metal needed to deposit ⅛ in. fillet

**0.035 in. Flux Cored Electrode
Estimated at $2 per lb Material Cost**

18,705 × $2 = $37,410
4765 × $2 = $9530

$37,410
− 9,530   Potential savings per boat in filler metal cost
$27,880

$27,880 × 2 = $55,760*   *Total potential savings* on the project in
filler metal costs by reducing overwelding

*These savings are calculated in terms of filler metal cost only. Substantial *additional* savings would also result from the reduction in welding labor and overhead expense.

Figure 2

Welding is performed in all positions with approximately 35% of it out of position. A 75% argon 25% CO_2 shielding gas is used. Recommended current settings for the 0.035-in. (0.89-mm) diameter flux cored wire range as high as 30 V and 280 A for flat welding, and as low as 17 V and 100 A for vertical up-welding. Inspection standards are the same as those required by the U.S. Navy.

Approximately 150 x-rays are taken on the welds of each vessel.

Overwelding is being kept to a minimum on this job using the 0.035-in. (0.89-mm) flux cored wire. The result is reduced cost, while quality is maintained.

(Adapted from information provided by Alloy Rods Corporation, Hanover, Pennsylvania 17331.)

REVIEW

1. What generates the heat for fusion in resistance welding?

2. What can be used to produce the force needed to hold the work together for RWs?

3. What are the basic resistance welding processes?

4. What steps can be included in RSW?

5. What are the three types of spot welding machines in common use?

6. What are some of the uses for a portable spot welder?

7. How is seam welding similar to spot welding?

8. What is the most common joint for seam welds?

9. What is RSEW-HF used for?

10. What metals are readily welded by the RPW process?

11. Why is FW not usually cost effective for short production runs?

12. What is the main difference between upset welding and flash welding?

13. What unusual metal combination can be welded using percussion welding?

14. What is the shape and characteristics of an EB weld?

15. What is the least amount of vacuum that a part must be subjected to in order to make an EB weld?

16. How is the beam focus changed in the EB welding process?

17. How can a misaligned seam be track automatically for EB welds?

18. What types of welds are most commonly made with the US welding process?

19. What equipment is needed to make US welds?

20. What are the typical applications for US welding?

21. List the steps of the inertia welding process.

22. Referring to **Table 27-1** (page 689), what metals can be joined by inertia welding
    a. to 1100 or 6061 aluminum?
    b. to pure titanium?

23. How does a laser drill holes?

24. What difficult metals can be welded with a laser?

25. How is the laser light released from the ruby rod?

26. How is the plasma arc stiffened?

27. How does stud welding work?

28. What types of wear can hardfacing protect against?

29. What base metals are used for most hardfacing alloys?

30. What are the advantages of oxyacetylene flame hardfacing?

31. Why should there be as little dilution as possible of the base metal when hardfacing?

32. How can wear provide a self-sharpening effect on some hardfaced parts?

33. What hardfacing allow is best applied with the gas tungsten arc welding process?

34. Why is THSP know as a cold buildup process?

35. Which thermal spray process can be used to apply ceramics?

36. Why should thermal spray coats be applied as thin coats?

37. What is the advantage of using an inert gas for plasma spraying?

38. What causes the coalescence during a cold weld?

Section Seven

OXYFUEL

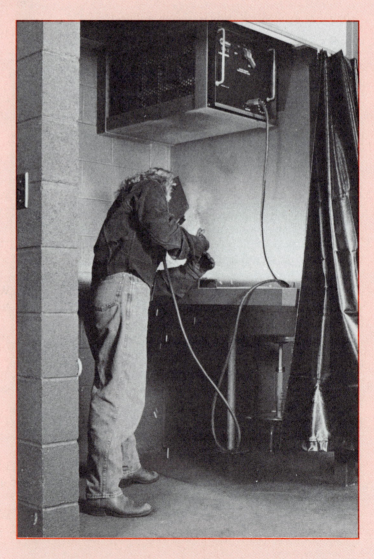

SUCCESS STORY

Courtney Lee Newcomer was only 14 years old when her father taught her to cut and weld for a school art project. She used an oxyacetylene torch to create a perfectly balanced mobile of horses, roughly cut from mild steel plate.

It was not until almost ten years later that Courtney had another opportunity to use a torch. She redeveloped her skills and created her first series of sculptures: a life-size horse, three hounds, and two foxes. Satisfied with her work, she was proud to lend the pieces to a local equestrian club as party decorations. Imagine her surprise when they were all sold, she received requests for more of her work, and she was invited to exhibit at a local art fair! This series of events opened Courtney's eyes to the real possibility of a career as a torch wielder and welder—as a steel sculptor.

Encouraged and enabled by her initial successes, Courtney purchased a plasma arc cutter and a wire welder. The capabilities of this new equipment revolutionized her work by allowing her to create pieces with much finer detail and to keep up with a rapidly expanding business.

Cutting and welding will always be an important part of Courtney's creative process. She insists on doing this work herself to control the character of each sculpture. The whimsical tilt of a cat's head, the amount of bend in a dog's floppy ear, and the flow of a horse's mane and tail as he runs, all are governed by Courtney's artistic eye.

Courtney has spent the last five years since CLN Metal Art, her own business, was established experimenting, creating new pieces, and expanding both her inventory and clientele. She participates in four to five art fairs and several gallery shows each year, and she displays her art permanently in galleries and retail shops across the Midwest. Her success can be attributed to her direct involvement with production and the enjoyment and personal satisfaction she finds in turning a lifeless sheet of steel plate into an animated creature.

Chapter 28

OXYFUEL WELDING AND CUTTING EQUIPMENT, SETUP, AND OPERATION

OBJECTIVES

After completing this chapter, the student should be able to:

- describe how to maintain the major components of oxyfuel welding equipment.
- explain the method of testing an oxyfuel system for leaks.
- demonstrate how to set up, light, adjust, extinguish, and disassemble oxyfuel welding equipment safely.

KEY TERMS

regulators	leak-detecting solution
diaphragm	oxyfuel gas torch
two-stage regulators	combination welding and cutting torch
regulator gauges	cutting torches
working pressure	mixing chamber
cylinder pressure	injector chamber
line drop	valve packing
Bourdon tube	flashback arrestor
atmospheric pressure	Siamese hose
gauge pressure	purge
absolute pressure	crack
safety release valve	sparklighter
safety disc	manifold system
seat	back pressure release
creep	

INTRODUCTION

Oxyfuel welding, cutting, brazing, hardsurfacing, heating, and other processes use the same basic equipment. When storing, handling, assembling, testing, adjusting, lighting, shutting off, and disassembling this basic equipment, the same safety procedures must be followed for each process. Improper or careless work habits can cause serious safety hazards. Proper attention to all details makes these processes safe.

Certain basic equipment is common to all gas welding. Cylinders, regulators, hoses, hose fittings, safety valves, torches, and tips are some of the basic equipment used. Although numerous manufacturers produce a large variety of gas equipment, it all works on the same principle. When welders are not sure how new equipment is operated, they should seek professional help. A welder should never experiment with any equipment.

All oxyfuel processes use a high-heat, high-temperature flame produced by burning a fuel gas mixed with pure oxygen. The gases are supplied in pressurized cylinders. The gas pressure from the cylinder must be reduced by using a regulator. The gas then flows through flexible hoses to the torch. The torch controls this flow and mixes the gases in the proper proportion for good combustion at the end of the tip.

Acetylene is the most widely used fuel gas, but about twenty-five other gases are available. The tip is usually the only equipment change required in order to use another fuel gas. The adjustment and skill required are often different, but the storage, handling, assembling, and testing are the same. When changing gases, make sure the tip can be used safely with the new gas.

PRESSURE REGULATORS

All **regulators** do the same type of job. That is, they all work on the same principle whether they are low or high pressure; single or multiple stage; cylinder, manifold, line, or station regulator; and regardless of the type of gas they regulate. The regulator reduces a high pressure to a lower, working pressure. The lower pressure must be held constant over a range of flow rates.

■ CAUTION

Although all regulators work the same way, they cannot be safely used interchangeably on different types of gas or for different pressure ranges without the possibility of a fire or an explosion.

Operation

A regulator works by holding the forces on both sides of a **diaphragm** in balance, Figure 28-1. As the adjusting screw is turned inward, it increases the force of a spring on the flexible diaphragm and bends the diaphragm away. As the diaphragm is moved, the small high-pressure valve is opened, allowing more gas to flow into the regulator. The gas pressure cancels the spring pressure, and the spring bends the diaphragm back to its original position, closing the high-pressure valve, **Figure 28-2** (page 708). When the regulator is used, the gas pressure on the back side of the diaphragm is reduced, the spring again forces the valve open, and gas flows. The drop in the internal pressure can be seen on the working pressure gauge, Figure 28-3 (page 708).

The size of a regulator determines its ability to hold the working pressure constant over a wider range of flow rates. **Two-stage regulators,** Figure 28-4 (page 709), are able to keep the pressure constant at very high flow rates as the cylinder empties. This type of regulator has two sets of springs, diaphragms, and valves. The first spring is preset to reduce the cylinder pressure to 225 psig (1,550 kPag). The second spring is adjusted like other regulators. Because the second high-pressure valve has to control a maximum pressure of only around 225 psig (1,550 kPag), it can be larger, thus allowing a greater flow.

Regulator Gauges

There may be one or two gauges on a regulator. One **regulator gauge** shows the working pressure, and the other indicates the cylinder pressure, Figure 28-5 (page 709). The **working pressure** gauge shows the pressure at the regulator and not at the torch. The torch pressure is always less than the working pressure shown on the gauge because of line drop. The smaller in diameter or longer a line is, the greater the drop will be, Table 28-1

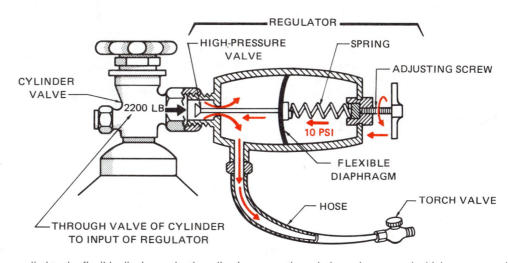

Figure 28-1 Force applied to the flexible diaphragm by the adjusting screw through the spring opens the high-pressure valve.

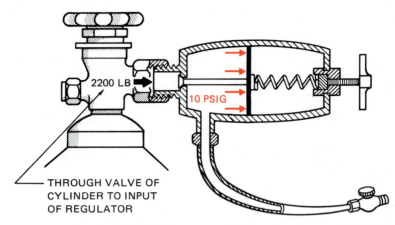

Figure 28-2 When the gas pressure against the flexible diaphragm equals the spring pressure, the high-pressure valve closes.

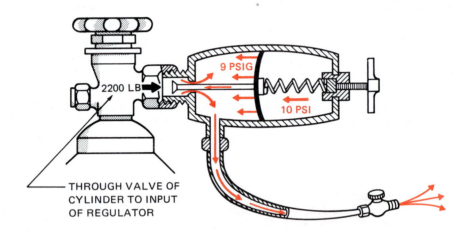

Figure 28-3 A drop in the working pressure occurs when the torch valve is opened and gas flows through the regulator at a constant pressure.

Tip Pressure psig (kg/cm²G)	Regulator Pressure* for Hose Lengths ft (m)				
	10 ft (3 m)	25 ft (7.6 m)	50 ft (15.2 m)	75 ft (22.9 m)	100 ft (30.5 m)
1　(0.1)	1　(0.1)	2.25　(0.15)	3.5　(0.27)	4.75　(0.35)	6　(0.4)
5　(0.35)	5　(0.35)	6.25　(0.4)	7.5　(0.52)	8.75　(0.6)	10　(0.7)
10　(0.7)	10　(0.7)	11.25　(0.75)	12.5　(0.85)	13.75　(0.95)	15　(1.0)

*These values are for hose with a diameter of 1/4 in (6 mm); larger or smaller hose diameters or high flow rates will change these pressures.

Table 28-1 Regulator Pressure for Various Lengths of Hose.

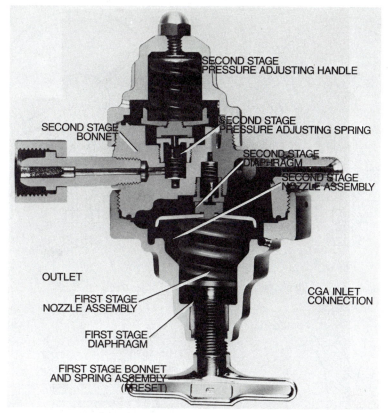

Figure 28-4 Two-stage oxygen regulator. (Courtesy of Alphagaz Liquid Air Corporation.)

Figure 28-5 Safety release valve on an oxygen regulator. (Courtesy of Victor Equipment Company.)

(page 708). The line drop is caused by the resistance of a gas as it flows through a line.

EXPERIMENT 28-1

Line Resistance

In this experiment, two pieces of the same diameter hose are required. One piece of hose is to be a short length, less than 10 feet (3 m) long. The other piece of hose is to be more than 25 feet (8 m) long. Blow through the short piece and then blow through the long piece. Observe the difference. ◆

The high-pressure gauge on a regulator shows cylinder pressure only. This gauge may be used to indicate the amount of gas that remains in a cylinder. However, cylinders containing liquified gases, such as CO_2, propane, and MPS, must be weighed to determine the amount of gas remaining.

Inside a regulator gauge there is a Bourdon tube. This tube is bent in the shape of the letter C, with one end attached solidly to the gauge body and the other end attached through a gear to a needle, **Figure 28-6** (page 710). As the pressure inside the tube increases, the tube tries to straighten out.

The pressure shown on a gauge is read as pounds per square inch gauge (psig) or kilopascals (kPag). The atmospheric pressure, 14.7 psi (101.35 kPa), must be

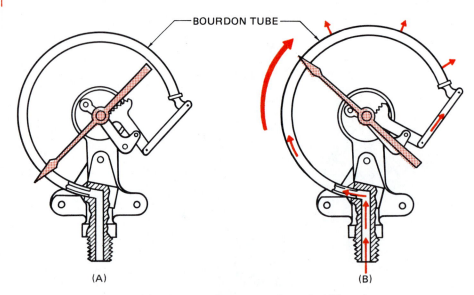

Figure 28-6 Gauge before pressure is applied (A) and gauge after pressure is applied (B).

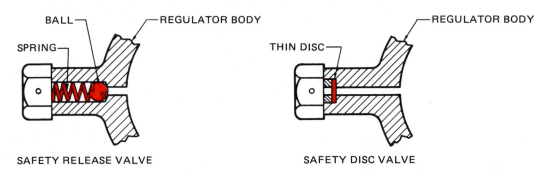

Figure 28-7 Pressure release valves.

added to the **gauge pressure** to find the **absolute pressure,** psia (kPaa). In welding, psig and psi (kPag and kPa) are used interchangeably.

Safety Release Device

Regulators may be equipped with either a **safety release valve** or a **safety disc** to prevent excessively high pressures from damaging the regulator. A safety release valve is made up of a small ball held tightly against a **seat** by a spring. The release valve will reseat itself after the excessive pressure has been released. A safety disc is a thin piece of metal held between two seals, **Figure 28-7.** The disc will burst to release excessive pressure. The disc must then be replaced before the regulator can be used again.

Fittings

A variety of inlet or cylinder fittings are available to ensure that the regulator cannot be connected to the wrong gas or pressure, **Figures 28-8(A)** through **28-8(D).** A few adapters are available that will allow some regulators to be attached to a different type of fitting. The two most common types: (1) adapt a left-hand male acetylene

cylinder fitting to a right-hand female regulator fitting, or vice versa, and (2) adapt an argon or mixed gas male fitting to a female flat washer-type CO_2 fitting, **Figure 28-9.**

The connections to the cylinder and to the hose must be kept free of dirt and oil. Fittings should screw together freely by hand and only require light wrench pressure to be leak tight. If the fitting does not tighten freely on the connection, both parts should be cleaned. If the joint leaks after it has been tightened with a wrench, the seat should be checked. Examine the seat and threads for damage. If the seat is damaged, it can be repaired with a reamer specially made for this purpose, **Figure 28-10.** The threads can be repaired by using a die or thread file, **Figure 28-11** (page 712). Care should be taken not to gouge or score the soft brass when repairing a connector. Severely damaged connections must be replaced.

Safety Precautions

If the adjusting screw is not loosened when the work is completed, and the cylinder valve is turned off, it is possible that long-term damage to the diaphragm and springs may result. In addition, when the cylinder valve is reopened, some high-pressure gas can pass by

Figure 28-8(A) Acetylene cylinder valve (left-hand thread).

Figure 28-8(B) Oxygen cylinder valve.

Figure 28-8(C) Argon cylinder valve.

Figure 28-8(D) Carbon dioxide (CO_2) cylinder valve.

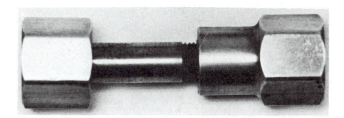

Figure 28-9 Carbon dioxide-to-argon adapter.

Figure 28-10 A reamer is used to repair damaged torch seats.

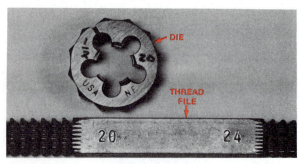

Figure 28-11 The tools used to repair damaged threads.

the open high-pressure valve before the diaphragm can close it. This condition may cause the diaphragm to rupture or the low-pressure gauge to explode, or both. High-pressure valve seats that leak will result in a creep or rising pressure on the working side of the regulator. If the leakage at the seat is severe, the maximum safe pressure can be exceeded on the working side, resulting in damage to the diaphragm, gauge, hoses, or other equipment.

■ *CAUTION*

Regulators that creep should not be used.

A diaphragm can be tested for leaks by first setting the regulator to 14 psig (95 kPag) for fuel gases or 45 psig (310 kPag) for oxygen and other gases. Once the pressure is set, place a finger over the vent hole and spray it with a leak-detecting solution, Figure 28-12. Slowly move the finger from the hole and watch for bubbles, which indicate a leak.

A gauge that gives a faulty reading or that is damaged can result in dangerous pressure settings. Gauges that do not read "0" (zero) pressure when the pressure is released, or those that have a damaged glass or case, must be repaired or replaced.

■ *CAUTION*

All work on regulators must be done by properly trained repair technicians.

■ *CAUTION*

Regulators should be located far enough from the actual work that flames or sparks cannot reach them.

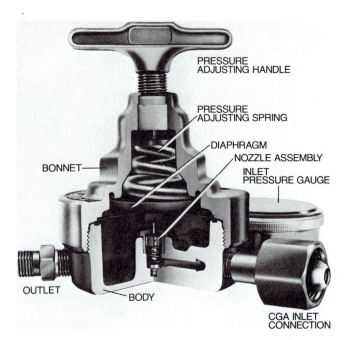

Figure 28-12 Single-stage oxygen regulator. (Courtesy of Alphagaz Liquid Air Corporation.)

The outlet connection on a regulator is either a right-hand fitting for oxygen or a left-hand fitting for fuel gases. A left-hand threaded fitting has a notched nut, **Figure 28-13**.

Regulator Care and Use

There are no internal or external moving parts on a regulator or a gauge that require oiling, **Figure 28-14**.

■ *CAUTION*

Oiling a regulator is unsafe and may cause a fire or an explosion.

Figure 28-13 Left-hand threaded fittings are identified with a notch.

Figure 28-14 Never oil a regulator. (Courtesy of Air Products and Chemicals, Inc., Material Sciences Division.)

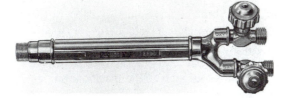

Figure 28-15 A torch body or handle used for welding or cutting. (Courtesy of Victor Equipment Company.)

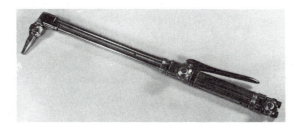

Figure 28-16 A torch used for cutting only.

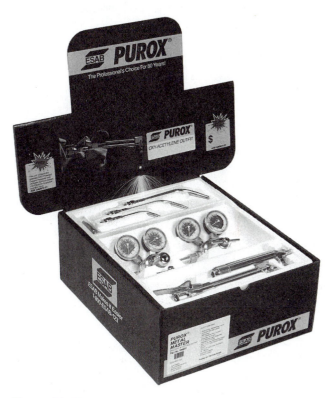

Figure 28-17 A combination welding and cutting torch kit. (Courtesy of ESAB Welding and Cutting Products.)

If the adjusting screw becomes tight and difficult to turn, it can be removed and cleaned with a dry, oil-free rag. When replacing the adjusting screw, be sure it does not become cross-threaded. Many regulators use a nylon nut in the regulator body, and the nylon is easily cross-threaded.

When welding is finished and the cylinders are turned off, the gas pressure must be released and the adjusting screw backed out. This procedure is done so that the diaphragm, gauges, and adjusting spring are not under load. A regulator that is left pressurized causes the diaphragm to stretch, the Bourdon tube to straighten, and the adjusting spring to compress. These changes result in a less accurate regulator with a shorter life expectancy.

WELDING AND CUTTING TORCHES DESIGN AND SERVICE

The oxyacetylene hand torch is the most common type of oxyfuel gas torch used in industry. The hand torch may be either a combination welding and cutting torch or a cutting torch only, **Figures 28-15** and **28-16**.

The combination welding and cutting torch offers more flexibility because a cutting head, welding tip, or heating tip can be attached quickly to the same torch body, **Figure 28-17**. Combination torch sets are often used in schools, automotive repair shops, auto body shops, small welding shops, or in any other situation where flexibility is needed. The combination torch sets usually are more practical for portable welding since the one unit can be used for both cutting and welding.

Straight cutting torches are usually longer than combination torches. The longer length helps keep the operator farther away from heat and sparks. In addition, thicker material can be cut with greater comfort.

Most manufacturers make torches in a variety of sizes for different types of work. There are small torches for jewelry work, **Figure 28-18** (page 714), and large torches for heavy plates. Specialty torches for heating, brazing, or soldering are also available. Some of these use a fuel-air mixture, **Figure 28-19** (page 714). Fuel-air torches are often used by plumbers and air-conditioning technicians for brazing and soldering copper pipe or tubing. There are no industrial standards for tip size identification, tip threads, or seats. Therefore, each style, size, and type of torch can be used only with the tips made by the same manufacturer to fit the specific torch.

Mixing the Gases

Two basic methods are used for mixing the oxygen and fuel gas to form a hot, uniform flame. The two gases must be mixed completely before they leave the tip and create the flame. If the gases are not mixed completely, the torch will have a greater tendency to backfire or flashback. One method uses equal or balanced pressures, and the gases are mixed in a **mixing chamber.** The other method uses a higher oxygen pressure, and the gases are mixed in an **injector chamber.**

The mixing chamber of the equal-pressure torch may be located in the torch body, attached to the tip, or in the tip, **Figure 28-20.** Both gases must enter the enlarged mixing chamber through small, separate openings. Because the mixing chamber is larger than the total size of

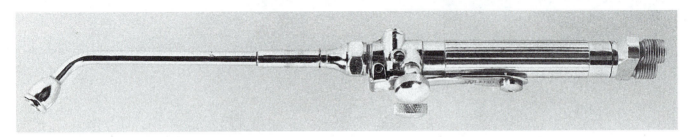

Figure 28-18 Lightweight torch for small, delicate jobs. (Courtesy of National Torch Tip Co.)

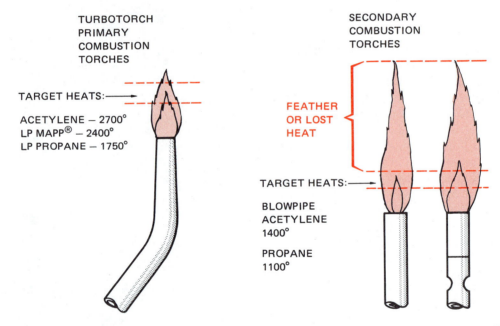

Figure 28-19 Some air/gas torches use a special tip that improves the combustion for a hotter, more effective flame. (Courtesy of Victor Equipment Company.)

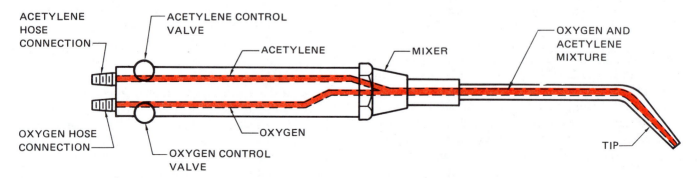

Figure 28-20 Schematic drawing of an oxyacetylene welding torch.

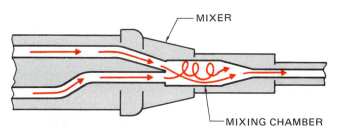

Figure 28-21 Equal-pressure mixing chamber.

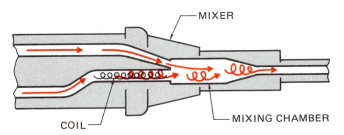

Figure 28-22 A metal coil in the oxygen tube spins the gas, ensuring a complete mixing of gases.

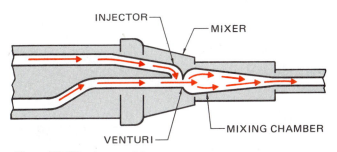

Figure 28-23 Injector mixing system.

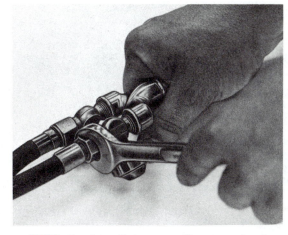

Figure 28-24 One hose-fitting nut will protect the threads when the other nut is loosened or tightened.

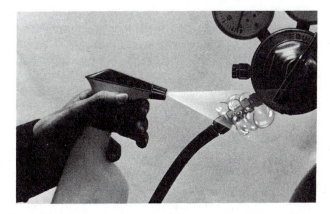

Figure 28-25 Check all connections for possible leaks and tighten if necessary.

both entrance holes and the exit hole, the gases experience a rapid drop in pressure as they enter the chamber. The rapid drop in pressure causes the gases to become turbulent and mix thoroughly, **Figure 28-21**. Some manufacturers spin the gases before they enter the chamber to ensure complete mixing, **Figure 28-22**.

The injector torches work both with equal gas pressures or low fuel-gas pressures, **Figure 28-23**. The injector allows oxygen at the higher pressure to draw the fuel gas into the chamber, even when the fuel gas pressure is as low as 6 oz/in² (26.3 g/cm²). The injector works by passing the oxygen through a venturi, which creates a vacuum to pull the fuel gas in and then mixes the gases together. An injector-type torch must be used if a low-pressure acetylene generator or low-pressure residential natural gas is used as a fuel gas supply.

Torch Care and Use

The torch body contains threaded connections for the hoses and tips. These connections must be protected from any damage. Most torch connections are external and made of soft brass that is easily damaged. Some connections, however, are more protected because they have either internal threads or stainless steel threads for the tips. The best protection against damage and dirt is to leave the tip and hoses connected when the torch is not in use.

Because the hose connections are close to each other, a wrench should never be used on one nut unless the other connection is protected with a hose-fitting nut, **Figure 28-24**.

The hose connections should not leak after they are tightened with a wrench. If leaks are present, the seat should be repaired or replaced. Some torches have removable hose connection fittings so that replacement is possible.

The valves should be easily turned on and off and should stop all gas flowing with minimum finger pressure. To find leaking valve seats, set the regulators to a working pressure. With the torch valves off, spray the tip with a leak-detecting solution. The presence of bubbles indicates a leaking valve seat, **Figure 28-25**. The gas should not leak past the valve stem packing when the valve is open or when it is closed. To test leaks around the valve stem, set the regulator to a working pressure. With the valves off, spray the valve stem with

Figure 28-26 The torch valves should be checked for leaks, and the valve packing nut should be tightened if necessary.

a leak-detecting solution and watch for bubbles, indicating a leaking valve packing. The valve stem packing can now be tested with the valve open. Place a finger over the hole in the tip and open the valve. Spray the stem and watch for bubbles, which would indicate a leaking valve packing, **Figure 28-26.** If either test indicates a leak, the valve stem packing nut can be tightened until the leak stops. After the leak stops, turn the valve knob. It should still turn freely. If it does not, or if the leak cannot be stopped, replace the valve packing.

The valve packing and valve seat can be easily repaired on most torches by following the instructions given in the repair kit. On some torches, the entire valve assembly can be replaced, if necessary.

WELDING AND HEATING TIPS

Because no industrial standard tip size identification system exists, the student must become familiar with the size of the orifice (hole) in the tip and the thickness range for which it can be used. Comparing the overall size of the tip can be done only for tips made by the same manufacturer for the same type and style of torch, **Figure 28-27.** Learning a specific manufacturer's system is not always the answer because on older, worn tips the orifice may have been enlarged by repeated cleaning.

Tip sizes can be compared to the numbered drill size used to make the hole, **Table 28-2.** The sizes of tip cleaners are given according to the drill size of the hole they fit. By knowing the tip cleaner size commonly used to clean a tip, the welder can find the same size tip made by a different manufacturer. The tip size can also be determined by trial and error.

On some torch sets, each tip has its own mixing chamber. On other torch sets, however, one mixing chamber may be used with a variety of tip sizes.

Tip Care and Use

Torch tips may have metal-to-metal seals, or they may have an O ring or gasket between the tip and the

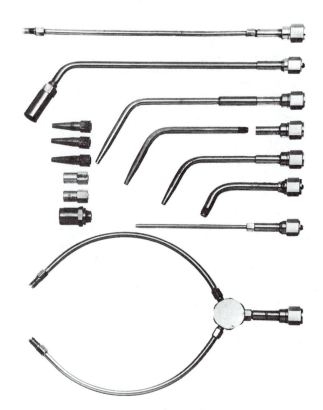

Figure 28-27 A variety of tip styles and sizes for one torch body. (Courtesy of Victor Equipment Company.)

Tip Cleaner Standard Set			
SMALLEST	Use Cleaner	For Drill	77 = .0160″ (0.4064 mm)
	1.	77-76	
	2.	75-74	
	3.	73-72-71	
	4.	70-69-68	
	5.	67-66-65	
	6.	64-63-62	
	7.	61-60	
	8.	59-58	
	9.	57	
	10.	56	
	11.	55-54	
LARGEST	12.	53-52	49 = .0730″ (1.8542 mm)
	13.	51-50-49	

Table 28-2 Tip Cleaner Size Compared to Drill Size Found on Most Standard Tip-Cleaning Sets.

torch seat. Metal-to-metal seal tips must be tightened with a wrench. Tips with an O ring or gasket may be tightened with a wrench or by hand, depending on the manufacturer's recommendations. Using the wrong method of tightening the tip fitting may result in damage to the torch body or the tip.

After a tip is installed, it should be checked for leaks. To test for leaks, set the regulator to a working pressure, turn the torch valve on, and put a finger over the holes in

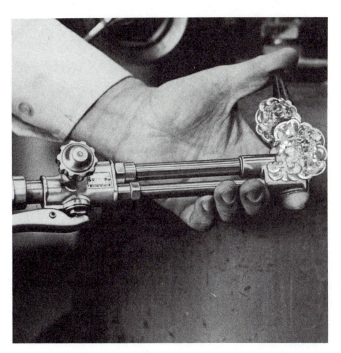

Figure 28-28 Checking a torch tip for a leaking seat. (Courtesy of D. Rhodes.)

Figure 28-29 Standard set of tip cleaners. (Courtesy of Uniweld Products, Inc.)

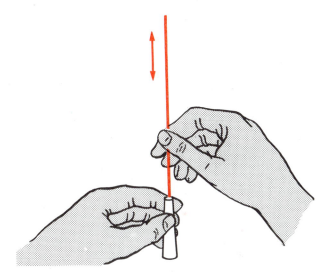

Figure 28-30 Cleaning a tip with a standard tip cleaner.

the end of the tip. Spray a leak-detecting solution around the base of the tip at the fitting and watch for bubbles, indicating a leaking seat, **Figure 28-28**.

If the leaking tip seat has an O ring or gasket, then the O ring or gasket is replaced with a new one. If the leaking tip seat is the metal-to-metal type, then a special reamer is used to smooth the seat.

Used, dirty tips can be cleaned using a set of tip cleaners. A little oxygen flow should be turned on so that dirt loosened during cleaning will be blown away. Using the file provided in the tip-cleaning set, **Figure 28-29**, file the end of the tip smooth and square. Next, select the size of tip cleaner that fits easily into the orifice. The tip cleaner is a small, round file and should only be moved in and out of the orifice a few times, **Figure 28-30**. Be sure the tip cleaner is straight and that it is held in a steady position to prevent it from bending or breaking off in the tip. Excessive use of the tip cleaner tends to ream the orifice, making it too large. Therefore, use the tip cleaner only as required.

Damaged tips or tips with cleaners broken in them can be reconditioned, but they require a good deal of work and some specialized tools, **Figure 28-31** (page 718). To remove broken tip cleaners, push them back out of the tip using a long tip cleaner inserted from the back. If this method does not work, the end of the tip can be filed back while taking care not to push the tip cleaner farther into the tip. When the tip is filed back approximately 1/8 in. (3 mm), a pair of pliers can be used to pull the broken cleaner from the tip, **Figure 28-32** (page 718).

Tips that have had the end burned off or are worn unevenly can be straightened using a tip end reamer.

Holding the reamer square and firmly against the tip, spin the reamer until the tip end is flat and square.

For tip orifices that are worn out of round, or have become tapered, or need a major cleaning, a small set of drills can be used. The proper size drill should be selected and tightened in the chuck. Guide the bit carefully into the tip hole while turning it between your fingers. If the orifice is made too large for its mixing chamber, the gases may not burn correctly, and the tip should therefore be discarded.

REVERSE FLOW AND FLASHBACK VALVES

The purpose of the reverse flow valve is to prevent gases from accidentally flowing through the torch and into the wrong hose. If the gases being used are allowed

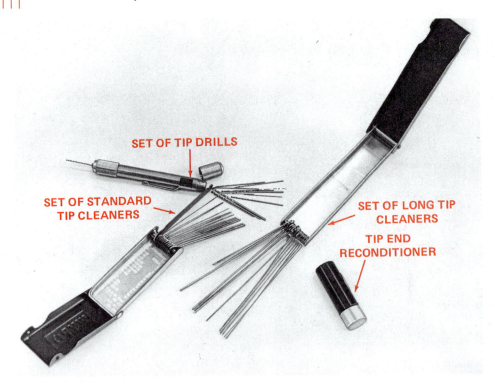

Figure 28-31 Tools used to repair tips.

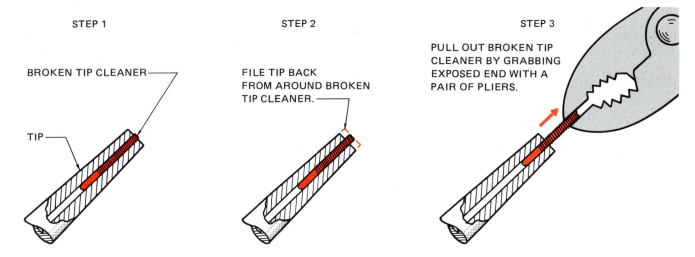

Figure 28-32 Steps that are used to remove a broken tip cleaner.

to mix in the hose or regulator, they might explode. The reverse flow valve is a spring-loaded check valve that closes when gas pressure from a back flow tries to occur through the torch valves, **Figure 28-33**. Some torches have reverse flow valves built into the torch body, but most torches must have these safety devices added. If the torch does not come with a reverse flow valve, they must be added to either the torch end or regulator end of the hose.

A reverse flow of gas will occur if the torch is not turned off or bled properly. The torch valves must be opened one at a time so that the gas pressure in that hose will be vented into the atmosphere and not into the other hose, **Figure 28-34** (page 720).

■ **CAUTION**

If both valves are opened at the same time, one gas may be pushed back up the hose of the other gas.

A reverse flow valve will not stop the flame from a flashback from continuing through the hoses. A flashback arrestor will do the job of a reverse flow valve, and it will stop the flame of a flashback, **Figure 28-35** (page 720). The flashback arrestor is designed to quickly stop the flow of gas during a flashback. These valves work on a similar principle to the gas valve at a service station. They are

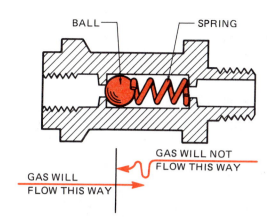

Figure 28-33 Reverse flow valve only. (Courtesy of Airco Welding Products.)

very sensitive to any back pressure in the hose and stop the flow if any back pressure is detected.

Care of the Reverse Flow Valve and Flashback Arrestor

Both devices must be checked on a regular basis to see that they are working correctly. The internal valves may become plugged with dirt or they may become sticky and not operate correctly. To test the check valve, you can try to blow air backwards through the valve. To test the flashback arrestor, follow the manufacturer's recommended procedure. If the safety device does not function correctly, it should be replaced.

HOSES AND FITTINGS

Most welding hoses used today are molded together as one piece and are referred to as Siamese hose. Hoses that are not of the Siamese type, or hose ends that have separated, may be taped together. When taping the hoses, they must not be wrapped solidly. The hoses should be wrapped for about 2 in. (51 mm) every 12 in. (305 mm), allowing the colors of the hose to be seen.

Fuel gas hoses must be red and have left-hand threaded fittings. Oxygen hoses must be green and have right-hand threaded fittings.

Hoses are available in four sizes: 3/16 in. (4.8 mm), 1/4 in. (6 mm), 5/16 in. (8 mm), and 3/8 in. (10 mm). The size given is the inside diameter of the hose. Larger sizes offer less resistance to gas flow and should be used where long hose lengths are required. The smaller sizes are more flexible and easier to handle for detailed work.

The three sizes of hose end fittings available are A (small), B (standard), and C (large). The three sizes are made to fit all hose sizes.

Hose Care and Use

When hoses are not in use, the gas should be turned off and the pressure bled off. Turning off the equipment and releasing the pressure prevents any undetected leaks from causing a fire or explosion. This action also eliminates a dangerous situation that would be created if a hose were cut by equipment or materials being handled by workers who were unfamiliar with welding equipment. In addition, hoses are permeable to gases (ability of the gas to dissolve into or through the hose walls). Thus, gases left under pressure for long periods of time can migrate through the hose walls and mix with each other, **Figure 28-36** (page 721). If the gases mix and the torch is lit without first purging the lines, the hoses can explode. For this reason, if the welder is not certain that the hoses were bled, it is recommended that they be purged before the torch is lit.

Hoses are resistant to burns, but they are not burnproof. They should be kept out of direct flame, sparks, and hot metal. You must be especially cautious when using a cutting torch. If they become damaged, the damaged section should be removed and the hose repaired with a splice. Damaged hoses should never be taped to stop leaks.

Hoses should be checked periodically for leaks. To test a hose for leaks, adjust the regulator to a working pressure with the torch valves closed. Wet the hose with a leak-detecting solution by rubbing it with a wet rag, spraying it, or dipping it in a bucket. Then watch for bubbles, which indicate that the hose leaks.

The hose fittings can be changed if the old ones become damaged. Several kits are available that have new nuts, nipples, ferrules, a ferrule crimping tool, and any other supplies required to replace the hose ends, **Figure 28-37** (page 721).

To replace the hose end, the hose is first cut square. The correct size ferrule is inserted. Then both the hose end and nipple are sprayed with a leak-detecting solution. This will help the nipple slide in more easily. Screw the nipple and nut on a torch body. This will hold the nipple deep inside the nut, and the body will act as a handle for leverage as the nipple is pushed inside the hose, **Figure 28-38** (page 721). After the hose is slid up tightly to the nut, crimp the ferrule until it is tight. The

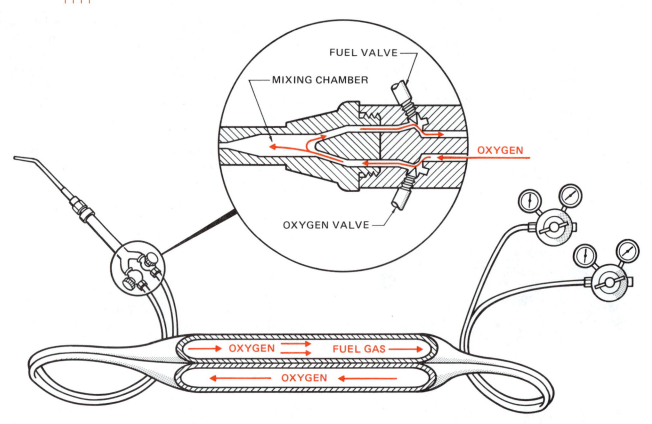

Figure 28-34 Gas may flow back up the hose if both valves are opened at the same time when the system is being bled down after use. Installing reverse flow valves on the torch can prevent this.

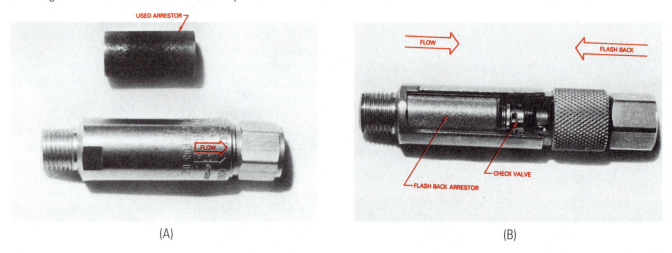

(A)   (B)

Figure 28-35 Flashback arrestor and reverse low valve combination. (A) Arrestor cartridge can be replaced after it has done its job. (B) Cutaway.

crimping tool should be squeezed twice, the second time at right angles to the first, **Figure 28-39.** When the crimping is complete, install the hose on a torch and regulator. Then adjust the regulator to a working pressure and spray the fitting with a leak-detecting solution. Watch for any bubbles, which indicate a leaking fitting.

BACKFIRES AND FLASHBACKS

A backfire occurs when a flame goes out with a loud snap or pop. A backfire may be caused by: (1) touching the tip against the workpiece, (2) over-heating the tip, (3) operating the torch when the flame settings are too low, (4) a loose tip, (5) damaged seats, or (6) dirt in the tip. The problem that caused the backfire must be corrected before relighting the torch. A backfire may cause a flashback.

A flashback occurs when the flame burns back inside of the tip, torch, hose, or regulator. If the torch does flashback, close the oxygen valve at once and then close the fuel valve. The order in which the valves are closed is not as important as the speed at which they are

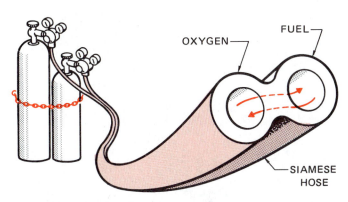

Figure 28-36 Gas left under pressure may migrate through the hose walls.

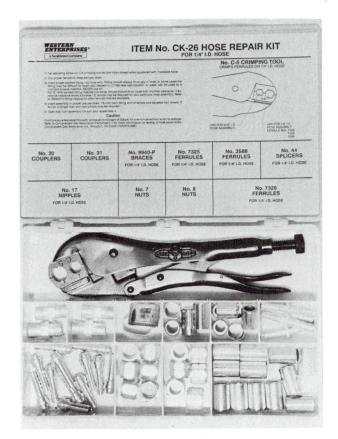

Figure 28-37 Hose repair kit.

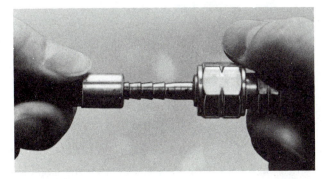

Figure 28-38 Screwing the hose nut onto a fitting will help when pushing the nipple into the hose.

Figure 28-39 (A) Crimping hose ferrule.

Figure 28-39 (B) 2-ear clamp. (Courtesy of Oetiker, Inc.)

closed. A flashback that reaches the cylinder may cause a fire or an explosion.

A flashback usually makes a high-pitched squealing or hissing sound. Closing the torch oxygen valve stops the flame inside at once. Then the fuel gas valve should be closed and the torch allowed to cool off before repairing the problem. When a flashback occurs, there is usually a serious problem with the equipment, and a qualified technician should be called. After locating and repairing the problem, gas should be blown through the tip for a few seconds to clear out any soot that may have accumulated in the passages. A flashback that burns in the hose leaves a carbon char inside that may explode and burn in a pressurized oxygen system. A fuel gas is not required to kindle a hot, severe fire inside such hose sections. Discard hose sections in which a flashback has occurred and obtain new hose.

TYPES OF FLAMES

There are three distinctly different oxyacetylene flame settings. A carburizing flame has an excess of fuel gas. See Color Plate 47a. This flame has the lowest temperature and may put extra carbon in the weld metal. A neutral flame has a balance of fuel gas and oxygen. See Color Plate 47b. It is the most commonly used flame because it adds nothing to the weld metal. An oxidizing flame has an excess of oxygen. See Color Plate 47c. This flame has the highest temperature and may put oxides in the weld metal.

LEAK DETECTION

A leak-detecting solution can be purchased pre-mixed and ready to use, or as a concentrate that must be mixed with water. It can also be mixed in the shop by using a small quantity of liquid dishwashing detergent in water. Use only enough detergent to produce bubbles; too much detergent will leave a soapy film.

A leak-detecting solution must be free flowing so that it can seep into small joints, cracks, and other areas that may have a leak. The solution must produce a good quantity of bubbles without leaving a film. The solution can be dipped, sprayed, or brushed on the joints.

■ CAUTION

Some detergents are not suitable for O_2 because of an oil base. Use only O_2 approved leak-detection solution on oxygen fittings.

PRACTICE 28-1

Setting up an Oxyfuel Torch Set

This practice requires a disassembled oxyfuel torch set, consisting of two regulators, two reverse flow valves, one set of hoses, a torch body, a welding tip, two cylinders, a portable cart or supporting wall, and a wrench. The student is to assemble the equipment in a safe manner.

1. Safety chain the cylinders in the cart or to a wall, Figure 28-40. Then remove the valve protection caps, Figure 28-41.
2. Crack the cylinder valve on each cylinder for a second to blow away dirt that may be in the valve, Figure 28-42.

■ CAUTION

If a fuel gas cylinder does not have a valve hand wheel permanently attached, you must use a nonadjustable wrench to open the cylinder valve. The wrench must stay with the cylinder as long as the cylinder is on, Figure 28-43.

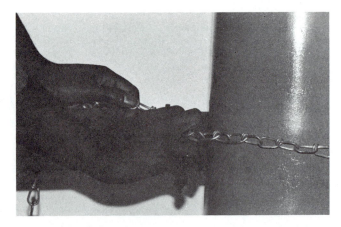

Figure 28-40 Safety chain cylinder.

Figure 28-41 Unscrew the valve protector caps. Put the caps in a safe place; they must be replaced on empty cylinders before they are returned.

Figure 28-42 "Cracking" the oxygen and fuel cylinder valves to blow out any dirt lodged in the valves. (Courtesy of Victor Equipment Company.)

Small combination wrench

Large combination wrench

T-Wrench

Figure 28-43 Nonadjustable wrenches for acetylene cylinders. (Courtesy of ESAB Welding and Cutting Products.)

■ *CAUTION*

Always stand to one side. Point the valve away from anyone in the area and be sure there are no sources of ignition when cracking the valve.

3. Attach the regulators to the cylinder valves, **Figure 28-44(A)** The nuts can be started by hand and then tightened with a wrench, **Figure 28-44(B)**.

4. Attach a reverse flow valve or flashback arrestor, if the torch does not have them built in, to the hose connection on the regulator or to the hose connection on the torch body, depending on the type of reverse flow valve in the set, **Figure 28-45** (page 724). Occasionally test each reverse flow valve by blowing through it to make sure it works properly.

5. Connect the hoses. The red hose has a left-hand nut and attaches to the fuel gas regulator. The green hose has a right-hand nut and attaches to the oxygen regulator.

6. Attach the torch to the hoses, **Figure 28-46** (page 724). Connect both hose nuts fingertight before using a wrench to tighten either one.

7. Check the tip seals for nicks or O rings, if used, for damage. Check the owner's manual, or a supplier, to determine if the torch tip should be tightened by hand only, or should be tightened with a wrench, **Figure 28-47** (page 724).

■ *CAUTION*

Tightening a tip the incorrect way may be dangerous and might damage the equipment.

(A)

(B)

Figure 28-44 Attach the oxygen regulator (A) to the oxygen cylinder valve. Using a wrench (B), tighten the nut.

Figure 28-45 Attach reverse flow valves.

Check all connections to be sure they are tight. The oxyfuel equipment is now assembled and ready for use. ◆

PRACTICE 28-2

Turning On and Testing a Torch

Using the oxyfuel equipment that was properly assembled in Practice 28-1, a nonadjustable tank wrench, and a leak-detecting solution, you will pressurize the system and test for leaks.

1. Back out the regulator pressure adjusting screws until they are loose, **Figure 28-48.**
2. Standing to one side of the regulator, open the cylinder valve SLOWLY so that the pressure rises on the gauge slowly, **Figure 28-49.**

■ CAUTION

If the valve is opened quickly, the regulator or gauge may be damaged, or the gauge may explode.

3. Open the oxygen valve all the way until it is sealed at the top, **Figure 28-50.**
4. Open the acetylene or other fuel gas valve 1/4 turn, or just enough to get gas pressure. **Figure 28-51.** If the cylinder valve does not have a handwheel, use a nonadjustable wrench and leave it in place on the valve stem while the gas is on.

■ CAUTION

The acetylene valve should never be opened more than 1 1/2 turns, so that in an emergency it can be turned off quickly.

Figure 28-46 Connect the free ends of the oxygen (green) and the acetylene (red) hoses to the welding torch. (Courtesy of Albany Calcium Light Co., Inc.)

Figure 28-47 Select the proper tip or nozzle and install it on the torch body. (Courtesy of Albany Calcium Light Co., Inc.)

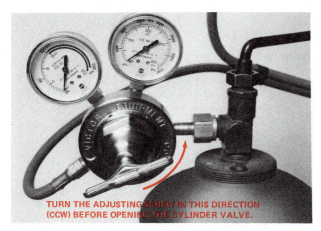

Figure 28-48 Back out both regulator adjusting screws before opening the cylinder valve. (Courtesy of Albany Calcium Light Co., Inc.)

Figure 28-49 Stand to one side when opening the cylinder valve. (Courtesy of Albany Calcium Light Co., Inc.)

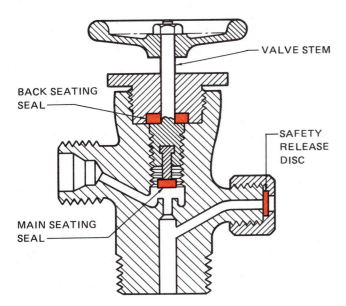

Figure 28-50 Cutaway of an oxygen cylinder valve showing the two separate seals. The back seating seal prevents leakage around the valve stem when the valve is open.

Figure 28-51 Open the cylinder valve slowly.

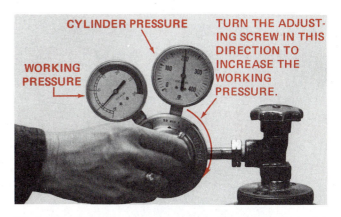

Figure 28-52 Adjust the regulator to read 5 psig (0.35 kg/cm²g) working pressure.

5. Open one torch valve, and point the tip away from any source of ignition. Slowly turn in the pressure adjusting screw until gas can be heard escaping from the torch. The gas should flow long enough to allow the hose to be completely purged (emptied) of air and replaced by the gas before the torch valve is closed. Repeat this process with the other gas.

6. After purging is completed, and with both torch valves off, adjust both regulators to read 5 psig (35 kPag), **Figure 28-52**.

7. Spray a leak-detecting solution on each hose and regulator connection, and on each valve stem on the torch and cylinders. Watch for bubbles, which indicate a leak. Turn off the cylinder valve before tightening any leaking connections, **Figure 28-53** (page 726).

■ CAUTION

Connections should not be overtightened. If they do not seal properly, repair or replace them.

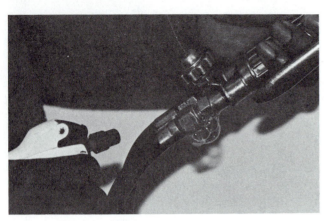

Figure 28-53 Spray fittings with a leak-detecting solution.

Figure 28-54 Identify any cylinder that has a problem by marking it.

Figure 28-55 Correct position to hold a sparklighter.

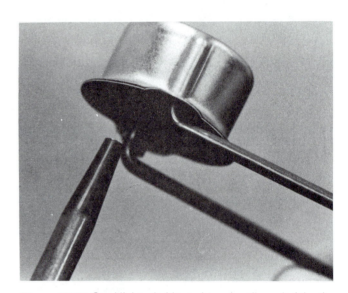

Figure 28-56 Sparklighter held too close over the end of the tip.

■ CAUTION

Leaking cylinder valve stems should not be repaired. Turn off the valve, disconnect the cylinder, mark the cylinder, and notify the supplier to come and pick up the bad cylinder, Figure 28-54. ◆

The assembled oxyfuel welding equipment is now tested and ready to be ignited and adjusted. ◆

PRACTICE 28-3

Lighting and Adjusting an Oxyacetylene Flame

Using the assembled and tested oxyfuel welding equipment from Practice 28-2, a sparklighter, gas welding goggles, gloves, and proper protective clothing, you will light and adjust an oxyacetylene torch for welding.

1. Wearing proper clothing, gloves, and gas welding goggles, turn both regulator adjusting screws in

until the working pressure gauges read 5 psig (35 kPag). If you mistakenly turn on more than 5 psig (35 kPag), open the torch valve to allow the pressure to drop as the adjusting screw is turned outward.

2. Turn on the torch fuel-gas valve just enough so that some gas escapes.

■ CAUTION

Be sure the torch is pointed away from any sources of ignition or any object or person that might be damaged or harmed by the flame when it is lit.

3. Using a sparklighter, light the torch. Hold the lighter near the end, **Figure 28-55**, of the tip but not covering the end, **Figure 28-56**.

4. With the torch lit, increase the flow of acetylene until the flame stops smoking. See Color Plate 47d.

5. Slowly turn on the oxygen and adjust the torch to a neutral flame.

This flame setting uses the minimum gas flow rate for this specific tip. The fuel flow should never be adjusted to a rate below the point where the smoke stops. This is the minimum flow rate at which the cool gases will pull the flame heat out of the tip, **Figure 28-57**. If excessive heat is allowed to build up in a tip, it can cause a backfire or flashback.

The maximum gas flow rate gives a flame that, when adjusted to the neutral setting, does not settle back on the tip. ◆

PRACTICE 28-4

Shutting Off and Disassembling Oxyfuel Welding Equipment

Using the properly lit and adjusted torch from Practice 28-3 and a wrench, you will extinguish the flame and disassemble the torch set.

1. First, quickly turn off the torch fuel-gas valve. This action blows the flame out and away from the tip, ensuring that the fire is out. In addition, it prevents the flame from burning back inside the torch. On large tips or hot tips, turning the fuel off first may cause the tip to pop. The pop is caused by a lean fuel mixture in the tip.

    If you find that the tip pops each time you turn the fuel off first, turn the oxygen off first to prevent the pop. Be sure that the flame is out before putting the torch down.

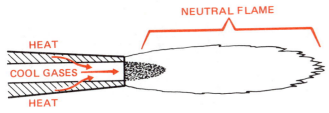

Figure 28-57 Enough cool gas flowing through the tip will help prevent popping.

2. After the flame is out, turn off the oxygen valve.

3. Turn off the cylinder valves.

4. Open one torch valve at a time to bleed off the pressure.

5. When all of the pressure is released from the system, back both regulator adjusting screws out until they are loose.

6. Loosen both ends of both hoses and unscrew them.

7. Loosen both regulators and unscrew them from the cylinder valves.

8. Replace the valve protection caps. ◆

MANIFOLD SYSTEMS

A **manifold system** can be used if there are a number of work stations or if a high volume of gas will be used. The manifold can also be used to keep the cylinders out of the work area and to reduce the number of cylinders needed at one time.

Manifolds must be located 20 ft (6 m) or more from the actual work, or they must be located so that sparks cannot reach them. Oxygen manifolds must be at least 20 ft (6 m) from fuel gas manifolds and flammable materials, or be separated from them by a noncombustible wall 5 ft (1.5 m) high, with a 1/2-hour, fire-resistant rating. Inert gas manifolds can be placed with either oxygen or fuel gas manifolds.

The rooms that are used for manifolds can also be used for cylinder storage, but full and empty cylinders must be kept separated within the room. Fuel gas manifold rooms must have good ventilation and explosion-proof lights, **Figure 28-58** (page 728). They must also have a sign on the door that reads "Danger: No Smoking, Matches, or Open Lights," or similar wording.

Piping for the high-pressure side of a manifold must be steel, stainless steel, or alloyed copper. Piping for the low-pressure side, except acetylene, can be stainless steel, copper (type L or K), brass, steel, or wrought iron. All acetylene piping must be steel or wrought iron. Unalloyed copper used with acetylene forms an unstable and explosive compound, copper acetylide.

The pipe joints in copper or brass lines can be welded, brazed, threaded, or flanged. Joints in steel or iron pipe can be welded, threaded, or flanged. Nitrogen should be purged through pipes being welded or brazed. All piping must be clean, oil-free, and should have a slight angle back toward the manifold so that any moisture will run back toward it, **Figure 28-59** (page 728).

Manifold systems should be tested for leaks at one-and-a-half (1 1/2) times the operating pressure. The fuel gas lines must be protected from oxygen flowing back into the lines by installing a reverse flow valve. In addition, gas fuel lines must have flashback protection and back pressure release. One device can be installed to satisfy all three requirements, **Figure 28-60** (page 728).

Figure 28-58 Explosion-proof light fixtures suitable for a manifold room. (Courtesy of Crouse-Hinds Company.)

Manifold Operation

The pipes should be cleaned with an oil-free, noncombustible fluid before the regulators are attached. Solutions of caustic soda or trisodium phosphate are good for this purpose. Once the system is cleaned, install the regulators and purge the system with nitrogen. Nitrogen will pick up any moisture in the system while providing a noncombustible gas in the system.

After the system has been filled with nitrogen, start filling the pipes with the oxygen or fuel gas. Allow the gas to escape freely from the station farthest away from the manifold. Continue purging until all the nitrogen is removed from the manifold.

■ *CAUTION*

Make sure there is good ventilation during the purging, and that there are no sources of ignition near the escaping gas.

Set the line pressure as low as possible, still ensuring that the type of work being done at the work stations can be performed satisfactorily, **Figure 28-61**. When

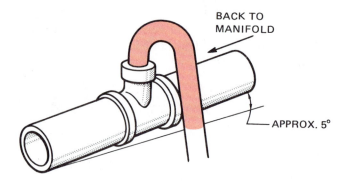

Figure 28-59 A slight angle will allow moisture to flow back away from the station regulators.

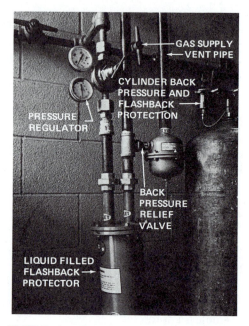

Figure 28-60 Fuel gas manifold system with flashback and back pressure safety equipment.

Figure 28-61 Dymo tape on gauge notes the maximum pressure to be used.

(A)

(B)

Figure 28-62 Manifold station regulators (A) and station valve safety cap (B).

work is completed, the system must be turned off and each station, manifold, and cylinder valve closed. If a station regulator is removed, a cap must be put on the line so that air cannot enter the system, **Figure 28-62(A)** and **(B)**. A complete list of operating, maintenance, and emergency procedures should be clearly posted for all manifolds.

■ *CAUTION*

In case of fire or severe threatening weather, such as a tornado, all cylinder valves must be turned off.

SAFETY

After you have developed your skills and found a welding job, you will be exposed to welders with many years of experience. These welders will have developed many good shortcuts through the years that can help you on the job. However, there are few safe shortcuts in equipment setup and maintenance. The safe way of setting up and testing a system should always be followed. Be careful to avoid questionable practices. Always refer to the manufacturer's operating instructions and safety recommendations for the type of equipment you are using.

Alcohol Processing Plants Provide Fuel for New Mexico

Figure 1

Everyone speaks well of alternative energy sources, such as alcohol, these days, but few people actually do anything about them.

In the state of New Mexico, however, alternative energy is regarded as the key to the area's future economic vitality. In fact, a special promotional program for alcohol fuels was established by the state government.

The state's alcohol fuel industries lead the nation in the construction of new processing and distribution plants. In the past year alone, one construction company has been responsible for the construction of thirteen alcohol processing plants.

Equipment used in the construction project included 16 diesel engine drives and a single 200-amp welder/power plant. At the beginning of the project, the engine drives were used for auxiliary power more than for welding. Then, as work progressed on each plant, the requirements gradually shifted to more welding and less auxiliary power.

The first eight plants were situated on a 46-acre plot of land. There was no utility power available on the site, so everything had to be powered by the welding power generators.

There is a great deal of piping, tank fabrication, and structural welding required for an alcohol plant. But first the forms must be built, concrete foundations poured, and other basic preparations completed.

Working conditions at the first and subsequent sites were less than ideal. Snow, frozen mud, and cold temperatures were problems in the winter, while dust and high temperatures prevailed in the summer.

A typical alcohol plant consists of packaged boiler, eight 12,000-gallon tanks, and two silo tanks with a capacity of 60,000 gallons apiece. A 100-foot tall distillation column completes the basic setup, and a network of schedule 40 pipe connects the elements to form an integrated processing system.

Welders relied on 6010 and 7018 rods for all-position welding of both the pipe and the 3/16-inch-thick tank walls, **Figure 1**. For the steel silo,

Figure 2

preformed sections were prepared first, **Figure 2**. Each section was then lifted into place with a crane and joined to the structure with horizontal and vertical welds. The systems operate at low pressure and have a life expectancy of twenty years.

The construction crews finished a plant in approximately sixty working days. A single plant can provide up to one million gallons of high-octane fuel per year and consumes approximately 400,000 bushels of grain in the process.

The alcohol made from the grain is just part of the benefit. The processing plants also produce nearly seven million pounds of stillage per year, which contains almost 30 percent protein by weight.

The stillage can be dried, pelletized, and sold as high-quality agricultural feed. The distillation process also yields carbon dioxide, which in the future may be collected to produce even more revenue from each plant.

(Courtesy of Miller Mfg. Co., Appleton, Wisconsin 54912.)

REVIEW

1. What is the purpose of a pressure regulator?

2. What may result if a pressure regulator is not used on the type of gas or pressure range for which it was designed?

3. Describe how a single stage regulator operates.

4. Is the torch pressure always the same as the working gauge pressure? Why or why not?

5. Why doesn't the high pressure gauge on a regulator always indicate the amount of gas in the cylinder?

6. How do the operation of a safety relief valve and the operation of a safety disc valve differ?

7. Describe the difference between an argon cylinder valve fitting and a carbon dioxide cylinder valve fitting.

8. What is meant by *regulator creep*?

9. Who can repair regulators?

10. Why must the pressure be released from a regulator when work is finished?

11. Why are combination welding and cutting torches considered to be more versatile?

12. What is the advantage of using an injector-type mixing chamber?

13. What should be done to the valve packing if the valve knob does not turn freely after it has been tightened to stop a leak?

14. What may happen to a tip seat if it is incorrectly tightened?

15. What can happen to a tip if it is excessively cleaned with a tip cleaner?

16. What is the difference between a reverse flow valve and a flashback arrestor?

17. What are Siamese hoses?

18. Why must the pressure be bled off hoses when work is complete?

19. What is the difference between a backfire and a flashback?

20. Why is a neutral flame the most commonly used oxyacetylene flame used?

21. What properties should a good leak-detecting solution have?

22. Why must the oxygen cylinder valve be opened all the way to the top?

23. How long should hoses be purged?

24. What should be done with cylinders that have leaking valve stems?

25. How should the spark lighter be held to light a torch?

26. Once the torch is lit why must the acetylene flow be increased until the flame stops smoking before the oxygen is turned on for adjustment?

27. What type of piping can be used for a manifold system?

Chapter 29

OXYFUEL GASES AND FILLER METALS

OBJECTIVES

After completing this chapter, the student should be able to:

- explain the chemical reaction that takes place in any oxyfuel flame.
- list the major advantages and disadvantages of the different fuel gases.
- demonstrate an ability to choose correct filler metals.
- explain what conditions affect the selection of filler metal.

KEY TERMS

oxyfuel flame	acetone
optical pyrometer	oxidizing flame
carbonizing (carburizing) flame	inner cone
hydrocarbons	outer envelope
molecules	liquefied fuel gases
atoms	methylacetylene-propadiene
acetylene (C_2H_2)	(MPS)
combustion	MAPP®
primary combustion	piccolo tube
secondary combustion	neutral flame
combustion rate	oxyhydrogen
heat energy	oxyhydrogen flame
flashback	filler metals
backfire	ferrous filler metal

INTRODUCTION

The general grouping of processes known as oxyfuel consists of a number of separate processes, all of which burn a fuel gas with oxygen. The oxyfuel flame was used for fusion welding as early as the first half of the 1800s when scientists developed the oxyhydrogen torch. Before that time, air fuel torches were used, but because the flame was not hot enough they had limited success. The early use of pure oxygen with hydrogen or acetylene as the fuel gas often resulted in flashbacks and explosions. The use of water traps helped prevent most flashbacks from becoming explosions. But until the early development of the torch mixing chamber, welding was a very dangerous occupation. The mixing chamber gave a more uniform flame that was less likely to flash back.

During the early 1900s, the oxyacetylene flame became more popular as the primary means of welding. Since 1900, when the first shielded metal arc welding (stick) electrodes were introduced by Strohmeyer in Britain, the use of the oxyacetylene flame for welding has declined. During its prime, plates 1 in.

(25 mm) thick or more were gas welded to build everything from large, sea-going ships to massive machines used during the Industrial Revolution. Today, because of improvements in other processes, the oxyacetylene flame is seldom used on metal thicker than l/16 in. (2 mm).

The advances in shielded metal arc welding, gas tungsten arc welding, plasma arc cutting, and gas metal arc welding have overshadowed oxyfuel welding. These processes are faster, cleaner, and cause less distortion than oxyfuel welding. At present, oxyfuel welding is used mainly for farm repairs, maintenance, and in smaller shops.

Uses of the Oxyfuel Flame

The usefulness of the oxyfuel flame as a primary means of cutting ferrous metals has increased. In 1887, the flame was used to melt through thin metal. Around 1900 the oxygen lance was introduced. The oxygen lance allowed high-pressure oxygen to be directed onto metal heated by a torch, resulting in much improved cuts. The later development of a cutting torch with preheating flames surrounding a central oxygen hole in the tip brought oxyfuel cutting into its own. The cutting torch, used by hand or as part of a machine, is used to rapidly cut out steel parts, **Figure 29-1**.

Today, a large number of manufactured items are touched in some way by the oxyfuel flame. The parts may have been cut out, heated, hardened, or joined with an oxyfuel flame. The expansion of the role of the oxyfuel flame in industry has lead to the introduction of new fuel gases and gas mixtures. Each of the new gases has certain advantages and disadvantages. The decision of which gas to use must be based on cost, availability, welder skills and skill changes required, equipment

changes, safety, handling, performance, and other concerns. The information supplied for a special gas often points out its strengths and may use comparisons to prove its advantages.

Characteristics of the Fuel Gas Flame

The data available for fuel gas flame characteristics is not gathered in a consistent manner. Temperature and heat, for example, are very basic physical facts about the flame of a specific gas. But even these facts can be misleading. The flame condition, such as neutral, oxidizing, or carbonizing (carburizing), and/or the purity of the gases being used will affect the temperature of the flame. The method by which the temperature is measured also affects its value. The highest potential temperature values are determined by chemical analysis and by the calculation of the theoretical energy released. However, no combustion is perfect, so this temperature is never attained. An infrared analysis or optical pyrometer of the flame gives the highest temperature in the flame. But this temperature is concentrated in a thin layer around the inner cone and is so small that it is not of practical use.

The optical pyrometer and the infrared analysis both give an accurate temperature reading, but where the temperature is measured makes a difference.

Differences in heat values may also be misleading, depending upon how they are obtained. The temperatures, heat, and other flame characteristics noted in this chapter are as close as possible to those that occur during use. They are not necessarily the highest or lowest possible values, and therefore should not be considered absolute. But they can, however, be compared with each other for the purpose of selecting a fuel gas.

FUEL GASES

Most fuel gases used for welding are hydrocarbons. The gases are made up of hydrogen (H) and carbon (C) atoms. The atoms are bound together tightly to form molecules. Each molecule of a specific gas has the same type, number, and arrangement of atoms. The number of atoms in a molecule of gas varies from one gas to another. A molecule of acetylene is made up of two hydrogen (H) and two carbon (C) atoms (C_2H_2). A molecule of propane is made up of eight hydrogen (H) and three carbon (C) atoms (C_3H_8). The chemical formulas for these two fuel gases are shown in **Figure 29-2**.

If a mixture of acetylene (C_2H_2) and oxygen (O_2) is ignited, the acetylene molecule bonds are broken, and new bonds between the hydrogen, carbon, and oxygen atoms are formed. The breaking down of acetylene molecules releases energy in the form of heat and light. The rate at which this reaction occurs results in a change in temperature. The formation of new bonds with the oxygen releases still more energy. This chemical reaction, known as combustion, is a form of rapid oxidation.

Figure 29-1 Modern hand-held cutting torch. (Courtesy of Victor Equipment Company.)

$$C_2H_2$$
ACETYLENE

$$CH_3\,CH_2\,CH_3$$
PROPANE

Figure 29-2 Chemical formulas for two hydrocarbons used as fuel gases.

	Cu ft/cu ft			
	Total O_2 required	Supplied O_2 through torch*	% total through torch	Cu ft O_2 per lb of fuel
Acetylene — 5589°F/1470 Btu	2.5	1.3	50.0	18.9
MAPP® gas — 5301°F/2406 Btu	4.0	2.5	62.5	22.1
Natural gas — 4600°F/1000 Btu	2.0	1.9	95.0	44.9
Propane — 4579°F/2498 Btu	5.0	4.3	85.0	37.2
Propylene — 5193°F/2371 Btu	4.5	3.5	77.0	31.0

*Balance of total oxygen demand is entrained in the fuel-gas flame from the atmosphere.

Source: MAPP® Products

Table 29-1 Oxygen Consumption for Fuels (neutral flame).

The number of oxygen atoms required to completely combust the fuel gas varies, depending upon its molecular makeup. **Table 29-1** lists some fuel gases and the amount of oxygen required to complete the reaction in the flame.

Combustion is divided into two separate chemical reactions. The first reaction is referred to as primary combustion, and the second reaction is referred to as secondary combustion, **Figure 29-3**.

In the primary reaction of acetylene (C_2H_2), the oxygen (O_2) atoms unite with the carbon (C) atoms. This reaction liberates energy and forms carbon monoxide (CO) and free hydrogen, **Figure 29-4**.

In the secondary reaction, Oxygen (O_2) unites with free hydrogen (H) to form water vapor (H_2O) and liberates more heat. Also, the carbon monoxide (CO) unites with additional oxygen (O_2) to form carbon dioxide (CO_2), **Figure 29-5** (page 736).

The final products of all clean-burning hydrocarbon flames are the same, water vapor and carbon dioxide. The temperature, rate of combustion, and quantity of heat release are the characteristics that make each flame different. These characteristics make some gases better than others for certain operations.

FLAME RATE OF BURNING

The combustion rate or rate of propagation of a flame is the rate or speed at which the flame burns. The combustion or burn rate of a fuel gas is given in feet per second (meters per second).

The combustion rate of a gas is determined by the amount of heat energy required to break the bonds

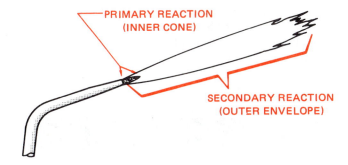

Figure 29-3 Parts of an oxyfuel flame.

PRIMARY REACTION (INNER CONE)

SECONDARY REACTION (OUTER ENVELOPE)

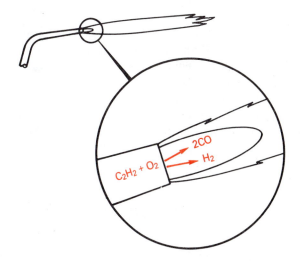

Figure 29-4 Primary flame reaction (with acetylene as the fuel gas).

$C_2H_2 + O_2$ 2CO H_2

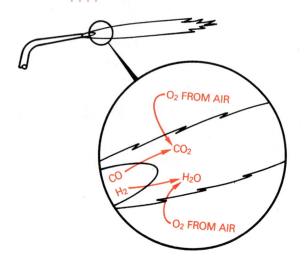

Figure 29-5 Secondary flame reaction.

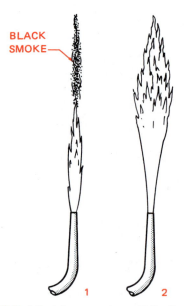

BLACK SMOKE

Figure 29-6 Adjust the gas valve until the flame stops smoking.

between the atoms of its molecules and by the amount of heat energy liberated as the bonds are broken. The ratio of fuel to oxygen and the homogeneity of their mixture also affect the propagation rate.

A homogeneous mixture of 50% acetylene and 50% oxygen has a burn rate of 22.7 feet per second (6.9 meters per second). As the percentage of oxygen in the mixture increases, the burn rate increases; as the percentage of oxygen decreases, the burn rate also decreases.

The higher the combustion rate of an oxygen fuel mixture, the more prone the mixture is to backfire or flashback. Table 29-2 lists the combustion rates for most fuel gases.

EXPERIMENT 29-1

Burn Rate

Using properly set up and adjusted oxyfuel welding equipment, striker, goggles, gloves, and any other required safety equipment, you are going to observe how changing the mixture ratios of oxygen and acetylene affects the combustion rate.

Set both regulators at 5 psig (35 kPag) and purge the lines as required. With only the acetylene gas torch valve on, use the striker to light the torch. Adjust the gas valve until the flame stops smoking, **Figure 29-6**. See Color Plate 47(d). Turn on the oxygen torch valve and note the immediate reduction in the flame size. Continue opening the valve until the flame is at a neutral setting, **Figure 29-7**. See Color Plate 47(c). With the flame at a neutral setting, the total flame size is smaller than it was with just

the acetylene even though a larger total volume of gas is coming from the tip. Now increase the oxygen flow by opening the valve more and notice that the size of the flame again decreases, **Figure 29-8**. See Color Plate 47(b).

In this experiment you have observed that as the percentage of oxygen in the fuel gas increases, the flame propagation or burning rate also increases. This increase is indicated because, even with the increase in total volume of gas, the actual size of the flame decreases. ◆

■ CAUTION

An extremely oxygen-rich fuel gas mixture may have a combustion rate that is greater than the gas flow rates of large tips, resulting in a backfire or flashback. If this occurs, close the oxygen valve immediately.

ACETYLENE (C₂H₂)

Acetylene (C_2H_2) is the most frequently used fuel gas. The mixture of oxygen and acetylene produces a high heat and high temperature flame that is widely used for welding, cutting, brazing, heating, metallizing, and hardsurfacing.

Acetylene is produced by mixing calcium carbide (CaC_2), often referred to as carbide, with water (H_2O).

Gas	Acetylene	MAPP®	Natural Gas	Propane	Propylene	Hydrogen
Burning Velocity in Oxygen, ft/sec (mm/s)	22.7 (6097)	15.4 (4694)	15.2 (4633)	12.2 (3718)	15.1 (4602)	36 (6540)

Table 29-2 Combustion Rates.

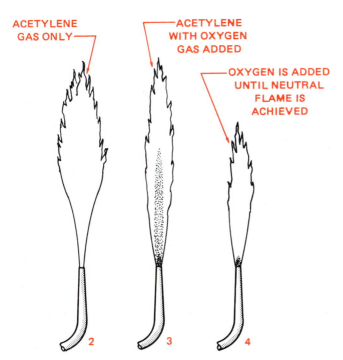

ACETYLENE
GAS ONLY

ACETYLENE
WITH OXYGEN
GAS ADDED

OXYGEN IS ADDED
UNTIL NEUTRAL
FLAME IS
ACHIEVED

2   3   4

Figure 29-7 Achieving the neutral setting of the flame.

Calcium carbide is produced by smelting coke (a coal by-product) and lime in electric arc furnaces. After smelting, the carbide is cooled, crushed, and packed in dry, airtight containers. Either a fixed or portable acetylene generator, **Figure 29-9**, is used to mix the carbide with water. In the generator, crushed carbide is dropped into a large tank of water. The carbon (C) in calcium carbide (CaC_2) unites with the hydrogen (H) in water (H_2O) to form two products. These two products are acetylene (C_2H_2), which bubbles out, and calcium hydroxide [$Ca(OH)_2$], which drops to the bottom, **Figure 29-10** (page 738). The acetylene gas is drawn off and may be used immediately or pumped into storage cylinders for later use.

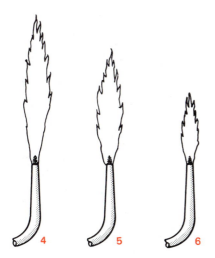

Figure 29-8 An increase in oxygen flow again decreases the size of the flame.

4   5   6

Acetylene is colorless, lighter than air, and has a strong garlic smell. Acetylene is also unstable at pressures above 30 psig (200 kPag), or at temperatures above 1,435°F (780°C). Above the critical pressure or temperature, an explosion may occur if acetylene rapidly decomposes. This decomposition can occur without the presence of oxygen and may occur as a result of electrical shock or extreme physical shock.

■ *CAUTION*

Because of the instability of acetylene, it must never be used at pressures above 15 psig (100 kPag) or subjected to any possible electrical shock, excessive heat, or rough handling. Acetylene must not be manifolded or distributed through copper lines because it forms copper acetylide, which is explosive.

Acetone is used inside acetylene cylinders to absorb the gas and make it more stable. The cylinder is filled with a porous material, and then acetone is added to the cylinder, where it absorbs about twenty-four times its own weight in acetylene. Acetone absorbs acetylene in the same manner as water absorbs carbon dioxide in a carbonated drink. As the acetylene is drawn off for use, it evaporates out of the acetone. Excessively high withdrawal rates will cause the acetylene to boil out of the acetone. This rapid boiling can cause a portion of the acetone to be carried out of the cylinder with the acetylene, **Figure 29-11** (page 738). Withdrawing the acetone-acetylene mixture and burning it in the flame will contaminate the weld and may damage the rubber seals and internal parts of the regulator and other equipment. In addition to the danger caused by damaged equipment, the cylinder itself may be damaged, resulting in another hazard.

The withdrawal rate of gas from a cylinder should not exceed one-seventh of the total cylinder capacity per hour. **Table 29-3** (page 738) lists the flow rates in cubic feet per hour (liters per minute) for various torch tip sizes. It also lists the minimum cylinder size to be used with a single torch. If more than one torch is to be used

Figure 29-9 Acetylene generator. (Courtesy of Rexarc, Inc.)

with a cylinder, the draw rates of the tips should be added together to obtain the minimum cylinder size or the number of cylinders that must be manifolded together. As the gas from the cylinder(s) is emptied, a new maximum withdrawal rate must be calculated based upon current cylinder capacity.

Heat and Temperature

The neutral oxyacetylene flame burns at a temperature of approximately 5,589°F (3,087°C). The maximum temperature of a strongly oxidizing flame is approximately 5,615°F (3,102°C). The flame burns in two parts, the inner cone and the outer envelope; refer to Figure 29-12. The heat produced by the flame can be divided into portions produced in the inner cone and outer envelope. The inner cone produces 507 Btu per cubic foot of gas (19 kilogram-calories per cubic meter), and the outer envelope produces 963 Btu per cubic foot of gas (36 kg-cal/m^3). The total heat produced by the flame is 1,470 Btu/ft^3 (55 kg-cal/m^3).

The high temperature produced by the oxyacetylene flame is concentrated around the inner cone, Figure 29-13. As the flame is moved back away from the work, the localized heating is reduced quickly. For welding, this highly concentrated temperature is the greatest advantage of the oxyacetylene flame over other oxyfuel gases. The molten weld pool is easily controlled by torch angle and position of the inner cone to the work.

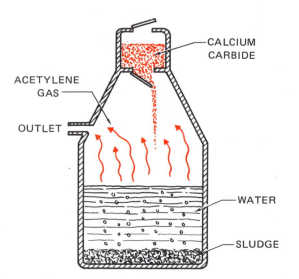

Figure 29-10 An acetylene generator is used to mix calcium carbide with water.

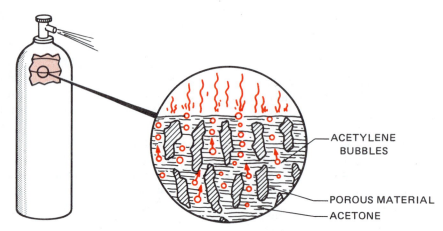

Figure 29-11 The result of too rapid withdrawal of acetylene.

Trip Orifice Size - Drill No.	Thickness of Metal - in (mm)		Oxygen and Acetylene Pressure - psi (kg/cm2)		Acetylene Flow Rate - cfh (L/min)*		Oxygen Flow Rate - cfh (L/min)	
70	1/64	(0.4)	1	(0.0703)	.1	(0.0471)	.1	(0.0471)
65	1/32	(0.8)	1	(0.0703)	.4	(0.188)	.4	(0.188)
60	1/16	(1.6)	1	(0.0703)	1	(0.4719)	1.1	(0.519)
59	3/32	(2.4)	2	(0.1406)	2	(0.943)	2.2	(1.038)
53	1/8	(3.2)	3	(0.2109)	8	(3.775)	8.8	(4.152)
49	3/16	(4.8)	4	(0.2812)	17	(8.022)	18	(8.494)
43	1/4	(6.4)	5	(0.3515)	25	(11.797)	27	(12.741)
36	5/16	(8.0)	6	(0.4218)	34	(16.044)	37	(17.460)
30	3/8	(9.5)	7	(0.4921)	43	(20.291)	47	(22.179)

*This flow rate must not exceed 1/7 of the cylinder's capacity per hour. For example, a large acetylene cylinder contains approximately 275 cu ft (7788 L) and its withdrawal rate must not exceed 39 cfh (18.40 L/min).

Table 29-3 Flow Rates for Various Tip Sizes.

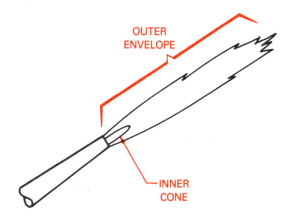

OUTER ENVELOPE

INNER CONE

Figure 29-12

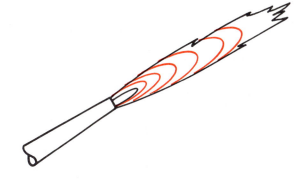

Figure 29-13 Thermal gradients of a flame.

Although more heat is produced in the secondary flame than in the primary flame, the temperature is much lower. The lower temperature and the relatively low concentration of the heat produced by the outer flame are distinct disadvantages of the oxyacetylene flame when it is used for heating. The flame must be held close to the work and moved constantly in order to obtain uniform heating of large parts.

LIQUEFIED FUEL GASES

Some fuel gases are available in liquefied form in pressurized cylinders. Liquefied fuel gases can be obtained in either individual cylinders or bulk tanks. The high-volume use of fuel gas often requires excessive cylinder inventory and handling. Setting up a bulk system will eliminate much of the cylinder inventory and handling.

PRESSURE

The pressure in a cylinder containing a liquefied gas is not an indication of the level of gas in the tank. The pressure is based on the type of gas and the gas temperature. The gas pressure of a liquid increases as the temperature increases and decreases as the temperature decreases. At high temperatures, approximately 200°F (93°C), the pressure may cause the release valve to open to release excessive pressures, over 375 psig (2,690 kPag). At extremely low temperatures, the cylinder may not have any working pressure. For example, at temperatures below 31°F (-1°C), butane has no pressure. Table 29-4 (page 740) lists the gas pressures for various gases at different temperatures.

High withdrawal rates of gas from liquefied gas cylinders will cause a drop in pressure, a lowering of the cylinder temperature, and the possibility of freezing the regulator. As a gas is drawn out of the cylinder, the liquid inside absorbs heat from the outside to produce more gas vapor, **Figure 29-14**. If the gas is drawn out faster than heat can be absorbed, the cylinder will begin to cool off. A ring of frost may appear around the bottom of the

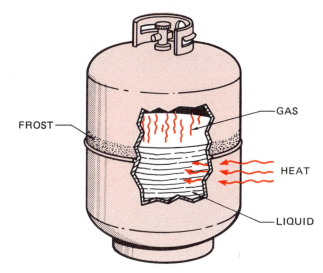

FROST

GAS

HEAT

LIQUID

Figure 29-14 The heat absorbed by the liquid propane causes it to change to a gas.

cylinder. If high withdrawal rates continue, the regulator may also start to ice up. For applications requiring high withdrawal rates, the cylinder should be placed in a warm area.

■ *CAUTION*

Heat should never be applied directly to a cylinder.

METHYLACETYLENE-PROPADIENE (MPS)

Many different methylacetylene-propadiene (MPS) gases are in use today as fuel gases for oxyfuel cutting, heating, brazing, metallizing, and to a limited extent for welding. MPS gases are mixtures of two or more of the following gases: propane (C_3H_4), butane (C_4H_{10}), butadiene (C_4H_6), methyl acetylene (C_3H_4), and propadiene (C_3H_6). The mixtures vary in composition and characteristics from one manufacturer to another. However, all manufacturers provide MPS gases as liquefied gases in pressurized cylinders.

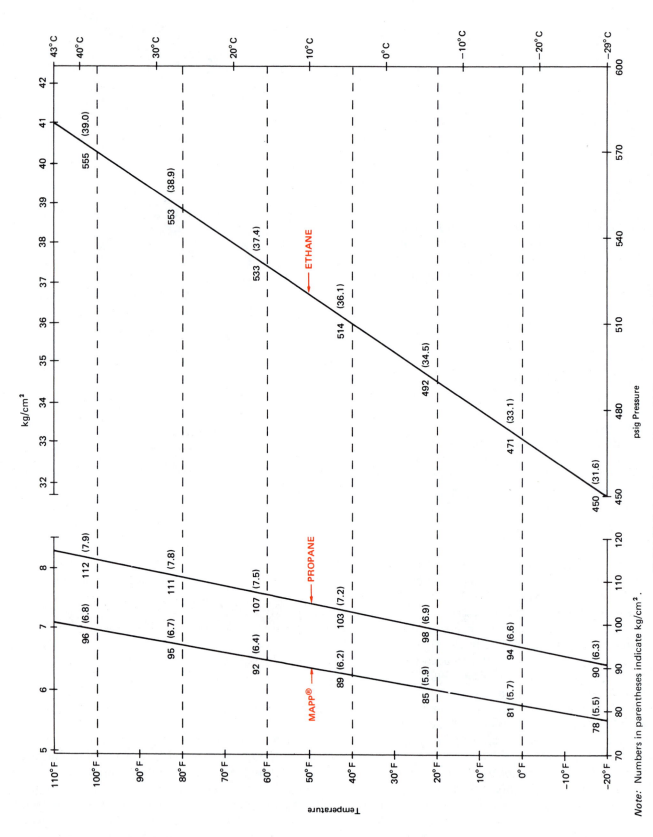

kg/cm²

555	(39.0)
553	(38.9)
533	(37.4)
514	(36.1)
492	(34.5)
471	(33.1)
450	(31.6)

ETHANE

43°C
40°C
30°C
20°C
10°C
0°C
−10°C
−20°C
−29°C

psig Pressure

PROPANE

MAPP®

112	(7.9)		
111	(7.8)		
107	(7.5)		
103	(7.2)		
98	(6.9)		
94	(6.6)		
90	(6.3)		
96	(6.8)		
95	(6.7)		
92	(6.4)		
89	(6.2)		
85	(5.9)		
81	(5.7)		
78	(5.5)		

110°F
100°F
90°F
80°F
70°F
60°F
50°F
40°F
30°F
20°F
10°F
0°F
−10°F
−20°F

Temperature

Table 29-4 Gas Pressures Vary with Temperature Changes.

Note: Numbers in parentheses indicate kg/cm².

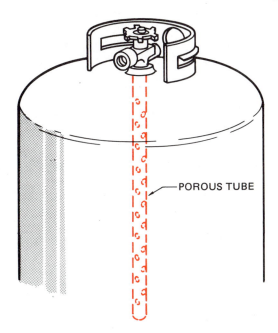

Figure 29-15 A piccolo tube helps keep gases mixed during use.

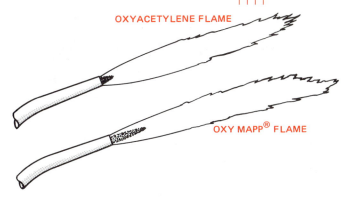

Figure 29-16 The inner cone and outer envelope of a MAPP® flame are longer than an equal-volume flame of acetylene because of the slower burn rate of MAPP®.

METHYLACETYLENE   PROPADIENE

Figure 29-17 MAPP® gas molecules.

Production

Approximately twenty-six MPS gases are marketed: MAPP®, Chem-O-Lean® Apachigas®, FG-2®, Gulf HP®, Flamex®, and Hy Temp®, among others. The gases may be mixed by a local supplier as the cylinders are filled, or the supplier may receive the gas premixed. In cylinders that are stored for a long time, some mixtures tend to separate. Before using the cylinder, it should be moved enough to remix the gases inside. In some cylinders a piccolo tube is used, Figure 29-15, to improve the mixing of the gas by causing the liquid to be agitated as it is used. Without a means of stirring the gas, some MPS gases may burn differently when the cylinder is full and when it is nearly empty.

Temperature and Heat

MPS gases have a neutral oxyfuel flame temperature of about 5,301°F (2,927°C). The exact temperature varies with the specific mixture. The flame temperature and heat are high enough to be used for welding, but MPS gases are seldom used for this purpose.

The heat produced by the primary flame, approximately 570 Btu/ft³ (21 kg-cal/m³), is about the same as acetylene. The secondary flame produces approximately 1,889 Btu/ft³ (70 kg-cal/m³) of heat, or twice the heat of acetylene. These values make MPS gases much better than acetylene for heating, brazing, and some types of cutting. The slower burn rate of these gases results in a poorly concentrated flame, which is difficult to use for welding.

MAPP®

MAPP® gas is the trade name for the stabilized liquefied mixture of methylacetylene (CH₃:C:CH) and propa-diene (CH₂:C:CH₂) gases. Oxy MAPP® combusts with a high-heat, high-temperature flame that works well for cutting, heating, brazing, and metallizing, Figure 29-16.

The gases mixed to produce MAPP® have the same atomic composition. This means that three carbon and four hydrogen atoms are present in each molecule of gas. Molecules of each gas have the same mass and size even though they are shaped differently, Figure 29-17. Because they are the same mass and size, the gas molecules form a stable mixture that remains uniformly mixed in the cylinder. The stabilization of mixture assures a uniform and consistent flame for easier quality control.

Oxy MAPP® produces a neutral flame temperature of 5,301°F (2,927°C) that yields a heat value of 517 Btu/ft³ (19 kg-cal/m³) in the primary flame and 1,889 Btu/ft³ (70 kg-cal/m³) in the secondary flame. The total heat value is 2,406 Btu/ft³ (90 kg-cal/m³). Table 29-5 (page 742) compares the temperatures and heat produced by five different fuel gases. The difference between the temperature produced by a neutral oxyacetylene flame and a neutral oxy MAPP® flame is only 288°F (142°C). Both flames are well above the approximate 2,800°F (1,536°C) temperature required to melt mild steel. Although MAPP® is not normally recommended for use in gas welding, it can be used successfully. Because the secondary flame of MAPP® produces almost twice the heat of the acetylene flame, distortion is more of a problem with MAPP® than with acetylene. Also, it is more difficult to melt the root of a the joint using MAPP®.

Because of the higher heat, the sides of the joint melt first, Figure 29-18 (page 742). This is not as much of a problem with other joint designs.

All gases used as alternatives to acetylene are safer to use, store, and handle. MAPP® has each one of the

Fuel	Neutral Flame Temp °F (°C)	Primary Flame Btu/ft³ (kg-cal/m³)	Secondary Flame Btu/ft³ (kg-cal/m³)	Total Heat Btu/ft³ (kg-cal/m³)
Acetylene	5589 (3087)	507 (4510)	963 (8570)	1470 (13,090)
MAPP® gas	5301 (2927)	517 (4600)	1889 (16,820)	2406 (21,420)
Natural gas	4600 (2538)	11 (98)	989 (8810)	1000 (8900)
Propane	4579 (2526)	255 (2270)	2243 (19,970)	2498 (22,240)
Propylene	5193 (2867)	438 (3900)	1962 (17,470)	2371 (21,110)

Source: Courtesy of BOC Gases.

Table 29-5 Heating Value of Major Industrial Fuel Gases (combusted with pure oxygen).

safety features of the other fuel gases. These safety features include shock stability, narrow explosive limits in air, no pressure limitation, and slow burning velocities, Figure 29-19. Another safety advantage of MAPP® is its smell. The odor of MAPP® can be detected when there is as little as 100 parts per million (ppm) of the gas or 1/340 of its lower explosive limit in air. By law, propane, natural gas, and propylene must have an odor added to them so that they can be detected at a concentration of 1/15 of their lower explosive limit in air. The ability to detect even small leaks can save gas and avoid the possibility of explosions. The foul odor of MAPP® allows leaks to be found more than twenty-two times faster than acetylene leaks can be found.

Table 29-6 compares acetylene, MAPP® gas, and propylene for various procedures. The higher the number given, the better the performance of the gas for that procedure.

PROPANE AND NATURAL GAS

Propane, butane, and natural gas find limited use in the welding industry. Because of their relatively low-temperature, low-heat flames, they are seldom used for purposes other than heating and cutting.

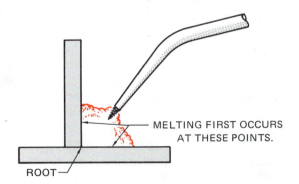

Figure 29-18 Secondary heat causes the sides of a tee joint to melt before the root of the joint melts.

The major advantage of propane and natural gas is that they are often used for heating the shop. Therefore, a supply of these gases is readily available. The handling of cylinders is reduced or eliminated because natural gas is piped directly to the shop and propane can be delivered in bulk tanks. Both gases are easily piped through the shop.

Propane and natural gas are both obtained from the petroleum industry. Natural gas comes from gas wells, and propane is produced from oil and gas at refineries. An artificial odor, a mercaptan chemical, must be added so that leaks can be detected for safety purposes.

Chemically, propane is C_3H_8, and natural gas is mostly methane (CH_4) and ethane (C_2H_6). The major disadvantage is that both gases consume a large amount of oxygen. Propane requires 85% of the flame's oxygen from the cylinder, and natural gas requires 95% of its oxygen from the cylinder, compared with an oxygen consumption of as little as 50% for the oxyacetylene flame.

HYDROGEN

Oxyhydrogen produces only a primary combustion flame, unlike hydrocarbon gases, which have both primary and secondary combustion. The hydrogen flame is almost colorless and can only be seen when dirt, dust, and other contaminants from the air glow while burning in the flame. Hydrogen is not widely used in welding because of its expense, limited availability, and some myths about its safety.

Hydrogen has the fastest burning velocity of any of the fuel gases at 36 ft/s (10.9 m/s). Acetylene has a burn rate of less than one-half that of hydrogen. Hydrogen has a very slight tendency to backfire, yet it does not flashback. Unlike acetylene, which can explosively decompose without oxygen, hydrogen cannot be made to react without the presence of sufficient oxygen.

Hydrogen is much lighter than air. Therefore, when it is released, it diffuses quickly, reducing the possibility of accidental combustion. If a large quantity of hydrogen is allowed to burn uncontrolled, the gas rises into the flame. This means it burns in an upward direction, away from people in an area. Most other gases burn in a downward direction, which can trap people in an area. The chance of large quantities of hydrogen exploding is limited. For example, when the hydrogen-filled airship, the *Hindenburg*, caught fire and burned in 1931, no explosion occurred, and most of the people on board the airship survived.

The low-flame temperature restricts the use of the oxyhydrogen flame to cutting, usually underwater, and to gas welding and brazing on low temperature metals such

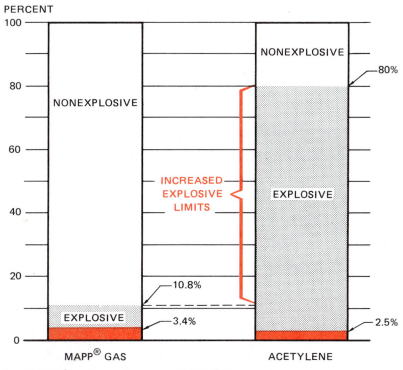

Figure 29-19 Explosive limits of MAPP® gas in air. (Courtesy of MAPP® Products.)

Application	Acetylene	MAPP® Gas	Propylene
Cutting			
Under 3/8 in thick	100	95	90
5/8 in to 5 in thick	95	100	95
Over 5 in thick	80	100	95
Cutting dirty or scaled surfaces	100	95	80
Repetitive cutting	100	100	80
Stack cutting	90	100	95
Cutting low-alloy specialty steels	100	90	80
Beveling	100	100	85
Cutting rounds	95	100	85
Piercing	100	100	85
Blind-hole piercing	100	90	80
Rivet washing	100	95	80
Gouging	100	100	85
Wire metallizing	80	100	90
Powder metallizing	100	0	0
Heating, stress relief, bending	70	100	90
Deep flame hardening	90	100	90
Shallow flame hardening	95	100	80
Cobalt-base hardsurfacing	100	0	0
Other alloy hardsurfacing	100	85	70
Welding	100	70	0
Braze welding	100	90	70
Brazing	100	100	90

Source: Courtesy of BOC Gases.

Table 29-6 Average Performance Ratings of Some Oxyfuel Flames.

as aluminum. The flame can be made reducing (needing oxygen) to help protect the aluminum from oxidation, without having excessive carbon to contaminate the weld. The finished flame product is water, H_2O. Only one-quarter of the flame oxygen comes from the cylinder.

Two major safety problems exist when hydrogen is used as a fuel gas. First, hydrogen has no smell, which makes it difficult to detect leaks. Second, the molecule is extremely small so that it leaks easily. When using hydrogen, an active leak-checking schedule must be followed to find small problems before they develop into disasters. It is possible for a leak to be on fire and not be noticed, because the hydrogen flame is almost invisible.

EXPERIMENT 29-2

Oxyfuel Flames

Using an identical torch set with each available fuel gas, you are going to observe the flame as each fuel gas is safely lit, adjusted, and extinguished.

Set all fuel and oxygen regulators at approximately 5 psig (35 kPag). Each torch should have the same size tip. The tip should have an orifice equal to a number 53 to 60 drill. Place the torches on a table with the tips pointed up, **Figure 29-20**.

Starting with the oxyacetylene torch, turn on the fuel gas valve slightly. Using a flint lighter, light the torch and adjust the gas valve so that the flame is not smoking. After securely placing the lit torch back on the table, repeat the process with all of the other torches. Adjust the flame of each torch to the same size as the acetylene flame, **Figure 29-21**.

Pick up the acetylene torch and move it back and forth, **Figure 29-22(A)**. The flame should be stable, with only the top deflecting as the torch is moved. Replace the torch carefully on the table.

One at a time, pick up each of the other torches and move them just as you did with the acetylene torch. The flames on these torches will deflect more, and some flames may go out, **Figure 29-22 (B)**. The reason for the difference is that less gas is flowing with each of the other gases. The acetylene flame is more stable because it has the highest flow rate and the highest burn rate. Thus, the flame is more compact.

Turn on the oxygen valve slowly until the acetylene flame is adjusted to a neutral setting. Repeat this procedure with each of the other flames. The flames may blow out as the oxygen is turned on. If this happens, direct the flame against the firebrick top of the welding table and readjust the oxygen until a neutral flame is reached with each torch. The flames may blow out because of the slow burning velocities of the gases. The flames should all look nearly the same in color, but the inner cone on the acetylene flame will be the shortest, **Figure 29-23** (page 746).

Line up the torches and hold the end of a gas welding rod with a 1/6-in. (3-mm) diameter in each flame. The welding rods should all be put in the flames at the same time. They should all be held the same height above the inner cone, about 1/4 in. (6 mm), **Figure 29-24** (page 746). Watch as the welding rods melt, each one at a different rate. Cool the welding rods and repeat the experiment with the ends of the welding rods 1/4 in. (6 mm) higher than before. After the welding rods melt, cool and raise the welding rods another 1/4 in. (6 mm). Keep repeating this step until the welding rods stop melting or turning red.

The acetylene flame should melt the low welding rod fastest, but one of the other gases should heat the rod faster as the distance above the inner cone increases. ◆

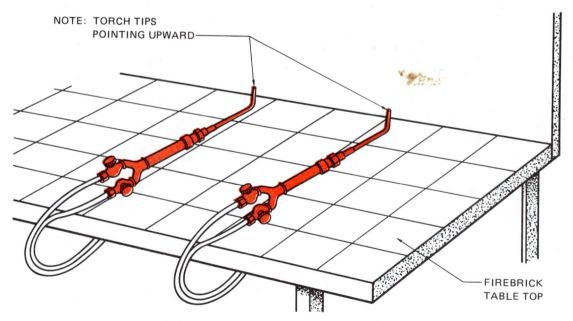

Figure 29-20 Torches set up to compare fuel gases.

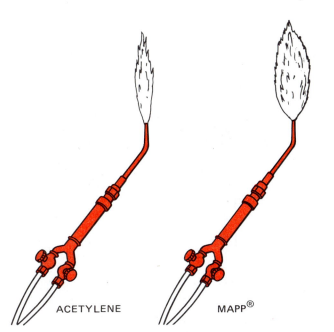

Figure 29-21 The second fuel gas flame may be slightly off the tip.

FILLER METALS

Filler metals specifically designed to be used with an oxyfuel torch are generally divided into three groups. One group of welding rods used for welding is designated with the prefix letter R. Another group of rods used for brazing is designated with the prefix letter B. A third group is used for buildup, wear resistance surfacing, or both. Welding rods in this group may also be classified in one of the other groups, may be patented and use a trade name, or may be tubular with a granular material in the center. The tubular welding rods are designated with an RWC prefix. Some filler metals, for example BRCuZn, are classified both as a braze welding rod (R) and a brazing rod (B) because they can be used either way.

Ferrous Metals

Ferrous filler metals are welding rods that are mainly iron. They may have other elements added to change their strength, corrosion resistance, weldability, or another physical property. There are three major AWS specifications for ferrous filler metals. These three specifications are A5.2 low-carbon, low-alloy steel, A5.15 cast iron, and A5.9 stainless steel. Within each specification, there are classes; for example, in group A5.2 the classes are RG45, RG60, and RG65.

The specifications and classes have minimum and maximum limits for the alloys that are added to provide the required physical properties of the weld they produce. Each manufacturer is free to make changes in the wire composition within the specified limits. The changes generally are concerned with weldability, tensile strength, ductility, cracking, appearance, porosity, impact strength, and hardness.

Physical changes are most often affected by changes in the percentages of alloys of carbon (C), silicon (Si), manganese (Mn), chromium (Cr), vanadium (V), nickel (Ni), and molybdenum (Mo). Contaminants from fuels used in the production of iron, such as phosphorus (P) and sulfur (S), have a negative effect on many of the desired properties and should be kept as low as possible.

Small shops sometimes use other types of wire for weld filler metals. The most popular substitution is often coat-hanger wire. Using such substitutes can cause weld failure. Coat-hanger wires were not manufactured for welding purposes, and their chemistry varies greatly. Porosity inside the weld deposit is common due to higher than acceptable levels of phosphorus (P) and sulfur (S). The painted finish of the wire will

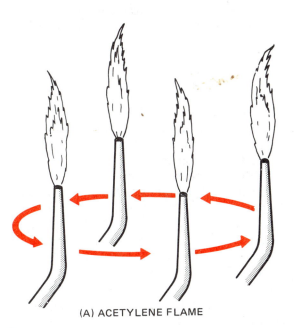

(A) ACETYLENE FLAME

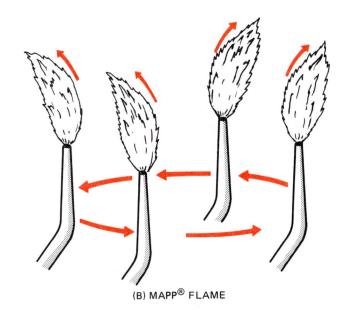

(B) MAPP® FLAME

Figure 29-22 Move the torches back and forth and watch the flame.

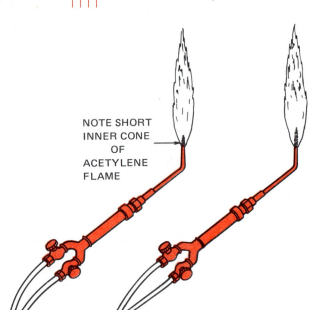

NOTE SHORT
INNER CONE
OF
ACETYLENE
FLAME

Figure 29-23 Compare the flames.

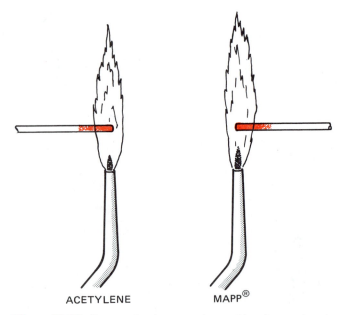

ACETYLENE   MAPP®

Figure 29-24 Observe the degree and rate of heating produced by each flame.

burn, causing further weld contamination and fumes that can be hazardous to the welder. Another type of substitute wire is often cadmium (Cd), plated to prevent rust. However, when cadmium is burned or vaporized, poisonous fumes are produced. The only safe filler metals to use are the ones specifically designed for welding.

MILD STEEL

Ferrous metals are generally classified by the American Welding Society (AWS) as mild steel, low-alloy steel, or cast iron. Mild steel and low-alloy steel are the materials that are most frequently gas welded. They are easily welded without a flux. Cast iron and stainless steels require fluxes and special techniques.

Mild steel and low-alloy gas welding rods are classified by the AWS as RG45, RG60, and RG65. The R refers to the welding rod, and the G refers to the ability to use gas for welding. The two digits indicate the tensile strength range of the weld deposited.

Class RG45 is a general-purpose gas welding rod that is often used for training welders. It has a smooth, shiny, molten weld pool that leaves a nice looking bead. The low carbon content helps make low-strength 45,000 psi to 55,000 psi (3,163 kg/cm^2 to 3,866 kg/cm^2) welds in the tensile strength range that are ductile. Automotive, auto body, wrought iron, and general welding shops use this welding rod for most or all of their gas welding.

Class RG60 welding rods, compared to RG45 welding rods, have a slightly higher carbon content and produce a higher weld strength in the range of 50,000 psi to 65,000 psi (3,515 kg/cm^2 to 4,569 kg/cm^2). The addition of silicon, manganese, and other metallic elements may improve the retention of carbon or other easily oxidized material. The molten weld pool is not as clear or shiny as

that obtained with RG45 welding rods, and the weld bead also may not look as nice. The RG60 welding rod can be used on low-alloy steels requiring good strength and ductility. It is frequently used for mild steel pipe welds, structural shapes, chrome-moly aircraft tubing, American Society of Mechanical Engineers (ASME) code welds, and gas tungsten arc welding.

Class RG65 welding rods are low-alloy, high-strength, 65,000 psi to 75,000 psi (4,569 kg/cm^2 to 5,272 kg/cm^2), low-creep, corrosion-resistant welding rods. The high content of silicon and manganese helps to retain carbon, but may cause a thick, crusty layer of manganese silicate flux (MnSiO$_3$) to float on top of the molten weld pool. Although this molten weld pool has the roughest looks, it is the purest and strongest. The RG65 welding rod is used in ASME high-pressure piping, tubing, and gas tungsten arc welding.

Cast Iron

Cast iron filler rods for gas welding are small, round, or square iron castings. The prefix R, which refers to the welding rod, is used in front of Cl, which stands for cast iron. A high-temperature, borax-based flux must be used to prevent the carbon from burning out.

Class RCI is the lower strength filler metal, which has a tensile strength of 20,000 psi to 26,000 psi (1,406 kg/cm^2 to 1,757 kg/cm^2). This class is recommended for most general cast iron repair and for the buildup and fill-in of damaged castings. The weld can be remachined if necessary.

As compared to RCI welding rods, class RCI-A welding rods have a higher tensile strength, in the range of 35,000 psi to 40,000 psi (2,460 kg/cm^2 to 2,612 kg/cm^2). An RCI-A welding rod is used on alloy cast irons or where higher strengths are required.

What's Being Done To Weld Metal-Matrix Composites?

If composites are going to be the main materials of construction for the 21st century, then the welding industry has its work cut out for it.

From the designers of metal-matrix composite mountain bicycles to those responsible for drawing up the blueprints of future exotic space platforms made out of polymer-matrix composite tubing, the cry goes out. How are we supposed to weld these materials anyway? Many of the answers aren't in the textbooks yet.

According to one of the world's leading authorities on composites, who asked not to be identified because of the sensitivity of his comment, the so-called joining problem is the excuse companies use when they don't want to get involved with these materials in the first place.

It is no picnic, the experts say, to try to weld, or even join by some other means, a material having a matrix of aluminum or titanium or thermoplastic and a reinforcement of glass or graphite or some complicated ceramic.

Companies that are into composites on a large scale are dealing with this because of the outstanding properties that can be gained in using them.

Those who sneer at composites, or at any other advanced material for that matter, contend that composites are nothing more than expensive aerospace materials. You'll never find them in any practical application, they say.

This is not true. Welded metal-matrix composites are starting to be used for the seamless tubing in drive shafts for light trucks. This tubing has to be welded to forged aluminum yokes, in production. Also, the frames for a new all-terrain "mountain" bicycle are being produced from aluminum-matrix composites. In this design, many of the sections have to be welded together.

But the procedure is not the same. The welder does not use his gas metal arc welding gun as he did when welding neat 6061 aluminum. In fact, he might not even be using gas metal arc welding at all.

Figure 1

Metal-matrix composite (MMC) extrusions are furnished by Duralcan USA of San Diego to Specialized Bicycle Components, Inc., Morgan Hill, California, for the frame in that company's new line of Stumpjumper M2 mountain bikes (Figure 1). The frames, which are guaranteed for life, consist of aluminum oxide-reinforced 6061 aluminum that is welded by the gas tungsten arc process using ER5356 filler metal.

The literature from Specialized Bicycle describes the MMC on the Stumpjumper M2 as "one-third the weight of steel, stronger and stiffer than aluminum alloys, but has unsurpassed shock-dampening abilities. In short, it's the ultimate material for a mountain bike frame."

ER5356 filler metal can be used in gas metal arc welding. It is also recommended that a power supply with a steep volt/ampere characteristic be used. Postweld heat treatment should be used in order to obtain the best mechanical properties.

Chemical Compositions of Weld Beads, % by weight							
Weld Pass	Cu	Fe	Mg	Mn	Si	Cr	Al_2O_3
2nd	0.06	0.31	3.86	0.06	0.31	0.1	6.6
3rd	0.06	0.33	3.82	0.06	0.24	0.1	7.9
4th	0.06	0.32	4.10	0.06	0.26	0.1	5.9

Source: Alcan International Ltd.

Table 1

Mechanical Properties of Welds			
Conditions	Ultimate Tensile Strength, MPa	*Yield Strength, MPa	Elongation, %
Weld Reinforcement Beads On			
As Welded	230 (2)	138 (1)	4.7 (0.9)
T5 Temper	252 (2)	169 (1)	3.6 (0.1)
T6 Temper	265 (12)	204 (9)	1.8 (0.6)
Weld Reinforcement Beads Off			
As Welded	228 (4)	132 (1)	6.6 (0.9)
T5 Temper	283 (1)	189 (10)	3.9 (0.2)
T6 Temper	283 (1)	189 (10)	3.9 (0.2)

*Yield Strength — 0.2% offset in 2 inch gauge length
Numbers in ( ) — standard deviation
Source: Alcan International Ltd.

1 Ksi = 6.896 MPa

Table 2

Welding studies have been performed involving the welding of 6061 aluminum reinforced by 20% by-volume aluminum oxide. The results highlighted the chemical composition of the weld beads (**Table 1**) and the mechanical properties of the same welds (**Table 2**).

The trick to welding any of these materials is to know what processes and procedures are being used successfully in industry, then move in that same direction.

(Adapted from an article by Bob Irving appearing in the Welding Journal. Courtesy of the American Welding Society, Inc.)

REVIEW

1. What elements make up all hydrocarbons?
2. What are the separate parts that make up an oxyacetylene flame?
3. Use **Table 29-1** (page 735) to determine which fuel gas requires the largest amount of oxygen from the torch.
4. Approximately how long would it take a 50/50 mixture of oxygen and acetylene to flash back through a 25-foot (7.62-m) -long hose?
5. Use **Table 29-2** (page 736) to determine which fuel gas has the highest burning velocity when mixed with oxygen.
6. How is acetylene produced?
7. Why is it not safe to use acetylene above 15 psig (100 kPag)?
8. Use **Table 6-3** to determine the largest tip orifice size (drill number) that could be used with an acetylene cylinder containing 175 cubic feet (4956 L).
9. Where is the highest temperature and where is the greatest heat produced in a neutral oxyacetylene flame?
10. Use **Table 6-4** to determine what would be the pressures in cylinders containing (a) MAPP, (b) propane, and (c) ethane on a 100°F (38°C) day.
11. What are methylactylene-propadiene fuel gases used for?
12. Use **Table 29-5** (page 736) to determine which oxygen fuel–gas mixture produces the highest total heat.
13. Which fuel gas has the strongest odor and is easiest to detect?
14. What is the major advantage of using propane or natural gas?
15. Use **Table 29-6** (page 736) to determine which fuel gas would be best for (a) cutting material thicker than 5 in. (125 mm), (b) powdered metallizing, (c) stack cutting.
16. What two major safety problems does hydrogen present?
17. Why should coat hangers not be used as gas welding filler metal?
18. Explain the significance of the AWS filler metal classification RG45.

Chapter 30

OXYACETYLENE WELDING

OBJECTIVES

After completing this chapter, the student should be able to:

- explain how to set up and weld mild steel.
- make a variety of welded joints in any position on thin-gauge, mild steel sheet.
- make a satisfactory weld on small-diameter pipe and tubing in any position.
- explain the effects of torch angle, flame height, filler metal size, and welding speed on gas welds.

KEY TERMS

penetration	vertical welds
torch angle	trailing edge
torch manipulation	flashing
molten weld pool	horizontal welds
weld crater	shelf
burnthrough	undercut
kindling temperature	overlap
heat sink	overhead welds
outside corner joint	1G position
flat butt joint	5G position
lap joint	2G position
tee joint	6G position
out-of-position welding	key hole

INTRODUCTION

Oxyacetylene welding is limited to thin metal sections or to times when portability is important. During the early years of welding, oxyacetylene was used to weld thick plate, 1 in. (25 mm) and thicker. Today it is used almost exclusively on thin metal, 11 gauge or thinner. One of the arc welding processes is most often used today for welding metal thicker than 16 gauge. Some of the arc welding processes, such as GMAW, are replacing the gas welding processes on metals as thin as 28 gauge, **Figure 30-1** (page 750). Because of the expanded use of arc welding processes on thinner sections, we will concentrate on the use of gas welding on metal having a thickness of 16 gauge (approximately 1/16 in. (2 mm)) or thinner.

Figure 30-1 Gas metal arc welded (GMAW) on 16-gauge mild steel.

MILD STEEL WELDS

Mild steel is the easiest metal to gas weld. With this metal, it is possible to make welds with 100% integrity (lack of porosity, oxides, or other defects) and that have excellent strength, ductility, and other positive characteristics. The secondary flame shields the molten weld pool from the air, which would cause oxidation. The atmospheric oxygen combines with the carbon monoxide (CO) from the outer flame envelope to produce carbon dioxide (CO_2). The carbon dioxide will not react with the molten weld pool. In addition, the carbon dioxide forces the surrounding atmosphere away from the weld.

Factors Affecting the Weld

Torch Tip Size. The torch tip size should be used to control the weld bead width, penetration, and speed.

Penetration is the depth into the base metal that the weld fusion or melting extends from the surface, excluding any reinforcement. Because each tip size has a limited operating range in which it can be used, tip sizes must be changed to suit the thickness and size of the metal being welded. Never lower the size of the torch flame when the correct tip size is unavailable. Other factors that can be changed to control the weld size are the torch angle, the flame-to-metal distance, the welding rod size, or the way the torch is manipulated.

Torch Angle. The torch angle and the angle between the inner cone and the metal have a great effect on the speed of melting and size of the molten weld pool. The ideal angle for the welding torch is 45°. As this angle increases toward 90°, the rate of heating increases. As the angle decreases toward 0°, the rate of heating decreases, as illustrated in **Figure 30-2**. The distance between the inner cone and the metal ideally should be 1/8 in. to 1/4 in. (3 mm to 6 mm). As this distance increases, the rate of heating decreases; as it decreases, the heating rate increases, **Figure 30-3**.

Welding Rod Size. Welding rod size and torch manipulation can be used to control the weld bead characteristics. A larger size welding rod can be used to cool the molten weld pool, increase buildup, and reduce penetration, **Figure 30-4(A), (B),** and **(C).** The torch can be manipulated so that the direct heat from the flame is flashed off the molten weld pool for a moment to allow it to cool, **Figure 30-5.**

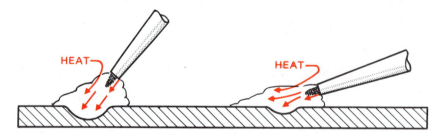

Figure 30-2 Changing the torch angle changes the percentage of heat that is transferred into the metal.

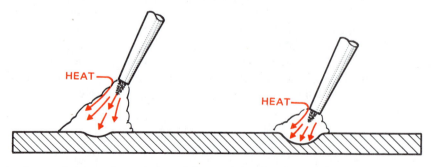

Figure 30-3 Changing the distance between the torch tip and the metal changes the percentage of heat input into the metal.

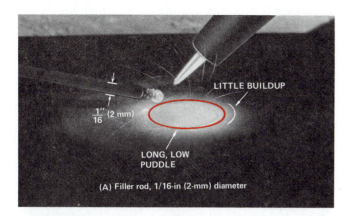

(A) Filler rod, 1/16-in (2-mm) diameter

LITTLE BUILDUP

1/16" (2 mm)

LONG, LOW PUDDLE

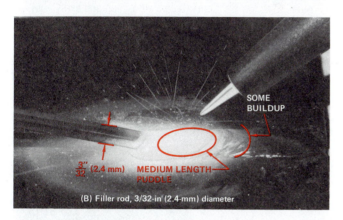

SOME BUILDUP

3/32" (2.4 mm)   MEDIUM LENGTH PUDDLE

(B) Filler rod, 3/32-in (2.4-mm) diameter

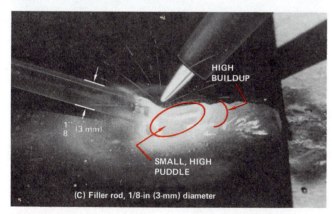

HIGH BUILDUP

1/8" (3 mm)

SMALL, HIGH PUDDLE

(C) Filler rod, 1/8-in (3-mm) diameter

Figure 30-4 If all other conditions remain the same, changing the size of the filler rod will affect the weld as shown in (A), (B), and (C).

Characteristics of the Weld

The molten weld pool must be protected by the secondary flame to prevent the atmosphere from contaminating the metal. If the secondary flame is suddenly moved away from a molten weld pool, the pool will throw off a large number of sparks. These sparks are caused by the rapid burning of the metal and its alloys as they come into contact with oxygen in the air. This is particularly a problem when a weld is stopped. The weld crater is especially susceptible to cracking. This tendency is greatly increased if the molten weld pool is allowed to burn out,

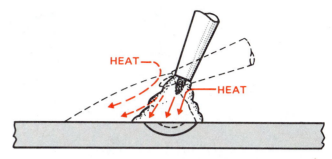

HEAT

HEAT

Figure 30-5 Flashing the flame off the metal will allow the molten weld pool to cool and reduce in size.

Figure 30-6 Building up the molten weld pool before it is ended will help prevent crater cracking.

Figure 30-6. To prevent burnout, the torch should be raised or tilted, keeping the outer flame envelope over the molten weld pool until it solidifies. As the molten weld pool is being cooled, the molten weld pool should also be filled with welding rod so that it is a uniform height compared to the surrounding weld bead.

The sparks that occur as the weld progresses are due to metal components that are being burned out of the weldment. Silicon oxides make up most of the sparks, and extra silicon can be added by the filler metal so that the weldment retains its desired soundness. A change in the number of sparks given off by the weld as it progresses can be used as an indication of changes in weld temperature. An increase in sparks on clean metal means an increase in weld temperature. A decrease in sparks indicates a decrease in weld temperature. Often the number of sparks in the air increases just before a burnthrough takes place; that is, burning out the molten metal that appears on the back side of the plate. This burnout does not happen to molten metal until it reaches the kindling temperature (the temperature that must be attained before something begins to burn). Small amounts of total penetration usually will not cause a burnout. When the sparks increase quickly, the torch should be pulled back to allow the metal to cool and prevent a burnthrough.

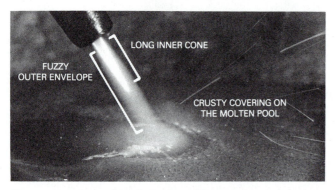

Figure 30-7 Carbonizing flame (excessive acetylene).

LONG INNER CONE

FUZZY OUTER ENVELOPE

CRUSTY COVERING ON THE MOLTEN POOL

SMOOTH MOLTEN POOL

Figure 30-8 Neutral flame (balanced oxygen and acetylene).

EXPERIMENT 30-1

Flame Effect on Metal

The first experiment examines how the flame affects mild steel. Use a piece of 16-gauge mild steel and the proper size torch tip. Light and adjust the flame by turning down the oxygen so that the flame has excessive acetylene. Hold the flame on the metal until it melts, and observe what happens, **Figure 30-7**. See Color Plate 49. Now adjust the flame by turning up the oxygen so that the flame is neutral. Hold this flame on the metal until it melts, and observe what takes place, **Figure 30-8**. See Color Plate 48. Next, adjust the flame by turning up the oxygen so that the flame has excessive oxygen. Hold this flame on the metal until it melts, and observe what happens, **Figure 30-9**. See Color Plate 50. Repeat this experiment until you can easily identify each of the three flame settings by the flame, molten weld pool, sound, and sparks.

Before actual welding is started, it is a good idea to find a comfortable position. The more comfortable or relaxed you are, the easier it will be for you to make uniform welds. The angle of the plate to you and the direction of travel are important.

Place a plate on the table in front of you and, with the torch off, practice moving the torch in one of the suggested patterns along a straight line, as illustrated in **Figure 30-10**. Turn the plate and repeat this step until you determine the most comfortable direction of travel. Later, when you have mastered several joints, you should change this angle and try to weld in a less comfortable position. Welding in the field or shop must often be done in positions that are less than comfortable, so the welder needs to be somewhat versatile.

It is important to feed the welding wire into the molten weld pool at a uniform rate. **Figure 30-11(A)** and **(B)** shows some suggested methods of feeding the wire by hand. It is also suggested that you not cut the welding wire in two pieces for welding. Short lengths are easy to use, but this practice is not widely accepted in industry. The end of the welding wire may be rested on your shoulder so that it is easy to handle.

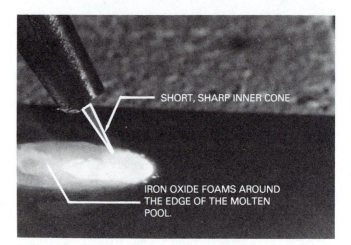

SHORT, SHARP INNER CONE

IRON OXIDE FOAMS AROUND THE EDGE OF THE MOLTEN POOL.

Figure 30-9 Oxidizing flame (excessive oxygen).

■ CAUTION

The end of the filler rod should have a hook bent in it so that you can readily tell which end may be hot, and so that the sharp end will not be a hazard to a welder who may be working next to you, **Figure 30-12** (page 754).

The torch hoses may be stiff and may therefore cause the torch to twist. Before you start welding, move the hoses so that there is no twisting of the torch. This will make the torch easy to manipulate and will be more relaxing for the welder. ◆

PRACTICE 30-1

Pushing a Molten Weld Pool

Using a clean piece of 16-gauge mild steel sheet, approximately 6 in. (152 mm) long, and a torch that is lit

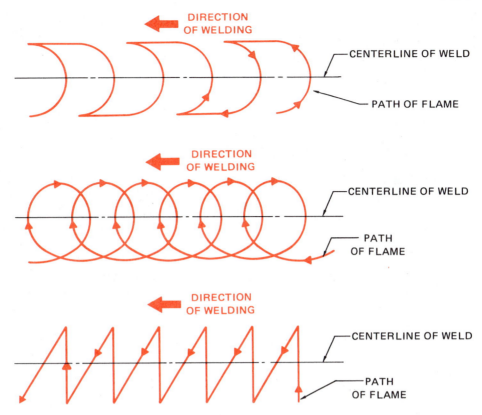

Figure 30-10 A few torch patterns.

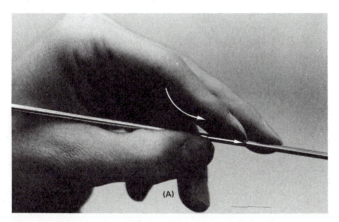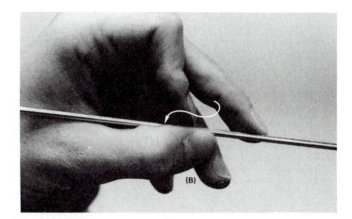

Figure 30-11 Feed the filler rod by using your index finger.

and adjusted to a neutral flame, push a molten weld pool in a straight line down the sheet. Start at one end and hold the torch at a 45° angle in the direction of the weld, **Figure 30-13** (page 754). When the metal starts to melt, move the torch in a circular pattern down the sheet toward the other end. If the size of the molten weld pool changes, speed up or slow down to keep it the same size all the way down the sheet, **Figure 30-14** (page 754). Repeat this practice until you can keep the width of the molten weld pool uniform and the direction of travel in a straight line.

Uniformity in width shows that you have control of the molten weld pool. A straight line indicates that you can see more than the molten weld pool itself. Students just learning to weld usually see only the molten weld pool. As you master the technique of welding, your visual range will increase. A broad visual range is important so that later you will be able to follow the joint, watch for distortion or see if other adjustments are needed, and relax as you weld. Turn off the cylinders, bleed the hoses, back out the regulator adjusting screw, and clean up your work area when you are finished. ◆

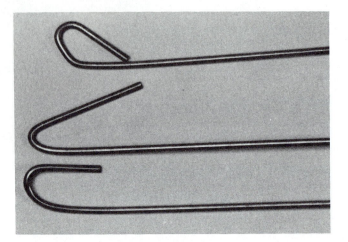

Figure 30-12 The end of the filler rod should be bent for safety and easy identification.

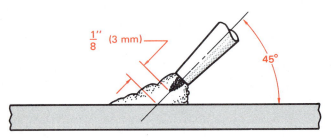

Figure 30-13 Hold the torch at a 45° angle in the direction of the weld.

Effect of Torch Angle and Torch Height Changes

Use a clean piece of 16-gauge mild steel and a torch adjusted to a neutral flame. You will be experimenting with different torch angles and torch heights to change the size of the molten weld pool, **Figure 30-15**.

To start this experiment, hold the torch at a 4° angle to the metal with the inner cone about 1/8 in. (3 mm) above the metal surface, **Figure 30-16(A), (B),** and **(C).** As the metal starts to melt, move the torch slowly down the sheet. As you move, increase and decrease the angle of the torch to the sheet and observe the change in the size of the molten weld pool. Repeat the experiment, but this time as you move down the sheet, raise and lower the torch and observe the effect on the size of the molten weld pool. ◆

Beading

Repeat Experiment 30-2 until you can control the change in the size of the molten weld pool, as shown in **Figure 30-17.** After the control of the molten weld pool has been mastered, you are ready to start adding filler rod. See Color Plate 51. When selecting and adding filler metal, the following are some facts to remember.

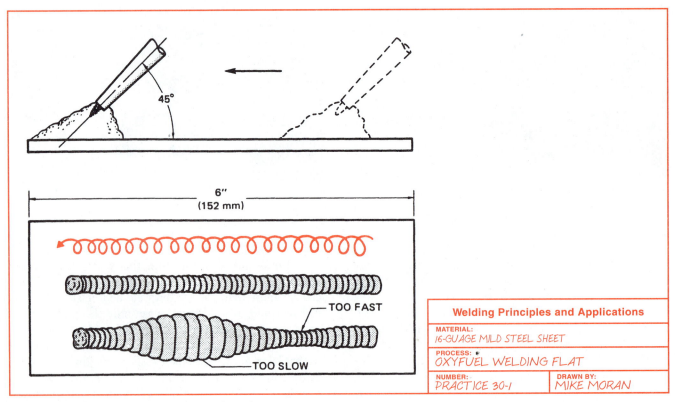

Figure 30-14 Effect of changing the rate of travel.

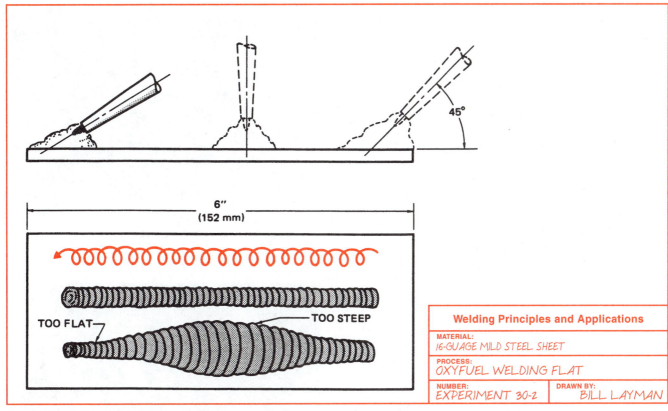

Figure 30-15 Effect of changing torch angle.

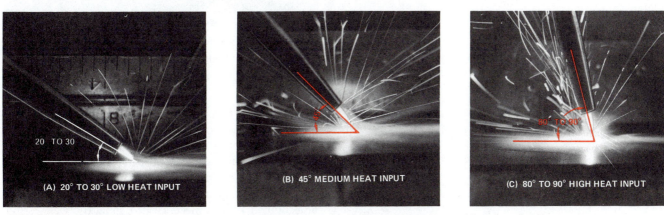

Figure 30-16 Changes in the angle between the torch and the work will change the molten weld pool produced.

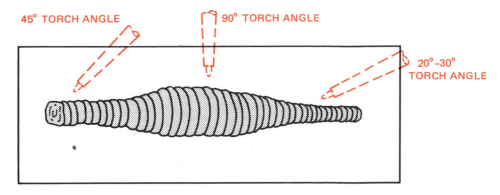

Figure 30-17 Changing the weld bead size by changing the torch angle.

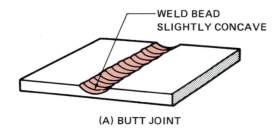

(A) BUTT JOINT

WELD BEAD
SLIGHTLY CONCAVE

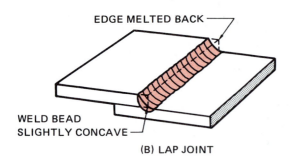

EDGE MELTED BACK

WELD BEAD
SLIGHTLY CONCAVE

(B) LAP JOINT

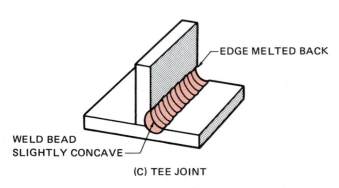

EDGE MELTED BACK

WELD BEAD
SLIGHTLY CONCAVE

(C) TEE JOINT

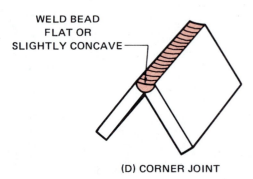

WELD BEAD
FLAT OR
SLIGHTLY CONCAVE

(D) CORNER JOINT

Figure 30-18 Indications that more filler metal should have been added.

Selecting the proper size or correct diameter of filler metal will help control the weld bead width, buildup, and penetration. A large-diameter filler rod can be used as a heat sink (something to draw off excessive heat) to keep the molten weld pool narrow, with little penetration and high buildup. As the diameter of the filler wire is decreased, the heat absorbing effect decreases, and the molten weld pool becomes wider, with deeper penetration and reduced buildup. On thin metal, it may be necessary to use a large-diameter filler rod to control burnthrough. On thick metal, a small-diameter filler rod may be used to increase penetration.

Figure 30-19 The hot end of the filler rod is protected from the atmosphere by the outer flame envelope.

Figure 30-20 Filler metal being correctly added by dipping the rod into the leading edge of the molten weld pool.

Another thing to remember when adding filler metal to a weld is to keep a smooth and uniform rhythm. Each weld joint has its own indicator that will tell you when to add welding rod, **Figure 30-18**. As a new welder, watch for these indicators and try to anticipate adding the rod so that the indicators do not appear. At the end of a joint, it may be necessary to apply the welding rod faster and cool the molten weld pool to keep the weld bead uniform.

The end of the welding rod should always be kept inside the protective envelope of the flame, **Figure 30-19**. The hot end of the welding rod oxidizes each time it is removed from the protection of the flame. This oxide is deposited in the molten weld pool, causing sparks and a weak or brittle weld.

When the rod is added to the molten weld pool, the flame should be moved back as the end of the rod is dipped into the leading edge of the molten weld pool, **Figure 30-20**. See Color Plate 52. If the torch is not moved back, the rod may melt and drip into the molten weld pool, **Figure 30-21**. The major problems with adding rod in this manner are: (1) the drop of metal tends to overheat, resulting in important alloys being

Figure 30-21 Filler metal being incorrectly added by allowing the rod to melt and drip into the molten weld pool.

burned out; (2) the metal cannot always be added where it is needed; and (3) the method works only in the flat position. When dipping the rod into the molten weld pool, if the rod touches the hot metal around the molten weld pool, it will stick. When this happens, move the flame directly to the end of the rod to melt and free it. Turn off the cylinders, bleed the hoses, back out the regulator adjusting screw, and clean up your work area when you are finished. ◆

EXPERIMENT 30-3

Effect of Rod Size on the Molten Weld Pool

Use a properly lit and adjusted torch, 6 in. (152 mm) of 16-gauge mild steel, and three different diameters of RG45 gas welding rods, 1/8 in. (3 mm), 3/32 in. (2.4 mm), and 1/16 in. (2 mm), by 36 in. (914 mm) long. In this experiment, you will observe the effect on the molten weld pool of changing the size of filler metal. You also will practice adding the filler metal to the molten weld pool.

Starting with the 1/8-in. (3-mm) filler metal, make a weld 6 in. (152 mm) long. Next to this weld, make another one with the 3/32-in. (2.4-mm) rod, and then one with the 1/16-in. (2-mm) rod. Try to keep the angle, speed, height, and pattern of the torch the same during each of the welds. Observe the differences in each of the weld sizes. ◆

PRACTICE 30-3

Stringer Bead, Flat Position

Repeat Experiment 30-3 until you have mastered a straight and uniform weld with any of the three sizes of filler rod.

Joining two or more clean pieces of metal to form a welded joint is the next step in learning to weld. The joints must be uniform in width and reinforcement so that they will have maximum strength. For each type of joint, the amount of penetration required to give maximum strength will vary and may not be 100%, **Figure 30-22**. For example, in thin sheet metal there is usually enough reinforcement on the weld to give the weld adequate strength. But if that reinforcement has to be removed, then 100% penetration is important, **Figure 30-23**. Some joints, such as the lap joint, never need 100% penetration. However, they do need 100% fusion. Turn off the cylinders, bleed the hoses, back out the regulator adjusting screw, and clean up your work area when you are finished. ◆

OUTSIDE CORNER JOINT

The flat outside corner joint can be made with or without the addition of filler metal. This joint is one of the easiest welded joints to make. If the sheets are tacked together properly, the addition of filler metal is not needed. However, if filler metal is added, it should be added uniformly, as in the stringer beads in Practice 30-3.

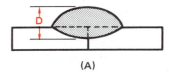
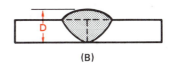

(A)   (B)

Figure 30-22 Welds (A) and (B) both have approximately the same strength — but only if the reinforcement does not have to be removed.

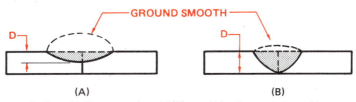

(A)   (B)

Figure 30-23 If the reinforcement on both welds is removed, weld (B) would be the stronger weld.

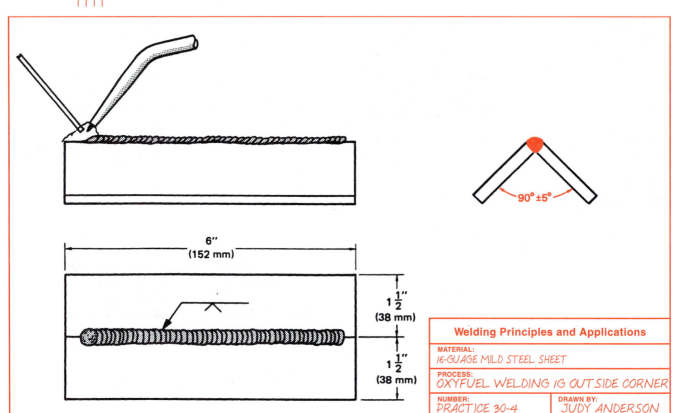

Figure 30-24 Outside corner.

PRACTICE 30-4

Outside Corner Joint Flat Position

Using a properly lit and adjusted torch, two clean pieces of 16-gauge mild steel, 6 in. (152 mm) long, and filler metal, you will make a flat outside corner welded joint, **Figure 30-24.**

Place one of the pieces of metal in a jig or on a firebrick and hold or brace the other piece of metal vertically on it, as shown in **Figures 30-25** and **30-26.** Tack the ends of the two sheets together. Then set it upright and put two or three more tacks on the joint, **Figure 30-27.** Holding the torch as shown in **Figure 30-28,** make a uniform weld along the joint. Repeat this until the weld can be made without defects. Turn off the cylinders, bleed

the hoses, back out the regulator adjusting screw, and clean up your work area when you are finished. ◆

BUTT JOINT

The flat butt joint is a welded joint and one of the easiest to make. To make the butt joint, place two clean pieces of metal flat on the table and tack weld both ends together as illustrated in **Figure 30-29.** Tack welds may also be placed along the joint before welding begins. Point the torch so that the flame is distributed equally on both sheets. The flame is to be in the direction that the weld is to progress. If the sheets to be welded are of different sizes or thicknesses, the torch should be pointed so that both pieces melt at the same time, **Figure 30-30.**

When both sheet edges have melted, add the filler rod in the same manner as in Practice 30-3.

PRACTICE 30-5

Butt Joint Flat Position

Using a properly lit and adjusted torch, two clean pieces of 16-gauge mild steel, 6 in. (152 mm) long, and filler metal, you will make a welded butt joint, **Figure 30-31** (page 760).

Place the two pieces of metal in a jig or on a firebrick and tack weld both ends together. The tack on the ends can be made by simply heating the ends and allowing

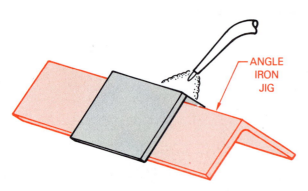

Figure 30-25 Angle iron jig for holding metal so it can be tack welded.

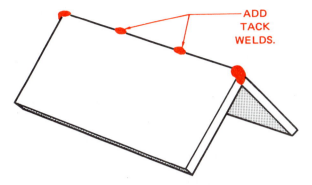

Figure 30-27

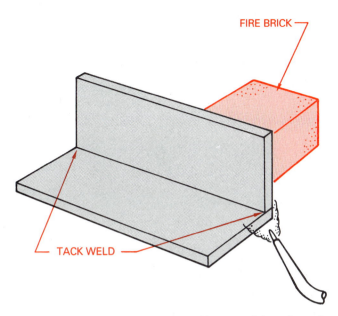

Figure 30-26 Tack welding the outside corner joint using a fire brick to support the metal.

Figure 30-28 Outside corner joint.

them to fuse together or by placing a small drop of filler metal on the sheet and heating the filler metal until it fuses to the sheet. The latter method is especially convenient if you have to use one hand to hold the sheets together and the other to hold the torch. After both ends are tacked together, place one or two small tacks along the joint to prevent warping during welding.

With the sheets tacked together, start welding from one end to the other using the technique learned in Practice 30-3. Repeat this weld until you can make a welded butt joint that is uniform in width and reinforcement and has no visual defects. The penetration of this practice weld may vary. Turn off the cylinders, bleed the hoses, back out the regulator adjusting screw, and clean up your work area when you are finished. ◆

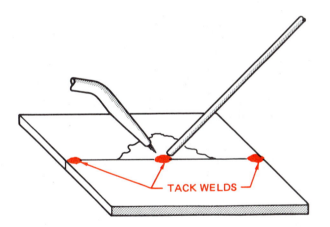

Figure 30-29 Making a tack weld.

PRACTICE 30-6

Butt Joint with 100% Penetration

Using the same equipment, materials, and setup as described in Practice 30-5, make a welded butt joint with

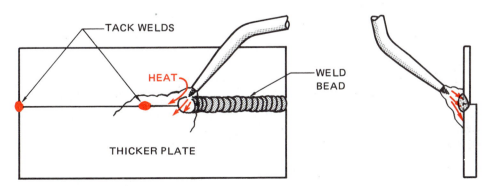

Figure 30-30 Direct the flame on the thicker plate.

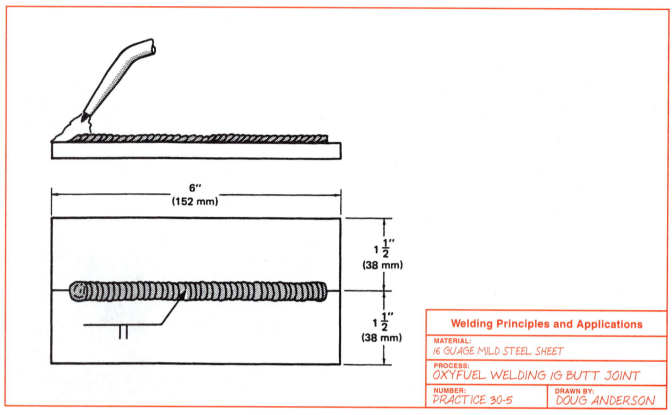

Welding Principles and Applications

MATERIAL:
16 GUAGE MILD STEEL SHEET

PROCESS:
OXYFUEL WELDING 1G BUTT JOINT

NUMBER: *PRACTICE 30-5*   DRAWN BY: *DOUG ANDERSON*

Figure 30-31 Butt joint.

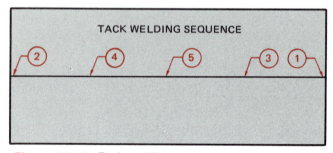

TACK WELDING SEQUENCE

Figure 30-32 Tack welding sequence used to minimize distortion.

100% penetration along the entire 6 in. (152 mm) of welded joint, and then visually inspect (VT) the root (back) to see if it has complete penetration. Turn off the cylinders, bleed the hoses, back out the regulator adjusting screw, and clean up your work area when you are finished. ◆

PRACTICE 30-7

Butt Joint with Minimum Distortion

Using a properly lit and adjusted torch, two clean pieces of 16-gauge mild steel, 6 in. (152 mm) long, and filler metal, you will make a welded butt joint while controlling distortion and penetration.

Distortion can be controlled by back stepping a weld, proper tacking, and clamping. For this weld, back

stepping and proper tacking will be used to control distortion. The tacking sequence to be used is shown in **Figure 30-32**. The back-stepping method to be used is illustrated in **Figure 30-33**. Back stepping will also eliminate the problem of burning away the end of the sheet. Practice this weld until you can pass a visual inspection (VT) for distortion. Turn off the cylinders, bleed the hoses, back out the regulator adjusting screw, and clean up your work area when you are finished. ◆

LAP JOINT

The flat lap joint can be easily welded if some basic manipulations are used. When heating the two clean sheets, caution must be exercised to ensure that both sheets start melting at the same time. Heat is not distributed uniformly in the lap joint, **Figure 30-34**. Because of this difference in heating rate, the flame must be directed on the bottom sheet and away from the metal top sheet, **Figure 30-35**. The filler rod should be added to the top sheet. Gravity will pull the molten weld pool down to the bottom sheet, so it is therefore not necessary to put metal on the bottom sheet. If the filler metal is not added to the top sheet or if it is not added fast enough, surface tension will pull the molten weld pool back from the joint, **Figure 30-36** (page 762). When this happens, the rod should be added directly into this notch, and it will close. The weld appearance and strength will not be affected.

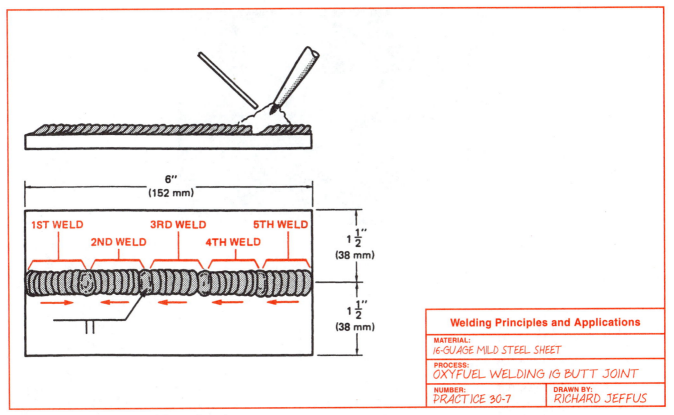

Figure 30-33 Welding with minimum distortion.

PRACTICE 30-8

Lap Joint Flat Position

Using a properly lit and adjusted torch, two clean pieces of 16-gauge mild steel, 6 in. (152 mm) long, and filler metal, you will make a welded lap joint, **Figure 30-37** (page 762). Place the two pieces of metal on a firebrick and tack both ends, as shown in **Figures 30-38** and 30-39 (page 762). Make two or three more tack welds. Starting at one end, make a uniform weld along the joint. Both sides of the joint can be welded. Repeat this practice until the weld can be made without defects. If you want to test your skill, shear out a strip 1 in. (25 mm) wide, as shown in **Figure 30-40** (page 763), and test it for 100% root penetration, **Figure 30-41** (page 763). Turn off the cylinders, bleed the hoses, back out the regulator adjusting screw, and clean up your work area when you are finished. ◆

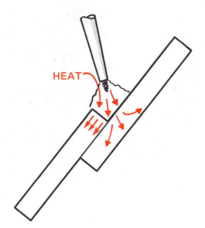

Figure 30-34 Heat is conducted away more quickly in the bottom plate, resulting in the top plate's melting more quickly.

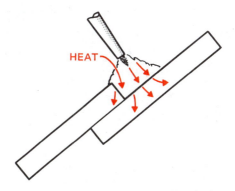

Figure 30-35 Flame heat should be directed at the bottom plate to compensate for thermal conductivity.

NOTE: THE TOP PLATE MELTS
BACK FORMING A NOTCH,
INDICATING THE NEED
FOR MORE FILLER METAL.

Figure 30-36

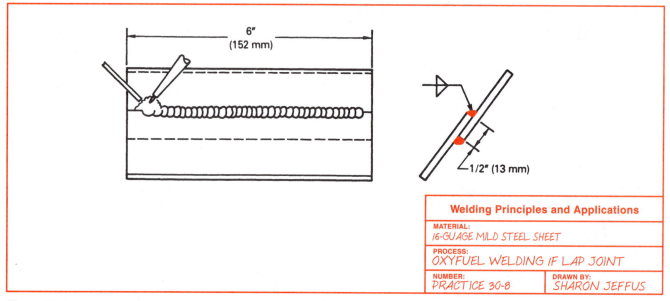

6″
(152 mm)

1/2″ (13 mm)

Welding Principles and Applications

MATERIAL:
16-GUAGE MILD STEEL SHEET

PROCESS:
OXYFUEL WELDING IF LAP JOINT

NUMBER:
PRACTICE 30-8

DRAWN BY:
SHARON JEFFUS

Figure 30-37

BOTH PLATES ARE
HEATED EQUALLY

Figure 30-38 Heating the joint before tacking.

NOTE: THE FLAME IS
DIRECTED MORE TOWARD
THE BOTTOM PLATE

Figure 30-39 Filler rod is added after both pieces are heated to a melt.

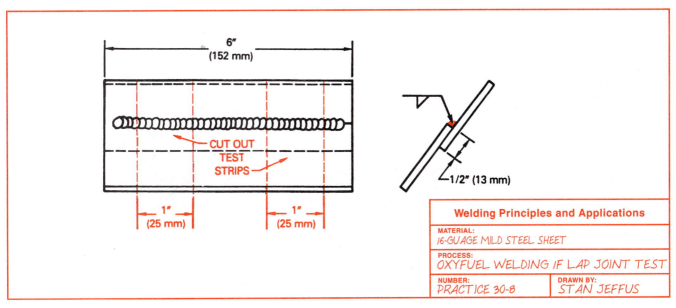

Welding Principles and Applications

MATERIAL:
16-GUAGE MILD STEEL SHEET

PROCESS:
OXYFUEL WELDING IF LAP JOINT TEST

NUMBER: DRAWN BY:
PRACTICE 30-8   *STAN JEFFUS*

Figure 30-40 Cut out test strips.

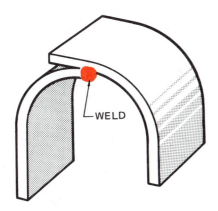

Figure 30-41 180° bend to test lap weld quality.

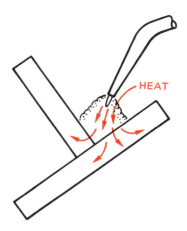

Figure 30-42 Direct the heat on the bottom plate to equalize the heating rates.

TEE JOINT

The flat **tee joint** is more difficult to make than the butt or lap joints. The tee joint has the same problem with uneven heating as the lap joint does. It is important to hold the flame so that both sheets melt at the same time, **Figure 30-42**. Another problem that is unique to the tee joint is that a large percentage of the welding heat is reflected back on the torch. This reflected heat can cause even a properly cleaned and adjusted torch to backfire or pop. To help prevent this from happening, angle the torch more in the direction of the weld travel. Because of the slightly restricted atmosphere of the tee joint, it may be necessary to adjust the flame so that it is somewhat oxidizing. The beginning student should not be overly concerned with this.

PRACTICE 30-9

Tee Joint, Flat Position

Using a properly lit and adjusted torch, two clean pieces of 16-gauge mild steel, 6 in. (152 mm) long, and filler metal, you will weld a flat tee joint, **Figure 30-43**.

Place the first piece of metal flat on a firebrick and hold or brace the second piece vertically on the first piece. The vertical piece should be within 5° of square to the bottom sheet. Tack the two sheets at the ends. Then put two or three more tacks along the joint and brace the tee joint in position. If you want to test your skill, cut out a strip 1 in. (25 mm) wide, as shown in **Figure 30-44**, and test it for 100% root penetration, **Figure 30-45** (page 765). Turn off the cylinders, bleed the hoses, back out the regulator adjusting screw, and clean up your work area when you are finished. ◆

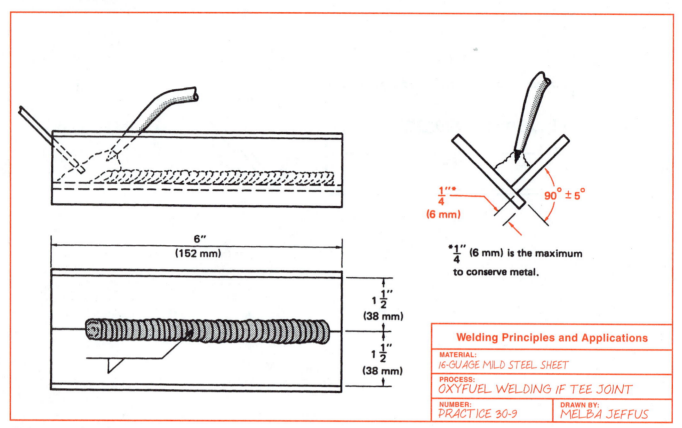

Figure 30-43 Tee joint.

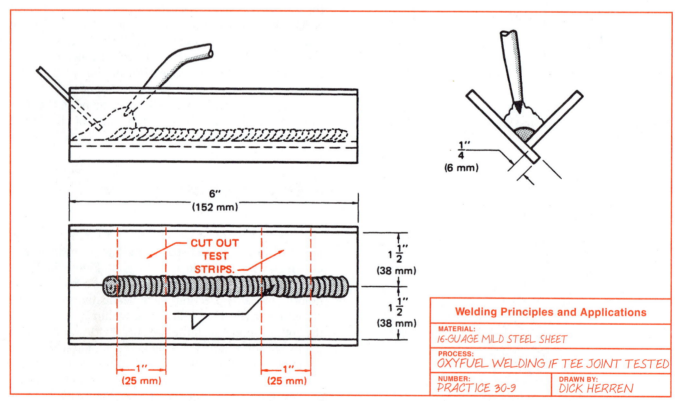

Figure 30-44 Cut out strips for testing.

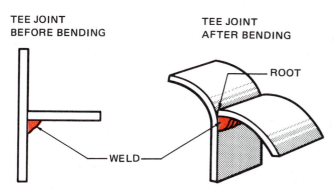

Figure 30-45 Bend the test strip to be sure the weld had good root fusion.

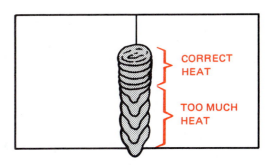

Figure 30-46 Vertical weld showing effect of too much heat.

OUT-OF-POSITION WELDING

A part to be welded cannot always be positioned so that it can be welded in the flat position. Whenever a weld is performed in a position other than flat, it is said to be out-of-position. Welds made in the vertical, horizontal, or overhead positions are out-of-position and somewhat more difficult than flat welds.

VERTICAL WELDS

A vertical weld is the most common out-of-position weld that a welder is required to perform. When making a vertical weld, it is important to control the size of the molten weld pool. If the molten weld pool size increases beyond that which the shelf will support, **Figure 30-46**, the molten weld pool will overflow and drip down the weld. These drops, when cooled, look like the drips of wax on a candle. To prevent the molten weld pool from dripping, the trailing edge of the molten weld pool must be watched. The trailing edge will constantly be solidifying, forming a new shelf to support the molten weld pool as the weld progresses upward, **Figure 30-47**. Small molten weld pools are less likely then large ones to drip.

The less vertical the sheet, the easier the weld is to make, but the type of manipulation required is the same. Welding on a sheet at a 45° angle requires the same manipulation and skill as welding on a vertical sheet. However, the speed of manipulation is slower, and the skill is less critical than at 90° vertical. This welding technique should be mastered at a 45° angle. Then the angle of the sheet is increased until it is possible to make totally vertical welds. Each practice weld in this section should be started on an incline, and, as skill is gained, the angle should be increased until the sheets are vertical.

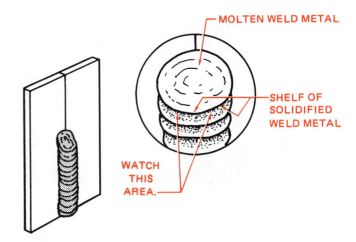

Figure 30-47 Watch the trailing edge to see that the molten pool stays properly supported on the shelf.

filler metal, you will make a bead at a 45° angle.

The filler metal should be added as you did in Practice 30-3. It may be necessary to flash the torch off the molten weld pool to allow it to cool, **Figure 30-48(A)**, **(B)**, and **(C)** (page 766). Flashing the torch off allows the molten weld pool to cool by moving the hotter inner cone away from the molten weld pool itself. While still protecting the molten metal with the outer flame envelope, a rhythm of moving the torch and adding the rod should be established. This rhythm helps make the bead uniform. Repeat this practice until the weld can be made without defects. Turn off the cylinders, bleed the hoses, back out the regulator adjusting screw, and clean up your work area when you are finished. ◆

PRACTICE 30-11

Stringer Bead, Vertical Position

Repeat Practice 30-10 until you have mastered a straight and uniform weld bead in a vertical position. Turn off the cylinders, bleed the hoses, back out the regulator adjusting screw, and clean up your work area when you are finished. ◆

PRACTICE 30-10

Stringer Bead at a 45° Angle

Using a properly lit and adjusted torch, two clean pieces of 16-gauge mild steel, 6 in. (152 mm) long, and

BUTT JOINT

PRACTICE 30-12

Butt Joint at a 45° Angle

Using a properly lit and adjusted torch, two clean pieces of 16-gauge mild steel, 6 in. (152 mm) long, and filler metal, you will make a welded butt joint at a 45° angle, **Figure 30-49**.

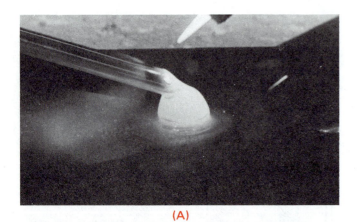

(A)

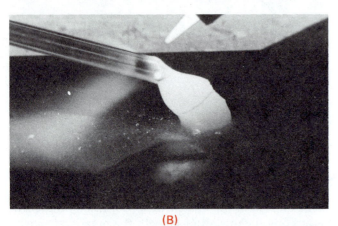

(B)

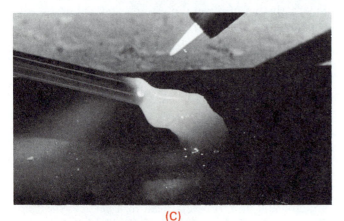

(C)

Figure 30-48 By flashing the flame off and controlling the pool size, a weld can be built up (A), and up (B), and over (C).

Tack the sheets together and support them at a 45° angle. Weld using the method of flashing the torch off the molten weld pool to control penetration and weld contour. Make a weld that has uniform width and reinforcement. Repeat this practice until the weld can be made without defects. Turn off the cylinders, bleed the hoses, back out the regulator adjusting screw, and clean up your work area when you are finished. ◆

PRACTICE 30-13

Butt Joint, Vertical Position

Using the same equipment, materials, and setup as described in Practice 30-12, make a welded butt joint in the vertical position. Make a weld that is uniform in width and reinforcement and has no visual detects. The penetration of this practice weld may vary. Turn off the cylinders, bleed the hoses, back out the regulator adjusting screw, and clean up your work area when you are finished. ◆

PRACTICE 30-14

Butt Joint, Vertical Position, with 100% Penetration

Using the same equipment, materials, and setup as listed in Practice 30-12, weld a butt joint in the vertical position with 100% penetration along the entire 6 in. (152 mm) of welded joint. Repeat this practice until this weld can be made without defects. If you want to test your skill, shear out a strip 1 in. (25 mm) wide, and test it for 100% root penetration, **Figure 30-50**. Turn off the cylinders, bleed the hoses, back out the regulator adjusting screw, and clean up your work area when you are finished. ◆

LAP JOINT

PRACTICE 30-15

Lap Joint at a 45° Angle

Using a properly lit and adjusted torch, two clean pieces of 16-gauge mild steel, 6 in. (152 mm) long, and filler metal, you will weld a lap joint at a 45° angle.

After tacking the sheets together and supporting them at a 45° angle, use the same method of adding rod as you did for the flat lap joint. Again, flash off the torch as needed to control the molten weld pool.

Repeat this weld until you can make a weld that is uniform in width and reinforcement and has no visual defects. Both sides of the joint can be welded. Turn off the cylinders, bleed the hoses, back out the regulator adjusting screw, and clean up your work area when you are finished. ◆

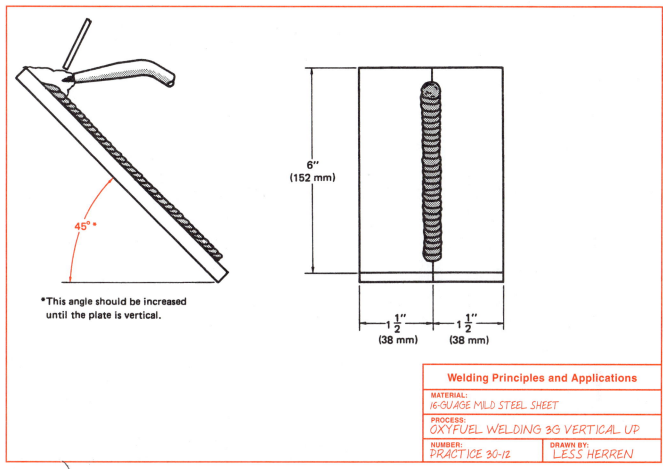

Welding Principles and Applications

MATERIAL:
16-GUAGE MILD STEEL SHEET

PROCESS:
OXYFUEL WELDING 3G VERTICAL UP

NUMBER:
PRACTICE 30-12

DRAWN BY:
LESS HERREN

Figure 30-49 Butt joint at a 45° angle.

PRACTICE 30-16

Lap Joint Vertical Position

Using the same equipment, materials, and setup as listed in Practice 30-15, weld a lap joint in the vertical position. Make a weld that is uniform in width and reinforcement and has no visual defects. Both sides of the joint can be welded. Repeat this practice until the weld can be made without defects. If you want to test your skill, shear out a strip 1 in. (25 mm) wide, and test it for 100% root penetration. Turn off the cylinders, bleed the hoses, back out the regulator adjusting screw, and clean up your work area when you are finished. ◆

TEE JOINT

The vertical tee joint has a right and a left side. Figure 30-51 (page 768) shows the best way to place the sheets depending upon whether the welder is right-handed or left-handed. It is a good idea to try both the right-hand and left-hand joints because in the field you may not be able to change the joint direction. Use the same method of adjusting the torch and torch angle that was practiced for the flat tee joint. In addition, use the method of flashing the torch off for molten weld pool con-

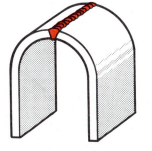
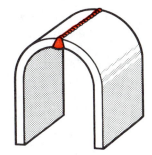

Figure 30-50 Bend strips to check a weld.

trol. Surface tension in the molten weld pool of a tee joint enables a larger weld to be made than either the butt or lap joints without as severe a problem with dripping.

PRACTICE 30-17

Tee Joint at a 45° Angle

Using a properly lit and adjusted torch, two clean pieces of 16-gauge mild steel, 6 in. (152 mm) long, and filler metal, you will weld a tee joint at a 45° angle.

After tacking the sheets together and supporting them at a 45° angle, make a fillet weld that has uniform

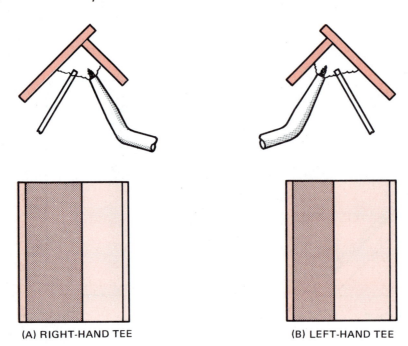

(A) RIGHT-HAND TEE   (B) LEFT-HAND TEE

Figure 30-51 Some vertical tee joints are easier for right-handed or left-handed welders.

Figure 30-52 A "J" weave pattern for horizontal welds.

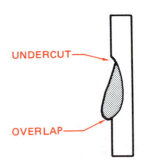

Figure 30-53 Too large a molten weld pool.

width and reinforcement and no visual defects. It is often best to weld only one side of the practice tee joint unless the oxides can be easily removed from the back side of the previous weld. Repeat this practice until the weld can be made without defects. Turn off the cylinders, bleed the hoses, back out the regulator adjusting screw, and clean up your work area when you are finished. ◆

Tee Joint, Vertical Position

Using the same equipment, materials, and setup as listed in Practice 30-17, make a fillet weld. Cut out a strip 1 in. (25 mm) wide (refer to Figure 30-44, page 764) and test it for 100% root penetration. Repeat this practice until the weld passes the test. Turn off the cylinders, bleed the hoses, back out the regulator adjusting screw, and clean up your work area when you are finished. ◆

HORIZONTAL WELDS

Horizontal welds, like vertical welds, must rely on some part of the weld bead to support the molten weld pool as the weld is made. The shelf that supports a horizontal weld must be built up under the molten weld pool and at the same time keep the weld bead uniform. The weave pattern required for a horizontal weld is completely different from that of any of the other positions. The pattern, **Figure 30-52**, builds a shelf on the bottom side of the bead to support the molten weld pool, which is elongated across the top. The sheet may be tipped back at a 45° angle for the stringer bead. Doing this allows the student to acquire the needed skills before proceeding to the more difficult, fully horizontal position. As with the vertically inclined sheet, the skills required are the same.

HORIZONTAL STRINGER BEAD

When starting a horizontal bead, it is important to start with a small bead and build it to the desired size. If too large a molten weld pool is started, the shelf does not

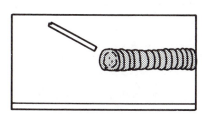
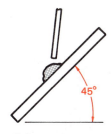

Figure 30-54 Horizontal stringer bead at a reclining angle.

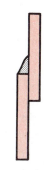

Figure 30-55 Horizontal lap joint.

have time to form properly. The weld bead will tend to sag downward and not be uniform. As a result, there may be an undercut of the top edge and an overlap on the bottom edge, **Figure 30-53**.

PRACTICE 30-19

Horizontal Stringer Bead at a 45° Sheet Angle

Using a properly lit and adjusted torch, one clean piece of 16-gauge mild steel, 6 in. (152 mm) long, and filler metal, you will make a horizontal bead at a 45° reclining angle, **Figure 30-54**.

Add the filler metal along the top leading edge of the molten weld pool. Surface tension will help hold it on the top. The weld should be uniform in width and reinforcement and have no visual defects. Repeat this practice until the weld can be made without defects. Turn off the cylinders, bleed the hoses, back out the regulator adjusting screw, and clean up your work area when you are finished. ◆

PRACTICE 30-20

Stringer Bead, Horizontal Position

Using the same equipment, materials, and setup as listed in Practice 30-19, make a stringer bead in the horizontal position. The stringer bead should be uniform in width and reinforcement and have no visual defects. Turn off the cylinders, bleed the hoses, back out the regulator adjusting screw, and clean up your work area when you are finished. ◆

BUTT JOINT

PRACTICE 30-21

Butt Joint, Horizontal Position

Using a properly lit and adjusted torch, two clean pieces of 16-gauge mild steel, 6 in. (152 mm) long, and filler metal, you will weld a butt joint in the horizontal position.

After tacking the sheets together and supporting them in the horizontal position, make a weld using the same technique as practiced in the horizontal beading, Practice 30-20. The weld must be uniform in width and reinforcement and have no visual defects. If you want to test your skill, shear out a strip 1 in. (25 mm) wide, and test it for 100% root penetration. Repeat this practice until the weld can be made without defects. Turn off the cylinders, bleed the hoses, back out the regulator adjusting screw, and clean up your work area when you are finished. ◆

LAP JOINT

PRACTICE 30-22

Lap Joint, Horizontal Position

Using a properly lit and adjusted torch, two clean pieces of 16-gauge mild steel, 6 in. (152 mm) long, and filler metal, you will weld a lap joint in the horizontal position.

After tacking the sheets together, support the assembly as illustrated in **Figure 30-55**. The weld must be uniform in width and reinforcement and have no visual defects. The sheet can be turned over, and the other side can be welded. If you want to test your skill, shear out a strip 1 in. (25 mm) wide, and test it for 100% root penetration. Repeat this practice until the weld can be made without defects. Turn off the cylinders, bleed the hoses, back out the regulator adjusting screw, and clean up your work area when you are finished. ◆

TEE JOINT

The horizontal and flat tee joints are very similar in relation to the types of skills required to preform metal welds. The horizontal fillet weld tends to flow down toward the horizontal sheet from the vertical sheet. To correct this problem, the filler metal should be added to the top edge of the molten weld pool.

Tee Joint, Horizontal Position

Using a properly lit and adjusted torch, two clean pieces of 16-gauge mild steel, 6 in. (162 mm) long, and filler metal, you will weld one side of a tee joint in the horizontal position. After tacking the sheets together, make a fillet weld that is uniform in width and reinforcement and has no visual defects. If you want to test your skill, shear out a strip 1 in. (25 mm) wide, and test it for 100% root penetration. Repeat this practice until the weld can be made without defects. Turn off the cylinders, bleed the hoses, back out the regulator adjusting screw, and clean up your work area when you are finished. ◆

OVERHEAD WELDS

When welding in the overhead position, it is important to wear the proper personal protection, including leather gloves, leather sleeves, a leather apron, and a cap. The possibility of being burned increases greatly when welding in the overhead position. However, with the proper protective clothing you should avoid being burned.

With the overhead weld, the molten weld pool is held to the sheet by surface tension in the same manner that a drop of water is held to the bottom of a glass sheet. If the molten weld pool gets too large, big drops of metal may fall. If the welding rod is not dipped into the molten weld pool, but is allowed to melt in the flame, it also may drip. As long as the molten weld pool is controlled and the rod is added properly, overhead welding is safe.

The direction that you choose to weld in the overhead position is one of personal preference. It is a good idea to try several directions before deciding on one. The height of the sheet also affects your skill and progress. Welders often prefer to stand while overhead welding so that sparks do not land in their laps. If you decide to stand, you need to somehow brace yourself to help your stability.

STRINGER BEAD

Place the metal at a height recommended by your instructor. With the torch off, your goggles down, and a rod in your hand, try to progress across the sheet in a straight line. Use several directions until you find the direction that best suits you. Change the height of the

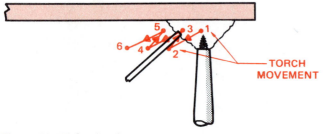

Figure 30-56 Overhead.

sheet up and down to determine the height at which welding is most comfortable.

Stringer Bead, Overhead Position

Using a properly lit and adjusted torch, and one clean piece of 16-gauge mild steel, 6 in. (152 mm) long, you will make a bead in the overhead position.

Heat the sheet until it melts and forms a molten weld pool. Put the welding rod into the molten weld pool as the torch tip is moved away from the molten weld pool. Return the flame to the molten weld pool as you remove the rod, **Figure 30-56**. Continue repeating this sequence as you move along the sheet. The weld bead should be uniform in width and reinforcement and have no visual defects. Repeat this practice until the weld can be made without defects. Turn off the cylinders, bleed the hoses, back out the regulator adjusting screw, and clean up your work area when you are finished. ◆

Butt Joint, Overhead Position

Using a properly lit and adjusted torch, two clean pieces of 16-gauge mild steel, 6 in. (152 mm) long, and filler metal, you will weld a butt joint in the overhead position.

After tacking the sheets together, put them in the overhead position. Following the sequence used in Practice 30-24 for the overhead stringer bead, make a weld along the joint. The weld should be uniform in width and reinforcement and have no visual defects. If you want to test your skill, shear out a strip 1 in. (25 mm) wide, and test it for 100% root penetration. Repeat this practice until the weld can be made without defects. Turn off the cylinders, bleed the hoses, back out the regulator adjusting screw, and clean up your work area when you are finished. ◆

Lap Joint, Overhead Position

Using a properly lit and adjusted torch, two clean pieces of 16-gauge mild steel, 6 in. (152 mm) long, and filler metal, you will weld a lap joint in the overhead position.

After tacking the sheets together, put them in the overhead position. Using the sequence for the overhead stringer bead, make a weld down the joint. The filler metal should be added to the leading edge of the molten weld pool on the top sheet. The weld should be uniform in width and reinforcement and have no visual defects. If you want to test your skill, shear out a strip 1 in. (25 mm) wide, and test it for 100% root penetration. Repeat this practice until the weld can be made without defects. Turn off the cylinders, bleed the hoses, back out the regulator adjusting screw, and clean up your work area when you are finished. ◆

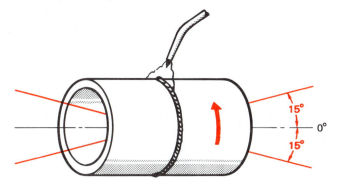

Figure 30-58 1G position. The pipe is rolled horizontally. The weld is made in the flat position (approximately 12 o'clock as the pipe is rolled).

MILD STEEL PIPE AND TUBING

Mild steel pipe and tubing, both small diameter and thin wall, can be gas welded. The welding process for both pipe and tubing are usually the same. Thin-wall material does not require a grooved preparation. Gas welding is very seldom used to manufacture piping systems. It is used on both pipe and tubing to make structures, such as bicycle and motorcycle frames, gates, works of art, hand rails, and light aircraft frames, **Figure 30-57.**

HORIZONTAL ROLLED POSITION 1G

The experiments and practices that follow (through Practice 30-29) will give the student the opportunity to gain skill in making welds in the 1G position, **Figure 30-58.**

EXPERIMENT 30-4

Effect of Changing Angle on Molten Weld Pool

This experiment will show how the molten weld pool is affected by changing the surface angle. With a piece of pipe having a diameter of approximately 2 in. (51 mm), you will push a molten weld pool across the top of the pipe. The pipe is in the 1G position, **Figure 30-59.** Starting at the 2 o'clock position, weld upward and across to the 11 o'clock position. Use the same torch

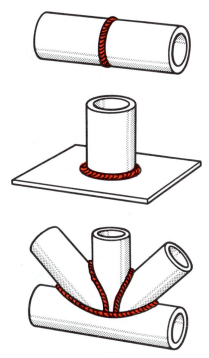

Figure 30-57 Examples of tube joints commonly used in industry.

PRACTICE 30-27

Tee Joint, Overhead Position

Using a properly lit and adjusted torch, two clean pieces of 16-gauge mild steel, 6 in. (152 mm) long, and filler metal, you will weld one side of a tee joint in the overhead position.

After tacking the sheets together, put them in the overhead position and make a fillet weld. The filler metal should be added to the top sheet. The weld should be uniform in width and reinforcement and have no visual defects. If you want to test your skill, cut out a strip 1 in. (25 mm) wide, and test it for 100% root penetration. Repeat this practice until the weld can be made without defects. Turn off the cylinders, bleed the hoses, back out the regulator adjusting screw, and clean up your work area when you are finished. ◆

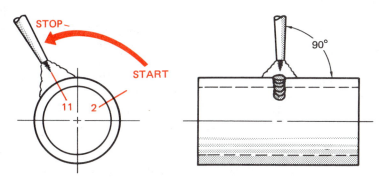

Figure 30-59 When the weld is finished, stop and roll the pipe so that the end of the weld is at the 2 o'clock position.

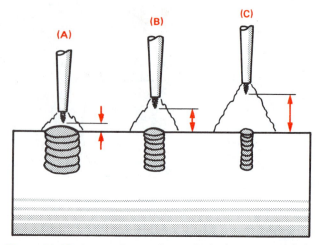

Figure 30-60 As the distance between the inner core and pipe changes, the size of the molten weld pool changes.

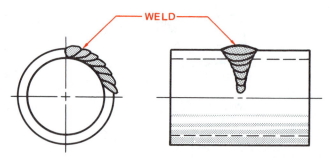

Figure 30-61 The weld bead shape is affected by its relative position on the pipe.

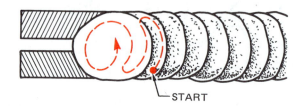

Figure 30-62 When restarting a weld pool, be sure the entire pool area is remelted before starting to add weld metal.

angle and distances that were learned in Practice 30-2 and Experiment 30-3. Repeat this experiment until you can make a straight weld bead that is uniform in width.

Keeping the torch at the correct angle to the pipe surface requires some practice. Changes in the relative torch angle will greatly affect the size of the molten weld pool. The distance of the inner cone from the pipe surface is also important, **Figure 30-60.** You must be comfortable and able to freely move around. Your free hand should not be used for supporting or steadying the torch or yourself. Later, when you need this hand to add filler metal, you will not have to learn how to be steady without the use of your hand. ◆

EXPERIMENT 30-5

Stringer Bead, 1G Position

Using a properly lit and adjusted torch, one clean piece of pipe approximately 2 in. (51 mm) in diameter, and three different sizes of filler metal, you will make a stringer bead on the pipe.

With the pipe in the 1G position, start a molten weld pool at the 2 o'clock position and weld to the 11 o'clock position. When you are at the 2 o'clock position, gravity will pull the molten metal outward, making the bead high and narrow, **Figure 30-61.** As the weld progresses toward the 12 o'clock position, if you keep using the same technique, the weld metal is pulled down, making the bead flat and wide. Repeat this experiment until you can make a straight weld bead that is uniform in width and reinforcement.

To keep the contour and width of the weld bead uniform, you must adjust your technique as the weld progresses. The following are some methods of keeping the buildup uniform:

- ■ Move the flame farther from the surface of the pipe.
- ■ Decrease the torch angle relative to the pipe surface.
- ■ Travel at an increasing rate of speed.

A combination of these methods can be used to control the spreading weld bead. The spreading weld bead should be treated as if it is becoming too hot and you must cool it down. However, gravity rather than temperature is the problem. But decreasing the temperature can solve the problem. ◆

EXPERIMENT 30-6

Stops and Starts

Using a properly lit and adjusted torch, one clean piece of pipe having a diameter of about 2 in. (51 mm), and filler metal, you will learn how to make good starts and stops. When welding pipe, you will frequently need to stop and restart. With practice and the proper technique, it is possible to make uniform stops and starts that are as strong as the surrounding weld bead.

To make a proper stop, you should slightly taper down the molten weld pool by flashing the torch off. This allows the molten weld pool to solidify before the flame is totally removed from the weld pool. If the flame is removed too soon from the molten weld pool, it will rapidly oxidize, throwing out a burst of sparks. The pocket of oxides will greatly weaken the weld at this point.

Restarting the weld requires that a molten weld pool equal in size to the original one be re-established. To restart the molten weld pool, point the flame slightly ahead of the crater at the end of the weld bead, **Figure 30-62.** When this metal starts to melt, move the flame back to the weld crater and melt the entire crater. Once

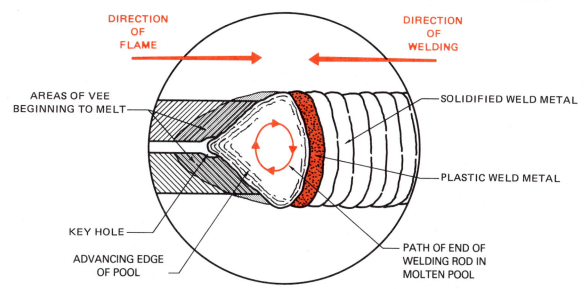

Figure 30-63

the crater melts, start adding filler metal and continue with the weld. If the metal ahead of the crater is not heated up first when the weld metal is added to the crater, it may form a cold lap over the base metal.

Practice stops and starts until you can make them so that they are uniform and unnoticeable on the weld bead. ◆

PRACTICE 30-28

Stringer Bead, 1G Position

Using a properly lit and adjusted torch and one clean piece of pipe approximately 2 in. (51 mm) in

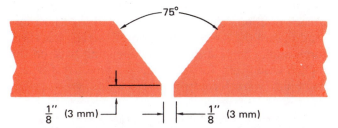

Figure 30-64   The bevel on the pipe may be oxyfuel flamecut, ground, or machined.

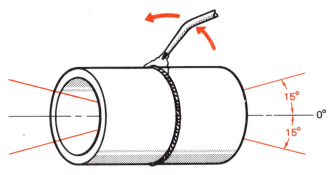

Figure 30-65   5G position. The pipe is fixed horizontally.

diameter, you will make a stringer bead around the pipe.

With the pipe in the 1G position, start a welding bead at the 2 o'clock position and weld toward the 12 o'clock position. When you must stop to change positions, roll the pipe so the ending crater is at the 2 o'clock position. Start the weld bead as practiced in Experiment 30-6 and proceed with the weld as before **Figure 30-63**. Repeat this process until the weld bead extends all the way around the pipe. Repeat this practice until you can produce a straight weld bead that is uniform in width and reinforcement and has no visual defects. Turn off the cylinders, bleed the hoses, back out the regulator adjusting screw, and clean up your work area when you are finished. ◆

PRACTICE 30-29

Butt Joint, 1G Position

Using a properly lit and adjusted torch, two clean pieces of schedule 40 pipe approximately 2 in. (51 mm) in diameter, and filler metal, you will weld a butt joint in the 1G position.

The pipe ends should be prepared as shown in **Figure 30-64**. The weld will be made with one pass. Tack weld the pipe together and place it on a firebrick. Using the principles and applications you learned in Practice 30-28, make this weld. Repeat this weld until it can be made without defects. Turn off the cylinders, bleed the hoses, back out the regulator adjusting screw, and clean up your work area when you are finished. ◆

HORIZONTAL FIXED POSITION 5G

The horizontal fixed position 5G weld, **Figure 30-65**, requires little skill development after completing the horizontal rolled position 1G welds. The torch height and

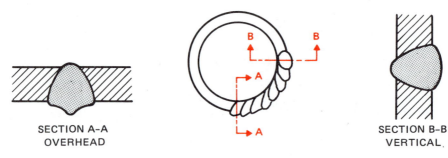

Figure 30-66 Changes in weld bead shape at different locations.

angle skills you have developed will help you master the 5G position.

As the weld changes from overhead (6 o'clock) to vertical (3 o'clock), the weld contour changes. At the overhead position, the weld bead tends to be shaped as shown in **Figure 30-66, Section A-A.** Note that the bead

is high in the center and recessed and possibly undercut on the sides. The vertical position has a high and narrow bead shape, **Figure 30-66, Section B-B.**

The overhead bead shape can be controlled by stepping the molten weld pool and moving the flame and rod back and forth at the same time. This process will deposit the metal and allow it to cool before it can sag. It allows the surface tension to hold the metal in place.

As the weld progresses toward the vertical section, the need to step the weld decreases. When the weld reaches the vertical section, the bead shape should be controlled by torch angle and flame distance.

Figure 30-67 Pipe stand.

THE BASE MAY BE ATTACHED TO A TABLE OR FLOOR PLATE.

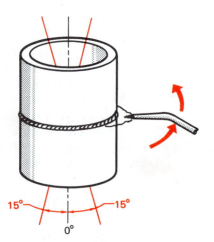

Figure 30-68 2G position. The pipe is fixed vertically and welded horizontally.

EXPERIMENT 30-7

5G Position

In this experiment, you are going to push a molten weld pool from the bottom of a fixed pipe up to the side of the pipe to see how torch manipulation affects the bead shape.

Using a properly lit and adjusted torch and one clean piece of pipe approximately 2 in. (51 mm) in diameter, start by establishing a molten weld pool at the 6 o'clock position. Then move the molten weld pool forward toward the 3 o'clock position. Observe what effect torch angle, flame distance, and stepping have on the molten weld pool. Turn the pipe and repeat this experiment until you can control the bead width. ◆

PRACTICE 30-30

Stringer Bead, 5G Position

Using a properly lit and adjusted torch, one clean piece of pipe having a diameter of about 2 in. (51 mm), and filler metal, you will make a stringer bead upward ground one side of the pipe.

With the pipe in the 5G position, start a welding bead at the 6 o'clock position and weld toward the 12 o'clock position. Use all procedures necessary to keep the weld bead uniform. Repeat this practice until you can produce a straight weld bead that is uniform in width and reinforcement and has no visual defects. Turn off the cylinders, bleed the hoses, back out the regulator adjusting screw, and clean up your work area when you are finished. ◆

PRACTICE 30-31

Butt Joint, 5G Position

Using a properly lit and adjusted torch, two clean pieces of pipe having a diameter of approximately 2 in. (51 mm), and filler metal, you will weld a butt joint in the 5G position.

The pipe ends should be beveled as shown in Figure 30-64 and tack welded together. Secure the pipe on a stand, such as the one shown in **Figure 30-67**, and start welding at the 6 o'clock position, moving toward the 12 o'clock position. When you reach the top, stop, restart back at the 6 o'clock position, and continue up the other side. Repeat this weld until it can be made without defects. Turn off the cylinders, bleed the hoses, back out the regulator adjusting screw, and clean up your work area when you are finished. ◆

VERTICAL FIXED POSITION 2G

The vertically fixed pipe requires a horizontal weld, **Figure 30-68**. The welding manipulative skill required for the 2G position is similar to that for a horizontal butt joint. The pipe may be rotated around its vertical axis but may not be turned end for end after you have started welding, **Figure 30-69**.

PRACTICE 30-32

Stringer Bead, 2G Position

Using a properly lit and adjusted torch and one clean piece of pipe having a diameter of approximately 2 in. (51 mm), you will make a stringer bead around the pipe.

With the pipe in the 2G position, start with a small bead, as illustrated in **Figure 30-70**, and then increase the bead to the desired size. Starting small will allow you to build a shelf to support the molten weld pool. It also will let you tie the end of the weld into the start of the weld so that a stronger weld is obtained. Repeat this weld until it can be made without defects. Turn off the cylinders, bleed the hoses, back out the regulator adjusting screw, and clean up your work area when you are finished. ◆

PRACTICE 30-33

Butt Joint, 2G Position

Using a properly lit and adjusted torch, two clean pieces of pipe having a diameter of approximately 2 in. (51 mm), and filler metal, you will weld a butt joint in the 2G position. The pipe ends should be beveled and tack welded together. Secure the pipe in the vertical

position and start welding. Repeat this weld until it can be made without defects. Turn off the cylinders, bleed the hoses, back out the regulator adjusting screw, and clean up your work area when you are finished. ◆

45° FIXED POSITION 6G

The 45° fixed pipe position, **Figure 30-71**, requires careful manipulation of the molten weld pool to ensure a uniform and satisfactory weld. The weld progresses

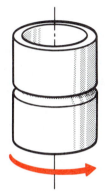

Figure 30-69 2G vertical fixed position.

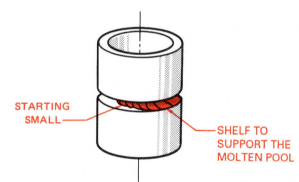

STARTING SMALL

SHELF TO SUPPORT THE MOLTEN POOL

Figure 30-70 The proper starting technique will aid with tying in the weld when it is completed around the pipe.

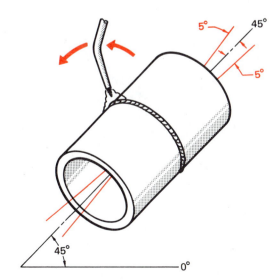

Figure 30-71 6G position. The pipe is inclined at a 45° angle.

around the pipe, changing from vertical to horizontal to overhead to flat and not completely in any one position. It is the combination of compound angles that makes the 6G position particularly difficult.

Stringer Bead, 6G Position

Using a properly lit and adjusted torch, one clean piece of pipe having a diameter of approximately 2 in. (51 mm), and filler metal, you will make a stringer bead around the pipe with the pipe in the 6G position.

Start at the bottom and weld upward toward the top of the pipe, Figure 30-72. The weld bead shape will change as you move around the pipe. To prevent the bottom from pulling to one side and slight movement to the high side, the side movement will create a shelf to hold the metal in place.

The side of the bead will pull to one side, but not so severely as at the bottom. To keep the side from being pulled out of shape, simply add the filler metal on the high side of the joint. Some side movement may be required but not so much as for the bottom.

The top will pull down to one side and tend to be flatter than the other parts of the bead, especially along the side of the pipe. To prevent one-sidedness, the filler metal is added to the top side. To add to the buildup of the bead, change the torch angle or the flame height.

Repeat this weld until it can be made straight and uniform in width and reinforcement. Turn off the cylinders, bleed the hoses, back out the regulator adjust-ing screw, and clean up your work area when you are finished. ◆

Butt Joint, 6G Position

Using a properly lit and adjusted torch, one clean piece of pipe approximately 2 in. (51 mm) in diameter, and filler metal, you will weld a butt joint in the 6G position. The pipe ends should be beveled and tack welded together. Secure the pipe at a 45° angle and start welding. Repeat this weld until it can be made without defects. Turn off the cylinders, bleed the hoses, back out the regulator adjusting screw, and clean up your work area when you are finished. ◆

THIN-WALL TUBING

Welding thin-wall tubing requires a technique similar to welding a stringer bead around pipe. If penetration is not a concern, the welding proceeds as if you are making a stringer bead on pipe. However, if penetration is required, then the weld will probably have a key hole to ensure 100% penetration, Figure 30-73. This will be similar to welding the root pass.

If you want to practice on tubing, and it is not readily available, then 16-gauge sheet metal can be rolled up and tack welded. Since 16-gauge is a standard wall thickness for tubing, fabricating your own will give you a realistic experience.

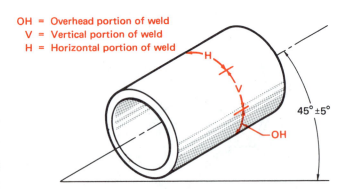

OH = Overhead portion of weld
V = Vertical portion of weld
H = Horizontal portion of weld

45° ±5°

Figure 30-72 6G position.

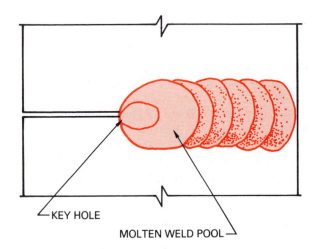

KEY HOLE

MOLTEN WELD POOL

Figure 30-73

The Alton Dam — Taming the Great Mississippi

Figure 1

One of the largest civil engineering projects undertaken in recent years is the replacement of a set of locks and a dam on the Mississippi River north of St. Louis.

The dam is owned by the U.S. Army Corps of Engineers and is strategically located at the center of the 25,000-mile inland waterway system. Designed and built in the 1930s, the existing locks and dam have seen an increase in tonnage transported of 10.5% per year since their completion in 1938. Many of the tows that travel the Mississippi today are too large to pass through the old locks; therefore they had to be double-locked: separated, locked in sections, and then reassembled following lockage. This created an expensive bottleneck, with average waiting times of twenty hours for valuable cargos of grain, chemicals, and energy products that pass through the locks each day. Traffic projections made by the Army Corps of Engineers demonstrated that the problem would grow more severe, with demand soon exceeding the maximum 63 million ton annual locking capacity of the existing structure.

The new lock and dam, planned for completion in three phases, is being built two miles downstream from the existing structures. The first phase involves the following: dewatering a 25-acre work area (containing 250 million gallons of water) extending 1,000 feet across the Mississippi River channel; driving nearly 4,800 foundation H-Piles; pacing 120,000 cubic yards of concrete; consumption of over 13,000 pounds of weld metal for field erection of six prefabricated Tainter gates, the largest ever constructed, **Figure 1**; erecting-in-place eighteen new cofferdam cells with connecting arcs, **Figure 2**; and removing the existing cofferdam. To complete Phase 1, the Army Corps of Engineers allotted a total of 900 days, in keeping with its schedule for all three phases of this major civil engineering project.

All contractors involved in the project soon discovered that the Mississippi River had little respect for man-made schedules. During one six-month period alone, the project experienced three high-river emergencies, one of which required actual rewatering of the cofferdam and subsequent cleanup. This was followed by a runaway barge that punched a hole in the outside of the cofferdam. On a project in which the proper sequence is a crucial determinant of success, procedures planned for execution in summer and fall weather were unavoidably delayed, forcing their completion in the bitterest of winter weather. In some cases, crews braved wind-chill factors of -60°F to keep the project on schedule.

Approaching the end of Phase 1, the contractor faced the on-site fabrication and erection of the six 512-ton Tainter gates, each measuring 110 feet wide by 60 feet tall. **Figure 3** shows the completed gates in position. This work was a major challenge, even with skilled ironworkers. Semiautomatic welding equipment was used on this project to achieve cost and time savings, as well as to maintain weld quality and control distortion. Extremely high quality welds were required at production

Figure 2

rates nearly double those possible with manual metal arc welding.

For the final field assembly of the giant Tainter gates, a number of equipment/electrode combinations were used to best meet the specific requirements of each step in the process. Portable wire welders were used in conjunction with electrodes for all-position work as well as electrodes designed for all-position welding and open gap root pass welding. In all, each Tainter gate required 2,358 pounds of weld metal — approximately 41 miles of semiautomatic flux-cored wire, melted into more than 2,000 linear feet of weld joint. The iron-workers who did the welding often worked under adverse conditions. At times they had to weld overhead up to a 45 degree incline (a very difficult weld to make), or they had to make a 36-in.-long vertical up weld in an area of 18 inches by 12 inches with minimum preheat of 150°F and maximum of 400°F, depending on the material. In many places the welders used air hoses attached to themselves to withstand the heat.

Phase 1 of the project is complete after 3-1/2 years. Phase 2, the construction of a 1,200-foot-

Figure 3

long lock, is underway with completion scheduled for 44 months after startup.

(Courtesy of The Lincoln Electric Company, Cleveland, Ohio 44117.)

REVIEW

1. What protects the molten weld pool from oxidation during a gas weld?

2. How does tip size affect a gas weld?

3. How does torch angle affect a gas weld?

4. How does welding rod size affect a gas weld?

5. What is a good indication that the molten weld pool is not being protected from oxidation?

6. Why should the end of the gas filler rod be bent?

7. What is the purpose of a heat sink?

8. Why should the end of the welding rod be kept inside the flame envelope?

9. Does a weld have to have 100% penetration to be strong?

10. What technique can be used to minimize weld distortion on a flat butt joint?

11. Why does the edge of the top plate on a lap joint tend to melt more easily?

12. What is meant by the term *out-of-position*?

13. What holds the molten weld pool in place on a vertical weld?

14. What additional personal safety protection should be used for overhead welds?

15. What force holds the molten weld pool in place on an overhead weld?

16. What is the welding sequence for a 1G weld?

17. What is the proper method for stopping a weld pool so it can be easily restarted?

18. What welding positions are required when making a 5G weld on pipe?

19. What welding positions are required when making a 6G weld on pipe?

20. How can 100% penetration be ensured when making welds on thin wall tubing?

Chapter 31

SOLDERING, BRAZING, AND BRAZE WELDING

OBJECTIVES

After completing this chapter, the student should be able to:

■ define the terms *soldering, brazing,* and *braze welding.*
■ explain the advantages and disadvantages of liquid solid phase bonding.
■ demonstrate an ability to properly clean, assemble, and perform required practice joints.
■ describe the functions of fluxes in making proper liquid-solid phase bonded joints.

KEY TERMS

liquid-solid phase
   bonding processes
phase
soldering
brazing
braze welding
capillary action
tensile strength
shear strength
ductility
fatigue resistance
elastic limit
fatigue failures
corrosion resistance

fluxes
torch soldering/brazing
furnace soldering/brazing
induction soldering/brazing
dip soldering/brazing
resistance soldering/brazing
paste range
eutectic composition
soldering alloys
brazing alloys
low-fuming alloy
silver braze
braze buildup

INTRODUCTION

Soldering and brazing are both classified by the American Welding Society as liquid-solid phase bonding processes. Liquid means that the filler metal is melted; solid means that the base material or materials are not melted. The phase is the temperature at which bonding takes place between the solid base material and the liquid filler metal. The bond between the base material and filler metal is a metallurgical bond because no melting or alloying of the base metal occurs. If done correctly, this bond results in a joint having four or five times the tensile strength of that of the filler metal itself.

Soldering and brazing differ only in that soldering takes place at a temperature below 840°F (450°C) and brazing occurs at a temperature above 840°F (450°C). Because only the temperature separates the two processes, it is possible to do both soldering and brazing using different mixtures of the same metals, depending upon the alloys used and their melting temperatures.

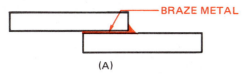

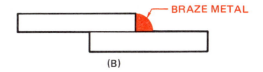

Figure 31-1 A brazed lap joint (A) and a braze welded lap joint (B).

Brazing is divided into two major categories, brazing and braze welding. In brazing, the parts being joined must be fitted so that the joint spacing is very small, approximately .025 in. (0.6 mm) to .002 in. (0.06 mm), Figure 31-1. This small spacing allows capillary action to

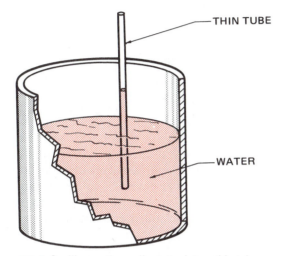

Figure 31-2 Capillary action pulls water into a thin tube.

draw the filler metal into the joint when the parts reach the proper phase temperature.

Capillary action is the force that pulls water up into a paper towel, or pulls a liquid into a very fine straw, Figure 31-2. Braze welding does not need capillary action to pull filler metal into the joint. Examples of brazing and braze welding joint designs are shown in Figure 31-3.

ADVANTAGES OF SOLDERING AND BRAZING

Some advantages of soldering and brazing as compared to other methods of joining include

- Low temperature — Since the base metal does not have to melt, a low-temperature heat source can be used.

- May be permanently or temporarily joined — Since the base metal is not damaged, parts may be disassembled at a later time by simply reapplying heat. The parts then can be reused. However, the joint is solid enough to be permanent, Figure 31-4 (page 782).

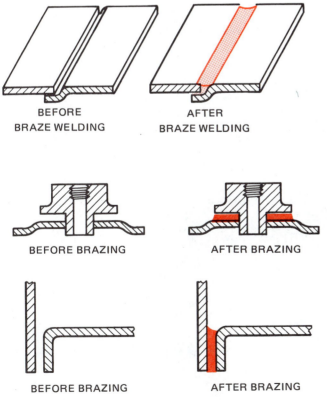

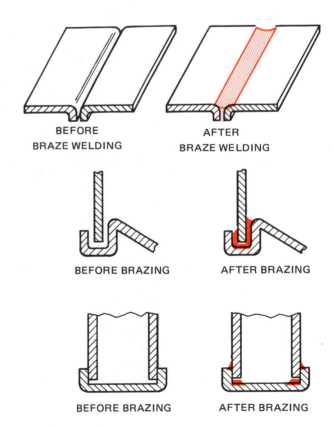

Figure 31-3 Examples of brazing and braze welded joints.

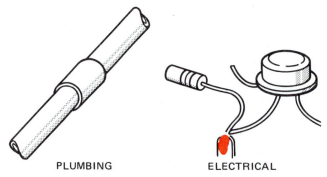

PLUMBING   ELECTRICAL

Figure 31-4 Examples of permanent joints that can easily be disassembled and the parts reused.

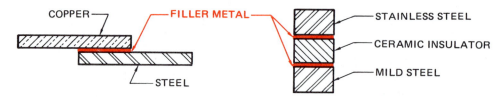

COPPER — FILLER METAL — STAINLESS STEEL
— CERAMIC INSULATOR
— MILD STEEL
— STEEL

Figure 31-5 Dissimilar materials joined.

Figure 31-6 Furnace brazed part.

■ Dissimilar materials can be joined — It is easy to join dissimilar metals, such as copper to steel, aluminum to brass, and cast iron to stainless steel, **Figure 31-5.** It is also possible to join nonmetals to each other or nonmetals to metals. Ceramics are easily brazed to each other or to metals.

■ Speed of joining —

a. Parts can be preassembled and dipped or furnace soldered or brazed in large quantities, **Figure 31-6.**

b. A lower temperature means less time in heating.

■ Less chance of damaging parts — A heat source can be used that has a maximum temperature below the

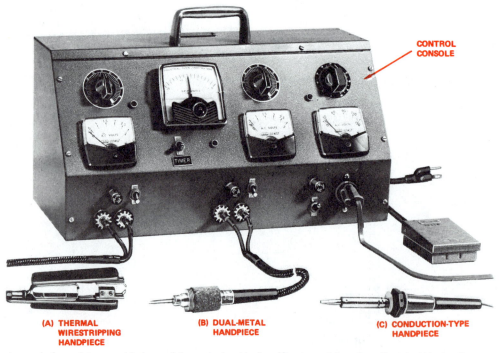

CONTROL CONSOLE

(A) THERMAL WIRESTRIPPING HANDPIECE   (B) DUAL-METAL HANDPIECE   (C) CONDUCTION-TYPE HANDPIECE

Figure 31-7 Control console for resistance soldering and thermal wirestripping. (Courtesy of American Electrical Heater Company.)

Figure 31-8 Joint in tension.

temperature that may cause damage to the parts. With the controlled temperature sufficiently low, even damage from unskilled or semiskilled workers can be eliminated, **Figure 31-7**.

■ Slow rate of heating and cooling — Because it is not necessary to heat a small area to its melting temperature and then allow it to cool quickly to a solid, the internal stresses caused by rapid temperature changes can be reduced.

■ Parts of varying thicknesses can be joined — Very thin parts or a thin part and a thick part can be joined without burning or overheating them.

■ Easy realignment — Parts can easily be realigned by reheating the joint and then repositioning the part.

PHYSICAL PROPERTIES OF THE JOINT

Tensile Strength

The tensile strength of a joint is its ability to withstand being pulled apart, **Figure 31-8**. A brazed joint can be made that has a tensile strength four to five times higher than the filler metal itself. If a few drops of water are placed between two smooth and flat panes of glass and the panes are pressed together, a tensile load is required to pull the panes of glass apart. The water, which has no tensile strength itself, has added tensile strength to the glass joint.

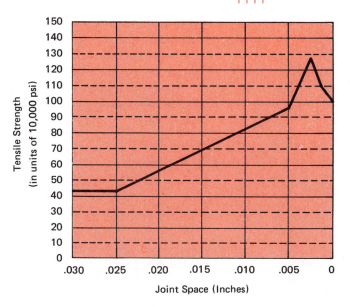

Table 31-1 Tensile Strength of Brazed Joint Increases as Joint Space Decreases.

The glass is being held together by the surface tension of the water. As the space between the pieces of glass decreases, the tensile strength increases. The same action takes place with a soldered or brazed joint. As the joint spacing decreases, the surface tension increases the tensile strength of the joint, **Table 31-1**.

Shear Strength

The shear strength of a joint is its ability to withstand a force parallel to the joint, **Figure 31-9**. For a solder or braze joint, the shear strength depends upon the amount of overlapping area of the base parts. The

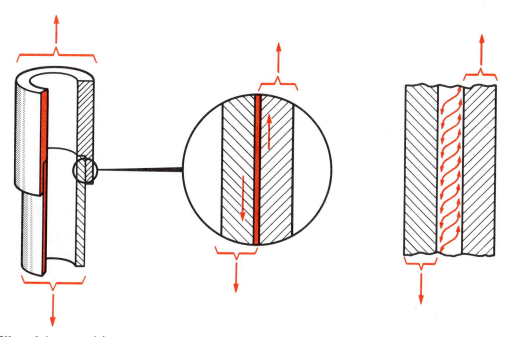

Figure 31-9 Effect of shear on a joint.

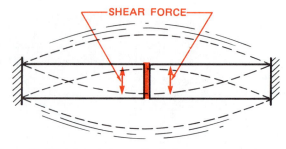

Figure 31-10 A joint being cyclically bent.

greater the area that is overlapped, the greater is the strength.

Ductility

Ductility of a joint is its ability to bend without failing. Most soldering and brazing alloys are ductile metals, so the joint made with these alloys is also ductile.

Fatigue Resistance

The fatigue resistance of a metal is its ability to be bent repeatedly without exceeding its elastic limit and without failure. For most soldered or brazed joints, fatigue resistance is usually fairly low. As a joint is bent, the less ductile base materials cause a shear force to be applied to the filler metal, **Figure 31-10**, resulting in joint failure. Fatigue failures may also occur as a result of vibration.

Corrosion Resistance

Corrosion resistance of a joint is its ability to resist chemical attack. The compatibility of the base materials to the filler metal will determine the corrosion resistance. Using the proper filler metal with the base materials that are listed in this chapter will result in corrosion-free joints. However, using filler metals on base materials that are not recommended in this chapter may result in a joint that looks good when completed but will eventually corrode. For example, brass (BCuZn) will make a nice-looking joint on stainless steel, but the zinc in the brass will combine with the nickel in the stainless steel if the part is kept hot for too long. As a result, an embrittled structure is formed in the joint, reducing strength.

FLUXES

General

Fluxes used in soldering and brazing have three major functions:

- They must remove any oxides that form as a result of heating the parts.
- They must promote wetting.
- They should aid in capillary action.

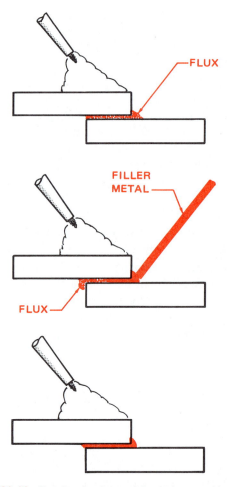

Figure 31-11 Flux flowing into a joint reduces oxides to clean the surfaces and gives rise to a capillary action that causes the filler metal to flow behind it.

The flux, when heated to its reacting temperature, must be thin and flow through the gap provided at the joint. As it flows through the joint, the flux absorbs and dissolves oxides, allowing the molten filler metal to be pulled in behind it, **Figure 31-11**. After the joint is complete, the flux residue should be easily removable.

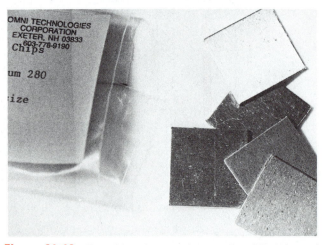

Figure 31-12 Flux chips that can be preplaced in a braze/solder joint.

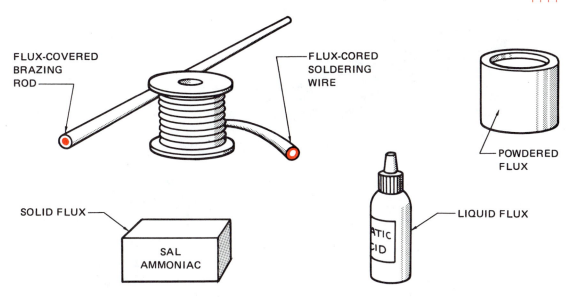

FLUX-COVERED BRAZING ROD

FLUX-CORED SOLDERING WIRE

POWDERED FLUX

SOLID FLUX

SAL AMMONIAC

ATIC CID

LIQUID FLUX

Figure 31-13 Flux can be purchased with the filler metal or separately.

Fluxes are available in many forms, such as solids, powders, pastes, liquids, sheets, rings, and washers, **Figure 31-12.** They are also available mixed with the filler metal, inside the filler metal, or on the outside of the filler metal, **Figure 31-13.** Sheets, rings, and washers may be placed within the joints of an assembly before heating so that a good bond inside the joint can be assured. Paste and liquids can be injected into a joint from tubes using a special gun, **Figure 31-14.** Paste, powders, and liquids may be brushed on the joint before or after the material is heated. Paste and powders may also be applied to the end of the rod by heating the rod and dipping it in the flux. Most powders can be made into a paste, or a paste can be thinned by adding distilled water or alcohol; see manufacturers' specifications for details. If water is used, it should be distilled because tap water may contain minerals that will weaken the flux.

Some liquid fluxes may also be added to the gas when using an oxyfuel gas torch for soldering or brazing. The flux is picked up by the fuel gas as it is bubbled through the flux container and is then carried to the torch where it becomes part of the flame.

Flux and filler metal combinations are most convenient and easy to use, **Figure 31-15.** It may be necessary to stock more than one type of flux-filler metal combination for different jobs. These combinations are more expensive than buying the filler and flux separately. In cases where the flux covers the outside of the filler metal, it may be damaged by humidity or chipped off during storage.

Using excessive flux in a joint may result in flux being trapped in the joint, weakening the joint, or causing the joint to leak or fail.

Fluxing Action

Soldering and brazing fluxes will remove light surface oxides, promote wetting, and aid in capillary

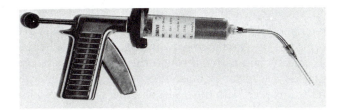

Figure 31-14 Gun for injecting flux into joint.

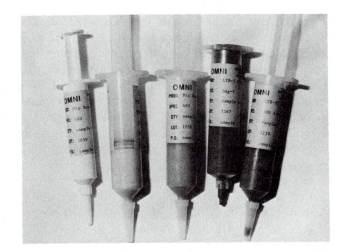

Figure 31-15 Tubes that contain flux filler metal mixtures.

action. The use of fluxes does not eliminate the need for good joint cleaning. Fluxes will not remove oil, dirt, paint, glues, heavy oxide layers, or other surface contaminants.

Soldering fluxes are chemical compounds such as muriatic acid (hydrochloric acid), sal ammoniac (ammonium chloride), or rosin. Brazing fluxes are chemical compounds such as fluorides, chlorides, boric acids, and alkalies. These compounds react to dissolve, absorb, or

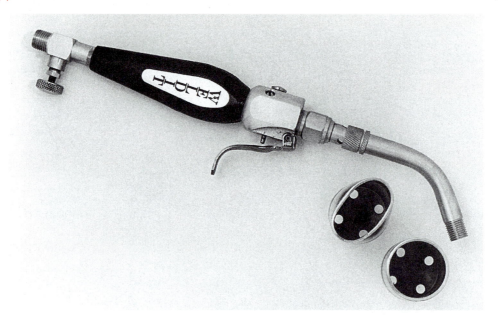

Figure 31-16 An air propane torch can be used in soldering joints. (Courtesy of National Torch Tip Co.)

mechanically break up thin surface oxides that are formed as the parts are being heated. They must be stable and remain active through the entire temperature range of the solder or braze filler metal. The chemicals in the flux react with the oxides as either acids or alkaline (bases). Some dip fluxes are salts.

The reactivity of a flux is greatly affected by temperature. As the parts are heated to the soldering or brazing temperature, the flux becomes more active. Some fluxes are completely inactive at room temperature. Most fluxes have a temperature range within which they are most effective. Care should be taken to avoid overheating fluxes. If they become overheated or burned, they will stop working as fluxes, and they become a contamination in the joint. If overheating has occurred, the welder must stop and clean off the damaged flux before continuing.

Fluxes that are active at room temperature must be neutralized (made inactive) or washed off after the job is complete. If these fluxes are left on the joint, premature failure may result due to flux-induced corrosion. Fluxes that are inactive at room temperature do not have to be cleaned off the part. However, if the part is to be painted or auto body plastic is to be applied, fluxes must be removed.

SOLDERING AND BRAZING METHODS

General

Soldering and brazing methods are grouped according to the method with which heat is applied: torch, furnace, induction, dip, or resistance.

Torch Soldering and Brazing

Oxyfuel or air fuel torches can be used either manually or automatically, **Figure 31-16**. Acetylene is often

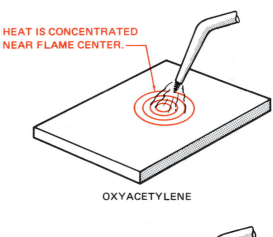

HEAT IS CONCENTRATED NEAR FLAME CENTER.

OXYACETYLENE

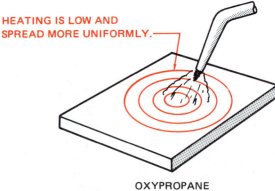

HEATING IS LOW AND SPREAD MORE UNIFORMLY.

OXYPROPANE

Figure 31-17 The high temperature of an oxyacetylene flame may cause localized overheating.

used as the fuel gas, but it is preferable to use one of the other fuel gases having a higher heat level in the secondary flame, **Figure 31-17**. The oxyacetylene flame has a very high temperature near the inner cone, but it has little heat in the outer flame. This often results in the parts being overheated in a localized area. Such fuel

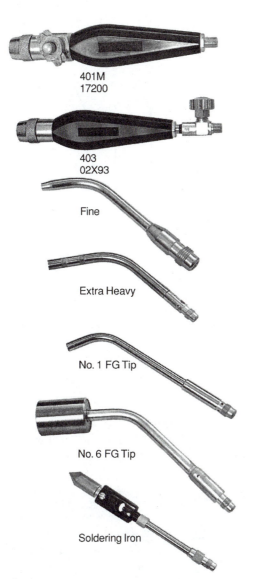

Figure 31-18 Examples of torch tips and handles that use air–fuel mixtures for brazing. (Courtesy of ESAB Welding and Cutting Products.)

gases as MAPP®, propane, butane, and natural gas have a flame that will heat parts more uniformly. Often torches are used that mix air with the fuel gas in a swirling or turbulent manner to increase the flame's temperature, **Figure 31-18**. The flame may even completely surround a small-diameter pipe, heating it from all sides at once, **Figure 31-19**.

Some advantages of using a torch include the following:

- ■ Versatility — Using a torch is the most versatile method. Both small and large parts in a wide variety of materials can be joined with the same torch.

- ■ Portability — A torch is very portable. Anyplace a set of cylinders can be taken or anywhere the hoses can be pulled into can be soldered or brazed with a torch.

- ■ Speed — The flame of the torch is one of the quickest ways of heating the material to be joined, especially on thicker sections.

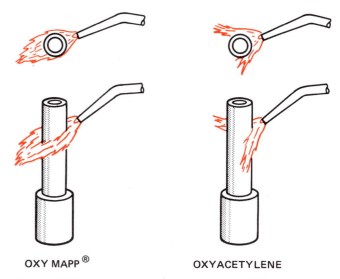

Figure 31-19 Heating characteristics of oxy MAPP® compared with oxyacetylene on round materials.

Some of the disadvantages of using a torch include the following:

- ■ Overheating — When using a torch, it is easy to overheat or burn the parts, flux, or filler metal.

- ■ Skill — A high level of skill with a torch is required to produce consistently good joints.

- ■ Fires — It is easy to start a fire if a torch is used around combustible (flammable) materials.

Furnace Soldering and Brazing

In this method, the parts are heated to their soldering or brazing temperature by passing them through or putting them into a furnace. The furnace may be heated by electricity, oil, natural gas, or by any other locally available fuel. The parts may be passed through the furnace on a conveyor belt in trays or placed on the belt itself, **Figure 31-20** (page 788). The parts also may be loaded in trays to be placed in a furnace that does not use a conveyor belt, **Figure 31-21** (page 788).

Some of the advantages of using a furnace are:

- ■ Temperature control — The furnace temperature can be accurately controlled to ensure that the parts will not overheat.

- ■ Controlled atmosphere — The furnace can be filled with an inert gas to prevent oxides from forming on the parts.

- ■ Uniform heating — The uniform heating of the parts reduces stresses and distortion.

- ■ Mass production — By using a furnace, it is easy to produce quality parts consistently.

Some of the disadvantages of using a furnace are:

- ■ Size — Unless parts are small, the length of time required to heat them is extremely long.

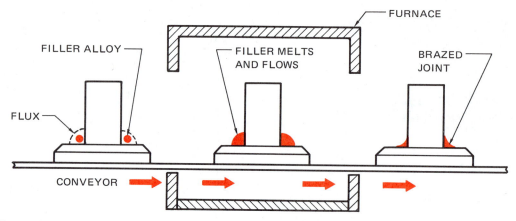

Figure 31-20 Furnace brazing permits the rapid joining of parts on a production basis.

■ Heat damage — The entire part must be able to withstand heating without burning.

Induction Soldering and Brazing

The induction method of heating uses a high-frequency electrical current to establish a corresponding current on the surface of the part, Figure 31-22. The current on the part causes rapid and very localized heating of the surface only. There is little, if any, internal heating of the part except by conductivity of heat from the surface.

The advantage of the induction method is:

■ Speed — Very little time is required for the part to reach the desired temperature.

Some of the disadvantages of the induction method include:

■ Distortion — The very localized heating may result in some distortion.

■ Lack of temperature control — The electrical resistance of the part increases as the part heats up. This, in turn, increases the rate of heating.

Figure 31-21 Small furnace brazed part.

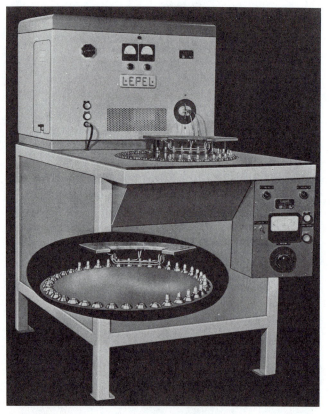

Figure 31-22 Induction brazing and soldering machine. (Courtesy of Lepel Corp.)

■ Incomplete penetration — Because the inside of the part is not directly heated, it may be too cool to permit the filler metal to flow fully through the joint.

Dip Soldering and Brazing

Two types of dip soldering or brazing are used: molten flux bath and molten metal bath. With the molten flux method, the soldering or brazing filler metal in a suitable form is preplaced in the joint, and the assembly is immersed in a bath of molten flux, as shown in **Figure 31-23**. The bath supplies the heat needed to preheat the joint and fuse the solder or braze metal, and also provides protection from oxidation.

With the molten metal method, the prefluxed parts are immersed in a bath of fused solder or braze metal, which is protected by a cover of molten flux. This method is confined to wires and other small parts. Once they are removed from the bath, the ends of the wires and parts must not be allowed to move until the solder or braze metal has solidified. As with all soldering or brazing operations, any movement of the parts as they cool from a liquid through the paste range to become a solid will result in micro fractures in the filler metal. In electronic parts these micro fractures cause resistance to the electron flow and may render the part unfit for service. Reheating can be used to refuse the joint only if reheating will not damage the part beyond use.

Some of the advantages of dip processing include the following:

■ Mass production — It is possible to dip many small parts at one time.

■ Corrosion protection — The entire surface of the part can be covered with the filler metal at the same time that it is being joined. If a corrosion-resistant

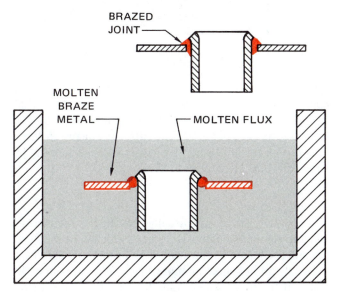

Figure 31-23 Dip brazing eliminates the need for a separate fluxing operation. (Courtesy of Reynolds Metals Company.)

filler metal is used, the thin layer provided will help protect the part from corrosion.

■ Distortion minimized — The entire part is heated uniformly, which reduces distortion.

Some of the disadvantages of dip processing include the following:

■ Steam explosions — Moisture trapped in the joint may cause a steam explosion that can scatter molten metal.

■ Corrosion — If any of the salt is trapped in the joint or is left on the surface, corrosion may cause the part to fail at some time in the future.

■ Size — Parts must be small to be effectively joined.

■ Quantity — Only a large quantity of parts can justify heating the large amount of molten filler metal or salt required for dipping.

Resistance Soldering and Brazing

The resistance method of heating uses an electric current that is passed through the part. The resistance of the part to the current flow results in the heat needed to produce the bond. The flux is usually preplaced, and the material must have sufficient electrical resistance to produce the desired heating. The machine used in this method resembles a spot welder.

Some of the advantages of the resistance heating method include:

■ Localized heating — The heat can be localized so that the entire part may not get hot.

■ Speed — A wide variety of spots can be made on the same machine without having to make major adjustments on the machine.

■ Multiple spots — Many spots can be joined in a small area without disturbing joints that are already made.

Some of the disadvantages of the resistance heating method include:

■ Distortion — Localized heating may result in distortion.

■ Conductors — Parts must be able to conduct electricity.

■ Joint design — Lap joints in plate are the only joint designs that can be made.

Special Methods

A few other methods of producing soldered or brazed parts are used that do not entirely depend upon heat to produce the joint. Ultrasonic uses high-frequency sound waves to produce the bond or to aid with heat in the bonding, **Figure 31-24** (page 790). Another method is diffusion, which uses pressure and may also use heat or ultrasonic to form a bond. Still another process uses infrared light to heat the part for soldering or brazing.

FILLER METALS

General

The type of filler metal used for any specific joint should be selected by considering as many as possible of the criteria listed in **Figure 31-25**. It would be impossible to consider each of these items with the same importance. Welders must decide which things they feel are the most important and then base their selection on that decision.

Soldering and brazing metals are alloys, that is, a mixture of two or more metals. Each alloy is available in a variety of different percentage mixtures. Some mixtures are stronger, and some melt at lower temperatures than other mixtures. Each one has specific properties. Almost all of the alloys used for soldering or brazing have a paste range. A **paste range** is the temperature range in which a metal is partly solid and partly liquid as it is heated or cooled. As the joined part cools through the paste range, it is important that the part not be moved. If the part is moved, the solder or braze metal may crumble like dry clay, destroying the bond.

EXPERIMENT 31-1

Paste Range

This experiment shows the effect on bonding of moving a part as the filler metal cools through its paste range. The experiment also shows how metal can be "worked" using its paste range. You will need tin-lead solder composed of 20% to 50% tin, with the remaining percentage being lead. You also will need a properly lit and adjusted torch, a short piece of brazing rod, and a piece of sheet metal. Using a hammer make a dent in the sheet metal about the size of a quarter (25¢), **Figure 31-26**.

Using the torch, melt a small amount of the solder into the dent and allow it to harden. Remelt the solder slowly, frequently flashing the torch off and touching the solder with the brazing rod until it is evident the solder has all melted. Once it has melted, stick the brazing rod in the solder and remove the torch. As the solder cools, move the brazing rod in the metal and observe what happens, **Figure 31-27**.

As the solder cools to the uppermost temperature of its paste range, it will have a rough surface appearance as the rod is moved. When the solder cools more, it will start to break up around the rod. Finally, as it becomes a solid, it will be completely broken away from the rod.

Now slowly reheat the solder and work the surface with the rod until it can be shaped like clay. If the surface is slightly rough, a quick touch of the flame will smooth it. This is the same way in which "lead" is

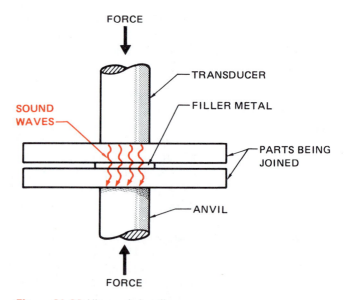

Figure 31-24 Ultrasonic bonding.

- Material being joined
- Strength required
- Joint design
- Availability and cost
- Appearance
- Service (corrosion)
- Heating process to be used
- Cost

Figure 31-25 Criteria for selecting filler metal.

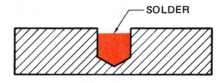

Figure 31-26 Partially fill the drilled hole with solder.

applied to some body panel joints on a new car so that the joints are not seen on the car when it is finished. The lead used is actually a tin-lead alloy or solder. A large area can be made as smooth as glass without sanding by simply flashing the area with the flame. ◆

Soldering Alloys

Soldering alloys are usually identified by their major alloying elements. **Table 31-2** lists the major types of solder and the materials they will join. In many cases, a base material can be joined by more than one solder alloy. In addition to the considerations for selecting filler metal listed in Figure 31-24, specific factors are listed in the following sections for the major soldering alloys.

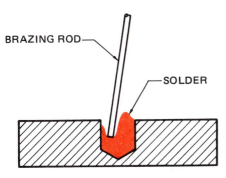

Figure 31-27 Solder being shaped as it cools to its paste range.

Tin-Lead	Copper and copper alloys Mild steel Galvanized metal
Tin-Antimony	Copper and copper alloys Mild steel
Cadmium-Silver	High strength for copper and copper alloys Mild steel Stainless steel
Cadmium-Zinc	Aluminum and aluminum alloys

Table 31-2 Soldering Alloys.

Tin-Lead. This is the most popular solder and is the least expensive one. An alloy of 61.9% tin and 38.1% lead melts at 362°F (183°C) and has no paste range. This is the eutectic composition (lowest possible melting point of an alloy) of the tin-lead solder. An alloy of 60% tin and 40% lead is commercially available and is close enough to the eutectic alloy to have the same low melting point with only a 12°F (7.8°C) paste range. The widest paste range is 173°F (78°C) for a mixture of 19.5% tin and 80.5% lead. This mixture begins to solidify at 535°F (289°C) and is totally solid at 362°F (193°C). The closest mixture that is commercially available is a 20% tin and 80% lead alloy. Table 31-3 lists the percentages, temperatures, and paste ranges for tin-lead solders. Tin-lead solders are most commonly used on electrical connections, but must never be used for water piping. Most health and construction codes will not allow tin-lead solders for use on water or food-handling equipment.

■ *CAUTION*

Tin-lead solders must not be used where lead could become a health hazard in things such as food or water.

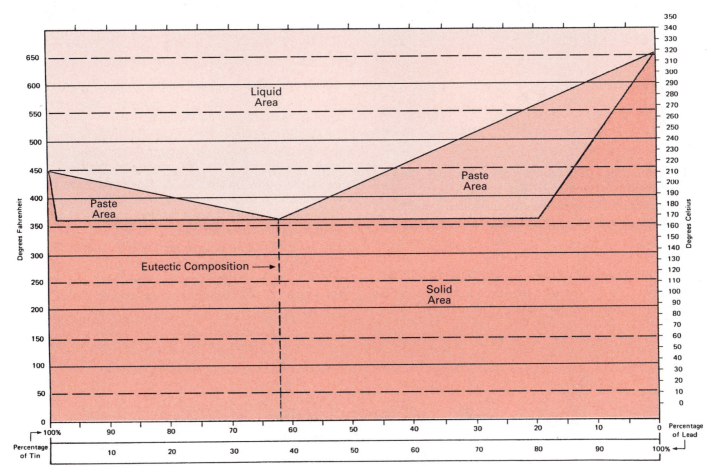

Table 31-3 Melting, Solidification, and Paste Range Temperatures for Tin-Lead Solders.

Tin-Antimony. This family of solder alloys has a higher tensile strength and lower creep than the tin-lead solders. The most common alloy is 95/5, 95% tin and 5% antimony. This is often referred to as "hard solder." This is the most common solder used in plumbing because it is lead free. The use of "C" flux, which is a mixture of flux and small flakes of solder, makes it easier to fabricate quality joints much easier. This mixture of flux and solder will draw additional solder into the joint as it is added.

Cadmium-Silver. These solder alloys have excellent wetting, flow, and strength characteristics, but they are expensive. The silver in this solder helps improve wetting and strength. Cadmium-silver alloys melt at a temperature of around 740°F (393°C); they are called high-temperature solders because they retain their strength at temperatures above most other solders. These solder alloys can be used to join aluminum to itself or other metals, for example, to piping that is used in air-conditioning equipment.

■ CAUTION

When silver soldering on food handling equipment, use a cadmium-free silver solder.

■ CAUTION

If the cadmium is overheated, the fumes can be hazardous unless the area is properly ventilated.

Cadmium-Zinc. Cadmium-zinc alloys have good wetting action and corrosion resistance on aluminum and aluminum alloys. The melting temperature is high, and some alloys have a wide paste range, Table 31-4.

Brazing Alloys

The American Welding Society's classification system for brazing alloys uses the letter B to indicate that the alloy is to be used for brazing. The next series of let-

ters in the classification indicates the atomic symbol of metals used to make up the alloy, such as CuZn (copper and zinc). There may be a dash followed by a letter or number to indicate a specific alloyed percentage. The letter R may be added to indicate that the braze metal is in rod form. An example of a filler metal designation is BRCuZn-A, which indicates a copper-zinc brazing rod with 59.25% copper, 40% zinc, and 0.75% tin; Table 31-5 is a list of the base metals and the most common alloys used to join the base metals. Not all of the available brazing alloys have an AWS classification. Some special alloys are known by registered trade names.

Copper-Zinc. Copper-zinc alloys are the most popular brazing alloys. They are available as regular and low-fuming alloys. The zinc in this braze metal has a tendency to burn out if it is overheated. Overheating is indicated by a red glow on the molten pool, which gives off a white smoke. The white smoke is zinc oxide. If zinc oxide is breathed in, it can cause zinc poisoning. Using a low-fuming alloy will help eliminate this problem. Examples of low-fuming alloys are RCuZn-B and RCuZn-C.

■ CAUTION

Breathing zinc oxide can cause zinc poisoning. If you think you have zinc poisoning, get medical treatment immediately.

Copper-Zinc and Copper-Phosphorus A5.8. The copper-zinc filler rods are often grouped together and known as brazing rods. The copper-phosphorus rods are referred to as phos-copper. Both terms do not adequately describe the metals in this group. There are vast differences among each of the five major classifications of the copper-zinc filler metals, as well as among the five major classifications of the copper-phosphorus filler metals. The following material describes the five major classifications of copper-zinc filler rods.

Class BRCuZn is used for the same application as BCu fillers. The addition of 40% zinc (Zn) and 60% copper (Cu) improves the corrosion resistance and aids in this rod's use with silicon-bronze, copper-nickel, and stainless steel.

Cadmium	Zinc	Completely Liquid	Completely Solid	Paste Range
82.5%	17.5%	509°F (265°C)	509°F (265°C)	No paste range
40.0%	60.0%	635°F (335°C)	509°F (265°C)	126°F (52°C)
10.0%	90.0%	750°F (399°C)	509°F (265°C)	241°F (116°C)

Table 31-4 Cadmium-Zinc Alloys.

Base Metal	Brazing Filler Metal
Aluminum	BAlSi, aluminum silicon
Carbon Steel	BCuZn, brass (copper-zinc) BCu, copper alloy BAg, silver alloy
Alloy Steel	BAg, silver alloy BNi, nickel alloy
Stainless Steel	BAg, silver alloy BAu, gold base alloy BNi, nickel alloy
Cast Iron	BCuZn, brass (copper-zinc)
Galvanized Iron	BCuZn, brass (copper-zinc)
Nickel	BAu, gold base alloy BAg, silver alloy BNi, nickel alloy
Nickel-copper Alloy	BNi, nickel alloy BAg, silver alloy BCuZn, brass (copper-zinc)
Copper	BCuZn, brass (copper-zinc) BAg, silver alloy BCuP, copper-phosphorus
Silicon Bronze	BCuZn, brass (copper-zinc) BAg, silver alloy BCuP, copper-phosphorus
Tungsten	BCuP, copper-phosphorus

Table 31-5 Base Metals and Common Brazing Filler Metals Used to Join the Base Metals.

■ **CAUTION**

Care must be exercised in order to prevent overheating this alloy, as the zinc will vaporize, causing porosity and poisonous zinc fumes.

Class BRCuZn-A is commonly referred to as naval brass and can be used to fuse weld naval brass. The addition of 17% tin (Sn) to the alloy adds strength and corrosion resistance. The same types of metal can be joined with this rod as could be joined with BRCuZn.

Class BRCuZn-B is a manganese-bronze filler metal. It has a relatively low melting point and is free flowing. This rod can be used to braze weld steel, cast iron, brass, and bronze. The deposited metal is higher than BRCuZn or BRCuZn-A in strength, hardness, and corrosion resistance.

Class BRCuZn-C is a low-fuming, high silicon (Si) bronze rod. It is especially good for general-purpose work due to the low-fuming characteristic of the silicon on the zinc.

Class BRCuZn-D is a nickel-bronze rod with enough silicon to be low fuming. The nickel gives the deposit a silver-white appearance and is referred to as white brass. This rod is used to braze and braze weld

steel, malleable iron, cast iron, and for building up wear surfaces on bearings.

Copper-Phosphorus. This alloy is sometimes referred to as phos-copper. It is a good alloy to consider for joints where silver braze alloys may have been used in the past. Phos-copper has good fluidity and wettability on copper and copper alloys. The joint spacing should be from .001 in. (0.03 mm) to .005 in. (0.12 mm) for the strongest joints. Heavy buildup of this alloy may cause brittleness in the joint. Phosphorus forms brittle iron phosphide at brazing temperatures on steel. Copper-phos or copper-phos-silver should not be used on copper-clad fittings with ferrous substrates because the copper can easily be burned off, exposing the underlying metal to phosphorus embrittlement.

The copper-phosphorus (BCuP group) rods are used in air-conditioning applications and in plumbing to join copper piping. The phosphorus makes the rod self-fluxing on copper. This feature is one of the major advantages of copper-phosphorus rods. The addition of a small amount of silver, approximately 2%, helps with wetting and flow into joints.

Class BCuP-1 has a low wetting characteristic and a lower flow rate than the other phos-copper alloys. This type of filler metal should be preplaced in the joint. The major advantage of this type of filler metal is its increased ductility.

Classes BCuP-2 and BCuP-4 both have good flow into the joint. The high phosphorus content of the rods makes them self-fluxing on copper. Both of these classes are used often for plumbing installations.

Classes BCuP-3 and BCuP-5 both have high surface tension and low flow so that they are used when close fit-ups are not available.

Copper-Phosphorus-Silver. This alloy is sometimes referred to as sil-phos. Its characteristics are similar to copper-phosphorus except the silver gives this alloy a little better wetting and flow characteristic. Often it is not necessary to use flux with alloys containing 5% or more of silver when joining copper pipe. This is the most common brazing alloy used in air conditioning and refrigeration work. When sil-phos is used on air-conditioning compressor fittings that are copper-clad steel, care must be taken to make the braze quickly. If the fitting is heated too much or for too long, the copper cladding can be burned off. With this burn-off, the phosphorus can make the steel fitting very brittle, and embrittlement can cause the fitting to crack and leak sometime later.

Silver-Copper. Silver-copper alloys can be used to join almost any metal, ferrous or nonferrous, except aluminum, magnesium, zinc, and a few other low-melting metals. This alloy is often referred to as silver braze and is the most versatile. It is among the most expensive alloys, except for the gold alloys.

Nickel. Nickel alloys are used for joining materials that need high strength and corrosion resistance at an elevated temperature. Some applications of these alloys include joining turbine blades in jet engines, torch parts, furnace parts, and nuclear reactor tubing. Nickel will wet and flow acceptably on most metals. When used on copper-based alloys, nickel may diffuse into the copper, stopping its capillary flow.

Nickel and Nickel Alloys A5.14. Nickel and nickel alloys are increasingly used as a substitute for silver-based alloys. Nickel is generally more difficult than silver to use because it has lower wetting and flow characteristics. However, nickel has much higher strength then silver.

Class BNi-1 is a high-strength, heat-resistant alloy that is ideal for brazing jet engine parts and for other similar applications.

Class BNi-2 is similar to BNi-1, but has a lower melting point and a better flow characteristic.

Class BNi-3 has a high flow rate that is excellent for large areas and close-fitted joints.

Class BNi-4 has a higher surface tension than the other nickel filler rods, which allows larger fillets and poor-fitted joints to be filled.

Class BNi-5 has a high oxidation resistance and high strength at elevated temperatures, and can be used for nuclear applications.

Class BNi-6 is extremely free flowing and has good wetting characteristics. The high corrosion resistance gives this class an advantage when joining low-chromium steels in corrosive applications.

Class BNi-7 has a high resistance to erosion and can be used for thin or honeycomb structures.

Aluminum-silicon (BAlSi) brazing filler metals can be used to join most aluminum sheet and cast alloys. The AWS type number 1 flux must be used when brazing aluminum. It is very easy to overheat the joint. If the flux is burned by overheating, it will obstruct wetting. Use standard torch brazing practices but guard against overheating.

Copper and Copper Alloys A5.7. Although pure copper (Cu) can be gas fusion welded successfully using a neutral oxyfuel flame without a flux, most copper filler metals are used to join other metals in a brazing process.

Class BCu-1 can be used to join ferrous, nickel, and copper-nickel metals with or without a flux. BCu-1 is also available as a powder that is classified as BCu-1a. This material has the same applications as Bcu-1. The AWS type number 3B flux must be used with metals that are prone to rapid oxidation or with heavy oxides such as chromium, titanium, manganese, and others.

Class BCu-2 has applications similar to those for BCu-1. However, BCu-2 contains copper oxide suspended in an organic compound. Since copper oxides can cause porosity, tying up the oxides with the organic compounds reduces the porosity.

Silver and Gold. Silver and gold are both used in small quantities when joining metals that will be used under corrosive conditions, when high joint ductility is needed, or when low electrical resistance is important. Because of the ever-increasing price and reduced availability of these precious metals, other filler metals should first be considered. In many cases, other alloys can be used with great success. When substituting a different filler metal for one that has been used successfully, the new metal and joint should first be extensively tested.

JOINT DESIGN

General

The spacing between the parts being joined greatly affects the tensile strength of the finished part. Table 31-6 lists the spacing requirements at the joining temperature for the most common alloys. As the parts are heated, the initial space may increase or decrease, depending upon the joint design and fixturing. The changes due to expansion can be calculated, but trial and error also works.

The strongest joints are obtained when the parts use lap or scarf joints where the joining area is equal to three times the thickness of the thinnest joint member, **Figure 31-28**. The strength of a butt joint can be increased if the area being joined can be increased. Parts that are 1/4 in. (6 mm) thick should not be considered for brazing or soldering if another process will work successfully.

Some joints can be designed so that the flux and filler metal may be preplaced. When this is possible, visual checking for filler metal around the outside of the joint is easy. Evidence of filler metal around the outside is a good indication of an acceptable joint.

Joint preparation is also very important to a successful soldered or brazed part. The surface must be cleaned of all oil, dirt, paint, oxides, or any other contaminants. The surface can be mechanically cleaned by using a wire brush or by sanding, sand-blasting, grounding, scraping, or filing, or it can be cleaned chemically with an acid, alkaline, or salt bath. Soldering or brazing should start as soon as possible after the parts are cleaned to prevent any additional contamination of the joint.

Filler Metal	Joint Spacing	
	in	mm
BAlSi	.006–.025	(0.15–0.61)
BAg	.002–.005	(0.05–0.12)
BAu	.002–.005	(0.05–0.12)
BCuP	.001–.005	(0.03–0.12)
BCuZn	.002–.005	(0.05–0.12)
BNi	.002–.005	(0.05–0.12)

Table 31-6

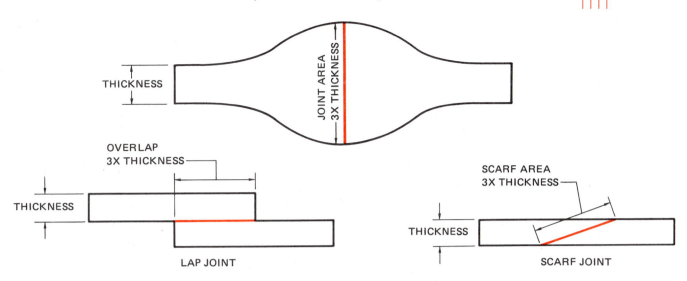

Figure 31-28 The joining area should be three times the thickness of the thinnest joint member.

EXPERIMENT 31-2

Fluxing Action

In this experiment, you will observe oxide removal by a flux as the flux reaches its effective temperature. For this experiment, you will need a piece of copper, either tubing or sheet, rosin or C flux, and a properly lit and adjusted torch.

Any paint, oil, or dirt must first be removed from the copper. Do not remove the oxide layer unless it is blue-black in color. Put some flux on the copper and start heating it with the torch. When the flux becomes active, the copper that is covered by the flux will suddenly change to a bright coppery color. The copper that is not covered by the flux will become darker and possibly turn blue-black, **Figure 31-29.** Continue heating the copper until the flux is burned off and the once clean spot quickly builds an oxide layer.

Repeat this experiment, but this time hold the torch further from the metal's surface. When the flux begins to clean the copper flash the torch off the metal (quickly move the flame off and back onto the same spot). Try to control the heat so that the flux does not burn off. ◆

EXPERIMENT 31-3

Uniform Heating

In this experiment, you will learn how to control the flame direction so that two pieces of metal of unequal size are heated at the same rate to the same temperature. You will need, for this experiment, two pieces of mild steel, one 16 gauge and the other 1/8 in. (3 mm) thick, and a properly lit and adjusted torch.

Place the two pieces of metal on a firebrick to form a butt joint. Then take the torch and point the flame toward the thicker piece of metal, moving it as needed so that both plates turn red at the same time. Now move the torch so that the red area is equal in size on both plates. Keep the spot red but do not allow it to melt. Repeat this experiment until you can control the area and rate of heating of both plates at the same time. ◆

EXPERIMENT 31-4

Tinning or Phase Temperature

In this experiment, you will observe the wetting of a piece of metal by a filler metal. For this experiment, you will need one piece of 16-gauge mild steel, BRCuZn filler metal rod, powdered flux, and a properly lit and adjusted torch.

Place the sheet flat on a firebrick. Heat the end of the rod and dip it in the flux so that some flux sticks on the rod, **Figures 31-30(A), (B), and (C) (page 796).** Direct the flame onto the plate. When the sheet gets hot, hold the brazing rod in contact with the sheet, directing the flame so that a large area of the sheet is dull red and

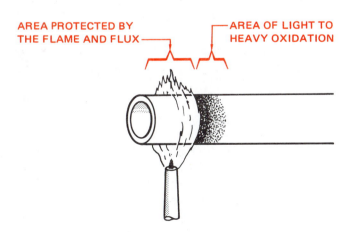

AREA PROTECTED BY THE FLAME AND FLUX   **AREA OF LIGHT TO HEAVY OXIDATION**

Figure 31-29 Copper pipe fluxed and exposed to heat.

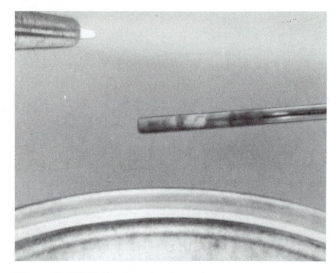

Figure 31-30(A) Heating a brazing rod.

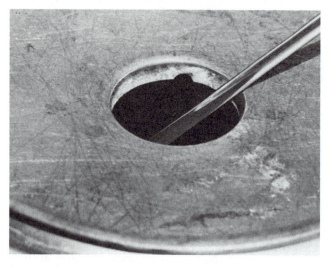

Figure 31-30(B) Dipping the heated rod into the flux.

Figure 31-30(C) Flux stuck to rod ready for brazing.

the rod starts to melt, **Figure 31-31**. After a molten pool of braze metal is deposited on the sheet, remove the rod and continue heating the sheet and molten pool until the braze metal flows out. Repeat this experiment until you can get the braze metal to flow out in all directions equally at the same time. ◆

PRACTICE 31-1

Brazed Stringer Bead

Using a properly lit and adjusted torch, 6 in. (152 mm) of clean 16-gauge mild steel, brazing flux, and BRCuZn brazing rod, you will make a straight bead the length of the sheet.

Place the sheet flat on a firebrick and hold the flame at one end until the metal reaches the proper temperature. Then touch the flux-covered rod to the sheet and allow a small amount of brazing rod to melt onto the hot sheet, **Figure 31-32**. See Color Plate 53. Once the molten brazing metal wets the sheet, start moving the torch in a circular pattern while dipping the rod into the molten braze pool as you move along the sheet. See Color Plate 54. If the size of the molten pool increases, you can control it by reducing the torch angle, raising the torch, traveling at a faster rate, or flashing the flame off the molten braze pool, **Figures 31-33(A), (B), and (C)**. Flashing the torch off a braze joint will not cause oxidation problems as it does when welding, because the molten metal is protected by a layer of flux.

As the braze bead progresses across the sheet, dip the end of the rod back in the flux, if a powdered flux is used, as often as needed to keep a small molten pool of flux ahead of the bead, **Figure 31-34** (page 798).

The object of this practice is to learn how to control the size and direction of the braze bead. Controlling the width, buildup, and shape shows that you have a good understanding and control of the process. Keeping the braze bead in a straight line indicates that you have mastered the bead well enough to watch the bead and the direction at the same time. Turn off the cylinders, bleed the hoses, back out the regulator adjusting screw, and clean up your work area when you are finished. ◆

PRACTICE 31-2

Brazed Butt Joint

Using the same equipment and setup as listed in Practice 31-1, make a braze butt joint on two pieces of 16-gauge mild steel sheet, 6 in. (152 mm) long.

Place the metal flat on a firebrick, hold the plates tightly together, and make a tack braze at both ends of the joint. If the plates become distorted, they can be bent back into shape with a hammer before making another tack weld in the center. Align the sheets so that you can comfortably make a braze bead along the joint.

Figure 31-31 Deposit a spot of braze on the plate and continue heating the plate until the braze flows onto the surface.

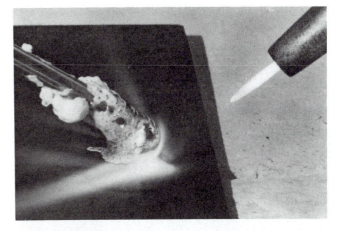

Figure 31-33(A) Once the plate is up to temperature, start adding more filler.

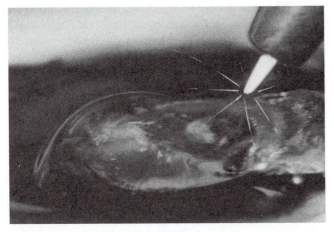

Figure 31-32 Checking the surface temperature with a spot of braze metal.

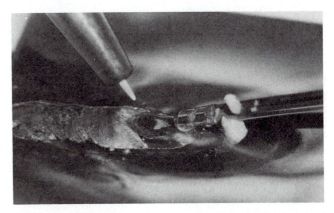

Figure 31-33(B) Dip the brazing rod into the leading edge of the molten weld pool.

Starting as you did in Practice 31-1, make a uniform braze along the joint. Repeat this practice until a uniform braze can be made without defects. Turn off the cylinders, bleed the hoses, back out the regulator adjusting screw, and clean up your work area when you are finished. ◆

Figure 31-33(C) Remove the rod from the flame area when it is not being added to the molten weld pool.

PRACTICE 31-3

Brazed Butt Joint with 100% Penetration

Using the same equipment, material, and setup as described in Practice 31-2, make a braze butt joint with 100% penetration, **Figure 31-35** (page 798). To ensure that 100% penetration is obtained, a little additional heat is required to flow the braze metal through the joint. Apply the additional heat just ahead of the bead on the base sheets. After the braze is completed, turn the plate over and look for a small amount of braze showing along the entire joint. Repeat this practice until it can be made without defects. Turn off the cylinders, bleed the hoses, back out the regulator adjusting screw, and clean up your work area when you are finished. ◆

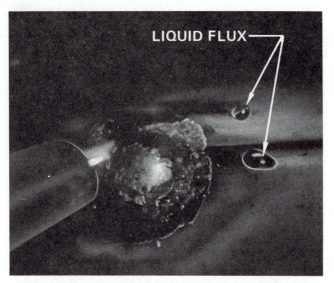

Figure 31-34 Observe the molten flux flowing ahead of the molten weld pool.

Brazed Tee Joint

Using the same equipment, material, and setup as listed in Practice 31-2, you will make a brazed tee joint with 100% root penetration, **Figure 31-36**.

Tack the pieces of metal into a tee joint as you did in Practice 31-2. To obtain 100% penetration, direct the flame on the sheets just ahead of the braze bead, being careful not to overheat the braze metal. If the bead has a notch, the root of the joint is still not hot enough to allow the braze metal to flow properly. If the braze metal appears to have flowed properly after you have completed the joint, look at the back of the joint for a line of braze metal that flowed through. Repeat this practice until the joint can be made without defects. Turn off the cylinders, bleed the hoses, back out the regulator adjusting screw, and clean up your work area when you are finished. ◆

Brazed Lap Joint

Using the same equipment, material, and setup as listed in Practice 31-2, you will make a brazed lap joint in the flat position.

Place the pieces of sheet metal on a firebrick so that they overlap each other by approximately 1/2 in. (13 mm). It is important that the pieces be held flat relative to each other, **Figure 31-37**. Make a small tack braze on both ends, and then one or two tack brazes along the joint. Hold the torch so the flame moves along the joint and heats up both pieces at the same time. When the sheets are hot, touch the rod to the sheets and make a bead similar to the butt joint. After completing the

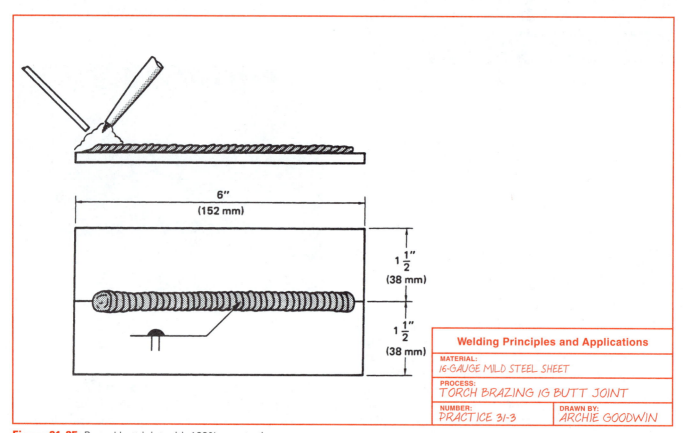

Welding Principles and Applications	
MATERIAL: 16-GAUGE MILD STEEL SHEET	
PROCESS: TORCH BRAZING 1G BUTT JOINT	
NUMBER: PRACTICE 31-3	DRAWN BY: ARCHIE GOODWIN

Figure 31-35 Brazed butt joint with 100% penetration.

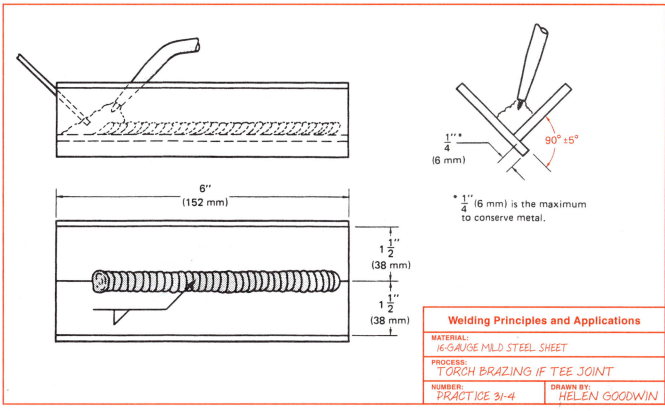

Figure 31-36 Brazed tee joint.

brazed joint, it should be uniform in width and appearance. Repeat this practice until the joint can be made without defects. Turn off the cylinders, bleed the hoses, back out the regulator adjusting screw, and clean up your work area when you are finished. ◆

PRACTICE 31-6

Brazed Lap Joint with 100% Penetration

Using the same equipment, material, and setup as listed in Practice 31-2, you will make a lap joint having 100% penetration in the flat position, **Figure 31-38** (page 800).

Tack the plates together and start the bead the same way as you did in Practice 31-5. Next move the torch back onto the top plate, as you did for Experiment 31-5, so that the braze metal will be drawn into the joint. After the joint is completed, check the back side of the joint for braze metal showing along the joint. Repeat this practice until the joint can be made without defects. Turn off the cylinders, bleed the hoses, back out the regulator adjusting screw, and clean up your work area when you are finished. ◆

PRACTICE 31-7

Brazed Tee Joint, Thin to Thick Metal

Using a properly lit and adjusted torch, two pieces of mild steel 6 in. (152 mm) long, one 16-gauge and

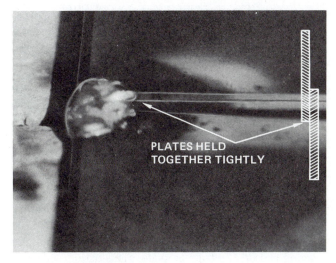

Figure 31-37 Tack braze lap joint.

the other 1/4 in. (6 mm) thick, brazing flux, and BRCuZn brazing rod, you will make a tee joint in the flat position.

Hold the 16-gauge metal vertically on the 1/4 in. (6 mm) plate and tack braze both ends. The vertical member of a tee joint heats up faster than the flat member because the heat on the vertical member can only be conducted, **Figure 31-39** (page 800). The thin plate will heat up faster than the thick plate because there is less mass (metal). For the braze bead to be equal, both plates must be heated equally. Direct the flame on the

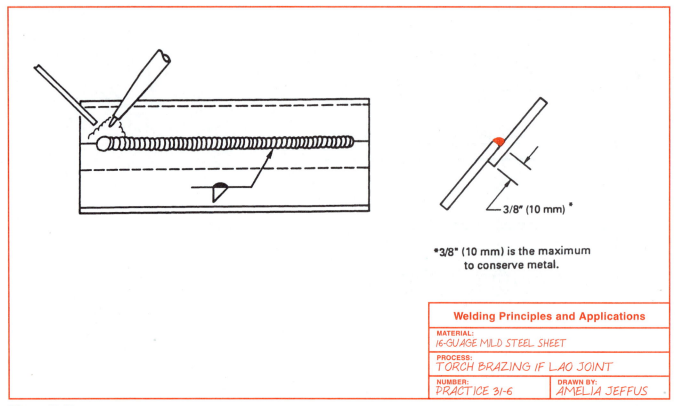

Welding Principles and Applications

MATERIAL:
16-GUAGE MILD STEEL SHEET

PROCESS:
TORCH BRAZING IF LAO JOINT

NUMBER: DRAWN BY:
PRACTICE 31-6   AMELIA JEFFUS

Figure 31-38

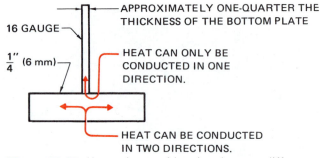

Figure 31-39 Unequal rate of heating due to a difference in mass.

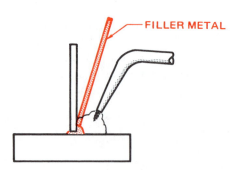

Figure 31-40 Torch and rod positions to balance heating between parts of unequal thickness.

thicker plate, as shown in **Figure 31-40**, and add the brazing rod on the thinner plate. This action will keep the thin plate from overheating. Make a braze along the joint that is uniform in appearance. Repeat this practice until the joint can be made without defects. Turn off the cylinders, bleed the hoses, back out the regulator adjusting screw, and clean up your work area when you are finished. ◆

PRACTICE 31-8

Brazed Lap Joint, Thin to Thick Metal

Using the same equipment, materials, and setup as described in Practice 31-7, make a lap joint in the horizontal position, **Figure 31-41**.

Tack braze the pieces together, being sure that they are held tightly together. Place the metal on a firebrick with the thin metal up. Apply heat to the exposed thick metal and more slowly to the overlapping thin metal go that conduction from the thin metal will heat the thick metal at the lap, **Figure 31-42**. If the braze is started before the thick metal is sufficiently heated, the filler metal will be chilled, and a bond will not occur. After the joint is completed and cooled, tap the joint with a hammer to see if there is a good bonded joint. **Figure 31-43** shows a broken braze joint that bonded properly only in certain areas. Repeat this practice until the joint can be made without defects. Turn off the cylinders, bleed the hoses, back out the regulator adjusting screw, and clean up your work area when you are finished. ◆

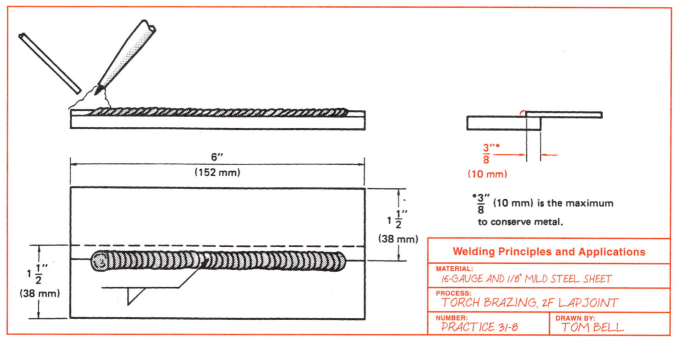

Figure 31-41 Brazed lap joint, thin to thick metal.

PRACTICE 31-9

Braze Welded Butt Joint, Thick Metal

Using a properly lit and adjusted torch, 6 in. (152 mm) of mild steel plate, 1/4 in. (6 mm) thick, flux, and BRCuZn filler metal, you will make a flat braze welded butt joint.

Grind the edges of the plate so that they are slightly rounded, **Figure 31-44** (page 802). The rounded edges are better than the shape used for brazing because they distribute the strain on the braze more uniformly. After the plates have been prepared, tack braze both ends together. Because of the mass of the plates, it may be necessary to preheat the plates. This helps the penetration and also eliminates cold lap at the root. The flame should be moved in a triangular motion so that the root is heated, as well as the top of the bead, **Figure 31-45** (page 802). When the joint is complete and cool, bend the joint to check for complete root bonding, **Figure 31-46** (page 802). Repeat this practice until the joint can be made without defects. Turn off the cylinders, bleed the hoses, back out the regulator adjusting screw, and clean up your work area when you are finished. ◆

PRACTICE 31-10

Braze Welded Tee Joint, Thick Metal

Using the same equipment, materials, and setup as listed in Practice 31-9, you will make a braze welded tee joint in the flat position.

As with the butt joint, the edge of the plate is to be slightly rounded, and the metal may have to be preheated

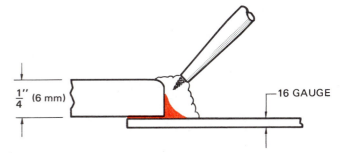

Figure 31-42 Direct the heat on the thicker section to ensure proper bonding.

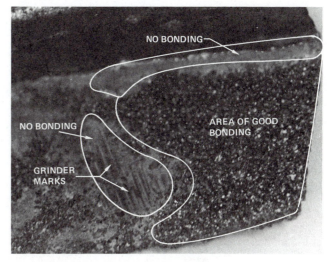

Figure 31-43 Only part of this braze joint is bonded properly. Grinder marks can be seen casted into the braze. The base metal was too cool for proper bonding to occur.

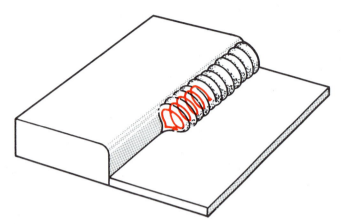

Figure 31-44 Joint preparation.

Figure 31-45 Flame movement to ensure a 100% root penetration.

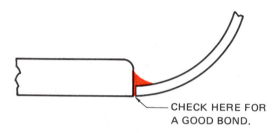

CHECK HERE FOR A GOOD BOND.

Figure 31-46 Bend the braze joint to check for a good bond.

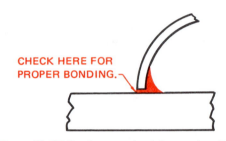

CHECK HERE FOR PROPER BONDING.

Figure 31-47 Bend test to check for root bonding.

to get a good bond at the root. Direct the flame toward the bottom plate and into the root, and add the filler metal back into the root. Watch for a notch, which indicates lack of bonding at the root. When the braze is complete and cool, bend it with a hammer to check for a bond at the root, **Figure 31-47**. Repeat this practice until the joint can be made without defects. Turn off the cylinders, bleed the hoses, back out the regulator adjusting screw, and clean up your work area when you are finished. ◆

BUILDING UP SURFACES AND FILLING HOLES

Surfaces on worn parts are often built up again with braze metal. Braze metal is ideal for parts that receive limited abrasive wear because the buildup is easily machinable. Unlike welding or hardsurfacing, **braze buildup** has no hard spots that make remachining difficult. Braze buildups are good both for flat and round stock. The lower temperature used in brazing does not tend to harden the base metal as much as in welding.

Holes in light-gauge metal can be filled using braze metal. The filled hole can be ground flush if it is required for clearance, leaving a strong patch with minimum distortion.

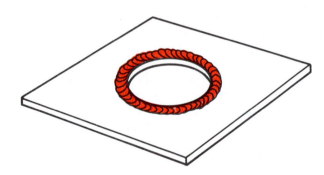

Figure 31-48 Filling a hole with braze. First run a bead around the outside of the hole.

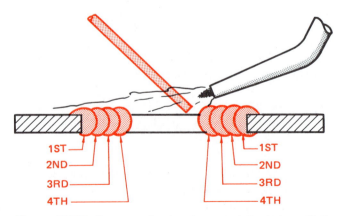

1ST
2ND
3RD
4TH

1ST
2ND
3RD
4TH

Figure 31-49 Keep running beads around the hole until it is closed.

PRACTICE 31-11

Braze Welding to Fill a Hole

Using a properly lit and adjusted torch, one piece of 16-gauge mild steel, flux, and BRCuZn filler rod, you

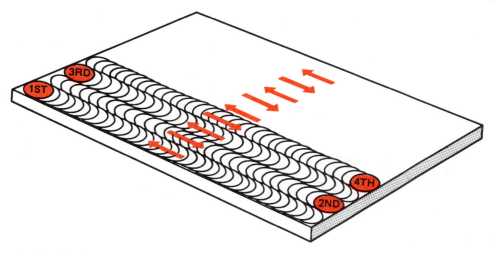

Figure 31-50 Braze buildup, first layer.

will fill a 1-in. (25-mm) hole. Place the piece of metal on two firebricks so that the hole is between them. Start by running a stringer bead around the hole, **Figure 31-48.** Once the bead is complete, turn the torch at a very sharp angle and point it at the edge of the hole nearest the torch. Hold the end of the filler rod in the flame so that both the bead around the hole and the rod meet at the same time, **Figure 31-49.** Put the rod in the molten bead and flash the torch off to allow the molten braze pool to cool. When it has cooled, repeat this process. Surface tension will hold a small piece of molten metal in place. If the piece of molten metal becomes too large, it will drop through. Progress around the hole as many times as needed to fill the hole. When the braze weld is complete, it should be fairly flat with the surrounding metal. Repeat this practice until it can be made without defects. Turn off the cylinders, bleed the hoses, back out the regulator adjusting screw, and clean up your work area when you are finished. ◆

Flat Surface Buildup

Using a properly lit and adjusted torch, a 3-in. (76-mm) square of 1/4-in. (6-mm) mild steel, flux, and BRCuZn filler rod, you will build up a surface.

Place the square plate flat on a firebrick. Start along one side of the plate and make a braze weld down that side. When you get to the end, turn the plate 180° and braze back alongside the first braze, **Figure 31-50,** covering about one-half of the first braze bead. Repeat this procedure until the side is covered with braze metal.

Turn the plate 90° and repeat the process, **Figure 31-51.** Be sure that you are getting good fusion with the first layer and that there are no slag deposits trapped under the braze. Be sure to build up the edges so that they could be cut back square. This process should be repeated until there is at least 1/4 in. (6 mm) of buildup

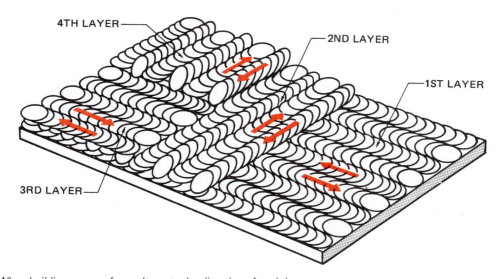

Figure 31-51 When building up a surface, alternate the direction of each layer.

on the surface. The surfacing can be checked visually or by machining the square to a 3-in. (76-mm) x 1/2-in. (13-mm) block and checking for slag inclusions. The plate will warp as a result of this buildup. If the plate were going to be used, it should be clamped down to prevent distortion. Turn off the cylinders, bleed the hoses, back out the regulator adjusting screw, and clean up your work area when you are finished. ◆

PRACTICE 31-13

Round Surface Buildup

Using a properly lit and adjusted torch, one piece of mild steel rod, 1/2 in. (13 mm) in diameter x 3 in. (76 mm) long, flux, and BRCuZn brazing rod, you will build up a round surface.

In the flat position, start at one end and make a braze weld bead, 1 1/2 in. (38 mm) long, along the side of the steel rod. Turn the rod and make another bead next to the first bead, covering about one-half of the first, **Figures 31-52** and **31-53**. Repeat this procedure until the rod is 1 in. (25 mm) in diameter. It may be necessary to make a braze bead around both ends of the buildup to keep it square. The buildup can be visually inspected, or it can be turned down in a lathe. Repeat this practice until it can be made without defects. Turn off the cylinders, bleed the

Figure 31-52 Round shaft built up with braze.

Figure 31-53 Shaft turned down to check for slag inclusions or poor bonding.

hoses, back out the regulator adjusting screw, and clean up your work area when you are finished. ◆

SOLDERING

The soldering practices that follow will use tin-lead or tin-antimony solders. Both solders have low melting temperature. If an oxyacetylene torch is used, its very easy to overheat the solder. Caution is necessary because most of the fluxes used with this type of solder are easily over-heated. The best type of flame to use for this type of soldering is air acetylene, air MAPP®, air propane, or any air fuel-gas mixture. The most popular types are air acetylene or air propane. If galvanized metal is used, additional ventilation should be used to prevent zinc oxide poisoning.

PRACTICE 31-14

Soldered Tee Joint

Using a properly lit and adjusted torch, two pieces of 18-gauge to 24-gauge mild steel sheet, 6 in. (152 mm) long, flux, and tin-lead or tin-antimony solder wire, you will solder a flat tee joint.

Hold one piece of metal vertical on the other piece and spot solder both ends. If flux-cored wire is not being used, paint the flux on the joint at this time. Hold the torch flame so it moves down the joint in the same direction you will be soldering. Continue flashing the torch off and touching the solder wire to the joint until the solder begins to melt. Keeping the molten pool small enough to work with is a major problem with soldering. The flame must be flashed off frequently to prevent over-heating. When the joint is completed, the solder should be uniform. Repeat this practice until it can be made without defects. Turn off the cylinders, bleed the hoses, back out the regulator adjusting screw, and clean up your work area when you are finished. ◆

PRACTICE 31-15

Soldered Lap Joint

Using the same equipment, materials, and setup as listed in Practice 31-14, you will solder a lap joint in the horizontal position.

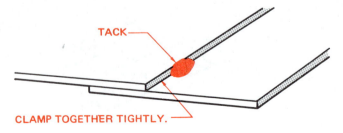

TACK

CLAMP TOGETHER TIGHTLY.

Figure 31-54 Tacking metal together for soldering.

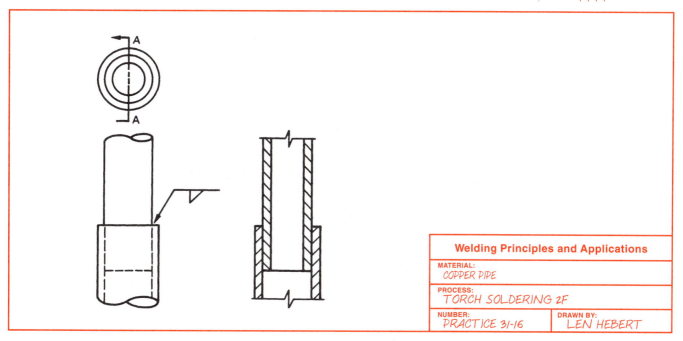

Welding Principles and Applications

MATERIAL:
COPPER PIPE

PROCESS:
TORCH SOLDERING 2F

NUMBER:
PRACTICE 31-16

DRAWN BY:
LEN HEBERT

Figure 31-55

Tack the pieces of metal together as shown in Figure 31-54. Apply the flux and heat the metal slowly, checking the temperature by touching the solder wire to the metal often. When the work gets hot enough, flash the flame off frequently to prevent overheating and proceed along the joint. When the joint is completed, the solder should be uniform. Repeat this practice until it can be made without defects. Turn off the cylinders, bleed the hoses, back out the regulator adjusting screw, and clean up your work area when you are finished. ◆

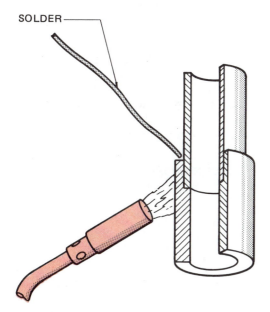

SOLDER

Figure 31-56 Soldering copper fitting to copper pipe. (Courtesy of Praxair, Inc.)

PRACTICE 31-16

Soldering Copper Pipe 2F Vertical Down Position

Using a properly lit and adjusted torch, a piece of 1/2 in. to 1 in. (13 mm to 25 mm) copper pipe, a copper pipe fitting, steel wool, flux, and tin-lead or tin-antimony solder wire, you will solder a pipe joint in the vertical position, **Figure 31-55**.

Clean the pipe and fitting using steel wool and apply the flux to both parts. Slide the fitting onto the pipe and twist the fitting to insure that the flux is applied completely around the inside of the joint.

Make a bend in the solder wire about 3/4 in. (19 mm) from the end. This will give you a gauge so that you don't put too much solder in the joint. Excessive solder will flow inside the pipe, and it may cause problems to the system later.

Heat the pipe and the fitting with the torch. As the parts become hot, keep checking the parts with the solder wire so that you know when they reach the correct temperature. When the solder starts to wet, remove the flame and wipe the joint with the end of the solder wire, **Figure 31-56**. Next, heat the fitting more than the pipe so that the solder will be drawn into the joint. Rewipe the joint with the solder as needed. There should be a small fillet of solder around the joint. This fillet adds very little strength to the joint, but it is an easy way to make sure that the joint is leak free.

After the pipe has cooled, notch it diagonally with a hacksaw, put a screw driver in the cut, and twist the joint apart, **Figure 31-57** (page 806). With the joint separated,

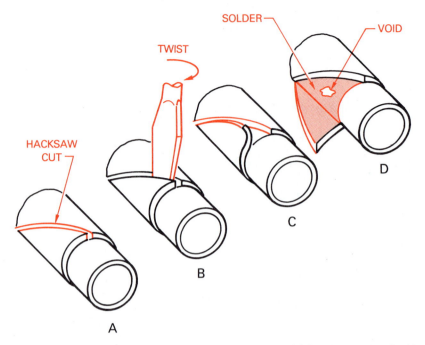

Figure 31-57 (A) Use a hacksaw to cut a groove through the outside copper pipe. (B) Carefully push a flat blade screwdriver into the saw cut groove. (C) Twist the blade to open up the groove. (D) Unwrap the outer copper pipe to reveal the solder in the joint.

check for (1) complete penetration, (2) small porosity caused by overheating the solder, (3) drops of solder inside of the pipe, or other defects. Repeat this practice until it can be made without defects. Turn off the cylinders, bleed the hoses, back out the regulator adjusting screw, and clean up your work area when you are finished. ◆

PRACTICE 31-17

Soldering Copper Pipe 1G Position

Using the same equipment and setup as listed in Practice 31-16, make a soldered joint with the pipe held horizontal, **Figure 31-58**. Test the joint as before and repeat as necessary until it passes the test. Turn off the cylinders, bleed the hoses, back out the regulator adjusting screw, and clean up your work area when you are finished. ◆

PRACTICE 31-18

Soldering Copper Pipe 4F Vertical Up Position

Using the same equipment and setup as listed in Practice 31-16, make a soldered joint with the pipe held in the vertical position with the solder flowing up hill. Test the joint as before and repeat as necessary until it can be made without defects. Turn off the cylinders, bleed the hoses, back out the regulator adjusting screw, and clean up your work area when you are finished. ◆

PRACTICE 31-19

Soldering Aluminum to Copper

Using a properly lit and adjusted torch, a piece of aluminum plate, a copper penny, steel wool, flux, and tin-lead or tin-antimony solder wire, you will tin both the

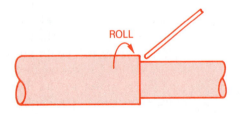

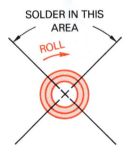

Figure 31-58 1G soldering.

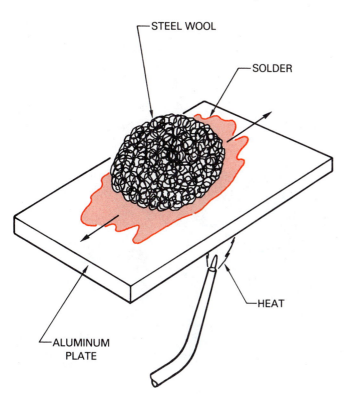

Figure 31-59 Tinning aluminum with solder.

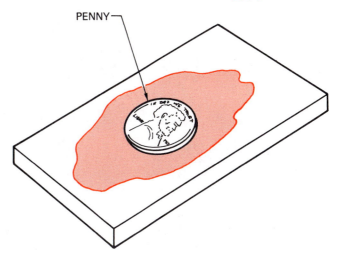

Figure 31-60 Copper penny soldered to aluminum using solder.

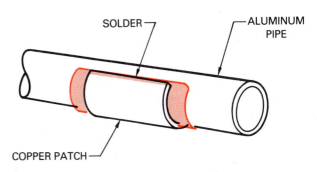

Figure 31-61

aluminum and copper with solder and then join both together.

The surface of the aluminum must be clean and free of paint, oils, dirt, and coatings such as anodizes. Hold the flame on the aluminum until it warms up slightly. Hold the solder in the flame and allow a small amount to melt and drop on the aluminum plate; don't add flux, **Figure 31-59**. Move the flame off the plate and rub the liquid solder with the steel wool. Be careful not to burn your fingers or allow the flame to touch the steel wool. The solder should be stuck in the steel wool when it is lifted off the plate. Alternately heat the plate and rub it with the steel wool-solder. When the plate becomes hot enough it will melt the solder, and the solder will tin the aluminum surface, **Figure 31-60**.

Use some flux and solder and tin the penny. Place the penny on the aluminum plate so the solder on both the penny and plate are touching each other. Heat the two until the solder melts and flows out from between the pennies and plate. When the parts cool, to check the bond, try to break the joint apart.

This process will work on other types of metals that have a strong oxide layer that prevents the solder from bonding. By breaking the oxide layer free with the mechanical action of the steel wool the metals can join. This process can allow a copper patch over an aluminum tube such as those used in air conditioning or refrigeration, **Figure 31-61**. Turn off the cylinders, bleed the hoses, back out the regulator adjusting screw, and clean up your work area when you are finished. ◆

Metal Powder Continuous Electrode Saves Production Time

Figure 1

Figure 2

One of the nation's leading designers and fabricators of industrial-size washing machines recently produced over 800 machines in one year in their modern, air-conditioned, 100,100 sq ft (9,300 m²) facility.

Although the company fabricates a variety of washing machines, ranging in dry weight load capacity from 65 to 700 lb (29.5 to 247.5 kg), a representative machine for a large industrial user has a dry weight load capacity of 450 lb (181.4 kg) with a cylinder 64 in. (160 cm) in diameter and 40 in. (100 cm) deep. Overall dimensions of the finished machine are 88 in. (220 cm) wide, 95 in. (237 cm) high, and 94 1/2 in. (236 cm) deep with a total shipping weight of 19,600 lb (8,890 kg).

Welding plays a major role at the company. All major structural components of the washers are welded with the GMAW, FCAW, or SMAW processes. The continuous wire processes are emphasized for their time- and cost-saving advantages. The base and housing of the machines are constructed of mild steel, while internal components in contact with water during the washing cycle are stainless steel.

The welding supervisor recently looked for a better electrode for welding the rear head assembly of the washing cylinder. This assembly is an integral part of the machine, as the driveshaft that rotates the washing cylinder is attached to it. Three motors connected to the driveshaft accelerate the cylinder through four speeds ranging from 25 rpm to 530 rpm.

It was desired to use a welding wire that would achieve greater efficiency, less distortion, and better weld appearance than the previously used wire provided. The wire selected is a metal powder composite electrode that combines the high deposition rate of a flux cored wire with the high efficiency of a solid wire. The fluxing ingredients are virtually eliminated from the core of the wire and are replaced with metal powder. The only slag formed by this wire consists of small islands of silicates, similar to those found on a solid wire deposit. Multiple passes are made without deslagging.

The rear head weldment consists of a plate 1 in. (2.5 cm) thick and 64 in. (160 cm) in diameter with a 12-in. (30-cm) diameter center opening, a 12-in. (30-cm) solid round hub bored to receive the driveshaft, and a 1/2-in. (1.2-cm) thick plate 30 in. (76 cm) in diameter with an opening cut in the center. The 1/2-in. (1.2-cm) plate is pressed into a bowl shape, allowing it to fit over and enclose the protruding hub.

The welding on the head assembly is done automatically and in the flat position. A custom-designed arm holds the welding gun in a fixed position with the only adjustments being gun angle and arc length. The assembly rests on a rotating table. The electrode is fed at a constant speed, with

Figure 3

adjustment for travel rate being made by varying the rotational speed of the table. A 1/16-in. (1.6-mm) electrode operated at 29 volts and 330 amperes is used for a smooth transfer, good wetting, and minimal spatter, and a shielding mixture of 90% argon/10% CO_2 proved best.

The first step in welding the head is the attachment of the 12-in. (30-cm) round hub to the center of the 64-in. (160-cm) diameter, 1 in. (2.5 cm) thick plate, **Figure 1**. This joint is filled in three passes on the top and three passes on the underside. The 1/2-in. (1.2-cm) bowl-shaped plate, which is fitted over the hub, is welded to the 1-in. (2.5-cm) plate in two passes and to the top of the 12-in. (30-cm) round hub in four passes, **Figure 2**.

Every head is magnetic particle inspected, **Figure 3**, before the driveshaft is attached to the hub. Once the driveshaft is attached, the whole assembly is joined to the stainless steel washing cylinder, forming its rear wall. This wall is clad with stainless steel sheet.

Due to the greater deposition rates and better penetration of the composite electrode, four passes were eliminated from the previous welding procedure and, as a result, the total job is performed 25 to 30% faster.

(Adapted from information provided by Alloy Rods Corporation, Hanover, Pennsylvania 17331.)

REVIEW

1. Explain the difference between brazing and soldering.

2. How does capillary action separate brazing?

3. Why can brazing be both a permanent and a temporary joining method?

4. Why is it less likely that a semiskilled worker would damage a part with brazing than with welding?

5. What is the effect of joint spacing on joint tensile strength?

6. Why are braze joints subject to fatigue failure?

7. Will all braze joints resist corrosion? Give an example.

8. What are the three primary functions that a flux must perform?

9. In what forms are fluxes available?

10. How can liquid fluxes be delivered to the joint through the torch?

11. How do fluxes react with the base metal?

12. How are soldering and brazing methods grouped?

13. What are the advantages of torch soldering?

14. What are the advantages of furnace brazing?

15. How does the induction brazing method heat the part being brazed?

16. What soldering process can be used to join parts and provide a protective coating to the part at the same time?

17. What soldering or brazing process uses a machine similar to a spot welder to produce the heat required to make a joint?

18. What is a metal alloy?

19. Why must parts not be moved as they cool through the paste range?

20. Why must you not use tin-lead solders on water piping?

21. Using Table 31-3 (page 791), determine the approximate temperature that a mixture of 90% tin with 10% lead would become 100% solder.

22. What is a major use of tin-antimony solder?

23. What solder alloy can retain its strength at elevated temperatures?

24. Why do brazing alloys use letters such as "CuZn" to identify the metal?

25. What is the white smoke that can be given off from copper-zinc brazing alloy?

26. Which copper-zinc brazing alloy could be used to join each of the following metals?
    a. cast iron
    b. build up wear surfaces
    c. malleable iron
    d. naval brass

27. Why should copper-phosphorus not be used on fittings with ferrous substrates?

28. Silver-copper alloys can be used to join which metals?

29. Which nickel alloy would be best for joining the following?
    a. poor fitting joints
    b. honeycomb structures
    c. jet engine parts
    d. corrosive applications

30. Why does BCu-2 brazing alloy use an organic compound mixed with copper?

31. Using Table 31-6 (page 794), determine the ideal joint spacings for the following braze alloys.
    a. BCuZn
    b. BAlSi
    c. BCuP

32. What indicates that you have overheated the solder flux on copper?

33. How can the size of a braze bead be controlled?

34. Why is brazing a better process to rebuild some surfaces than welding?

Appendix

I. CONVERSION OF DECIMAL INCHES TO MILLIMETERS AND FRACTIONAL INCHES TO DECIMAL INCHES AND MILLIMETERS

Inches dec	mm	Inches dec	mm	Inches frac	dec	mm	Inches frac	dec	mm
0.01	0.2540	0.51	12.9540	1/64	0.015625	0.3969	33/64	0.515625	13.0969
0.02	0.5080	0.52	13.2080						
0.03	0.7620	0.53	13.4620	1/32	0.031250	0.7938	17/32	0.531250	13.4938
0.04	1.0160	0.54	13.7160	3/64	0.046875	1.1906	35/64	0.546875	13.8906
0.05	1.2700	0.55	13.9700						
0.06	1.5240	0.56	14.2240	1/16	0.062500	1.5875	9/16	0.562500	14.2875
0.07	1.7780	0.57	14.4780	5/64	0.078125	1.9844	37/64	0.578125	14.6844
0.08	2.0320	0.58	14.7320						
0.09	2.2860	0.59	14.9860	3/32	0.093750	2.3812	19/32	0.593750	15.0812
0.10	2.5400	0.60	15.2400						
0.11	2.7940	0.61	15.4940	7/64	0.109375	2.7781	39/64	0.609375	15.4781
0.12	3.0480	0.62	15.7480	1/8	0.125000	3.1750	5/8	0.625000	15.8750
0.13	3.3020	0.63	16.0020						
0.14	3.5560	0.64	16.2560	9/64	0.140625	3.5719	41/64	0.640625	16.2719
0.15	3.8100	0.65	16.5100	5/32	0.156250	3.9688	21/32	0.656250	16.6688
0.16	4.0640	0.66	16.7640						
0.17	4.3180	0.67	17.0180	11/64	0.171875	4.3656	43/64	0.671875	17.0656
0.18	4.5720	0.68	17.2720	3/16	0.187500	4.7625	11/16	0.687500	17.4625
0.19	4.8260	0.69	17.5260						
0.20	5.0800	0.70	17.7800	13/64	0.203125	5.1594	45/64	0.703125	17.8594
0.21	5.3340	0.71	18.0340						
0.22	5.5880	0.72	18.2880	7/32	0.218750	5.5562	23/32	0.718750	18.2562
0.23	5.8420	0.73	18.5420	15/64	0.234375	5.9531	47/64	0.734375	18.6531
0.24	6.0960	0.74	18.7960						
0.25	6.3500	0.75	19.0500	1/4	0.250000	6.3500	3/4	0.750000	19.0500
0.26	6.6040	0.76	19.3040						
0.27	6.8580	0.77	19.5580	17/64	0.265625	6.7469	49/64	0.765625	19.4469
0.28	7.1120	0.78	19.8120	9/32	0.281250	7.1438	25/32	0.781250	19.8437
0.29	7.3660	0.79	20.0660						
0.30	7.6200	0.80	20.3200	19/64	0.296875	7.5406	51/64	0.796875	20.2406
0.31	7.8740	0.81	20.5740	5/16	0.312500	7.9375	13/16	0.812500	20.6375
0.32	8.1280	0.82	20.8280						
0.33	8.3820	0.83	21.0820	21/64	0.328125	8.3344	53/64	0.828125	21.0344
0.34	8.6360	0.84	21.3360						
0.35	8.8900	0.85	21.5900	11/32	0.343750	8.7312	27/32	0.843750	21.4312
0.36	9.1440	0.86	21.8440	23/64	0.359375	9.1281	55/64	0.859375	21.8281
0.37	9.3980	0.87	22.0980						
0.38	9.6520	0.88	22.3520	3/8	0.375000	9.5250	7/8	0.875000	22.2250
0.39	9.9060	0.89	22.6060	25/64	0.390625	9.9219	57/64	0.890625	22.6219
0.40	10.1600	0.90	22.8600						
0.41	10.4140	0.91	23.1140	13/32	0.406250	10.3188	29/32	0.906250	23.0188
0.42	10.6680	0.92	23.3680						
0.43	10.9220	0.93	23.6220	27/64	0.421875	10.7156	59/64	0.921875	23.4156
0.44	11.1760	0.94	23.8760	7/16	0.437500	11.1125	15/16	0.937500	23.8125
0.45	11.4300	0.95	24.1300						
0.46	11.6840	0.96	24.3840	29/64	0.453125	11.5094	61/64	0.953125	24.2094
0.47	11.9380	0.97	24.6380	15/32	0.468750	11.9062	31/32	0.968750	24.6062
0.48	12.1920	0.98	24.8920						
0.49	12.4460	0.99	25.1460	31/64	0.484375	12.3031	62/64	0.984375	25.0031
0.50	12.7000	1.00	25.4000	1/2	0.500000	12.7000	1	1.000000	25.4000

For converting decimal-inches in "thousandths," move decimal point in both columns to left.

II. CONVERSION FACTORS: U.S. CUSTOMARY (STANDARD) UNITS AND METRIC UNITS (SI)

TEMPERATURE
Units

°F (each 1° change)	= 0.555°C (change)
°C (each 1° change)	= 1.8°F (change)
32°F (ice freezing)	= 0°Celsius
212°F (boiling water)	= 100°Celsius
−460°F (absolute zero)	= 0° Rankine
−273°C (absolute zero)	= 0° Kelvin

Conversions

°F to °C _____ °F − 32 = _____ × .555 = _____ °C

°C to °F _____ °C × 1.8 = _____ + 32 = _____ °F

LINEAR MEASUREMENT
Units

1 inch	= 25.4 millimeters
1 inch	= 2.54 centimeters
1 millimeter	= 0.0394 inch
1 centimeter	= 0.3937 inch
12 inches	= 1 foot
3 feet	= 1 yard
5280 feet	= 1 mile
10 millimeters	= 1 centimeter
10 centimeters	= 1 decimeter
10 decimeters	= 1 meter
1,000 meters	= 1 kilometer

Conversions

in. to mm _____ in. × 25.4 = _____ mm

in. to cm _____ in. × 2.54 = _____ cm

ft to mm _____ ft × 304.8 = _____ mm

ft to m _____ ft × 0.3048 = _____ m

mm to in. _____ mm × 0.0394 = _____ in.

cm to in. _____ cm × 0.3937 = _____ in.

mm to ft _____ mm × 0.00328 = _____ ft

m to ft _____ m × 32.8 = _____ ft

AREA MEASUREMENT
Units

1 sq in.	= 0.0069 sq ft
1 sq ft	= 144 sq in.
1 sq ft	= 0.111 sq yd
1 sq yd	= 9 sq ft
1 sq in.	= 645.16 sq mm
1 sq mm	= 0.00155 sq in.
1 sq cm	= 100 sq mm
1 sq m	= 1,000 sq cm

Conversions

sq in. to sq mm _____ sq in. × 645.16 = _____ sq mm

sq mm to sq in. _____ sq mm × 0.00155 = _____ sq in.

VOLUME MEASUREMENT
Units

1 cu in.	= 0.000578 cu ft
1 cu ft	= 1728 cu in.
1 cu ft	= 0.03704 cu yd
1 cu ft	= 28.32 L
1 cu ft	= 7.48 gal (U.S.)
1 gal (U.S.)	= 3.737 L
1 cu yd	= 27 cu ft
1 gal	= 0.1336 cu ft
1 cu in.	= 16.39 cu cm
1 L	= 1,000 cu cm

II. CONVERSION FACTORS: U.S. CUSTOMARY (STANDARD) UNITS AND METRIC UNITS (SI) (continued)

1 L	= 61.02 cu in.
1 L	= 0.03531 cu ft
1 L	= 0.2642 gal (U.S.)
1 cu yd	= 0.769 cu m
1 cu m	= 1.3 cu yd

Conversions

cu in. to L	_____ cu in.	× 0.01638	=	_____ L
L to cu in.	_____ L	× 61.02	=	_____ cu in.
cu ft to L	_____ cu ft	× 28.32	=	_____ L
L to cu ft	_____ L	× 0.03531	=	_____ cu ft
L to gal	_____ L	× 0.2642	=	_____ gal
gal to L	_____ gal	× 3.737	=	_____ L

WEIGHT (MASS) MEASUREMENT
Units

1 oz	= 0.0625 lb
1 lb	= 16 oz
1 oz	= 28.35 g
1 g	= 0.03527 oz
1 lb	= 0.0005 ton
1 ton	= 2,000 lb
1 oz	= 0.283 kg
1 lb	= 0.4535 kg
1 kg	= 35.27 oz
1 kg	= 2.205 lb
1 kg	= 1,000 g

Conversions

lb to kg	_____ lb	× 0.4535	=	_____ kg
kg to lb	_____ kg	× 2.205	=	_____ lb
oz to g	_____ oz	× 0.03527	=	_____ g
g to oz	_____ g	× 28.35	=	_____ oz

PRESSURE and FORCE MEASUREMENTS
Units

1 psig	= 6.8948 kPa
1 kPa	= 0.145 psig
1 psig	= 0.000703 kg/sq mm
1 kg/sq mm	= 6894 psig
1 lb (force)	= 4.448 N
1 N (force)	= 0.2248 lb

Conversions

psig to kPa	_____ psig	× 6.8948	=	_____ kPa
kPa to psig	_____ kPa	× 0.145	=	_____ psig
lb to N	_____ lb	× 4.448	=	_____ N
N to lb	_____ N	× 0.2248	=	_____ psig

VELOCITY MEASUREMENTS
Units

1 in./sec	= 0.0833 ft/sec
1 ft/sec	= 12 in/sec
1 ft/min	= 720 in./sec
1 in./sec	= 0.4233 mm/sec
1 mm/sec	= 2.362 in./sec
1 cfm	= 0.4719 L/min
1 L/min	= 2.119 cfm

Conversions

ft/min to in./sec	_____ ft/min	× 720	=	_____ in./sec
in./min to mm/sec	_____ in./min	× 0.4233	=	_____ mm/sec
mm/sec to in./min	_____ mm/sec	× 2.362	=	_____ in./min
cfm to L/min	_____ cfm	× 0.4719	=	_____ L/min
L/min to cfm	_____ L/min	× 2.119	=	_____ cfm

III. ABBREVIATIONS AND SYMBOLS

U.S. Customary (Standard) Units

°F	= degrees Fahrenheit
°R	= degrees Rankine = degrees absolute F
lb	= pound
psi	= pounds per square inch = lb per sq in.
psia	= pounds per square inch absolute = psi + atmospheric pressure
in.	= inches = i = "
ft	= foot or feet = f = '
sq in.	= square inch = in.
sq ft	= square foot = ft
cu in.	= cubic inch = in.
cu ft	= cubic foot = ft
ft-lb	= foot-pound
ton	= ton of refrigeration effect
qt	= quart

Metric Units (SI)

°C	= degress Celsius
K	= kelvin
mm	= millimeter
cm	= centimeter
cm	= centimeter squared
cm	= centimeter cubed
dm	= decimeter
dm	= decimeter squared
dm	= decimeter cubed
m	= meter
m	= meter squared
m	= meter cubed
L	= liter
g	= gram
kg	= kilogram
J	= joule
kJ	= kilojoule
N	= newton
Pa	= pascal
kPa	= kilopascal
W	= watt
kW	= kilowatt
MW	= megawatt

Miscellaneous Abbreviations

P	= pressure
h	= hours
sec	= seconds
D	= diameter
r	= radius of circle
A	= area
π	= 3.1416 (a constant used in determining the area of a circle)
V	= volume
∞	= infinity

IV. METRIC CONVERSIONS APPROXIMATIONS

Metric Conversions Approximations
1/4 inch = 6 mm
1/2 inch = 13 mm
3/4 inch = 18 mm
1 inch = 25 mm
2 inches = 50 mm
1/2 gal = 2 L
1 gal = 4 L
1 lb = 1/2 K
2 lb = 1 K
1 psig = 7 kPa
1°F = 2°C

V. PRESSURE CONVERSION

psi	kPa	psi	kPa
1	7 (6.9)	100	690 (689)
2	14 (13.7)	110	760 (758)
3	20 (20.6)	120	820 (827)
4	30 (27.5)	130	900 (896)
5	35 (34.4)	140	970 (965)
6	40 (41.3)	150	1030 (1034)
7	50 (48.2)	160	1100 (1103)
8	55 (55.1)	170	1170 (1172)
9	60 (62.0)	180	1240 (1241)
10	70 (69.9)	190	1310 (1310)
15	100 (103)	200	1380 (1379)
20	140 (137)	225	1550 (1551)
25	170 (172)	250	1720 (1723)
30	200 (206)	275	1900 (1896)
35	240 (241)	300	2070 (2068)
40	280 (275)	325	2240 (2240)
45	310 (310)	350	2410 (2413)
50	340 (344)	375	2590 (2585)
55	380 (379)	400	2760 (2757)
60	410 (413)	450	3100 (3102)
65	450 (448)	500	3450 (3447)
70	480 (482)	550	3790 (3792)
75	520 (517)	600	4140 (4136)
80	550 (551)	650	4480 (4481)
85	590 (586)	700	4830 (4826)
90	620 (620)	750	5170 (5171)
95	660 (655)	800	5520 (5515)
		850	5860 (5860)
		900	6210 (6205)
		950	6550 (6550)
		1000	6890 (6894)

Pounds per square inch (psi) converted to kilopascals (kPa). One psi equals 6.8948 kPa. In most applications the conversion from standard units of pressure to SI units can be rounded to an even number. The number in the (XXX) is the value before it's rounded off.

VI. STRUCTURAL METAL SHAPE DESIGNATIONS

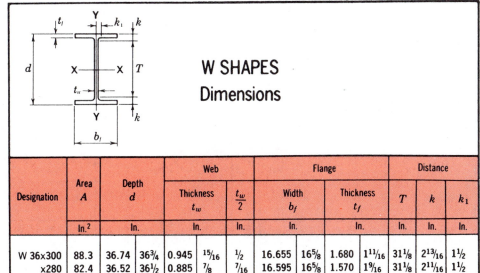

W SHAPES
Dimensions

Designation	Area A	Depth d		Web Thickness t_w		$\frac{t_w}{2}$	Flange Width b_f		Flange Thickness t_f		T	k	k_1
	In.²	In.		In.		In.	In.		In.		In.	In.	In.
W 36×300	88.3	36.74	36¾	0.945	15/16	½	16.655	16⅝	1.680	1 11/16	31⅛	2 13/16	1½
x280	82.4	36.52	36½	0.885	7/8	7/16	16.595	16⅝	1.570	1 9/16	31⅛	2 11/16	1½
x260	76.5	36.26	36¼	0.840	13/16	7/16	16.550	16½	1.440	1 7/16	31⅛	2 9/16	1½
x245	72.1	36.08	36⅛	0.800	13/16	7/16	16.510	16½	1.350	1⅜	31⅛	2½	1 7/16
x230	67.6	35.90	35⅞	0.760	¾	3/8	16.470	16½	1.260	1¼	31⅛	2⅜	1 7/16
W 36×210	61.8	36.69	36¾	0.830	13/16	7/16	12.180	12⅛	1.360	1⅜	32⅛	2 5/16	1¼
x194	57.0	36.49	36½	0.765	¾	3/8	12.115	12⅛	1.260	1¼	32⅛	2 3/16	1 3/16
x182	53.6	36.33	36⅜	0.725	¾	3/8	12.075	12⅛	1.180	1 3/16	32⅛	2⅛	1 3/16
x170	50.0	36.17	36⅛	0.680	11/16	3/8	12.030	12	1.100	1⅛	32⅛	2	1 3/16
x160	47.0	36.01	36	0.650	5/8	5/16	12.000	12	1.020	1	32⅛	1 15/16	1⅛
x150	44.2	35.85	35⅞	0.625	5/8	5/16	11.975	12	0.940	15/16	32⅛	1⅞	1⅛
x135	39.7	35.55	35½	0.600	5/8	5/16	11.950	12	0.790	13/16	32⅛	1 11/16	1⅛
W 33×241	70.9	34.18	34⅛	0.830	13/16	7/16	15.860	15⅞	1.400	1⅜	29¾	2 3/16	1 3/16
x221	65.0	33.93	33⅞	0.775	¾	3/8	15.805	15¾	1.275	1¼	29¾	2 1/16	1 3/16
x201	59.1	33.68	33⅝	0.715	11/16	3/8	15.745	15¾	1.150	1⅛	29¾	1 15/16	1⅛
W 33×152	44.7	33.49	33½	0.635	5/8	5/16	11.565	11⅝	1.055	1 1/16	29¾	1⅞	1⅛
x141	41.6	33.30	33¼	0.605	5/8	5/16	11.535	11½	0.960	15/16	29¾	1¾	1 1/16
x130	38.3	33.09	33⅛	0.580	9/16	5/16	11.510	11½	0.855	7/8	29¾	1 11/16	1 1/16
x118	34.7	32.86	32⅞	0.550	9/16	5/16	11.480	11½	0.740	¾	29¾	1 9/16	1 1/16
W 30×211	62.0	30.94	31	0.775	¾	3/8	15.105	15⅛	1.315	1 5/16	26¾	2⅛	1⅛
x191	56.1	30.68	30⅝	0.710	11/16	3/8	15.040	15	1.185	1 3/16	26¾	1 15/16	1 1/16
x173	50.8	30.44	30½	0.655	5/8	5/16	14.985	15	1.065	1 1/16	26¾	1⅞	1 1/16
W 30×132	38.9	30.31	30¼	0.615	5/8	5/16	10.545	10½	1.000	1	26¾	1¾	1 1/16
x124	36.5	30.17	30⅛	0.585	9/16	5/16	10.515	10½	0.930	15/16	26¾	1 11/16	1
x116	34.2	30.01	30	0.565	9/16	5/16	10.495	10½	0.850	7/8	26¾	1⅝	1
x108	31.7	29.83	29⅞	0.545	9/16	5/16	10.475	10½	0.760	¾	26¾	1 9/16	1
x 99	29.1	29.65	29⅝	0.520	½	¼	10.450	10½	0.670	11/16	26¾	1 7/16	1

VI. STRUCTURAL METAL SHAPE DESIGNATIONS (continued)

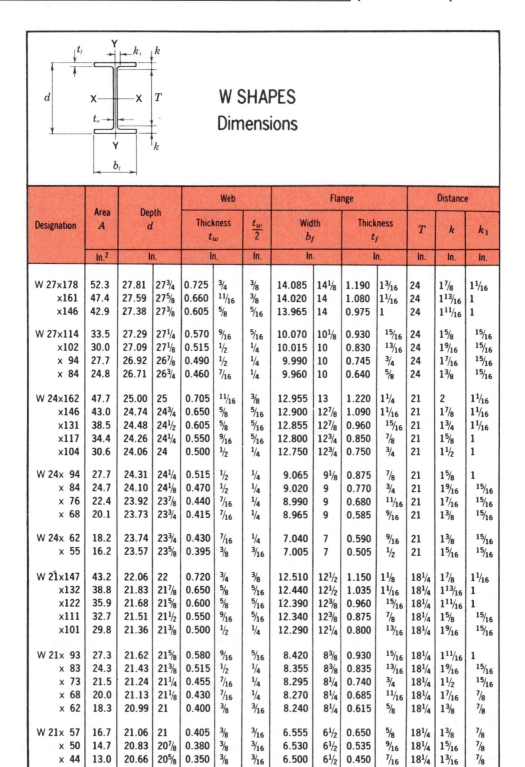

W SHAPES
Dimensions

Designation	Area A	Depth d		Web Thickness t_w		$\dfrac{t_w}{2}$	Flange Width b_f		Flange Thickness t_f		T	k	k_1
	In.²	In.		In.		In.	In.		In.		In.	In.	In.
W 27×178	52.3	27.81	27¾	0.725	¾	⅜	14.085	14⅛	1.190	1³/₁₆	24	1⅞	1¹/₁₆
x161	47.4	27.59	27⅝	0.660	¹¹/₁₆	⅜	14.020	14	1.080	1¹/₁₆	24	1¹³/₁₆	1
x146	42.9	27.38	27⅜	0.605	⅝	⁵/₁₆	13.965	14	0.975	1	24	1¹¹/₁₆	1
W 27×114	33.5	27.29	27¼	0.570	⁹/₁₆	⁵/₁₆	10.070	10⅛	0.930	¹⁵/₁₆	24	1⅝	¹⁵/₁₆
x102	30.0	27.09	27⅛	0.515	½	¼	10.015	10	0.830	¹³/₁₆	24	1⁹/₁₆	¹⁵/₁₆
x 94	27.7	26.92	26⅞	0.490	½	¼	9.990	10	0.745	¾	24	1⁷/₁₆	¹⁵/₁₆
x 84	24.8	26.71	26¾	0.460	⁷/₁₆	¼	9.960	10	0.640	⅝	24	1⅜	¹⁵/₁₆
W 24×162	47.7	25.00	25	0.705	¹¹/₁₆	⅜	12.955	13	1.220	1¼	21	2	1¹/₁₆
x146	43.0	24.74	24¾	0.650	⅝	⁵/₁₆	12.900	12⅞	1.090	1¹/₁₆	21	1⅞	1¹/₁₆
x131	38.5	24.48	24½	0.605	⅝	⁵/₁₆	12.855	12⅞	0.960	¹⁵/₁₆	21	1¾	1¹/₁₆
x117	34.4	24.26	24¼	0.550	⁹/₁₆	⁵/₁₆	12.800	12¾	0.850	⅞	21	1⅝	1
x104	30.6	24.06	24	0.500	½	¼	12.750	12¾	0.750	¾	21	1½	1
W 24x 94	27.7	24.31	24¼	0.515	½	¼	9.065	9⅛	0.875	⅞	21	1⅝	1
x 84	24.7	24.10	24⅛	0.470	½	¼	9.020	9	0.770	¾	21	1⁹/₁₆	¹⁵/₁₆
x 76	22.4	23.92	23⅞	0.440	⁷/₁₆	¼	8.990	9	0.680	¹¹/₁₆	21	1⁷/₁₆	¹⁵/₁₆
x 68	20.1	23.73	23¾	0.415	⁷/₁₆	¼	8.965	9	0.585	⁹/₁₆	21	1⅜	¹⁵/₁₆
W 24x 62	18.2	23.74	23¾	0.430	⁷/₁₆	¼	7.040	7	0.590	⁹/₁₆	21	1⅜	¹⁵/₁₆
x 55	16.2	23.57	23⅝	0.395	⅜	³/₁₆	7.005	7	0.505	½	21	1⁵/₁₆	¹⁵/₁₆
W 21×147	43.2	22.06	22	0.720	¾	⅜	12.510	12½	1.150	1⅛	18¼	1⅞	1¹/₁₆
x132	38.8	21.83	21⅞	0.650	⅝	⁵/₁₆	12.440	12½	1.035	1¹/₁₆	18¼	1¹³/₁₆	1
x122	35.9	21.68	21⅝	0.600	⅝	⁵/₁₆	12.390	12⅜	0.960	¹⁵/₁₆	18¼	1¹¹/₁₆	1
x111	32.7	21.51	21½	0.550	⁹/₁₆	⁵/₁₆	12.340	12⅜	0.875	⅞	18¼	1⅝	¹⁵/₁₆
x101	29.8	21.36	21⅜	0.500	½	¼	12.290	12¼	0.800	¹³/₁₆	18¼	1⁹/₁₆	¹⁵/₁₆
W 21x 93	27.3	21.62	21⅝	0.580	⁹/₁₆	⁵/₁₆	8.420	8⅜	0.930	¹⁵/₁₆	18¼	1¹¹/₁₆	1
x 83	24.3	21.43	21⅜	0.515	½	¼	8.355	8⅜	0.835	¹³/₁₆	18¼	1⁹/₁₆	¹⁵/₁₆
x 73	21.5	21.24	21¼	0.455	⁷/₁₆	¼	8.295	8¼	0.740	¾	18¼	1½	¹⁵/₁₆
x 68	20.0	21.13	21⅛	0.430	⁷/₁₆	¼	8.270	8¼	0.685	¹¹/₁₆	18¼	1⁷/₁₆	⅞
x 62	18.3	20.99	21	0.400	⅜	³/₁₆	8.240	8¼	0.615	⅝	18¼	1⅜	⅞
W 21x 57	16.7	21.06	21	0.405	⅜	³/₁₆	6.555	6½	0.650	⅝	18¼	1⅜	⅞
x 50	14.7	20.83	20⅞	0.380	⅜	³/₁₆	6.530	6½	0.535	⁹/₁₆	18¼	1⁵/₁₆	⅞
x 44	13.0	20.66	20⅝	0.350	⅜	³/₁₆	6.500	6½	0.450	⁷/₁₆	18¼	1³/₁₆	⅞

VI. <u>STRUCTURAL METAL SHAPE DESIGNATIONS</u> (continued)

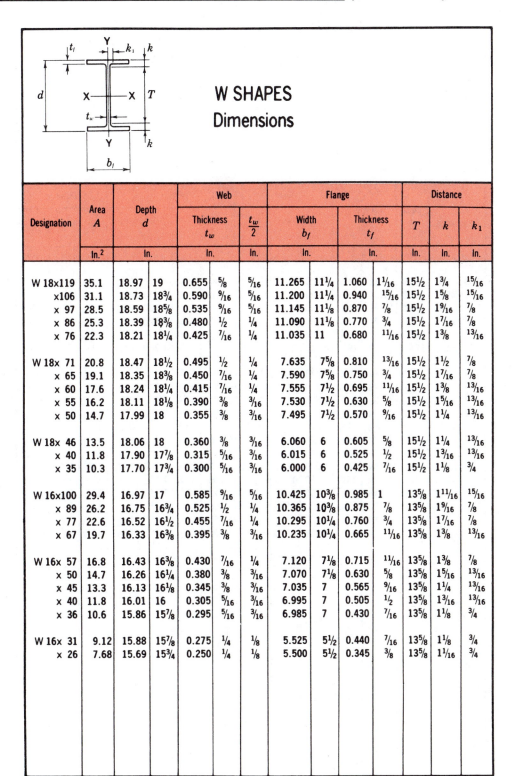

W SHAPES
Dimensions

Designation	Area A	Depth d		Web			Flange				Distance		
				Thickness t_w		$\frac{t_w}{2}$	Width b_f		Thickness t_f		T	k	k_1
	In.²	In.		In.		In.	In.		In.		In.	In.	In.
W 18×119	35.1	18.97	19	0.655	5/8	5/16	11.265	11¼	1.060	1 1/16	15½	1¾	15/16
×106	31.1	18.73	18¾	0.590	9/16	5/16	11.200	11¼	0.940	15/16	15½	1 5/8	15/16
× 97	28.5	18.59	18 5/8	0.535	9/16	5/16	11.145	11 1/8	0.870	7/8	15½	1 9/16	7/8
× 86	25.3	18.39	18 3/8	0.480	½	¼	11.090	11 1/8	0.770	¾	15½	1 7/16	7/8
× 76	22.3	18.21	18¼	0.425	7/16	¼	11.035	11	0.680	11/16	15½	1 3/8	13/16
W 18× 71	20.8	18.47	18½	0.495	½	¼	7.635	7 5/8	0.810	13/16	15½	1½	7/8
× 65	19.1	18.35	18 3/8	0.450	7/16	¼	7.590	7 5/8	0.750	¾	15½	1 7/16	7/8
× 60	17.6	18.24	18¼	0.415	7/16	¼	7.555	7½	0.695	11/16	15½	1 3/8	13/16
× 55	16.2	18.11	18 1/8	0.390	3/8	3/16	7.530	7½	0.630	5/8	15½	1 5/16	13/16
× 50	14.7	17.99	18	0.355	3/8	3/16	7.495	7½	0.570	9/16	15½	1¼	13/16
W 18× 46	13.5	18.06	18	0.360	3/8	3/16	6.060	6	0.605	5/8	15½	1¼	13/16
× 40	11.8	17.90	17 7/8	0.315	5/16	3/16	6.015	6	0.525	½	15½	1 3/16	13/16
× 35	10.3	17.70	17¾	0.300	5/16	3/16	6.000	6	0.425	7/16	15½	1 1/8	¾
W 16×100	29.4	16.97	17	0.585	9/16	5/16	10.425	10 3/8	0.985	1	13 5/8	1 11/16	15/16
× 89	26.2	16.75	16¾	0.525	½	¼	10.365	10 3/8	0.875	7/8	13 5/8	1 9/16	7/8
× 77	22.6	16.52	16½	0.455	7/16	¼	10.295	10¼	0.760	¾	13 5/8	1 7/16	7/8
× 67	19.7	16.33	16 3/8	0.395	3/8	3/16	10.235	10¼	0.665	11/16	13 5/8	1 3/8	13/16
W 16× 57	16.8	16.43	16 3/8	0.430	7/16	¼	7.120	7 1/8	0.715	11/16	13 5/8	1 3/8	7/8
× 50	14.7	16.26	16¼	0.380	3/8	3/16	7.070	7 1/8	0.630	5/8	13 5/8	1 5/16	13/16
× 45	13.3	16.13	16 1/8	0.345	3/8	3/16	7.035	7	0.565	9/16	13 5/8	1¼	13/16
× 40	11.8	16.01	16	0.305	5/16	3/16	6.995	7	0.505	½	13 5/8	1 3/16	13/16
× 36	10.6	15.86	15 7/8	0.295	5/16	3/16	6.985	7	0.430	7/16	13 5/8	1 1/8	¾
W 16× 31	9.12	15.88	15 7/8	0.275	¼	1/8	5.525	5½	0.440	7/16	13 5/8	1 1/8	¾
× 26	7.68	15.69	15¾	0.250	¼	1/8	5.500	5½	0.345	3/8	13 5/8	1 1/16	¾

VI. <u>STRUCTURAL METAL SHAPE DESIGNATIONS</u> (continued)

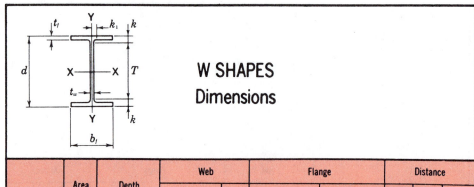

W SHAPES
Dimensions

Designation	Area A	Depth d		Web Thickness t_w		Web $\frac{t_w}{2}$	Flange Width b_f		Flange Thickness t_f		Distance T	Distance k	Distance k_1
	In.²	In.		In.		In.	In.		In.		In.	In.	In.
W 14×730	215.0	22.42	22⅜	3.070	3 1/16	1 9/16	17.890	17⅞	4.910	4 15/16	11¼	5 9/16	2 3/16
×665	196.0	21.64	21⅝	2.830	2 13/16	1 7/16	17.650	17⅝	4.520	4½	11¼	5 3/16	2 1/16
×605	178.0	20.92	20⅞	2.595	2⅝	1 5/16	17.415	17⅜	4.160	4 3/16	11¼	4 13/16	1 15/16
×550	162.0	20.24	20¼	2.380	2⅜	1 3/16	17.200	17¼	3.820	3 13/16	11¼	4½	1 13/16
×500	147.0	19.60	19⅝	2.190	2 3/16	1⅛	17.010	17	3.500	3½	11¼	4 3/16	1¾
×455	134.0	19.02	19	2.015	2	1	16.835	16⅞	3.210	3 3/16	11¼	3⅞	1⅝
W 14×426	125.0	18.67	18⅝	1.875	1⅞	15/16	16.695	16¾	3.035	3 1/16	11¼	3 11/16	1 9/16
×398	117.0	18.29	18¼	1.770	1¾	⅞	16.590	16⅝	2.845	2⅞	11¼	3½	1½
×370	109.0	17.92	17⅞	1.655	1⅝	13/16	16.475	16½	2.660	2 11/16	11¼	3 5/16	1 7/16
×342	101.0	17.54	17½	1.540	1 9/16	13/16	16.360	16⅜	2.470	2½	11¼	3⅛	1⅜
×311	91.4	17.12	17⅛	1.410	1 7/16	¾	16.230	16¼	2.260	2¼	11¼	2 15/16	1 5/16
×283	83.3	16.74	16¾	1.290	1 5/16	11/16	16.110	16⅛	2.070	2 1/16	11¼	2¾	1¼
×257	75.6	16.38	16⅜	1.175	1 3/16	⅝	15.995	16	1.890	1⅞	11¼	2 9/16	1 3/16
×233	68.5	16.04	16	1.070	1 1/16	9/16	15.890	15⅞	1.720	1¾	11¼	2⅜	1 3/16
×211	62.0	15.72	15¾	0.980	1	½	15.800	15¾	1.560	1 9/16	11¼	2¼	1⅛
×193	56.8	15.48	15½	0.890	⅞	7/16	15.710	15¾	1.440	1 7/16	11¼	2⅛	1 1/16
×176	51.8	15.22	15¼	0.830	13/16	7/16	15.650	15⅝	1.310	1 5/16	11¼	2	1 1/16
×159	46.7	14.98	15	0.745	¾	⅜	15.565	15⅝	1.190	1 3/16	11¼	1⅞	1
×145	42.7	14.78	14¾	0.680	11/16	⅜	15.500	15½	1.090	1 1/16	11¼	1¾	1

AMERICAN INSTITUTE OF STEEL CONSTRUCTION

VI. STRUCTURAL METAL SHAPE DESIGNATIONS (continued)

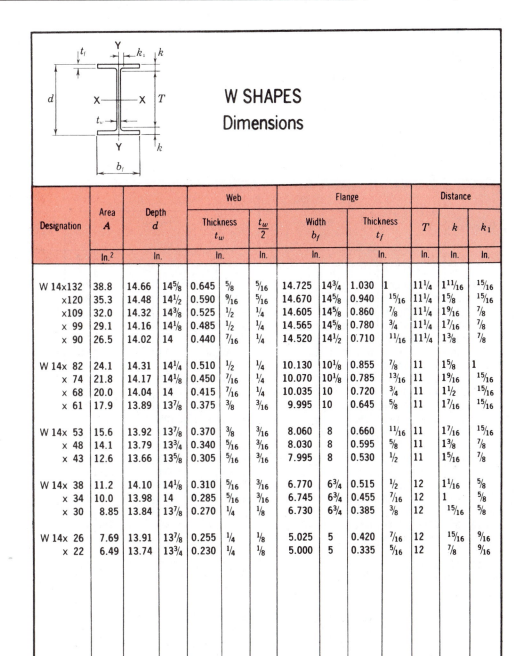

W SHAPES
Dimensions

Designation	Area A	Depth d		Web Thickness t_w	$\frac{t_w}{2}$	Flange Width b_f		Thickness t_f		Distance T	k	k_1
	In.²	In.		In.	In.	In.		In.		In.	In.	In.
W 14x132	38.8	14.66	14⅝	0.645	⅝	5/16	14.725	14¾	1.030	1	11¼	1 11/16	15/16
x120	35.3	14.48	14½	0.590	9/16	5/16	14.670	14⅝	0.940	15/16	11¼	1⅝	15/16
x109	32.0	14.32	14⅜	0.525	½	¼	14.605	14⅝	0.860	⅞	11¼	1 9/16	⅞
x 99	29.1	14.16	14⅛	0.485	½	¼	14.565	14⅝	0.780	¾	11¼	1 7/16	⅞
x 90	26.5	14.02	14	0.440	7/16	¼	14.520	14½	0.710	11/16	11¼	1⅜	⅞
W 14x 82	24.1	14.31	14¼	0.510	½	¼	10.130	10⅛	0.855	⅞	11	1⅝	1
x 74	21.8	14.17	14⅛	0.450	7/16	¼	10.070	10⅛	0.785	13/16	11	1 9/16	15/16
x 68	20.0	14.04	14	0.415	7/16	¼	10.035	10	0.720	¾	11	1½	15/16
x 61	17.9	13.89	13⅞	0.375	⅜	3/16	9.995	10	0.645	⅝	11	1 7/16	15/16
W 14x 53	15.6	13.92	13⅞	0.370	⅜	3/16	8.060	8	0.660	11/16	11	1 7/16	15/16
x 48	14.1	13.79	13¾	0.340	5/16	3/16	8.030	8	0.595	⅝	11	1⅜	⅞
x 43	12.6	13.66	13⅝	0.305	5/16	3/16	7.995	8	0.530	½	11	1 5/16	⅞
W 14x 38	11.2	14.10	14⅛	0.310	5/16	3/16	6.770	6¾	0.515	½	12	1 1/16	⅝
x 34	10.0	13.98	14	0.285	5/16	3/16	6.745	6¾	0.455	7/16	12	1	⅝
x 30	8.85	13.84	13⅞	0.270	¼	⅛	6.730	6¾	0.385	⅜	12	15/16	⅝
W 14x 26	7.69	13.91	13⅞	0.255	¼	⅛	5.025	5	0.420	7/16	12	15/16	9/16
x 22	6.49	13.74	13¾	0.230	¼	⅛	5.000	5	0.335	5/16	12	⅞	9/16

VI. STRUCTURAL METAL SHAPE DESIGNATIONS (continued)

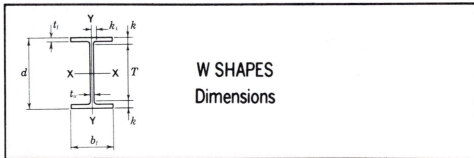

W SHAPES
Dimensions

Designation	Area A (In.²)	Depth d (In.)	Web Thickness t_w (In.)	Web $t_w/2$ (In.)	Flange Width b_f (In.)	Flange Thickness t_f (In.)	T (In.)	k (In.)	k_1 (In.)
W 12×336	98.8	16.82 16 7/8	1.775 1 3/4	7/8	13.385 13 3/8	2.955 2 15/16	9 1/2	3 11/16	1 1/2
×305	89.6	16.32 16 3/8	1.625 1 5/8	13/16	13.235 13 1/4	2.705 2 11/16	9 1/2	3 7/16	1 7/16
×279	81.9	15.85 15 7/8	1.530 1 1/2	3/4	13.140 13 1/8	2.470 2 1/2	9 1/2	3 3/16	1 3/8
×252	74.1	15.41 15 3/8	1.395 1 3/8	11/16	13.005 13	2.250 2 1/4	9 1/2	2 15/16	1 5/16
×230	67.7	15.05 15	1.285 1 5/16	11/16	12.895 12 7/8	2.070 2 1/16	9 1/2	2 3/4	1 1/4
×210	61.8	14.71 14 3/4	1.180 1 3/16	5/8	12.790 12 3/4	1.900 1 7/8	9 1/2	2 5/8	1 1/4
×190	55.8	14.38 14 3/8	1.060 1 1/16	9/16	12.670 12 5/8	1.735 1 3/4	9 1/2	2 7/16	1 3/16
×170	50.0	14.03 14	0.960 15/16	1/2	12.570 12 5/8	1.560 1 9/16	9 1/2	2 1/4	1 1/8
×152	44.7	13.71 13 3/4	0.870 7/8	7/16	12.480 12 1/2	1.400 1 3/8	9 1/2	2 1/8	1 1/16
×136	39.9	13.41 13 3/8	0.790 13/16	7/16	12.400 12 3/8	1.250 1 1/4	9 1/2	1 15/16	1
×120	35.3	13.12 13 1/8	0.710 11/16	3/8	12.320 12 3/8	1.105 1 1/8	9 1/2	1 13/16	1
×106	31.2	12.89 12 7/8	0.610 5/8	5/16	12.220 12 1/4	0.990 1	9 1/2	1 11/16	15/16
× 96	28.2	12.71 12 3/4	0.550 9/16	5/16	12.160 12 1/8	0.900 7/8	9 1/2	1 5/8	7/8
× 87	25.6	12.53 12 1/2	0.515 1/2	1/4	12.125 12 1/8	0.810 13/16	9 1/2	1 1/2	7/8
× 79	23.2	12.38 12 3/8	0.470 1/2	1/4	12.080 12 1/8	0.735 3/4	9 1/2	1 7/16	7/8
× 72	21.1	12.25 12 1/4	0.430 7/16	1/4	12.040 12	0.670 11/16	9 1/2	1 3/8	7/8
× 65	19.1	12.12 12 1/8	0.390 3/8	3/16	12.000 12	0.605 5/8	9 1/2	1 5/16	13/16
W 12× 58	17.0	12.19 12 1/4	0.360 3/8	3/16	10.010 10	0.640 5/8	9 1/2	1 3/8	13/16
× 53	15.6	12.06 12	0.345 3/8	3/16	9.995 10	0.575 9/16	9 1/2	1 1/4	13/16
W 12× 50	14.7	12.19 12 1/4	0.370 3/8	3/16	8.080 8 1/8	0.640 5/8	9 1/2	1 3/8	13/16
× 45	13.2	12.06 12	0.335 5/16	3/16	8.045 8	0.575 9/16	9 1/2	1 1/4	13/16
× 40	11.8	11.94 12	0.295 5/16	3/16	8.005 8	0.515 1/2	9 1/2	1 1/4	3/4
W 12× 35	10.3	12.50 12 1/2	0.300 5/16	3/16	6.560 6 1/2	0.520 1/2	10 1/2	1	9/16
× 30	8.79	12.34 12 3/8	0.260 1/4	1/8	6.520 6 1/2	0.440 7/16	10 1/2	15/16	1/2
× 26	7.65	12.22 12 1/4	0.230 1/4	1/8	6.490 6 1/2	0.380 3/8	10 1/2	7/8	1/2
W 12× 22	6.48	12.31 12 1/4	0.260 1/4	1/8	4.030 4	0.425 7/16	10 1/2	7/8	1/2
× 19	5.57	12.16 12 1/8	0.235 1/4	1/8	4.005 4	0.350 3/8	10 1/2	13/16	1/2
×·16	4.71	11.99 12	0.220 1/4	1/8	3.990 4	0.265 1/4	10 1/2	3/4	1/2
× 14	4.16	11.91 11 7/8	0.200 3/16	1/8	3.970 4	0.225 1/4	10 1/2	11/16	1/2

VI. STRUCTURAL METAL SHAPE DESIGNATIONS (continued)

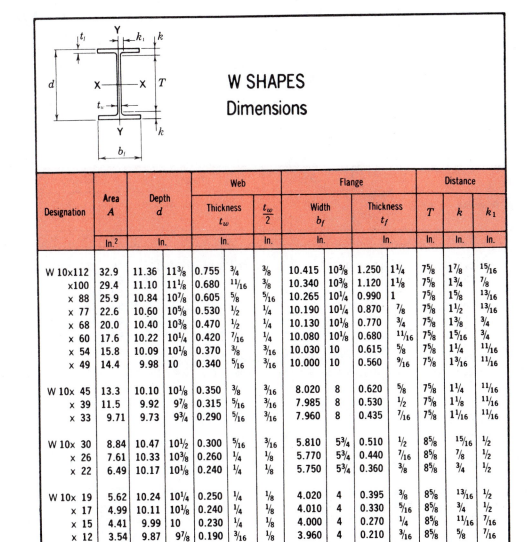

W SHAPES
Dimensions

Designation	Area A	Depth d	Web			Flange				Distance		
			Thickness t_w		$\frac{t_w}{2}$	Width b_f		Thickness t_f		T	k	k_1
	In.²	In.	In.		In.	In.		In.		In.	In.	In.
W 10x112	32.9	11.36	11³/₈	0.755	³/₄	³/₈	10.415	10³/₈	1.250	1¹/₄	7⁵/₈	1⁷/₈	15/₁₆
x100	29.4	11.10	11¹/₈	0.680	11/₁₆	³/₈	10.340	10³/₈	1.120	1¹/₈	7⁵/₈	1³/₄	⁷/₈
x 88	25.9	10.84	10⁷/₈	0.605	⁵/₈	⁵/₁₆	10.265	10¹/₄	0.990	1	7⁵/₈	1⁵/₈	13/₁₆
x 77	22.6	10.60	10⁵/₈	0.530	¹/₂	¹/₄	10.190	10¹/₄	0.870	⁷/₈	7⁵/₈	1¹/₂	13/₁₆
x 68	20.0	10.40	10³/₈	0.470	¹/₂	¹/₄	10.130	10¹/₈	0.770	³/₄	7⁵/₈	1³/₈	³/₄
x 60	17.6	10.22	10¹/₄	0.420	⁷/₁₆	¹/₄	10.080	10¹/₈	0.680	11/₁₆	7⁵/₈	1⁵/₁₆	³/₄
x 54	15.8	10.09	10¹/₈	0.370	³/₈	³/₁₆	10.030	10	0.615	⁵/₈	7⁵/₈	1¹/₄	11/₁₆
x 49	14.4	9.98	10	0.340	⁵/₁₆	³/₁₆	10.000	10	0.560	⁹/₁₆	7⁵/₈	1³/₁₆	11/₁₆
W 10x 45	13.3	10.10	10¹/₈	0.350	³/₈	³/₁₆	8.020	8	0.620	⁵/₈	7⁵/₈	1¹/₄	11/₁₆
x 39	11.5	9.92	9⁷/₈	0.315	⁵/₁₆	³/₁₆	7.985	8	0.530	¹/₂	7⁵/₈	1¹/₈	11/₁₆
x 33	9.71	9.73	9³/₄	0.290	⁵/₁₆	³/₁₆	7.960	8	0.435	⁷/₁₆	7⁵/₈	1¹/₁₆	11/₁₆
W 10x 30	8.84	10.47	10¹/₂	0.300	⁵/₁₆	³/₁₆	5.810	5³/₄	0.510	¹/₂	8⁵/₈	15/₁₆	¹/₂
x 26	7.61	10.33	10³/₈	0.260	¹/₄	¹/₈	5.770	5³/₄	0.440	⁷/₁₆	8⁵/₈	⁷/₈	¹/₂
x 22	6.49	10.17	10¹/₈	0.240	¹/₄	¹/₈	5.750	5³/₄	0.360	³/₈	8⁵/₈	³/₄	¹/₂
W 10x 19	5.62	10.24	10¹/₄	0.250	¹/₄	¹/₈	4.020	4	0.395	³/₈	8⁵/₈	13/₁₆	¹/₂
x 17	4.99	10.11	10¹/₈	0.240	¹/₄	¹/₈	4.010	4	0.330	⁵/₁₆	8⁵/₈	³/₄	¹/₂
x 15	4.41	9.99	10	0.230	¹/₄	¹/₈	4.000	4	0.270	¹/₄	8⁵/₈	11/₁₆	⁷/₁₆
x 12	3.54	9.87	9⁷/₈	0.190	³/₁₆	¹/₈	3.960	4	0.210	³/₁₆	8⁵/₈	⁵/₈	⁷/₁₆

AMERICAN INSTITUTE OF STEEL CONSTRUCTION

VI. <u>STRUCTURAL METAL SHAPE DESIGNATIONS</u> (continued)

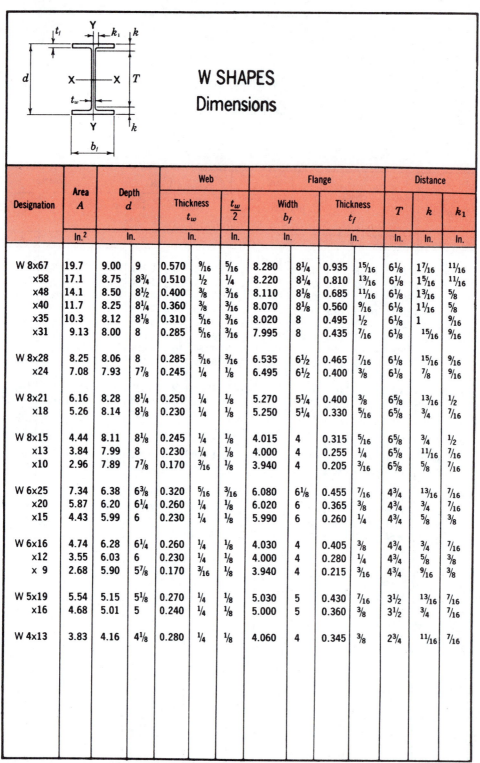

W SHAPES
Dimensions

Designation	Area A	Depth d		Web Thickness t_w		$\dfrac{t_w}{2}$	Flange Width b_f		Thickness t_f		Distance T	k	k_1
	In.²	In.		In.		In.	In.		In.		In.	In.	In.
W 8x67	19.7	9.00	9	0.570	9/16	5/16	8.280	8¼	0.935	15/16	6⅛	1 7/16	11/16
x58	17.1	8.75	8¾	0.510	½	¼	8.220	8¼	0.810	13/16	6⅛	1 5/16	11/16
x48	14.1	8.50	8½	0.400	⅜	3/16	8.110	8⅛	0.685	11/16	6⅛	1 3/16	⅝
x40	11.7	8.25	8¼	0.360	⅜	3/16	8.070	8⅛	0.560	9/16	6⅛	1 1/16	⅝
x35	10.3	8.12	8⅛	0.310	5/16	3/16	8.020	8	0.495	½	6⅛	1	9/16
x31	9.13	8.00	8	0.285	5/16	3/16	7.995	8	0.435	7/16	6⅛	15/16	9/16
W 8x28	8.25	8.06	8	0.285	5/16	3/16	6.535	6½	0.465	7/16	6⅛	15/16	9/16
x24	7.08	7.93	7⅞	0.245	¼	⅛	6.495	6½	0.400	⅜	6⅛	⅞	9/16
W 8x21	6.16	8.28	8¼	0.250	¼	⅛	5.270	5¼	0.400	⅜	6⅝	13/16	½
x18	5.26	8.14	8⅛	0.230	¼	⅛	5.250	5¼	0.330	5/16	6⅝	¾	7/16
W 8x15	4.44	8.11	8⅛	0.245	¼	⅛	4.015	4	0.315	5/16	6⅝	¾	½
x13	3.84	7.99	8	0.230	¼	⅛	4.000	4	0.255	¼	6⅝	11/16	7/16
x10	2.96	7.89	7⅞	0.170	3/16	⅛	3.940	4	0.205	3/16	6⅝	⅝	7/16
W 6x25	7.34	6.38	6⅜	0.320	5/16	3/16	6.080	6⅛	0.455	7/16	4¾	13/16	7/16
x20	5.87	6.20	6¼	0.260	¼	⅛	6.020	6	0.365	⅜	4¾	¾	7/16
x15	4.43	5.99	6	0.230	¼	⅛	5.990	6	0.260	¼	4¾	⅝	⅜
W 6x16	4.74	6.28	6¼	0.260	¼	⅛	4.030	4	0.405	⅜	4¾	¾	7/16
x12	3.55	6.03	6	0.230	¼	⅛	4.000	4	0.280	¼	4¾	⅝	⅜
x 9	2.68	5.90	5⅞	0.170	3/16	⅛	3.940	4	0.215	3/16	4¾	9/16	⅜
W 5x19	5.54	5.15	5⅛	0.270	¼	⅛	5.030	5	0.430	7/16	3½	13/16	7/16
x16	4.68	5.01	5	0.240	¼	⅛	5.000	5	0.360	⅜	3½	¾	7/16
W 4x13	3.83	4.16	4⅛	0.280	¼	⅛	4.060	4	0.345	⅜	2¾	11/16	7/16

AMERICAN INSTITUTE OF STEEL CONSTRUCTION

VI. <u>STRUCTURAL METAL SHAPE DESIGNATIONS</u> (continued)

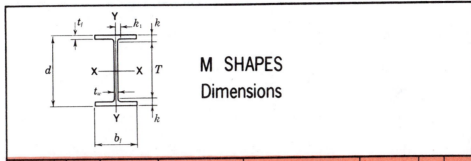

M SHAPES
Dimensions

Designation	Area A	Depth d		Web Thickness t_w		$\frac{t_w}{2}$	Flange Width b_f		Flange Thickness t_f		Distance T	Distance k	Grip	Max. Flge. Fastener
	In.²	In.		In.		In.	In.		In.		In.	In.	In.	In.
M 14x18	5.10	14.00	14	0.215	3/16	1/8	4.000	4	0.270	1/4	12¾	5/8	1/4	3/4
M 12x11.8	3.47	12.00	12	0.177	3/16	1/8	3.065	3⅛	0.225	1/4	10⅞	9/16	1/4	—
M 10x9	2.65	10.00	10	0.157	3/16	1/8	2.690	2¾	0.206	3/16	8⅞	9/16	3/16	—
M 8x6.5	1.92	8.00	8	0.135	1/8	1/16	2.281	2¼	0.189	3/16	7	1/2	3/16	—
M 6x20	5.89	6.00	6	0.250	1/4	1/8	5.938	6	0.379	3/8	4¼	7/8	3/8	7/8
M 6x4.4	1.29	6.00	6	0.114	1/8	1/16	1.844	1⅞	0.171	3/16	5⅛	7/16	3/16	—
M 5x18.9	5.55	5.00	5	0.316	5/16	3/16	5.003	5	0.416	7/16	3¼	7/8	7/16	7/8
M 4x13	3.81	4.00	4	0.254	1/4	1/8	3.940	4	0.371	3/8	2⅜	13/16	3/8	3/4

AMERICAN INSTITUTE OF STEEL CONSTRUCTION

VI. <u>STRUCTURAL METAL SHAPE DESIGNATIONS</u> (continued)

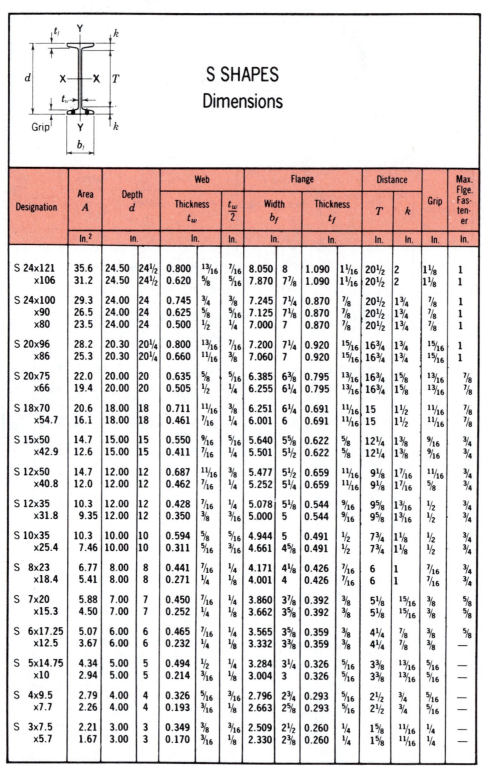

S SHAPES
Dimensions

Designation	Area A	Depth d		Web Thickness t_w		$\frac{t_w}{2}$	Flange Width b_f		Thickness t_f	Distance T	k	Grip	Max. Flge. Fastener
	In.²	In.		In.		In.	In.		In.	In.	In.	In.	In.
S 24x121	35.6	24.50	24½	0.800	13/16	7/16	8.050	8	1.090	1 1/16	20½	2	1 1/8	1
x106	31.2	24.50	24½	0.620	5/8	5/16	7.870	7 7/8	1.090	1 1/16	20½	2	1 1/8	1
S 24x100	29.3	24.00	24	0.745	3/4	3/8	7.245	7 1/4	0.870	7/8	20½	1 3/4	7/8	1
x90	26.5	24.00	24	0.625	5/8	5/16	7.125	7 1/8	0.870	7/8	20½	1 3/4	7/8	1
x80	23.5	24.00	24	0.500	1/2	1/4	7.000	7	0.870	7/8	20½	1 3/4	7/8	1
S 20x96	28.2	20.30	20¼	0.800	13/16	7/16	7.200	7 1/4	0.920	15/16	16 3/4	1 3/4	15/16	1
x86	25.3	20.30	20¼	0.660	11/16	3/8	7.060	7	0.920	15/16	16 3/4	1 3/4	15/16	1
S 20x75	22.0	20.00	20	0.635	5/8	5/16	6.385	6 3/8	0.795	13/16	16 3/4	1 5/8	13/16	7/8
x66	19.4	20.00	20	0.505	1/2	1/4	6.255	6 1/4	0.795	13/16	16 3/4	1 5/8	13/16	7/8
S 18x70	20.6	18.00	18	0.711	11/16	3/8	6.251	6 1/4	0.691	11/16	15	1 1/2	11/16	7/8
x54.7	16.1	18.00	18	0.461	7/16	1/4	6.001	6	0.691	11/16	15	1 1/2	11/16	7/8
S 15x50	14.7	15.00	15	0.550	9/16	5/16	5.640	5 5/8	0.622	5/8	12 1/4	1 3/8	9/16	3/4
x42.9	12.6	15.00	15	0.411	7/16	1/4	5.501	5 1/2	0.622	5/8	12 1/4	1 3/8	9/16	3/4
S 12x50	14.7	12.00	12	0.687	11/16	3/8	5.477	5 1/2	0.659	11/16	9 1/8	1 7/16	11/16	3/4
x40.8	12.0	12.00	12	0.462	7/16	1/4	5.252	5 1/4	0.659	11/16	9 1/8	1 7/16	5/8	3/4
S 12x35	10.3	12.00	12	0.428	7/16	1/4	5.078	5 1/8	0.544	9/16	9 5/8	1 3/16	1/2	3/4
x31.8	9.35	12.00	12	0.350	3/8	3/16	5.000	5	0.544	9/16	9 5/8	1 3/16	1/2	3/4
S 10x35	10.3	10.00	10	0.594	5/8	5/16	4.944	5	0.491	1/2	7 3/4	1 1/8	1/2	3/4
x25.4	7.46	10.00	10	0.311	5/16	3/16	4.661	4 5/8	0.491	1/2	7 3/4	1 1/8	1/2	3/4
S 8x23	6.77	8.00	8	0.441	7/16	1/4	4.171	4 1/8	0.426	7/16	6	1	7/16	3/4
x18.4	5.41	8.00	8	0.271	1/4	1/8	4.001	4	0.426	7/16	6	1	7/16	3/4
S 7x20	5.88	7.00	7	0.450	7/16	1/4	3.860	3 7/8	0.392	3/8	5 1/8	15/16	3/8	5/8
x15.3	4.50	7.00	7	0.252	1/4	1/8	3.662	3 5/8	0.392	3/8	5 1/8	15/16	3/8	5/8
S 6x17.25	5.07	6.00	6	0.465	7/16	1/4	3.565	3 5/8	0.359	3/8	4 1/4	7/8	3/8	5/8
x12.5	3.67	6.00	6	0.232	1/4	1/8	3.332	3 3/8	0.359	3/8	4 1/4	7/8	3/8	—
S 5x14.75	4.34	5.00	5	0.494	1/2	1/4	3.284	3 1/4	0.326	5/16	3 3/8	13/16	5/16	—
x10	2.94	5.00	5	0.214	3/16	1/8	3.004	3	0.326	5/16	3 3/8	13/16	5/16	—
S 4x9.5	2.79	4.00	4	0.326	5/16	3/16	2.796	2 3/4	0.293	5/16	2 1/2	3/4	5/16	—
x7.7	2.26	4.00	4	0.193	3/16	1/8	2.663	2 5/8	0.293	5/16	2 1/2	3/4	5/16	—
S 3x7.5	2.21	3.00	3	0.349	3/8	3/16	2.509	2 1/2	0.260	1/4	1 5/8	11/16	1/4	—
x5.7	1.67	3.00	3	0.170	3/16	1/8	2.330	2 3/8	0.260	1/4	1 5/8	11/16	1/4	—

AMERICAN INSTITUTE OF STEEL CONSTRUCTION

VI. <u>STRUCTURAL METAL SHAPE DESIGNATIONS</u> (continued)

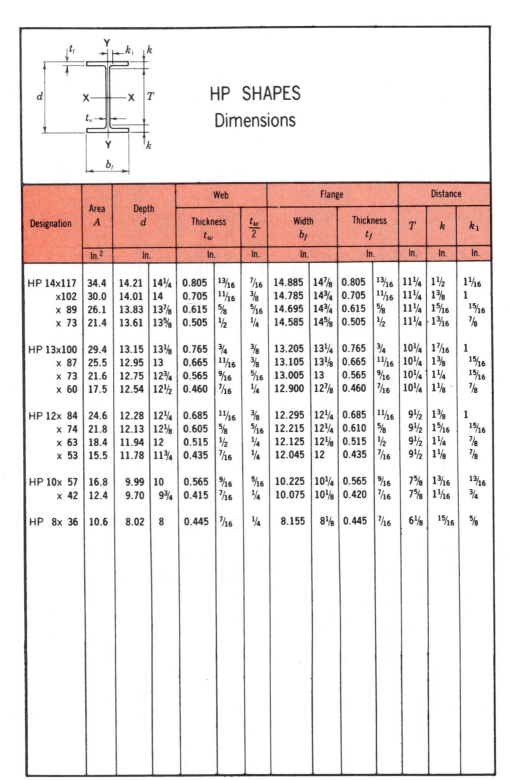

HP SHAPES
Dimensions

Designation	Area A	Depth d		Web Thickness t_w		$\frac{t_w}{2}$	Flange Width b_f		Thickness t_f		Distance T	k	k_1
	In.²	In.		In.		In.	In.		In.		In.	In.	In.
HP 14x117	34.4	14.21	14¼	0.805	¹³/₁₆	⁷/₁₆	14.885	14⅞	0.805	¹³/₁₆	11¼	1½	1¹/₁₆
x102	30.0	14.01	14	0.705	¹¹/₁₆	⅜	14.785	14¾	0.705	¹¹/₁₆	11¼	1⅜	1
x 89	26.1	13.83	13⅞	0.615	⅝	⁵/₁₆	14.695	14¾	0.615	⅝	11¼	1⁵/₁₆	¹⁵/₁₆
x 73	21.4	13.61	13⅝	0.505	½	¼	14.585	14⅝	0.505	½	11¼	1³/₁₆	⅞
HP 13x100	29.4	13.15	13⅛	0.765	¾	⅜	13.205	13¼	0.765	¾	10¼	1⁷/₁₆	1
x 87	25.5	12.95	13	0.665	¹¹/₁₆	⅜	13.105	13⅛	0.665	¹¹/₁₆	10¼	1⅜	¹⁵/₁₆
x 73	21.6	12.75	12¾	0.565	⁹/₁₆	⁵/₁₆	13.005	13	0.565	⁹/₁₆	10¼	1¼	¹⁵/₁₆
x 60	17.5	12.54	12½	0.460	⁷/₁₆	¼	12.900	12⅞	0.460	⁷/₁₆	10¼	1⅛	⅞
HP 12x 84	24.6	12.28	12¼	0.685	¹¹/₁₆	⅜	12.295	12¼	0.685	¹¹/₁₆	9½	1⅜	1
x 74	21.8	12.13	12⅛	0.605	⅝	⁵/₁₆	12.215	12¼	0.610	⅝	9½	1⁵/₁₆	¹⁵/₁₆
x 63	18.4	11.94	12	0.515	½	¼	12.125	12⅛	0.515	½	9½	1¼	⅞
x 53	15.5	11.78	11¾	0.435	⁷/₁₆	¼	12.045	12	0.435	⁷/₁₆	9½	1⅛	⅞
HP 10x 57	16.8	9.99	10	0.565	⁹/₁₆	⁵/₁₆	10.225	10¼	0.565	⁹/₁₆	7⅝	1³/₁₆	¹³/₁₆
x 42	12.4	9.70	9¾	0.415	⁷/₁₆	¼	10.075	10⅛	0.420	⁷/₁₆	7⅝	1¹/₁₆	¾
HP 8x 36	10.6	8.02	8	0.445	⁷/₁₆	¼	8.155	8⅛	0.445	⁷/₁₆	6⅛	¹⁵/₁₆	⅝

AMERICAN INSTITUTE OF STEEL CONSTRUCTION

VI. <u>STRUCTURAL METAL SHAPE DESIGNATIONS</u> (continued)

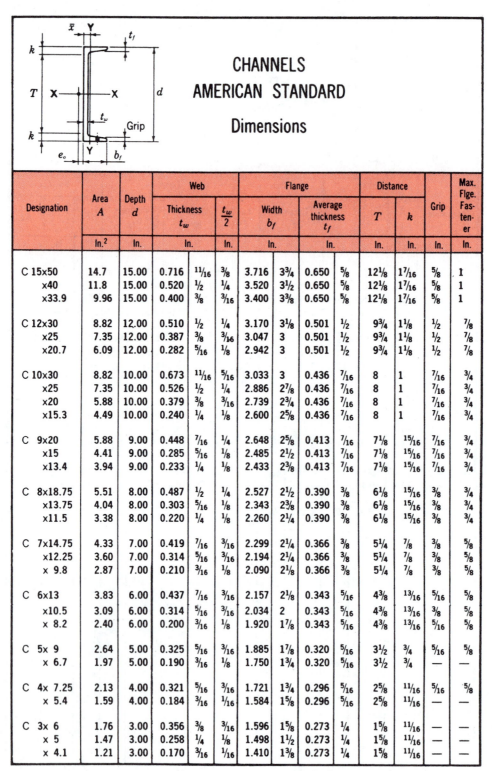

CHANNELS
AMERICAN STANDARD
Dimensions

Designation	Area A	Depth d	Web Thickness t_w	$\frac{t_w}{2}$	Flange Width b_f	Flange Average thickness t_f	Distance T	k	Grip	Max. Flge. Fastener
	In.²	In.	In.	In.	In.	In.	In.	In.	In.	In.
C 15x50	14.7	15.00	0.716 ¹¹⁄₁₆	³⁄₈	3.716 3³⁄₄	0.650 ⁵⁄₈	12¹⁄₈	1⁷⁄₁₆	⁵⁄₈	1
x40	11.8	15.00	0.520 ¹⁄₂	¹⁄₄	3.520 3¹⁄₂	0.650 ⁵⁄₈	12¹⁄₈	1⁷⁄₁₆	⁵⁄₈	1
x33.9	9.96	15.00	0.400 ³⁄₈	³⁄₁₆	3.400 3³⁄₈	0.650 ⁵⁄₈	12¹⁄₈	1⁷⁄₁₆	⁵⁄₈	1
C 12x30	8.82	12.00	0.510 ¹⁄₂	¹⁄₄	3.170 3¹⁄₈	0.501 ¹⁄₂	9³⁄₄	1¹⁄₈	¹⁄₂	⁷⁄₈
x25	7.35	12.00	0.387 ³⁄₈	³⁄₁₆	3.047 3	0.501 ¹⁄₂	9³⁄₄	1¹⁄₈	¹⁄₂	⁷⁄₈
x20.7	6.09	12.00	0.282 ⁵⁄₁₆	¹⁄₈	2.942 3	0.501 ¹⁄₂	9³⁄₄	1¹⁄₈	¹⁄₂	⁷⁄₈
C 10x30	8.82	10.00	0.673 ¹¹⁄₁₆	⁵⁄₁₆	3.033 3	0.436 ⁷⁄₁₆	8	1	⁷⁄₁₆	³⁄₄
x25	7.35	10.00	0.526 ¹⁄₂	¹⁄₄	2.886 2⁷⁄₈	0.436 ⁷⁄₁₆	8	1	⁷⁄₁₆	³⁄₄
x20	5.88	10.00	0.379 ³⁄₈	³⁄₁₆	2.739 2³⁄₄	0.436 ⁷⁄₁₆	8	1	⁷⁄₁₆	³⁄₄
x15.3	4.49	10.00	0.240 ¹⁄₄	¹⁄₈	2.600 2⁵⁄₈	0.436 ⁷⁄₁₆	8	1	⁷⁄₁₆	³⁄₄
C 9x20	5.88	9.00	0.448 ⁷⁄₁₆	¹⁄₄	2.648 2⁵⁄₈	0.413 ⁷⁄₁₆	7¹⁄₈	¹⁵⁄₁₆	⁷⁄₁₆	³⁄₄
x15	4.41	9.00	0.285 ⁵⁄₁₆	¹⁄₈	2.485 2¹⁄₂	0.413 ⁷⁄₁₆	7¹⁄₈	¹⁵⁄₁₆	⁷⁄₁₆	³⁄₄
x13.4	3.94	9.00	0.233 ¹⁄₄	¹⁄₈	2.433 2³⁄₈	0.413 ⁷⁄₁₆	7¹⁄₈	¹⁵⁄₁₆	⁷⁄₁₆	³⁄₄
C 8x18.75	5.51	8.00	0.487 ¹⁄₂	¹⁄₄	2.527 2¹⁄₂	0.390 ³⁄₈	6¹⁄₈	¹⁵⁄₁₆	³⁄₈	³⁄₄
x13.75	4.04	8.00	0.303 ⁵⁄₁₆	¹⁄₈	2.343 2³⁄₈	0.390 ³⁄₈	6¹⁄₈	¹⁵⁄₁₆	³⁄₈	³⁄₄
x11.5	3.38	8.00	0.220 ¹⁄₄	¹⁄₈	2.260 2¹⁄₄	0.390 ³⁄₈	6¹⁄₈	¹⁵⁄₁₆	³⁄₈	³⁄₄
C 7x14.75	4.33	7.00	0.419 ⁷⁄₁₆	³⁄₁₆	2.299 2¹⁄₄	0.366 ³⁄₈	5¹⁄₄	⁷⁄₈	³⁄₈	⁵⁄₈
x12.25	3.60	7.00	0.314 ⁵⁄₁₆	³⁄₁₆	2.194 2¹⁄₄	0.366 ³⁄₈	5¹⁄₄	⁷⁄₈	³⁄₈	⁵⁄₈
x 9.8	2.87	7.00	0.210 ³⁄₁₆	¹⁄₈	2.090 2¹⁄₈	0.366 ³⁄₈	5¹⁄₄	⁷⁄₈	³⁄₈	⁵⁄₈
C 6x13	3.83	6.00	0.437 ⁷⁄₁₆	³⁄₁₆	2.157 2¹⁄₈	0.343 ⁵⁄₁₆	4³⁄₈	¹³⁄₁₆	⁵⁄₁₆	⁵⁄₈
x10.5	3.09	6.00	0.314 ⁵⁄₁₆	³⁄₁₆	2.034 2	0.343 ⁵⁄₁₆	4³⁄₈	¹³⁄₁₆	³⁄₈	⁵⁄₈
x 8.2	2.40	6.00	0.200 ³⁄₁₆	¹⁄₈	1.920 1⁷⁄₈	0.343 ⁵⁄₁₆	4³⁄₈	¹³⁄₁₆	⁵⁄₁₆	⁵⁄₈
C 5x9	2.64	5.00	0.325 ⁵⁄₁₆	³⁄₁₆	1.885 1⁷⁄₈	0.320 ⁵⁄₁₆	3¹⁄₂	³⁄₄	⁵⁄₁₆	⁵⁄₈
x 6.7	1.97	5.00	0.190 ³⁄₁₆	¹⁄₈	1.750 1³⁄₄	0.320 ⁵⁄₁₆	3¹⁄₂	³⁄₄	—	—
C 4x 7.25	2.13	4.00	0.321 ⁵⁄₁₆	³⁄₁₆	1.721 1³⁄₄	0.296 ⁵⁄₁₆	2⁵⁄₈	¹¹⁄₁₆	⁵⁄₁₆	⁵⁄₈
x 5.4	1.59	4.00	0.184 ³⁄₁₆	¹⁄₁₆	1.584 1⁵⁄₈	0.296 ⁵⁄₁₆	2⁵⁄₈	¹¹⁄₁₆	—	—
C 3x 6	1.76	3.00	0.356 ³⁄₈	³⁄₁₆	1.596 1⁵⁄₈	0.273 ¹⁄₄	1⁵⁄₈	¹¹⁄₁₆	—	—
x 5	1.47	3.00	0.258 ¹⁄₄	¹⁄₈	1.498 1¹⁄₂	0.273 ¹⁄₄	1⁵⁄₈	¹¹⁄₁₆	—	—
x 4.1	1.21	3.00	0.170 ³⁄₁₆	¹⁄₁₆	1.410 1³⁄₈	0.273 ¹⁄₄	1⁵⁄₈	¹¹⁄₁₆	—	—

AMERICAN INSTITUTE OF STEEL CONSTRUCTION

VI. STRUCTURAL METAL SHAPE DESIGNATIONS (continued)

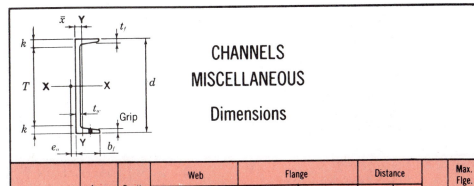

CHANNELS
MISCELLANEOUS
Dimensions

Designation	Area A	Depth d	Web			Flange				Distance		Grip	Max. Flge. Fastener
			Thickness t_w		$\frac{t_w}{2}$	Width b_f		Average thickness t_f		T	k		
	In.²	In.	In.		In.	In.		In.		In.	In.	In.	In.
MC 18x58	17.1	18.00	0.700	¹¹/₁₆	³/₈	4.200	4¹/₄	0.625	⁵/₈	15¹/₄	1³/₈	⁵/₈	1
x51.9	15.3	18.00	0.600	⁵/₈	⁵/₁₆	4.100	4¹/₈	0.625	⁵/₈	15¹/₄	1³/₈	⁵/₈	1
x45.8	13.5	18.00	0.500	¹/₂	¹/₄	4.000	4	0.625	⁵/₈	15¹/₄	1³/₈	⁵/₈	1
x42.7	12.6	18.00	0.450	⁷/₁₆	¹/₄	3.950	4	0.625	⁵/₈	15¹/₄	1³/₈	⁵/₈	1
MC 13x50	14.7	13.00	0.787	¹³/₁₆	³/₈	4.412	4³/₈	0.610	⁵/₈	10¹/₄	1³/₈	⁵/₈	1
x40	11.8	13.00	0.560	⁹/₁₆	¹/₄	4.185	4¹/₈	0.610	⁵/₈	10¹/₄	1³/₈	⁹/₁₆	1
x35	10.3	13.00	0.447	⁷/₁₆	¹/₄	4.072	4¹/₈	0.610	⁵/₈	10¹/₄	1³/₈	⁹/₁₆	1
x31.8	9.35	13.00	0.375	³/₈	³/₁₆	4.000	4	0.610	⁵/₈	10¹/₄	1³/₈	⁹/₁₆	1
MC 12x50	14.7	12.00	0.835	¹³/₁₆	⁷/₁₆	4.135	4¹/₈	0.700	¹¹/₁₆	9³/₈	1⁵/₁₆	¹¹/₁₆	1
x45	13.2	12.00	0.712	¹¹/₁₆	³/₈	4.012	4	0.700	¹¹/₁₆	9³/₈	1⁵/₁₆	¹¹/₁₆	1
x40	11.8	12.00	0.590	⁹/₁₆	⁵/₁₆	3.890	3⁷/₈	0.700	¹¹/₁₆	9³/₈	1⁵/₁₆	¹¹/₁₆	1
x35	10.3	12.00	0.467	⁷/₁₆	¹/₄	3.767	3³/₄	0.700	¹¹/₁₆	9³/₈	1⁵/₁₆	¹¹/₁₆	1
MC 12x37	10.9	12.00	0.600	⁵/₈	⁵/₁₆	3.600	3⁵/₈	0.600	⁵/₈	9³/₈	1⁵/₁₆	⁵/₈	⁷/₈
x32.9	9.67	12.00	0.500	¹/₂	¹/₄	3.500	3¹/₂	0.600	⁵/₈	9³/₈	1⁵/₁₆	⁹/₁₆	⁷/₈
x30.9	9.07	12.00	0.450	⁷/₁₆	¹/₄	3.450	3¹/₂	0.600	⁵/₈	9³/₈	1⁵/₁₆	⁹/₁₆	⁷/₈
MC 12x10.6	3.10	12.00	0.190	³/₁₆	¹/₈	1.500	1¹/₂	0.309	⁵/₁₆	10⁵/₈	¹¹/₁₆	—	—
MC 10x41.1	12.1	10.00	0.796	¹³/₁₆	³/₈	4.321	4³/₈	0.575	⁹/₁₆	7¹/₂	1¹/₄	⁹/₁₆	⁷/₈
x33.6	9.87	10.00	0.575	⁹/₁₆	⁵/₁₆	4.100	4¹/₈	0.575	⁹/₁₆	7¹/₂	1¹/₄	⁹/₁₆	⁷/₈
x28.5	8.37	10.00	0.425	⁷/₁₆	³/₁₆	3.950	4	0.575	⁹/₁₆	7¹/₂	1¹/₄	⁹/₁₆	⁷/₈
MC 10x28.3	8.32	10.00	0.477	¹/₂	¹/₄	3.502	3¹/₂	0.575	⁹/₁₆	7¹/₂	1¹/₄	⁹/₁₆	⁷/₈
x25.3	7.43	10.00	0.425	⁷/₁₆	³/₁₆	3.550	3¹/₂	0.500	¹/₂	7³/₄	1¹/₈	¹/₂	⁷/₈
x24.9	7.32	10.00	0.377	³/₈	³/₁₆	3.402	3³/₈	0.575	⁹/₁₆	7¹/₂	1¹/₄	⁹/₁₆	⁷/₈
x21.9	6.43	10.00	0.325	⁵/₁₆	³/₁₆	3.450	3¹/₂	0.500	¹/₂	7³/₄	1¹/₈	¹/₂	⁷/₈
MC 10x 8.4	2.46	10.00	0.170	³/₁₆	¹/₁₆	1.500	1¹/₂	0.280	¹/₄	8⁵/₈	¹¹/₁₆	—	—
MC 10x 6.5	1.91	10.00	0.152	¹/₈	¹/₁₆	1.127	1¹/₈	0.202	³/₁₆	9¹/₈	⁷/₁₆	—	—

AMERICAN INSTITUTE OF STEEL CONSTRUCTION

VI. <u>STRUCTURAL METAL SHAPE DESIGNATIONS</u> (continued)

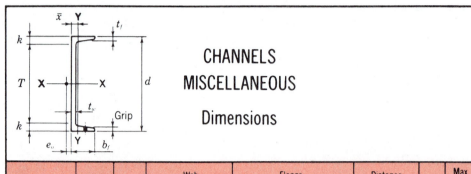

CHANNELS
MISCELLANEOUS
Dimensions

Designation	Area A	Depth d	Web Thickness t_w		$\frac{t_w}{2}$	Flange Width b_f		Average thickness t_f		Distance T	k	Grip	Max. Flge. Fastener
	In.²	In.	In.		In.	In.		In.		In.	In.	In.	In.
MC 9x25.4	7.47	9.00	0.450	$7/16$	$1/4$	3.500	$3\frac{1}{2}$	0.550	$9/16$	$6\frac{5}{8}$	$1\frac{3}{16}$	$9/16$	$7/8$
x23.9	7.02	9.00	0.400	$3/8$	$3/16$	3.450	$3\frac{1}{2}$	0.550	$9/16$	$6\frac{5}{8}$	$1\frac{3}{16}$	$9/16$	$7/8$
MC 8x22.8	6.70	8.00	0.427	$7/16$	$3/16$	3.502	$3\frac{1}{2}$	0.525	$1/2$	$5\frac{5}{8}$	$1\frac{3}{16}$	$1/2$	$7/8$
x21.4	6.28	8.00	0.375	$3/8$	$3/16$	3.450	$3\frac{1}{2}$	0.525	$1/2$	$5\frac{5}{8}$	$1\frac{3}{16}$	$1/2$	$7/8$
MC 8x20	5.88	8.00	0.400	$3/8$	$3/16$	3.025	3	0.500	$1/2$	$5\frac{3}{4}$	$1\frac{1}{8}$	$1/2$	$7/8$
x18.7	5.50	8.00	0.353	$3/8$	$3/16$	2.978	3	0.500	$1/2$	$5\frac{3}{4}$	$1\frac{1}{8}$	$1/2$	$7/8$
MC 8x 8.5	2.50	8.00	0.179	$3/16$	$1/16$	1.874	$1\frac{7}{8}$	0.311	$5/16$	$6\frac{1}{2}$	$3/4$	$5/16$	$5/8$
MC 7x22.7	6.67	7.00	0.503	$1/2$	$1/4$	3.603	$3\frac{5}{8}$	0.500	$1/2$	$4\frac{3}{4}$	$1\frac{1}{8}$	$1/2$	$7/8$
x19.1	5.61	7.00	0.352	$3/8$	$3/16$	3.452	$3\frac{1}{2}$	0.500	$1/2$	$4\frac{3}{4}$	$1\frac{1}{8}$	$1/2$	$7/8$
MC 7x17.6	5.17	7.00	0.375	$3/8$	$3/16$	3.000	3	0.475	$1/2$	$4\frac{7}{8}$	$1\frac{1}{16}$	$1/2$	$3/4$
MC 6x18	5.29	6.00	0.379	$3/8$	$3/16$	3.504	$3\frac{1}{2}$	0.475	$1/2$	$3\frac{7}{8}$	$1\frac{1}{16}$	$1/2$	$7/8$
x15.3	4.50	6.00	0.340	$5/16$	$3/16$	3.500	$3\frac{1}{2}$	0.385	$3/8$	$4\frac{1}{4}$	$7/8$	$3/8$	$7/8$
MC 6x16.3	4.79	6.00	0.375	$3/8$	$3/16$	3.000	3	0.475	$1/2$	$3\frac{7}{8}$	$1\frac{1}{16}$	$1/2$	$3/4$
x15.1	4.44	6.00	0.316	$5/16$	$3/16$	2.941	3	0.475	$1/2$	$3\frac{7}{8}$	$1\frac{1}{16}$	$1/2$	$3/4$
MC 6x12	3.53	6.00	0.310	$5/16$	$1/8$	2.497	$2\frac{1}{2}$	0.375	$3/8$	$4\frac{3}{8}$	$13/16$	$3/8$	$5/8$

VII. WELDING CODES AND SPECIFICATIONS

A *welding code* is a detailed listing of the rules or principles which are to be applied to a specific classification or type of product.

A *welding specification* is a detailed statement of the legal requirements for a specific classification or type of product. Products manufactured to code or specification requirements commonly must be inspected and tested to ensure compliance.

There are a number of agencies and organizations that publish welding codes and specifications. The application of the particular code or specification to a weldment can be the result of one or more of the following requirements:

- Local, state, or federal government regulations
- Bonding or insuring company
- End user (customer) requirements
- Standard industrial practices

The three most popular codes are:

#1104, American Petroleum Institute — used for pipelines

Section IX, American Society of Mechanical Engineers — used for pressure vessels

D1.1, American Welding Society — used for bridges and buildings

The following organizations publish welding codes and/or specifications.

AASHT

American Association of State Highway and Transportation Officials
444 North Capitol Street, NW
Washington, DC 20001

AISC

American Institute of Steel Construction
1 East Wacker Drive
Chicago, IL 60601

ANSI

American National Standards Institute
11 W. 42nd Street
New York, NY 10036

API

American Petroleum Institute
2101 L Street, NW
Washington, DC 20005

AREA

American Railway Engineering Association
50 F. Street, NW
Washington, DC 20001

ASME

American Society of Mechanical Engineers
345 East 47th Street
New York, NY 10017

AWWA

American Water Works Association
6666 West Quincy Avenue
Denver, CO 80235

AWS

American Welding Society
550 NW LeJeune Road
Miami, FL 33126

AAR

Association of American Railroads
50 F. Street, NW
Washington, DC 20001

MIL

Department of Defense
Washington, DC 20301

SAE

Society of Automotive Engineers
400 Commonwealth Drive
Warrendale, PA 15096

VIII. WELDING ASSOCIATIONS AND ORGANIZATIONS

AA

Aluminum Association
900 19th Street, NW
Washington, DC 20006

AASHTO

American Association of State Highway
and Transportation Officials
444 North Capital Street, NW
Washington, DC 20001

ABS

American Bureau of Shipping
2 World Trade Center
New York, NY 10048

AGA

American Gas Association
1515 Wilson Blvd.
Arlington, VA 22209

AIME

American Institute of Mining, Metallurgical
and Petroleum Engineering
345 East 47th Street
New York, NY 10017

AISC

American Institute of Steel Construction
61 East Wacker Drive
Chicago, IL 60601

AISI

American Iron and Steel Institute
1101 17th Street, NW
Washington, DC 20036

ANS

American Nuclear Society
555 North Kensington Avenue
La Grange Park, IL 60526

ANSI

American National Standards Institute
11 West 42nd Street
New York, NY 10036

API

American Petroleum Institute
2101 L Street, NW
Washington, DC 20005

ASCE

American Society of Civil Engineers
1801 Alexander Bell Drive
Reston, VA 20191

ASME

American Society of Mechanical Engineers
345 East 47th Street
New York, NY 10017

ASNT

American Society for Nondestructive Testing
1711 Arlingate Lane
Columbus, OH 43228

ASTM

American Society for Testing and Materials
100 Barr Harbor Drive
West Conshohocken, PA 19428

AWI

American Welding Institute
10628 Dutchtown Road
Knoxville, TN 37932

AWS

American Welding Society
550 NW LeJeune Road
Miami, FL 33126

CWB

Canadian Welding Bureau
7250 West Credit Avenue
Mississauga, Ont. L5N 5N1
Canada

EWI

Edison Welding Institute
1250 Arthur E. Adams Drive
Columbus, OH 43221

IIW

International Institute of Welding
550 NW LeJeune Road
Miami, FL 33126

NSC

National Safety Council
1121 Spring Lake Business Park
Itaska, IL 60143

VIII. <u>WELDING ASSOCIATIONS AND ORGANIZATIONS</u> (continued)

PFI

Pipe Fabrication Institute
612 Lenore Avenue
Springdale, PA 15144

SAE

Society of Automotive Engineers
400 Commonwealth Drive
Warrendale, PA 15096

SME

Society of Manufacturing Engineers
One SME Drive
P.O. Box 930
Dearborn, MI 48121

TWI

The Welding Institute
Abington Hall
Abington, Cambridge CB1 6AL
United Kingdom

WRC

Welding Research Council
345 East 47th, 14th Floor
New York, NY 10017

Glossary

The terms and definitions in this glossary are extracted from the American Welding Society publication AWS A3.0-80 Welding Terms and Definitions. The terms with an asterisk are from a source other than the American Welding Society. Note: The English term and definition is given first, followed by the same term and definition in Spanish.

A

*absolute pressure. The sum of the gauge pressure and the atmospheric pressure.
presión absoluta. La suma de la presión manómetro y la presión atmosférica.

absorptive lens. A filter lens designed to attenuate the effects of glare and reflected and stray light.
lente absorbente. Un lente de filtro diseñado para disminuir los efectos de la luz y la reflexión de la luz extraviada.

acceptable criteria. Agreed upon standards that must be satisfactorily met.
criterios aceptables. Las normas sobre las que se ha llegado a un acuerdo y que deben cumplirse en forma satisfactoria.

acceptable weld. A weld that meets all the requirements and the acceptance criteria prescribed by welding specifications.
soldadura aceptable. Una soldadura que satisface los requisitos y el criterio aceptable prescribida por las especificaciónes de la soldadura.

*acetone. A fragrant liquid chemical used in acetylene cylinders. The cylinder is filled with a porous material and acetone is then added to fill. Acetylene is then added and absorbed by the acetone, which can absorb up to 28 times its own volume of the gas.
acetona. Un liquido fragante químico que se usa en los cilindros del acetileno. El cilindro se llena de un material poroso y luego se le agrega la acetona hasta que se llene. El acetileno es absorbido por la acetona, la cual puede absorber 28 veces el propio volumen del gas.

*acetylene. A fuel gas used for welding and cutting. It is produced as a result of the chemical reaction between calcium carbide and water. The chemical formula for acetylene is C_2H_2. It is colorless, lighter than air, and has a strong garlic-like smell. Acetylene is unstable above pressures of 15 psig (1.05 kg/cm^2 g). When burned in the presence of oxygen, acetylene produces one of the highest flame temperatures available.
acetileno. Un gas combustible que se usa para soldar y cortar. Es producido a consecuencia de una reacción química de agua y calcio y carburo. La fórmula química para el acetileno es C_2H_2. No tiene color, es más ligero que el aire, y tiene un olor fuerte como a ajo. El acetileno es inestable en presiones más altas de 15 psig (1.05 kg/cm^2 g). Cuando se quema en presencia del oxígeno, el acetileno produce una de las llamás con una temperatura más alta que la que se utiliza.

actual throat. See throat of a fillet weld.
garganta actual. Vea garganta de soldadura filete.

*adaptable. Capable of making self-directed corrections in a robot, this is often accomplished with visual, force, or tactile sensors.
adaptable. Capaz de hacer correcciones por instrucción propia de un robot, esto se lleva a cabo con sensores tangibles visuales, o de fuerza.

air acetylene welding (AAW). An oxyfuel gas welding process that uses an air-acetylene flame. The process is used without the application of pressure. This is an obsolete or seldom-used process.
soldadura de aire acetileno. Un proceso de soldar con gas (oxi/combustible) que usa aire-acetileno sin aplicarse presión. Un proceso anticuado que es una rareza.

air carbon arc cutting (CAC-A). A carbon arc cutting process variation that removes molten metal with a jet of air.
arco de carbón con aire. Un proceso de cortar con arco de carbón variante que quita el metal derretido con un chorro de aire.

allotropic metals. Metals that have a specific lattice or crystal structures that form when the metal is cool and change within the solid metal as it is heated and before it melts.
metales alotrópicos. metales que tienen un determinado enrejado o estructuras cristalinas que se forman cuando el metal está frío y cambian dentro del metal sólido mientras se lo calienta y antes de que éste se derrita.

allotropic transformation. A change in the crystalline lattice pattern of a metal due to a change in temperature.
transformación alotropico. Un cambio en el modelo cristalino enrejado del metal debido a un cambio en la temperatura.

*alloy. A metal with one or more elements added to it, resulting in a significant change in the metal's properties.
aleación. Un metal en que se le agrega uno o más elementos resultando en un cambio significativo en las propiedades del metal.

*alloying elements. Elements in the flux that mix with the filler metal and become part of the weld metal. Major alloying elements are molydenum, nickel, chromium, manganese, and vanadium.
elementos de mezcla. Elementos en el flujo que se mezclan con el metal para rellenar y formar parte del metal soldado. Los elementos principales de mezcla son molibdeno, niquel, cromo, manganeso y vanadio.

all-weld-metal test specimen. A test specimen with the reduced section composed wholly of weld metal.
prueba de metal soldado. Una prueba con una sección reducida compuesta totalmente del metal de la soldadura.

*amperage. A measurement of the rate of flow of electrons; amperage controls the size of the arc.

amperaje. Una medida de la proporción de la corriente de electrones; el amperaje controla el tamaño del arco.

amperage range. The lower and upper limits of welding power, in amperage, that can be produced by a welding machine or used with an electrode or by a process.

rango de amperaje. Los límites máximos y mínimos de poder de soldadura (en amperaje) que puede tener una máquina para soldar o que pueden usarse con un electrodo o a través de un proceso.

angle of bevel. See preferred term **bevel angle.**

ángulo del bisel. Es preferible que vea el término **ángulo del bisel.**

*anode. Material with a lack of electrons; thus has a positive charge.

ánodo. Un material que carece electrones; por eso tiene una carga positiva.

arc blow. The deflection of arc from its normal path because of magnetic forces.

soplo del arco. Desviación de un arco eléctrico de su senda normal a causa de fuerzas magnéticas.

arc brazing (AB). A brazing process in which the heat required is obtained from an electric arc.

soldadura fuerte aplicada por arco. Un proceso de soldadura fuerte donde el calor requerido es obtenido de un arco eléctrico.

arc cutting (AC). A group of thermal cutting processes that severs or removes metal by melting with the heat of an arc between an electrode and the workpiece.

corte con arco. Un grupo de procesos termales para cortar que desúne o quita el metal derretido con el calor del arco en medio del electrodo y la pieza de trabajo.

arc force. The axial force developed by a plasma.

fuerza del arco. La fuerza axial desarrollada por la plasma.

arc gouging. Thermal gouging that uses an arc cutting process variation to form a bevel or groove.

gubiadura con arco. Gubiadura termal que usa un proceso variante de corte con arco para formar un bisel o ranura.

arc length. The length from the tip of the welding electrode to the adjacent surface of the weld pool.

largura del arco. La distancia de la punta del electrodo a la superficie que colinda con el charco de la soldadura.

arc plasma. A state of matter found in the region of an electrical discharge (arc). (See **plasma.**)

arco de plasma. Un estado de la materia encontrado en la región de una descarga eléctrica (arco). (Vea **plasma.**)

arc spot weld. A spot weld made by an arc welding process.

soldadura de puntos por arco. Una soldadura de punto hecha por un proceso de soldadura de arco.

arc strike. A discontinuity consisting of any localized remelted metal, heat-affected metal, or change in the surface profile of any part of a weld or base metal resulting from an arc.

golpe del arco. Una discontinuidad que consiste de cualquier rederretimiento del metal localizado, metal afectado por el calor, o cambio en el perfil de la superficie de cualquier parte de la soldadura o metal base resultante de un arco.

arc time. The time during which an arc is maintained in making an arc weld.

tiempo del arco. El tiempo durante el que el arco se mantiene al hacer una soldadura de arco.

arc voltage. The voltage across the welding arc.

voltaje del arco. El voltaje a través del arco de soldar.

arc welding (AW). A group of welding processes that produces coalescence of workpieces by heating them with an arc. The processes are used with or without the application of pressure and with or without filler metal.

soldadura de arco. Un grupo de procesos de soldadura que producen una unión de piezas de trabajo calentándolas con un arco. Los procesos se usan con o sin la aplicación de presión y con o sin metal para rellenar.

arc welding deposition efficiency. The ratio of the weight of filler metal deposited in the weld metal to the weight of filler metal melted, expressed in percent.

eficiencia de deposición de soldadura de arco. La relación del peso del metal depositado en la soldadura al peso del metal para rellenar, y el metal derretido, expresado en por ciento.

arc welding electrode. A component of the welding circuit through which current is conducted between the electrode holder and the arc. (See **arc welding.**)

electrodo para soldadura de arco. Un componente del circuito de soldadura que conduce la corriente a través del porta-electrodo y el arco. (Vea **soldadura de arco.**)

arc welding gun. A device used to transfer current to a continuously fed consumable electrode, guide the electrode, and direct the shielding gas.

pistola de soldadura de arco. Aparato que se usa para transferir corriente eléctrica continuamente a un alimentador de electrodo consumible. También se usa para guiar al electrodo y dirigir el gas de protección.

*arm. An interconnected set of links and powered joints comprising a manipulator, which supports or moves a wrist and hand or end effector.

brazo. Una entreconexión que une un juego de eslabones y coyunturas de potencia conteniendo un manipulador, que apolla o mueve una muñeca o mano o el que efectúa al final.

as-welded. Pertaining to the weld metal, welded joints, and weldments after welding but prior to any subsequent thermal, mechanical, or chemical treatments.

como-soldado. Pertenece a metal soldado, juntas soldadas, o soldaduras ya soldadas pero antes de que se les hagan tratamientos termales, mecánicos, o químicos.

*atmospheric pressure. The pressure at sea level resulting from the weight of a column of air on a specified area; expressed for an area of one square inch or square centimeter; normally given as 14.7 psi (1.05 kg/cm^2).

presion atmosferica. La presión al nivel del mar que resulta del peso de una columna de aire en una area especificada; expresada para una área de una pulgada cuadrada o un centímetro cuadrado. Normalmente dado como 14.7 psi (1.05 kg/cm²).

atomic hydrogen. A single, free, unbounded hydrogen atom (H) usually formed when molecular hydrogen is exposed to an arc.

hidrógeno atómico. un solo átomo de hidrógeno (H) libre, que normalmente se forma cuando se expone hidrógeno molecular a un arco.

atomic hydrogen welding (AHW). An arc welding process that used an arc between two metal electrodes in a shielding atmosphere of hydrogen and without the application of pressure. This is an obsolete or seldom-used process.

soldadura atómica hidrógena. Un proceso de soldadura de arco que usa un arco entre dos electrodos de metal en una atmósfera de protección de hidrógeno y sin la aplicación de presión. Esto es un proceso anticuado y es rara la vez que se use.

austenitic manganese steel. A steel alloy with a high carbon content containing 10% or more manganese that is very tough and that will harden when cold worked.

acero al manganeso austenítico. Aleación de acero con un alto contenido de carbono, que contiene un 10% o más de manganeso, y que es muy tenaz y endurece cuando se lo trabaja en frío.

autogenous weld. A fusion weld made without filler metal.

soldadura autógena. Una soldadura fundida sin metal de rellenar.

*automated operation. Welding operations are performed repeatedly by a robot or other machine that is programmed to perform a variety of processes.

operación automatizada. Operaciones de soldaduras que se ejecutan repetidamente por un robot u otra maquina que está programada para hacer una variedad de procesos.

automatic arc welding downslope time. The time during which the current is changed continuously from final taper current or welding current to final current.

tiempo del pendiente con descenso en soldadura de arco automático. El tiempo durante en que la corriente cambia continuamente disminuyendo la corriente final o la corriente de la soldadura a la corriente final.

automatic arc welding upslope time. The time during which the current changes continuously from the initial current to the welding current.

tiempo de pendiente con ascenso en soldadura de arco automático. El tiempo durante en que la corriente cambia continuamente de donde se inició la corriente a la corriente de la soldadura.

*automatic operation. Welding operations are performed repeatedly by a machine that has been programmed to do an entire operation without the interaction of the operator.

operación automática. Operaciones de soldadura que se ejecutan repetidamente por una maquina que ha sido programada para hacer una operación entera sin influencia del operador.

automatic oxygen cutting. Oxygen cutting with equipment that performs the cutting operation without constant observation and adjustment of the controls by an operator. The equipment may or may not perform loading and unloading of the work.

corte del oxígeno automático. Cortadura del oxígeno con equipo que hace la operación de cortar sin observación y ajuste de los controles por el operador. El equipo puede que ejecute o no el trabajo de cargar o descargar las piezas.

automatic welding. Welding with equipment that requires only occasional or no observation of the welding, and no manual adjustment of the equipment controls. Variations of this term are automatic brazing, automatic soldering, automatic thermal cutting, and automatic thermal spraying.

soldadura automática. Soldadura con equipos que requieren ocasional o ninguna observación, y ningun ajuste manual de los controles de los equipos. Variaciones de éste término son soldadura automática, corte termal automático, rociadura termal automático.

axis of a weld. A line through the length of a weld, perpendicular to and at the geometric center of its cross section.

eje de la soldadura. Una linea a lo largo de la soldadura perpendicula a y al centro geométrico de su corte transversal.

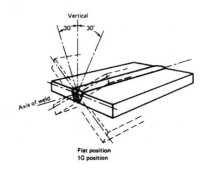

Flat position
1G position

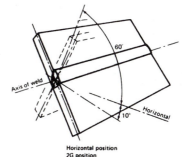

Horizontal position
2G position

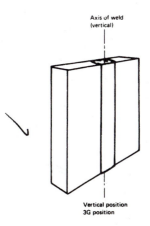

Vertical position
3G position

Positions of welding — groove welds

(figure continued on page 838)

(figure continued from page 837)

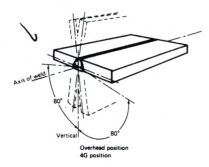

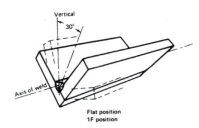

Positions of welding — groove welds

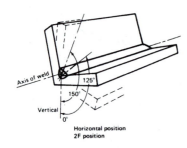

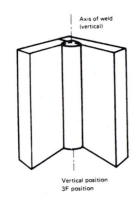

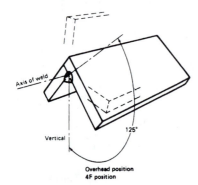

Positions of welding — fillet welds

B

backfire. The momentary recession of the flame into the welding tip, cutting tip, or flame spraying gun, followed by immediate reappearance or complete extinction of the flame.

llama de retroceso. El retroceso momentaneo de la llama dentro de la punta para soldar, punta para cortar, o pistola para rociar con llama. La llama puede reaparecer inmediatamente o apagarse completamente.

backgouging. The removal of weld metal and base metal from the weld root side of a welded joint to facilitate complete fusion and complete joint penetration upon subsequent welding from that side.

gubia trasera. Quitar el metal soldado y el metal base del lado de la raíz de una junta soldada para facilitar una fusión completa y penetración completa de la junta soldada subsecuente a soldar de ese lado.

backhand welding. A welding technique in which the welding torch or gun is directed opposite to the progress of welding. Sometimes referred to as the ``pull gun technique'' in GMAW and FCAW. See travel angle, work angle, and drag angle.

soldadura en revés. Una técnica de soldar la cual el soplete o pistola es guiada en la dirección contraria al adelantamiento de la soldadura. A veces se refiere como una tecnica de "estirar la pistola" en GMAW y FCAW. Vea ángulo de avance, ángulo de trabajo y ángulo del tiro.

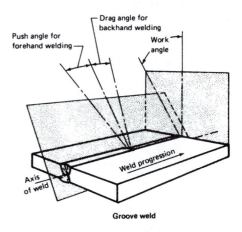

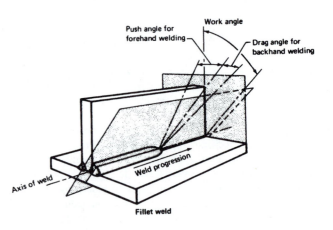

Position of electrode or torch

(figure continued on page 839)

(figure continued from page 838)

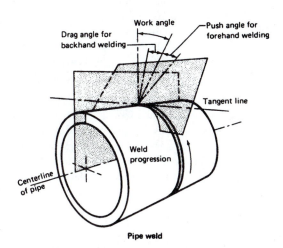

Pipe weld

Position of electrode or torch

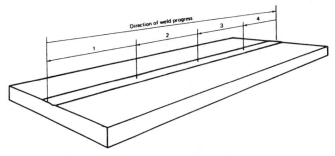

Backstep sequence

backing. A material (base metal, weld metal, carbon, or granular material) placed at the root of a weld joint for the purpose of supporting molten weld metal.
respaldo. Un material (metal base, metal de soldadura, carbón o material granulado) puesto en la raíz de la junta soldada con el proposito de sostener el metal de la soldadura que está derretido.

backing pass. A pass made to deposit a backing weld.
pasada de respaldo. Una pasada hecha para depositar la pasada del respaldo.

backing ring. Backing in the form of a ring, generally used in the welding of piping.
anillo o argolla de respaldo. Respaldo en forma de argolla, generalmente se usa en soldaduras de tubos.

backing strip. Backing in the form of a strip.
tira de respaldo. Un respaldo en la forma de una tira.

backing weld. Backing in the form of a weld.
soldadura de respaldo. Respaldo en la forma de soldadura.

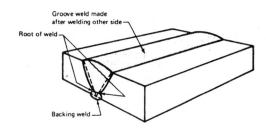

backstep sequence. A longitudinal sequence in which the weld bead increments are deposited in the direction opposite to the progress of welding the joint.
secuencia a la inversa. Una serie de soldaduras en secuencia longitudinal hechas en la dirección opuesta del progreso de la soldadura.

back weld. A weld deposited at the back of a single groove weld.
soldadura de atrás. Una soldadura que se deposita en la parte de atrás de una soldadura de ranura sencilla.

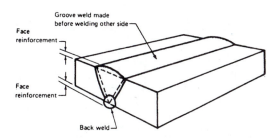

bare electrode. A filler metal electrode that has been produced as a wire, strip, or bar with no coating or covering other than that which is incidental to its manufacture or preservation.
electrodo descubierto. Un electrodo de metal para rellenar que se ha producido como alambre, tira, o barra sin revestimiento o cubierto con solo lo necesario para su fabricación y conservación.

base material. The material that is welded, brazed, soldered, or cut. See also **base metal** and **substrate**.
material base. El material que está soldado, soldado con soldadura fuerte, soldado con soldadura blanda, o cortado. Vea también **metal base** y **substrato**.

base metal. The metal or alloy that is welded, brazed, soldered, or cut. See also **base material** and **substrate**.
metal base. El metal que está soldado con soldadura fuerte, soldado con soldadura blanda, o cortado. Vea también **material base** y **substrato**.

base metal test specimen. A test specimen composed wholly of base metal.
probeta para metal base. Una probeta totalmente compuesta de metal base.

bend test. A test in which a specimen is bent to a specified bend radius. See also **face bend test, root bend test,** and **side bend test**.
prueba de dobléz. Una prueba donde la probeta se dobla a una vuelta con un radio especificado. Vea también **prueba de dobléz de cara, prueba de dobléz de raíz, y prueba de dobléz de lado**.

bevel. An angular type of edge preparation.
bisel. Una preparación de tipo angular con filo.

bevel angle. The angle formed between the prepared edge of a member and a plane perpendicular to the surface of the member. Refer to drawings for **bevel**.

ángulo del bisel. El ángulo formado entre el corte preparado de un miembro y la plana perpendicular a la superficie del miembro. Refiera a los dibujos del bisel.

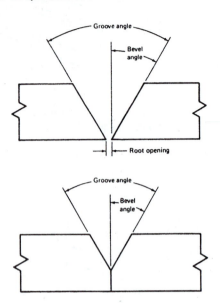

blind joint. A joint, no portion of which is visible.

junta ciega. Una junta en que no hay porción visible.

block sequence. A combined longitudinal and cross-sectional sequence for a continuous multiple pass weld in which separated increments are completely or partially welded before intervening increments are welded.

secuencia de bloques. Una soldadura continua de pasadas multiples en sucesión combinadas longitudinal y sección transversa donde los incrementos separados son completamente o parcialmente soldados antes que los incrementos sean soldados.

bonding force. The force that holds two atoms together; it results from a decrease in energy as two atoms are brought closer to one another.

fuerza ligamentosa. La fuerza que detiene dos átomos juntos; es el resultado del decrecimiento en la energía cuando dos átomos son traídos cerca del uno al otro.

braze. A weld produced by heating an assembly to suitable temperatures and by using a filler metal having a liquidus above 450°C (840°F) and below the solidus of the base metal. The filler metal is distributed between the closely fitted faying surfaces of the joint by capillary action.

soldadura de latón. Una soldadura producida cuando se calienta un montaje a una temperatura conveniente usando un metal de relleno que se liquida arriba de 450°C (840°F) y abajo del estado sólido del metal base. El metal de relleno es distribuido por acción capilar en una junta entre las superficies empalmadas montadas muy cerca.

*braze buildup. Braze metal added to the surface of a part to repair wear.

formación con bronce. Reparación de partes gastadas donde se agrega bronce.

braze metal. That portion of a braze that has been melted during brazing.

latón. La porción del bronce que se derrite cuando se solda.

braze welding. A welding process variation that uses a filler metal with a liquidus above 450°C (840°F) and below the solidus of the base metal. Unlike brazing, in braze welding the filler metal is not distributed in the joint by capillary action.

soldadura con bronce. Es un proceso de soldar variante que usa un metal de relleno con un liquido arriba de 450°C (840°F) y abajo del estado del metal base. Distinto a la soldadura fuerte, el metal de relleno no es distribuido por acción capilar.

brazeability. The capacity of a metal to be braced under the fabrication conditions imposed into a specific suitably designed structure and to perform satisfactorily in the intended service.

soldabilidad fuerte. La capacidad de un metal de refuerzo bajo las condiciones impuestas en la fabricación de una estructura diseñada especificamente para funcionar satisfactoriamente en los servicios intentados.

brazement. An assembly whose component parts are joined by brazing.

montaje de soldadura fuerte. Un montaje donde las partes son unidas por soldadura fuerte.

brazing (B). A group of welding processes that produces coalescence of materials by heating them to the brazing temperature in the presence of a filler metal with a liquidus above 450°C (840°F) and below the solidus of the base metal. The filler metal is distributed between the closely fitted faying surfaces of the joint by capillary action.

soldadura fuerte (B). Un grupo de procesos de soldadura que produce coalescencia de materiales calentándolos a una temperatura de soldar en la presencia de un material de relleno el cual se derrite a una temperatura de 450°C (840°F) y bajo del estado sólido del metal base. El metal de relleno se distribuye por acción capilar de una junta entre las superficies empalmadas montadas muy cerca.

brazing filler metal. The metal that fills the capillary joint clearance and has a liquidus above 450°C (840°F), but below the solidus of the base metal.

metal de relleno para soldadura fuerte. El metal que rellena el espacio libre en la junta capilar y se derrite a una temperatura de 450°C(840°F) y bajo del estado sólido del metal base.

brazing procedure qualification record (BPQR). A record of brazing variables used to produce an acceptable test brazement and the results of tests conducted on the brazement to qualify a brazing procedure specification.

registro del procedimiento de calificación de la soldadura fuerte (BPQR). Un registro de variables de la soldadura fuerte que se usan para producir una probeta bronceada aceptable y los resultados de la prueba conducidas en el bronceamiento para calificar la especificación del procedimiento de la soldadura fuerte.

brazing procedure specification (BPS). A document specifying the required brazing variables for a specific application.

especificación del procedimiento de la soldadura fuerte (BPS). Un documento especificando los variables requeridos de la soldadura fuerte para una aplicación especificada.

brazing rod. Filler metal used in the brazing process is supplied in the form of rods; the filler metal is usually an alloy of two or more metals; the percentages and types of metals used in the alloy impart different characteristics to the braze being made.

Several classification systems are in use by manufacturers of filler metals. See **brazing filler metal**.

varilla de latón. Metal de relleno usado en procesos de soldadura fuerte es surtida en forma de varillas; el metal para rellenar es usualmente un aleación de dos or más metales; los porcentajes y tipos de metales usados en la aleación imparten diferentes características a la soldadura que se está haciendo. Varios sistemás de clasificación se están usando por los fabricantes de metales de relleno. Vea **metal de relleno para soldadura fuerte**.

brazing temperature. The temperature to which the base metal is heated to enable the filler metal to wet the base metal and form a brazed joint.

temperatura de soldadura fuerte. La temperatura a la cual se calienta el metal base para permitir que el metal de relleno moje al metal base y forme la soldadura fuerte.

buildup. A surfacing variation in which surfacing material is deposited to achieve the required dimensions. See also **buttering**, **cladding**, and **hardfacing**.

recubrimiento. Una variación en la superficie donde el metal es depositado para que pueda obtener las dimensiones requeridas. Vea **recubrimiento antes de terminar una soldadura**, **capa de revestimiento**, y **endurecimiento de caras**.

*buried arc transfer. In gas metal arc welding, a method of transfer in which the wire tip is driven below the surface of the weld pool due to the force of the carbon dioxide shielding gas. The shorter arc reduces the size of the drop, and any spatter is trapped in the cavity produced by the arc.

traslado de arco enterrado. En soldaduras de arco metalico con gas, un método de transferir en la cual la punta del alambre es enterrado debajo de la superficie del metal derretido debido a la fuerza del carbón bióxido del gas protector. Lo corto del arco reduce el tamaño de la gota y la salpicadura es atrapada en el hueco producido por el arco.

*burnthrough. Burning out of molten metal on the back side of the plate.

metal quemado que pasa al otro lado. Metal derretido que se quema en el lado de atrás del plato.

buttering. A surfacing variation in which one or more layers of weld metal are deposited on the groove face of one member (for example, a high alloy weld deposit on steel base metal that is to be welded to a dissimilar base metal). The buttering provides a suitable transition weld deposit for subsequent completion of the butt joint.

recubrimiento antes de terminar una soldadura. Una variación de la superficie donde se deposita una o más capas de metal soldado en la ranura de un miembro (por ejemplo, un deposito de soldadura de aleación alta en un metal base de acero la cual será soldada a un metal base diferente). El recubrimiento proporciona una transición conveniente a la soldadura depositada para el acabamiento subsiguiente de una junta tope.

butt joint. A joint between two members aligned approximately in the same plane.

junta a tope. Una junta entre dos miembros alineados aproximadamente en el mismo plano.

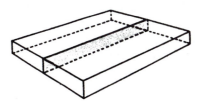

Applicable welds

Square-Groove   J-Groove
V-Groove   Flare-V-Groove
Bevel-Groove   Flare-Bevel-Groove
U-Groove   Edge-Flange
    Braze

Butt joint

C

capillary action. The force by which liquid, in contact with a solid, is distributed between closely fitted faying surfaces of the joint to be brazed or soldered.

acción capilar. La fuerza por la que el liquido, en contacto con un sólido, es distribuido entre el empalme de las juntas del superficie para soldadura fuerte o blanda.

carbon arc cutting (CAC). An arc cutting process that uses a carbon electrode. See also **air carbon arc cutting**.

corte con arco y carbón. Un proceso de corte con arco que usa un electrodo de carbón. Vea **arco de carbón con aire**.

carbon arc welding (CAW). An arc welding process that uses an arc between a carbon electrode and the weld pool. The process is used with or without shielding and without the application of pressure.

soldadura con arco de carbón. Un proceso de soldar de arco en que se usa un arco entre el electrodo de carbón y el metal derretido. El proceso se usa con o sin protección y sin aplicación de presión.

carbon electrode. A nonfiller metal electrode used in arc welding and cutting, consisting of a carbon or graphite rod, which may be coated with copper or other materials.

electrodo de carbón. Un electrodo de metal que no se rellena usado en soldaduras de arco y para cortes consistiendo de varillas de carbón o grafito que pueden ser cubiertas de cobre u otros materiales.

*carbon steel. Steel whose physical properties are primarily the result of the percentage of carbon contained within the alloy. Carbon content ranges from 0.04 to 1.4%, often referred to as plain carbon steel, low carbon steel, or straight carbon steel.

acero al carbono. Acero cuyas propiedades físicas son primariamente el resultado del porcentaje de carbón que es contenido dentro de la aleación. El contenido del carbón es clasificado entre 0.04 a 1.4%, frecuentemente es referido como el carbono de acero liso, acero de bajo carbón, o carbon de acero recto.

carbonizing (carburizing) flame. A reducing oxyfuel gas flame in which there is an excess of fuel gas, resulting in a carbon-rich zone extending around and beyond the cone.

llama carburante. Una llama minorada de gas combustible a oxígeno donde hay un exceso de gas combustible, resultando en una zona rica de carboón extendiendose alrededor y al otro lado del cono.

*cast. The natural curve in the electrode wire for gas metal arc welding as it is removed from the spool; cast is measured by the diameter of the circle that the wire makes when it is placed on a flat surface without any restraint.

distancia. La curva natural en el alambre electrodo para soldadura de arco metálico para gas cuando se aparta del carrete; la distancia es medida en el círculo que hace el alambre cuando es puesto en una superficie plana sin restricción.

*cast iron. A combination of iron and carbon. The carbon may range from 2% to 4%. Approximately 0.8% of the carbon is combined with the iron. The remaining free carbon is found as graphite mixed throughout the metal. Gray cast iron is the most common form of cast iron.

acero vaciado. Una combinación de acero y carbono. El carbono puede ser clasificado de 2% a 4%. Aproximadamente 0.8% del carbono es combinado con el hierro. El resto de carbono libre es encontrado como grafito mezclado por todo el metal. El acero fundido gris es la forma más común.

cathode. A natural curve material with an excess of electrons, thus has a negative charge.

cátodo. Un material de curva natural con un exceso de electrones, por eso tiene una carga negativa.

*cell. A manufacturing unit consisting of two or more work stations and the material transport mechanisms and storage buffers that interconnect them.

celda. Una unidad manufacturera la cual consiste de dos o más estaciones de trabajo y mecanismos para trasladar el material y los amortiguadores del almacén que los entreconecta.

cellulose-based electrode fluxes. Fluxes that use an organic-based cellulose ($C_6H_{10}O_5$) (a material commonly used to make paper) held together with a lime binder. When this flux is exposed to the heat of the arc, it burns and forms a rapidly expanding gaseous cloud of CO_2 that protects the molten weld pool from oxidation. Most of the fluxing material is burned, and little slag is deposited on the weld. E6010 is an example of an electrode that uses this type of flux.

fundentes para electrodos celulósicos. Fundentes que usan celulosa de base orgánica (C6H10O5) (un material normalmente utilizado para fabricar papel), y que se mantienen unidos con un aglomerante de cal. Cuando a este fundente se lo expone al calor del arco, se consume y forma una nube gaseosa de Co2 que se expande rápidamente y protege de la oxidación al charco de soldadura derretido. La mayor parte del material del fundente se consume, y se deposita poca escoria en la soldadura. El E6010 es un ejemplo de un electrodo que utiliza este tipo de fundente.

*center. A manufacturing unit consisting of two or more cells and the materials transport and storage buffers that interconnect them.

centro. Una unidad manufacturera la cual consiste de dos o más celdas y el traslado de materiales y los amortiguadores del almacén que los entreconecta.

certified welder. An individual who has demonstrated his/her welding skills for a process by passing a specific welding test.

soldador certificado. Persona que ha demostrado, mediante una prueba específica de soldadura, su habilidad para soldar en un proceso.

chain intermittent welds. Intermittent welds on both sides of a joint in which the weld increments on one side are approximately opposite those on the other side.

soldadura intermitente de cadena. Soldadura intermitente en los dos lados de una junta en cual los incrementos de soldadura están aproximadamente opuestos a los del otro lado.

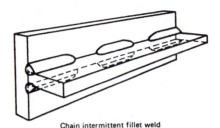

Chain intermittent fillet weld

*chill plate. A large piece of metal used in welding to correct overheating.

plato desalentador. Una pieza de metal grande que se usa para corregir el sobrecalentamiento.

cladding. A relatively thick layer (> 1 mm [0.04 in]) of material applied by surfacing for the purpose of improved corrosion resistance or other properties. See coating, surfacing, and hardfacing.

capa de revestimiento. Una capa de material relativamente grueso (> 1 mm [0.04 pulgadas]) aplicada por la superficie con el objeto de mejorar la resistencia a la corrosión u otras propiedades. Vea revestimiento, recubrimiento superficial y endurecimiento de caras.

coalescence. The growing together or growth into one body of the materials being welded.

coelescencia. El crecimiento o desarrollo de un cuerpo de los materiales los cuales se están soldando.

coating. A relatively thin layer (< 1 mm [0.04 in]) of material applied by surfacing for the purpose of corrosion prevention, resistance to high temperature scaling, wear resistance, lubrication, or other purposes. See cladding, surfacing, and hardfacing.

revestimiento. Una capa de material relativamente delgado (< 1 mm [0.04 pulgadas]) aplicada por la superficie con el propósito de prevenir corrosión, resistencia a las altas temperaturas, resistencia a la deterioración, lubricación, o para otros propósitos. Vea capa de revestimiento, recubrimiento superficial y endurecimiento de caras.

cold soldered joint. A joint with incomplete coalescence caused by insufficient application of heat to the base metal during soldering.

junta soldada fría. Una junta con coalescencia incompleta causada por no haber aplicado suficiente calor al metal base durante la soldadura.

*combustion rate. Also known as rate of propagation of a flame; this is the speed at which the fuel gas burns, in ft/sec (m/sec). The ratio of fuel gas to oxygen affects the rate of burning: a higher percentage of oxygen increases the burn rate.

velocidad de combustión. También es conocida como velocidad de propagación de una llama; está es la velocidad en la cual se quema el gas combustible, en pies/sec (m/sec). La proporción del gas combustible al oxígeno afecta la proporción de quemadura: un porcentaje más alto del oxígeno aumenta la proporción de quemarse.

complete fusion. Fusion over the entire fusion faces and between all adjoining weld beads.

fusión completa. Fusión sobre todas las caras de fusión y en medio de todos los cordónes de soldadura inmediatos.

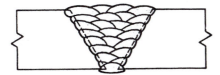

composite electrode. A generic term for multicomponent filler metal electrodes in various physical forms, such as stranded wires, tubes, and covered wire. See also **covered electrode, flux cored electrode,** and **stranded electrode.**

electrodo compuesto. Un término genérico para componentes múltiples para electrodos de metal de aporte en varias formas físicas, como cable de alambre, tubos, y alambre cubierto. Vea **electrodo cubierto, electrodo de núcleo de fundente** y **electrodo cable.**

***computer control.** Control involving one or more electronic digital computers.

control de computadora. Un control que incluye una o más computadoras electrónicas dactilares.

concave fillet weld. A fillet weld with a concave face.
soldadura de filete cóncava. Soldadura de filete con cara cóncava.

concave root surface. A root surface that is concave.
superficie raíz cóncava. La superficie del cordón raíz con cara cóncava.

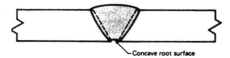

concavity. The maximum distance from the face of a concave fillet weld perpendicular to a line joining the toes.

concavidad. La distancia máxima de la cara de una soldadura de filete cóncava perpendicular a una linea que une con los pies.

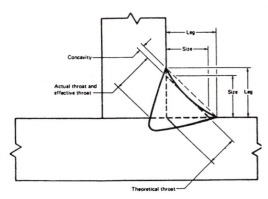

Concave fillet weld

conduit liner. A flexible steel tube that guides the welding wire from the feed rollers through the welding lead to the gun used for GMAW and FCAW welding. The steel conduit liner may

have a nylon or Teflon inner surface for use with soft metals such as aluminum.

revestimiento de conducto. Un tubo flexible de acero que guía el alambre para soldar desde los rodillos de alimentación, a través de los cables para soldar, hasta la pistola, usado en soldaduras de tipo GMAW y FCAW. El revestimiento del conducto de acero puede tener una superficie interior de Teflon o nylon para su uso con metales blandos como el aluminio.

cone. The conical part of an oxyfuel gas flame adjacent to the orifice of the tip.

cono. La parte cónica de la llama del gas de oxígeno combustible que colinda con la abertura de la punta.

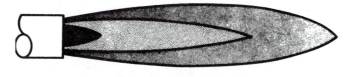

	Neutral flame	Oxidizing flame	Reducing flame
	White blue cone	White cone	Intense white cone
	Nearly colorless	Orange to purplish	White or colorless
	Bluish to orange		Orange to bluish

Oxyacetylene flame

constricted arc. A plasma arc column that is shaped by the constricting orifice in the nozzle of the plasma arch torch or plasma spraying gun.

arco constreñido. Una columna de arco plasma que está formada por el constreñimiento del orificio en la lanza de la antorcha del arco plasma o pistola de rociado plasma.

constricting nozzle. A device at the exit end of a plasma arc torchor plasma spraying gun containing the constricting orifice.

boquilla de constreñimiento. Un aparato a la salida de la antorcha de un arco plasma o la pistola de rociado plasma que contiene la boquilla de constreñimiento.

constricting orifice. The hole in the constricting nozzle of the plasma arc torch or plasma spraying gun through which the arc plasma passes.

orificio de constreñimiento. El agujero en la boquilla del constreñimiento en la antorcha de arco plasma o de la pistola de rociado plasma por donde pasa el arco de plasma.

consumable electrode. An electrode that provides filler metal.
electrodo consumible. Un electrodo que surte el metal de relleno.

consumable insert. Preplaced filler metal that is completely fused into the root of the joint and becomes part of the weld.

inserción consumible. Metal de relleno antepuesto que se funde completamente en la raíz de la junta y se hace parte de la soldadura.

contact tube. A device that transfers current to a continuous electrode.

tubo de contacto. Un aparato que traslada corriente continua a un electrodo.

continuous weld. A weld that extends continuously from one end of a joint to the other. Where the joint is essentially circular, it extends completely around the joint.

soldadura continua. Una soldadura que se extiende continuamente de una punta de la junta a la otra. Donde la junta es esencialmente circular, se extiende completamente alrededor de la junta.

convex fillet weld. A fillet weld with a convex face.

soldadura de filete convexa. Una soldadura de filete con una cara convexa.

convex root surface. A root surface that is convex.

raíz superficie convexa. La raíz que es convexa.

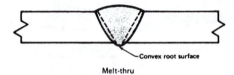

Melt-thru

convexity. The maximum distance from the face of a convex fillet weld perpendicular to a line joining the toes.

convexidad. La distancia máxima de la cara de la soldadura convexa filete perpendicula a la linea que une los pies.

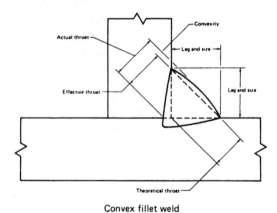

Convex fillet weld

cored solder. A solder wire or bar containing flux as a core.

soldadura de núcleo. Un alambre o barra para soldar que contiene fundente en el núcleo.

*core wire. The wire portion of the coated electrode for shielded metal arc welding. The wire carries the welding current and adds most of the filler metal required in the finished weld. The composition of the core wire depends upon the metals to be welded.

alambre del centro. La porción del alambre del electrodo forrado para proteger el metal de la soldadura de arco. El alambre lleva la corriente de la soldadura y añade casi todo el metal para rellenar que es requerido para terminar la soldadura. La composición del alambre del centro depende de los metales que se van a usar para soldar.

corner joint. A joint between two members located approximately at right angles to each other.

junta de esquina. Una junta dentro de dos miembros localizados aproximadamente a ángulos rectos de unos a otros.

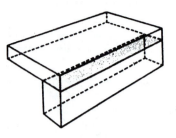

Applicable welds

Fillet	Flare-Bevel-Groove
Square-Groove	Edge-Flange
V-Groove	Corner-Flange
Bevel-Groove	Spot
U-Groove	Projection
J-Groove	Seam
Flare-V-Groove	Braze

Corner joint

*corrosion resistance. The ability of the joint to withstand chemical attack; determined by the compatibility of the base materials to the filler metal.

resistencia a la corrosión. La abilidad de una junta de resistir ataques químicos; determinado por la compatibilidad de los materiales bases al metal de relleno.

corrosive flux. A flux with a residue that chemically attacks the base metal. It may be composed of inorganic salts and acids, organic salts and acids, or activated rosins or resins.

fundente corrosivo. Un fundente con un residuo que ataca químicamente al metal base. Puede estar compuesto de sales y ácidos inorgánicos, sales y ácidos orgánicos, o abelinotes o resinas activados.

cosmetic pass. A weld pass made primarily to enhance appearance.

pasada cosmética. Una pasada que se le hace a la soldadura para mejorar la apariencia.

cover lens. A round cover plate.

lente para cubrir. Un plato redondo de vidrio para cubrir el lente obscuro.

*cover pass. The last layer of weld beads on a multipass weld. The final bead should be uniform in width and reinforcement, not excessively wide, and free of any visual defects.

pasada para cubrir. La última capa de cordónes soldadura de pasadas múltiples. La pasada final debe ser uniforme en anchura y refuerzo, no excesivamente ancha, y libre de defectos visuales.

cover plate. A removable pane of colorless glass, plastic-coated-glass, or plastic that covers the filter plate and protects it from weld spatter, pitting, or scratching.

plato para cubrir. Una hoja removible de vidrio claro, vidrio cubierto con plástico o plástico que cubre el plato filtrado y lo protege de salpicadura, picaduras o de que se rayen.

covered electrode. A composite filler metal electrode consisting of a core of a bare electrode or metal cored electrode to which a covering sufficient to provide a slag layer on the weld metal has been applied. The covering may contain materials providing such functions as shielding from the atmosphere, deoxidation, and arc stabilization and can serve as a source of metallic additions to the weld.

electrodo cubierto. Un electrodo compuesto de metal para rellenar que consiste de un núcleo de un electrodo liso o electrodo con núcleo de metal el cual se le agrega cubrimiento suficiente para proveer una capa de escoria sobre el metal de la soldadura que se le aplicó. El cubierto puede contener materiales que pueden proveer funciones como protección de la atmósfera, deoxidación, y estabilización del arco, y también puede servir como fuente para añadir metales adicionales a la soldadura.

*coupling distance. The distance to be maintained between the inner cones of the cutting flame and the surface of the metal being cut, in the range of 1/8 in. (3 mm) to 3/8 in. (10 mm).

distancia de acoplamiento. La distancia que debe de mantenerse entre los conos internos de la llama y la superficie del metal que se está cortando, varía de 1/8 pulgadas (3 mm) a 3/8 pulgadas (10 mm).

crack. A fracture type discontinuity characterized by a sharp tip and high ratio of length and width to opening displacement.

grieta. Una desunión discontinuidada de tipo fractura caracterizada por una punta filoza y proporción alta de lo largo y de lo ancho al desplazamiento de la abertura.

crater. A depression in the weld face at the termination of a weld bead.

crater. Una depresión en la superficie de la soldadura a donde se termina el cordón de soldadura.

crater crack. A crack in the crater of a weld bead.

grieta de crater. Una grieta en el crater del cordón de soldar.

critical weld. A weld so important to the soundness of the weldment that its failure could result in the loss or destruction of the weldment and injury or death.

soldadura crítica. Una soldadura tan importante para la calidad del conjunto de partes soldadas, que su fracaso podría ocasionar la pérdida o destrucción de dicho conjunto, así como también lesiones o muerte.

cross-sectional sequence. The order in which the weld passes of a multiple pass weld are made with respect to the cross-section of the weld. See also **block sequence**.

secuencia del corte transversal. La orden en la cual se hacen las pasadas de la soldadura en una soldadura de pasadas múltiples hechas al respecto al corte transversal de la soldadura. Vea también **secuencia de bloque**.

crucible. A high-temperature container that holds the thermite welding mixture as it begins its thermal reaction before the molten metal is released into the mold.

crisol. un recipiente de alta temperatura que contiene la mezcla de la soldadura con termita en el momento en que comienza su reacción térmica antes de que el metal fundido se vierta en el molde.

*crystalline structure. The orderly arrangement of atoms in a solid in a specific geometric pattern. Sometimes called **lattice**.

estructura espacial cristalina. Un arreglo metódico de átomos en un modelo geométrico preciso. A veces lo llaman **celosia**.

cup. A nonstandard term for **gas nozzle**.

tazón. Un término que no es la norma para de **boquilla de gas**.

cutting attachment. A device for converting an oxyfuel gas welding torch into an oxygen cutting torch.

equipo para cortar. Un aparato para convertir una antorcha para soldar en una antorcha para cortar con oxígeno.

cutting head. The part of a cutting machine in which a cutting torch or tip is incorporated.

cabeza de la antorcha para cortar. La parte de una maquina para cortar en donde una antorcha para cortar o una punta es incorporada.

cutting tip. That part of an oxygen cutting torch from which the gases issue.

punta para cortar. Esa parte de la antorcha para cortar con oxígeno por donde salen los gases.

cutting tip, high speed. Designed to provide higher oxygen pressure, thus allowing the torch to travel faster.

punta para cortar a alta velocidad. Diseñada para proveer presión más alta de oxígeno, asi puede caminar la antorcha más rápidamente.

cutting torch. (plasma arc). A device used for plasma arc cutting to control the position of the electrode, to transfer current to the arc, and to direct the flow of plasma and shielding gas.

antorcha para cortar. (arco de plasma). Un aparato que se usa para cortes de arco de plasma para el control de la posición del electrodo.

*cycle time. The period of time from starting one machine operation to starting another (in a pattern of continuous repetition).

tiempo del ciclo. El período de tiempo de cuando se empieza la operación de una maquina y cuando se empieza otra. (en una norma de repetición continua).

cylinder. A portable container used for transportation and storage of a compressed gas.

cilindro. Un recipiente portátil que se usa para transportar y guardar un gas comprimido.

cylinder manifold. A multiple header for interconnection of gas sources with distribution points.

conexión de cilindros múltiple. Una tuberia con conexiones múltiples que sirve como fuente de gas con puntos de distribución.

*cylinder pressure. The pressure at which a gas is stored in approximately 2,200 pounds per square inch (psi), and acetylene is stored at approximately 225 psi.

presión del cilindro. La presión del cilindro en el cual un gas se guarda en aproximadamente 2,200 libras por pulgada cuadrada (psi), y el acetilino se guarda a aproximadamente 225 psi.

D

defect. A discontinuity or discontinuities that by nature or accumulated effect (for example, total crack length) render a part or product unable to meet minimum applicable acceptance standards or specifications. This term designated resectability and flaw. See **discontinuity** and **flaw**.

defecto. Una desunión o desuniónes que por la naturaleza o efectos acumulados (por ejemplo, distancia total de una grieta) hace que una parte o producto no esté de acuerdo con las normas o especificaciones mínimas para aceptarse. Este término designado resectabilidad y falta. Vea **discontinuidad** y **falta**.

defective weld. A weld containing one or more defects.

soldadura defectuosa. Una soldadura que contiene uno o más defectos.

deposited metal. Filler metal that has been added during a welding operation.

metal depositado. Metal de relleno que se ha agregado durante una operación de soldadura.

deposition efficiency (arc welding). The ratio of the weight of deposited metal to the net weight of filler metal consumed, exclusive of stubs.

eficiencia de deposición. La relación del peso del metal depositado al peso neto del metal de relleno consumido, excluyendo los tacones.

deposition rate. The weight of material deposited in a unit of time. It is usually expressed as kilograms per hour (kg/hr) (pounds per hour [lb/h]).

relación de deposición. El peso del material depositado en una unidad de tiempo. Es regularmente expresado en kilogramos por hora (kg/hora) (libras por hora {lb/hora}).

depth of fusion. The distance that fusion extends into the base metal or previous bead from the surface melted during welding.

grueso de fusión. La distancia en que la fusión se extiende dentro del metal base o del cordón anterior de la superficie que se derritió durante la soldadura.

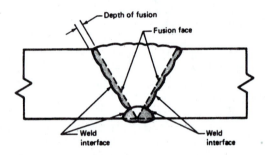

*destructive testing. Mechanical testing of weld specimens to measure strength and other properties. The tests are made on specimens that duplicate the material and weld procedures required for the job.

prueba destructiva. Pruebas mecánicas de probetas de soldadura para medir la fuerza y otras propiedades. Las pruebas se hacen en probetas que duplican el material y los procedimientos de la soldadura requeridos para el trabajo.

dimensioning. The measurements of an object such as its length, width, and height, or the measurements for locating such things as parts, holes, or surfaces.

acotación. Las medidas de un objeto, tal como su longitud, ancho y altura; o las medidas para ubicar cosas como piezas, agujeros o superficies.

dip brazing (DB). A brazing process that uses heat from a molten salt or metal bath. When a molten salt is used, the bath may act as a flux. When a molten metal is used, the bath provides the filler material.

soldadura fuerte por inmersión. Es un proceso de soldadura fuerte que usa el calor de una sal fundida o un baño de metal. Cuando se usa la sal fundida el baño puede actuar como flujo. Cuando se usa el metal fundido, el baño proporciona el metal de relleno.

dip soldering (DS). A soldering process using the heat furnished by a molten metal bath that provides the solder filler metal.

soldadura blanda de bajo punto de fusión por inmersión. Un proceso que usa calor proporcionado por un baño de metal derretido que provee el metal de relleno para soldar.

direct current electrode negative (DCEN). The arrangement of direct current arc welding leads on which the electrode is the negative pole and the workpiece is the positive pole of the welding arc.

corriente directa con electrodo negativo. El arreglo de los cables para soldar con la soldadura de arco donde el electrodo es polo negativo y la pieza de trabajo es polo positivo de la soldadura de arco.

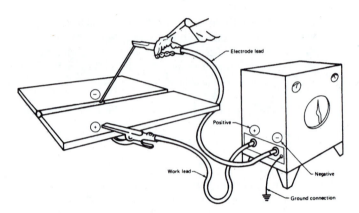

Direct current electrode negative
(straight polarity)

direct current electrode positive (DCEP). The arrangement of direct current arc welding leads on which the electrode is the positive pole and the workpiece is the negative pole of the welding arc.

corriente directa con el electrodo positivo. El arreglo de los cables para soldar con la soldadura de arco con el electrodo es el polo positivo y la pieza de trabajo es el polo negativo de la soldadura de arco.

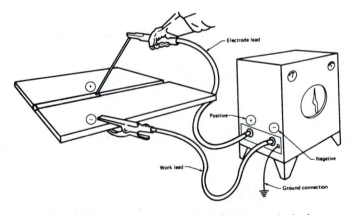

Direct current electrode positive (reverse polarity)

discontinuity. An interruption of the typical structure of a material, such as a lack of homogeneity in its mechanical, metallurgical, or physical characteristics. A discontinuity is not necessarily a defect. See **defect, flaw.**

discontinuidad. Una interrupción de la estructura típica de un material, el que falta de homogenidad en sus características mecánicas, metalúrgicas, o física. Vea **defecto** o **falta.**

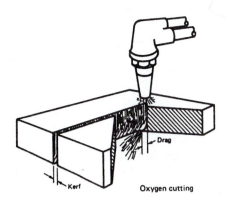

Oxygen cutting

*distortion. Movement or warping of parts being welded, from the pre-welding position and condition compared to the post-welding condition and position.

deformacion. Movimiento o torcimiento de las partes que se están soldando, comparando la posición antes de soldar a la posicion despues de soldar.

double-bevel-groove weld. A type of groove weld. Refer to drawing for groove weld.

soldadura de ranura con doble bisel. Es un tipo de soldadura de ranura. Refiérase al dibujo para soldadura de ranura.

double-flare-bevel-groove weld. A type of groove weld. Refer to drawing for groove weld.

soldadura de ranura con doble bisel acampanada. Un tipo de soldadura de ranura. Refiérase al dibujo para soldadura de ranura.

double-flare-V-groove weld. A weld in grooves formed by two members with curved surfaces. Refer to drawing for groove weld.

soldadura de ranura de doble V acampanada. Una soldadura de ranura formada por dos miembros con superficies curvados. Refiérase al dibujo para soldadura de ranura.

double-J-groove weld. A type of groove weld. Refer to drawing for groove weld.

soldadura de ranura de doble -J-. Un tipo de soldadura de ranura. Refiérase al dibujo para soldadura de ranura.

double-U-groove weld. A type of groove weld. Refer to drawing for groove weld.

soldadura de ranura de doble-U-. Un tipo de soldadura de ranura. Refiérase al dibujo para soldadura de ranura.

double-V-groove weld. A type of groove weld. Refer to drawing for groove weld.

soldadura de ranura de doble-V-. Un tipo de soldadura de ranura. Refiérase al dibujo para soldadura de ranura.

downslope time. See automatic arc welding downslope time and resistance welding downslope time.

tiempo de caída del pendiente. Vea tiempo del pendiente con descenso en soldadura de arco automático y tiempo del pendiente descenso en soldadura de resistencia.

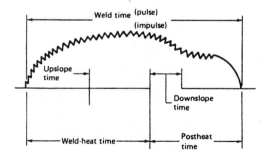

Electronic frequency converter (1/2 cycle operation) or metallic rectifier control

drag (thermal cutting). The offset distance between the actual and straight line exit points of the gas stream or cutting beam measured on the exit surface of the base metal.

tiro (Corte termal). La distancia desalineada entre la actual y la linea recta del punto de salida del chorro de gas o el rayo de cortar medido a la salida de la superficie del metal base.

drag angle. The travel angle when the electrode is pointing in a direction opposite to the progression of welding. This angle can also be used to partially define the position of guns, torches, rods, and beams.

ángulo del tiro. El ángulo de avance cuando el electrodo está apuntando en una dirección opuesta del progreso de la soldadura. Este ángulo también se puede usar para parcialmente definir la posición de pistolas, antorchas, varillas, y rayos.

*drag lines. High-pressure oxygen flow during cutting forms lines on the cut faces. A correctly made cut has up and down drag lines (zero drag); any deviation from the pattern indicates a change in one of the variables affecting the cutting process; with experience the welder can interpret the drag lines to determine how to correct the cut by adjusting one or more variables.

lineas del tiro. La salida del oxígeno a presión elevada durante el corte forma lineas en las caras del corte. Un corte hecho correctamente tiene lineas hacia arriba y hacia abajo (zero tiro); cualquier desviación de la norma indica un cambio en uno de los variables que afectan el proceso de cortar; con experiencia el soldador puede interpretar las lineas de tiro y determinar como corregir el corte ajustando uno o más variables.

*drift. The tendency of a system's response to gradually move away from the desired response.

deriva. La tendencia de la respuesta del sistema de retirarse gradualmente de la respuesta deseada.

*drooping output. Volt-ampere characteristic of the shielded metal arc process power supply where the voltage output decreases as increasing current is required of the power supply. This characteristic provides a reasonably high voltage at a constant current.

reducción de potencia de salida. Característica de voltio-amperios de la alimentación de poder de un proceso de soldadura de arco protegido donde la salida del voltaje disminuye mientras un aumento de la corriente es requerida de la alimentación de poder. Está característica proporciona un voltaje razonable alto con corriente constante.

*ductility. As applied to a soldered or brazed joint, it is the ability of the joint to bend without failing.

ductilidad. Como aplicada a junta de soldadura fuerte o soldadura blanda, es la abilidad de la junta de doblarse sin fallar.

duty cycle. The percentage of time during an arbitrary test period that a power source or its accessories can be operated at rated output without overheating.

ciclo de trabajo. El porcentaje de tiempo durante un período a prueba arbitraria de una fuente de poder y sus accesorios que pueden operarse a la capacidad de carga de salida sin sobrecalentarse.

E

edge joint. A joint between the edges of two or more parallel or nearly parallel members.

junta de orilla. Una junta en medio de las orillas de dos o más miembros paralelos o casi paralelos.

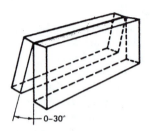

0–30°

Applicable welds

Square-Groove	Edge-Flange
Bevel-Groove	Corner-Flange
V-Groove	Seam
U-Groove	Edge
J-Groove	

edge preparation. The surface prepared on the edge of a member for welding.

preparación de orilla. La superficie preparada en la orilla de un miembro que se va a soldar.

effective length of weld. The length of weld throughout which the correctly proportioned cross section exists. In a curved weld, it shall be measured along the axis of the weld.

distancia efectiva de soldadura. La distancia de una sección transversal correctamente proporcionada que existe por toda la soldadura. En una soldadura en curva, debe medirse por el axis de la soldadura.

effective throat. The minimum distance from the root of a weld to its face, less any reinforcement. See also **joint penetration**. Refer to drawing for **convexity**.

garganta efectiva. La distancia mínima de la raíz a la cara de una soldadura, menos el refuerzo. Vea **penetración de junta**. Refiérase al dibujo para **convexidad**.

*elastic limit. The maximum force that can be applied to a material or joint without causing permanent deformation or failure.

limite elástico. La fuerza máxima que se le puede aplicar a un material o junta sin causar deformación o falta permanente.

*elbow. The joint that connects a robot's upper arm and forearm.

codo. La junta que conecta al brazo de arriba con el brazo de enfrente en un robot.

electrode. A component of the electrical circuit that terminates at the arc, molten conductive slag, or base metal.

electrodo. Un componente del circuito eléctrico que termina al arco, escoria derretida conductiva, o metal base.

*electrode angle. The angle between the electrode and the surface of the metal; also known as the direction of travel (leading angle or trailing angle); leading angle pushes molten metal and slag ahead of the weld; trailing angle pushes the molten metal away from the leading edge of the molten weld pool toward the back, where it solidifies.

ángulo del electrodo. El ángulo en medio del electrodo y la superficie del metal; también conocido como la dirección de avance (apuntado hacia adelante o apuntado hacia atras); el ángulo apuntado empuja el metal derretido y la escoria enfrente de la soldadura; y el ángulo apuntado hacia atrás empuja el metal derretido lejos de la orilla delantera del charco del metal derretido hacia atrás, donde se solidifica.

*electrode classification. Any of several systems developed to identify shielded metal arc welding electrodes. The most widely used identification system was developed by the American Welding Society (AWS). The information represented by the classification generally includes the minimum tensile strength of a good weld, the position(s) in which the electrode can be used, the type of flux coating, and the type(s) of welding currents with which the electrode can be used.

clasificación de electrodo. Cualquiera de los varios sistemas desarrollados para identificar electrodos protegidos para soldadura de arco. El sistema de identificación que se usa mucho más fue desarrollado por la Sociedad de Soldadura Americana (AWS). La información representada por la clasificación generalmente incluye la fuerza tensible mínima de una soldadura, la posición(es) donde se puede usar el electrodo, el tipo de recubrimiento de fundente y los tipo(s) de corrientes para soldar con la cual se puede usar el electrodo.

electrode extension (GMAW, FCAW, SAW). The length of unmelted electrode extending beyond the end of the contact tube during welding.

extensión del electrodo (GMAW, FCAW, SAW). La distancia de extensión del electrodo que no está derretido más allá de la punta del tubo de contacto durante la soldadura.

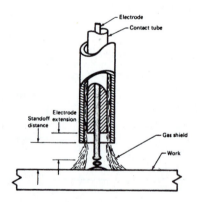

electrode holder. A device used for mechanically holding and conducting current to an electrode during welding.

porta electrodo. Un aparato usado para detener mecánicamente y conducir corriente a un electrodo durante la soldadura.

electrode lead. The electrical conductor between the source of arc welding current and the electrode holder. Refer to drawing for **direct current electrode negative**.

cable de electrodo. Un conductor eléctrico en medio de la fuente para la corriente de soldar con arco y el portaelectrodo. Refiérase al dibujo de **corriente directa con electrodo negativo.**

electrode setback. The distance the electrode is recessed behind the constricting orifice of the plasma arc torch or thermal spraying gun, measured from the outer face of the nozzle.

retroceso del electrodo. La distancia del hueco del electrodo que está detrás del orificio constringente de la antorcha de arco plasma o pistola de rocío termal, se mide de la cara de afuera a la boquilla.

electron beam cutting (EBC). A cutting process that uses the heat obtained from a concentrated beam composed primarily of high velocity electrons which impinge upon the workpieces to be cut; it may or may not use an externally supplied gas.

cortes a rayo de electron. Un proceso de cortar que usa calor obtenido de un rayo concentrado primeramente de electrones de alta velocidad que choca sobre la pieza de trabajo la cual se va a cortar; puede o no usar un gas surtido externamente.

electron beam welding (EBW). A welding process that produces coalescence with a concentrated beam, composed primarily of high velocity electrons, impinging on the joint. The process is used without shielding gas and without the application of pressure.

soldadura a rayo de electron. Un proceso de soldadura la cual produce coalescencia de un rayo concentrado, compuesto primeramente de electrones de alta velocidad al chocar con la junta. Este proceso no usa gas de protección y sin la aplicación de presión.

electroslag welding electrode. A filler metal component of the welding circuit through which current is conducted from the electrode guiding member to the molten slag.

electrodo para soldadura de electroescoria. Un componente de metal de relleno del circuito para soldar por donde la corriente es conducida del miembro que guía el electrodo a la escoria derretida.

emissive electrode. A filler metal electrode consisting of a core of a bare electrode or a composite electrode to which a very light coating has been applied to produce a stable arc.

electrodo emisivo. Un electrodo de metal de relleno consistiendo de un electrodo liso o un electrodo compuesto de una capa ligera que se le aplica para producir un arco estable.

*end effector. An actuator, gripper, or mechanical device attached to the wrist of a manipulator by which objects can be grasped or acted upon.

punta que efectúa. Un movedor, el que agarra o un aparato mecánico fijo a la muñeca de un manipulador por el cual objectos se pueden agarrar u obrar en impulso.

entry level welder. A person just entering the welding profession.

soldador principiante. Una persona que acaba de comenzar en la profesión de la soldadura.

*error signal. The difference between desired response and actual response.

señal equivocada. La diferencia entre la respuesta deseada y la respuesta efectiva.

*eutectic composition. The composition of an alloy that has the lowest possible melting point for that mixture of metals.

composición de tipo eutectico. La composición de un aleado que tenga el punto de fusión lo más bajo posible para esa mezcla de metales.

*exhaust pickup. A component of a forced ventilation system that has sufficient suction to pick up fumes, ozone, and smoke from the welding area and carry the fumes, etc. outside of the area.

recogedor de extracción. Un componente de un sistema de ventilación forzada que tiene suficiente succión para recoger vaho, ozono, y humo de la área de soldadura y lleva al vaho, etc. a fuera de la area.

F

face bend test. A test in which the weld face is on the convex surface of a specified bend radius.

prueba de dobléz de cara. Una prueba donde la cara de la soldadura está en la superficie convexa al radio de dobléz especificado.

face of weld. The exposed surface of a weld on the side from which welding was done.

cara de la soldadura. La superficie expuesta de una soldadura del lado de donde se hizo la soldadura.

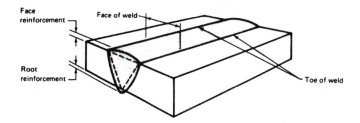

face reinforcement. Reinforcement of weld at the side of the joint from which welding was done. See also **root reinforcement.** Refer to drawing for **face of weld.**

refuerzo de cara. Refuerzo de una soldadura en el lado de la junta de donde se hizo la soldadura. Vea también **refuerzo de raíz.** Refiérase al dibujo para **cara de la soldadura.**

face shield (eye protection). A device positioned in front of the eyes and a portion of, or all of, the face, whose predominant function is protection of the eyes and face. See also **helmet.**

protector de cara sostenido a mano (protección del ojo). Un aparato puesto en frente de los ojos y una porción, o en toda la cara, cuya función predominante es de proteger los ojos y la cara. Vea también **casco.**

*fast freezing electrode. An electrode whose flux forms a high-temperature slag that solidifies before the weld metal solidifies, thus holding the molten metal in place. This is an advantage for vertical, horizontal, and overhead welding positions.

electrodo de congelación rápida. Un electrodo cuyo flujo forma una escoria a temperaturas altas que se puede solidificar antes de que el metal de soldadura se pueda solidificar, asi detiene el metal derretido en su lugar. Está es una ventaja en soldaduras de posiciones vertical, horizontal y sobrecabeza.

*fatigue resistance. As applied to a soldered or brazed joint, it is the ability of the joint to be bent repeatedly without exceeding its elastic limit and without failure. Generally, fatigue resistance is low for most soldered and brazed joints.

resistencia a la fatiga. Como aplicada a una junta de soldadura con metales de bajo punto de fusión y soldadura fuerte, es la abilidad de una junta de ser doblada repetidamente sin exceder los limites elásticos y sin fracaso. Generalmente, la resistencia a la fatiga es muy baja para la mayoría de las juntas de soldadura con bajo punto de fusión y soldadura fuerte.

faying surface. The mating surface of a member that is in contact with or in close proximity to another member to which it is to be joined.

superficie de unión. La superficie de apareamiento de un miembro del que está en contacto con otro miembro o está en proximidad cercana a otro miembro que está para ser unido.

feed rollers. A set of two or four individual rollers which, when pressed tightly against the filler wire and powered up, feed the wire through the conduit liner to the gun for GMAW and FCAW welding.

rodillos de alimentación. Un conjunto de dos o cuatro rodillos individuales que al ser presionados fuertemente contra el alambre de relleno y ser accionados alimentan al alambre a través del revestimiento de canal hasta la pistola, en soldaduras tipo GMAW y FCAW.

filler metal. The metal or alloy to be added in making a welded, brazed, or soldered joint.

metal de aporte. El metal o aleado que se agrega cuando se hace una soldadura blanda o soldadura fuerte.

*filler pass. One or more weld beads used to fill the groove with weld metal. The bead must be cleaned after each pass to prevent slag inclusions.

pasada para rellenar. Uno o más cordones de soldadura usados para llenar la ranura con el metal de soldadura. El cordón debe ser limpiado después de cada pasada para prevenir inclusiones de escoria.

fillet weld. A weld of approximately triangular cross section joining two surfaces approximately at right angles to each other in a lap joint, T-joint, or corner joint. Refer to drawing for **convexity**.

soldadura de filete. Una soldadura de filete de sección transversa aproximadamente triangular que une dos superficies aproximademente en ángulos rectos de uno al otro en junta de traslape, junta en- T- o junta de esquina. Refiérase al dibujo para **convexidad**.

fillet weld break test. A test in which the specimen is loaded so that the weld root is in tension.

prueba de rotura en soldadura de filete. Una prueba en donde la probeta es cargada de manera en que la tensión esté sobre la soldadura.

fillet weld leg. The distance from the joint root to the toe of the fillet weld.

pierna de soldadura filete. La distancia de la raíz de la junta al pie de la soldadura filete.

fillet weld size. For equal leg fillet welds, the leg lengths of the-largest isosceles right triangle that can be inscribed within the fillet weld cross section. For unequal leg fillet welds, the leg lengths of the largest right triangle that can be inscribed within the fillet weld cross section.

tamaño de soldadura filete. Para soldaduras de filete que tienen piernas iguales, lo largo de las piernas del isósceles más grande del triángulo recto que puede ser inscribido dentro de la sección. Para soldaduras de filete con piernas desiguales, lo largo de las piernas del triángulo recto más grande puede inscribirse dentro de la sección transversal.

filter plate. An optical material that protects the eyes against excessive ultraviolet, infrared, and visible radiation.

lente filtrante. Un material óptico que protege los ojos contra ultravioleta excesiva, infrarrojo, y radiación visible.

final current. The current after downslope but prior to current shut-off.

corriente final. La corriente después del pendiente en descenso pero antes de que la corriente sea cerrada.

fisheye. A discontinuity found on the fracture surface of a weld in steel that consists of a small pore or inclusion surrounded by an approximately round, bright area.

ojo de pescado. Una discontinuidad que se encuentra en una fractura de superficie en una soldadura de acero que consiste de poros pequeños o inclusiones rodeadas de áreas aproximadamente redondas y brillantes.

fissure. A small, crack-like discontinuity with only slight separation (opening displacement) of the fracture surfaces. The prefixes macro or micro indicate relative size.

hendemiento. Una pequeña, discontinuidad como una grieta con solamente una separación (abertura desalojada) de las superficies fracturadas. El prefijo marco o micro indica el tamaño relativo.

*fixed inclined (6G) position. For pipe welding, the pipe is fixed at a 45° angle to the work surface. The effective welding angle changes as the weld progresses around the pipe.

posición (6G) inclinado fijo. Para soldadura de tubo, el tubo se fija a un ángulo de 45° de la superficie del trabajo. El ángulo efectivo de la soldadura cambia cuando la soldadura progresa alrededor del tubo.

fixture. A device designed to hold parts to be joined in proper relation to each other.

fijación. Una devisa diseñada para detener partes que se van a unir en relación propia de una a la otra.

flame propagation rate. The speed at which flame travels through a mixture of gases.

cantidad de propagación de la llama. La rapidez en que la llama camina a través de una mezcla de gas.

flame spraying (FLSP). A thermal spraying process in which an oxyfuel gas flame is the source of heat for melting the surfacing material. Compressed gas may or may not be used for atomizing the propellant and surfacing material to the substrate.

rociado a llama. Un proceso de rociado termal en donde la llama del gas oxicombustible es la fuente de calor para derretir el material de revestimiento. El gas comprimido se puede o no se puede usar para automizar el propulsor y el material de revestimiento al substrato.

flange weld. A weld made on the edges of two or more joint members, at least one of which is flanged.

soldadura de reborde. Una soldadura que se hace en las orillas de dos o más miembros de junta, donde por lo menos uno tiene reborde.

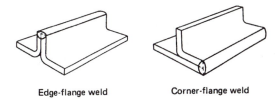

Edge-flange weld   Corner-flange weld

flash. The material that is expelled or squeezed out of a weld joint and that forms around the weld.

ráfaga. El material que es despedido o exprimido fuera de una junta de soldadura y se forma alrededor de la soldadura.

flashback. A recession of the flame into or back of the mixing chamber of the oxyfuel gas torch or flame spraying gun.

llamarada de retroceso. Una recesión de la llama adentro o atrás de la cámara mezcladora de una antorcha de gas oxicombustible o pistola de rociar a llama.

flashback arrester. A device to limit damage from a flashback by preventing propagation of the flame front beyond the point at which the arrester is installed.

válvula de retención. Un aparato para limitar el daño de una llamarada de retroceso para prevenir la propagación del frente de la llama más allá del punto donde se instala la válvula de retención.

flash welding (FW). A resistance welding process that produces a weld at the faying surfaces of a butt joint by a flashing action and by the application of pressure after heating is substantially completed. The flashing action, caused by the very high current densities at small contact points between the workpieces, forcibly expels the material from the joint as the workpieces are slowly moved together. The weld is completed by a rapid upsetting of the workpieces.

soldadura de relámpago. Un proceso de soldadura de resistencia que produce una soldadura en el empalme de la superficie de una junta tope por una acción de relampagueo y por la aplicación de presión después que el calentamiento este substancialmente acabado. La acción del relampagueo, causado por densidades de corrientes altas a unos puntos de contacto pequeños en medio de las piezas de trabajo, despiden fuertemente el material de la junta cuando las piezas de trabajo se mueven despacio. La soldadura es terminada por un acortamiento rápido de las piezas de trabajo.

flat position. The welding position used to weld from the upper side of the joint; the weld face is approximately horizontal.

posición plana. La posición de soldadura que se usa para soldar del lado de arriba de una junta; la cara de la soldadura está aproximadamente horizontal.

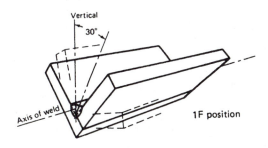

1F position

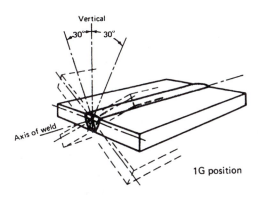

1G position

flaw. An undesirable **discontinuity.**

falta. Una **discontinuidad** indeseable.

flow rate. The rate at which a given volume of shielding gas is delivered to the weld zone. The units used for welding are cubic feet, inches, meters, or centimeters.

caudal. velocidad a la cual llega un determinado volumen de gas protector a la zona de soldadura. Las unidades usadas para la soldadura son pies cúbicos, pulgadas, metros o centímetros.

flux. A material used to hinder or prevent the formation of oxides and other undesirable substances in molten metal and on solid metal surfaces, and to dissolve or otherwise facilitate the removal of such substances.

flujo. Un material que se usa para impedir o prevenir la formación de óxidos y otras substancias indeseables en el metal derretido y en las superficies del metal sólido, y para desolver o de otra manera facilitar el removimiento de dichas substancias.

flux cored arc welding (FCAW). An arc welding process that uses an arc between a continuous filler metal electrode and the weld pool. The process is used with shielding gas from a flux contained within the tubular electrode, with or without additional shielding from an externally supplied gas, and without the application of pressure.

soldadura de arco con núcleo de fundente. Un proceso de soldadura de arco que usa un arco entre medio de un electrodo de metal rellenado continuo y el charco de la soldadura. El proceso es usado con gas de protección del flujo contenido dentro del electrodo tubular, y sin usarse protección adicional de abastecimiento de gas externo, y sin aplicarse presión.

flux cored electrode. A composite tubular filler metal electrode consisting of a metal sheath and a core of various powered materials, producing an extensive slag cover on the face of a weld bead. External shielding may be required.

electrodo de núcleo de fundente. Un electrodo tubular de metal para rellenar con una compostura que consiste de una envoltura de metal y un núcleo con varios materiales de polvo, que producen un forro extensivo de escoria en la superficie del cordón de soldadura. Protección externa puede ser requerida.

flux cover. In metal bath dip brazing and dip soldering, a cover of flux over the molten filler metal bath.

tapa de fundente. En metal de baño soldadura fuerte y soldadura blanda por inmersión, una tapa de fundente sobre el baño del metal de relleno derretido.

***forced ventilation.** To remove excessive fumes, ozone, or smoke from a welding area, a ventilation system may be required to supplement natural ventilation. Where forced ventilation of the welding area is required, the rate of 200 cu ft (56m³) or more per welder is needed.

ventilación forzada. Para quitar excesivo vaho, ozono y humo de la área donde se solda, un sistema de ventilación puede ser requerido para suplementar la ventilación natural. Donde la ventilación forzada de la área de la soldadura es requerida, la rázon de 200 pies cúbicos (56m³) o más es requerido por cada soldador.

forehand welding. A welding technique in which the welding torch or gun is directed toward the progress of welding. See also **travel angle, work angle,** and **push angle.** Refer to drawing for **backhand welding.**

soldadura directa. Una técnica de soldar en cual la pistola o la antorcha para soldar es dirigida hacia al progreso de la soldadura. Vea también **ángulo de avance, ángulo de trabajo, ángulo de empuje.** Refiérase al dibujo **soldadura en revés.**

forge welding (FOW). A solid-state welding process that produces a weld by heating the workpieces to welding temperature and applying blows sufficient to cause permanent deformation at the faying surfaces.

soldadura por forjado. Un proceso de soldadura de estado sólido que produce una soldadura calentando las piezas de trabajo a una temperatura de soldadura y aplicando golpes suficientes para causar una deformación permanente en las superficies del empalme.

frog. The large rail structures, usually castings, that form the center of a crossing or the rail crossing point of a switch.

corazón de cruzamiento. Las grandes estructuras de rieles, normalmente piezas moldeadas, que forman el centro de un cruzamiento o el punto de cruzamiento de rieles en el sitio de convergencia.

fuel gases. A gas such as acetylene, natural gas, hydrogen, propane, stabilized methylacetylene propadiene, and other fuels normally used with oxygen in one of the oxyfuel processes and for heating.

gases combustibles. Gases como acetileno, gas natural, hidrógeno, propano, metilacetileno propodieno estabilizado, y otros combustibles normalmente usados con oxígeno en uno de los procesos de oxicombustible y para calentar.

full fillet weld. A fillet weld whose size is equal to the thickness of the thinner member joined.

soldadura de filete llena. Una soldadura de filete cuyo tamaño es igual de grueso como el miembro más delgado de la junta.

full penetration. A nonstandard term for complete **joint penetration.**

penetración llena. Un término fuera de la norma en vez de la **penetración de junta.**

furnace brazing (FB). A brazing process in which the parts to be joined are placed in a furnace heated to a suitable temperature.

soldadura fuerte en horno. Un proceso de soldadura fuerte en donde las partes que se van a unir se ponen en un horno calentado a una temperatura adecuada.

furnace soldering (FS). A soldering process in which the parts to be joined are placed in a furnace heated to a suitable temperature.

soldadura blanda en horno. Un proceso de soldadura blanda en donde las partes que se van a unir se ponen en un horno calentado a una temperatura adecuada.

fusion. The melting together of filler metal and base metal, or of base metal only, to produce a weld. (See **depth of fusion.**)

fusión. El derretir el metal de relleno y el metal base juntos o el metal base solamente, para producir una soldadura. (Vea **grueso de fusión.**)

fusion welding. Any welding process or method that uses fusion to complete the weld.

soldadura de fusión. Cualquier proceso de soldadura o método que usa fusión para completar la soldadura.

fusion zone. The area of base metal melted as determined on the cross section of a weld. Refer to drawing for **depth of fusion.**

zona de fusión. La área del metal base que se derritió como determinada en la sección transversa de la soldadura. Refiérase al dibujo **grueso de fusion.**

G

gap. A nonstandard term when used for **arc length, joint clearance,** and **root opening.**

abertura. Un término fuera de norma cuando se usa en **lugar del arco, despejo de junta,** y **abertura de raíz.**

gas cup. A nonstandard term for **gas nozzle.**

tazón de gas. Un término fuera de norma en vez de **boquilla de gas.**

gas cylinder. A portable container used for transportation and storage of compressed gas.

cilindro de gas. Un recipiente portátil que se usa para transportación y deposito de gas comprimido.

gas lens. One or more fine mesh screens located in the torch nozzle to produce a stable stream of shielding gas. Primarily used for gas tungsten arc welding.

lente para gas. Uno o más cedazos de malla fina localizados en la lanza de la antorcha para producir un chorro estable de gas de protección primeramente usada para soldaduras de arco tungsteno y gas.

gas metal arc cutting (GMAC). An arc cutting process that uses a continuous consumable electrode and a shielding gas.

cortes de arco metálico con gas. Un proceso de corte con arco que usa un alambre consumible continuo y un gas de protección.

gas metal arc welding (GMAW). An arc welding process that uses an arc between a continuous filler metal electrode and the weld pool. The process is used with shielding from an externally supplied gas and without the application of pressure.

soldadura de arco metálico con gas. Un proceso de soldar con arco que usa un arco en medio de un electrodo de metal para rellenar continuo y el charco de soldadura. El proceso usa protección de un abastecedor externo de gas y sin la aplicación de presión.

gas metal arc welding @ pulsed arc (GMAW-P). A gas metal arc welding process variation in which the current is pulsed.

soldadura con arco metálico con gas arco pulsado. Un proceso de soldadura de arco metálico con gas con variación en cual la corriente es de pulsación.

gas metal arc welding @ short circuit arc (GMAW-S). A gas metal arc welding process variation in which the consumable electrode is deposited during repeated short circuits. Sometimes this process is referred to as MIG or CO_2 welding (nonpreferred terms).

soldadura con arco metálico con gas @ arco de corto circuito. Un proceso de soldadura de arco metálico con gas con variación en cual el electrodo consumible es depositado durante los cortos circuitos repetidos. A veces el proceso es referido como soldadura MIG o $C0_2$ (términos que no son preferidos).

gas nozzle. A device at the exit end of the torch or gun that directs shielding gas.

boquilla de gas. Un aparato a la salida de la punta de la antorcha o pistola que dirige el gas protector.

gas regulator. A device for controlling the delivery of gas at some substantially constant pressure.

regulador de gas. Un aparato para controlar la salida de gas a una presión substancialmente constante.

gas tungsten arc cutting (GTAC). An arc cutting process that uses a single tungsten electrode with gas shielding.

corte de arco con tungsteno y gas. Un proceso de corte de arco que usa un electrodo de tungsteno sencillo con gas de protección.

gas tungsten arc welding (GTAW). An arc welding process that uses an arc between a tungsten electrode (nonconsumable) and the weld pool. The process is used with shielding gas and without the application of pressure.

soldadura de arco de tungsteno con gas. Un proceso de soldadura de arco que usa un arco en medio del electrodo tungsteno (no consumible) y el charco de la soldadura. El proceso es usado con gas de protección y sin aplicación de presión.

gas tungsten arc welding @ pulsed arc (GTAW-P). A gas tungsten arc welding process variation in which the current is pulsed.

soldadura de arco de tungsteno con gas @ arco pulsado. Un proceso de soldadura de arco tungsteno con variación en cual la corriente es de pulsación.

***gauge pressure.** The actual pressure shown on the gauge; does not take into account atmospheric pressure.

manómetro para presión. La presión actual que se enseña en el manómetro; no toma en cuenta la presión atmosférica.

***gauge (regulator).** A device mounted on a regulator to indicate the pressure of the gas passing into the gauge. A regulator is provided with two gauges; one (high-pressure gauge) indicates the pressure of the gas in the cylinder; the second gauge (low-pressure gauge) shows the pressure of the gas at the torch.

manómetro (regulador). Un aparato montado en un regulador para indicar la presión del gas que está pasando por el manómetro. El regulador tiene dos manómetros; uno (manómetro de alta presión) indica la presión del gas en el cilindro; el segundo manómetro (manómetro de presión baja) enseña la presión del gas en la antorcha.

globular transfer. The transfer of molten metal in large drops from a consumable electrode across the arc.

traslado globular. El traslado del metal derretido en gotas grandes de un electrodo consumible a través del arco.

gouging. The forming of a bevel or groove by material removal. See also **back gouging, arc gouging,** and **oxygen gouging.**

escopleando con gubia. Formando un bisel o ranura removiendo el material. Vea también **gubia trasera, gubia dura con arco** y **escopleando con la gubia con oxígeno.**

groove. An opening or channel in the surface of a part or between two components, which provides space to contain a weld.

ranura. Una abertura o canal en la superficie de una parte o en medio de dos componentes, la cual provee espacio para contener una soldadura.

groove angle. The total included angle of the groove between parts to be joined by a groove weld.

ángulo de ranura. El ángulo total incluido de la ranura entre partes para unirse por una soldadura de ranura.

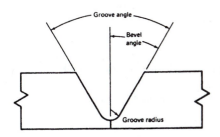

groove face. The surface of a joint member included in the groove.

cara de ranura. La superficie de un miembro de una junta incluido en la ranura.

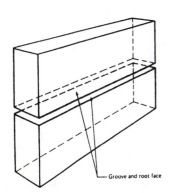
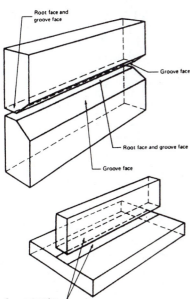

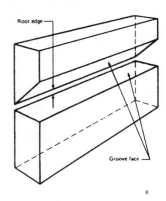

Groove face, root edge, and root face

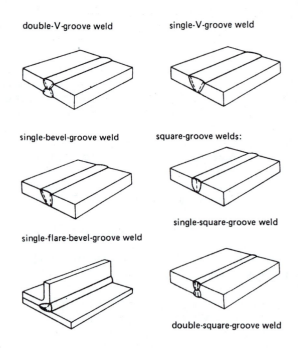

double-V-groove weld   single-V-groove weld

single-bevel-groove weld   square-groove welds:

single-square-groove weld

single-flare-bevel-groove weld

double-square-groove weld

groove radius. The radius used to form the shape of a J- or U-groove weld joint. Refer to drawings for **bevel**.

radio de ranura. La radio que se usa para formar una junta de una soldadura con una ranura de forma U o J. Refiérase a los dibujos para **bisel**.

groove weld. A weld made in the groove between two members to be joined. The standard types of groove welds are as follows:

soldadura de ranura. Una soldadura hecha en la ranura dentro de dos miembros que se unen. Los tipos normales de soldadura de ranura son los siguientes:

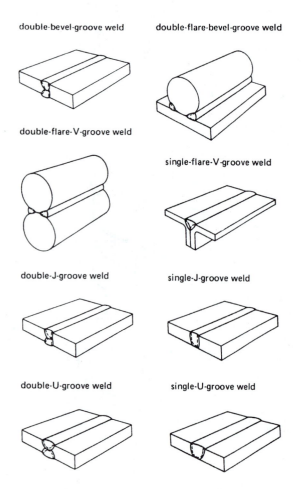

double-bevel-groove weld   double-flare-bevel-groove weld

double-flare-V-groove weld

single-flare-V-groove weld

double-J-groove weld   single-J-groove weld

double-U-groove weld   single-U-groove weld

ground connection. An electrical connection of the welding machine frame to the earth for safety.

conexión a tierra. Una conexión eléctrica del marco de la máquina de soldar a la tierra para seguridad.

ground lead. A nonstandard and incorrect term for **workpiece lead**.

cable de tierra. Un término fuera de norma e incorrecto que se usa en vez de **cable de pieza de trabajo**.

*group technology. A system for coding parts based on similarities in geometrical shape or other characteristics of parts. The grouping of parts into families based on similarities in their production so that parts of a particular family could be processed together.

codificador. Un sistema para codificar partes basadas en similaridades en forma geométrica y otras características de las partes. La agrupación de partes en familias basadas en similaridades en su producción para que partes de una familia en particular puedan ser procesadas juntas.

*guided bend specimen. Any bend specimen that will be bend tested in a fixture that controls the bend radii, such as the AWS bend test fixture.

probeta de dobléz guiada. Cualquier probeta de dobléz en la cual se va a hacer un dobléz guiado en una máquina que controla el radio del dobléz, como la máquina de dobléz guiado del AWS.

H

hardfacing. A surfacing variation in which surfacing material is deposited to reduce wear.

endurecimiento de caras. Una variación superficial donde el material superficial es depositado para reducir el desgastamiento.

heat-affected zone. The portion of the base metal whose mechanical properties or microstructure have been altered by the heat of welding, brazing, soldering, or thermal cutting.

zona afectada por el calor. La porción del metal base cuya propiedad mecánica o microestructura ha sido alterada por el calor de soldadura, soldadura fuerte, soldadura blanda, o corte termal.

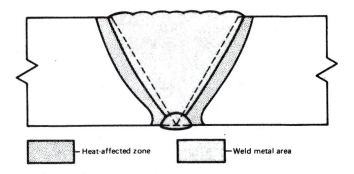

- Heat-affected zone   - Weld metal area

heating torch. A device for directing the heating flame produced by the controlled combustion of fuel gases.

antorcha de calentamiento. Un aparato para dirigir la llama de calentamiento que es producida por una combustión controlada de gases de combustión.

helmet. A device designed to be worn on the head to protect eyes, face, and neck from arc radiation, radiated heat, spatter, or other harmful matter expelled during arc welding, arc cutting, and thermal spraying.

casco. Un aparato diseñado para usarse sobre la cabeza para proteger ojos, cara y cuello de radiación del arco, calor radiado, salpicadura, u otra materia dañosa despedida durante la soldadura de arco, corte por arco, y rociado termal.

horizontal fixed position (pipe welding). The position of a pipe joint in which the axis of the pipe is approximately horizontal, and the pipe is not rotated during welding.

posición fija horizontal (soldadura de tubos). La posición de una junta de tubo la cual el axis del tubo es aproximadamente horizontal, y el tubo no da vueltas durante la soldadura.

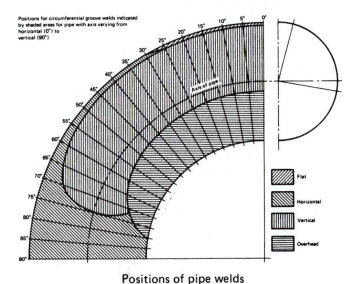

Positions of pipe welds

Horizontal fixed position

***horizontal fixed (5G) position weld.** For pipe welding, the pipe is fixed horizontally (cannot be rolled). The weld progresses from overhead, to vertical, to flat position around the pipe.

soldadura de posición fija horizontal (5G). Para soldadura de tubos, el tubo está fijo horizontalmente (no se pueder rodar). La soldadura progresa de sobre cabeza, a vertical, a la posición plana alrededor del tubo.

horizontal position (fillet weld). The position in which welding is performed on the upper side of an approximately horizontal surface and against an approximately vertical surface.

posición horizontal (soldadura de filete). La posición de la soldadura la cual es hecha en el lado de arriba de una superficie horizontal aproximadamente y junto a una superficie vertical aproximadamente.

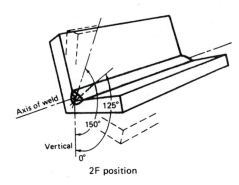

2F position

horizontal position (groove weld). The position of welding in which the weld axis lies in an approximately horizontal plane and the weld face lies in an approximately vertical plane.

posición horizontal (de ranura). La posición para soldar en la cual el axis de la soldadura está en una plana horizontal aproximadamente, y la cara de la soldadura está en una plana vertical aproximadamente.

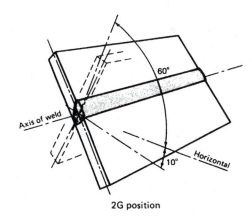

2G position

horizontal rolled position (pipe welding). The position of a pipe joint in which the axis of the pipe is approximately horizontal, and welding is performed in the flat position by rotating the pipe.

posición horizontal rodada (soldadura de tubo). La posición de una junta de tubo en la cual el axis del tubo es horizontal aproximadamente, y la soldadura se hace en la posición plana con rotación del tubo.

(continued on page 856)

(continued from page 855)

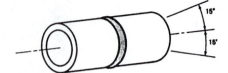

*horizontal rolled (1G) position. For pipe welding, this position yields high-quality and high-quantity welds. Pipe to be welded is placed horizontally on the welding table in a fixture to hold it steady and permit each rolling. The weld proceeds in steps, with the pipe being rolled between each step, until the weld is complete. For plate, see **axis of a weld.**
posición (1G) horizontal rodada. Para soldadura de tubo, está posición produce soldaduras de alta calidad y alta cantidad. El tubo que se va a soldar se pone horizontalmente sobre la mesa de soldadura en una instalación fija que lo detiene seguro y permite cada rodadura. La soldadura procede en pasos, con el tubo siendo rodado entre cada paso, hasta que la soldadura esté completa. Para plato, vea **eje de la soldadura.**

hot crack. A crack formed at temperatures near the completion of solidification.
grieta caliente. Una grieta formada a temperaturas cerca de la terminación de la solidificación.

*hot pass. The welding electrode is passed over the root pass at a higher than normal amperage setting and travel rate to reshape an irregular bead and turn out trapped slag. A small amount of metal is deposited during the hot pass so the weld bead is convex, promoting easier cleaning.
pasada caliente. El electrodo de soldadura se pasa sobre la pasada de raíz poniendo el amperaje más alto que lo normal y proporción de avance para reformar un cordón irregular y sacar la escoria atrapada. Una cantidad pequeña de metal es depositada durante la pasada caliente para que el cordón soldado sea convexo, promoviendo más fácil la limpieza.

hot start current. A very brief current pulse at arc initiation to stabilize the arc quickly. Refer to drawing for **upslope time.**
corriente caliente para empezar. Un pulso muy breve de corriente a iniciación de arco para estabilizar el arco aprisa. Refiérase al dibujo **tiempo del pendiente en ascenso.**

hydrogen embrittlement. The delayed cracking in steel that may occur hours, days, or weeks following welding. It is a result of hydrogen atoms that dissolved in the molten weld pool during welding.
fragilidad causada por el hidrógeno. el fisuramiento retardado en el acero que puede ocurrir horas, días o semanas después de la soldadura. Es el resultado de la disolución de átomos de hidrógeno en el charco de soldadura derretido durante la soldadura.

I

inclined position. The position of a pipe joint in which the axis of the pipe is at an angle of approximately 45 degrees to the horizontal, and the pipe is not rotated during welding.
posición inclinada. Una posición de junta de tubo en la cual el axis del tubo está a un ángulo de aproximadamente 45 grados a la horizontal, y no se le da vueltas al tubo durante la soldadura.

inclined position, with restriction ring. The position of a pipe joint in which the axis of the pipe is at an angle of approximately 45 degrees

to the horizontal, and a restriction ring is located near the joint. The pipe is not rotated during welding.
posición inclinada con argolla de restricción. La posición de una junta de tubo en la cual el axis del tubo está a un ángulo de aproximadamente 45 grados a la horizontal, y una argolla de restricción está localizada cerca de la junta. No se le da vuelta al tubo durante la soldadura.

included angle. A nonstandard term for **groove angle.**
ángulo incluido. Un término fuera de norma para **ángulo de ranura.**

inclusion. Entrapped foreign solid material, such as slag, flux, tungsten, or oxide.
inclusión. Material extraño atrapado sólido, como escoria, flujo, tungsteno u óxido.

incomplete fusion. A weld discontinuity in which fusion did not occur between weld metal and fusion faces or adjoining weld beads.
fusión incompleta. Una discontinuidad en la soldadura en la cual no ocurrió fusión entre el metal soldado y caras de fusión o cordones soldados inmediatos.

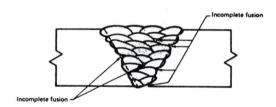

incomplete joint penetration. Joint penetration that is unintentionally less than the thickness of the weld joint.
penetración de junta incompleta. Penetración de la junta que no es intencionalmente menos de lo grueso de la junta de soldar.

induction brazing (IB). A brazing process that uses heat from the resistance of the workpieces to induced electric current.
soldadura fuerte por inducción. Un proceso de soldadura blanda que usa calor de la resistencia de las piezas de trabajo para inducir la corriente eléctrica.

induction soldering (IS). A soldering process in which the heat required is obtained from the resistance of the workpieces to induced electric current.
soldadura blanda por inducción. Un proceso de soldadura blanda en el cual el calor requerido es obtenido de la resistencia de las piezas de trabajo a la corriente eléctrica inducida.

inert gas. A gas that normally does not combine chemically with materials.
gas inerte. Un gas que normalmente no se combina químicamente con materiales.

*inertia welding. A welding process in which one workpiece revolves rapidly and one is stationary. At a predetermined speed the power is cut, the rotating part is thrust against the stationary part, and frictional heating occurs. The weld bond is formed when rotation stops.
soldadura inercia. Un proceso en el cual una pieza de trabajo voltea rápidamente y la otra está fija. A una velocidad predeterminada se le corta la potencia, la parte que está volteando es empujada en contra de la parte fija, y el calor de fricción ocurre. La unión soldada es formada cuando se para la rotación.

infrared radiation. Electromagnetic energy with wavelengths from 770 to 12,000 nonometers.

radiación infrarrojo. Energía electromagnética con longitud de ondas de 770 a 12,000 nonometros.

*injector chamber. One method of completely mixing the fuel gas and oxygen to form a flame. High-pressure oxygen is passed through a narrowed opening (venturi) to the mixing chamber. This action creates a vacuum, which pulls the fuel gas into the chamber and ensures thorough mixing. Used for equal gas pressures and is particularly useful for low-pressure fuel gases.

cámara de inyector. Un método de mezclar completamente el gas de combustión y el oxígeno para formar una llama. Oxígeno a alta presión es pasado por una abertura angosta (venturi) a la cámara de mezcla. Está acción hace un vacuo, la cual estira el gas combustible para dentro de la cámara y asegura una mezcla completa . Usada para presiones de gas que son iguales y es particularmente útil para gases combustibles de presión baja.

*inner cone. The portion of the oxyacetylene flame closest to the welding tip. The primary combustion reaction occurs in the inner cone. The size and color of the cone serves as an indicator of the type of flame (carburizing, oxidizing, neutral).

cono interno. La porción de la llama de oxiacetileno más cerca de la punta para soldar. La reacción de combustión principal ocurre en el cono interno. El tamaño y el calor del cono sirve como un indicador del tipo de la llama (carburante, oxidante, neutral).

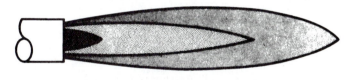

	Neutral flame	Oxidizing flame	Reducing flame
	White blue cone	White cone	Intense white cone
	Nearly colorless	Orange to purplish	White or colorless
	Bluish to orange		Orange to bluish

Oxyacetylene flame

*intelligent robot. A robot that can be programmed to make performance choices contingent on sensory inputs.

robot inteligente. Un robot que puede ser programado para hacer preferencias contigentes en sensorios de entrada.

*interface. A shared boundary. An interface might be mechanical or electrical connection between two devices; it might be a portion of computer storage accessed by two or more programs; or it might be a device for communication to or from a human operator.

interface. Un limite repartido. Un interface puede ser una conexión mecánica o eléctrica entre dos aparatos; puede ser una porción de deposito con accesión a dos o más programas; o puede ser un aparato para comunicarse a o con un operador humano.

interpass temperature. In multipass weld, the temperature of the weld area between weld passes.

temperatura de pasada interna. En soldaduras de pasadas multiples, la temperatura en la área de la soldadura entre pasadas de soldaduras.

*iron. An element. Very seldom used in its pure form. The most common element alloyed with iron is carbon.

hierro. Un elemento. Es muy raro que se use en forma pura. El elemento más común del aleado con hierro es carbón.

J

J-groove weld. A type of groove weld.

soldadura con ranura-J. Es un tipo de soldadura de ranura.

joint. The junction of members or the edges of members that are to be joined or have been joined.

junta. El punto en que se unen dos miembros o las orillas de los miembros que están para unirse o han sido unidos.

Types of joints

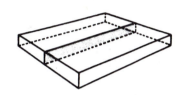

Applicable welds

Square-Groove   J-Groove
V-Groove   Flare-V-Groove
Bevel-Groove   Flare-Bevel-Groove
U-Groove   Edge-Flange
                 Braze

Butt joint

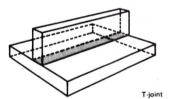

Applicable welds

Fillet   Flare-Bevel-Groove
Square-Groove   Edge-Flange
V-Groove   Corner-Flange
Bevel-Groove   Spot
U-Groove   Projection
J-Groove   Seam
Flare-V-Groove   Braze

Corner joint

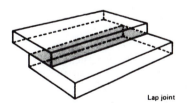

Applicable welds

Fillet   J-Groove
Plug   Flare-Bevel-Groove
Slot   Spot
Square-Groove   Projection
Bevel-Groove   Seam
                 Braze

T-joint

Applicable welds

Fillet   J-Groove
Plug   Flare-Bevel-Groove
Slot   Spot
Bevel-Groove   Projection
                 Seam
                 Braze

Lap joint

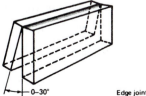

Applicable welds

Square-Groove   Edge-Flange
Bevel-Groove   Corner-Flange
V-Groove   Seam
U-Groove   Edge
J-Groove

0–30°   Edge joint

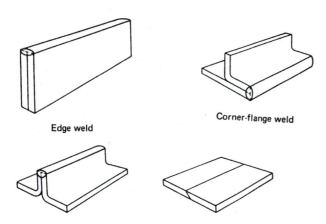

Edge weld

Corner-flange weld

joint buildup sequence. The order in which the weld beads of a multiple-pass weld are deposited with respect to the cross section of the joint.

secuencia de formación de una junta. La orden en la cual los cordones de soldadura en una soldadura de pasadas múltiples son depositadas con respecto a la sección transversa de la junta.

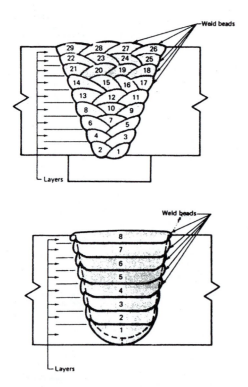

joint clearance. The distance between the faying surfaces of a joint.

despejo de junta. La distancia entre las superficies del empalme de una junta.

joint design. The joint geometry together with the required dimensions of the welded joint.

diseño de junta. La geometría de la junta junto con las dimensiones requeridas de la junta de la soldadura.

joint efficiency. The ratio of strength of a joint to the strength of the base metal, expressed in percent.

eficiencia de junta. La razón de fuerza de una junta a la fuerza del metal base, expresada en por ciento.

joint geometry. The shape and dimensions of a joint in cross section prior to welding.

geometría de junta. La figura y dimensión de una junta en sección transversa antes de soldarse.

joint penetration. The distance the weld metal extends from its face into a joint, exclusive of weld reinforcement.

penetración de junta. La distancia del metal soldado que se extiende de su cara hacia adentro de la junta, exclusiva de la soldadura de refuerzo.

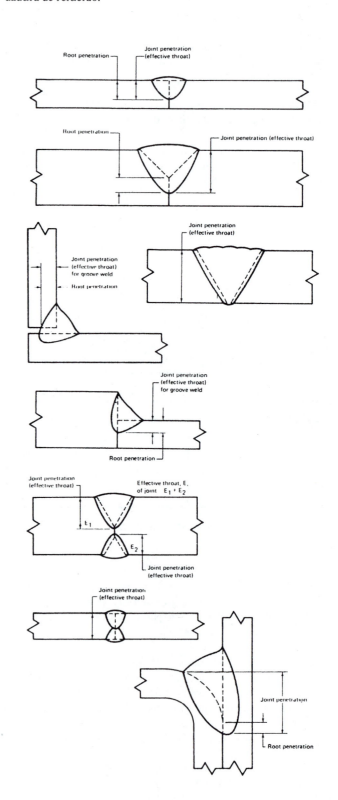

joint root. That portion of a joint to be welded where the members approach closest to each other. In cross section, the joint root may be either a point, a line, or an area.

raíz de junta. Esa porción de una junta que está para soldarse donde los miembros están más cercanos uno del otro. En la sección transversa, la raíz de la junta puede ser una punta, una línea, o una área.

joint type. A weld joint classification based on the five basic arrangements of the component parts such as a butt joint, corner joint, edge joint, lap joint, and T-joint.

tipo de junta. Una clasificación de una junta de soldadura basada en los cinco arreglos del componente de partes como junta a tope, junta en esquina, junta de orilla, junta de solape, y junta en T.

joint welding sequence. See preferred term **joint buildup sequence.**

secuencia para soldar una junta. Vea el término preferido **secuencia de formación de una junta.**

K

kerf. The width of the cut produced during a cutting process. Refer to drawing for **drag.**

cortadura. La anchura del corte producido durante un proceso de cortar. Refiérase al dibujo de **tiro.**

keyhole welding. A technique in which a concentrated heat source penetrates completely through a workpiece, forming a hole at the leading edge of the weld pool. As the heat source progresses, the molten metal fills in behind the hole to form the weld bead.

soldadura con pocillo. Una técnica en la cual una fuente de calor concentrado se penetra completamente a través de la pieza de trabajo, formando un agujero en la orilla del frente del charco de la soldadura. Asi como progresa la potencia de calor, el metal derretido rellena detrás del agujero para formar un cordón de soldadura.

*kindling point. The lowest temperature at which a material will burn.

punto de ignición. La temperatura más baja la cual un material se puede quemar.

L

lack of fusion. A nonstandard term for **incomplete fusion.**
falta de fusión. Un término fuera de norma para **fusión incompleta.**

lack of penetration. A nonstandard term for **incomplete joint penetration.**
falta de penetración. Un término fuera de norma para **penetración de junta incompleta.**

lamellar tear. A subsurface terrace and step-like crack in the base metal with a basic orientation parallel to the wrought surface. It is caused by stresses in the through-thickness direction of the base metals weakened by the presence of small, dispersed, planar shaped, non-metallic inclusions parallel to the metal surface.

rasgadura laminar. Una terraza subsuperficie y una grieta como un escalón en el metal base con una orientación paralela a la superficie forjada. Es causada por tensión en la dirección de lo grueso-continuo de los metales de base debilitados por la presencia

de pequeños, dispersados, formados como plano, inclusiones no metálicas paralelas a la superficie del metal.

land. See preferred term **root face.**
hombro. Vea el término preferido **cara de raíz.**

lap joint. A joint between two overlapping members.
junta de solape. Una junta entre dos miembros traslapadas.

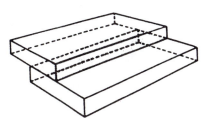

Applicable welds

Fillet	J-Groove
Plug	Flare-Bevel-Groove
Slot	Spot
Bevel-Groove	Projection
	Seam
	Braze

laser beam cutting (LBC). A thermal cutting process that severs metal by locally melting or vaporizing with the heat from a laser beam. The process is used with or without assist gas to aid the removal of molten and vaporized material.

cortes con rayo laser. Un proceso de cortes termal que separa al metal vaporizado o derretido localmente con el calor de un rayo laser. El proceso es usado sin gas que asiste a remover el material vaporizado o derretido.

laser beam welding (LBW). A welding process that produces coalescence with the heat from a laser beam impinging on the joint. The process is used without a shielding gas and without the application of pressure.

soldadura con rayo laser. Un proceso de soldar que produce coalescencia con calor de un rayo laser al golpear contra la junta.

lattice. An orderly geometric pattern of atoms within a solid metal. The lattice structure is responsible for many of the mechanical properties of the metal.

enrejado. Es una forma geométrica bien arreglada de átomos dentro de un metal sólido. La estructura de enrejado es responsable por muchas propiedades mecánicas del metal.

layer. A stratum of weld metal or surfacing material. The layer may consist of one or more weld beads laid side by side. Refer to drawing for **joint buildup sequence.**

capa. Un estrato de metal de soldadura o material de superficie. La capa puede consistir de uno o más cordones de soldadura depositados o puestos de lado. Refiérase al dibujo **secuencia de formación de una junta.**

*leak-detecting solution. A solution, usually soapy water, that is brushed or sprayed on the hose fittings at the regulator and torch to detect gas leaks. If a small leak exists, soap bubbles form.

solución para descubrir escape. Una solución, por lo regular de agua enjabonada, que se acepilla o se rocía sobre las conexiones de las mangueras y los reguladores y antorcha para detectar escape de gas. Si existe un escape pequeño, se forman burbujas de jabón.

leg of a fillet weld. See fillet weld leg.

pierna de soldadura filete. Vea **pierna de soldadura filete.**

lightly coated electrode. A filler metal electrode consisting of a metal wire with a light coating applied subsequent to the drawing operation, primarily for stabilizing the arc.

electrodo con recubrimiento ligero. Un electrodo de metal de aporte consistiendo de un alambre de metal con un recubrimiento ligero aplicado subsecuente a la operación del dibujo, principalmente para estabilizar el arco.

*line drop. The difference between the pressure at the low-pressure gauge and the pressure at the torch; results from the resistance to gas flow offered by the hose, and how it is affected by the diameter and length of the hose. The smaller the hose diameter, or the longer the hose is, the greater is the line drop.

descenso de línea. La diferencia de la presión en el manómetro de baja presión y la presión en la antorcha; resultados de la resistencia de la corriente del gas causada por la manguera, y como es afectada por el diámetro, y lo largo de la manguera. Si el diámetro de la manguera es más chica o es más larga la manguera, más grande es el descenso.

*liquid-solid phase bonding process. Soldering or brazing where the filler metal is melted (liquid) and the base material does not melt (solid); the phase is the state at which bonding takes place between the solid base material and liquid filler metal. There is no alloying of the base metal.

proceso de ligación de fase líquido-sólido. Soldando con soldadura blanda o soldadura fuerte donde el metal de relleno se derrite (líquido) y el material base no se derrite (sólido); la fase es el estado la cual el ligamento se lleva a cabo entre el material base sólido y el metal de relleno (líquido). No se mezcla con el metal base.

liquidus. The lowest temperature at which a metal or an alloy is completely liquid.

liquidus. La temperatura más baja en la cual un metal o un aleado es completamente líquido.

local preheating. Preheating a specific portion of a structure.

precalentamiento local. El precalentamiento de una porción especificada de un estructura.

local stress relief heat treatment. Stress relief heat treatment of a specific portion of a structure.

tratamiento de calor para relevar la tensión local. Un tratamiento de calor el cual releva la tensión de una porción especificada de una estructura.

longitudinal sequence. The order in which the weld passes of a continuous weld are made in respect to its length. See also **back-step sequence.**

secuencia longitudinal. La orden en que las pasadas de un soldadura continua son hechas en respecto a su longitud. Vea también **secuencia a la inversa.**

M

*machine operation. Welding operations are performed automatically under the observation and correction of the operator.

operación de máquina. Operaciones de soldadura son ejecutadas automáticamente bajo la observación y corrección del operador.

machine welding. Welding with equipment that performs the welding operation under the constant observation and control of a welding operator. The equipment may or may not perform the loading and unloading of the work. See **automatic welding.**

máquina para soldadura. Soldadura con equipo que ejecutan la operación de soldadura bajo la observación constante de un operador de soldadura. El equipo pueda o no ejecutar el cargar o descargar del trabajo. Vea **soldadura automática.**

macroetch test. A test in which a specimen is prepared with a fine finish, etched, and examined under low magnification.

prueba con grabado al agua fuerte y examinado por magnificación. Una prueba en una probeta preparada con acabado fino, grabada al agua fuerte, y examinado debajo de un amplificador de aumento bajo.

*macro structure. A structure large enough to be seen with the naked eye or low magnification, usually under 30 power.

estructura macro. Una estructura suficientemente grande que puede verse con el puro ojo o con un amplificador de aumento bajo, regularmente abajo de poder 30.

*magnetic flux lines. Parallel lines of force that always go from the north pole to the south pole in a magnet, and surround a dc current-carrying wire.

líneas magnéticas de flujo. Líneas paralelas de fuerza que siempre van del polo norte al polo sur en un magneto, y rodea un alambre que lleva corriente dc.

manifold. A multiple header for interconnection of gas or fluid sources with distribution points.

conexión múltiple. Una tuberia con conexiones múltiples que sirve como fuente de gas o flúido con puntos de distribución.

*manifold system. Used when there are a number of work stations or a high volume of gas is required. A piping system that allows several oxygen and fuel gas cylinders to be connected to several welding stations. Normally regulators are provided at the manifold and at the stations to provide control of the oxygen and fuel gas pressures. Safety features such as reverse flow valves, flashback arrestors, and back pressure release must be provided at the manifold.

sistema de conexiones múltiples. Usado cuando hay un número de estaciones de trabajo o cuando se requiere un alto volumen de gas. Un sistema de tubos que permite que se conecten varios cilindros de oxígeno y gas combustible a varias estaciones de soldadura. Normalmente se usan reguladores en el tubo de conexiones múltiples y en las estaciones para mantener el control de la presión del oxígeno y el gas combustible. Normas de seguridad como válvulas de retención, protector de agua contra retroceso de llama, y escape de presión deben usarse en el tubo de conexiones múltiples.

*manipulator. A mechanism, usually consisting of a series of segments, joined or sliding relative to one another for the purpose of grasping and moving objects, usually in several degrees of freedom. It may be remotely controlled by a computer or by a human.

manipulador. Un mecanismo, que consiste regularmente de una serie de segmentos unidos o corredizos con relación del uno al otro con el propósito de que agarre y mueva objetos, por lo regular en varios grados de libertad. Puede ser controlado remotamente por una computadora o un humano.

*manual operation. The entire welding process is manipulated by the welding operator.

operación manual. Todo el proceso de soldadura es manipulado por un operador de soldadura.

manual welding. Welding with the torch, gun, or electrode holder held and manipulated by hand. Variations of this term are manual brazing, manual soldering, manual thermal cutting, and manual thermal spraying.

soldadura manual. Soldando con la antorcha, pistola, porta electrodo detenido y manipulado por la mano. Variaciones de este término son soldadura fuerte manual, soldadura blanda manual, cortes termal manual, rociado termal manual.

*MAPP®. One manufacturer's trade name for a specific stabilized, liquefied, MPS mixture. MAPP® has a distinctive odor, which makes it easy to detect; used for welding and cutting. (See methylacetylene propadiene.)

MAPP®. Un nombre comercial de un fabricante para una específica estabilizada, licuada, mezcla MPS. MAPP® tiene un olor distintivo, el cual es muy fácil de descubrir; es usado para cortes y soldaduras. (Vea metilacetileno y propadieno.)

*martensite. A very hard and brittle solid-solution phase that is found in medium- and high-carbon steels.

martensita. Una solución sólida con un aspecto muy duro y quebradizo que se encuentra en aceros medianos y de alto carbón.

meltback time. The time interval at the end of crater fill time to arc outage during which electrode feed is stopped. Arc voltage and arc length increase and current decreases to zero to prevent the electrode from freezing in the weld deposit.

tiempo de refundición. El tiempo de intervalo al fin del tiempo en que se llena el crater hasta que se apaga el arco durante el cual el alimento del electrodo se detiene. El voltaje del arco y lo largo del arco aumenta y la corriente empieza a desminuir hasta llegar a cero para prevenir la congelación del electrodo en el deposito de la soldadura.

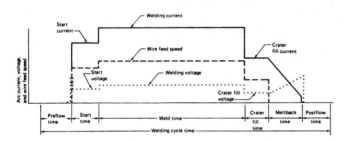

Typical GMAW, FCAW, and SAW program for automatic welding

melting range. The temperature range between solidus and liquidus.

variación de derretimiento. La variación de temperatura entre solidus y liquidus.

melting rate. The weight or length of electrode, wire, rod, or powder, melted in a unit of time.

cantidad de derretimiento. El peso o lo largo de un electrodo, alambre, varilla, o polvo derretido en una unidad de tiempo.

melt-through. Complete joint penetration for a joint welded from one side. Visible root reinforcement is produced.

derretir de un lado a otro. Una junta con penetración completa para una junta que está soldada de un lado. Refuerzo de raíz visible es producido.

metal. An opaque, lustrous, elemental, chemical substance that is a good conductor of heat and electricity, usually malleable, ductile, and more dense than other elemental substance.

metal. Una opaca, brillante, elemental, substancia química que es una buena conductora de calor y electricidad, por lo regular es maleable, ductil, y es más densa que otras substancias elementales.

metal arc cutting (MAC). Any of a group of arc cutting processes that sever metals by melting them with the heat of an arc between a metal electrode and the base metal. See shielded metal arc cutting and gas metal arc cutting.

cortes de metal con arco. Cualquiera de un grupo de procesos de cortes con arco que corta metales derritiéndolos con el calor de un arco entre un electrodo de metal y el metal base. Vea cortes de arco metálico protegido y cortes de arco metálico con gas.

metal cored electrode. A composite tubular filler metal electrode consisting of a metal sheath and a core of various powdered materials, producing no more than slag islands on the face of a weld bead. External shielding may be required.

electrodo de metal de núcleo. Un electrodo de metal para rellenar tubular compuesto consistiendo de una envoltura de metal y núcleo de varios materiales en polvo, que producen nada más que islas de escoria en la cara del cordón de soldadura. Protección externa puede ser requerida.

metal electrode. A filler or non-filler metal electrode used in arc welding or cutting, which consists of a metal wire or rod that has been manufactured by any method and that is either bare or covered with a suitable covering or coating.

electrodo de metal. Un electrodo de metal que se usa para rellenar o para no rellenar la soldadura de arco o para cortar, que consiste de un alambre de metal o varilla que ha sido fabricada por cualquier método ya sea liso o cubierto con un cubierto o revestimiento propio.

*methylacetylene propadiene (MPS). A family of fuel gases that are mixtures of two or more gases (propane, butane, butadiene, methyacetylene, and propadiene). The neutral flame temperature is approximately 5,031°F (2,927°C), depending upon the actual gas mixture. MPS is used for oxyfuel cutting, heating, brazing and metallizing; rarely used for welding.

metilacetileno y propadieno. Una familia de gases de combustión que son mezclas de dos o más gases (propano, butano, butadiano, metilacetileno, propadieno). La temperatura de la llama natural es aproximadamente 5,031°F (2927°C), dependiendo de la mezcla actual del gas. MPS es usado como gas de combustión para cortar, calentar, soldadura fuerte, y metalizar; es muy raro que se use para soldar.

*micro structure. A structure that is visible only with high magnification or with the aid of a microscope.

estructura micronesia. Una estructura que es visible solamente con un amplificador de poder muy alto o con la ayuda de un microscopio.

*microcomputer. A computer that uses a microprocessor as its basic element.

computadora micronesia. Una computadora que usa un procesor micronesio como su elemento básico.

microetch test. A test in which the specimen is prepared with a polished finish, etched, and examined under high magnification.

prueba con grabado al agua fuerte y examinada por un amplificador de alto poder. Una prueba en una probeta preparada con acabado fino, grabada al agua fuerte y examinado bajo un amplíficador de alto poder.

*microprocessor. The principal processing element of a microcomputer, made as a single, integrated circuit.

procesor micronesio. El elemento principal de un procedimiento de una computadora micronesia, hecha con un solo circuito integrado.

mineral-based electrode fluxes. Fluxes that use inorganic compounds such as the rutile-based flux (titanium dioxide TiO^2). These mineral compounds do not contain hydrogen, and electrodes that use these fluxes are often referred to as low-hydrogen electrodes. Less smoke is generated with this welding electrode than with cellulose-based fluxes, but a thicker slag layer is deposited on the weld. E7018 is an example of an electrode that uses this type of flux.

fundentes para electrodos de base mineral. Fundentes que usan compuestos inorgánicos, como por ejemplo, el fundente a base de rutilo (bióxido de titanio TI02). Estos compuestos minerales no contienen hidrógeno, y a los electrodos que usan estos fundentes se los llama con frecuencia electrodos de bajo hidrógeno. En la soldadura con electrodos se producen menos humos que en la que se realiza con fundentes celulósicos, pero se deposita una capa de escoria más gruesa en la soldadura. El F7018 es un ejemplo de un electrodo que usa este tipo de fundente.

mixing chamber. That part of a welding or cutting torch in which a fuel gas and oxygen are mixed.

cámara mezcladora. Esa parte de una antorcha para soldar y cortar por la cual el gas combustible y el oxígeno son mezclados.

mold. A high-temperature container into which liquid metal from the thermite welding process is poured and held until it cools and hardens into the container's interior shape.

molde. un contenedor de alta temperatura en el cual se vierte y se mantiene metal líquido del proceso de soldadura con termita hasta que éste se enfríe y se solidifique tomando la forma interior del contenedor.

molecular hydrogen. A bonded pair of hydrogen atoms (H$_2$). This is the configuration that all hydrogen atoms try to form.

hidrógeno molecular. Un par de átomos de hidrógeno (H2) unidos. Ésta es la configuración que tratan de formar todos los átomos de hidrógeno.

molten weld pool. The liquid state of a weld prior to solidification as weld material.

charco de soldadura derretido. El estado líquido de una soldadura antes de solidificarse como material de soldadura.

*multipass weld. Weld requiring more than one pass to ensure complete and satisfactory joining of the metal pieces.

soldadura de pasadas múltiples. Soldadura que requiere más de una pasada para asegurar una completa y satisfactoria unión de las piezas de metal.

N

neutral flame. An oxyfuel gas flame that has characteristics neither oxidizing nor reducing. Refer to drawing for cone.

llama neutral. Una llama de gas oxicombustible que no tiene características de oxidación ni de reducción. Refiérase al dibujo para cono.

nonconsumable electrode. An electrode that does not provide filler metal.

electrodo no consumible. Un electrodo que no provee metal de relleno.

noncorrosive flux. A soldering flux that in neither its original nor residual form chemically attacks the base metal. It usually is composed of rosin or resin-base materials.

flujo no corrosivo. Un flujo para soldadura blanda que ni en su forma original ni en su forma restante químicamente ataca el metal base. Regularmente es compuesto de materiales de colofonia o resino de base.

*nondestructive testing. Methods that do not alter or damage the weld being examined; used to locate both surface and internal defects. Methods include visual inspection, penetrant inspection, magnetic particle inspection, radiographic inspection, and ultrasonic inspection.

pruebas no destructivas. Métodos que no alteran ni dañan la soldadura que se está examinando. Se usa para encontrar ambos defectos internos y de superficie. Incluye métodos como inspección visual, inspección penetrante, inspección de partículas magnéticas, inspección de radiografía, inspección ultrasónica.

nontransferred arc. An arc established between the electrode and the constricting nozzle of the plasma arc torch or thermal spraying. The workpiece is not in the electrical circuit. Also see transferred arc.

arco no transferible. Un arco establecido entre el electrodo y la boquilla constrictiva de la antorcha del arco de plasma o pistola termal para rociar. La pieza de trabajo no está en el circuito eléctrico. Vea también arco transferido.

nozzle. A device that directs shielding media.

boquilla. Un aparato que dirige el medio de protección.

nugget. The weld metal joining the workpieces in spot, seam, or projection welds.

botón. El metal de soldadura que une a las piezas de trabajo en soldadura de puntos, costura, o proyección de soldaduras.

nugget size. The diameter or width of the nugget measured in the plane of the interface between the pieces joined.

tamaño del botón. El diámetro o lo ancho del botón medido en el plano del interfaze entre las piezas unidas.

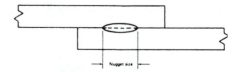

O

open circuit voltage. The voltage between the output terminals of the power source when no current is flowing to the torch or gun.

voltaje de circuito abierto. El voltaje entre los terminales de salida de una fuente de poder cuando la corriente no está corriendo a la antorcha o pistola.

open root joint. An unwelded joint without backing or consumable insert.

junta de raíz abierta. Una junta que no está para soldarse sin respaldo o inserto consumible.

orifice. See constricting orifice.

orifice. Vea orifice de constreñimiento.

orifice gas. The gas that is directed into the plasma arc torch or thermal spraying gun to surround the electrode. It becomes ionized in the arc to form the arc plasma and issues from the constricting orifice of the nozzle as a plasma jet.

gas para orifice. El gas que es dirigido dentro de la antorcha de plasma o la pistola de rociado termal para rodear el electrodo. Se vuelve ionizado dentro del arco para formar el arco de plasma y sale de la orifice de constreñimiento a la boquilla como chorro de plasma.

orifice throat length. The length of the constricting orifice in the plasma arc torch or thermal spraying gun.

largo de garganta del orifice. Lo largo de la orifice constreñida en la antorcha de plasma o en la pistola de plasma para rociar.

*outer envelope. The outer boundary of the oxyacetylene flame. The secondary combustion reaction occurs in the outer envelope.

envoltura externa. El límite de afuera de la llama de oxiacetileno. La reacción de la combustión secundaria ocurre en la envoltura externa.

*out of position welding. Any welding position other than the flat position; includes vertical, horizontal, or overhead positions.

soldadura fuera de posición. Cualquier posición de soldadura menos la de la posición plana; incluye vertical, horizontal, o posiciones de sobrecabeza.

overhead position. The position in which welding is performed from the underside of the joint.

posición de sobrecabeza. La posición en la cual se hace la soldadura por el lado de abajo de la junta.

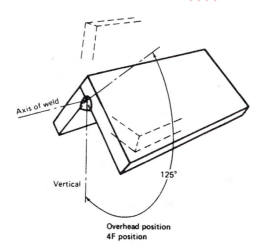

Overhead position
4F position

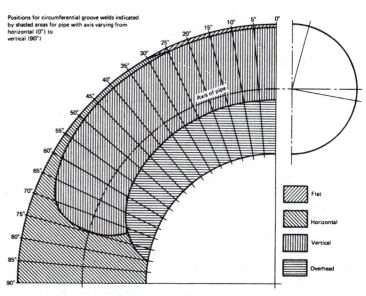

Positions of pipe welds

overlap. The protrusion of weld metal beyond the toe, face, or root of the weld; in resistance seam welding, the area in the preceding weld remelted by the succeeding weld.

traslapo. El metal de la soldadura que sobresale más allá del pie, cara, o de la raíz de una soldadura; en soldaduras de costuras por resistencia, la área de la soldadura anterior se rederrite por la soldadura subsiguiente.

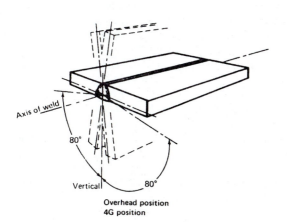

Overhead position
4G position

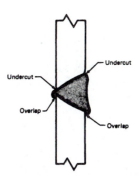

oxidizing flame. An oxyfuel gas flame in which there is an excess of oxygen, resulting in an oxygen-rich zone extending around and beyond the cone.

llama oxidante. Una llama de gas oxicombustible en la cual hay un exceso de oxígeno, resultando en una zona rica de oxígeno extendiéndose alrededor y más allá del cono.

oxyacetylene cutting (OFC-A). An oxyfuel cutting process variation that uses acetylene as the gas.

cortes con oxiacetileno. Un proceso de cortes con gas con variación que usa acetileno como gas de combustión.

*oxyacetylene hand torch. Most commonly used oxyfuel gas cutting torch; may be a cutting torch only or a combination welding and cutting torch set. On the combination set different tips can be attached to the same torch body. The torch mixes the oxygen and fuel gas and directs the mixture to the tip. The torch can be an equal pressure type (equal pressures of oxygen and fuel gas) or an injector type (equal pressures of high-pressure oxygen and low-pressure fuel gas).

antorcha de mano oxiacetileno. La antorcha de gas oxicombustible es la que se usa más frecuentemente; puede ser una antorcha para hacer cortes solamente o una combinación de un juego de antorcha para soldar y hacer cortes. En el juego de combinación diferentes puntas pueden ser conectadas al mismo mango de la antorcha. La antorcha mezcla el oxígeno y gas combustible y dirige la mezcla a la punta. La antorcha puede ser de tipo de presión igual (presiones iguales de oxígeno y gas combustible) o de tipo inyector (presiones iguales de oxígeno de alta presión y baja presión de gas combustible).

oxyacetylene welding (OAW). An oxyfuel gas welding process that uses acetylene as the fuel gas. The process is used without the application of pressure.

soldadura con oxiacetileno (OAW). Un proceso de soldadura de gas oxicombustible que usa acetileno con gas de combustión. El proceso se usa sin aplicación de presión.

*oxyfuel flame. A flame resulting from the combustion of oxygen mixed with a fuel gas. This intense flame is applied to two pieces of metal to cause them to melt to form weld pools. When the edges of the weld pools run together and fuse, the two pieces of metal are joined.

llama oxicombustible. Una llama que resulta de una combustión de oxígeno mezclado con gas combustible. Está llama intensa es aplicada a dos piezas de metal para hacer que se derritan para formar charcos de soldadura. Cuando las orillas de los charcos de soldadura se juntan y se derriten, las dos piezas de metal se unen.

oxyfuel gas cutting (OFC). A group of oxygen cutting processes that uses heat from an oxyfuel gas flame. See oxygen cutting, oxyacetylene cutting, oxyhydrogen cutting, and oxypropane cutting.

gas para cortar oxicombustible. Un grupo de procesos para cortar con oxígeno que usa calor de una llama de gas oxicombustible. Vea cortes con oxigeno, cortes con oxiacetileno, cortes con oxihidrógeno, y cortes con oxipropano.

oxyfuel gas cutting torch. A device used for directing the preheating flame produced by the controlled combustion of fuel gases and to direct and control the cutting oxygen.

antorcha para cortes de gas oxicombustible. Un aparato que se usa para dirigir la llama precalentada producida por la combustión controlada de los gases de combustión y para dirigir y controlar el oxígeno para cortar.

oxyfuel gas welding (OFW). A group of welding processes that produces coalescence of workpieces by heating them with an oxyfuel gas flame. The processes are used with or without the application of pressure and with or without filler metal.

soldadura con gas oxicombustible (OFW). Un grupo de procesos de soldadura que produce coalescencia de las piezas de trabajo calentándolas con una llama de gas oxicombustible. Los procesos son usados sin la aplicación de presión y con o sin el metal para rellenar.

oxyfuel gas welding torch. A device used in oxyfuel gas welding, torch brazing, and torch soldering for directing the heating flame produced by the controlled combustion of fuel gases.

antorcha para soldar con gas oxicombustible. Un aparato que se usa para soldar con gas oxicombustible, soldadura fuerte con antorcha, soldadura blanda con antorcha y para dirigir la llama calentada producida por combustión controlada de gases de combustión.

oxygen arc cutting (AOC). An oxygen cutting process that uses an arc between the workpiece and a consumable tubular electrode, through which oxygen is directed to the workpiece.

cortes de oxígeno con arco (AOC). Es un proceso de cortar con oxígeno que usa un arco entre la pieza de trabajo y un electrodo tubular consumible, por el cual el oxígeno es dirigido a la pieza de trabajo.

oxygen cutting (OC). A group of thermal cutting processes that severs or removes metal by means of the chemical reaction between oxygen and the base metal at elevated temperature. The necessary temperature is maintained by the heat from an arc, an oxyfuel gas maintained by the heat from an arc, an oxyfuel gas flame, or other sources. See oxyfuel gas cutting.

cortes con oxígeno. Un grupo de procesos termales que corta y quita el metal por medio de una reacción química entre el oxígeno y el metal base a una temperatura elevada. La temperatura necesaria es mantenida por el calor del arco, un gas oxicombustible mantenido por el calor del arco, una llama de gas oxicombustible, o de otras fuentes. Vea gas para cortar oxicombustible.

oxygen gouging. Thermal gouging that uses an oxygen cutting process variation to form a bevel or groove.

escopleando con la gubia con oxígeno. Gubia termal que usa un proceso de variación de corte con oxígeno para formar un bisel o ranura.

oxygen lance. A length of pipe used to convey oxygen to the point of cutting in oxygen lance cutting.

lanza de oxígeno. Un tramo de tubo usado para conducir oxígeno al punto de cortar en cortes con lanza de oxígeno.

oxygen lance cutting (LOC). An oxygen cutting process that uses oxygen supplied through a consumable lance. The preheat to start the cutting is obtained by other means.

cortes con lanza de oxígeno. Un proceso de cortar con oxígeno que usa oxígeno surtido por una lanza consumible. El precalentamiento para empezar a cortar es obtenido por otros medios.

oxyhydrogen cutting (OFC-H). An oxyfuel gas cutting process variation that uses hydrogen as the fuel gas.

cortes con oxihidrógeno. Un proceso de cortar de gas oxicombustible con variación que usa hidrógeno como gas combustible.

*oxyhydrogen flame. A specific flame resulting from the combustion of oxygen and hydrogen: consists of primary combustion region only; used for welding and cutting.

llama oxihidrógeno. Una llama específica que resulta de la combustión del oxígeno e hidrógeno; consiste solamente de la región de combustión primaria; usada para cortar y soldar.

oxyhydrogen welding (OHW). An oxyfuel gas welding process that uses hydrogen as the fuel gas. The process is used without the application of pressure.

soldadura oxihidrógeno (OHW). Un proceso de soldar con gas oxicombustible que usa hidrógeno como gas de combustible. El proceso se usa sin la aplicación de presión.

oxypropane cutting (OFC-P). An oxyfuel gas cutting process variation that uses propane as the fuel gas.

cortes con oxipropano. Un proceso de cortar con gas combustible con variación que usa propano como gas combustible.

P

parent metal. See preferred term base metal.

metal de origen. Vea término preferido metal base.

*paste range. The temperature range of soldering and brazing filler metal alloys in which the metal is partly solid and partly liquid as it is heated or cooled.

grados de la pasta. Los grados de la temperatura del metal para rellenar aleados para soldadura blanda o soldadura fuerte cuando se calienta o se enfria.

peel test. A destructive method of inspection that mechanically separates a lap joint by peeling.

prueba por pelar. Un método de inspección destructivo de pelar que separa mecánicamente una junta de solape.

peening. The mechanical working of metals using impact blows.

martillazos (con martillo de bola). Metales que se trabajan mecánicamente con golpes de impacto.

*penetration. The depth into the base metal (from the surface) that the weld metal extends, excluding any reinforcement.

penetración. La profundidad de adentro del metal base (de la superficie) que el metal de soldadura se extiende, excluyendo cualquier refuerzo.

percussion welding (PEW). A welding process that produces coalescence with an arc resulting from a rapid discharge of electrical energy. Pressure is applied percussively during or immediately following the electrical discharge.

soldadura a percusión (PEW). Un proceso de soldadura que produce coalescencia con un arco resultando de una descarga rápida de energia. Presión es aplicada a percusión durante o inmediatamente después de la descarga eléctrica.

*phase diagram. Provides information on the crystalline constituents of metal alloys at different temperatures in three different phases: pure metal, solid solutions of two or more metals, and intermetallic compounds.

diagrama de equilibrio. Proporciona información en los constituyentes cristalinos de metales aleados a diferentes temperaturas en tres aspectos: metal puro, soluciones sólidas de dos o más metales, y mezclas intermetálicas.

*pick-and-lace robot. A simple robot, often with only two or three degrees of freedom, which transfers items from place to place by means of point-to-point moves. Little or no trajectory control is available. Often referred to as a "bank-bank" robot.

robot de escoger y atar. Un robot simple, frecuentemente con solo dos o tres grados de libertad, el cual traslada artículos de un lugar a otro por medio de movidas de punto a punto. Un poco o nada de control trayectoria es utilizado. A veces es referido como un"banco-banco" robot.

pilot arc. A low current arc between the electrode and the constricting nozzle of the plasma arc torch to ionize the gas and facilitate the start of the welding arc.

piloto del arco. Un arco de corriente baja en medio del electrodo y la boquilla constreñida de la antorcha de arco de plasma para ionizar el gas y facilitar el arranque del arco para soldar.

*pitch. The angular rotation of a moving body about an axis perpendicular to its direction of motion and in the same plane as its top side.

grado de inclinación. La rotación angular de un cuerpo en movimiento alrededor de un eje perpendicular a su dirección y en el mismo plano como el del lado de arriba.

plasma. A gas that has been heated to an at least partially ionized condition, enabling it to conduct an electric current.

plasma. Un gas que ha sido calentado a lo menos parcialmente a una condicón ionizada permitiendo que conduzca una corriente eléctrica.

plasma arc cutting (PAC). An arc cutting process that uses a constricted arc and removes the molten metal with a high velocity jet of ionized gas issuing from the constricting orifice.

cortes con arco de plasma. Un proceso de cortar con el arco que usa un arco constreñido y quita el metal derretido con un chorro de alta velocidad de gas ionizado que sale de la orifice constringente.

plasma arc welding (PAW). An arc welding process that uses a constricted arc between a nonconsumable electrode and the weld pool (transferred arc) or between the electrode and the constricting nozzle (nontransferred arc). Shielding is obtained from the ionized gas issuing from the torch, which may be supplemented by an auxiliary source of shielding gas. The process is used without the application of pressure.

soldadura con arco de plasma (PAW). Un proceso de soldadura de arco que usa un arco constreñido entre un electrodo que no se consume y el charco de la soldadura (arco transferido) o entre el electrodo y la lanza constreñida (arco no transferido). La protección es obtenida del gas ionizado que sale de la antorcha, el cual puede ser suplementado por una fuente auxiliar de gas para protección. El proceso es usado sin la aplicación de presión.

plug weld. A weld made in a circular hole in one member of a joint fusing that member to another member. A fillet-welded hole should not be construed as conforming to this definition.

soldadura de tapón. Una soldadura que se hace en un agujero circular en un miembro de una junta uniendo ese miembro con otro miembro. Un agujero de soldadura de filete no debe ser interpretado como confirmación de está definición.

(continued on page 866)

(continued from page 865)

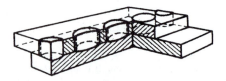

*point-to point control. A control scheme whereby the inputs or commands specify only a limited number of points along a desired path of motion. The control system determines the intervening path segments.

control de punto a punto. Una esquema de control con que las entradas o las ordenes especifican solamente un número limitado de puntos a lo largo de la senda de moción deseada. El sistema de control determina el intervenio de los segmentos de la senda.

porosity. Cavity type discontinuities formed by gas entrapment during solidification or in a thermal spray deposit.

porosidad. Un tipo de cavidad de desuniones formadas por gas atrapado durante la solidificación o en un deposito rociado termal.

postflow time. The time interval from current shut-off to shielding gas and/or cooling water shut-off.

tiempo de poscorriente. El intervalo de tiempo de cuando se cierra la corriente a cuando se cierra el gas de protección y o cuando se cierra el agua para enfriar.

postheat current (resistance welding). The current through the welding circuit during postheat time in resistance welding.

corriente de poscalentamiento (soldadura de resistencia). La corriente que va de un lado a otro del circuito durante el tiempo de poscalentamiento en la soldadura de resistencia.

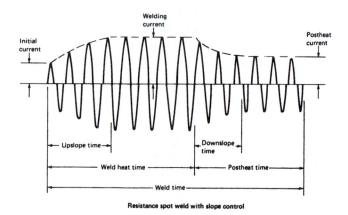

Resistance welding current characteristics

postheat time (resistance welding). The time from the end of weld heat time to the end of weld time. Refer to drawing for downslope time.

tiempo de poscalentamiento (soldadura de resistencia). El tiempo del fin del calor de la soldadura al tiempo al fin del tiempo de la soldadura. Refiérase al dibujo de **tiempo de cadía del pendiente.**

postheating. The application of heat to an assembly after welding, brazing, soldering, thermal spraying, or thermal cutting. See **postweld heat treatment.**

poscalentamiento. La aplicación de calor a una asamblea después de la soldadura, soldadura fuerte, soldadura blanda, rociado termal o corte termal. Vea **tratamiento de calor postsoldadura.**

*postpurge. Once welding current has stopped in gas tungsten arc welding, this is the time during which the gas continues to flow to protect the molten pool and the tungsten electrode as cooling takes place to a temperature at which they will not oxidize rapidly.

pospurgante. Cuando la corriente de soldar se ha dentenido en la soldadura de arco gas tungsteno, este es el tiempo durante en que el gas continua a salir para proteger el charco de soldadura derretido y el electrodo de tungsteno se enfrian a una temperatura donde no se oxidan rápidamente.

postweld heat treatment. Any heat treatment subsequent to welding.

tratamiento de calor postsoldadura. Cualquier tratamiento de calor subsiguiente a la soldadura.

powder flame spraying. A thermal spraying process variation in which the material to be sprayed is in powder form. See **flame spraying (FLSP).**

rociado de polvo con llama. Un proceso termal para rociar con variación el cual el material que está para rociar se está en forma de polvo. Vea **rociado a llama .**

power source. An apparatus for supplying current and voltage suitable for welding, thermal cutting, or thermal spraying.

fuente de poder. Un aparato para surtir corriente y voltaje conveniente para soldar, para hacer cortes termales, o rociado termal.

preheat. The heat applied to the base metal or substrate to attain and maintain preheat temperature.

precalentamiento. El calor aplicado al metal base o substrato para obtener y mantener temperatura de precalentamiento.

*preheat flame. Brings the temperature of the metal to be cut above its kindling point, after which the high-pressure oxygen stream causes rapid oxidation of the metal to perform the cutting.

llama para precalentamiento. Sube la temperatura del metal que está para cortarse a una temperatura de encendimiento, después que la corriente del oxígeno de alta presión cause una oxidación rápida del metal para hacer el corte.

*preheat holes. The cutting tip has a central hole through which the oxygen flows. Surrounding this central hole are a number of other holes called preheat holes. The differences in the type or number of preheat holes determine the type of fuel gas to be used in the tip.

agujeros para precalentamiento. La boquilla para cortar tiene un agujero central por donde corre el oxígeno. Rodeando este agujero central hay un numero de otros agujeros que se llaman agujeros para precalentar. Las diferencias en el tipo o número de agujeros percalentados determina el tipo de gas combustible que se usará en la boquilla.

preheat temperature. The temperature of the base metal or substrate in the welding, brazing, soldering, thermal spraying, or thermal cutting area immediately before these operations are performed. In a multipass operation, it is also the temperature in the area immediately before the second and subsequent passes are started.

temperatura de precalentamiento. La temperatura del metal base o substrato en la soldadura, soldadura fuerte, soldadura blanda, rociado termal, o en la área de los cortes termal inmediatamente antes de que estas operaciones sean ejecutadas. En una operación multipasada, es también la temperatura en la área inmediatamente antes de empezar la segunda pasada y pasadas subsiguientes.

preheating. The application of heat to the base metal immediately before welding, brazing, soldering, thermal spraying, or cutting.

precalentamiento. La aplicación de calor al metal base inmediatamente antes de la soldadura, soldadura fuerte, soldadura blanda, rociado termal o cortes.

*prepurge. In gas tungsten arc welding, the time during which gas flows through the torch to clear out any air in the cup or surrounding the weld zone. Prepurge time is set by the operator and is completed before the welding current is started.

prepurgar. En soldadura de arco de tungsteno con gas, el tiempo durante el cual el gas corre por la antorcha para quitar el aire en la boquilla o la zona de soldadura. El tiempo de prepurgar es determinado por el operador y es acabado antes de que la corriente de soldadura es empezada.

*primary combustion. The first reaction in the chemical reaction resulting when a mixture of acetylene and oxygen is ignited. This reaction frees energy and forms carbon monoxide (CO) and free hydrogen.

combustión primaria. La primera reacción en una reacción química resulta cuando una mezcla de oxígeno y acetileno es encendida. Está reacción libra la energia y forma carbón monóxido (CO) e hidrógeno libre.

procedure qualification. The demonstration that welds made by a specific procedure can meet prescribed standards.

calificación de procedimiento. La demostración en que las soldaduras hechas por un procedimiento específico conformen con las normas prescritas.

projection weld. A weld made by projection welding.

soldadura de proyección. Una soldadura hecha con soldadura de proyección.

protective atmosphere. A gas envelope surrounding the part to be brazed, welded, or thermal sprayed, with the gas composition controlled with respect to chemical composition, dew point, pressure, flow rate, etc. Examples are inert gases, combusted fuel gases, hydrogen, and vacuum.

atmósfera protectora. Una envoltura de gas que está alrededor de la parte que está para soldarse con soldadura fuerte, soldadura o rociada termal, con la composición del gas controlado con respecto a la química compuesta, punto de rocío, presión, cantidad de corriente, etc. Ejemplos son gas inerto, gases de combustión que ya están encendidos, hidrógeno, y vacuo.

*proximity sensor. A device that senses that an object is only a short distance (e.g., a few inches or feet) away, and/or measures how far away it is. Proximity sensors work on the principles of triangulation of reflected light, lapsed time for reflected sound, or intensity induced eddy currents, magnetic fields, back pressure from air jets, and others.

sensor de proximidad. Un aparato que siente que un objeto está solamente a una corta distancia (e.g., unas pulgadas o pies) afuera, y/o mide que tan lejos está. Sensores de proximidad trabajan en los fundamentos de triangulación de luz reflejada, tiempo lapso del sonido reflejado, o en las corrientes de Fancault inducidas con intensidad, campos magnéticos, contrapresión trasera del chorro de aire, y otras.

puddle. See preferred term **weld pool.**

charco. Vea término preferido **charco de soldadura.**

*pulsed arc metal transfer. In gas metal arc welding, pulsing the current from a level below the transition current to a level above the transition current to achieve a controlled spray transfer at lower average currents; spray transfer occurs at the higher current level.

transferir el metal por arco pulsado. En la soldadura de arco metálico con gas, se pulsa la corriente de un nivel más alto de la corriente de transición para lograr un traslado de rocío controlado a una corriente media baja; el traslado del rocío ocurre al nivel más alto de la corriente.

pulse start delay time. The time interval from current initiation to the beginning of current pulsation, if pulsation is used. Refer to drawing for **upslope time.**

tiempo de dilación en empezar la pulsación. El tiempo del intervalo de donde se inicia la corriente al principio de la pulsación de la corriente, si es que se use pulsación. Refiérase al dibujo de **tiempo del pendiente en ascenso.**

*purge. The process of opening first one cylinder valve and then the other to replace all air in the hoses with the appropriate gas prior to welding.

limpiar. El proceso de abrir primero una válvula de un cilindro y luego el otro para reemplazar todo el aire en las mangueras con un gas apropiado antes de empezar a soldar.

push angle. The travel angle when the electrode is pointing in the direction of weld progression. This angle can also be used to partially define the position of guns, torches, rods, and beams.

ángulo de empuje. El ángulo de avance cuando el electrodo apunta en la dirección en que la soldadura progresa. Este ángulo también puede ser usado para parcialmente definir la posición de pistolas, antorchas, varillas, y rayos.

push weld (resistance welding). A spot or projection weld made by push welding.

soldadura de empuje (soldadura de resistencia). Una soldadura de botón o proyección hecha por soldadura de empuje.

Q

qualification. See preferred terms **welder performance qualification** and **procedure qualification.**

calificación. Vea el término preferido **calificación de ejecución del soldador** y **calificación de procedimiento.**

R

reactor (arc welding). A device used in arc welding circuits for the purpose of minimizing irregularities in the flow of welding current.

reactor (soldadura de arco). Un aparato usado en los circuitos de la soldadura de arco con el propósito de reducir a lo mínimo las irregularidades en la manera que corre la corriente de soldadura de arco.

reduced section tension test. A test in which a transverse section of the weld is located in the center of the reduced section of the specimen.

prueba de tensión de sección reducida. Una prueba en la cual la sección transversa de la soldadura está ubicada en el centro de la sección reducida de la probeta.

reducing atmosphere. A chemically active protective atmosphere, which at elevated temperature will reduce metal oxides to their metallic state. (Reducing atmosphere is a relative term, and such an atmosphere may be reducing to one oxide but not to another oxide.)

atmósfera de reducción. Una atmósfera protectiva activa, la cual a una temperatura elevada reduce los óxidos del metal a sus estados metalicos. (Atmósfera de reducción es un término relativo, y cierta atmósfera puede reducir a un óxido pero no al otro óxido.)

reducing flame. An oxyfuel gas flame with an excess of fuel gas.

llama de reducción. Una llama de gas oxicombustible con un exceso de gas combustible.

regulator. A device for controlling the delivery of gas at some substantially constant pressure.

regulador. Un aparato para controlar la expedición de gas a una presión substancialmente constante.

residual stress. Stress present in a joint member or material that is free of external forces or thermal gradients.

fuerza residual. Fuerza presente en un miembro de una junta o material que está libre de fuerzas externas o ambulantes termales.

resistance brazing (RB). A brazing process that uses heat from the resistance to electric current flow in a circuit of which the workpieces are a part.

soldadura fuerte por resistencia. Un proceso de soldadura fuerte que usa calor de la resistencia al correr de la corriente eléctrica en un circuito en las cuales las piezas de trabajo forman parte.

resistance soldering (RS). A soldering process that uses heat from the resistance to electric current flow in a circuit of which the workpieces are a part.

soldadura blanda por resistencia. Un proceso de soldadura blanda que usa calor de la resistencia al correr de la corriente eléctrica en un circuito en las cuales las piezas de trabajo forman parte.

resistance spot welding (RSW). A resistance welding process that produces a weld at the faying surfaces of a joint by the heat obtained from resistance to the flow of welding current through the workpieces from electrodes that serve to concentrate the welding current and pressure at the weld area.

soldadura de puntos por resistencia. Un proceso de soldar por resistencia que produce una soldadura en los empalmes de la superficie de una junta por el calor obtenido de la resistencia al correr la corriente a través de las piezas de trabajo de los electrodos que sirven para concentrar la corriente para soldar y la presión en la área de la soldadura.

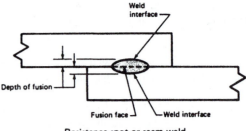

Resistance spot or seam weld

resistance welding (RW). A group of welding processes that produces coalescence of the faying surfaces with the heat obtained from resistance of the workpieces to the flow of the welding current in a circuit of which the workpieces are a part, and by the application of pressure.

soldadura por resistencia. Un grupo de procesos para soldar que producen coalescencia de las superficies empalmadas con el calor obtenido de la resistencia de las piezas de trabajo al correr la corriente de soldadura en un circuito en las cuales las piezas de trabajo forman parte, y por la aplicación de presión.

resistance welding downslope time. The time during which the welding current is continuously decreasing.

tiempo del pendiente de descenso en soldadura de resistencia. El tiempo durante el cual la corriente está continuamente disminuyendo.

resistance welding electrode. The part of a resistance welding machine through which the welding current and, in most cases, force are applied directly to the workpiece. The electrode may be in the form of a rotating wheel, rotating roll, bar, cylinder, plate, clamp, chuck, or modification thereof.

electrodo para soldadura por resistencia. La parte de una máquina para soldar por resistencia por cual la corriente de soldar y, en muchos casos, la fuerza es aplicada directamente a la pieza de trabajo. El electrodo puede ser en la forma de una rueda que da vueltas, rollo rotativo, barra, cilindro, plato, empalme, calzo, o modificación de ello.

reverse polarity. The arrangement of direct current arc welding leads with the work as the negative pole and the electrode as the positive pole of the welding arc. A synonym for **direct current electrode.** Refer to drawing for **direct current electrode positive.**

polaridad invertida. El arreglo de los cables para soldar con el arco con corriente directa con el cable de tierra como el polo negativo y el electrodo como polo positivo del arco para soldar. Un sinónimo para **corriente directa electrodo.** Refiérase al dibujo para **corriente directa con el electrodo positivo.**

*robot. A reprogrammable, multifunctional manipulator designed to move material, parts, tools, or specialized devices through variable programmed motions for the performance of a variety of tasks.

robot. Un manipulador reprogramable, multifuncional diseñado para mover material, partes, herramienta, o aparatos especializados por medio de mociones programadas variables para la ejecución de una variedad de tareas.

*robot programming language. A computer language especially designed for writing programs for controlling robots.

lenguaje para programación del robot. Un lenguaje para computadoras con un diseño especial para escribir programas para el control de los robots.

root. See preferred terms of **root of joint** and **root of weld**.

raíz. Vea el término preferido de **raíz de junta** y **raíz de soldadura**.

root bead. A weld bead that extends into, or includes part or all of the joint root.

cordón de raíz. Un cordón de soldadura que se extiende adentro, o incluye parte o toda la junta de raíz.

root bend test. A test in which the weld root is on the convex surface of a specified bend radius.

prueba de dobléz de raíz. Una prueba en la cual la raíz de la soldadura está en una superficie convexa de un radio especificado para el dobléz.

root crack. A crack in the weld or heat-affected zone occurring at the root of a weld.

grieta de raíz. Una grieta en la soldadura o en la zona afectada por el calor que ocurre en la raíz de la soldadura.

root edge. A root face of zero width. See **root face**. Refer to drawing for **groove face**.

orilla de raíz. Una cara de raíz con una anchura de cero. Vea **cara de raíz**. Refiérase al dibujo para **cara de ranura**.

root face. That portion of the groove face adjacent to the root of the joint. Refer to drawing for **groove face**.

cara de raíz. La porción de la cara de la ranura adyacente a la raíz de la junta. Refiérase al dibujo para **cara de ranura**.

root gap. See preferred term **root opening**.

rendija de raíz. Vea el término preferido **abertura de ráiz**.

root of joint. That portion of a joint to be welded where the members approach closest to each other. In cross section, the root of the joint may be either a point, a line, or an area.

raíz de junta. Esa porción de una junta que está para soldarse donde los miembros se acercan muy cerca del uno al otro. En sección transversa, la raíz de una junta puede ser una punta, una línea, o una área.

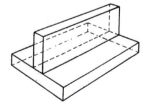 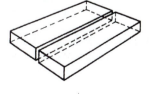

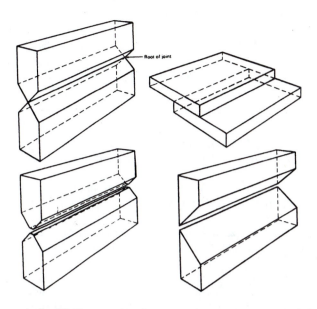

root of weld. The points, as shown in cross section, at which the back of the weld intersects the base metal surfaces.

raíz de soldadura. Las puntas, como ensena la sección transversa, donde la parte de atrás cruza con la superficie del metal base.

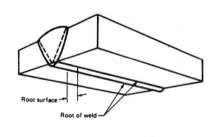

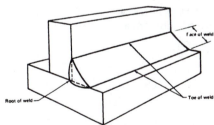

root opening. The separation between the members to be joined at the root of the joint. Refer to drawings for **bevel**.

abertura de raíz. La separación entre los miembros que están para unirse a la raíz de la junta. Refiérase al dibujo para **bisel**.

*root pass. The first weld of a multipass weld. The root pass fuses the two pieces together and establishes the depth of weld metal penetration.

pasada de raíz. La primera soldadura de una soldadura de pasadas múltiples. La pasada de raíz funde las dos piezas juntas y establece la profundidad de la penetración del metal soldado.

root penetration. The distance the weld metal extends into the joint root.

penetración de raíz. La distancia que se extiende el metal de soldadura adentro de la junta de raíz.

(continued on page 870)

(continued from page 869)

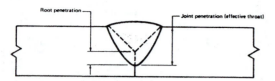

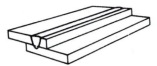

Arc seam weld

Resistance seam weld

Electron beam seam weld

root radius. See preferred term **groove radius.**
radio de raíz. Vea el término preferido **radio de ranura.**

root reinforcement. Reinforcement of weld at the side other than that from which welding was done. Refer to drawing for **face of weld.**
refuerzo de raíz. Refuerzo de soldadura en el lado opuesto de donde se hizo la soldadura. Refiérase al dibujo para **cara de la soldadura.**

root surface. The exposed surface of a weld on the side other than that from which welding was done. Refer to drawings for **root of weld.**
superficie de raíz. La superficie expuesta de una soldadura en el lado opuesto de donde se hizo la soldadura. Refiérase a los dibujos para **raíz de soldadura.**

runoff weld tab. Additional material that extends beyond the end of the joint, on which the weld is terminated.
solera de carrera final de soldadura. Material adicional que se extiende más allá de donde se acaba la junta, en la cual la soldadura es terminada.

S

scarf joint. A form of butt joint.
junta de echarpe. Una forma de junta a tope.

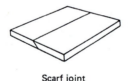

Scarf joint

*scavenger. Elements in the flux that pick up contaminants in the molten weld pool and float them to the surface where they become part of the slag.
limpiadores o (expulsadores). Elementos en el flujo que levantan los contaminantes en el charco de soldadura derretida y los flotan a la superficie donde se forman parte de la escoria.

seam weld. A continuous weld made between or upon overlapping members, in which coalescence may start and occur on the faying surfaces, or may have proceeded from the surface of one member. The continuous weld may consist of a single weld bead or a series of overlapping spot welds.
soldadura de costura. Una soldadura continua hecha en medio o encima de los miembros traslapados, en la cual la coalescencia puede empezar y ocurrir en la superficie del empalme, o puede haber procedido de la superficie de un miembro. La soldadura continua puede consistir de un solo cordón de soldadura o una serie de puntos traslapados en las soldaduras.

*secondary combustion. In the combustion of acetylene and oxygen, the secondary reaction unites oxygen and the free hydrogen to form water vapor (H_2O) and liberate more heat. The carbon monoxide unites with more oxygen to form carbon dioxide (CO_2).
combustión secundaria. En la combustión de acetileno y oxígeno, la reacción secundaria une el oxígeno y el hidrógeno libre para formar vapor de agua (H_2O) y liberar más calor. El carbón monóxido se une con más oxígeno para formar carbón bióxido (CO_2).

semiautomatic arc welding. Arc welding with equipment that controls only the filler metal feed. The advance of the welding is manually controlled.
soldadura de arco semiautomático. La soldadura de arco con equipo que controla solamente la alimentación del metal de relleno. El avance de la soldadura es controlado manualmente.

*semiautomatic operation. During the welding process, the filler metal is added automatically, and all other manipulation is performed manually by the operator.
operación semiautomática. Durante el proceso de la soldadura, el metal de relleno es añadido automáticamente, y todas las otras manipulaciones son ejecutadas manualmente por el operador.

*sensor. A transducer whose input is a physical phenomenon and whose output is a quantitative measure of the physical phenomenon.
sensor. Un transducor cuya entrada es un fenómeno físico y cuya medida es una medida cuantitativa del fenómeno físico.

sequence time (automatic arc welding). See preferred term **welding cycle.**
tiempo de secuencia (soldadura de arco automático). Vea el término preferido **ciclo de soldadura.**

*shear strength. As applied to a soldered or brazed joint, it is the ability of the joint to withstand a force applied parallel to the joint.
fuerza cizallada. Asi como es aplicada a una junta de soldadura fuerte o soldadura blanda, es la habilidad de la junta de resistir una fuerza aplicada al paralelo de la junta.

shielded metal arc cutting (SMAC). An arc cutting process that uses a covered electrode.
cortes de arco métalico protegido (SMAC). Un proceso de cortar con arco que usa un electrodo cubierto.

shielded metal arc welding (SMAW). An arc welding process with an arc between a covered electrode and the weld pool. The process is used with shielding from the decomposition of the electrode covering, without the application of pressure, and with filler metal from the electrode.

soldadura de arco metálico protegido (SMAW). Un proceso de soldadura de arco con un arco en medio de un electrodo cubierto y el charco de soldadura. El proceso se usa con protección de descomposición del cubrimiento del electrodo sin la aplicación de presión, y con el metal de relleno del electrodo.

shielding gas. Protective gas used to prevent or reduce atmospheric contamination.

gas protector. El gas protector se usa para prevenir o reducir la contaminación atmosférica.

short arc. A nonstandard term for **short circuiting transfer** arc welding.

arco corto. Un término fuera de la norma para **transferir por corto circuito** (soldadura de arco).

short circuiting arc welding. A nonstandard term for **short circuiting transfer** (arc welding).

soldadura de arco con corto circuito. Un término fuera de la norma para **transferir por corto circuito** (soldadura de arco).

short circuiting transfer (arc welding). Metal transfer in which molten metal from a consumable electrode is deposited during repeated short circuits.

transferir por corto circuito (soldadura de arco). Transferir metal el cual el metal derretido del electrodo consumible es depositado durante repetidos cortos circuitos.

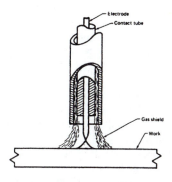

shoulder. See preferred term **root face.**
hombro. Vea término preferido **cara de raíz.**

shrinkage void. A cavity type discontinuity normally formed by shrinkage during solidification.
vacío de encogimiento. Una discontinuidad tipo cavidad normalmente formada por encogimiento durante solidificación.

side bend test. A test in which the side of a transverse section of the weld is on the convex surface of a specified bend radius.
prueba de dobléz de lado. Una prueba en la cual el lado de una sección transversa de la soldadura está en la superficie convexa de un radio de dobléz especificado.

silver soldering, silver alloy brazing. Nonpreferred terms used to denote brazing with a silver-base filler metal. See preferred term **furnace brazing.**
soldadura blanda con plata, soldadura fuerte con aleación de plata. Términos no preferidos que se usan para denotar soldadura

fuerte con metal para rellenar con base de plata. Vea el término preferido **soldadura fuerte en horno.**

single-bevel-groove weld. A type of groove weld. Refer to drawing for **groove weld.**
soldadura de ranura de un solo bisel. Tipo de soldadura de ranura. Refiérase al dibujo para **soldadura de ranura.**

single-flare-bevel-groove weld. A type of groove weld. Refer to drawing for **groove weld.**
soldadura de ranura de un solo bisel acampanado. Un tipo de soldadura de ranura. Refiérase al dibujo para **soldadura de ranura.**

single-flare-V-groove weld. A type of groove weld. Refer to drawing for **groove weld.**
soldadura de ranura de una sola V acampanada. Un tipo de soldadura de ranura. Refiérase al dibujo para **soldadura de ranura.**

single-J-groove weld. A type of groove weld. Refer to drawing for **groove weld.**
soldadura de ranura de una sola J. Un tipo de soldadura de ranura. Refiérase al dibujo para **soldadura de ranura.**

single-port nozzle. A constricting nozzle of the plasma arc torch that contains one orifice, located below and concentric with the electrode.
boquilla de una sola abertura. Es una boquilla constreñida de la antorcha de arco de plasma que contiene un orificio, situado debajo y concéntrico al electrodo.

single-square-groove weld. A type of groove weld. Refer to drawing for **groove weld.**
soldadura de ranura de una sola escuadra. Un tipo de soldadura de ranura. Refiérase al dibujo de **soldadura de ranura.**

single-U-groove weld. A type of groove weld. Refer to drawing for **groove weld.**
soldadura de ranura de una sola U. Un tipo de soldadura de ranura. Refiérase al dibujo para **soldadura de ranura.**

single-V-groove weld. A type of groove weld. Refer to drawing for **groove weld.**
soldadura de ranura de una sola V. Un tipo de soldadura de ranura. Refiérase al dibujo de **soldadura de ranura.**

size of weld.
groove weld. The joint penetration (depth of bevel plus the root penetration when specified). The size of a groove weld and its effective throat are one and the same.
fillet weld. For equal leg fillet welds, the leg lengths of the largest isosceles right triangle that can be inscribed within the fillet weld cross section. Refer to drawings for **concavity** and **convexity.** For unequal leg fillet welds, the leg lengths of the largest right triangle that can be inscribed within the fillet weld cross section.
Note: When one member makes an angle with the other member greater than 105 degrees, the leg length (size) is of less significance than the effective throat, which is the controlling factor for the strength of a weld.
flange weld. The weld metal thickness measured at the root of the weld.
tamaño de la soldadura.

soldadura de ranura. La penetración de la junta (profundidad del bisel más la penetración de la raíz cuando está especificada). El tamaño de la soldadura de ranura y la garganta efectiva son una y la misma.

soldadura filete. Para soldaduras con piernas iguales de filete, lo largo de las piernas del triángulo recto con el isosceles más grande que puede ser inscrito dentro de la sección transversa de la soldadura de filete. Refiérase al dibujo para **concavidad** y **convexidad.** Para piernas de soldadura de filete desiguales, lo largo de las piernas del triángulo recto más grande que puede ser inscrito dentro de la sección transversa de la soldadura de filete.

Nota: Cuando un miembro hace un ángulo con otro miembro más grande de 105 grados, lo largo de la pierna (tamaño) es de menor significado que la garganta efectiva, la cual es el factor de control para la fuerza de una soldadura.

soldadura de brida. Lo grueso del metal de soldadura se mide a la raíz de la soldadura.

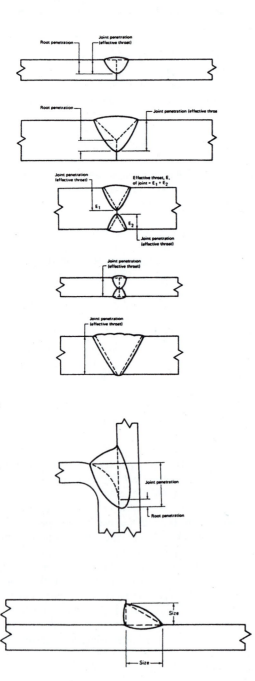

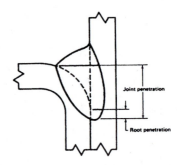

slag. A nonmetallic product resulting from the mutual dissolution of flux and nonmetallic impurities in some welding and brazing processes.

escoria. Un producto que no es metálico resultando de una disolución mutual del flujo y las impuridades no metálicas en unos procesos de soldadura y soldadura fuerte.

slag inclusion. Nonmetallic solid material entrapped in weld metal or between weld metal and base metal.

inclusion de escoria. Un material sólido que no es metálico atrapado en el metal de soldadura en medio del metal base.

slag pan. A high-temperature container that holds the slag from a thermite weld until it cools.

bandeja para escoria. un recipiente de alta temperatura que contiene la escoria de una soldadura con termita hasta que se enfría.

*slope. For gas metal arc welding, the volt-ampere curve of the power supply indicates that there is a slight decrease in voltage as the amperage increases; the rate of voltage decrease in the slope.

pendiente. Para soldadura de arco de metal con gas, la curva voltio-amperio de la fuente de poder indica que si hay un ligero decremento en voltaje cuando los amerios aumentan; la proporción del voltaje decrementa en el pendiente.

slot weld. A weld made in an elongated hole in one member of a lap or T-joint joining that member to that portion of the surface of the other member that is exposed through the hole. The hole may be open at one end and may be partially or completely filled with weld metal. (A fillet welded slot should not be construed as conforming to this definition.)

soldadura de ranura alargada. Una soldadura hecha en un agujero alargado en un miembro de una junta en solape o T uniendo ese miembro a esa porción de la superficie del otro miembro que está expuesto a través del agujero. El agujero puede ser abierto en una punta y puede ser parcialmente o completamente rellenado con metal de soldadura. (Una ranura alargada con soldadura de filete no debe de interpretarse como conforme a está definición.)

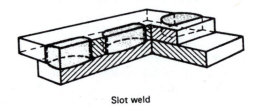

Slot weld

slugging. The act of adding a separate piece or pieces of material in a joint before or during welding that results in a welded joint not complying with design, drawing, or specification requirements.

usar trozos de metal. El acto de agregar una pieza o piezas separadas de material en una junta antes o durante la soldadura que resulta en una junta soldada que no cumple con diseño, dibujo, o las especificaciones requeridas.

solder. A filler metal used in soldering that has a liquidus not exceeding 450°C (840°F).

soldadura (material para soldar). Un metal de relleno usado para soldadura blanda que tiene un liquidus que no excede de 450°C (840°F).

soldering (S). A group of welding processes that produces coalescence of materials by heating them to the soldering temperature and by using a filler metal with a liquidus not exceeding 450°C (840°F) and below the solidus of the base metals. The filler metal is distributed between closely fitted faying surfaces of the joint by capillary action.

soldadura blanda. Un grupo de procesos de soldadura que produce coalescencia de materiales calentándolos a una temperatura de soldar y usando un metal para rellenar con un liquidus que no exceda de 450°C (840°F) y más abajo del solidus de los metales base. El metal para rellenar es distribuido en medio de las superficies empalmadas acopladas muy cerca de la junta por acción capilar.

soldering gun. An electrical soldering iron with a pistol grip and a quick heating, relatively small bit.

pistola de soldar. Un fierro eléctrico para soldar con mango de pistola, rápido para calentarse, y tiene una punta relativamente pequeña.

solid-state welding (SSW). A group of welding processes that produces coalescence by the application of pressure at a welding temperature below the melting temperatures of the base metal and the filler metal.

soldadura de estado sólido. Un grupo de procesos para soldar que produce coalescencia cuando se le aplica presión a una temperatura de soldadura más baja que las temperatures que se usan para derretir el metal base y el metal de relleno.

solidus. The highest temperature at which a metal or an alloy is completely solid.

solidus. La temperatura más alta cuando un metal o una aleación está completamente sólido.

spatter. The metal particles expelled during welding and that do not form a part of the weld.

salpicadura. Las partículas de metal que se despidan cuando se está soldando y que no forman parte de la soldadura.

spatter loss. Metal lost due to spatter.

pérdida causa salpicadura. El metal perdido debido a la salpicadura.

spool. A type of filter metal package consisting of a continuous length of electrode wound on a cylinder (called the barrel), which is flanged at both ends. The flange extends below the inside diameter of the barrel and contains a spindle hole.

carrete. Un paquete de metal tipo filtro consistiendo de una extensión continua de un electrodo enrollado en un cilindro (llamado el barril), el cual tiene una brida en los dos extremos. La brida se extiende debajo del diámetro de adentro del barril y contiene un agujero huso.

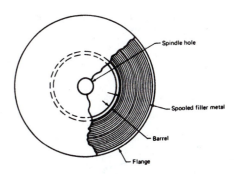

spot weld. A weld made between or upon overlapping members in which coalescence may start and occur on the faying surfaces or may proceed from the surface of one member. The weld cross section (plan view) is approximately circular. See also **arc spot weld** and **resistance spot welding.**

soldadura de puntos. Una soldadura hecha en medio o sobre miembros traslapados en la cual la coalescencia puede empezar y ocurrir en las superficies empalmadas o puede continuar en la superficie de un miembro. La sección transversa (plan de vista) es aproximadamente circular. Vea también **soldadura de puntos por arco** y **soldadura de puntos por resistencia.**

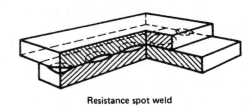

Resistance spot weld

spray arc. A nonstandard term for **spray transfer.**

arco para rociar. Un término fuera de norma para **traslado rociado.**

spray transfer (arc welding). Metal transfer in which molten metal from a consumable electrode is propelled axially across the arc in small droplets.

traslado rociado (soldadura de arco). Transferir el metal el cual el metal derretido de un electrodo consumible es propelado axialmente a traves del arco en gotitas pequeñas.

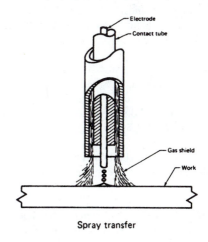

Spray transfer

*square butt joint. Joint made when two flat pieces of metal face each other with no edge preparation. (See **square groove weld.**)

junta escuadra de tope. Una junta hecha cuando dos piezas planas de metal se enfrentan una a la otra sin preparación de orilla. (Vea **soldadura de ranura escuadra.**)

square butt weld. (See butt joint.)
junta escuadra de tope. (Vea junta a tope.)

square-groove weld. A type of groove weld.
soldadura de ranura escuadra. Un tipo de soldadura de ranura.

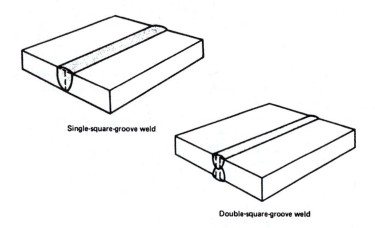

Single-square-groove weld

Double-square-groove weld

stack cutting. Thermal cutting of stacked metal plates arranged so that all the plates are severed by a single cut.
corte de metal apilado. Un corte termal de hojas de metal apilados arregladas para que todas las hojas sean cortadas por un solo corte.

staggered intermittent welds. Intermittent welds on both sides of a joint in which the weld increments on one side are alternated with respect to those on the other side.
soldadura intermitente de cadena. Soldaduras intermitentes en los dos lados de una junta en cual los incrementos de soldadura son alternados de un lado con respecto a los del otro lado.

standoff distance. The distance between a nozzle and the workpiece.
distancia de alejamiento. La distancia entre la boquilla y la pieza de trabajo.

starting weld tab. Additional material that extends beyond the beginning of the joint, on which the weld is started.
solera para empezar a soldar. Material adicional que se extiende más allá del principio de la junta, en donde la soldadura es empezada.

*steel. An alloy consisting primarily of iron and carbon. The carbon content may be as high as 2.2% but is usually less than 1.5%.
acero. Una aleación que consiste primeramente de hierro y carbón. El contenido del carbón puede ser tan alto como 2.2% pero es regularmente menos de 1.5%.

stick electrode. A nonstandard term for covered electrode.
electrodo de varilla. Un término fuera de norma por electrodo cubierto.

stickout. See preferred term electrode extension.
sobresalga. Vea término preferido extensión del electrodo.

straight polarity. The arrangement of direct current arc welding leads in which the work is the positive pole and the electrode is the negative pole of the welding arc. A synonym for **direct current electrode negative.** Refer to drawing for **direct current electrode negative.**
polaridad directa. El arreglo de los cables de soldadura de arco con corriente directa donde el cable de la tierra es el polo positivo y el porta electrodo es el polo negativo del arco de soldadura. Un sinónimo para **corriente directa con electrodo negativo.** Refiérase al dibujo para **corriente directa con electrodo negativo.**

stranded electrode. A composite filler metal electrode consisting of stranded wires that may mechanically enclose materials to improve properties, stabilize the arc, or provide shielding.
electrodo cable. Electrodo de metal para rellenar compuesto que consiste de cable de alambres que pueden encerrar materiales mecánicamente para mejorar propiedades, estabilizar el arco, o proveer protección.

*stress point. Any point in a weld where incomplete fusion of the weld on one or both sides of the root gives rise to stress, which can result in premature cracking or failure of the weld at a load well under the expected strength of the weld.
punto de tensión. Cualquier punto en una soldadura donde la fusión incompleta en la soldadura en uno o en los dos lados de la raíz le aumenta la tensión, la cual puede resultar en una grieta o falta prematura en la soldadura con una carga mucho menos que la fuerza de la soldadura que se esperaba.

stress relief heat treatment. Uniform heating of a structure or a portion thereof to a sufficient temperature to relieve the major portion of the residual stresses, followed by uniform cooling.
tratamiento de calor para relevar la tensión. Calentamiento uniforme de una estructura o una porción a una temperatura suficiente para relevar la mayor porción de las tensiones restantes, seguido por enfriamiento uniforme.

stringer bead. A type of weld bead made without appreciable weaving motion. See also weave bead.
cordón encordador. Un tipo de cordón de soldadura sin movimiento del tejido apreciable. Vea también cordón tejido.

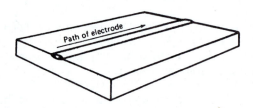

Path of electrode

stud arc welding (SW). An arc welding process that uses an arc between a metal stud, or similar part, and the other workpiece. The process is used with or without shielding gas or flux, with or without partial shielding from a ceramic ferrule surrounding the stud, with the application of pressure after the faying surfaces are sufficiently heated, and without filler metal.
esparrago (tachón) para soldadura de arco. Un proceso de soldadura de arco que usa un arco entre un esparrago (tachón) de metal, o parte similar, y la otra pieza de trabajo. El proceso es usado con o sin flujo o protección de gas, con o sin protección parcial del casquillo cerámico que rodea el esparrago (tachón), con la aplicación de presión después que las superficies empalmadas tengan suficiente calor, y sin metal de relleno.

stud welding. A general term for the joining of a metal stud or similar part to a workpiece. Welding may be accomplished by arc, resistance, friction, or other suitable process with or without external gas shielding.

soldadura de esparrago (tachón). Un término general para la unión de esparragos (tachones) o parte similar a una pieza de trabajo. La soldadura se puede efectuar por arco, resistencia, fricción, u otro proceso conveniente con o sin gas externo para protección.

submerged arc welding (SAW). An arc welding process that uses an arc or arcs between a bare metal electrode or electrodes and the weld pool. The arc and molten metal are shielded by a blanket of granular flux on the workpieces. The process is used without pressure and with filler metal from the electrode and sometimes from a supplemental source (welding rod, flux, or metal granules).

soldadura por arco sumergido. Un proceso de soldar con arco que usa un arco o arcos entre un electrodo de metal liso o electrodos y el charco de la soldadura. El arco y el metal derretido son protegidos por una capa de flujo granular sobre la pieza de trabajo. El proceso es usado sin presión y con metal para rellenar del electrodo y a veces de una fuente suplementaria (varilla para soldar, flujo, o gránulos de metal).

substrate. Any base material to which a thermal sprayed coating or surfacing weld is applied.

substrato. Cualquier material base al cual se le aplica una capa termal o una soldadura de superficie.

suck-back. See preferred term **concave root surface.**

succión del cordón de raíz. Vea el término preferido **superficie raíz concavo.**

surface preparation. The operations necessary to produce a desired or specified surface condition.

preparación de la superficie. Las operaciones necesarias para producir una deseada o una especificada condición de la superficie.

surfacing. The application by welding, brazing, or thermal spraying of a layer of material to a surface to obtain desired properties or dimensions, as opposed to making a joint. See also **buttering, cladding, coating,** and **hardfacing.**

recubrimiento superficial. La aplicación a la soldadura, a la soldadura fuerte o rociado termal de una capa de material a la superficie para obtener las deseadas propiedades o dimensiones, contrario a la hechura de una junta. Vea también **recubrimiento antes de terminar una soldadura, capa de revestimiento, revestimiento y endurecimiento de caras.**

*synergic system. Pulsed arc metal transfer system in which the power supply and wire feed settings are made by adjusting a single knob.

sistema sinérgico. Sistema de transferir metal por arco pulsado en cual la fuente de poder y los ajustes del alimentador de alambre son hechos por el ajuste de un botón solamente.

T

tab. See **runoff weld tab, starting weld tab,** and **weld tab.**

solera. Vea **solera de carrera final de soldadura, solera para empezar a soldar** y **solera para soldar.**

tack weld. A weld made to hold parts of a weldment in proper alignment until the final welds are made.

soldadura de puntos aislados. Una soldadura hecha para detener las partes en su propio alineamiento hasta que se hagan las soldaduras finales.

*tactile sensor. A transducer that is sensitive to touch.

sensor táctil. Un transducor que es sensitivo al tocar.

taps. Connections to a transformer winding that are used to vary the transformer turns ratio, thereby controlling welding voltage and current.

grifo. Conexiones al arrollamiento de un transformador que se usan para variar la proporción de vueltas del transformador, asi se puede controlar la corriente y el voltaje para soldar.

*teach. To program a manipulator arm by guiding it through a series of points or in a motion pattern that is recorded for subsequent automatic action by the manipulator.

enseñar. Para programar un manipulador de brazo guiándolo por una serie de puntos o en una muestra de movimiento que está registrada para acción automática subsiguiente por el manipulador.

*teaching interface. The mechanisms or devices by which a human operator teaches a machine.

enseñanza de interfaze. Los mecanismos o aparatos por los cuales un operador humano enseña a la máquina.

tee joint. (See **T-joint.**)

junta en Te. (Vea **junta en T.**)

*tempering. Reheating hardened metal before it cools to room temperature to make it tough, not brittle.

templar. Recalentando un metal endurecido antes de que se enfrie a la temperatura del ambiente para hacerlo duro, no frágil.

*tensile strength. As applied to a brazed or soldered joint, it is the ability of the joint to withstand being pulled apart.

resistencia a la tensión. Como es aplicada a una junta de soldadura fuerte o soldadura blanda, es la habilidad de una junta que resista ser estirada hasta que se rompa en dos pedazos.

tension test. A test in which a specimen is loaded in tension until failure occurs. See also **reduced section tension test.**

prueba de tensión. Una prueba en la cual la probeta está cargada de tensión hasta que ocurra el fracaso. Vea también **prueba de tensión de sección reducida.**

thermal cutting (TC). A group of cutting processes that severs or removes metal by localized melting, burning, or vaporizing of the workpiece. See also **arc cutting, electron beam cutting, laser beam cutting,** and **oxygen cutting.**

corte termal (TC). Un grupo de procesos para cortar que desúne o quita el metal para derretir o quemar o vaporizar localmente las piezas de trabajo. Vea también **corte con arco, cortes rayo de electron, corte de rayo laser, cortes de oxígeno.**

thermal spraying (THSP). A group of processes in which finely divided metallic or nonmetallic surfacing materials are deposited in a molten or semimolten condition on a substrate to form a thermal spray deposit. The surfacing material may be in the form of powder, rod, cord, or wire.

rociado termal. Un grupo de procesos el cual los materiales metálicos o no metálicos de superficie que son depositados en una condición derretida o semiderretida sobre el substrato para formar un depósito de rociado termal. El material de superficie puede ser en forma de polvo, varilla, cordón, o alambre.

thermal stresses. Stresses in metal resulting from nonuniform temperature distributions.

tensión termal. Tensiones en el metal resultando cuando la distribución de la temperatura no está uniforme.

thermit welding (TW). A welding process that produces coalescence of metals by heating them with superheated liquid metal from a chemical reaction between a metal oxide and aluminum, with or without the application of pressure. Filler metal, when used, is obtained from the liquid metal.

soldadura termita. Un proceso de soldadura que produce coalescencia del metal calentándolos con un metal supercalentado líquido da una reacción química entre un metal óxido y el aluminio, con o sin la aplicación de presión. El metal de relleno, cuando es usado, es obtenido del metal líquido.

throat area. The area bounded by the physical parts of the secondary circuit in a resistance spot, seam, or projection welding machine. Used to determine the dimensions of a part that can be welded and determine, in part , the secondary impedance of the equipment.

área de garganta. La área limitada por las partes físicas del circuito secundario en un punto de resistencia, costura, o una máquina de soldar de proyección. Se usa para determinar las dimensiones de una parte que puede ser soldada y determinar, en parte, la impedancia secundaria del equipo.

throat depth. In a resistance spot, seam, or projection welding machine, the distance from the centerline of the electrodes or platens to the nearest point of interference for flat sheets.

profundidad de garganta. En una punta de resistencia, costura, o máquina de soldar de proyección, la distancia de la línea del centro del electrodo o platinas al punto más cercano de interferencia para las hojas planas.

throat of a fillet weld.
actual throat. The shortest distance from the root of weld to its face. Refer to drawing for **convexity.**
effective throat. The minimum distance minus any reinforcement from the root of weld to its face. Refer to drawing for **convexity.**
theoretical throat. The distance from the beginning of the root of the joint perpendicular to the hypotenuse of the largest right triangle that can be inscribed within the fillet weld cross section. This dimension is based on the assumption that the root opening is equal to zero. Refer to drawing for **convexity.**
garganta de soldadura filete.
garganta actual. La distancia más corta de la raíz de una soldadura a su cara. Refiérase al dibujo para **convexidad.**
garganta efectiva. La distancia mínima menos cualquier refuerzo de la raíz de la soldadura a su cara. Refiérase al dibujo para **convexidad.**
garganta teórica. La distancia de donde empieza la raíz de la junta perpendicular a la hipotenusa del triángulo recto más grande que puede ser inscrito adentro de la sección transversa de una soldadura de filete. Esta dimensión está basada en la proposición que la abertura de la raíz es igual a cero. Refiérase al dibujo para **convexidad.**

TIG welding. A nonstandard term when used for **gas tungsten arc welding.**

soldadura TIG. Un término fuera de norma cuando es usado por **soldadura de arco de tungsteno con gas.**

T-joint. A joint between two members located approximately at right angles to each other in the form of a T.

junta en T. Una junta en medio de dos miembros que están localizados aproximadamente a ángulos rectos de uno al otro en la forma de T.

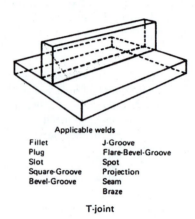

Applicable welds

Fillet	J-Groove
Plug	Flare-Bevel-Groove
Slot	Spot
Square-Groove	Projection
Bevel-Groove	Seam
	Braze

T-joint

toe crack. A crack in the base metal occurring at the toe of a weld.

grieta de pie. Una grieta en el metal base que ocurre al pie de la soldadura.

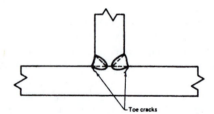

Toe cracks

toe of weld. The junction between the face of a weld and the base metal. Refer to drawing for **face of weld.**

pie de la soldadura. La unión entre la cara de la soldadura y el metal base. Refiérase al dibujo para **cara de la soldadura.**

tolerances. The allowable deviation in accuracy or precision between the measurement specified and the part as laid out or produced.

tolerancias. desviación permitida en la precisión entre la medida especificada y la pieza instalada o producida.

torch. See preferred terms **cutting torch** and **welding torch.**

antorcha. Vea el término preferido **antorcha para cortar y antorcha para soldar.**

*torch angle. The angle between the centerline of the torch and the work surface; the ideal torch angle is 45°. The torch angle affects the percentage of heat input into the metal, thus affecting the speed of melting and the size of the molten weld pool.

ángulo de antorcha. El ángulo en medio de la línea del centro de la antorcha y la superficie del trabajo; el ángulo ideal de la antorcha es 45°. El ángulo de la antorcha afecta el por ciento de calor que entra dentro del metal, asi afectando la rapidez de derretimiento y el tamaño del charco del metal derretido de la soldadura.

torch brazing (TB). A brazing process that uses heat from a fuel gas flame.

soldadura fuerte con antorcha. Un proceso de soldadura fuerte que usa calor de una llama de gas combustible.

*torch manipulation. The movement of the torch by the operator to control the weld bead characteristics.

manipulacion de la antorcha. El movimiento de la antorcha por el operador para el control de las características del cordón de la soldadura.

torch soldering (TS). A soldering process that uses heat from a fuel gas flame.

soldadura blanda con antorcha. Un proceso de soldadura blanda que usa calor de una llama de gas combustible.

transducer. A device that transforms one form of energy into another.

transducor. Un aparato que convierte una forma de energía a otra.

transferred arc. A plasma arc established between the electrode of the plasma arc torch and the workpiece.

arco transferido. Un arco de plasma establecido entre el electrodo de la antorcha de arco de plasma y la pieza de trabajo.

*transition current. In gas metal arc welding, current above a critical level to permit spray transfer; the rate at which drops are transferred changes in relationship to the current. Transition current depends upon the alloy bearing welded and is proportional to the wire diameter.

corriente de transición. En soldadura de arco y metal con gas, corriente arriba de un nivel crítico para permitir el traslado del rociado; la proporción en la cual las gotas son transferidas cambia en relación a la corriente. La corriente de transición depende del aleado que se está soldando y es proporcional al diámetro del alambre.

transverse face-bend. (See face bend.)
doblez de cara transversal. (Vea doblez de cara.)

transverse root-bend. (See root bend.)
cordón de raíz transversal. (Vea cordón de raíz.)

travel angle. The angle less than 90 degrees between the electrode axis and a line perpendicular to the weld axis, in a plane determined by the electrode axis and the weld axis. The angle can also be used to partially define the position of guns, torches, rods, and beams. See also drag angle and push angle. Refer to drawing for backhand welding.

ángulo de avance. El ángulo menos de 90 grados entre el eje del electrodo y una línea perpendicular al eje de la soldadura, en un plano determinado por el eje del electrodo y el eje de la soldadura. El ángulo también puede ser usado para parcialmente definir la posición de las pistolas, antorchas, varillas, y rayos. Vea también ángulo del tiro, ángulo de empuje. Refiérase al dibujo para soldadura en revés.

travel angle (pipe). The angle less than 90 degrees between the electrode axis and a line perpendicular to the weld axis at its point of intersection with the extension of the electrode axis, in a plane

determined by the electrode axis and a line tangent to the pipe surface of the same point. This angle can also be used to partially define the position of guns, torches, rods, and beams. Refer to drawing for **backhand welding.**

ángulo de avance (tubo). El ángulo menos de 90 grados entre el eje del electrodo y la línea perpendicular al eje a su punto de intersección con la extensión del eje del electrodo, en un plano determinado por el eje del electrodo y una línea tangente a la superficie del tubo del mismo punto. Este ángulo también puede ser usado para parcialmente definir la posición de las pistolas, varillas, y rayos. Refiérase al dibujo para **soldadura en revés.**

tungsten electrode. A nonfiller metal electrode used in arc welding, arc cutting, and plasma spraying, made principally of tungsten.

electrodo de tungsteno. Un electrodo de metal que no se rellena que se usa para soldadura de arco, cortes por arco, rociado por plasma, y hecho principalmente de tungsteno.

*type A fire extinguisher. Extinguisher used for combustible solids, such as paper, wood, and cloth. Identifying symbol is a green triangle enclosing the letter A.

extinguidor para incendios tipo A. Extinguidor que se usa para combustibles sólidos como papel, madera, y tela. El símbolo de identificación es un triángulo verde con la letra A adentro.

*type B fire extinguisher. Extinguisher used for combustible liquids, such as oil and gas. Identifying symbol is a red square enclosing the letter B.

extinguidor para incendios tipo B. Extinguidor que se usa para liquidos combustibles, como aceite y gas. El símbolo de identificación es un cuadro rojo con la letra B adentro.

*type C fire extinguisher. Extinguisher used for electrical fires. Identifying symbol is a blue circle enclosing the letter C.

extinguidor para incendios tipo C. Extinguidor que se usa para incendios eléctricos. El símbolo de identificación es un círculo azul con la letra C adentro.

*type D fire extinguisher. Extinguisher used on fires involving combustible metals, such as zinc, magnesium, and titanium. Identifying symbol is a yellow star enclosing the letter D.

extinguidor para incendios tipo D. Extinguidor que se usa para incendios de metales combustibles, como zinc, magnesio, y titanio. El símbolo de identificación es una estrella amarilla con una letra D adentro.

U

U-groove weld. A type of groove weld.
soldadura de ranura en U. Un tipo de soldadura de ranura.

ultrasonic coupler (ultrasonic soldering) and ultrasonic welding. Elements through which ultrasonic vibration is transmitted from the transducer to the tip.

acoplador ultrasónico (soldadura blanda ultrasónica y soldadura ultrasónica). Los elementos por los cuales la vibración ultrasónica es transmitida del transducor a la punta.

ultrasonic welding (USW). A solid-state welding process that produces a weld by the local application of high-frequency vibratory energy as the workpieces are held together under pressure.

soldadura ultrasónica. Un proceso de soldadura de estado sólido que produce una soldadura por la aplicación local de energía vibratoria de alta frecuencia asi cuando las piezas de trabajo están agarradas juntas bajo presión.

underbead crack. A crack in the heat-affected zone, generally not extending to the surface of the base metal.

grieta entre o bajo cordones. Una grieta en la zona afectada por el calor, generalmente no se extiende a la superficie del metal base.

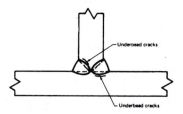

undercut. A groove melted into the base metal adjacent to the toe or root of a weld and left unfilled by weld metal. Refer to drawing for **overlap.**

socavación. Una ranura dentro del metal base adyacente al pie o raíz de la soldadura y se deja sin rellenar con el metal de soldadura. Refiérase al dibujo para **traslapo.**

underfill. A depression on the face of the weld or root surface extending below the surface of the adjacent base metal.

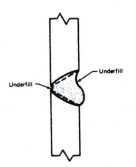

valle. Una depresión en la cara de la soldadura o la superficie de la raíz extendiéndose más abajo de la superficie del adyacente metal base.

uphill. Welding with an upward progression.
soldando hacia arriba. Solando con progresión hacia arriba.

*****upper arm** (robot). That portion of a jointed arm that is connected to the shoulder.
brazo de arriba (robot). La porción que úne al brazo que conecta con el hombro.

upset. Bulk deformation resulting from the application of pressure in welding. The upset may be measured as a percent increase in interfacial for area, a reduction in length, or a percent reduction in thickness (for lap joints).
recalada. Deformación en bulto resultado de la aplicación de presión en la soldadura. El recalado puede ser medido como un aumento en porciento en una área interfacial, una reducción en lo largo, o un porciento de reducción en lo grueso (para juntas en solape).

upset welding (UW). A resistance welding that produces coalescence over the entire area of faying surfaces or progressively along a butt joint by the heat obtained from the resistance to the flow of welding current through the area where those surfaces are in contact. Pressure is used to complete the weld.
soldadura recalada. Un proceso de soldadura por resistencia que produce coalescencia sobre toda la área de superficie empalmada o progresivamente sobre una junta a tope con el calor obtenido de la resistencia del flujo de la corriente de la soldadura por la área donde esas superficies están en contacto. Presión es usada para completar la soldadura.

upslope time (automatic arc welding). The time during which the current changes continuously from initial current valve to the welding value.
tiempo del pendiente en ascenso (soldadura automática de arco). El tiempo durante el cual la corriente cambia continuamente el valor de la corriente inicial al valor de la soldadura.

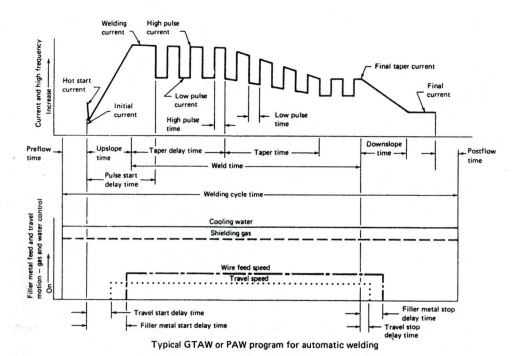

Typical GTAW or PAW program for automatic welding

V

vertical position. The position of welding in which the axis of the weld is approximately vertical.

posición vertical. La posición de la soldadura en la cual el eje para soldarse es aproximadamente vertical.

regulador de voltaje. Un aparato de control eléctrico automático para mantener y proporcionar un voltaje constante a la primaria de un transformador de una soldadura.

V-groove weld. A type of groove weld.

soldadura de ranura V. Un tipo de soldadura de ranura.

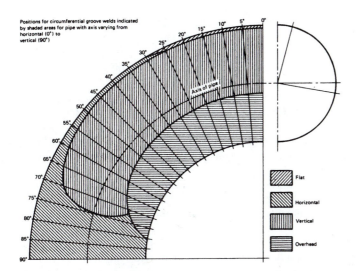

Positions for circumferential groove welds indicated by shaded areas for pipe with axis varying from horizontal (0°) to vertical (90°)

Axis of pipe

Flat
Horizontal
Vertical
Overhead

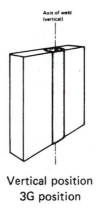

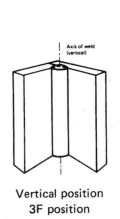

Axis of weld (vertical)

Vertical position
3G position

Axis of weld (vertical)

Vertical position
3F position

vertical position (pipe welding). The position of a pipe joint in which welding is performed in the horizontal position and the pipe may or may not be rotated.

posición vertical (soldadura de tubo). La posición de una junta de tubo en la cual la soldadura se hace en la posición horizontal y el tubo puede o no dar vueltas.

W

*wagon tracks. A pattern of trapped slag inclusions in the weld that show up as discontinuities in x-rays of the weld.

huellas de carreta. Una muestra de inclusiones de escoria atrapadas en la soldadura que enseña que hay discontinuidades en los rayos-x de la soldadura.

*water-arc plasma cutting (PAC). In this process, nitrogen is used as the plasma gas. The plasma is created at a high temperature in an arc between the electrode and the orifice. A tap water spray applied to the plasma causes it to constrict and accelerate. This column causes the cutting action and melts a very narrow kerf in the material.

cortes de arco-agua plasma. En este proceso, nitrogeno es usado como el gas de plasma. La plasma es creada a una temperatura alta en un arco en medio del electrodo y la orifice. La aplicación de agua del grifo a la plasma la hace que se constriñe y acelera. Está columna causa la acción de cortar y derritir un corte que es muy estrecho en el material.

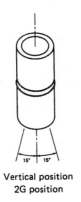

Vertical position
2G position

*wattage. A measurement of the amount of power in the arc; the wattage of the arc controls the width and depth of the weld bead.

número de vatios. Una medida de la cantidad de poder en el arco; el número de vatios del arco controla lo ancho y hondo del cordón de la soldadura.

voltage range. The lower and upper limits of welding power, in volts, that can be produced by a welding machine or used with an electrode or by a process.

rango de voltaje. Los límites máximos y mínimos de poder de soldadura (en voltios) que puede tener una máquina para soldar o que pueden usarse con un electrodo o a través de un proceso.

wax pattern (thermit welding). Wax molded around the workpieces to the form desired for the completed weld.

soldadura termita (molde de cera). Un molde de cera alrededor de las piezas de trabajo a la forma deseada para la soldadura terminada.

voltage regulator. An automatic electrical control device for maintaining a constant voltage supply to the primary of a welding transformer.

weave bead. A type of weld bead made with transverse oscillation.

cordón tejido. Un tipo de cordón de soldadura hecha con oscilación transversa.

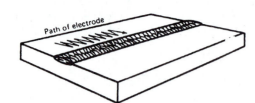

Path of electrode

*weave pattern. The movement of the welding electrode as the weld progresses; common weave patterns include circular, square, zigzag, stepped, C, J, T and Figure 8.

muestra de tejido. El movimiento del electrodo para soldar a como progresa la soldadura; las muestras de tejidos comunes incluyen circular, de cuadro, zigzag, de pasos, C, J, T y la figura 8.

weld. A localized coalescence of metals or nonmetals produced either by heating the materials to suitable temperatures, with or without the application of pressure, or by the application of pressure alone and with or without the use of the filler material.

soldar. Una coalescencia localizada de metales o metaloides producida al calentar los materiales a una temperatura adecuada, con o sin la aplicación de presión, o por la aplicación de presión solamente y con o sin el uso del material de relleno.

weld axis. A line through the length of the weld, perpendicular to and at the geometric center of its cross section.

eje de la soldadura. Una línea a través de lo largo de la soldadura, perpendicular a y al centro geométrico de su sección transversa.

weld bead. weld resulting from a pass. See also stringer bead and weave bead.

cordón de soldadura. Una soldadura que resulta de una pasada. Vea también cordón encordador o cordón tejido.

weld brazing. A joining method that combines resistance welding with brazing.

soldadura y soldadura fuerte. Un método de unir que combina soldadura de resistencia con soldadura fuerte.

weld crack. crack in weld metal.

grieta en la soldadura. Una grieta en el metal de soldadura.

weld face. The exposed surface of a weld on the side from which welding was done.

cara de la soldadura. La superficie expuesta de una soldadura en el lado de donde se hizo la soldadura.

weld gauge. A device designed for checking the shape and size of welds.

instrumento para medir la soldadura. Un aparato diseñado para comprobar la forma y tamaño de las soldaduras.

weld groove. A channel in the surface of a workpiece or an opening between two joint members that provides space to contain a weld.

soldadura de ranura. Un canal en la superficie de una pieza de trabajo o una abertura entre dos miembros de junta que provee espacio para contener una soldadura.

weld interface. The interface between weld metal and base metal in a fusion weld, between base metals in a solid-state weld without filler metal, or between filler metal and base metal in a solid-state weld with filler metal. Refer to drawing for depth of fusion.

interfase de la soldadura. La interfase entre el metal de soldadura y el metal base en una soldadura de fusión, entre metales de base en una soldadura de estado sólido sin metal para rellenar, o entre metal de rellenar y metal base en una soldadura de estado sólido con metal para rellenar. Refiérase al dibujo grueso de fusión.

weld length. See effective length of weld.

largura de la soldadura. Vea distancia efectiva de soldadura.

weld metal. The portion of a fusion weld that has been completely melted during welding.

metal de soldadura. La porción de una soldadura de fusión que se ha derretido completamente durante la soldadura.

weld metal area. The area of the weld metal as measured on the cross section of a weld. Refer to drawing for heat-affected zone.

área de metal de soldadura. La área del metal de la soldadura la cual fue medida en la sección transversa de la soldadura. Refiérase al dibujo para zona affectada por el calor.

weld pass. A single progression of welding along a joint. The result of a pass is a weld bead or layer.

pasada de soldadura. Una progresión singular de la soldadura a lo largo de una junta. El resultado de una pasada es un cordón o una capa.

weld pass sequence. The order in which the weld passes are made. See longitudinal sequence and cross-sectional sequence.

secuencia de pasadas de soldadura. La orden en que las pasadas de soldadura se hacen. Vea secuencia longitudinal y secuencia del corte transversal.

weld penetration. A nonstandard term for joint penetration and root penetration.

penetracion de soldadura. Un término fuera de norma para penetración de junta y penetración de raíz.

weld pool. The localized volume of molten metal in a weld prior to its solidification as weld metal.

charco de soldadura. El volumen localizado del metal derretido en una soldadura antes de su solidificación como metal de soldadura.

weld puddle. A nonstandard term for weld pool.

charco de soldadura. Un término fuera de norma para charco de soldadura.

weld reinforcement. Weld metal in excess of the quantity required to fill a joint. See also face reinforcement and root reinforcement.

refuerzo de soldadura. Metal de soldar en exceso de la cantidad requerida para llenar una junta. Vea también refuerzo de cara y refuerzo de raíz.

weld root. The points, shown in a cross section, at which the root surface intersects the base metal surfaces.

raíz de soldadura. Los puntos, enseñados en una sección transversa, la cual la superficie de la raíz se interseca con las superficies del metal base.

weld size. See preferred term **size of weld.**

tamaño de soldadura. Vea el término preferido **tamaño de la soldadura.**

***weld specimen.** A sample removed from a welded plate according to AWS specifications, which detail the preparation of the plate, the cutting of the plate, and the size of the specimen to be tested.

probeta de soldadura. Una prueba apartada del plato soldado de acuerdo con las especificaciones del AWS, las cuales detallan la preparación del plato, el corte del plato, y el tamaño de la probeta que se va a probar.

weld symbol. A graphical character connected to the welding symbol indicating the type of weld.

símbolo de soldadura. Un signo gráfico conectado al símbolo de soldadura indicando el tipo de soldadura.

weld tab. Additional material that extends beyond either end of the joint, on which the weld is started or terminated.

solera para soldar. Material adicional que se extiende más allá de cualquier punto de la junta en la cual la soldadura es empezada o terminada.

weld test. A welding performance test to a specific code or standard.

prueba de soldadura. Una prueba de ejecución de soldadura según una norma o código específico.

weld time (automatic arc welding). The time interval from the end of start time or end of upslope to beginning of crater fill time or beginning of downslope. Refer to the drawings for **upslope time.**

tiempo de soldadura (soldadura de arco automática). El intervalo de tiempo del fin del tiempo de arranque o el fin del pendiente en ascenso al principio del tiempo de llenar el crater o el principio del tiempo del pendiente en descenso. Refiérase al dibujo para **tiempo del pendiente en ascenso.**

weld timer. A device that controls only the weld time in resistance welding.

contador de tiempo para soldadura. Un aparato que controla solamente el tiempo de soldar en soldaduras de resistencia.

weld toe. The junction of the weld face and the base metal.

pie de la soldadura. La unión de la cara de la soldadura y el metal base.

weldability. The capacity of a material to be welded under the fabrication conditions imposed into a specific, suitably designed structure and to perform satisfactorily in the intended service.

soldabilidad. La capacidad de un material para soldarse bajo las condiciones de fabricación impuestas en un específico, en un diseño de estructura adecuada y para ejecutar satisfactoriamente los servicios intentados.

welder. One who performs manual or semiautomatic welding.

soldador. Uno que ejecuta soldadura manual o semiautomática.

welder certification. Written verification that a welder has produced welds meeting a prescribed standard of welder performance.

certificación del soldador. Verificación escrita de que un soldador ha producido soldaduras que cumplen con la norma prescrita de la ejecución del soldador.

welder performance qualification. The demonstration of a welder's ability to produce welds meeting prescribed standards.

calificación de ejecución del soldador. La demostración de la habilidad del soldador de producir soldaduras que cumplen con las normas prescritas.

welder registration. The act of registering a welder certification or a photostatic copy thereof.

registración del soldador. El acto de registrar una certificación del soldador o una copia fotostata de ello.

welding. A joining process that produces coalescence of materials by heating them to the welding temperature, with or without the application of pressure or by the application of pressure alone, and with or without the use of filler metal.

soldadura. Un proceso de unión que produce coalescencia de materiales calentándolos a la temperatura de soldadura, con o sin la aplicación de presión o por la aplicación de presión solamente, y con o sin el uso del metal de relleno.

welding arc. A controlled electrical discharge between the electrode and the workpiece that is formed and sustained by the establishment of a gaseous conductive medium, called an arc plasma.

arco de soldadura. Una descarga eléctrica controlada entre el electrodo y la pieza de trabajo que es formada y sostenida por el establecimiento de un medio conductivo gaseoso, llamado un arco de plasma.

welding current. The current in the welding circuit during the making of a weld.

corriente para soldadura. La corriente en el circuito de soldar durante la hechura de una soldadura.

welding current (automatic arc welding). The current in the welding circuit during the making of a weld, but excluding upslope, downslope, start, and crater fill current. Refer to drawing for **upslope time.**

corriente de soldadura (soldadura de arco automático). La corriente en el circuito de soldar durante la hechura de una soldadura, pero excluyendo el pendiente en ascenso, pendiente en descenso, empiezo, y corriente par llenar el crater. Refiérase al dibujo para **tiempo del pendiente en ascenso.**

welding cycle. The complete series of events involved in the making of a weld. Refer to drawings for **downslope time** and **upslope time.**

ciclo de soldadura. Una serie completa de eventos envueltos en hacer una soldadura. Refiérase al dibujo para **tiempo de cáida del pendiente** y **tiempo del pendiente en descenso.**

welding electrode. A component of the welding circuit through which current is conducted and that terminates at the arc, molten conductive slag, or base metal. See also **arc welding electrode, bare electrode, carbon electrode, composite electrode, covered electrode, electroslag welding electrode, emissive electrode, flux cored electrode, lightly coated electrode, metal cored electrode, metal electrode, resistance welding electrode, stranded electrode,** and **tungsten electrode.**

soldadura con electrodo. Un componente del circuito de soldar por donde la corriente es conducida y que termina en el arco, en la escoria derretida conductiva, o en el metal base. Vea también electro para soldar de arco, electrodo descubierto, electrodo de carbón, electrodo compuesto, electrodo cubierto, electrodo para soldadura de electroescoria, electrodo emisivo, electrodo de núcleo de fundente, electrodo con recubrimiento ligero, electrodo de metal de núcleo , electrodo de metal, electrodo para soldadura por resistencia, electrodo cable, y electrodo de tungsteno.

welding filler metal. The metal or alloy to be added in making a weld joint that alloys with the base metal to form weld metal in a fusion welded joint.

metal de soldadura para rellenar. El metal o aleación que se va a agregar en la hechura de una junta de soldadura que se mezcla con el metal base para formar metal de soldadura en una junta de fusión de soldadura.

welding generator. A generator used for supplying current for welding.

generador para soldar. Un generador que se usa para proporcionar la corriente para la soldadura.

welding ground. A nonstandard and incorrect term for **workpiece connection.**

tierra de soldadura. Un término fuera de norma e incorrecto para **conexión de pieza de trabajo.**

welding head. The part of a welding machine in which a welding gun or torch is incorporated.

cabeza de soldar. La parte de una máquina para soldar la cual una pistola de soldadura o una antorcha se puede incorporar.

welding leads. The work lead and electrode lead of an arc welding circuit. Refer to drawing for **direct current electrode positive.**

cables para soldar. Los cables de pieza de trabajo y el portelectrodo de un circuito de soldadura de arco. Refiérase al dibujo **corriente directa con el electrodo positivo.**

welding machine. Equipment used to perform the welding operation. For example, spot welding machine, arc welding machine, seam welding machine, etc.

máquina para soldar. El equipo que se usa para ejecutar la operación de soldadura. Por ejemplo, máquina de soldadura por puntos, máquina de soldadura de arco, máquina de soldadura de costura, etc.

welding operator. One who operates adaptive control, automatic, mechanized, or robotic welding equipment.

operador de soldadura. Uno que opera control adaptivo, automático, mecanizado, o equipo robótico para soldar.

welding position. See flat position, horizontal position, horizontal fixed position, horizontal rolled position, overhead position, and vertical position.

posición de soldadura. Vea posición plana, posición horizontal, posición fija horizontal, posición horizontal rodada, posición de sobrecabeza, posición vertical.

POSITION OF WELDING

Flat. See flat position.
Horizontal. See horizontal position, horizontal fixed position, and horizontal rolled position.
Vertical. See vertical position.
Overhead. See overhead position.

POSITION FOR QUALIFICATION

Plate welds

Groove welds

1G. See flat position.
2G. See horizontal position.
3G. See vertical position.
4G. See overhead position.

Fillet welds

1F. See flat position.
2F. See horizontal position.
3F. See vertical position.
4F. See overhead position.

Pipe welds

Groove welds

1G. See horizontal rolled position.
2G. See vertical position.
5G. See horizontal fixed position.
6G. Inclined position.
6GR. Inclined position

Position

welding power source. An apparatus for supplying current and voltage suitable for welding. See also **welding generator, welding rectifier,** and **welding transformer.**

fuente de poder para soldar. Un aparato para surtir corriente y voltaje adecuado para soldar. Vea también **generador para soldar, rectificador para soldar,** y **transformador para soldar.**

welding procedure qualification record (WPQR). A record of welding variables used to produce an acceptable test weldment and the results of tests conducted on the weldment to qualify a welding procedure specification.

registro de calificación de procedimiento de la soldadura. Un registro de los variables usados para producir una probeta aceptable y los resultados de la prueba conducida en la probeta para calificar el procedimiento de especificación.

welding procedure specification (WPS). A document providing in detail the required variables for specific application to assure repeatability by properly trained welders and welding operators.

calificación de procedimiento de soldadura. Un documento que provee en detalle los variables requeridos para la aplicación específica para asegurar la habilidad de repetir el procedimiento por soldadores y operadores que estén propiamente preparados.

welding process. A materials joining process that produces coalescence of materials by heating them to suitable temperatures, with or without the application of pressure or by the application of pressure alone, and with or without the use of filler metal.

proceso para soldar. Un proceso para unir materiales que produce coalescencia calentándolos a una temperatura adecuada con o sin la aplicación de presión solamente y con o sin usarse material para rellenar.

welding rectifier. A device in a welding machine for converting alternating current to direct current.

rectificador para soldar. Un aparato en una máquina para soldar para convertir la corriente alterna a corriente directa.

welding rod. A form of welding filler metal, normally packaged in straight lengths, that does not conduct the welding current.

varilla para soldar. Una forma de metal de soldadura para rellenar, normalmente empaquetada en piezas derechas, que no conduce la corriente para soldar.

welding sequence. The order of making the welds in a weldment.

orden de sucesión (para soldar). La orden de hacer las pasadas de soldar en una soldadura.

welding symbol. A graphical representation of a weld.

símbolo de soldadura. Una representación gráfica de una soldadura.

welding technique. The details of a welding procedure that are controlled by the welder or welding operator.

ejecución de soldadura. Los detalles del procedimiento que son controlados por el soldador u operador de soldadura.

welding tip. A welding torch tip designed for welding.

boquilla (punta) para soldar. Una boquilla en la antorcha de soldadura que está diseñada para soldar.

welding torch (arc). A device used in the gas tungsten and plasma arc welding processes to control the position of the electrode, to transfer current to the arc, and to direct the flow of shielding and plasma gas.

antorcha para soldar (arco). Un aparato usado en los procesos de soldadura del gas tungsteno y arco plasma para controlar la posición del electrodo, para transferir corriente al arco, y para dirigir la corriente del gas protector y gas de la plasma.

welding torch (oxyfuel gas). A device used in oxyfuel gas welding, torch brazing, and torch soldering for directing the heating flame produced by the controlled combustion of fuel gases.

antorcha para soldar (gas oxicombustible). Un aparato usado en soldadura de gas oxicombustible, soldadura blanda con antorcha y soldadura fuerte con antorcha y para dirigir la llama para calentar producida por la combustión controlada de gases de combustión.

welding transformer. A transformer used to supplying current for welding. See also **reactor** (arc welding).

transformador para soldar. Un transformador que se usa para dar corriente para la soldadura. Vea también **reactor** (soldadura de arco).

welding voltage. See **arc voltage**.

voltaje para soldar. Vea **voltaje del arco**.

welding wire. A form of welding filler metal, normally packaged as coils or spools, that may or may not conduct electrical current, depending upon the filler metal and base metal in a solid-state weld with filler metal.

alambre para soldar. Una forma de metal para rellenar con soldadura, normalmente empaquetado en rollos o en carretes que pueda o pueda que no conducir corriente eléctrica, dependiendo en el metal de relleno y el metal base en una soldadura que está en estado sólido con metal de relleno.

weldment. An assembly whose component parts are joined by welding.

conjunto de partes soldadas. Una asamblea cuyas partes componentes están unidas por la soldadura.

weldor. See preferred term **welder**.
soldador. Vea el término preferido **soldador**.

wetting. The phenomenon whereby a liquid filler metal or flux spreads and adheres in a thin, continuous layer on a solid base metal.

exudación. El fenómeno de que un metal para rellenar líquido o un flujo se puede desparramar y adherirse en una capa delgada, capa continua en un sólido metal base.

wire feed speed. The rate at which wire is consumed in arc cutting, thermal spraying, or welding.

velocidad de alimentador de alambre. La velocidad que el alambre es consumido en cortes de arco, rociado termal o soldadura.

work angle. The angle less than 90 degrees between a line perpendicular to the major workpiece surface and a plane determined by the electrode axis and the weld axis. In a T-joint or a corner joint, the line is perpendicular to the nonbutting member. This angle can also be used to partially define the position of guns, torches, rods, and beams.

ángulo de trabajo. El ángulo menos de 90 grados entre una línea perpendicular a la superficie de pieza de trabajo mayor y una plana determinada por el eje del electrodo y el eje de la soldadura. En una junta-T o en una junta de esquina, la línea es perpendicular a un miembro que no topa. Este ángulo puede ser usado también para parcialmente definir la posición de pistolas, antorchas, varillas y rayos.

work angle (pipe). The angle less than 90 degrees between a line, which is perpendicular to the cylindrical pipe surface at the point of intersection of the weld axis and the extension of the electrode axis, and a plane determined by the electrode axis and a line tangent to the pipe at the same point. In a T-joint, the line is perpendicular to the nonbutting member. This angle can also be used to partially define the position of guns, torches, rods, and beams.

ángulo de trabajo (tubo). El ángulo menos de 90 grados entre una línea, la cual es perpendicular a la superficie de un tubo cilíndrico al punto de intersección del eje de la soldadura y la extensión del eje del electrodo, y un plano determinado por el eje del electrodo y una línea tangente al tubo al mismo punto. En una junta-T, la línea es perpendicular a un miembro que no topa. Este ángulo puede también usarse para definir parcialmente la posición de pistolas, antorchas, varillas, y rayos.

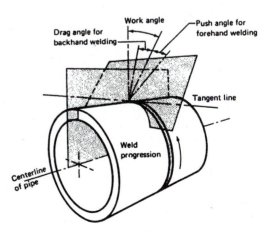

Pipe weld

work connection. The connection of the work lead to the work. Refer to drawing for **direct current electrode negative.**

pinza de tierra. La conexión del cable de trabajo (tierra) al trabajo. Refiérase al dibujo para **corriente directa con electrodo negativo.**

work lead. The electric conductor between the source of arc welding current and the work. Refer to drawing for **direct current electrode negative.**

cable de tierra. Un conductor eléctrico entre la fuente de la corriente del arco y la pieza de trabajo. Refiérase al dibujo **corriente directa con electrodo negativo.**

work station. A manufacturing unit consisting of one or more numerically controlled machine tools serviced by a robot.

estación de trabajo. Una unidad manufacturera de una o más herramienta numerada que es controlada por una máquina y abatecida por un robot.

working envelope. The set of points representing the maximum extent or reach of the robot hand or working tool in all directions.

alcance de operación. Un juego de puntos que representan la máxima extensión o alcance de la mano del robot o la herramienta del trabajo en todas las direcciones.

*working pressure. The pressure at the low-pressure gauge, ranging from 0 to 45 psi (depending on the type of gas), used for welding and cutting.

presión de trabajo. La presión en el manómetro de baja presión, con escala de 0 a 45 psi (dependiendo en el tipo de gas), usado para cortar y soldar.

working range. All positions within the working envelope. The range of any variable within which the system normally operates.

extensión de trabajo. Todas las posiciones dentro del alcance del trabajo. El alcance de cualquier variable dentro del sistema que opera normalmente.

workpiece. The part that is welded, brazed, soldered, thermal cut, or thermal sprayed.

pieza de trabajo. La parte que está soldada, con soldadura fuerte, soldadura blanda, corte termal, o rociado termal.

workpiece connection. The connection of the workpiece lead to the workpiece.

conexion de pieza de trabajo (pinzas). La conexión del cable de la pieza de trabajo a la pieza de trabajo.

workpiece lead. The electrical conductor between the arc welding current source and workpiece connection.

cable de pieza de trabajo. El conductor eléctrico entre la fuente de corriente de soldadura de arco y la conexión de la pieza de trabajo.

wrist. A set of rotary joints between the arm and hand that allow the hand to be oriented to the workpiece.

muñeca. Un juego de coyunturas rotatorias entre el brazo y la mano que permite a la mano ser orientada a la pieza de trabajo.

Y

yaw. The angular displacement of a moving body about an axis that is perpendicular to the line of motion and to the top side of the body.

guiñada. El desalojamiento de un cuerpo en movimiento alrededor de un eje que está perpendicular a la línea de movimiento y al lado más alto del cuerpo.

Index

Note: Page numbers in **bold** reference non-text material.